Laser Theory

Laser Theory

Edited by

Frank S. Barnes

Chairman
Department of Electrical Engineering
University of Colorado, Boulder

A volume in the IEEE PRESS Selected Reprint Series, prepared under the sponsorship of the IEEE Electron Devices Group and the IEEE Microwave Theory and Techniques Group.

IEEE PRESS

The Institute of Electrical and Electronics Engineers, Inc. New York

$14.95

International Standard Book Numbers:
Clothbound: 0-87942-019-7
Paperbound: 0-87942-020-0

Library of Congress Catalog Card Number 72-91997

PRINTED IN THE UNITED STATES OF AMERICA

Acknowledgment

It is the editor's pleasure to express his deep appreciation to Tony Siegman, Steve Harris, Dietrich Marcuse, Gene Gordon, and Bill Louisell for their very constructive comments in selecting titles to be included in this reprint volume.

Contents

Part 2: Optical Resonators

Part 3: Laser Oscillator Theory

Part 4: Laser Amplifiers

Part 5: Internal Modulation and Mode Locking

Introduction

This reprint volume has brought together a collection of papers in an attempt to serve a portion of the needs of the serious student and laser research worker who have the need to refer to some aspects of laser theory in depth. The criteria for the selection of papers for this volume is to some degree arbitrary as there are many more important papers than may be collected in a single volume of reasonable length. In addition to important resource papers, two review papers are included to improve the readability of parts of the material and to provide extensive references. The papers in this collection have been broken down into six parts.

Part 1 consists of twelve papers that were chosen primarily on the basis of their historical significance. In teaching a graduate course, the editor has found that these are papers that serious students are most likely to want to refer to, either out of curiosity about the way important discoveries were first reported or because the explanations of important ideas are particularly clearly stated in them. The paper by Wang and Townes [1] also represents a useful contribution to the early development of maser theory but was omitted for lack of space.

The three papers in Part 2 deal with the theory of optical resonators. This subject is much more extensive than can be covered in this reprint volume without making its length excessive. Thus the editor has chosen the review paper by Kogelnik and Li as a way of summarizing important ideas and of providing a bibliography to much of the pertinent literature. The papers by Boyd and Gordon and by Fox and Li are historically significant, and they provide starting points for most of the analyses of optical resonators that have been done since then.

Part 3 deals with extensions of laser oscillator theory. The paper by Jaynes and Cummings provides an introduction to semiclassical maser theory and a beginning to full quantum theory. The paper by Bennett introduces the important concept of hole burning and its influence on multimode laser oscillations. The semiclassical theory of laser operation is brought to a high degree of completeness in the paper by Lamb. Although this paper will be difficult reading for many electrical engineers, the detailed characteristics of lasers are needed by an experimentalist for work on such things as laser frequency standards. Some of the results from this paper are reviewed in [2] and [3]. The papers by Scully and Lamb and by Gordon represent two different approaches to a full quantum theory for the laser oscillator. The Scully and Lamb paper is the first of a sequence of five papers for one approach to the development of a quantum theory of lasers (references to other papers in this series can be found in [4]). The Gordon paper is closely related to a sequence of at least thirteen papers by Lax and co-authors (references to other papers in this sequence and related work can be found in [5]) and develops a quantum theory of maser oscillators from a slightly different point of view. A third school for the development of the quantum theory for laser oscillators has been developed by Haken and his associates in Germany [6]. This work, along with individual papers by a wide variety of authors, is important to the understanding of the noise properties of lasers and such things as the power spectrum of the oscillator output.

Part 4 of this volume deals with transient phenomena in laser amplifiers and begins with the paper by Statz and deMars that initiated the rate equation treatment of the transient build-up of oscillation in masers. This treatment is extended in the paper by Wittke and Warter to include a steady-state pulse propagating in long laser amplifiers. The paper by Iesvgi and Lamb provides detailed treatment of the light pulse propagating in a laser amplifier with a wide variety of atomic parameters for characterizing various laser systems. (Other papers of interest in the development of the theory for the amplification of the laser pulse include [7] and [8].) The final two papers by Rigrod deal with the influence of saturation and the coupling of the output of the mirror on the operating characteristics of the laser.

Part 5 has to do with the problems associated with internal modulation and mode locking. The paper by Smith is a review paper that both summarizes much of the recent work on mode locking and provides a good bibliography to most of the original work. The two papers by Harris and McDuff present a theory for the internal modulation of a laser from the point of a multimode laser oscillator. The papers by Kuizenga and Siegman describe the same phenomenon from the point of view of a pulse that circulates through the laser cavity. The point of view that is most useful depends on the application. The last two papers in Part 5 introduce the problems of a parametric conversion of optical energy in the laser internal to the cavity resonator. The field of optical parametrics is too large to be covered in the depth that it deserves in the space available in this book. However, these two papers have been selected to improve the completeness of the coverage of process that may occur in the laser operation internal to the cavity. Other papers contributing to this field include [9]-[13].

The final section of the book, Part 6, deals briefly with the problems of noise in maser amplifiers. The two papers by Muller and Pound represent a portion of the early development of the theory that showed that microwave masers could be useful as a very low noise amplifier for radio signals. The paper by Lax develops the basic equations from which many of the detailed quantum characteristics have been derived in later papers [5]. A few other noise papers that provide access to the literature or contribute to the early development of the subject are [14]-[17].

References

[1] T. C. Wang and C. H. Townes, "Further aspects of the theory of the maser," *Phys. Rev.*, vol. 102, pp. 1308-1321, June 1956.

[2] D. C. Sinclair and W. E. Bell, *Gas Laser Technology.* New York: Holt, Rhinehart, and Winston, 1969, ch. 3.

[3] A. E. Siegman, *An Introduction to Lasers and Masers.* New York: McGraw-Hill, 1971.

[4] D. M. Kim, M. O. Scully, and W. E. Lamb, Jr., "Quantum theory of an optical maser V—atomic motion and recoil," *Phys. Rev.*, vol. 2, pp. 2534-2541, Dec. 1970.

[5] M. Lax and W. H. Louisell, "Quantum noise XII density—operator treatment of field and population fluctuations," *Phys. Rev.*, vol. 185, pp. 568-591, Sept. 1969.

[6] H. Haken and W. Weidlich, "A theorem on the calculations of multi-time correlation functions by the single-time density matrix," *Z. Phys.*, vol. 205, pp. 96-102, 1967.

[7] E. O. Schultz-DuBois, "Pulse sharpening and gain saturation in traveling wave masers," *Bell Syst. Tech. J.*, vol. 43, pp. 625-658, Jan. 1964.

[8] F. T. Arecchi and R. Bonifaco, "Theory of optical maser amplifiers," *IEEE J. Quantum Electron.*, vol. QE-1, pp. 169-178, July 1965.

[9] A. Yariv, "Parametric interactions of optical modes," *IEEE J. Quantum Electron.*, vol. QE-2, pp. 30-37, Feb. 1966.

[10] A. Yariv and W. H. Louisell, "Theory of the optical parametric oscillator," *IEEE J. Quantum Electron.*, vol. QE-2, pp. 418-424, Sept. 1966.

[11] G. D. Boyd and D. A. Kleinman, "Parametric interactions of focused Gaussian light beams," *J. Appl. Phys.*, vol. 39, pp. 3597-3639, 1968.

[12] J. A. Giordmaine, "Mixing of light beams in crystals," *Phys. Rev. Lett.*, vol. 8, pp. 19-20, Jan. 1962.

[13] P. A. Franken and J. F. Ward, "Optical harmonics and non-linear phenomena," *Rev. Mod. Phys.*, vol. 35, pp. 23-39, 1963.

[14] K. Shimoda, H. Takahashi, and C. H. Townes, "Fluctuations in amplifications of quanti with applications of maser amplifiers," *J. Phys. Soc. Jap.*, vol. 12, pp. 686-700, 1957.

[15] J. P. Gordon and L. D. White, "Noise in maser amplifiers—theory and experiment," *Proc. IRE*, vol. 46, pp. 1588-1594, Sept. 1958.

[16] E. I. Gordon, "Optical maser oscillators and noise," *Bell Syst. Tech. J.*, vol. 43, pp. 507-539, Jan. 1964.

[17] C. Freed and H. Haus, "Photoelectron statistics produced by a laser operating below and above the threshold of oscillation," *IEEE J. Quantum Electron.*, vol. QE-2, pp. 190-195, Aug. 1966.

Part 1
Historical Papers

On the Quantum Theory of Radiation†

A. EINSTEIN

THE formal similarity of the curve of the chromatic distribution of black-body radiation and the Maxwell velocity-distribution is too striking to be hidden for long. Indeed, already Wien in his important theoretical paper in which he derived his displacement law

$$\rho = \nu^3 f(\nu/T) \tag{1}$$

was led by this similarity to a further determination of the radiation formula. It is well known that he then found the formula

$$\rho = \alpha\nu^3\, e^{-h\nu/kT}, \tag{2}$$

which is also nowadays accepted as being correct as a limiting law for large values of ν/T (Wien's radiation law). We know nowadays that no considerations based on classical mechanics and electrodynamics can give us a usable radiation formula, and that classical theory necessarily leads to the Rayleigh formula

$$\rho = \frac{k\alpha}{h}\nu^2 T. \tag{3}$$

As soon as Planck in his classical investigation based his radiation formula

$$\rho = \alpha\nu^3 \frac{1}{e^{h\nu/kT} - 1} \tag{4}$$

on the assumption of discrete elements of energy, from which very

† *Physikalische Zeitschrift* **18**, 121 (1917).

Reprinted with permission from *The Old Quantum Theory* (Elmsford, N.Y.: Pergamon, 1967, pp. 167–183).

quickly quantum theory developed, it was natural that Wien's discussion which led to equation (2) became forgotten.

Recently[1]† I found a derivation of Planck's radiation formula which is based upon the basic assumption of quantum theory and which is related to Wien's original considerations; in this derivation, the relationship between the Maxwell distribution and the chromatic black-body distribution plays a role. This derivation is of interest not only because it is simple, but especially because it seems to clarify somewhat the at present unexplained phenomena of emission and absorption of radiation by matter. I have shown, on the basis of a few assumptions about the emission and absorption of radiation by molecules, which are closely related to quantum theory, that molecules distributed in temperature equilibrium over states in a way which is compatible with quantum theory are in dynamic equilibrium with the Planck radiation. In this way, I deduced in a remarkably simple and general manner Planck's formula (4). It was a consequence of the condition that the distribution of the molecules over the states of their internal energy, which is required by quantum theory, must be established solely through the absorption and emission of radiation.

If the assumptions about the interaction between radiation and matter which we have introduced are essentially correct, they must, however, yield more than the correct statistical distribution of the internal energy of the molecules. In fact, in absorption and emission of radiation, momentum is transferred to the molecules; this entails that merely through the interaction of radiation and molecules the velocities of the molecules will acquire a certain distribution. This must clearly be the same velocity distribution as the one which the molecules attain through the action of their mutual collisions alone, that is, it must be the same as the Maxwell distribution. We must require that the average kinetic energy (per degree of freedom) which a molecule acquires in the Planck radiation field of temperature T is equal to $\frac{1}{2}kT$; this must be true independent of the nature of the molecules considered and independent of the frequencies of the light emitted or absorbed by

† The considerations given in that paper are repeated in the present one.

them. In the present paper, we want to show that our simple hypotheses about the elementary processes of emission and absorption obtain another support.

In order to obtain the above-mentioned result we must, however, complete to some extent the hypotheses upon which our earlier work was based, as the earlier hypotheses were concerned only with the exchange of energy. The question arises: does the molecule receive an impulse when it absorbs or emits the energy ε? Let us, for instance, consider the emission from the point of view of classical electrodynamics. If a body emits the energy ε, it receives a recoil (momentum) ε/c if all of the radiation ε is emitted in the same direction. If, however, the emission takes place as an isotropic process, for instance, in the form of spherical waves, no recoil at all occurs. This alternative also plays a role in the quantum theory of radiation. When a molecule during a transition from one quantum-theoretically possible state to another absorbs or emits energy ε in the form of radiation, such an elementary process can be thought of either as being a partially or completely directed or as being a symmetrical (non-directional) process. *It now turns out that we arrive at a consistent theory only, if we assume each elementary process to be completely directional.* This is the main result of the following considerations.

1. Basic Hypothesis of Quantum Theory. Canonical Distribution over States

According to quantum theory, a molecule of a given kind can take up—apart from its orientation and its translational motion—only a discrete set of states $Z_1, Z_2, \ldots, Z_n, \ldots$ with (internal) energies $\varepsilon_1, \varepsilon_2, \ldots, \varepsilon_n, \ldots$. If molecules of this kind form a gas of temperature T, the relative occurrence W_n of these states Z_n is given by the formula giving the canonical distribution of statistical mechanics:

$$W_n = p_n e^{-\varepsilon_n/kT}. \qquad (5)$$

In this equation $k = R/N$ is the well-known Boltzmann constant, and p_n a number which is characteristic for the molecule and its *n*th quantum state and which is independent of T; it can be called the statistical "weight" of the state. One can derive equation (5) either from Boltzmann's principle or by purely thermodynamic means. Equation (5) expresses the greatest generalisation of Maxwell's velocity distribution law.

Recent important progress in quantum theory relates to the theoretical determination of quantum theoretically possible states Z_n and their weight p_n. For our considerations of the principles involved in radiation, we do not need a detailed determination of the quantum states.

2. Hypotheses about Energy Exchange through Radiation

Let Z_n and Z_m be two quantum-theoretically possible states of the gas molecule, and let their energies ε_n and ε_m satisfy the inequality $\varepsilon_m > \varepsilon_n$. Let the molecule be able to make a transition from the state Z_n to the state Z_m by absorbing radiative energy $\varepsilon_m - \varepsilon_n$; similarly let a transition from Z_m to Z_n be possible in which this radiative energy is emitted. Let the frequency of the radiation absorbed or emitted by the molecule in such transitions be ν; it is characteristic for the combination (m, n) of the indices.

We make a few hypotheses about the laws valid for this transition; these are obtained by using the relations known from classical theory for a Planck resonator, as the quantum-theoretical relations which are as yet unknown.

(a) *Spontaneous emission.*† It is well known that a vibrating Planck resonator emits according to Hertz energy independent of whether it is excited by an external field or not. Accordingly, let it be possible for a molecule to make without external stimulation a transition from the state Z_m to the state Z_n while emitting the radiation energy $\varepsilon_m - \varepsilon_n$ of frequency ν. Let the probability dW that this will in fact take place in the time interval dt be

$$dW = A_m^n \, dt, \tag{A}$$

† Einstein uses *Ausstrahlung* and *Einstrahlung* for spontaneous emission and induced radiation [D. t. H.].

where A^n_m denotes a constant which is characteristic for the combination of indices considered.

The statistical law assumed here corresponds to the law of a radioactive reaction, and the elementary process assumed here corresponds to a reaction in which only γ-rays are emitted. It is not necessary to assume that this process takes place instantaneously; it is only necessary that the time this process takes is negligible compared with the time during which the molecule is in the state $Z_1, \ldots$.

(b) *Induced radiation processes.* If a Planck resonator is in a radiation field, the energy of the resonator can be changed by the transfer of energy from the electromagnetic field to the resonator; this energy can be positive or negative depending on the phases of the resonator and of the oscillating field. Accordingly we introduce the following quantum-theoretical hypothesis. Under the influence of a radiation density ρ of frequency ν a molecule can make a transition from the state Z_n to the state Z_m by absorbing the radiative energy $\varepsilon_m - \varepsilon_n$ and the probability law for this process is

$$dW = B^m_n \rho \, dt. \tag{B}$$

Similarly, a transition $Z_m \rightarrow Z_n$ may also be possible under the influence of the radiation; in this process the radiative energy $\varepsilon_m - \varepsilon_n$ will be freed according to the probability law

$$dW = B^n_m \rho \, dt. \tag{B'}$$

The B^m_n and B^n_m are constants. These two processes we shall call "changes in state, induced by radiation".

The question now arises: what is the momentum transferred to the molecule in these changes in state? Let us begin with the induced processes. If a radiation beam with a well-defined direction does work on a Planck resonator, the corresponding energy is taken from the beam. According to the law of conservation of momentum, this energy transfer corresponds also to a momentum transfer from the beam to the resonator. The resonator is thus subject to the action of a force in the direction of the beam. If the energy transferred is negative, the action of the force on the resonator is also in the opposite direction. This means clearly the following in the case of the quantum hypothesis. If through the irradiation by a beam of light a transition $Z_n \to Z_m$ is induced, the momentum $(\varepsilon_m - \varepsilon_n)/c$ is transferred to the molecule in the direction of propagation of the beam. In the induced transition $Z_m \to Z_n$ the transferred momentum has the same magnitude but is in the opposite direction. We assume that in the case where the molecule is simultaneously subjected to several radiation beams, the total energy $\varepsilon_m - \varepsilon_n$ of an elementary process is absorbed from or added to *one* of these beams, so that also in that case the momentum $(\varepsilon_m - \varepsilon_n)/c$ is transferred to the molecule.

In the case of a Planck resonator, when the energy is emitted through a spontaneous emission process, no momentum is transferred to the resonator, since according to classical theory the emission is in the form of a spherical wave. We have, however, already noted that we can only obtain a consistent quantum theory by assuming that the spontaneous emission process is also a directed one. In that case, in each spontaneous emission elementary process $(Z_m \to Z_n)$ momentum of magnitude $(\varepsilon_m - \varepsilon_n)/c$ is transferred to the molecule. If the molecule is isotropic, we must assume that all directions of emission are equally probable. If the molecule is not isotropic, we arrive at the same statement if the orientation changes in a random fashion in time. We must, of course, make a similar assumption for the statistical laws (B) and (B′) for the induced processes, as otherwise the constants should depend on direction, but we can avoid this through the assumption of isotropy or pseudo-isotropy (through time-averaging) of the molecule.

3. Derivation of the Planck Radiation Law

We now ask for that radiation density ρ which must be present in order that the exchange of energy between radiation and molecules according to the statistical laws (A), (B), and (B′) does not perturb the distribution (5) of the molecules. For this it is necessary and sufficient that on the average per unit time as many elementary processes of type (B) take place as of types (A) and (B′) combined. This combination leads, because of (5), (A), (B), and (B′), to the following equation for the elementary processes corresponding to the index combination (m, n):

$$p_n e^{-\varepsilon_n/kT} B_n^m \rho = p_m e^{-\varepsilon_m/kT} (B_m^n \rho + A_m^n).$$

If, furthermore, ρ will increase to infinity with T, as we shall assume, the following relation must exist between the constants B_n^m and B_m^n:

$$p_n B_n^m = p_m B_m^n. \tag{6}$$

We then obtain from our equation the following condition for dynamic equilibrium:

$$\rho = \frac{A_m^n / B_m^n}{e^{(\varepsilon_m - \varepsilon_n)/kT} - 1}. \tag{7}$$

This is the temperature-dependence of the radiation density of the Planck law. From Wien's displacement law (1) it follows from this immediately that

$$\frac{A_m^n}{B_m^n} = \alpha \nu^3, \tag{8}$$

and

$$\varepsilon_m - \varepsilon_n = h\nu, \tag{9}$$

where α and h are constants. To find the numerical value of the constant α, we should have an exact theory of the electrodynamic and mechanic processes; for the time being we must use the Rayleigh limit of high temperatures, for which the classical theory is valid as a limiting case.

Equation (9) is, of course, the second main hypothesis of Bohr's theory of spectra of which we can now state after Sommerfeld's and Epstein's extensions that it belongs to those parts of our science which are sure. It contains implicitly, as I have shown, also the photochemical equivalence rule.

4. Method of Calculating the Motion of Molecules in the Radiation Field

We now turn to the investigation of the motions which our molecules execute under the influence of the radiation. To do this, we use a method which is well known from the theory of Brownian motion, and which I have used already many times for numerical calculations of motion in radiation. To simplify the calculations, we only perform them for the case where the motion takes place only in one direction, the X-direction of our system or coordinates. We shall moreover restrict ourselves to calculating the average value of the kinetic energy of the translational motion, and thus do not give the proof that these velocities v are distributed according to Maxwell's law. Let the mass M of the molecule be sufficiently large that we can neglect higher powers of v/c in comparison with lower ones; we can then apply ordinary mechanics to the molecule. Moreover, without any real loss of generality, we can perform our calculations as if the states with indices m and n were the only ones which the molecule can take on.

The momentum Mv of a molecule is changed in two ways in the short time τ. Although the radiation is the same in all directions, because of its motion the molecule will feel a force acting in the opposite direction of its motion which comes from the radiation. Let this force be Rv, where R is a constant to be evaluated later on. This force would bring the molecule to rest, if the irregularity of the action of the radiation did not have as a consequence that during the time τ a momentum Δ of varying sign and varying magnitude is transferred to the molecule; this unsystematic influence will against the earlier mentioned force maintain a certain motion of the molecule. At the end of the short time τ, which we are considering, the momentum of the molecule will have the value

$$Mv - Rv\tau + \Delta.$$

As the velocity distribution must remain the same in time, this quantity must have the same average absolute magnitude as Mv; therefore, the average squares of those two quantities, averaged over a long period or over a large number of molecules, must be equal to one another:

$$\overline{(Mv - Rv\tau + \Delta)^2} = \overline{(Mv)^2}.$$

As we have separately taken into account the systematic influence of v on the momentum of the molecule, we must neglect the average $\overline{\Delta v}$. Expanding the left-hand side of the equation, we get then

$$\overline{\Delta^2} = 2RM\overline{v^2}\tau. \tag{10}$$

The average $\overline{v^2}$ for our molecules, which is caused by radiation of temperature T through its interaction with the molecules must be equal to the average value $\overline{v^2}$, which according to the kinetic theory of gases a molecule in the gas would have according to the gas laws at the temperature T. Otherwise, the presence of our molecules would disturb the thermodynamic equilibrium between black-body radiation and any gas of the same temperature. We must then have

$$\tfrac{1}{2}M\overline{v^2} = \tfrac{1}{2}kT. \tag{11}$$

Equation (10) thus becomes

$$\frac{\overline{\Delta^2}}{\tau} = 2RkT. \tag{12}$$

The investigation must now proceed as follows. For given radiation $[\rho(\nu)]$ we can calculate $\overline{\Delta^2}$ and R with our hypotheses about the interaction between radiation and molecules. Inserting these results into (12), this equation must be satisfied identically if ρ as function of ν and T is expressed by the Planck equation (4).

5. Calculation of R

Let a molecule of the kind considered move uniformly with velocity v along the X-axis of the system of coordinates K. We ask for the average momentum transferred by the radiation to the molecule per unit time. To be able to evaluate this, we must consider the radiation in a system of coordinates K' which is at rest relative to the molecule under consideration, because we have only formulated our hypotheses about emission and absorption for molecules at rest. The transformation to the system K' has often been given in the literature and especially accurately in Mosengeil's Berlin thesis.[2] For the sake of completeness, I shall, however, repeat the simple considerations.

In K the radiation is isotropic, that is, we have for the radiation per unit volume in a frequency range $d\nu$ and propagating in a direction within a given infinitesimal solid angle $d\kappa$:

$$\rho \, d\nu \frac{d\kappa}{4\pi}, \tag{13}$$

where ρ depends only on the frequency ν, but not on the direction. This particular radiation corresponds in the coordinate system K' to a particular radiation, which is also characterised by a frequency range $d\nu'$ and a certain solid angle $d\kappa'$. The volume density of this particular radiation is

$$\rho'(\nu', \phi') \, d\nu' \frac{d\kappa'}{4\pi}. \tag{13'}$$

This defines ρ'. It depends on the direction which is defined in the usual way by the angle ϕ' with the X'-axis and the angle ψ' between the projection in the $Y'Z'$-plane with the Y'-axis. These angles correspond to the angles ϕ and ψ which in a similar manner fix the direction of $d\kappa$ with respect to K.

First of all it is clear that the same transformation law must be valid between (13) and (13') as between the squares of the amplitude A^2 and A'^2 of a plane wave of the appropriate direction of propagation. Therefore in the approximation we want, we have

$$\frac{\rho'(\nu', \phi') \, d\nu' \, d\kappa'}{\rho(\nu) \, d\nu \, d\kappa} = 1 - 2\frac{v}{c}\cos\phi, \tag{14}$$

or

$$\rho'(\nu', \phi') = \rho(\nu) \frac{d\nu}{d\nu'} \frac{d\kappa}{d\kappa'} \left(1 - 2\frac{v}{c}\cos\phi\right). \tag{14'}$$

The theory of relativity further gives the following formulae, valid in the approximation needed here,

$$\nu' = \nu\left(1-\frac{v}{c}\cos\phi\right), \tag{15}$$

$$\cos\phi' = \cos\phi-\frac{v}{c}+\frac{v}{c}\cos^2\phi, \tag{16}$$

$$\psi' = \psi. \tag{17}$$

From (15) in the same approximation it follows that

$$\nu = \nu'\left(1+\frac{v}{c}\cos\phi'\right).$$

Therefore, also in the same approximation,

$$\rho(\nu) = \rho\left(\nu'+\frac{v}{c}\nu'\cos\phi'\right),$$

or

$$\rho(\nu) = \rho(\nu')+\frac{\partial\rho(\nu')}{\partial\nu}\frac{v}{c}\nu'\cos\phi'. \tag{18}$$

Furthermore from (15), (16), and (17) we have

$$\frac{d\nu}{d\nu'} = 1+\frac{v}{c}\cos\phi',$$

$$\frac{d\kappa}{d\kappa'} = \frac{\sin\phi\,d\phi\,d\psi}{\sin\phi'\,d\phi'\,d\psi'} = \frac{d(\cos\phi)}{d(\cos\phi')} = 1-2\frac{v}{c}\cos\phi'.$$

Using these two relations and (18), we get from (14′)

$$\rho(\nu',\phi') = \left[\rho(\nu)+\frac{v}{c}\nu'\cos\phi'\frac{\partial\rho(\nu)}{\partial\nu}\right]\left(1-3\frac{v}{c}\cos\phi'\right). \tag{19}$$

From (19) and our hypothesis about the spontaneous emission and the induced processes of the molecule, we can easily calculate the average momentum transferred per unit time to the molecule. Before doing this we must, however, say something to justify the method used. One can object that the equations (14), (15), and (16) are based upon Maxwell's theory of the electromagnetic field which is incompatible with quantum theory. This objection, however, touches the form rather than the essence of the matter. Whatever the form of the theory of electromagnetic processes, surely in any case the Doppler principle and the aberration law will remain valid, and thus also equations (15) and (16). Furthermore, the validity of the energy relation (14) certainly extends beyond that of the wave theory; this transformation law is, for instance, also valid according to the theory of relativity for the energy density of a mass having an infinitesimal rest mass and moving with (quasi-) light-velocity. We can thus claim the validity of equation (19) for any theory of radiation.

The radiation corresponding to the spatial angle $d\kappa'$ will according to (B) lead per second to

$$B_n^m \rho'(\nu', \phi') \frac{d\kappa'}{4\pi}$$

induced elementary processes of the type $Z_n \to Z_m$, provided the molecule is brought back to the state Z_n immediately after each such elementary process. In reality, however, the time spent per second in the state Z_n is according to (5) equal to

$$\frac{1}{S} p_n e^{-\varepsilon_n/kT},$$

where we used the abbreviation

$$S = p_n e^{-\varepsilon_n/kT} + p_m e^{-\varepsilon_m/kT}. \tag{20}$$

In actual fact the number of these processes per second is thus

$$\frac{1}{S} p_n e^{-\varepsilon_n/kT} B_n^m \rho'(\nu', \phi') \frac{d\kappa'}{4\pi}.$$

In each process the momentum

$$\frac{\varepsilon_m - \varepsilon_n}{c} \cos \phi'$$

is transferred to the molecule in the direction of the positive X'-axis. Similarly, we find, using (B′) that the corresponding number of induced elementary processes of the kind $Z_m \rightarrow Z_n$ per second is

$$\frac{1}{S} p_m e^{-\varepsilon_m/kT} B_m^n \rho'(\nu', \phi') \frac{d\kappa'}{4\pi},$$

and in each such elementary process the momentum

$$-\frac{\varepsilon_m - \varepsilon_n}{c} \cos \phi'$$

is transferred to the molecule. The total momentum transferred per unit time to the molecule through induced processes is thus, taking (6) and (9) into account,

$$\frac{h\nu'}{cS} p_n B_n^m (e^{-\varepsilon_n/kT} - e^{-\varepsilon_m/kT}) \int \rho'(\nu', \phi') \cos \phi' \frac{d\kappa'}{4\pi},$$

where the integration is over all elements of solid angle. Performing the integration we get, using (19), the value

$$-\frac{h\nu}{c^2 S}\left(\rho - \tfrac{1}{3}\nu \frac{\partial \rho}{\partial \nu}\right) p_n B_n^m (e^{-\varepsilon_n/kT} - e^{-\varepsilon_m/kT})\, v.$$

Here we have denoted the frequency involved by ν (instead of ν').

This expression represents, however, the total average momentum transferred per unit time to a molecule moving with a velocity v; because it is clear that the spontaneous emission processes which take place without the action of radiation do not have a preferential direction, considered in the system K', so that they can on average not transfer any momentum to the molecule. We obtain thus as the final result of our considerations:

$$R = \frac{h\nu}{c^2 S}\left(\rho - \tfrac{1}{3}\nu \frac{\partial \rho}{\partial \nu}\right) p_n B_n^m e^{-\varepsilon_n/kT} (1 - e^{-h\nu/kT}). \tag{21}$$

6. Calculation of $\overline{\Delta^2}$

It is much simpler to calculate the influence of the irregularity of the elementary processes on the mechanical behaviour of the molecule, as we can base this calculation on a molecule at rest in the approximation which we have used from the start.

Let some event lead to the transfer of a momentum λ in the X-direction to a molecule. Let this momentum have varying sign and varying magnitude in different cases, but let there be such a statistical law for λ that the average value of λ vanishes. Let now $\lambda_1, \lambda_2, \ldots$ be the values of the momentum in the X-direction transferred to the molecule through several, independently acting causes so that the resultant transfer of momentum Δ is given by

$$\Delta = \Sigma \lambda_v.$$

As the average value $\overline{\lambda_v}$ vanishes for the separate λ_v, we must have

$$\overline{\Delta^2} = \Sigma \overline{\lambda_v^2}. \tag{22}$$

If the averages $\overline{\lambda_v^2}$ of the separate momenta are equal to one another $(=\overline{\lambda^2})$, and if l is the total number of momentum transferring processes, we have the relation

$$\overline{\Delta^2} = l\overline{\lambda^2}. \tag{22a}$$

According to our hypothesis in each elementary process, induced or spontaneous, the momentum

$$\lambda = \frac{h\nu}{c}\cos\phi$$

is transferred to the molecule. Here ϕ is the angle between the X-axis and a direction chosen randomly. Therefore we have

$$\overline{\lambda^2} = \frac{1}{3}\left(\frac{h\nu}{c}\right)^2. \tag{23}$$

As we assume that we may take all elementary processes which take place to be independent of one another, we may apply (22a). In that case, l is the number of all elementary processes taking place during the time τ. This is twice the number of the number of induced processes $Z_n \rightarrow Z_m$ during the time τ. We have thus

$$l = \frac{2}{S} p_n B_n^m e^{-\varepsilon_n/kT} \rho\tau. \tag{24}$$

We get from (23), (24), and (22)

$$\frac{\overline{\Delta^2}}{\tau} = \frac{2}{3S}\left(\frac{h\nu}{c}\right)^2 p_n B_n^m e^{-\varepsilon_n/kT} \rho. \tag{25}$$

7. Results

To prove now that the momenta transferred by the radiation to the molecules in accordance with our hypotheses never disturb the thermodynamic equilibrium, we only need to substitute the values (25) and (21) for $\overline{\Delta^2}/\tau$ and R which we have calculated after we have used (4) to replace in (21) the quantity

$$\left(\rho - \tfrac{1}{3}\nu\frac{\partial\rho}{\partial\nu}\right)(1 - e^{-h\nu/kT})$$

by $\rho h\nu/3kT$. It then turns out immediately that our fundamental equation (12) is identically satisfied.

The considerations which are now finished give strong support for the hypotheses given in Section 2 about the interaction between matter and radiation through absorption and emission processes, that is, through spontaneous and induced radiation processes. I was led to these hypotheses by my attempt to postulate as simply as possible a quantum theoretical behaviour of the molecules which would be similar to the behaviour of a Planck resonator of the classical theory. I obtained then in a natural fashion from the general quantum assumption for matter the second Bohr rule (equation (9)) as well as the Planck radiation formula.

The most important result seems to me, however, to be the one about the momentum transferred to the molecule in spontaneous or induced radiation processes. If one of our assumptions about this momentum transfer is changed, this would lead to a violation of equation (12); it seems hardly possible to remain in agreement with this relation which is required by the theory of heat otherwise than on the basis of our assumptions. We can thus consider the following as rather certainly proved.

If a ray of light causes a molecule hit by it to absorb or emit through an elementary process an amount of energy $h\nu$ in the form of radiation (induced radiation process), the momentum $h\nu/c$ is always transferred to the molecule, and in such a way that the momentum is directed along the direction of propagation of the ray if the energy is absorbed, and directed in the opposite direction, if the energy is emitted. If the molecule is subjected to the action of several directed rays of light, always only one of them will participate in an induced elementary process; this ray alone defines then the direction of the momentum transferred to the molecule.

If the molecule undergoes a loss of energy of magnitude $h\nu$ without external influence, by emitting this energy in the form of radiation (spontaneous emission), this process is also a *directed one*. There is no emission in spherical waves. The molecule suffers in the spontaneous elementary process a recoil of magnitude $h\nu/c$ in a direction which is in the present state of the theory determined only by "chance".

These properties of the elementary processes required by equation (12) make it seem practically unavoidable that one must construct an essentially quantum theoretical theory of radiation. The weakness of the theory lies, on the one hand, in the fact that it does not bring any nearer the connexion with the wave theory and, on the other hand, in the fact that it leaves moment and direction of the elementary processes to "chance"; all the same, I have complete confidence in the reliability of the method used here.

Still one more general remark may be made here. Practically

all theories of black-body radiation are based on a consideration of the interaction between radiation and molecules. However, in general one restricts oneself to considering energy-exchange, without taking momentum-exchange into account. One feels easily justified in this as the smallness of the momenta transferred by the radiation entails that these momenta are practically always in reality negligible compared to other processes causing a change in motion. However, for the *theoretical* discussion, these small actions must be considered to be completely as important as the obvious actions of the *energy*-exchange through radiation, as energy and momentum are closely connected; one can, therefore, consider a theory to be justified only when it is shown that according to it the momenta transferred by the radiation to the matter lead to such motion as is required by the theory of heat.

References

1. A. Einstein, *Verh. Dtsch. Phys. Ges.* **18**, 318 (1916).
2. K. von Mosengeil, *Ann. Physik* **22**, 867 (1907).

Molecular Microwave Oscillator and New Hyperfine Structure in the Microwave Spectrum of NH_3†

J. P. Gordon, H. J. Zeiger,* and C. H. Townes
Department of Physics, Columbia University, New York, New York
(Received May 5, 1954)

An experimental device, which can be used as a very high resolution microwave spectrometer, a microwave amplifier, or a very stable oscillator, has been built and operated. The device, as used on the ammonia inversion spectrum, depends on the emission of energy inside a high-Q cavity by a beam of ammonia molecules. Lines whose total width at half-maximum is six to eight kilocycles have been observed with the device operated as a spectrometer. As an oscillator, the apparatus promises to be a rather simple source of a very stable frequency.

A block diagram of the apparatus is shown in Fig. 1. A beam of ammonia molecules emerges from the source and enters a system of focusing electrodes. These electrodes establish a quadrupolar cylindrical electrostatic field whose axis is in the direction of the beam. Of the inversion levels, the upper states experience a radial inward (focusing) force, while the lower states see a radial outward force. The molecules arriving at the cavity are then virtually all in the upper states. Transitions are induced in the cavity, resulting in a change in the cavity power level when the beam of molecules is present. Power of varying frequency is transmitted through the cavity, and an emission line is seen when the klystron frequency goes through the molecular transition frequency.

If the power emitted from the beam is enough to maintain the field strength in the cavity at a sufficiently high level to induce transitions in the following beam, then self-sustained oscillations will result. Such oscillations have been produced. Although the power level has not yet been directly measured, it is estimated at about 10^{-8} watt. The frequency stability of the oscillation promises to compare favorably with that of other possible varieties of "atomic clocks."

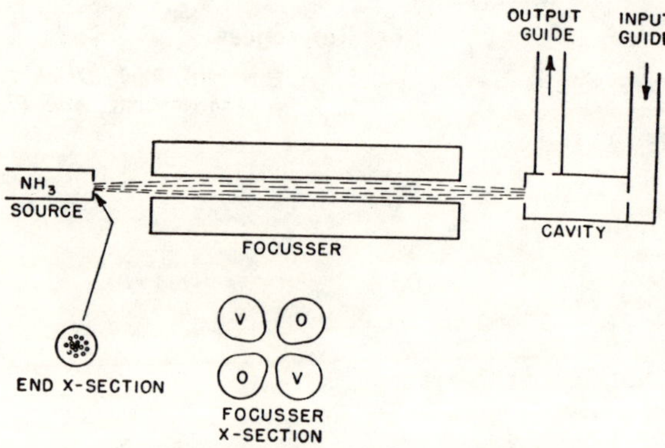

Fig. 1. Block diagram of the molecular beam spectrometer and oscillator.

Under conditions such that oscillations are not maintained, the device acts like an amplifier of microwave power near a molecular resonance. Such an amplifier may have a noise figure very near unity.

High resolution is obtained with the apparatus by utilizing the directivity of the molecules in the beam. A cylindrical copper cavity was used, operating in the $TE011$ mode. The molecules, which travel parallel to the axis of the cylinder, then see a field which varies in amplitude as $\sin(\pi x/L)$, where x varies from 0 to L. In particular, a molecule traveling with a velocity v sees a field varying with time as $\sin(\pi vt/L)\sin(\Omega t)$, where Ω is the frequency of the rf field in the cavity. A Fourier analysis of this field, which the molecule

Reprinted with permission from *Phys. Rev.*, vol. 95, pp. 282–284, July 1, 1954.

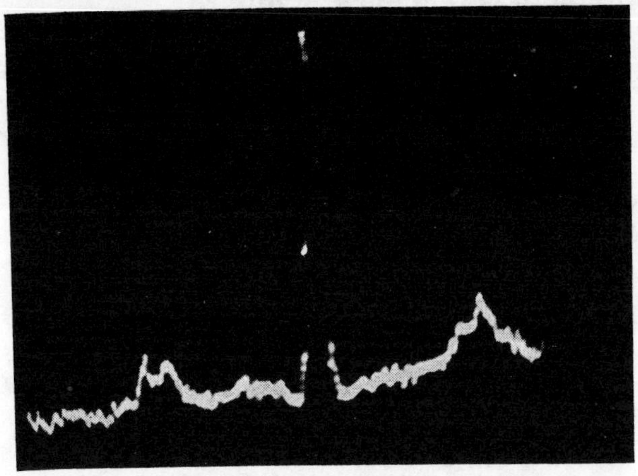

FIG. 2. A typical oscilloscope photograph of the NH_3, $J=K=3$ inversion line at 23 870 Mc/sec, showing the resolved magnetic satellites. Frequency increases to the left.

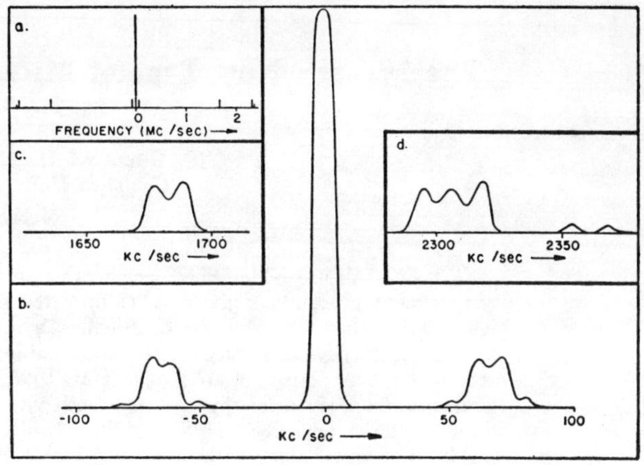

FIG. 3. The observed hyperfine spectrum of the 3,3 inversion line. (a) Complete spectrum, showing the spacings of the quadrupole satellites. (b) Main line with magnetic satellites. (c) Structure of the inner quadrupole satellites. (d) Structure of the outer quadrupole satellites. The quadrupole satellites on the low-frequency side of the main line are the mirror images of those shown, which are the ones on the high-frequency side.

sees from $t=0$ to $t=L/v$, gives a frequency distribution whose amplitude drops to 0.707 of its maximum at points separated by a $\Delta\nu$ of $1.2v/L$. The cavity used was twelve centimeters long, and the most probable velocity of ammonia molecules in a beam at room temperature is 4×10^4 cm/sec. Since the transition probability is proportional to the square of the field amplitude, the resulting line should have a total width at half-maximum given by the above expression, which in the present case is 4 kc/sec. The observed line width of 6–8 kc/sec is close to this value.

The hyperfine structure of the ammonia inversion transitions for $J=K=2$ and $J=K=3$ has been examined, and previously unresolved structure due to the reorientation of the hydrogen spins has been observed. Figure 2 is a typical scope photograph of these new magnetic satellites on the 3,3 line. The observed spectra for the 3,3 line is shown in Fig. 3, which contains all the observed hyperfine structure components, including the quadrupole reorientation transitions of the nitrogen nucleus, which have been previously observed as single lines.

Within the resolution of the apparatus, the hyperfine structures of the upper and lower inversion levels are identical, as evidenced by the fact that the main line is not split. Symmetry considerations require that the hydrogen spins be in a symmetric state under 120-degree rotations about the molecular axis. Thus for the 3,3 state, $I_H=3/2$, and one expects each of the quadrupole levels to be further split into four components by the interaction of the hydrogen magnetic moments with the various magnetic fields of the molecule. At the present writing, the finer details of the expected magnetic splittings have not been worked out.

This type of apparatus has considerable potentialities as a more general spectrometer. Since the effective dipole moments of molecules depend on their rotational state, some selection of rotational states could be effected by such a focuser. Similarly, a focuser using magnetic fields would allow spectroscopy of atoms. Sizable dipole moments are required for a strong focusing action, but within this limitation, the device may prove to have a fairly general applicability for the detection of transitions in the microwave region.

The authors would like to acknowledge the expert help of Mr. T. C. Wang during the latter stages of this experiment.

† Work supported jointly by the Signal Corps, the U. S. Office of Naval Research, and the Air Force.
* Carbide and Carbon post-doctoral Fellow in Physics, now at Project Lincoln, Massachusetts Institute of Technology, Cambridge, Massachusetts.

The Maser—New Type of Microwave Amplifier, Frequency Standard, and Spectrometer*†

J. P. GORDON,‡ H. J. ZEIGER,§ AND C. H. TOWNES
Columbia University, New York, New York

(Received May 4, 1955)

A type of device is described which can be used as a microwave amplifier, spectrometer, or oscillator. Experimental results are given. When operated as a spectrometer, the device has good sensitivity, and, by eliminating the usual Doppler broadening, a resolution of 7 kc/sec has been achieved. Operated as an oscillator, the device produced a frequency stable to at least 4 parts in 10^{12} in times of the order of a second, and stable over periods of an hour or more to at least a part in 10^{10}. The device is examined theoretically, and results are given for the expected sensitivity of the spectrometer, the stability and purity of the oscillation, and the noise figure of the amplifier. Under certain conditions a noise figure approaching the theoretical limit of unity, along with reasonably high gain, should be attainable.

INTRODUCTION

A TYPE of device is described below can be used as a microwave spectrometer, a microwave amplifier, or as an oscillator. As a spectrometer, it has good sensitivity and very high resolution since it can virtually eliminate the Doppler effect. As an amplifier of microwaves, it should have a narrow band width, a very low noise figure and the general properties of a feedback amplifier which can produce sustained oscillations. Power output of the amplifier or oscillator is small, but sufficiently large for many purposes.

The device utilizes a molecular beam in which molecules in the excited state of a microwave transition are selected. Interaction between these excited molecules and a microwave field produces additional radiation and hence amplification by stimulated emission. We call an apparatus utilizing this technique a "maser," which is an acronym for "microwave amplification by stimulated emission of radiation."

Some results obtained with this device have already been briefly reported.[1] An independent proposal for a system of this general type has also been published.[2] We shall here examine in some detail the general behavior and characteristics of the maser and compare experimental results with theoretical expectations. Particular attention is given to its operation with ammonia molecules. The preceding paper,[3] which will hereafter be referred to as (I), discusses an investigation of the hyperfine structure of the microwave spectrum of $N^{14}H_3$ with this apparatus. Certain of its properties which are necessary for an understanding of the relative intensities of the hyperfine structure components are also discussed there.

BRIEF DESCRIPTION OF OPERATION

A molecular beam of ammonia is produced by allowing ammonia molecules to diffuse out a directional source consisting of many fine tubes. The beam then transverses a region in which a highly nonuniform electrostatic field forms a selective lens, focusing those molecules which are in upper inversion states while defocusing those in lower inversion states. The upper inversion state molecules emerge from the focusing field and enter a resonant cavity in which downward transitions to the lower inversion states are induced. A simplified block diagram of this apparatus is given in Fig. 1. The source, focuser, and resonant cavity are all enclosed in a vacuum chamber.

For operation of the maser as a spectrometer, power of varying frequency is introduced into the cavity from an external source. The molecular resonances are then observed as sharp increases in the power level in the cavity when the external oscillator frequency passes the molecular resonance frequencies.

At the frequencies of the molecular transitions, the beam amplifies the power input to the cavity. Thus the maser may be used as a narrow-band amplifier. Since the molecules are uncharged, the usual shot noise existing in an electronic amplifier is missing, and essentially no noise in addition to fundamental thermal noise is present in the amplifier.

If the number of molecules in the beam is increased beyond a certain critical value the maser oscillates. At the critical beam strength a high microwave energy density can be maintained in the cavity by the beam alone since the power emitted from the beam compensates for the power lost to the cavity walls and coupled wave guides. This oscillation is shown both experimentally and theoretically to be extremely monochromatic.

---

* Work supported jointly by the Signal Corps, the Office of Naval Research, and the Air Research and Development Command.

† Submitted by J. P. Gordon in partial fulfillment of the requirements of the degree of Doctor of Philosophy at Columbia University.

‡ Now at the Bell Telephone Laboratories, Inc., Murray Hill, New Jersey.

§ Carbide and Carbon Postdoctoral Fellow in Physics, now at Project Lincoln, Massachusetts Institute of Technology, Cambridge, Massachusetts.

[1] Gordon, Zeiger, and Townes, Phys. Rev. **95**, 282 (1954).

[2] N. G. Bassov and A. M. Prokhorov, J. Exptl. Theoret. Phys. (U.S.S.R.) **27**, 431 (1954). Also N. G. Bassov and A. M. Prokhorov, Proc. Acad. of Sciences (U.S.S.R.) **101**, 47 (1945).

[3] J. P. Gordon, preceding paper [Phys. Rev. **99**, 1253 (1955)].

Reprinted with permission from *Phys. Rev.*, vol. 99, pp. 1264–1274, Aug. 15, 1955.

APPARATUS

The geometrical details of the apparatus are not at all critical, and so only a brief description of them will be made. Two ammonia masers have been constructed with somewhat different focusers. Both have operated satisfactorily.

A source designed to create a directional beam of the ammonia molecules was used. An array of fine tubes is produced in accordance with a technique described by Zacharias,[4] which is as follows. A $\frac{1}{4}$ in. wide strip of 0.001-in. metal foil (stainless steel or nickel, for example) is corregated by rolling it between two fine-toothed gears. This strip is laid beside a similar uncorregated strip. The corregations then form channels leading from one edge of the pair of strips to the other. Many such pairs can then be stacked together to create a two-dimensional array of channels, or, as was done in this work, one pair of strips can be rolled up on a thin spindle. The channels so produced were about 0.002 in. by 0.006 in. in cross section. The area covered by the array of channels was a circle of radius about 0.2 in., which was about equal to the opening into the focuser. Gas from a tank of anhydrous ammonia was maintained behind this source at a pressure of a few millimeters of mercury.

This type of source should produce a strong but directed beam of molecules flowing in the direction of the channels. It proved experimentally to be several times more effective than a source consisting of one annular ring a few mils wide at a radius of 0.12 in., which was also tried.

The electrodes of the focuser were arranged as shown in Fig. 1. High voltage is applied to the two electrodes marked V, while the other two are kept at ground. Paul *et al.*[5,6] have used similar magnetic pole arrangements for the focusing of atomic beams.

In the first maser which was constructed the inner faces of the electrodes were shaped to form hyperbolas with 0.4-in. separating opposing electrodes. The distance of closest approach between adjacent electrodes was 0.08 in., and the focuser was about 22 in. long. Voltages up to 15 kv could be applied to these electrodes before sparking occurred. In the second maser the electrodes were shaped in the same way, but were separated from each other by 0.16 in. This allowed voltages up to almost 30 kv to be applied, and somewhat more satisfactory operation was obtained since higher field gradients could be achieved in the region between the electrodes. This second focuser was only 8 in. long. Teflon spacers were used to keep the electrodes in place. To provide more adequate pumping of the large amount of ammonia released into the vacuum system from the source the focuser electrodes were hollow and were filled with liquid nitrogen.

The resonant cavities used in most of this work were circular in cross section, about 0.6 in. in diameter by 4.5 in. long, and were resonant in the TE_{011} mode at the frequency of interest (about 24 kMc/sec). Each cavity could be turned over a range of about 50 Mc/sec by means of a short section of enlarged diameter and variable length at one end. A hole 0.4 in. in diameter in the other end allowed the beam to enter. The beam traversed the length of the cavity. The cavities were made long to provide a considerable time for the molecules to interact with the microwave field. Only one-half wavelength of the microwave field in the cavity in the axial direction was allowed for reasons which will appear later in the paper. Since the free space wavelength of 24-kMc/sec microwaves is only about 0.5 in., and an axial wavelength of about 9 in. was required in the cavity, the diameter of the cavity had to be very close to the cut-off diameter for the TE_{01} mode in circular wave guide. The diameter of the beam entrance hole was well beyond cutoff for this mode and so very little loss of microwave power from it was encountered. The cavities were machined and mechanically polished. They were made of copper or silver-plated Invar, and had values of Q near 12 000. Some work was also done with cavities in the TM_{01} mode which has some advantages over the TE_{01} mode. However, the measurements described here all apply to the TE_{011} cavities.

Microwave power was coupled into and out from the cavities in several ways. Some cavities had separate input and output wave guides, power being coupled into the cavity through a two-hole input in the end of the cavity furthest from the source and coupled out through a hole in the sidewall of the cavity. In other cavities the sidewall hole served as both input and output, and the end-wall coupling was eliminated. About the same spectroscopic sensitivity was obtained with both types of cavities.

Three MCF 300 diffusion pumps (Consolidated Vacuum Company, Inc.) were used to maintain the necessary vacuum of less than 10^{-5} mm Hg. Nevertheless, due to the large volume of gas released into the system through the source, satisfactory operation has

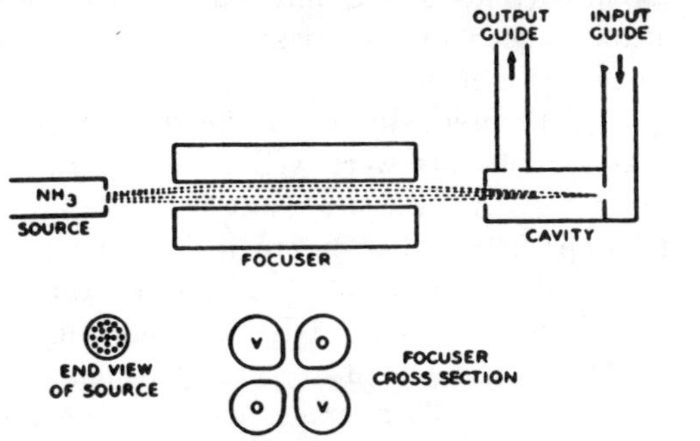

FIG. 1. Simplified diagram of the essential parts of the maser.

[4] J. R. Zacharias and R. D. Haun Jr., Quarterly Progress Report, Massachusetts Institute of Technology Research Laboratory of Electronics, 34, October, 1954 (unpublished).
[5] H. Friedberg and W. Paul, Naturwiss. 38, 159 (1951).
[6] H. G. Bennewitz and W. Paul, Z. Physik 139, 489 (1954).

not yet been attained without cooling the focuser electrodes with liquid nitrogen. At 78°K the vapor pressure of ammonia is considerably less than 10^{-6} mm Hg and so the cold electrode surfaces provide a large trapping area which helps maintain a sufficiently low pressure in the vacuum chamber. The pumping could undoubtedly be accomplished by liquid air traps alone; however the diffusión pumps alone have so far proven insufficient. The solidified ammonia which builds up on the focuser electrodes is somewhat of a nuisance as electrostatic charges which distort the focusing field tend to build up on it, and crystals form which can eventually impede the flow of gas. For the relatively short runs, however, which are required for spectroscopic work, this arrangement has been fairly satisfactory.

EXPERIMENTAL RESULTS

Experimental results have been obtained with the maser as a spectrometer and as an oscillator. Although it has been operated as an amplifier, there has as yet been no measurement of its characteristics in this role. Its properties as an amplifier are examined theoretically below.

The reader is referred to (I) for the results obtained from an examination of the hyperfine structure of the $N^{14}H_3$ inversion spectrum with the maser. Resolution of about seven kc/sec was obtained, which is a considerable improvement over the limit of about[7] 65 kc/sec imposed by Doppler broadening in the usual absorption-cell type of microwave spectrometer. This resolution can be improved still further by appropriate cavity design. The sensitivity of the maser was considerably better than that of other spectrometers which have had comparably high resolution.[8–10]

The factors which determine the sensitivity and resolution of the maser spectrometer are discussed in detail below, but we may make a general comment here. The sensitivity of the maser depends in part on the physical separation of quantum states by the focuser and thus on the forces exerted by the focuser on molecules in the various quantum states. For this reason its sensitivity is not simply related to the gas absorption coefficient for a given molecular transition. Each individual case must be examined in detail. Due to the focuser, for example, the sensitivity of the maser varies more rapidly with the dipole moment of the molecule to be studied than does that of the ordinary absorption spectrometer.

The experimental results obtained with the maser in its role as an oscillator agree with the theory given below and show that its oscillation is indeed extremely monochromatic, in fact more monochromatic than any other known source of waves. Oscillations have been produced at the frequencies of the 3–3 and 2–2 inversion lines of the ammonia spectrum, those for the 3–3 line being the stronger. Tests of the oscillator stability were made using the 3–3 line, so we shall limit the discussion to oscillation at this frequency. Other ammonia transitions, or transitions of other molecules could, of course, be used to operate a maser oscillator.

The frequency of the $N^{14}H_3$ 3–3 inversion transition is 23 870 mc/sec. The maser oscillation at this frequency was sufficiently stable in an experimental test so that a clean audio-frequency beat note between the two masers could be obtained. This beat note, which was typically at about 30 cycles per second, appeared on an oscilloscope as a perfect sine wave, with no random phase variations observable above the noise in the detecting system. The power emitted from the beams during this test was not measured directly, but is estimated to be about 5×10^{-10} watt.

The test of the oscillators was made by combining signals from the two maser oscillators together in a 1N26 crystal detector. A heterodyne detection scheme was used, with a 2K50 klystron as a local oscillator and a 30-Mc/sec intermediate-frequency (IF) amplifier. The amplified intermediate frequency signals from the two maser oscillators were then beat together in a diode detector, and their difference, which was then a direct beat between the two maser oscillator frequencies, displayed on an oscilloscope. The over-all band width of this detecting system was about 2×10^4 cps, and the beat note appeared on the oscilloscope with a signal to noise ratio of about 20 to 1.

It was found that the frequency of oscillation of each maser could be varied one or two kc/sec on either side of the molecular transition frequency by varying the cavity resonance frequency about the transition frequency. If the cavity was detuned too far, the oscillation ceased. The ratio of the frequency shift of the oscillation to the frequency shift of the cavity was almost exactly equal to the ratio of the frequency width of the molecular response (that is, the line width of the molecular transition as seen by the maser spectrometer) to the frequency width of the cavity mode. This behavior is to be expected theoretically as will be shown below. The two maser oscillators were well enough isolated from one another so that the beat note could be lowered to about 20 cps before they began to lock together. The appearance of this beat note has been noted above. As perhaps $\frac{1}{10}$-cycle phase variation could have been easily detected in a time of a second (which is about the time the eye normally averages what it observes), the appearance of the beat indicates a spectral purity of each oscillator of at least 0.1 part in 2.4×10^{10}, or 4 parts in 10^{12} in a time of the order of a second.

By using Invar cavities maintained in contact with ice water to control thermal shifts in their resonant frequencies, the oscillators were kept in operation for

[7] Gunther-Mohr, White, Schawlow, Good, and Coles, Phys. Rev. 94, 1184 (1954).
[8] G. Newell and R. H. Dicke, Phys. Rev. 83, 1064 (1951).
[9] R. H. Romer and R. H. Dicke, Phys. Rev. 98, 1160(A) (1955).
[10] M. W. P. Strandberg and H. Dreicer, Phys. Rev. 94, 1393 (1954).

periods of an hour or so with maximum variations in the beat frequency of about 5 cps or 2 parts in 10^{10} and an average variation of about one part in 10^{10}. Even these small variations seemed to be connected with temperature changes such as those associated with replenishing the liquid nitrogen supply in the focusers. Theory indicates that variations of about 0.1°C in temperature, which was about the accuracy of the temperature control, would cause frequency deviations of just this amount.

It was found that the oscillation frequency was slightly dependent on the source pressure and the focuser voltage, both of which affect the strength of the beam. These often produced frequency changes of the order of 20 cycles per second when either voltage or pressure was changed by about 25%. As the cavity was tuned, however, both these effects changed direction, and the null points for the two masers coincide to within about 30 cps. The frequency at which these effects disappear is probably very near the center frequency of the molecular response, so this may provide a very convenient way of resetting the frequency of a maser oscillator without reference to any other external standard of frequency.

THE FOCUSER

In (I) it was shown that forces are exerted by the nonuniform electric field of the focuser on the ammonia molecules, the force being radially inward toward the focuser axis for molecules in upper inversion states and radially outward for molecules in lower inversion states. Molecules in upper inversion states are therefore focused by the field, and only these molecules reach the cavity. Moreover, the quadrupole hyperfine splitting of the upper inversion state was shown to affect the focusing since the flight of the molecules through the focuser is adiabatic with respect to transitions between the different quadrupole levels. As a result the higher energy quadrupole levels are focused considerably more strongly than the lower energy ones. The further slight splitting of the various quadrupole states by the magnetic hyperfine interactions of the hydrogen nuclei has little effect since the molecules make many transitions between these closely spaced levels as they enter and leave the focuser. In regions of high field strength where hyperfine effects are unimportant and can be neglected the energy of the molecules in an electric field may be written as

$$W = W_{\text{rotation}}(J,K) \pm \left\{ \left(\frac{h\nu_0}{2}\right)^2 + \left(\frac{M_J K}{J(J+1)}\mu\mathcal{E}\right)^2 \right\}^{\frac{1}{2}}, \quad (1)$$

where ν_0 is the zero-field inversion frequency, J, K, and M_J specify the rotational state of the molecule relative to the direction of the field, μ is the molecular dipole moment, and $\mathcal{E}$ is the magnitude of the electric field.

With these considerations in mind, an approximate calculation of the total number of molecules in the upper inversion state which are trapped by the potential well of the focuser and delivered to the cavity is fairly straightforward. It involves some computation, since the line used for the oscillation (the main line of the $J=K=3$ inversion transition) is composed of three different but unresolved component transitions between quadrupole sublevels of the inversion states, and therefore the number of molecules trapped by the focuser must be calculated for each of these three sublevels and the results added. This calculation is outlined below. We shall consider in detail the properties of the first maser oscillator, with which the work reported in (I) was done.

The focuser electrodes form approximate equipotentials of the potential $V = V_0 r^2 \cos 2\theta$, where r and θ are cylindrical coordinates of a system whose z axis coincides with the axis of the focuser. 15 kv applied to the high-voltage focuser electrodes establishes an electric field whose magnitude is given by

$$\mathcal{E} = 200r, \quad (2)$$

where $\mathcal{E}$ is measured in esu and r is in cm. For simplicity we shall assume that the source is small in area and is located on the axis of the focuser. We shall also assume that all molecules which can travel farther than 0.5 cm from the focuser axis collide with the focuser electrodes and are removed from the beam. From (1) and (2) it is seen that the force ($\mathbf{f} = -\text{grad}W$) on the molecules is radial, and for small field strength is proportional to r. Furthermore it can be seen from energy considerations that all molecules which emerge from the source with radial velocity v_r less than $v_{\max}$, where $\frac{1}{2}mv_{\max}^2 = W(r=0.5\text{ cm}) - W(r=0)$ are held within the focuser by the electric field, while all molecules whose radial velocity is greater than $v_{\max}$ collide with the electrodes. Since $v_{\max}$ is a function of M_J (M_J is the projection of $\mathbf{J}$ on the direction of the electric field of the focuser) the number of molecules focused from a given zero-field quadrupole level depends on the high field distribution of these molecules among the various possible M states.

From kinetic theory, the number of molecules per second emerging from a thin-walled source of area S with radial velocity less than $v_{\max}$ is given by

$$N = PSv_0\Omega/(2\pi)^{\frac{1}{2}}kT, \quad (3)$$

where P is the source pressure, $v_0 = (kT/m)^{\frac{1}{2}}$ is the most probable velocity of molecules in the beam, T is the absolute temperature, and Ω is a solid angle defined by $\Omega = \pi(v_{\max}/v_0)^2$. The number of molecules per second in a given quadrupole level which are focused is therefore

$$N(F_1) = \frac{PSv_0}{(2\pi)^{\frac{1}{2}}kT} f(JKF_1) \sum_{M_J} \varphi(F_1 M_J)\Omega(M_J), \quad (4)$$

where $f(JKF_1)$ is the fraction of molecules emerging from the source in the quadrupole state characterized by J, K, and F_1 ($\mathbf{F}_1=\mathbf{J}+\mathbf{I}_N$, where I_N is the spin of the nitrogen nucleus), and $\varphi(F_1, M_J)$ is that fraction of these molecules which, according to the discussion in (I), go adiabatically into the state characterized by the quantum number M_J as they enter the high electric field of the focuser. The total number of molecules per second in the upper inversion state which are delivered to the cavity by the focuser is then just the sum of the $N(F_1)$ for the three quadrupole levels, and so is

$$N(J,K)=\frac{PSv_0}{(2\pi)^{\frac{1}{2}}kT}f(JK)\Omega(JK), \qquad (5)$$

where

$$\Omega(JK)=\sum_{F_1}\frac{f(JKF_1)}{f(JK)}\sum_{M_J}\varphi(F_1M_J)\Omega(M_J)$$

is an average solid angle for the upper inversion state, and $f(JK)$ is the fraction of molecules emerging from the source in the upper inversion state of the JK rotational level.

If each of these N molecules could be induced to make a transition to the lower inversion state while in the resonant cavity the total power delivered by the beam would be just $N(JK)h\nu_0$. Actually only about 50 to 75% average transition probability for the molecules in the beam can be obtained due to the variation of their velocities and spatial orientations. Assuming 50% transition probability, a source temperature of 300°K, and the geometry and voltage of the focuser given above, a calculation of the solid angle for the 3-3 line gives $\Omega(3\text{-}3)=4\times10^{-3}$ steradian, and available power of 1.5×10^{-9} watt per square millimeter of source area at 1 mm Hg source pressure.

It is estimated that the total number of molecules emerging from the source in the solid angle from which the upper inversion state molecules are selected is about 10^{15} per second. This estimate comes from knowledge of the number of molecules necessary to induce oscillation. This indicates that the present source is operating fairly inefficiently.

RESONANT CAVITY AND LINE WIDTH

The beam of molecules which enters the resonant cavity is almost completely composed of molecules in the upper inversion state. During their flight through the cavity the molecules are induced to make downward transitions by the rf electric field existing in the cavity. The transition probability for any particular molecule at low field strengths is given from first-order perturbation theory by

$$P_{ab}=\hbar^{-2}|\mu_{ab}|^2\left|\int_0^{L/v}\mathcal{E}(t)e^{-i2\pi\nu_0t}dt\right|^2, \qquad (6)$$

where μ_{ab} is the dipole matrix element for the transition, L is the length of the cavity, v is the velocity of the molecule, and $\mathcal{E}(t)$ is the rf electric field at the position of the molecule.

An average transition probability $\bar{P}_{ab}$ can be obtained for all molecules in the beam by averaging over the various velocities, trajectories, and values of $|\mu_{ab}|$ for the molecules in the several states which contribute to each spectral line. The power emitted from the beam is then just

$$P=Nh\nu\bar{P}_{ab}, \qquad (7)$$

so $\bar{P}_{ab}$ as a function of the frequency of the applied field determines the line width of the molecular response.

Under the simplifying assumptions that the molecules all travel axially down the length of the cavity, that their velocity is uniform and equal to v_0, and that the cavity is a perfect cylinder with only one-half wavelength of rf field in the axial direction, we find that the emitted power has a maximum at the natural transition frequency ν_B, and a total width at half-maximum of $1.2v_0/L$. If the field is assumed to be uniform along the axis rather than one-half of a sine wave, the corresponding total width at half-maximum is $0.9v_0/L$. This line width of about v_0/L can alternatively be obtained from the uncertainty principle and the finite time of interaction of the molecules with the rf field. Thus $\Delta\nu\approx1/\Delta t$, where Δt is the time of flight of the molecule in the cavity, or L/v_0. The identity of the "Doppler broadening" of the spectral line has essentially disappeared. The sharpness of the molecular response as opposed to that obtained in the usual spectrometer may alternatively be attributed to the long wavelength of the rf field in the cavity in the direction of travel of the beam. If the cavity is excited in a mode in which there is more than one-half wavelength in the direction of travel of the beam, then the molecular emission line, as given by Eq. (6), has two peaks symmetrically spaced about the transition frequency. The frequency separation of these peaks can be associated with the Doppler shift.

Equations (6) and (7) show that for small rf field strengths the emitted power P is proportional to $\mathcal{E}_{max}^2$ and thus to the energy stored in the cavity. For larger field strengths, of course, the molecular transitions begin to saturate and Eq. (6) is no longer sufficient for the calculation of P_{ab}. The effects of this saturation will be considered in detail in a later paper; however, we can say that P_{ab} must certainly be less than 1, and that if a high field strength is maintained in the cavity then the average transition probability $\bar{P}_{ab}(\nu_B)$ will be about 0.5. The total power available from the beam is therefore about $Nh\nu_B/2$. Power saturation in this case is rather similar to that in the usual molecular beam experiment, for which it has been considered by Torrey.[11]

Associated with the power emitted from the beam is an anomalous dispersion, that is, a sharp variation

[11] H. C. Torrey, Phys. Rev. 59, 293 (1941).

in the dielectric constant of the cavity medium due to the beam. These two effects can be considered at the same time by thinking of the beam as a polarizable medium introduced into the cavity, whose average electric susceptibility is given by $\chi=\chi'+i\chi''$. The power emitted from the beam can then be shown directly from Maxwell's equations to be[12]

$$P=8\pi^2\nu_B W\chi'', \tag{8}$$

where W is the energy stored in the cavity. Thus, from Eqs. (7) and (8), χ'' is related to $\bar{P}_{ab}$ by

$$\chi''=Nh\bar{P}_{ab}/8\pi^2W. \tag{9}$$

The value of χ' is given from χ'' by Kramer's relation,[13] which for a sharp resonance line can be approximated by[14]

$$\chi'(\nu)=\frac{1}{\pi}\int_0^\infty \frac{\chi''(\nu')d\nu'}{\nu'-\nu}. \tag{10}$$

Figure 2 shows the form of χ' and χ'', calculated with the assumptions that all molecules are traveling parallel to the axis of the cavity with uniform velocity, that the cavity is excited in the TE_{011} mode, and that there is a small field strength in the cavity so that $P_{ab}\ll 1$ and Eq. (6) is valid. χ' and χ'' can also be found directly by calculation of the induced dipole moments of the molecules as they traverse the cavity.

If the Q of the cavity is defined in terms of net power loss (i.e., $dW/dt=-2\pi\nu W/Q_C$) then the presence of the beam can be considered as causing a change in the effective Q_C given by $1/Q_{CB}=1/Q_C-4\pi\chi''$, where Q_{CB} and Q_C are respectively the cavity Q's with and without the beam, along with a shift in the resonant frequency of the cavity given by $\nu_{CB}=\nu_C(1-2\pi\chi')$ if $\chi'\ll 1$. These relations can also be easily derived directly from Maxwell's equations, and they will prove important in determining the properties of the maser.

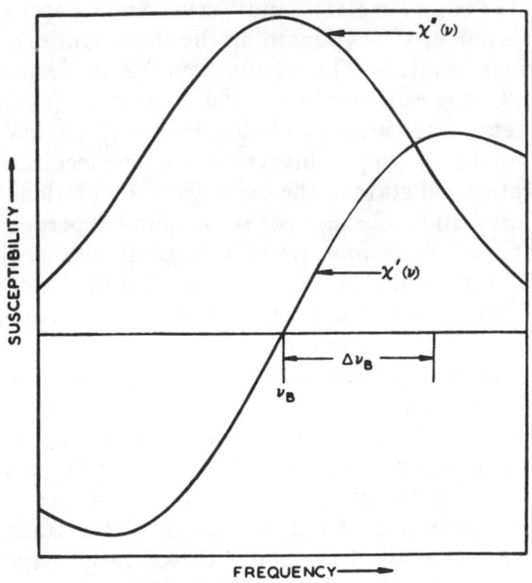

FIG. 2. Real and imaginary parts of the electric susceptibility χ of the molecular beam in the cavity. $\chi=\chi'+i\chi''$.

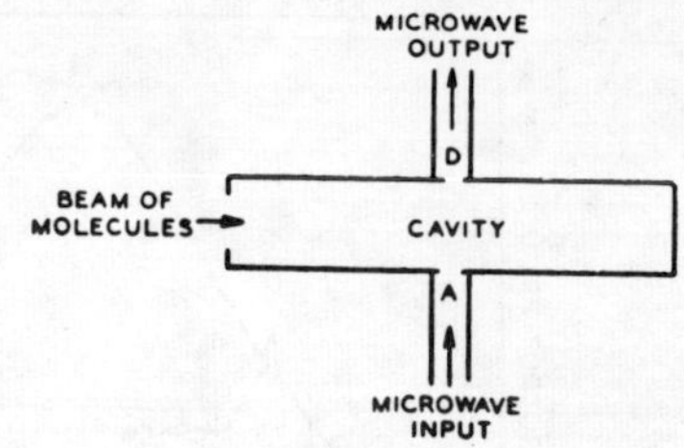

FIG. 3. Schematic diagram of the resonant cavity and molecular beam.

THE MASER SPECTROMETER

Observed Line Shape as a Function of the Cavity Resonant Frequency

Consider the situation shown in Fig. 3. Power P_0 is incident on the cavity from wave guide A, and the power transmitted on out through wave guide D is detected as a function of the frequency of the input power. The power transmitted through the cavity in the absence of the beam is given by[12]

$$P_D(\nu)=\frac{P_0}{Q_AQ_D}\bigg/\left[\left(\frac{1}{2Q_L}\right)^2+\left(\frac{\nu-\nu_C}{\nu_C}\right)^2\right], \tag{11}$$

where Q_A and Q_D are defined in terms of the power losses from the cavity to wave guides A and D respectively, and Q_L is the loaded Q of the cavity, given by $1/Q_L=1/Q_C+1/Q_A+1/Q_D$. As was shown in the last section, the change in $P_D(\nu)$ caused by the presence of the beam can be described through variations in Q_C and ν_C near the transition frequency ν_B. Thus in the presence of the beam we find P_D modified to

$$P_{DB}(\nu)=\frac{P_0}{Q_AQ_D}\bigg/\left[\left(\frac{1}{2Q_L}-2\pi\chi''(\nu)\right)^2+\left(\frac{\nu-\nu_C+2\pi\nu_C\chi'(\nu)}{\nu_C}\right)^2\right]. \tag{12}$$

As long as the power output P_{DB} is not so high that nonlinearities in the molecular response are important, (12) gives the output power as a function of frequency in the presence of the beam and represents the spectrum which may be observed. For most spectroscopic applications we are interested in the case for which $\chi''(\nu)\ll 1/4\pi Q_L$ for all ν. For this case, an appropriate expansion of Eq. (12) shows that if $\nu_C=\nu_B$, where ν_B is the center frequency of the molecular response, then the presence of the beam shows up as a pip of the shape

[12] J. C. Slater, *Microwave Electronics* (D. Van Nostrand Company, Inc., New York, 1950).
[13] J. H. Van Vleck, *Massachusetts Institute of Technology Radiation Laboratory Report*, 735, see also *Radiation Laboratory Series* (McGraw-Hill Book Company, Inc., New York, 1948), Vol. 13, Chap. 8.
[14] G. E. Pake and E. M. Purcell, Phys. Rev. 74, 1184 (1948).

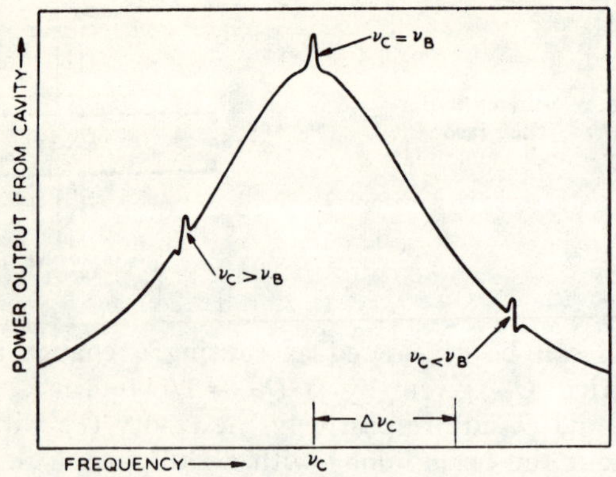

FIG. 4. Spectral line shapes as observed in the maser.

of χ'' superimposed on the cavity mode. The half-width $\Delta\nu_B$ of the molecular response is therefore defined as half the frequency separation of the half-maximum points of $\chi''(\nu)$. If the cavity frequency is altered so that the molecular line appears on the wings of the cavity mode, then the pip due to the beam assumes the shape of $\pm\chi'(\nu)$, the + or − sign depending on whether the line appears on the low or high frequency side of the cavity mode. These three situations are illustrated in Fig. 4.

Sensitivity of the Maser Spectrometer

We have here the usual problem occurring in microwave spectroscopy of detecting a small change in output power P_0 caused by the presence of the molecules. Assume that $\nu_C=\nu_B=\nu$ in Eq. (12). Then

$$P_{DB}=\frac{P_0}{Q_AQ_D}\bigg/\left(\frac{1}{2Q_L}-2\pi\chi''(\nu_B)\right)^2. \tag{13}$$

The change in output power which must be detected is then

$$\delta P_D=P_{DB}-P_D\cong P_0(4Q_L^2/Q_AQ_D)\times[8\pi Q_L\chi''(\nu_B)]. \tag{14}$$

Consider now the change in output voltage δV_D, which is proportional to $(P_D+\delta P_D)^{\frac{1}{2}}-P_D^{\frac{1}{2}}$ or to $\delta P_D/2P_D^{\frac{1}{2}}$ when $\delta P_D\ll P_D$. The noise voltage at the output of a linear detector is just $(FkT\Delta\nu)^{\frac{1}{2}}$, where $\Delta\nu$ is the band width of the detector and F is its over-all noise figure. Thus the voltage signal-to-noise ratio is given by

$$\delta V_D/V_N=[(\delta P_D)^2/4FkTP_D\Delta\nu]^{\frac{1}{2}}. \tag{15}$$

As long as the change in power δP_D due to the beam increases linearly with the power level P_D, the signal to noise ratio from (15) continues to increase. Hence the power input to the cavity should be increased until the transition begins to saturate. For saturation and a given Q_C, it can be shown that maximum sensitivity is achieved by using a small input coupling ($Q_A\gg Q_C$) and a matched output coupling ($Q_D=Q_C$). If we approximate the saturation condition by setting W equal to the level at which the power emitted by the beam is just $\frac{1}{2}Nh\nu_B$, then with a little algebra we find, from Eqs. (8), (14), and the known relationship of W to P_0,[12] that the change in output power δP_D is just $\frac{1}{2}Nh\nu_B$ (the beam emits $\frac{1}{2}Nh\nu_B$ of power, and due to the change in the input match caused by the beam, $\frac{1}{2}Nh\nu_B$ more power enters the cavity through the input coupling hole A. Thus the increase in power input to the cavity is twice the power emitted from the beam. Half of this increase emerges into wave guide D since it was assumed to be matched to the cavity. The power level P_D is now determined by the required energy W from the relation

$$P_D=2\pi\nu_BW/Q_C. \tag{16}$$

This, with (8), gives

$$P_D=P/4\pi\chi''(\nu_B)Q_C=Nh\nu_B/8\pi\chi''(\nu_B)Q_C. \tag{17}$$

Inserting these values for P_D and δP_D in (15) yields

$$\delta V_D/V_N=[\pi Q_CNh\nu_B\chi''(\nu_B)/2FkT\Delta\nu]^{\frac{1}{2}}. \tag{18}$$

This relation gives the sensitivity of the maser, once the value of $\chi''(\nu_B)$ for a given molecule is calculated. χ'' is, of course, related to the average transition probability $\bar{P}_{ab}$ by Eq. (9), so that Eq. (18) can easily be rewritten in terms of the transition probability.

For the ammonia 3–3 line, a calculation of the number of molecules in the 3–3 state necessary to make $\delta V_D/V_N=1$ was done, assuming $F=100$, $\Delta\nu=1$ cps, $Q_C=12\,000$, $T=300°$ and using an approximate calculation of $\chi''(\nu_B)$ based on the considerations of the previous section. The result was 10^9 molecules per second. It is estimated from the value of $\chi''(\nu_B)$ which is necessary to cause oscillation (see next section) that the number of upper inversion state molecules in the 3–3 rotational state in the beam when oscillations occur is at least 10^{13} molecules per second, and experimentally a number about four times this great was achieved. Thus, for ammonia, the maser should have good sensitivity, and the results described in paper (I) show that this is indeed the case.

In the case of the ammonia inversion spectrum, the focuser can effect an almost complete separation of the upper states from the lower states of the transitions. For some other transitions, this ideal state of affairs may not be attainable, but yet the focuser may preferentially focus one of the two states of the transition. In such a case all of the above considerations apply so long as one uses for N just the excess number of molecules in one of the two states. It is, of course, unimportant for spectroscopic purposes whether the more highly focused state is the upper or lower state of the transition. The high sensitivity attained in the observation of the ammonia spectrum with the maser gives promise that it may be generally useful as a microwave spectrometer of very high resolution.

THE MASER OSCILLATOR AND AMPLIFIER

By extending the considerations of the previous section to include amplification of the thermal noise which exists in the cavity, we can discuss the properties of the maser as an oscillator or amplifier. The results of this analysis, which is made below, are as follows:

(1) The center frequency ν_0 of the oscillation is given to a good approximation by the equation

$$\nu_0 = \nu_B + \frac{\Delta\nu_B}{\Delta\nu_C}(\nu_C - \nu_B), \tag{19}$$

where $\Delta\nu_C$ and $\Delta\nu_B$ are respectively the half-widths of the cavity mode and of the molecular emission line; and $\nu_C - \nu_B$ is the difference between the cavity resonant frequency ν_C and the line frequency ν_B.

(2) The total width at half-power of the spectral distribution of the oscillation is approximately

$$2\delta\nu = 8\pi kT(\Delta\nu_B)^2/P_B \tag{20}$$

where T is the temperature and P_B is the power emitted from the beam. Inserting in (20) values which approximate the experimental conditions, $T = 300°K$, $\Delta\nu_B = 3\times10^3$ cps, $P_B = 10^{-10}$ watt, we find $2\delta\nu \approx 10^{-2}$ cps, or $\nu_B/\delta\nu = 5\times10^{12}$.

(3) If the beam is sufficiently strong, the maser may be used as an amplifier with a gain greater than unity and a noise figure very close to unity.

The argument goes along the following lines. Consider the situation of Fig. 3, the cavity with two wave guides. The whole system will be assumed to be in thermal equilibrium in the absence of the beam. Noise power of amount kT per unit band width is incident on the cavity from each wave guide, and the cavity walls emit noise power within the cavity. Of the noise power incident on the cavity from wave guide A, a certain amount within the frequency range of the cavity mode enters the cavity; part of this power is then absorbed by the cavity walls and part is transmitted on out through wave guide D. A similar situation holds for noise power incident on the cavity from wave guide D; some is absorbed in the cavity and some is transmitted through to wave guide A. The cavity walls emit noise power in the region of the cavity mode and some of this power goes out through each wave guide. When the beam is not present we have assumed the system to be in thermal equilibrium, so there must be kT per unit band width of noise power flowing away from the cavity down each wave guide and there must be kT of noise energy in the cavity mode, as required by the equipartition theorem.

In the presence of the beam thermal equilibrium is upset. The beam, since it is composed solely of upper inversion state molecules, and since the probability for spontaneous decay of these molecules to the lower states is negligible during the time they take to traverse the cavity, contributes no random noise of its own to the rf field of the cavity. What it does is merely to amplify, in a way described by its effect on the loaded Q and resonant frequency of the cavity, all of the noise signals which exist in the cavity. Thus to the noise sources in the wave guides the intrinsic Q of the cavity seems to have been altered; whereas to the noise source within the cavity, the loading on the cavity seems to have changed. In fact, the presence of the beam can be duplicated in the imagination by attaching to the cavity a third wave guide, with a negative Q equal to $\frac{1}{4}\pi\chi''(\nu)$ describing its coupling to the cavity, and by simultaneously shifting the resonant frequency of the cavity by an amount $-2\pi\nu_C\chi'(\nu)$.

From these considerations we will show that in the presence of the beam more than kT of power per unit bandwidth travels down each wave guide away from the cavity. The extra power, of course, comes from the beam. At a certain critical beam intensity this power suddenly becomes large, corresponding to sustained oscillations.

Let $\Delta\nu$ be some arbitrarily small element of the frequency spectrum at frequency ν. Within this range noise power of magnitude $kT\Delta\nu$ is incident on the cavity from each wave guide, independent of ν. Let $P_A\Delta\nu$ be the amount of noise power which enters the cavity from the incident power in wave guide A, and let $P_A'\Delta\nu$ be the total noise power re-emitted into wave guide A from inside the cavity. The presence of the beam will be indicated by an added subscript; i.e., P_{AB} will represent the value of P_A when the beam is present, etc. Similar definitions apply to the output guide D. Since the noise powers generated in the wave guides and in the cavity are completely incoherent with one another, we can simply add power coming from various sources to obtain the total power in any element of the system. Thus the total energy $W\Delta\nu$ stored in the cavity is merely the sum $\sum_i W_i\Delta\nu$ of all the energies due to power coming from the various different power sources. (W is energy per unit band width.)

Consider now the flow of power when no beam is present. The noise power entering the cavity from wave guide A is given by[12]

$$P_A = \frac{kT}{Q_A Q_L^A}\Bigg/\left[\left(\frac{1}{2Q_L}\right)^2 + \left(\frac{\nu-\nu_C}{\nu_C}\right)^2\right], \tag{21}$$

where $1/Q_L^x = \sum_{m\neq x} 1/Q_m$ is just proportional to all the losses from the cavity except that due to Q_x. Of this power P_A, some is absorbed in the cavity walls, and the rest is transmitted on out through wave guide D. The energy per unit band width stored in the cavity due to this power input is

$$W_A = P_A Q_L^A/2\pi\nu$$
$$= \frac{kT}{2\pi\nu Q_A}\Bigg/\left[\left(\frac{1}{2Q_L}\right)^2 + \left(\frac{\nu-\nu_C}{\nu_C}\right)^2\right], \tag{22}$$

and the power transmitted on to the output wave guide D is $2\pi\nu W_A/Q_D$, or $P_A Q_L^A/Q_D$. Similar expressions hold for the noise power incident on the cavity from wave guide D. Furthermore, the cavity loss associated with Q_C may be assumed to be due to a third wave guide with coupling characterized by Q_C to a perfectly conducting cavity, so that the energy W_C emitted into the actual cavity from its wall has the same form as (22). The total energy stored in the cavity per unit frequency interval is hence

$$W=W_A+W_D+W_C$$

$$=\frac{kT}{2\pi\nu Q_L}\Big/\left[\left(\frac{1}{2Q_L}\right)^2+\left(\frac{\nu-\nu_C}{\nu_C}\right)^2\right], \quad (23)$$

where

$$1/Q_L=1/Q_A+1/Q_D+1/Q_C.$$

The total energy stored in the cavity, given by $\int_0^\infty W d\nu$, is easily shown to be equal to kT (we make the assumption that $Q_L \gg 1$, so that in the integration the approximation $\nu \approx \nu_C$ may be made) as required by the equipartition theorem. The net noise power flowing in the wave guides A or D is also easily shown to be zero, so that the system is indeed seen to be in thermal equilibrium.

Consider now the case when the beam is present. The noise power incident from each wave guide sees a cavity whose rates of internal loss has been reduced by an amount $4\pi\chi''(\nu)$ by the energy emitted from the beam and whose resonant frequency has been shifted by an amount $-2\pi\nu_C\chi'(\nu)$. Corresponding to Eq. (21), the power entering the cavity from wave guide A in the presence of the beam is

$$P_{AB}=\frac{kT}{Q_{LB}^A Q_A}\Big/\left[\left(\frac{1}{2Q_{LB}}\right)^2+\left(\frac{\nu-\nu_{CB}}{\nu_{CB}}\right)^2\right], \quad (24)$$

where $1/Q_{xB}=1/Q_x-4\pi\chi''(\nu)$ for any x and $\nu_{CB}=\nu_C[1-2\pi\chi'(\nu)]$, while that entering from wave guide D is similarly

$$P_{DB}=\frac{kT}{Q_{LB}^D Q_D}\Big/\left[\left(\frac{1}{2Q_{LB}}\right)^2+\left(\frac{\nu-\nu_{CB}}{\nu_{CB}}\right)^2\right]. \quad (25)$$

The noise energy stored in the cavity due to these two sources is

$$W_{AB}=\frac{P_{AB}Q_{LB}^A}{2\pi\nu}, \text{ and } W_{DB}=P_{DB}Q_{LB}^D/2\pi\nu. \quad (26)$$

At the same time, the energy stored in the cavity due to its own internal noise source is changed as though the loading on the cavity has been altered while its internal loss was unaffected. This energy is therefore given by

$$W_{CB}=\frac{kT}{2\pi\nu Q_C}\Big/\left[\left(\frac{1}{2Q_{LB}}\right)^2+\left(\frac{\nu-\nu_{CB}}{\nu_{CB}}\right)^2\right]. \quad (27)$$

Due to the presence of the beam the net noise power emitted from the cavity into the output wave guide D is now no longer zero. The power emerging from the cavity is now

$$P_{DB}'=\frac{2\pi\nu}{Q_D}[W_{CB}+W_{AB}]$$

$$=\frac{kT}{Q_D Q_L^D}\Big/\left[\left(\frac{1}{2Q_{LB}}\right)^2+\left(\frac{\nu-\nu_{CB}}{\nu_{CB}}\right)^2\right]. \quad (28)$$

Thus the additional noise output in wave guide D due to the beam is, from (25) and (28),

$$P_{DN}\equiv P_{DB}'-P_{DB}$$

$$=\frac{kT}{Q_D}4\pi\chi''(\nu)\Big/\left[\left(\frac{1}{2Q_{LB}}\right)^2+\left(\frac{\nu-\nu_{CB}}{\nu_{CB}}\right)^2\right]. \quad (29)$$

The power which must be emitted from the beam to give this amount of power in wave guide D is just

$$P_N\equiv P_{DN}Q_D/Q_L=\frac{kT}{Q_L}4\pi\chi''(\nu)\Big/$$

$$\left[\left(\frac{1}{2Q_L}-2\pi\chi''(\nu)\right)^2+\left(\frac{\nu-\nu_C+2\pi\nu_C\chi'(\nu)}{\nu_C}\right)^2\right], \quad (30)$$

where ν_C has replaced ν_{CB} in the denominator of the last term in the denominator of this expression since $\nu_{CB}\approx\nu_C$. Note that (30) is just equivalent to (24) if the beam is thought of as a wave guide coupled to the cavity with a Q of $-1/4\pi\chi''$.

Expression (30) gives the complete spectrum of the power emitted from the beam due to amplification of the noise signals which are always present in the cavity. The necessary condition for the existence of oscillations as some cavity frequency is evidently that $\chi''(\nu_B)\approx\frac{1}{4}\pi Q_L$.

Assume that the cavity is tuned so that $|\nu_C-\nu_B| \ll \Delta\nu_C$, and then let the beam strength slowly increase so that χ'' increases. Then at the critical beam strength where $\chi''(\nu_B)\rightarrow 1/4\pi Q_L$, the total power $\int P_N d\nu$ emitted from the beam approaches infinity accordingly to (30). Obviously, the total power emitted from the beam cannot go to infinity, but is limited to about $\frac{1}{2}Nh\nu_B$. When the power level in the cavity reaches the point at which the molecular transition begins to saturate, χ'' and χ' become functions of the power level, and, of course, vary in such a way that $\int P_N d\nu$ is always less than $\frac{1}{2}Nh\nu_B$. We can, for simplicity, avoid the problem of dealing with this saturation merely by increasing χ'' until $\int P_N d\nu=\frac{1}{2}Nh\nu_B$ and examining the frequency spectrum of the power emitted from the beam at this level of output. Although χ'' is not independent of the electric field strength when saturation occurs, it varies much more slowly with time than does the oscillation, so that it may be considered constant in treating the short-term behavior of the microwave field.

As the critical number of molecules is reached, P_N becomes very large at frequencies very close to ν_B. Hence it is appropriate to expand χ' and χ'' about the center frequency ν_B. This gives approximately

$$\chi'=\chi_0\left(\frac{\nu-\nu_B}{\Delta\nu_B}\right)+\cdots,$$
$$\chi''=\chi_0\left[1-\frac{1}{2}\left(\frac{\nu-\nu_B}{\Delta\nu_B}\right)^2+\cdots\right]. \tag{31}$$

Writing Eq. (30) in terms of (31), and setting $\int P_N d\nu$ equal to P_B, where $P_B=\frac{1}{2}Nh\nu_B$, one obtains

$$P_N\approx 4kT(\Delta\nu_B)^2\bigg/\left[(\nu-\nu_0)^2+\left(\frac{4\pi kT}{P_B}(\Delta\nu_B)^2\right)^2\right], \tag{32}$$

where ν_0, the oscillation frequency, is given by the equation

$$\frac{\nu_0-\nu_B}{\nu_0-\nu_C}=-\frac{\Delta\nu_B}{\Delta\nu_C}, \tag{33}$$

or, as in (19),

$$\nu_0=\nu_B+(\nu_B-\nu_C)\Delta\nu_B/\Delta\nu_C \quad \text{if} \quad \Delta\nu_B/\Delta\nu_C\ll 1.$$

The total width $2\delta\nu$ at half-maximum power of this "noise" output is, from (32),

$$2\delta\nu=(8\pi kT/P_B)(\Delta\nu_B)^2 \tag{34}$$

as already stated in (20).

It should be remembered that (32) involves the assumption that the maser is a linear noise amplifier of very high gain. Actually, the noise properties of an oscillator depend to a considerable extent on the nonlinearities in its response, or the overload, and (32) does not accurately represent the precise noise spectrum of the maser as an oscillator. However, the approximate width of its noise spectrum is properly given by (34). As in the more usual types of oscillators, this oscillator actually maintains a nearly fixed amplitude of oscillation, but its phase slowly varies with time in a random way, corresponding to a noise spectrum of a width given approximately by (34). A more detailed discussion of noise will be given in a later publication.

The half-width of this oscillation signal is not to be confused with the half-width of the molecular response $\Delta\nu_B$. The latter represents the band width of the maser amplifier at low gain, whereas the former gives the band width of the oscillation signal. The oscillation frequency ν_0 can be varied throughout the range over which the molecules will amplify in accordance with (33) or (19). Hence care must be taken to keep the cavity frequency ν_C constant if it is desired to keep the oscillation frequency constant for any extended period of time.

Noise Figure and Band Width of the Amplifier

The noise figure of the maser amplifier may be easily found from the results of the foregoing sections. Assume that $\nu_C=\nu_B=\nu$, where ν is the frequency of the signal to be amplified. Also assume that the detector has a band width $\Delta\nu_{\text{det}}$ such that $\Delta\nu_{\text{det}}\ll\Delta\nu_B$. Equation (13) gives the signal power at the cavity output, while Eq. (29) gives the noise at the output in excess of kT. Thus we see that the signal-to-noise ratio at the output is just

$$\frac{P_0\bigg/\left[Q_AQ_D\left(\frac{1}{2Q_L}-2\pi\chi_0\right)^2\right]}{kT\Delta\nu_{\text{det}}\left[1+\frac{4\pi\chi_0}{Q_D}\bigg/\left(\frac{1}{2Q_L}-2\pi\chi_0\right)^2\right]}, \tag{35}$$

where $\chi''(\nu_B)=\chi_0$. At the input to the cavity, the signal to noise ratio is $P_0/kT\Delta\nu_{\text{det}}$. Therefore the noise figure F, which is just the ratio of these two quantities, is

$$F=Q_AQ_D\left[\frac{4\pi\chi_0}{Q_D}+\left(\frac{1}{2Q_L}-2\pi\chi_0\right)^2\right]. \tag{36}$$

At the same time, the power amplification available is, from (13), given simply by

$$\mu=P_{DB}/P_0=\left[Q_AQ_D\left(\frac{1}{2Q_L}-2\pi\chi_0\right)^2\right]^{-1}. \tag{37}$$

It can be shown from (37) that $\mu<1$ if $4\pi\chi_0<1/Q_C$, i.e., if there is a net loss of power within the cavity itself. Thus unless it is possible to produce oscillation by putting lossless reflections in all the wave guides so that $Q_L\to Q_C$, it is also impossible to create an amplifier with a gain greater than unity. In order to obtain a large gain μ, one must have $1/Q_L\approx 4\pi\chi_0$. If the gain is large, then a noise figure approaching unity is attainable by making $1/Q_A\approx +4\pi\chi_0\approx 1/Q_L$. This shows that for high amplification and at the same time a low noise figure, a fairly large input coupling to the cavity and a small output coupling is needed. Furthermore a sufficiently strong beam is required so that the maser is not too far from oscillation.

The maser acts as a regenerative amplifier, as can be seen from (12). Thus under conditions such that $4\pi\chi_0\approx 1/Q_L$ so that the midband gain is high, the band width becomes substantially smaller than $2\Delta\nu_B$.

It might also be noted that a certain amount of modulation of the amplified output is to be expected due to random variations of the number of molecules in the cavity at any time. These effects, however, are proportional to the input signal strength, and so are quite different from thermal noise signals which have no dependence on input power. Furthermore, they represent a modulation of only about one part in 10^6 since there are 10^{12} or more molecules in the cavity at any time. This type of modulation can be neglected when small input signals are considered and is not important under most circumstances. This shot effect and also the effect of power flow through the cavity on the frequency dependence of the amplification will be discussed in more detail in a subsequent paper.

Amplification may also be accomplished using one wave guide as both input and output, and the noise figure of such an amplifier can also approach unity. The amplified output signal might be coupled out and detected through a directional coupler, which would have to have a fairly small coupling so that little of the input power was lost to it. Then so long as the amplified input noise appearing at the detector was large compared to kT, the noise figure of this amplifier would be small.

The maser amplifier may be useful in a restricted range of applications in spite of its narrow band width because of its potentially low noise figure. For example, suppose that the signal to be amplified came from outer space, where the temperature is only a few degrees absolute. Then by making the coupling through the cavity fairly large so that little noise is contributed by the cavity itself, amplification should be attainable while keeping the noise figure, *based on the temperature of the signal source*, fairly low. This might prove to have a considerable advantage over electronic amplifiers. It might also be possible to tune the frequency of a maser amplifier through the use of the Stark or Zeeman effects on the molecular transition frequencies.

ACKNOWLEDGMENTS

The authors would like to express their gratitude to the personnel of the Columbia Radiation Laboratory who assisted in the construction of the experimental apparatus. They would also like to thank Dr. T. C. Wang and Dr. Koichi Shimoda for their assistance and suggestions during the later stages of this work.

Electronic Structure of F Centers: Saturation of the Electron Spin Resonance

A. M. Portis
Department of Physics, University of California, Berkeley, California
(Received May 18, 1953)

It is shown that the unusual observed saturation behavior of the microwave electron spin resonance associated with F centers in KCl, NaCl, and KBr crystals can be accounted for if the overall width is ascribed to interaction between the F-center electrons and the nuclear magnetic moments of the ions adjacent to the F centers. The measured saturation factor gives for F centers in KCl a spin-lattice relaxation time of 2.5×10^{-6} sec at room temperature. The observed saturation behavior in which only the absorption saturates is in marked disagreement with the Kramers-Kronig relations. However it is shown, that the Kramers-Kronig relations are not applicable to saturated systems. Expressions which avoid the use of these relations are presented for saturable systems.

I. INTRODUCTION

IT is shown in this paper that the unusual saturation behavior of the microwave spin resonance absorption and dispersion associated with F centers in alkali halide crystals can be accounted for if the over-all width is caused by hyperfine interaction.[1] The details of the saturation behavior of a system depend markedly on the nature of the broadening mechanism. If the broadening arises from dipolar interaction between like spins or from interaction with the radiation field, then the thermal equilibrium of the spin system will be preserved during resonance absorption. This will also be true if the line width comes from some mechanism which is external to the spin system but is fluctuating rapidly compared with the time associated with a spin transition. This first case we call the ***homogeneous*** case. The consequence of homogeneous broadening is that the energy absorbed from the microwave field is distributed to all the spins and thermal equilibrium of the spin system is maintained through resonance. In an effort to understand the observed saturation results, our measurements were first compared with the behavior expected for this kind of broadening. This comparison is shown in Fig. 1. The observed and the theoretical behavior for a simple line of the observed shape were fitted so as to have the same slope at low microwave fields. It can be seen that the simple theory completely fails to account for our saturation results.

In attempting to account for our results we realized that if the line width were to come from variations in the local magnetic fields the physical response of the system would be markedly different. For this second case, which we call the inhomogeneous case, energy will be transferred only to those spins whose local fields satisfy the resonance condition. Further, the processes for spin-spin interaction will be slow as compared with the direct interaction of the spins with the lattice, since in order for spins in different local fields to come to equilibrium, energy will have to be transferred to the lattice. It is useful for this case to think of spin packets

[1] A preliminary account of this work was given by A. M. Portis and A. F. Kip, Bull. Am. Phys. Soc. 28, No. 2, 9 (1953).

Reprinted with permission from *Phys. Rev.*, vol. 91, pp. 1071–1078, Sept. 1, 1953.

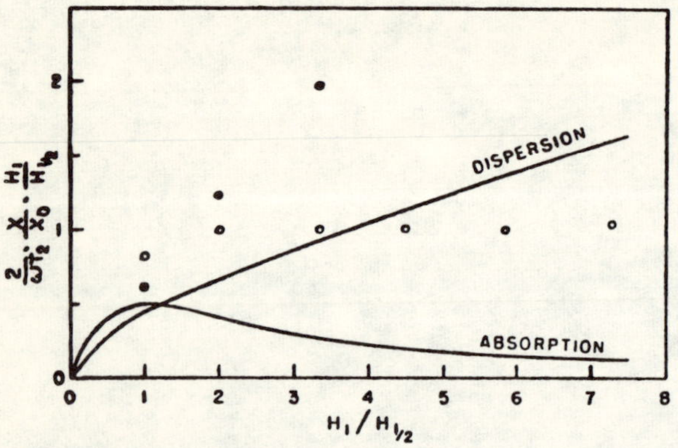

FIG. 1. Comparison between the observed saturation behavior of *F* centers in KCl and the simple theory of Sec. II.

having little or no interaction with each other and of width given by the simple dipole-dipole interaction. Then the over-all response of the spin system will be a superposition of the individual responses of the spin packets. It is clear that the broadening of the spin resonance in alkali halides where the width is from the distribution in hyperfine fields is of the inhomogeneous type. The excellent agreement between our results and the predicted saturation behavior for this case is shown in Sec. V.

By studying the way in which the system saturates at high-microwave fields it is possible to determine the spin-lattice relaxation time. Of course, for a steady-state experiment, one does not obtain directly the spin lattice relaxation time, T_1, but rather the product T_1T_2, where T_2 is the time associated with the width of the spin packets. This case is closely related to one considered by Bloembergen, Purcell, and Pound[2] where the over-all width comes from inhomogeneities in the applied magnetic field. At high-microwave fields the relation between absorption and dispersion deviated from the relationship developed by Kramers and Kronig.[3] This deviation has led here to a re-examination of the derivation of these relations. It has been found that certain assumptions made in the derivation regarding the behavior of the complex susceptibility are not valid for saturable systems. This problem is discussed in Sec. VI and Appendix I, where a scheme which introduces the Kramers-Kronig relations only in a restricted way is presented.

II. THE ELEMENTARY THEORY OF SATURATION

We consider here the saturation behavior of a system for which the broadening is of the *homogeneous* type. For simplicity we consider single electron spins so that only on the application of an external magnetic field is the spin degeneracy lifted. We then have two states per electron of frequency separation:

$$\omega_0 = g(e/2mc)H_0, \tag{1}$$

where g is the spectroscopic splitting factor and H_0 the applied magnetic field. These spin levels are not perfectly defined but are somewhat broadened. Sources of *homogeneous* broadening include:

(*a*) Dipolar interaction between like spins;
(*b*) Spin-lattice relaxation;
(*c*) Interaction with the radiation field;
(*d*) Motion of carriers in the microwave field;
(*e*) Diffusion of excitation through the sample;
(*f*) Motionally narrowed fluctuations in the local field.

We define N^+ to be the number of spins in the higher-energy spin state and N^- the number in the lower-energy state. For the entire spin system in thermal equilibrium we have

$$N^+/N^- = \exp(-\hbar\omega_0/kT_s), \tag{2}$$

where T_s is the effective spin temperature. If the spin system is in thermal equilibrium with the lattice at some temperature T_l,

$$N^+ = N_0^+, \quad N^- = N_0^-,$$

and

$$N_0^+/N_0^- = \exp(-\hbar\omega_0/kT_l). \tag{3}$$

We consider now the deviations of the spin system from equilibrium with the lattice under the interaction of the spin system with the radiation field and with the lattice. If we let $n = N^- - N^+$:

$$dn/dt = (dn/dt)_{rf} + (dn/dt)_{sl}, \tag{4}$$

where $(dn/dt)_{rf}$ and $(dn/dt)_{sl}$ give the rate of change of the difference in populations from interaction with the radiation field and spin-lattice interaction respectively.

The rate of change of the difference in populations from interaction with the radiation field at a frequency ω is given by

$$(dn/dt)_{rf} = -\tfrac{1}{4}\pi\gamma^2 H_1^2 g(\omega-\omega_0)n, \tag{5}$$

where γ is the magnetomechanical ratio, H_1 the maximum amplitude of the microwave magnetic field, and $g(\omega-\omega_0)$ the normalized shape factor for the transition. Following Bloembergen,[2] we take for the interaction with the lattice:

$$(dn/dt)_{sl} = (n_0 - n)/T_1, \tag{6}$$

where $n_0 = N_0^- - N_0^+ \simeq (\hbar\omega_0/2kT_l)N$ and T_1 is the spin-lattice relaxation time.

If the microwave power incident on the sample is varying slowly as compared with T_1, we obtain a quasi-steady state condition for which

$$dn/dt \cong 0,$$

and

$$n = n_0/[1 + \tfrac{1}{4}\pi\gamma^2 H_1^2 T_1 g(\omega-\omega_0)]. \tag{7}$$

For the case where the microwave power is amplitude modulated at a frequency high compared with $1/T_1$,

[2] Bloembergen, Purcell, and Pound, Phys. Rev. 73, 679 (1948).
[3] R. de L. Kronig, J. Opt. Soc. Am. 12, 547 (1926); H. A. Kramers, Atti congr. intern. fis. Como 2, 545 (1927).

n will not be able to follow the variations in microwave power over a modulation cycle but will respond to the average power. These two cases are examined in detail in Appendix II.

We are concerned here with the rate at which energy is absorbed from the radiation field:

$$P_a = -\tfrac{1}{2}\hbar\omega(dn/dt)_{rf}. \tag{8}$$

Substituting from Eqs. (5) and (7), we find that

$$P_a = \tfrac{1}{2}\omega\left\{\tfrac{1}{2}\omega_0\left(\frac{\gamma^2\hbar n_0}{2\omega_0}\right)\cdot\frac{\pi g(\omega-\omega_0)}{1+\frac{1}{4}\pi\gamma^2H_1^2T_1g(\omega-\omega_0)}\right\}. \tag{9}$$

Noting that the complex part of the rf susceptibility is defined by

$$P_a = \tfrac{1}{2}\omega\chi''H_1^2, \tag{10}$$

and that the static spin susceptibility,

$$\chi_0 = \gamma^2\hbar n_0/2\omega_0, \tag{11}$$

we have

$$\chi''(\omega) = \tfrac{1}{2}\chi_0\omega_0\frac{\pi g(\omega-\omega_0)}{1+\frac{1}{4}\pi\gamma^2H_1^2T_1g(\omega-\omega_0)}. \tag{12}$$

We define T_2 such that at the line center

$$g(0) = T_2/\pi. \tag{13}$$

Then

$$\chi''(\omega_0) = \tfrac{1}{2}\chi_0\omega_0T_2\frac{1}{1+\frac{1}{4}\gamma^2H_1^2T_1T_2}. \tag{14}$$

The expression derived in Appendix I for the real part of the rf susceptibility is

$$\chi'(\omega) = \tfrac{1}{2}\chi_0\omega_0\frac{1}{1+\frac{1}{4}\pi\gamma^2H_1^2T_1g(\omega-\omega_0)} \times\int_0^\infty\frac{2\omega'g(\omega'-\omega_0)}{\omega'^2-\omega^2}d\omega'. \tag{15}$$

III. THE GENERAL THEORY OF SATURATION

We considered in Sec. II the saturation behavior for *homogeneous* broadening of the spin levels. However, as in spin resonance from *F* centers in alkali halides, the over-all width may be caused by factors which do not maintain the spin system in equilibrium. Here one must take the results of Sec. II as applying to a spin packet of width approximately $1/T_2$. These lines must be then assembled under an envelope the shape of which is given by the details of the inhomogeneous broadening as illustrated in Fig. 2. The distinction between homogeneous and inhomogeneous broadening is that the inhomogeneous broadening must come from interactions outside the spin system and must be slowly varying over the time required for a spin transition. Examples of inhomogeneous broadening include:

(*a*) Hyperfine interaction,
(*b*) Anisotropy broadening,
(*c*) Dipolar interaction between spins with different Larmor frequencies,
(*d*) Inhomogeneities in the applied magnetic field.

The distribution in local fields is given by $h(\omega-\omega_0)$, which is normalized so that

$$\int_0^\infty h(\omega-\omega_0)d\omega = 1.$$

We define T_2^* so that at the center of the distribution

$$h(0) = T_2^*/\pi. \tag{16}$$

Then we obtain in the general case for the absorption

$$\chi''(\omega) = \tfrac{1}{2}\chi_0\int_0^\infty\frac{\pi\omega'g(\omega-\omega')}{1+\frac{1}{4}\pi\gamma^2H_1^2T_1g(\omega-\omega')}h(\omega'-\omega_0)d\omega'. \tag{17}$$

The dispersion is given by

$$\chi'(\omega) = \tfrac{1}{2}\chi_0\int_0^\infty\frac{\omega'h(\omega'-\omega_0)d\omega'}{1+\frac{1}{4}\pi\gamma^2H_1^2T_1g(\omega-\omega')} \times\int_0^\infty\frac{2\omega''g(\omega''-\omega')d\omega''}{\omega''^2-\omega^2}. \tag{18}$$

As is shown in Appendix I, if the over-all broadening is large compared with the width of the spin packets, these expressions simplify to

$$\chi''(\omega) = \tfrac{1}{2}\chi_0\omega h(\omega-\omega_0)\int_0^\infty\frac{\pi g(\omega-\omega')}{1+\frac{1}{4}\pi\gamma^2H_1^2T_1g(\omega-\omega')}d\omega', \tag{19}$$

$$\chi'(\omega) = \tfrac{1}{2}\chi_0\int_0^\infty\frac{2\omega'^2h(\omega'-\omega_0)}{\omega'^2-\omega^2}d\omega'. \tag{20}$$

We note that the absorption line does not change shape on saturation since the integral in Eq. (19) will not be a function of ω. The details of the way in which

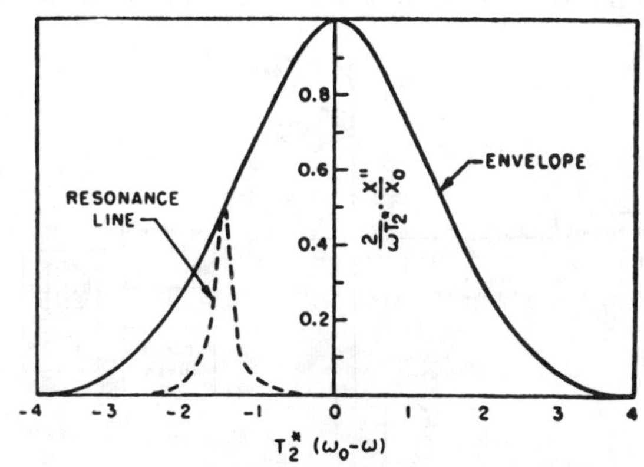

FIG. 2. Diagram illustrating the distinction between the resonance line for a spin packet and the envelope caused by inhomogeneous broadening.

$\chi''(\omega)$ changes in amplitude depend on the form of the shape function $g(\omega-\omega')$. As will be shown in Sec. V the observed behavior of the spin resonance is in good agreement with these results and allows further the determination of the shape of $g(\omega-\omega')$ even though it is masked by the over-all broadening.

IV. EXPERIMENTAL PROCEDURE

The experiments described in this and the preceding paper were performed at a frequency of approximately 8590 Mc/sec corresponding to a resonance field of about 3200 oersteds. A 2K39 reflex klystron nominally delivering 250 mw of microwave power was employed as the microwave source. The power was 100 percent modulated at 6 kc/sec by a microwave gyrator.[4] A simplified block diagram of the resonance equipment is shown in Fig. 3. The equipment for Pound stabilization, power monitoring, and power modulation of the microwave oscillator is not shown. The signal detected by a 1N23 crystal in the arm to the right of the magic-tee bridge is proportional to the difference in power reflected from the variable probe and the resonant cavity. In practice the bridge was first balanced by adjusting the position and depth of the variable probe until no microwave power reached the detecting crystal. The bridge was then unbalanced until about 1 μw of microwave power reached the detecting crystal.

One can unbalance the bridge in one of two ways, either by translating the probe along the guide or by decreasing the depth of its penetration into the guide. Translating the probe makes the bridge sensitive to variations in cavity frequency giving a signal at the crystal proportional to the dispersion. Raising the probe makes the bridge sensitive to variations in cavity Q giving a signal proportional to the absorption.

After unbalancing the bridge by raising the probe, the position of the probe along the guide was always readjusted for minimum power to the crystal in order to avoid a mixture of absorption and dispersion. In this way the amplitudes of the absorption and dispersion signals were compared over a range of 20 db in microwave power. After each power change the unbalance of the bridge was readjusted in order to maintain the same power on the detecting crystal. Using this procedure, it can easily be shown that the detected absorption and dispersion signals are proportional to $H_1\chi''$ and $H_1\chi'$ respectively.

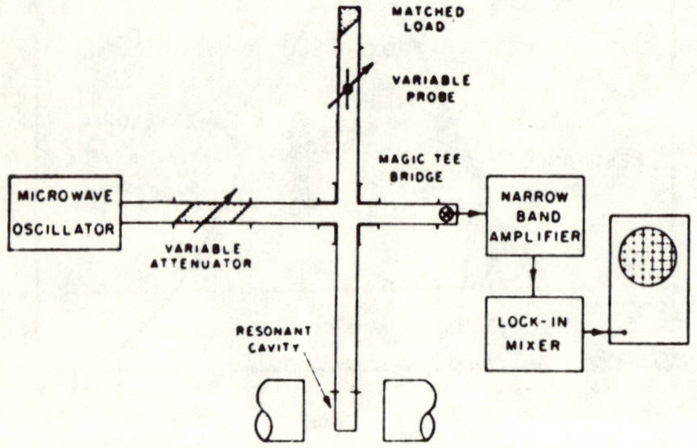

FIG. 3. Microwave apparatus.

[4] C. L. Hogan, Bell System Tech. J. 31, 1 (1952).

The microwave field in the cavity was calculated from measurements of the loaded Q of the cavity, the voltage standing wave ratio in the cavity arm, and the power incident on the cavity. The power reflection coefficient of the cavity,

$$|\Gamma|^2=[(\beta-1)/(\beta+1)]^2, \qquad (21)$$

where β is the voltage standing wave ratio and is equal to Q_e/Q_u, the ratio of the external Q to the unloaded Q of the cavity. It can be shown that the condition for optimum sensitivity is

$$\delta^2|\Gamma|^2/\delta Q_e\delta Q_u=0. \qquad (22)$$

From differentiation of Eq. (21) it follows that for optimum sensitivity, assuming square law detection, the coupling iris should be chosen so that one-third of the incident power is reflected from the cavity. Since

$$1/Q_l=1/Q_u+1/Q_e, \qquad (23)$$

where Q_l is the loaded Q, a measurement of Q_l and β will suffice to determine the unloaded Q of the cavity. If P is the power incident on the cavity it can be shown that

$$H_1^2=32\pi Q_u(1-|\Gamma|^2)P/\omega V, \qquad (24)$$

where V is the volume of the resonant cavity.

V. ANALYSIS OF RESULTS IN KCl

The saturation results in crystals of KCl, KBr, and NaCl were qualitatively similar and further work is in progress. We will describe here only the results on KCl, for which a detailed study has been made at room temperature. Since at ordinary power levels the dispersion signal is nearly five times as strong as the absorption signal, the former was used for the analysis of line shape. Figure 4 presents a comparison of the results obtained for KCl with the theoretical Gaussian and Lorentz line shapes. As shown the line is essentially Gaussian in shape, in good agreement with the theory of hyperfine interaction as presented in the preceding paper. It was found further that the line width and line shape of both the absorption and dispersion were independent of the microwave power level.

Figure 5 presents the saturation results for KCl as a function of H_1. The parameters plotted are the peak absorption and dispersion signals, suitably normalized. $H_{\frac{1}{2}}$ is the value of the microwave field such that the saturation parameter at the line center,

$$S(\omega_0, H_1)=1/(1+\tfrac{1}{4}\gamma^2H_1^2T_1T_2)=\tfrac{1}{2}. \qquad (25)$$

The results are compared with the theory of Sec. III with $g(\omega-\omega')$ taken to be Lorentz in shape.

It is clear from Fig. 1 that the results cannot be explained in any way on the basis of the simple theory of Sec. II. Equation (12) of that section gives for the absorption signal at the line center:

$$\frac{2}{\omega_0 T_2}\frac{\chi''(\omega_0)}{\chi_0}\frac{H_1}{H_{\frac{1}{2}}}=\frac{1}{1+(H_1/H_{\frac{1}{2}})^2}\frac{H_1}{H_{\frac{1}{2}}}. \tag{26}$$

For large H_1, on this basis, the absorption signal should go as $H_{\frac{1}{2}}/H_1$, as contrasted with the signal obtained, which is independent of H_1. Further, the theory of Sec. II predicts for a Gaussian line a flattening of the central portion of the line and a nearly twofold increase in line width over the range investigated. Neither of these effects was observed.

However, based on the theory of Sec. III, no change in line width or line shape should be expected. The saturation broadens and reduces the intensity of the absorption of the individual spin packets, but as long as the packets are narrow as compared with the inhomogeneous broadening of the resonance, this effect will be reflected only in a reduction in the intensity of the absorption. As shown in Appendix I, the dispersion is not affected by the saturation of the spin packets, in agreement with the observations on KCl.

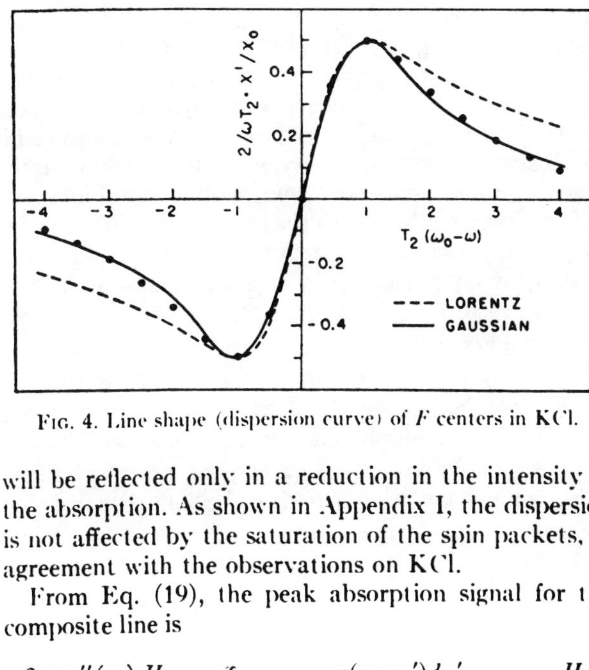

Fig. 4. Line shape (dispersion curve) of *F* centers in KCl.

From Eq. (19), the peak absorption signal for the composite line is

$$\frac{2}{\omega T_2^*}\frac{\chi''(\omega_0)}{\chi_0}\frac{H_1}{H_{\frac{1}{2}}}=\int_0^\infty\frac{g(\omega-\omega')d\omega'}{1+(H_1/H_{\frac{1}{2}})^2\pi[g(\omega-\omega')/T_2]}\frac{H_1}{H_{\frac{1}{2}}}. \tag{27}$$

We wish to determine the form of $g(\omega-\omega')$ such that $\chi''(\omega_0)$ will behave in the observed way as a function of H_1. We require that for $H_1>H_{\frac{1}{2}}$

$$\int_0^\infty\frac{g(\omega-\omega')d\omega'}{1+(H_1H_{\frac{1}{2}})^2\pi[g(\omega-\omega')/T_2]}\frac{H_1}{H_{\frac{1}{2}}}\cong 1. \tag{28}$$

If $g(\omega-\omega')$ represents a Lorentz line, that is

$$g(\omega-\omega')=(T_2/\pi)[1/1+T_2^2(\omega-\omega')^2], \tag{29}$$

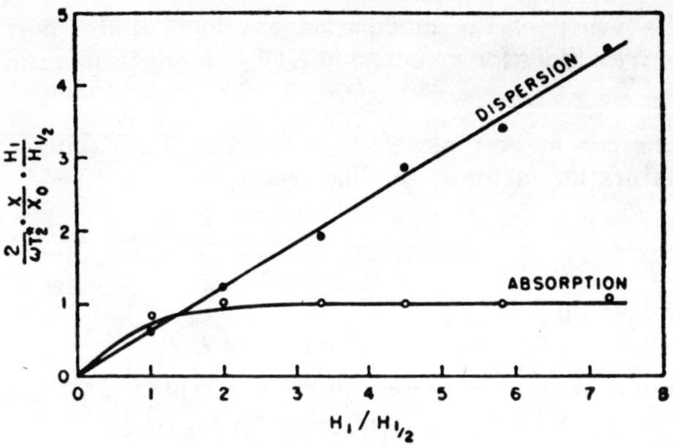

Fig. 5. Saturation behavior of *F* centers in KCl.

then the left side of Eq. (28) is equal to

$$\{1/[1+(H_1/H_{\frac{1}{2}})^2]^{\frac{1}{2}}\}(H_1/H_{\frac{1}{2}}), \tag{30}$$

which for large H_1 satisfies the required condition. If $g(\omega-\omega')$ were taken as cutting off more sharply in the wings than a Lorentz line, then the left side of Eq. (28) would be decreasing for $H_1>H_{\frac{1}{2}}$, in contrast with the observed behavior. On this basis we must conclude that the spin packets are very closely Lorentz in shape. This result is in good agreement with the arguments of Anderson[5] and the rigorous calculations of Kittel and Abrahams[5] of the frequency moments for dipolar broadening from paramagnetic ions randomly distributed on a simple cubic lattice. The result of Kittel and Abrahams is that, for a fractional magnetic population of less than 0.01, the resonance line is very nearly Lorentz with a half-width at half-maximum intensity,

$$\Delta H=5.3g\mu_B N, \tag{31}$$

where N is the concentration of paramagnetic ions. Substituting for N, the concentration of *F* centers in the sample studied as determined from the strength of the optical absorption in the *F* band, the value $2.9\times10^{17}/\text{cm}^3$, we obtain for the half-width of the spin packets:

$$\Delta H=0.028 \text{ oersted}.$$

Comparing this with the half-width at half-maximum intensity for the over-all line of 27 oersteds, it is clear that the assumption made that the packets be narrow as compared with the over-all line width is well justified.

From a study of the saturation behavior it is possible to estimate the spin-lattice relaxation time. The saturation parameter at the line center is defined as

$$S(\omega_0, H_1)=\chi''(\omega_0, H_1)/\chi''(\omega_0, 0). \tag{32}$$

For $H_1>H_{\frac{1}{2}}$ we find

$$S(\omega_0, H_1)\cong 0.012/H_1.$$

However, as shown in Appendix II, the form of the saturation factor will depend to some extent on whether $\omega_m T_1$ is less than or greater than unity, where ω_m is the

[5] P. W. Anderson, Phys. Rev. **82**, 342 (1952); C. Kittel and E. Abrahams, Phys. Rev. **90**, 238 (1953).

frequency of the modulation envelope of the microwaves. Since for F centers in KCl at room temperature,

$$\omega_m T_1 \sim 1,$$

one can at best make an estimate of T_1. Taking the saturation factor at the line center,

$$S(\omega_0, H_1) \simeq \frac{1.6}{\gamma(T_1T_2)^{\frac{1}{2}}} \frac{1}{H_1},$$

we obtain

$$T_1T_2 = 5.7\times 10^{-11}\ \text{sec}^2.$$

From the known concentration of F centers

$$T_2 = 1/\gamma\Delta H = 2.3\times 10^{-6}\ \text{sec.}$$

This gives for the spin-lattice relaxation time at room temperature

$$T_1 = 2.5\times 10^{-5}\ \text{sec.}^{6}$$

VI. ON THE KRAMERS-KRONIG RELATIONS

The Kramers-Kronig relations[3] give the form of the dispersion if the absorption is known for all frequencies and, conversely, the absorption if the dispersion is known,

$$\chi'(\omega) = \frac{2}{\pi}\int_0^\infty \frac{\omega'\chi''(\omega')}{\omega'^2-\omega^2}d\omega', \tag{33}$$

$$\chi''(\omega) = -\frac{2}{\pi}\omega\int_0^\infty \frac{\chi'(\omega')}{\omega'^2-\omega^2}d\omega'. \tag{34}$$

Our experimental results are in strong contradiction with the above relations. We find that at high power levels χ' is independent of power level but χ'' decreases in amplitude with increasing microwave power. Neither the absorption nor the dispersion vary in shape or width. This strong discrepancy led to a re-examination of the validity of these relations. Their derivation is presented clearly by Van Vleck,[7] and it is his presentation which will be examined here. The usual procedure is to define a complex susceptibility,

$$\chi(\omega') = \chi'(\omega') - i\chi''(\omega'), \tag{35}$$

where ω' is a complex frequency. One then integrates the function

$$\chi(\omega')/(\omega'-\omega),$$

where ω is real, along the real frequency axis closing the contour around the lower half of the complex frequency plane. Since for physically realizable systems the absorption is an even function of the frequency and the dispersion an odd function, and further stipulating that there be no singularities in the lower half of the complex plane, the relations Eqs. (33) and (34) are obtained.

It is in particular the stipulation that the complex susceptibility be analytic over the lower half of the complex frequency plane which we wish to examine here. Van Vleck establishes this condition on the basis of the linearity of the system. In considering the behavior of the complex dielectric constant, he expresses the electric polarization of the system in terms of a Fourier integral over the spectrum of the electrical disturbance by introducing a dielectric constant which is a function of frequency. It is concluded then that the dielectric constant can have no singularities in the lower half of the complex frequency plane, since if this were not the case one would have a response of the system preceding the arrival of the electrical disturbance. This situation is closely related to the impossibility of having exponentially increasing solutions for the response of a physical system in the absence of any excitation. We show here that if the system is nonlinear the treatment of Van Vleck does not apply, nor are his conclusions about the behavior of the system in the absence of any excitation valid. The susceptibility will in fact have singularities in the lower half of the complex frequency plane. The existence of these singularities is contingent on the presence of excitation, and it is in this important respect that linear and nonlinear systems differ. It does not follow then that for nonlinear systems the presence of singularities in the lower half of the complex frequency plane has as a consequence that there will be exponentially increasing solutions in the absence of excitation. Further, the response cannot precede the disturbance since it is the disturbance which creates the singularities.

Taking the results of Appendix I for the complex susceptibility in the simple case,

$$\chi(\omega) = \tfrac{1}{2}\chi_0\omega_0 \frac{\left\{\int_0^\infty [2\omega' g(\omega'-\omega_0)/(\omega'^2-\omega^2)]d\omega'\right\} - i\pi g(\omega-\omega_0)}{1+(H_1/H_{\frac{1}{2}})^2[\pi g(\omega-\omega_0)/T_2]}, \tag{36}$$

the complex susceptibility will have simple poles in the complex frequency plane for

$$[\pi g(\omega-\omega_0)/T_2]^{\frac{1}{2}}(H_1/H_{\frac{1}{2}}) = \pm i. \tag{37}$$

These poles must occur in conjugate complex pairs, from which it follows that if H_1 is nonvanishing there will be poles in the lower half of the complex frequency plane. The residues at these poles will be of order $(H_1/H_{\frac{1}{2}})^2$, so that for low power levels they can be disregarded. However, as has been seen at high power levels, the presence of these poles may profoundly alter the relation between χ' and χ'' so that their presence cannot be neglected. It does not seem possible in

[6] An estimate of the spin-lattice relaxation time from the hyperfine interaction, using the results of Waller for Raman processes [Z. Physik 79, 370 (1932)], yields a value about ten times greater than this. However, because of the roughness of the estimate a more exact calculation might still be profitable.

[7] J. H. Van Vleck, Massachusetts Institute of Technology Radiation Laboratory Technical Report 735, 1945 (unpublished).

general to obtain an analytical expression, for example, for a saturated disperion line from a knowledge of the form of the saturated absorption line. It is possible on physical grounds to circumvent this problem as is done in Appendix I. It is required, however, that the form of the absorption in the absence of saturation be known.

As an example of the above considerations, we examine the solutions of the Bloch equations[8] for the behavior of a paramagnetic system in crossed constant and rotating magnetic fields. The importance of the Bloch equations here is that they yield expressions for the absorption and dispersion for any H_1 without the need for introducing the Kramers-Kronig relations. We obtain for a rotating microwave field of amplitude H_1

$$\chi'(\omega)=\tfrac{1}{2}\chi_0\omega_0T_2\{T_2(\omega_0-\omega)/[1+s+T_2^2(\omega_0-\omega)^2]\}, \quad (38)$$

$$\chi''(\omega)=\tfrac{1}{2}\chi_0\omega_0T_2\{1/[1+s+T_2^2(\omega_0-\omega)^2]\}, \quad (39)$$

where $s=\gamma^2H_1^2T_1T_2$.

$$\chi(\omega)=\chi'(\omega)-i\chi''(\omega)=-\tfrac{1}{2}\chi_0\omega_0\frac{\omega-[\omega_0-(i/T_2)]}{\{\omega-[\omega_0-(i/T_2)(1+s)^{\frac{1}{2}}]\}\{\omega-[\omega_0+(i/T_2)(1+s)^{\frac{1}{2}}]\}}. \quad (40)$$

The complex susceptibility will have poles at

$$\omega=\omega_0\pm(i/T_2)(1+s)^{\frac{1}{2}}.$$

The residue at the pole in the lower half of the complex plane is

$$-\tfrac{1}{4}\chi_0\omega_0[1-(1+s)^{-\frac{1}{2}}],$$

which is of order $(H_1/H_{\frac{1}{2}})^2$.

If one substitutes the expression of Eq. (39) for the absorption into Eq. (33), one does not obtain the expression of Eq. (38) for a nonvanishing microwave field. The solutions of the Bloch equation are then not in agreement with the Kramers-Kronig relations. They are, however, completely consistent with our Eqs. (12) and (15), which were derived using the Kramers-Kronig relations in only a very restricted way.

I wish to acknowledge my indebtesness to Professor A. F. Kip and Professor C. Kittel for their stimulating direction and invaluable contributions during the course of this work. I have benefited in addition from discussions with Professor W. D. Knight and Professor W. A. Nierenberg. This work was supported in part by the U. S. Office of Naval Research and was carried out during the tenure of a fellowship from the U. S. Atomic Energy Commission.

APPENDIX I. DERIVATION OF THE SUSCEPTIBILITY EXPRESSIONS

In the absence of saturation we have, from Eq. (12),

$$\chi''(\omega)=\tfrac{1}{2}\pi\chi_0\omega_0g(\omega-\omega_0).$$

Since we are dealing here with an unsaturated system, we can obtain the unsaturated dispersion from Eq. (33), the first of the Kramers-Kronig relations. Then

$$\chi'(\omega)=\tfrac{1}{2}\chi_0\omega_0\int_0^\infty\frac{2\omega'g(\omega'-\omega_0)}{\omega'^2-\omega^2}d\omega'.$$

The saturation is introduced through Eq. (7), which gives the way in which the difference in populations of the spin levels decreases with increasing microwave power. Then, for arbitrary microwave field strength

$$\chi''(\omega)=\tfrac{1}{2}\chi_0\omega_0\frac{\pi g(\omega-\omega_0)}{1+\frac{1}{4}\pi\gamma^2H_1^2T_1g(\omega-\omega_0)}, \quad (12)$$

$$\chi'(\omega)=\tfrac{1}{2}\chi_0\omega_0\frac{\int_0^\infty[2\omega'g(\omega'-\omega_0)/(\omega'^2-\omega^2)]d\omega'}{1+\frac{1}{4}\pi\gamma^2H_1^2T_1g(\omega-\omega_0)}. \quad (15)$$

In the general case with both homogeneous and inhomogeneous broadening, we assemble the response of the spin packets under an envelope of shape $h(\omega-\omega_0)$. We obtain

$$\chi''(\omega)=\tfrac{1}{2}\chi_0\int_0^\infty\frac{\pi\omega'g(\omega-\omega')}{1+\frac{1}{4}\pi\gamma^2H_1^2T_1g(\omega-\omega')}h(\omega'-\omega_0)d\omega', \quad (17)$$

$$\chi'(\omega)=\tfrac{1}{2}\chi_0\int_0^\infty\frac{\omega'h(\omega'-\omega_0)d\omega'}{1+\frac{1}{4}\pi\gamma^2H_1^2T_1g(\omega-\omega')}\times\int_0^\infty\frac{2\omega''g(\omega''-\omega')}{\omega''^2-\omega^2}d\omega''. \quad (18)$$

For the case where the spin packets are narrow compared with the over-all line width, $g(\omega-\omega')$ will be appreciably different from zero only for $\omega'\sim\omega$. Since $h(\omega'-\omega_0)$ is slowly varying over this region, we obtain from Eq. (17)

$$\chi''(\omega)=\tfrac{1}{2}\chi_0\omega h(\omega-\omega_0)\int_0^\infty\frac{\pi g(\omega-\omega')d\omega'}{1+\frac{1}{4}\pi\gamma^2H_1^2T_1g(\omega-\omega')}. \quad (19)$$

To obtain Eq. (20), we first reverse the order of integration of Eq. (18)

$$\chi'(\omega)=\tfrac{1}{2}\chi_0\int_0^\infty\frac{2\omega''d\omega''}{\omega''^2-\omega^2}\times\int_0^\infty\frac{\omega'h(\omega'-\omega_0)g(\omega''-\omega')}{1+\frac{1}{4}\pi\gamma^2H_1^2T_1g(\omega-\omega')}d\omega'. \quad (41)$$

[8] F. Bloch, Phys. Rev. 70, 460 (1946).

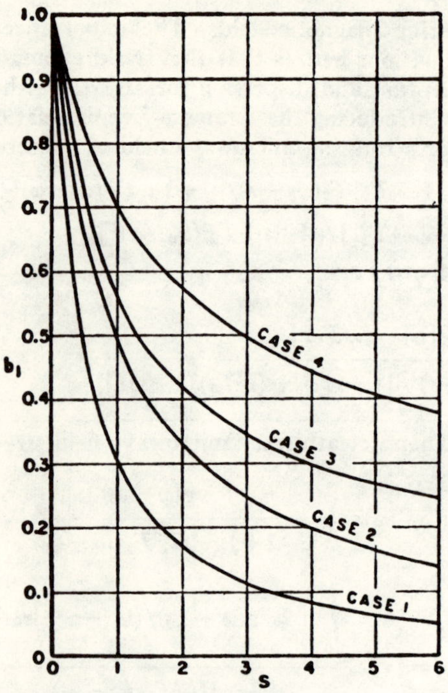

FIG. 6. Variation of the saturation parameter as a function of microwave power.

The contribution of the denominator in the integral over ω' will be of order T_2^*/T_2, so that we neglect it under the assumptions being made in this treatment. Then

$$\chi'(\omega)=\tfrac{1}{2}\chi_0\int_0^\infty \frac{2\omega''d\omega''}{\omega''^2-\omega^2}\int_0^\infty \omega' h(\omega'-\omega_0)g(\omega''-\omega')d\omega'.$$

Since $h(\omega'-\omega_0)$ is slowly varying in the vicinity of $\omega''\sim\omega'$, where $g(\omega''-\omega')$ is appreciable, and since

$$\int_0^\infty g(\omega''-\omega')d\omega'=1,$$

we obtain finally

$$\chi'(\omega)=\tfrac{1}{2}\chi_0\int_0^\infty \frac{2\omega''^2 h(\omega''-\omega_0)}{\omega''^2-\omega^2}d\omega''. \tag{20}$$

It should be observed that this is just the dispersion that one obtains from the Kramers-Kronig relations for an absorption of shape $h(\omega-\omega_0)$.

APPENDIX II

Saturation Behavior for Power Modulated Microwaves

We have treated the microwave power level as varying slowly enough compared with T_1 so that we could consider the spin system as being in a quasi-steady state. However if ω_m, the frequency of power modulation of the microwaves, is of the order of or greater than $1/T_1$, the saturation behavior is somewhat altered. We consider here four cases:

Case 1. Homogeneous broadening and $\omega_m T_1<1$. The system will follow the variations in microwave power. Or if we take the saturation factor

$$S(\omega,H_1)=\chi''(\omega,H_1)/\chi''(\omega,0),$$

the saturation factor will be periodic at the frequency ω_m.

Case 2. Homogeneous broadening and $\omega_m T_1>1$. For this case the system will not follow the variations in microwave power but will saturate at the average power level.

Case 3. Inhomogeneous broadening and $\omega_m T_1<1$. Here we will saturate spin packets individually rather than transferring power at once to the whole spin system. However, the spin packets will individually follow the periodic variations in power level.

Case 4. Inhomogeneous broadening and $\omega_m T_1>1$.

If the microwave field is power modulated, the power in the signal at the detecting crystal is given by

$$P_s=F\chi_0\omega_0 T_2 S(\omega_0,H_1)[1+\sin(\omega_m t)]P,$$

where P is the power incident on the microwave cavity and F is a factor which depends on the Q of the cavity, the filling factor for the sample, and the conversion loss of the detecting crystal. Since the detected signal is amplified by a narrow band amplifier and detected by a lock-in mixer, in the Fourier expansion of P_s:

$$P_s=F\chi_0\omega_0 T_2 P\sum_n a_n\cos(n\omega_m t)+b_n\sin(n\omega_m t),$$

the observed signal will be proportional to b_1. Our result is that for

Case 1: $b_1=(2/s^2)[(1+s)/(1+2s)^{\frac{1}{2}}]$;

Case 2: $b_1=1/(1+s)$;

Case 3: $b_1\simeq(1+0.60s)/(1+s)^{\frac{3}{2}}$;

Case 4: $b_1=1/(1+s)^{\frac{1}{2}}$,

where $s=\frac{1}{4}\gamma^2H_1{}^2T_1T_2$.

These functions are plotted in Fig. 6 and behave in the expected way, with Case 1 saturating the most rapidly and Case 4 the most slowly.

Proposal for a New Type Solid State Maser*

N. Bloembergen
Cruft Laboratory, Harvard University, Cambridge, Massachusetts
(Received July 6, 1956)

The Overhauser effect may be used in the spin multiplet of certain paramagnetic ions to obtain a negative absorption or stimulated emission at microwave frequencies. The use of nickel fluosilicate or gadolinium ethyl sulfate at liquid helium temperature is suggested to obtain a low noise microwave amplifier or frequency converter. The operation of a solid state maser based on this principle is discussed.

TOWNES and co-workers[1,2] have shown that microwave amplification can be obtained by stimulated emission of radiation from systems in which a higher energy level is more densely populated than a lower one. In paramagnetic systems an inversion of the population of the spin levels may be obtained in a variety of ways. The "180° pulse" and the "adiabatic rapid passage" have been extensively applied in nuclear magnetic resonance. Combrisson and Honig[2] applied the fast passage technique to the two electron spin levels of a P donor in silicon, and obtained a noticeable power amplification.

Attention is called to the usefulness of power saturation of one transition in a multiple energy level system to obtain a change of sign of the population difference between another pair of levels. A variation in level populations obtained in this manner has been demonstrated by Pound.[3] Such effects have since acquired wide recognition through the work of Overhauser.[4]

Consider for example a system with three unequally spaced energy levels, $E_3 > E_2 > E_1$. Introduce the notation,

$$h\nu_{31} = E_3 - E_1 \quad h\nu_{32} = E_3 - E_2 \quad h\nu_{21} = E_2 - E_1.$$

Denote the transition probabilities between these spin levels under the influence of the thermal motion of the heat reservoir (lattice) by

$$w_{12} = w_{21} \exp(-h\nu_{21}/kT), \quad w_{13} = w_{31} \exp(-h\nu_{31}/kT),$$
$$w_{23} = w_{32} \exp(-h\nu_{32}/kT).$$

The w's correspond to the inverse of spin lattice relaxation times. Denote the transition probability caused by a large saturating field $H(\nu_{31})$ of frequency

* Supported by the Joint Services.

[1] Gordon, Zeiger, and Townes, Phys. Rev. **99**, 1264 (1955).

[2] Combrisson, Honig, and Townes, Compt. rend. **242**, 2451 (1956).

[3] R. V. Pound, Phys. Rev. **79**, 685 (1950).

[4] A. W. Overhauser, Phys. Rev. **92**, 411 (1953).

Reprinted with permission from *Phys. Rev.*, vol. 104, pp. 324–327, Oct. 15, 1956.

ν_{31} by W_{13}. Let a relatively small signal of frequency ν_{32} cause transitions between levels two and three at a rate W_{32}. The numbers of spins occupying the three levels n_1, n_2, and n_3, satisfy the conservation law

$$n_1+n_2+n_3=N.$$

For $h\nu_{32}/kT\ll 1$ the populations obey the equations[5]:

$$\frac{dn_3}{dt}=w_{13}\left(n_1-n_3-\frac{N}{3}\frac{h\nu_{31}}{kT}\right)+w_{23}\left(n_2-n_3-\frac{N}{3}\frac{h\nu_{32}}{kT}\right)+W_{31}(n_1-n_3)+W_{32}(n_2-n_3),$$

$$\frac{dn_2}{dt}=w_{23}\left(n_3-n_2+\frac{N}{3}\frac{h\nu_{32}}{kT}\right)+w_{21}\left(n_1-n_2-\frac{N}{3}\frac{h\nu_{21}}{kT}\right)+W_{32}(n_3-n_2), \tag{1}$$

$$\frac{dn_1}{dt}=w_{13}\left(n_3-n_1+\frac{N}{3}\frac{h\nu_{31}}{kT}\right)+w_{21}\left(n_2-n_1+\frac{N}{3}\frac{h\nu_{21}}{kT}\right)-W_{31}(n_1-n_3).$$

In the steady state the left-hand sides are zero. If the saturating field at frequency ν_{31} is very large, $W_{31}\gg W_{32}$ and w's, the solution is obtained

$$n_1-n_2=n_3-n_2=-\frac{1}{3}\frac{hN}{kT}\frac{-w_{23}\nu_{32}+w_{21}\nu_{21}}{w_{23}+w_{12}+W_{32}}. \tag{2}$$

This population difference will be positive, corresponding to negative absorption or stimulated emission at the frequency ν_{32}, if

$$w_{21}\nu_{21}>w_{32}\nu_{32}. \tag{3}$$

If the opposite is true, stimulated emission will occur at the frequency ν_{21}. The following discussion could easily be adapted to this situation. The power emitted by the magnetic specimen is

$$P_{\mathrm{magn}}=\frac{Nh^2\nu_{32}}{3kT}\frac{(w_{21}\nu_{21}-w_{32}\nu_{32})W_{32}}{w_{23}+w_{12}+W_{32}}. \tag{4}$$

For a magnetic resonance line with a normalized response curve $g(\nu)$ and $g(\nu_{\max})=T_2$, the transition probability at resonance is given by

$$W_{32}=\hbar^{-2}|(2|M_x|3)|^2H_x^2(\nu_{32})T_2. \tag{5}$$

For simplicity it has been assumed that the signal field $H(\nu_{32})$ is uniform in the x direction over the volume of the sample. A similar expression holds for W_{31}.

For the moment we shall restrict ourselves to the important case that the signal excitation at frequency ν_{32} is small, $W_{32}\ll w_{23}+w_{31}$. No saturation effects at this transition occur and a magnetic quality factor can be defined by

$$-1/Q_{\mathrm{magn}}=\frac{4P_{\mathrm{magn}}}{\nu_{23}\langle H^2(\nu_{32})\rangle_{\mathrm{Av}}V_c}. \tag{6}$$

Q_{magn} is negative for stimulated emission, $P_{\mathrm{magn}}>0$. V_c is the volume of the cavity, and $\langle H^2\rangle_{\mathrm{Av}}$ represents a volume average over the cavity. The losses in the cavity, exclusive of the magnetic losses or gains in the sample, are described by the unloaded quality factor Q_0. The external losses from the coupling to a wave guide or coaxial line are described by Q_e. Introduce the voltage standing wave ratio β for the cavity tuned to resonance,

$$\beta=(Q_e/Q_0)+(Q_e/Q_{\mathrm{magn}}).$$

The ratio of reflected to incident power is

$$\frac{P_r}{P_i}=\frac{(1-\beta)^2}{(1+\beta)^2}.$$

There is a power gain or amplification, when β is negative or, $-Q_{\mathrm{magn}}^{-1}>Q_0^{-1}$. Oscillation will occur when

$$-Q_{\mathrm{magn}}^{-1}>Q_0^{-1}+Q_e^{-1}=Q_L^{-1},$$

where Q_L is the "loaded Q." The amplitude of the oscillation will be limited by the saturation effect, embodied by the W_{32} in the denominator of Eq. (4). The absolute value of $1/Q_{\mathrm{magn}}$ decreases as the power level increases. In the oscillating region the device will act as a microwave frequency converter. Power input is at the frequency ν_{13}, a smaller power output at the frequency ν_{23}. The balance of power is dissipated in the form of heat through the spin-lattice relaxation and through conduction losses in the cavity walls. For $-Q_{\mathrm{magn}}=Q_L$, $\beta=-1$, and the amplification factor would be infinite. The device will act as a stable c.w. amplifier at frequency ν_{23}, if

$$Q_0^{-1}+Q_e^{-1}>-Q_{\mathrm{magn}}^{-1}>Q_0^{-1}. \tag{7}$$

The choice of paramagnetic substance is largely dependent on the existance of suitable energy levels and the existence of matrix elements of the magnet moment operator between the various spin levels. The absorption and stimulated emission process depend directly on this operator, but the relaxation terms (w) also depend on the spin angular momentum operator via spin-orbit coupling terms. It is essential that all off-diagonal elements between the three spin levels under consideration be nonvanishing. This can be achieved by putting a paramagnetic salt with a crystalline field splitting δ in a magnetic field, which makes an angle with the crystalline field axis. The magnitude of the field is such that the Zeeman energy is comparable to the crystalline field splitting. In this case the states with magnetic quantum numbers m_s are all scrambled. This situation is usually avoided to unravel paramagnetic resonance spectra, but occasionally "forbidden lines" have been observed, indicating mixing of the m_s states. For our

[5] In case $h\nu_{31}\sim kT$, the Boltzmann exponential factors cannot be approximated by the linear terms. The algebra becomes more involved without changing the character of the effect.

purposes the mixing up of the spin states by Zeeman and crystalline field interactions of comparable magnitude is essential. The energy levels and matrix elements of the spin angular momentum operator can be obtained by a numerical solution of the determinantal problem of the spin Hamiltonian.[6] The number of electron spin levels may be larger than three. One may choose the three levels between which the operation will take place. The analysis will be similar, but algebraically more complicated. One has a considerable amount of freedom by the choice of the external dc magnetic field, to adjust the frequencies ν_{23} and ν_{13} and to vary the values of the inverse relaxation times w. It is advisable—although perhaps not absolutely necessary—to operate at liquid helium temperature. This will give relatively long relaxation times (between 10^{-2} and 10^{-4} sec), and thus keep the power requirements for saturation down. The factor T in the denominator of Eq. (4) will also increase the emission at low temperature. Although the order of magnitude of the w's is known through the work of Leiden school,[7] there is only one instance where w's have been measured for some individual transitions.[8] Van Vleck's[9] theory of paramagnetic relaxation should be extended to the geometries envisioned in this paper. If a Debye spectrum of the lattice vibrations is assumed, the relaxation times will increase with decreasing frequency at liquid helium temperature, where Raman processes are negligible. This implies that the condition (3) should be easily realizable when $\nu_{32}<\nu_{21}$.

Important applications as a microwave amplifier could, e.g., be obtained for $\nu_{32}=1420$ Mc/sec, corresponding to the interstellar hydrogen line, or to another relatively low microwave frequency used in radar systems. The frequency ν_{31} could be chosen in the X band, $\nu_{31}=10^{10}$ cps. To obtain well scrambled states with these frequency splittings one should have crystalline field splittings between 0.03 cm^{-1} and 0.3 cm^{-1}. Paramagnetic crystals which are suggested by these considerations are nickel fluosilicate[10] and gadolinium ethyl sulfate.[11] These crystals have the additional advantage that all magnetic ions have the same crystalline field and nuclear hyperfine splitting is absent, thus keeping the total number of possible transitions down. The use of magnetically dilute salts is indicated to reduce the line width, increase the value of T_2 in Eq. (5) and to separate the individual resonance transitions.

A single crystal 5% Ni 95% Zn Si $F_6\cdot 6H_2O$ has a line width of 50 oersted ($T_2=1.2\times10^{-9}$ sec) and an average crystalline field splitting $\delta=0.12$ cm^{-1} for the Ni^{++} ions. With an effective spin value $S=1$ there are indeed three energy levels of importance. The spin lattice relaxation time is about 10^{-4} sec at 2°K as measured in a saturation experiment by Meyer.[12] Further dilution does not decrease the line width, as there is a distribution of crystalline fields in the diluted salt.

A single crystal of 1% Gd 99% La $(C_2H_5SO_4)_3\cdot 9H_2O$ has an effective spin $S=7/2$. In zero field there are four doublets separated respectively by $\delta=0.113$ cm^{-1}, 0.083 cm^{-1}, and 0.046 cm^{-1} as measured at 20°K. These splittings are practically independent of temperature. The line width is 7 oersteds due to the distribution of local fields arising from the proton magnetic moments. This width could be reduced by a factor three by using the deuterated salt. The relaxation time is not known, but should be about the same as in other Gd salts,[7] which give $T_1\sim10^{-2}$ sec at 2°K.

In the absence of detailed calculations for the relaxation mechanism, we shall take $w_{12}=w_{13}=w_{32}=10^4$ sec^{-1} for the nickel salt and equal to 10^2 sec^{-1} for the gadolinium salt. The matrix elements $(2|M_x|3)$, etc., can be calculated exactly by solving the spin determinant. For the purpose of judging the operation of the maser using these salts, we shall take the off-diagonal elements of magnetic moment operator simply equal to $g\beta_0$, where $g=2$ is the Landé spin factor and β_0 is the Bohr magneton. For the higher spin value of the Gd^{+++} some elements will be larger but this effect is offset by the distribution of the ions over eight rather than three spin levels. Take $T=2$°K and $Q_0=10^4$, which is readily obtained in a cavity of pure metal at this temperature. A coaxial cavity may be used which has a fundamental mode resonating at the frequency $\nu_{32}=1.42\times10^9$ cps and a higher mode resonating at $\nu_{31}\approx10^{10}$ cps. Take the volume of the cavity $V=60$ cm^3 and $H_s^2=6\langle H^2\rangle_{Av}$. If these values are substituted in Eqs. (4)–(6), the condition (7) for amplification is satisfied if $N>3\times10^{18}$ for nickel fluosilicate and $N>3\times10^{17}$ for gadolinium ethyl sulfate ($N>10^{17}$ for the deuterated salt). The minimum required number of Ni^{++} ions are contained in 0.02 cm^3 of the diluted nickel salt. The gadolinium salt, diluted to 1% Gd, contains the required number in about the same volume. The critical volume is only 0.006 cm^3 for the deuterated salt. Crystals of appreciably larger size can still be fitted conveniently in the cavity. A c.w. amplifier or frequency converter should therefore be realizable with these substances. A larger amount of power can be handled by these crystals than by the P impurities in silicon which have a very long relaxation time, and require an intermittant operation, and where it is harder to get the required number of spins in the cavity.

So far we have assumed that the width corresponds to the inverse of a true transverse relaxation time T_2.

[6] See, e.g., Bleaney and K. H. W. Stevens, Repts. Progr. in Phys. **16**, 108 (1953).
[7] See, e.g., C. J. Gorter, *Paramagnetic Relaxation* (Elsevier Publishing Company, Amsterdam, 1948).
[8] A. H. Eschenfelder and R. T. Weidner **92**, 869 (1953).
[9] J. H. Van Vleck, Phys. Rev. **57**, 426 (1940).
[10] R. P. Penrose and K. H. W. Stevens, Proc. Phys. Soc. (London) **A63**, 29 (1949).
[11] Bleaney, Scovil, and Trenam, Proc. Roy. Soc. (London) **A223**, 15 (1954).
[12] J. W. Meyer, Lincoln Laboratory Report 1955 (unpublished).

Actually the width $1/T_2^*$ is due to an internal inhomogeneity broadening with normalized distribution $h(\nu)$ and $h(\nu_{max})\approx T_2^*$ in both cases. The response curve for a single magnetic ion is probably very narrow indeed, $g(\nu_{max})=T_2\approx T_1$, and $T_1=10^{-4}$ should be used in Eq. (5) rather than $T_2^*=1.2\times10^{-9}$ sec. The response to a weak threshold signal at ν_{32} now originates, however, from a small fraction of the magnetic ions. If $\gamma H(\nu_{32})<1/T_1\approx10^4$ cps, then only T_2^*/T_1 of the ions contribute to the stimulated emission and the net result is the same as calculated above. In most applications the incoming signal will be so weak that this situation will apply, even with a power amplification of 30 or 40 db.

For use as an oscillator or high level amplifier with a field $H(\nu_{32})$ in the cavity larger than $1/\gamma T_1$, one has essentially complete saturation ($W_{32}\gg w_{23}+w_{13}$) in Eq. (4) for those magnetic ions lying in a width $2\pi\Delta\nu=\gamma H(\nu_{32})$ in the distribution $h(\nu)$. One has then for the power emitted instead of Eqs. (4) and (5)

$$P_{\text{magn}}=\frac{h^2\nu_{32}}{3kT}N(-w_{32}\nu_{32}+w_{21}\nu_{21})\gamma H(\nu_{31})T_2^*. \quad (8)$$

The power is proportional to the amplitude of the radio frequency field rather than its square. This effect has been discussed in more detail by Portis.[13] It will limit the oscillation or amplification to an amplitude which can be calculated by using Eq. (8) in conjunction with Eqs. (6) and (7).

The driving field $H(\nu_{31})$ will necessarily have to satisfy the condition $\gamma H(\nu_{31})>w_{31}=T_1^{-1}$ to obtain saturation between levels 1 and 3. The power absorbed in the crystal will be proportional to the amplitude $H(\nu_{31})$, and is in order of magnitude given by

$$P_{\text{abs}}\sim N\frac{h^2\nu_{13}^2}{3kT}w_{13}\gamma H(\nu_{31})T_2^*. \quad (9)$$

This equation looses its validity if $\gamma H(\nu_{31})>T_2^{*-1}$. In this case the whole line would be saturated, but such excessive power levels will not be used. For $T_1^{-1}<\gamma H(\nu_{31})<T_2^{*-1}$, the effective band width of the amplifier is determined by $H(\nu_{31})$. It is about 0.5 Mc/sec for $H(\nu_{31})=0.2$ oersted. The power dissipated in a specimen of fluosilicate ten times the critical size is 0.5 milliwatt under these circumstances. For the gadolinium salt, also ten times the critical size, either deuterated or not, the dissipation is only 0.005 milliwatt. There should be no difficulty in carrying this amount out of the paramagnetic crystal without excessive heating. The power dissipation in the walls under these conditions will be 5 milliwatts. Liquid helium will boil off at the rate of only 0.01 cc/min due to heating in the cavity. Since helium is superfluid at 2°K, troublesome vapor bubbles in the cavity are eliminated.

The noise power generated in this type of amplifier should be very low. The cavity with the paramagnetic salt can be represented by two resonant coupled circuits as discussed by Bloembergen and Pound.[14] Noise generators are associated with the losses in the cavity walls, kept at 2°K, and with the paramagnetic spin abosrption which is described by an effective spin temperature, associated with the distribution of the spin population. The absolute value of this effective temperature also has the order of magnitude of 1°K. The input is from an antenna, which sees essentailly the radiation temperature of interstellar space. Reflected power is channeled by a circulating nonreciprocal element[15] into a heterodyne receiver, or, if necessary, into a second stage Maser cavity. The circulator makes the connection: antenna→maser cavity→heterodyne receiver→dummy load→antenna. If the antenna is not well matched, the dummy load may be a matched termination kept at liquid helium temperature to prevent extra power from entering the cavity. The input arm of the cavity at frequency ν_{31} will be beyond cutoff for the frequency ν_{32}. The coaxial line passing the signal at ν_{32} between cavity and circulator will contain a rejection filter at frequency ν_{31} to prevent overloading and noise mixing at the mixer crystal of the super heterodyne receiver.

It may be concluded that the realization of a low-noise c.w. microwave amplifier by saturation of a spin level system at a higher frequency seems promising. The device should be particularly suited for detection of weak signals at relatively long wavelength, e.g., the 21-cm interstellar hydrogen radiation. It may also be operated as a microwave frequency converter, capable of handling milliwatt power. More detailed calculations and design of the cavity are in progress.

[13] A. M. Portis, Phys. Rev. **91**, 1071 (1953).

[14] N. Bloembergen and R. V. Pound, Phys. Rev. **95**, 8 (1954).

[15] C. L. Hogan, Bell System Tech. J. **31**, 1 (1952).

Infrared and Optical Masers

A. L. Schawlow and C. H. Townes*
Bell Telephone Laboratories, Murray Hill, New Jersey
(Received August 26, 1958)

The extension of maser techniques to the infrared and optical region is considered. It is shown that by using a resonant cavity of centimeter dimensions, having many resonant modes, maser oscillation at these wavelengths can be achieved by pumping with reasonable amounts of incoherent light. For wavelengths much shorter than those of the ultraviolet region, maser-type amplification appears to be quite impractical. Although use of a multimode cavity is suggested, a single mode may be selected by making only the end walls highly reflecting, and defining a suitably small angular aperture. Then extremely monochromatic and coherent light is produced. The design principles are illustrated by reference to a system using potassium vapor.

INTRODUCTION

AMPLIFIERS and oscillators using atomic and molecular processes, as do the various varieties of masers,[1–4] may in principle be extended far beyond the range of frequencies which have been generated electronically, and into the infrared, the optical region, or beyond. Such techniques give the attractive promise of coherent amplification at these high frequencies and of generation of very monochromatic radiation. In the infrared region in particular, the generation of reasonably intense and monochromatic radiation would allow the possibility of spectroscopy at very much higher resolution than is now possible. As one attempts to extend maser operation towards very short wavelengths, a number of new aspects and problems arise, which require a quantitative reorientation of theoretical discussions and considerable modification of the experimental techniques used. Our purpose is to discuss theoretical aspects of maser-like devices for wavelengths considerably shorter than one centimeter, to examine the short-wavelength limit for practical devices of this type, and to outline design considerations for an example of a maser oscillator for producing radiation in the infrared region. In the general discussion, roughly reasonable values of design parameters will be used. They will be justified later by more detailed examination of one particular atomic system.

* Permanent address: Columbia University, New York, New York.

1 Gordon, Zeiger, and Townes, Phys. Rev. **99**, 1264 (1955).

2 Combrisson, Honig, and Townes, Compt. rend. **242**, 2451 (1956).

3 N. Bloembergen, Phys. Rev. **104**, 329 (1956).

4 E. Allais, Compt. rend. **245**, 157 (1957).

CHARACTERISTICS OF MASERS FOR MICROWAVE FREQUENCIES

For comparison, we shall consider first the characteristics of masers operating in the normal microwave range. Here an unstable ensemble of atomic or molecular systems is introduced into a cavity which would normally have one resonant mode near the frequency which corresponds to radiative transitions of these systems. In some cases, such an ensemble may be located in a wave guide rather than in a cavity but again there would be characteristically one or a very few modes of propagation allowed by the wave guide in the frequency range of interest. The condition of oscillation for n atomic systems excited with random phase and located in a cavity of appropriate frequency may be written (see references 1 and 2)

$$n \geq hV\Delta\nu/(4\pi\mu^2 Q_c), \tag{1}$$

where n is more precisely the difference $n_1 - n_2$ in number of systems in the upper and lower states, V is the volume of the cavity, $\Delta\nu$ is the half-width of the atomic resonance at half-maximum intensity, assuming a Lorentzian line shape, μ is the matrix element involved in the transition, and Q_c is the quality factor of the cavity.

The energy emitted by such a maser oscillator is usually in an extremely monochromatic wave, since the energy produced by stimulated emission is very much larger than that due to spontaneous emission or to the normal background of thermal radiation. The frequency range over which appreciable energy is distributed is given approximately by[1]

$$\delta\nu = 4\pi kT(\Delta\nu)^2/P, \tag{2}$$

where $\Delta\nu$ is the half-width at half-maximum of the resonant response of a single atomic system, P is the total power emitted, k is Boltzmann's constant, and T the absolute temperature of the cavity walls and wave guide. Since in all maser oscillators at microwave frequencies which have so far been considered, $P \gg kT\Delta\nu$, the radiation is largely emitted over a region very much smaller than $\Delta\nu$, or $\delta\nu \ll \Delta\nu$.

As amplifiers of microwave or radio-frequency energy, masers have the capability of very high sensitivity, approaching in the limit the possibility of detecting one or a few quanta. This corresponds to a noise temperature of $h\nu/k$, which for microwave frequencies is of the order of 1°K.

USE OF MULTIMODE CAVITIES AT HIGH FREQUENCIES

Consider now some of the modifications necessary to operate a maser at frequencies as high as that of infrared radiation. To maintain a single isolated mode in a cavity at infrared frequencies, the linear dimension of the cavity would need to be of the order of one wavelength which, at least in the higher frequency part of the infrared spectrum, would be too small to be practical. Hence, one is led to consider cavities which are large compared to a wavelength, and which may support a large number of modes within the frequency range of interest. For very short wavelengths, it is perhaps more usual to consider a plane wave reflected many times from the walls of such a cavity, rather than the field of a standing wave which would correspond to a mode.

The condition for oscillation may be obtained by requiring that the power produced by stimulated emission is as great as that lost to the cavity walls or other types of absorption. That is,

$$\left(\frac{\mu' E}{\hbar}\right)^2 \frac{h\nu n}{4\pi\Delta\nu} \geq \frac{E^2}{8\pi}\frac{V}{t}, \tag{3}$$

where μ' is the matrix element for the emissive transition, E^2 is the mean square of the electric field. (For a multiresonant cavity, E^2 may be considered identical in all parts of the cavity.) n is the excess number of atoms in the upper state over those in the lower state, V is the volume of the cavity, t is the time constant for the rate of decay of the energy, $\Delta\nu$ is the half-width of the resonance at half maximum intensity, if a Lorentzian shape is assumed. The decay time t may be written as $2\pi\nu/Q$, but is perhaps more naturally expressed in terms of the reflection coefficient α of the cavity walls.

$$t = 6V/(1-\alpha)Ac, \tag{4}$$

where A is the wall area and c the velocity of light. For a cube of dimension L, $t = L/(1-\alpha)c$. The condition for oscillation from (3) is then

$$n \geq \frac{3hV}{8\pi^2\mu^2 t}\frac{\Delta\nu}{\nu}, \tag{5}$$

or

$$n \geq \frac{\Delta\nu}{\nu}\frac{h(1-\alpha)Ac}{16\pi^2\mu^2}. \tag{6}$$

Here μ'^2 has been replaced by $\mu^2/3$, since μ'^2 is the square of the matrix element for the transition which, when averaged over all orientations of the system, is just one-third of the quantity μ^2 which is usually taken as the square of the matrix elment.

In a gas at low pressure, most infrared or optical transitions will have a width $\Delta\nu$ determined by Doppler effects. Then the resonance half-width at half-maximum intensity is

$$\Delta\nu = \frac{\nu}{c}\left(\frac{2kT}{m}\ln 2\right)^{\frac{1}{2}}, \tag{7}$$

where m is the molecular mass, k is Boltzmann's constant, and T the temperature. Because of the Gaussian line shape in this case, expression (6) becomes

$$n \geq \frac{\Delta\nu}{\nu}\frac{h(1-\alpha)Ac}{16\pi^2\mu^2(\pi\ln 2)^{\frac{1}{2}}}, \tag{8}$$

or

$$n \geq \frac{h(1-\alpha)A}{16\pi^2\mu^2}\left(\frac{2kT}{\pi m}\right)^{\frac{1}{2}}. \tag{9}$$

It may be noted that expression (9) for the number of excited systems required for oscillation is independent of the frequency. Furthermore, this number n is not impractically large. Assuming the cavity is a cube of 1 cm dimension and that $\alpha = 0.98$, $\mu = 5\times10^{-18}$ esu, $T = 400$°K, and $m = 100$ amu, one obtains $n = 5\times10^8$.

The condition for oscillation, indicated in (5), may be conveniently related to the lifetime τ of the state due to spontaneous emission of radiation by a transition between the two levels in question. This lifetime is given, by well-known theory, as

$$\tau = 3hc^3/(64\pi^4\nu^3\mu^2). \tag{10}$$

Now the rate of stimulated emission due to a single quantum in a single mode is just equal to the rate of spontaneous emission into the same single mode. Hence, $1/\tau$ is this rate multiplied by the number of modes p which are effective in producing spontaneous emission. Assuming a single quantum present in a mode at the resonant frequency, the condition for instability can then be written

$$nh\nu/p\tau \geq h\nu/t_1,$$

or

$$n \geq p\tau/t. \tag{11}$$

This gives a simple expression which may sometimes

be useful, and which is equivalent to (5), since

$$p=\int p(\nu)\frac{(\Delta\nu)^2 d\nu}{(\nu-\nu_0)^2+(\Delta\nu)^2}, \tag{12}$$

where $p(\nu)d\nu$ is the number of modes between ν and $\nu+d\nu$, which is well known to be

$$p(\nu)d\nu=8\pi\nu^2 V d\nu/c^3. \tag{13}$$

From (12) and (13), one obtains for a Lorentzian line shape,

$$p=8\pi^2\nu^2 V\Delta\nu/c^3. \tag{14}$$

Or, for a line broadened by Doppler effects, the corresponding number of effective modes is

$$p=8\pi^2\nu^2 V\Delta\nu/(\pi \ln 2)^{\frac{1}{2}}c^3. \tag{15}$$

If τ and p are inserted into (11) from expressions (10) and (14), respectively, it becomes identical with (5), as one must expect.

The minimum power which must be supplied in order to maintain n systems in excited states is

$$P=nh\nu/\tau=ph\nu/t. \tag{16}$$

This expression is independent of the lifetime or matrix element. However, if there are alternate modes of decay of each system, as by collisions or other transitions, the necessary power may be larger than that given by (16) and dependent on details of the system involved. Furthermore, some quantum of higher frequency than that emitted will normally be required to excite the system, which will increase the power somewhat above the value given by (16). Assuming the case considered above, i.e., a cube of 1-cm dimension with $\alpha=0.98$, $\lambda=10^4$ A, and broadening due to Doppler effect, (16) gives $P=0.8\times10^{-3}$ watt. Supply of this much power in a spectral line does not seem to be extremely difficult.

The power generated in the coherent oscillation of the maser may be extremely small, if the condition of instability is fulfilled in a very marginal way, and hence can be much less than the total power, which would be of the order of 10^{-3} watt, radiated spontaneously. However, if the number of excited systems exceeds the critical number appreciably, e.g., by a factor of two, then the power of stimulated radiation is given roughly by $h\nu$ times the rate at which excited systems are supplied, assuming the excitation is not lost by some process not yet considered, such as by collisions. The electromagnetic field then builds up so that the stimulated emission may be appreciably greater than the total spontaneous emission. For values even slightly above the critical number, the stimulated power is of the order of the power $nh\nu/\tau$ supplied, or hence of the order of one milliwatt under the conditions assumed above.

The most obvious and apparently most convenient method for supplying excited atoms is excitation at a higher frequency, as in optical pumping or a three-level maser system. The power supplied must, of course, be appreciably greater than the emitted power in expression (9). There is no requirement that the pumping frequency be much higher than the frequency emitted, as long as the difference in frequency is much greater than kT/h, which can assure the possibility of negative temperatures. Since, for the high frequencies required, an incoherent source of pumping power must be used, a desirable operating frequency would be near the point where the maximum number of quanta are emitted by a given transition from a discharge or some other source of high effective temperature. This maximum will occur somewhere near the maximum of the blackbody radiation at the effective temperature of such a source, or hence in the visible or ultraviolet region. The number of quanta required per second would probably be about one order of magnitude greater than the number emitted at the oscillating frequency, so that the input power required would be about ten times the output given by (16), or 10 milliwatts. This amount of energy in an individual spectroscopic line is, fortunately, obtainable in electrical discharges.

Very desirable features of a maser oscillator at infrared or optical frequencies would be a high order of monochromaticity and tunability. In the microwave range, a maser oscillator is almost inherently a very monochromatic device. However, a solid state maser can also normally be tuned over a rather large fractional variation in frequency. Both of these features are much more difficult to obtain in the infrared or optical regions. Frequencies of atomic or molecular resonances can in principle be tuned by Stark or Zeeman effects, as they would be in the radio-frequency or microwave range. However, such tuning is usually limited to a few wave numbers (or a few times 10 000 Mc/sec), which represents a large fractional change in the microwave range and only a small fractional change in the optical region. Certain optical and infrared transitions of atoms in solids are strongly affected by neighboring atoms. This may be the result of Stark effects due to internal electric fields or, as in the case of antiferromagnetic resonances, internal magnetic fields may vary enough with temperature to provide tuning over a few tens of wave numbers. Hence variation of temperature or pressure can produce some tuning. However, it appears unreasonable to expect more than a small fractional amount of tuning in an infrared or optical maser using discrete levels.

SPECTRUM OF A MASER OSCILLATOR

Monochromaticity of a maser oscillator is very closely connected with noise properties of the device as an amplifier. Consider first a maser cavity for optical or infrared frequencies which supports a single isolated mode. As in the microwave case, it is capable of detecting, in the limit, one or a few quanta, corresponding to a noise temperature of $h\nu/k$. However, at a wavelength of 10 000 A, this noise temperature is about 14 000°K, and hence not remarkably low. Furthermore, other well-known photon detectors, such as a photoelectric tube, are capable of detecting a single quantum. At such frequencies, a maser has no great advantage over well-known techniques in detecting small numbers of quanta. It does offer the new possibility of coherent amplification. However, if many modes rather than a single one are present in the cavity, a rather large background of noise can occur, the noise temperature being proportional to the number of modes which are confused within the resonance width of the atomic or molecular system. A method for isolation of an individual mode which avoids this severe difficulty will be discussed below.

Let us examine now the extent to which the normal line width of the emission spectrum of an atomic system will be narrowed by maser action, or hence how monochromatic the emission from an infrared or optical maser would be. Considerations were given above concerning the number of excited systems required to produce stimulated power which would be as large as spontaneous emission due to all modes of a multimode cavity which lie within the resonance width of the system. Assume for the moment that a single mode can be isolated. Spontaneous emission into this mode adds waves of random phase to the electromagnetic oscillations, and hence produces a finite frequency width which may be obtained by analogy with expression (2) as

$$\Delta\nu_{osc} = (4\pi h\nu/P)(\Delta\nu)^2, \tag{17}$$

where $\Delta\nu$ is the half-width of the resonance at half-maximum intensity, and P the power in the oscillating field. Note that kT, the energy due to thermal agitation, has been replaced in expression (15) by $h\nu$, the energy in one quantum. Usually at these high frequencies, $h\nu \gg kT$, and there is essentially no "thermal" noise. There remains, however, "zero-point fluctuations" which produce random noise through spontaneous emission, or an effective temperature of $h\nu/k$.

For the case considered numerically above, $4\pi h\nu\Delta\nu/P$ is near 10^{-6} when P is given by expression (16), so that $\Delta\nu_{osc} \sim 10^{-6}\Delta\nu$. This corresponds to a remarkably monochromatic emission. However, for a multimode cavity, this very monochromatic emission is superimposed on a background of stimulated emission which has width $\Delta\nu$, and which, for the power P assumed, is of intensity equal to that of the stimulated emission. Only if the power is increased by some additional factor of about ten, or if the desired mode is separated from the large number of undesired ones, would the rather monochromatic radiation stand out clearly against the much wider frequency distribution of spontaneous emission.

Another problem of masers using multimode cavities which is perhaps not fundamental, but may involve considerable practical difficulty, is the possibility of oscillations being set up first in one mode, then in another—or perhaps of continual change of modes which would represent many sudden jumps in frequency. If the cavity dimensions, density distribution of gas and distribution of excited states remains precisely constant, it seems unlikely that oscillations will build up on more than one mode because of the usual nonlinearities which would allow the most favored mode to suppress oscillations in those which are less favored. However, if many nearby modes are present, a very small change in cavity dimensions or other characteristics may produce a shift of the oscillations from one mode to another, with a concomitant variation in frequency.

SELECTION OF MODES FOR AMPLIFICATION

We shall consider now methods which deviate from those which are obvious extensions of the microwave or radio-frequency techniques for obtaining maser action. The large number of modes at infrared or optical frequencies which are present in any cavity of reasonable size poses problems because of the large amount of spontaneous emission which they imply. The number of modes per frequency interval per unit volume cannot very well be reduced for a cavity with dimensions which are very large compared to a wavelength. However, radiation from these various modes can be almost completely isolated by using the directional properties of wave propagation when the wavelength is short compared with important dimensions of the region in which the wave is propagated.

Consider first a rectangular cavity of length D and with two square end walls of dimension L which are slightly transparent, its other surfaces being perfectly reflecting. Transparency of the end walls provides coupling to external space by a continuously distributed excitation which corresponds to the distribution of field strength at these walls. The resulting radiation produces a diffraction pattern which can be easily calculated at a large distance from the cavity, and which is effectively separated from the diffraction pattern due to any other mode of the cavity at essentially the same frequency.

The field distribution along the end wall, taken as the xy plane, may be proportional, for example, to $\sin(\pi rx/L)\cos(\pi sy/L)$. The resonant wavelength is of the form

$$\lambda = \frac{2}{[(q/D)^2 + (r/L)^2 + (s/L)^2]^{\frac{1}{2}}}, \tag{18}$$

where q is the number of half-wavelengths along the z direction. If L is not much smaller than D, and if $q \gg r$ or s, the resonant wavelength is approximately

$$\lambda = \frac{2D}{q}\left[1 - \tfrac{1}{2}\left(\frac{Dr}{Lq}\right)^2 - \tfrac{1}{2}\left(\frac{Ds}{Lq}\right)^2\right], \tag{19}$$

which is primarily dependent on q and insensitive to r or s. The direction of radiation from the end walls, however, is critically dependent on r and s. The Fraunhofer diffraction pattern of the radiation has an intensity variation in the x direction given by

$$I \propto (2\pi r)^2 \sin^2\left(\frac{\pi L \sin\theta}{\lambda}+\frac{\pi r}{2}\right) \Bigg/ \left(\pi r+\frac{2\pi L \sin\theta}{\lambda}\right)^2\left(\pi r-\frac{2\pi L \sin\theta}{\lambda}\right)^2, \tag{20}$$

where θ is the angle between the direction of observation and the perpendicular to the end walls. For a given value of r, the strongest diffraction maxima occur at

$$\sin\theta = \pm r\lambda/2L,$$

and the first minima on either side of the maxima at

$$\sin\theta = \pm r\lambda/2L \pm \lambda/L.$$

Thus the maximum of the radiation from a mode designated by $r+1$ falls approximately at the half-intensity point of the diffraction pattern from the mode designated by r, which is just sufficient for significant resolution of their individual beams of radiation. This provides a method for separately coupling into or out of one or a few individual modes in the multimode cavity. A practical experimental technique for selecting one or a few modes is to focus radiation from the end walls by means of a lens onto a black screen in the focal plane. A suitable small hole in the screen will accept only radiation from the desired mode or modes.

There may, of course, be more than one mode which has similar values of r and s but different values of q, and which radiate in essentially identical directions. However, the frequencies of such modes are appreciably different, and may be sufficiently separated from each other by an appropriate choice of the distance D between plates. Thus if only one mode with a particular value of r and s is wanted within the range of response $2\Delta\nu$ of the material used to produce oscillations, D should be less than $c/4\Delta\nu$. Or, if it is undesirable to adjust D precisely for a particular mode, and approximately one mode of this type is wanted in the range $2\Delta\nu$, one may choose

$$D \approx c/4\Delta\nu. \tag{21}$$

For the conditions assumed above, the value of D given by (21) has the very practical magnitude of about 10 cm.

It is desirable not only to be able to select radiation from a single mode, but also to make all but one or a few modes of the multimode cavity lossy in order to suppress oscillations in unwanted modes. This again can be done by making use of directional properties. Loss may be introduced perhaps most simply by removing the perfectly reflecting walls of the cavity. The "cavity" is then reduced to partially transparent end plates and nonexistent (or lossy) perfectly-matched side walls.

Suppose now that one of the modes of such a cavity is excited by suddenly introducing the appropriate field distribution on one of the end walls. This will radiate a wave into the cavity having directional properties such as those indicated by the diffraction pattern (20). If r and s have their minimum values, the maximum energy occurs near $\theta=0$, and the wave travels more or less straight back and forth between the two plates, except for a gradual spreading due to diffraction. If r or s are larger, the maximum energy occurs at an appreciable angle θ, and the wave packet will wander off the reflecting plates and be lost, perhaps after a number of reflections. Those modes for which θ is large are highly damped and merge into a continuum, since energy radiated into them travels immediately to the walls and is lost from the cavity. However, modes for which θ is quite small may have relatively high Q and hence be essentially discrete.

For estimates of damping, consider first two end plates of infinite extent, but excited only over a square area of dimension L by a distribution which corresponds to one of our original modes. The radiated wave will be reflected back and forth many times, gradually spreading out in the diffraction pattern indicated by (20). If a mode with small values of r and s is used, the wave undergoes reflection every time it travels a distance D, and the rate of loss of energy W is given by the equations

$$dW/dt = -c(1-\alpha)W/D, \qquad W = W_0 e^{-c(1-\alpha)t/D}. \tag{22}$$

The decay time t is then $D/c(1-\alpha)$ rather than that given by (4) for the multimode case, or the effective distance traveled is $D/(1-\alpha)$.

Since the wavelength for modes with small r and s is given by (19), the frequency separation between modes with successive values of q is given by the usual Fabry-Perot condition

$$\delta\nu = c/2D. \tag{23}$$

Thus $\delta\nu \gg 1/t$ and the modes with successive values of q are discrete if $1-\alpha$, the loss on reflection, is much less than unity. On the other hand, the various modes given by small values of r or s and the same value of q are nearly degenerate, since according to (19) their frequency difference is less than $\delta\nu$ given in (23) by the factor r/q, which is of the order 10^{-4} for a typical case. These modes must be separated purely by their directional properties, rather than by their differences in frequency.

After traveling a distance $D/(1-\alpha)$, the radiation resulting from the excitation discussed above will have moved sideways in the x direction along the infinite parallel plates a distance of approximately $D\theta/(1-\alpha)$, where θ is the angle of one of the two large diffraction maxima given by (20). This distance is then

$$x = D\lambda r / [2(1-\alpha)L]. \tag{24}$$

Consider now the case of finite end plates of dimension L without their infinite extension which was assumed immediately above. After a number of reflections, the diffraction pattern would no longer be precisely that given by (20). However, expression (24) would still give a reasonable approximation to the distance of sideways motion, and if this distance is larger than the end-wall dimension L, the radiation will have been lost to the cavity, and the decay time for the mode in question is appreciably shorter than that indicated by (22). This condition occurs when

$$D\lambda r/[2(1-\alpha)L] \gtrsim L, \quad \text{or} \quad r \gtrsim 2(1-\alpha)L^2/D\lambda. \tag{25}$$

Thus to damp out modes with $r \gtrsim 10$, when $L = \frac{1}{2}$ cm, $\alpha = 0.98$, and $\lambda = 10^{-4}$ cm, the separation D between plates needs to be as large as a few centimeters. By choosing L sufficiently small, it is possible to discriminate by such losses between the lowest mode ($r=1$), and any higher modes. Too small a ratio $L^2/D\lambda$ will, however, begin to appreciably add to the losses from the lowest mode, and hence is undesirable if the longest possible delay times are needed.

The precise distribution of radiation intensity in the plane of the end walls which will give minimum loss, or which will occur during maser oscillation, cannot be very easily evaluated. It must, however, be somewhat like the lowest mode, $r=1$, $s=0$. A normal and straightforward method for exciting a Fabry-Perot interferometer is to use a plane wave moving perpendicular to the reflecting plates, and screened so that it illuminates uniformly all but the edge of the plates. Such a distribution may be expressed in terms of the nearly degenerate modes of the "cavity" with various r and s, and the considerable majority of its energy will be found in the lowest mode $r=1$, $s=0$, if it is polarized in the y direction. There is, of course, an exactly degenerate mode of the same type which is polarized in the x direction. Any much more complicated distribution than some approximation to uniform illumination or to the lowest mode $r=1$, $s=0$ of our rectangular cavity will produce a wider diffraction pattern which would be lost to a detector arranged to accept a very small angle θ near zero, and which would also be subject to greater losses when $L^2/D\lambda$ is small. However, nonuniform distribution of excited atoms between the reflecting plates could compensate for the larger diffraction losses, and in some cases induce oscillations with rather complex distributions of energy.

The above discussion in terms of modes of a rectangular cavity illustrates relations between the arrangement using a Fabry-Perot interferometer and the usual microwave resonant cavity.† An alternative approach which uses the approximation of geometrical optics more directly may also be helpful and clarifying. An atom radiating spontaneously in any direction has a decay time τ given by expression (10). The probability per unit time of emission of a quantum within a given solid angle $\Delta\Omega$ is then

$$\frac{1}{t'} = \frac{16\pi^3\nu^3}{3hc^3}\mu^2\Delta\Omega. \tag{26}$$

Hence if a sufficiently small solid angle is selected from the radiation, the amount of spontaneous emission can be made arbitrarily small. However, if essentially all the stimulated emission emitted from the end-wall of the interferometer is to be collected in a receiver or detector, allowance must be made for diffraction and a solid angle as large as about $(\lambda/L)^2$ must be used, so that the rate of spontaneous emission into the detector is

$$\frac{1}{t'} = \frac{16\pi^3\mu^2\nu}{3hcL^2}. \tag{27}$$

The rate of spontaneous emission (27) within the diffraction angle may be compared with the rate of induced transitions produced by one photon reflected back and forth in the volume L^2D. This rate is, as in (3), $(\mu'E/\hbar)^2(4\pi\Delta\nu)^{-1}$, where $E^2L^2D/8\pi = h\nu$. That is, since $\mu'^2 = \mu^2/3$,

$$\frac{1}{t''} = \frac{8\pi^2\mu^2\nu}{3\Delta\nu L^2 Dh}. \tag{28}$$

If D is $c/4\Delta\nu$ as in expression (21), so that there is approximately one and only one interference maximum of the interferometer in a particular direction within the range $2\Delta\nu$ of emission, then (28) becomes

$$\frac{1}{t''} = \frac{32\pi^2\mu^2\nu}{3hcL^2}. \tag{29}$$

Except for a small numerical factor of the order of the accuracy of the approximations used here, $1/t''$ given by (29) may be seen to equal $1/t'$. That is, use of the limiting amount of directional selection reduces the background of spontaneous emission to the same rate as that of stimulated emission due to a single photon. This is similar to the situation of a single mode in a cavity at microwave frequencies. It affords the limit of sensitivity which can be obtained by the usual maser amplifier, and the smallest possible noise for such a system as an oscillator.

† *Note added in proof.*—Use of two parallel plates for a maser operating at short wavelengths has also recently been suggested by A. M. Prokhorov [JETP 34, 1658 (1958)] and by R. H. Dicke [U. S. Patent 2,851,652 (September 9, 1958)]. These sources do not, however, discuss the reduction of excess modes or spontaneous emission.

Consider now the rate of loss of energy from a beam being reflected back and forth between the two end plates in the approximation of geometric optics. If the angle of deviation from the direction perpendicular to the plates is θ, then the additional rate of energy loss from a plane wave due to its spilling off the edges of the reflecting surfaces is

$$dW'/dt = -c\theta W/L. \qquad (30)$$

Expression (30) assumes, to be precise, that the deviation is parallel to one edge of the end plate. Thus, when $\theta=(1-\alpha)L/D$, the decay time is one-half that for $\theta=0$. Because of nonlinearities when oscillations set in, it may be seen from expression (3) that only those modes with the largest decay times will fulfill the condition for oscillation. The fraction ϵ of all modes of the "cavity" which have decay times greater than one-half that of the maximum decay time is approximately $(2\theta)^2/2\pi$, or

$$\epsilon = 2(1-\alpha)^2 L^2/\pi D^2. \qquad (31)$$

Letting $(1-\alpha)=1/20$, $L=1$ cm, and $D=10$ cm, one obtains $\epsilon=1.6\times10^{-5}$. This enormously reduces the number of modes which are likely to produce oscillations. Since the total number of modes is, from (14), $(8\pi^2\nu^2\Delta\nu/c^3)L^2D$, this number which may produce oscillations is

$$p' = \frac{16\pi(1-\alpha)^2\nu^2\Delta\nu L^4}{c^3 D}. \qquad (32)$$

Under the assumptions used above, p' may be found to be approximately 10^5, which is very much smaller than the total number of modes in the multimode cavity, but still may be an inconveniently large number. By using limiting values of the solid angle $(2\theta)^2$ set by diffraction, the number of modes can be further reduced to approximately unity, as was seen above.

FURTHER DISCUSSION OF PROPERTIES OF MASERS USING LARGE DIMENSIONS

It is important to notice that in the parallel plate case a very large amount of spontaneous emission may be radiating in all directions, even though only the very small amount indicated above is accepted in the detector, or is confused with the amplified wave. This property is quite different from the normal case in the microwave or radio-frequency range, and requires a rather rapid rate of supply of excited systems in order to maintain enough for maser action. Furthermore, great care must be taken to avoid scattering of light from undesired modes into the one which is desired. The fraction of spontaneous light which is scattered into the detector must be typically as low as about 10^{-6} or 10^{-7} in order to approach genuine isolation of a single mode.

Admission of a signal into the region between the two parallel plates is very similar to the process involved in a microwave cavity. The partially reflecting surfaces are analogous to coupling holes. If a monochromatic plane wave strikes the outside surface of one of the partially reflecting planes, energy will build up with the region between the planes, and the relations between input wave, energy in the "cavity", and output waves are just analogous to those for a microwave impinging on an appropriate cavity with input and output coupling holes.

Another interesting property of optical or infrared maser action which is associated with directional selection is that a beam of light may be passed through an ensemble of excited states with resulting amplification, but no important change in the wave front or phase. This amplification is just the inverse of an absorption, where it is well known that the wave front and phase are not distorted. Suppose, for example, that parallel light is focused by a lens. If an amplifying medium of excited gas is interposed between the lens and its focal point, the image will be intensified, but not otherwise changed except for some more or less normal effects which may be attributed to the dielectric constant of the gas. The same situation can, in principle, occur for maser amplification of microwaves. However, at these lower frequencies the amplification per unit length is usually so small that an impractically large volume of excited material would be required for amplification of a wave in free space to be evident. There may be a considerable amount of spontaneous emission in all directions, but only a very small fraction of the total spontaneous emission will fall at the focal point of the lens and be superimposed as noise on the intensified image. Noise from spontaneous emission decreases, for example, with the inverse square of distance from the emitting material, whereas the intensity of the focused beam increases as one approaches the focal point.

A SPECIFIC EXAMPLE

As an example of a particular system for an infrared maser, let us consider potassium. Atomic potassium is easily vaporized and has a simple spectrum as indicated by the energy levels shown in Fig. 1. Absorption transitions can occur from the $4s\ {}^2S_{\frac{1}{2}}$ ground state only to the various p levels. In particular, the atoms can be excited to the $5p\ {}^2P_{\frac{1}{2},\frac{3}{2}}$ by radiation of wavelength 4047 A. Just the right exciting frequency can be obtained from another potassium lamp, whose light is filtered to remove the red radiation at 7700 A. These excited atoms will decay to the $5s$ or $3d$ states in about 2×10^{-7} sec, or more slowly to the $4s$ ground state. However, if excited atoms are supplied fast enough, a sizable population can be maintained in the $5p$ state.

The minimum number of excited atoms required for maser-type oscillation may be found from (6), if the dipole matrix element were known, or from (11) if the lifetime were known. Although the wave functions necessary for obtaining the matrix elements have been calculated,[5,6] only estimates of the matrix element or lifetime can be made at present. The rate at which atoms must be supplied may, however, be obtained without detailed knowledge of the matrix elements by a small modification of expression (11). If τ is the mean life for spontaneous radiation of the desired wavelength, and φ is the fraction of the all decay processes from the upper level which occur in this manner, then the actual mean life in the excited state is $\tau'=\varphi\tau$. The number of atoms needed per second can be obtained from (11) as

$$\frac{n}{\tau'}=\frac{p}{\varphi t}=\frac{8\pi^2 V}{(\pi \ln 2)^{\frac{1}{2}}\varphi t\lambda^3}\frac{\Delta\nu}{\nu}=8\pi(2\pi)^{\frac{1}{2}}\frac{V}{\varphi\lambda^3 ct}\left(\frac{kT}{m}\right)^{\frac{1}{2}}, \qquad (33)$$

where λ is the wavelength. Thus, if the fraction φ is known, no other detailed properties of the atomic transition are required to evaluate the rate at which excited atoms must be supplied.

For the particular case of a gas (such as potassium vapor) at sufficiently low pressure that collisions are not too frequent, we can obtain φ from the relative intensities of the various radiative transitions out of the excited state. Observed relative intensities[7] show that for potassium the $5p\rightarrow 3d$ transitions are about 4 times more intense than the $5p\rightarrow 5s$ transitions. Within the $5p\rightarrow 3d$ transitions it follows from elementary angular momentum theory that the $5p_{\frac{3}{2}}\rightarrow 3d_{\frac{5}{2}}$ is the most intense, accounting for 9/15 of the radiation

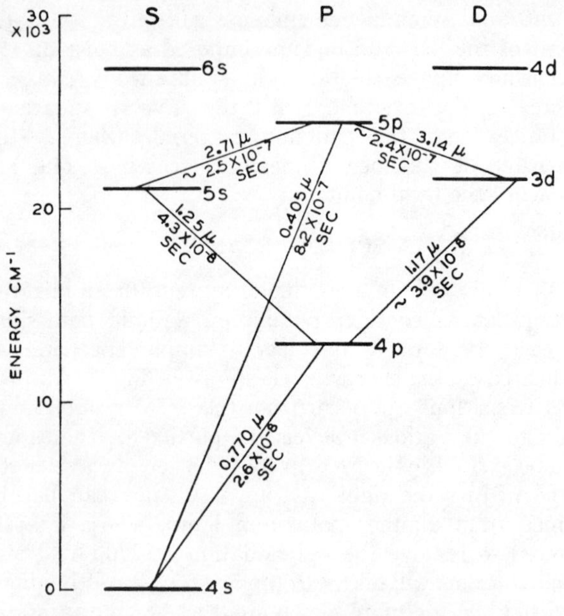

FIG. 1. Low-lying energy levels and transitions of atomic potassium.

[5] L. Biermann and K. Lubeck, Z. Astrophys. **25**, 325 (1948).
[6] D. S. Villars, J. Opt. Soc. Am. **42**, 552 (1952).
[7] *Viz.*, Tabulation in the *Handbook of Chemistry and Physics*, edited by D. Hodgman (Chemical Rubber Publishing Company, Cleveland, 1957), thirty-ninth edition.

emitted. Using the observed intensity ratio to allow for transitions to the $5s$ level, we conclude that about $9/18=\frac{1}{2}$ of those atoms excited to the $5p_{\frac{3}{2}}$ level decay to the $3d_{\frac{5}{2}}$ level. Decay to the $4s$ ground state is almost certainly less likely since we do know that this matrix element is not very large ($f=0.010$,[5] so that $\mu=0.65\times 10^{-18}$ esu). Thus $\varphi=\frac{1}{2}$ for the transition $5p_{\frac{3}{2}}-3d_{\frac{5}{2}}$ at 31 391 A.

Assume now two parallel plates of area 1 cm² and 10 cm apart, having a reflectivity α of 0.98. The decay time t for radiation in the space between the plates is $(10/3\times 10^{10})\times 50$ sec and $V=10$ cm³. For potassium vapor of suitable pressure, $T=435°$K and $m=39$ amu. Hence, from (29), the number of excited atoms needed per second is $dn/dt\geq 2.5\times 10^{15}$.

The energy needed per second is $d/dt(nh\nu)$, where ν is the frequency of the exciting radiation. Its value is 1.2×10^{-3} watt. This energy requirement is quite attainable. Incomplete absorption of the exciting radiation, reflection losses and multiplicity of the atomic states might raise this requirement somewhat. The absorption of the existing radiation is easily calculable and can be adjusted by controlling the density of the vapor:

$$k_0=\frac{1}{\Delta\nu_D}\left(\frac{\ln 2}{\pi}\right)^{\frac{1}{2}}\frac{\pi e^2}{mc}Nf, \qquad (34)$$

where k_0 is the absorption coefficient at the peak of the line $\Delta\nu_D$ is the (Doppler) line half-width, e is the electron charge, m is the electron mass, c is the velocity of light, N is the number of initial state atoms per cc, and f is the oscillator strength of the transition; i.e.,

$$k_0=1.25\times 10^{-2}(Nf/\Delta\nu_D).$$

For the exciting transition, 4046 A, in potassium, $\nu_0=7.42\times 10^{14}$ cycles/sec and at 435°K, $\Delta\nu_D=0.84\times 10^9$ cycles/sec. At this temperature the vapor pressure is 10^{-3} mm of mercury, so that in saturated vapor $N=2.5\times 10^{13}$/cc. Since $f=0.10$ for the $4s_{\frac{1}{2}}\rightarrow 5p_{\frac{3}{2}}$ transition,[6] $k_0=3.72$. This is high enough that the exciting radiation would be absorbed in a thin layer; if necessary it can be reduced by changing the pressure or temperature.

The light power for excitation is proportional to

$$\frac{V}{t}=\frac{AD}{(1-\alpha D/c)}=\frac{Ac}{1-\alpha}, \qquad (35)$$

where V is the volume of the cavity, t is the decay time for light in the cavity, D is the length of the cavity, A is the cross-section area of the cavity, α is the reflectivity of the end plates, and c is the velocity of light. This is independent of length, so that for a given cross-sectional area the light density needed can be reduced by increasing the length.

LIGHT SOURCES FOR EXCITATION

A small commercial potassium lamp (Osram) was operated with an input of 15 watts, 60 cycles, and its output was measured. In the red lines (7664–7699 A), the total light output was 28 mw from about 5 cc volume. At the same time, the total output in the violet lines (4044–4047 A) was 0.12 mw,[8] so that the output in $4s-5p_{\frac{3}{2}}$ was 0.08 mw. By increasing the current from 1.5 to 6 amp, with forced air cooling (the outer jacket being removed), the total violet output was increased to 0.6 mw. These outputs are somewhat short of the power level needed, but they may be considerably increased by adjusting discharge conditions to favor production of the violet line, and by using microwave excitation. With a long maser cell, the lamp area can be greatly increased. If necessary, very high peak light powers could be obtained in pulsed operation, although one would have to be careful not to broaden the line excessively.

Another possibility for excitation is to find an accidental coincidence with a strong line of some other element. The $8p$ level of cesium is an example of this type, since it can be excited very well by a helium line. The 4047 A line of mercury is 5 cm^{-1} from the potassium line, and is probably too far away to be useful even when pressure broadened and shifted.

Different modes correspond to different directions of propagation, and we only want to produce one or a few modes. Thus the cavity need only have two good reflecting walls opposite each other. The side walls need not reflect at all, nor do they need to transmit infrared radiation.

Unfortunately, most elements which have simple spectra, are quite reactive. Sapphire has good chemical inertness and excellent infrared transmission, being almost completely transparent as far as about 4 microns wavelength.[9] With such good transmission, the principal reflecting surfaces can be put outside the cell, and hence chosen for good reflectivity without regard to chemical inertness. Thus, one could use gold which has less than 2% absorption in this region, and attain a reflectivity of ~97% with 1% transmission. Even better reflectivity might be obtained with multiple dielectric layers of alternately high- and low-dielectric constant. The inner walls of the sapphire cell would reflect about 5% of the infrared light, and the thickness should be chosen so that the reflections from the two surfaces are in phase. The phase angle between reflections from the two surfaces depends on the thickness and the refractive index. Since sapphire is crystalline and the index is different for ordinary and extraordinary rays, the thickness could be chosen to give constructive interference for one polarization, and destructive interference for the perpendicular polarization. Thus, one could discriminate, if desired, between modes traveling in the same direction with different polarization.

To select just one from among the very many modes possible within the line width, the stimulated emission of radiation with one chosen direction of propagation must be favored. Thus the cell should be made long in the desired direction and fitted with highly reflecting end plates. The desired wave then has a long path as it travels back and forth, and so has a good chance to pick up energy from the excited atoms. A large width decreases the angular discrimination, and increases the pumping power needed.

For the potassium radiation at 3.14×10^{-4} cm wavelength, and $\Delta\nu$ being the Doppler width at 435°K, i.e., $\Delta\nu/\nu_0=1.2\times10^{-6}$, the number of modes is $2.0\times10^6\ V$ from expression (15). If we consider a cavity 1 cm square by 10 cm long, this number is 2.0×10^7, or 3.2×10^6 modes per steradian (forward and backward directions are taken as equivalent for standing waves). The angular separation between modes is then $(32\times10^6/2)^{-\frac{1}{2}}=2.5\times10^{-4}$ radian, where the 2 in the denominator removes the polarization degeneracy. The angular aperture accepted by this cavity is 1/10, but if the end plates had 98% reflectivity, the effective length would be increased by a factor of 50, and the angular aperture reduced to 2×10^{-3} radian. Thus there would be only 8 modes of each polarization within the effective aperture of the cell. Obviously this type of mode selection could be pushed further by making the cavity longer or narrower or more reflecting but this should not be necessary. Furthermore, the emission line does not have constant intensity over the width $\Delta\nu$, and the mode nearest the center frequency would be the first to oscillate at the threshold of emission.

SOLID-STATE DEVICES

There are a good many crystals, notably rare earth salts, which have spectra with sharp absorption lines, some of them having appeared also in fluorescence. In a solid, a concentration of atoms as large as 10^{19} per cc may be obtained. The oscillator strengths of the sharp lines are characteristically low, perhaps 10^{-6}. If the f value is low, radiative lifetimes are long, and in some cases lifetimes are as long as 10^{-3} sec or even more.

If the lifetime is primarily governed by radiation in the desired line, the pumping power required for the onset of stimulated oscillation is independent of the f value, as was shown above. For the atomic potassium level considered earlier, there are several alternative radiative decay paths (to the $4s$ and $3d$ states). In a solid there may also be rapid decay by nonradiative processes. If the storage time is long, because of a small f value, there is more time for competing processes to occur. Even lines which are sharp for solids are likely to be broader than those obtainable in gases. This larger width makes the attainment of maser oscillation more difficult, and it adds greatly to the difficulty of

[8] We are indebted to R. J. Collins for making these measurements.

[9] R. W. Kebler, *Optical Properties of Synthetic Sapphire* (Linde Company, New York).

selecting a single mode. However, there may very well be suitable transitions among the very many compounds.

The problem of populating the upper state does not have as obvious a solution in the solid case as in the gas. Lamps do not exist which give just the right radiation for pumping. However, there may be even more elegant solutions. Thus it may be feasible to pump to a state above one which is metastable. Atoms will then decay to the metastable state (possibly by nonradiative processes involving the crystal lattice) and accumulate until there are enough for maser action. This kind of accumulation is most likely to occur when there is a substantial empty gap below the excited level.

SUMMARY AND HIGH-FREQUENCY LIMITS

The prospect is favorable for masers which produce oscillations in the infrared or optical regions. However, operation of this type of device at frequencies which are still very much higher seems difficult. It does not appear practical to surround an atomic system with cavity walls which would very much affect its rate of spontaneous emission at very short wavelengths. Hence any ensemble of excited systems which is capable of producing coherent amplification at very high frequencies must also be expected to emit the usual amount of spontaneous emission. The power in this spontaneous emission, from expressions (14) and (16), increases very rapidly with frequency—as ν^4 if the width $\Delta\nu$ is due to Doppler effects, or as ν^6 if the width is produced by spontaneous emission. By choice of small matrix elements, $\Delta\nu$ can, in principle, be limited to that associated with Doppler effects, but the increase in spontaneously emitted power as fast as ν^4 is unavoidable.

For a wavelength $\lambda = 10^4$ A, it was seen above that spontaneous emission produced a few milliwatts of power in a maser system of dimensions near one centimeter, assuming reflectivities which seem attainable at this wavelength. Thus in the ultraviolet region at $\lambda = 1000$ A, one may expect spontaneous emissions of intensities near ten watts. This is so large that supply of this much power by excitation in some other spectral line becomes very difficult. Another decrease of a factor of 10 in λ would bring the spontaneous emission to the clearly prohibitive value of 100 kilowatts. These figures show that maser systems can be expected to operate successfully in the infrared, optical, and perhaps in the ultraviolet regions, but that, unless some radically new approach is found, they cannot be pushed to wavelengths much shorter than those in the ultraviolet region.

For reasonably favorable maser design in the short wavelength regions, highly reflecting surfaces and means of efficient focusing of radiation must be used. If good reflecting surfaces are not available, the number of excited systems used must, from (6), be very much increased with a resulting increase in spontaneous emission and difficulty in supply of excited systems. If focusing is not possible, the directional selection of radiation can in principle be achieved by detection at a sufficiently large distance from the parallel plates. However, without focusing the directional selection is much more difficult, and the background of spontaneous emission may give serious interference as noise superimposed on the desired radiation.

Finally, it must be emphasized that, as masers are pushed to higher frequencies, the fractional range of tunability must be expected to decrease more or less inversely as the frequency. The absolute range of variation can be at least as large as the width of an individual spectral line, or as the few wave numbers shift which can be obtained by Zeeman effects. However, continuous tuning over larger ranges of frequency will require materials with very special properties.

ACKNOWLEDGMENTS

The authors wish to thank W. S. Boyle, M. Peter, A. M. Clogston, and R. J. Collins for several stimulating discussions.

Stimulated Optical Emission in Fluorescent Solids. I. Theoretical Considerations

T. H. MAIMAN*

Hughes Research Laboratories, A Division of Hughes Aircraft Company, Malibu, California

(Received January 27, 1961; revised manuscript received May 17, 1961)

An analysis of stimulated emission processes in fluorescent solids is presented. The kinetic equations are discussed and expressions for pumping power and effective temperature of the exciting source are given in terms of the material parameters. A comparison of excitation intensity for three- and four-level systems is given. The spectral width of the stimulated radiation is discussed with particular attention to imperfect crystals.

INTRODUCTION

EXPERIMENTS culminating in the achievement of stimulated optical radiation from Cr^{3+} in Al_2O_3 (ruby) were recently reported by the author[1,2]; in addition, stimulated emission from other solids has since been reported.[3] A discussion of stimulated emission processes in the infrared-optical spectral region was presented previously by Schawlow and Townes,[4] with particular emphasis on an alkali vapor system; however, the modifications of both the analytical and experimental problems encountered with solids are considerable. Some of these extensions and revisions are presented here.

The general class of materials to be considered are fluorescent solids whose emission spectra consist of one or more sharp spectral lines. Excitation is normally supplied to these solids by radiation of frequencies which produce absorption into one or more bands. Some of this excitation energy is lost by a combination of spontaneous emission and thermal relaxation to lower-lying states; however, if the solid has a relatively high fluorescent efficiency[1] most of the energy is transferred to the sharp fluorescent levels by means of a non-radiative process. Subsequently, by a combination of spontaneous emission and thermal relaxation, the excited atoms (ions) return either to the ground state, or another low-lying state. The spontaneous emission from these sharp levels is the observed fluorescent radiation. If the exciting radiation is sufficiently intense it is possible to obtain a population density in one of the fluorescent levels greater than that of the lower-lying terminal state. In this situation, spontaneously emitted (fluorescent) photons traveling through the crystal stimulate upper state atoms to radiate, and a net component of induced emission is superimposed on the spontaneous emission.

We distinguish at this point between two useful solid-state systems: a three-level scheme typified by the level diagram shown in Fig. 1, and a four-level arrangement as shown in Fig. 2. Other configurations are, of course, possible, such as one in which all of the levels are relatively sharp and therefore more analogous to the gaseous situation. However, one of the features of the solid system is the possible use of a broad absorption band for the pump transition. This situation allows a relatively high pumping efficiency to be realized since most high-power optical sources have very broad spectral distributions in their radiant energy.

PUMPING POWER CONSIDERATIONS

Three-Level System

In the configuration shown in Fig. 1, the ground state is the terminal level for the spontaneous emission transition $2 \rightarrow 1$. Therefore, to produce a net stimulated emission component it is necessary to have an incident pump radiation intensity large enough to excite at least one-half of the total number of ground-state atoms into level 2. The extreme difficulty of achieving such a radiation intensity led early investigators to ignore the three-level system here described.[5]

A more quantitative description of this scheme can be obtained from a solution of the steady-state rate equations:

$$dN_3/dt = W_{13}N_1 - (W_{31}+A_{31}+S_{32})N_3 = 0, \quad (1)$$

$$dN_2/dt = W_{12}N_1 - (A_{21}+W_{21})N_2 + S_{32}N_3 = 0, \quad (2)$$

$$N_1 + N_2 + N_3 = N_0. \quad (3)$$

These processes are indicated in Fig. 1, where W_{13} is the induced transition probability per unit time for the transition $(1 \rightarrow 3)$ caused by the exciting radiation of frequency ν_{13}, W_{21} is the induced probability $(2 \rightarrow 1)$ due to the presence of radiation of frequency ν_{21}, A_{31} and A_{21} are Einstein A coefficients to account for the spontaneous radiation, and S_{32} is the transition probability for the nonradiative process $(3 \rightarrow 2)$. N_1, N_2, and N_3 are the respective level population densities, and N_0 is the total active ion density in the crystal. It has been assumed that the thermal processes for both $(2 \rightarrow 1)$ and $(3 \rightarrow 1)$ are negligible compared to the correspond-

---

* Now at Quantatron, Inc., 2520 Colorado Avenue, Santa Monica, California.

1 T. H. Maiman, Phys. Rev. Letters 4, 564 (1960).

2 T. H. Maiman, Nature 187, 493 (1960); Brit. Comm. and Electronics 7, 674 (1960); Program of the 45th Annual Meeting of the Optical Society of America, October 12–14, 1960 (unpublished), p. 14.

3 P. P. Sorokin and M. J. Stevenson, Phys. Rev. Letters 5, 557 (1960); IBM J. Research Develop. 5 (1961).

4 A. L. Schawlow and C. H. Townes, Phys. Rev. 11?, 1940 (1958).

5 A. L. Schawlow, in *Quantum Electronics*, edited by C. H. Townes (Columbia University Press, New York, 1960), pp. 553–563.

Reprinted with permission from *Phys. Rev.*, vol. 123 pp. 1145–1150, Aug. 15, 1961.

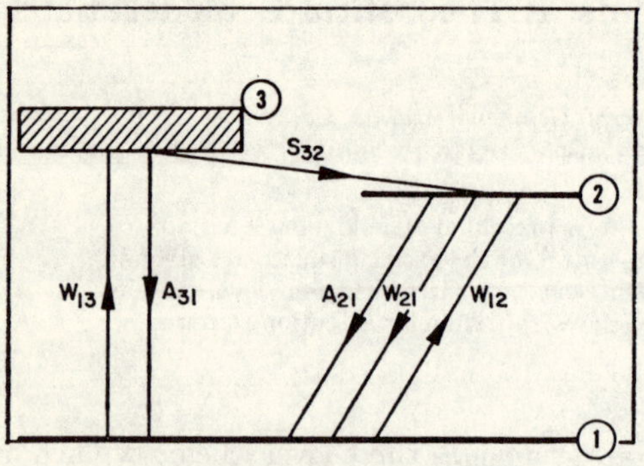

FIG. 1. Optical energy-level diagram for a three-level fluorescent solid.

ing radiative processes but that the reverse is true for the transition $(3 \rightarrow 2)$.[1] It is also assumed that each of the energy-level separations is large compared to kT so that the Boltzmann factors are negligible.

A solution to the above equations is

$$\frac{N_2}{N_1}=\left[W_{13}\left(\frac{S_{32}}{W_{31}+A_{31}+S_{32}}\right)+W_{12}\right]\Big/ (A_{21}+W_{21}). \quad (4)$$

If the solid has a high fluorescent quantum efficiency, e.g., ruby, $A_{31} \ll S_{32}$. Also, it will be found that even for very high pumping powers $W_{31} \ll S_{32}$. These approximations imply $N_3 \ll N_1$, $N_3 \ll N_2$ and (4) simplifies to

$$N_2/N_1 \cong (W_{13}+W_{12})/(A_{21}+W_{21}), \quad (5)$$

alternatively,

$$(N_2-N_1)/N_0 \cong (W_{13}-A_{21})/(W_{13}+A_{21}+2W_{12}). \quad (6)$$

In order to obtain stimulated emission at the frequency ν_{12}, it is necessary that $N_2 > N_1$ and therefore, from (6) we must have $W_{13} > A_{21}$. This is a minimum condition. In order to achieve a useful stimulated-emission amplifier or oscillator, it is further required that the excess population (N_2-N_1) be sufficient to overcome circuit losses.[4] The degree of population inversion needed can be conveniently represented for the optical case as follows.

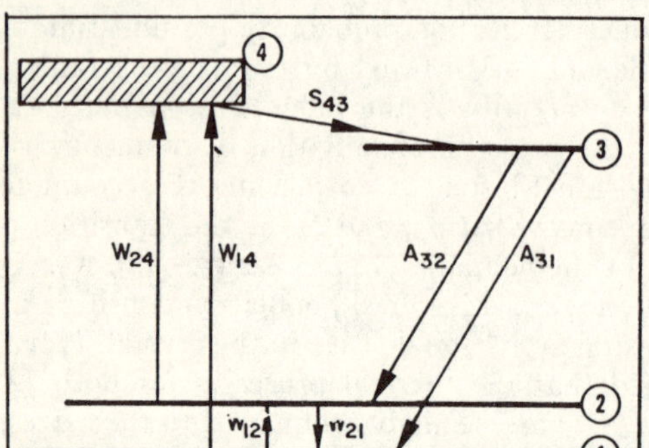

FIG. 2. Energy-level diagram in a fluorescent solid to illustrate a four-level method of producing stimulated optical radiation.

The energy in a wave propagating through a material with an absorption coefficient α is attenuated by the factor $e^{-\alpha l}$, if l is the length of material. The absorption coefficient is related to the population difference by

$$\alpha_{12}=(N_1-N_2)\sigma_{12}, \quad (7)$$

where σ_{12} is the absorption cross section for the pertinent transition. If $N_2 > N_1$, α is negative and $e^{-\alpha l} > 1$, i.e., amplification instead of absorption ensues. However, if $|\alpha|$ is small, a useful degree of amplification will be obtained only with an impractically long crystal. In such cases the interaction of radiation with the material can be increased by placing it between reflecting plates and advantage taken of the resulting regeneration due to waves reflected back and forth many times through the crystal. A wave that has traveled through the crystal once and reflected once has changed in energy content by the factor $e^{-\alpha l}r$, where r is the reflection coefficient of the end plate. In order to achieve a high amplification, or in the limit a coherent oscillation, the requirement is $e^{-\alpha l}r \cong 1$. In most cases $-\alpha l \ll 1$ so that the oscillation condition reduces to

$$-\alpha l \cong 1-r \quad (8)$$

or

$$(N_2-N_1)=(1-r)/\sigma_{12}l. \quad (9)$$

A different approach is to recognize that the parallel plate material system is an optical cavity, although not quite the same as a Fabry-Perot etalon because of the presence of the dielectric material; this latter point is discussed under "Spectral Width." The "Q" of the axial modes in this cavity can be obtained by solving for the mean decay time τ of photons moving perpendicular to the end plates.[4] When $(1-r) \ll 1$,

$$\tau \cong l\epsilon^{\frac{1}{2}}/[c(1-r)], \quad (10)$$

where l is the distance between the plates and ϵ is the dielectric constant of the material between the plates. Therefore,

$$Q_c=\omega\tau=2\pi l\epsilon^{\frac{1}{2}}/[\lambda(1-r)]. \quad (11)$$

To account for the absorption (or emission) of energy within the material itself, we can define a material Q and find by similar reasoning to that above

$$Q_M=2\pi\epsilon^{\frac{1}{2}}/\alpha\lambda. \quad (12)$$

In these terms the oscillation condition is $Q_c \cong -Q_M$ and is seen to be the same as Eq. (8). If we denote as α_0 the normally measured absorption coefficient for the transition $1 \rightarrow 2$ under low-power excitation, then, at the point of oscillation,

$$(N_2-N_1)/N_0 \cong (1-r)/\alpha_0 l \quad (13)$$

for a typical situation $(1-r) \ll \alpha_0 l$; therefore,

$$N_2/N_1 \cong 1+2(1-r)/\alpha_0 l. \quad (14)$$

and only a small excess over equal populations is required.

As the pumping power is increased above the point where $W_{13}=A_{21}$ (equal populations), the crystal begins to emit stimulated as well as spontaneous radiation. The spontaneous emission power is $N_2 h\nu_{12} A_{21}$, and since N_2 normally increases by only a small amount over the equal population value, as indicated above, almost all additional power developed at the frequency ν_{21} appears as stimulated emission. Since only this latter component constitutes useful output it would be desirable to pump at levels where $W_{13} \gg A_{21}$; unfortunately, however, it is difficult in practice even to reach the condition $W_{13}=A_{21}$.

Consider a parallel beam of light incident on the surface of an optically thin sample of the material. The power in the beam is

$$P_0 = nh\nu_p, \tag{15}$$

where n is the number of incident photons per second in the spectral region of the absorption band and ν_p is the center frequency of the pumping light. From the definition of transition probability

$$W_{13} = n\sigma_p / A \tag{16}$$

where σ_p is the absorption cross section for the pumping transition $(1 \rightarrow 3)$ and A is the area of the crystal perpendicular to the incident beam. Therefore, the flux density to produce equal populations $(W_{13}=A_{21})$ is

$$F = P_0/A = A_{21}(h\nu_p/\sigma_p). \tag{17}$$

Thus, in the case of an optically thin sample $(\alpha_p l \ll 1)$ the required flux density is independent of crystal concentration. Of course, in this case, only a small fraction of the incident power is absorbed, but increasing the ion density in the crystal results in nonuniform pumping so that in practice a compromise is made. The problem is alleviated somewhat if the back surface of the crystal is coated with reflecting material or if the entire crystal surface is illuminated uniformly.

The energy density U_p in the crystal is related to the flux density by $U_p = F\epsilon^{\frac{1}{2}}/c$ in the case of a parallel beam; therefore, using (17),

$$U_p = W_{13} h \epsilon^{\frac{1}{2}} / \lambda_p \sigma_p. \tag{18}$$

Although derived for the case of a parallel beam, the energy density is the fundamental quantity that determines transition probabilities, and, therefore, (18) is correct for other pumping geometries as long as the crystal is optically thin. The flux density, on the other hand, will depend on the particular configuration used, [e.g., $F=(c/4\epsilon^{\frac{1}{2}})U_p$ in the case of uniform illumination].

A configuration more likely to be realized in practice is one in which the material is more or less uniformly illuminated over most of its surface. Such an arrangement might be comprised of an appropriate reflector, lens, or reflector-lens combination. An alternative technique would be to use an extended source and place the active material in close proximity to it. For example, the source may be in the form of a hollow cylinder or helix with the active material placed inside. A disadvantage of the latter arrangement is that unused radiation from the exciting source aggravates heat dissipation problems; however, the over-all structure is much more compact.

The type of source most likely to be used is a high-pressure gaseous or vapor arc lamp. The spectral lines characteristic of the particular gas or vapor are practically nonexistent in such a lamp because of a combination of pressure and Stark broadening. The resultant output has some vague peaks, but is otherwise a more or less continuous spectrum with an effective radiation temperature typically between 4000°K and 10 000°K. Therefore, it is convenient and not too unrealistic, to consider that the active material is immersed in isotropic blackbody radiation.

We again assume an optically thin crystal so that the radiation density is uniform throughout its volume. Equation (18) is applicable to monochromatic radiation; in the case of broad-band radiation,

$$W_{13} = \int \frac{c\sigma(\nu)\rho(\nu)d\nu}{h\nu_p \epsilon^{\frac{1}{2}}}, \tag{19}$$

where $\rho(\nu)$ is the energy density per unit volume per unit frequency interval. Taking $\sigma(\nu)$ as Lorentzian, and using

$$\rho(\nu) = \frac{8\pi h\nu^3 \epsilon^{\frac{3}{2}}}{c^3(e^{h\nu/kT}-1)}, \tag{20}$$

we get

$$W_{13} \cong \frac{4\pi^2 \epsilon \Delta\nu_p \sigma_p \nu_p^2}{c^2(e^{h\nu/kT}-1)}, \tag{21}$$

where $\Delta\nu_p$ is the half-width of the absorption band. From radiation theory,

$$A_{31} = 4\pi^2 \epsilon \Delta\nu_p \sigma_p \nu_p^2 / c^2 \tag{22}$$

for a Lorentzian line; therefore,

$$W_{31} = A_{31}/(e^{h\nu_p/kT_s}-1). \tag{23}$$

The source temperature required for equal populations $(N_2=N_1,\ W_{13}=A_{21})$ is therefore

$$T_s = \frac{h\nu_p}{k \ln[1+(A_{31}/A_{21})]}. \tag{24}$$

If the lifetime of the fluorescent level is not wholly radiative, $1/A_{21}$ should be replaced by τ_2, the actual lifetime of this state.

The important material parameters connected with pumping power requirements for the three-level system are indicated in (24). The effective source temperature is a measure of the power per unit area per unit frequency interval and is, therefore, germane to the pumping problem. Since the necessary temperature is proportional to the pump frequency ν_p, operation at shorter wavelengths becomes increasingly difficult. The main consideration at a given wavelength is to maximize the ratio $A_{31}:A_{21}$.

One of the main problems in obtaining stimulated emission from the three-level solid is the very high power density required for the pump source with its associated heat-dissipation problem. A possible solution is to use a pulsed rather than a continuous source, i.e., a high-power electronic flash lamp. If the pulse is short compared to the life of the fluorescent level ($\tau_s \ll \tau_2$) the atoms excited to the fluorescent level do not appreciably decay during such an exciting pulse. The total energy absorbed when equal populations are attained is $E=\frac{1}{2}N_0 h\nu_p V$. If the crystal is isotropically irradiated, the absorbed energy is also (from the definition of Q_M at the pump frequency)

$$E=(2\pi\nu_p U_p V/Q_M)\tau_s. \tag{25}$$

Combining with previously derived relations, we find that the required energy per unit area radiated in one pulse from the flash tube is

$$J=F\tau_s \cong h\nu_p/4\sigma_p. \tag{26}$$

If, on the other hand, the pulse is long compared to the fluorescent decay time then the necessary *peak power* developed by the source is the same as that derived previously for a continuous source. The *average* power, however, can be made arbitrarily small by keeping the flash rate low. If large average powers are desired, larger flash tubes and consequently larger crystals can be employed. Although it requires about the same average input power to obtain a given average output power, the power dissipation problem is much less severe with pulsed operation than with continuous operation because of the greater volume of material involved.

Four-Level System

The level scheme depicted in Fig. 2 requires, in principle, lower exciting intensity than the three-level system discussed above.[5] If the low-lying level 2 is high enough above the ground state so that $e^{-h\nu_{12}/kT} \ll 1$, then level 2 will be relatively unpopulated. It will be as difficult as before to obtain an excess population with respect to the ground state, but negative temperatures for the transition $3 \rightarrow 2$ will be obtained at much lower intensities. The rate equations for this system are:

$$dN_3/dt=W_{14}N_1+W_{24}N_2-(A_{32}+A_{31})N_3=0, \tag{27}$$

$$dN_2/dt=w_{12}N_1-(W_{24}+w_{21})N_2+A_{32}N_3=0, \tag{28}$$

$$N_1+N_2+N_3=N_0. \tag{29}$$

We have neglected the induced rates $W_{32}=W_{23}$ which do not become appreciable until inversion is produced and have assumed as previously that the nonradiative process S_{43} is much greater than either A_{42}, A_{41}, W_{14}, or W_{24}. The solution is then

$$\frac{N_3}{N_2} \cong \frac{W_{14}(W_{24}+w_{21})+(W_{24}+w_{12})W_{24}}{W_{14}A_{32}+w_{12}(A_{32}+A_{31})}. \tag{30}$$

The quantities w_{12} and w_{21} are thermal transition probabilities, and the other quantities have the same meaning as previously stated. Since at room temperature thermal relaxation between levels 1 and 2 is apt to be very fast ($\sim 10^{-6}$ sec), $w_{12} \gg W_{14}$ and $w_{21} \gg W_{24}$; also the thermal probabilities are related by $w_{12}=w_{21}\exp[-h\nu_{12}/kT]$. Therefore,

$$\left(\frac{N_3}{N_2}\right) \cong \frac{W_{14}e^{h\nu_{12}/kT}+W_{24}}{A_{32}+A_{31}}. \tag{31}$$

and

$$N_2/N_1 \cong e^{-h\nu_{12}/kT}.$$

If we make the further assumption $W_{14} \approx W_{24}$, and define $A_3=A_{31}+A_{32}$, we find that

$$\frac{N_3-N_2}{N_0} \cong \frac{W_{14}-A_3/(1+e^{h\nu_{12}/kT})}{W_{14}+A_3}. \tag{32}$$

In order to obtain high gain or oscillation the excess population must be

$$N_3-N_2 \cong (1-r)/\sigma_{32}l. \tag{33}$$

Combining (32) and (33) and assuming $N_0\sigma_{32}l \gg 1-r$, we find

$$W_{14}\tau \cong \frac{1-r}{N_0\sigma_{32}l}+\frac{1}{1+e^{h\nu_{12}/kT}}, \tag{34}$$

where $\tau=1/A_3$ is the lifetime of the fluorescent level. This result is to be compared with the equivalent requirement $W\tau \approx 1$ in the three-level system.

SPECTRAL WIDTH

The spectral width of the radiation emitted from an optical maser is discussed by Schawlow and Townes in the ideal limiting case in which the oscillations excite only a single mode in a Fabry-Perot type of resonator. This situation is possible with extremely precise fabrication in a gaseous or vapor system but has not been realized in solid state devices.[2,3,6] The effect of the solid dielectric is twofold. The first is the introduction of "light pipe" modes of propagation, due to internal reflection; the second is distortion of the mode patterns because of strains, inhomogeneities, and deviations from single crystallinity. The first effect is not so serious since it can, in principle, be circumvented, for example, by using a roughened cylindrical surface or immersing the maser material in a liquid (or solid) having nearly the same or higher refractive index; however, this is an additional problem to deal with.

---

[6] R. J. Collins, D. F. Nelson, A. L. Schawlow, W. Bond, C. G. B. Garrett, and W. Kaiser, Phys. Rev. Letters **5**, 303 (1960).

Temperature effects in a solid are apt to be more serious than in a gaseous system. Generally speaking solids require a much higher power density because of the much larger density of atoms. For this reason, the magnitude of temperature shifts will be larger, and in addition the same temperature shifts will usually be of more consequence. The chamber for a gaseous system can be made of quartz or other special low temperature material whereas the plate separation in the solid situation is determined by the maser material itself which in turn may have a relatively high expansion coefficient. Finally, the energy levels are temperature dependent in a solid [about 3 (kMc/sec)/°C near room temperature for the ruby R_1 line].

Crystal imperfections, light-pipe modes, and end-plate misalignments generally introduce ambiguity in the mode selection process; i.e., there may be a very large number of modes which are of equal status in contrast to the perfect situation where the axial (or nearly axial) cavity mode closest to the atomic resonance peak is clearly the preferred one. Temperature shifts give rise to frequency sweeping and mode hopping.

As discussed by Schawlow and Townes,[4] multimoding introduces additional noise into the system due to spontaneous emission considerations; in addition, as shown in any text on oscillator theory, this same noise input reduces the monochromaticity of the output radiation from the system when it is oscillating.

We consider a system such as described above in which a large number of cavity modes enter into the oscillation because of poor mode selection. For purposes of illustration, we consider a cylindrical rod of three-level fluorescent material (such as ruby) with reflecting end plates to form an optical resonator. The power dissipated in this cavity is the sum of spontaneous and stimulated emission components. The stimulated emission power results primarily from the energy density built up in the cavity by standing waves and so essentially all of the stimulated emission power radiated by the material is coupled to the cavity. On the other hand, only a fraction (denoted by f) of the spontaneous emission power is dissipated in and excites the cavity. When regeneration ensues due to excess upper-state population, only the highest Q (axial and near axial) modes are excited appreciably. Although in practice only a small fraction of the total number of modes may be excited (limited principally by crystal perfection), the absolute number might be very large. (In a ruby of one-cm dimensions, the total number of modes is $\sim 2\times 10^{11}$.)

The power per unit volume coupled to the cavity is

$$\begin{aligned} p_c/V &= 2\pi\nu_{12}U/Q_c \\ &= fA_{21}N_2h\nu_{12}+W_{21}(N_2-N_1)h\nu_{12}. \end{aligned} \tag{35}$$

Here U is the energy density in the cavity, and Q_c is the loaded cavity Q (not including the material but including output coupling). Because we are dealing with a multimode cavity with dimensions very large compared to a wavelength, the spectral distribution of the spontaneous radiation is essentially the same as the fluorescence radiating into free space. If we assume a Lorentzian line, the frequency distribution of spontaneous power is $N_2A_{21}h\nu g(\nu)$, where

$$g(\nu)=\frac{\Delta\nu}{2\pi}\frac{1}{(\nu-\nu_0)^2+(\Delta\nu/2)^2} \tag{36}$$

and $\Delta\nu$ is the half-power width of the atomic transition $(2\rightarrow 1)$. Analogous to Eq. (19), we have

$$W_{21}=\int\frac{c\sigma(\nu)\rho(\nu)d\nu}{h\nu\epsilon^{\frac{1}{2}}}, \tag{37}$$

where $\sigma(\nu)=\frac{1}{2}\pi\Delta\nu\sigma_0 g(\nu)$. The power per unit volume per unit frequency interval is then

$$\frac{2\pi\nu\rho(\nu)}{Q_c}=fN_2A_{21}h\nu g(\nu)+\frac{c}{\epsilon^{\frac{1}{2}}}\sigma(\nu)\rho(\nu)(N_2-N_1), \tag{38}$$

therefore, using (12)

$$\rho(\nu)=\frac{1}{4\pi^2}\frac{fN_2A_{21}h\Delta\nu Q_c}{(\nu-\nu_0)^2+(\Delta\nu/2)^2(1+Q_c/Q_M)}. \tag{39}$$

Thus, the energy density is Lorentzian with a half-amplitude width

$$B=\Delta\nu(1+Q_c/Q_M)^{\frac{1}{2}}. \tag{40}$$

The total power delivered to the cavity is

$$P_c=\frac{2\pi\nu V}{Q_c}\int\rho(\nu)d\nu,$$

therefore, using (36) and (37),

$$B=fN_2A_{21}h\nu\Delta\nu V/P_c. \tag{41}$$

The fraction f in terms of cavity modes is well known to be

$$f=\frac{\bar{n}}{4\pi^2\nu^2\Delta\nu V\epsilon^{\frac{3}{2}}/c^3}, \tag{42}$$

where the denominator is the total number of modes within the integrated width $(\frac{1}{2}\pi\Delta\nu)$ of the atomic transition and $\bar{n}$ is the actual number of modes participating in the oscillation; also since

$$A_{21}=4\pi^2\epsilon\Delta\nu\sigma\nu^2/c^2,$$

we find

$$B=\bar{n}cN_2\sigma h\nu\Delta\nu/\epsilon^{\frac{1}{2}}P_c. \tag{43}$$

Then, using $N_2/N_1=e^{-h\nu/kT_e}$ (where T_e is the effective temperature of the two-level system composed of the fluorescent level and the ground state), $|Q_M|=2\pi\epsilon^{\frac{1}{2}}/|\alpha|\lambda\cong Q_c$ (when the system is oscillating), and $Q_c=\nu/\delta\nu$ (where $\delta\nu$ is the width of a single cavity mode)

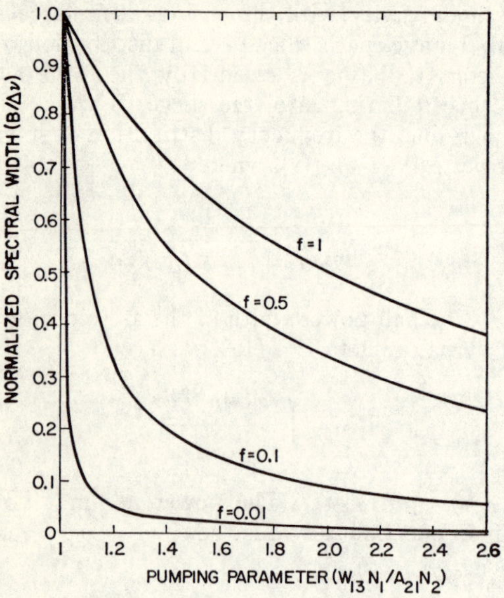

FIG. 3. Spectral bandwidth of the stimulated emission as a function of pumping power and the fractional number of cavity modes entering into the oscillation.

we obtain from (43)

$$B=2\pi\bar{n}h\nu\Delta\nu\delta\nu/P_c(1-e^{h\nu/kT_e}) \tag{44}$$

(T_e is a negative quantity).

An alternative expression for the spectral bandwidth is obtained by considering the total power developed at the frequency ν_{12},

$$P_t=N_1W_{13}h\nu_{12}V=P_c+(1-f)N_2A_{21}h\nu V, \tag{45}$$

then using (41), we find

$$B=\frac{f\Delta\nu}{(W_{13}/A_{21})(N_1/N_2)-(1-f)}. \tag{46}$$

When the pumping power is increased to the point where $W_{13}=A_{21}(N_1=N_2)$, from (46) $B=\Delta\nu$ independent of f. Any further increase in pumping power results in the production of stimulated emission and narrowing of the spectral width. Equation (46) is plotted in this region for various values of the parameter f and shown in Fig. 3. Thus, with a fairly good crystal, where good mode selection is feasible ($f\ll 1$), an abrupt change in the spectral width occurs with a fairly well-defined threshold pump power, but with a badly strained crystal ($f\sim 1$), the energy is scattered into many modes and no well-defined threshold can be found. Both types of behavior have been observed (see following paper).

The validity of the foregoing analysis breaks down when the spectral width of the "mode packet" is comparable to, or less than, that of the atomic resonance. In this case account must be taken of the modified spectral distribution of the spontaneous radiation as discussed by Purcell[7] and Pound.[8] In the extreme case of a single mode oscillation one can use a wire circuit analysis[8,9] and arrive at an equation analogous to (44), i.e.,

$$B=\frac{2\pi h\nu(\delta\nu)^2}{P_c(1-e^{h\nu/kT_e})}. \tag{47}$$

This expression is similar to the one presented by Schawlow and Townes,[4] but more general in that account is taken of a finite lower state population (their result implies $T_e\to -0$). Also the width of the cavity mode appears here, as it should, since usually (for both solids and gases) $\delta\nu<\Delta\nu$ and the narrower of these two is the dominant element in determining the spectral width.

CONCLUSIONS

Optical excitation into a broad absorption band is possible with many fluorescent solids; thus, stimulated emission may be produced with relatively efficient utilization of typical broad band pumping sources. A three-level solid system as described here (e.g., ruby), uses the ground state as the terminal level for the stimulated emission transition. The required pumping power for this scheme is quite high; however, it has been shown that the important pumping parameter is the effective temperature of the source. A four-level fluorescent solid may also utilize a broad absorption band; however, in some cases, depending on the material parameters, much less pumping power is required than for the three-level scheme.

Solid-state systems pose additional problems in mode selection not present in a gaseous system. A dielectric rod constitutes a waveguide (light pipe) and thus additional modes of propagation (which may be low loss) are introduced. Even more important are strains, inhomogeneities, and deviations from single crystallinity which tend to scatter energy into undesired modes. Under the assumption that the mode selection is poor, i.e., many modes participate in the oscillation, an analysis was presented for the case of badly strained crystals. The general features of this formulation have been observed in the laboratory as discussed in the following paper.

No attempt has been made here to treat the transient behavior or multiple spectrum observed in pulsed excitation of ruby and other solids. A quantitative description would require an analytical expression for the variation of Q of the various mode possibilities. This in turn is a function of the particular crystal and its history of fabrication (see following paper).

---

[7] E. M. Purcell, Phys. Rev. **89**, 681(A) (1946).
[8] R. V. Pound, Ann. Phys. (N. Y.) **1**, 24 (1957).
[9] T. H. Maiman, in *Quantum Electronics*, edited by C. H. Townes (Columbia University Press, New York, 1960), pp. 324–332.

Stimulated Optical Emission in Fluorescent Solids. II. Spectroscopy and Stimulated Emission in Ruby

T. H. Maiman,* R. H. Hoskins,* I. J. D'Haenens, C. K. Asawa, and V. Evtuhov
Hughes Research Laboratories, A Division of Hughes Aircraft Company, Malibu, California
(Received January 27, 1961)

Optical absorption cross sections and the fluorescent quantum efficiency in ruby have been determined. This data has been used to correlate calculations with the analysis of the preceding paper. Stimulated emission from ruby under pulsed excitation has been studied in some detail; the observations are found to depend strongly on the perfection of the particular crystal under study. A peak power output of approximately 5 kw, total output energy of near 1 joule, beam collimation of less than 10^{-2} rad, and a spectral width of individual components in the output radiation of about 6×10^{-4} A at 6943 A have been measured. It is suggested that mode instabilities due to temperature shifts and a time-varying magnetic field are contributing to an oscillatory behavior of the output pulse.

INTRODUCTION

The purpose of this paper is to describe the results of spectroscopic and stimulated emission experiments on Cr^{3+} in Al_2O_3 (ruby), and to apply the analysis of the preceding paper[1] to these results.

The energy-level diagram for ruby (taken from the work of Sugano and Tanabe[2]) is shown in Fig. 1. Fluorescence in this crystal is easily demonstrated by irradiating it with green light to excite the ${}^4A_2\rightarrow{}^4F_2$ transition, violet light to excite the ${}^4A_2\rightarrow{}^4F_1$ transition, or ultraviolet to excite a high-lying charge transfer band (not shown in the diagram[3]). The emission spectrum consists of a sharp doublet (${}^2E\rightarrow{}^4A_2$) in the red whose components at room temperature are at about 6943 A (R_1) and 6929 A (R_2) with respective half-power spectral widths of 4 and 3 A.

ABSORPTION SPECTRA

The absorption spectrum of ruby taken with a Cary spectrophotometer is shown in Fig. 2. The sample was a 1-cm cube cut with the c axis perpendicular to one pair of faces. The chromium concentration, determined by chemical analysis, was 0.0515 ± 0.0005 weight percent of Cr_2O_3:Al_2O_3 corresponding to a chromium ion density $N_0=1.62\times10^{19}/cm^3$. Absorption cross sections were calculated using $\sigma=\alpha/N_0$, where α is the linear absorption coefficient.

The absorption data for the R lines (Fig. 2) is inaccurate because of insufficient resolution; therefore, an independent experiment was performed using a ruby optical-maser source to measure transmission through the cube. It was found that with light propagating parallel to the c axis the absorption coefficient for the R_1 line $\alpha\cong0.4$ cm^{-1} and therefore $\sigma\cong2.5\times10^{-20}$ cm^2.

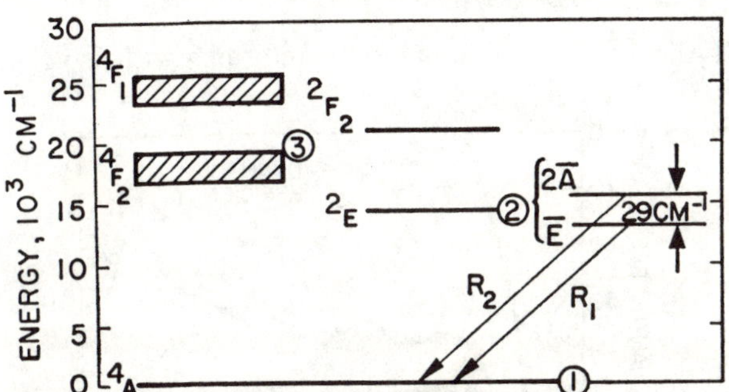

Fig. 1. Energy-level diagram for ruby.

FLUORESCENT EFFICIENCY

A parameter of particular interest in the study of fluorescent solids is the fluorescent quantum efficiency, defined as the ratio of the number of fluorescent photons emitted to the number of exciting photons *absorbed*. Wieder[4] reported 10^{-2} for this quantity, whereas Maiman[5] obtained a value near unity for excitation into the center of the green band. These measurements were repeated and extended to cover excitation in the spectral range of 3500–6000 A.

The sample used was the ruby cube referred to above

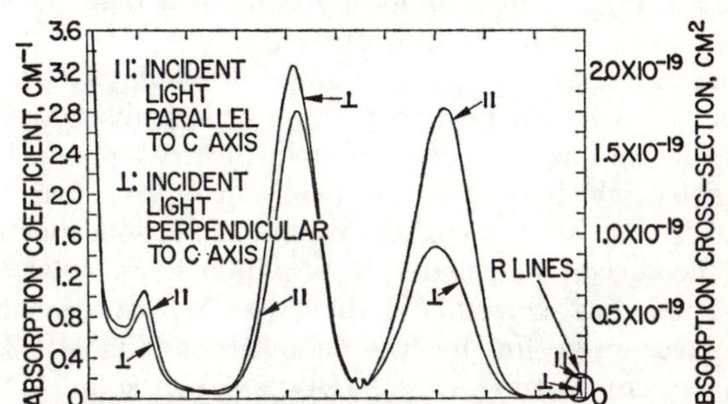

Fig. 2. Spectrophotometric absorption spectrum of ruby. The absorption cross section σ (right scale) is given by $\sigma=\alpha/N_0$, where α is the linear absorption coefficient and N_0 the number of absorbing centers per cm^3.

* Now at Quantatron, Inc., 2520 Colorado Avenue, Santa Monica, California.

[1] T. H. Maimam, preceding paper, Phys. Rev. **123**, 1145 (1961). Hereafter referred to as Part I.

[2] S. Sugano and Y. Tanabe, J. Phys. Soc. (Japan) **13**, 880 (1958).

[3] D. McClure, in *Solid-State Physics*, edited by F. Seitz and D. Turnbull (Academic Press, Inc., New York, 1960), Vol. 9.

[4] I. Wieder, Rev. Sci. Instr. **30**, 995 (1959).

[5] T. H. Maiman, Phys. Rev. Letters **4**, 564 (1960)

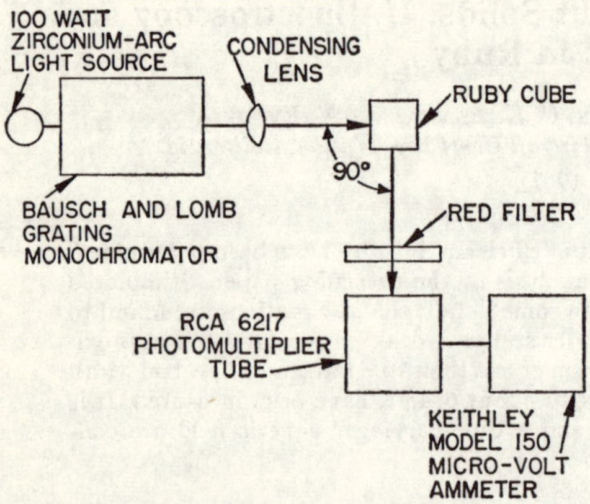

FIG. 3. Block diagram of the experiment setup for the determination of fluorescent efficiency in ruby.

A block diagram of the experiment is shown in Fig. 3. Radiation from a Bausch & Lomb monochromator at a power level of about 25 μw/cm^2 and of 60 A spectral width was incident on a side of the cube normal to the c axis. The total R-line fluorescence radiation was measured both along and normal to the c axis with a photomultiplier tube placed 30 cm from the sample. This light was monitored through a red-transmitting filter and a 0.08-cm^2 aperture. The total radiated power was computed assuming the radiation pattern to be an ellipsoid with an axis of revolution along the crystal c axis.

The light source and monochromator combination was calibrated by replacing the ruby sample by an Eppley silver-cadmium thermopile. The photomultiplier was calibrated by replacing the ruby by a mirror and comparing the resultant photomultiplier reading with that of the thermopile in the same position. In making the latter measurement a calibrated neutral-density (2.0) filter was used with the photomultiplier.

The fluorescent quantum and power efficiencies were calculated from these measurements making use of the data of Fig. 2 and accounting for multiple reflections in the crystal. The results are shown in Fig. 4. The accuracy of the dotted portions of the curves is lower than in the solid portion because of low absorptivity, and hence low absorbed power in the crystal in these regions, which reduces the precision of the measurements. The accuracy of the results in the solid portion of the curves is estimated to be within 10%.

The indicated results of about 70±5% can probably be accounted for by the following argument. The fluorescent lifetime of $^2E \rightarrow {}^4A_2$ is 4.3 msec at 77°K[6] but is reduced to 3.0 msec at 300°K.[5] Estimates of the variation of the matrix elements with temperature indicate a much smaller change than this; therefore, we assume that the reduced lifetime at 300°K is due to thermal relaxation and this alone would restrict the fluorescent quantum efficiency to a maximum of 70% at this temperature.

[6] F. Varsanyi, D. A. Wood, and A. L. Schawlow, Phys. Rev. Letters 3, 544 (1959).

STIMULATED EMISSION

Pumping Power Considerations

We now consider the amount of input power required to produce oscillations. The thermal relaxation time between the excited states $2\bar{A}(^2E)$ and $\bar{E}(^2E)$ at room temperature is estimated to be the order of one μsec or less, whereas the lifetime of ^2E at this temperature is 3.0 msec.[5] For this reason it can be assumed that the components of ^2E are nearly in thermal equilibrium in the experiments to be described. Because of a somewhat greater population in $\bar{E}(^2E)$ with respect to $2\bar{A}(^2E)$, due to the Boltzmann factor and because the matrix element for the transition $\bar{E}(^2E) \rightarrow {}^4A_2$ is larger than that for $2\bar{A}(^2E) \rightarrow {}^4A_2$, the former transition (R_1) will be favored in stimulated emission experiments.

It was shown in Part I that the excess population needed to produce oscillation is

$$(N_2 - N_1)/N_0 \cong (1-r)/\alpha_0 l.$$

In the case considered here, $N_2 - N_1$ is the population excess of $\bar{E}(^2E)$ with respect to $^4A_2$. If we use $\alpha_0 \sim 0.4$ cm^{-1}, $l = 2$ cm, and $(1-r) = 0.015$ (evaporated silver at 6943 A), we find $(N_2 - N_1)/N_0 \approx 0.019$. Thus, only a small excess over equal populations is required.

A simple extension of the discussion in Part I indicates that equal populations will be produced when

$$W_{13} \cong (1 + e^{-h\nu_a/kT}) A_{21}, \tag{1}$$

where ν_a is the frequency of the transition

$$2A(^2E) \rightarrow E(^2E).$$

This implies an energy density of pump radiation

$$U_p = (1 + e^{-h\nu_a/kT}) A_{21} h(\sqrt{\epsilon})/\lambda_p \sigma_p. \tag{2}$$

If the crystal is illuminated uniformly with isotropic

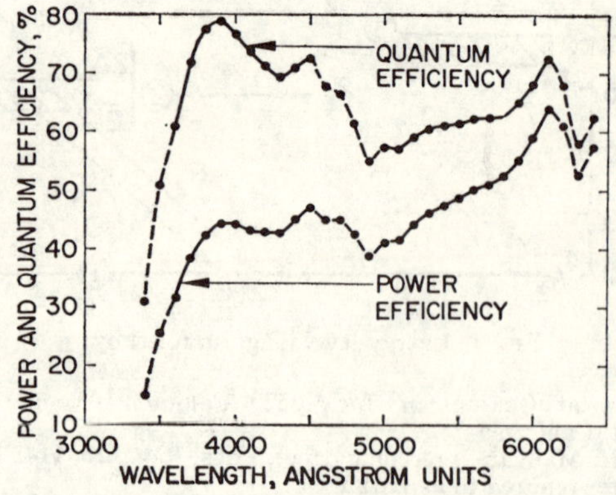

FIG. 4. Fluorescent quantum and power efficiencies for ruby: ratio of output fluorescence to input absorption as a function of wavelength.

radiation,

$$F=\frac{c}{4\sqrt{\epsilon}}U_p=\frac{(1+e^{-h\nu_a/kT})h\nu_pA_{21}}{4\sigma_p}. \tag{3}$$

Substituting the values applicable to ruby, i.e.,

$$\sigma_p=10^{-19}\ \text{cm}^2,\quad \nu_a=8.7\times10^{11}/\text{sec},\quad \nu_p=5.4\times10^{14}/\text{sec},\quad A_{31}=330/\text{sec};$$

we find that $F>555$ w/cm² is required to produce stimulated emission in this crystal.

Because of the broad spectral distribution of typical high-intensity exciting sources,[1] it is interesting to consider the brightness and spectral efficiency of a blackbody radiator in this connection. The flux density and relative spectral efficiency of such a radiator are given in Figs. 5 and 6. Radiation centered at both 5500 and 4100 A and 1000 A wide are considered, corresponding approximately to the green and violet bands in ruby. If we consider excitation into the green band only, for the moment, it is seen that a radiation temperature of about 5250°K is required for the onset of stimulated emission and that reasonable spectral efficiencies would be realized up to perhaps 15 000°K. Excitation into the violet band is not as efficient from a power standpoint (see Fig. 4) as the green band. Also, a side effect is possible from violet radiation. Maiman[5] has shown that appropriate wavelengths can stimulate transitions from ^2E to the charge transfer band; therefore, at high power densities, where the population of ^2E is appreciable, the net effect of the blue radiation could easily be to *de*populate ^2E.

Due to the need for high source intensities to produce stimulated emission in ruby and because of associated heat dissipation problems, these experiments were performed using a pulsed light source. For the case in which the exciting light pulses are short compared to the fluorescent lifetime, the requirement on the flash tube is that the energy per unit area[1] is

$$J=(1+e^{-h\nu_a/kT})h\nu_p/4\sigma_p; \tag{4}$$

for ruby, this requirement is $J\cong1.67$ joules/cm².

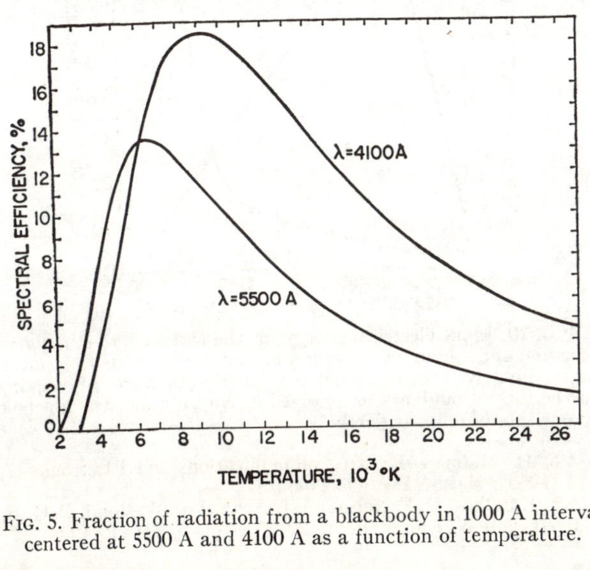

Fig. 5. Fraction of radiation from a blackbody in 1000 A interval centered at 5500 A and 4100 A as a function of temperature.

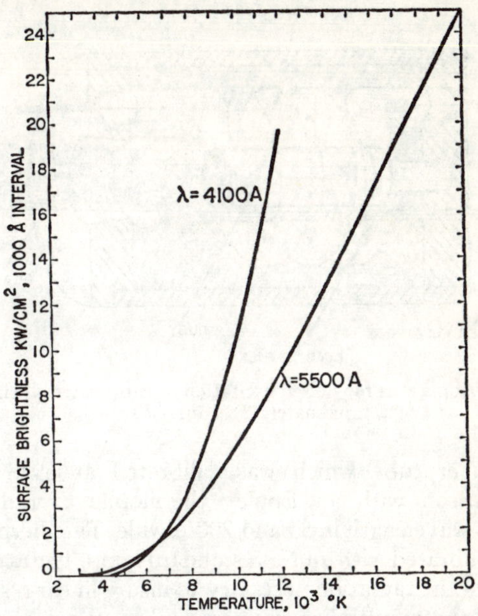

Fig. 6. Surface brightness from a blackbody in 1000 A interval centered at 5500 A and 4100 A as a function of temperature.

The source which was used was a General Electric type FT-506 xenon-filled quartz flash tube; this tube has a luminous efficiency of 40 lumens per watt. By coincidence, the green absorption band of ruby is almost identical to the sensitivity response of the human eye; therefore, lumens are meaningful units in this experiment corresponding to 0.0016 w of radiant energy and the spectral efficiency of the lamp is therefore about 0.064. The radiating area of this source is approximately 25 cm² so that an electrical input energy to the lamp of 650 joules would be required to produce stimulated emission in ruby on the basis of the previous considerations.

Apparatus

The material samples were ruby cylinders about $\frac{3}{8}$ in. in diameter and $\frac{3}{4}$ in. long with the ends flat and parallel to within $\lambda/3$ at 6943 A. The rubies were supported inside the helix of the flash tube, which in turn was enclosed in a polished aluminum cylinder (see Fig. 7); provision was made for forced air cooling. The ruby cylinders were coated with evaporated silver at each end; one end was opaque and the other was either semitransparent or opaque with a small hole in the center.

A block diagram of the experiment is shown in Fig. 8. The energy to the flashtube was obtained by discharging a 1350-μf capacitor bank and the input energy was varied by changing the charging potential. The R_1 output radiation was monitored with a type 6217 photo-

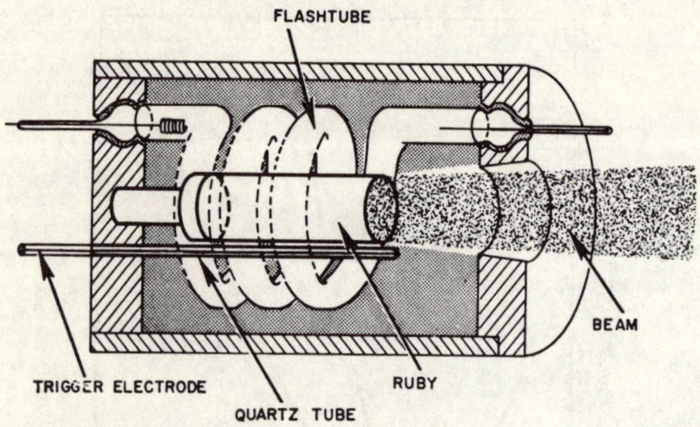

FIG. 7. Apparatus for pulsed excitation of ruby. (Actual size, approximately 2×1 in. o.d.)

multiplier tube which was calibrated at 6943 A by comparison with an Eppley thermopile to radiation at this wavelength in a band 200 A wide. The thermopile was calibrated with an NBS standard lamp. The attentuation of the radiation necessary to insure linear response of the photomultiplier was obtained by the use of calibrated neutral-density gelatin filters. Peak output power and details of the output pulse were obtained from the phototube output across a 1000-ohm resistor on an oscilloscope; the total instrumental response time was about 0.1 μsec. The total energy in the output pulse was obtained by integrating the phototube current with a 0.1-μf mica capacitor.

Experimental Results

1. Summary

It was found with high-intensity excitation that the nature of the output radiation from the various ruby samples which were tried could be divided into two categories:

A. Crystals which exhibited R_1 line narrowing of only 4 or 5 times, a faster but smooth time decay of the output (compared to the fluorescence), an output beam angle of about 1 rad, and no clear-cut evidence of a threshold excitation. This type of behavior was reported and discussed by Maiman.[7]

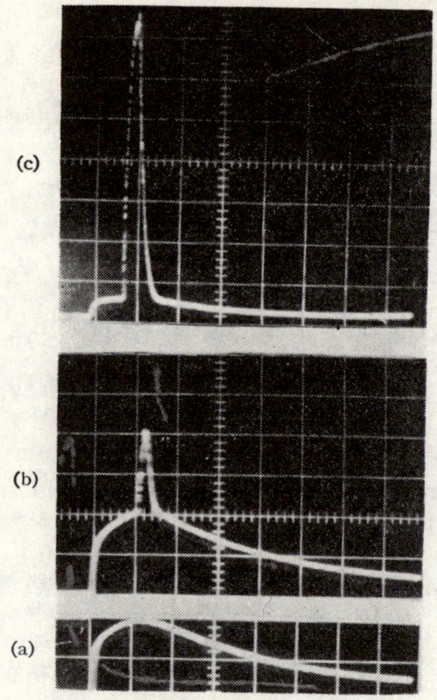

FIG. 9. R_1 line output pulse with various input energies: (a) below threshold, (b) threshold, (c) above threshold. The instrumental response time is too long to resolve the oscillatory structure of the stimulated emission pulse.

B. Crystals which exhibited a pronounced line narrowing of nearly four orders of magnitude, an oscillatory behavior of the output pulse, and a beam angle of about 10^{-2} rad; these crystals were particularly characterized by a very clear-cut threshold input energy where the pronounced line and beam narrowing occurred. This second category of behavior was reported by Collins *et al.*,[8] and is the subject of further study reported here.

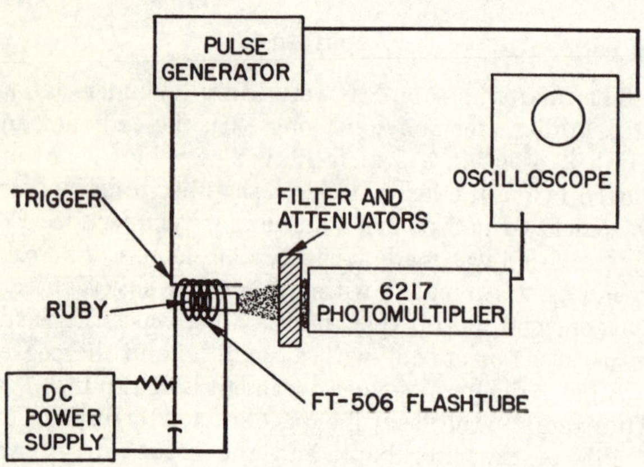

FIG. 8. Block diagram of experimental setup for the observation of stimulated emission in ruby.

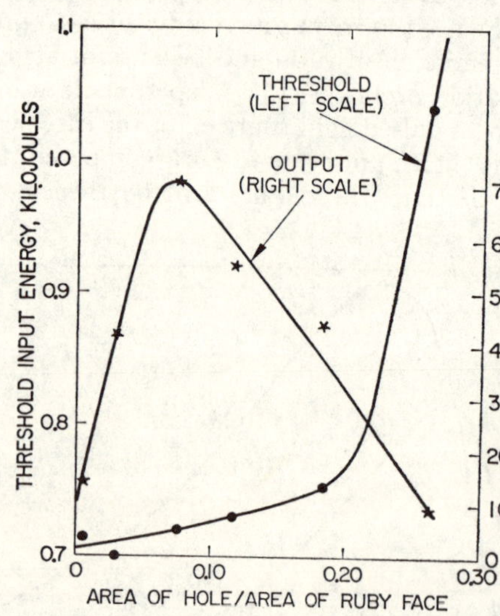

FIG. 10. Input electrical energy for the threshold of stimulated emission and stimulated emission output energy with an input of approximately twice threshold as a function of degree of coupling to the ruby. Coupling was changed by varying the size of a hole in one totally reflecting end.

[7] T. H. Maiman, British Communications and Electronics 7, 674 (1960); Nature 187, 493 (1960).

[8] R. J. Collins, D. F. Nelson, A. L. Schawlow, W. Bond, C. G. B. Garrett, and W. Kaiser, Phys. Rev. Letters 5, 303 (1960).

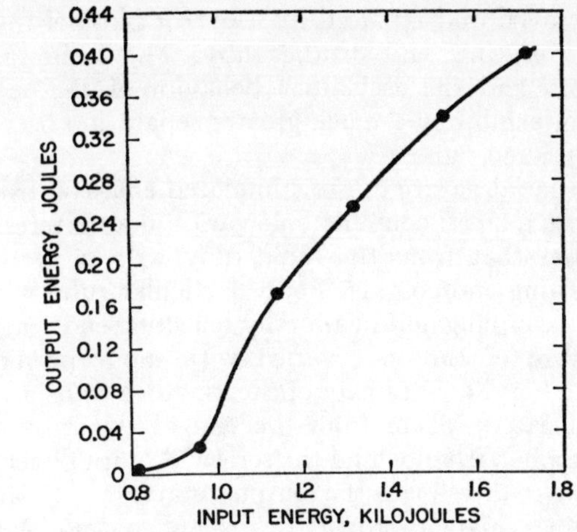

FIG. 11. Stimulated emission output energy from a ruby with a semitransparent (7% transmitting) end face as a function of electrical energy input to the exciting lamp source.

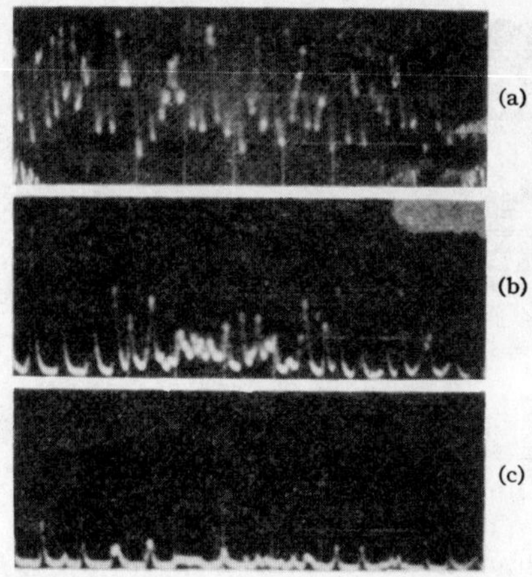

FIG. 13. Output pulse from ruby on an expanded time scale (10 μsec/division). (a), (b), and (c) represent the output approximately 600, 1000, and 1200 μsec after the onset of oscillation. The vertical sensitivity and the base line are the same in each case.

2. *Threshold*

The R_1 output light at various levels of input illumination is shown in Fig. 9. The spontaneous fluorescent radiation is shown in (a). In (b), the input energy has been increased by approximately one part in 10^4 and the onset of stimulated emission can clearly be seen. We define this condition as threshold. In (c) (different vertical scale), the input energy has been increased still further and the stimulated emission is seen to increase rapidly over the spontaneous fluorescent radiation. The instrumental response time for these particular photographs was long; hence, the structure of the stimulated emission pulse (discussed below) is not seen.

The variation in input energy required for the onset of stimulated emission as a function of degree of coupling to the ruby is shown in Fig. 10. Coupling was varied by gradually enlarging a small hole in the center of one of the silvered ends of the ruby. The abscissa of the curve is the ratio of the area of the unsilvered hole to the total area of the face. Variation in coupling was also obtained by changing the thickness of a partially transmitting silver film on one end of the ruby. The latter mathod is more suitable to analytic treatment of the data; however, in practice, it is much more difficult to take the ruby from its holder, remove a silver film, and resilver in a controlled manner than simply to enlarge a hole without changing any other experimental conditions.

The qualitative behavior shown in the figure is expected, since increasing the hole size reduces the effective reflection coefficient of the end face; a larger population excess is then required, which in turn requires larger input energy to the system. When the hole size becomes sufficiently large, the necessary excess population to overcome the cavity losses is not attainable at any value of pumping energy.

3. *Output Energy and Peak Power*

The variation of the total energy output with size of coupling hole at a fixed energy input (approximately twice threshold energy for small coupling) is also shown in Fig. 10. An optimum hole size for maximum energy output (which varies with available input energy and from sample to sample) is clearly indicated. This vari-

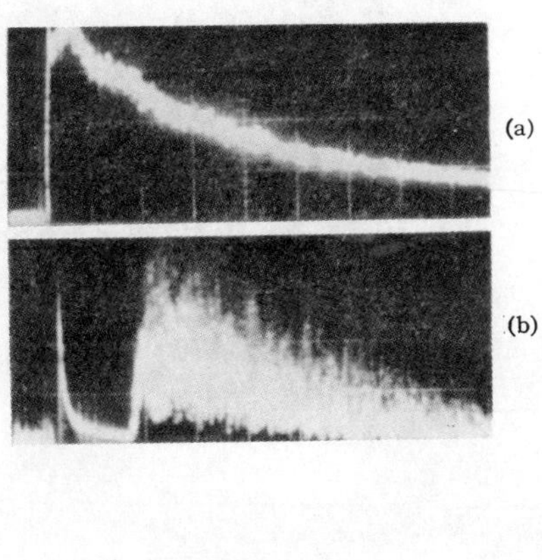

FIG. 12. Light output pulse from the exciting lamp in the green (a) and stimulated emission output pulse from ruby (b) (time scale 200 μsec/division). The excitation energy for (b) was approximately twice threshold. The initial spike in (b) is an electrical transient arising from the trigger circuitry.

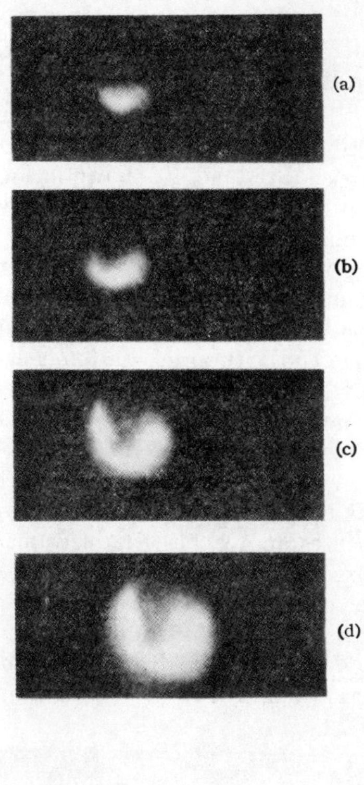

FIG. 14. Change in stimulated emission output beam with input energy for a ruby with a small hole in one totally reflecting end. The angular width of the largest pattern is about 1°.

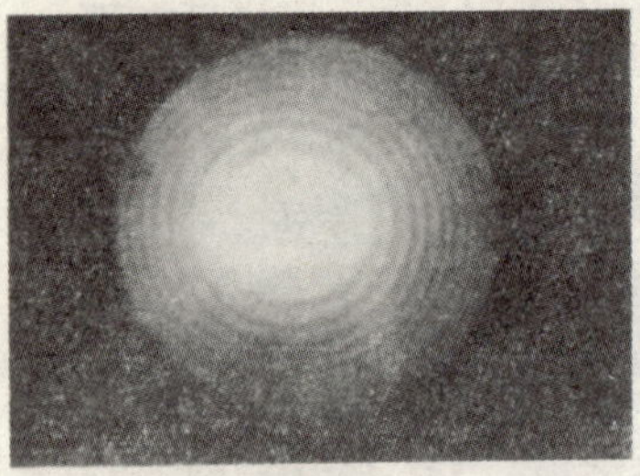

FIG. 15. Output beam pattern for a partially silvered ruby.

ation is also readily understood in terms of coupling to a resonant cavity. The peak power output follows a similar curve.

The energy output from a ruby with a partially silvered end face (7% transmitting) as a function of electrical energy input to the flash tube is given in Fig. 11. The slow initial rise of the curve is due to integration of spontaneous fluorescent radiation which makes up a relatively large portion of the light reaching the phototube detector at input energy levels near threshold. At higher input energies, the curve rises more rapidly as the spontaneous emission radiation becomes smaller in relation to the total output. Since the efficiency of the flash tube increases as the input energy increases, the curve does not represent light output as a function of *light* input. The peak power output, difficult to determine closely because of the pulsed nature of the radiation, again follows the same general curve, but with a faster initial rise, since the spontaneous emission is not integrated. Under the conditions of Fig. 11, the peak output power with 1730 joule input was more than 3 kw and the total output energy was 0.4 joule (using a different ruby sample, a peak power output in excess of 5 kw and an integrated output energy of near 1 joule was measured).

Characteristics of Emitted Radiation

The exciting flash tube output in the green and the stimulated emission pulse with an input approximately twice threshold are shown in Fig. 12, the time scale for both traces being 200 μsec/division. The stimulated emission pulse (b) is seen to lag the input pulse (a) by approximately 300 μsec, the time required for the buildup of excess population in the ^2E level. The output pulse from a ruby with a 3% transmitting silver film on one end with an excitation energy approximately twice threshold is shown in Fig. 13 on an expanded time scale. The amplitude of the individual pulses making up the output are seen to vary in an erratic manner and the average pulse repetition rate is seen to decrease markedly as the input light intensity decays. The rise time of the pulses approaches the instrumental rise time of ~0.1 μsec and does not appear to depend upon the excitation level. No apparent variation of pulse repetition rate or rise time as a function of coupling was found.

Mode selection by means of the following experiment was tried. The silver coating from one end of the crystal was removed and replaced by a mirror parallel to and 5 cm from this end of the ruby. The beam angle decreased and the oscillatory behavior of the output changed, exhibiting a much greater separation between the individual pulses.

The spatial nature of the stimulated emission from a ruby with a small coupling hole varies in an interesting way from that from the same ruby with a partially transmitting end coating. For a particular ruby with a 0.125-in. coupling hole in a totally reflecting end coating, the nature of the spot varied with input power as shown in Fig. 14. The image of the spot on a transluscent screen located 75 cm from the end of the ruby was photographed from behind the screen. At input energies just above threshold, the output starts as a barely discernible, small, approximately round spot not shown in the figure. As the input energy increases, the spot spreads in two directions to form, at input levels approximately twice threshold, a horseshoe shaped pattern of approximately one degree angular width.

The same ruby with a partially transmitting end coating gives, slightly above threshold, a small round spot (of smaller linear dimension than the ruby face at close distances) surrounded by barely discernible rings. As the input energy is increased, the central spot appears to grow and fill in the space between the rings. Such behavior might arise if only a portion of the ruby were active. As the input energy is increased, the active portion of the ruby increases in size. The ring pattern from another ruby with a partially transmitting end is shown in Fig. 15. The size of the central spot projects to about 3 mm in diameter at the face of the ruby (the diameter of the ruby is about 9 mm) and has an angular width of 10^{-2} rad. The angular diameter of the rings varies as $(n+a)^{\frac{1}{2}}$, where n is an integer and a is a constant. The appearance of these rings is discussed by Wagner and Birnbaum[9] and explained on the basis of a Fabry-Perot type phenomena within the ruby.

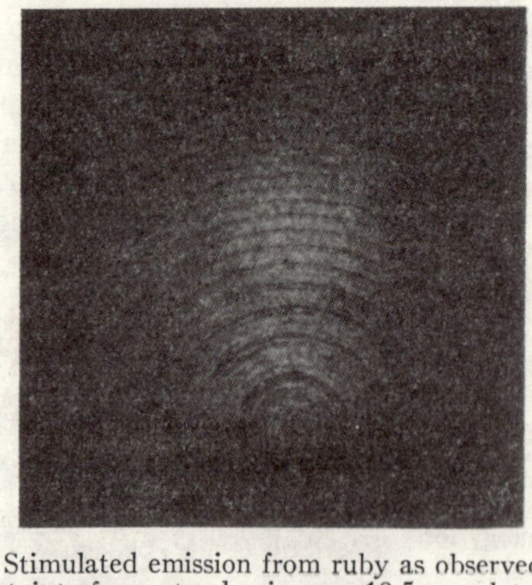

FIG. 16. Stimulated emission from ruby as observed through a Fabry-Perot interferometer having an 18.5-cm plate separation (interorder spacing 0.0135 A). Bright fringes are the white areas of the figure.

[9] W. G. Wagner and G. Birnbaum (to be published).

The spectral width of the emitted radiation was investigated with a Fabry-Perot interferometer. In general, one or more sharp components were observed of spectral width $\sim 6\times 10^{-4}$ A superimposed on a broader background (see Fig. 16). The fringe patterns were not, however, reproducible; that is, the number and relative intensity of the components changed in each photographic exposure.

DISCUSSION

The variation in the behavior of stimulated emission in ruby can be explained on the basis of the discussion in Part I. It was asserted that badly strained crystals scatter the energy into many cavity modes and that from the curves presented there it was expected that a clearly defined threshold would not be present in such cases. This is corroborated by the fact that the rubies which exhibit the pronounced beam and spectral narrowing, when viewed with polarized light, appear to be less strained than the others.

Several theories to account for the oscillatory nature of the output based upon relaxation behavior have been advanced.[8,10] There is, however, a possibility that some type of mode-hopping process is also taking place, since the frequency of the inverted transition is certainly being swept in time during the oscillation pulse due to temperature changes and also due to a time-varying magnetic field produced by the current flow in the helical flash tube. Moreover, it is difficult to explain the appearance of several extremely narrowed lines observed with the Fabry-Perot interferometer unless some sort of mode sweeping is invoked. Further experimental work is indicated before the characteristics of the emitted light can be fully understood.

ACKNOWLEDGMENTS

The authors are indebted to R. C. Pastor and H. Kimura for the chemical analysis of the rubies, to J. Bernath and M. Himber for the frabrication of the ruby samples, and to C. R. Duncan for the electronic instrumentation. We have benefited from helpful discussions with G. Birnbaum, R. W. Hellwarth, and W. G. Wagner.

[10] R. W. Hellwarth, Phys. Rev. Letters 6, 9 (1961).

Coherence in Spontaneous Radiation Processes

R. H. Dicke
Palmer Physical Laboratory, Princeton University, Princeton, New Jersey
(Received August 25, 1953)

By considering a radiating gas as a single quantum-mechanical system, energy levels corresponding to certain correlations between individual molecules are described. Spontaneous emission of radiation in a transition between two such levels leads to the emission of coherent radiation. The discussion is limited first to a gas of dimension small compared with a wavelength. Spontaneous radiation rates and natural line breadths are calculated. For a gas of large extent the effect of photon recoil momentum on coherence is calculated. The effect of a radiation pulse in exciting "super-radiant" states is discussed. The angular correlation between successive photons spontaneously emitted by a gas initially in thermal equilibrium is calculated.

IN the usual treatment of spontaneous radiation by a gas, the radiation process is calculated as though the separate molecules radiate independently of each other. To justify this assumption it might be argued that, as a result of the large distance between molecules and subsequent weak interactions, the probability of a given molecule emitting a photon should be independent of the states of other molecules. It is clear that this model is incapable of describing a coherent spontaneous radiation process since the radiation rate is proportional to the molecular concentration rather than to the square of the concentration. This simplified picture overlooks the fact that all the molecules are interacting with a common radiation field and hence cannot be treated as independent. The model is wrong in principle and many of the results obtained from it are incorrect.

A simple example will be used to illustrate the inadequacy of this description. Assume that a neutron is placed in a uniform magnetic field in the higher energy of the two spin states. In due course the neutron will spontaneously radiate a photon via a magnetic dipole transition and drop to the lower energy state. The probability of finding the neutron in its upper energy state falls exponentially to zero.[1,2]

If, now, a neutron in its ground state is placed near the first excited neutron (a distance small compared with a radiation wavelength but large compared with a particle wavelength and such that the dipole-dipole interaction is negligible), the radiation process would, according to the above hypothesis of independence, be unaffected. Actually, the radiation process would be strongly affected. The initial transition probability would be the same as before but the probability of finding an excited neutron would fall exponentially to one-half rather than to zero.

The justification for these assertions is the following: The initial state of the neutron system finds neutron 1 excited and neutron 2 unexcited. (It is assumed that the particles have nonoverlapping space functions, so that particle symmetry plays no role.) This initial state may be considered to be a superposition of the triplet and singlet states of the particles. The triplet state is capable of radiating to the ground state (triplet) but the singlet state will not couple with the triplet system. Consequently, only the triplet part is modified by the coupling with the field. After a long time there is still a probability of one-half that a photon has not been emitted. If, after a long period of time, no photon has been emitted, the neutrons are in a singlet state and it is impossible to predict which neutron is the excited one.

On the other hand, if the initial state of the two neutrons were triplet with $s=1$, $m_s=0$ namely a state with one excited neutron, a photon would be certain to be emitted and the transition probability would be just double that for a lone excited neutron. Thus, the presence of the unexcited neutron in this case doubles the radiation rate.

In recent years the excitation of correlated states of atomic radiating systems with the subsequent emission of spontaneous coherent radiation has become an important technique for nuclear magnetic resonance research.[3] The description usually given of this process is a classical one based on a spin system in a magnetic field. The purpose of this note is to generalize these results to any system of radiators with a magnetic or electric dipole transition and to see what effects, if any, result from a quantum mechanical treatment of the radiation process. Most of the previous work[4] was quite early and not concerned with the problems being considered here. In a subsequent article to be published in the *Review of Scientific Instruments* some of these results will be applied to the problem of instrumentation for microwave spectroscopy.

In this treatment the gas as a whole will be considered as a single quantum-mechanical system. The problem will be one of finding those energy states representing correlated motions in the system. The spontaneous emission of coherent radiation will accompany transitions between such levels. In the first problem to be considered the gas volumes will be assumed to have

[1] W. Heitler, *The Quantum Theory of Radiation* (Clarendon Press, Oxford, 1936), first edition, p. 112.
[2] E. P. Wigner and V. Weisskopf, Z. Physik **63**, 54 (1930).
[3] E. L. Hahn, Phys. Rev. **77**, 297 (1950); **80**, 580 (1950).
[4] E.g., W. Pauli, *Handbuch der Physik* (Springer, Berlin, 1933), Vol. 24, Part I, p. 210; G. Wentzel, *Handbuch der Physik* (Springer, Berlin, 1933), Vol. 24, Part I, p. 758.

Reprinted with permission from *Phys. Rev.*, vol. 93, pp. 99–110, Jan. 1, 1954.

dimensions small compared with a radiation wavelength. This case, which is of particular importance for nuclear magnetic resonance experiments and some microwave spectroscopic applications, is treated first quantum mechanically and then semiclassically, the radiation process being treated classically. A classical model is also described. In the next case to be considered the gas is assumed to be of large extent. The effect of molecular motion on coherence and the effect on coherence of the recoil momentum accompanying the emission of a photon are discussed. Finally, the two principal methods of exciting coherent states by the absorption of photons from an intense radiation pulse or the emission of photons by the gas are discussed. Calculations of these two effects are made for the gas system initially in thermal equilibrium. The effect of photon emission on inducing coherence is discussed as a problem in the angular correlation of the emitted photons.

DIPOLE APPROXIMATION

The first problem to be considered is that of a gas confined to a container the dimensions of which are small compared with a wavelength. It is assumed that the walls of the container are transparent to the radiation field. In order to avoid difficulties arising from collision broadening it will be assumed that collisions do not affect the internal states of the molecules. It will be assumed that the transition under question takes place between two nondegenerate states of the molecule. The assumption of nondegeneracy is made in order to limit the scope of the problem to its bare essentials. It might be assumed that nondegenerate states are present as a result of a uniform static electric or magnetic field acting on the gas. Actually, for many of the questions being discussed it is not essential that the degeneracies be split. Also, it will be assumed that there is insufficient overlap in the wave functions of separate molecules to require that the wave functions be symmetrized.

Since it is assumed that internal coordinates of the individual molecules are unaffected by collisions and but two internal states are involved for each molecule, the wave function for the gas may be written conveniently in a representation diagonal in the center-of-mass coordinates and the internal energies of the molecules. The internal energy coordinate takes on only two values. Omitting for the moment the radiation field, the Hamiltonian for an n molecule gas can be written

$$H=H_0+E\sum_{j=1}^{n} R_{j3}, \tag{1}$$

where $E=\hbar\omega=$ molecular excitation energy. Here H_0 acts on the center-of-mass coordinates and represents the translational and intermolecular interaction energies of the gas. ER_{j3} is the internal energy of the jth molecule and has eigenvalues $\pm\frac{1}{2}E$. H_0 and all the R_{j3} commute with each other. Consequently, energy eigenfunctions may be chosen to be simultaneous eigenfunctions of H_0, R_{13}, R_{23}, $\cdots$, R_{n3}.

Let a typical energy state be written as

$$\psi_{gm}=U_g(\mathbf{r}_1\cdots\mathbf{r}_n)[++-+\cdots]. \tag{2}$$

Here $\mathbf{r}_1\cdots\mathbf{r}_n$ designates the center-of-mass coordinates of the n molecules, and $+$ and $-$ symbols represent the internal energies of the various molecules. If the number of $+$ and $-$ symbols are denoted by n_+ and n_-, respectively, then m is defined as

$$m=\tfrac{1}{2}(n_+-n_-),$$
$$n=n_++n_-=\text{number of gaseous molecules.} \tag{3}$$

If the energy of motion and mutual interaction of the molecules is denoted by E_g, then the total energy of the system is

$$E_{gm}=E_g+mE. \tag{4}$$

It is evident that the index m is integral or half-integral depending upon whether n is even or odd. Because of the various orders in which the $+$ and $-$ symbols can be arranged, the energy E_{gm} has a degeneracy

$$\frac{n!}{(\frac{1}{2}n+m)!(\frac{1}{2}n-m)!}. \tag{5}$$

This degeneracy has its origin in the internal coordinates only.

In addition, the wave function may have additional degeneracy from the center-of-mass coordinates. It should be noted in this connection that the degeneracy of the total wave function will depend upon whether or not the molecules are regarded as distinguishable or not.

If the molecules are indistinguishable, the symmetry of U_g will depend upon the symmetries of the wave function under interchanges of internal coordinates. For example, the states with all molecules excited are symmetric under an interchange of the internal coordinates of any two molecules. Consequently, for these states U_g must be symmetric for Bose molecules and antisymmetric for Fermi molecules. The limitations of symmetry are normally without physical significance as it is assumed that the gas is of such low density that the various molecules have nonoverlapping wave functions.

Of the Hamiltonian equation (1), H_0 operates on the center-of-mass coordinates only and gives

$$H_0U_g=E_gU_g, \tag{6}$$

whereas R_{j3} operates on the plus or minus symbol in the jth place corresponding to the internal energy of the jth molecule. Except for the factor $\frac{1}{2}$, it is analogous to one of the Pauli spin operators. As operators similar to the other two Pauli operators are also needed in this development, the properties of all three are listed here.

$$\begin{aligned} &\qquad\quad\ j\\ &\qquad\quad\ \downarrow\\ R_{j1}&[\cdots\pm\cdots]=\tfrac{1}{2}[\cdots\mp\cdots],\\ R_{j2}&[\cdots\pm\cdots]=\pm\tfrac{1}{2}i[\cdots\mp\cdots],\\ R_{j3}&[\cdots\pm\cdots]=\pm\tfrac{1}{2}[\cdots\pm\cdots]. \end{aligned} \tag{7}$$

It is also convenient to define the operators

$$R_k = \sum_{j=1}^{n} R_{jk}, \quad k=1, 2, 3, \tag{8}$$

and the operator

$$R^2 = R_1^2 + R_2^2 + R_3^2. \tag{9}$$

In this notation the Hamiltonian becomes

$$H = H_0 + ER_3, \tag{10}$$

and

$$R_3\psi_{gm} = m\psi_{gm}. \tag{11}$$

To complete the description of the dynamical system, there must be added to the Hamiltonian that of the radiation field and the interaction term between field and the molecular system.

For the purpose of definiteness the ineraction of a molecule with the electromagnetic field will be assumed to be electric dipole. The main results are actually independent of the type of coupling. The interaction energy of the jth molecule with the electromagnetic field can be written as

$$-\mathbf{A}(\mathbf{r}_j)\cdot \sum_{k=1}^{N-1} \frac{e_k}{m_k c}\mathbf{P}_k. \tag{12}$$

Here the configuration coordinates of the molecule are taken to be the center-of-mass coordinates and the coordinates relative to the center of mass of any $N-1$ of the N particles which constitute the jth molecule. e_k and m_k are the charge and mass of the kth particle, and $\mathbf{P}_k$ is the momentum conjugate to the position of the kth particle relative to the center of mass. The molecule is assumed electrically neutral.

Since $\mathbf{P}_k$ is an odd operator, it has only off-diagonal elements in a representation with internal energy diagonal. Hence the general form of Eq. (12) is

$$-\mathbf{A}(\mathbf{r}_j)\cdot(\mathbf{e}_1 R_{j1} + \mathbf{e}_2 R_{j2}). \tag{13}$$

$\mathbf{e}_1$ and $\mathbf{e}_2$ are constant real vectors the same for all molecules. The total interaction energy then becomes

$$H_1 = -\textstyle\sum_j \mathbf{A}(\mathbf{r}_j)\cdot(\mathbf{e}_1 R_{j1} + \mathbf{e}_2 R_{j2}). \tag{14}$$

Since the dimensions of the gas cell are small compared with a wavelength, the dependence of the vector potential on the center of mass of the molecules can be omitted and the interaction energy (12) becomes

$$H_1 = -\mathbf{A}(0)\cdot(\mathbf{e}_1 R_1 + \mathbf{e}_2 R_2). \tag{15}$$

Since the interaction term Eq. (15) does not contain the center-of-mass coordinates, the selection rule on the molecular motion quantum number g is $\Delta g = 0$. Consequently there is no Doppler broadening of the transition frequency. This results solely from the small size of the gas container.[5]

The operators R_1, R_2, and R_3, apart from a factor of $\hbar$, obey the same commutation relations as the three components of angular momentum. Consequently, the interaction operator Eq. (15) obeys the selection rule $\Delta m = \pm 1$. In general, it has nonvanishing matrix elements between a given state Eq. (2) and a large number of states with $\Delta m = \pm 1$. In order to simplify the calculation of spontaneous radiation transitions, it is desirable that a set of stationary states be selected in such a way that the interaction term has matrix elements joining a given state with, at most, one state of higher and lower energy, respectively. Because of the very close analogy between this formalism and that of a system of particles of spin $\frac{1}{2}$, known results can be taken over from the spin formalism.

In a manner similar to an angular momentum formalism,[6] the operations H and R^2 commute; consequently, stationary states can be chosen to be eigenstates of R^2. These new states are linear combinations of the states of Eq. (2). The operator R^2 has eigenvalues $r(r+1)$. r is integral or half-integral and positive, such that

$$|m| \leqslant r \leqslant \tfrac{1}{2}n. \tag{16}$$

The eigenvalue r will be called the "cooperation number" of the gas. Denote the new eigenstates by

$$\psi_{gmr}. \tag{17}$$

Here

$$H\psi_{gmr} = (E_g + mE)\psi_{gmr}, \tag{18}$$

$$R^2\psi_{gmr} = r(r+1)\psi_{gmr}. \tag{19}$$

The degeneracy of the stationary states is not completely removed by introducing R^2. The state (g, m, r) has a degeneracy

$$\frac{n!(2r+1)}{(\frac{1}{2}n+r+1)!(\frac{1}{2}n-r)!}. \tag{20}$$

The complete set of eigenstates ψ_{gmr} may be specified in the following way: the largest value of m and r is

$$r = m = \tfrac{1}{2}n.$$

This state is nondegenerate in the internal coordinates and may be written as

$$\psi_{g,\frac{1}{2}n,\frac{1}{2}n} = U_g\cdot[++\cdots+]. \tag{20a}$$

All the states with this same value of $r=\frac{1}{2}n$, but with different values of m, are nondegenerate also and may be generated as[7]

$$\psi_{gmr} = [(R^2 - R_3^2 - R_3)^{-\frac{1}{2}}(R_1 - iR_2)]^{r-m}\psi_{grr}. \tag{21}$$

The operator $R_1 - iR_2$ reduces the m index by unity every time it is applied and the fractional power operator is to preserve the normalization of the wave function.[8] The fractional power operator is defined as having positive eigenvalues only.

[5] R. H. Dicke, Phys. Rev. **89**, 472 (1953).

[6] E. U. Condon and G. H. Shortley, *The Theory of Atomic Spectra* (Cambridge University Press, Cambridge, 1935), pp. 45–49.

[7] See reference 6, p. 48, Eq. (3).

[8] See reference 6, p. 48.

The state $\psi_{g,\frac{1}{2}n-1,\frac{1}{2}n}$ is one of n states with this value of m. The remaining $n-1$ states should be chosen to be orthogonal to this state, orthogonal to each other, and normalized. Since these remaining $n-1$ states are not states of $r=\frac{1}{2}n$, they must be states of $r=\frac{1}{2}n-1$, the only other possibility. Again the complete set of states with this value of r can be generated using Eq. (21), where now $r=\frac{1}{2}n-1$, and the operator in Eq. (21) is applied to each of the $n-1$ orthogonal states of $r=m=\frac{1}{2}n-1$. This procedure can be repeated until all possible values of r are exhausted, in which case all the stationary states have been defined.

With this definition of the stationary states, the interaction energy operator has matrix elements joining a given state of the gas to but two other states. Aside from the factor involving the radiation field operator, the matrix elements of the interaction energy may be written[8]

$$(g,r,m|\mathbf{e}_1R_1+\mathbf{e}_2R_2|g,r,m\mp 1)$$
$$=\tfrac{1}{2}(\mathbf{e}_1\pm i\mathbf{e}_2)[(r\pm m)(r\mp m+1)]^{\frac{1}{2}}. \quad (23)$$

Transition probabilities will be proportional to the square of the matrix elements. In particular, the spontaneous radiation probabilities will be

$$I=I_0(r+m)(r-m+1). \quad (24)$$

Here, by setting $r=m=\frac{1}{2}$, it is evident that I_0 is the radiation rate of a gas composed of one molecule in its excited state. I_0 has the value[9]

$$I_0=\frac{4}{3}\frac{\omega^2}{c}\left|\left(\sum_k\frac{e_k\mathbf{P}_k}{m_kc}\right)_{+-}\right|^2=\frac{1}{3}\frac{\omega^2}{c}|\mathbf{e}_1-i\mathbf{e}_2|^2$$
$$=\frac{1}{3}\frac{\omega^2}{c}(e_1^2+e_2^2). \quad (25)$$

If $m=r=\frac{1}{2}n$ (i.e., all n molecules excited),

$$I=nI_0. \quad (26)$$

Coherent radiation is emitted when r is large but $|m|$ small. For example, for even n let

$$r=\tfrac{1}{2}n,\quad m=0;\quad I=\tfrac{1}{2}n(\tfrac{1}{2}n+1)I_0. \quad (27)$$

This is the largest rate at which a gas with an even number of molecules can radiate spontaneously. It should be noted that for large n it is proportional to the square of the number of molecules.

Because of the fact that with the choice of stationary states given by Eq. (21) a given state couples with but one state of lower energy, this radiation rate [Eq. (27)], is an absolute maximum. Any superposition state will radiate at the rate

$$I=I_0\sum_{r,m}P_{r,m}(r+m)(r-m+1)$$
$$=I_0\langle(R_1+iR_2)(R_1-iR_2)\rangle, \quad (28)$$

where $P_{r,m}$ is the probability of being in the state r, m.

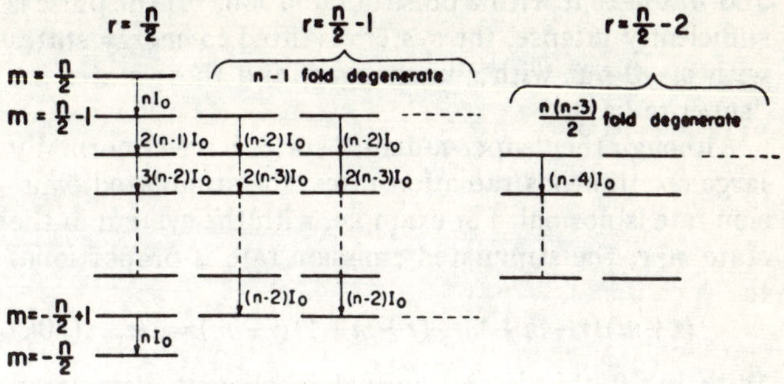

FIG. 1. Energy level diagram of an n-molecule gas, each molecule having 2 nondegenerate energy levels. Spontaneous radiation rates are indicated. $E_m=mE$.

There are no interference terms. Consequently, no superposition state can radiate more strongly than Eq. (27). An energy level diagram which shows the relative magnitudes of the various radiation probabilities is given in Fig. 1.

States with a low "cooperation number" are also highly correlated but in such a way as to have abnormally low radiation rates. For example, a gas in the state $r=m=0$ does not radiate at all. This state, which exists only for an even number of molecules, is analogous to a classical system of an even number of oscillators swinging in pairs oppositely phased.

The energy trapping which results from the internal scattering of photons by the gas appears naturally in the formalism. As an example, consider an initial state of the gas for which one definite molecule, and only this molecule, is excited. The gas at first radiates at the normal incoherent rate for a short time and thereafter fails to radiate. The probability of a photon's being emitted during the radiating period is $1/n$. These results follow from the fact that the assumed state is a linear superposition of the various states with $m=1-n/2$, and that $1/n$ is the probability of being in the state $r=\frac{1}{2}n$. The probability that the energy will be "trapped" is $(n-1)/n$. This is analogous to the radiation by a classical oscillator when $n-1$ similar unexcited oscillators are near. The solution of this classical problem shows that only $1/n$ of the excitation energy is radiated. The remainder appears in nonradiating normal modes of the system.

For want of a better term, a gas which is radiating strongly because of coherence will be called "super-radiant." There are two obvious ways in which a "super-radiant" state may be excited. First, if all the molecules be excited, the gas is in the state characterized by

$$r=m=\tfrac{1}{2}n. \quad (29)$$

As the system radiates it passes to states of lower m with r unchanged. This will take the system to the "super-radiant" region $m\sim 0$.

Another way in which such a state can be excited is to start with the gas in its ground state,

$$r=-m=\tfrac{1}{2}n, \quad (30)$$

[9] Reference 1, p. 106.

and irradiate it with a pulse of radiation.[9a] If the pulse is sufficiently intense, the system is lifted to energy states with $m\sim 0$ but with r unchanged, and these states are "super-radiant."

Although the "super-radiant" states have abnormally large spontaneous radiation rates, the stimulated emission rate is normal. For example, with the system in the state m, r, the stimulated emission rate is proportional to

$$(r+m)(r-m+1)-(r+m+1)(r-m)=2m. \quad (30a)$$

With $m>0$ this is the normal incoherent stimulated emission rate. For $m<0$ this becomes the negative of the incoherent absorption rate.

As has been pointed out, the pulse technique for exciting "super-radiant" states is commonly used in nuclear magnetic resonance experiments. Here there is one important point that needs clarification, however. Instead of starting in the highly organized state given by Eq. (30) the pulse is applied to a system that is in thermal equilibrium at high temperatures. For example, if the system be a set of proton spins, the energy necessary to turn a spin over in the magnetic field may be about

$$E\sim 10^{-5}kT. \quad (31)$$

Under these conditions the two spin states of the proton are very nearly equally populated and it might be expected that thermal equilibrium would imply a badly disorganized system. The randomness in the initial state does not imply, however, complete randomness in m and r. For a gas with n, large states of low r have a high degeneracy. These states have a high statistical weight and are favored. However, Eq. (16) sets a lower bound on r for any m. The result is a relatively small range of values of m and r. For a system with n molecules in thermal equilibrium the mean square deviation from the mean of m is

$$n/4-\bar{m}^2/n. \quad (32)$$

Here $\bar{m}$ is the mean of m and is for high temperatures equal to

$$\bar{m}=-\tfrac{1}{4}nE/kT. \quad (33)$$

For a definite value of m the mean value of $r(r+1)$ is

$$m^2+\tfrac{1}{2}n, \quad (34)$$

and the mean square deviation is

$$\tfrac{1}{4}n^2-m^2. \quad (35)$$

The expression (32)–(35) may be easily derived using the density matrix formalism assuming the appropriate statistical ensemble.

It is hence clear that if

$$\bar{m}^2\gg n\gg 1, \quad (36)$$

the percentage deviation from the mean of m is small, that the percent deviation from the mean of $r(r+1)$ is small, and that the mean of $r(r+1)$ is approximately the smallest value compatible with the mean value of m. Thus, in the case of a gas system at high temperature, for sufficiently large n, values of m and r cluster to such an extent that the system may be considered as approximately in a state of definite $r=m=-nE/4kT$. If this gas is excited by a pulse of the proper intensity to excite states $m\sim 0$, the radiation rate after the pulse is approximately

$$I\cong I_0 r(r+1)\cong I_0 n^2(E/4kT)^2, \quad (37)$$

which is proportional to n^2 and hence coherent. A better calculation good for all temperatures gives the result [see Eq. (78) with $\theta=90°$]

$$I=\tfrac{1}{4}I_0 n(n-1)\tanh^2(E/2kT)+\tfrac{1}{2}nI_0. \quad (37a)$$

SEMICLASSICAL TREATMENT

For the spontaneous radiation from super-radiant states ($m\sim 0$) a semiclassical treatment is generally adequate. This method, which is a generalization of the well-known picture used in describing radiation from a nuclear spin system,[10] treats the molecular systems quantum mechanically but calculates the radiation process classically. In the following calculation the gas system will be assumed to be excited by a radiation pulse, which excites it from thermal equilibrium to a set of super-radiant states. To calculate the radiation rate, the expectation value of the electric dipole moment is treated as a classical dipole. When the gas contains a large number of molecules the dipole moment of the gas as a whole should be given by the sum of the expectation values of the individual dipole moments.

In thermal equilibrium the gas may be considered as having n_- molecules in the ground state and n_+ molecules in the excited state. A molecule which is initially in its ground state is assumed to be thrown into a superposition state of + and − by the radiation pulse. It is assumed that there is a unity probability ratio. The internal part of the wave function of the molecules after the pulse is given by

$$\psi_+=\frac{1}{\sqrt{2}}\left\{[+]\exp\left(-i\frac{\omega}{2}t\right)+[-]\exp i\left(\frac{\omega}{2}t+\delta\right)\right\}. \quad (38)$$

This is the most general form for ψ_+ apart from a possible multiplication phase factor. Here δ is a phase given by the phase of the exciting pulse. In a similar way a molecule in the excited state has its wave function converted to

$$\psi_-=\frac{1}{\sqrt{2}}\left\{[-]\exp i\frac{\omega}{2}t-[+]\exp\left(-i\frac{\omega}{2}t-i\delta\right)\right\}. \quad (39)$$

Instead of calculating the expectation value of the electric dipole moment it is more convenient to calculate the expectation value of the polarization current of the

[9a] See F. Bloch and I. I. Rabi, Revs. Modern Phys. 17, 237 (1945), for a discussion of the effect of a pulse on the analogous spin-½ system.

[10] F. Bloch, Phys. Rev. 70, 460 (1946).

jth molecule given by

$$\left(\sum_{k=1}^{N-1}\frac{e_k\mathbf{P}_k}{m_k}\right)=c\langle\mathbf{e}_1R_{j1}+\mathbf{e}_2R_{j2}\rangle$$

$$=\pm\tfrac{1}{2}c[\mathbf{e}_1\cos(\omega t+\delta)+\mathbf{e}_2\sin(\omega t+\delta)]. \quad (40)$$

The plus sign is obtained from the plus state, Eq. (38), and the negative sign from Eq. (39). Note the oscillating time dependence which results from the states being energy-superposition states. The polarization current for the gas as a whole is then

$$\mathbf{j}=(n_+-n_-)(c/2)[\mathbf{e}_1\cos(\omega t+\delta)+\mathbf{e}_2\sin(\omega t+\delta)]. \quad (41)$$

The radiation rate calculated classically is then[11]

$$I=\frac{2}{3}\frac{\omega^2}{c_3}|\mathbf{J}^2|=\frac{1}{12}\frac{\omega^2}{c}(n_+-n_-)^2(e_1^2+e_2^2). \quad (42)$$

In thermal equilibrium $n_+/n_-=\exp(-E/kT)$, from which

$$n_+-n_-=n\tanh(E/2kT). \quad (43)$$

Substituting into Eq. (42) gives the classical radiation rate

$$I=\frac{1}{12}\frac{\omega^2}{c}n^2(e_1^2+e_2^2)\tanh^2\left(\frac{E}{2kT}\right). \quad (44)$$

This may be compared with the quantum-mechanical result [Eq. (37a) and Eq. (25)]. For large n the two results are equal.

CLASSICAL MODEL

When the gas is in a state of definite "cooperation number" r which has a very large value, it is possible to represent it in its interaction with the electromagnetic field by a simple classical model. The energy-level spacing and the matrix elements joining adjacent levels are similar to those of a rotating top of large angular momentum and carrying an electric dipole moment. The details depend upon $\mathbf{e}_1$ and $\mathbf{e}_2$, which in turn depend on the nature of the original states. Let us consider a specific example. Assume that the radiators are atoms having a $^1P_1$ excited state and a $^1S_0$ ground state. Assume that the degeneracy of the excited state is split by a magnetic field in the z direction and that the $m_l=1$ excited level is being used. Under these conditions $\mathbf{e}_1$ and $\mathbf{e}_2$ are orthogonal to each other and the z axis, and the system has energy levels and interactions with the field identical with those of a spinning top having an electric dipole moment along its axis and precessing about the z axis as a result of an interaction with a static *electric* field in that direction. Consequently, since large quantum numbers are involved, to a good approximation the gas can be replaced by this classical model, which consists of a spinning top, in calculating both the interaction of the field on the gas and *vice versa*.

[11] Reference 1, p. 26.

RADIATION LINE BREADTH AND SHAPE

Under conditions for which the above "classical model" is valid, it is easy to calculate the natural line breadth and shape factor. This is of considerable importance in microwave spectroscopy. It has been customary to regard the natural line breadth as too small to be of any practical importance. However, as will be seen below, when coherence is properly taken into account the natural radiation breadth of the line may be far from negligible.

Using the above classical model, the angle between the spin axis and the z axis (the polar angle) will be designated as φ. In this approximation the quantum number m may be replaced by

$$m=r\cos\varphi, \quad (44a)$$

from which, using Eq. (24), the radiation rate becomes

$$I=I_0r^2\sin^2\varphi. \quad (44b)$$

Also, the internal energy of the gas is

$$mE=rE\cos\varphi. \quad (44c)$$

Balancing the radiation rate to the energy loss of the gas gives

$$\dot{\varphi}=(I_0r/E)\sin\varphi,$$

from which, assuming $\varphi=90°$ if $t=0$,

$$\sin\varphi=\operatorname{sech}(\alpha t),$$

where $\alpha=I_0r/E$. The radiated wave has the following form as a function of time:

$$A(t)=\begin{cases}e^{i\omega t}\sin\varphi, & t>0,\\ 0, & t<0,\end{cases}\qquad \hbar\omega=E.$$

The Fourier transform gives the line shape and has the value

$$\mathfrak{a}(\beta)=\left(\frac{\pi}{2}\right)^{\frac{1}{2}}\frac{1}{\alpha}\operatorname{sech}\left(\frac{\pi}{2}\frac{\beta-\omega}{\alpha}\right). \quad (44d)$$

It should be noted that this is not of the usual Lorentz form. The line width at half-intensity points is

$$\Delta\omega=1.12I_0r/E=1.12\gamma r. \quad (44e)$$

Here γ is the line width at half-intensity points for the radiation from isolated single molecules. Putting in the maximum value of r gives a line breadth of $\Delta\omega=1.12\gamma n/2$, which is generally very substantially larger than γ.

RADIATION FROM A GAS OF LARGE EXTENT

A classical system of simple harmonic oscillators distributed over a large region of space can be so phased relative to each other that coherent radiation is obtained in a particular direction. It might be expected also that the radiating gas under consideration would have energy levels such that spontaneous radiation occurs coherently in one direction.

It will be assumed that the gas occupies a region having dimensions generally larger than radiation wavelength but small compared with the reciprocal of the natural line width,

$$\Delta k = \Delta\omega / c.$$

It is necessary to turn again to the general expression for the interaction term in the Hamiltonian equation (13). The vector potential operator can be expanded in plane waves:

$$\mathbf{A}(\mathbf{r}) = \sum_{\mathbf{k}'} [\mathbf{v}_{\mathbf{k}'} \exp(i\mathbf{k}' \cdot \mathbf{r}) + \mathbf{v}_{\mathbf{k}'}^* \exp(-i\mathbf{k}' \cdot \mathbf{r})]. \quad (45)$$

$\mathbf{v}_{\mathbf{k}'}$ and its Hermitian adjoint $\mathbf{v}_{\mathbf{k}'}^*$ are photon destruction and creation operators, respectively. After substituting Eq. (45) into (13), the interaction term becomes

$$H_1 = -\tfrac{1}{2} \sum_{\mathbf{k}'} \mathbf{v}_{\mathbf{k}'} \cdot (\mathbf{e}_1 - i\mathbf{e}_2) \sum_{j=1}^{n} R_{j+} \exp(i\mathbf{k}' \cdot \mathbf{r}_j)$$

$$-\tfrac{1}{2} \sum_{\mathbf{k}'} \mathbf{v}_{\mathbf{k}'}^* \cdot (\mathbf{e}_1 + i\mathbf{e}_2) \sum_{j=1}^{n} R_{j-} \exp(-i\mathbf{k}' \cdot \mathbf{r}_j), \quad (46)$$

where $R_{j\pm} = R_{j1} \pm iR_{j2}$. In this expression, terms involving the product of the photon creation operator and the "excitation operator" R_{j+}, etc., have been dropped as these terms do not lead to first-order transitions for which energy is conserved. The form of Eq. (46) suggests defining the operators:

$$\begin{aligned} R_{\mathbf{k}1} &= \sum_j (R_{j1} \cos\mathbf{k} \cdot \mathbf{r}_j - R_{j2} \sin\mathbf{k} \cdot \mathbf{r}_j), \\ R_{\mathbf{k}2} &= \sum_j (R_{j1} \sin\mathbf{k} \cdot \mathbf{r}_j + R_{j2} \cos\mathbf{k} \cdot \mathbf{r}_j). \end{aligned} \quad (47)$$

In terms of these operators the interaction energy becomes

$$H_1 = -\tfrac{1}{2} \sum_{\mathbf{k}'} (\mathbf{v}_{\mathbf{k}'} \cdot \mathbf{e} R_{\mathbf{k}'+} + \mathbf{v}_{\mathbf{k}'}^* \cdot \mathbf{e}^* R_{\mathbf{k}-'}), \quad (48)$$

where

$$R_{\mathbf{k}'\pm} = R_{\mathbf{k}'1} \pm iR_{\mathbf{k}'2} = \sum_{j=1}^{n} R_{j\pm} \exp(\pm i\mathbf{k}' \cdot \mathbf{r}_j),$$

$$\mathbf{e} = \mathbf{e}_1 - i\mathbf{e}_2.$$

For every direction of propagation $\mathbf{k}$ there are two orthogonal polarizations $\mathbf{v}_{\mathbf{k}}$ of $\mathbf{A}$. By a proper choice of polarization basis, the dot product of one of the basic polarizations with $\mathbf{e}$ can be assumed zero. This radiation oscillator is never excited and can be ignored. The orthogonal polarization is the one which couples with the gas. The polarization of emitted or absorbed radiation is uniquely given by the direction of propagation and need not be explicitly indicated.

The operators of Eq. (47), together with R_3, obey the angular momentum commutation relations. The operator

$$R_{\mathbf{k}}^2 = R_{\mathbf{k}1}^2 + R_{\mathbf{k}2}^2 + R_3^2 \quad (49)$$

commutes with the operators of Eq. (47) and with R_3. In Eq. (49) $\mathbf{k}$ is regarded as a fixed index. This operator does not commute with another one of the same type having a different index. Omitting for a moment the translational part of the wave function, wave functions may be so chosen as to be simultaneous eigenfunctions of the internal energy ER_3 and $R_{\mathbf{k}}^2$. They may be written as ψ_{mr} and are generated by an expression analogous to Eq. (21):

$$R_{\mathbf{k}}^2 \psi_{mr} = r(r+1)\psi_{mr}, \quad ER_3\psi_{mr} = mE\psi_{mr}. \quad (50)$$

By analogy with the development leading to Eq. (24) it is clear that these states represent correlated states of the gas for which radiation emitted in the $\mathbf{k}$ direction is coherent. Thus, coherence is limited to a particular direction only, provided the initial state of the gas is given by a function of the same type as Eq. (50). The selection rules for the absorption or emission of a photon with momentum $\mathbf{k}$ are

$$\Delta r = 0, \quad \Delta m = \pm 1. \quad (51)$$

The spontaneous radiation rate in the direction $\mathbf{k}$ is given by Eq. (24), where I and I_0 are now to be interpreted as radiation rates per unit solid angle in the direction $\mathbf{k}$. This may be written as

$$I(\mathbf{k}) = I_0(\mathbf{k})[(r+m)(r-m+1)]. \quad (51a)$$

If a photon is emitted or absorbed having a momentum $\mathbf{k}' \neq \mathbf{k}$, the selection rules are

$$\Delta r = \pm 1, 0; \quad \Delta m = \pm 1. \quad (52)$$

To prove this, it may be noted that the commutation relations of the $2n$ operators

$$\begin{aligned} R_{j1}' &= R_{j1} \cos(\mathbf{k} \cdot \mathbf{r}_j) - R_{j2} \sin(\mathbf{k} \cdot \mathbf{r}_j), \\ R_{j2}' &= R_{j1} \sin(\mathbf{k} \cdot \mathbf{r}_j) + R_{j2} \cos(\mathbf{k} \cdot \mathbf{r}_j), \end{aligned} \quad (53)$$

with those of Eq. (47) are of the same type as denoted by Condon and Shortley[12] as **T**. The selection rules satisfied by these operators are of the type given by Eq. (52).[13] The operators of Eq. (47), with $\mathbf{k} = \mathbf{k}'$, may be expressed as linear combinations of those of Eq. (53). Hence the operators of Eq. (47), with $\mathbf{k}$ replaced by $\mathbf{k}'$, satisfy the selection rules given by Eq. (52).

As was discussed previously in the dipole approximation, super-radiant states may be excited by irradiating the gas with radiation until states in the vicinity of $m = 0$ are excited. In the present case the incident radiation is assumed to be plane with a propagation vector $\mathbf{k}$. After excitation the gas radiates coherently in the $\mathbf{k}$ direction. Because of the selection rules Eq. (52), radiation in directions other than $\mathbf{k}$ tends to destroy the coherence with respect to the direction $\mathbf{k}$ by causing transitions generally to states of lower r.

DOPPLER EFFECT

Because of the occurrence of the center-of-mass coordinates in the "cooperation" operator Eq. (49), it fails to commute with H_0 [Eq. (1)]; hence eigenstates of $R_{\mathbf{k}}^2$ are generally not stationary. This is equivalent to the fact that relative motion of classical oscillators will gradually destroy the coherence of the emitted radiation. If, on the other hand, a set of classical oscillators all move with the same velocity, the state of coherence

[12] Reference 6, p. 59.
[13] Reference 6, pp. 60–61.

is stationary. The corresponding question in the case of the quantum mechanical system is whether there exist simultaneous eigenstates of H and $R_{\mathbf{k}}^2$ such that coherent radiation is emitted in a transition from one state to another. By starting with the state defined by

$$\psi_{srr}=(\exp i\mathbf{s}\cdot\textstyle\sum_j \mathbf{r}_j)\cdot[+++\cdots+],\quad r=n/2, \tag{54}$$

and using the method leading to Eq. (21), there is obtained the set of states

$$\psi_{smr}=[(R_{\mathbf{k}}^2-R_3^2-R_3)^{-\frac{1}{2}}(R_{\mathbf{k}1}-iR_{\mathbf{k}2})]^{r-m}\psi_{srr}. \tag{55}$$

If it is assumed that the gas is free, the functions Eq. (55) are simultaneous eigenfunctions of H and $R_{\mathbf{k}}^2$. Consequently, the coherence in the $\mathbf{k}$ direction is stationary.

These states are analogous to the classical oscillators all moving with the same speed. Note one important difference, however; from Eq. (55) the momentum of an excited molecule is always

$$\mathbf{p}_+=\hbar\mathbf{s}, \tag{56}$$

whereas if a molecule is in its ground state the momentum, as given by Eq. (55), is

$$\mathbf{p}_-=\hbar(\mathbf{s}-\mathbf{k}), \tag{57}$$

the difference being the recoil momentum of the photon. Thus, the coherent states Eq. (55) are always a superposition of states such that the excited molecules have one momentum and the unexcited have another. Hence it is clear that the recoil momentum given to a molecule when it radiates in the $\mathbf{k}$ direction does not produce a molecular motion which destroys the coherence but rather is required to preserve the coherence.

The gain or loss in photon energy which has its origin in the Doppler effect is equal to the loss or gain in the kinetic energy of a radiator which results from the photon-induced recoil. Expressed as a fractional shift in photon frequency, this is

$$\frac{\Delta\omega}{\omega}=\frac{\hbar(\mathbf{S}-\frac{1}{2}\mathbf{k})\cdot\mathbf{k}}{Mck}. \tag{58}$$

Here M is the molecular mass. For energy states such that $|m|\ll n/2$, Eq. (58) can be written as

$$\frac{\Delta\omega}{\omega}=\frac{\mathbf{v}\cdot\mathbf{k}}{ck}. \tag{59}$$

Where $\mathbf{v}$ is the total momentum of the gas divided by its total mass. Equation (59) is the usual classical expression for the Doppler shift for a radiator moving with a velocity $\mathbf{v}$. Consequently, for the highly correlated states $|m|\sim 0$ the Doppler effect can be described in classical terms.

The stationary states Eq. (55) do not form a complete set. In particular, the final state, a photon being emitted or absorbed with a momentum not $\mathbf{k}$, is not one of these states. The set of stationary states may be made complete by adding all the other possible orthogonal plane wave states, each being characterized by a definite momentum and internal energy for each molecule. With this set of orthogonal states, matrix elements can be easily calculated for transitions from the states given by Eq. (55) to states in which photons appear having momenta not equal to $\mathbf{k}$. These matrix elements are found to have a magnitude characteristic of the incoherent radiation process. It should be noted that only for one magnitude of $\mathbf{k}$ as well as for direction are the matrix elements of a coherent transition obtained.

PULSE-INDUCED COHERENCE RADIATION

It will be assumed in this section that a gas initially in thermal equilibrium is illuminated for a short time by an intense radiation pulse. The intensity and angular dependence of the spontaneous radiation emitted after the pulse will be calculated. In order to avoid the difficulties associated with motional effects, the molecules will be assumed so massive that their center-of-mass coordinates can be represented by small stationary wave packets. The center-of-mass coordinates will be then treated as time-independent parameters in the equation. It is assumed that the intensity of the exciting radiation pulse is so great that the fields acting on the gas during the pulse can be considered as described classically. The spontaneous radiation rate after the exciting pulse will be calculated quantum mechanically.

Because the initial state of the gas is a mixed state describing thermodynamic equilibrium, it is convenient to use the density matrix formalism.[14] It will be assumed that one has an ensemble of gas systems statistically identical and that what one is calculating is certain ensemble averages.

For a pure state, Eq. (28) shows that the spontaneous radiation rate in the $\mathbf{k}'$ direction can be written as the expectation value

$$I(\mathbf{k}')=I_0(\mathbf{k}')\langle R_{\mathbf{k}'+}R_{\mathbf{k}'-}\rangle. \tag{60}$$

For a state which may be mixed or pure using the density matrix formalism this becomes the trace

$$I(\mathbf{k}')=I_0(\mathbf{k}')\ \mathrm{tr}R_{\mathbf{k}'-}\rho R_{\mathbf{k}'+}. \tag{62}$$

Here the density matrix is defined as the ensemble mean

$$\rho=[\psi\psi^*]_{\mathrm{Av}}. \tag{63}$$

In Eq. (63) the wave function ψ is interpreted as a column vector and the * is the Hermitian adjoint. The symbol $[\ ]_{\mathrm{Av}}$ signifies an ensemble mean.

Assume that the exciting radiation pulse is in the form of a plane wave in the $\mathbf{k}$ direction. The fields which act on the various molecules differ only in their arrival time. The Hamiltonian of the system can be written

$$H=\hbar\omega R_3-\textstyle\sum_j \mathbf{A}_j(t)\cdot(\mathbf{e}_1R_{j1}+\mathbf{e}_2R_{j2}). \tag{64}$$

Here $\mathbf{A}_j(t)$ is a classical field quantity and

$$\mathbf{A}_j(t)=0,\quad \begin{matrix} t<t_j, \\ t>t_j+\tau \end{matrix} \tag{65}$$

[14] R. C. Tolman, *The Principles of Statistical Mechanics* (Clarendon Press, Oxford, 1938), p. 325.

where t_j is the arrival time of the radiation pulse at the jth molecule. Neglecting for the moment the interaction term, the time dependence of the wave function can be given by the unitary transformation

$$\psi(t)=\exp(-i\omega t R_3)\cdot\psi(0). \quad (66)$$

In general, the wave function after the interaction with the electromagnetic field can be obtained through a unitary transformation on the wave function prior to the pulse. The wave function of the gas after the radiation pulse has passed completely over the gas can be related to that before by

$$\psi'(t)=\exp(-i\omega t R_3)T\psi(0). \quad (67)$$

Here T is a unitary matrix which represents the effect of the pulse on the gas. To find the most general form of T it is convenient to consider the effect of the pulse on a particular molecule. Since this molecule has only two internal states of interest, its wave function can be regarded as a spinor in a pseudo "spin space." Then, apart from a multiplicative phase factor which has no physical significance, any unitary transformation can be represented as a rotation in "spin space." Any arbitrary rotation can be represented as a rotation about the No. 3 axis followed by a rotation about an axis perpendicular to No. 3. Except for the arrival time the radiation pulse is identical in its effect on each molecule of the gas. The operator T can be written then as the product

$$T=\exp[i\omega \textstyle\sum_j t_j R_{j3}] \cdot\prod_l \exp i\left[\frac{\theta}{2}(R_{l+}\alpha+R_{l-}\alpha^*)+\theta' R_{l3}\right] \cdot\exp[-i\omega \textstyle\sum_j t_j R_{j3}]. \quad (67a)$$

The first and second rotations are through angles of θ' and θ, respectively, and the phase of α determines the direction of the 2nd rotation axis. It is assumed that $|\alpha|=1$ and that the arrival time at the jth molecule is

$$t_j=(1/\omega)\mathbf{k}\cdot\mathbf{r}_j. \quad (67b)$$

Equation (67a) becomes Eq. (68) after making use of (67b):

$$T=\exp i\frac{\theta}{2}(R_{\mathbf{k}+}\alpha+R_{\mathbf{k}-}\alpha^*)\cdot\exp i\theta' R_3. \quad (68)$$

It should be noted that the effect of the different times of arrival of the pulse at the various molecules is contained in $\mathbf{k}\cdot\mathbf{r}_j$ which appears in $R_{\mathbf{k}\pm}$ in Eq. (68).

The reason for choosing this transformation to be a rotation about No. 3 followed by a perpendicular rotation is that the rotation about No. 3 is the same as a time displacement and has no effect since the initial state is assumed to be one of thermal equilibrium.

Assume that the initial density matrix can be written as

$$\rho_0=\frac{\exp(-ER_3/kT)}{\text{tr}\exp(-ER_3/kT)}=2^{-n}\prod_j(1-\gamma R_{j3}), \quad (69)$$
$$\gamma=2\tanh(E/2kT).$$

The density matrix after the radiation pulse is

$$\rho(t)=\exp(-i\omega t R_3)\cdot T\rho_0 T^{-1}\exp(i\omega t R_3). \quad (70)$$

The spontaneous radiation rate after the exciting pulse is given by Eq. (62) which becomes

$$I(\mathbf{k}')=I_0(\mathbf{k}')\,\text{tr}T\rho_0T^{-1}R_{\mathbf{k}'+}R_{\mathbf{k}'-}, \quad (71)$$

since R_3 commutes with $R_{\mathbf{k}'+}R_{\mathbf{k}'-}$. The radiation rate is thus independent of the time after the exciting pulse. This is because the effect of the radiated field on the gas has been neglected. Equation (71) is to be interpreted as the radiation rate immediately after the exciting pulse. Since ρ_0 and R_3 commute, Eq. (71) can be written as

$$I(\mathbf{k}')=I_0(\mathbf{k}')\,\text{tr}\exp[\tfrac{1}{2}i\theta(R_{\mathbf{k}+}\alpha+R_{\mathbf{k}-}\alpha^*)]\cdot\rho_0 \cdot\exp[-\tfrac{1}{2}i\theta(R_{\mathbf{k}+}\alpha+R_{\mathbf{k}-}\alpha^*)]\cdot R_{\mathbf{k}'+}R_{\mathbf{k}'-}. \quad (72)$$

It is desirable to transform ρ_0 before evaluating the trace

$$\rho'=\exp[\tfrac{1}{2}i\theta(R_{\mathbf{k}+}\alpha+R_{\mathbf{k}-}\alpha^*)] \cdot\rho_0\exp[-\tfrac{1}{2}i\theta(R_{\mathbf{k}+}\alpha+R_{\mathbf{k}-}\alpha^*)] =2^{-n}\textstyle\prod_j(1-\gamma R_{j3}^{\dagger}), \quad (73)$$

where

$$R_{j3}^{\dagger}=R_{j3}\cos\theta-\tfrac{1}{2}i(R_{j+}'\alpha-R_{j-}'\alpha^*)\sin\theta. \quad (74)$$

The primed operators are obtained from Eq. (53) as

$$R_{j\pm}'=R_{j1}'\pm iR_{j2}'=R_{j\pm}\exp(\pm i\mathbf{k}\cdot\mathbf{r}_j). \quad (75)$$

The trace in Eq. (72) can now be evaluated to give

$$I(\mathbf{k}')=I_0(\mathbf{k}')\sum_{jl}\text{tr}2^{-n}\prod_s(1-\gamma R_{s3}^{\dagger})R_{j+}''R_{l-}''. \quad (76)$$

The double prime is Eq. (75) referred to the $\mathbf{k}'$ direction. To evaluate the trace the following relations are needed: For A_i and B_j functions of the R's of molecules i and j,

$$\text{tr}A_iB_j=2^{-n}\,\text{tr}A_i\,\text{tr}B_j,$$
$$\text{tr}R_{j3}=\text{tr}R_{j\pm}=0,\quad \text{tr}R_{j3}^2=2^{n-2}, \quad (77)$$
$$\text{tr}R_{j+}R_{j-}=\text{tr}R_{j-}R_{j+}=2^{n-1}.$$

The final result is

$$I(\mathbf{k}')=I_0(\mathbf{k}')\cdot\tfrac{1}{2}n[1-\cos\theta\cdot\tanh(E/2kT) +\tfrac{1}{2}\sin^2\theta\cdot\tanh^2(E/2kT) \cdot(n|[\exp i(\mathbf{k}-\mathbf{k}')\cdot\mathbf{r}]_{\text{Av}}|^2-1)]. \quad (78)$$

Here the symbol $[\ ]_{\text{Av}}$ signifies a mean over all the molecules of the gas. For the example considered in Eq. (37a) this mean is unity, and Eq. (37a) follows by integrating over all directions of the emitted radiation. Aside from the factor $I_0(\mathbf{k}')$, the directional dependence of the emitted radiation is given by this mean. This factor is identical with the distribution factor for radiation about a set of classical isotropic radiators which have been excited by a plane wave. Consequently, for a θ of 90° and $n\tanh^2(E/kT)$ large compared with unity, the angular distribution of radiation is just the classical one.

The physical significance of the angle θ is that $\sin^2\frac{1}{2}\theta$

is the probability of the pulse exciting a molecule in its ground state. Also, if the exciting pulse is a constant amplitude wave of frequency ω during the duration of the pulse, the angle θ is proportional to the product of pulse amplitude and duration.

If the radiating system consists of a set of particles of spin $\frac{1}{2}$ in a uniform magnetic field, the angle θ has a geometrical significance. The initial state of a particle will have spin parallel or antiparallel to the field. The radiofrequency pulse will change its state such that its spin axis will be tipped through an angle θ. Note that if $\theta=180°$ the populations of the + and − populations have been just interchanged, corresponding to a transition from a positive temperature T to the negative temperature $-T$.[15] $\theta=90°$ corresponds to the excitation of molecules to energy superposition states Eqs. (38) and (39) for which the gas is radiating coherently.

ANGULAR CORRELATION OF SUCCESSIVE PHOTONS

The system to be considered here is assumed to be initially in thermal equilibrium. It is allowed to radiate spontaneously. The angular correlation between successive photons is calculated. This correlation was implicit in some of the earlier development, for example in Eq. (51a). As an example, consider a gas composed of widely separated molecules, all excited. Assume that a photon is emitted in the $\mathbf{k}$ direction. The radiation rate for the second photon in this direction is by Eq. (51a).

$$I(\mathbf{k})=I_0(\mathbf{k})2(n-1). \tag{79}$$

This is twice the incoherent rate. It is not hard to show that for an intermolecular spacing large compared with a radiation wavelength the radiation rate averaged over all directions is the incoherent rate. Hence from Eq. (79) the radiation probability in the direction $\mathbf{k}$ has twice the probability averaged over all directions.

In the problem to be considered, the system will consist initially of the gas in thermal equilibrium having a temperature T (possibly negative) and a photonless field. The molecules will be assumed fixed in position and with intermolecular distances large compared with a radiation wavelength. Photons are observed to be emitted in the directions $\mathbf{k}_1, \mathbf{k}_2, \cdots, \mathbf{k}_{s-1}$ and only these photons are emitted. The problem is one of finding the radiation rate in the $\mathbf{k}_s$ direction for the next photon.

Stated more exactly, it is assumed that there is an ensemble of gaseous systems, each with its own external radiation field. Every member of the ensemble which is capable of radiating will eventually radiate a photon. Those members which radiate their first photon into a small solid angle in the direction $\mathbf{k}_1$, are selected to form a new ensemble. For this second ensemble the time zero is taken to be the time that a photon was detected for each member of the ensemble.

It is convenient to calculate correlations for the gas systems forming a microcanonical distribution having an energy per gas system of m_0E. The results for a canonical distribution with a temperature T can subsequently be determined as an average over the microcanonical distributions.

Since the initial state of the system is assumed photonless, it is sufficient to give the explicit dependence of the initial density matrix on the molecular coordinates. Except for normalization this can be written as a projection operator for states of molecular energy m_0E. A particularly useful form for this density matrix is

$$\rho_0=\frac{\sum_{q=1}^{n}\exp 2\pi i\frac{q}{n}(R_3-m_0)}{\operatorname{tr}\sum_{q=1}^{n}\exp 2\pi i\frac{q}{n}(R_3-m_0)}. \tag{80}$$

This is a convenient way to write the density matrix because of the relation

$$\exp\left(2\pi i\frac{q}{n}R_3\right)=\prod_j \exp\left(2\pi i\frac{q}{n}R_{j3}\right)$$
$$=\prod_j\left[\cos\left(\pi\frac{q}{n}\right)+2iR_{j3}\sin\left(\pi\frac{q}{n}\right)\right]. \tag{81}$$

Here the product is over $j=1, \cdots, n$. To illustrate the importance of Eq. (81) the trace appearing in the denominator D of Eq. (80) will be calculated using the relations Eq. (77).

$$D=\sum_{q=1}^{n}\exp\left(-2\pi i\frac{q}{n}m_0\right)\cdot\operatorname{tr}\prod_j\left[\cos\left(\pi\frac{q}{n}\right)+2iR_{j3}\sin\left(\pi\frac{q}{n}\right)\right]$$
$$=\sum_{q=1}^{n}2^n\exp\left(-2\pi i\frac{q}{n}m_0\right)\cdot\cos^n\left(\pi\frac{q}{n}\right)$$
$$=\frac{n!n}{(\frac{1}{2}n+m_0)!(\frac{1}{2}n-m_0)!},\quad |m_0|<\frac{n}{2}$$
$$=2n\quad\text{for } |m_0|=n/2. \tag{82}$$

After one photon has been emitted and absorbed in the photon detector, the system is again photonless and its density matrix is (see Appendix 1)

$$\rho_1=(R_{\mathbf{k}_1-}\rho_0R_{\mathbf{k}_1+})/(\operatorname{tr}R_{\mathbf{k}_1-}\rho_0R_{\mathbf{k}_1+}). \tag{83}$$

After $s-1$ photons it is

$$\rho_{s-1}=\frac{R_{\mathbf{k}_{s-1}-}\cdots R_{\mathbf{k}_1-}\rho_0R_{\mathbf{k}_1+}\cdots R_{\mathbf{k}_{s-1}+}}{\operatorname{tr}R_{\mathbf{k}_{s-1}-}\cdots R_{\mathbf{k}_1-}\rho_0R_{\mathbf{k}_1+}\cdots R_{\mathbf{k}_{s-1}+}}. \tag{84}$$

The R's are defined in Eqs. (48) and (47) or (46). The radiation rate in the $\mathbf{k}_s$ direction immediately after the $s-1$ photon is from Eq. (62)

$$I(\mathbf{k}_s)=I_0(\mathbf{k}_s)\operatorname{tr}R_{\mathbf{k}_s-}\rho_{s-1}R_{\mathbf{k}_s+}. \tag{85}$$

Note that $s\leq\frac{1}{2}n+m_0$. For any l, $R_{l\pm}{}^2=0$. Consequently,

[15] E. M. Purcell and R. V. Pound, Phys. Rev. 81, 279 (1951).

the numerator of Eq. (84) can be written

$$\frac{1}{(s-1)!}\sum_{u,v\cdots=1}^{s-1}\sum_{u',v'\cdots=1}^{s-1}\sum_{j,l\cdots=1}^{n}$$
$$\times\exp i[(\mathbf{k}_u-\mathbf{k}_{u'})\cdot\mathbf{r}_j+(\mathbf{k}_v-\mathbf{k}_{v'})\cdot\mathbf{r}_l+\cdots]. \quad (86)$$
$$R_{j-}R_{l-}\cdots\rho_0\cdots R_{l+}R_{j+}.$$

Each of the above sums is over $s-1$ indices, including only terms for which all $s-1$ indices take on different values. The trace of the expression appears in the denominator of Eq. (84). In order to evaluate this trace it is necessary first to evaluate

$$\mathrm{tr}R_{j-}R_{l-}\cdots\rho_0\cdots R_{l+}R_{j+}=\mathrm{tr}\rho_0\cdots R_{l+}R_{l-}R_{j+}R_{j-}$$
$$=\mathrm{tr}\rho_0\cdots(\tfrac{1}{2}+R_{l3})(\tfrac{1}{2}+R_{j3}). \quad (87)$$

If Eqs. (80), (81), and (82) are substituted into Eq. (87), and use is made of Eq. (77) and the equality

$$\mathrm{tr}[\cos(\pi q/n)+2iR_{j3}\sin(\pi q/n)](\tfrac{1}{2}+R_{j3})$$
$$=2^{-1}\exp(i\pi q/n), \quad (87a)$$

Eq. (87) becomes

$$=\frac{2^{n-s+1}}{D}\sum_{q=1}^{n}\exp\left[i\pi\frac{q}{n}(s-1-2m_0)\right]\cdot\cos^{n-s+1}\left(\pi\frac{q}{n}\right)$$
$$=\frac{(n-s+1)!(\tfrac{1}{2}n+m_0)!}{n!(\tfrac{1}{2}n+m_0-s+1)!},\quad |m_0|<\tfrac{1}{2}n \text{ or } |m_0|=\tfrac{1}{2}n,\ s=1$$
$$=\tfrac{1}{2},\quad |m_0|=\tfrac{1}{2}n,\ s>1. \quad (88)$$

Making use of Eq. (88) the denominator of Eq. (84) can be written as

$$=P_{s-1}\frac{(n-s+1)!(\tfrac{1}{2}n+m_0)!}{n!(\tfrac{1}{2}n+m_0-s+1)!},\quad |m_0|<\tfrac{1}{2}n \text{ or}$$
$$|m_0|=\tfrac{1}{2}n,\quad s=1$$
$$=\tfrac{1}{2}P_{s-1},\quad m_0=\tfrac{1}{2}n,\ s>1, \quad (89)$$

where

$$P_{s-1}=\frac{1}{(s-1)!}\sum_{u,v\cdots=1}^{s-1}\sum_{u',v'\cdots=1}^{s-1}\sum_{j,l\cdots=1}^{n}$$
$$\times\exp i[(\mathbf{k}_u-\mathbf{k}_{u'})\cdot\mathbf{r}_j+(\mathbf{k}_v-\mathbf{k}_{v'})\cdot\mathbf{r}_l+\cdots],\quad s>1$$
$$P_0=1. \quad (90)$$

Here, as before, each of the above sums is over $s-1$ indices, including only terms for which all $s-1$ indices take on different values. If Eq. (84) is substituted into Eq. (85), the numerator is Eq. (89) with s increased by one unit. Consequently, substituting Eq. (89) into Eq. (85),

$$I(\mathbf{k}_s)=I_0(\mathbf{k}_s)\frac{P_s(\tfrac{1}{2}n+m_0-s+1)}{P_{s-1}(n-s+1)}. \quad (91)$$

To restate the meaning of this equation, $I(\mathbf{k}_s)$ is the radiation probability per unit time per unit solid angle in the direction $\mathbf{k}_s$; $I_0(\mathbf{k}_s)$ is the corresponding radiation probability for a single isolated excited molecule. It has been assumed that the gas was initially in the energy state m_0E [see Eq. (3)] with a random distribution over the degeneracy of this state. The gas was observed to radiate photons $\mathbf{k}_1$, $\mathbf{k}_2$, $\cdots\mathbf{k}_{s-1}$ previously to $\mathbf{k}_s$. Equation (91) is the radiation rate immediately after the $\mathbf{k}_{s-1}$ photon was observed. As a check on the correctness of this expression, note that the incoherent rate is obtained if $s=1$. Also, for $m_0=\frac{1}{2}n$ and $\mathbf{k}_1=\mathbf{k}_2=\cdots=\mathbf{k}_s=\mathbf{k}$, the radiation rate Eq. (91) agrees with Eq. (51a).

It should be noted that Eq. (91) is independent of the ordering of the subscripts 1, $\cdots$, $s-1$. Consequently, the angular distribution of the s photon is dependent upon the direction of a previous photon but is independent of the previous photon's position in the sequence of prior photons.

For a gas which contains a large number of randomly positioned molecules and for which previous photons have either been emitted in the direction $\mathbf{k}_3$ or in quite different directions, the radiation rate [Eq. (91)] is approximately equal to the incoherent rate times the number of photons previously emitted in this direction plus one.

Perhaps the case of most physical interest is where $s=2$. In this case Eq. (91) becomes

$$I(\mathbf{k}_2)=I_0(\mathbf{k}_2)\frac{\tfrac{1}{2}n+m_0-1}{n-1}[n|[\exp i\Delta\mathbf{k}\cdot\mathbf{r}]_{Av}|^2+n-2],$$
$$\Delta\mathbf{k}=\mathbf{k}_2-\mathbf{k}_1. \quad (92)$$

The symbol $[\ ]_{Av}$ signifies an average over all the molecular positions.

In case of a gas system at a temperature T, Eq. (91) must be averaged over all possible values of m_0 to give

$$\bar{I}(\mathbf{k}_s)=I_0(\mathbf{k}_s)\frac{P_s\displaystyle\sum_{m_0=s-\frac{1}{2}n-1}^{\frac{1}{2}n}(\tfrac{1}{2}n+m_0+1-s)\frac{n!}{(\tfrac{1}{2}n+m_0)!(\tfrac{1}{2}n-m_0)!}\exp\left(-\frac{m_0E}{kT}\right)}{(n-s+1)P_{s-1}\displaystyle\sum_{m_0=s-\frac{1}{2}n-1}^{\frac{1}{2}n}\frac{n!\exp(-m_0E/kT)}{(\tfrac{1}{2}n+m_0)!(\tfrac{1}{2}n-m_0)!}}. \quad (93)$$

For $|E/kT|\ll1$ and $s\ll n$, Eq. (93) can be approximated by

$$\bar{I}(\mathbf{k}_s)=I_0(\mathbf{k}_s)\frac{(\tfrac{1}{2}n+\bar{m}_0+1-s)P_s}{(n-s+1)P_{s-1}}, \quad (94)$$

where

$$\bar{m}_0=-\tfrac{1}{4}nE/kT.$$

It is a pleasure to acknowledge the assistance of the author's colleague, Professor A. S. Wightman, who read the manuscript and made a number of helpful suggestions.

APPENDIX I

It is assumed that the system consists initially of a gas with an energy m_0E and a photonless radiation field. A photon and only one photon is observed to be emitted. The effect of the photon emission on the state of the system is required.

There are two separate effects to be considered. First there is the effect on the state of the system which has its origin in the interaction between the field and gas. Second there is the effect of the observation which determines that a photon and one photon only has been emitted, that this photon was emitted in the **k** direction, and that the photon was absorbed in the detector. The first part of the problem is solved using Schrödinger's equation. The Hamiltonian of the system is

$$H=\hbar\omega R_3+H_0+H',\quad H_0=\textstyle\sum_{\mathbf{k}'}H_{\mathbf{k}'},$$
$$H'=-\tfrac{1}{2}\textstyle\sum_{\mathbf{k}'}[\mathbf{v}_{\mathbf{k}'}\cdot\mathbf{e}R_{\mathbf{k}'+}+\mathbf{v}_{\mathbf{k}'}^*\cdot\mathbf{e}^*R_{\mathbf{k}'-}]. \tag{95}$$

Here $H_{\mathbf{k}'}$ is the energy of the $\mathbf{k}'$ radiation oscillator. Assume a pure state represented by a wave function ψ_0 at a time $t=0$. Assume that ψ_0 is an eigenstate of R_3 and is photonless. At some later time it is

$$\psi(t)=\exp(-iHt/\hbar)\psi_0=\left(1-\frac{i}{\hbar}Ht-\frac{H^2}{2\hbar^2}t^2+\cdots\right)\psi_0. \tag{96}$$

For the quadratic and higher powers of t each term will be a sum of products of H' and $(H_0+\hbar\omega R_3)$. However, the interaction term H' consists of sums of terms of the type

$$U_{\mathbf{k}'}=\mathbf{v}_{\mathbf{k}'}\cdot\mathbf{e}R_{\mathbf{k}'+} \tag{97}$$

and its Hermitian adjoint. The operator $U_{\mathbf{k}'}$ consists of the product of a photon annihilation operator and a gas excitation operator. It converts an eigenstate of R_3 and H_0 into another such or it gives zero. The most general term operating on ψ_0 in Eq. (96) is therefore a product of powers of $H_0+\hbar\omega R_3$ and terms of the type $U_{\mathbf{k}'}$ and $U_{\mathbf{k}'}^*$ taken in various orders. In each of these terms $H_0+\hbar\omega R_3$ always operates on an eigenfunction and consequently can be moved to the end of the product as a number, the eigenvalue. Consequently $\psi(t)$ becomes

$$\psi(t)=[1+\textstyle\sum_{\mathbf{k}'}g_{\mathbf{k}'}(t)U_{\mathbf{k}'}^*+\sum_{\mathbf{k}'}h_{\mathbf{k}'}(t)U_{\mathbf{k}'}U_{\mathbf{k}'}^*$$
$$+\sum_{\mathbf{k}'\mathbf{k}''}g_{\mathbf{k}'\mathbf{k}''}(t)U_{\mathbf{k}'}^*U_{\mathbf{k}''}^*+\cdots]\psi_0. \tag{98}$$

The g's and h's are numbers, functions of the time. It may be noted that since ψ_0 represents a photonless state, an annihilation operator for a given radiation oscillator $\mathbf{k}'$ appears only if preceded by the corresponding creation operator.

Assuming that at the time t a photon measurement is made which indicates the presence of photon **k** and no other photons, the wave function after the measurement is

$$\psi'=P_{\mathbf{k}}\psi, \tag{99}$$

where the operator $P_{\mathbf{k}}$ is a projection operator for the **k** photon state.

$$P_{\mathbf{k}}=\frac{H_{\mathbf{k}}}{\hbar\omega_k}\prod_{\mathbf{k}'}{}'\left(\frac{\hbar\omega_{k'}-H_{\mathbf{k}'}}{\hbar\omega_{k'}}\right). \tag{100}$$

The product is over all $\mathbf{k}'\neq\mathbf{k}$. Two-photon excitation of one radiation oscillator has been neglected.

$$\psi'=[g_{\mathbf{k}}(t)U_{\mathbf{k}}^*+\textstyle\sum_{\mathbf{k}'}H_{\mathbf{k}\mathbf{k}'}(t)U_{\mathbf{k}}^*U_{\mathbf{k}'}U_{\mathbf{k}'}^*$$
$$+\textstyle\sum_{\mathbf{k}}I_{\mathbf{k}\mathbf{k}'}(t)U_{\mathbf{k}'}U_{\mathbf{k}'}^*U_{\mathbf{k}}^*+\cdots]\psi_0. \tag{101}$$

In summing over the direction of $\mathbf{k}'$ in the second and third terms above, the expression

$$R_{\mathbf{k}'+}R_{\mathbf{k}'-}=\textstyle\sum_{ab}\exp[i\mathbf{k}'\cdot(\mathbf{r}_a-\mathbf{r}_b)]\cdot R_{a+}R_{b-} \tag{102}$$

appears under the integral. By expanding the exponential in spherical harmonics it can be seen that for $a\neq b$ this integral vanishes, as it has been assumed that

$$\mathbf{k}'\cdot(\mathbf{r}_a-\mathbf{r}_b)\gg1 \text{ for } a\neq b.$$

It should be indicated that the angular dependence is not wholly in the exponential in Eq. (102) but exists in part in the square of the dot product of **e** and $\mathbf{v}_{\mathbf{k}'}$. However, this contribution to the angular dependence includes only spherical harmonics of finite degree in fact with $l<3$. As the only terms which need to be included in Eq. (102) are $a=b$, Eq. (102) becomes

$$R_{\mathbf{k}'+}R_{\mathbf{k}'-}=\tfrac{1}{2}+R_3+(\text{terms from } a\neq b). \tag{103}$$

Independent of its position in a series of products of U's the expression on the right side of Eq. (103) will operate on an eigenfunction and becomes an eigenvalue which can be removed as a number. In the higher-order terms in Eq. (101) $U_{\mathbf{k}'}$ and $U_{\mathbf{k}'}^*$ may not appear adjacent to each other, but if they do not, some other pair such as $U_{\mathbf{k}''}U_{\mathbf{k}''}^*$ will appear, and after removing this as an eigenvalue another such pair will occur, and eventually the $\mathbf{k}'$ pair will be adjacent. Consequently, to all orders in the expansion

$$\psi'=f(t)U_{\mathbf{k}}^*\psi_0, \tag{104}$$

where f is a function of the time of observation. As the photon detector also absorbs the photon, the wave function must be multiplied by the annihilation operator $\mathbf{e}\cdot\mathbf{v}_{\mathbf{k}}$. This gives, except for the time factor,

$$\psi''\sim R_{\mathbf{k}-}\psi_0, \tag{105}$$

which is another photonless state but with one quantum less energy.

If the initial density matrix ρ_0 contains only photonless states of the same energy m_0E, then from Eqs. (63) and (105) it is transformed to

$$\rho_1=R_{\mathbf{k}-}\rho_0R_{\mathbf{k}+}/\mathrm{tr}(R_{\mathbf{k}-}\rho_0R_{\mathbf{k}+}), \tag{106}$$

representing the photonless state of the ensemble of systems after the emission, detection, and absorption of photon described by **k**.

Radiation Damping in Magnetic Resonance Experiments

N. Bloembergen and R. V. Pound
Harvard University, Cambridge, Massachusetts
(Received March 22, 1954)

Magnetic resonance experiments can be described by analogy to a coupled pair of circuits, one of which is the ordinary electrical resonant circuit. The other circuit is formed by the rotating magnetization. For transient phenomena, such as occur, e.g., in the pulse techniques of free nuclear induction, the coupling gives rise to a damping of the magnetic resonance by the electric circuit. Such damping can also be considered as spontaneous radiation damping. It is shown that in certain cases of nuclear induction this radiation damping is more important than the damping from the spin-spin and the spin-lattice relaxation mechanisms usually considered. For ferromagnetic materials at microwave frequencies the radiation damping can become very large.

I. FREE MAGNETIC INDUCTION

By way of introduction, consider a system of volume V with a uniform macroscopic magnetization per unit volume $\mathbf{M}_0$, which precesses with an angular frequency

$$\omega_0=\gamma H_0=g\beta\hbar^{-1}H_0 \tag{1}$$

around a constant magnetic field H_0, parallel to the z axis. Here g is the gyromagnetic ratio, β the Bohr magneton. The angle between $\mathbf{M}_0$ and $\mathbf{H}_0$ is θ_0. Assume that the magnetization can precess freely, $M_x=M_0\cos\theta\sin\omega t$, $M_y=M_0\cos\theta\cos\omega t$, and $M_z=M_0\sin\theta$ and neglect for the time being any internal damping of the magnetic system. When we introduce a pickup coil, tuned by a condenser to the frequency ω_0, in the x-y plane, a periodic voltage is induced in this coil by the precessing magnetization. If the coil has n turns of cross section A, this induction signal is given by

$$V_s=-nA\eta\frac{4\pi}{c}\frac{dM_x}{dt}=nA\frac{4\pi\omega}{c}\eta M_0\sin\theta\cos\omega_0 t, \tag{2}$$

where η is a filling factor, equal to unity if the coil is completely immersed in the magnetic material, or otherwise

$$\eta=\frac{\int \mathbf{M}\cdot(\mathbf{H}_c/i_c)dV}{\mathbf{M}_0\cdot\int(\mathbf{H}_c/i_c)dV}, \tag{3}$$

where $\mathbf{M}$ is the magnetic moment per unit volume and $\mathbf{M}_0$ its average value, allowing the extension to a sample of nonuniform magnetization. Each point in the integration over the space coordinates is weighted by the field $\mathbf{H}_c/i_c$ which a unit current through the coil would produce at that point.

A current in phase with the induced voltage will flow in the tuned circuit, and the Joule heat dissipated per unit time is $V_s^2/2R$. The energy source which generates this power is the magnetic energy of the magnetized sample in the field H_0. The energy of the permanent magnet $\mathbf{M}_0V$ is

$$W=M_0VH_0(1-\cos\theta). \tag{4}$$

As the current flows the angle θ will gradually decrease to zero. The equation of motion can be written as the requirement of conservation of energy,

$$dW/dt=M_0VH_0\sin\theta(d\theta/dt)=V_s^2/2R. \tag{5}$$

The motion can also be found by considering the torque exerted on the magnetization by the field connected with the induced current. In a long coil of n turns over a length l, the induced field is

$$H_x=4\pi V_sn/clR=4\pi\eta QM_0\sin\theta\cos\omega_0 t.$$

The quality factor Q of the circuit is given by

$$Q=\omega L/R=\omega 4\pi n^2A/Rlc^2.$$

The torque produced by this field gives rise to the equation of motion:

$$dM_z/dt=-\gamma M_yH_x,$$

or

$$d\theta/dt=-2\pi\eta M_0Q\gamma\sin\theta. \tag{6}$$

Equation (5) can easily be reduced to this same form. The solution of Eq. (6) for the special case that $\theta=\pi/2$ at $t=0$ is

$$\tan(\tfrac{1}{2}\theta)=\exp(-2\pi M_0Q\eta\gamma t), \tag{7}$$

and the amplitude of the induction signal Eq. (2) decreases proportional to $\operatorname{sech}(2\pi M_0Q\eta\gamma t)$. A damping time constant due to the reaction of the induced field on the magnetization can be defined as

$$\tau_R=(2\pi\eta M_0Q\gamma)^{-1}. \tag{8}$$

Suryan[1] has first called attention to the importance of this type of damping without giving a detailed quantitative discussion. We shall give the pertinent circuit equations, from which the above result can be derived as a special case, in the next section. The relation of this damping to magnetic relaxation processes will be discussed in a final section, where, also, a comparison will be made with the natural line width of optical spectral lines. This radiation damping has an

[1] G. Suryan, Current Sci. (India) 18, 203 (1949).

Reprinted with permission from *Phys. Rev.*, vol. 95, pp. 8–12, July 1, 1954.

appreciable magnitude because of the coherence between the individual spins producing the magnetization. These coherence effects have been discussed from a fundamental quantum mechanical point of view by Dicke.[2] The fact that it is possible by radiospectroscopy techniques to produce states with definite coherent phase relations between the individual elements, allows for a classical discussion of the problem in terms of one macroscopic magnet as first introduced by Bloch[3] in his classical theory of nuclear induction. We shall illustrate now with a few examples that the damping described by Eq. (8) is frequently not negligible.

a. The Proton Resonance in Water in a Field of 7000 gauss

In thermal equilibrium the protons acquire a magnetization per unit volume in this field given by

$$M_0=\chi_0 H_0=[N_0\gamma^2\hbar^2 I(I+1)/3kT]H_0,$$

where χ_0 is the nuclear paramagnetic volume susceptibility. At room temperature the static magnetization is about 2×10^{-6} oersted in a field of 7000 oersteds. A short radiofrequency pulse[4,5] at the resonance frequency and of amplitude H_{rf} and duration t such that $\gamma H_{rf}t=\frac{1}{2}\pi$, will turn this magnetization into the plane perpendicular to H_0. A coherent "superradiant" state is thus produced. At the end of the pulse free nuclear induction will occur which is damped in a time τ_R given by Eq. (8). Assuming that the resonant circuit has a $Q=100$ and $\eta=1$, we find $\tau_R=0.03$ sec. This is much shorter than the damping from spin-spin and spin-lattice relaxation mechanisms in pure water. The fact that spin-echo pulses can be obtained after times much longer than τ_R is explained by the absence of this radiation damping when the nuclear spins are out of phase in the inhomogeneous field. To make the radiation per echo negligible the inhomogeneity ΔH should satisfy the inequality $\gamma\Delta H\tau_R\gg1$. This allows detection of fine structures with separations much smaller than the inverse damping time given by Eq. (8).

When the initial magnetization is reversed by a 180° pulse, an unstable radiationless state is established. The magnetization will change only due to the spin-lattice relaxation mechanism, the angle θ remaining 180°.

b. The Proton Resonance in Water in the Earth's Magnetic Field

Recently Packard and Varian have observed the proton resonance in the earth's field at 2185 cps.[6] The radiant state was produced by magnetizing the sample in a field H_0' of about 100 gauss perpendicular to the earth's field. The field H_0' is reduced to a field H_0'' of a few gauss in a time short compared to the spin lattice relaxation time. During this adiabatic demagnetization the magnetization $M_0=\chi_0 H_0'$ remains unchanged. Then the field H_0'' is reduced to zero in a time t' such that $\gamma H_0'' t'\ll1$. During this process, which is nonadiabatic in the Ehrenfest sense, the magnetization vector has no time to reorient. Thus, a radiant state is constructed with a magnetization $M_0\approx3\times10^{-8}$ oersteds precessing with a frequency of 2185 cps in the plane perpendicular to the earth's field. The radiation damping time τ_R, assuming $\eta=1$, is about $200Q^{-1}$ sec. For sufficiently high Q, this could explain the observation that the duration of the signal was shorter than the known relaxation times in pure water. It is interesting to note that in this sense a spectral line in the audiofrequency range of the electromagnetic spectrum has a "natural width."

c. A Small Ferrite Sphere in a Microwave Cavity

Assume that a ferrite with a volume magnetization of 300 gauss cm^{-3} has a g value equal to that of a free electron and is placed in a cavity with $Q=2000$. For a sphere of 0.2 mm in diameter in a cavity of 10 cc, the filling factor η is of the order of 10^{-6}. Then we find that τ_R 10^{-8} sec. This time is inversely proportional to the filling factor or roughly to the volume of the ferrite sample.

It seems as if the radiation damping would be able to broaden the line so much as to wipe out the resonance completely for larger spheres. When the time τ_R becomes shorter than the characteristic time of the electrical circuit Q/ω, the analysis of this introductory paragraph is not valid. We shall therefore proceed with a more accurate description of the coupling.

II. THE COUPLING BETWEEN THE RESONANT CIRCUIT AND THE MAGNETIZATION

The equations of motion describing the complete system consisting of the magnetic material in a constant field H_0 in the z direction and two crossed coils parallel to the x and y direction, respectively, are

$$\frac{dM_{x,y}}{dt}=\gamma(\mathbf{M}\times\mathbf{H})_{x,y}-\frac{M_{x,y}}{T_2},$$

$$\frac{dM_z}{dt}=\gamma(\mathbf{M}\times\mathbf{H})_z-\frac{M_z-M_0}{T_1},$$

$$-K_x\frac{dM_x}{dt}+L_x\frac{di_x}{dt}+R_x i_x+\frac{1}{C_x}\int i_x dt=V_{x,\,\mathrm{app}},$$

$$-K_y\frac{dM_y}{dt}+L_y\frac{di_y}{dt}+R_y i_y+\frac{1}{C_y}\int i_y dt=V_{y,\,\mathrm{app}},$$

$$H_x=K_x' i_x,\quad H_y=K_y' i_y,\quad H_z=H_0.$$

[2] R. H. Dicke, Phys. Rev. **93**, 99 (1954).
[3] F. Bloch, Phys. Rev. **70**, 460 (1946).
[4] E. L. Hahn, Phys. Rev. **80**, 580 (1950); **88**, 1070 (1952).
[5] H. Y. Carr, and E. M. Purcell, Phys. Rev. **94**, 630 (1954).
[6] M. Packard and R. Varian, Phys. Rev. **93**, 941 (1954).

In general there are five simultaneous nonlinear differential equations connecting the three components of magnetization and the two currents. The first three equations are the familiar Bloch equations.[3] The transverse field components H_x and H_y are now, however, related to the magnetization by the last two equations and must not be considered as impressed constants. The phenomenological damping terms for the spin-spin and spin-lattice relaxation are responsible for the fact that the magnetization has now in general not a constant magnitude M_0. Frequently there will be only one coil, and the number of equations is consequently reduced by one. In writing the relations between the transverse components of the field and the currents we have assumed that demagnetizing and anisotropy fields are zero. The K and K' are geometrical factors. They are related to each other by $KK'=4\pi L\eta$. The symmetrical arrangement of two crossed coils with identical circuit constants has the advantage that rotating fields can be produced. If the x and y components of the driving force are the real and imaginary parts of one complex function $V_{\mathrm{app}}^{+}=V_{x,\mathrm{app}}+jV_{y,\mathrm{app}}$ the solutions for $M^{+}=M_x+jM_y$, $i^{+}=i_x+ji_y$ can be found from a set of only three differential equations:

$$-K\frac{dM^{+}}{dt}+L\frac{di^{+}}{dt}+Ri^{+}+\frac{1}{C}\int i^{+}dt=V_{\mathrm{app}}^{+}, \tag{9}$$

$$\frac{dM^{+}}{dt}=-j\gamma M^{+}H_0-\frac{M^{+}}{T_2}-j\gamma M_zK'i^{+}, \tag{10}$$

$$\frac{dM_z}{dt}=\mathrm{Im}(\gamma K'M^{+*}i^{+})-\frac{M_z-M_0}{T_1}. \tag{11}$$

We first give the steady-state solution, when the circuit is driven by a harmonic rotating potential $V_{x,\mathrm{app}}+jV_{y,\mathrm{app}}=V_{\mathrm{app}}\exp(-j\omega t)$. A solution with $dM_z/dt=0$ and the other components varying with the frequency ω is obtained, $M^{+}=M_1\exp(-j\omega t)$, $i^{+}=i\exp(-j\omega t)$.

$$M_z=\frac{1+(\Delta\omega T_2)^2}{1+(\Delta\omega T_2)^2+\gamma^2H_1^2T_1T_2}M_0, \tag{12}$$

$$M_1=\chi H_1=\chi K'i=-\frac{j|\gamma|M_zT_2}{1+j(\omega_0-\omega)T_2}K'i, \tag{13}$$

$$i=[j\omega L(1+4\pi\eta\chi)+R+(j\omega C)^{-1}]^{-1}V_{\mathrm{app}}. \tag{14}$$

This solution is identical with those usually given for the stationary state.[7] The effect of the rotating magnetization can be described by a complex susceptibility χ. The over-all impedance of each circuit is given by the expression between square brackets in Eq. (14). There is no additional radiation damping of the magnetic resonance in this case. The z component of the magnetization is constant. The generator supplies a current which exactly balances the effect of the radiation damping. The steady-state response of a circuit should be analyzed in terms of the impedance given by Eq. (14). Under suitable steady-state conditions it is possible to resolve fine structure lines of nuclear magnetic resonance in liquids which are only a few cycles apart, although these lines under conditions of free nuclear induction would be damped in a time short compared to their inverse spacing. The same situation prevails in ferromagnetic resonance experiments. Ordinary resonance lines are observed under steady-state conditions, although the freely precessing magnetization would radiate with a time short compared to the inverse line width.

The expression for the impedance for small H_1, allowing M_z to be taken equal to M_0, can be seen to be identical to that of a double tuned coupled circuit. It is therefore apparent that a condition of critical or more than critical coupling can be established and that a resonance detector that uses the electrical circuit as the frequency determining element of oscillation[8] is subject to frequency pulling in a discontinuous manner, sometimes called a "drag link" effect.

For transient phenomena the Eqs. (10) and (11) for the magnetization cannot be considered independently from the circuit Eq. (9). The solution in the introductory section is obtained as an approximation by dropping the last two terms in Eq. (10) and the last term in Eq. (11). The damping time τ_R is found to be half as long as given by Eq. (8) because of the presence of two crossed coils creating a rotating reaction field.

A general solution cannot be obtained since nonlinear terms containing products M_zi^{+} and $M^{+*}i^{+}$ occur. For small values of θ, we can, however, replace M_z in Eq. (10) by M_0. Equations (9) and (10) then represent two coupled linear circuits. In the absence of a driving force we obtain a third degree equation for the proper frequencies of the system. If we assume that the electric circuit is tuned to the precession frequency of the magnetization $\omega_0^2LC=\gamma^2H_0^2LC=1$ and neglect terms of the order $\Delta\omega/\omega_0$, we obtain a quadratic equation for the complex proper frequencies $\omega=\omega_0+\Delta\omega$ of the system, with the solution

$$\Delta\omega=\frac{j}{2}\left(\frac{\omega_0}{2Q}+\frac{1}{T_2}\right)\pm\left\{-\frac{1}{4}\left(\frac{\omega_0}{2Q}+\frac{1}{T_2}\right)^2+\frac{\omega_0}{Q}\left(\frac{1}{\tau_R}+\frac{1}{T_2}\right)\right\}^{\frac{1}{2}},$$

where τ_R is given by Eq. (8).

For negligible spin-spin damping, we can consider the following two limiting cases:

Case I, $1/\tau_R\ll\omega_0/Q$. The solutions are $\Delta\omega=j\omega_0/2Q$ or $\Delta\omega=2j/\tau_R$. The first solution corresponds to the damping of the mode with the electric circuit excited,

[7] N. Bloembergen, thesis, Leiden, 1948.

[8] R. V. Pound and W. D. Knight, Rev. Sci. Instr. 21, 219 (1950).

the second to the damping of the magnetization. This last solution corresponds to the examples of nuclear induction of Sec. I. In example **b** there is no current flowing at $t=0$. We must take a linear combination of the two solutions, and initially no damping. Only after a time Q/ω_0 when the electric circuit has been excited, is the damping given by the characteristic time τ_R.

Case II, $1/\tau_R \gg \omega_0/Q$. The solutions are now

$$\Delta\omega \approx j\omega_0/4Q \pm (\omega_0/Q\tau_R)^{\frac{1}{2}}.$$

The energy swings back and forth between the electric circuit and the magnet system with a frequency $(\omega_0/Q\tau_R)^{\frac{1}{2}}$ until the energy is eventually dissipated in the resistance of the electric circuit. This situation can be realized by ferrites in a microwave cavity. The energy is never dissipated faster than is permitted by the Q of the cavity,[9] so long as $1/T_2$ is negligible. It is quite possible, however, that the magnetization of the ferrite at the end of a microwave pulse will return to its position parallel to H_0 in a time short compared to both relaxation times T_1 and T_2. This observation does not alter the interpretation of the data in reference 9 since the relaxation times were evaluated from the power absorbed in the steady-state existing for the duration of the microwave pulse. The relaxation times thus obtained are independent of the size of the sample or the filling factor.

In the third mode of the system $\omega \approx -\omega_0$, the electrical circuits are excited at the same frequency but produce a field rotating in the opposite direction. This mode is only very slightly perturbed by the presence of the magnetization.

III. SPONTANEOUS EMISSION AND THERMAL RELAXATION MECHANISMS

The transition probability for spontaneous emission of a quantum $h\nu=\gamma\hbar H_0$ by a single spin $I=\frac{1}{2}$ is given by the well-known formula of radiation theory,[10]

$$W=16\pi^2\gamma^2 h\nu^3 c^{-3}. \tag{15}$$

Substituting numerical values, we find for the lifetime of a single free proton in a field of 10^4 oersted the astronomical value $T_1=10^{25}$ sec. In the earth's magnetic field the lifetime against spontaneous radiation would be $T_1 \approx 10^{38}$ sec. How can this result be consistent with the radiation damping of the order of one second calculated in Sec. I? This discrepancy is resolved by considering two factors. One is the coherence which exists between the individual proton spins, the other is the increase in the density of the radiation field in the tuned circuit over that in free space.[11]

In the situation of interest, the spins of protons have such phase relations that they create a macroscopic magnetic dipole which is N times as large as that of a single one. Since the emitted radiation is proportional to the square of the dipole moment, the transition probability is increased by a factor N^2. Dicke[2] has given a quantum mechanical analysis of the coherent state and also arrives at this result. The damping is, however, not increased simply by a factor N^2, but by a factor N. The quantum levels of our macroscopic system are all equally spaced. We must not consider a single transition but the damping of an N-fold excitation with coherence between all possible transitions. The same problem arises in calculating the lifetime of an excited harmonic oscillator.[12] Classically, the factor N is immediately obvious, as the radiation rate is proportional to N^2 and the stored energy proportional to N. The factor N is related to the magnetization introduced earlier by

$$N=M_0V_m/\gamma\hbar. \tag{16}$$

Here V_m is the volume of sample.

The magnetic radiation density in the coil of volume V_c of a resonant circuit is increased over the density in free space by a factor

$$Q\lambda^3/8\pi^2V_c. \tag{17}$$

Note that this factor increases with increasing wavelength, cancelling the decrease predicted by Eq. (15). Multiplying Eq. (15) by Eqs. (16) and (17) we find a radiation damping time τ_R which is identical with Eq. (8), provided we multiply by a factor $\frac{1}{2}$ to take account of the fact that only one of the two circularly polarized modes in the coil is effective.

The interaction of a single spin with a thermal radiation field leads to induced emission and absorption. The transition probability for these processes is obtained by multiplying Eq. (15) by the Bose-Einstein factor,

$$[\exp(h\nu/kT)-1]^{-1} \approx kT/h\nu \gg 1, \tag{18}$$

for the frequency range of interest in magnetic resonance. The damping for the radiation from a single spin is consequently increased by this rather large factor.[13]

This factor should, however, not be added to Eq. (8). The influence of the thermal radiation field in the case of coherent spins in a resonant circuit can best be discussed in a classical manner. The classical analog of the magnetic thermal radiation field is the field produced in the coil by the thermal noise current $i(t)$. This current produces a fluctuating torque on the magnetization, leading to the equation of motion:

$$d\theta/dt=\gamma H_{th}(t). \tag{19}$$

Since $\langle H_{th}(t)\rangle_{Av}=0$, there is on the average no change in θ, due to the noise current. There are only fluctuations

[9] N. Bloembergen and S. Wang, Phys. Rev. **93**, 72 (1954).
[10] W. Heitler, *Quantum Theory of Radiation* (Oxford University Press, London, 1943).
[11] E. M. Purcell, Phys. Rev. **69**, 681 (1946).
[12] V. Weisskopf and E. Wigner, Z. Physik **65**, 18 (1931).
[13] V. Weisskopf, Z. Physik **34**, 1 (1933).

in θ. The mean square deviation of θ after a time t' is

$$\langle\theta_{t'}^2\rangle_{\text{Av}}=\gamma^2t'\left\langle\int_0^\infty H_{th}(t)H_{th}(t+t'')dt''\right\rangle_{\text{Av}}$$

$$=\gamma^2\langle H_{th}^2\rangle_{\text{Av}}\frac{L}{R}t'=\gamma^2\frac{L}{R}\frac{4\pi kT}{V_c}t'.$$

The average rate of change in the stored energy due to the interaction with the thermal noise field is

$$\langle(dW/dt)_{\text{noise}}\rangle_{\text{Av}}=\tfrac{1}{2}M_0VH_0\cos\theta(d\langle\theta^2\rangle_{\text{Av}}/dt)$$
$$=2\pi kT\gamma Q\eta M_0\cos\theta,\quad(20)$$

and the average thermal damping is consequently

$$\left\langle\frac{1}{\tau_{th}}\right\rangle_{\text{Av}}=2\pi\gamma QM_0\frac{kT}{M_0H_0V_c}\frac{\cos\theta}{(1-\cos\theta)}.\quad(21)$$

Apart from the angular dependence, this is smaller than Eq. (8) by a factor equal to the thermal energy kT over the total magnetic energy of the system. It is very small for a large coherent magnetization, but reduces properly to the result of increased damping for an individual spin. One might put the result Eq. (21) in words by saying that there is no coherence between the rotating magnetization and the thermal field, while there is between the rotating magnetization and its own reaction field.

The relaxation or damping discussed here refers only to the approach to equilibrium energy of orientation of the classical precessing magnet that corresponds to the given coherent state. Changes in the over-all magnetization of the system that lead toward a true equilibrium state of magnetization can only occur through the mechanisms leading to T_2 and thus cannot occur in times shorter than T_2. Thus, a spin temperature is defined only after the time T_2, the spin-spin relaxation time, has elapsed. It is not correct to say that after a radiofrequency pulse which has turned the magnetization through an angle between 90° and 180°, the spin system has assumed a negative temperature. Immediately after the pulse there are certain phase relations between the spins, which change to a situation of internal thermal equilibrium[14] only after a time T_2. The radiation processes which do not change the coherence relations among the spins, do not contribute to the spin-spin relaxation mechanism.

A significant remark may be made about the conditions under which τ_R may be expected to be as short as T_2. In a rigid lattice, T_2 is of the order of magnitude of $r^3/\gamma^2\hbar$ where r is the interspin distance. Since r^3 is approximately $1/N_0$, for magnetization described by Curie's law, $(\tau_R/T_2)\sim kT/\omega\hbar\eta Q$. Thus, unless ηQ is quite large, only under the condition $\omega\hbar\gg kT$, leading to saturation magnetization, would $\tau_R\ll T_2$. The liquid narrowing, in the example of proton resonance, and the exchange effect, in the ferromagnetic case, are essential to the presence of a significant radiative effect.

Finally, we wish to stress the fact that a coherently radiating state can only be produced after a certain macroscopic magnetization has been established. It is still necessary that a relaxation mechanism produces a thermal distribution over the various energy levels of the individual spins, which are initially incoherent. A homogeneous magnetic field, such as exists in the coil and in general in a system which has dimensions small compared to the wavelength of the radiation, will never produce the initial thermal relaxation. The Hamiltonian for the interaction energy with a homogeneous magnetic field commutes with the total angular momentum or magnetization of the sample, which is therefore a constant of the motion. Radiation damping discussed in this paper does therefore not represent a spin-lattice relaxation mechanism in the usual sense. To establish a magnetization in thermal equilibrium with the surroundings, the inhomogeneous internal or local fields[7] are essential. The inhomogeneity of the rf field in the coil was explicitly taken into account in the definition of the filling factor η. This results in a complication that the total magnetization is not rigorously a constant of the motion. The gist of the statement, however, that the radiation field in the coil does not provide a microscopic thermal relaxation mechanism is still valid.

[14] E. M. Purcell and R. V. Pound, Phys. Rev. **81**, 279 (1951).

Molecular Ringing

STANLEY BLOOM
RCA Laboratories, Princeton, New Jersey
(Received February 20, 1956)

Semiclassical radiation theory is used to describe the response of an assemblage of two-state molecules driven by an electromagnetic field. When the field is suddenly removed, the assemblage does not immediately become quiescent; it continues to radiate in diminishing amount. This coherent molecular-ringing radiation persists until the molecular populations return to the values they had at the beginning of the driving pulse. Depending upon the strength and duration of the driving pulse, the ringing radiation may exhibit a delayed peak.

INTRODUCTION

WE inquire: What happens when an electromagnetic field applied to a molecular assemblage is suddenly switched off? This problem has been treated by Dicke[1] in a detailed and rigorous fashion. Taking the whole assemblage to be a single quantum-mechanical system, he finds energy levels corresponding to correlations between individual molecules. Spontaneous transitions between highly correlated states lead to coherent emission. These coherent states may be excited by either the emission of photons by the gas or the absorption of photons from an externally applied radiation pulse.

If, however, there is a large difference between the populations of two particular single-molecule energy states, then the molecular assemblage need not be considered as a single quantum system. For with a large population imbalance, a net macroscopic polarization exists which oscillates in time and whose radiation rate can be computed classically.

A different attack is used here. Instead of dealing with a macroscopic polarization, we employ, in Sec. I, the simple and well-known methods of time-dependent perturbation theory to compute the population variation of an assemblage of two-state molecules driven by a classical electromagnetic field. This driving field orders the state phases of the various molecules, which before the application of the field had random phases. In Sec. II, the primary driving field is cut off and the pulse-induced coherent radiation—or briefly, "molecular ringing"—is investigated. Collisional effects are at first ignored. Because the induced emission does not stop instantaneously at cutoff, the molecules are now driven by their own radiation. The magnitude of this ringing field is proportional to the rate of change of the state populations, and this rate of change is itself due to the ringing field. Thus by a self-consistent calculation we can follow the system's behavior in time as the populations slowly return to the values they had at the onset of the primary driving field. Furthermore, and this is one point of the present paper, it is shown that the ringing power may, depending upon the strength and duration of the driving pulse, exhibit a *delayed* peak.

[1] R. H. Dicke, Phys. Rev. **93**, 99 (1954).

The effects of molecular impacts, which tend to destroy the phase coherency of the radiating molecules produced by the driving field, are calculated in Sec. III.

I. RESPONSE TO DRIVING FIELD

The well-known procedure of time-dependent perturbation theory[2] starts with the Schrödinger equation

$$i\hbar\partial\Psi/\partial t=[H_0+V(t)]\Psi$$

and leads to the transition probabilities between the unperturbed eigenstates

$$H_0\phi_\lambda=\epsilon_\lambda\phi_\lambda$$

of the isolated molecule as induced by the perturbation $V(t)$. Only two states of energy difference

$$\epsilon_2-\epsilon_1=\hbar\omega_0$$

are assumed to be of importance for the transition of interest. The substitution of

$$\Psi=a_1\phi_1\exp(-i\epsilon_1 t/\hbar)+a_2\phi_2\exp(-i\epsilon_2 t/\hbar)$$

into the Schrödinger equation shows the probability amplitudes, a_λ, to grow in time as

$$\begin{aligned}i\hbar\dot{a}_1&=V_{11}a_1+V_{12}a_2\exp(-i\omega_0 t),\\ i\hbar\dot{a}_2&=V_{12}{}^*a_1\exp(i\omega_0 t)+V_{22}a_2.\end{aligned}$$

Many approximate schemes have been described for solving these two coupled equations for particular types of perturbations $V(t)$. The following method, however, gives a solution which, though implicit, is perfectly general.

The transformation

$$a_\lambda\equiv c_\lambda\exp\left(-i\int_0^t dtV_{\lambda\lambda}/\hbar\right),$$

puts the coupled equations into the form

$$\begin{aligned}\dot{c}_1&=fc_2,\\ \dot{c}_2&=-f^*c_1,\end{aligned}\tag{1}$$

[2] For example see L. I. Schiff, *Quantum Mechanics* (McGraw-Hill Book Company, Inc., New York, 1949).

Reprinted with permission from *J. Appl. Phys.*, vol. 27, pp. 785–788, July 1956.

where the driving function is

$$f(t)=-i\hbar^{-1}V_{12}\exp\left[-i\omega_0 t-i\int_0^t dt(V_{22}-V_{11})/\hbar\right].$$

Elimination of c_1 or c_2 from Eqs. (1) leads to

$$d(\dot{c}_1/f)/dt=-f^*c_1,$$
$$d(\dot{c}_2^*/f)/dt=-f^*c_2^*,$$

showing that c_1 and c_2^* satisfy the same differential equation, namely,

$$d(\dot{z}/f)/dt=-f^*z$$

or

$$\ddot{z}-\dot{z}\dot{f}/f+z|f|^2=0. \tag{2}$$

Therefore if z is any solution of Eq. (2), so is $(\dot{z}/f)^*$ and the general solution of Eqs. (1) is

$$c_1(t)=c_1(0)z-c_2(0)(\dot{z}/f)^*,$$
$$c_2(t)=c_1(0)\dot{z}/f+c_2(0)z^*, \tag{3}$$

with z subject to the initial conditions

$$z(0)=1, \quad \dot{z}(0)=0, \tag{4}$$

and fulfilling the normalization requirement

$$|z|^2+|\dot{z}/f|^2=1. \tag{5}$$

The probabilities of state occupancy, $|a_\lambda(t)|^2=|c_\lambda(t)|^2$, contain the cross term $c_1(0)c_2^*(0)=|c_1(0)c_2(0)| \times \exp i(\alpha_1-\alpha_2)$. These vanish for a molecular assemblage when we average over the initial phase angles, taking the α's to be distributed uniformly and independently. If, then, N is the total population of the two states, there are

$$N_1(t)=N\langle|c_1(t)|^2\rangle_\alpha$$
$$=N_2(0)+[N_1(0)-N_2(0)]|z(t)|^2$$

molecules in the lower state and $N_2(t)=N-N_1(t)$ in the upper. In terms of the excess number in the lower state,

$$D(t)=N_1(t)-N_2(t)=D(0)(2|z(t)|^2-1), \tag{6}$$

where $D(0)$ is the excess population at the instant of independent phases; its value is determined by the Boltzmann factor.

The perturbation of interest is that due to a classical monochromatic electromagnetic field of frequency ω and amplitude $\mathbf{E}$ acting on the dipole moment $\boldsymbol{\mu}$ and adding the energy

$$V(t)=-\mathbf{E}\cdot\boldsymbol{\mu}\cos\omega t$$

to the molecule. In this case the driving function is

$$f(t)=ip\cos\omega t\exp\left[-i\omega_0 t+i\int_0^t dt(p_{22}-p_{11})\cos\omega t\right],$$

where

$$p\equiv\mathbf{E}\cdot|\boldsymbol{\mu}_{12}|/\hbar, \tag{7}$$

since only the magnitude of the matrix element matters.

Two approximations on $f(t)$ are now in order. First, the integral term in the exponent will be neglected; it is usually zero by virtue of $p_{11}=p_{22}$, or in any case it is of the order of $(p_{22}-p_{11})/\omega$ which is small compared to $\omega_0 t$. Secondly, we neglect the sum-frequency term in the expansion of $\cos\omega t\exp(-i\omega_0 t)$ since we are interested mainly in frequencies not too far from resonance, where terms in $\omega+\omega_0$ produce only small effects. Consequently, the driving function becomes simply

$$f(t)=\tfrac{1}{2}ip\exp(i\delta t), \quad \delta\equiv\omega-\omega_0, \tag{8}$$

and Eq. (2) reads

$$\ddot{z}-(i\delta+\dot{p}/p)\dot{z}+(p/2)^2z=0. \tag{9}$$

For a constant-amplitude driving field, $p=p_c=$ const, Eqs. (9) and (4) give

$$z(t)=(\cos\Omega t/2-i\delta\Omega^{-1}\sin\Omega t/2)\exp(i\delta t/2),$$

where $\Omega^2\equiv p_c^2+\delta^2$. The excess population in the lower state at time t is therefore

$$D(t)=D(0)[(\delta/\Omega)^2+(p_c/\Omega)^2\cos\Omega t], \tag{10}$$

which is a familiar result,[3] though differently obtained.

If, on the other hand, the primary driving field is of arbitrary amplitude but tuned to resonance ($\delta=0$), the solution of Eq. (9) is simply

$$z(t)=\cos\theta/2,$$

so that

$$D(t)=D(0)\cos\theta, \tag{11}$$

where

$$\theta\equiv\int_0^t p(t)dt.$$

Figure 1(a) shows D *vs* the driving phase angle θ, together with the state populations N_1 and N_2. It is seen that, in contrast to a nonresonant drive, a resonant driving field is capable of completely inverting the state populations.

II. AFTER CUTOFF

Consider now what happens in the molecular assemblage when the primary driving field is suddenly removed. Collisions, which tend to destroy the phase memory of the molecules produced by the driving field, will first be ignored. Following cutoff, the system radiates power at the resonant frequency and in an amount proportional to the instantaneous rate of change of the population excess:

$$P=\hbar\omega_0\dot{D}/2. \tag{12}$$

This molecular-ringing power is related to its own electric field, $\mathfrak{E}$, filling a lossless wave guide of area A, by

$$P=cA\mathfrak{E}^2/4\pi,$$

---

[3] For example see H. S. Snyder and P. I. Richards, Phys. Rev. **73**, 1178 (1948).

or, for a cavity of volume V and loaded quality factor Q_L, by

$$P=\omega_0 V \mathfrak{E}^2/8\pi Q_L.$$

Thus, because $p=\mathfrak{E}\cdot|\mathbf{u}_{12}|/\hbar$, the value of p following cutoff is

$$p^2=k\dot{D}, \tag{13}$$

where

$$\begin{aligned} k&=2\pi\omega_0|\mu_{12}|^2/3c\hbar A \quad \text{(wave guide)}\\ k&=4\pi Q_L|\mu_{12}|^2/3\hbar V \quad \text{(cavity)}. \end{aligned} \tag{14}$$

(The factor $\frac{1}{3}$ comes from averaging over the angles between $\mathfrak{E}$ and $\mathbf{u}$.)

Although it is not necessary to do so, we shall assume that the primary driving field is tuned to the molecular resonance frequency ω_0. Not only is this the simplext case to treat, but it is also the one of greatest practical importance. Equation (11) then applies. If the phase duration of the primary drive is θ, the post-cutoff excess population in the lower state is

$$D(\tau)=D(0)\cos\left(\theta+\int_0^\tau p d\tau\right), \tag{15}$$

where τ is the time measured from cutoff. Taking $\dot{D}$, squaring, and using Eq. (13) we obtain the differential equation

$$\dot{D}=k[D^2(0)-D^2],$$

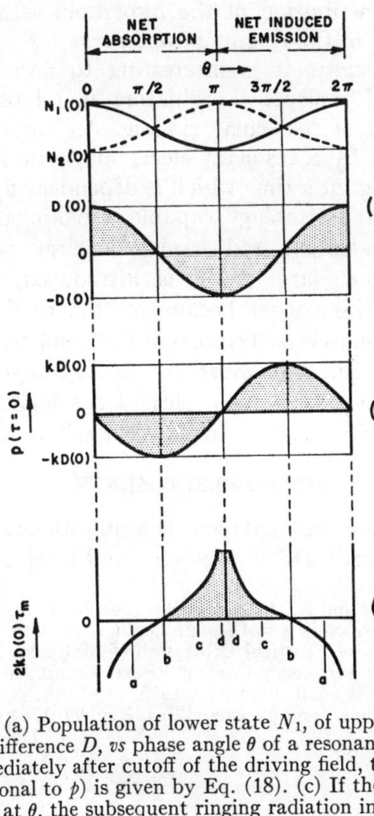

FIG. 1. (a) Population of lower state N_1, of upper state N_2 and of their difference D, *vs* phase angle θ of a resonant driving pulse. (b) Immediately after cutoff of the driving field, the ringing field (proportional to p) is given by Eq. (18). (c) If the primary drive is cut off at θ, the subsequent ringing radiation in the absence of impacts reaches its maximum at the time τ_m, when the two state populations become instantaneously equal.

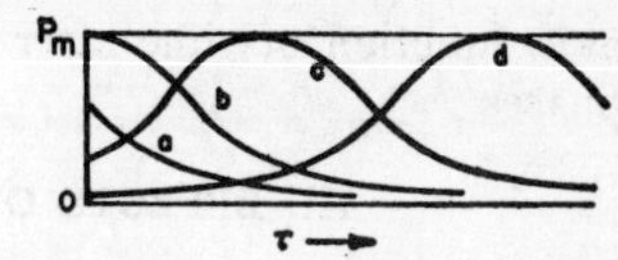

FIG. 2. Molecular-ringing power *vs* time τ after cutoff, for various driving phase angles θ. The behavior is symmetrical about $\theta=\pi$.

which has the solution

$$D(\tau)=D(0)\tanh[kD(0)(\tau-\tau_m)], \tag{16}$$

where we have called

$$\exp[2kD(0)\tau_m]\equiv(1-\cos\theta)/(1+\cos\theta). \tag{17}$$

At the time $\tau=\tau_m$ the two state populations are instantaneously equal.

The value of the ringing field $\mathfrak{E}$ is proportional to p, which from Eq. (13) is found to be

$$p(\tau)=-kD(0)\sin\left(\theta+\int_0^\tau p d\tau\right),$$

having the value immediately after cutoff of

$$p(\tau=0)=-kD(0)\sin\theta. \tag{18}$$

This is shown in Fig. 1(b). We see that if $D(0)$ is positive, corresponding to more molecules in the lower than in the upper state at the onset of the primary driving field, then p is negative if cutoff lies in $0\leqslant\theta\leqslant\pi$, and p is positive if $\pi\leqslant\theta\leqslant2\pi$. In other words, if at cutoff the gas was exhibiting a net absorption from the primary field, the sudden removal of that field produces a 180° phase reversal in the field detected at the matched termination of the wave guide.

The ringing power $P(\tau)$ is obtainable from Eqs. (12) and (16); it is

$$P(\tau)=P_m\,\text{sech}^2[kD(0)(\tau-\tau_m)], \tag{19}$$

where

$$P_m=\tfrac{1}{2}\hbar\omega_0 kD^2(0). \tag{20}$$

The ringing field is proportional to the $D(0)$ molecules and this field acts on the $D(0)$ molecules to give an emitted power proportional to $D^2(0)$; the square denotes coherency. Also, because p^2 is proportional to $\dot{D}$, a resonant driving field is more effective in producing ringing than is a nonresonant drive; the maximum value of $\dot{D}$ from Eq. (10) is smaller than that from Eq. (11).

The quantity $2kD(0)\tau_m$ of Eq. (17) is shown schematically in Fig. 1(c). An interesting fact emerges from this figure and from Eq. (19). The coherent radiation following cutoff may exhibit a delayed "surge" of power occuring at the time τ_m. These delayed peaks, of magnitude P_m, arise whenever cutoff finds the system with a greater population in the upper state than in the lower, that is, whenever $D(\tau=0)$ is negative. The peak occurs at the instant τ_m when the ringing field has driven the populations to an instantaneous equality, at which point the gas disgorges its stored energy most rapidly. The behavior of the ringing power is sketched in Fig. 2

as a function of time after cutoff, for various driving phases.

III. EFFECTS OF COLLISIONS

In the absence of collisions, the excess population in the lower state following cutoff was given by Eq. (15). With collisions, of mean free time T, this equation is changed to

$$D(\tau)=D(0)e^{-\tau/T}\times\cos\left(\theta+\int_0^\tau p\,d\tau\right)+D(0)(1-e^{-\tau/T}),\quad (21)$$

which reduces to Eq. (15) when $T=\infty$. The derivative of D is

$$\dot{D}=-pD(0)e^{-\tau/T}\times\sin\left(\theta+\int_0^\tau p\,d\tau\right)+[D(0)-D]/T;\quad (22)$$

the first term describes the change in D due to the ringing of those molecules which have not yet collided, the second term describes the change in D due to collisions alone which tend to return D to its thermal value $D(0)$. In determinging p (i.e., the strength of the ringing field) from Eq. (13), it is the first term of Eq. (22) which applies; thus

$$p(\tau)=-kD(0)\sin\left(\theta+\int_0^\tau p\,d\tau\right)e^{-\tau/T}.\quad (23)$$

This equation is solved for p by means of the substitution

$$\phi(\tau)=\int_0^\tau p\,d\tau,$$

which gives

$$p=d\phi/d\tau=-kD(0)e^{-\tau/T}\sin(\theta+\phi).$$

Integration yields

$$\ln\left(\tan\frac{\theta}{2}\bigg/\tan\frac{\theta+\phi}{2}\right)=kD(0)T(1-e^{-\tau/T}).$$

Solving for $\sin(\theta+\phi)$ and inserting it into Eq. (23) we obtain, after using Eq. (17) for τ_m,

$$p(\tau)=-kD(0)e^{-\tau/T}\operatorname{sech}\{kD(0)[T(1-e^{-\tau/T})-\tau_m]\}.$$

Thus, by Eqs. (12) and (13), the ringing power in the presence of impacts is

$$P(\tau)=P_m e^{-2\tau/T}\operatorname{sech}^2\{kD(0)[T(1-e^{-\tau/T})-\tau_m]\},\quad (24)$$

which reduces to Eq. (19) when $T=\infty$.

Under what conditions and at what time does the ringing power in the presence of collisions exhibit a delayed peak? Differentiating $P(\tau)$ we find that if the power has a maximum it occurs at the time τ_M satisfying the equation

$$1=kD(0)Te^{-\tau_M/T}\times\tanh\{kD(0)[\tau_m-T(1-e^{-\tau_M/T})]\}.\quad (25)$$

At time τ_M the ringing power is

$$P_M=P_m\{e^{-2\tau_M/T}-[kD(0)T]^{-2}\},\quad (26)$$

where P_m is the maximum power in the absence of impacts, as given by Eq. (20). However, in order for a delayed peak to exist, it is necessary, as seen from Eq. (25), that

$$kD(0)T>1.\quad (27)$$

This quantity, which, assuming collision times inversely proportional to density, is independent of density, can, for a given molecular system, be made large by increasing the Q of the cavity or the length of the wave guide.

The molecular ringing induced by a driving pulse of $\theta=\pi/2$—leading to maximum ringing immediately after cutoff—has been observed by Dicke and Romer.[4] Norton, of this Laboratory, in 1953 also observed these effects[5] and is presently refining new measurements in an attempt to observe either the delayed peaks or at least the contribution of the hyperbolic-secant factor to the shape of the ringing power curve.

Parenthetically, it is interesting to note that the operation of molecular oscillators is related to the phenomenon of molecular ringing. An empty cavity, when struck by a resonant electromagnetic field, continues to ring for a time which is dependent upon its Q. If a molecular assemblage, capable of making a resonant radiative transition and having a large population imbalance, i.e., large $D(0)$, is introduced, then the cavity will ring longer because of the ringing of the molecules themselves. To prevent this composite ringing from dying out, one must continually replenish the supply of large-$D(0)$ molecules to the high-Q cavity. This is just what is done in the "maser" oscillator.[6]

ACKNOWLEDGMENTS

The author has enjoyed helpful discussions with Dr. D. O. North, Dr. R. W. Peter, and Dr. J. P. Wittke.

[4] R. H. Dicke and R. H. Romer, Rev. Sci. Instr. **26**, 915 (1955).
[5] To be described in a forthcoming article. See also "Investigation and Study of Practical Utilization of Molecular Absorption Phenomena for Frequency Control," Final Report No. 10, Signal Corps Project 33-142B, February, 1955.
[6] Gordon, Zeiger, and Townes, Phys. Rev. **99**, 1264 (1955).

Geometrical Representation of the Schrödinger Equation for Solving Maser Problems

RICHARD P. FEYNMAN AND FRANK L. VERNON, JR., *California Institute of Technology, Pasadena, California*

AND

ROBERT W. HELLWARTH, *Microwave Laboratory, Hughes Aircraft Company, Culver City, California*

(Received September 18, 1956)

A simple, rigorous geometrical representation for the Schrödinger equation is developed to describe the behavior of an ensemble of two quantum-level, noninteracting systems which are under the influence of a perturbation. In this case the Schrödinger equation may be written, after a suitable transformation, in the form of the real three-dimensional vector equation $d\mathbf{r}/dt=\boldsymbol{\omega}\times\mathbf{r}$, where the components of the vector $\mathbf{r}$ uniquely determine ψ of a given system and the components of $\boldsymbol{\omega}$ represent the perturbation. When magnetic interaction with a spin $\frac{1}{2}$ system is under consideration, "r" space reduces to physical space. By analogy the techniques developed for analyzing the magnetic resonance precession model can be adapted for use in any two-level problems. The quantum-mechanical behavior of the state of a system under various different conditions is easily visualized by simply observing how $\mathbf{r}$ varies under the action of different types of $\boldsymbol{\omega}$. Such a picture can be used to advantage in analyzing various MASER-type devices such as amplifiers and oscillators. In the two illustrative examples given (the beam-type MASER and radiation damping) the application of the picture in determining the effect of the perturbing field on the molecules is shown and its interpretation for use in the complex Maxwell's equations to determine the reaction of the molecules back on the field is given.

INTRODUCTION

ELECTROMAGNETIC resonances in matter have become a fundamental tool for studying the structure of matter. Moreover, recently it has become of interest to use such resonances for radio and microwave frequency circuit components, such as highly stable oscillators, high Q filters, isolators, and amplifiers. The purpose of this paper will be to aid in the understanding of simple resonances and especially in the conception and design of microwave "atomic" devices (now commonly called MASER-type devices) which involve these simple resonances. In this paper we propose to do the following things: (a) To develop a simple but rigorous and complete geometrical picture of the Schrödinger equation describing the resonance behavior of a quantum system when only a pair of energy levels is involved (the resulting picture has the same form as the well-known three-dimensional classical precession of a gyromagnet in a magnetic field); (b) To note further properties of the model which permit its direct interpretation in terms of the physical properties which couple the quantum systems to the electromagnetic fields, and to state these explicitly for dipole transitions; (c) To illustrate the use of the picture by solving the particular cases of the beam MASER oscillator characteristics and "radiation damping."

Although the approach does not obtain results inaccessible to straight-forward calculation, the simplicity of the pictorial representation enables one to gain physical insight and to obtain results quickly which display the main features of interest.

FORMULATION

We will be concerned with an ensemble of spacially non-overlapping systems, e.g., molecules in a molecular beam, such that the wave function for any one individual system may be written

$$\psi(t)=a(t)\psi_a+b(t)\psi_b \tag{1}$$

during some time of interest. ψ_a and ψ_b are the two eigenstates of interest of the Hamiltonian for the single system corresponding to the energies $W+\hbar\omega_0/2$ and $W-\hbar\omega_0/2$ respectively. W is the mean energy of the two levels determined by velocities and internal interactions which remain unchanged. W will be taken as the zero of energy for each system. ω_0 is the resonant angular frequency associated with a transition between the two levels and is always taken positive.

It is usual to solve Schrödinger's equation with some perturbation V for the complex coefficients $a(t)$ and $b(t)$, and from them calculate the physical properties of the system. However, the mathematics is not always transparent and the complex coefficients do not give directly the values of real physical observables. Neither is it sufficient to know only the real magnitudes of a and b, i.e., the level populations and transition probabilities, when coherent processes are involved. We propose instead to take advantage of the fact that the phase of $\psi(t)$ has no influence so that only three real numbers are needed to completely specify $\psi(t)$. We construct three real functions (r_1,r_2,r_3) of a and b which have direct physical meaning and which define a 3-vector $\mathbf{r}$ whose time dependance is easily pictured:

$$\begin{aligned} r_1&\equiv ab^*+ba^*\\ r_2&\equiv i(ab^*-ba^*)\\ r_3&\equiv aa^*-bb^*. \end{aligned} \tag{2}$$

(*) always indicates complex conjugate. The time dependence of $\mathbf{r}$ can be obtained from Schrödinger's

Reprinted with permission from *J. Appl. Phys.*, vol. 28, pp. 49–52, Jan. 1957.

equation which gives

$$i\hbar da/dt = a[(\hbar\omega_0/2)+V_{aa}]+bV_{ab} \qquad (3)$$

and similar equations for db/dt, da^*/dt, db^*/dt. The subscripts on V indicate the usual matrix elements. $V_{aa}=V_{bb}=0$ for most all cases of interest, and whenever these can be neglected compared to $\hbar\omega_0/2$, V need be neither small nor of short duration for the results to be exact. Using Eqs. (3) to find the differential equation for $\mathbf{r}$ gives

$$d\mathbf{r}/dt=\boldsymbol{\omega}\times\mathbf{r} \qquad (4)$$

where $\boldsymbol{\omega}$ is also a three vector in "r" space defined by the three real components:

$$\omega_1 \equiv (V_{ab}+V_{ba})/\hbar$$
$$\omega_2 \equiv i(V_{ab}-V_{ba})/\hbar \qquad (5)$$
$$\omega_3 \equiv \omega_0.$$

The $\times$ symbol has the usual vector product meaning. It is easily shown that the remaining real combination aa^*+bb^* is just equal to the length of the $\mathbf{r}$ vector, $(r_1^2+r_2^2+r_3^2)^{\frac{1}{2}}$, and is constant in time. It equals one when ψ is normalized to unity. The motion described by Eq. (4) is of the form for the precession of a classical gyromagnet in a magnetic field. Therefore, it is not surprising that in the case of transitions between the two magnetic levels of a spin $\frac{1}{2}$ particle, this mathematical $\mathbf{r}$ space will be equivalent to physical space with r_1, r_2, r_3 proportional to the expectation values of μ_x, μ_y, μ_z, and ω_1, ω_2, ω_3 proportional to the components of the magnetic field H_x, H_y, H_z respectively. Although in general the formalism does not represent physical space, by analogy any transitions under the stated conditions may be thought of rigorously in terms of the well-known classical vector model for spin precession. The extensive and explicit use of rotating coordinate procedures, as was introduced by Bloch, Ramsey, Rabi, and Schwinger[1,2] for special kinds of magnetic transitions, is generally applicable in dealing with the r space.

INTERPRETATION

The effect of the presence of the quantum systems on the surrounding electromagnetic field is observed in many resonance experiments or devices, so it is of interest to deduce such quantities as the energy given up by the systems and effective polarization densities which, in general, are not linear in the impressed fields. The internal energy, or expectation value of the unperturbed Hamiltonian H at any time t is

$$\langle H\rangle = \int \psi^* H\psi d(\mathrm{Vol}) = (aa^*-bb^*)\hbar\omega_0/2 = r_3\hbar\omega_0/2 \qquad (6)$$

or just r_3 in units of $\hbar\omega_0/2$. The total internal energy in any ensemble of these systems is of course the sum of the r_3 values (in units of $\hbar\omega_0/2$) in the region, or the projection on the 3 axis of the vector sum $\mathbf{R}=\sum_i \mathbf{r}^i$ over the region. In fact, any operator x such as the dipole moment operator, which is separable in the systems, has an expectation value of the form

$$x_{ab}\sum_i (a^i)^* b^i + x_{ba}\sum_i (b^i)^* a^i + x_{aa}\sum_i (a^i)^* a^i + x_{bb}\sum_i (b^i)^* b^i$$

and is therefore a linear combination of the r_1's, r_2's, and r_3's, or R_1, R_2, and R_3; it is proportional to a projection of $\mathbf{R}$ on some axis, plus perhaps a constant.

It remains to determine the proper projections for particular cases and also state explicitly the values of $\boldsymbol{\omega}$. Since all common microwave transitions such as hyperfine structure, spin flip, molecular rotational and inversion transitions are dipole transitions, we will examine only these cases.

For electric dipole $\Delta m=0$ transitions,

$$V_{ab}=-\mu_{ab}E \qquad (7)$$

where μ_{ab} is the matrix element between the two states for the component of the dipole moment along the electric field E. If μ_{ab} is made real by proper choice of the phases of ψ_a and ψ_b, then

$$\omega_1 = (V_{ab}+V_{ba})/\hbar = -(2\mu_{ab}/\hbar)E$$
$$\omega_2 = i(V_{ab}-V_{ba})/\hbar = 0 \qquad (8)$$
$$\omega_3=\omega_0$$

ω_1 is the electric field strength in units of $-2\mu_{ab}/\hbar$. In this case

$$\langle\mu\rangle = a^*b\mu_{ab}+b^*a\mu_{ba}=r_1\mu_{ab}. \qquad (9)$$

This means that the component of the polarization density P along the electric field will equal the average projection of $\mathbf{r}$ on the 1 axis in some small region of space and given in units of $\rho\mu_{ab}$ where ρ is the particle density.

In the case of magnetic dipole $\Delta m=0$ transitions, the same formulas apply substituting H for E and the appropriate magnetic dipole for μ.

In the case of electric or magnetic $\Delta m=\pm 1$ dipole transitions, considering E_x and E_y to be the relevant spacial components of either the electric or magnetic fields,

$$V=(-1/2)(\mu^+E^-+\mu^-E^+) \qquad (10)$$

where $E^\pm \equiv E_x \pm iE_y$ and $\mu^\pm \equiv \mu_x \pm i\mu_y$. By the well-known properties of the $\mu^\pm$ operators:

$$V_{ab}=-(1/2)\mu_{ab}^+(E_x-iE_y)$$
$$V_{ba}=-(1/2)\mu_{ba}^-(E_x+iE_y). \qquad (11)$$

Choosing the phases of ψ_a and ψ_b such that μ_{ab}^+ is a real number γ, then $\mu_{ab}^+=\mu_{ba}^-$ by their definitions, and:

$$\omega_1=-(\gamma/\hbar)E_x$$
$$\omega_2=-(\gamma/\hbar)E_y; \qquad (12)$$

thus $\boldsymbol{\omega}$ behaves in the 1–2 plane exactly as does E in

[1] Rabi, Ramsey, and Schwinger, Revs. Modern Phys. 26, 167 (1954).

[2] R. K. Wangsness, Am. J. Phys. 24, 60 (1956).

the $x-y$ plane of space. By noting that $\langle\mu^+\rangle=\gamma a^*b$ and $\langle\mu^-\rangle=\gamma b^*a$, we find:

$$\begin{aligned}\langle\mu_x\rangle&=(\gamma/2)r_1\\ \langle\mu_y\rangle&=(\gamma/2)r_2.\end{aligned}\tag{13}$$

If there exists a component μ_z such that $-\mu_z E_z=H$, then it can be seen that the mathematical "r" space reduces to physical space, as in the case of free spin $\frac{1}{2}$ Zeeman transitions. By similar procedures any kind of perturbation affecting only two levels can be thought of in terms of the familiar behavior of vectors rotating in space, according to $d\mathbf{r}/dt=\boldsymbol{\omega}\times\mathbf{r}$.

SAMPLE APPLICATIONS

Beam Type Maser Oscillator[3]

To examine how this viewpoint leads to the solution of a particular problem, we first solve the effect of a given field on the particles involved; secondly, we formulate the classical field equations in a way suitable to the experimental situation, and using the proper projections of the **r** vector we find the conditions which satisfy both Schrödinger's and Maxwell's equations simultaneously. Consider a beam of molecules which enters a microwave cavity which is near resonance with a $\Delta m=0$ transition of the molecule. The molecules have been prepared so that only those in the higher energy state enter the cavity. Assume for simplicity that the cavity mode shape is such that the molecules see an oscillatory field of constant amplitude and phase as they pass through the cavity. The oscillating ω_1 can be separated into two counter-rotating components in the 1–2 plane. For coherent perturbations such as this it is convenient to transform to a coordinate frame in which the appropriate component of ω_1 appears stationary, and neglect the other counter-rotating component. The rotating axes will be designated the I, II, and III axes. We take the I axis in the plane of the stationary driving torque which now has the following constant components (see Fig. 1):

$$\begin{aligned}\omega_I&=1/2|\omega_1|\\ \omega_{II}&=0\\ \omega_{III}&=\omega_0-\omega.\end{aligned}$$

ω is the frequency of the perturbation. The molecules enter the cavity with $\mathbf{r}=\mathbf{III}$ and at a time t later the components r_I and r_{II} can be seen by inspection of Fig. 1 to be

$$\begin{aligned}r_I&=\frac{\omega_I(\omega_0-\omega)}{\Omega^2}[1-\cos(\Omega t)]\\ r_{II}&=-\frac{\omega_I}{\Omega}\sin(\Omega t).\end{aligned}\tag{14}$$

Ω is the magnitude $[\omega_I^2+(\omega_0-\omega)^2]^{\frac{1}{2}}$ of the driving torque as seen in the rotating frame.

[3] Gordon, Zeiger, and Townes, Phys. Rev. **95**, 282 (1954).

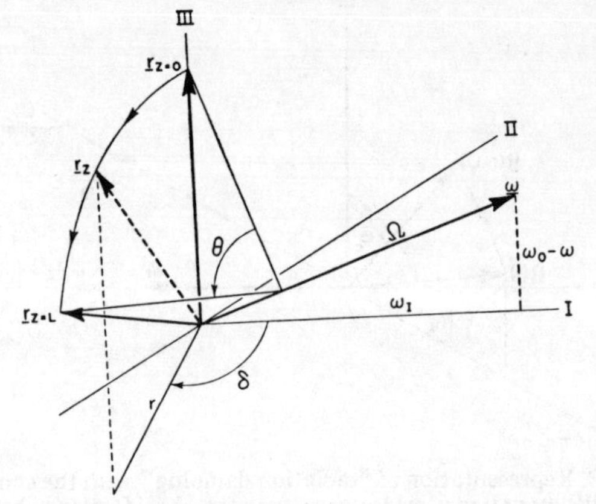

FIG. 1. MASER oscillator diagram in rotating coordinates.

To reduce these results to the stationary frame we choose the time reference such that $\omega_1=2\omega_I\cos(\omega t)$. Then $r_1=r(t)\cos[\omega t+\delta(t)]$ where $r(t)$ is the magnitude of the projection of **r** on the $1-2$ plane and $\delta(t)=\tan^{-1}r_{II}/r_I$. If we use complex quantities to represent time dependence at frequency ω, it is evident if ω_1 is represented by ω_I then r_1 is represented by (r_I+ir_{II}). Assuming all the molecules to have a velocity v then the complex polarization density P at a distance z along the cavity is the simple expression $\rho\mu_{ab}(r_I+ir_{II})$ with $t=z/v$.[4] In a thin beam, P_z, the polarization per unit length of beam is $(n/v)\mu_{ab}\times(r_I+ir_{II})$. n is number per second entering the cavity. Thus in practice one obtains the quantities of interest directly from the rotating frame.

The electric field configuration in the cavity has been assumed to be the normal configuration $\mathbf{E}_c(x,y,z)$ of the nondegenerate mode employed, where the normalization is taken such that $\int|\mathbf{E}_c|^2d\mathcal{V}=1$. $|\mathbf{E}_c|$ at the beam is taken to be the constant $f\mathcal{V}^{-\frac{1}{2}}$. $\mathcal{V}$ is the volume of the cavity. f is a form factor which would be unity were the field uniform throughout. The electric field may be written $\mathbf{E}=\mathbf{E}_c(x,y,z)\mathcal{E}(t)e^{i\omega t}$ where $\mathcal{E}$ is a real amplitude, constant in the steady state of oscillation. Then Maxwell's equations in complex form give

$$-\omega^2[\mathcal{E}\mathbf{E}_c+(4\pi\mathbf{E}_c/|\mathbf{E}_c|)P]+i(\omega\omega_c/Q)\mathbf{E}_c\mathcal{E}+\omega_c^2\mathbf{E}_c\mathcal{E}=0.\tag{15}$$

ω_c is the resonant frequency of the cavity and Q is quality factor of the cavity. Integrating Eq. (15) by $\cdot\mathbf{E}_c$ over the cavity volume gives in the case of a very thin beam

$$-\omega^2\left[\mathcal{E}+(4\pi n/v)\mu_{ab}\int_0^L f\mathcal{V}^{-\frac{1}{2}}(r_I+ir_{II})dz\right]+i(\omega\omega_c/Q)\mathcal{E}+\omega_c^2\mathcal{E}=0.\tag{16}$$

Performing the indicated integration, the imaginary

[4] J. Helmer, M. L. Report No. 311, Signal Corps Contract DA 36-039 SC-71178, Stanford University.

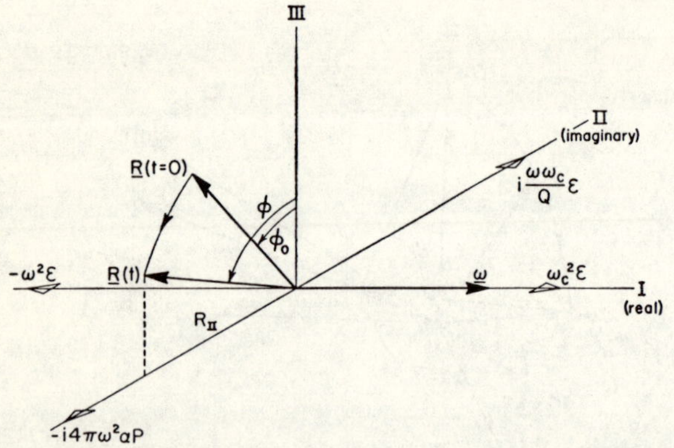

FIG. 2. Representation of "radiation damping" with the complex Maxwell's equations represented on the $I-II$ plane by the hollow arrows. $\mathbf{R}=\Sigma i\mathbf{r}^i$, $\alpha=fV^{-\frac{1}{2}}\mu_{ab}$, $P=$total polarization (or magnetization).

part of Eq. (16) gives

$$\frac{n}{n_{th}}=\frac{\theta^2}{2(1-\cos\theta)}. \tag{17}$$

$n_{th}\equiv\hbar\mathcal{V}v^2/2\pi f^2\mu_{ab}{}^2L^2Q$ and is the threshold number per second required to sustain oscillation. θ is the total angle $\Omega L/v$ through which each r precesses about the effective $\boldsymbol{\omega}$.

Equation (17) gives θ if n is known and thus the spread of frequencies at which oscillation is possible. To determine the magnitude of the electric field and the frequency of oscillation for a particular ω_c and cavity Q consider the real part of Eq. (16).

This may be written as

$$\frac{\omega_0-\omega}{\omega-\omega_c}=\frac{Q}{\pi Q_B}\frac{1-\cos\theta}{1-(\sin\theta)/\theta}\approx\frac{\omega_0-\omega}{\omega_0-\omega_c} \tag{18}$$

where $Q_B\equiv 2\pi\omega_0L/v\approx\omega_0/\Delta\omega$ is a parameter describing the natural molecular resonance line width $\Delta\omega$. Given the amount of cavity detuning $\omega_0-\omega_c$ and θ from Eq. (17), Eq. (18) enables one to determine the frequency of oscillation ω and then $\mathcal{E}\sim\omega_I$ by using the definition of θ. These are essentially the results of Shimoda, Wang, and Townes,[5] though it appears here that no restrictions need be placed on ω to obtain them. Since the parameters θ, $\omega-\omega_0$, $\omega_I\sim\mathcal{E}$, the internal energy, and the dipole moment all appear as geometrical quantities in Fig. 1, it is easy to visualize the effects of changing any of them. Also, it is often easy to visualize, if not to solve, more complicated situations such as those which involve cavities with nonuniform modes, multiple cavities, or externally-driven cavities.

To picture the coupling of the molecules, governed by the Schrödinger equation, with the field, governed by Maxwell's equation, it is useful to think of the $I-II$ plane in the rotating frame as a complex plane representing relative time phase, with the II axis as the imaginary axis. Then the complex Maxwell's equation (16) can be drawn on the $I-II$ plane and the way in which the various quantities must vary to balance the equation to zero (or to some other driving force, if present) can be visualized. Imagining the $I-II$ plane as complex is especially useful when the $\mathbf{r}$ vectors throughout the cavity have all seen the same perturbation for the same length of time, in which case the integrals are just proportional at any time to the resultant $\mathbf{R}=\sum_i\mathbf{r}^i$ which behaves in the same manner as the individual $\mathbf{r}$'s, i.e., $d\mathbf{R}/dt=\boldsymbol{\omega}\times\mathbf{R}$. This picture is easily applied to the phenomenon of "radiation damping."[6,7]

Radiation Damping

To examine the spontaneous behavior of an ensemble of dipoles in an arbitrary state (represented by an $\mathbf{R}$) and enclosed in some small portion of a microwave cavity, we may write Maxwell's equations for the cavity as before. When the ensemble is in thermal equilibrium $\mathbf{R}$ is $-\mathbf{III}R_0$ where R_0 is given by the number present and Boltzmann statistics. Assume some other $\mathbf{R}$ state is obtained (this can be done by applying a short intense rf pulse at ω_0) and $\mathbf{R}$ is left tipped at an angle ϕ_0 to the III axis in the II, III plane ($R_I=0$). Further, we assume that the cavity is tuned to the molecular resonant frequency so that in this case $\omega=\omega_0=\omega_c$. Figure 2 is drawn for this case. $R_{II}=R_0\sin\phi$ is proportional to $\mathcal{E}$ from balancing imaginary parts of the diagram. We must now assume that $d\mathcal{E}/dt\ll(\omega_0/Q)\mathcal{E}$ and $(\omega_I/\omega_0)^2\ll 1$ as we have replaced time derivatives by $i\omega$ only. Now $d\mathbf{R}/dt=\boldsymbol{\omega}\times\mathbf{R}$ means that $d\phi/dt\sim\sin\phi$. So the radiation damping obeys $d\phi/dt\sim\sin\phi$ at resonance. The solution with constants evaluated is

$$\tan(\phi/2)=\tan(\phi_0/2)e^{t/\tau}. \tag{19}$$

$\tau=\mathcal{V}\hbar/4\pi f^2\mu_{ab}{}^2QR_0$ for $\Delta m=0$ transitions, and $\tau=\mathcal{V}\hbar/\pi f^2\gamma^2R_0Q$ for the case of $\Delta m\pm 1$ transitions in a linearly polarized field (that is, a nondegenerate cavity mode). The case of a circularly polarized field involving two cavity modes and $\Delta m=\pm 1$ transitions is more complicated and involves both $\langle\mu_x\rangle$ and $\langle\mu_y\rangle$ each coupling to a separate mode.

In conclusion, we wish to emphasize the usefulness of the geometrical model in visualizing and solving problems involving transitions between two levels. However, the way in which this model would be interpreted and used in a given situation depends upon the particular problem as is indicated by the two examples given. This technique of using the geometrical model does not make solutions of problems possible which were not solvable previously. However, even in many of these insoluble cases one can gain considerable insight into the behavior of the processes being investigated by observing how the parameters in the model vary.

---

[5] Shimoda, Wang, and Townes, Phys. Rev. **102**, 1308 (1956).

[6] N. Bloembergen and R. V. Pound, Phys. Rev. **95**, 8 (1954).

[7] R. H. Dicke, Phys. Rev. **93**, 99 (1954).

Part 2
Optical Resonators

Laser Beams and Resonators

H. KOGELNIK AND T. LI

Abstract—**This paper is a review of the theory of laser beams and resonators. It is meant to be tutorial in nature and useful in scope. No attempt is made to be exhaustive in the treatment. Rather, emphasis is placed on formulations and derivations which lead to basic understanding and on results which bear practical significance.**

1. Introduction

THE COHERENT radiation generated by lasers or masers operating in the optical or infrared wavelength regions usually appears as a beam whose transverse extent is large compared to the wavelength. The resonant properties of such a beam in the resonator structure, its propagation characteristics in free space, and its interaction behavior with various optical elements and devices have been studied extensively in recent years. This paper is a review of the theory of laser beams and resonators. Emphasis is placed on formulations and derivations which lead to basic understanding and on results which are of practical value.

Historically, the subject of laser resonators had its origin when Dicke [1], Prokhorov [2], and Schawlow and Townes [3] independently proposed to use the Fabry-Perot interferometer as a laser resonator. The modes in such a structure, as determined by diffraction effects, were first calculated by Fox and Li [4]. Boyd and Gordon [5], and Boyd and Kogelnik [6] developed a theory for resonators with spherical mirrors and approximated the modes by wave beams. The concept of electromagnetic wave beams was also introduced by Goubau and Schwering [7], who investigated the properties of sequences of lenses for the guided transmission of electromagnetic waves. Another treatment of wave beams was given by Pierce [8]. The behavior of Gaussian laser beams as they interact with various optical structures has been analyzed by Goubau [9], Kogelnik [10], [11], and others.

The present paper summarizes the various theories and is divided into three parts. The first part treats the passage of paraxial rays through optical structures and is based on geometrical optics. The second part is an analysis of laser beams and resonators, taking into account the wave nature of the beams but ignoring diffraction effects due to the finite size of the apertures. The third part treats the resonator modes, taking into account aperture diffraction effects. Whenever applicable, useful results are presented in the forms of formulas, tables, charts, and graphs.

Manuscript received July 12, 1966.
H. Kogelnik is with Bell Telephone Laboratories, Inc., Murray Hill, N. J.
T. Li is with Bell Telephone Laboratories, Inc., Holmdel, N. J.

2. Paraxial Ray Analysis

A study of the passage of paraxial rays through optical resonators, transmission lines, and similar structures can reveal many important properties of these systems. One such "geometrical" property is the stability of the structure [6], another is the loss of unstable resonators [12]. The propagation of paraxial rays through various optical structures can be described by ray transfer matrices. Knowledge of these matrices is particularly useful as they also describe the propagation of Gaussian beams through these structures; this will be discussed in Section 3. The present section describes briefly some ray concepts which are useful in understanding laser beams and resonators, and lists the ray matrices of several optical systems of interest. A more detailed treatment of ray propagation can be found in textbooks [13] and in the literature on laser resonators [14].

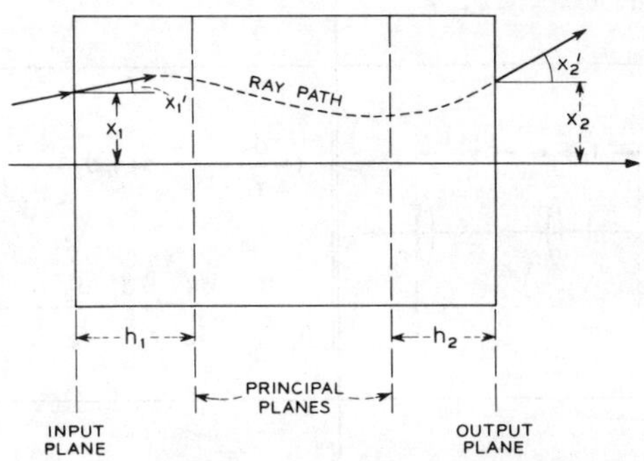

Fig. 1. Reference planes of an optical system. A typical ray path is indicated.

2.1 *Ray Transfer Matrix*

A paraxial ray in a given cross section (z=const) of an optical system is characterized by its distance x from the optic (z) axis and by its angle or slope x' with respect to that axis. A typical ray path through an optical structure is shown in Fig. 1. The slope x' of paraxial rays is assumed to be small. The ray path through a given structure depends on the optical properties of the structure and on the input conditions, i.e., the position x_1 and the slope x_1' of the ray in the input plane of the system. For paraxial rays the corresponding output quantities x_2 and x_2' are linearly dependent on the input quantities. This is conveniently written in the matrix form

$$\begin{vmatrix} x_2 \\ x_2' \end{vmatrix} = \begin{vmatrix} A & B \\ C & D \end{vmatrix} \begin{vmatrix} x_1 \\ x_1' \end{vmatrix} \tag{1}$$

Reprinted from *Proc. IEEE*, vol. 54, pp. 1312–1329, Oct. 1966.

TABLE I

RAY TRANSFER MATRICES OF SIX ELEMENTARY OPTICAL STRUCTURES

NO.	OPTICAL SYSTEM	RAY TRANSFER MATRIX
1	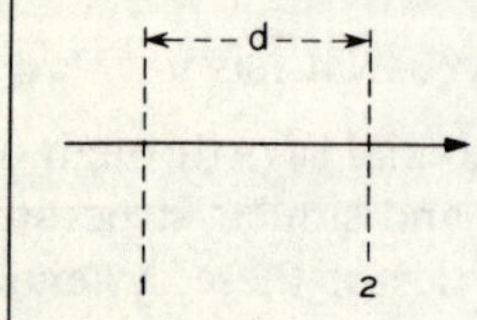	$\begin{vmatrix} 1 & d \\ 0 & 1 \end{vmatrix}$
2	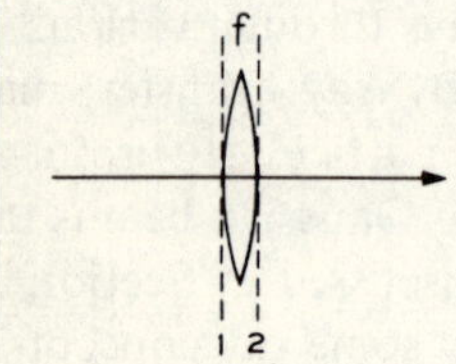	$\begin{vmatrix} 1 & 0 \\ -\frac{1}{f} & 1 \end{vmatrix}$
3	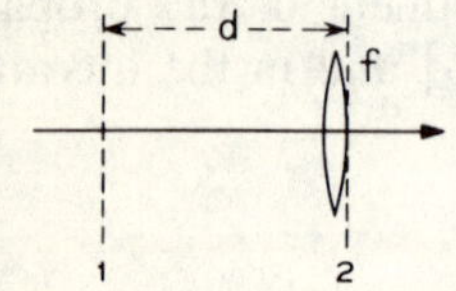	$\begin{vmatrix} 1 & d \\ -\frac{1}{f} & 1-\frac{d}{f} \end{vmatrix}$
4	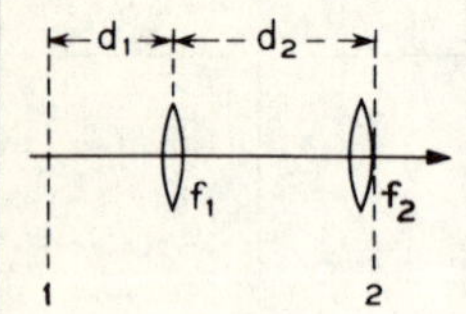	$\begin{vmatrix} 1-\frac{d_2}{f_1} & d_1+d_2-\frac{d_1 d_2}{f_1} \\ -\frac{1}{f_1}-\frac{1}{f_2}+\frac{d_2}{f_1 f_2} & 1-\frac{d_1}{f_1}-\frac{d_2}{f_2}-\frac{d_1}{f_2}+\frac{d_1 d_2}{f_1 f_2} \end{vmatrix}$
5	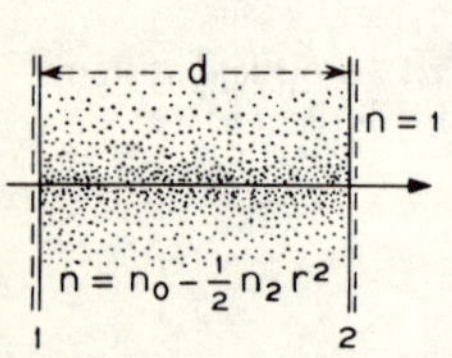	$\begin{vmatrix} \cos d\sqrt{\frac{n_2}{n_0}} & \frac{1}{\sqrt{n_0 n_2}} \sin d\sqrt{\frac{n_2}{n_0}} \\ -\sqrt{n_0 n_2}\, \sin d\sqrt{\frac{n_2}{n_0}} & \cos d\sqrt{\frac{n_2}{n_0}} \end{vmatrix}$
6	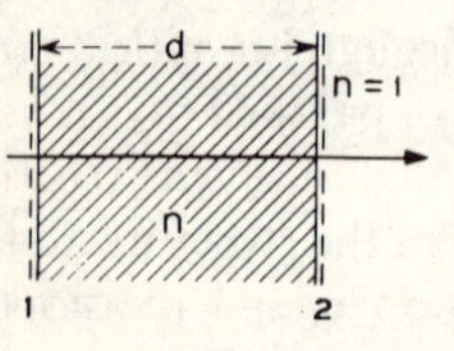	$\begin{vmatrix} 1 & d/n \\ 0 & 1 \end{vmatrix}$

where the slopes are measured positive as indicated in the figure. The *ABCD* matrix is called the ray transfer matrix. Its determinant is generally unity

$$AD - BC = 1. \tag{2}$$

The matrix elements are related to the focal length f of the system and to the location of the principal planes by

$$f = -\frac{1}{C}$$

$$h_1 = \frac{D-1}{C} \tag{3}$$

$$h_2 = \frac{A-1}{C}$$

where h_1 and h_2 are the distances of the principal planes from the input and output planes as shown in Fig. 1.

In Table I there are listed the ray transfer matrices of six elementary optical structures. The matrix of No. 1 describes the ray transfer over a distance d. No. 2 describes the transfer of rays through a thin lens of focal length f. Here the input and output planes are immediately to the left and right of the lens. No. 3 is a combination of the first two. It governs rays passing first over a distance d and then through a thin lens. If the sequence is reversed the diagonal elements are interchanged. The matrix of No. 4 describes the rays passing through two structures of the No. 3 type. It is obtained by matrix multiplication. The ray transfer matrix for a lenslike medium of length d is given in No. 5. In this medium the refractive index varies quadratically with the distance r from the optic axis.

$$n = n_0 - \tfrac{1}{2} n_2 r^2. \tag{4}$$

An index variation of this kind can occur in laser crystals and in gas lenses. The matrix of a dielectric material of index n and length d is given in No. 6. The matrix is referred to the surrounding medium of index 1 and is computed by means of Snell's law. Comparison with No. 1 shows that for paraxial rays the effective distance is *shortened* by the optically denser material, while, as is well known, the "optical distance" is lengthened.

2.2 *Periodic Sequences*

Light rays that bounce back and forth between the spherical mirrors of a laser resonator experience a periodic focusing action. The effect on the rays is the same as in a periodic sequence of lenses [15] which can be used as an optical transmission line. A periodic sequence of identical optical systems is schematically indicated in Fig. 2. A single element of the sequence is characterized by its *ABCD* matrix. The ray transfer through n consecutive elements of the sequence is described by the nth power of this matrix. This can be evaluated by means of Sylvester's theorem

$$\begin{vmatrix} A & B \\ C & D \end{vmatrix}^n = \frac{1}{\sin\Theta} \tag{5}$$

$$\cdot \begin{vmatrix} A \sin n\Theta - \sin(n-1)\Theta & B \sin n\Theta \\ C \sin n\Theta & D \sin n\Theta - \sin(n-1)\Theta \end{vmatrix}$$

where

$$\cos \Theta = \tfrac{1}{2}(A + D). \tag{6}$$

Periodic sequences can be classified as either *stable* or *unstable*. Sequences are stable when the trace $(A+D)$ obeys the inequality

$$-1 < \tfrac{1}{2}(A + D) < 1. \tag{7}$$

Inspection of (5) shows that rays passing through a stable sequence are periodically refocused. For unstable systems, the trigonometric functions in that equation become hyperbolic functions, which indicates that the rays become more and more dispersed the further they pass through the sequence.

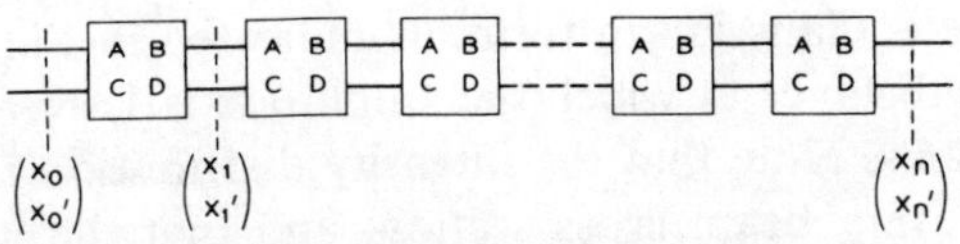

Fig. 2. Periodic sequence of identical systems, each characterized by its *ABCD* matrix.

2.3 *Stability of Laser Resonators*

A laser resonator with spherical mirrors of unequal curvature is a typical example of a periodic sequence that can be either stable or unstable [6]. In Fig. 3 such a resonator is shown together with its dual, which is a sequence of lenses. The ray paths through the two structures are the same, except that the ray pattern is folded in the resonator and unfolded in the lens sequence. The focal lengths f_1 and f_2 of the lenses are the same as the focal lengths of the mirrors, i.e., they are determined by the radii of curvature R_1 and R_2 of the mirrors ($f_1 = R_1/2$, $f_2 = R_2/2$). The lens spacings are the same as the mirror spacing d. One can choose, as an element of the periodic sequence, a spacing followed by one lens plus another spacing followed by the second lens. The *ABCD* matrix of such an element is given in No. 4 of Table I. From this one can obtain the trace, and write the stability condition (7) in the form

$$0 < \left(1 - \frac{d}{R_1}\right)\left(1 - \frac{d}{R_2}\right) < 1. \tag{8}$$

To show graphically which type of resonator is stable and which is unstable, it is useful to plot a stability diagram on which each resonator type is represented by a point. This is shown in Fig. 4 where the parameters d/R_1 and d/R_2 are drawn as the coordinate axes; unstable systems are represented by points in the shaded areas. Various resonator types, as characterized by the relative positions of the centers of curvature of the mirrors, are indicated in the appropriate regions of the diagram. Also entered as alternate coordinate axes are the parameters g_1 and g_2 which play an important role in the diffraction theory of resonators (see Section 4).

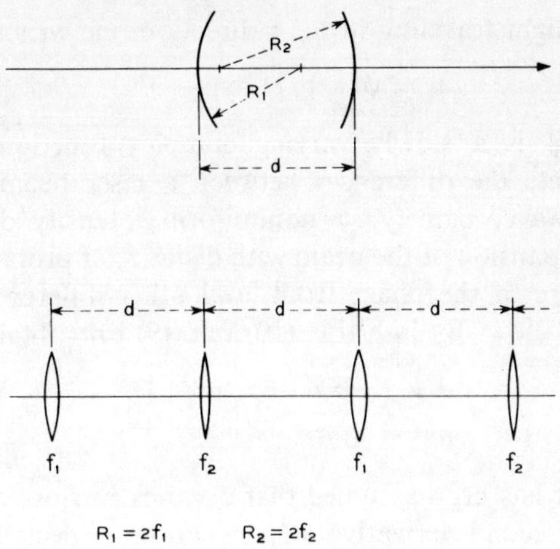

Fig. 3. Spherical-mirror resonator and the equivalent sequence of lenses.

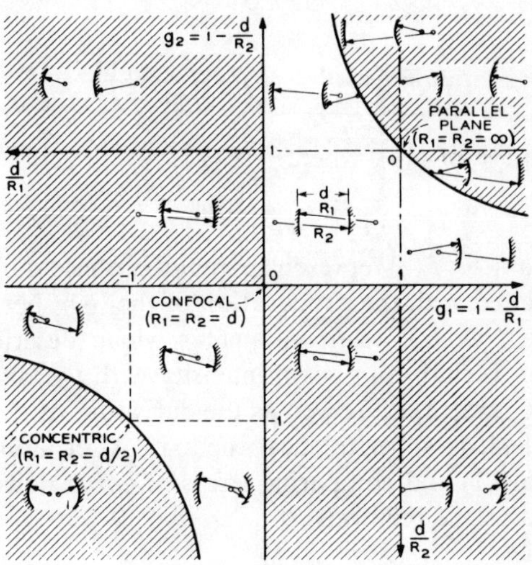

Fig. 4. Stability diagram. Unstable resonator systems lie in shaded regions.

3. Wave Analysis of Beams and Resonators

In this section the wave nature of laser beams is taken into account, but diffraction effects due to the finite size of apertures are neglected. The latter will be discussed in Section 4. The results derived here are applicable to optical systems with "large apertures," i.e., with apertures that intercept only a negligible portion of the beam power. A theory of light beams or "beam waves" of this kind was first given by Boyd and Gordon [5] and by Goubau and Schwering [7]. The present discussion follows an analysis given in [11].

3.1 *Approximate Solution of the Wave Equation*

Laser beams are similar in many respects to plane waves; however, their intensity distributions are not uniform, but are concentrated near the axis of propagation and their phase fronts are slightly curved. A field component or potential u of the coherent light satisfies the scalar wave equation

$$\nabla^2 u + k^2 u = 0 \tag{9}$$

where $k = 2\pi/\lambda$ is the propagation constant in the medium.

For light traveling in the z direction one writes

$$u = \psi(x, y, z) \exp(-jkz) \tag{10}$$

where ψ is a slowly varying complex function which represents the differences between a laser beam and a plane wave, namely: a nonuniform intensity distribution, expansion of the beam with distance of propagation, curvature of the phase front, and other differences discussed below. By inserting (10) into (9) one obtains

$$\frac{\partial^2 \psi}{\partial x^2} + \frac{\partial^2 \psi}{\partial y^2} - 2jk \frac{\partial \psi}{\partial z} = 0 \tag{11}$$

where it has been assumed that ψ varies so slowly with z that its second derivative $\partial^2\psi/\partial z^2$ can be neglected.

The differential equation (11) for ψ has a form similar to the time dependent Schrödinger equation. It is easy to see that

$$\psi = \exp\left\{-j\left(P + \frac{k}{2q} r^2\right)\right\} \tag{12}$$

is a solution of (11), where

$$r^2 = x^2 + y^2. \tag{13}$$

The parameter $P(z)$ represents a *complex* phase shift which is associated with the propagation of the light beam, and $q(z)$ is a *complex* beam parameter which describes the Gaussian variation in beam intensity with the distance r from the optic axis, as well as the curvature of the phase front which is spherical near the axis. After insertion of (12) into (11) and comparing terms of equal powers in r one obtains the relations

$$q' = 1 \tag{14}$$

and

$$P' = -\frac{j}{q} \tag{15}$$

where the prime indicates differentiation with respect to z. The integration of (14) yields

$$q_2 = q_1 + z \tag{16}$$

which relates the beam parameter q_2 in one plane (output plane) to the parameter q_1 in a second plane (input plane) separated from the first by a distance z.

3.2 *Propagation Laws for the Fundamental Mode*

A coherent light beam with a Gaussian intensity profile as obtained above is not the only solution of (11), but is perhaps the most important one. This beam is often called the "fundamental mode" as compared to the higher order modes to be discussed later. Because of its importance it is discussed here in greater detail.

For convenience one introduces two *real* beam parameters R and w related to the complex parameter q by

$$\frac{1}{q} = \frac{1}{R} - j\frac{\lambda}{\pi w^2}. \tag{17}$$

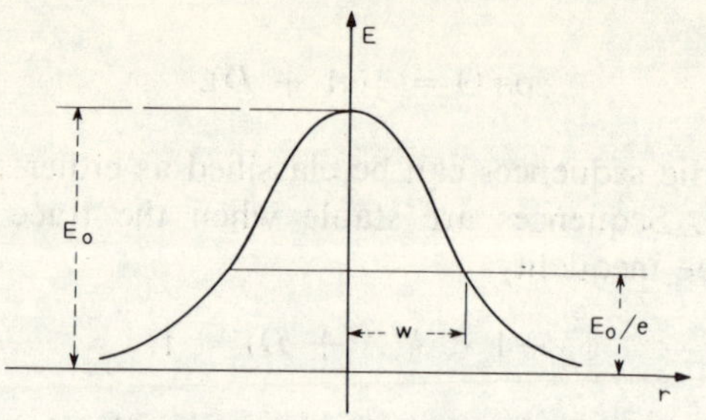

Fig. 5. Amplitude distribution of the fundamental beam.

When (17) is inserted in (12) the physical meaning of these two parameters becomes clear. One sees that $R(z)$ is the radius of curvature of the wavefront that intersects the axis at z, and $w(z)$ is a measure of the decrease of the field amplitude E with the distance from the axis. This decrease is Gaussian in form, as indicated in Fig. 5, and w is the distance at which the amplitude is $1/e$ times that on the axis. Note that the intensity distribution is Gaussian in every beam cross section, and that the width of that Gaussian intensity profile changes along the axis. The parameter w is often called the beam radius or "spot size," and $2w$, the beam diameter.

The Gaussian beam contracts to a minimum diameter $2w_0$ at the *beam waist* where the phase front is plane. If one measures z from this waist, the expansion laws for the beam assume a simple form. The complex beam parameter at the waist is purely imaginary

$$q_0 = j\frac{\pi w_0^2}{\lambda} \tag{18}$$

and a distance z away from the waist the parameter is

$$q = q_0 + z = j\frac{\pi w_0^2}{\lambda} + z. \tag{19}$$

After combining (19) and (17) one equates the real and imaginary parts to obtain

$$w^2(z) = w_0^2\left[1 + \left(\frac{\lambda z}{\pi w_0^2}\right)^2\right] \tag{20}$$

and

$$R(z) = z\left[1 + \left(\frac{\pi w_0^2}{\lambda z}\right)^2\right]. \tag{21}$$

Figure 6 shows the expansion of the beam according to (20). The beam contour $w(z)$ is a hyperbola with asymptotes inclined to the axis at an angle

$$\theta = \frac{\lambda}{\pi w_0}. \tag{22}$$

This is the far-field diffraction angle of the fundamental mode.

Dividing (21) by (20), one obtains the useful relation

$$\frac{\lambda z}{\pi w_0^2} = \frac{\pi w^2}{\lambda R} \tag{23}$$

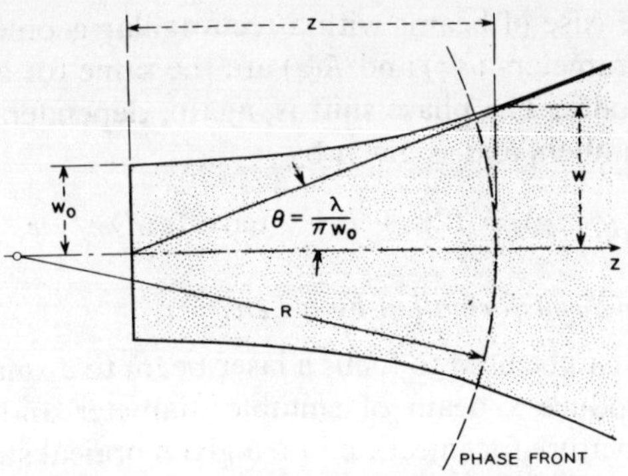

Fig. 6. Contour of a Gaussian beam.

which can be used to express w_0 and z in terms of w and R:

$$w_0{}^2 = w^2 \Big/ \left[1 + \left(\frac{\pi w^2}{\lambda R}\right)^2\right] \tag{24}$$

$$z = R \Big/ \left[1 + \left(\frac{\lambda R}{\pi w^2}\right)^2\right]. \tag{25}$$

To calculate the complex phase shift a distance z away from the waist, one inserts (19) into (15) to get

$$P' = -\frac{j}{q} = -\frac{j}{z + j(\pi w_0{}^2/\lambda)}. \tag{26}$$

Integration of (26) yields the result

$$\begin{aligned} jP(z) &= \ln[1 - j(\lambda z/\pi w_0{}^2)] \\ &= \ln\sqrt{1 + (\lambda z/\pi w_0{}^2)^2} - j \arctan(\lambda z/\pi w_0{}^2). \end{aligned} \tag{27}$$

The real part of P represents a phase shift difference Φ between the Gaussian beam and an ideal plane wave, while the imaginary part produces an amplitude factor w_0/w which gives the expected intensity decrease on the axis due to the expansion of the beam. With these results for the fundamental Gaussian beam, (10) can be written in the form

$$u(r, z) = \frac{w_0}{w} \cdot \exp\left\{-j(kz - \Phi) - r^2\left(\frac{1}{w^2} + \frac{jk}{2R}\right)\right\} \tag{28}$$

where

$$\Phi = \arctan(\lambda z/\pi w_0{}^2). \tag{29}$$

It will be seen in Section 3.5 that Gaussian beams of this kind are produced by many lasers that oscillate in the fundamental mode.

3.3 *Higher Order Modes*

In the preceding section only one solution of (11) was discussed, i.e., a light beam with the property that its intensity profile in every beam cross section is given by the same function, namely, a Gaussian. The width of this Gaussian distribution changes as the beam propagates along its axis. There are other solutions of (11) with similar properties, and they are discussed in this section. These solutions form a complete and orthogonal set of functions and are called the "modes of propagation." Every arbitrary distribution of monochromatic light can be expanded in terms of these modes. Because of space limitations the derivation of these modes can only be sketched here.

a) Modes in Cartesian Coordinates: For a system with a rectangular (x, y, z) geometry one can try a solution for (11) of the form

$$\psi = g\left(\frac{x}{w}\right) \cdot h\left(\frac{y}{w}\right) \cdot \exp\left\{-j\left[P + \frac{k}{2q}(x^2 + y^2)\right]\right\} \tag{30}$$

where g is a function of x and z, and h is a function of y and z. For real g and h this postulates mode beams whose intensity patterns scale according to the width $2w(z)$ of a Gaussian beam. After inserting this trial solution into (11) one arrives at differential equations for g and h of the form

$$\frac{d^2 H_m}{dx^2} - 2x\frac{dH_m}{dx} + 2mH_m = 0. \tag{31}$$

This is the differential equation for the Hermite polynomial $H_m(x)$ of order m. Equation (11) is satisfied if

$$g \cdot h = H_m\left(\sqrt{2}\,\frac{x}{w}\right) H_n\left(\sqrt{2}\,\frac{y}{w}\right) \tag{32}$$

where m and n are the (transverse) mode numbers. Note that the same pattern scaling parameter $w(z)$ applies to modes of all orders.

Some Hermite polynomials of low order are

$$\begin{aligned} H_0(x) &= 1 \\ H_1(x) &= x \\ H_2(x) &= 4x^2 - 2 \\ H_3(x) &= 8x^3 - 12x. \end{aligned} \tag{33}$$

Expression (28) can be used as a mathematical description of higher order light beams, if one inserts the product $g \cdot h$ as a factor on the right-hand side. The intensity pattern in a cross section of a higher order beam is, thus, described by the product of Hermite and Gaussian functions. Photographs of such mode patterns are shown in Fig. 7. They were produced as modes of oscillation in a gas laser oscillator [16]. Note that the number of zeros in a mode pattern is equal to the corresponding mode number, and that the area occupied by a mode increases with the mode number.

The parameter $R(z)$ in (28) is the same for all modes, implying that the phase-front curvature is the same and changes in the same way for modes of all orders. The phase shift Φ, however, is a function of the mode numbers. One obtains

$$\Phi(m, n; z) = (m + n + 1) \arctan(\lambda z/\pi w_0{}^2). \tag{34}$$

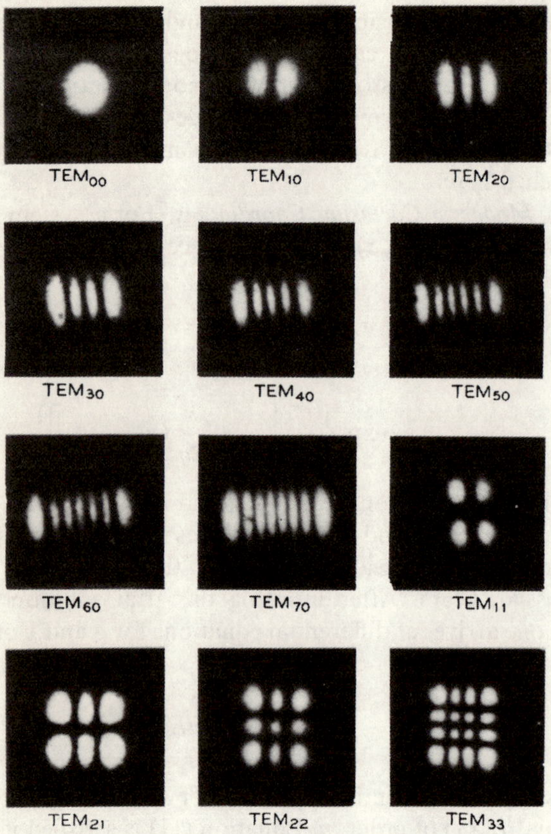

Fig. 7. Mode patterns of a gas laser oscillator (rectangular symmetry).

This means that the phase velocity increases with increasing mode number. In resonators this leads to differences in the resonant frequencies of the various modes of oscillation.

b) Modes in Cylindrical Coordinates: For a system with a cylindrical (r, ϕ, z) geometry one uses a trial solution for (11) of the form

$$\psi = g\left(\frac{r}{w}\right) \cdot \exp\left\{-j\left(P + \frac{k}{2q} r^2 + l\phi\right)\right\}. \tag{35}$$

After some calculation one finds

$$g = \left(\sqrt{2}\,\frac{r}{w}\right)^l \cdot L_p{}^l\left(2\,\frac{r^2}{w^2}\right) \tag{36}$$

where $L_p{}^l$ is a generalized Laguerre polynomial, and p and l are the radial and angular mode numbers. $L_p{}^l(x)$ obeys the differential equation

$$x\,\frac{d^2 L_p{}^l}{dx^2} + (l + 1 - x)\,\frac{dL_p{}^l}{dx} + pL_p{}^l = 0. \tag{37}$$

Some polynomials of low order are

$$L_0{}^l(x) = 1$$

$$L_1{}^l(x) = l + 1 - x$$

$$L_2{}^l(x) = \tfrac{1}{2}(l+1)(l+2) - (l+2)x + \tfrac{1}{2}x^2. \tag{38}$$

As in the case of beams with a rectangular geometry, the beam parameters $w(z)$ and $R(z)$ are the same for all cylindrical modes. The phase shift is, again, dependent on the mode numbers and is given by

$$\Phi(p, l; z) = (2p + l + 1) \text{ arc } \tan(\lambda z/\pi w_0{}^2). \tag{39}$$

3.4 *Beam Transformation by a Lens*

A lens can be used to focus a laser beam to a small spot, or to produce a beam of suitable diameter and phase-front curvature for injection into a given optical structure. An ideal lens leaves the transverse field distribution of a beam mode unchanged, i.e., an incoming fundamental Gaussian beam will emerge from the lens as a fundamental beam, and a higher order mode remains a mode of the same order after passing through the lens. However, a lens does change the beam parameters $R(z)$ and $w(z)$. As these two parameters are the same for modes of all orders, the following discussion is valid for all orders; the relationship between the parameters of an incoming beam (labeled here with the index 1) and the parameters of the corresponding outgoing beam (index 2) is studied in detail.

An ideal thin lens of focal length f transforms an incoming spherical wave with a radius R_1 immediately to the left of the lens into a spherical wave with the radius R_2 immediately to the right of it, where

$$\frac{1}{R_2} = \frac{1}{R_1} - \frac{1}{f}. \tag{40}$$

Figure 8 illustrates this situation. The radius of curvature is taken to be positive if the wavefront is convex as viewed from $z = \infty$. The lens transforms the phase fronts of laser beams in eactly the same way as those of spherical waves. As the diameter of a beam is the same immediately to the left and to the right of a *thin* lens, the q-parameters of the incoming and outgoing beams are related by

$$\frac{1}{q_2} = \frac{1}{q_1} - \frac{1}{f}, \tag{41}$$

where the q's are measured at the lens. If q_1 and q_2 are measured at distances d_1 and d_2 from the lens as indicated in Fig. 9, the relation between them becomes

$$q_2 = \frac{(1 - d_2/f)q_1 + (d_1 + d_2 - d_1 d_2/f)}{-(q_1/f) + (1 - d_1/f)}. \tag{42}$$

This formula is derived using (16) and (41).

More complicated optical structures, such as gas lenses, combinations of lenses, or thick lenses, can be thought of as composed of a series of thin lenses at various spacings. Repeated application of (16) and (41) is, therefore, sufficient to calculate the effect of complicated structures on the propagation of laser beams. If the *ABCD* matrix for the transfer of paraxial rays through the structure is known, the q parameter of the output beam can be calculated from

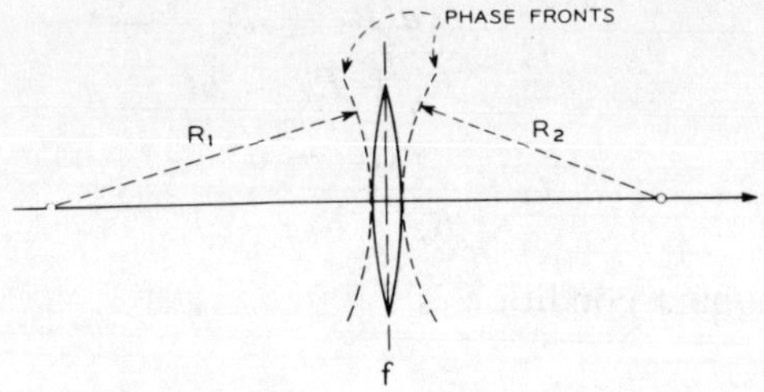

Fig. 8. Transformation of wavefronts by a thin lens.

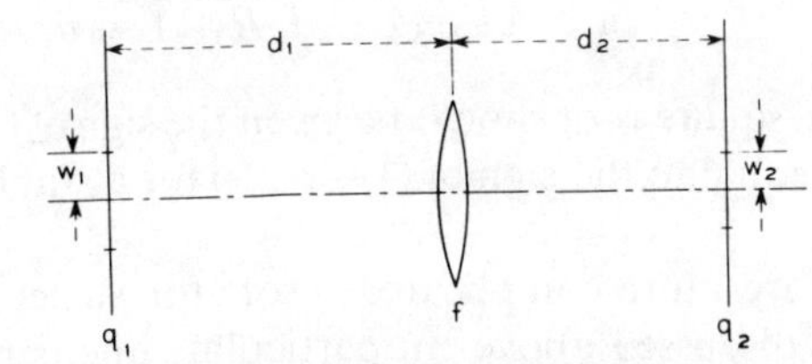

Fig. 9. Distances and parameters for a beam transformed by a thin lens.

$$q_2 = \frac{Aq_1 + B}{Cq_1 + D}. \tag{43}$$

This is a generalized form of (42) and has been called the *ABCD* law [10]. The matrices of several optical structures are given in Section II. The *ABCD* law follows from the analogy between the laws for laser beams and the laws obeyed by the spherical waves in geometrical optics. The radius of the spherical waves R obeys laws of the same form as (16) and (41) for the complex beam parameter q. A more detailed discussion of this analogy is given in [11].

3.5 *Laser Resonators* (*Infinite Aperture*)

The most commonly used laser resonators are composed of two spherical (or flat) mirrors facing each other. The stability of such "open" resonators has been discussed in Section 2 in terms of paraxial rays. To study the *modes* of laser resonators one has to take account of their wave nature, and this is done here by studying wave beams of the kind discussed above as they propagate back and forth between the mirrors. As aperture diffraction effects are neglected throughout this section, the present discussion applies only to stable resonators with mirror apertures that are large compared to the spot size of the beams.

A mode of a resonator is defined as a self-consistent field configuragion. If a mode can be represented by a wave beam propagating back and forth between the mirrors, the beam parameters must be the same after one complete return trip of the beam. This condition is used to calculate the mode parameters. As the beam that represents a mode travels in both directions between the mirrors it forms the axial standing-wave pattern that is expected for a resonator mode.

A laser resonator with mirrors of equal curvature is shown in Fig. 10 together with the equivalent unfolded system, a sequence of lenses. For this symmetrical structure it is sufficient to postulate self-consistency for one transit of the resonator (which is equivalent to one full period of the lens sequence), instead of a complete return trip. If the complex beam parameter is given by q_1, immediately to the right of a particular lens, the beam parameter q_2, immediately to the right of the next lens, can be calculated by means of (16) and (41) as

$$\frac{1}{q_2} = \frac{1}{q_1 + d} - \frac{1}{f}. \tag{44}$$

Self-consistency requires that $q_1 = q_2 = q$, which leads to a quadratic equation for the beam parameter q at the lenses (or at the mirrors of the resonator):

$$\frac{1}{q^2} + \frac{1}{fq} + \frac{1}{fd} = 0. \tag{45}$$

The roots of this equation are

$$\frac{1}{q} = -\frac{1}{2f} \overset{-}{(+)}\, j \sqrt{\frac{1}{fd} - \frac{1}{4f^2}} \tag{46}$$

where only the root that yields a real beamwidth is used. (Note that one gets a real beamwidth for stable resonators only.)

From (46) one obtains immediately the real beam parameters defined in (17). One sees that R is equal to the radius of curvature of the mirrors, which means that the mirror surfaces are coincident with the phase fronts of the resonator modes. The width $2w$ of the fundamental mode is given by

$$w^2 = \left(\frac{\lambda R}{\pi}\right) \Big/ \sqrt{2\frac{R}{d} - 1}. \tag{47}$$

To calculate the beam radius w_0 in the center of the resonator where the phase front is plane, one uses (23) with $z = d/2$ and gets

$$w_0^2 = \frac{\lambda}{2\pi}\sqrt{d(2R - d)}. \tag{48}$$

The beam parameters R and w describe the modes of all orders. But the phase velocities are different for the different orders, so that the resonant conditions depend on the mode numbers. Resonance occurs when the phase shift from one mirror to the other is a multiple of π. Using (28) and (34) this condition can be written as

$$kd - 2(m + n + 1) \arctan(\lambda d / 2\pi w_0^2) = \pi(q + 1) \tag{49}$$

where q is the number of nodes of the axial standing-wave pattern (the number of half wavelengths is $q+1$),[1] and m and n are the rectangular mode numbers defined in Section 3.3. For the modes of circular geometry one obtains a similar condition where $(2p+l+1)$ replaces $(m+n+1)$.

The fundamental beat frequency ν_0, i.e., the frequency spacing between successive longitudinal resonances, is given by

$$\nu_0 = c/2d \tag{50}$$

[1] This q is not to be confused with the complex beam parameter.

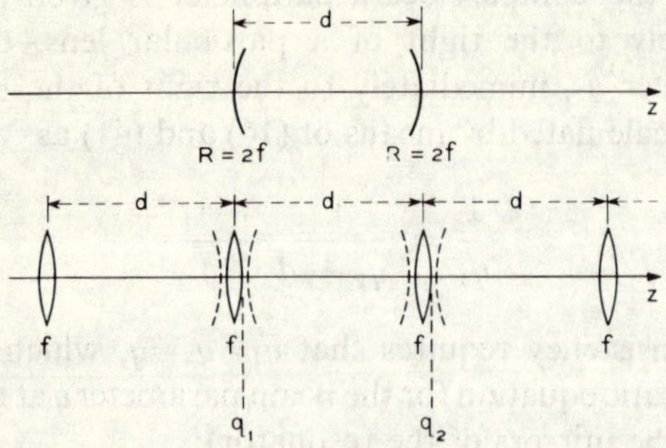

Fig. 10. Symmetrical laser resonator and the equivalent sequence of lenses. The beam parameters, q_1 and q_2, are indicated.

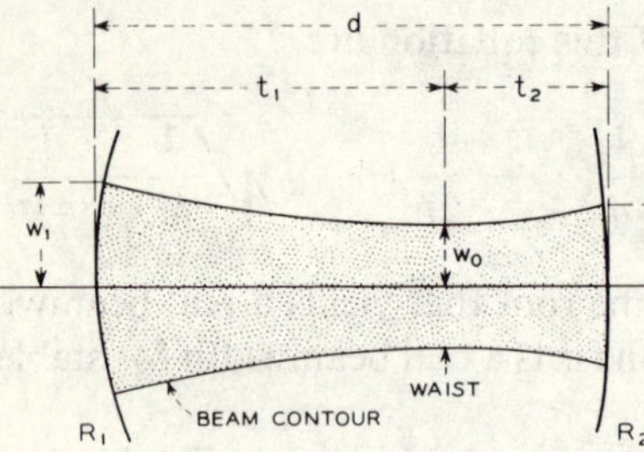

Fig. 11. Mode parameters of interest for a resonator with mirrors of unequal curvature.

where c is the velocity of light. After some algebraic manipulations one obtains from (49) the following formula for the resonant frequency ν of a mode

$$\nu/\nu_0 = (q+1) + \frac{1}{\pi}(m+n+1) \text{ arc cos}(1-d/R). \tag{51}$$

For the special case of the confocal resonator ($d=R=b$), the above relations become

$$w^2 = \lambda b/\pi, \qquad w_0{}^2 = \lambda b/2\pi;$$

$$\nu/\nu_0 = (q+1) + \tfrac{1}{2}(m+n+1). \tag{52}$$

The parameter b is known as the confocal parameter.

Resonators with mirrors of unequal curvature can be treated in a similar manner. The geometry of such a resonator where the radii of curvature of the mirrors are R_1 and R_2 is shown in Fig. 11. The diameters of the beam at the mirrors of a stable resonator, $2w_1$ and $2w_2$, are given by

$$w_1{}^4 = (\lambda R_1/\pi)^2 \frac{R_2-d}{R_1-d}\,\frac{d}{R_1+R_2-d}$$

$$w_2{}^4 = (\lambda R_2/\pi)^2 \frac{R_1-d}{R_2-d}\,\frac{d}{R_1+R_2-d}. \tag{53}$$

The diameter of the beam waist $2w_0$, which is formed either inside or outside the resonator, is given by

$$w_0{}^4 = \left(\frac{\lambda}{\pi}\right)^2 \frac{d(R_1-d)(R_2-d)(R_1+R_2-d)}{(R_1+R_2-2d)^2}. \tag{54}$$

The distances t_1 and t_2 between the waist and the mirrors, measured positive as shown in the figure, are

$$t_1 = \frac{d(R_2-d)}{R_1+R_2-2d}$$

$$t_2 = \frac{d(R_1-d)}{R_1+R_2-2d}. \tag{55}$$

The resonant condition is

$$\nu/\nu_0 = (q+1) + \frac{1}{\pi}(m+n+1) \text{ arc cos}\sqrt{(1-d/R_1)(1-d/R_2)} \tag{56}$$

where the square root should be given the sign of $(1-d/R_1)$, which is equal to the sign of $(1-d/R_2)$ for a stable resonator.

There are more complicated resonator structures than the ones discussed above. In particular, one can insert a lens or several lenses between the mirrors. But in every case, the unfolded resonator is equivalent to a periodic sequence of identical optical systems as shown in Fig. 2. The elements of the $ABCD$ matrix of this system can be used to calculate the mode parameters of the resonator. One uses the $ABCD$ law (43) and postulates self-consistency by putting $q_1 = q_2 = q$. The roots of the resulting quadratic equation are

$$\frac{1}{q} = \frac{D-A}{2B} \overset{-}{(+)} \frac{j}{2B}\sqrt{4-(A+D)^2}, \tag{57}$$

which yields, for the corresponding beam radius w,

$$w^2 = (2\lambda B/\pi)/\sqrt{4-(A+D)^2}. \tag{58}$$

3.6 *Mode Matching*

It was shown in the preceding section that the modes of laser resonators can be characterized by light beams with certain properties and parameters which are defined by the resonator geometry. These beams are often injected into other optical structures with different sets of beam parameters. These optical structures can assume various physical forms, such as resonators used in scanning Fabry-Perot interferometers or regenerative amplifiers, sequences of dielectric or gas lenses used as optical transmission lines, or crystals of nonlinear dielectric material employed in parametric optics experiments. To match the modes of one structure to those of another one must transform a given Gaussian beam (or higher order mode) into another beam with prescribed properties. This transformation is usually accomplished with a thin lens, but other more complex optical systems can be used. Although the present discussion is devoted to the simple case of the thin lens, it is also applicable to more complex systems, provided one measures the distances from the principal planes and uses the combined focal length f of the more complex system.

The location of the waists of the two beams to be transformed into each other and the beam diameters at the waists are usually known or can be computed. To match the beams one has to choose a lens of a focal length

f that is larger than a characteristic length f_0 defined by the two beams, and one has to adjust the distances between the lens and the two beam waists according to rules derived below.

In Fig. 9 the two beam waists are assumed to be located at distances d_1 and d_2 from the lens. The complex beam parameters at the waists are purely imaginary; they are

$$q_1 = j\pi w_1^2/\lambda, \qquad q_2 = j\pi w_2^2/\lambda \tag{59}$$

where $2w_1$ and $2w_2$ are the diameters of the two beams at their waists. If one inserts these expressions for q_1 and q_2 into (42) and equates the imaginary parts, one obtains

$$\frac{d_1 - f}{d_2 - f} = \frac{w_1^2}{w_2^2}. \tag{60}$$

Equating the real parts results in

$$(d_1 - f)(d_2 - f) = f^2 - f_0^2 \tag{61}$$

where

$$f_0 = \pi w_1 w_2/\lambda. \tag{62}$$

Note that the characteristic length f_0 is defined by the waist diameters of the beams to be matched. Except for the term f_0^2, which goes to zero for infinitely small wavelengths, (61) resembles Newton's imaging formula of geometrical optics.

Any lens with a focal length $f > f_0$ can be used to perform the matching transformation. Once f is chosen, the distances d_1 and d_2 have to be adjusted to satisfy the matching formulas [10]

$$d_1 = f \pm \frac{w_1}{w_2}\sqrt{f^2 - f_0^2},$$

$$d_2 = f \pm \frac{w_2}{w_1}\sqrt{f^2 - f_0^2}. \tag{63}$$

These relations are derived by combining (60) and (61). In (63) one can choose either both plus signs or both minus signs for matching.

It is often useful to introduce the confocal parameters b_1 and b_2 into the matching formulas. They are defined by the waist diameters of the two systems to be matched

$$b_1 = 2\pi w_1^2/\lambda, \qquad b_2 = 2\pi w_2^2/\lambda. \tag{64}$$

Using these parameters one gets for the characteristic length f_0

$$f_0^2 = \tfrac{1}{4} b_1 b_2, \tag{65}$$

and for the matching distances

$$d_1 = f \pm \tfrac{1}{2} b_1 \sqrt{(f^2/f_0^2) - 1},$$

$$d_2 = f \pm \tfrac{1}{2} b_2 \sqrt{(f^2/f_0^2) - 1}. \tag{66}$$

Note that in this form of the matching formulas, the wavelength does not appear explicitly.

Table II lists, for quick reference, formulas for the two important parameters of beams that *emerge* from various optical structures commonly encountered. They are the confocal parameter b and the distance t which gives the waist location of the emerging beam. System No. 1 is a resonator formed by a flat mirror and a spherical mirror of radius R. System No. 2 is a resonator formed by two equal spherical mirrors. System No. 3 is a resonator formed by mirrors of unequal curvature. System No. 4

TABLE II

FORMULAS FOR THE CONFOCAL PARAMETER AND THE LOCATION OF BEAM WAIST FOR VARIOUS OPTICAL STRUCTURES

NO.	OPTICAL SYSTEM	$\frac{1}{2}b = \pi w_0^2/\lambda$	t
1	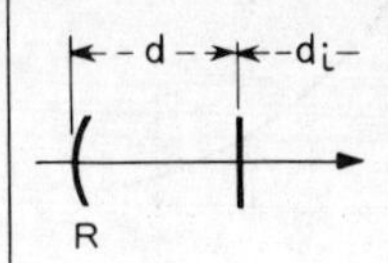	$\sqrt{d(R-d)}$	—
2	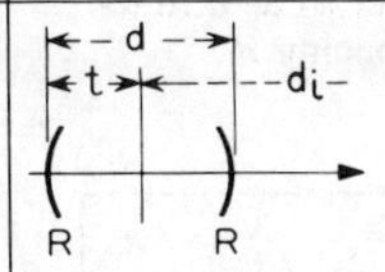	$\frac{1}{2}\sqrt{d(2R-d)}$	$\frac{1}{2}d$
3	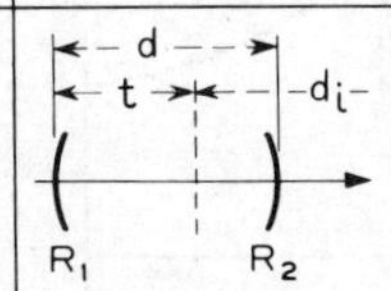	$\frac{\sqrt{d(R_1-d)(R_2-d)(R_1+R_2-d)}}{R_1+R_2-2d}$	$\frac{d(R_2-d)}{R_1+R_2-2d}$
4	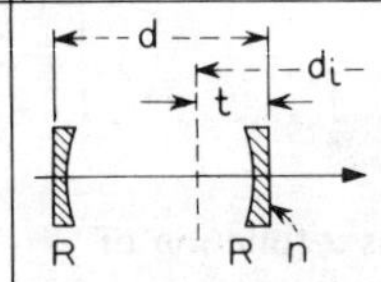	$\frac{R\sqrt{d(2R-d)}}{2R+d(n^2-1)}$	$\frac{ndR}{2R+d(n^2-1)}$
5	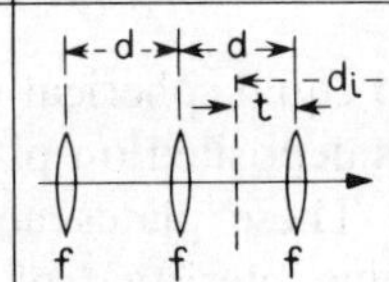	$\frac{1}{2}\sqrt{d(4f-d)}$	$\frac{1}{2}d$
6	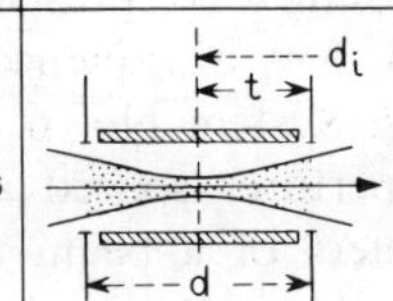	$\frac{1}{2}d$	$\frac{1}{2}d$
7	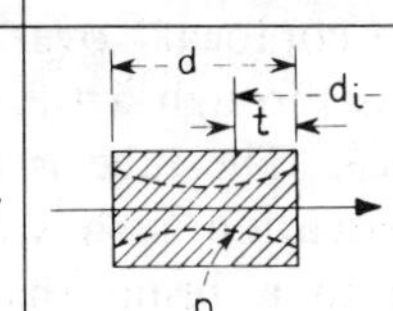	$\frac{d}{2n}$	$\frac{d}{2n}$
8	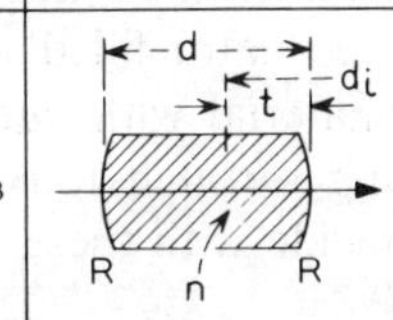	$\frac{nR\sqrt{d(2R-d)}}{2n^2R-d(n^2-1)}$	$\frac{dR}{2n^2R-d(n^2-1)}$

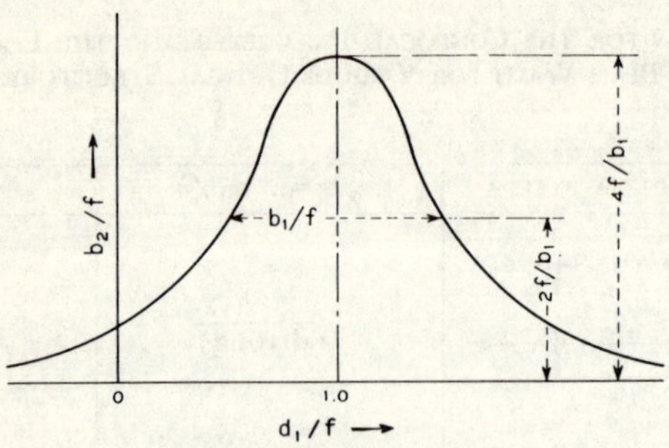

Fig. 12. The confocal parameter b_2 as a function of the lens-waist spacing d_1.

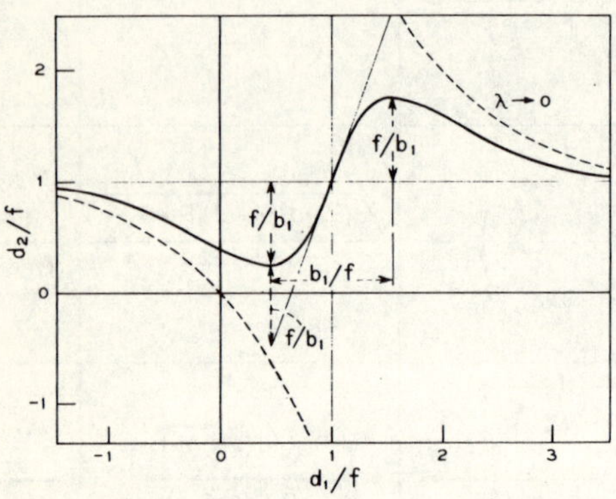

Fig. 13. The waist spacing d_2 as a function of the lens-waist spacing d_1.

is, again, a resonator formed by two equal spherical mirrors, but with the reflecting surfaces deposited on plano-concave optical plates of index n. These plates act as negative lenses and change the characteristics of the emerging beam. This lens effect is assumed not present in Systems Nos. 2 and 3. System No. 5 is a sequence of thin lenses of equal focal lengths f. System No. 6 is a system of two irises with equal apertures spaced at a distance d. Shown are the parameters of a beam that will pass through both irises with the least possible beam diameter. This is a beam which is "confocal" over the distance d. This beam will also pass through a tube of length d with the optimum clearance. (The tube is also indicated in the figure.) A similar situation is shown in System No. 7, which corresponds to a beam that is confocal over the length d of optical material of index n. System No. 8 is a spherical mirror resonator filled with material of index n, or an optical material with curved end surfaces where the beam passing through it is assumed to have phase fronts that coincide with these surfaces.

When one designs a matching system, it is useful to know the accuracy required of the distance adjustments. The discussion below indicates how the parameters b_2 and d_2 change when b_1 and f are fixed and the lens spacing d_1 to the waist of the input beam is varied. Equations (60) and (61) can be solved for b_2 with the result [9]

$$b_2/f = \frac{b_1/f}{(1 - d_1/f)^2 + (b_1/2f)^2}. \tag{67}$$

This means that the parameter b_2 of the beam emerging from the lens changes with d_1 according to a Lorentzian functional form as shown in Fig. 12. The Lorentzian is centered at $d_1 = f$ and has a width of b_1. The maximum value of b_2 is $4f^2/b_1$.

If one inserts (67) into (60) one gets

$$1 - d_2/f = \frac{1 - d_1/f}{(1 - d_1/f)^2 + (b_1/2f)^2} \tag{68}$$

which shows the change of d_2 with d_1. The change is reminiscent of a dispersion curve associated with a Lorentzian as shown in Fig. 13. The extrema of this curve occur at the halfpower points of the Lorentzian. The slope of the curve at $d_1 = f$ is $(2f/b_1)^2$. The dashed curves in the figure correspond to the geometrical optics imaging relation between d_1, d_2, and f [20].

3.7 *Circle Diagrams*

The propagation of Gaussian laser beams can be represented graphically on a circle diagram. On such a diagram one can follow a beam as it propagates in free space or passes through lenses, thereby affording a graphic solution of the mode matching problem. The circle diagrams for beams are similar to the impedance charts, such as the Smith chart. In fact there is a close analogy between transmission-line and laser-beam problems, and there are analog electric networks for every optical system [17].

The first circle diagram for beams was proposed by Collins [18]. A dual chart was discussed in [19]. The basis for the derivation of these charts are the beam propagation laws discussed in Section 3.2. One combines (17) and (19) and eliminates q to obtain

$$\left(\frac{\lambda}{\pi w^2} + j\frac{1}{R}\right)\left(\frac{\pi w_0^2}{\lambda} - jz\right) = 1. \tag{69}$$

This relation contains the four quantities w, R, w_0, and z which were used to describe the propagation of Gaussian beams in Section 3.2. Each pair of these quantities can be expressed in complex variables W and Z:

$$W = \frac{\lambda}{\pi w^2} + j\frac{1}{R}$$

$$Z = \frac{\pi w_0^2}{\lambda} - jz = b/2 - jz, \tag{70}$$

where b is the confocal parameter of the beam. For these variables (69) defines a conformal transformation

$$W = 1/Z. \tag{71}$$

The two dual circle diagrams are plotted in the complex planes of W and Z, respectively. The W-plane diagram [18] is shown in Fig. 14 where the variables $\lambda/\pi w^2$ and $1/R$ are plotted as axes. In this plane the lines of constant $b/2 = \pi w_0^2/\lambda$ and the lines of constant z of the Z plane appear as circles through the origin. A beam is represented by a circle of constant b, and the beam parameters w and R at a distance z from the beam waist can be easily read

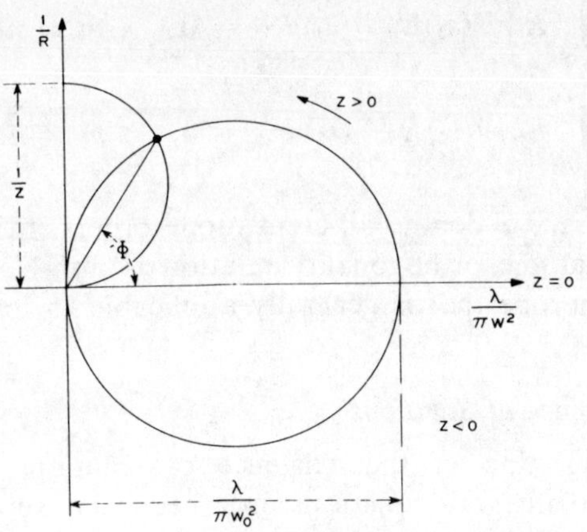

Fig. 14. Geometry for the W-plane circle diagram.

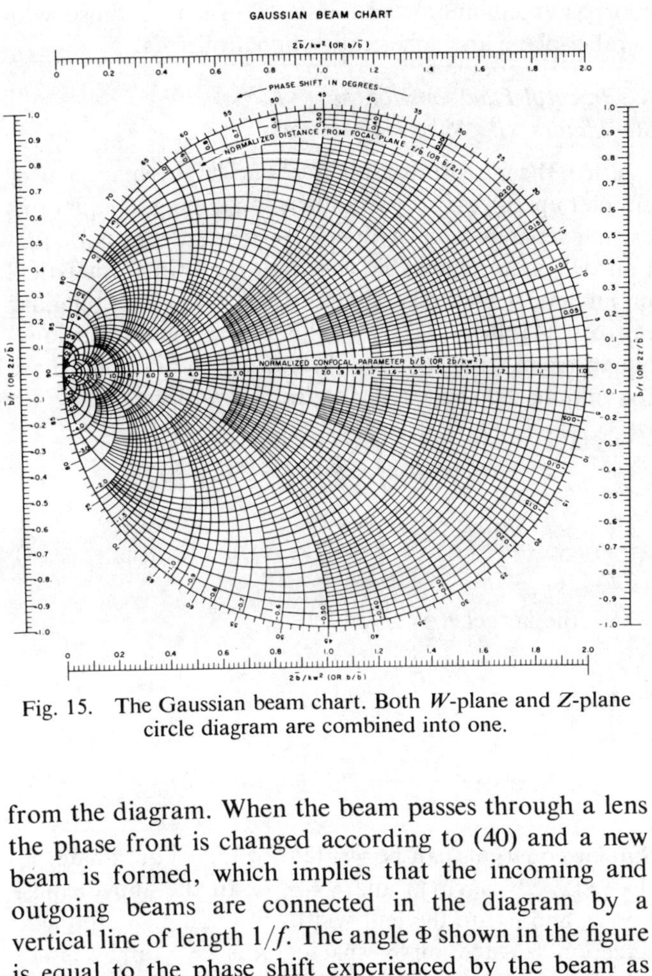

Fig. 15. The Gaussian beam chart. Both W-plane and Z-plane circle diagram are combined into one.

from the diagram. When the beam passes through a lens the phase front is changed according to (40) and a new beam is formed, which implies that the incoming and outgoing beams are connected in the diagram by a vertical line of length $1/f$. The angle Φ shown in the figure is equal to the phase shift experienced by the beam as given by (29); this is easily shown using (23).

The dual diagram [19] is plotted in the Z plane. The sets of circles in both diagrams have the same form, and only the labeling of the axes and circles is different. In Fig. 15 both diagrams are unified in one chart. The labels in parentheses correspond to the Z-plane diagram, and $\bar{b}$ is a normalizing parameter which can be arbitrarily chosen for convenience.

One can plot various other circle diagrams which are related to the above by conformal transformations. One such transformation makes it possible to use the Smith chart for determining complex mismatch coefficients for Gaussian beams [20]. Other circle diagrams include those for optical resonators [21] which allow the graphic determination of certain parameters of the resonator modes.

4. Laser Resonators (Finite Aperture)

4.1 *General Mathematical Formulation*

In this section aperture diffraction effects due to the finite size of the mirrors are taken into account; these effects were neglected in the preceding sections. There, it was mentioned that resonators used in laser oscillators usually take the form of an open structure consisting of a pair of mirrors facing each other. Such a structure with finite mirror apertures is intrinsically lossy and, unless energy is supplied to it continuously, the electromagnetic field in it will decay. In this case a mode of the resonator is a *slowly decaying* field configuration whose relative distribution does not change with time [4]. In a laser oscillator the active medium supplies enough energy to overcome the losses so that a steady-state field can exist. However, because of nonlinear gain saturation the medium will exhibit less gain in those regions where the field is high than in those where the field is low, and so the oscillating modes of an active resonator are expected to be somewhat different from the decaying modes of the passive resonator. The problem of an active resonator filled with a saturable-gain medium has been solved recently [22], [23], and the computed results show that if the gain is not too large the resonator modes are essentially unperturbed by saturation effects. This is fortunate as the results which have been obtained for the passive resonator can also be used to describe the active modes of laser oscillators.

The problem of the open resonator is a difficult one and a rigorous solution is yet to be found. However, if certain simplifying assumptions are made, the problem becomes tractable and physically meaningful results can be obtained. The simplifying assumptions involve essentially the quasi-optic nature of the problem; specifically, they are 1) that the dimensions of the resonator are large compared to the wavelength and 2) that the field in the resonator is substantially transverse electromagnetic (TEM). So long as those assumptions are valid, the Fresnel-Kirchhoff formulation of Huygens' principle can be invoked to obtain a pair of integral equations which relate the fields of the two opposing mirrors. Furthermore, if the mirror separation is large compared to mirror dimensions and if the mirrors are only slightly curved, the two orthogonal Cartesian components of the vector field are essentially uncoupled, so that separate scalar equations can be written for each component. The solutions of these scalar equations yield resonator modes which are uniformly polarized in one direction. Other polarization configurations can be constructed from the uniformly polarized modes by linear superposition.

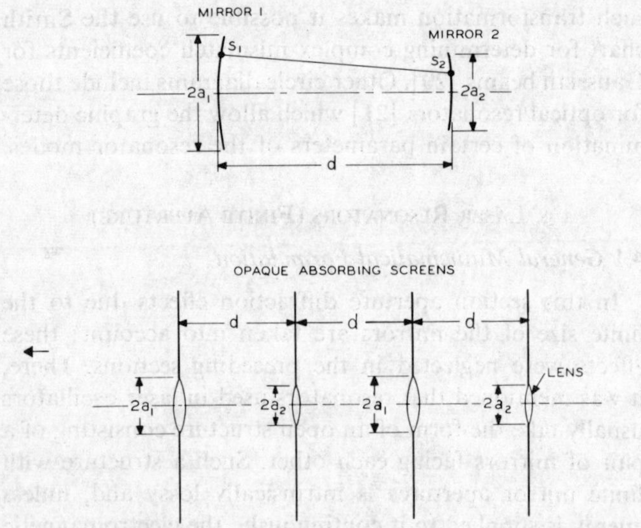

Fig. 16. Geometry of a spherical-mirror resonator with finite mirror apertures and the equivalent sequence of lenses set in opaque absorbing screens.

In deriving the integral equations, it is assumed that a traveling TEM wave is reflected back and forth between the mirrors. The resonator is thus analogous to a transmission medium consisting of apertures or lenses set in opaque absorbing screens (see Fig. 16). The fields at the two mirrors are related by the equations [24]

$$\gamma^{(1)}E^{(1)}(s_1) = \int_{S_2} K^{(2)}(s_1, s_2)E^{(2)}(s_2)dS_2$$

$$\gamma^{(2)}E^{(2)}(s_2) = \int_{S_1} K^{(1)}(s_2, s_1)E^{(1)}(s_1)dS_1 \qquad (72)$$

where the integrations are taken over the mirror surfaces S_2 and S_1, respectively. In the above equations the subscripts and superscripts one and two denote mirrors one and two; s_1 and s_2 are symbolic notations for transverse coordinates on the mirror surface, e.g., $s_1=(x_1, y_1)$ and $s_2=(x_2, y_2)$ or $s_1=(r_1, \phi_1)$ and $s_2=(r_2, \phi_2)$; $E^{(1)}$ and $E^{(2)}$ are the relative field distribution functions over the mirrors; $\gamma^{(1)}$ and $\gamma^{(2)}$ give the attenuation and phase shift suffered by the wave in transit from one mirror to the other; the kernels $K^{(1)}$ and $K^{(2)}$ are functions of the distance between s_1 and s_2 and, therefore, depend on the mirror geometry; they are equal $[K^{(1)}(s_2, s_1)=K^{(2)}(s_1, s_2)]$ but, in general, are not symmetric $[K^{(1)}(s_2, s_1)\neq K^{(1)}(s_1, s_2)$, $K^{(2)}(s_1, s_2)\neq K^{(2)}(s_2, s_1)]$.

The integral equations given by (72) express the field at each mirror in terms of the reflected field at the other; that is, they are single-transit equations. By substituting one into the other, one obtains the double-transit or round-trip equations, which state that the field at each mirror must reproduce itself after a round trip. Since the kernel for each of the double-transit equations is symmetric [24], it follows [25] that the field distribution functions corresponding to the different mode orders are orthogonal over their respective mirror surfaces; that is

$$\int_{S_1} E_m^{(1)}(s_1)E_n^{(1)}(s_1)dS_1 = 0, \qquad m \neq n$$

$$\int_{S_2} E_m^{(2)}(s_2)E_n^{(2)}(s_2)dS_2 = 0, \qquad m \neq n \qquad (73)$$

where m and n denote different mode orders. It is to be noted that the orthogonality relation is non-Hermitian and is the one that is generally applicable to lossy systems.

4.2 *Existence of Solutions*

The question of the existence of solutions to the resonator integral equations has been the subject of investigation by several authors [26]–[28]. They have given rigorous proofs of the existence of eigenvalues and eigenfunctions for kernels which belong to resonator geometries commonly encountered, such as those with parallel-plane and spherically curved mirrors.

4.3 *Integral Equations for Resonators with Spherical Mirrors*

When the mirrors are spherical and have rectangular or circular apertures, the two-dimensional integral equations can be separated and reduced to one-dimensional equations which are amenable to solution by either analytical or numerical methods. Thus, in the case of rectangular mirrors [4]–[6], [24], [29], [30], the one-dimensional equations in Cartesian coordinates are the same as those for infinite-strip mirrors; for the x coordinate, they are

$$\gamma_x^{(1)}u^{(1)}(x_1) = \int_{-a_2}^{a_2} K(x_1, x_2)u^{(2)}(x_2)dx_2$$

$$\gamma_x^{(2)}u^{(2)}(x_2) = \int_{-a_1}^{a_1} K(x_1, x_2)u^{(1)}(x_1)dx_1 \qquad (74)$$

where the kernel K is given by

$$K(x_1, x_2) = \sqrt{\frac{j}{\lambda d}} \cdot \exp\left\{-\frac{jk}{2d}(g_1x_1^2 + g_2x_2^2 - 2x_1x_2)\right\}. \qquad (75)$$

Similar equations can be written for the y coordinate, so that $E(x, y)=u(x)v(y)$ and $\gamma=\gamma_x\gamma_y$. In the above equation a_1 and a_2 are the half-widths of the mirrors in the x direction, d is the mirror spacing, k is $2\pi/\lambda$, and λ is the wavelength. The radii of curvature of the mirrors R_1 and R_2 are contained in the factors

$$g_1 = 1 - \frac{d}{R_1}$$

$$g_2 = 1 - \frac{d}{R_2} \qquad (76)$$

For the case of circular mirrors [4], [31], [32] the equations are reduced to the one-dimensional form by using cylindrical coordinates and by assuming a sinusoidal azimuthal variation of the field; that is, $E(r, \phi) = R_l(r)e^{-jl\phi}$. The radial distribution functions $R_l^{(1)}$ and $R_l^{(2)}$ satisfy the one-dimensional integral equations:

$$\gamma_l^{(1)} R_l^{(1)}(r_1)\sqrt{r_1} = \int_0^{a_2} K_l(r_1, r_2) R_l^{(2)}(r_2)\sqrt{r_2}\, dr_2$$

$$\gamma_l^{(2)} R_l^{(2)}(r_2)\sqrt{r_2} = \int_0^{a_1} K_l(r_1, r_2) R_l^{(1)}(r_1)\sqrt{r_1}\, dr_1 \quad (77)$$

where the kernel K_l is given by

$$K_l(r_1, r_2) = \frac{j^{l+1}}{d} J_l\left(k\frac{r_1 r_2}{d}\right)\sqrt{r_1 r_2} \cdot \exp\left\{-\frac{jk}{2d}(g_1 r_1^2 + g_2 r_2^2)\right\} \quad (78)$$

and J_l is a Bessel function of the first kind and lth order. In (77), a_1 and a_2 are the radii of the mirror apertures and d is the mirror spacing; the factors g_1 and g_2 are given by (76).

Except for the special case of the confocal resonator [5] ($g_1 = g_2 = 0$), no exact analytical solution has been found for either (74) or (77), but approximate methods and numerical techniques have been employed with success for their solutions. Before presenting results, it is appropriate to discuss two important properties which apply in general to resonators with spherical mirrors; these are the properties of "equivalence" and "stability."

4.4 *Equivalent Resonator Systems*

The equivalence properties [24], [33] of spherical-mirror resonators are obtained by simple algebraic manipulations of the integral equations. First, it is obvious that the mirrors can be interchanged without affecting the results; that is, the subscripts and superscripts one and two can be interchanged. Second, the diffraction loss and the intensity pattern of the mode remain invariant if both g_1 and g_2 are reversed in sign; the eigenfunctions E and the eigenvalues γ merely take on complex conjugate values. An example of such equivalent systems is that of parallel-plane ($g_1 = g_2 = 1$) and concentric ($g_1 = g_2 = -1$) resonator systems.

The third equivalence property involves the Fresnel number N and the stability factors G_1 and G_2, where

$$N = \frac{a_1 a_2}{\lambda d}$$

$$G_1 = g_1 \frac{a_1}{a_2}$$

$$G_2 = g_2 \frac{a_2}{a_1} \cdot \quad (79)$$

If these three parameters are the same for any two resonators, then they would have the same diffraction loss, the same resonant frequency, and mode patterns that are scaled versions of each other. Thus, the equivalence relations reduce greatly the number of calculations which are necessary for obtaining the solutions for the various resonator geometries.

4.5 *Stability Condition and Diagram*

Stability of optical resonators has been discussed in Section 2 in terms of geometrical optics. The stability condition is given by (8). In terms of the stability factors G_1 and G_2, it is

$$0 < G_1 G_2 < 1$$

or

$$0 < g_1 g_2 < 1. \quad (80)$$

Resonators are stable if this condition is satisfied and unstable otherwise.

A stability diagram [6], [24] for the various resonator geometries is shown in Fig. 4 where g_1 and g_2 are the coordinate axes and each point on the diagram represents a particular resonator geometry. The boundaries between stable and unstable (shaded) regions are determined by (80), which is based on geometrical optics. The fields of the modes in stable resonators are more concentrated near the resonator axes than those in unstable resonators and, therefore, the diffraction losses of unstable resonators are much higher than those of stable resonators. The transition, which occurs near the boundaries, is gradual for resonators with small Fresnel numbers and more abrupt for those with large Fresnel numbers. The origin of the diagram represents the confocal system with mirrors of equal curvature ($R_1 = R_2 = d$) and is a point of lowest diffraction loss for a given Fresnel number. The fact that a system with minor deviations from the ideal confocal system may become unstable should be borne in mind when designing laser resonators.

4.6 *Modes of the Resonator*

The transverse field distributions of the resonator modes are given by the eigenfunctions of the integral equations. As yet, no exact analytical solution has been found for the general case of arbitrary G_1 and G_2, but *approximate* analytical expressions have been obtained to describe the fields in *stable* spherical-mirror resonators [5], [6]. These approximate eigenfunctions are the same as those of the optical beam modes which are discussed in Section 2; that is, the field distributions are given *approximately* by Hermite-Gaussian functions for rectangular mirrors [5], [6], [34], and by Laguerre-Gaussian functions for circular mirrors [6], [7]. The designation of the resonator modes is given in Section 3.5. (The modes are designated as TEM_{mnq} for rectangular mirrors and TEM_{plq} for circular mirrors.) Figure 7 shows photographs of some of the rectangular mode patterns of a

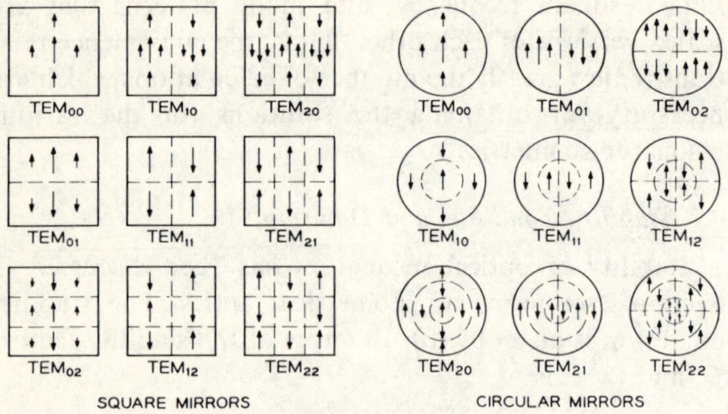

Fig. 17. Linearly polarized resonator mode configurations for square and circular mirrors.

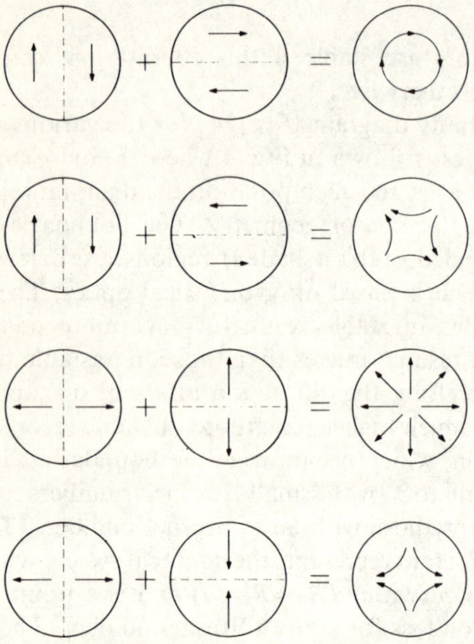

Fig. 18. Synthesis of different polarization configurations from the linearly polarized TEM_{01} mode.

laser. Linearly polarized mode configurations for square mirrors and for circular mirrors are shown in Fig. 17. By combining two orthogonally polarized modes of the same order, it is possible to synthesize other polarization configurations; this is shown in Fig. 18 for the TEM_{01} mode.

Field distributions of the resonator modes for any value of G could be obtained numerically by solving the integral equations either by the method of successive approximations [4], [24], [31] or by the method of kernel expansion [30], [32]. The former method of solution is equivalent to calculating the transient behavior of the resonator when it is excited initially with a wave of arbitrary distribution. This wave is assumed to travel back and forth between the mirrors of the resonator, undergoing changes from transit to transit and losing energy by diffraction. After many transits a quasi steady-state condition is attained where the fields for successive transits differ only by a constant multiplicative factor. This steady-state *relative* field distribution is then an eigenfunction of the integral equations and is, in fact, the field distribution of the mode that has the lowest diffraction loss for the symmetry assumed (e.g., for even or odd symmetry in the case of infinite-strip mirrors, or for a given azimuthal mode index number l in the case of circular mirrors); the constant multiplicative factor is the eigenvalue associated with the eigenfunction and gives the diffraction loss and the phase shift of the mode. Although this simple form of the iterative method gives only the lower order solutions, it can, nevertheless, be modified to yield higher order ones [24], [35]. The method of kernel expansion, however, is capable of yielding both low-order and high-order solutions.

Figures 19 and 20 show the relative field distributions of the TEM_{00} and TEM_{01} modes for a resonator with a pair of identical, circular mirrors ($N=1$, $a_1=a_2$, $g_1=g_2=g$) as obtained by the numerical iterative method. Several curves are shown for different values of g, ranging from zero (confocal) through one (parallel-plane) to 1.2 (convex, unstable). By virtue of the equivalence property discussed in Section 4.4, the curves are also applicable to resonators with their g values reversed in sign, provided the sign of the ordinate for the phase distribution is also reversed. It is seen that the field is most concentrated near the resonator axis for $g=0$ and tends to spread out as $|g|$ increases. Therefore, the diffraction loss is expected to be the least for confocal resonators.

Figure 21 shows the relative field distributions of some of the low order modes of a Fabry-Perot resonator with (parallel-plane) circular mirrors ($N=10$, $a_1=a_2$, $g_1=g_2=1$) as obtained by a modified numerical iterative method [35]. It is interesting to note that these curves are not very smooth but have small wiggles on them, the number of which are related to the Fresnel number. These wiggles are entirely absent for the confocal resonator and appear when the resonator geometry is unstable or nearly unstable. Approximate expressions for the field distributions of the Fabry-Perot resonator modes have also been obtained by various analytical techniques [36], [37]. They are represented to first order, by sine and cosine functions for infinite-strip mirrors and by Bessel functions for circular mirrors.

For the special case of the confocal resonator ($g_1=g_2=0$), the eigenfunctions are self-reciprocal under the *finite* Fourier (infinite-strip mirrors) or Hankel (circular mirrors) transformation and exact analytical solutions exist [5], [38]–[40]. The eigenfunctions for infinite-strip mirrors are given by the prolate spheroidal wave functions and, for circular mirrors, by the generalized prolate spheroidal or hyperspheroidal wave functions. For large Fresnel numbers these functions can be closely approximated by Hermite-Gaussian and Laguerre-Gaussian functions which are the eigenfunctions for the beam modes.

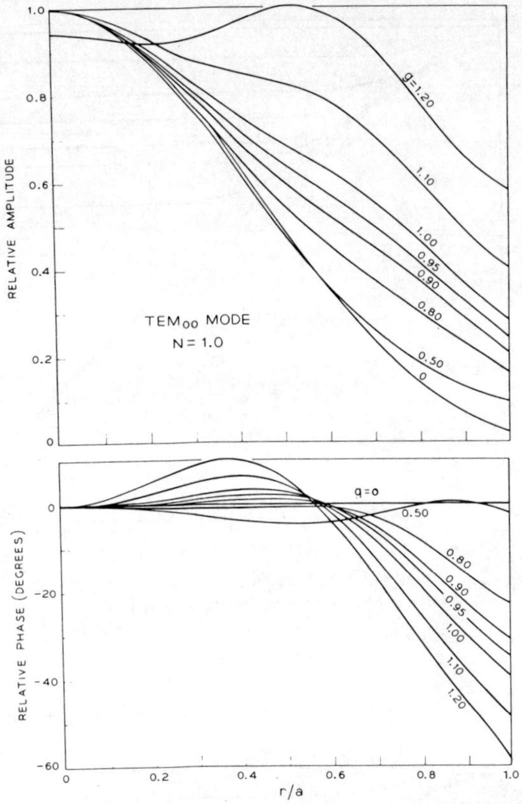

Fig. 19. Relative field distributions of the TEM_{00} mode for a resonator with circular mirrors ($N=1$).

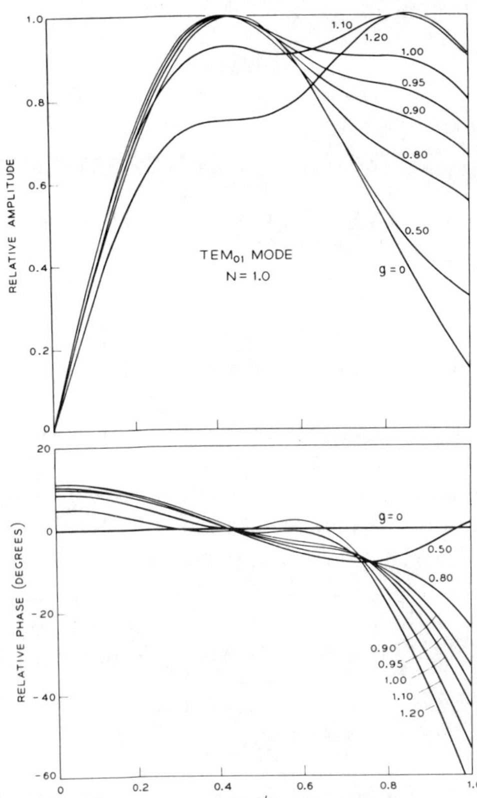

Fig. 20. Relative field distributions of the TEM_{01} mode for a resonator with circular mirrors ($N=1$).

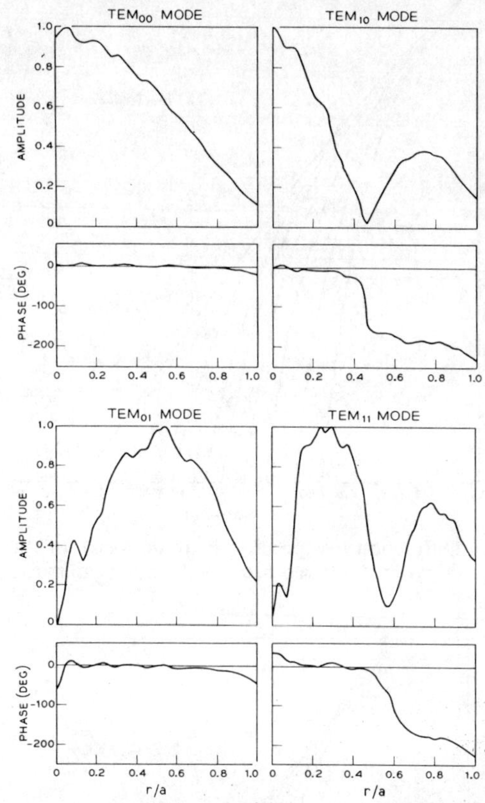

Fig. 21. Relative field distributions of four of the low order modes of a Fabry-Perot resonator with (parallel-plane) circular mirrors ($N=10$).

4.7 *Diffraction Losses and Phase Shifts*

The diffraction loss α and the phase shift β for a particular mode are important quantities in that they determine the Q and the resonant frequency of the resonator for that mode. The diffraction loss is given by

$$\alpha = 1 - |\gamma|^2 \tag{81}$$

which is the fractional energy lost per transit due to diffraction effects at the mirrors. The phase shift is given by

$$\beta = \text{angle of } \gamma \tag{82}$$

which is the phase shift suffered (or enjoyed) by the wave in transit from one mirror to the other, in addition to the geometrical phase shift which is given by $2\pi d/\lambda$. The eigenvalue γ in (81) and (82) is the appropriate γ for the mode under consideration. If the total resonator loss is small, the Q of the resonator can be approximated by

$$Q = \frac{2\pi d}{\lambda \alpha_t} \tag{83}$$

where α_t, the total resonator loss, includes losses due to diffraction, output coupling, absorption, scattering, and other effects. The resonant frequency ν is given by

$$\nu/\nu_0 = (q + 1) + \beta/\pi \tag{84}$$

where q, the longitudinal mode order, and ν_0, the fundamental beat frequency, are defined in Section 3.5.

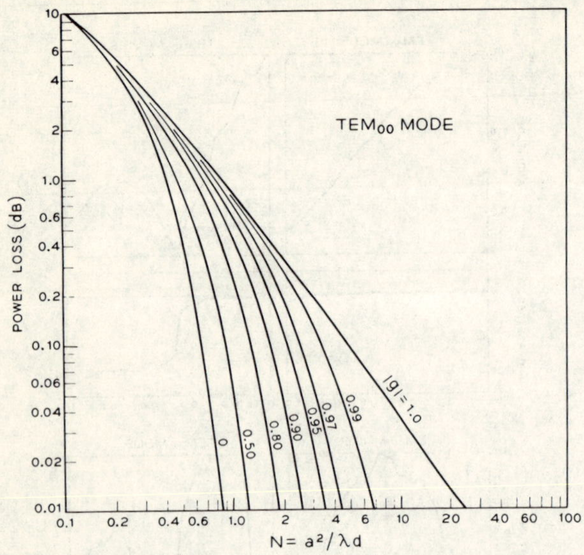

Fig. 22. Diffraction loss per transit (in decibels) for the TEM$_{00}$ mode of a stable resonator with circular mirrors.

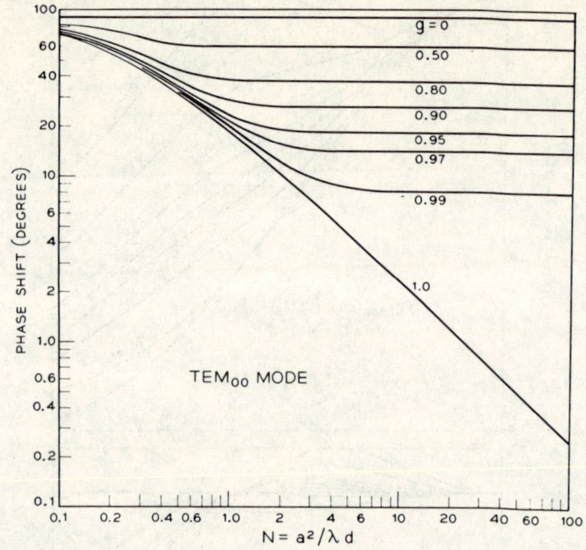

Fig. 24. Phase shift per transit for the TEM$_{01}$ mode of a stable resonator with circular mirrors.

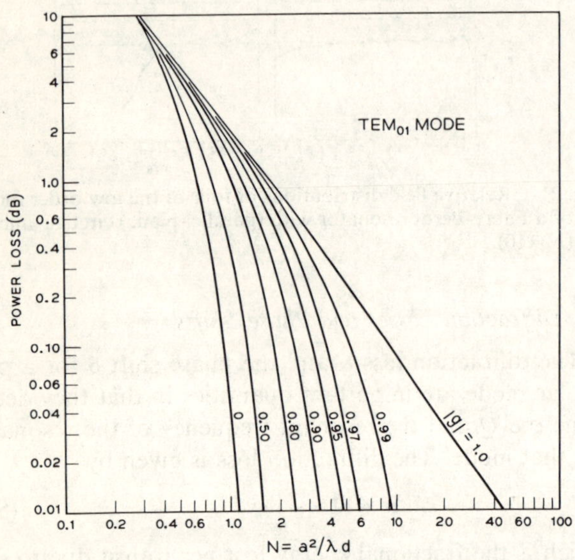

Fig. 23. Diffraction loss per transit (in decibels) for the TEM$_{01}$ mode of a stable resonator with circular mirrors.

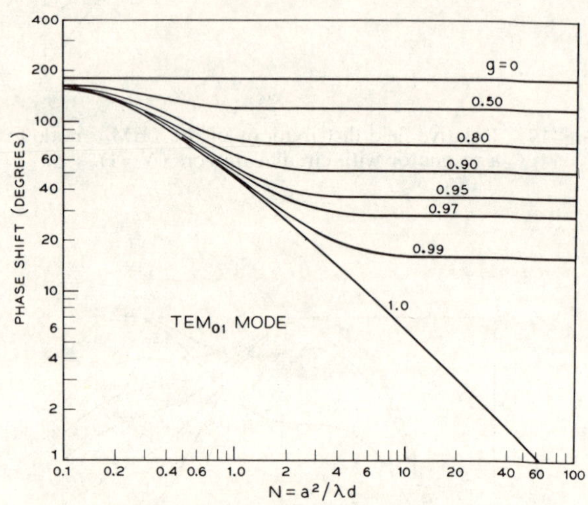

Fig. 25. Phase shift per transit for the TEM$_{01}$ mode of a stable resonator with circular mirrors.

The diffraction losses for the two lowest order (TEM$_{00}$ and TEM$_{01}$) modes of a stable resonator with a pair of identical, circular mirrors ($a_1=a_2$, $g_1=g_2=g$) are given in Figs. 22 and 23 as functions of the Fresnel number N and for various values of g. The curves are obtained by solving (77) numerically using the method of successive approximations [31]. Corresponding curves for the phase shifts are shown in Figs. 24 and 25. The horizontal portions of the phase shift curves can be calculated from the formula

$$\begin{aligned}\beta &= (2p + l + 1) \text{ arc cos } \sqrt{g_1 g_2} \\ &= (2p + l + 1) \text{ arc cos } g, \qquad \text{for } g_1 = g_2 \end{aligned} \tag{85}$$

which is equal to the phase shift for the beam modes derived in Section 3.5. It is to be noted that the loss curves are applicable to both positive and negative values of g while the phase-shift curves are for positive g only; the phase shift for negative g is equal to 180 degrees minus that for positive g.

Analytical expressions for the diffraction loss and the phase shift have been obtained for the special cases of parallel-plane ($g=1.0$) and confocal ($g=0$) geometries when the Fresnel number is either very large (small diffraction loss) or very small (large diffraction loss) [36], [38], [39], [41], [42]. In the case of the parallel-plane resonator with circular mirrors, the approximate expressions valid for large N, as derived by Vainshtein [36], are

$$\alpha = 8\kappa_{pl} \frac{\delta(M+\delta)}{[(M+\delta)^2+\delta^2]^2} \tag{86}$$

$$\beta = \left(\frac{M}{4\delta}\right)\alpha \tag{87}$$

where $\delta=0.824$, $M=\sqrt{8\pi N}$, and κ_{pl} is the $(p+1)$th zero of the Bessel function of order l. For the confocal resonator with circular mirrors, the corresponding expressions are [39]

$$\alpha = \frac{2\pi(8\pi N)^{2p+l+1}e^{-4\pi N}}{p!(p+l+1)!}\left[1+0\left(\frac{1}{2\pi N}\right)\right] \quad (88)$$

$$\beta = (2p+l+1)\frac{\pi}{2}. \quad (89)$$

Similar expressions exist for resonators with infinite-strip or rectangular mirrors [36], [39]. The agreement between the values obtained from the above formulas and those from numerical methods is excellent.

The loss of the lowest order (TEM_{00}) mode of an *unstable* resonator is, to first order, independent of the mirror size or shape. The formula for the loss, which is based on geometrical optics, is [12]

$$\alpha = 1 \pm \frac{1-\sqrt{1-(g_1g_2)^{-1}}}{1+\sqrt{1-(g_1g_2)^{-1}}} \quad (90)$$

where the plus sign in front of the fraction applies for g values lying in the first and third quadrants of the stability diagram, and the minus sign applies in the other two quadrants. Loss curves (plotted vs. N) obtained by solving the integral equations numerically have a ripply behavior which is attributable to diffraction effects [24], [43]. However, the average values agree well with those obtained from (90).

5. Concluding Remarks

Space limitations made it necessary to concentrate the discussion of this article on the basic aspects of laser beams and resonators. It was not possible to include such interesting topics as perturbations of resonators, resonators with tilted mirrors, or to consider in detail the effect of nonlinear, saturating host media. Also omitted was a discussion of various resonator structures other than those formed of spherical mirrors, e.g., resonators with corner cube reflectors, resonators with output holes, or fiber resonators. Another important, but omitted, field is that of mode selection where much research work is currently in progress. A brief survey of some of these topics is given in [44].

References

[1] R. H. Dicke, "Molecular amplification and generation systems and methods," U. S. Patent 2 851 652, September 9, 1958.

[2] A. M. Prokhorov, "Molecular amplifier and generator for submillimeter waves," *JETP (USSR)*, vol. 34, pp. 1658–1659, June 1958; *Sov. Phys. JETP*, vol. 7, pp. 1140–1141, December 1958.

[3] A. L. Schawlow and C. H. Townes, "Infrared and optical masers," *Phys. Rev.*, vol. 29, pp. 1940–1949, December 1958.

[4] A. G. Fox and T. Li, "Resonant modes in an optical maser," *Proc. IRE (Correspondence)*, vol. 48, pp. 1904–1905, November 1960; "Resonant modes in a maser interferometer," *Bell Sys. Tech. J.*, vol. 40, pp. 453–488, March 1961.

[5] G. D. Boyd and J. P. Gordon, "Confocal multimode resonator for millimeter through optical wavelength masers," *Bell Sys. Tech. J.*, vol. 40, pp. 489–508, March 1961.

[6] G. D. Boyd and H. Kogelnik, "Generalized confocal resonator theory," *Bell Sys. Tech. J.*, vol. 41, pp. 1347–1369, July 1962.

[7] G. Goubau and F. Schwering, "On the guided propagation of electromagnetic wave beams," *IRE Trans. on Antennas and Propagation*, vol. AP-9, pp. 248–256, May 1961.

[8] J. R. Pierce, "Modes in sequences of lenses," *Proc. Nat'l Acad. Sci.*, vol. 47, pp. 1808–1813, November 1961.

[9] G. Goubau, "Optical relations for coherent wave beams," in *Electromagnetic Theory and Antennas.* New York: Macmillan, 1963, pp. 907–918.

[10] H. Kogelnik, "Imaging of optical mode—Resonators with internal lenses," *Bell Sys. Tech. J.*, vol. 44, pp. 455–494, March 1965.

[11] ——, "On the propagation of Gaussian beams of light through lenslike media including those with a loss or gain variation," *Appl. Opt.*, vol. 4, pp. 1562–1569, December 1965.

[12] A. E. Siegman, "Unstable optical resonators for laser applications," *Proc. IEEE*, vol. 53, pp. 277–287, March 1965.

[13] W. Brower, *Matrix Methods in Optical Instrument Design.* New York: Benjamin, 1964. E. L. O'Neill, *Introduction to Statistical Optics.* Reading, Mass.: Addison-Wesley, 1963.

[14] M. Bertolotti, "Matrix representation of geometrical properties of laser cavities," *Nuovo Cimento*, vol. 32, pp. 1242–1257, June 1964. V. P. Bykov and L. A. Vainshtein, "Geometrical optics of open resonators," *JETP (USSR)*, vol. 47, pp. 508–517, August 1964. B. Macke, "Laser cavities in geometrical optics approximation," *J. Phys. (Paris)*, vol. 26, pp. 104A–112A, March 1965. W. K. Kahn, "Geometric optical derivation of formula for the variation of the spot size in a spherical mirror resonator," *Appl. Opt.*, vol. 4, pp. 758–759, June 1965.

[15] J. R. Pierce, *Theory and Design of Electron Beams.* New York: Van Nostrand, 1954, p. 194.

[16] H. Kogelnik and W. W. Rigrod, "Visual display of isolated optical-resonator modes," *Proc. IRE (Correspondence)*, vol. 50, p. 220, February 1962.

[17] G. A. Deschamps and P. E. Mast, "Beam tracing and applications," in *Proc. Symposium on Quasi-Optics.* New York: Polytechnic Press, 1964, pp. 379–395.

[18] S. A. Collins, "Analysis of optical resonators involving focusing elements," *Appl. Opt.*, vol. 3, pp. 1263–1275, November 1964.

[19] T. Li, "Dual forms of the Gaussian beam chart," *Appl. Opt.*, vol. 3, pp. 1315–1317, November 1964.

[20] T. S. Chu, "Geometrical representation of Gaussian beam propagation," *Bell Sys. Tech. J.*, vol. 45, pp. 287–299, February 1966.

[21] J. P. Gordon, "A circle diagram for optical resonators," *Bell Sys. Tech. J.*, vol. 43, pp. 1826–1827, July 1964. M. J. Offerhaus, "Geometry of the radiation field for a laser interferometer," *Philips Res. Rept.*, vol. 19, pp. 520–523, December 1964.

[22] H. Statz and C. L. Tang, "Problem of mode deformation in optical masers," *J. Appl. Phys.*, vol. 36, pp. 1816–1819, June 1965.

[23] A. G. Fox and T. Li, "Effect of gain saturation on the oscillating modes of optical masers," *IEEE J. of Quantum Electronics*, vol. QE-2, p. lxii, April 1966.

[24] ——, "Modes in a maser interferometer with curved and tilted mirrors," *Proc. IEEE*, vol. 51, pp. 80–89, January 1963.

[25] F. B. Hildebrand, *Methods of Applied Mathematics.* Englewood Cliffs, N. J.: Prentice Hall, 1952, pp. 412–413.

[26] D. J. Newman and S. P. Morgan, "Existence of eigenvalues of a class of integral equations arising in laser theory," *Bell Sys. Tech. J.*, vol. 43, pp. 113–126, January 1964.

[27] J. A. Cochran, "The existence of eigenvalues for the integral equations of laser theory," *Bell Sys. Tech. J.*, vol. 44, pp. 77–88, January 1965.

[28] H. Hochstadt, "On the eigenvalue of a class of integral equations arising in laser theory," *SIAM Rev.*, vol. 8, pp. 62–65, January 1966.

[29] D. Gloge, "Calculations of Fabry-Perot laser resonators by scattering matrices," *Arch. Elect. Ubertrag.*, vol. 18, pp. 197–203, March 1964.

[30] W. Streifer, "Optical resonator modes—rectangular reflectors of spherical curvature," *J. Opt. Soc. Am.*, vol. 55, pp. 868–877, July 1965

[31] T. Li, "Diffraction loss and selection of modes in maser resonators with circular mirrors," *Bell Sys. Tech. J.*, vol. 44, pp. 917–932, May–June, 1965.

[32] J. C. Heurtley and W. Streifer, "Optical resonator modes—

circular reflectors of spherical curvature," *J. Opt. Soc. Am.*, vol. 55, pp. 1472–1479, November 1965.
[33] J. P. Gordon and H. Kogelnik, "Equivalence relations among spherical mirror optical resonators," *Bell Sys. Tech. J.*, vol. 43, pp. 2873–2886, November 1964.
[34] F. Schwering, "Reiterative wave beams of rectangular symmetry," *Arch. Elect. Übertrag.*, vol. 15, pp. 555–564, December 1961
[35] A. G. Fox and T. Li, to be published.
[36] L. A. Vainshtein, "Open resonators for lasers," *JETP (USSR)*, vol. 44, pp. 1050–1067, March 1963; *Sov. Phys. JETP*, vol. 17, pp. 709–719, September 1963.
[37] S. R. Barone, "Resonances of the Fabry-Perot laser," *J. Appl. Phys.*, vol. 34, pp. 831–843, April 1963.
[38] D. Slepian and H. O. Pollak, "Prolate spheroidal wave functions, Fourier analysis and uncertainty—I," *Bell Sys. Tech. J.*, vol. 40, pp. 43–64, January 1961.
[39] D. Slepian, "Prolate spheroidal wave functions, Fourier analysis and uncertainty—IV: Extensions to many dimensions; generalized prolate spheroidal functions," *Bell Sys. Tech. J.*, vol. 43, pp. 3009–3057, November 1964.
[40] J. C. Heurtley, "Hyperspheroidal functions—optical resonators with circular mirrors," in *Proc. Symposium on Quasi-Optics.* New York: Polytechnic Press, 1964, pp. 367–375.
[41] S. R. Barone and M. C. Newstein, "Fabry-Perot resonances at small Fresnel numbers," *Appl. Opt.*, vol. 3, p. 1194, October 1964.
[42] L. Bergstein and H. Schachter, "Resonant modes of optic cavities of small Fresnel numbers," *J. Opt. Soc. Am.*, vol. 55, pp. 1226–1233, October 1965.
[43] A. G. Fox and T. Li, "Modes in a maser interferometer with curved mirrors," in *Proc. Third International Congress on Quantum Electronics.* New York: Columbia University Press, 1964, pp. 1263–1270.
[44] H. Kogelnik, "Modes in optical resonators," in *Lasers*, A. K. Levine, Ed. New York: Dekker, 1966.

Confocal Multimode Resonator for Millimeter Through Optical Wavelength Masers

By G. D. BOYD and J. P. GORDON

(Manuscript received September 12, 1960)

Multimode resonators of high quality factor will very likely play a significant role in the development of devices, such as the maser, which operate in the millimeter through optical wavelength range. It has been suggested that a plane-parallel Fabry-Perot interferometer could act as a suitable resonator. In this paper a resonator consisting of two identical concave spherical reflectors, separated by any distance up to twice their common radius of curvature, is considered.

Mode patterns and diffraction losses for the low-loss modes of such a resonator are obtained analytically, using an approximate method which was suggested by W. D. Lewis. The results show that the diffraction losses are generally considerably lower for the curved surfaces than for the plane surfaces. Diffraction losses and mode volume are a minimum when the reflector spacing equals the common radius of curvature of the reflectors. For this case the resonator may be termed confocal. A further property of the concave spherical resonator is that the optical alignment is not extremely critical.

I. INTRODUCTION

Schawlow and Townes[1] proposed that coherent amplification could be achieved in the infrared through optical regions of the frequency spectrum by maser techniques. At such frequencies multimode resonators are necessary to achieve reasonable dimensions and high Q. They and Prokhorov[2] and Dicke[3] have suggested as a resonator two plane-parallel reflecting planes, known as a Fabry-Perot interferometer, or etalon.[4]

In Fabry-Perot resonators the major factors contributing to the Q (i.e., resolving power) are reflection losses and diffraction losses. Reflection losses result from absorption in the reflectors, and from transmission through them. At optical frequencies a very good layered dielectric reflector[5] can have a $99\frac{1}{2}$ per cent reflection coefficient. Diffraction losses result from the finite aperture of the reflectors and from imperfections in their "flatness."

Fox and Li have shown in the accompanying paper[6] that modes, in the sense of a self-reproducing field pattern, exist for an open structure such as a Fabry-Perot interferometer. They also have recognized that the diffraction losses of a plane-parallel Fabry-Perot are very much less than those obtained by assuming a uniform intensity distribution over the reflector and the Fraunhofer far field diffraction angle. They have made numerical self-consistent field calculations based on Huygens' principle to determine the actual diffraction losses and mode patterns.

In interferometry using a Fabry-Perot resonator, one normally excites a system of plane waves traveling at certain discrete angles to the axis. Constructive interference at each of these discrete angles, as is appropriate to ring order, wavelength and spacing, results in a pattern of concentric bright rings. Schawlow and Townes indicated that each ring of the interference pattern is *not* a pure mode of the resonator but an infinite sum of such modes, each representing a different field pattern over the reflector. This idea has been given much substance by the work of Fox and Li.

The plane-parallel Fabry-Perot is not necessarily ideal, however, as a high-frequency multimode resonator. A resonator formed by two spherical reflectors of equal curvature separated by their common radius of curvature is considered in detail in this paper. The focal length of a spherical mirror is one-half of its radius of curvature. Therefore the focal points of the reflectors are coincident and the resonator is termed confocal. G. W. Series, Fox and Li[6] and Lewis[7] have also suggested the confocal resonator. Lewis has recognized that it would have lower diffraction losses than the plane-parallel Fabry-Perot and has described the analytic solution presented here.

The use of confocal reflectors as an interferometer has been described by Connes.[8] The adjustment of the spherical Connes interferometer is trivial compared to the Fabry-Perot. Parallelism between the reflectors is not a strict requirement, the only fine adjustment therefore being the spacing between the surfaces. Parabolic surfaces may also be used, but they have an axis and thus lose the advantage of ease of adjustment.

II. RESONATOR QUALITY FACTOR

Resonator quality factor, or Q, is defined as

$$Q = \omega \frac{\text{energy stored}}{\text{energy lost per second}}. \qquad (1)$$

Consider an interferometer consisting of two reflecting surfaces separated by a distance d which is large compared to the wavelength in the medium λ. By considering waves bouncing back and forth between the surfaces, one may derive an approximate Q as

$$Q = \frac{2\pi d}{\alpha\lambda}, \tag{2}$$

where α is the fractional power loss per bounce from a reflector and is the sum of diffraction and reflection losses. This is to be compared to the resolving power derived in optics[9] as

$$R = \frac{2\pi d\sqrt{r}}{\lambda(1 - r)}, \tag{3}$$

where the power reflection coefficient per bounce is $r \equiv 1 - \alpha$. Resolving power is thus synonymous with Q within the small loss approximation of (2).

If diffraction losses are small compared with reflection losses, then resonator Q is proportional to the spacing between the reflecting surfaces. For a given reflector aperture size, the resonator Q will continue to increase with the spacing d between the reflectors until the diffraction losses become roughly comparable with the reflection losses. Further increase in spacing then decreases the Q because of increasing diffraction losses.

III. MODES AND DIFFRACTION LOSSES OF A CONFOCAL RESONATOR

All resonator dimensions are assumed large compared to a wavelength; the modes and diffraction losses of the confocal resonator are therefore obtainable from a self-consistent field analysis using Huygens' principle.[10] A confocal resonator is considered, with identical spherical reflectors of radius b, as shown in Fig. 1. Assume the field to be linearly polarized over the P' surface in the y direction and given by $E_0 f_m(x')g_n(y')$, where E_0 is a constant amplitude factor and $f_m(x')$ and $g_n(y')$ are the field variations over the aperture. At point $P(x,y)$ on the other surface, one computes the electric field by summing over contributions from the differential Huygens sources at all points $P'(x',y')$. The result is

$$E_y = \int_{S'} \frac{ik(1 + \cos\theta)}{4\pi\rho} e^{-ik\rho} E_0 f_m(x')g_n(y')\, dS'. \tag{4}$$

Here ρ is the distance between P and P', θ is the angle between the line $P'P$ and the normal to the reflector surface at P', and k is the propagation constant of the medium between the reflectors. Note that $k =$

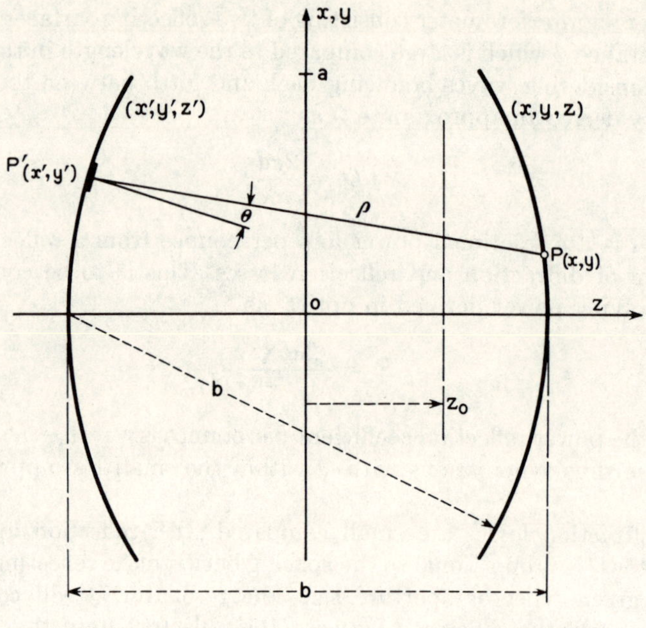

Fig. 1 — Confocal resonator with spherical reflectors.

$2\pi/\lambda$, where λ is the wavelength in the medium. The electric field in the xz plane is approximately zero. The reflector is assumed *square* and of dimension $2a$, which is small compared to the spacing b (since the confocal spacing is under consideration $d = b$), and thus θ is very nearly zero. The medium is assumed to fill all space.

The normal modes or eigenfunctions of the confocal resonator are obtained by requiring that the field distribution over $x'y'$ reproduce itself within a constant over the xy aperture, and thus $E_y = E_1 f_m(x) g_n(y)$, where $E_1 = \sigma_m \sigma_n E_0$. The proportionality factor $\sigma_m \sigma_n$ is generally complex, giving both amplitude and phase changes. The resulting integral equation is

$$\sigma_m \sigma_n f_m(x) g_n(y) = \iint_{-a}^{+a} \frac{ik}{2\pi\rho}\, e^{-ik\rho} f_m(x') g_n(y')\, dx'dy' . \tag{5}$$

The distance ρ varies only a small amount for small apertures and thus may be replaced by the separation b except in the exponential phase term. For x and y small compared to b one can show that

$$\frac{\rho}{b} = 1 - \frac{xx' + yy'}{b^2} + \frac{w^2 w'^2}{4b^4} + \cdots , \tag{6}$$

where $w^2 = x^2 + y^2$. The third term makes a negligible contribution to the phase when $a^2/b\lambda \ll b^2/a^2$. Note that in this approximation one cannot distinguish between spherical and parabolic surfaces. In terms of some dimensionless variables

$$c \equiv \frac{a^2 k}{b} = 2\pi\left(\frac{a^2}{b\lambda}\right) \qquad X \equiv \frac{x\sqrt{c}}{a}, \qquad Y \equiv \frac{y\sqrt{c}}{a}, \tag{7}$$

and with $F_m(X) \equiv f_m(x)$ etc., (5) becomes

$$\sigma_m\sigma_n F_m(X)G_n(Y) = \frac{ie^{-ikb}}{2\pi}\int_{-\sqrt{c}}^{+\sqrt{c}} F_m(X')e^{+iXX'}\,dX' \cdot \int_{-\sqrt{c}}^{+\sqrt{c}} G_n(Y')e^{+iYY'}\,dY'. \tag{8}$$

Slepian and Pollak[11] have considered the following integral equation:

$$F_m(X) = \frac{1}{\sqrt{2\pi}\chi_m}\int_{-\sqrt{c}}^{+\sqrt{c}} F_m(X')e^{+iXX'}\,dX'. \tag{9}$$

This is a homogeneous Fredholm equation of the second kind with $e^{iXX'}$ as the kernel. It is often referred to as a finite Fourier transform. They have shown solutions to be

$$F_m(c,\eta) \propto S_{0m}(c,\eta), \tag{10}$$

$$\chi_m = \sqrt{\frac{2c}{\pi}}\, i^m R_{0m}{}^{(1)}(c,1), \qquad m = 0, 1, 2, \cdots, \tag{11}$$

where $S_{0m}(c,\eta)$ and $R_{0m}{}^{(1)}(c,1)$ are respectively the angular and radial wave functions in prolate spheroidal coordinates as defined by Flammer,[12] and where $\eta = X/\sqrt{c} = x/a$ and $\eta = Y/\sqrt{c} = y/a$ respectively for $F_m(X)$ and $G_n(Y)$. There is an infinite number of eigenfunctions and corresponding eigenvalue solutions to (9) for any value of c. Flammer[12] gives values of these functions for $c \leqq 5$ and Slepian and Pollak[11,13] have computed the eigenvalues χ_m for the important region of $c > 5$.

The eigenfunction solutions of (8) are thus the spheroidal wave functions $S_{0m}(c,x/a)S_{0n}(c,y/a)$. The eigenfunctions are real; therefore, the reflecting surfaces are of constant phase. The eigenvalues are

$$\sigma_m\sigma_n = \chi_m\chi_n i\, e^{-ikb}. \tag{12}$$

The phase shift between the two reflecting confocal surfaces equals the phase angle of $\sigma_m\sigma_n$. For resonance the round-trip phase shift must equal an integer q times 2π. From (11) and (12), one finds therefore

$$2\pi q = 2\left|\frac{\pi}{2} - kb + (m+n)\frac{\pi}{2}\right|. \tag{13}$$

Since $k = 2\pi/\lambda$, one obtains for the condition of resonance

$$\frac{4b}{\lambda} = 2q + (1 + m + n). \tag{14}$$

The confocal resonator is seen to have resonances only for integer values of the quantity $4b/\lambda$. If $4b/\lambda$ is odd, $(m + n)$ must be even, likewise if $4b/\lambda$ is even, $(m + n)$ must be odd. Note that considerable degeneracy exists in the spectrum; increasing $(m + n)$ by two and decreasing q by unity gives the same frequency. The degenerate modes are orthogonal over the reflector surface since they satisfy the integral (5) with different eigenvalues. The modes have negligible axial electric and magnetic fields and thus will be designated by TEM_{mnq}, where m and n equal 0, 1, 2, $\cdots$, and refer to variations in the x and y directions, while q equals the number of half-guide wavelength variations in the z direction between reflectors.

The fractional energy loss per reflection due to diffraction effects is given by

$$\alpha_D = 1 - |\sigma_m\sigma_n|^2 = 1 - |\chi_m\chi_n|^2. \tag{15}$$

The function $1 - |\chi_m|^2$ versus c is shown in Fig. 2 for $m = 0, 1, 2$. It can be shown that Fig. 2 also gives the diffraction losses for an infinite cylindrical reflector strip of width $2a$ and radius of curvature b. The diffraction losses for various TEM_{mnq} modes are shown in Fig. 3. Note that TEM_{uvq} and TEM_{vuq} $(u \neq v)$ have the same diffraction losses; also that the diffraction losses of the TEM_{02q} and TEM_{12q} are so nearly equal that they can be plotted as one curve. As indicated previously, these last two types of modes cannot both be resonant at the same frequency. Note that the losses are primarily determined by the higher of the transverse mode numbers m, n, regardless of the field polarization.

In Fig. 3 the results of Fox and Li[6] for the plane-parallel resonator with circular reflectors are also shown. The diffraction losses for the confocal resonator are seen to be orders of magnitude *smaller* than for the plane parallel resonator. Fox and Li have also obtained numerical results for the confocal resonator with circular cross section of radius a. These are in good agreement with the results presented here, allowing for the fact that in this paper the reflectors have a square cross section of width $2a$.

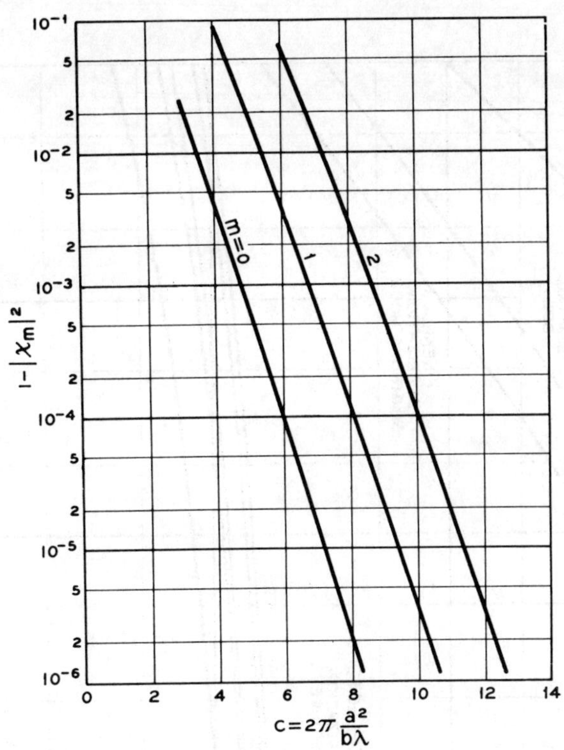

Fig. 2 — Eigenvalues of integral equation; also the diffraction losses of an infinitely long cylindrical reflector of width $2a$.

If one approximates the diffraction loss curve by a function $\alpha_D = A \times 10^{-B(a^2/b\lambda)}$, one may then show for a given reflection loss and reflector radius a that the resonator Q is a maximum as a function of the confocal spacing b when the reflection loss equals $[2.30B(a^2/b\lambda) - 1]$ times the diffraction loss. For the TEM_{00q} mode, $A = 10.9$ and $B = 4.94$; thus, if $a^2/b\lambda = 0.8$, then the diffraction loss is approximately one-eighth of the reflection loss.

The diffraction loss for the plane-parallel case assuming a uniform field and phase distribution and a diffraction angle of $\theta = \lambda/2a$ is also shown. This diffraction angle corresponds to the first Fraunhofer minimum in far field theory. For a square (or circular) reflector of side $2a$ the diffraction loss is approximately

$$\alpha_D \approx \left(\frac{a^2}{b\lambda}\right)^{-1}. \tag{16}$$

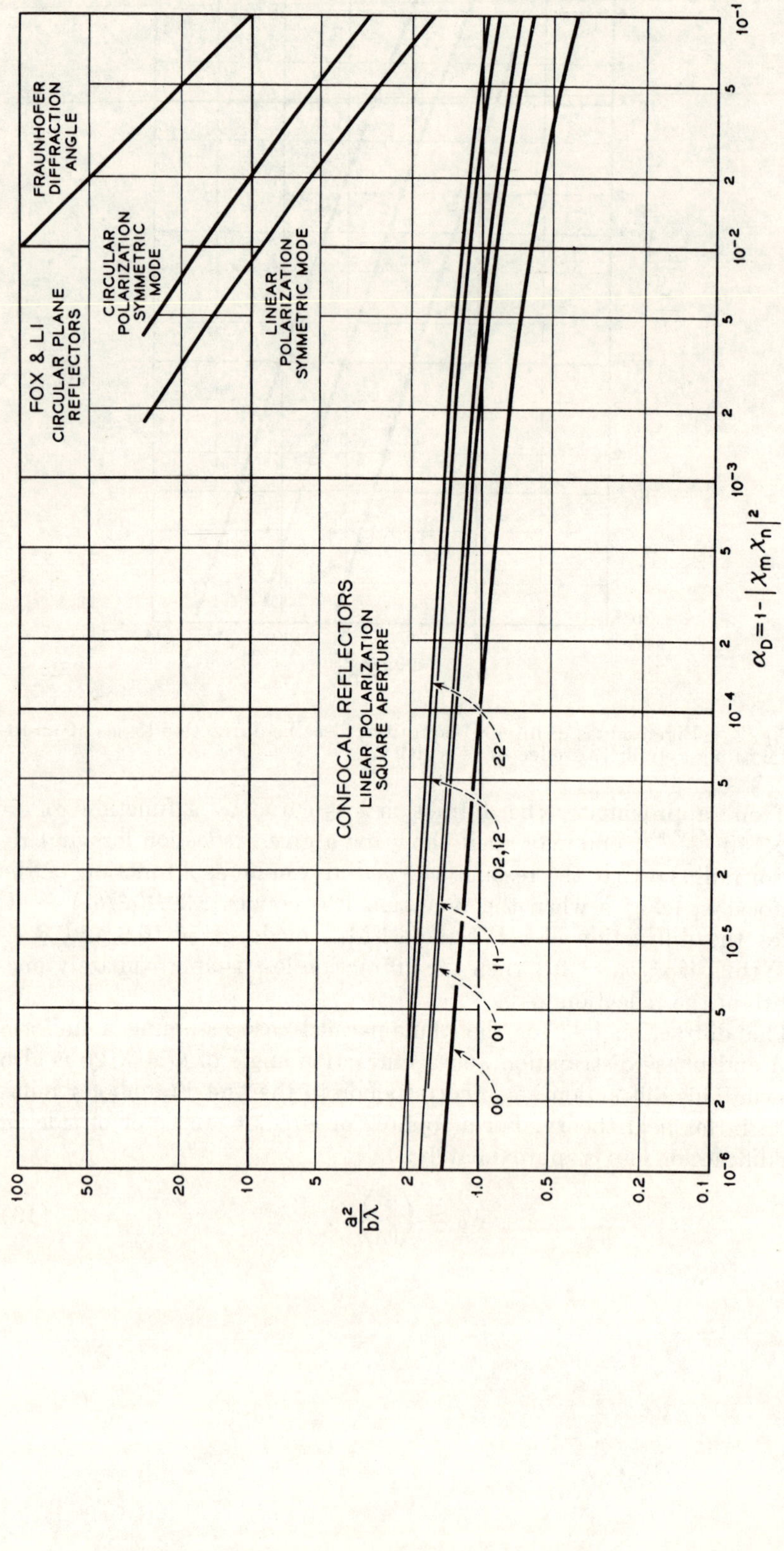

Fig. 3 — Diffraction losses for confocal and plane-parallel[6] resonators.

Fig. 3 clearly demonstrates the inadequacy of the assumption of uniform intensity distribution.

Though the eigenvalues given by (12) must be known accurately the eigenfunctions are only of approximate interest. Flammer[12] shows that, in the approximation of $\eta^2 \ll 1$ (near the center of the reflector), (10) becomes

$$F_m(X) \approx \frac{\Gamma\left(\frac{m}{2}+1\right)}{\Gamma(m+1)} H_m(X) e^{-\frac{1}{2}X^2}$$
$$= \frac{\Gamma\left(\frac{m}{2}+1\right)}{\Gamma(m+1)} (-1)^m e^{+\frac{1}{2}X^2} \frac{d^m}{dX^m} e^{-X^2}. \tag{17}$$

The mode shape is thus approximately a Gaussian times a Hermite polynomial $H_m(X)$. The gamma function is arbitrarily chosen as normalization such that $F_m(X = 0) = \pm 1$ for m even:

$$\begin{aligned} F_0(c,\eta) &= e^{-\frac{1}{2}c\eta^2}, \\ F_1(c,\eta) &= \sqrt{\pi c}\,\eta\, e^{-\frac{1}{2}c\eta^2}, \\ F_2(c,\eta) &= (2c\eta^2 - 1)e^{-\frac{1}{2}c\eta^2}. \end{aligned} \tag{18}$$

The approximation involved in (17) fails away from the center of the reflector. For reasonably large values of c, however, the field is weak there, and of little interest. The diffraction losses were previously obtained from (15). Curves representing (18) for various values of c are shown in Fig. 4. The dotted curves for $c = 5$ are the true eigenfunctions $S_{0m}(c,\eta)$ as obtained from Flammer.[12]

The exponential dependence of the electric field on $c\eta^2$, which is independent of the reflector half-width a, leads one to define a "spot size" at the reflector of radius $w = w_s$, where $w^2 = x^2 + y^2$, at which the exponential term falls to e^{-1}:

$$w_s = \sqrt{\frac{b\lambda}{\pi}}. \tag{19}$$

The only effect of increasing the reflector width $2a$ is to reduce the diffraction losses; the spot size is unaffected.

If one allows the reflectors to be somewhat lossy or partially transparent, then the resonator Q is reduced over that implied by diffraction losses alone. The field distribution, i.e., the mode pattern, is *not* seriously affected so long as the losses are small and fairly uniform over the plates.

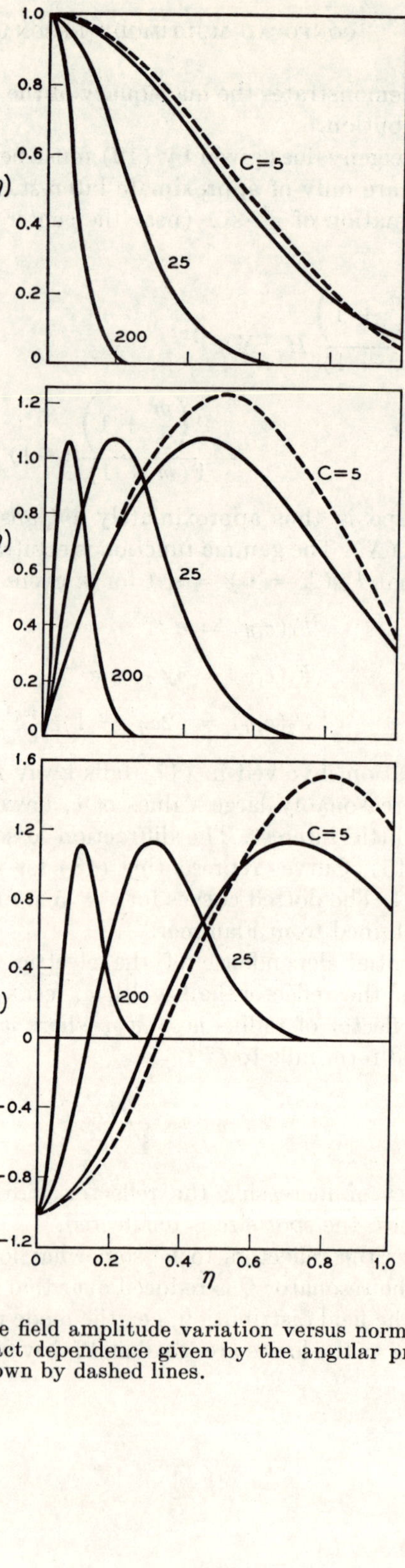

Fig. 4 — Approximate field amplitude variation versus normalized radius for various modes. The exact dependence given by the angular prolate spheroidal function $S_{0m}(c, \eta)$ is shown by dashed lines.

The electric field patterns derived thus far have all been linearly polarized. Fox and Li[6] have recognized that, by superimposing the TEM_{01q} mode linearly polarized in the x direction and the TEM_{10q} mode linearly polarized in the y direction, the lowest-order circular electric mode can result, and it has the same diffraction losses as the linearly polarized TEM_{01q} mode. Many other polarization configurations can be obtained in this manner.

IV. FIELDS OF THE CONFOCAL RESONATOR

The field over the confocal aperture has been obtained in the preceding section. The field over an arbitrary plane $z = z_0$, as in Fig. 1, is also obtainable by a straightforward application of Huygens' principle as stated in (4). The arbitrary plane z_0 may be placed outside the confocal geometry as well as inside provided one takes into account the transmission loss of the reflector. The field distribution over the confocal surface is given by $F_m(c,x/a)G_n(c,y/a)$. For large c the spheroidal functions may be approximated by the Gaussian-Hermite functions. The integral can be evaluated in the limit of $c \to \infty$.

Within these approximations, the traveling wave field of the confocal resonator resulting from the field at one of the reflectors is given by

$$\frac{E(x,y,z_0)}{E_0} = \sqrt{\frac{2}{1+\xi^2}} \frac{\Gamma\left(\frac{m}{2}+1\right)\Gamma\left(\frac{n}{2}+1\right)}{\Gamma(m+1)\Gamma(n+1)} H_m\left(X\sqrt{\frac{2}{1+\xi^2}}\right)$$
$$\cdot H_n\left(Y\sqrt{\frac{2}{1+\xi^2}}\right) \exp\left[-\frac{kw^2}{b(1+\xi^2)}\right] \tag{20}$$
$$\cdot \exp\left(-i\left\{k\left[\frac{b}{2}(1+\xi)+\frac{\xi}{1+\xi^2}\frac{w^2}{b}\right]-(1+m+n)\left(\frac{\pi}{2}-\varphi\right)\right\}\right),$$

where

$$w^2 = x^2 + y^2,$$
$$\xi = \frac{2z_0}{b}, \tag{21}$$
$$\tan\varphi = \frac{1-\xi}{1+\xi}.$$

When the reflecting surface is made partially transparent, as will be the case with optical or infrared masers, the field of the transmitted wave will be a traveling wave as given in (20) reduced by the transmission coefficient of the reflector. Within the resonator, the field will be a standing wave. The transverse standing wave is as given in (20) except that the exponential phase function is replaced by the sine function.

The surface of constant phase which intersects the axis at z_0 as obtained from (20) is given approximately by

$$z - z_0 \approx -\frac{\xi}{1+\xi^2}\frac{w^2}{b}, \tag{22}$$

neglecting the small variation in φ due to variation in z. This surface is spherical, within the approximations of this paper, and has a radius of curvature b' given by

$$b' = \left|\frac{1+\xi^2}{2\xi}\right| b. \tag{23}$$

At $\xi = \pm 1$ it coincides with the spherical reflector as expected. Also note that the symmetry or focal plane ($\xi = 0$) is a surface of constant phase.

The field distribution throughout the resonator is given by the modulus of (20). The complete field distribution within the confocal resonator is shown schematically in Fig. 5 for the low-loss TEM_{00q} mode.

The field distribution over the focal plane is less spread out than over the spherical reflectors. The field spot size over the spherical reflectors was defined by (19). In any arbitrary plane z_0 the exponential term in the field distribution falls to e^{-1} at a radius

$$w_s = \sqrt{\frac{b\lambda(1+\xi^2)}{2\pi}}. \tag{24}$$

The smallest achievable spot size is in the focal plane at $\xi = 0$.

To obtain the radiation pattern angular beam width of the TEM_{00q} mode spherical wave, one takes the ratio of the spot diameter from (20) or (24), as $\xi \to \infty$, to the distance from the center of the resonator. The beam width between the *half-power points* is given by

$$\theta = 2\sqrt{\frac{\ln 2}{\pi}}\sqrt{\frac{\lambda}{b}} = 0.939\sqrt{\frac{\lambda}{b}} \text{ radians.} \tag{25}$$

V. RESONATOR WITH NONCONFOCAL SPACING

Since the surfaces of constant phase of the confocal resonator are spherical, it is apparent that (20) also represents approximately the field distribution between two spherical reflectors of arbitrary spacing. That is, any two surfaces of constant phase may be replaced by reflectors. The frequencies at which such a resonator will be resonant will of course be determined by satisfaction of the phase condition.

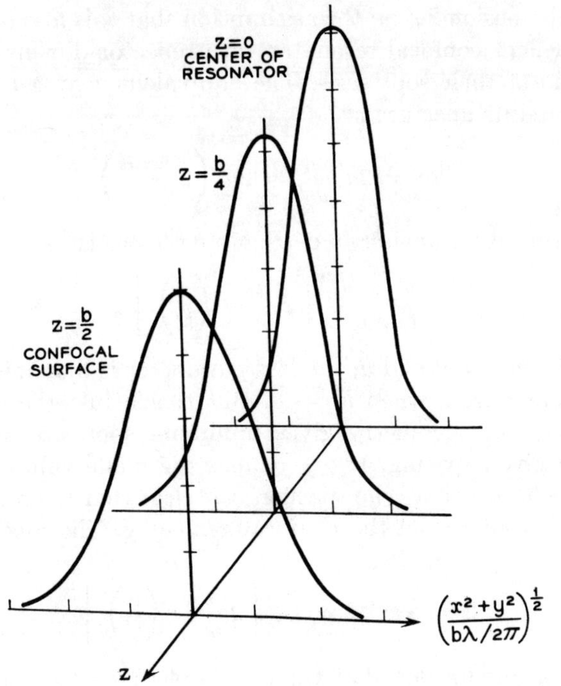

Fig. 5 — Field strength distribution within the confocal resonator for the TEM_{00q} mode.

Consider two identical spherical reflectors of radius of curvature b' spaced a distance d. The only restriction is that $b' \geqq d/2$. The confocal geometry of spacing b of which this resonator is a part is [set $\xi = d/b$ in (23)]:

$$b^2 = 2db' - d^2, \qquad b' \geqq \frac{d}{2}. \tag{26}$$

The spot size at the reflectors in the nonconfocal resonator may be immediately obtained from (24) with $\xi = \pm d/b$. It is

$$w_s' = \left(\frac{d\lambda}{\pi}\right)^{\frac{1}{2}} \left[2\frac{d}{b'} - \left(\frac{d}{b'}\right)^2\right]^{-\frac{1}{4}} \tag{27}$$

Note that the factor $[2(d/b') - (d/b')^2]$ achieves a maximum of unity, as a function of b', when $b' = d$. Thus, for a given spacing between reflectors, the spot size is a minimum for the confocal resonator.

One may estimate the loss of a nonconfocal resonator of square cross section of dimension $2a'$ on the assumption that this loss is equal to that of its equivalent confocal resonator with reflector dimensions scaled up by the ratio of their spot sizes. The equivalent confocal resonator has spacing b, and its aperture is

$$2a = 2a' \frac{w_s}{w_s'} = 2a' \left(2 - \frac{d}{b'}\right)^{\frac{1}{2}}. \tag{28}$$

The important parameter in determining losses is

$$\frac{a^2}{b\lambda} = \frac{a'^2}{d\lambda} \left[2 \frac{d}{b'} - \left(\frac{d}{b'}\right)^2\right]^{\frac{1}{2}}. \tag{29}$$

For given values of a' and d, the loss parameter is maximized, and thus losses are minimized, when $b' = d$. But this is just the confocal case. Thus the confocal geometry gives minimum spot size and minimum losses for a given spacing. If one defines the mode volume as the spot size at the reflector times the spacing, it is clear that the minimum mode volume also results from the confocal geometry. The mode volume, so defined, is

$$V = \pi w_s'^2 d = \lambda d^2 \left[2 \frac{d}{b'} - \left(\frac{d}{b'}\right)^2\right]^{-\frac{1}{2}}. \tag{30}$$

It is important to note that the results of this section are valid only when the diffraction losses derived from the "equivalent" confocal geometry are small, that is, when the reflector dimension a' is somewhat larger than the spot size. In an exact solution for the nonconfocal case one should again start from the integral (4), and clearly the field distribution and losses so derived will depart from that obtained from the equivalent confocal case if the confocal field is not substantially all intercepted by the nonconfocal reflectors. Conversely, so long as the spot size is small compared to the reflector dimension a', one expects the field distribution and losses to be very nearly correctly given by the equivalent confocal solution.

The phase shift between the two reflecting nonconfocal surfaces may be obtained from (20). The condition of resonance may then be shown to be

$$\frac{4d}{\lambda} = 2q + (1 + m + n)\left(1 - \frac{4}{\pi} \tan^{-1} \frac{b - d}{b + d}\right). \tag{31}$$

In the nonconfocal case $4d/\lambda$ is no longer necessarily an integer at resonance. It is more important, though, that the modes are no longer degenerate in $m + n$. The spectral range or mode separation for the nonconfocal resonator is given by

$$\Delta\left(\frac{1}{\lambda}\right) = \frac{1}{4d}\left[2\Delta q + \left(1 - \frac{4}{\pi}\tan^{-1}\frac{b-d}{b+d}\right)\Delta(m+n)\right]. \quad (32)$$

Note that in the confocal case the set of modes $mnq = 00q$, $01q$ are maximally split in frequency, whereas if the parameter in parenthesis equals $\frac{1}{2}$ (when $d/b = 0.414$) then the $mnq = 00q$, $01q$, $11q$, $12q$ modes are maximally split in frequency.

When $b \approx d$, (31) becomes

$$\frac{4d}{\lambda} \approx 2q + (1 + m + n)\left[1 - \frac{2}{\pi}\left(1 - \frac{d}{b}\right)\right], \quad (33)$$

where m and n are small integers and q a large integer. In the confocal case ($b = d$) note that equations (31) and (33) reduce to (14).

The theory of this section does not extend to the limit of plane-parallel reflectors, i.e., infinite radii of curvature. Let the spacing d remain fixed while b', and consequently [by (26)] the confocal radius b, approaches infinity. The spot size, as seen from (27), keeps increasing with b', and, as has been noted above, this results eventually in the breakdown of the whole idea of an equivalent confocal resonator. The relations for the nonconfocal resonator are valid as long as the reflector aperture radius a' is somewhat larger than the field spot size radius given by (27). That is, one must require

$$\frac{a'^2}{d\lambda} > \frac{1}{\pi}\left[2\frac{d}{b'} - \left(\frac{d}{b'}\right)^2\right]^{-\frac{1}{2}}. \quad (34)$$

VI. RESONANT MODES OF THE PLANE-PARALLEL RESONATOR

For comparison purposes, consider the resonances of a rectangular conducting box in the manner of Schawlow and Townes.[1] Let the dimensions be $2a \times 2a \times b$:

$$\left(\frac{2}{\lambda}\right)^2 = \left(\frac{q}{b}\right)^2 + \left(\frac{r}{2a}\right)^2 + \left(\frac{s}{2a}\right)^2, \quad (35)$$

where q, r, and s are integers. Modes where $q \gg r,s$ can be thought of physically as waves bouncing predominantly back and forth between the reflecting end plates of the rectangular box. The spectral range or mode separation is given by

$$\Delta\left(\frac{1}{\lambda}\right) = \frac{1}{2b}\left[\Delta q + \frac{1}{16}\left(\frac{b\lambda}{a^2}\right)(2r\Delta r + \Delta r^2 + 2s\Delta s + \Delta s^2)\right], \quad (36)$$

where $q \approx 2b/\lambda$ for $r,\ s = 1,2,3,\cdots$.

Removing the conducting side walls causes large diffraction losses for the $r = 0$ or $s = 0$ modes since they have a strong field at the edge of the reflectors. Large r or s modes represent waves traveling at a considerable angle to the normal between the reflectors and thus these modes have such large diffraction losses that they are eliminated as resolvable resonant modes. Modes with $r, s = 1,2,\cdots$, have small diffraction losses, and are approximations to the actual modes which can exist in the resonator without the conducting side walls. Fox and Li's[6] work shows that for $a^2/b\lambda$ greater than unity the mode separations of a plane-parallel Fabry-Perot are given approximately by (36), the approximation improving rapidly with increasing $a^2/b\lambda$.

The mode separation corresponding to Δr or $\Delta s = 1$ has, to the writers' knowledge, never been resolved at optical frequencies due to the large values of $a^2/b\lambda$ and low values of reflectance used. Calculations show, though, that for reflectance coefficients of about 0.99 and $a^2/b\lambda \approx 4$, such that diffraction losses are comparable with reflector losses, the resonances should be resolvable.

The mode separation due to $\Delta q = 1$ is easily resolvable and is given by $\Delta(1/\lambda) = 1/2b$. This is the spectral range as normally stated for the plane parallel Fabry-Perot interferometer. It corresponds to changing the number of half wavelengths between the reflecting surfaces by one.

The confocal resonator is resonant for integer values of $4b/\lambda$. The mode separation due to $\Delta q = 1$ is $\Delta(1/\lambda) = 1/2b$. The modes are degenerate in frequency in that for a given integer $4b/\lambda$ all TEM_{mnq} modes are resonant such that $m + n$ remains even or odd according to whether $4b/\lambda$ is odd or even. The modes of the plane-parallel Fabry-Perot are not degenerate, except for rsq and srq. A possible advantage of this degeneracy of the confocal modes will be discussed in the next section.

VII. CONFOCAL RESONATOR APPLIED TO OPTICAL MASERS

A type of solid state optical maser has recently been demonstrated by Maiman[14] and by Collins et al.[15] It consists of a fluorescent crystal material (ruby) a few centimeters in length and a few millimeters in diameter. The crystal material should be optically homogeneous. The ends of the crystal are optically flat and parallel. The ends are silver-coated for high reflectance. One of the reflecting surfaces must be slightly transparent, as the output of the optical maser is obtained through the reflecting surfaces. Thus far, silver has been used to provide the reflection, but for ultimate performance multiple-layer dielectrics[5] should be used to obtain low transmission loss as well as high reflectance. The pump power enters the fluorescent crystal from the side.

It is seen in (2) that, if diffraction losses are small compared to reflection losses, then the resonator Q is proportional to the spacing between the reflecting surfaces. Consider a confocal resonator and a plane-parallel resonator each of spacing b and of equal Q. The energy distribution in the former is more concentrated on the axis and thus the confocal resonator has a smaller effective mode volume. The volume of maser material required will thus be less for the confocal than for the plane parallel resonator. For maser oscillation the required excess density of excited states depends only on the cavity Q and in no other way upon the resonator shape.[1] The pump power is proportional to the volume of maser material times the density of excited states divided by the natural lifetime of the excited state. Thus, assuming equal Q, the confocal resonator with its smaller volume of material requires less pump power than the plane parallel resonator by the ratio of their cross-sectional areas. Snitzer[16] recently pointed out this relation between mode volume and pump power with regard to the use of optical fibers in maser applications.

The minimum volume of maser material is limited by diffraction losses. If diffraction losses are to be considerably less than 1 per cent for the lowest-order mode so as to be small compared to achievable reflection losses, then $a^2/b\lambda \gtrsim 1$. The minimum volume of maser material is then

$$V_m = \pi a^2 b \approx \pi b^2 \lambda. \tag{37}$$

If $b = 4$ cm and $\lambda = 10^{-4}$ cm, then the rod of maser material should be approximately 0.4 mm in diameter. A rod of larger diameter would waste pump power in that the field of the confocal resonator would be very weak outside this minimum diameter of material.

The analysis of the confocal resonator assumes a uniform dielectric material between the spherical reflectors. For reasons of minimizing the pump power, it is necessary to use a small diameter of maser material. Therefore, to prevent internal reflection of energy from the sides of the maser material, it may be advisable to grind rough the sides of the rod of maser material or to immerse it in a surrounding medium of equal dielectric constant. If this is not done, the energy assumed lost due to diffraction effects would not escape and the electric field pattern will not be as computed herein. A more important effect of internal reflection from the side walls would be to increase the Q of the transverse modes which would increase the spontaneous and stimulated emission power to these undesired modes, and thus increase the over-all pump power required.

The natural linewidth of the material used in an optical maser will, for reflector spacing d of a few centimeters, be large compared to the mode separation determined by integer changes in r,s for a plane-parallel resonator. Hopefully, the natural linewidth of the maser material will be less than the mode separation corresponding to integer changes in q. Thus, there is the possibility that a plane-parallel resonator optical maser may frequency wander between low-order r,s modes.

If the diffraction losses are comparable to or exceed the reflection losses for the lowest mode then, as can be seen from Fig. 3, the ratio of the Q's of the lowest two modes of the confocal resonator exceeds considerably the ratio of the Q's of the lowest two modes of the plane-parallel resonator. By the lowest order mode is meant $m = n = 0$, and $r = s = 1$, respectively, for the confocal and plane-parallel resonator. Therefore, maser oscillation is more likely to take place in *only* the lowest-order mode of the confocal than of the plane-parallel resonator. This greater loss discrimination between modes may be one of the significant advantages of the confocal resonator.

In the confocal resonator optical maser, if the maser oscillation wanders between modes the output beam pattern will change, just as in the plane-parallel resonator, but the frequency will remain fixed due to the mode degeneracy. Thus, the observed linewidth of the maser output may be narrower for the confocal resonator.

The required accuracy on the confocal condition to achieve degeneracy may be estimated from (33) and (36). It can be shown that if

$$|d/b - 1| \approx 0.03 \quad \text{and} \quad (a^2/b\lambda) \approx 10,$$

the mode splitting of the near-confocal resonator equals the mode separation of the plane parallel resonator. To achieve a significantly smaller mode separation in the near-confocal resonator than the plane parallel resonator would require proportionately greater accuracy in the radius of curvature and spacing of the curved surfaces.

The plane Fabry-Perot requires accurately parallel reflecting surfaces. The confocal resonator requires only that the axis of the confocal resonator approximately coincide with the axis of the rod of maser material. The axis of the confocal resonator is the line passing through the two centers of curvature. The resonator axis must intersect the two reflecting surfaces near their center. Define the effective aperture radius as the distance from the point of intersection of the axis of the confocal resonator with the reflector surface to the nearest edge of the aperture. The diffraction losses will be approximately determined by this distance.

If the minimum diameter of maser material is used, then the axis of the confocal resonator must coincide with the material axis. Increasing the diameter of the maser material wastes pump power but relaxes the tolerance on the resonator axis.

It is well to note that a single spherical reflecting surface and a plane reflecting surface spaced by approximately half the radius of curvature will have similar properties to the confocal resonator and may be advantageous if it is desired to bring the output through a plane surface.

VIII. CONCLUSIONS

A confocal multimode resonator formed by two spherical reflectors spaced by their common radii of curvature has been considered. The mode patterns and diffraction losses have been obtained. The confocal spacing of the reflectors is found to be optimum in the sense of minimum diffraction losses and minimum mode volume.

The diffraction losses are found to be orders of magnitude smaller than those of the plane-parallel Fabry-Perot, as obtained by Fox and Li.[6] It is more important, though, that a greater diffraction loss discrimination between modes occurs, and thus oscillation in other than the lowest-order mode is less likely for the confocal resonator, assuming that diffraction losses are comparable to reflection losses.

The modes of the confocal resonator are degenerate, in that one-half of all the possible field pattern variations over the aperture are resonant at any one time. This degeneracy is split if the resonator is nonconfocal. The splitting is comparable with that of the plane-parallel resonator (with $a^2/b\lambda \approx 10$) if the spacing of the reflectors is about 3 per cent different from the common radius. The mode volume and diffraction losses are insensitive to the confocal condition.

The required volume of maser material is smaller for the confocal resonator than for the plane-parallel resonator, and thus the required pump power is less. The confocal resonator is relatively easy to adjust in that no strict parallelism is required between the reflectors. The only requirement is that the axis of the confocal resonator intersect each reflector sufficiently far from its edge so that the diffraction losses are not excessive.

The example of a confocal resonator mentioned here was taken at infrared-optical wavelengths; however, such resonators may be useful down to the millimeter wave range by virtue of their low loss. In this connection, recent work of Culshaw[17] on the plane-parallel Fabry-Perot at millimeter wavelengths is of importance.

The writers have been informed that Goubau and Schwering[18] have recently investigated diffraction losses of parabolic reflectors and that their results agree with the work presented here.

IX. ACKNOWLEDGMENT

Fruitful discussions with A. G. Fox, W. D. Lewis, T. Li, D. Marcuse, S. P. Morgan and G. W. Series are sincerely appreciated. Mrs. F. J. MacWilliams performed the computations.

REFERENCES

1. Schawlow, A. L. and Townes, C. H., Phys. Rev., **112,** 1958, p. 1940.
2. Prokhorov, A. M., J.E.T.P., **34,** 1958, p. 1658.
3. Dicke, R. H., U.S. Patent 2,851,652, September 9, 1958.
4. Meissner, K. W., J. Opt. Soc. Am., **31,** 1941, p. 405; **32,** 1942, p. 185.
5. Heavens, O. S., *Optical Properties of Thin Films*, Butterworths, London, 1955.
6. Fox, A. G. and Li, T., this issue, p. 453; Proc. I.R.E., **48,** 1960, p. 1904.
7. Lewis, W. D., private communication.
8. Connes, P., Revue d'Optique, **35,** 1956, p. 37; J. Phys. Radium, **19,** 1958, p. 262.
9. Jenkins, F. A. and White, H. E., *Fundamentals of Optics*, 3rd ed., McGraw-Hill, New York, 1957.
10. Silver, S., *Microwave Antenna Theory and Design*, M.I.T. Radiation Laboratory Series, Vol. 12, McGraw-Hill, New York, 1949.
11. Slepian, D. and Pollak, H. O., B.S.T.J., **40,** 1961, p. 43.
12. Flammer, C., *Spheroidal Wave Functions*, Stanford Univ. Press, Palo Alto, Calif., 1957.
13. Slepian, D. and Pollak, H. O., private communication.
14. Maiman, T. H., Nature, **187,** 1960, p. 493.
15. Collins, R. J., Nelson, D. F., Schawlow, A. L., Bond, W., Garrett, C. G. B. and Kaiser, W., Phys. Rev. Letters, **5,** 1960, p. 303.
16. Snitzer, E., J. Appl. Phys., **32,** 1961, p. 36.
17. Culshaw, W., I.R.E. Trans., **MTT-7**, 1959, p. 221; **MTT-8**, 1960, p. 182.
18. Goubau, G. and Schwering, F., U.R.S.I.-I.R.E. Spring Meeting, Washington, May 1960.

Resonant Modes in a Maser Interferometer

By A. G. FOX and TINGYE LI

(Manuscript received October 20, 1960)

A theoretical investigation has been undertaken to study diffraction of electromagnetic waves in Fabry-Perot interferometers when they are used as resonators in optical masers. An electronic digital computer was programmed to compute the electromagnetic field across the mirrors of the interferometer where an initially launched wave is reflected back and forth between the mirrors.

It was found that after many reflections a state is reached in which the relative field distribution does not vary from transit to transit and the amplitude of the field decays at an exponential rate. This steady-state field distribution is regarded as a normal mode of the interferometer. Many such normal modes are possible depending upon the initial wave distribution. The lowest-order mode, which has the lowest diffraction loss, has a high intensity at the middle of the mirror and rather low intensities at the edges. Therefore, the diffraction loss is much lower than would be predicted for a uniform plane wave. Curves for field distribution and diffraction loss are given for different mirror geometries and different modes.

Since each mode has a characteristic loss and phase shift per transit, a uniform plane wave which can be resolved into many modes cannot, properly speaking, be resonated in an interferometer. In the usual optical interferometers, the resolution is too poor to resolve the individual mode resonances and the uniform plane wave distribution may be maintained approximately. However, in an oscillating maser, the lowest-order mode should dominate if the mirror spacing is correct for resonance.

A confocal spherical system has also been investigated and the losses are shown to be orders of magnitude less than for plane mirrors.

I. INTRODUCTION

Schawlow and Townes[1] have proposed infrared and optical masers using Fabry-Perot interferometers as resonators. Very recently, Maiman[2] and Collins et al.[3] have demonstrated experimentally the feasibility of stimulated optical radiation in ruby. In these experiments two parallel faces of the ruby sample were polished and silvered so as to form an interferometer. The radiation due to stimulated emission resonates in the interferometer and emerges from a partially silvered face as a coherent beam of light.

In a maser using an interferometer for a resonator, a wave leaving one mirror and traveling toward the other will be amplified as it travels through the active medium. At the same time it will lose some power due to scattering by inhomogeneities in the medium. When the wave arrives at the second mirror some power will be lost in reflection due to the finite conductivity of the mirror and some power will be lost by radiation around the edges of the mirror. For oscillation to occur, the total loss in power due to density scattering, diffractive spillover and reflection loss must be less than the power gained by travel through the active medium. Thus diffraction loss is expected to be an important factor, both in determining the start-oscillation condition, and in determining the distribution of energy in the interferometer during oscillation.

While it is common practice to regard a Fabry-Perot interferometer as being simultaneously resonant for uniform plane waves traveling parallel to the axis and at certain discrete angles from the axis, this picture is not adequate for the computation of diffraction loss in a maser. It is true that, when the interferometer is operated as a passive instrument with uniform plane waves continuously supplied from an external source, the internal fields may be essentially those of uniform plane waves. In an oscillating maser where power is supplied only from within the interferometer, the recurring loss of power from the edges of a wave due to diffraction causes a marked departure from uniform amplitude and phase across the mirror.

The purpose of our study is to investigate the effects of diffraction on the electromagnetic field in a Fabry-Perot interferometer in free space. The conclusions can be applied equally well to gaseous or solid state masers provided the interferometer is immersed in the active medium, i.e., there are no side-wall discontinuities.

II. FORMULATION OF THE PROBLEM

2.1 *General Formulation*

Our approach is to consider a propagating wave which is reflected back and forth by two parallel plane mirrors, as shown in Fig. 1(a). [This is equivalent to the case of a transmission medium comprising a series of collinear identical apertures cut into parallel and equally spaced black (perfectly absorbing) partitions of infinite extent, as in Fig. 1(b).] We assume at first an arbitrary initial field distribution at the first mirror and proceed to compute the field produced at the second mirror as a result of the first transit. The newly calculated field distribution is then used to compute the field produced at the first mirror as a result of the second transit. This computation is repeated over and over again for subsequent successive transits. The questions we have in mind are: (a) whether, after many transits, the relative field distribution approaches a steady state; (b) whether, if a steady-state distribution results, there are any other steady-state solutions; and (c) what the losses associated with these solutions would be. While it is by no means obvious that steady-state solutions (corresponding to normal modes) exist for a system which has no side-wall boundaries, it will be shown that such solutions do indeed exist.*

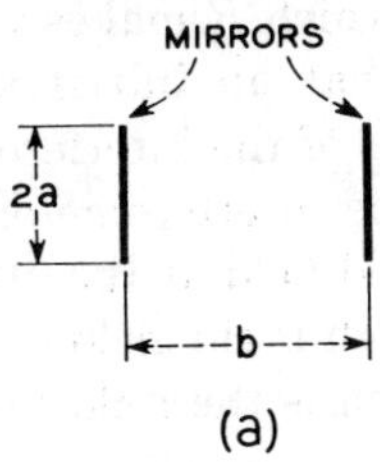

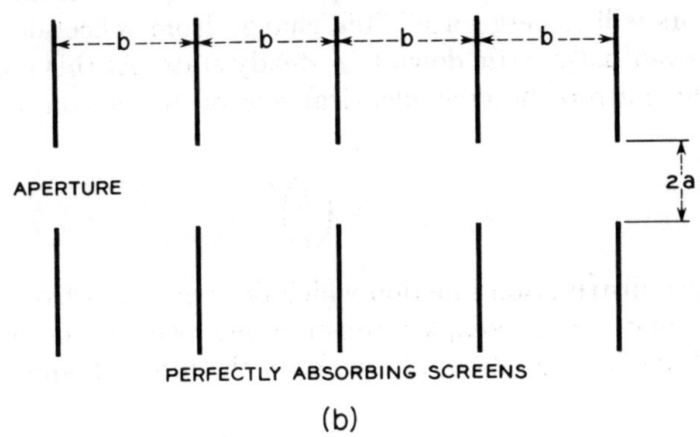

Fig. 1 — The Fabry-Perot interferometer and the transmission medium analog.

* Schawlow and Townes[1] suggested the possibility that resonant modes for a parallel plate interferometer might be similar in form to those for a totally enclosed cavity.

We shall use the scalar formulation of Huygens' principle to compute the electromagnetic field at one of the mirrors in terms of an integral of the field at the other. This is permissible if the dimensions of the mirror are large in terms of wavelength and if the field is very nearly transverse electromagnetic and is uniformly polarized in one direction. Later, we shall show that these assumptions are consistent with the results of our solutions and therefore are justifiable. We shall also show that other polarization configurations can be constructed from the solutions of the scalar problem by linear superposition.

The Fresnel field u_p due to an illuminated aperture A is given by the surface integral[4]

$$u_p = \frac{jk}{4\pi} \int_A u_a \frac{e^{-jkR}}{R} (1 + \cos \theta) \, dS, \tag{1}$$

where u_a is the aperture field, k is the propagation constant of the medium, R is the distance from a point on the aperture to the point of observation and θ is the angle which R makes with the unit normal to the aperture. We now assume that an initial wave of distribution u_p is launched at one of the mirrors of the interferometer and is allowed to be reflected back and forth in the interferometer. After q transits the field at a mirror due to the reflected field at the other is simply given by (1) with u_p replaced by u_{q+1}, which is the field across the mirror under consideration and u_a by u_q, which is the reflected field across the opposite mirror giving rise to u_{q+1}.

It is conceivable that after many transits the distribution of field at the mirrors will undergo negligible change from reflection to reflection and will eventually settle down to a steady state. At this point the fields across the mirrors become identical except for a complex constant; that is,

$$u_q = \left(\frac{1}{\gamma}\right)^q v, \tag{2}$$

where v is a distribution function which does not vary from reflection to reflection and γ is a complex constant independent of position coordinates. Substituting (2) in (1) we have the integral equation

$$v = \gamma \int_A K v \, dS \tag{3}$$

in which the kernel of the integral equation, K, is equal to $(jk/4\pi R) \cdot (1 + \cos \theta) e^{-jkR}$. The distribution function v, which satisfies (3), can be regarded as a normal mode of the interferometer defined at the mirror surface, and the logarithm of γ, which specifies the attenuation and the phase shift the wave suffers during each transit, can be regarded as the propagation constant associated with the normal mode.

The integral equation (3) can be solved numerically by the method of successive approximations (Ref. 5, p. 421). It is interesting to note that this iterative method of solution is analogous to the physical process of launching an initial distribution of wavefront in the interferometer and letting it bounce back and forth between the mirrors as described in the foregoing paragraphs.

We have studied and obtained numerical solutions for several geometric configurations of the interferometer. These are (a) rectangular plane mirrors, (b) circular plane mirrors and (c) confocal spherical or paraboloidal mirrors.

2.2 *Rectangular Plane Mirrors*

When the mirror separation is very much larger than the mirror dimensions the problem of the rectangular mirrors reduces to a two-dimensional problem of infinite strip mirrors. This is shown in Appendix A. The integral equation for the problem of infinite strip mirrors, when $a^2/b\lambda$ is much less than $(b/a)^2$, is

$$v(x_2) = \gamma \int_{-a}^{a} K(x_2, x_1)v(x_1)\, dx_1 \tag{4}$$

with

$$K(x_2, x_1) = \frac{e^{j(\pi/4)}}{\sqrt{\lambda b}}\, e^{-jk(x_1-x_2)^2/2b} \tag{4a}$$

The various symbols are defined in Fig. 2 and Appendix A.

Equation (4) is a homogeneous linear integral equation of the second kind. Since the kernel is continuous and symmetric $[K(x_2, x_1) = K(x_1, x_2)]$, its eigenfunctions v_n corresponding to distinct eigenvalues γ_n are orthogonal in the interval $(-a,a)$; that is (Ref. 5, p. 413),

$$\int_{-a}^{a} v_m(x)v_n(x)\, dx = 0, \qquad m \neq n. \tag{5}$$

It should be noted that the eigenfunctions are in general complex and are defined over the surface of the mirrors only. They are not orthogonal in the power (Hermitian) sense as commonly encountered in lossless systems. Here, the system is basically a lossy one and the orthogonality relation is

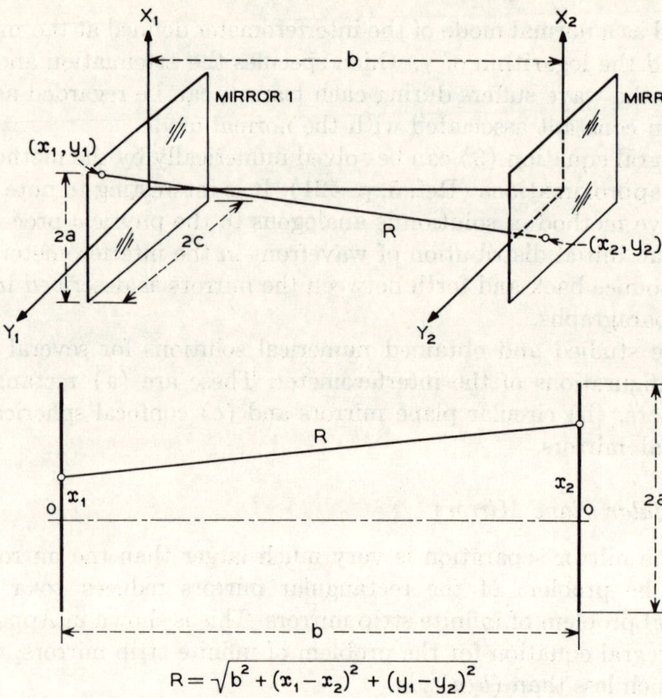

Fig. 2 — Geometry of rectangular plane mirrors.

one which is generally applicable to lossy systems, such as lossy-wall waveguides.

The eigenfunctions are distribution functions of the field over mirror surfaces and represent the various normal modes of the system. The normal modes for rectangular plane mirrors are obtained by taking the products of the normal modes for infinite strip mirrors in x and y directions; that is,

$$v_{mn}(x,y) = v_{x,m}(x)v_{y,n}(y). \tag{6}$$

We designate this as the TEM_{mn} mode for the rectangular plane-mirror interferometer. In view of (5) we see that the normal mode distribution functions v_{mn} are orthogonal over the surface of the rectangular mirror.

The logarithms of the eigenvalues represent propagation constants associated with the normal modes. The propagation constant for the TEM_{mn} mode of rectangular plane mirrors is given by

$$\log \gamma_{mn} = \log \gamma_{x,m} + \log \gamma_{y,n}. \tag{7}$$

The real part of the propagation constant specifies the loss per transit and the imaginary part the phase shift per transit, in addition to the geometrical phase shift, for the normal modes.

2.3 *Circular Plane Mirrors*

It is shown in Appendix B that the solutions to the integral equation for circular plane mirrors (Fig. 3) when $a^2/b\lambda$ is much less than $(b/a)^2$, are given by

$$v(r,\varphi) = R_n(r)e^{-jn\varphi} \qquad (n = \text{integer}), \tag{8}$$

where $R_n(r)$ satisfies the reduced integral equation

$$R_n(r_2)\sqrt{r_2} = \gamma_n \int_0^a K_n(r_2, r_1)R_n(r_1)\sqrt{r_1}\, dr_1, \tag{9}$$

with

$$K_n(r_2, r_1) = j^{n+1}\frac{k}{b} J_n\left(k\frac{r_1 r_2}{b}\right)\sqrt{r_1 r_2}\, e^{-jk(r_1^2+r_2^2)/2b}, \tag{9a}$$

where J_n is a Bessel function of the first kind and nth order. As in the

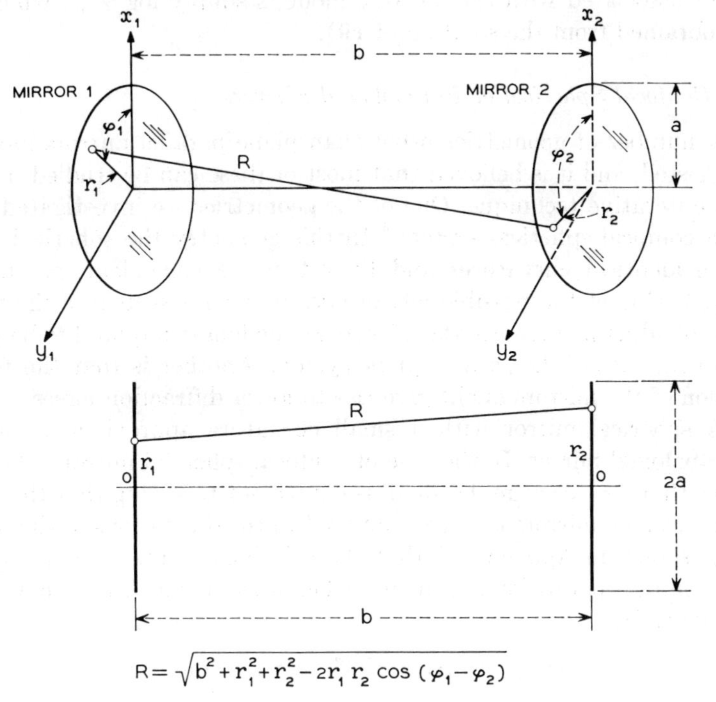

Fig. 3 — Geometry of circular plane mirrors.

problem of infinite strip mirrors, (9) is a homogeneous linear integral equation of the second kind with a continuous and symmetric kernel. Its eigenfunctions corresponding to distinct eigenvalues are orthogonal in the interval $(0,a)$; that is,

$$\int_0^a R_{nl}(r)R_{nm}(r)r\,dr = 0, \qquad (l \neq m). \tag{10}$$

Therefore, we see that the distribution functions $v_{nm}(r,\varphi) = R_{nm}(r)e^{-jn\varphi}$ corresponding to distinct eigenvalues γ_{nm} are orthogonal over the surface of the mirror; that is,

$$\int_0^{2\pi}\int_0^a v_{nm}(r,\varphi)v_{kl}(r,\varphi)r\,dr\,d\varphi = 0 \qquad (\text{either } n \neq k \text{ or } m \neq l). \tag{11}$$

The set of eigenfunctions R_{nm} describes the radial variations of field intensity on the circular mirrors, and the angular variations are sinusoidal in form. We designate a normal mode of the circular plane mirrors as the TEM_{nm} mode, with n denoting the order of angular variation and m denoting the order of radial variation. The propagation constant associated with the TEM_{nm} mode is simply $\log \gamma_{nm}$, which must be obtained from the solution of (9).

2.4 *Confocal Spherical or Paraboloidal Mirrors*

A number of geometries other than plane parallel mirrors have been suggested, and it is believed that most of these can be studied using the same iterative technique. One of the geometries we investigated is that of a confocal spherical system.[6] In this geometry the spherical mirrors have identical curvatures and their foci are coincident, as shown in Fig. 4. One of the possible advantages of such a system is the relative ease of adjustment, since the mirrors are no longer required to be parallel as in the case of the parallel plane system. Another is that the focusing action of the mirrors might give rise to lower diffraction losses.

A spherical mirror with a small curvature approximates closely a paraboloidal mirror. In the case of confocal spherical mirrors, the conditions that its curvature be small is equivalent to saying that the separation between mirrors is large compared to the dimensions of the mirrors. It is shown in Appendix C that the solutions to the integral equation for confocal paraboloidal mirrors, when $a^2/b\lambda$ is much less than $(b/a)^2$, are given by

$$v(r,\varphi) = S_n(r)e^{-jn\varphi} \qquad (n = \text{integer}), \tag{12}$$

where $S_n(r)$ satisfies the reduced integral equation

$$S_n(r_2)\sqrt{r_2} = \gamma_n \int_0^a K_n(r_2, r_1) S_n(r_1) \sqrt{r_1}\, dr_1, \tag{13}$$

with

$$K_n(r_2, r_1) = j^{n+1} \frac{k}{b} J_n\left(k \frac{r_1 r_2}{b}\right) \sqrt{r_1 r_2}. \tag{13a}$$

Again, we see that (13) is a homogeneous linear integral equation of the second kind with a continuous and symmetric kernel. Therefore, general remarks concerning the normal modes of circular plane mirrors given in the foregoing section are also applicable to confocal spherical or paraboloidal mirrors.

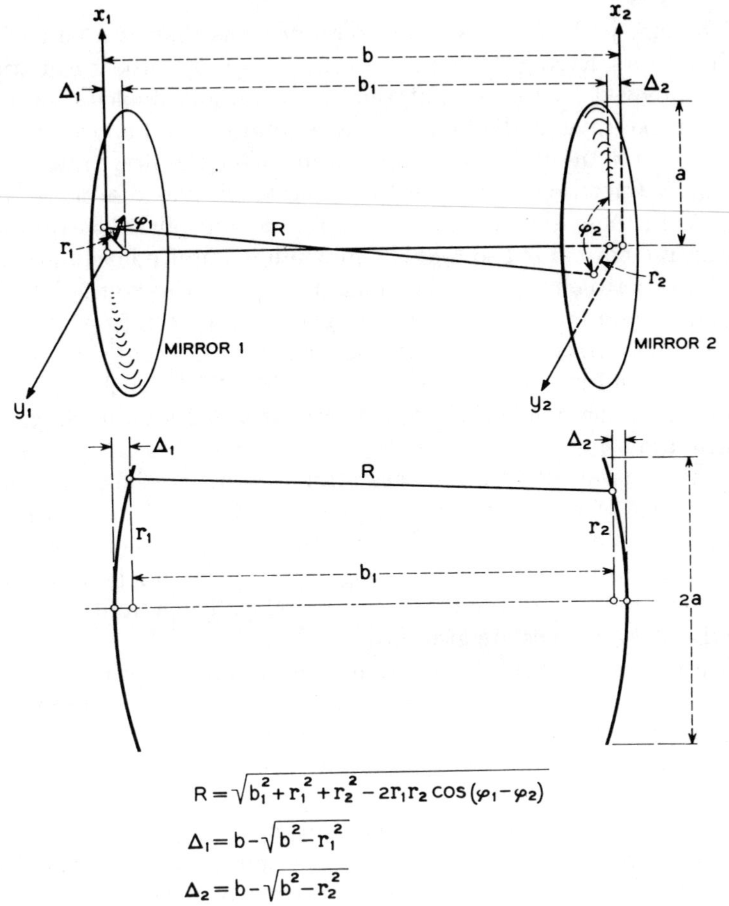

Fig. 4 — Geometry of confocal spherical mirrors.

III. COMPUTER SOLUTIONS

3.1 *General*

An IBM 704 computer was programmed to solve the integral equations for the various geometries of the interferometer by the method of successive approximations. As mentioned previously, this is analogous to the physical process of launching an initial distribution of wavefront in the interferometer and letting it bounce to and fro between the mirrors.

3.2 *Infinite Strip Mirrors*

The first problem put on the computer was that of a pair of infinite strip mirrors, having the dimensions $2a = 50\lambda$, $b = 100\lambda$. Equation (26) was employed for the computation, using an initial excitation of a uniform plane wave at the first mirror. A total of one hundred increments was used for the numerical integration. After the first transit the field intensity (electric or magnetic) had the amplitude and phase shown in Fig. 5. In these and subsequent amplitude and phase distributions the curves are normalized so that the maximum amplitude is unity, and the phase at that point is zero. The large ripples are due to the fact that the initial wave front contains 6.25 Fresnel zones as seen from the center of the second mirror. Therefore, in passing from the center to the edge of the second mirror there is a change of 3×6.25 Fresnel zones, and this agrees with the number of reversals in curvature seen in the amplitude distribution.

With subsequent transits, these ripples grow smaller, the amplitude at the edge of the mirror decreases, and the relative field distributions approach a steady state. By the time the wave had made three hundred bounces, the fluctuations occurring from bounce to bounce were less than 0.03 per cent of the final average value. The amplitude and phase for the 300th bounce are also shown in Fig. 5.

We regard this field distribution as an iterative normal mode of the interferometer. In other words, if this distribution is introduced as an initial wave at one mirror it will reproduce the same distribution at the other mirror. Indeed, this is what the computer is verifying when we compute the 301st bounce.

Once the solutions have reached a steady state, we can pick any point on the wavefront, say the center of the mirror, and examine how the absolute phase and amplitude change from bounce to bounce. In this way we determined that the power loss of this mode is 0.688 per cent per transit and the phase shift per transit has a lead of 1.59 degrees.

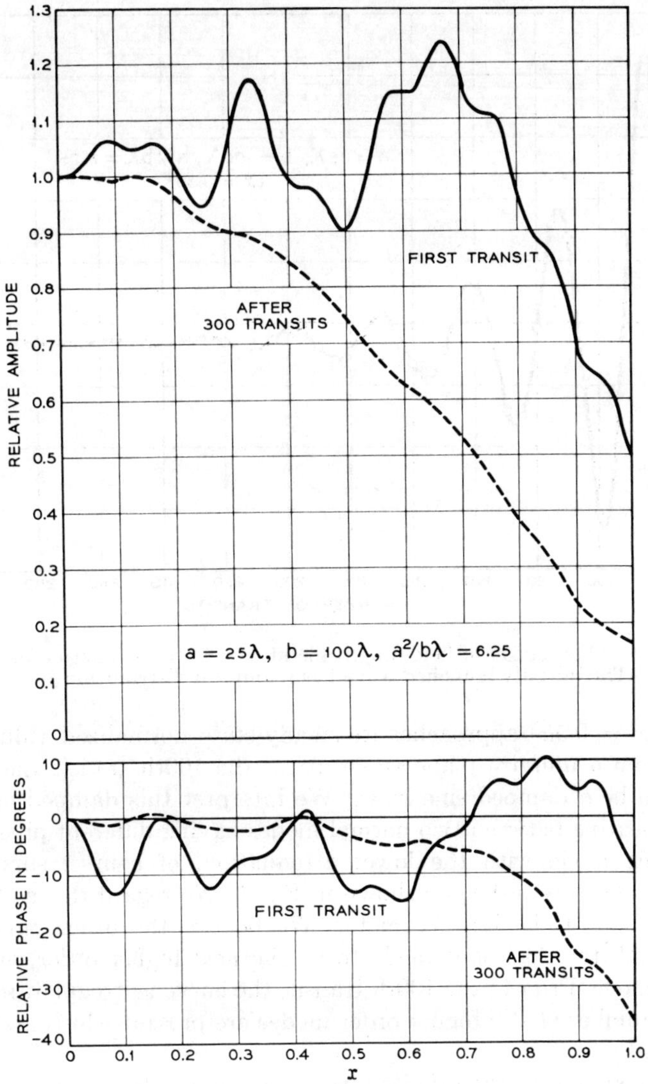

Fig. 5 — Relative amplitude and phase distributions of field intensity for infinite strip mirrors. (The initially launched wave has a uniform distribution.)

Since phase shift is measured relative to the free-space electrical length between the mirrors (360 b/λ degrees), this means that the mode has an effective phase velocity which is slightly greater than the speed of light, just as for a metal tube waveguide.

In Fig. 6 is shown how the field intensity at an arbitrary off-center

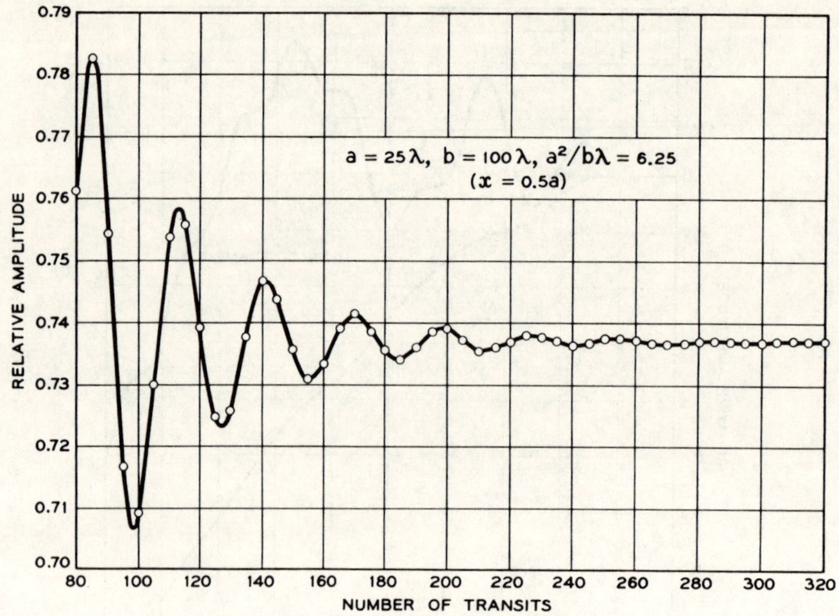

Fig. 6 — Fluctuation of field amplitude at $x = 0.5$ as a function of number of transits. (The initially launched wave has a uniform distribution.)

point ($x = 0.5a$) approaches its steady-state normalized value after a start from a uniform plane wave. After the 100th transit the plot appears to be a damped sine wave. We interpret this damped oscillation as the beating between two normal modes having different phase velocities. The mode with the lower attenuation, of course, survives the longest, and this is the one shown in Fig. 5. We regard this as the dominant mode of the interferometer. We believe the other mode which beats with the dominant mode to be the next-higher order, even-symmetric mode. Prior to the 100th transit, the curve is irregular, indicating that a number of still higher order modes are present which are damped out rapidly.

The next step in the infinite strip problem was to repeat solutions of the above type for other sets of dimensions. However, if $a^2/b\lambda$ is very small compared to $(b/a)^2$, the actual dimensions of the mirrors and their spacing are no longer important, the only parameter of importance being the Fresnel number $N = a^2/b\lambda$. This is approximately equal to the number of Fresnel zones seen in one mirror from the center of the other mirror, and as pointed out earlier, it determines the number of ripples in the field distributions. Amplitude and phase distributions for the dominant mode obtained by solving (27) are shown in Fig. 7 for different values of N. The larger the N, the weaker is the field intensity at the edge of the mirror, and the smaller is the power loss due to spill-over. The plot of power loss per transit as a function of N is approximately a straight line on log-log paper and is shown as the lowest line in Fig. 8. The phase shift per transit as a function of N is given by the lowest line in Fig. 9.

A uniform plane wave excitation can never give rise to a mode with odd symmetry. In order to investigate the possibility of modes of this

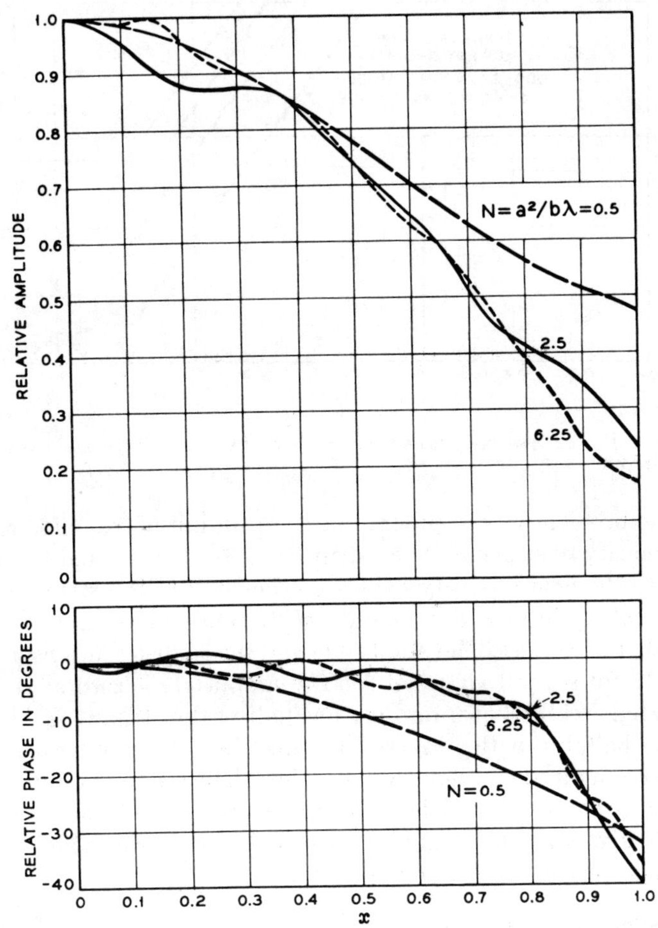

Fig. 7 — Relative amplitude and phase distributions of field intensity of the lowest order even-symmetric mode for infinite strip mirrors.

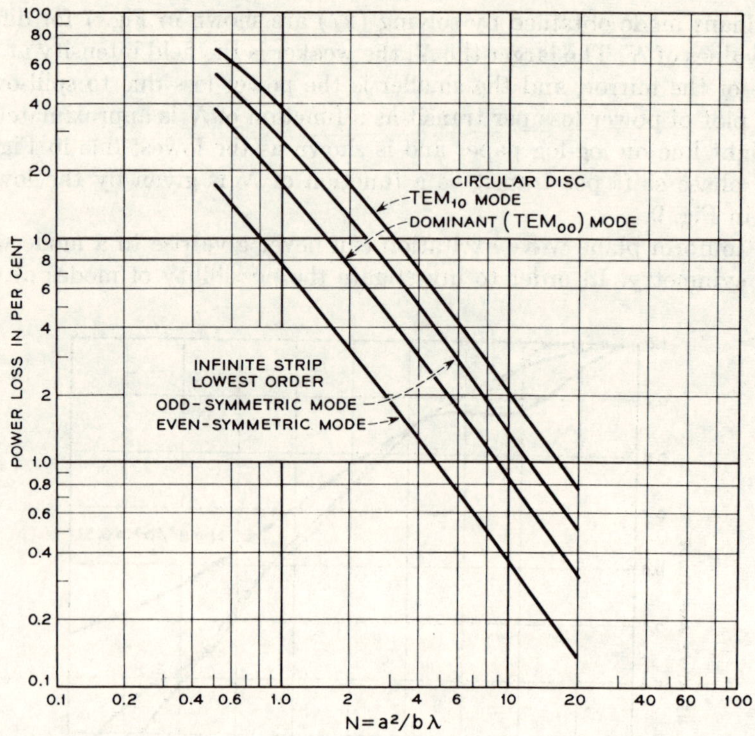

Fig. 8 — Power loss per transit vs. $N = a^2/b\lambda$ for infinite strip and circular plane mirrors.

type, the problem was re-programmed for an initial wave for which the field intensity over one-half the strip (0 to $+a$) was equal but opposite in sign to the field intensity over the other half of the strip (0 to $-a$). Steady-state solutions did indeed result, and odd-symmetric normal modes therefore exist. The amplitude and phase distributions are shown in Fig. 10 for several values of N. The amplitude is zero at the center, as expected. While shown for only one half of the strip, it is the same in the other half, but with a reversal in sign. Note that for the same values of N, the amplitude at the edge is higher than for the dominant mode. The spill-over loss should be higher and this is confirmed by the loss curve in Fig. 8 labeled "infinite strip odd-symmetric mode." The corresponding phase shift curve is shown in Fig. 9.

3.3 *Circular Plane Mirrors*

The feasibility of obtaining the normal mode solutions for the infinite strip mirrors having been established, programs were next set up to investigate the modes for plane circular mirrors. The first case considered was that for uniform plane wave excitation of the system. Once again, the polarization was assumed to be everywhere parallel to the same axis, and this results in a scalar wave solution having circular symmetry [(9) with $n = 0$]. That is, the amplitude and phase of the field intensity is the same for all points at the same radius from the center. The transverse field distributions for the lowest order mode of this type are shown in Fig. 11 for various values of N. The loss and phase shift are shown in Figs. 8 and 9 under the title "circular disc (dominant mode, TEM_{00})." One hundred increments along the radius were used for the numerical integrations involved.

Next we examined modes of the odd-symmetric type for circular plane mirrors. The equation we used was (9) with $n = 1$. Fig. 12 shows amplitude and phase distributions for the lowest order mode of the odd-symmetric type for circular plane mirrors. Again the loss and phase

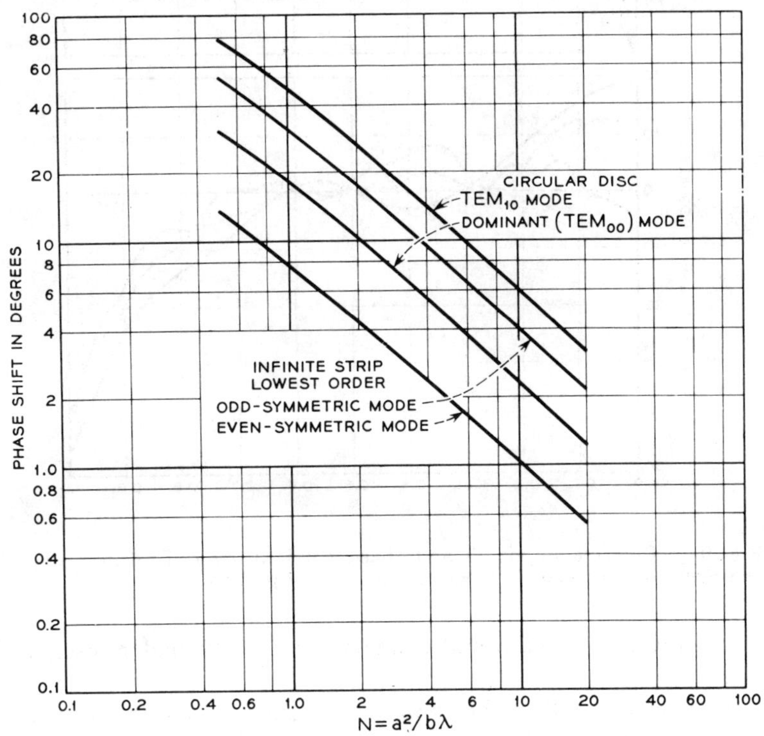

Fig. 9 — Phase shift per transit (leading relative to geometrical phase shift) vs. $N = a^2/b\lambda$ for infinite strip and circular plane mirrors.

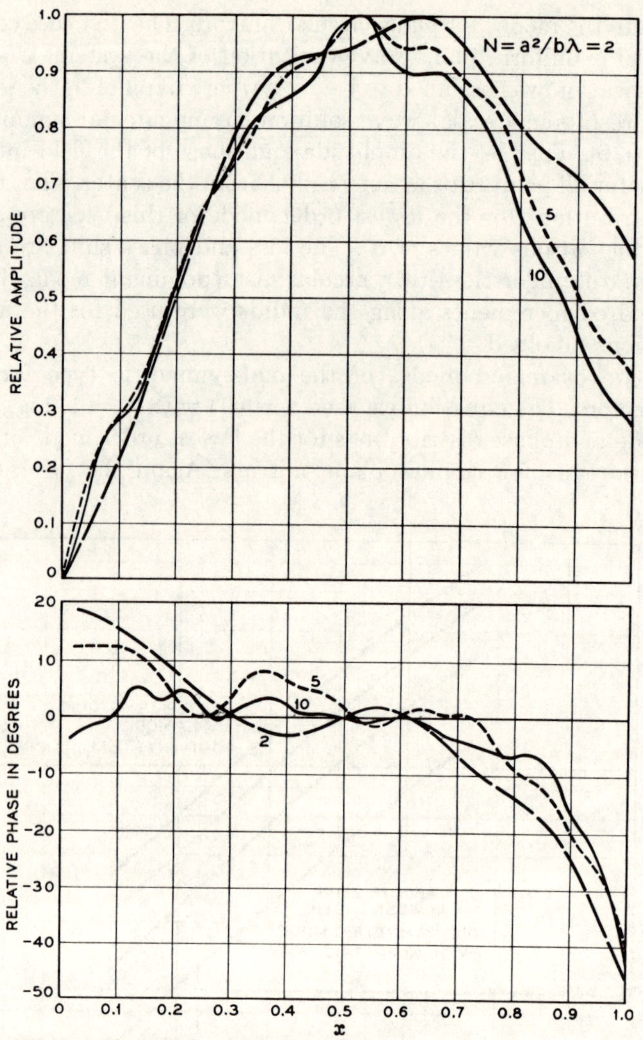

Fig. 10 — Relative amplitude and phase distributions of field intensity of the lowest-order odd-symmetric mode for infinite strip mirrors.

shift are given in Figs. 8 and 9 under the title "circular disc, TEM_{10} mode."

Normal modes with higher orders of angular variation ($n \geqq 2$) and radial variation ($m \geqq 1$) have greater losses and phase shifts than those of TEM_{00} and TEM_{10} modes. The mode with the least attenuation is therefore the lowest order, of TEM_{00} mode, which we designate as the dominant mode for circular plane mirrors.

3.4 *Confocal Spherical Mirrors*

Before (13) was programmed for solutions on the computer a more general method for solving the problem of the confocal spherical mirrors

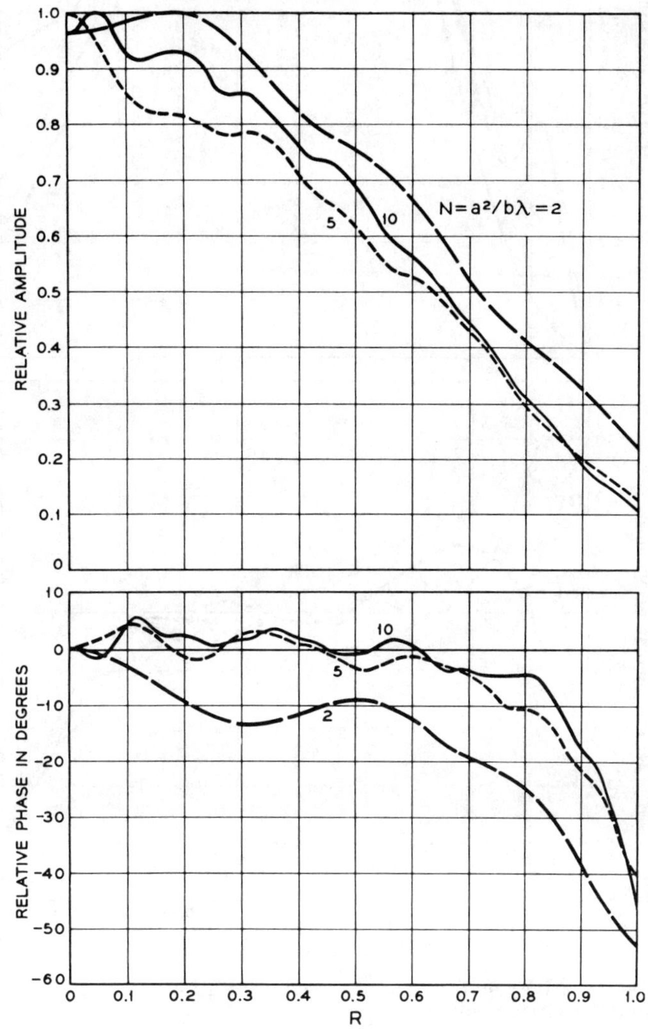

Fig. 11 — Relative amplitude and phase distributions of field intensity of the dominant (TEM_{00}) mode for circular plane mirrors.

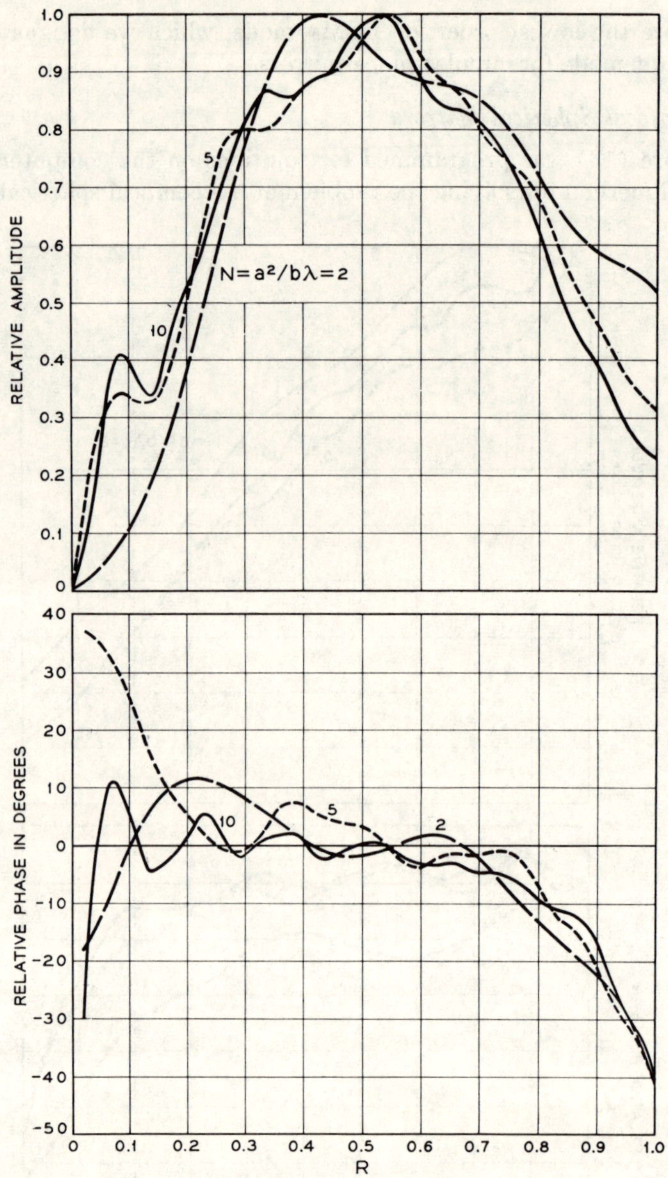

Fig. 12 — Relative amplitude and phase distributions of field intensity of the TEM_{10} mode for circular plane mirrors.

was tried — a procedure that can be used to solve problems involving mirrors with rather arbitrary but small curvatures. In this method the field at each mirror is calculated using the equation for circular plane mirrors and then a phase distribution corresponding to the curvature of the mirror is added to this field before it is used in the next iterative computation. The results from this general method of solution and from solving (13) are in perfect agreement.

The problem of confocal spherical mirrors has also been solved by Goubau[8] and Boyd and Gordon.[9] The results of their analyses are in good agreement with our computed results.

Amplitude distributions of the field intensity for TEM_{00} and TEM_{10} modes are shown in Figs. 13 and 14. The phase distributions are all uniform over the surface of the mirrors and therefore are not plotted. The loss and phase shift per transit are given in Figs. 15 and 16. We note some rather remarkable differences between these solutions and those obtained for circular plane mirrors. First, the field is much more tightly concentrated near the axis of the reflector and falls to a much lower value at the edge than is true for plane mirrors; also the amplitude distribution does not have ripples in it, but is smooth. Second,

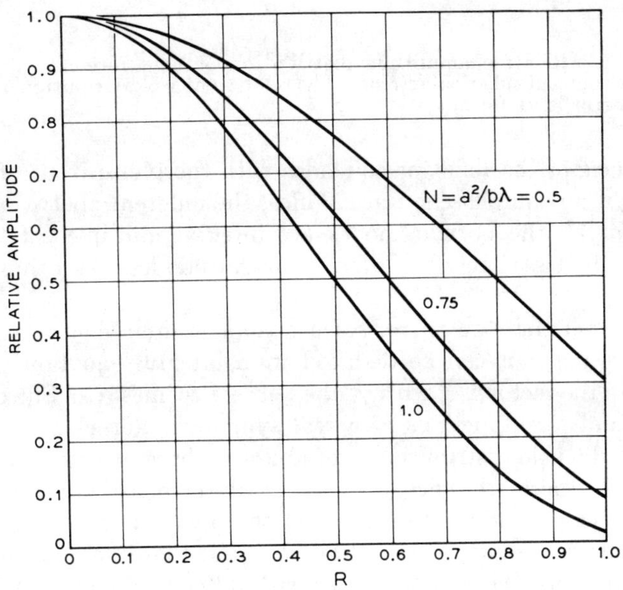

Fig. 13 — Relative amplitude distribution of field intensity of the dominant (TEM_{00}) mode for confocal spherical mirrors. The relative phase distribution on the surface of the mirror is uniform.

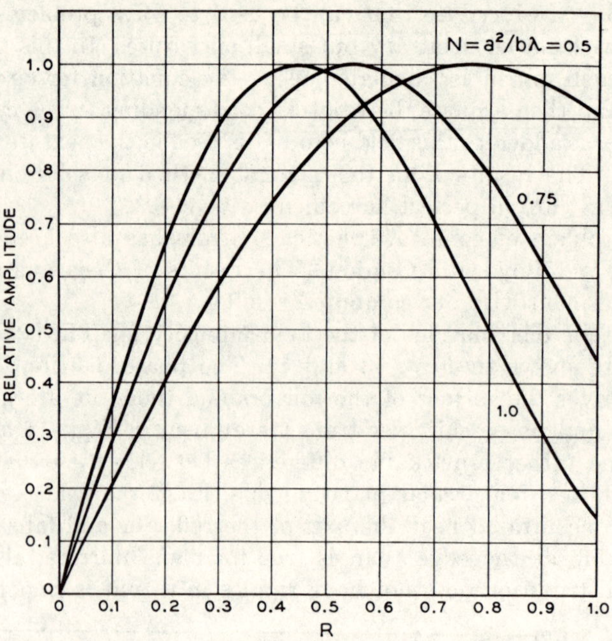

Fig. 14 — Relative amplitude distribution of field intensity of the TEM_{10} mode for confocal spherical mirrors. The relative phase distribution on the surface of the mirror is uniform.

the surface of the reflector coincides with the phase front of the wave, making it an equiphase surface. Third, the difference between the phase shifts for all the normal modes are integral multiples of 90 degrees. Fourth, the losses may be orders of magnitude less than those for plane mirrors.

The result that the mirror surface is an equiphase surface should not be surprising, but can be deduced from integral equation (13). If we associate the factor j^{n+1} with γ_n the kernel becomes real. Since the eigenvalues and eigenfunctions of a real symmetric kernel are all real,[7] we see that the field distribution is of uniform phase over the surface of the mirror. Furthermore, since $(j^{n+1}\gamma_n)$ is real, the phase shift for the normal modes belonging to a set of modes with a given angular variation must be an integral multiple of 180 degrees and the difference between the phase shifts for the normal modes with different angular variations but the same radial variation is an integral multiple of 90 degrees; that is, the phase shift is equal to $[180m + 90(n + 1)]$ degrees. Therefore, if the mirrors are adjusted for the resonance of a particular normal mode, half of the totality of all the modes are also resonant. However, the resonant mode with the lowest loss would persist longest in the resonator. Just as in the case of plane parallel mirrors, the mode with the lowest loss is the TEM_{00} mode.

IV. DISCUSSION OF RESULTS

The results of machine computation have shown that a two-mirror interferometer, whether of the plane or concave mirror type, can have

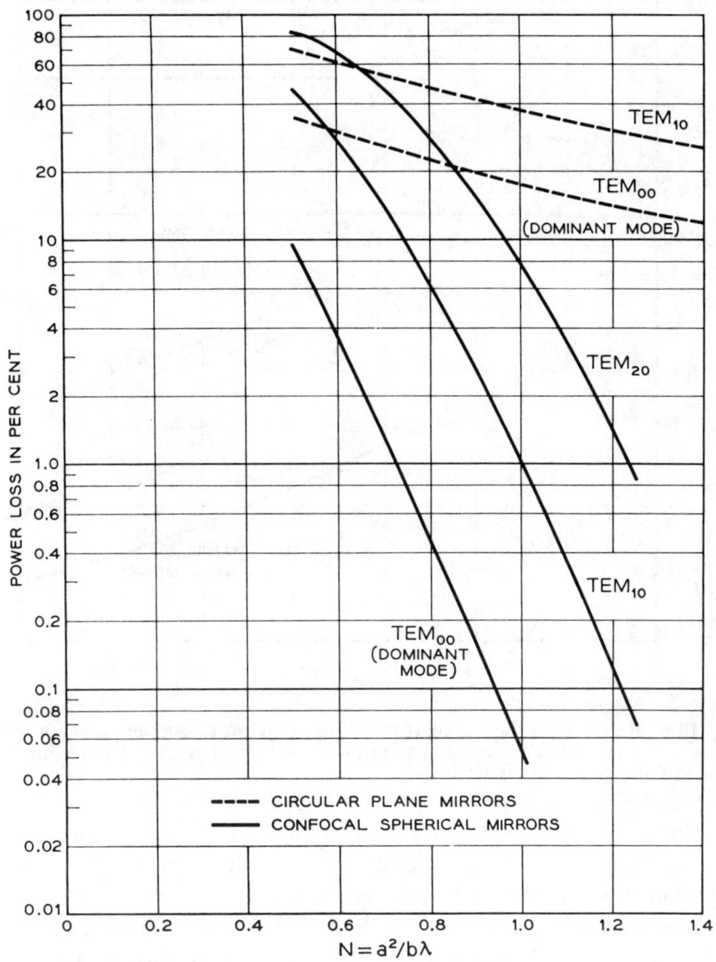

Fig. 15 — Power loss per transit vs. $N = a^2/b\lambda$ for confocal spherical mirrors. (Dashed curves for circular plane mirrors are shown for comparison).

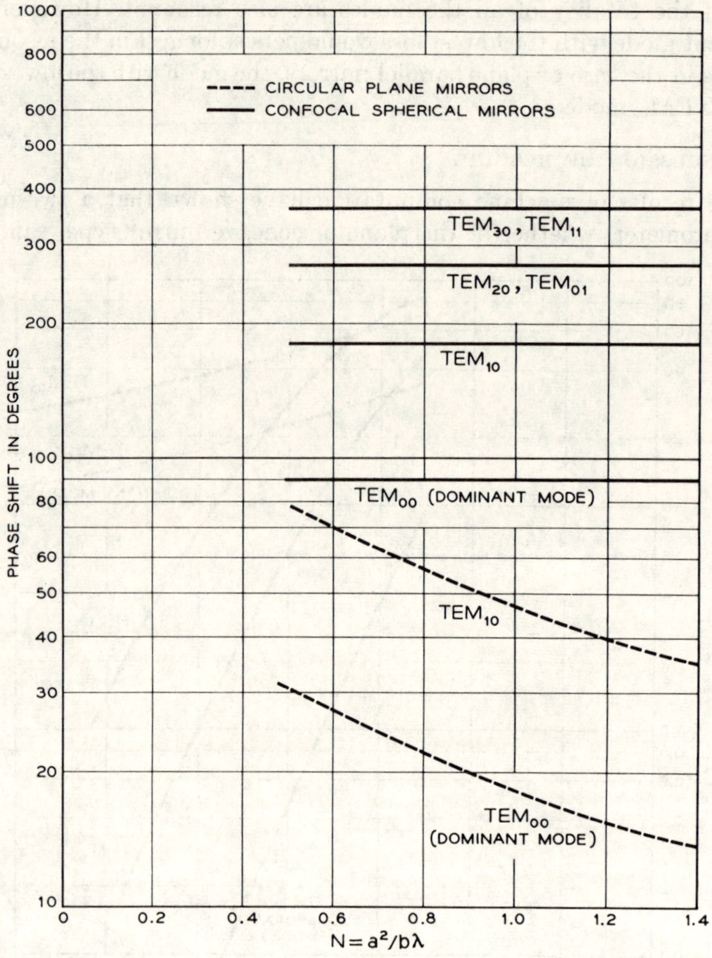

Fig. 16 — Phase shift per transit (leading relative to geometrical phase shift) vs. $N = a^2/b\lambda$ for confocal spherical mirrors. (Dashed curves for circular plane mirrors are shown for comparison.)

normal modes of propagation which are self-perpetuating or self-reproducing in the distance of one transit. We use the term mode of propagation rather than mode of resonance to emphasize the fact that these steady-state solutions are the result of multiple transits whether or not the plate separation happens to be adjusted for resonance. An analog of the plane mirror interferometer is a transmission medium consisting of a series of periodic collinear apertures, as was shown in Fig. 1. The same solutions apply, and here it is clear that the reproduction of a normal mode field at successive apertures does not depend on any critical relation between b and λ.

In Fig. 17 is shown the way in which a number of square-plate modes can be synthesized from the infinite-strip modes. Diagram A shows schematically the field distribution for the dominant square-plate mode obtained as the product of the field distributions of two even-symmetric strip modes crossed at right angles and with polarization as shown. Since the eigenvalue for the square plate is the product of the eigenvalues for the two strips, the phase shift per transit is the sum of the

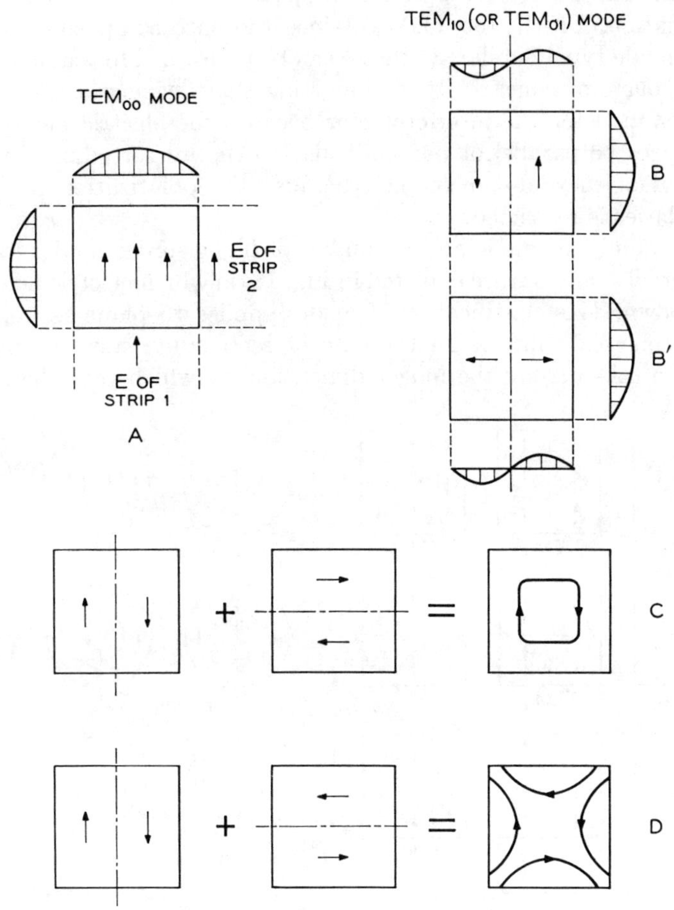

Fig. 17 — Synthesis of normal modes for square mirrors.

phase shifts for the two strips and, if the loss is small, the loss per transit is essentially the sum of the losses for the two strips. Diagram B represents an odd-symmetric square-plate mode formed by taking the product of an even- and an odd-symmetric strip mode; B′ is the same mode but with the polarization rotated 90°; C is a circular electric type of mode formed by *adding* two modes of the type B. This addition is permitted because the two components are degenerate. It follows that the circular electric mode C is degenerate with B and has the same loss and phase shift per transit. By taking the difference between the same two B modes as shown, the mode D is obtained, resembling the TE_{21} mode in circular waveguide. We give all the patterns B, B′, C and D the same designation, TEM_{10} (or TEM_{01}), since they are composites of the one basic mode type. Similar syntheses can be performed for circular mirrors, either plane or concave. It is interesting that degeneracies of this type are common for the interferometer because the electric vector E is at liberty to be parallel or perpendicular to the mirror edges. In a metal waveguide they are uncommon because the polarization of E at the boundaries is restricted.

The dominant mode and a number of higher-order modes for square and circular mirrors are depicted in Fig. 18, in which electric field vectors are shown. This classification of modes applies to plane as well as confocal spherical mirrors. In the case of rectangular mirrors, the x axis may be taken along the longer dimension, in which case the first sub-

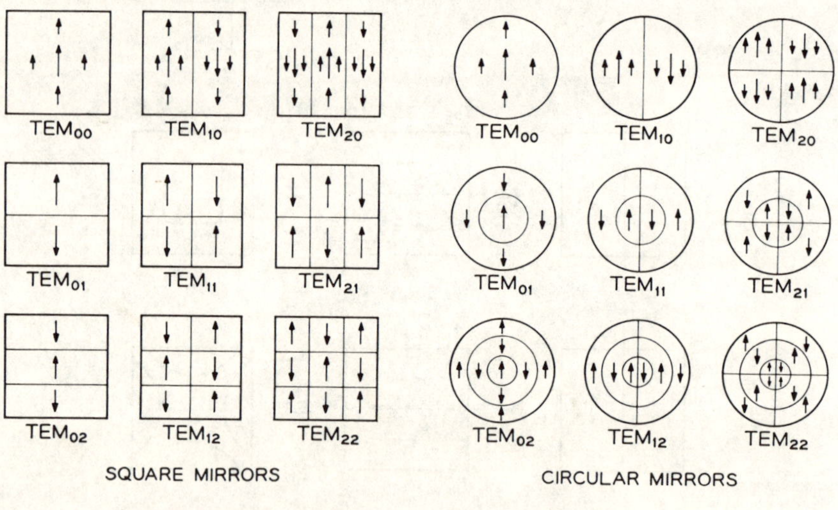

Fig. 18 — Field configuration of normal modes for square and circular mirrors.

script always denotes the number of field reversals along the longer dimension.

In formulating the problem we have assumed that the waves were almost transverse electromagnetic. The solutions for the flat mirror are consistent with this assumption. At the edges of the mirror there is a phase lag of approximately 45 degrees relative to the center, but this is only one-eighth of a wavelength out of many wavelengths for the mirror diameter. Thus the curvature of the wavefront away from the transverse plane is exceedingly small, and the assumption appears justified. For higher-order modes such as B′ of Fig. 17, it is clear that the field lines must have longitudinal components. This is illustrated by an edge view in Fig. 19. However, provided the width of a cell c is much greater than a half-wavelength, the longitudinal field intensity should be negligible compared to the transverse. Only for very high-order modes should this approximation begin to fail. Because the low-order modes of importance are essentially transverse electromagnetic, they are designated as TEM modes.

The plane mirror modes have a phase which is not constant over the mirror. This does not mean that it is impossible to space the mirrors for resonance of the entire field pattern. Actually, the phase delay for one transit is the same for every point on the wavefront. Therefore, if the plates are separated by the distance b plus an additional amount for the phase shift per transit of the mode desired, that mode should resonate in

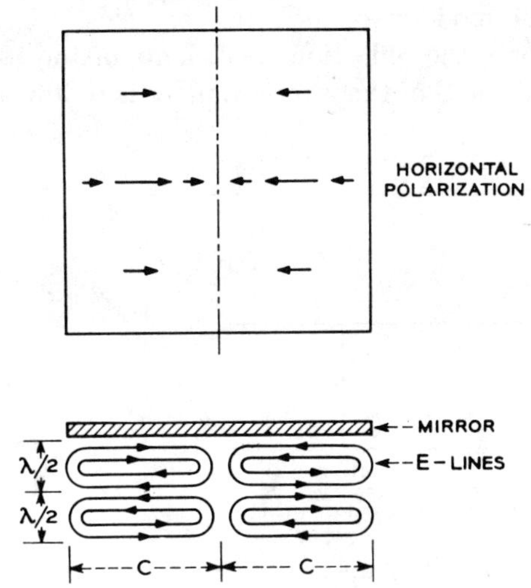

Fig. 19 — Field configuration of the TEM_{10} mode for square mirrors.

the interferometer. Other modes should not be resonant for this separation because they have different phase shifts per transit.

Since the field configurations of many of the normal modes of the interferometer are very similar to those of metal tube and parallel-plane waveguides, it is not surprising to find that simple waveguide theory can be used to predict certain characteristics of the interferometer modes. One of these characteristics is phase shift per transit. For instance, the field distributions of the normal modes for infinite strip mirrors are very similar to those of the TE modes of parallel-plane waveguide; also, by adding two orthogonally polarized TEM_{10} modes for circular plane mirrors, one obtains a field configuration which is very similar to that of the circular electric (TE_{01}) mode of circular waveguide (Fig. 20). Thus the amount of phase shift per transit computed for these modes of the interferometer agrees well with the phase shifts obtained for TE modes of parallel-plane waveguide and TE_{01} mode of circular waveguide. This is illustrated in Fig. 21. We see that agreement becomes better for larger values of N. This is because the similarity between field configurations becomes closer for larger values of N.

If we regard a uniform plane wave as being resolvable into a set of normal modes, there can be no such thing as a resonance for a uniform plane wave. Why then does it appear that there is such a resonance in passive optical interferometers? It is because for the usual optical case $a^2/b\lambda$ is in the thousands. The phase shifts per transit are extremely small, hence the mode resonances lie very close together in frequency. At the same time, the reflection coefficients of the best optical mirrors are so poor, and the Q of the interferometer is so low, that the resonance

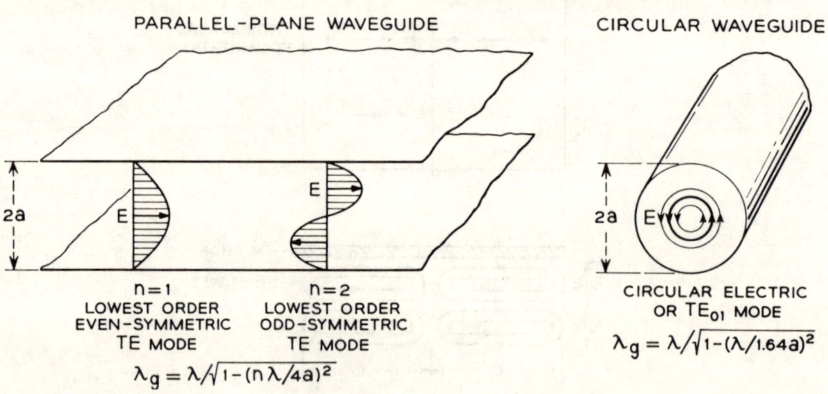

Fig. 20 — TE modes in a parallel-plane waveguide and circular electric mode in a circular waveguide.

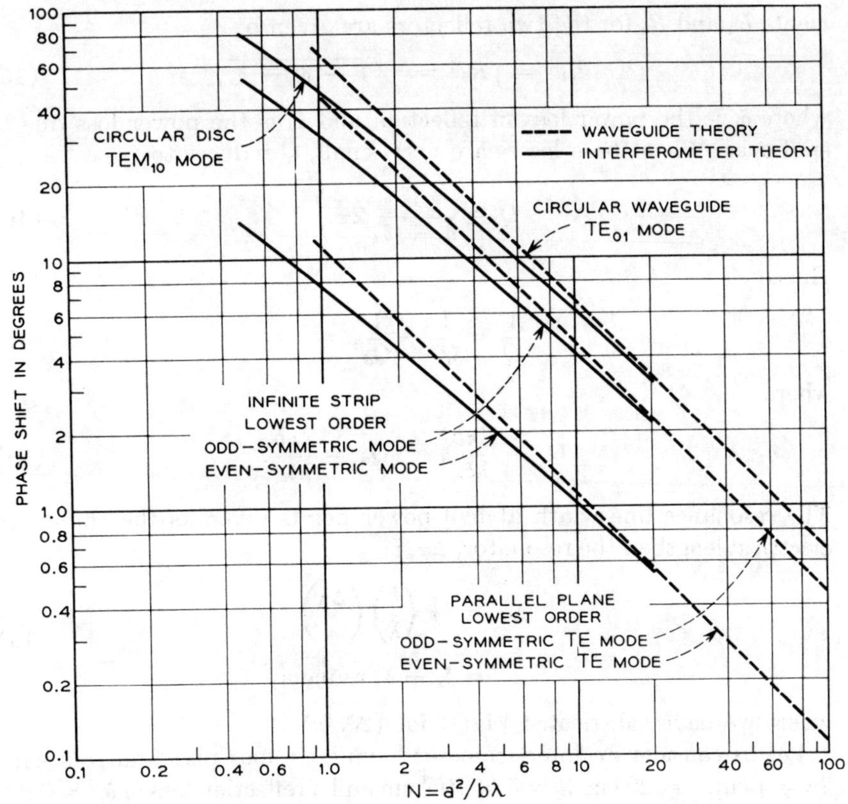

Fig. 21 — Comparison of computed phase shifts based on waveguide theory and on interferometer theory.

line width contains hundreds of normal mode resonances. Thus the uniform plane wave undergoes very little decomposition when resonated. Nevertheless, in the case of an active interferometer, the decomposition may be complete.

We now make use of the formula for the Q of a resonant waveguide cavity to compute the Q of an interferometer system. The Q of a resonant waveguide cavity is given by

$$Q = \frac{|R_1R_2e^{-2\alpha b}|}{1 - |R_1R_2e^{-2\alpha b}|}\left(\frac{2\pi b}{\lambda_g}\right)\left(\frac{\lambda_g}{\lambda}\right)^2, \tag{14}$$

where α is the attenuation constant of the waveguide and λ_g is the guide wavelength. For the interferometer we assume that α is zero and that λ_g is equal to λ, the free-space wavelength. The voltage reflection coefficients R_1 and R_2 for the two reflectors are given by

$$|R_1| = |R_2| = \sqrt{1 - \delta_r - \delta_d}, \tag{15}$$

where δ_r is the power loss in reflection and δ_d is the power loss due to spill-over. When these losses are very small, Q reduces to

$$Q \cong \frac{1}{\delta_r + \delta_d} 2\pi \frac{b}{\lambda}. \tag{16}$$

Hence

$$\frac{1}{Q} = \frac{1}{Q_r} + \frac{1}{Q_d}, \tag{17}$$

where

$$Q_r = \frac{2\pi b}{\lambda \delta_r}, \qquad Q_d = \frac{2\pi b}{\lambda \delta_d}. \tag{18}$$

The resonance line width at half-power points given as the change in electrical length of the resonator, $\Delta\varphi$, is

$$\begin{aligned} \Delta\varphi &= 2\pi \left(\frac{b}{\lambda}\right)\left(\frac{\Delta\lambda}{\lambda}\right) \\ &= \delta_r + \delta_d \text{ radians}, \end{aligned} \tag{19}$$

where we have substituted $(1/Q)$ for $(\Delta\lambda/\lambda)$.

Let us consider an interferometer having circular plane mirrors with $2a = 1$ cm, $b = 20$ cm, $\lambda = 5 \times 10^{-5}$ cm and a reflection loss of $\delta_r = 0.02$. In this case $N = a^2/b\lambda = 250$. Extrapolating the loss and phase shift curves of Figs. 8 and 9, we obtain diffraction loss $\delta_d = 9 \times 10^{-5}$ and phase shift for the dominant (TEM_{00}) mode $\varphi_d = 0.11$ degree. The diffraction loss is thus negligible compared to reflection loss, which limits the Q to a value of 1.25×10^8. The phase shift for the next higher order (TEM_{10}) mode is 0.30 degree and therefore it is separated from the dominant (TEM_{00}) mode by 0.19 degree or 0.0033 radian. The resonance line width, as given by (19), is 0.02 radian. Thus we see that TEM_{00} and TEM_{10} modes are not resolved. As the mirror separation is reduced or mirror size increased, more and more normal modes will become unresolved and a uniform plane wave will suffer less decomposition when resonated.

When an interferometer is filled with an active medium, the medium can compensate for the mirror losses and yield an enormously increased Q. Under these circumstances, the modes may be clearly resolved, and their Q's will be determined by the diffraction losses. If the gain of the medium is increased until it compensates for mirror losses plus the diffraction loss of the lowest order mode, that mode will become unstable and oscillation can result. All higher-order modes will be stable and have positive net loss. If the gain of the medium is further increased, then many modes may become unstable. In starting from a quiescent condition, spontaneous emission can initiate a large number of characteristic waves in the interferometer. These may then start to grow, but the dominant mode will always grow faster and should saturate first. At saturation the steady-state field distribution will be considerably altered. The relative field at the edges of the mirrors should increase, thereby increasing the relative power loss. This can be described as a coupling of power into other modes as a result of the nonlinearity of the medium. No attempt has yet been made to analyze this situation. The linear theory is at present of most interest because it allows the computation of the starting conditions for oscillation.

With the development of the normal mode picture of interferometer operation and the computation of the losses for these modes, we may now ask if there is an optimum geometry for a maser interferometer which will permit oscillation for the lowest possible gain in the medium. We know that the power gained from the medium can be increased by increasing length. For very great lengths corresponding to the far-field region ($N < 0.1$), the power gained from the medium increases more rapidly than transmission loss as length is increased, and there must always be some length beyond which oscillations can occur. However, these lengths are too great to be of practical interest. In the near-field region ($N > 1$), represented by the curves of Fig. 8, the diffraction loss increases more rapidly than the medium gain. Therefore, if the reflection loss is sufficiently small, an optimum length may exist which is most favorable for oscillation.

To be more specific, let us consider a circular plane mirror interferometer. From Fig. 8 we find that the loss for the dominant mode may be represented by the expression

$$\delta_d = 0.207 \left(\frac{b\lambda}{a^2}\right)^{1.4}. \tag{20}$$

In order to find the optimum value of b to give a maximum Q, (20) and (18) are substituted in (17) and the resulting equation is differentiated with respect to b. For the optimum b, the diffraction loss is 2.5 times the reflection loss, and not equal to it, as might be supposed. Moreover, this result is general and holds for all modes and all shapes of plane mirrors represented in Fig. 8, provided the optimum falls on the straight-line portions of the loss curves. Since the power supplied by the medium is proportional to the stored energy in the interferometer, while the power loss of the passive interferometer is just ω/Q times the stored energy, oscillation is most likely to occur when Q is a maximum. Fig. 22 illustrates the way the interferometer dimensions affect Q. If a given mirror diameter is chosen (as represented by the dashed line A), there is clearly an optimum distance b which will produce a maximum Q (intersection of lines A and B). However, if the distance b is held constant, there is no optimum value for a. The larger a, the higher will be Q, although it will approach a limiting value beyond which there is nothing to be gained by further increase of a.

As an example, let us assume a case where

$$\lambda = 10^{-4} \text{ cm},$$

$$2a = \text{plate diameter} = 2 \text{ cm},$$

$$\delta_r = \text{power reflection loss} = 0.001.$$

The optimum proportions require that δ_d be 0.0025, and for this, b is 435 cm and the resulting Q is 7.8×10^9. The length of 435 cm is probably impractically large for a maser. If b is reduced to a more reasonable value of 50 cm, the Q will drop to 3.14×10^9, which is the limiting value due to reflection loss. (The value assumed here for δ_r is already much lower than can be obtained from evaporated metal films and would require the technique of multilayered dielectric films.) In order to oscillate, the active medium would have to have a power amplification factor in excess of 1.00002 per centimeter of path.

In the case of confocal paraboloidal mirrors of 2 cm diameter, the optimum length turns out to be 8900 cm. If the diameter is reduced to 0.5 cm, the optimum length is still 530 cm, and for these proportions Q is 3.1×10^{10}. It is clear that with confocal mirrors the diffraction losses are negligible for any reasonable proportions of the interferometer.

One question of importance is whether there is an optimum set of dimensions which will discriminate against unwanted modes. It has sometimes been suggested that by making the mirror diameter small relative to the mirror spacing, "slant rays" will be more rapidly lost from the system. However, from Fig. 8 it can be seen that the ratios of the losses for the several modes is independent of N provided N is greater than 1. Thus, if diffraction losses predominate, there is no way of discriminating against unwanted modes by juggling dimensions. The limiting amount of discrimination is merely governed by the ratio of the losses for the different modes, which is independent of the dimensions. However, if reflection losses predominate, the discrimination between

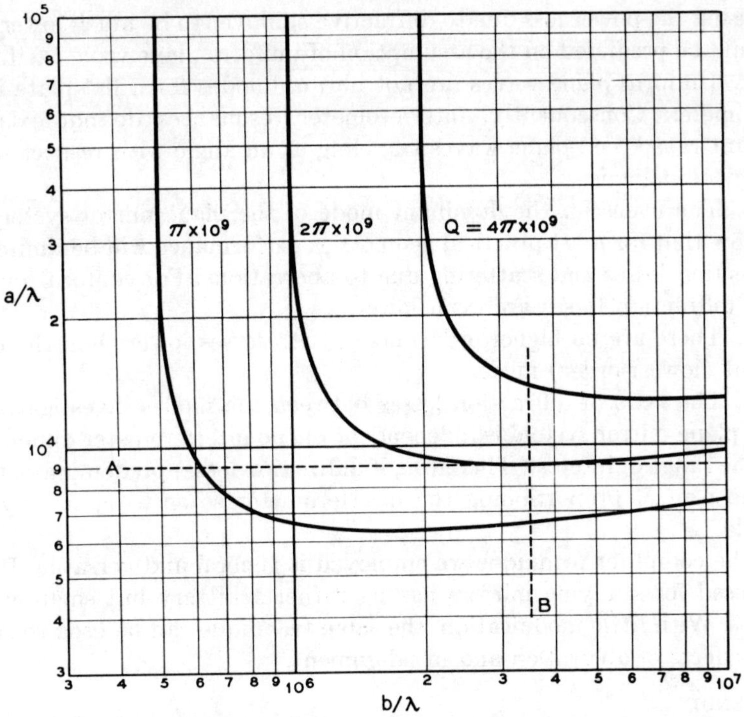

Fig. 22 — Interferometer dimensions for constant Q. (Circular plane mirrors, reflection loss = δ_r = 0.001.)

lower-order modes would be almost nonexistent and it would be advantageous to increase mirror separation and/or decrease mirror dimensions so as to make diffraction losses predominate. In the case of the confocal mirrors, the loss ratios between modes are not constant (Fig. 15) although, for values of N larger than those shown, they may become so. At any rate, for values of N close to unity, a small amount of increased discrimination against higher order modes can be obtained by making the mirrors *larger*.

V. CONCLUSIONS

Diffraction studies carried out on the IBM computer have led to the following conclusions:

1. Fabry-Perot interferometers, whether of the plane or concave mirror type, are characterized by a discrete set of normal modes which can be defined on an iterative basis. The dominant mode has a field intensity which falls to low values at the edges of the mirrors, thereby

causing the power loss due to diffractive spillover to be much lower than would be predicted on the assumption of uniform plane wave excitation.

2. Uniform plane waves are not normal modes for a flat-plate interferometer. Consequently, interferometer resonances do not exist for "slant rays," i.e., plane waves traveling at an angle with respect to the longitudinal axis.

3. The losses for the dominant mode of the plane mirror system are so low that for most practical geometries performance will be limited by reflection losses and scattering due to aberrations. For confocal mirrors the diffraction losses are even lower.

4. There are no higher-order modes with losses lower than the dominant (lowest-order) mode.

5. The ratio of diffraction losses between the modes investigated for the plane mirror system is independent of the interferometer dimensions in the range of interest. Therefore, if diffraction losses predominate, there is no way of proportioning the interferometer so as to favor any one mode.

The computer technique we employed is general and versatile. It can be used for studying mirrors having rather arbitrary but small curvatures. With little modification, the same technique can be used to study the effects of aberration and misalignment.

APPENDIX A

Rectangular Plane Mirrors

The geometry for rectangular plane mirrors parallel to the xy plane is shown in Fig. 2. According to (1), the iterative equation for computing the field at the surface of mirrors is

$$u_{q+1}(x_2, y_2) = \frac{j}{2\lambda} \int_{-c}^{c} \int_{-a}^{a} u_q(x_1, y_1) \frac{e^{-jkR}}{R} \left(1 + \frac{b}{R}\right) dx_1 \, dy_1, \quad (21)$$

where

$$R = \sqrt{b^2 + (x_1 - x_2)^2 + (y_1 - y_2)^2}.$$

If b/a and b/c are large, (21) can be reduced to

$$u_{q+1}(x_2, y_2) = \frac{je^{-jkb}}{\lambda b} \int_{-c}^{c} \int_{-a}^{a} u_q(x_1, y_1) e^{-jk[(x_1-x_2)^2+(y_1-y_2)^2]/2b} \, dx_1 \, dy_1, \quad (22)$$

which is valid for $(a^2/b\lambda) \ll (b/a)^2$ and $(c^2/b\lambda) \ll (b/c)^2$.* The corresponding integral equation is

* Actually, the stringency of this requirement can be relaxed somewhat for lower-order modes in which field intensities near the edges of the mirror are rather low. We have made check computations for the case $a^2/b\lambda = 5$ and $(b/a)^2 = 25$ and have found that the results based on the exact equation and on the approximate equation are in essential agreement.

$$v(x_2, y_2) = \gamma \int_{-c}^{c} \int_{-a}^{a} K(x_2, x_1; y_2, y_1) v(x_1, y_1)\, dx_1 dy_1, \tag{23}$$

where

$$K(x_2, x_1; y_2, y_1) = \frac{j}{\lambda b} e^{-jk[(x_1-x_2)^2+(y_1-y_2)^2]/2b} \tag{23a}$$

and the factor e^{-jkb} is absorbed in γ.

Here, the kernel of the integral equation is separable in x and y. If the distribution function v is assumed to be of the form

$$v(x,y) = v_x(x) v_y(y) \tag{24}$$

it is possible to separate (23) into two equations, one involving x only and the other involving y only; that is,

$$v_x(x_2) = \gamma_x \int_{-a}^{a} K_x(x_2, x_1) v_x(x_1)\, dx_1, \tag{25a}$$

$$v_y(y_2) = \gamma_y \int_{-c}^{c} K_y(y_2, y_1) v_y(y_1)\, dy_1, \tag{25b}$$

with

$$K_x = \frac{e^{j(\pi/4)}}{\sqrt{\lambda b}} e^{-jk(x_1-x_2)^2/2b} \tag{25c}$$

and

$$K_y = \frac{e^{j(\pi/4)}}{\sqrt{\lambda b}} e^{-jk(y_1-y_2)^2/2b}. \tag{25d}$$

The product of the eigenvalues γ_x and γ_y is equal to the eigenvalue γ in (23).

It remains to be shown that (25a) through (25d) represent integral equations for infinite strip mirrors. Let us consider a pair of infinite strip mirrors of width $2a$ and separated by b. The iterative equation for computing the field at the mirrors can be derived from (1). It is

$$u_{q+1}(x_2) = \frac{e^{j(\pi/4)}}{2\sqrt{\lambda}} \int_{-a}^{a} u_q(x_1) \frac{e^{-jk\rho}}{\sqrt{\rho}} \left(1 + \frac{b}{\rho}\right) dx_1, \tag{26}$$

where

$$\rho = \sqrt{b^2 + (x_1 - x_2)^2}.$$

For $(a^2/b\lambda) \ll (b/a)^2$, (26) reduces to

$$u_{q+1}(x_2) = \frac{e^{j[(\pi/4)-kb]}}{\sqrt{\lambda b}} \int_{-a}^{a} u_q(x_1)\, e^{-jk(x_1-x_2)^2/2b}\, dx_1. \tag{27}$$

The corresponding integral equation is

$$v(x_2) = \gamma \int_{-a}^{a} K(x_2, x_1)\, v(x_1)\, dx_1, \tag{28}$$

where

$$K(x_2, x_1) = \frac{e^{j(\pi/4)}}{\sqrt{\lambda b}}\, e^{-jk(x_1-x_2)^2/2b} \tag{28a}$$

and the factor e^{-jkb} is absorbed in γ. We see that (25) and (28) are identical in form.

APPENDIX B

Circular Plane Mirrors

Assuming approximately plane waves propagating normally to the circular plane mirrors (Fig. 3), the iterative equation for computing the steady-state field distribution can be written as

$$u_{q+1}(r_2, \varphi_2) = \frac{j}{2\lambda} \int_0^a \int_0^{2\pi} u_q(r_1, \varphi_1)\, \frac{e^{-jkR}}{R} \left(1 + \frac{b}{R}\right) r_1\, d\varphi_1\, dr_1, \tag{29}$$

where

$$R = \sqrt{b^2 + r_1^2 + r_2^2 - 2r_1 r_2 \cos(\varphi_1 - \varphi_2)}.$$

If b/a is large, (29) simplifies to

$$u_{q+1}(r_2, \varphi_2) = \frac{je^{-jkb}}{\lambda b} \int_0^a \int_0^{2\pi} u_q(r_1, \varphi_1) \cdot e^{-jk[(r_1^2+r_2^2)/2b-(r_1 r_2/b)\cos(\varphi_1-\varphi_2)]} r_1\, d\varphi_1\, dr_1, \tag{30}$$

which is valid for $(a^2/b\lambda) \ll (b/a)^2$.*

The integral equation corresponding to (30) is

$$v(r_2, \varphi_2) = \gamma \int_0^a \int_0^{2\pi} K(r_2, \varphi_2; r_1, \varphi_1) v(r_1, \varphi_1) r_1\, d\varphi_1\, dr_1, \tag{31}$$

with

$$K(r_2, \varphi_2; r_1, \varphi_1) = \frac{j}{\lambda b}\, e^{-jk[(r_1^2+r_2^2)/2b-(r_1 r_2/b)\cos(\varphi_1-\varphi_2)]} \tag{31a}$$

---

* Comments in Appendix A regarding the stringency of this requirement are also applicable herein.

and where the factor e^{-jkb} is absorbed in γ. Making use of the relation[10]

$$e^{jn[(\pi/2)-\varphi_2]} J_n\left(k\frac{r_1r_2}{b}\right) = \frac{1}{2\pi}\int_0^{2\pi} e^{jk(r_1r_2/b)\cos(\varphi_1-\varphi_2)-jn\varphi_1}\, d\varphi_1 \qquad (32)$$

and integrating (31) with respect to φ_1, it is seen that

$$v(r,\varphi) = R_n(r)e^{-jn\varphi}, \qquad (n = \text{integer}) \qquad (33)$$

satisfies (31). The function $R_n(r)$ satisfies the reduced integral equation

$$R_n(r_2)\sqrt{r_2} = \gamma_n \int_0^a K_n(r_2, r_1)R_n(r_1)\sqrt{r_1}\, dr_1, \qquad (34)$$

with

$$K_n(r_2, r_1) = \frac{j^{n+1}k}{b} J_n\left(k\frac{r_1r_2}{b}\right)\sqrt{r_1r_2}\, e^{-jk(r_1^2+r_2^2)/2b}, \qquad (34a)$$

where J_n is a Bessel function of the first kind and nth order.

APPENDIX C

Confocal Spherical or Paraboloidal Mirrors

For confocal spherical mirrors of circular cross section (Fig. 4), the iterative equation corresponding to (29) is

$$u_{q+1}(r_2, \varphi_2) = \frac{j}{2\lambda}\int_0^a\int_0^{2\pi} u_q(r_1, \varphi_1)\frac{e^{-jkR}}{R}\left(1 + \frac{b_1}{R}\right) r_1\, d\varphi_1\, dr_1 \qquad (35)$$

where

$$R = \sqrt{b_1^2 + r_1^2 + r_2^2 - 2r_1r_2\cos(\varphi_1 - \varphi_2)}.$$

The distance b_1 is given by

$$b_1 = b - \Delta_1 - \Delta_2 \qquad (36)$$

where, for confocal spherical mirrors,

$$\Delta_i = b - \sqrt{b^2 - r_i^2} \qquad i = 1,2. \qquad (36a)$$

If b/a is large, the distance Δ_i is given approximately by

$$\Delta_i \cong r_i^2/2b \qquad i = 1,2, \qquad (37)$$

which is exact for confocal paraboloids. In this case (35) simplifies to

$$u_{q+1}(r_2, \varphi_2) = \frac{je^{-jkb}}{\lambda b} \int_0^a \int_0^{2\pi} u_q(r_1, \varphi_1)\, e^{jk(r_1 r_2/b)\cos(\varphi_1 - \varphi_2)} r_1\, d\varphi_1\, dr_1, \quad (38)$$

which is valid for $(a^2/b\lambda) \ll (b/a)^2$.

The integral equation corresponding to (38) is

$$v(r_2, \varphi_2) = \gamma \int_0^a \int_0^{2\pi} K(r_2, \varphi_2; r_1, \varphi_1) v(r_1, \varphi_1) r_1\, d\varphi_1\, dr_1, \quad (39)$$

with

$$K(r_2, \varphi_2; r_1, \varphi_1) = \frac{j}{\lambda b} e^{jk(r_1 r_2/b)\cos(\varphi_1 - \varphi_2)} \quad (40)$$

and where the factor e^{-jkb} is absorbed in γ. Just as in the case of circular plane mirrors, it can be shown that

$$v(r,\varphi) = S_n(r) e^{-jn\varphi} \qquad (n = \text{integer}) \quad (41)$$

satisfies (39). The function $S_n(r)$ satisfies the reduced integral equation

$$S_n(r_2)\sqrt{r_2} = \gamma_n \int_0^a K_n(r_2, r_1) S_n(r_1) \sqrt{r_1}\, dr_1, \quad (42)$$

with

$$K_n(r_2, r_1) = \frac{j^{n+1} k}{b} J_n\left(k \frac{r_1 r_2}{b}\right) \sqrt{r_1 r_2}. \quad (42a)$$

REFERENCES

1. Schawlow, A. L. and Townes, C. H., Infrared and Optical Masers, Phys. Rev., **112,** 1958, p. 1940.
2. Maiman, T. H., Stimulated Optical Radiation in Ruby, Nature, **187,** 1960, p. 493.
3. Collins, R. J., Nelson, D. F., Schawlow, A. L., Bond, W., Garrett, C. G. B. and Kaiser, W., Coherence, Narrowing, Directionality and Relaxation Oscillations in the Light Emission from Ruby, Phys. Rev. Letters, **5,** 1960, p. 303.
4. Silver, S., *Microwave Antenna Theory and Design,* McGraw-Hill, New York, 1949, p. 167.
5. Hildebrand, F. B., *Methods of Applied Mathematics,* Prentice-Hall, Englewood Cliffs, N. J., 1952.
6. Connes, P., Increase of the Product of Luminosity and Resolving Power of Interferometers by Using a Path Difference Independent of the Angle of Incidence, Revue d'Optique, **35,** 1956, p. 37.
7. Lovitt, W. V., *Linear Integral Equations,* Dover Publications, New York, 1950, pp. 129; 137.
8. Goubau, G., and Schwering, F., On the Guided Propagation of Electromagnetic Wave Beams, URSI-IRE Spring Meeting, May 1960, Washington, D. C.
9. Boyd, G. D., and Gordon, J. P., Confocal Multimode Resonator for Millimeter through Optical Wavelength Masers, this issue, p. 489.
10. Stratton, J. A., *Electromagnetic Theory,* McGraw-Hill, New York, 1941, p. 372.

Part 3
Laser Oscillator Theory

Comparison of Quantum and Semiclassical Radiation Theories with Application to the Beam Maser*

E. T. JAYNES† AND F. W. CUMMINGS‡

Summary—**This paper has two purposes: 1) to clarify the relationship between the quantum theory of radiation, where the electromagnetic field-expansion coefficients satisfy commutation relations, and the semiclassical theory, where the electromagnetic field is considered as a definite function of time rather than as an operator; and 2) to apply some of the results in a study of amplitude and frequency stability in a molecular beam maser.**

In 1), it is shown that the semiclassical theory, when extended to take into account both the effect of the field on the molecules and the effect of the molecules on the field, reproduces almost quantitatively the same laws of energy exchange and coherence properties as the quantized field theory, even in the limit of one or a few quanta in the field mode. In particular, the semiclassical theory is shown to lead to a prediction of spontaneous emission, with the same decay rate as given by quantum electrodynamics, described by the Einstein A coefficients.

In 2), the semiclassical theory is applied to the molecular beam maser. Equilibrium amplitude and frequency of oscillation are obtained for an arbitrary velocity distribution of focused molecules, generalizing the results obtained previously by Gordon, Zeiger, and Townes for a singel-velocity beam, and by Lamb and Helmer for a Maxwellian beam. A somewhat surprising result is obtained; which is that the measurable properties of the maser, such as starting current, effective molecular Q, etc., depend mostly on the slowest 5 to 10 per cent of the molecules.

Next we calculate the effect of amplitude and frequency of oscillation, of small systematic perturbations. We obtain a prediction

* Received September 28, 1962.
† Washington University, St. Louis, Mo.
‡ Aeronutronic, Division of Ford Motor Co., Newport Beach, Calif.

Reprinted from *Proc. IEEE*, vol. 51, pp. 89–109, Jan. 1963.

that stability can be improved by adjusting the system so that the molecules emit all their energy $\hbar\Omega$ to the field, then reabsorb part of it, before leaving the cavity. In general, the most stable operation is obtained when the molecules are in the process of absorbing energy from the radiation as they leave the cavity, most unstable when they are still emitting energy at that time.

Finally, we consider the response of an oscillating maser to randomly time-varying perturbations. Graphs are given showing predicted response to a small superimposed signal of a frequency near the oscillation frequency. The existence of "noise enhancing" and "noise quieting" modes of operation found here is a general property of any oscillating system in which amplitude is limited by nonlinearity.

I. Introduction

THIS PAPER has two purposes: 1) to clarify the relationship between the quantum theory of radiation where the electromagnetic field expansion coefficients satisfy commutation relations, and the semiclassical theory where the electromagnetic field is considered as a definite function of time rather than as an operator, and 2) to apply some of the results thus obtained in a study of amplitude and frequency stability of the ammonia beam maser.

In 1), the relation between quantum electrodynamics and the semiclassical theory is shown to be quite different from that usually assumed. The semiclassical theory, when extended to take into account both the effect of the molecules on the field and the effect of the field on the molecules, reproduces almost quantitatively the same laws of energy exchange and coherence properties as the quantized field theory, even in the limit of one or a few quanta in the field cavity mode. In particular, the semiclassical theory is shown to lead to a prediction of spontaneous emission, with exactly the same decay rate as given by quantum electrodynamics, as described by the Einstein A coefficients.

There remain, however, several fundamental differences in the two theories. For example, quantum electrodynamics allows the possibility that the combined system (molecules plus field) may be in states which have properties qualitatively different than any that can be described in classical terms, even in the limit of arbitrarily high photon occupation numbers. Thus the common statement that quantum electrodynamics goes over into classical electrodynamics in the case of high quantum numbers for the field oscillators, needs to be somewhat qualified.

Having shown the essential equivalence of quantum electrodynamics and the semiclassical approach for the problems of interest, we turn to detailed calculations applying the semiclassical theory to the ammonia beam maser. Equilibrium amplitude and frequency of oscillation are obtained for an arbitrary velocity distribution of focused molecules, generalizing the results obtained previously by Gordon, Zeiger and Townes [1] for a single-velocity beam and by Lamb and Helmer [6] for a Maxwellian beam. A rather surprising result is obtained, namely that the measurable properties of the maser, such as starting current, effective molecular Q, etc., depend mostly on the slowest 5 to 10 per cent of the molecules.

Next we calculate the effect on amplitude and frequency of oscillation of small systematic perturbations. We obtain a prediction that stability can be improved by adjusting the system so that the molecules emit all their energy $\hbar\Omega$ to the field, then reabsorb part of it, before leaving the cavity. In general, the most stable operation is obtained when the molecules are in the process of absorbing energy from the radiation as they leave the cavity, the most unstable, when they are still emitting energy at that time.

Finally, we consider this response of an oscillating maser to random time-varying perturbations. Graphs are given showing predicted response to a small superimposed signal of a frequency near to the oscillation frequency. The results show a quite complicated variation as a function of the frequency difference and beam current, and resemble some results of Wiener, concerning nonlinear random phenomena.

Broadly speaking, there are two different levels of approximation used in gas maser theories published as of this writing:

1) The most common and also the crudest of these theories is the one wherein one treats the emission process of radiation from molecules as if the transition probabilities were proportional to the time. Such theories contain little that was not already contained in Einstein's 1917 paper which introduced the A and B coefficients. According to quantum mechanics, the idea of time proportional transition probabilities is an approximation, valid only when the correlation time of the radiation is short compared to the time required to accumulate an appreciable transition probability; that is, the radiation responsible for the transition must be random, with a spectrum wide compared to the line width. In an ammonia beam device the correlation time of the radiation may be of the order of 10^3 to 10^6 times the flight time of a molecule through the cavity, and thus any attempt to describe maser operation in terms of "Fermi golden rule" type of equations for the transition probabilities, *i.e.*,

$$W_{1\to 2} = \frac{2\pi}{\hbar^2} \mid H_{12} \mid^2 \rho(\omega)$$

may lead to conclusions qualitatively as well as quantitatively wrong. Most of the existing noise figure calculations are based on a treatment of this type [1], [2], and hence one cannot assess their worth until the calculation has been checked by a more rigorous theory.

2) The second method of treating the maser theoretically is that based on solving Schrödinger's time-dependent equation for a molecule as perturbed by a classically described field and finding then the expectation value of the dipole moment of the molecule and using the time derivative of this expectation value as the current source of the classical electromagnetic field.

This is essentially the calculation of Shimoda, Wang and Townes [4], and Basov and Prokhorov [5], and Lamb and Helmer [6], Feynman, Vernon and Hellwarth [7]. While this is clearly superior to the first method outline above, there are still several important approximations involved. In principle, the molecular beam should be treated as a single quantum-mechanical system, by a formalism like that of Dicke's "super-radiant gas" [8]. In the theories quoted above, the molecules were ascribed independent wave functions. Also, the electromagnetic field should be quantized and the problem treated as one of quantum electrodynamics. Although the theories above lead to definite predictions for saturation and frequency pulling, it is not at all clear that they can lead to reliable predictions of fluctuation effects involved in noise figure and frequency stability. It is generally thought that the semiclassical theory should be adequate for any effects at microwave frequencies due to the smallness of the Einstein A coefficient compared to the B coefficient. However, quantization of the electromagnetic field introduces many changes in addition to the appearance of A coefficients; for instance, quantization can lead to states qualitatively different from any describable in classical terms, even in the limit of arbitrarily high photon occupation numbers per field normal mode. Such states will be shown, in the calculations to follow, to actually be the ones produced in the maser under certain idealized conditions. Thus until these calculations based on these approximations are checked in some other way, our degree of confidence in them cannot be too great.

Our approach in this paper will be, stated briefly, to first treat simple problems in which we can talk of transition probabilities with all coherence properties retained, within the formalism of quantum electrodynamics. Then we will investigate the relationship between the "modified" semiclassical ("neoclassical" theory), as employed by Shimoda, Wang, and Townes [4], and quantum electrodynamics. The relationship is not at all that which is usually assumed, *i.e.*, that quantum electrodynamics goes into semiclassical theory only in the limit of high photon occupation numbers per field-normal mode. Rather the neoclassical theory, in which expectation values of quantum mechanical operators are interpreted as *actual* values of sources in the classical Maxwell equations, and both the effect of the radiation field on the molecule and the effect of the molecule back on the field are taken into account, *does* lead to a prediction of spontaneous emission, and to only very small quantitative differences in the decay rate for the case of a few microwave photons in the cavity.

In Section IV the neoclassical theory is applied to the problem of the ammonia beam device in which the indirect "coupling" between molecules via the field is treated and a steady-state solution is obtained under the assumption of an arbitrary velocity distribution, wall losses and/or external energy coupling. The solution is obtained for the frequency stability as a function of the mean values of the square and cube of the flight time and the Q of the cavity. This solution is found to agree with that of Gordon, Zeiger and Townes [1] in the univelocity case and with the analysis of Lamb and Helmer [6] in the case of a Maxwellian distribution of velocities. Then this solution is made the basis, or unperturbed solution, in a perturbation treatment of flucutation effects to the first order in the small departure from the steady-state solution. These problems in this case become linear, so that we can analyze the effect of small periodic perturbations proportional to exp $(i\beta t)$ and superpose the solutions to give solutions for the transient response to an arbitrary small perturbation. This can represent an extra signal fed in intentionally, or it might be a randomly varying function representing thermal noise in the cavity and/or load. A "noise quieting" phenomena is seen to occur for proper values of the flight time, and graphs are drawn which exhibit the power spectrum of the thermal noise as affected by the molecular beam.

II. Quantum Electrodynamic Solutions

We approach the theory of maser operation in several stages, starting with simple, special cases for which all details of the mathematics can be worked out, then adding various features which tend in the direction of more realistic models. The mathematical form of the theory is quite similar to what one encounters in the statistical mechanics of irreversible processes. Of particular interest, however, is the extent to which the semiclassical theory is derivable from quantum electrodynamics, and we are most interested in comparing the results of this section with those obtained in Section III. Also, the effect of different statistical assumptions concerning the initial states of the molecules is interesting in this same regard.

A. Field Quantization

We first develop the formalism of field quantization in a form suitable for microwave applications. There is, of course, no need for elegant covariant formulations here; the simple approach to electrodynamics given by Fermi [9] is quite adequate for our purposes. Here the usual plane-wave expansion is not appropriate and in its place we need to use the expansion of electromagnetic fields in terms of resonant modes of the particular cavity under consideration. We use the cavity normal mode functions as defined by Slater [10]. The cavity is represented by a volume V, bounded by a closed surface S. Let $\boldsymbol{E}_a(\mathbf{x})$, $k_a^2 = \omega_a^2/C^2$ be the eigenfunctions and eigenvalues of the boundary-value problem.

$$\begin{aligned} \nabla \times \nabla \times \boldsymbol{E} - k^2 \boldsymbol{E} &= 0 \qquad \text{in } V \\ \boldsymbol{n} \times \boldsymbol{E} &= 0 \qquad \text{on } S \end{aligned} \tag{1}$$

where $\boldsymbol{n}$ is a unit vector normal to S. The $\boldsymbol{E}_a(\boldsymbol{x})$ are so normalized that

$$\int_v (\boldsymbol{E}_a \cdot \boldsymbol{E}_b) dV = \delta_{ab}. \tag{2}$$

The vector functions $\boldsymbol{H}_a(\boldsymbol{x})$, related to $\boldsymbol{E}_a$ by

$$\nabla \times \boldsymbol{E}_a = k_a \boldsymbol{H}_a \qquad \nabla \times \boldsymbol{H}_a = k_a \boldsymbol{E}_a \tag{3}$$

are also orthonormal in V as follows:

$$\int_v (\boldsymbol{H}_a \cdot \boldsymbol{H}_b) dV = \delta_{ab}. \tag{4}$$

The electric and magnetic fields can be expanded in the following forms:

$$\boldsymbol{E}(x\ t) = -\sqrt{4\pi} \sum_a p_a(t) \boldsymbol{E}_a(x) \tag{5}$$

and

$$\boldsymbol{H}(x\ t) = \sqrt{4\pi} \sum_a \omega_a q_a(t) \boldsymbol{H}_a(x). \tag{6}$$

From these relations, we find for the total field energy

$$\mathcal{H} = \int \frac{E^2 + H^2}{8\pi} dV = \frac{1}{2} \sum_a (p_a{}^2 + \omega_a{}^2 q_a{}^2), \tag{7}$$

and the Maxwell equations,

$$\nabla \times \boldsymbol{E} = -\frac{1}{c} \frac{\partial \boldsymbol{H}}{\partial t} \tag{8}$$

and

$$\nabla \times \boldsymbol{H} = \frac{1}{c} \frac{\partial \boldsymbol{E}}{\partial t}, \tag{9}$$

then reduce to the Hamiltonian equations of motion,

$$\dot{q}_a = \frac{\partial \mathcal{H}}{\partial p_a} = p_a, \tag{8a}$$

$$\dot{p}_a = -\frac{\partial \mathcal{H}}{\partial q_a} = -\omega_a{}^2 q_a, \tag{8b}$$

respectively.

On quantization of the field the canonically conjugate coordinates and momenta satisfy the commutation rules,

$$[q_a, q_b] = [p_a, p_b] = 0 \tag{10}$$

and

$$[q_a, p_b] = i\hbar\delta_{ab}. \tag{11}$$

The operators $C_a{}^*$, C_a which create or annihilate a photon in the ath cavity mode are then

$$C_a{}^* = \frac{p_a + i\omega_a q_a}{\sqrt{2\hbar\omega_a}} \qquad C_a = \frac{p_a - i\omega_a q_a}{\sqrt{2\hbar\omega_a}} \tag{12}$$

with the commutation rule

$$[C_a\ C_b^*] = \delta_{ab}. \tag{13}$$

Denote by $\phi(n_1, n_2, \cdots)$ the state vector of the field for which there are n_1 quanta in mode 1, n_2 in mode 2, etc. The C_a operators have the properties

$$C_a \phi(\cdots, n_a, \cdots) = \sqrt{n_a}\, \phi(\cdots, n_a - 1, \cdots) \tag{14}$$

and

$$C_a{}^* \phi(\cdots, n_a, \cdots) = \sqrt{n_a + 1}\, \phi(\cdots, n_a + 1, \cdots) \tag{15}$$

from which we easily verify (13), and obtain the matrix elements in the n_a representation,

$$(n_a | C_a | n_a') = (n_a' | C_a{}^* | n_a) = \sqrt{n_a + 1}\, \delta(n_a'\ n_a + 1). \tag{16}$$

The Hamiltonian, with zero point energy removed, then reduces to

$$\mathcal{H} = \sum_a \hbar\omega_a C_a{}^* C_a = \sum_a \hbar\omega_a n_a. \tag{17}$$

Finally, we work out for later purposes the matrix elements of the electric field in the case of a cylindrical cavity with only the lowest TM mode excited. In this mode, the only nonvanishing component of $\boldsymbol{E}_a$ is $E_{az} = (\text{constant}) \times J_0(k_a r)$, independent of z and θ. The normalizing constant is obtained from evaluating the integral (2), with the result that on the axis of the cylinder (along which the molecules travel in an ammonia maser) the function E_{az} reduces to

$$E_{az} = \frac{1}{J_1\sqrt{V}} \cdot \tag{18}$$

Here $J_1 = J(u) = 0.5191$, and $u = 2.405$ is the first root of $J_0(u) = 0$. V is the volume of the cavity. The operator P_a involved in the electric field expansion is, from (12),

$$p_a = \sqrt{\frac{\hbar\omega_a}{2}}\, (C_a + C_a{}^*). \tag{19}$$

Combining (5), (16), (18), and (19), we obtain the matrix elements

$$(n | E | n') = -\left(\frac{2\pi\hbar\omega}{J_1{}^2 V}\right)^{1/2} [\sqrt{n}\, \delta_{n,n'+1} + \sqrt{n+1}\, \delta_{n+1,n'}] \tag{20}$$

in which we have dropped the subscript a, it being understood that (20) refers to the case where only the lowest TM mode is taken into account. For the matrix elements of electric field at points off the axis of the cylinder, this expression should be multiplied by $J_0(Kr)$.

B. Interaction with a Single Molecule

The simplest possible situation is one where we consider a lossless cavity, which has only a single resonant mode near the natural line frequency of the molecule, and a uniform field (electric or magnetic, whichever is the one effective in field-molecule interaction) along the path of the molecules. Suppose further that only a single molecule, which has only two possible energy levels, is in the cavity. With the molecule-field interaction in the usual $(\boldsymbol{J}\cdot\boldsymbol{A})$ form, it appears that even this problem cannot be solved exactly. However, because of the simplicity of the model, we will be able to treat it more accurately than is usually done in more difficult problems, where one resorts to an expansion in powers of $(e^2/\hbar c)$. The stationary states of the system (molecule plus field) can be found to an accuracy of perhaps one part in 10^7 for radiation energy densities up to the order of those encountered in masers, by a calculation which involves nothing worse than solving quadratic equations. By use of perturbation theory still better accuracy would be feasible, but this is not done here.

Let the two possible energy levels of the molecule be denoted by E_m, and the corresponding states by $\psi_m (m=1, 2)$. Similarly, the number of quanta in the field oscillator will be n, and the corresponding state of the field by $\phi(n=0, 1, 2, \cdots)$. The state vectors $\psi_m\phi_n$ then form a basis for the system (molecule plus field). In this representation, the total Hamiltonian is

$$(mn \mid H \mid m'n') = (E_m + n\hbar\omega)\delta_{mm'}\delta_{nn'} + (mn \mid H_{\text{int}} \mid m'n'). \quad (21)$$

The interaction Hamiltonian between molecule and field is taken of the form

$$H_{\text{int}} = -\boldsymbol{\mu}\cdot\boldsymbol{E} \quad (22)$$

where $\boldsymbol{\mu}$ is the electric dipole moment of the molecule, whose component along $\boldsymbol{E}$ shall have the matrix elements

$$(mn \mid \mu_z \mid m'n') = \mu(1 - \delta_{mm'})\delta_{nn'}. \quad (23)$$

Combining this with (20), we obtain the matrix elements for the interaction energy

$$(mn \mid H_{\text{int}} \mid m'n') = \hbar\alpha(1 - \delta_{mm'})[\sqrt{n}\,\delta_{n,n'+1} + \sqrt{n+1}\,\delta_{n+1,n'}] \quad (24)$$

where

$$\alpha = \frac{\mu}{J_1}\sqrt{\frac{2\pi\omega}{\hbar V}} \quad (25)$$

is the interaction constant. Using the value [11] $\mu = 1.47\times10^{-18}$ esu for ammonia, and a cavity 10 cm long, we find $(\alpha/\omega) = 2.08\times10^{-10}$ or, $\alpha\approx 5.0$ cps.

The interaction Hamiltonian has matrix elements of two different types: $H_{\text{int}} = V + W$, where

$$V_n = (1, n+1 \mid V \mid 2, n) = (2, n \mid V \mid 1, n+1) = \hbar\alpha\sqrt{n+1}$$

and

$$W_n = (1, n \mid W \mid 2, n+1) = (2, n+1 \mid W \mid 1, n) = \hbar\alpha\sqrt{n+1} \quad (26)$$

all other elements being zero. The term V cannot be treated as a perturbation, for its matrix elements connect "unperturbed" states with an energy separation $(E_2 - E_1 - \hbar\omega)$ which goes through zero as the cavity is tuned exactly on the natural line frequency. On the other hand, elements of W connect states with unperturbed energy separation $(E_2 - E_1 + \hbar\omega)\approx 2\hbar\omega$. Since in typical operation conditions $(n\approx 10^6)$ we have $W_n 2\hbar\omega < 10^{-7}$, we may treat W as a small perturbation, or even neglect it entirely. We thus write the Hamiltonian as

$$H = H_0 + W$$

in which the term $H_0 = (H_{\text{mol}} + H_{\text{field}} + V)$ must be diagonalized exactly. This is readily done, since H_0 has a "block form" consisting of many (2×2) matrices along the main diagonal. The eigenvalues and eigenfunctions of H_0, defined by $H_0\Phi_n^{\pm} = E_n^{\pm}\Phi_n^{\pm}$, are the ground state

$$E_0 = E_1 = \hbar\omega_0 \qquad \Phi_0 = \psi_1\phi_0 \quad (27)$$

and for $n > 0$,

$$E_n^{\pm} = \hbar\omega_n^{\pm} = \tfrac{1}{2}[E_1 + E_2 + (2n-1)\hbar\omega] \pm \tfrac{1}{2}[(E_2 - E_1 - \hbar\omega)^2 + 4n\hbar^2\alpha^2]^{1/2}. \quad (28)$$

We find it convenient now to define our zero molecular energy midway between the levels E_1 and E_2 such that

$$E_1 + E_2 = 0 \qquad E_2 - E_1 = \hbar\Omega$$

so that (28) now reads

$$E_n^{\pm} = \hbar\omega_n^{\pm} = (n - \tfrac{1}{2})\hbar\omega \pm \frac{\hbar}{2}[(\Omega - \omega)^2 + 4n\alpha^2]^{1/2}. \quad (28a)$$

Now

$$\Phi_n^{+} = \psi_2\phi_{n-1}\cos\theta_n + \psi_1\phi_n\sin\theta_n$$
$$\Phi_n^{-} = -\psi_2\phi_{n-1}\sin\theta_n + \psi_1\phi_n\cos\theta_n \quad (29)$$

where

$$\tan 2\theta_n = \frac{2\alpha\sqrt{n}}{\omega - \Omega}\cdot \quad (30)$$

We now require the time-development matrix (in units with $\hbar = 1$)

$$U(t, t') = U(t - t') = \exp[-iH(t - t')] \quad (31)$$

for which the perturbation expansion is

$$U(t) = e^{-iH_0 t} - i\int_0^t e^{i(t-t')H_0} W e^{it'H_0} dt' + \cdots. \quad (32)$$

The major term $U_0 = \exp(-iH_0t)$ has the matrix elements, for $n > 0$,

$$(2, n-1 | U_0 | 2, n-1) = a_n = \cos^2\theta_n e^{-i\omega_n^+ t} + \sin^2\theta_n e^{-i\omega_n^- t}$$

$$(2, n-1 | U_0 | 1, n) = b_n = \sin\theta_n \cos\theta_n (e^{-i\omega_n^- t} - e^{-i\omega_n^+ t})$$

$$(2, n | U_0 | 2, n-1) = b_n$$

$$(1, n | U_0 | 1, n) = c_n = \cos^2\theta_n e^{-i\omega_n^- t} + \sin^2\theta_n e^{-i\omega_n^+ t} \tag{33}$$

and, for $n = 0$,

$$(1, 0 | U_0 | 1, 0) = e^{-i\omega_0 t} \tag{34}$$

where now $\omega_0 = -\Omega/2$ all other elements vanish. The transition probability for emission or absorption of one photon during time t is therefore, neglecting terms in W,

$$|b_n|^2 = \sin^2 2\theta_n \sin^2(\omega_n^+ - \omega_n^-)t/2 = \frac{n\alpha^2 \sin^2 \beta t}{\beta^2} \tag{35}$$

where

$$4\beta^2 = (\omega - \Omega)^2 + 4n\alpha^2. \tag{36}$$

The above notation has been chosen in such a way that the block form of U_0 consists of the symmetric, (2×2) unitary matrices

$$\begin{bmatrix} a_n & b_n \\ b_n & c_n \end{bmatrix} \qquad n = 1, 2, \cdots$$

along the main diagonal. The first row and column, however, contain only the single term (34).

We now consider the effect on the field of passing a single molecule through the cavity, with flight time τ. At the instant $(t = 0)$ when the molecule enters the cavity, let its state be described by the density matrix $\rho_1(0)$, and the state of the field by the density matrix $\rho_f(0)$. The initial density matrix fo the entire system is thus the direct product $\rho(0) = \rho_1(0) \times \rho_f(0)$, with matrix elements

$$(mn | \rho(0) | m'n') = (m | \rho_1(0) | m')(n | \rho_f(0) | n'). \tag{37}$$

During the interaction, ρ undergoes a unitary transformation

$$\rho(t) = U(t, 0)\rho(0)U^{-1}(t, 0) \tag{38}$$

and the density matrix $\rho_f(t)$, which describes the state of the field only, is the projection[1] of (38) onto the space of the field variables

$$(n | \rho_f(t) | n') = \sum_m (mn | \rho(t) | mn'). \tag{39}$$

[1] This formalism is developed in detail by Jaynes [12].

The net change in the state of the field thus consists of a linear transformation,

$$(n | \rho_f(t) | n') = \sum_{k,k'} (nn' | G | kk')(k | \rho_f(0) | k') \tag{40}$$

or

$$\rho_f(\tau) = G\rho_f(0) \tag{41}$$

$$(nn' | G | kk') = \sum_{m,m',m''} (m''n | U | mk)(m'k' | U^{-1} | m''n')\sigma_{mm'} \tag{42}$$

where we have written for brevity

$$\sigma_{mm'} \equiv (m | \rho_1(0) | m'). \tag{43}$$

The sums (42) are readily evaluated with the use of (33), with the result that the only nonvanishing elements of G are

$$(nn' | G |'n, n') = a_{n+1}a^*_{n'+1}\sigma_{22} + c_n c^*_{n'}\sigma_{11} \tag{44a}$$

$$(nn' | G |'n+1, n') = b_{n+1}a^*_{n'+1}\sigma_{12} \tag{44b}$$

$$(n, n' | G | n, n'+1) = a_{n+1}b^*_{n'+1}\sigma_{21}, \tag{44c}$$

$$(n, n' | G | n, n'-1) = c_n b^*_{n'}\sigma_{12}, \tag{44d}$$

$$(n, n' | G | n-1, n') = b_n c^*_{n'}\sigma_{12}, \tag{44e}$$

$$(n, n' | G | n+1, n'+1) = b_{n+1}b^*_{n'+1}\sigma_{11}, \tag{44f}$$

$$(n, n' | G | n-1, n'-1) = b_n b^*_{n'}\sigma_{22}. \tag{44g}$$

These relations hold for all quantum numbers n if we understand that c_0 is not defined by (33) but by $c_0 = \exp(-i\omega_0 t)$.

To illustrate the use of this formalism, we discuss a few simple problems using (44). Consider first the case where the field is initially in its lowest state; $(0 | \rho_f(0) | 0) = 1$, all other elements of $\rho_f(0)$ vanish. Then according to (44), after a molecule with initial density matrix σ has passed through, the field density matrix has elements

$$\begin{aligned} (0 | \rho_f(\tau) | 0) &= |a_1|^2\sigma_{22} + \sigma_{11} \\ (0 | \rho_f(\tau) | 1) &= (1 | \rho_f(\tau) | 0)^* = c_0 b_1^* \sigma_{12} \\ (1 | \rho_f(\tau) | 1) &= |b_1|^2\sigma_{22}, \end{aligned} \tag{45}$$

all other elements still vanishing. If the molecule were initially in its lowest state then nothing happens, and the field remains in its ground state. If the molecule was initially in the upper state $[\sigma_{22} = 1,\ \sigma_{11} = \sigma_{12} = 0]$ we have a simple transition probability of $|b_1|^2$ for the molecule to emit one photon in passing through. If there was initially no coherence relation between upper and lower states of the molecule, then $\sigma_{12} = 0$, and ρ_f remains diagonal; no coherence between states $n = 0$ and $n = 1$ can be set up by the molecule unless there was some coherence initially between upper and lower states of the molecule.

The expectation value of electric field along the axis of the cavity, as obtained from (20), is

$$\langle E\rangle = \text{Trace}\,(\rho_f E)$$

$$= -\frac{\hbar\alpha}{\mu}\sum_n \sqrt{n+1}\,[(n\,|\,\rho_f\,|\,n+1) + (n+1\,|\,\rho_f\,|\,n)]$$

$$= -\frac{2\alpha\hbar}{\mu}\,\text{Re}\sum_n \sqrt{n+1}\,(n\,|\,\rho_f\,|\,n+1). \tag{46}$$

This remains zero as long as there is no coherence among adjacent levels, even though the energy stored in the field may be large. In the case (45), we obtain for $\langle E\rangle$,

$$\langle E\rangle = -\frac{2\alpha\hbar}{\mu}\,\text{Re}\,(c_0 b_1{}^* \sigma_{12})$$

$$= 2\hbar\alpha^2 \frac{\sin \beta t}{\mu\beta}\,\text{Re}\left[i\sigma_{12}e^{i(\Omega+\omega)t/2}\right] \tag{47}$$

where β is defined by (36) with $n=1$. Suppose now the cavity is so tuned that its resonant frequency ω is equal to Ω, then $\beta=\alpha$ and we obtain simply

$$\langle E\rangle = \frac{2\hbar\alpha}{\mu}\sin \alpha t\;\text{Re}\left[i\sigma_{12}e^{i\omega t}\right]. \tag{47a}$$

Since $\alpha \approx 5$ cps, the term $\sin(\alpha t)$ reaches its first maximum in a quarter cycle, or about 1/20 of a second. This is the interaction time required for a molecule to emit a photon, with probability one, into a lossless cavity initially in its ground state. This shows the great enhancement of spontaneous emission probability due to the presence of the resonant cavity, for the same molecule in empty space would emit with a natural line width (full width at half-maximum intensity),

$$\Delta\omega = \frac{8\omega^3\mu^2}{3\hbar c^3} \approx 10^{-7}\ \text{sec}^{-1}, \tag{48}$$

which leads to spontaneous emission times of the order of months at the frequencies here considered.

If the molecule and field are in arbitrary initial states, the general transformation of the field caused by passage of the molecule is, from (44),

$$(n\,|\rho_f(t)\,|n')$$
$$= \sigma_{11}[b_{n+1}\overset{*}{b}_{n'+1}(n+1\,|\,\rho_f(0)\,|\,n'+1) + c_n\overset{*}{c}_n(n\,|\,\rho_f(0)\,|\,n')]$$
$$+\sigma_{12}[b_{n+1}\overset{*}{a}_{n+1}(n+1\,|\,\rho_f(0)\,|\,n') + c_n\overset{*}{b}_{n'}(n\,|\,\rho_f(0)\,|\,n'-1)]$$
$$+\sigma_{12}[a_{n+1}\overset{*}{b}_{n'+1}(n\,|\,\rho_f(0)\,|\,n'+1) + b_n\overset{*}{c}_{n'}(n-1\,|\,\rho_f(0)\,|\,n')]$$
$$+\sigma_{22}[a_{n+1}\overset{*}{a}_{n'+1}(n\,|\,\rho_f(0)\,|\,n') \quad *$$
$$+ b_n\overset{*}{b}_{n'}(n-1\,|\,\rho_f(0)\,|\,n'-1)]. \tag{49}$$

If the field density matrix is initially diagonal,

$$(n\,|\,\rho_f(0)\,|\,n') = \rho_n\delta_{nn'}. \tag{50}$$

The only nonvanishing components of $\rho_f(t)$ are

$$(n\,|\,\rho_f(t)\,|\,n+1) = \sigma_{11}[\,|\,b_{n+1}\,|^2\rho_{n+1} + |\,c_n\,|^2\rho_n]$$
$$+ \sigma_{22}[\,|\,a_{n+1}\,|^2\rho_n + |\,b_n\,|^2\rho_{n-1}] \tag{51}$$

and

$$(n\,|\,\rho_f(t)\,|\,n+1) = (n+1\,|\,\rho_f(t)\,|\,n)^*$$
$$= \sigma_{12}[b_{n+1}\overset{*}{a}_{n+2}\rho_{n+1} + c_n\overset{*}{b}_{n+1}\rho_n]. \tag{52}$$

These relations will be used in the next section.

C. Successive Single-Molecule Interactions

If several molecules pass through the cavity in succession, the Nth entering as the $(N-1)$th leaves, all with the same initial state, this generates a Markov chain,

$$\rho_f(N\tau) = G^N\rho_f(0) = G\rho_f(N\tau - \tau). \tag{53}$$

Of particular interest is the limit $N\to\infty$.

If the density matrices of field and molecule are initially diagonal,

$$\sigma_{12} = \sigma_{21} = 0 \quad (n\,|\,\rho_f(0)\,|\,n') = \rho_n\delta_{nn'}, \tag{54}$$

then ρ_f remains diagonal for all time. In this case the entering molecules can always be described by a temperature, defined by

$$\sigma_{22} = \sigma_{11}e^{-x} = (e^x + 1)^{-1}$$
$$x = \hbar\Omega/kT \tag{55}$$

and, using (51), (53) reduces to

$$\rho_n(N\tau) = (e^x+1)^{-1}[(\,|\,a_{n+1}\,|^2 + |\,c_n\,|^2e^x)\rho_n(N\tau-\tau)$$
$$+ |\,b_{n+1}\,|^2e^x\rho_{n+1}(N\tau-\tau) + |\,b_n\,|^2\rho_{n-1}(N\tau-\tau)]. \tag{56}$$

From this the limiting form of ρ_n may be found. Taking note of the fact that fact that $|a_n|^2+|b_n|^2=|b_n|^2+|c_n|^2=1$, we find that a necessary and sufficient condition for a steady state $\rho_n(N\tau)=\rho_n(N\tau-\tau)=\rho_n$, is that the quantities

$$B_n = |\,b_n\,|^2(\rho_{n-1} - e^x\rho_n)$$

be independent of n. Now $\Sigma_n\rho_n=1$, and so $\rho_n\to 0$ as $n\to\infty$. Consequently, $B_n\to 0$, since $|b_n|^2\le 1$. Thus B_n can be independent of n only if $B_n=0$, and the only steady-state solution is the Boltzmann distribution,

$$\rho_n = e^{-x}\rho_{n-1}, \tag{57}$$

for all n for which $|b_n|^2\ne 0$. From (35) it is seen that b_n could vanish only for isolated special values of n.

Note that (57) is not a Boltzmann distribution with the same temperature T as that of the molecules, except in the case where the cavity is tuned exactly to the natural line frequency. The temperature of (57) is $T_f=\omega T/\Omega$. This difference would never be seen in practice because as soon as one detunes the cavity appreciably the transition probability $|b_n|^2$ becomes extremely small, and the temperature of the radiation would be determined by its interaction with the walls of the cavity, here neglected.

Nevertheless, in principle the difference is there, and we have an example of an interaction between two systems which maintains them at different temperatures. The origin of the phenomena lies in the fact that we have described the state of the molecule in terms of a temperature, which is not wholly justified, since nothing was said about their kinetic energy of translational motion. It is this translational motion which supplies or absorbs the excess energy so as to remove the above apparent violation of energy conservation. When a molecule enters or leaves the cavity it passes through a region of inhomogeneous field, and experiences a net force which very slightly changes its velocity.

In the "negative temperature" case where the entering molecules are more likely to be in the upper state, $\sigma_{22}>\sigma_{11}$, and $x<0$, the solution $B_n=$constant is still formally the only stationary one. But it now represents an infinite amount of energy in the field and could never be reached by any finite number of molecules passing through the cavity. It is, of course, only our neglect of losses which leads to such a result, and in practice the operating level quickly reaches a steady value which can be predicted by adding a phenomenological damping term to $\dot{\rho}$ in a well-known way.

As long as the density matrix σ of the entering molecules is diagonal, the density matrix of the field alone also remains diagonal; the expectation value of the electric field remains zero in spite of the fact that the number of photons present may be very large. That is, $\langle E^2\rangle$ can be very large but $\langle E\rangle$ remains zero. This is more or less to be expected since the entering molecules do not "tell" the field what phase to have. This situation raises certain questions, however, regarding the relation between quantum theory and classical theory. It is usually supposed that the condition for validity of classical electromagnetic theory is simply that the number of photons in each normal mode is large, and that then one may identify the classical electromagnetic field with the quantum-mechanical expectation value. It is seen, however, that this is a necessary but not sufficient condition, for here we have a situation where the semiclassical theory of radiation could not describe such states.

The statement, found in most books on quantum theory, that in the limit of large quantum numbers, quantum theory goes over into classical theory is somewhat misleading. Actually it is possible by coherent superposition of quantum states to construct states which are not describable in terms of classical theory at all. Thus it is that we arrive at the conclusion that classical theory is but a special case of quantum theory in the case of large quantum numbers, *i.e.*, large quantum numbers are necessary but not sufficient to insure the transition from quantum to classical theory.

The case in which two molecules pass through the cavity with flight time τ and leave just as two others enter, etc., has been worked out. The mathematics is tedious, and the result is substantially the same as the successive single-molecule interaction case.

III. Relation Between Quantum Electrodynamics and Semiclassical Radiation Theory

A. Semiclassical Electrodynamics

Now one considers that the electric field $E(t)$ is classically describable, and introduces a wave function,

$$\psi(t) = a(t)\psi_1 + b(t)\psi_2, \tag{58}$$

for the molecule alone, which develops in time according to the Schrödinger equation

$$i\hbar\dot{\psi} = (H_{\text{mol}} + H_{\text{int}})\psi \tag{59}$$

where

$$(m \mid H_{\text{mol}} \mid m') = E_m\delta_{mm'} \tag{60}$$

and

$$\begin{aligned}(m \mid H_{\text{int}} \mid m') &= (m \mid -\boldsymbol{\mu}\cdot \boldsymbol{E}(t) \mid m') \\ &= -\mu(1-\delta_{mm'})E(t).\end{aligned} \tag{61}$$

Schrödinger's equation (59) then reduces to

$$\begin{aligned}i\hbar\dot{a} &= E_1 a - \mu E(t)b \\ i\hbar\dot{b} &= -\mu E(t)a + E_2 b.\end{aligned} \tag{62}$$

These equations describe the effect of the field on the molecule.

Semiclassical theory as usually treated does not consider the effect of the molecule on the field. To find the effect of the molecule on the field, one calculates the expectation value of the dipole moment of the molecule from the solution of (62),

$$\boldsymbol{M}(t) \equiv \langle\boldsymbol{\mu}\rangle(t) = \boldsymbol{\mu}(ab^* + ba^*), \tag{63}$$

and assumes that the field satisfies the classical equations of motion which would result from interaction with a dipole of moment $\boldsymbol{M}(t)$. This is obtained most easily from the Hamiltonian equations of motion by addition of the interaction energy

$$-\boldsymbol{M}\cdot\boldsymbol{E} = +\sqrt{4\pi}\sum_a p_a(t)\boldsymbol{E}_a(\boldsymbol{x})\cdot\boldsymbol{M}(t) \tag{64}$$

to $\mathcal{H}$ in (7) of Section II, where $\boldsymbol{x}$ denotes the position of the molecule. The classical equations of motion are now

$$\dot{p}_a = -\frac{\partial\mathcal{H}}{\partial q_a} = -\omega_a{}^2 q_a$$

and

$$\dot{q}_a = \frac{\partial\mathcal{H}}{\partial p_a} = p_a + \sqrt{4\pi}\,\boldsymbol{M}\cdot\boldsymbol{E}_a(\boldsymbol{x}). \tag{65}$$

Eliminating q_a,

$$\ddot{p}_a + \omega_a{}^2 p_a = -\sqrt{4\pi}\,\omega_a{}^2 \boldsymbol{M}\cdot\boldsymbol{E}_a(\mathbf{x}). \tag{66}$$

Assuming that we have only one normal mode excited, the electric field of this mode satisfies the differential equation

$$\ddot{E} + \omega^2 E = \frac{+4\pi\omega^2}{J_1{}^2 V} M \tag{67}$$

where again we drop the subscript a. If the cavity has a finite Q, due to wall losses and/or energy coupled out, this is taken into account by adding a phenomenological damping term to (67), giving us

$$\ddot{E} + \frac{\omega}{Q}\dot{E} + \omega^2 E = \frac{4\pi\omega^2 M}{J_1{}^2 V}\cdot \tag{68}$$

By the "semiclassical" theory we mean the system of equations (62), (63) and (68). They may be given a somewhat neater formal appearance by eliminating the amplitudes $a(t)$, $b(t)$. The result is the nonlinear system of coupled equations,

$$\ddot{M} + \Omega^2 M = -K^2 W E, \tag{69a}$$

$$\dot{W} = E\dot{M} \tag{69b}$$

and

$$\ddot{E} + \omega/Q\dot{E} + \omega^2 E = SM, \tag{69c}$$

where

$$K = 2\mu/\hbar \qquad S = 4\pi\omega^2/J_1{}^2 V \tag{70}$$

and

$$W = E_1 |a|^2 + E_2 |b|^2 - \tfrac{1}{2}(E_1 + E_2)$$
$$= \frac{\hbar\Omega}{2}(|b|^2 - |a|^2) \tag{71}$$

is the expectation value of energy of the molecule, referred to a zero lying midway between the levels E_1, E_2. In the form (69) we have an apparently classical nonlinear system, all reference to "quantum-mechanical" quantities having disappeared.

The first two equations of (69) admit a first integral,

$$\dot{M}^2 + \Omega^2 M^2 + K^2 W^2 = \text{const.} = \left(\frac{K\hbar\Omega}{2}\right)^2. \tag{72a}$$

This is readily verified by eliminating E between them. Eq. (72a) is a disguised form of the principle of conservation of probability, $|a|^2 + |b|^2 = 1$. Similarly, the last two equations of (69) can be combined, in the case $Q = \infty$, to yield the constant of the motion

$$\dot{E}^2 + \omega^2 E^2 + 2S(W - ME) = \text{constant}, \tag{72b}$$

which is easily identified as the conservation of energy statement for the system.

B. The Relation Between Semiclassical and Quantum Electrodynamic Equations of Motion

For the equation of motion of any quantum-mechanical operator we have $i\hbar\dot{F} = [F, H]$. Differentiating this, we have

$$\hbar^2 \ddot{F} + [H, [H, F]] = i\hbar[\dot{H}, F] \tag{73}$$

which is exact for any operator F which has no explicit time dependence. Let us apply this identity to the electric field operator $F = E$. The total Hamiltonian $H = (H_{\text{mol}} + H_{\text{field}} + H_{\text{int}})$ has no explicit time dependence, so the right-hand side of (73) will vanish. To evaluate the double commutator, we note that H_{int} commutes with E but not with $[H_f, E]$, while H_m commutes with both. Therefore,

$$[H, [H, E]] = [H_f, [H_f, E]] + [H_{\text{int}}, [H_f, E]]. \tag{74}$$

These commutators are easily worked out, and the result is

$$[H_f, [H_f, E]] = \hbar^2\omega^2 E \tag{75}$$

$$[H_{\text{int}}, [H_f, E]] = -\hbar^2 S\mu_{\text{op}}. \tag{76}$$

Thus a special case of (73) is the operator identity

$$\ddot{E} + \omega^2 E = S\mu_{\text{op}} \tag{77}$$

which is to be compared to (12c). If we interpret (12c) as the expectation value of (20), they are seen to be identical in the limit $Q \to \infty$, provided that the expectation value of μ_{op} be defined, not in terms of $a(t)$ and $b(t)$ by means of (6), but as the expectation value taken over the complete density matrix $(mn/\rho/m'n')$, *i.e.*,

$$\langle\mu_{\text{op}}\rangle = \text{Tr}\,(\rho\mu_{\text{op}}) = \sum_{nmm'} (mn\rho m'n)(m' \mid \mu_{\text{op}} \mid m). \tag{78}$$

With this change in interpretation (69c) is seen to be an exact consequence of quantum electrodynamics.

We now write out the identity (73) for the operator $F = \mu_{\text{op}}$. This time H_{int} commutes with μ_{op}, but not with $[H_m, \mu_{\text{op}}]$, while H_f commutes with both. Therefore,

$$[H, [H, \mu_{\text{op}}]] = [H_m, [H_m, \mu_{\text{op}}]] + [H_{\text{int}}, [H_m, \mu_{\text{op}}]]. \tag{79}$$

Proceeding as before, a short calculation yields the following results:

$$[H_m, [H_m, \mu_{\text{op}}]] = \hbar^2\Omega^2\mu_{\text{op}} \tag{80}$$

and

$$[H_{\text{int}}, [H_m, \mu_{\text{op}}]] = \hbar^2 K^2 H' E \tag{81}$$

where we have defined an operator

$$H' = H_{\text{mol}} - \tfrac{1}{2}(E_1 + E_2) \tag{82}$$

with matrix elements

$$(mn \mid H' \mid m'n') = \frac{\hbar\Omega}{2}(-1)^m \delta_{mm'}\delta_{nn'}, \tag{83}$$

which is just the energy of the molecule, referred to a zero lying midway between its levels E_1, E_2. Combining these relations, we find that another special case of (73) is the operator identity

$$\ddot{\mu}_{op} + \Omega^2 \mu_{op} = -K^2 H' E \tag{84}$$

which is to be compared to (69a). However, now when we take the expectation value of (84) we do not get (69a) in general, for in the semiclassical equation the "driving term" appears as $\langle H' \rangle \langle E \rangle$, while quantum electrodynamics yields $\langle H'E \rangle$. The difference between these terms arises from the possibility of having correlated states, a situation inherent in quantum electrodynamics but not in semiclassical theory. When the states of field and molecule are uncorrelated, the density matrix reduces to a direct product $\rho = \rho_m \rho_f$, or

$$(mn \mid \rho \mid m'n') = (m \mid \rho_m \mid m')(n \mid \rho_f \mid n') \tag{85}$$

when (85) holds, then $\langle H'E \rangle = \langle H' \rangle \langle E \rangle$. But in general, $\langle H'E \rangle \neq \langle H' \rangle \langle E \rangle$.

The possibility of obtaining "correlated states" can arise whenever two or more quantum-mechanical systems interact. Quantum electrodynamics allows the possibility of states of the combined system (molecule plus field) which are in a definite pure state, but nevertheless one cannot ascribe any definite quantum state to the molecule alone, or the field alone. This possibility forms the basis of one of Einstein's objections to quantum mechanics. The Einstein-Podolsky-Rosen [13] paradox consists of the fact that when such correlated states exist, one has the possibility of predicting with certainty either one of two noncommuting quantities of a system by making measurements which do not involve any physical interaction with it.

An interesting line of thought is based on the fact that the semiclassical theory and quantum electrodynamics predict different equations of motion for a molecule in the field, the difference arising just from those correlated states which cause the above conceptual difficulties. Thus if one could find any experimental situation in which the difference between $\langle H'E \rangle$ and $\langle H' \rangle \langle E \rangle$ leads to any observable difference in maser operations, this would constitute an indirect, but convincing, check on those aspects of quantum theory which lead to the Einstein-Podolsky-Rosen paradox. However, as will be shown, the prospects of detecting such a difference are extremely dubious, for we will see that the semiclassical theory actually reproduces many of the features which one commonly supposes can be found only with field quantization.

C. Solution of Nonlinear Semiclassical Equations

The simplest approximate solution of the coupled semiclassical equations is the one wherein we ignore the time variation of W, thereby converting the problem into a linear one, similar to the case of two coupled pendulums. The normal modes are found by assuming that E and M have a common time factor $\exp(i\nu t)$; if $W = \text{constant}$, then (69a) and (69c) reduce to

$$(\omega^2 - \nu^2)(\Omega^2 - \nu^2) + K^2 SW = 0 \tag{86}$$

or

$$\nu^2 = \frac{\omega^2 + \Omega^2}{2} \pm \frac{1}{2}\sqrt{(\omega^2 - \Omega^2)^2 - 4K^2 SW}. \tag{87}$$

We see here a new feature, not present in coupled pendulums. If $W > 0$ and the cavity is tuned so closely to the natural line frequency that

$$|\omega^2 - \Omega^2| < \sqrt{4K^2 SW}, \tag{88}$$

the square root in (87) becomes imaginary; one of the normal modes grows exponentially, the other decays. Now an oscillation of growing amplitude represents energy being transferred from molecule to field, and therefore we see that the semiclassical theory *does* lead to a prediction of spontaneous emission. Since W is just the energy of the molecule, we see that the condition of unstable growing oscillation is just that the molecule's wave function contains more of the upper state than the lower, $|b|^2 > |a|^2$.

Suppose that the cavity is tuned exactly to the natural line frequency, $\omega = \Omega$. Then (87) reduces to

$$\nu^2 = \omega^2 \pm i\sqrt{K^2 SW} \tag{89}$$

or to an extremely good approximation,

$$\nu = \omega \pm \frac{i\sqrt{K^2 SW}}{2\omega}. \tag{90}$$

If we start with the molecule nearly in the upper state then $W = \hbar\Omega/2$ and the amplitude of the field varies like

$$\exp\left(\frac{\sqrt{K^2 SW}}{2\omega} t\right) e^{i\omega t} = \exp \alpha t e^{i\omega t} \tag{91}$$

where α is the interaction constant defined in (25). This is to be compared to the result (47a) describing spontaneous emission according to quantum electrodynamics. It is seen that although the two approaches lead to equations of different functional form, they predict exactly the same characteristic time $1/\alpha$ for spontaneous emission. This shows that the relation between quantum electrodynamics and the semiclassical theory of radiation is quite different from what is usually supposed. Physically, it means that whenever the molecule has a dipole moment different from zero, the fields set up by this dipole react back on the molecule and change its state in such a way that energy is delivered to the field, as long as $W > 0$. These linear relations do not hold indefinitely, of course. From the conservation law (72a) it is clear that when the amplitude of the M

oscillation increases, the magnitude of W must decrease, and this will eventually put a stop to the emission process.

The change in time of the variable W, for any reasonable value of field strength, as we are concerned with here, is slow compared to the time variation of E or M, and for typical ammonia maser operating conditions, E and M go through the order of 10^7 cycles for each cycle of W. For a qualitative picture of the slow changes in the case $\omega=\Omega$, we may consider the orbits in the $(\dot{E}, \omega E)$ plane and in the $(\dot{M}, \omega M)$ plane, as in Fig. 1. Noting that the interaction energy is typically about 10^{-6} times smaller than the energy of the molecule W, the conservation of energy law (72b) reduces, in almost all cases, to

$$(\dot{E})^2 + \omega^2 E^2 + 2SW = \text{constant}, \tag{92}$$

which shows that as W increases, the orbit in the $(\dot{E}, \omega E)$ plane must shrink, and vice versa. Also, the conservation law (72a) shows that if $|W|$ increases, the M orbit must shrink, and vice versa. Therefore the direction of all secular changes is determined by the sign of W and $\dot{W}$. In the equation $\dot{W}=E\dot{M}$ we can for all practical purposes replace $E\dot{M}$ by its average over one cycle, $\overline{E\dot{M}}$, since we are interested in the trend of W over time scales of many cycles, rather than small rapid fluctuations whose effect averages to zero over a cycle. Secular changes in W depend, thus, only on the sign of $\overline{E\dot{M}}$.

Whenever the E motion is advanced in phase over the M motion, we have $\overline{E\dot{M}}\geq 0$. In this case, W will slowly increase and the E orbit will shrink. The M orbit will then grow if $W<0$, shrink if $W>0$. If the M motion is advanced in phase over the E motion, all these changes are reversed. The situation is summarized by the orbit diagrams of Fig. 2. Or again, let us assume that W is given as some periodic function of time, so that we can summarize these same conclusions graphically as in Fig. 3.

Whenever the E orbit is expanding, energy is being delivered from the molecule to the field, and the necessary and sufficient condition for this is that the M motion be advanced in phase over the E motion. Thus in order to understand the long time course of events, one must study the secular changes in relative phases of the E and M motion.

To this end introduce the slowly varying complex amplitudes X and Y, defined by

$$\dot{E} + i\omega E = X(t)e^{i\omega t} \tag{93}$$

and

$$\dot{M} + i\omega M = Y(t)e^{i\omega t}. \tag{94}$$

The quantities depicted in Fig. 2 are just the complex numbers (93) and (94). Noting the properties,

$$(\dot{E})^2 + \omega^2 E^2 = |X|^2 \tag{95}$$

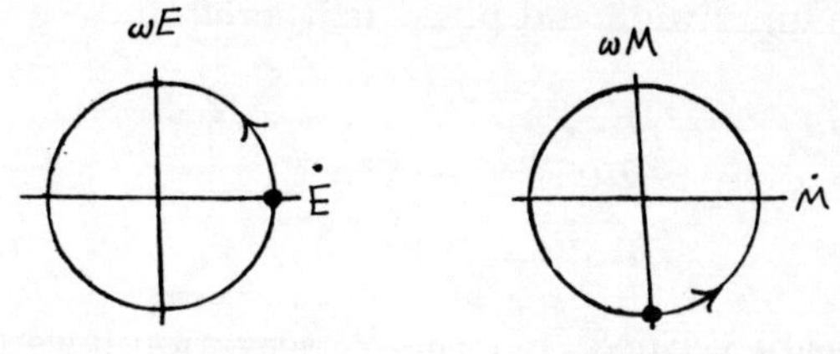

Fig. 1—Closed orbits in the phase space of the E and M oscillators. The dots indicate that the E motion is 90° ahead of the M motion in the phase.

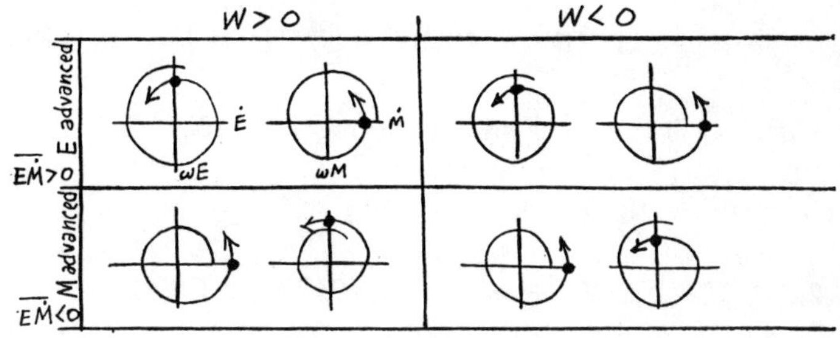

Fig. 2—Secular changes in orbits for the four combinations of signs of $\overline{E\dot{M}}$ and W.

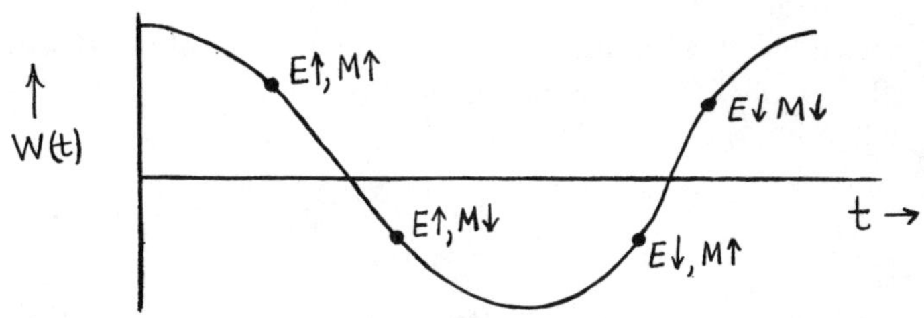

Fig. 3—Representation of the orbit diagrams in another way.

and

$$\ddot{E} + \omega^2 E = \dot{X}e^{i\omega t} \tag{96}$$

and similarly for M, we can write the equations of motion (69) in the form, for the case $\omega=\Omega$,

$$2i\omega\dot{X} = S(Y - Y^* e^{-2i\omega t}), \tag{97a}$$

$$2i\omega\dot{Y} = -K^2W(X - X^* e^{-2i\omega t}), \tag{97b}$$

$$4i\omega\dot{W} = XYe^{2i\omega t} + XY^* - X^*Y - X^*Y^*e^{-2i\omega t}. \tag{97c}$$

The conservation laws become

$$|Y|^2 + K^2W^2 = \text{constant} = \left(\frac{K\hbar\Omega}{2}\right)^2 \tag{98a}$$

and

$$|X|^2 + 2SW = \text{constant}. \tag{98b}$$

Now the functions X and Y are slowly varying functions of time, and again it is their average change over many cycles, rather than the very small rapid fluctuations at frequency 2ω, which are of interest. Thus the oscillating terms in (97) can be dropped, since their average over a cycle is negligible compared to their dc components.

The system of equations determining secular changes of both amplitude and phase is, therefore,

$$2i\omega\dot{X} = SY, \tag{99a}$$

$$2i\omega\dot{Y} = -K^2WX, \tag{99b}$$

$$4i\omega\dot{W} = XY^* - X^*Y. \tag{99c}$$

It is easily verified that the conservation laws (98) are exact consequences of (99). Differentiating (99c) once more and making use of the conservation laws, we can eliminate X and Y, obtaining the equation

$$4\omega^2\ddot{W} - 3SK^2W^2 + K^2CW + SK^2\left(\frac{\hbar\Omega}{2}\right)^2 = 0 \tag{100}$$

where C is the constant of the motion (98b). A first integral of (100) can be obtained immediately by multiplication with $\dot{W}$ and integrating

$$2\omega^2(\dot{W})^2 - SK^2W^3 + \frac{K^2CW^2}{2} + SK^2\left(\frac{\hbar\Omega}{2}\right)^2 W = \text{constant}. \tag{101}$$

This equation has the form of the Hamilton-Jacobi equation for motion of a particle in a particular potential well. For any motion in which either of the points $W = \pm(\hbar\Omega/2)$ is accessible, we have the constant on the right-hand side of (101) equal to

$$\frac{K^2C}{2}\left(\frac{\hbar\Omega}{2}\right)^2.$$

This is easily seen from (98a), for if $W = \pm(\hbar\Omega/2)$, then $Y=0$ and $\dot{W}=0$. For any such motion the cubic polynomial in (101) factors. To see this most easily, introduce the change of variable $(\hbar\Omega/2)z = W$. Then (101) takes the form,

$$b\dot{z}^2 - (z^3 - z - az^2 + a) = b\dot{z}^2 - [(z-1)(z+1)(z-a)] = 0, \tag{102}$$

where

$$b = \frac{4\omega^2}{SK^2\hbar\omega} = \frac{1}{2\alpha^2} \qquad a = \frac{C}{S\hbar\omega}\cdot \tag{103}$$

The solution is

$$\sqrt{2}\,\alpha t = \int_{z(0)}^{z(t)} \frac{dz}{\sqrt{(1-z)(1+z)(a-z)}}\cdot \tag{104}$$

The z motion is therefore periodic between turning points represented by singularities of the integrand. If $a>1$, these turning points are at $z=\pm 1$, while if $a<1$ they are at $z=-1$ and $z=a$. Now we consider evaluation of the constant $a = C/\hbar\omega S$. From (5), (9a) and (17) we have

$$\dot{E}^2 + \omega^2E^2 = \frac{4\pi\omega^2}{J_1{}^2V}(2n\hbar\omega)$$

where n is the number of photons stored in the cavity. Now, examination of (72b) with the small interaction term neglected gives us

$$\frac{4\pi\omega^2}{J_1{}^2V}(2n\hbar\omega) + S\hbar\omega = C = S\hbar\omega(2n+1)$$

if we assume there are n molecules in the field when the molecule is in its upper state, $W = +(\hbar\omega/2)$. Thus $a = (2n+1)$. There is in this theory of course no restriction on n to be an integer. The smallest value which a can attain is represented by zero energy in the field and the molecule in its ground state, $a=-1$, or $n=-1$. When n is negative this of course means that the total energy is insufficient for the molecule to get into its upper state, and this is the physical reason why the turning point of the z motion then occurs at $z=a$. The integral (104) is one of the standard forms defining elliptic functions. Using the standard notation sn (u, k), the solution for the case $n \geq 0$ is

$$z(t) = -1 + 2sn^2\left(\sqrt{n+1}\,\alpha t + Q, \frac{1}{\sqrt{n+1}}\right) \tag{105}$$

where

$$Q \equiv sn^{-1}\left(\sqrt{\frac{z(0)+1}{2}}, \frac{1}{\sqrt{n+1}}\right) \tag{106}$$

is the initial phase of the motion. In the limit of large n, the elliptic functions approach trigonometric functions, as is seen most easily from (104). If $a \gg 1$, then (104) reduces to

$$\sqrt{2}\,\alpha t \approx \frac{1}{\sqrt{a}}\int \frac{dz}{\sqrt{1-z^2}} = \frac{1}{\sqrt{a}}\sin^{-1} z(t) + \text{constant},$$

or

$$z(t) \approx \sin(2\sqrt{n}\,\alpha t + \theta). \tag{107}$$

The case $a=1$, $n=0$ is a special one, for the integrand of (104) then develops a first-order pole at $z=1$. The solution (105) is still valid but is no longer periodic; in fact sn $(u, 1)$ is equal to tanh u which approaches ± 1 asymptotically as $u \to \pm\infty$. This represents a case where the energy in the field exactly disappears just as the molecule gets into its upper state, and the final stages of the solution then represent the "shrinking normal mode" of (90), where E is 90° ahead of M. (This phase relation is in fact maintained throughout the part of the motion (105) in which z increases. Throughout the decreasing part, E is 90° behind M.)

The point $W = \hbar\Omega/2$, $z=1$ is a metastable point of the orbit in this case, for if we start out with exactly the initial comditions $z=1$, $E=M=0$ then nothing happens. All time derivatives remain zero and the molecule does not emit. However if there is the slightest change in this initial condition, the growing normal mode of (91) will be started up (unless the phase relations between M and E happen to be just the value for the pure shrinking mode), and eventually the energy of the

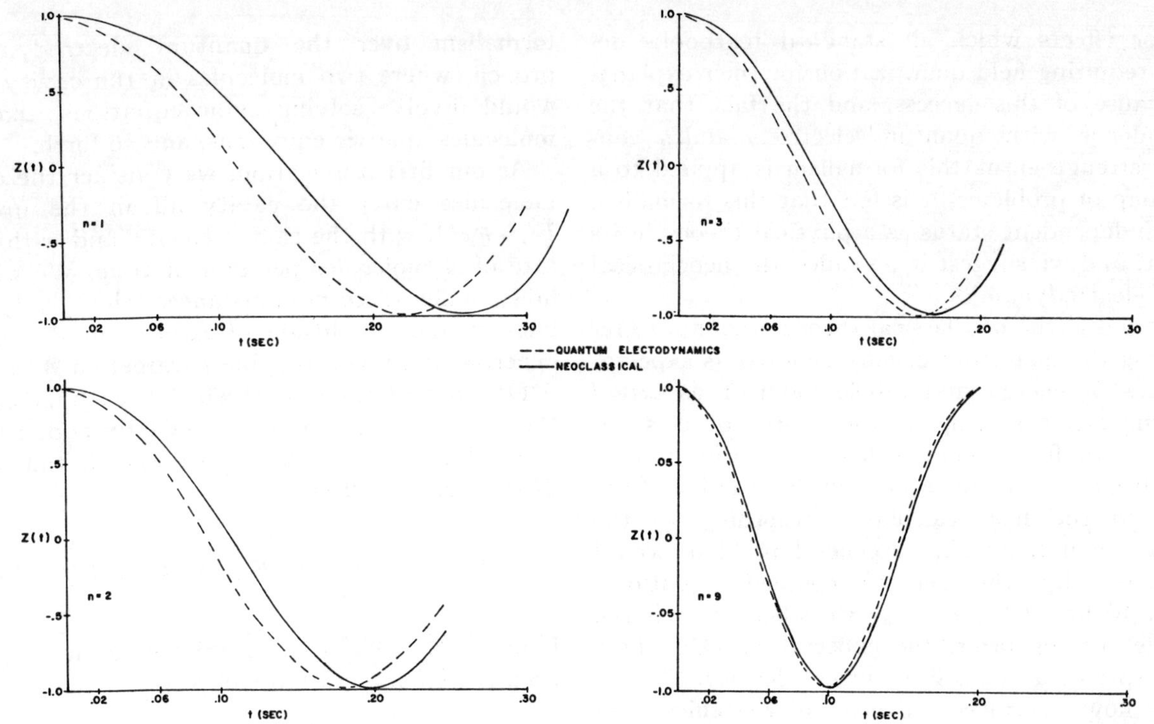

Fig. 4—Energy as a function of time for several n's.

molecule spills out entirely into the field when we reach the lower turning point $z=-1$. The molecule then re-absorbs the energy $\hbar\omega$ from the field, passing back to the metastable point $z=1$ but requiring an infinite time to do so. Fig. 4 shows $z(t)$ as a function of time for several values of the parameter n, the number of photons in the cavity, and these are compared with the corresponding quantum electrodyanmics curves. It is seen that for a few photons, the correspondence is almost exact. Even in the case of one or two quanta, the semi-classical theory gives solutions reproducing almost quantitatively everything that is found in the quantum electrodynamics analysis. Even the "quantum jumps" are still with us, but here they show up as perfectly continuous processes, where an instability develops in the solution of the nonlinear equations and an amount of energy $\hbar\omega$ is more or less rapidly transferred from molecule to field.

The semiclassical analysis gives a very interesting description of the process of spontaneous emission. Consider a large number of molecules, as nearly as possible in the upper state. In practice, of course, we cannot prepare them *exactly* in the upper state, but there will be a certain probability distribution of initial values of amplitude for the growing normal mode. A molecule with an initial value $M_+(0)$ will at time t have an M amplitude of $M_+(0)e^{\alpha t}=M_+(t)$.

If we agree to say that when this reaches the value K, the molecule is actively emitting energy, then, no matter what the probability distribution of initial values, provided only that this distribution is a continuous function in the neighborhood of the metastable point $W=\hbar\Omega/2$, we find that the number of molecules emitting at time t is proportional to $\exp(-2\alpha t)$. We can see this by a simple argument which runs as follows: We shall say that the molecule has reached stage K of the emission process when the amplitude of the oscillation reaches a value in the range, $K<M_+(0)e^{\alpha t}<(K+\delta K)$. We now ask, "How many atoms will be in stage K at time t?" Clearly, all those for which $M_+(0)$ lies in the range, $Ke^{-\alpha t}<M_+(0)<(K+\delta K)e^{-\alpha t}$. If the initial probability distribution in the phase space of the molecule is constant, this would be proportional to the area of the annular ring, $2\pi r dr=2\pi(Ke^{-\alpha t})(\delta Ke^{-\alpha t})\sim e^{-2\alpha t}$. Thus the "law of radioactive decay" or "time proportional transition probabilities" appears in this analysis as a consequence of the existence of metastable states. The time constant of the decay law is independent of the method of preparation of the molecules, and only depends on the interaction constant of the molecules with the electromagnetic field. The situation is exactly like that of a large number of pencils nearly perfectly balanced on their points. The time required for any one pencil to fall over depends on how close it was to vertical at $t=0$. If the probability distribution of initial states is continuous in the neighborhood of this metastable point, then we have a decay law with a time constant which depends only on the laws of mechanics, not on the method of preparation of initial states.

Mathematically, this semiclassical theory as expounded in this section is exactly the same as that already used by Shimoda, Wang and Townes [4]. The new feature is the realization that this formalism accounts for effects which all standard textbooks describe as requiring field quantization for their explanation. Because of this success, and the fact that the correspondence with quantum electrodynamics continues to strengthen as this formalism is applied to a larger group of problems, it is felt that this formalism deserves independent status as a physical theory in its own right, and we suggest it be called the neoclassical theory of electrodynamics.

Conceptually, the neoclassical theory amounts to reinterpreting the quantities usually denoted as expectation values of energy and dipole moment as *actual* values, the latter serving as the source of classical electromagnetic fields. These fields are then inserted in the Hamiltonian of the molecule and the reaction of the molecule to the field calculated according to the Schrödinger equation. Thus a general problem would be governed by the set of coupled equations, $i\hbar\dot{\psi}=H(A_u)\psi$ and $\Box A_u+4\pi\langle j_u\rangle=0$, where $\langle j_u\rangle$ is the current density operator, the expectation value here being interpreted as a classical current density.

Having now convinced ourselves of the efficacy of this method, we now turn to the application of these equations to the problem of the ammonia beam maser.

IV. Application of Neoclassical Radiation Theory to the Ammonia Maser

A. Ideal Steady-State Solution

Our starting point for obtaining the ideal steady-state solution for the ammonia maser will be (69), *i.e.*,

$$\ddot{M}_i + \Omega^2 M_i = -K^2 W_i E(t) \tag{108a}$$

$$\ddot{E} + \omega^2 E + \frac{\omega}{Q}\dot{E} = SM_i \tag{108b}$$

$$\dot{W}_i = E\dot{M}_i \tag{108c}$$

where now the subscript i refers to the ith molecule. If now all of the molecules are subjected to the same field $E(t)$ as in the Stanford [6] ammonia maser, we can simply define the total moment and energy

$$M(t) \equiv \sum_i M_i(t), \tag{109a}$$

$$W(t) \equiv \sum_i W_i(t). \tag{109b}$$

We see that (108a)–(108c) are still satisfied by these quantitites, in particular,

$$\ddot{E} + \omega^2 E + \frac{\omega}{Q}\dot{E} = SM = S\sum_i M_i. \tag{110}$$

The conservation law (72a) is still valid in the sense that the left-hand side is still constant; but, of course, the value of the constant now depends on the initial conditions. Since the problem of N molecules in the cavity is hardly any more difficult in this formalism than that of one molecule, we can see the advantage of this formalism over the quantum electrodynamics approach, where two molecules in the cavity at a time would involve solving cubic equations, that of three molecules, quartic equations, and so forth.

As our first application, we consider the case where molecules enter the cavity all in the upper state, $W_i=\hbar\Omega/2$, with the same velocity and with a uniform rate of A molecules per unit of time. We wish to find how, under these circumstances, the steady-state frequency and amplitude of oscillation depend on the experimentally controllable parameters A, ω, Q.

Denote by t_i the time at which the ith molecule enters the cavity. It is readily verified by substitution that the solution of (108a) with the initial conditions $M_i(t_i)=\dot{M}_i(t_i)=0$ is

$$M_i(t) = -\frac{K^2}{\Omega}\int_{t_i}^{t} W_i(t')E(t')\sin\Omega(t-t')dt'. \tag{111}$$

Using this, (108c) can be written as an integral equation. A time integration yields

$$W_i(t) - W_i(t_i) = \int_{t_i}^{t} dt''E(t'')\dot{M}(t'')$$

$$= -K^2\int_{ti}^{t} dt''E(t'')\int_{ti}^{t''} dt'W_i(t')E(t')\cos\Omega(t''-t'). \tag{112}$$

Interchanging the order of integration in (112), we find that $W_i(t)$ satisfies an integral equation of Volterra form,

$$W_i(t) - W_i(t_i) = \int_{t_i}^{t} G(t,t')W_i(t')dt', \tag{113}$$

with the kernel,

$$G(t,t') = -K^2\int_{t'}^{t} dt''E(t'')\cos\Omega(t''-t')E(t'). \tag{114}$$

We now assume the electric field is given by

$$E(t) = 2a\sin\nu t \tag{115}$$

where a and ν are parameters to be determined by the condition that (110), (111) and (113) be self-consistent. It is clear from (108c) that the exact solution of $W_i(t)$ contains terms oscillating at frequencies of the order of $(\Omega+\nu)$. While these terms may contribute appreciably to $\dot{W}_i$, their effect on W_i averages to zero in times of the order of one cycle of the RF. Since we are interested in the long time drift in W_i, rather than these small rapid fluctuations, we neglect terms in (114) of frequency $(\Omega+\nu)$. Their contribution to W_i is of the relative order of magnitude

$$\frac{(\Omega-\nu)}{(\Omega+\nu)} \lesssim 10^{-7}$$

in all cases of practical interest. With this approximation (114) reduces to

$$G(t, t') \approx -K^2a^2 \frac{\sin(\Omega - \nu)(t - t')}{(\Omega - \nu)} \quad (116)$$

and the slowly varying part of $W_i(t)$ satisfies the integral equation

$$W_i(t) = W_i(t_i) - K^2a^2 \int_{t_i}^{t} W_i(t') \frac{\sin(\Omega - \nu)(t - t')}{(\Omega - \nu)} dt'. \quad (117)$$

The exact solution of (117) with the initial condition $W_i(t_i) = \hbar\Omega/2$ is

$$W_i(t) = \frac{\hbar\Omega}{2\lambda^2} \left[(\Omega - \nu)^2 + a^2K^2 \cos \lambda(t - t_i)\right] \quad (118)$$

where

$$\lambda^2 \equiv (\Omega - \nu)^2 + (Ka)^2. \quad (119)$$

As a check, and an illustration of some of our previous remarks, we note that (118) and (119) agree with the results found from quantum electrodynamics, and (35) and (36), and also with the result of others who have treated the problem by direct integration of Schrödinger's equation.

The total dipole moment of all the moledules in the cavity is

$$M(t) = A \int_{t-\tau}^{t} M_i(t) dt_i = -\frac{AK^2}{\Omega} \int_{t-\tau}^{t} dt' E(t') \sin \Omega(t - t') \cdot \int_{t-\tau}^{t'} dt_i W_i(t') \quad (120)$$

where we have used (111) and inverted the order of integration. With the solution (118) for $W_i(t)$, this becomes

$$M(t) = -\frac{Aa\hbar\Omega K^2}{\Omega\lambda^2} \int_0^{\tau} \left[(\Omega - \nu)^2 q + \frac{a^2K^2}{\lambda} \sin \lambda q\right] \cdot \sin \nu(q + t - \tau) \sin \Omega(\tau - q) x dq \quad (121)$$

where $q = (\tau - t + t')$. As a function of q, the last factor of the integrand contains oscillating terms of frequencies $(\Omega \pm \nu)$, and again the relative contribution of the high-frequency term will be of the order of 10^{-7} or smaller under all conditions of interest. Neglecting this small term, (121) reduces to

$$M(t) = \frac{Aa\hbar\Omega K^2}{\Omega\lambda^2} \left[(1 - \cos \lambda\tau) \cos \nu t - \frac{(\Omega - \nu)}{\lambda} (\lambda\tau - \sin \lambda\tau) \sin \nu t\right]. \quad (122)$$

Details from this point on will be considered later, as a special case of a more general solution.

B. Velocity Distribution

In the preceding section, all the molecules were assumed to have the same flight time τ. To find the effect of a velocity distribution, we have only to note that the analysis leading to (122) is still valid, and it gives the contribution to total moment of those molecules with flight time in the range $d\tau$, provided that we replace A by $nf(\tau)d\tau$, where n is the total number of molecules entering the cavity per unit time, and $f(\tau)d\tau$ is the fraction of entering molecules with flight times in the range $d\tau$, normalized so that

$$\int_0^{\infty} f(\tau) d\tau = 1. \quad (123)$$

The total dipole moment of all molecules in the cavity is then obtained by one more integration of (122), as follows:

$$M(t) = n\gamma \left\{[1 - c(\lambda)] \cos \nu t - \frac{(\Omega - \nu)}{\lambda} [\lambda\bar{\tau} - S(\lambda)] \sin \nu t\right\} \quad (124)$$

where we have defined

$$\gamma = \frac{a\hbar\Omega K^2}{\Omega\lambda^2} \quad (125)$$

for convenience, and where

$$\bar{\tau} = \int_0^{\infty} \tau f(\tau) d\tau \quad (126)$$

is the mean flight time, and

$$c(\lambda) \equiv \int_0^{\infty} \cos \lambda\tau f(\tau) d\tau \quad (127)$$

and

$$s(\lambda) \equiv \int_0^{\infty} \sin \lambda\tau f(\tau) d\tau \quad (128)$$

are the Fourier transforms of the flight time distribution.

To obtain the conditions for a self-consistent solution, we substitute (124) into (110) and equate the coefficients of $\cos \nu t$, $\sin \nu t$. We obtain the relations

$$\frac{\omega\nu}{Q} = \frac{Sn\gamma}{2a} [1 - c(\lambda)] \quad (129)$$

and

$$\nu^2 - \omega^2 = \frac{Sn\gamma}{2a} \frac{\Omega - \nu}{\lambda} [\lambda\bar{\tau} - s(\lambda)]. \quad (130)$$

The starting current n_0 is determined by (129) for small λ. From (127) we have

$$\lim_{\lambda \to 0} \frac{1 - c(\lambda)}{\lambda^2} = \int_0^\infty \frac{1}{2} \tau^2 f(\tau) d\tau = \frac{\overline{\tau^2}}{2} \tag{131}$$

so that it is the mean-square flight time which determines the starting current, as follows:

$$n_0 = \frac{4a\omega\nu}{Q\lambda^2 j \overline{\tau^2} S} \approx \frac{\hbar J_1{}^2 V}{Q\overline{\tau^2} 4\pi\mu^2} \cdot \tag{132}$$

Similarly, we have from (128),

$$\lim_{\lambda \to 0} \frac{\lambda\bar{\tau} - S(\lambda)}{\lambda^3} = \frac{1}{6} \overline{\tau^3}, \tag{133}$$

so that if we define new functions,

$$F(\lambda) \equiv \frac{6[\lambda\bar{\tau} - s(\lambda)]}{\lambda^3 \overline{\tau^3}} \tag{134}$$

and

$$G(\lambda) \equiv \frac{2[1 - c(\lambda)]}{\lambda^2 \overline{\tau^2}}, \tag{135}$$

we have $F(0) = G(0) = 1$. Previous writers [1] have expressed their results in terms of an "effective Q" of the molecular beam. The appropriate definition here would be

$$Q_m \equiv \frac{\Omega \overline{\tau^3}}{6\overline{\tau^2}} \approx 10^6. \tag{136}$$

Our conditions (129) and (130) then assume the forms

$$n_0/n = G(\lambda) \tag{137}$$

and

$$(\nu - \Omega) \cong (\omega - \Omega) \frac{Q}{Q_m} \frac{G(\lambda)}{F(\lambda)} \tag{138}$$

if we neglect terms of order $(Q/Q_m)^2$. These relations are to be used graphically as follows: For a given velocity distribution, the functions $F(\lambda)$ and $G(\lambda)$ can be calculated and plotted. Then from (137) one determines λ as a function of the beam current. The frequency pulling factor (G/F) of (138) is then determined. Finally, the amplitude of oscillation a is determined for given beam current n and cavity tuning ω, by use of (138) and (119). For constant beam current, *i.e.*, constant λ, the graph of amplitude vs frequency of oscillation is a half ellipse, the amplitude reaching a maximum for perfect tuning of the cavity to the molecule line, in which case $Ka = \lambda$. Oscillations can persist over a frequency band $(\Omega - \lambda) < \nu < (\Omega + \lambda)$.

In the case of a single molecular velocity, $f(\tau) = \delta(\tau - \tau_0)$, these relations reduce to those of Shimoda, Wang, and Townes [4], and in the case of a Maxwellian velocity distribution, where the fraction of molecules per unit time with velocities in the range dv is proportional to $v^3 \exp(-v^2/v_0{}^2)\, dv$, or

$$f(\tau) \sim \exp(-L^2/v_0{}^2\tau^2)/\tau^5 \tag{139}$$

(L = length of cavity, $v_0{}^2 = 2KT/m$), they reduce to the theory of Lamb and Helmer [6]. In the Maxwellian case our dimensionless parameters F, G, Q_m become identical with those defined by Lamb and Helmer.

The assumption of a Maxwellian distribution is certainly a reasonable first approximation, but it is probably not very accurate because the quality of focusing is velocity dependent, the focuser having the property of focusing the small velocity molecules much more effectively than the high-speed ones; and, except for extremely strong focusing voltage, the distribution of flight times may be biased considerably more in favor of large τ than is indicated by (139). This could have an important effect on stability, as may be seen in the following.

In the case of a Maxwellian distribution, half of the mean-square flight time $\overline{\tau^2}$, which determines the starting current, is contributed by the slowest 12 per cent of the molecules. Half of the third moment $\overline{\tau^3}$, which determines the effective molecular Q, and hence the long time frequency stability, is due to the slowest 1.9 per cent. Any further biasing in favor of higher τ would have a considerable effect on $\overline{\tau^2}$, and a very large effect on $\overline{\tau^3}$. For this reason, effects of fluctuations in beam current may not be of the relative order of magnitude $1/\sqrt{N}$, where N is the total number of molecules in the cavity. Fluctuations in the experimentally significant quantities may be determined almost entirely by the slowest 5 per cent of the molecules, with corresponding greater relative variation.

The effect on the frequency-pulling function G/F of a "truncated" Maxwellian distribution has been worked out for the case where the velocity distribution is taken to be Maxwellian up to some $v_{\max}$ and zero thereafter, where $v_{\max}$ was taken arbitrarily to be $\frac{1}{2}v_0$. The results were compared to that of a Maxwellian distribution. A region of stability still appears, as in the analysis of Lamb and Helmer, but it occurs for smaller values of λ (about a factor of two) which corresponds to smaller values of beam flux.

V. Fluctuation Effects

The steady-state relations found in Section IV form the starting point for investigations of fluctuation effects, by perturbation methods in which we expand in powers of the small departure from the previous solutions. If we calculate only to the lowest nonvanishing order, these problems become linear. But in this case we can analyze the effect of small periodic perturbations, proportional to $\exp(i\beta t)$ and superpose the solutions to find the transient response to an arbitrary small perturbation. Thus, consider the effect of an additional signal $F(t)$ impressed on the cavity. This might be an extra signal intentionally fed in, or it might be a randomly varying function representing thermal noise generated in the cavity and/or load. The equation of motion for the electric field then becomes

$$\ddot{E} + \omega^2 E + \frac{\omega}{Q}\dot{E} = SM(t) + F(t) \tag{140}$$

and this $F(t)$ causes a change of E_1 in the electric field. Suppose now that $F(t)$ contains the time factor exp $(i\beta t)$; then if β is not too close to the oscillation frequency ν, the change in electric moment of the ith molecule will satisfy

$$\ddot{M}_{1_i} + \Omega^2 M_{1_i} = -K^2 W_i E_1 \tag{141}$$

where we have set

$$E = E_0 + E_1 = 2a \sin \nu t + E_1 \tag{142}$$

and

$$M_i = M_{0_i} + M_{1_i} \tag{143}$$

$$W_i = W_{0_i} + W_{1_i} \tag{144}$$

where the subscript "0" denotes the unperturbed or steady-state solutions of Section IV. Here we have dropped a term EW_{1_i} on the grounds that it will not have an appreciable component of frequency β. Under these conditions, the change in total moment of all molecules in the cavity will be simply

$$M_1 = \sum_i M_{1_i} = \frac{-K^2\overline{W}E_1}{\Omega^2 - \beta^2} \tag{145}$$

where $\overline{W}$ is the average energy of all molecules in the cavity. Combining these relations, we find the electric field fluctuation to be given by

$$E_1 = \frac{F(t)(\Omega^2 - \beta^2)}{\left[\left(\omega^2 - \beta^2 + \frac{i\omega\beta}{Q}\right)(\Omega^2 - \beta^2) + K^2 S\overline{W}\right]}. \tag{146}$$

If $\overline{W} = 0$, this reduces, as it must, to the response of the cavity alone. The effect of the molecules is, in this approximation, to suppress the magnitude of the electric field fluctuations E_1, for those frequency components which lie close to the natural line frequency.

If the period of the beat frequency $(\beta - \omega)$ is comparable to the flight time, however, then one should take into account the term W_{1_i}; generally, perturbations of any type can lead to the greatest effects when their frequency is related in this way to the flight time. Since the theory remains linear, flucutations due to any cause are readily calculated. We now proceed to the calculation of electric field fluctuation in which we retain the term W_{1_i}.

Keeping terms to only the first order in the perturbation, we have to solve the system of linear equations with time varying coefficients,

$$\ddot{E}_1 + \frac{\omega}{Q}\dot{E}_1 + \omega^2 E_1 = SM_1(t) + F(t)$$

$$= S\sum_i M_{1_i}(t) + F(t), \tag{147}$$

$$\ddot{M}_{1_i} + \Omega^2 M_{1_i} = -K^2 W_{1_i} E_0 - K^2 W_{0_i} E_1, \tag{148}$$

and

$$\dot{W}_{1_i} = E_1 \dot{M}_{0_i} + E_0 \dot{M}_{1_i}. \tag{149}$$

From (149) we have

$$W_{1_i}(t) = \int_{t_i}^{t} E_0(t')\left[-K^2 \int_{t_i}^{t'} (W_{1_i}(t'')E_0(t'') + W_{0_i}(t'')E_1(t'')) \cos \Omega(t' - t'')dt''\right] dt'$$

$$+ \int_{t_i}^{t} E_1(t')\left[-K^2 \int_{t_i}^{t'} W_{0_i}(t'')E_0(t'') \cos \Omega(t' - t'')dt''\right] dt' \tag{150}$$

where we have used

$$M_{1_i}(t) = -\frac{K^2}{\Omega}\int_{t_i}^{t} \sin \Omega(t - t') \cdot [W_{1_i}(t')E_0(t') + W_{0_i}(t')E_1(t')]dt'. \tag{151}$$

We specialize to the case of the tuned cavity, $\omega = \Omega$, since this will not significantly affect the results and renders the mathematics very much less tedious. Now, assuming a solution of the form

$$E_1(t) = a_1 e^{i\beta t} + a_2 e^{i(\beta - 2\Omega)t} \tag{152}$$

we are able to get self-consistent solutions for (147)–(149). Using (152) and the unperturbed solutions of the preceding chapter for the case of a single molecular flight time, we have for (150),

$$W_{1_i}(t) = \frac{K\hbar\Omega}{4\Omega'}(a_1 - a_2)(e^{-i\Omega' t} - e^{-i\Omega' t_i}) \sin \lambda(t - t_i). \tag{153}$$

We have dropped terms of frequency $\Omega + \beta$, as they do not contribute appreciably to W_1. Here $\Omega' \equiv \Omega - \beta$. Now we put this result into (151) and after integrating over all the molecules, *i.e.*,

$$M_1(t) = A\int_{t-\tau}^{t} M_{1_i}(t)dt_i \tag{154}$$

and, inverting the order of integration, we have

$$M_1(t) = -\frac{K^2A}{\Omega}\int_{t-\tau}^{t} dt' E_0(t') \sin \Omega(t-t') \int_{t-\tau}^{t} W_{1_i}(t')dt_i - \frac{K^2A}{\Omega}\int_{t-\tau}^{t} dt' E_1(t') \sin \Omega(t-t') \cdot \int_{t-\tau}^{t'} W_{0_i}(t')dt_i. \quad (155)$$

The result of the integrations is

$$SM_1(t) = a_1Be^{i\beta t} + a_1Ce^{i(\beta-2\Omega)t} - a_2Be^{i(\beta-2\Omega)t} - a_2Ce^{i\beta t} \quad (156)$$

where we have defined

$$C = \frac{AK^2S\hbar i}{4(\Omega'^2 - \lambda^2)} e^{i\Omega'\tau/2} \cdot \left[\cos \Omega'\tau/2(1-\cos\lambda\tau) - \frac{\lambda}{\Omega'}\sin \Omega'\tau/2 \sin \lambda\tau\right] \quad (157)$$

and

$$B = C + \frac{AK^2S\hbar i}{4(\Omega'^2 - \lambda^2)} \cdot \left[\cos \lambda\tau + i\frac{\Omega'}{\lambda}\sin\lambda\tau - \cos\Omega'\tau - i\sin\Omega'\tau\right]. \quad (158)$$

Inserting this expression into (147) we find

$$a_1 = \frac{F(\Omega_2 + B)}{[(\Omega_1 - B)(\Omega_2 + B) + C^2]} \quad (159)$$

and

$$a_2 = \frac{FC}{[(\Omega_1 - B)(\Omega_2 + B) + C^2]} \quad (160)$$

where we define

$$\Omega_1 = \Omega^2 - \beta^2 + i\Omega\beta/Q \quad (161)$$

and

$$\Omega_2 = \Omega^2 - (\beta - 2\Omega)^2 + i\frac{\Omega}{Q}(\beta - 2\Omega). \quad (162)$$

Note that $\Omega_1 \approx -\Omega_2$ if $\Omega - \beta \leq \pm 10\lambda$. The expressions (157) and (158) can be simplified somewhat by remembering that we can replace A by the starting current A_0 divided by $G(\lambda)$ for the univelocity case, *i.e.*,

$$A = A_0/G(\lambda) = \frac{4\Omega^3}{Q\tau^2(\hbar\Omega K^2 S)}\frac{\lambda^2\tau^2}{2(1-\cos\lambda\tau)} \quad (163)$$

from (132) and (135). Thus,

$$C = \frac{i\lambda^2(\Omega^2/Q)[\Omega' \cos \Omega'\tau/2 \sin \lambda\tau/2 - \lambda \cos\lambda\tau/2 \sin \Omega'\tau/2]}{2\Omega'(\Omega'^2 - \lambda^2)\sin\lambda\tau/2} \quad (164)$$

and

$$B = C + \frac{i\lambda^2\Omega^2/Q[(\cos\lambda\tau - \cos\Omega'\tau) + i/\lambda(\Omega'\sin\lambda\tau - \lambda\sin\Omega'\tau)]}{2\Omega'(\Omega'^2 - \lambda^2)(1-\cos\lambda\tau)}. \quad (165)$$

We see here the result of having time-varying coefficients in our linear equations (147)–(149). If we feed in a signal of frequency β, the field given back contains the frequencies β and $(2\Omega - \beta)$, *i.e.*, the frequency β and the reflection of β about Ω.

If we plot the amplitudes $|a_1|^2$ and $|a_2|^2$ of the frequencies β and $(2\Omega-\beta)$ as a function of the parameter k, such that $\Omega' = k\lambda$, for various values of the flight time τ, we observe a "quieting" effect for certain values of the flight time. For small values of the flight time, the effect of the molecular beam is to amplify any input signal, *e.g.*, thermal noise. The region of greatest stability, that at which "noise quieting" would be the greatest, appears at $\lambda\tau = 3\pi/2$. See Fig. 5, pp. 107–108.

This analysis was carried out under the assumption of a uniform velocity distribution and a tuned cavity. It is clear enough that these restrictions do not greatly impair the generality of the results obtained. The analysis for an untuned cavity would merely have the effect of shifting the axis in the plots of Fig. 5 by the amount $\Omega - \nu$, so that one would have plots symmetrical about ν instead of Ω. Also, very plausible speculation leads one to conclude that the only effect of a velocity distribution, which mathematically would appear as integrals over τ in B and C in the denominators of (159) and (160), would be to smooth out the "wiggles" which appear near the "bare" cavity response value $FQ/\Omega^2 = 1$.

One can recast the electric field,

$$E(t) = 2a\sin\Omega t + \int_{-\infty}^{+\infty} a_1 e^{i\beta t}d\beta + \int_{-\infty}^{+\infty} a_2 e^{i(2\Omega-\beta)^t}d\beta + cc \quad (166)$$

in the form

$$E(t) = A(t)\sin[\Omega t + \phi(t)] \quad (167)$$

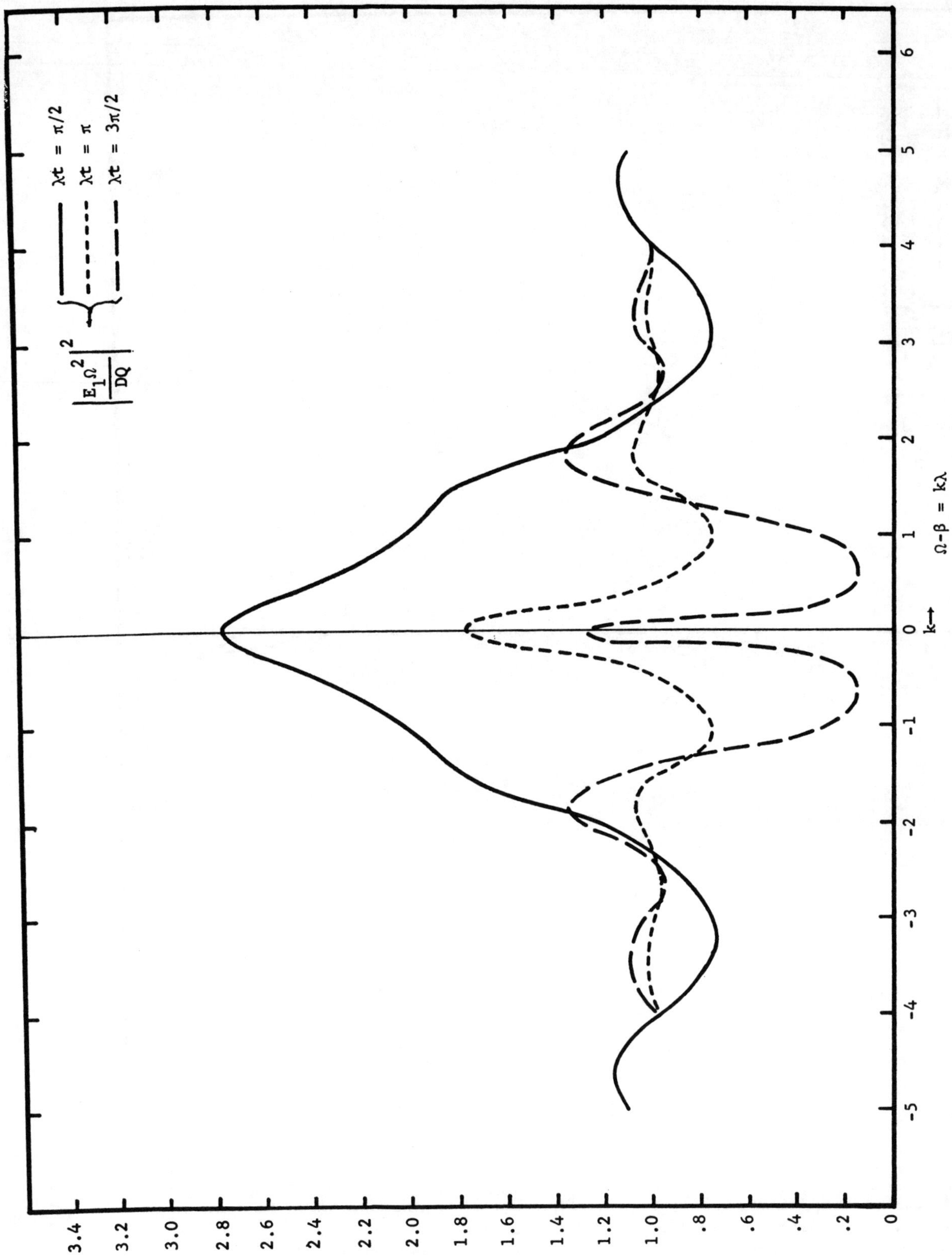

$\left|\frac{E_1\Omega^2}{DQ}\right|^2$
$\lambda t = \pi/2$
$\lambda t = \pi$
$\lambda t = 3\pi/2$
k⟶
$\Omega-\beta = k\lambda$
-5
-4
-3
-2
-1
0
1
2
3
4
5
6
0
.2
.4
.6
.8
1.0
1.2
1.4
1.6
1.8
2.0
2.2
2.4
2.6
2.8
3.0
3.2
3.4

(a)

Fig. 5.

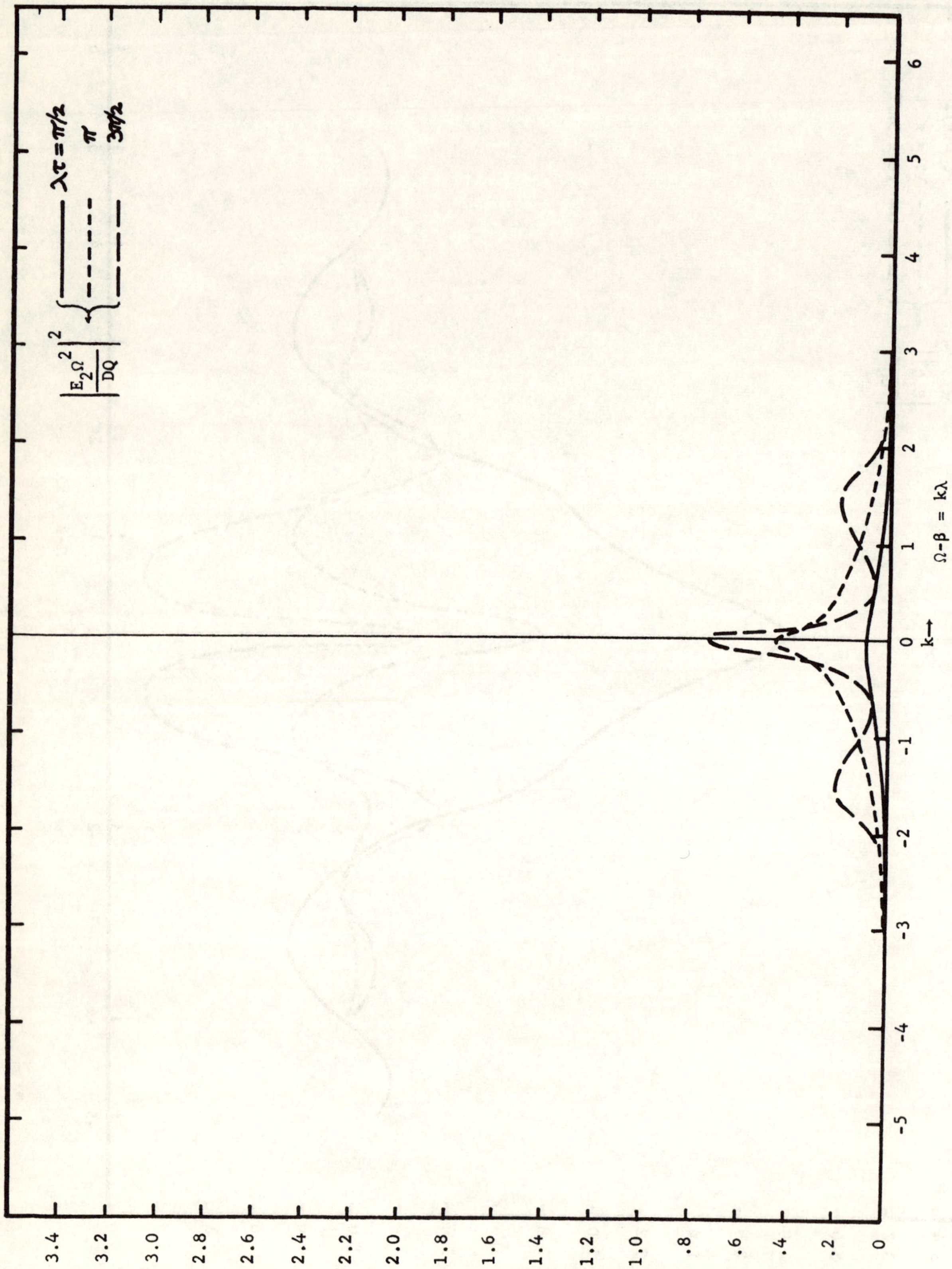
$\left|\frac{E_2\Omega^2}{DQ}\right|^2$
λτ = π/2
π
3π/2
Ω-β = kλ
k→
-5 -4 -3 -2 -1 0 1 2 3 4 5 6
0 .2 .4 .6 .8 1.0 1.2 1.4 1.6 1.8 2.0 2.2 2.4 2.6 2.8 3.0 3.2 3.4

(b)

Fig. 5.

where the power spectra of $A(t)$ and $\phi(t)$ are known from the knowledge of a_1 and a_2. One might then argue that we can obtain $\langle\phi^2\rangle$ and thus have an answer for the frequency stability; but we must remember that we have used a first-order perturbation method, which is valid only for small momentary departures from the steady-state solution—the frequency instability due to thermal noise in the walls probably causes the phase to wander in a manner closely akin to Brownian motion phenomena; and we cannot expect to predict cumulative phenomena by use of perturbations. However, since there is "restoring force" on the amplitude, it can never wander very far from the steady-state value, and we would expect that this analysis could lead to a good prediction of the amplitude stability due to a random perturbation such as thermal noise generated in the cavity or walls.

The criterion for the accuracy of any frequency standard, or "clock," is not so much how far the oscillation frequency drifts from its nominal value in a given time, but rather how well we are able to *predict* how it will wander. Here one would be concerned with the wander of the phase of the output under the perturbation of 1) the thermal radiation from the cavity walls and 2) the unavoidable random fluctuations in beam composition. The analysis given above indicates that under conditions likely to be realized in the foreseeable future, the former effect will be by far the most important.

By analogy with the classical Einstein treatment of Brownian motion, one expects that the phase ϕ will be uncertain in the following sense: While $\langle\phi(t)\rangle=0$, $\langle\phi^2(t)\rangle=Kt$. The constant K would then be a reasonable measure of the oscillation stability.

The analysis given above is not, however, adapted to answering questions of this type. In assuming that the actual output could be represented in the form

$$E(t) = 2a \sin \nu t + E_1(t)$$

with $|E_1|\ll a$, we have in effect restricted the theory to cases (or time intervals) such that the phase wandering ϕ is at most of order E_1/a. A more general theory of random processes is therefore needed before questions about very long time behavior can be answered.

Wiener [15] has given an interesting discussion of the response of nonlinear systems to a random perturbation. His curves are very similar to those of Fig. 5, which arise, for instance, in the analysis of the alpha rhythm of brain waves. It appears that the phenomena predicted by Fig. 5 is a general property of any nonlinear system which is attempting to stabilize itself.

References

[1] J. P. Gordon, H. J. Zeiger, and C. H. Townes, "Molecular microwave oscillator and new hyperfine structure in the microwave spectrum of NH_3," *Phys. Rev.*, vol. 95, pp. 282–284; July 1, 1954.
——, "The maser—new type of microwave amplifier, frequency standard, and spectrometer," *Phys. Rev.*, vol. 99, pp. 1264–1274; August 15, 1955.

[2] M. W. Muller, "Noise in a molecular amplifier," *Phys. Rev.*, vol. 106, pp. 8–12; April 1, 1957.

[3] J. P. Gordon and L. D. White, "Noise in maser amplifiers—theory and experiment," Proc. IRE, vol. 46, pp. 1599–1594; September, 1958.

[4] K. Shimoda, T. C. Wang, and C. H. Townes, "Further aspects of maser theory," *Phys. Rev.*, vol. 102, pp. 1308–1321; June 1, 1956.

[5] N. G. Basov and A. M. Prokhorov, "Applications of molecular beams to radio spectroscopic studies of rotation spectra of molecules," *J. Exp. Theoret. Phys.* (*USSR*), vol. 27, pp. 431–438; 1954. (In Russian.)
——, "On possible methods of producing active molecules for a molecular generator," *J. Exp. Theoret. Phys.* (*USSR*), vol. 28, (Letter), p. 249; 1955.

[6] J. C. Helmer, "Maser oscillators," *J. Appl. Phys.*, vol. 28, pp. 212–215; February, 1957.

[7] R. P. Feynman, F. L. Vernon, Jr., and R. W. Hellwarth, "Geometrical representation of the Schrödinger equation for solving maser problems," *J. Appl. Phys.*, vol. 28, pp. 49–52; January, 1957.

[8] R. H. Dicke, "Coherence in spontaneous radiation processes," *Phys. Rev.*, vol. 93, pp. 99–110; January, 1954.

[9] E. Fermi, "Quantum theory of radiation," *Rev. Mod. Phys.*, vol. 4, pp. 87–102; January, 1932.

[10] J. C. Slater, "Microwave Electronics," D. van Nostrand Co., Inc., New York, N. Y., ch. 4; 1950.

[11] D. K. Coles, et al., "Stark effect of the ammonia inversion spectrum," *Phys. Rev.*, vol. 82, pp. 877–879; June 15, 1951.

[12] E. T. Jaynes, "Information theory and statistical mechanics," *Phys. Rev.*, vol. 108, pp. 171–190; October, 15, 1957.

[13] A. Einstein, B. Podolsky, and N. Rosen, *Phys. Rev.*, vol. 47, p. 777; 1935.

[14] N. Wiener, "Extrapolation, Interpolation and Smoothing of Stationary Time Series," John Wiley and Sons, Inc., New York, N. Y.; 1950.

[15] ——, "Nonlinear Problems in Random Theory," John Wiley and Sons, Inc., New York, N. Y.; 1958.

On Maser Rate Equations and Transient Oscillations

C. L. Tang
Research Division, Raytheon Company, Waltham, Massachusetts
(Received 4 April 1963; in final form 16 May 1963)

Masers exhibit interesting transient behavior that cannot be completely understood on the basis of the rate equations. The use of the rate equations in most transient analyses is usually justified on a more or less intuitive basis and the implied assumptions are not always clear. In this paper, the macroscopic maser rate equations are derived systematically from the Boltzmann equation for the density matrix of the atomic systems and Maxwell's equations for the radiation fields. When the coherence linewidth (T_2^{-1}) of the atomic systems is much larger than the cavity linewidth and the natural linewidth (T_1^{-1}) of the atomic emission, and with a WKB approximation, in the lowest order of approximation one obtains the two widely used, coupled first-order nonlinear rate equations of Statz and deMars. On the other hand, if the cavity linewidth is much larger than the atomic linewidths (T_1^{-1} and T_2^{-1}), one can use the so-called "reaction-field principle" of Anderson and obtain, again, two coupled first-order rate equations; however, only one of the equations is nonlinear. The ranges of validity of both approaches are discussed in some detail.

I. INTRODUCTION

MASERS exhibit interesting transient behaviors that have not been completely explained.[1] In the microwave case, although the normally observed damped relaxation oscillation output of most masers can be satisfactorily understood on the basis of relatively simple phenomenological rate equations,[2] the reasons for the occasionally observed continuous undamped spiking output are still not clear.[3] The transient behavior of optical masers is even less well understood. On the basis of the widely used rate equations of Statz and deMars,[2] one would expect the transient oscillations of the maser output to damp out eventually. It can be said, however, that undamped random spiking output can be produced in almost all solid-state optical masers.[1] Regular damped spikes are observed in some,[4,5] but in ruby masers only under somewhat extraordinary circumstances have regular damped spikes as predicted by the simple rate equations been observed.[6]

Further extension of the original rate equations of Ref. 2 to take into account the spatially nonuniform coherent deexcitation of the inverted population and the multimode nature of the optical masers have also been made,[7] but the results of the detailed analyses appear to indicate that one would still expect the transient oscillations to damp out. The transient behavior of solid-state optical masers is, therefore, far from well understood. This question of the stability of the steady-state oscillations of solid-state optical masers is very important from both practical and theoretical viewpoints.

It is quite possible that physical processes completely overlooked in the usual analyses are responsible for this apparent paradox. However, in view of the fact that such a behavior is observed in solid-state masers of quite different materials and designs, and in quite different frequency ranges, it is not completely clear that the explanation has a purely physical origin. Furthermore, many useful analyses of masers are based on the simple rate equations, but the use of the rate equations is usually justified on a more or less intuitive basis and the assumptions implied are not always clear. It is, therefore, of interest to give a systematic derivation of these equations and to establish more precisely their ranges of validity.

In masers, the radiation field intensities of the oscillating mode are usually quite high, so that one does not have to quantize the radiation field, and a classical description of the radiation field in terms of Maxwell's equations is usually quite adequate. The active medium, on the other hand, must be treated quantum mechanically. Lamb, Jr.,[8,9] and others[10–12] have given detailed treatment of various aspects of masers using the density matrix formalism. Most of these studies, however, are concerned with either the threshold condition for oscillation or the steady-state solutions. Some of the approximations made in these cases are not so easily justifiable when one is also interested in the stability of the steady-state solutions or in the tran-

[1] C. G. B. Garret, "Review of Solid State Lasers," Third International Symposium on Quantum Electronics, February 1963.
[2] H. Statz and G. A. deMars, *Quantum Electronics*, edited by C. H. Townes (Columbia University Press, New York, 1960), p. 530.
[3] H. Statz, C. Luck, C. Shafer, and M. Cifton, *Quantum Electronics*, edited by J. R. Singer (Columbia University Press, New York, 1961), p. 342.
[4] P. P. Sorokin and M. J. Stevenson, *Quantum Electronics*, edited by J. R. Singer (Columbia University Press, New York, 1961), p. 65.
[5] E. Snitzer, "High Power and Fibre Neodymium Glass Lasers," Third International Symposium on Quantum Electronics, February 1963.
[6] K. Shimoda (private communication); R. J. Collins and J. A. Giordamaine, Bull. Am. Phys. Soc. **7**, 446 (1962); K. Gurs, Third International Symposium on Quantum Electronics, February 1963.
[7] C. L. Tang, H. Statz, and G. A. deMars, J. Appl. Phys. **34**, 2289 (1963); H. Statz, C. L. Tang, and G. deMars, Bull. Am. Phys. Soc. **8**, 87 (1963).
[8] W. E. Lamb, Jr., *Lectures in Theoretical Physics* (Interscience Publishers, Inc., New York, 1960), Vol. II, p. 435.
[9] W. E. Lamb, Jr., "Quantum Theory of Optical Masers," Third International Symposium on Quantum Electronics, February 1963.
[10] P. W. Anderson, J. Appl. Phys. **28**, 1049 (1957).
[11] S. Yatsiv, Phys. Rev. **113**, 1538 (1959).
[12] C. L. Tang and H. Statz, Phys. Rev. **128**, 1013 (1962).

Reprinted with permission from *J. Appl. Phys.*, vol. 34, pp. 2935–2940, Oct. 1963.

sient behavior of the maser as it approaches the steady state. In the transient period, the amplitude and frequency of the radiation field, the populations of the maser states, and the amplitude and frequency of the induced polarization of the active medium are all slowly varying with time. One must take care to determine which time derivative terms to keep in order to arrive at the correctly simplified rate equations that are amenable to further analysis. Depending upon the relative linewidths of the cavity radiation field and the atomic emission line, one obtains different simplified rate equations.

II. EQUATIONS OF MOTION FOR THE DENSITY MATRIX AND THE ELECTROMAGNETIC FIELD

We follow closely the formulation given by Lamb[8] for the single mode maser and consider for the present purpose the case where the active medium is dilute enough so that the induced macroscopic polarization and the radiation field of the only oscillating mode always have essentially the same spatial distribution as that of the unperturbed cavity mode. Consequently, one can ignore the spatial dependence of the density matrix characterizing the active medium and the radiation field, and act as though these are uniform. For optical masers, we consider electric dipole transitions. It follows from Maxwell's equations that the electric field $\mathbf{E}(t)$ satisfies the following equation of motion:

$$(d^2/dt^2)\mathbf{E}(t)+\gamma_0(d/dt)\mathbf{E}(t)+\omega_0^2\mathbf{E}(t) = (-d^2/dt^2)\mathbf{P}(t), \quad (2.1)$$

where $\mathbf{P}(t)$ is the macroscopic electric polarization of the active medium, ω_0 is the resonant frequency of the unperturbed cavity mode, and γ_0 characterizes its damping. The macroscopic polarization, $\mathbf{P}(t)$, is determined from the density matrix $\rho(t)$:

$$\mathbf{P}(t)=N \text{ trace } [\mathbf{p}\rho(t)], \quad (2.2)$$

where N is the density of the active atoms coupled to the radiation field and $\mathbf{p}$ is the electric dipole operator. With slight modification of Lamb's notation,[8] the quantum mechanical Boltzmann equation for the density matrix is

$$(d\rho/dt)=(-i/\hbar)[H,\rho] -\tfrac{1}{2}[\Gamma(\rho-\rho^{(0)})+(\rho-\rho^{(0)})\Gamma], \quad (2.3)$$

where the total Hamiltonian

$$H=H_0+V=H_0-\mathbf{p}\cdot\mathbf{E}(t) \quad (2.4)$$

is the sum of the unperturbed Hamiltonian of the atom H_0 and the interaction energy of the atom and the electromagnetic field. If we focus our attention only on the two atomic levels with energies E_1 and E_2 involved in the maser action and label the upper and lower maser states by 1 and 2 in a representation in which H_0 is diagonal, then Γ, which characterizes phenomenologically the relaxation processes, is a diagonal matrix $\Gamma_{\alpha\beta}=\gamma_\alpha\delta_{\alpha\beta}$. The matrix $\rho^{(0)}$ in Eq. (2.3) depends upon the pumping conditions in the maser. In the absence of the perturbation V, it can be seen from (2.3) and (2.4) that ρ approaches $\rho^{(0)}$ in the steady state. If the pumping process is incoherent, as in most present day optical masers, then $\rho^{(0)}$ is purely diagonal, and the diagonal elements $\rho_{11}^{(0)}$ and $\rho_{22}^{(0)}$ are proportional, respectively, to the thermal equilibrium populations of the upper and lower maser states with the pump on but in the absence of the cavity field, $\mathbf{E}(t)$. If the pumping process is not incoherent, $\rho^{(0)}$ would not be purely diagonal.[13,14] In the present study, we consider only incoherently pumped masers so that $\rho^{(0)}$ in the Boltzmann equation (2.3) is always diagonal.

In terms of the elements of the density matrices ρ and $\rho^{(0)}$, which are Hermitian, we have four independent equations of motion from Eqs. (2.1) and (2.3):

$$(d^2/dt^2)E+\gamma_0(d/dt)E+\omega_0^2E = -N(d^2/dt^2)[p_{12}\rho_{21}+p_{21}\rho_{12}], \quad (2.5)$$

$$(d/dt)(\rho_{11}+\rho_{22})=-\gamma_1(\rho_{11}-\rho_{11}^{(0)})-\gamma_2(\rho_{22}-\rho_{22}^{(0)}), \quad (2.6)$$

$$(d/dt)(\rho_{11}-\rho_{22})=(2i/\hbar)[p_{12}\rho_{21}-p_{21}\rho_{12}]E -\gamma_1(\rho_{11}-\rho_{11}^{(0)})+\gamma_2(\rho_{22}-\rho_{22}^{(0)}), \quad (2.7)$$

$$(d/dt)\rho_{12}+i\omega_{12}\rho_{12} = (-i/\hbar)(\rho_{11}-\rho_{22})p_{12}E -[(\gamma_1+\gamma_2)/2]\rho_{12}, \quad (2.8)$$

where $\omega_{12}=1/\hbar(E_1-E_2)$ and use is made of the fact that $\mathbf{E}//\mathbf{P}$ so that one can drop the vector notations. In deriving Eqs. (2.5)–(2.8), we have assumed that the maser levels are not accidentally degenerate and parity considerations exclude diagonal matrix elements of the dipole operator $\mathbf{p}$.

In most masers, and quite likely in three level masers such as the ruby maser, even in the transient period, the total population of the upper and lower states are essentially constant in time so that, from (2.6):

$$(d/dt)(\rho_{11}+\rho_{22})=0 =-\gamma_1(\rho_{11}-\rho_{11}^{(0)})-\gamma_2(\rho_{22}-\rho_{22}^{(0)}). \quad (2.9)$$

With this condition of conservation of $(\rho_{11}+\rho_{22})$ as expressed in (2.9), the remaining two equations for the elements of the density matrix, (2.7) and (2.8), can be further simplified:

$$(d/dt)(\rho_{11}-\rho_{22})+(1/T_1)[(\rho_{11}-\rho_{22})-(\rho_{11}^{(0)}-\rho_{22}^{(0)})] = (2i/\hbar)[p_{12}\rho_{21}-p_{21}\rho_{12}]E, \quad (2.10)$$

$$(d/dt)\rho_{12}+i\omega_{12}\rho_{12}+(1/T_2)\rho_{12} = (-i/\hbar)(\rho_{11}-\rho_{22})\rho_{12}E. \quad (2.11)$$

In analogy with magnetic resonance work, two re-

[13] J. Brossel, *Quantum Electronics*, edited by J. R. Singer (Columbia University Press, New York, 1961), p. 93.
[14] C. Cohen-Tanoudji, *Quantum Electronics*, edited by J. R. Singer (Columbia University Press, New York, 1961), p. 95.

laxation times $T_1=(\gamma_1+\gamma_2)/2\gamma_1\gamma_2$ and $T_2=2/(\gamma_1+\gamma_2)$, which appeared naturally in (2.10) and (2.11), have been introduced. Equations (2.5), (2.10), and (2.11) are the basic equations of motion for the idealized model of maser[8] under consideration.

The dynamical behavior of the maser at all times is now described in terms of 5 independent variables: the amplitudes and phases of the cavity radiation field, $\mathbf{E}(t)$, the macroscopic polarization of the active medium, $\mathbf{P}(t)$ or $\rho_{12}(t)$, and the excess population of the upper and lower maser states, $\rho_{11}(t)-\rho_{22}(t)$. These 5 variables satisfy three complex equations of motion: one second-order differential equation, (2.5), and two first-order equations (2.10) and (2.11). The last two equations are nonlinear in the sense that they involve products of two independent variables.

In the usual rate equations approach,[2] the maser is described by only two independent variables, the electromagnetic power and the excess population, which satisfy two coupled first-order nonlinear rate equations. Such a further contraction of description is possible only if the relaxation times satisfy certain conditions.

III. THE RATE EQUATIONS

A. Case 1: $T_2^{-1}\gg T_1^{-1}$ and γ_0

We consider first the case where the linewidth corresponding to the inverse of the coherence time T_2^{-1} is much larger than the natural linewidth T_1^{-1} and the cavity linewidth γ_0:

$$T_2^{-1}\gg T_1^{-1} \quad \text{and} \quad \gamma_0. \tag{3.1}$$

This is the case for most solid-state optical masers.

For convenience, we split $E(t)$ and $\rho_{12}(t)$ into a rapidly oscillating factor and a slowly varying amplitude factor:

$$E(t)=\tfrac{1}{2}[\bar{E}(t)e^{-i\omega t}+\bar{E}^*(t)e^{i\omega t}], \tag{3.2}$$

$$\rho_{12}(t)=\bar{\rho}_{12}(t)e^{-i\omega t}, \tag{3.3}$$

where ω is a constant to be specified and is assumed to be of the same order of magnitude as ω_0 or $|\omega-\omega_0|\approx 1/T_1$ and $\gamma_0\ll 1/T_2$.

Using (3.2) and (3.3), Eq. (2.11) can be recast in an integral form, neglecting the counter resonant term:

$$\bar{\rho}_{12}(t)=-\frac{ip_{12}}{\hbar}\int_0^t [\rho_{11}(t')-\rho_{22}(t')]\bar{E}(t') \times\exp\left\{\left[i(\omega_{12}-\omega)+\frac{1}{T_2}\right](t'-t)\right\}dt'. \tag{3.4}$$

Due to the factor $\exp[(t'-t)/T_2]$ in the integrand, the major contribution to the integral in (3.4) comes from the range $t-t'\lesssim T_2$. If, in addition to condition (3.1), we assume that the rate of pumping is not too large, so that the rate of change of $[\rho_{11}(t')-\rho_{22}(t')]$ and $\bar{E}(t')$ are small compared with T_2^{-1}, or:

$$\begin{aligned} &|1/\bar{E}(t)[d\bar{E}(t)/dt]|\,T_2\ll 1\\ &1/[\rho_{11}(t)-\rho_{22}(t)]\cdot d/dt[\rho_{11}(t)-\rho_{22}(t)]\cdot T_2\ll 1 \end{aligned} \tag{3.5}$$

then these factors can be taken out of the integral and evaluated at $t'=t$:

$$\begin{aligned}\bar{\rho}_{12}(t)&=-\frac{ip_{12}}{\hbar}[\rho_{11}(t)-\rho_{22}(t)]\bar{E}(t)\times\int_0^t \exp\left\{\left[i(\omega_{12}-\omega)+\frac{1}{T_2}\right](t'-t)\right\}dt'\\ &=-\frac{1}{p_{21}\omega}\chi[\rho_{11}(t)-\rho_{22}(t)]\bar{E}(t)\quad\text{for}\quad t\gg T_2,\end{aligned} \tag{3.6}$$

where

$$\chi=i|p_{12}|^2\omega T_2/\hbar[1+i(\omega_{12}-\omega)T_2]\equiv\chi'+i\chi''. \tag{3.6a}$$

Note that on account of (3.1), small variations in ω of the order of γ_0 have negligible effects on χ.

At this stage, one does not know *a priori* whether condition (3.5) is satisfied or not: so it can only be considered as an *a posteriori* consistency check. However, in general, so long as the rate of pumping is not too high, the transient oscillation period of $\bar{E}(t)$ and $[\rho_{11}(t)-\rho_{22}(t)]$ is governed by T_1 and γ_0. When condition (3.1) is satisfied, it is always safe to assume that (3.5) is satisfied. In any case, higher order correction terms to this can, in principle, be obtained by expanding $\bar{E}(t')$ and $[\rho_{11}(t')-\rho_{22}(t')]$ around t and then carrying out the integrations. This leads to an asymptotic series in powers of T_2. Such a procedure can be justified rigorously and is well known in the theory of asymptotic expansions[15] as the Laplace method of evaluating integrals of the type (3.4) in the limit where T_2 is very small.

Equations (3.6) and (3.6a) imply that for most of the time duration and time scales of interest, the induced macroscopic polarization of the active medium is always proportional to the amplitude of the applied field and the excess population even though these are time varying. This approximation is possible only in the asymptotic limit $T_2\to 0$, or, for all practical purposes, when T_2 is much shorter than all other relaxation times and the transient periods of the electromagnetic field and the excess population, so that one can neglect the transient effects of Eq. (2.11). The proportionality constant χ is now known explicitly and its linewidth is determined by T_2. This is an essential assumption implied in the rate equations of Statz and deMars, as we see shortly. For most solid-state optical masers, this is a good approximation. For example, in typical optical masers of good ruby crystals at room temperature, $T_2\approx 10^{-10}$ sec, $T_1\approx 10^{-3}$ sec, $\gamma_0\approx 10^7$ sec^{-1}, and the observed spiking period is of the order of 10^{-6} sec.

[15] See, for example, N. G. deBruijn, *Asymptotic Methods in Analysis* (Interscience Publishers, Inc., New York, 1958), Chap. 4, in particular p. 65.

With (3.6) and (3.6a), we can now obtain the equations of motion which govern the slow time variations of $\bar{E}(t)$ and $[\rho_{11}(t)-\rho_{22}(t)]$ from Eqs. (2.5) and (2.10). Substituting (3.6), (3.3), and (3.2) into (2.10) and omitting terms varying rapidly at 2ω, we obtain

$$(d/dt)(\rho_{11}-\rho_{22}) + (1/T_1)[(\rho_{11}-\rho_{22})-(\rho_{11}^{(0)}-\rho_{22}^{(0)})] = -[(\rho_{11}-\rho_{22})/\hbar\omega]\chi''|\bar{E}|^2, \quad (3.7)$$

which is the rate equation for the excess population proposed in Ref. 2.

Substituting (3.5) into (2.5) and keeping in mind that we are interested here in the time variations of $\bar{E}(t)$ and $[\rho_{11}(t)-\rho_{22}(t)]$, we obtain

$$[1-N(\rho_{11}-\rho_{22})\chi](d^2\bar{E}/dt^2) + [\gamma_0-2i\omega-2N\chi(d/dt)(\rho_{11}-\rho_{22})-2i\omega N\chi(\rho_{11}-\rho_{22})] \times(d\bar{E}/dt)+[\omega_0^2-\omega^2-i\gamma_0\omega+2i\omega N\chi(d/dt)(\rho_{11}-\rho_{22}) + N\chi\omega^2(\rho_{11}-\rho_{22})-N\chi(d^2/dt^2)(\rho_{11}-\rho_{22})]\bar{E}=0, \quad (3.8)$$

which is a good deal more complicated than the first-order rate equation for the electromagnetic power used in the usual rate equations approach.

Equation (3.8) is a second-order differential equation for $\bar{E}(t)$ with slowly varying coefficients; it can be further systematically approximated using a WKB type of approximation. We leave the detailed calculations to the Appendix. Briefly, the general solution of (3.8) can be written in the form:

$$\bar{E}(t)=Ae^{-\phi(t)}, \quad (3.9)$$

where $\phi(0)=0$ and $A=\bar{E}(0)$ is a constant depending on the initial condition of $\bar{E}(t)$. Substituting (3.9) into (3.8) gives a nonlinear differential equation in $\phi(t)$. If the perturbation terms involving γ_0 and χ are small so that they can be regarded as second-order quantities and if, furthermore, the second derivative of ϕ is a second-order quantity compared with the first derivative of ϕ, then we can expand

$$\phi=\phi_1+\phi_2+\phi_3+\cdots \quad (3.10)$$

in successive orders of approximation. Choosing $\omega=\omega_0$, for equation (3.8) as well as (3.7), (3.6), (3.22), and (3.23), then application of the WKB approximation procedure as given in the Appendix gives the lowest order equation:

$$(d\phi_1/dt)^2+2i\omega_0(d\phi_1/dt)=0. \quad (3.11)$$

or

$$d\phi_1/dt=0 \quad \text{or} \quad -2i\omega_0. \quad (3.12)$$

Substituting (3.12) into (3.9) and (3.2), we see immediately that the lowest order term in (3.10) just reproduces the positive and negative resonant frequencies of the unperturbed cavity mode: $\pm\omega_0$. The second-order equation corresponding to the $d\phi_1/dt=0$ solution is:

$$(d/dt)\phi_2=\gamma_0/2+(i/2)N\omega_0\chi'(\rho_{11}-\rho_{22}) -(N/2)\omega_0\chi''(\rho_{11}-\rho_{22}), \quad (3.13)$$

and, as shown in the Appendix, the higher-order equations can also be determined systematically. If, however, we stop at the second-order approximation, and consider the resonant component $\bar{E}(t)e^{-i\omega_0 t}$ of $E(t)$, we have from (3.9), (3.10), and (3.12)

$$\bar{E}(t)=Ae^{-\phi_2} \quad (3.14)$$

where $d\phi_2/dt$ is given by (3.13). It follows immediately from (3.14) and (3.13) that in this approximation, $|\bar{E}|^2$ satisfies the following equation of motion:

$$(d/dt)|\bar{E}|^2=-|\bar{E}|^2[2Re(d\phi_2/dt)] =-\gamma_0|\bar{E}|^2+N\chi''\omega_0(\rho_{11}-\rho_{22})|\bar{E}|^2, \quad (3.15)$$

which is precisely the first-order rate equation for the electromagnetic power proposed by Statz and deMars. Equations (3.15) and (3.7) constitute the usual rate equations approach. We now see under what conditions one might use this simple approach instead of using the more rigorous density matrix formalism. Detailed studies of these rate equations can be found in the literature.

It may also be of interest to consider Eq. (3.13). In the steady state, the amplitude of $E(t)$ is constant, which means that the real part of the time derivative of the complex phase of $\bar{E}(t)$, $(d/dt)\phi_2$, is equal to zero. From (3.13) we find the steady-state excess population

$$\bar{\rho}_{11}-\bar{\rho}_{22}=\gamma_0/N\omega_0\chi'', \quad (3.16)$$

which is also the steady-state solution for $[\rho_{11}-\rho_{22}]$ one would find from Eq. (3.15). The imaginary part of $(d/dt)\phi_2$ gives the small steady-state frequency shift from ω_0 of the maser emission:

$$\mathrm{Im}(d/dt)\phi_2=\gamma_0\chi'/2\chi''=(\gamma_0/2T_2^{-1})(\omega_{12}-\omega_0) \quad (3.17)$$

with the help of (3.16) and (3.6a), which means that the "frequency pulling" of the maser emission from the cavity center frequency ω_0 is equal to the difference between the cavity frequency and the atomic resonance frequency reduced by a factor equal to the ratio of the cavity linewidth to the atomic emission linewidth, as found by Schawlow and Townes.[16]

B. Case 2: $\gamma_0 \gg T_2^{-1}$ and T_1^{-1}

It often happens in experimental situations that the cavity linewidth is much larger than the coherence linewidth T_2^{-1} and the natural linewidth of the atomic emission, T_1^{-1}. This is often the case in microwave masers such as the ammonia beam masers[17] or the hydrogen maser.[18]

[16] A. L. Schawlow and C. H. Townes, Phys. Rev. **112**, 1940 (1958).

[17] J. P. Gordon, H. J. Zeiger, and C. H. Townes, Phys. Rev. **99**, 1264 (1955).

[18] D. Klepner, M. Goldenberg, and N. F. Ramsey, Phys. Rev. **126**, 603 (1962).

Under these conditions

$$\gamma_0 \gg T_2^{-1} \text{ and } T_1^{-1} \tag{3.18}$$

following exactly the same arguments as those given in the previous Sec. IIIA, it would now be a good approximation to ignore the transient effects of Eq. (2.5) and use its steady-state solution, since, now, the transient period of the cavity response is extremely short compared with other time scales of interest. We, therefore, have:

$$\bar{E}(t) = (2N/p_{12})y\tilde{\rho}_{12}(t) \tag{3.19}$$

where

$$y = |p_{12}|^2\omega^2/(\omega_0^2-\omega^2-i\gamma_0\omega) \equiv y'+iy''. \tag{3.20}$$

Again, the implication is that the cavity response $\bar{E}(t)$ is always proportional to $\tilde{\rho}_{12}(t)$ even though these are time varying, but now the linewidth of the proportionality constant y is. determined by the cavity linewidth γ_0. Note that Eq. (3.19) is equivalent to the so-called "reaction-field principle" used by Anderson[10] in his treatment of the magnetic resonance amplifier; indeed, he has pointed out that the same principle is applicable to the electric case. In both cases, however, this approximation is valid only in the asymptotic limit $\gamma_0 \to \infty$, or, for all practical purposes, if the cavity linewidth is much larger than all other linewidths of interest. With (3.19) and (3.20) the remaining derivation of the rate equations is trivial. Substituting (3.19) and (3.20) into (2.10) and (2.11), we obtain two coupled rate equations.

$$\frac{d}{dt}(\rho_{11}-\rho_{22}) + \frac{(\rho_{11}-\rho_{22})-(\rho_{11}^{(0)}-\rho_{22}^{(0)})}{T_1} = -\frac{|p_{12}|^2y''}{\hbar N|y|^2}|\bar{E}|^2 \tag{3.21}$$

$$\frac{d}{dt}\tilde{\rho}_{12} + i(\omega_{12}-\omega)\tilde{\rho}_{12} + \frac{1}{T_2}\tilde{\rho}_{12} = -\frac{iN}{\hbar}(\rho_{11}-\rho_{22})y\tilde{\rho}_{12}. \tag{3.22}$$

Equations (3.19) and (3.22) also lead to

$$d/dt|\bar{E}|^2 + 2|\bar{E}|^2/T_2 = 2N/\hbar(\rho_{11}-\rho_{22})y''|\bar{E}|^2. \tag{3.23}$$

Equations (3.21) and (3.23) are the two rate equations for the time variations of the excess population and the electromagnetic power. When the cavity linewidth is large, these two rate equations should be used; when the coherence linewidth T_2^{-1} is large, the rate equations of Statz and deMars should be used. The physical significance of Eqs. (3.21) and (3.22) is discussed in some detail in Ref. 10.

We consider first the frequency of maser emission in the steady state. By virtue of (3.18) we can set $\omega = \omega_{12}$ in (3.20)–(3.23), and find the small frequency pulling from Eq. (3.22). Again, from the steady-state solution for $(\rho_{11}-\rho_{22})$, we find the negative of the imaginary part of the logarithmic derivative of $\tilde{\rho}_{12}$, or the imaginary part of the time derivative of the complex phase of $\tilde{\rho}_{12}$:

$$-\text{Im}\frac{d}{dt}(\ln\tilde{\rho}_{12}) = \frac{y'}{T_2y''} = \frac{\omega_0^2-\omega_{12}^2}{T_2\gamma_0\omega_{12}} \approx \frac{2T_2^{-1}(\omega_0-\omega_{12})}{\gamma_0} \tag{3.24}$$

with the help of (3.20) and using the fact that $\omega_0 \approx \omega_{12}$ or $\omega_0^2-\omega_{12}^2 \approx 2\omega_{12}\cdot(\omega_0-\omega_{12})$. This means that, when the cavity linewidth is much larger than the atomic linewidths, the frequency pulling of the maser emission is equal to the difference of the cavity frequency and the atomic transition frequency reduced by a factor equal to the ratio of the atomic linewidth to the cavity linewidth, as given in Ref. 17.

Consider now the pair of coupled rate equations, (3.21) and (3.23), for the excess population and the electromagnetic power. It is of particular interest to note the differences between these equations, valid for the case where the cavity linewidth is broad, and the rate equations of Statz and deMars, (3.7) and (3.15), which are valid for the case where the coherence linewidth T_2^{-1} is large. Equation (3.21) is linear and the rate of decay of the excess population contains a term directly proportional to the electromagnetic power in the cavity, not proportional to the product of the excess population and the electromagnetic power as in Eq. (3.7). As a result, in the latter case, the population of the upper maser state is always larger than that of the lower maser state during the transient period while in the former case, during the transient period the population of the upper maser state could become less than that of the lower maser state. Both approaches, however, predict that the steady-state maser oscillations are always stable.

IV. SUMMARY

Because of the fact that the important transient behavior of solid-state masers cannot be completely explained on the basis of the widely used maser rate equations, an attempt has been made in this paper to determine more clearly just what assumptions are implied in these rate equations. It is hoped that this work will be of help in improving the present theory. The necessary conditions under which it is possible to reduce the more complete description in terms of the Boltzmann equation for the density matrix of the active medium and Maxwell's equations for the electromagnetic fields to the simple rate equations are determined. Summarizing, if the linewidth corresponding to the inverse of the coherence time of the atomic systems (T_2^{-1} in magnetic resonance terminology) is much larger than the cavity linewidth and the natural linewidth of the atomic emission (T_1^{-1}), it has been shown here, making use of a WKB type of approximation procedure, that the first-order equations describing the radiation field and the atomic systems are the rate equations of Statz and deMars.[2,19] On the other hand, if the radiation cavity linewidth is much larger than the atomic linewidths, then one can use the so-called reaction-field principle of Anderson,[10] and obtain the rate equations given in Ref. 10. Both sets of rate equations have stable steady-state solutions. It is clear, however, that addition of certain small terms in these rate equations could lead to undamped transient solutions. Several such modifications have been suggested by various people, but the physical interpretations of the additional terms are not yet completely clear.

APPENDIX. APPLICATION OF THE WKB METHOD TO EQ. (3.8)

Consider Eq. (3.8)

$$\begin{aligned}&[1-N(\rho_{11}-\rho_{22})\chi](d^2/dt^2)\bar{E}\\&\quad+[\gamma_0-2i\omega-2N\chi(d/dt)(\rho_{11}-\rho_{22})-2i\omega N\chi(\rho_{11}-\rho_{22})]\\&\quad\times(d/dt)\bar{E}+[\omega_0^2-\omega^2-i\gamma_0\omega+2i\omega N\chi(d/dt)(\rho_{11}-\rho_{22})\\&\quad+N\chi\omega^2(\rho_{11}-\rho_{22})-N\chi(d^2/dt^2)(\rho_{11}-\rho_{22})]\bar{E}=0,\qquad\text{(A1)}\end{aligned}$$

in which E is the independent variable and the coefficients are time varying due to the time dependence of $(\rho_{11}-\rho_{22})$. If we assume that the terms involving γ_0 and χ are small perturbations and, furthermore, that $(\rho_{11}-\rho_{22})$ changes very little in the time duration $(2\pi/\omega_0)$, then (A1) can be simplified by the well known WKB approximation procedure (see, for example, Schiff[20]). Accordingly, we let $\omega=\omega_0$ and write the general solution of (A1) in the form

$$\bar{E}(t)=Ae^{-\phi(t)},\qquad\text{(A2)}$$

where $\phi(0)=0$ and $A=\bar{E}(0)$ is a constant depending upon the initial condition on $\bar{E}(t)$. Substituting (A2) into (A1) leads to a nonlinear differential equation in ϕ:

$$\begin{aligned}&[1-N(\rho_{11}-\rho_{22})\chi][(d\phi/dt)^2-(d^2\phi/dt^2)]\\&\quad-[\gamma_0-2i\omega_0-2N\chi(d/dt)(\rho_{11}-\rho_{22})-2i\omega_0N(\rho_{11}-\rho_{22})\chi]\\&\quad(d\phi/dt)+[-i\gamma_0\omega_0+2i\omega_0N\chi(d/dt)(\rho_{11}-\rho_{22})\\&\quad+N\chi\omega_0^2(\rho_{11}-\rho_{22})-N\chi(d^2/dt^2)(\rho_{11}-\rho_{22})]=0.\qquad\text{(A3)}\end{aligned}$$

In order to fix the order of magnitude of the various terms in (A3) more clearly, we introduce the artifice ϵ, which eventually can be set to 1, and identify all second-order terms by multiplying them by ϵ. Consequently we replace in (A3)

$$\begin{aligned}\gamma_0&\to\epsilon\gamma_0,\\\chi&\to\epsilon\chi,\end{aligned}\qquad\text{(A4)}$$

and $t\to(1/\epsilon)t$, $\phi\to(1/\epsilon)\phi$ so that $(d\phi/dt)\to(d\phi/dt)$, but $(d^2\phi/dt^2)\to\epsilon(d^2\phi/dt^2)$. Substituting (A4) into (A3) gives

$$\begin{aligned}&[1-\epsilon N\chi(\rho_{11}-\rho_{22})][(d\phi/dt)^2-\epsilon(d^2\phi/dt^2)]\\&\quad-[\epsilon\gamma_0-2i\omega_0-\epsilon^2 2N\chi(d/dt)(\rho_{11}-\rho_{22})\\&\quad-\epsilon 2i\omega_0N\chi(\rho_{11}-\rho_{22})](d\phi/dt)\\&\quad+[-\epsilon i\gamma_0\omega_0+\epsilon^2\chi 2i\omega_0N(d/dt)(\rho_{11}-\rho_{22})\\&\quad+\epsilon N\chi\omega_0^2(\rho_{11}-\rho_{22})-\epsilon^3N\chi(d^2/dt^2)(\rho_{11}-\rho_{22})]=0.\qquad\text{(A5)}\end{aligned}$$

(A5) can now be solved by successive approximations:

$$\phi(t)=\phi_1(t)+\epsilon\phi_2(t)+\epsilon^2\phi_3(t)+\cdots,\qquad\text{(A6)}$$

with the initial conditions

$$\phi_1(0)=\phi_2(0)=\phi_3(0)=\cdots=0.$$

Substituting (A6) into (A5) and collecting terms of like powers in ϵ, we obtain:

$$L_0+\epsilon L_1+\epsilon^2L_2+\cdots=0,\qquad\text{(A7)}$$

where

$$L_0=(d\phi_1/dt)^2+2i\omega_0(d\phi_1/dt),\qquad\text{(A8)}$$

$$\begin{aligned}L_1=&-(d^2\phi_1/dt^2)+2(d\phi_1/dt)(d\phi_2/dt)\\&-N(\rho_{11}-\rho_{22})\chi(d\phi_1/dt)^2-2i\omega_0N\chi(\rho_{11}-\rho_{22})(d\phi_1/dt)\\&-\gamma_0(d\phi_1/dt)+2i\omega_0(d\phi_2/dt)-i\omega_0\gamma_0\\&+N\chi\omega_0^2(\rho_{11}-\rho_{22})\qquad\text{(A9)}\end{aligned}$$

and all the higher-order coefficients L_n can also be obtained. If we now let $\epsilon\to 1$, the n equations $L_1=L_2=\cdots=L_n=0$ constitute the nth order approximation of (A3) or (A1), with the help of (A2) and (A6).

It may be of interest to consider the simple case where $\chi\to 0$ but $\gamma_0\neq 0$ so that the right-hand side of the original equation (2.5), for $E(t)$ vanishes and the resulting equation describes also a weakly damped harmonic oscillator. Then the lowest-order equation (A8) gives

$$(d\phi_1/dt)=0\quad\text{and}\quad -2i\omega_0,$$

or

$$\phi_1=0\quad\text{and}\quad -2i\omega_0t\qquad\text{(A11)}$$

which when substituted into (A2) and (3.2) just reproduces the positive and negative resonant frequencies $\pm\omega_0$ of the undamped harmonic oscillator. Consider now, for definitness, the $d\phi_1/dt=0$ component; the second-order equation (A9) gives

$$(d\phi_2/dt)=(\gamma_0/2),\qquad\text{(A12)}$$

which is the well-known damping factor of the amplitude of the oscillation.

ACKNOWLEDGMENTS

I should like to thank Dr. H. Statz for many helpful discussions, and Professor K. Shimoda for stimulating discussions and for showing me his interesting experimental data on the observed regular damped oscillation output of a ruby maser before publication of his results.

[19] See also the interesting study of J. L. Kaplan, and R. Zier, J. Appl. Phys. **33**, 2372 (1962), on the transient oscillations of optical masers, in particular on the question of the domain of validity of the rate equations of Statz and deMars.

[20] L. I. Schiff, *Quantum Mechanics* (McGraw-Hill Book Company, Inc., New York, 1949).

Semiclassical Treatment of the Optical Maser*

L. W. DAVIS†

Summary—**Using the semiclassical theory of radiation, the steady-state operation of the optical maser oscillator is studied in the case where a single "cavity" mode is excited. On introducing certain simplifying assumptions, a straightforward calculation leads to concise results for the frequency and amplitude of the field oscillations. These results are either well known or readily interpreted. For example, a frequency pulling effect is predicted which corresponds to the Bloch-Siegert shift, or alternatively is understandable as due to the Stark effect.**

Introduction

IN THIS STUDY we apply the semiclassical theory of radiation to the problem of predicting the operation of the optical maser oscillator.[1] Therefore, on the one hand, the time-dependent Schrödinger equation for the effect of the classical electromagnetic field on the active medium leads to a nonlinear electronic equation while, on the other hand, the expectation value of quantum mechanical operators are interpreted as actual values of sources in the classical Maxwell equations, whereupon a linear circuit equation is obtained. The requirement that the resulting electronic and circuit equations must simultaneously be satisfied then determines the frequency and amplitude for a given mode of oscillation.

In order to simplify the problem, we confine our attention here to the following case: the resonance line of the active medium is homogeneously broadened, the active medium is continuously pumped, the optical maser is in steady-state operation, and only one mode of the optical cavity is excited.

General Formalism

Let us consider first the case where a single molecule interacts with an applied electric field $\boldsymbol{E}(t)$. For simplicity, we shall suppose that the unperturbed molecule has only two possible energy levers, E_1 and E_2 (where $E_2 > E_1$), with corresponding eigenfunctions ψ_1, ψ_2. Thus, the wave function for the molecule can be expressed as

$$\psi(t) = a(t)\psi_1 + b(t)\psi_2. \tag{1}$$

The development of $\psi(t)$ in time is given by the Schrödinger equation

$$i\hbar\dot{\psi} = (H_0 + H')\psi, \tag{2}$$

where H_0 denotes the molecule Hamiltonian, with matrix elements

$$\langle\psi_m \mid H_0 \mid \psi_{m'}\rangle = E_m\delta_{mm'} \qquad (m, m' = 1, 2), \tag{3}$$

and H' denotes the electric dipole interaction Hamiltonian, with matrix elements

$$\begin{aligned}\langle\psi_m \mid H' \mid \psi_{m'}\rangle &= \langle\psi_m \mid -\boldsymbol{\mu}_{\text{op}}\cdot\boldsymbol{E}(t) \mid \psi_{m'}\rangle \\ &= -\boldsymbol{\mu}\cdot(1 - \delta_{mm'})\boldsymbol{E}(t).\end{aligned} \tag{4}$$

In (4), $\boldsymbol{\mu}_{\text{op}}$ is the electric dipole moment quantum mechanical operator, and

$$\boldsymbol{\mu} = \langle\psi_1 \mid \boldsymbol{\mu}_{\text{op}} \mid \psi_2\rangle. \tag{5}$$

On substituting (1) into (2), we find that the amplitudes $a(t)$ and $b(t)$ satisfy

$$i\hbar\dot{a} = E_1 a - \boldsymbol{\mu}\cdot\boldsymbol{E}(t)b \tag{6a}$$

$$i\hbar\dot{b} = -\boldsymbol{\mu}\cdot\boldsymbol{E}(t)a + E_2 b. \tag{6b}$$

To find the effect of the molecule on the electromagnetic field, in this formalism we assume that the field obeys the classical Maxwell equations, with the expectation value of the electric dipole moment quantum mechanical operator of the molecule

$$\begin{aligned}\boldsymbol{P}(t) &= \langle\psi \mid \boldsymbol{\mu}_{\text{op}} \mid \psi\rangle \\ &= \boldsymbol{\mu}(ab^* + a^*b)\end{aligned} \tag{7}$$

identified as the *actual* value of its electric dipole moment. Now it can be shown from (6a) and (6b) that $\boldsymbol{P}(t)$ must satisfy the system

$$\ddot{\boldsymbol{P}} + \omega_m^2\boldsymbol{P} = -k^2W\boldsymbol{E}(t) \tag{8a}$$

$$\dot{W} = \dot{\boldsymbol{P}}\cdot\boldsymbol{E}(t) \tag{8b}$$

where

$$k^2 = \frac{4\boldsymbol{\mu}^2}{\hbar^2}, \tag{9}$$

and

$$\begin{aligned}W(t) &= \langle\psi \mid H_0 \mid \psi\rangle - \tfrac{1}{2}(E_1 + E_2) \\ &= \frac{\hbar\omega_m}{2}(bb^* - aa^*)\end{aligned} \tag{10}$$

is the expectation value of the energy of the molecule, referred to $\frac{1}{2}(E_1+E_2)$ as zero. In deriving (8a), we have assumed, merely for simplicity, that $\boldsymbol{E}(t)$ is parallel to $\boldsymbol{\mu}$; and note that in (10)

$$\omega_m = \frac{E_2 - E_1}{\hbar} \tag{11}$$

is the natural frequency of the molecule.

* Received August 13, 1962.

† Western Development Laboratories, Philco Corporation, Palo Alto, Calif.

[1] The relation between the semiclassical theory and quantum electrodynamics recently has been re-examined in a note by E. T. Jaynes [1], and it is shown to be quite different than is often supposed. There Jaynes discusses essentially the same semiclassical equations on which the results of the present study are based.

Reprinted from *Proc. IEEE*, vol. 51, pp. 76–80, Jan. 1963.

Let us consider now the case where a *system* of molecules interacts with an applied electric field $\boldsymbol{E}(r, t)$. To account for the effect of relaxation mechanisms (*e.g.*, collisions, spontaneous emission, etc.), we insert phenomenological relaxation terms in the single molecule equations of motion [(8)] as follows:

$$\frac{\partial^2 \boldsymbol{P}}{\partial t^2} + \frac{2}{T_2}\frac{\partial \boldsymbol{P}}{\partial t} + \omega_m^2 \boldsymbol{P} = -k^2 W \boldsymbol{E}(r, t) \tag{12a}$$

$$\frac{\partial W}{\partial t} + \frac{W - W_e}{T_1} = \frac{\partial \boldsymbol{P}}{\partial t} \cdot \boldsymbol{E}(r, t). \tag{12b}$$

It must be understood, of course, that $\boldsymbol{P}$ and W now represent the predicted electric polarization and energy *per unit volume*, due to the molecules. In the optical frequency range it would appear that T_1 should nearly correspond to the radiation lifetime of the molecules, and T_2 (inverse) is a measure of the width of the resonance line for the system of molecules, it being assumed that the line is homogeneously broadened. Finally, W_e is the thermal equilibrium value of W.

If pumping supplies energy per unit volume to the molecules at a rate $\dot{W}_p$=constant, we shall suppose that (12b) then becomes

$$\frac{\partial W}{\partial t} + \frac{W - W_e}{T_1} = \frac{\partial \boldsymbol{P}}{\partial t} \cdot \boldsymbol{E}(r, t) + \dot{W}_p,$$

which can be written as

$$\frac{\partial W}{\partial t} + \frac{W - W_e'}{T_1} = \frac{\partial \boldsymbol{P}}{\partial t} \cdot \boldsymbol{E}(r, t), \tag{12b$'$}$$

where

$$W_e' \equiv W_e + \dot{W}_p T_1. \tag{13}$$

Thus, in this treatment, pumping at a constant rate $\dot{W}_p$ has the effect of increasing W_e by an amount $\dot{W}_p T_1$.

The circuit equation is obtained by first expanding the electro-magnetic field in terms of the normal modes, as defined, for example, by Slater [2], for the optical cavity under consideration.

$$\boldsymbol{E}(r, t) = \sum_a v_a(t) \boldsymbol{E}_a(r) \tag{14a}$$

$$\boldsymbol{H}(r, t) = \sum_a \omega_a q_a(t) \boldsymbol{H}_a(r). \tag{14b}$$

The vector functions $\boldsymbol{E}_a(r)$ and $\boldsymbol{H}_a(r)$ in (14a) and (14b) are normalized such that

$$\int (\boldsymbol{E}_a \cdot \boldsymbol{E}_b) dV = \delta_{ab}, \qquad \int (\boldsymbol{H}_a \cdot \boldsymbol{H}_b) dV = \delta_{ab}, \tag{15}$$

and they satisfy

$$\nabla \times \boldsymbol{E}_a = k_a \boldsymbol{H}_a, \qquad \nabla \times \boldsymbol{H}_a = k_a \boldsymbol{E}_a \tag{16}$$

where

$$k_a^2 = \epsilon \mu \omega_a^2. \tag{17}$$

In this study, ϵ and μ always denote the inductive capacities of the *passive* medium (*e.g.*, the host lattice). From the Maxwell equations, for a field which interacts with an electric polarization $\boldsymbol{P}(r, t)$,

$$\nabla \cdot \boldsymbol{D} = 0 \qquad \nabla \cdot \boldsymbol{B} = 0$$

$$\nabla \times \boldsymbol{E} = -\frac{\partial \boldsymbol{B}}{\partial t} \qquad \nabla \times \boldsymbol{H} = \frac{\partial \boldsymbol{D}}{\partial t}, \tag{18}$$

where

$$\boldsymbol{D} = \epsilon \boldsymbol{E} + \boldsymbol{P} \qquad \boldsymbol{B} = \mu \boldsymbol{H}, \tag{19}$$

it then is found that, for the case where only the ath mode is excited, the amplitude $v_a(t)$ obeys the equation of motion

$$\ddot{v}_a + \frac{\omega_a}{Q}\dot{v}_a + \omega_a^2 v_a = -\frac{1}{\epsilon}\int \left(\frac{\partial^2 \boldsymbol{P}}{\partial t^2} \cdot \boldsymbol{E}_a\right) dV. \tag{20}$$

In (20), Q denotes the quality factor of the optical cavity.

Predicted Steady-State Solution

Thus, the semiclassical theory as applied here consists of the nonlinear system of coupled equations (12a), (12b′), and (20), where in the former two equations $\boldsymbol{E}(r, t) = v_a(t)\boldsymbol{E}_a(r)$. We wish now to solve this system for the case where the motion of the polarization and field is periodic in time with period ω. Consider first the response of $\boldsymbol{P}(r, t)$ due to a monochromatic electric field

$$\boldsymbol{E}(r, t) = \boldsymbol{E}_0(r) \cos \omega t \tag{21}$$

in which

$$\boldsymbol{E}_0(r) = v_0 \boldsymbol{E}_a(r). \tag{22}$$

There then exists a steady-state solution of the electronic equations (12a) and (12b′) wherein the polarization is of the form

$$\boldsymbol{P}(r, t) = \frac{1}{2}\sum_{-\infty}^{\infty}{}_n \boldsymbol{P}_n(r) e^{in\omega t} \qquad (n \text{ odd}). \tag{23}$$

However, if $(k^2 E_0^2)/\omega^2 \ll 1$ within the cavity (which strongly holds for steady-state optical maser operation of today), then E_0^2 is sufficiently small that to a very good approximation (23) reduces to

$$\boldsymbol{P}(r, t) = \tfrac{1}{2}\{\boldsymbol{P}_1(r) e^{i\omega t} + \boldsymbol{P}_{-1}(r) e^{-i\omega t}\}. \tag{24}$$

On inserting (24) into (12a) and (12b′), we find that

$$\boldsymbol{P}_{\pm 1}(r) = -k^2 W_e' \boldsymbol{E}_0 \cdot \frac{\left(\omega_m^2 + \dfrac{k^2 E_0^2}{8} - \omega^2\right) \mp \dfrac{i2\omega}{T_2}}{\left(\omega_m^2 + \dfrac{k^2 E_0^2}{8} - \omega^2\right)^2 + \dfrac{4\omega^2}{T_2^2} + \dfrac{T_1}{T_2}\omega^2 k^2 E_0^2}. \tag{25}$$

This result is to be consistent with

$$\int (\boldsymbol{P}_{\pm 1}\cdot \boldsymbol{E}_a)dV = \epsilon v_0 \left\{\left(\frac{\omega_a^2}{\omega^2} - 1\right) \pm \frac{i\omega_a}{Q\omega}\right\} \tag{26}$$

which is obtained from the circuit equation (20).

For the purpose of bringing out the essential features of the problem, let us now suppose that E_0^2 is identical in all parts of the cavity; *i.e.*,

$$E_0^2 = \text{constant}. \tag{27}$$

Then we immediately see that (25) yields

$$\int (\boldsymbol{P}_{\pm 1}\cdot \boldsymbol{E}_a)dV = -\,k^2 W_e' v_0 \cdot \frac{\left(\omega_m^2 + \frac{k^2E_0^2}{8} - \omega^2\right) \mp \frac{i2\omega}{T_2}}{\left(\omega_m^2 + \frac{k^2E_0^2}{8} - \omega^2\right)^2 + \frac{4\omega^2}{T_2^2} + \frac{T_1}{T_2}\omega^2k^2E_0^2}. \tag{28}$$

From the requirement that (26) and (28) must be identical, it then follows that

$$\frac{1}{Q} = \alpha \frac{\omega_m^2\omega}{\omega_a} \cdot \frac{\frac{2\omega}{T_2}}{\left(\omega_m^2 + \frac{k^2E_0^2}{8} - \omega^2\right)^2 + \frac{4\omega^2}{T_2^2} + \frac{T_1}{T_2}\omega^2k^2E_0^2} \tag{29}$$

and

$$\left(\frac{\omega}{\omega_a} - \frac{\omega_a}{\omega}\right) = \alpha \frac{\omega_m^2\omega}{\omega_a} \cdot \frac{\left(\omega_m^2 + \frac{k^2E_0^2}{8} - \omega^2\right)}{\left(\omega_m^2 + \frac{k^2E_0^2}{8} - \omega^2\right)^2 + \frac{4\omega^2}{T_2^2} + \frac{T_1}{T_2}\omega^2k^2E_0^2} \tag{30}$$

where

$$\alpha = \frac{k^2W_e'}{\epsilon\,\omega_m^2}. \tag{31}$$

In order to predict the frequency of oscillation, we first note that (29) and (30) give

$$\omega^2 = \omega_a^2 + \frac{\omega_a T_2}{2Q}\left(\omega_m^2 + \frac{k^2E_0^2}{8} - \omega^2\right). \tag{32}$$

Letting

$$\frac{\omega_a}{Q} = \Delta\omega_a \tag{33a}$$

$$\frac{2}{T_2} = \Delta\omega_m \tag{33b}$$

where $\Delta\omega_a$ and $\Delta\omega_m$ denote the half widths of the resonance curves of the cavity and active medium, respectively, we see that (32) may be written as

$$\omega^2 = \omega_a^2 + \frac{\frac{\Delta\omega_a}{\Delta\omega_m}}{1 + \frac{\Delta\omega_a}{\Delta\omega_m}}\left\{\omega_m^2\left(1 + \frac{k^2E_0^2}{8\omega_m^2}\right) - \omega_a^2\right\}. \tag{34}$$

For the optical maser, typically $Q \cong 10^7$ (for the mode of highest Q), $T_2/2 \cong 10^{-11}$ sec^{-1}, and $\omega_a \cong 10^{15}$ sec^{-1}; thus,

$$\frac{\Delta\omega_a}{\Delta\omega_m} \cong 10^{-3}. \tag{35}$$

For $E_0^2 = 0$, (34) agrees with the usual expression [3], [4] given for frequency pulling of the cavity line by the molecular system. However, for $E_0^2 \neq 0$, we see that ω_m is effectively *increased* by the factor $(1 + k^2E_0^2/8\omega_m^2)$, where we recall that $k^2 = 4\boldsymbol{\mu}^2/\hbar^2$. This effect is merely the electric analog of the well-known Bloch-Siegert shift which was predicted [5] in connection with magnetic resonance work. Alternatively, this frequency shift may be regarded as the result of a time-dependent Stark effect due to the coherent electric field.

In order to predict the amplitude of oscillation, we now insert ω^2, as given by (34), into (29) and solve for E_0^2. For additional simplicity, let us now consider the case where $\omega_a = \omega_m = \Omega$. Thus, on assuming that $\Delta\omega_a/\Delta\omega_m \ll 1$, we obtain the approximate expression

$$4 + T_2^2k^2E_0^2\left(\frac{T_1}{T_2} + \frac{k^2E_0^2}{64\Omega^2}\right) = 2\alpha\Omega T_2 Q. \tag{36}$$

Now, presumably under all conditions of interest

$$\frac{T_1}{T_2} \gg \frac{k^2E_0^2}{64\Omega^2} \tag{37}$$

would apply, and hence (36) reduces to approximately

$$4 + T_1T_2k^2E_0^2 = 2\alpha\Omega T_2 Q \tag{38}$$

or

$$E_0^2 = \frac{2QW_e'}{\epsilon\Omega T_1} - \frac{4}{k^2T_1T_2}. \tag{39}$$

It may be noted from (39), that the power delivered to the coherent field per unit volume of active material then is

$$P = \frac{\Omega}{Q}\frac{\epsilon E_0^2}{2} = \frac{W_e'}{T_1} - \frac{2\epsilon\Omega}{k^2T_1T_2Q}$$

$$= \frac{W_e'}{T_1} - \frac{\epsilon\Omega\hbar^2}{2\boldsymbol{\mu}^2T_1T_2Q}. \tag{40}$$

One sees that the "start oscillation" condition, as given by (40), is

$$W_e' \geq \frac{2\epsilon\Omega}{k^2 T_2 Q} = \frac{\epsilon\Omega\hbar^2}{2\mathfrak{u}^2 T_2 Q}. \tag{41}$$

In the limit of $E_0^2 = 0$, however,

$$W_e' = W = \frac{\hbar\Omega}{2}(N_2 - N_1) \tag{42}$$

where $(N_2 - N_1)$ denotes the steady-state population difference per unit volume. Therefore, (41) becomes

$$(N_2 - N_1) \geq \frac{\epsilon\hbar}{\mathfrak{u}^2 T_2 Q}, \tag{43}$$

which is the well-known "start oscillation" criterion derived by previous investigators [6]. Finally, on recalling that

$$W_e' = W_e + \dot{W}_p T_1,$$

from (40) there results

$$P = \frac{\Omega}{Q}\frac{\epsilon E_0^2}{2} = \dot{W}_p - \left\{\frac{\epsilon\Omega\hbar_2}{2\mathfrak{u}^2 T_1 T_2 Q} - \frac{W_e}{T_1}\right\}, \tag{44}$$

where W_e, of course, is a negative quantity. Hence, our analysis leads to the sensible (approximate) result that the power delivered to the field by the active molecules equals the power supplied to the active medium minus the power required to fulfill the condition of instability.

In applying the formalism, it was assumed for simplicity that

$$\frac{k^2 E_0^2}{\Omega^2} \ll 1. \tag{45}$$

Now, from (44), it is seen that

$$P \ll \frac{\epsilon\Omega^3}{2k^2 Q} = \frac{\epsilon\hbar^2\Omega^3}{8\mathfrak{u}^2 Q} \tag{46}$$

must then prevail. As an illustrative example, if $\epsilon \cong 1/(4\pi)$, $\Omega \cong 10^{15}$ sec^{-1}, $|\mathfrak{u}| \cong 10^{-18}$ cgs units, and $Q \cong 10^7$, then (46) merely requires that

$$P \ll 10^{11} \text{ watts/cm}^3. \tag{47}$$

Discussion

Application of the semiclassical equations, with phenomenological terms added in order to account for relaxation mechanisms and pumping, has yielded definite expressions for the steady-state frequency and amplitude of oscillation for an optical maser mode, in the case where just one mode is excited.

It is seen from (34) that we obtain the same frequency pulling expression which appears in the previous maser literature, with the difference that we find a shift which depends on the magnitude $\boldsymbol{E}_0$ of the alternating electric field. This field-dependent effect, which, as noted above, is the well-known Bloch-Siegert shift, may be characterized as: due to the "counter-rotating" component [7] of the electric field, as seen from (28), the maximum of the resonance line for the active molecules is shifted (as a function of frequency) from the unperturbed natural frequency ω_m to the value

$$\omega_m' \cong \omega_m\left(1 + \frac{k^2 E_0^2}{16\omega_m^2}\right) = \omega_m\left(1 + \frac{\mathfrak{u}^2 E_0^2}{4\hbar^2\omega_m^2}\right).$$

Now, if we suppose that $|\mathfrak{u}| \cong 10^{-18}$ cgs units, and $\omega_m \cong 10^{15}$ sec^{-1}, then we see that

$$\omega_m' - \omega_m \cong 2.3 \times 10^2 E_0^2 \text{ sec}^{-1},$$

for $\boldsymbol{E}_0$ expressed in esu units. Thus, at the high power levels which can be produced by the optical maser oscillator, in some situations this shift could be important.

Furthermore, it also is seen that from (28) that for sufficiently large field strength, the term $(T_1/T_2)\omega^2 k^2 E_0^2$ will tend to saturate the resonance line for the active molecules. This effect, which is well known to be due to the "co-rotating" component [7] of the electric field, leads through (29) to our prediction for the steady-state amplitude $\boldsymbol{E}_0$ of the electric field.

Here, no attempt has been made to treat fluctuation effects which would exist in practice due to conditions (such as thermal radiation, etc.) over which one has no experimental control. In our equations, $\boldsymbol{P}(r, t)$ and $\boldsymbol{E}(r, t)$ are to be interpreted as the *expected* polarization and electric field, as determined on the basis of the information which is available in practice.

In a more rigorous treatment, one would not assume that pumping may be accounted for merely by adding a term $\dot{W}_p$ to the right-hand side of (12b) for a two-level system. Instead, for example, it would be of interest to analyze a three-level system, in the case where there is optical pumping between levels E_1 and E_3, and optical maser operation between levels E_1 and E_2.

In conclusion, let us discuss briefly the question of the validity of the semiclassical theory of radiation, as applied above and by previous investigators [1], [4], [8]–[10], in predicting maser operation. The modern view, of course, is that in order to obtain complete agreement between theoretical and experimental results for problems where matter interacts with radiation, quantum electrodynamics is required. In the quantum electrodynamics formalism, the radiation field also is quantized, and then a wave function is introduced for the system consisting of matter plus the electromagnetic field. Recently, however, it has been shown [1] that experimentally observable effects, as predicted in this semiclassical theory and quantum electrodynamics, correspond to a degree which often is not sufficiently recognized. For example, it is found that the semiclassical theory does lead to *spontaneous* emission of radiation by a molecule, with the same radiation decay time that is obtained on quantizing the radiation field.

References

[1] E. T. Jaynes, "Some Aspects of Maser Theory," Microwave Lab., Stanford University, Stanford, Calif., Rept. No. 502; 1958.

[2] J. C. Slater, "Microwave Electronics," D. Van Nostrand Co., Inc., New York, N. Y.; 1950.

[3] J. P. Gordon, H. J. Zeiger, and C. H. Townes, "The maser—new type of microwave amplifier, frequency standard, and spectrometer," *Phys. Rev.*, vol. 99, pp. 1264–1274; August, 1955.

[4] W. G. Wagner and G. Birnbaum, "Theory of quantum oscillators in a multimode cavity," *J. Appl. Phys.*, vol. 32, pp. 1185–1194; July, 1961.

[5] F. Bloch and A. Siegert, "Magnetic resonance for non-rotating fields," *Phys. Rev.*, vol. 57, pp. 522–527; March, 1940.

[6] A. L. Schawlow and C. H. Townes, "Infrared and optical masers," *Phys. Rev.*, vol. 112, pp. 1940–1949; December, 1958.

[7] F. Bloch, "Nuclear induction," *Phys. Rev.*, vol. 70, pp. 460–474; October, 1946.

[8] K. Shimoda, T. C. Wang, and C. H. Townes, "Further aspects of the theory of the maser," *Phys. Rev.*, vol. 102, pp. 1308–1321; June, 1956.

[9] J. C. Helmer, "Maser oscillators," *J. Appl. Phys.*, vol. 28, pp. 212–215; February, 1957.

[10] R. P. Feynman, F. L. Vernon, and R. W. Hellwarth, "Geometrical representation of the Schrödinger equation for solving maser problems," *J. Appl. Phys.*, vol. 28, pp. 49–52; January, 1957.

Hole Burning Effects in a He-Ne Optical Maser

W. R. Bennett, Jr.
Bell Telephone Laboratories, Murray Hill, New Jersey
(Received November 20, 1961)

A study has been made of pulling effects by the amplifying media on the TEM_{00} modes of a helium-neon maser using a circular plane mirror Fabry-Perot cavity in which the mirror separation was known with precision. Approximate expressions are derived for mode pulling in homogeneously and inhomogeneously broadened optical masers. The experimental results suggest that the losses in the maser are not determined entirely by the mirror reflectance coefficient. A power-dependent splitting of the beat frequencies between simultaneously oscillating modes is explained in terms of a nonlinear frequency-dependent pulling effect arising from inhomogeneous broadening. The case of Lorentzian holes burned in a Gaussian line is treated specifically. It is suggested that the anomalous variation of beat frequencies with pumping rate results from hole repulsion effects and that the hole widths required arise from the combined effects of small-angle elastic scattering and stimulated emission. It is a consequence of the interpretation that the number of simultaneously oscillating even-symmetric modes in the helium-neon maser may easily be determined from the Fourier spectrum of the beat frequencies. A method is described by which the central cavity resonance may be stabilized near the center of the Doppler line in the case of three oscillating symmetric modes.

1. INTRODUCTION

THE narrow linewidths inherent in the helium-neon optical maser[1,2] are suggestive of a number of basic experiments. However, the ultimate usefulness of this narrow source of radiation, both as a research tool and a communications device, is largely limited by the frequency stability that may be obtained. For the latter reason, the present investigation was undertaken. Since the oscillation frequencies are primarily determined by the cavity resonances, and therefore the dimensions of the maser, it was not obvious *a priori* that a long-term frequency stability of much better than half the separation between adjacent cavity resonances (about 80 Mc/sec) could actually be obtained. It became apparent during the course of this research, however, that a method existed by which one could both determine the number of simultaneously oscillating modes and, in the case of three oscillations, set the central mode near the center of the Doppler line. The absolute stability that may be achieved in this way is still unknown and is currently under investigation. It seems clear that the method may at least be used to obtain an extremely high degree of relative stability.

2. MASER CONSTRUCTION

For the purpose of making the present study, the He-Ne maser shown in Fig. 1 was constructed. The mechanical design of the interferometer was altered from that of the original He-Ne maser,[1] in order to minimize mechanical fluctuations and permit an accurate determination of the plate separation. Each of the Fabry-Perot plates was fastened by spring clips to internal three-point mountings which, in turn, were securely fastened to the large external flanges shown at either end of the maser. The large flanges were surface-ground and separated by four 1-in. diameter Nilvar rods, cut to lengths which were identical within 0.0005 in. The bellows incorporated at each end of the discharge tube were sufficiently flexible to permit changing the plate separation by about 4.5% through the insertion of accurately machined spacers at the ends of the 1-in. diameter rods. Of several methods tried for controlling the plate angular alignment, one based on the magnetostrictive effect in the Nilvar supporting rods was found to be the most satisfactory. The magnetostrictive effect was also found to present a convenient method for changing the plate separation; however, with the adjustments used, some coupling between the plate separation and angular alignment controls existed. A more detailed description of the alignment method and interferometer construction will be given elsewhere. The improved mechanical stability obtained with the present maser arises largely by virtue of the fact that the tuning is accomplished through the slight distortion of a highly rigid structure. In other respects, the maser had properties similar to those reported in the first He-Ne maser with the exception that the device oscillated only on the strongest ($2s_2$ to $2p_4$ in Paschen notation at 11 522.76 A)[3] of the five transitions of neon previously observed. The plates used were flat to $\cong\lambda/100$ over the beam diameter and had a reflectance of 99% at the wavelength used prior to insertion in the maser.[4]

3. EXPERIMENTAL RESULTS

The modes of oscillation were studied by observing the difference frequencies produced when light from the maser was focussed on the surface of a 7102 photomultiplier tube. The photosurface acts as a square-law detector and one may observe all possible difference frequencies up to the maximum frequency response characteristic of the tube. This effect was first observed

[1] A. Javan, W. R. Bennett, Jr., D. R. Herriott, Phys. Rev. Letters 6, 106 (1961).

[2] A. Javan, E. A. Ballik, and W. L. Bond (to be published) have recently remeasured the linewidths in the first maser and have found that they are inherently in the order of 2 cycles/sec or less.

[3] The author is indebted to D. L. Wood for his help in firmly establishing the identity of this line. The small possibility that the strongest maser transition might have been the $2s_4$–$2p_7$ line at 11 525.016 A had not previously been ruled out experimentally.

[4] The author is indebted to D. R. Herriott for making these measurements.

Reprinted with permission from *Phys. Rev.*, vol. 126, pp. 580–593, Apr. 15, 1962.

by Forrester, Gudmundsen, and Johnson[5] using incoherent light sources and more recently was used to investigate the linewidths obtained in the first helium-neon maser.[1,2] A previous report on the difference-frequency spectrum obtained in this way from a He-Ne maser has been given by Herriott[6] and, for the sake of clarity, will be summarized here. Since the single-pass gain obtainable from a He-Ne maser is small, one is forced to use a Fabry-Perot cavity with an enormously high Q in order to exceed the threshold requirement for oscillation. In practice this implies a cavity width which is much narrower than the Doppler width for one transition and that the frequencies of oscillation will be primarily determined by the cavity resonances. The dominant cavity resonance frequencies are determined by requiring that the cavity length L be a half-integral multiple of the wavelength. These resonant frequencies correspond to the first even-symmetric radial modes calculated by Fox and Li[7] and differ in frequency by $c/2L \cong 160$ Mc/sec for the present work, where c is the velocity of light in vacuum. Since these frequency separations are small compared to the Doppler width, the maser may oscillate on several of these frequencies simultaneously. In addition to the even-symmetric modes, Fox and Li have shown that the next longitudinal modes of importance are those possessing odd radial symmetry and that these differ from the former by frequencies of the order of a Mc/sec with the present geometry. Hence, as was reported by Herriott,[6] one obtains a difference frequency spectrum of the type shown in Fig. 2. That is, a peak at zero frequency corresponding to each line beating with itself, followed by a peak $\cong 1$ Mc/sec away corresponding to all possible differences between the first even- and odd-symmetric radial modes having the same value of $c/2L$; a peak at $c/2L$ corresponding to the differences between all even-symmetric modes separated by $c/2L$ and all odd-symmetric modes separated by $c/2L$, surrounded by two satellites corresponding to the possible even-odd and odd-even difference frequencies at approximately $c/2L$, etc. As indicated by the arrows in Fig. 2, the satellites obtained as a result of the beats between the odd-even

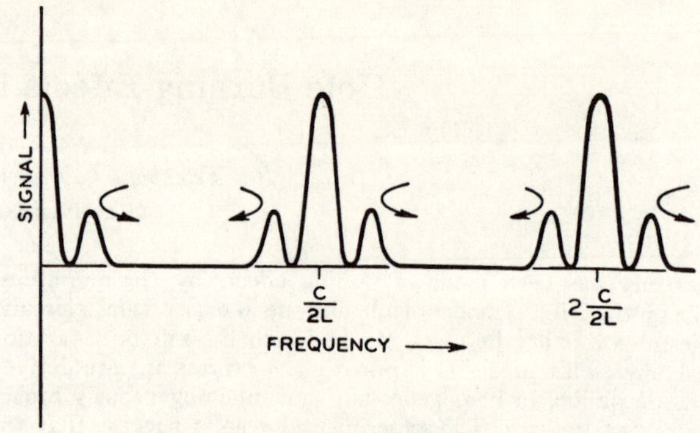

FIG. 2. Fourier spectrum from possible difference frequencies in the maser.

symmetric modes vary strongly with the interferometer plate alignment, reaching a minimum separation from the $c/2L$ beats at parallel alignment. For the latter reason, they are both easily identifiable, and unsuitable for the present study. The discussion which follows is concerned entirely with the central peaks obtained at multiples of $c/2L$ corresponding to the differences between the even-symmetric modes. That is, the low power data were taken under conditions where the satellites about the $c/2L$ and $2(c/2L)$ beats (and therefore the odd-symmetric modes) were absent.

The light output of the maser was found to have more or less random polarization. The output varied from the extreme case of linear polarization with arbitrary orientation of the electric field vector to nearly perfect circular polarization over time intervals in the order of seconds. This result would be expected from ideal cylindrical geometry and is in contrast with the behavior of the first He-Ne maser.[1,8] In addition, the various modes of oscillation in the present maser were also found to have unrelated polarizations. The latter was established by examining the behavior of the beat frequencies when a linear polaroid was inserted in the beam. Javan and Ballik[9] have observed that the central $c/2L$ beat may sometimes not be observable without the use of a polaroid. Their observation may be interpreted through the assumption that the two modes producing the beat are polarized linearly at right angles. The phototube, being a reasonably good square-law detector, fails to respond to the beat since the difference frequency is contained in the scalar product of the two electric fields. The insertion of a polaroid at 45° to the field components, however, restores the beat. That is, the polaroid transmits components of the two fields which are parallel. In order to avoid missing any of the beats in the present work, a polaroid was always inserted in the beam and rotated until a location was obtained in which all beats present were observed with comparable

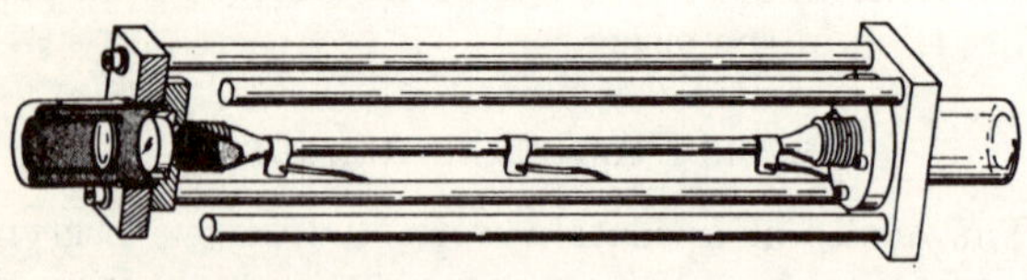

FIG. 1. He-Ne optical maser used. The "cut-away" section at the left shows the location of the exit window and Fabry-Perot plate. The discharge is driven by a 30-Mc/sec (20 to 50 watt) supply through the three external electrodes shown. The plate separation and angular alignment is controlled magneto-strictively by coils (not shown) placed about each of the four Nilvar supporting rods.

[5] A. T. Forrester, R. A. Gudmundsen, and P. O. Johnson, Phys. Rev. **99**, 1691 (1955).

[6] D. R. Herriott, in *Advances in Quantum Electronics*, edited by J. R. Singer (Columbia University Press, New York, 1961), p. 49.

[7] A. G. Fox and Tingye Li, Bell System Tech. J. **40**, 453 (1961).

[8] The first He-Ne maser was linearly polarized in a direction closely correlated with a striated pattern in the dielectric coatings. The absence of such striated patterns in the present maser may have resulted from the use of more modest bake-out temperatures.

[9] A. Javan and E. A. Ballik (private communication).

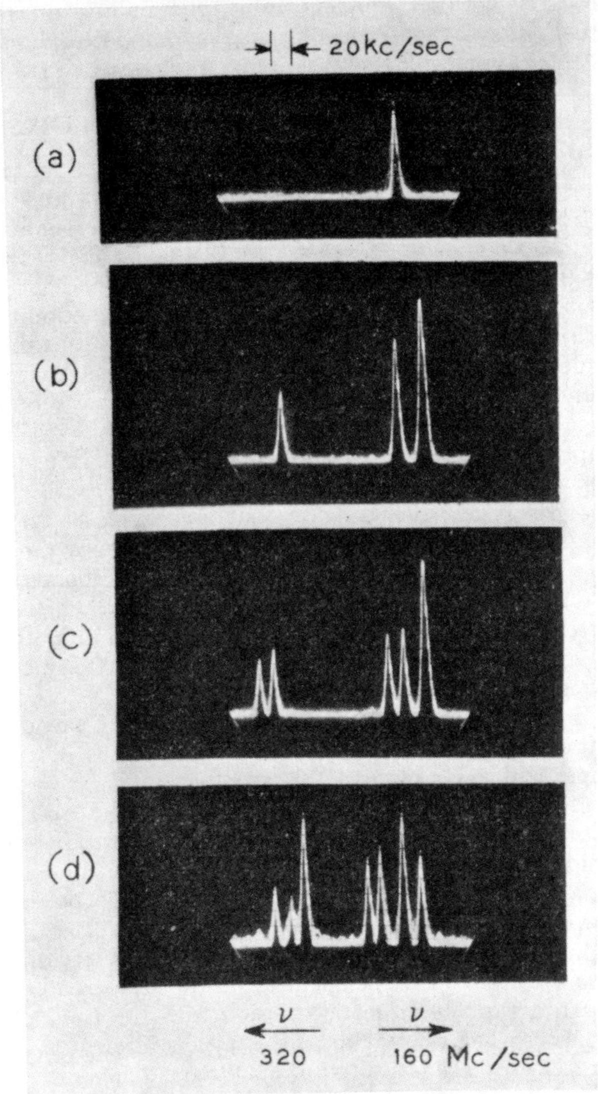

FIG. 3. Splitting of $c/2L$ and $2(c/2L)$ beats with power. Power increases from (a) to (d).

amplitude. For this reason, the relative amplitudes (for example of the split components in Fig. 3) are arbitrary.

The first measurements were made on the absolute frequency of the central peak corresponding to a difference frequency of $c/2L$. These measurements were made under the conditions of low power in which this beat was not split and the beat at $2c/2L$ was not observable, i.e., the conditions under which the data in Fig. 3(a) were taken. Measurements were made at three discrete interferometer lengths and these data are shown in Table I. The uncertainty in the absolute frequency measurement ($\cong$20 kc/sec) arises primarily from the variation of the beat frequency with power and plate separation as discussed below. As is obvious from Table I, the beat frequency is not equal to $c/2L$, but is less than $c/2L$ by about 1 part in 800. This effect may be readily explained on the basis of mode pulling by the Doppler line and is discussed in Sec. 13. It is a consequence of the analysis in Sec. 5 and the data in Table I that the loss in the interferometer exceeds the known reflectance loss by a factor of about two.

TABLE I. Comparsion of measured beat frequencies with the difference between calculated cavity resonance frequencies for three separate cavity lengths.

$c/2L$ (in Mc/sec)	Measured beat (in Mc/sec)	Fractional difference between $c/2L$ and measured beat
161.316 ± 0.022	161.107 ± 0.010	$-(1.3\pm0.2)\times10^{-3}$
158.713 ± 0.020	158.531 ± 0.010	$-(1.1\pm0.2)\times10^{-3}$
156.189 ± 0.018	155.982 ± 0.010	$-(1.3\pm0.2)\times10^{-3}$

There are two interesting aspects of the data obtained at increasing power levels:

(a) The frequency of the $c/2L$ beat was found to increase with increasing power. The magnitude of the increase before splitting occurred (see below) was dependent on the setting of the interferometer plates. It varied from nearly zero to a maximum of about 30 kc/sec. As will become evident from the analysis in Sec. 5, such an increase is anomalous. That is, for any singly-peaked line one would expect the pulling towards the line center to increase with the number of excited atoms and that the pulling would be less for a cavity resonance near the line center than for one farther away from it. Hence one would expect the frequency separation between adjacent cavity resonances to decrease with increasing power. The same anomalous behavior was encountered in the split components described below. The anomaly is explained through the hole repulsion effect in Sec. 13.

(b) A power-dependent splitting of the $c/2L$ and $2(c/2L)$ beats was encountered. Figs. 3(a) through 3(d) show data taken simultaneously on the central beats at both $c/2L$ and $2(c/2L)$ with increasing rf power. The data on the right show the $c/2L$ beat (160 Mc/sec) and those on the left the $2(c/2L)$ beat (320 Mc/sec). The frequency scale is the same in each case as indicated by the 20 kc/sec interval shown, with the exception that increasing frequency occurs in opposite directions for the two sets of data. These data were taken by adjusting the maser length so that the $c/2L$ beat occurred within the first if band of the spectrum analyzer used and by tuning the local oscillator in the analyzer to correspond to the $2(c/2L)$ beat. The data shown in Figs. 3(a) through 3(d) were taken at increasing power levels, where each figure represents about 10 oscilloscope traces. In Fig. 3(a) the $c/2L$ beat is present and the $2(c/2L)$ beat is not. As the power increases [Fig. 3(b)], the $c/2L$ beat splits by an amount which typically was in the order of 20 kc/sec and the $2(c/2L)$ beat appears. The separation between the two $(c/2L)$ components in Fig. 3(b) was dependent on the interferometer length. By changing the plate separation magnetostrictively, the splitting shown in Fig. 3(b) could be varied from a maximum of about 30 kc/sec to a minimum of $\leqslant 1$ kc/sec (i.e., 1 kc/sec was the minimum if resolution of the apparatus used). With a further increase in power [Fig. 3(c)], the $c/2L$ beat breaks into three components and the beat at $2(c/2L)$ splits. Again the frequency of the components was found to increase with power and the splitting was dependent on the interferometer setting. For some interferometer settings, the odd-symmetric modes (as evidenced by the appearance of the satellites in Fig. 2) were also present during conditions typical of Fig. 3(c). Their presence or absence seemed to have relatively little effect on the splitting of the central $c/2L$ and $2(c/2L)$ beats. Finally, at the highest power level shown in Fig 3, the $c/2L$ beat has broken into four components and the $2(c/2L)$ beat into three.

The data in Fig. 3 may be interpreted in the following way: Some nonlinear frequency-dependent pulling mechanism exists whereby the differences between adjacent oscillating modes are not identical In Fig. 3(a) the gain is only sufficient for the two modes nearest the line center to break into oscillation Consequently there is one component at $c/2L$ and none at $2(c/2L)$ In Fig. 3(b) the gain is adequate for three modes to oscillate; there exist two discrete frequencies at $c/2L$ and one at $2(c/2L)$. In Fig. 3(c) four modes are oscillating; there exist three different ways of obtaining the $c/2L$ beat and two for the $2(c/2L)$ beat. Finally, in Fig. 3(d) the gain is high enough for five modes to oscillate; there exist four ways of obtaining the 160-Mc/sec beat and three ways of obtaining the 320-Mc/sec beat. The expected components at frequencies higher than 320 Mc/sec have since been observed.[10]

There are two important consequences of the interpretation of the data in Fig. 3. The first is that one may deduce the number of even-symmetric modes which are oscillating from the beat splitting. The second is that in the case of three simultaneously oscillating modes [Fig. 3(b)] it seems probable that the frequency difference between the two $c/2L$ components should increase with the difference in frequency between the central cavity resonance and the center of the Doppler line. Hence, for example, the two $c/2L$ beats could be used to generate an error frequency (whose sharpness would be determined by the linewidth of the oscillation)[2,11] which could be used in a magnetostrictive tuning device to stabilize the central mode near the center of the Doppler line. The second consequence follows from a reasonably general symmetry argument. If we assume that the Doppler line shape is perfectly symmetric about its center frequency and that there are only three modes oscillating, any splitting in the $c/2L$ beats must arise from an asymmetric distribution of the three modes about the Doppler center. That is, if the central cavity mode is precisely on the Doppler center, the pulling on the other two resonances must be equal and opposite; hence in this case no splitting can occur.

[10] A. J. Rack (private communication).

[11] The linewidths implied by the data in Fig. 3 are limited by the resolution of the spectrum analyzer.

4. PARAMETERS IN THE PRESENT MASER

For convenience, we summarize the pertinent parameters characteristic of the present maser in this section.

The frequency separation of the cavity resonances ($c/2L$) is nominally $\cong$160 Mc/sec (Table I). The wavelength of the maser transition is 11 522.76 A and corresponds to the $2s_2$ to $2p_4$ line of neon in Paschen notation.[3] From direct gain measurements on this transition under similar discharge conditions, the fractional energy gain in a single pass through the interferometer should be about 6%. The known reflectance losses[4] ($\cong$1%) imply a cavity width ($\Delta\nu_c$) of about $\frac{1}{2}$ Mc/sec [Eq. (3) below]. However, the data in Table I taken with the analysis below, demonstrate that $\Delta\nu_c$ is closer to 1 Mc/sec and that the total energy loss per pass is about 2%. Hence the maximum gain available corresponds to about three times the threshold for oscillation. The latter is in good agreement with the maximum number of modes observed [Fig. 3(d)] and the 800-Mc/sec Doppler width assumed below.

In the limit of complete resonance trapping of vacuum ultraviolet photons, the effective decay rates of the upper and lower maser levels are about 10^7 sec^{-1} and 8×10^7 sec^{-1}, respectively.[12] The exact value of the natural linewidth ($\Delta\nu_n$) for the maser transition depends critically on the unknown, vacuum ultraviolet decay rate of the upper maser level. It is expected, however, that $15<\Delta\nu_n<80$ Mc/sec (see Sec. 10).

The full Doppler width at half-maximum intensity for neon atoms at the temperature of the maser is about 800 Mc/sec. The inelastic collision with the He($2\,^3S$) atom by which the excited neon state is formed[1] is, of course, exothermic by $\cong 1.5kT$ at room temperature. However, since the He atom has $\frac{1}{5}$ the mass of the neon atom, it will carry off most of the energy by which the reaction is exothermic. Although the Doppler width will increase with the gas temperature and hence with the power output in the maser, it does so as the square root of the absolute temperature. Hence a Doppler width much in excess of 800 Mc/sec is not to be expected. It is inconceivable, for example, that the Doppler width could increase sufficiently with power to explain the magnitude of the anomalous power dependence of the beat frequencies reported above.

5. MODE PULLING ANALYSIS

We are only concerned with the dominant modes of even radial symmetry corresponding to longitudinal propagation in a plane parallel Fabry-Perot interferometer. In what follows it is assumed that steady-

[12] W. R. Bennett, Jr., in *Advances in Quantum Electronics*, edited by J. R. Singer (Columbia University Press, New York, 1961), p. 28.

state oscillation has been obtained in the system and that no coupling effects exist between simultaneously oscillating modes through time-dependent nonlinearities in the medium. These assumptions are likely to break down at some point. However, what is sought are manageable, approximate solutions which will explain the dominant mode pulling effects.

The oscillation frequency in the maser is determined primarily by a condition on the phase of the electric field and the phase shifts of importance all arise from single-pass time delays. By definition, the phase shift resulting from a wave traveling once through an interferometer of length L at a phase velocity c/n (where n is the refractive index) is

$$\varphi = 2\pi\nu Ln/c. \qquad (1)$$

For a standing wave to build up in the cavity, the single-pass phase shift must be an integral multiple of π. Hence the evacuated cavity ($n=1$) has resonant frequencies which are separated by $c/2L$.

From Eq. (1), the dispersion $(\partial\varphi_c/\partial\nu)$ for the evacuated cavity ($n=1$) is a constant. It is convenient to express $\partial\varphi_c/\partial\nu$ in terms of the fractional energy loss per pass f and the full width of the cavity resonance at half-maximum intensity, $\Delta\nu_{0c}$. To accomplish the latter, we visualize a situation in which energy (U) is placed in the evacuated cavity in the mode of interest. This energy decays exponentially with time at the rate $(c/L)f$. As a result, the frequency response of the interferometer is not perfectly sharp and one may define a Q for the evacuated cavity given by

$$Q_{0c} = \frac{2\pi\nu_{0c}U}{(cf/L)U} = \frac{\nu_{0c}}{\Delta\nu_{0c}}. \qquad (2)$$

From (1) and (2),

$$\partial\varphi_c/\partial\nu = 2\pi L/c = f/\Delta\nu_{0c} = \text{const} \cong 2\times10^{-8}\ \text{rad-sec}. \qquad (3)$$

For the present system, a single-pass loss of one percent corresponds to $\Delta\nu_{0c} \cong (1/2)$Mc/sec.

There will, in general, be some entirely negligible contribution to $(\partial\varphi_c/\partial\nu)$ arising from the resonant nature of the mirror reflectance coefficient. For example, in the present case the mirror transmission loss varied by a factor of two over a total range of about 1000 A (or 2×10^{12} cycles/sec). If one then estimates the dispersion at maximum reflectance using a Lorentzian line shape [Eq. (9)], it is seen that the contribution from the mirrors is about 5×10^{-15} rad-sec. Hence the inclusion of this term alters Eq. (3) by about 2 parts in 10^7. The inclusion of this term in the analysis given below merely changes the final pulling term in Eq. (22) by the same fractional amount and has no important effect on the absolute frequency of the oscillation. We therefore ignore this contribution to $(\partial\varphi_c/\partial\nu)$.

The introduction of the amplifying medium changes the refractive index in the system, thereby altering the single-pass phase shift from that obtained in the evacuated case. Oscillation therefore occurs at another frequency $\bar{\nu}$, differing from the cavity resonance frequency ν_{0c}, such that the single-pass phase shift is still an integral multiple of π. Since the cavity dispersion is large compared to that for the amplifying medium, oscillation occurs close to ν_{0c} and the pulling is small. It is convenient to formulate the problem in terms of the difference in frequency from the cavity resonance. In particular, the maser oscillates at that frequency $\bar{\nu}$ such that

$$(\partial\varphi_c/\partial\nu)(\bar{\nu}-\nu_{0c}) + \Delta\Phi_m(\bar{\nu}) = 0. \qquad (4)$$

Here, $\Delta\Phi_m(\bar{\nu})$ is the total change in single-pass phase shift at the actual frequency of oscillation which is caused by the insertion of the medium. $\Delta\Phi_m(\bar{\nu})$ is composed of two parts and, from Eq. (1), may be expressed as

$$\begin{aligned}\Delta\Phi_m(\bar{\nu}) &= (2\pi L/c)[(n_0-1)+(n-1)]\bar{\nu}\\ &\equiv (2\pi L/c)(n_0-1)\bar{\nu} + \Delta\varphi_m(\nu).\end{aligned} \qquad (5)$$

The first term in Eq. (5), $(2\pi L/c)(n_0-1)\bar{\nu}$, arises from the density of ground-state atoms in the maser and from the density of excited atoms which may participate in neighboring transitions (e.g., the $2\,^3S-2\,^3P$ of He at 10 830 A). That is, this first term arises from a refractive index which is essentially independent of the frequency over the range of interest. Its main source in the helium-neon maser is the He(^1S) at 1 mm Hg for which $(n_0-1)\cong 5\times10^{-8}$.[13]

From Eqs. (3) and (5), Eq. (4) may be expressed,

$$\bar{\nu} = \left(\frac{\nu_{0c}}{n_0}\right) - \left(\frac{\Delta\nu_{0c}/n_0}{f}\right)\Delta\varphi_m(\bar{\nu}).$$

We include the effects of n_0 by defining

$$\bar{\nu}_c = \nu_{0c}/n_0 \quad\text{and}\quad \Delta\nu_c = \Delta\nu_{0c}/n_0. \qquad (6)$$

Hence the oscillator frequency ($\bar{\nu}$) is given by

$$\bar{\nu} = \nu_c - (\Delta\nu_c/f)\Delta\varphi_m(\bar{\nu}). \qquad (7)$$

For the helium-neon case, $\Delta\nu_c$ in Eq. (6) differs negligibly ($\cong$5 parts in 10^8) from $\Delta\nu_{0c}$ in Eq. (3). Similarly, the frequency separation between adjacent cavity resonances differs by the same negligible constant amount from $c/2L$. For an arbitrary setting of the interferometer, n_0 does introduce a pressure-dependent shift in the oscillator frequency of about 13 Mc/sec per mm Hg of He. However, in the special case that ν_c is tuned to the center of a symmetric line (see below), the oscillator frequency is independent of ν_c and this pressure-dependent shift vanishes.

The term $\Delta\varphi_m(\bar{\nu})$ in Eqs. (5) and (7) is a function of the fractional energy gain per pass $g(\bar{\nu})$. Since the latter varies with frequency over the transition responsible for the amplification $\Delta\varphi_m$ does also. Generally, $\Delta\varphi_m$ goes through zero at the line center (ν_m), is negative

[13] *American Institute of Physics Handbook* (McGraw-Hill Book Company, Inc., New York, 1957), pp. 6–21.

("anomalous dispersion") for frequencies less than ν_m, and is positive for frequencies greater than ν_m. Equation (6) therefore predicts a shift in the direction of the line center.

Threshold for oscillation occurs at that frequency $\bar{\nu}_T$, satisfying both Eq. (7) and the condition $g(\bar{\nu}_T)=f$. The latter generally results in a transcendental equation for $\bar{\nu}_T$. In the special case of a Lorentzian line, the transcendental nature of this equation is removed.

The existence of a steady-state oscillation above threshold obviously implies that the gain at the frequency of oscillation must saturate at

$$g(\bar{\nu})=f. \tag{8}$$

In the case of homogeneous or "natural" broadening, gain proportionality is always maintained over the line. That is, the reduction of the gain at frequency $\bar{\nu}$ necessary to satisfy Eq. (8) produces a proportionate reduction of the gain at all other frequencies. Hence in the case of homogeneous broadening, the oscillation frequency is always given by its value at threshold and there is no direct power-dependent pulling effect. In the case of inhomogeneous broadening (applying to the He-Ne maser), the interpretation of requirement (8) is more complicated.

6. ASSUMPTIONS ON LINE SHAPE

In general, $\Delta\varphi_m$ will be a function of the single-pass gain and for a given line shape could be calculated numerically from the Kramers-Kronig relations.[14]

For a Lorentzian line, it may be shown (see Appendix I) that

$$\Delta\varphi_m(\nu)\cong-\frac{g_m(\nu_m-\nu)}{\Delta\nu_m}\left[1+\frac{4(\nu_m-\nu)^2}{(\Delta\nu_m)^2}\right]^{-1}, \tag{9}$$

where the fractional energy gain per pass at the frequency ν is given by

$$g(\nu)\cong g_m[1+4(\nu_m-\nu)^2/(\Delta\nu_m)^2]^{-1}. \tag{10}$$

Here, as in the rest of this paper, we adopt the notation that g_m is the fractional energy gain per pass at the line center (ν_m), and that $\Delta\nu_m$ is the full width of the line at half-maximum fractional energy gain. The approximations in (9) and (10) arise from the assumption that $\nu+\nu_m=2\nu$ and that $g_m\ll 1$. Equations (9) and (10) may be expressed

$$\Delta\varphi_m(\nu)\cong-g(\nu)(\nu_m-\nu)/\Delta\nu_m. \tag{11}$$

The Lorentzian form will be used in this paper to obtain an accurate expression for the pulling in the case of natural broadening and to allow for the holes produced by inhomogeneous broadening in the case of the helium-neon maser.

What actually is involved in the helium-neon maser is a Gaussian distribution of Lorentzian lines resulting from the thermal motion of the excited neon atoms and the corresponding Doppler shift of the atoms' center frequencies. Hence, in the limit (holding here) that the natural linewidth is small compared to the Doppler width, the line much more closely resembles a Gaussian than a Lorentzian.

We are primarily concerned with effects taking place within the full width of the line at half-maximum intensity and in the limit of small energy gain. In the limit of small gain, it is apparent from the Kramers-Kronig relations[15] that the phase shift increases linearly with the gain. Hence for any typical symmetric line, the phase shift introduced by the amplifying medium may be expanded in the series

$$\Delta\varphi_m(\nu)\cong-ag_m\left(\frac{\nu_m-\nu}{\Delta\nu_m}\right)\left[1-b\left(\frac{\nu_m-\nu}{\Delta\nu_m}\right)^2+\cdots\right], \tag{12}$$

where the constants a and b will depend on the line shape. Since $\Delta\varphi_m$ must have odd symmetry about the line center, approximation (12) involves the assumption that $[(\nu_m-\nu)/\Delta\nu_m]^4\ll 1$. Hence, approximation (12) will be extremely good over a considerable range within the central portion of the line and will fail rapidly for frequencies occurring near the half-intensity points on the gain curve. Approximation (12), of course, consists of the first two terms in the expansion of the sine function and for manipulative reasons it is convenient to express $\Delta\varphi_m$ in the latter way. In particular, a "best fit" of numerical calculations made by Thomas[14] demonstrates that

$$\Delta\varphi_m(\nu)\cong-0.28_2 g_m \sin\left(\frac{\nu_m-\nu}{0.3\Delta\nu_m}\right) \tag{13}$$

is an exceptionally good approximation for a Gaussian. Both "angles" in Eq. (13) are expressed in radians. The errors introduced by a literal interpretation of (13) are less than 1% for $|\nu_m-\nu|\leq 0.4\Delta\nu_m$. Equation (13) is $\cong$6% low at the half-intensity points on the curve. For $|\nu_m-\nu|>0.5\Delta\nu_m$ the error increases exponentially. The corresponding fractional energy gain for the Gaussian is,

$$g(\nu)\cong g_m \exp\left[-\left(\frac{\nu_m-\nu}{0.6\Delta\nu_m}\right)^2\right]. \tag{14}$$

From (13) and (14)

$$\Delta\varphi_m(\nu)\cong-0.28_2 g(\nu)\exp\left[\left(\frac{\nu_m-\nu}{0.6\Delta\nu_m}\right)^2\right]\sin\left(\frac{\nu_m-\nu}{0.3\Delta\nu_m}\right) \tag{15}$$

holds for a Gaussian within the central region of the curve.

[14] D. E. Thomas (to be published) has extended his "Tables of Phase of a Semi-Infinite Unit Attenuation Slope" (Bell System Monograph 2550) to handle general problems of the present type numerically. His results for the special case of a Gaussian have been verified analytically by A. Javan.

[15] See, for example, J. H. Van Vleck, *Radiation Laboratory Series* (McGraw-Hill Book Company, Inc., New York, 1948), Vol. 13, Chap. 8.

7. HOMOGENEOUS BROADENING

For a homogeneously broadened Lorentzian line, (7), (8), and (11) yield

$$\bar{\nu}=(\nu_c\Delta\nu_m+\nu_m\Delta\nu_c)/(\Delta\nu_m+\Delta\nu_c) \tag{16}$$

for the oscillation frequency where definition (6) is assumed.[16] In the limit that $\Delta\nu_c<\Delta\nu_m$, (16) becomes

$$\bar{\nu}=\nu_c+(\nu_m-\nu_c)\left(\frac{\Delta\nu_c}{\Delta\nu_m}\right)\left[1-\frac{\Delta\nu_c}{\Delta\nu_m}+\cdots\right], \tag{17}$$

and in the Lorentzian case, the pulling is linearly dependent on the frequencies. The latter would, of course, not be true in the case of an inhomogeneously broadened Lorentzian line above threshold. As may be seen by comparing the expanded terms of Eqs. (9) and (13), the phase characteristic of the Lorentzian is actually about twice as nonlinear as that of the Gaussian near the line center.

For a Gaussian line, (7), (8), and (15) yield,

$$\bar{\nu}=\nu_c+0.28_2\Delta\nu_c\exp\left[\left(\frac{\nu_m-\bar{\nu}}{0.6\Delta\nu_m}\right)^2\right]\sin\left(\frac{\nu_m-\bar{\nu}}{0.3\Delta\nu_m}\right), \tag{18}$$

and the pulling is nonlinear even in the homogeneously broadened case. For frequencies very near the line center in the limit that $\Delta\nu_c<\Delta\nu_m$, Eq. (18) becomes

$$\bar{\nu}=\nu_c+(\nu_m-\nu_c)(0.94\Delta\nu_c/\Delta\nu_m)\times[1-0.94(\Delta\nu_c/\Delta\nu_m)+\cdots]. \tag{19}$$

Hence, for frequencies near the center of the line in the limit $\Delta\nu_c\ll\Delta\nu_m$, the main difference between the Gaussian and the Lorentzian is a 6% reduction in the pulling factor.

Equation (18) only applies to the helium-neon maser in the case of threshold for the first cavity resonance. For an inhomogeneously broadened line, an expansion of the threshold pulling term in powers of $\Delta\nu_c/\Delta\nu_m$ is misleading. The neglected nonlinear terms are generally much larger than $(\Delta\nu_c/\Delta\nu_m)^2$. The latter is apparent for Eq. (18), but is also true of Eq. (16). The inhomogeneously broadened case may be more realistically treated through the method given below.

8. GENERAL APPROXIMATE SOLUTION FOR OPTICAL MASERS

Generally for optical masers, $\Delta\nu_m\gg\Delta\nu_c$ and $\Delta\varphi_m(\nu)$ will be a slowly varying function of the frequency over the cavity resonance. We may therefore expand $\Delta\varphi_m$ in a Taylor series about the cavity resonance frequency, obtaining

$$\Delta\varphi_m=(\Delta\varphi_m)_{\nu_c}+(\partial\Delta\varphi_m/\partial\nu)_{\nu_c}(\nu-\nu_c)+\cdots. \tag{20}$$

[16] Equation (16) is equivalent to a result quoted by C. H. Townes, in *Advances in Quantum Electronics*, edited by J. R. Singer (Columbia University Press, New York, 1961), p. 10.

Substituting Eq. (20) in Eq. (7) yields

$$\bar{\nu}=\nu_c-(\Delta\nu_c/f)\Delta\varphi_m(\nu_c)\times[1-(\Delta\nu_c/f)(\partial\Delta\varphi_m/\partial\nu)_{\nu_c}+\cdots]. \tag{21}$$

As may be seen by comparison with Eqs. (17) and (19), this last approximation is equivalent to an expansion of the pulling terms in powers of $(\Delta\nu_c/\Delta\nu_m)$ which still retains the nonlinear properties of the phase characteristic. Typically, $(\Delta\nu_c/\Delta\nu_m)\cong10^{-3}$. Hence the oscillator frequency is given by

$$\bar{\nu}\cong\nu_c-(\Delta\nu_c/f)\Delta\varphi_m(\nu_c), \tag{22}$$

where errors in the second term of about a part in 10^3 (i.e., about 200 cycles/sec in the absolute frequency for the helium-neon maser) may be expected. $\Delta\varphi_m(\nu_c)$ represents the actual phase shift introduced by the amplifying transition at the cavity resonance in the presence of oscillation. Equation (22) may be solved numerically using the methods of Thomas[14] in the general case (e.g., asymmetric lines) for both homogeneous and inhomogeneous broadening.

9. EFFECT OF HOLES

It is in the interpretation of requirement (8) that an important difference arises between the helium-neon maser and, for example, the ammonia maser (or more generally, between the present oscillator and most other oscillators). In the present system, the line is primarily broadened by the Doppler effect. Consequently, reduction of the gain at the frequency $\bar{\nu}$ in order to satisfy Eq. (8) does not imply a proportionate reduction of the gain at other frequencies. If the converse were true (as in the case of homogeneous broadening), a second cavity mode would generally not go into oscillation; i.e., the saturation requirement (8) at the first resonance would prevent the second cavity resonance from reaching threshold. In the present case, (8) is satisfied by burning a hole in the line. $g(\bar{\nu})$ saturates at f, whereas the gain over the rest of the line continues to increase with pumping rate. In other words, the phase shift introduced at the cavity resonance continues to increase as the number of upper state atoms increases at frequencies well removed from the hole. We shall satisfy requirement (8) in the following analysis by subtracting the phase shift which would have been produced by atoms in the hole from the phase shift introduced by the entire distribution in the absence of oscillation. In evaluating these phase shifts, we make use of the same approximation implicit in Eq. (22)—namely, that $\bar{\nu}$ may be replaced by ν_c.

Before taking explicit account of the holes, we make the following two observations:

(a) Since atoms on opposite sides of the cavity resonance (ν_c) contribute to the gain at ν_c with the same sign, the total gain at ν_c is determined primarily by those atoms whose center frequencies fall within the resonance by about one natural linewidth. Hence, variations in the upper state density at large frequency separations from the cavity resonance do not affect the gain at ν_c appreciably, and a hole burned at one resonance does not have an important first-order effect on the gain at another resonance.

(b) Since atoms on opposing sides of the cavity resonance contribute to the phase shift at ν_c with opposite sign, the net phase shift at ν_c is determined primarily by atoms which are well removed from the cavity resonance. Hence the absence of atoms (i.e., a hole) at one resonance can have a large first-order effect on the phase shift at another resonance. The presence of a hole at one resonance reduces the pulling at another resonance which would have existed in the absence of the hole. That is, two holes always tend to repel each other. However, since a hole represents a symmetric removal of atoms about a cavity resonance to first order, a hole does not have a first-order effect on itself.

It is shown in Appendix II that the probability for stimulated emission is a Lorentzian function of the frequency separation between the mode of oscillation and the center frequency of the atoms involved. In addition, it is shown that the full width at half-maximum for this probability distribution is a function of the power in the mode and, hence, of the gain at the resonance. We therefore assume that the holes burned in the line will be Lorentzian in shape and have widths dependent on their location in respect to the line center. Although the Lorentzian hole shape would only hold strictly for a constant distribution of atoms, it is apparent that it will represent a good first-order approximation at low powers in the present case.

By analogy with Eq. (9), the phase shift introduced at frequency ν_2 by a Lorentzian hole centered at frequency ν_1, is

$$\Delta\varphi_H(\nu_2)=\frac{D_1(\nu_1-\nu_2)}{H_1}\left[1+\frac{4(\nu_1-\nu_2)^2}{H_1^2}\right]^{-1}, \qquad (23)$$

where D_1 is the depth of the hole at ν_1 and H_1 is the full width of this hole at half-maximum intensity. The sign on Eq. (23) is opposite to that for Eq. (9) since we are concerned with phase shifts introduced by the absence of atoms. Since we are only interested in the phase shift produced at one resonance by holes at different resonances [observation (b) above], $|\nu_1-\nu_2|\geqq(c/2L)$. In the low power limit, we may assume that the hole widths are less than the hole separations and consequently that $4(\nu_1-\nu_2)^2\gg H_1^2$. This approximation introduces an error of about 5% in the evaluation of $\Delta\varphi_H$ for the hole widths required below to explain the anomalous increase in beat frequency with pumping rate. Therefore,

$$\Delta\varphi_H(\nu_2)\cong\left(\frac{g(1)-f}{4}\right)\left(\frac{H_1}{\nu_1-\nu_2}\right), \qquad (24)$$

where we have defined $D_1=g(1)-f$ in conformance

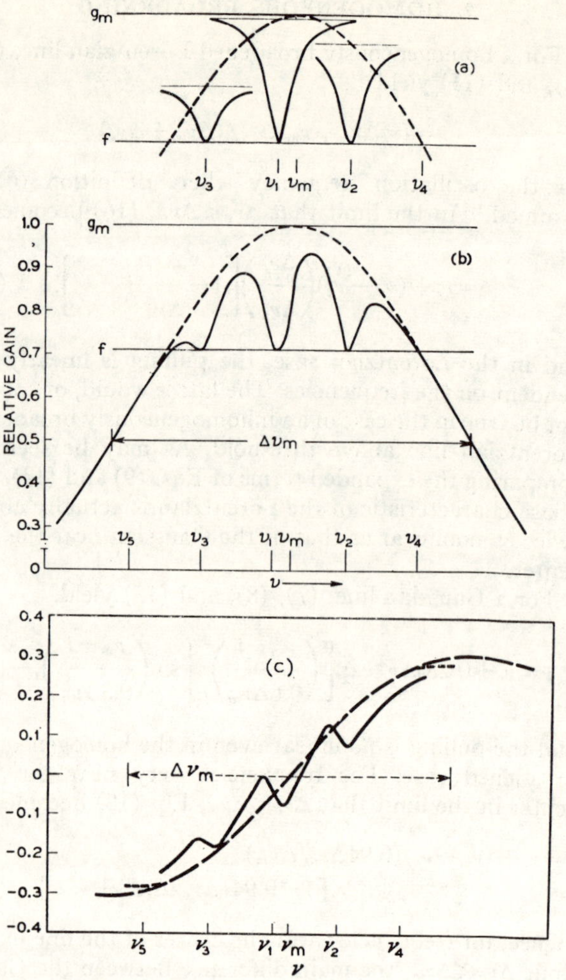

FIG. 4. "a" illustrates the choice of hole depth used to satisfy Eq. (8). "b" illustrates the resultant line shape at threshold for the fourth cavity resonance when $\nu_m-\nu_1=40$ Mc/sec. $\Delta\nu_m=800$ Mc/sec and $c/2L=160$ Mc/sec. Equal hole widths of 64 Mc/sec have arbitrarily been assumed. "c" (the solid curve) illustrates the resultant variation of the total phase shift for the condition in "b". The dashed curve represents the phase shift introduced by the original Gaussian. The dotted curve shows the departure of approximation (13) from the actual phase characteristic of the Gaussian.

with saturation condition (8). This choice of hole depth is illustrated in Fig. 4(a) for the case of three resonances.

The total phase shift introduced at frequency ν_j by N Lorentzian holes at frequencies ν_i is then

$$\Delta\varphi_H(j)\cong\sum_{i=1\ (i\neq j)}^{N}\left(\frac{g(i)-f}{4}\right)\left(\frac{H_i}{\nu_i-\nu_j}\right). \qquad (25)$$

From Eqs. (22) and (25) the oscillation frequency for the jth cavity resonance is given by

$$\bar{\nu}_j\cong\nu_j-\frac{\Delta\nu_c}{f}\Delta\varphi_m(\nu_j)-\frac{\Delta\nu_c}{4}\sum_{i=1\ (i\neq j)}^{N}\left(\frac{g(i)}{f}-1\right)\left(\frac{H_i}{\nu_i-\nu_j}\right), \qquad (26)$$

where by definition $g(i) \geqq f$ for a hole to occur at frequency ν_i. The quantities $g(i)$ are related by Eq. (14) and $\Delta\varphi_m$ is given by Eq. (13) for a Gaussian line. For a Lorentzian line, Eqs. (9) and (10) obtain.

A numerical evaluation of the total phase shift over the central portion of a Gaussian line is illustrated in Fig. 4(c) for three Lorentzian lines. The errors introduced by approximation (13) are also indicated in Fig. 4(c).

10. HOLE WIDTHS FROM STIMULATED EMISSION

From Eq. (II.11) of Appendix II, the full width (H) of a hole produced by stimulated emission is

$$H=\left(\frac{\gamma_a+\gamma_b}{2\pi}\right)\left(1+\frac{1}{\gamma_a\gamma_b}\left|\frac{2V}{\hbar}\right|^2\right)^{\frac{1}{2}}, \tag{27}$$

γ_a and γ_b represent the decay rates of the upper and lower maser levels, respectively, and the matrix element V is given by

$$|V|^2=|eE_0(z)_{a,b}/2|^2, \tag{28}$$

where E_0 is the amplitude of the electric field in the cavity mode. $E_0{}^2$ is proportional to the power (P) expended in the mode and may be readily expressed the latter way through the known cavity Q and volume ($v=108$ cm^3) of the discharge;

$$E_0{}^2=4P/(v\Delta\nu_c). \tag{29}$$

The upper maser level may decay by several transitions, hence

$$\gamma_a=\gamma_{ab}+\gamma_{ac}+\cdots, \tag{30}$$

where

$$\gamma_{ab}=\frac{4}{3}\frac{e^2}{\hbar\lambda^3}|r_{ab}|^2\cong\frac{4e^2}{\hbar\lambda^3}|z_{ab}|^2, \tag{31}$$

where λ and $\hbar$ are, respectively, the wavelength for the transition and Planck's constant divided by 2π. Combining Eqs. (27), (28), (29), and (31) yields

$$H\cong\frac{\gamma_a+\gamma_b}{2\pi}\left(1+\frac{\gamma_{ab}}{\gamma_a\gamma_b}\frac{\lambda^3P}{\hbar v\Delta\nu_c}\right)^{\frac{1}{2}}, \tag{32}$$

where cgs units are used and the coefficient outside the square root,

$$\Delta\nu_n=(\gamma_a+\gamma_b)/2\pi, \tag{33}$$

represents the natural linewidth for the transition.

If the decay rates in Eq. (32) were determined entirely by known spontaneous radiative values, the interpretation of Eq. (32) would be straightforward. The effective decay rates for the two maser levels have approximate values[12] of

$$\gamma_a\cong10^7\ \text{sec}^{-1},\qquad \gamma_b\cong8\times10^7\ \text{sec}^{-1}. \tag{34}$$

However, the vacuum ultraviolet portion of the decay rate of the upper maser level is unknown. The resonance trapping process [which determines the effective, long lifetime of the upper maser level given in Eq. (34)] will interrupt the phase of the wave function used to obtain Eq. (27) by some unknown, large extent. Consequently, an unambiguous interpretation of Eq. (32) cannot be given and we may only examine two limiting cases. The first limit is obtained by inserting the values (34) in Eq. (32). Here one obtains the minimum hole width (15 Mc/sec) at threshold and the maximum dependence of the hole width on power. From the measured power in the beam and estimates of the other parameters involved, the hole widths in this limiting case might extend to about 50 Mc/sec at the highest output powers obtained. The second limit is obtained through the marginal observation of a clean $c/2L$ beat at low powers in another He-Ne maser having twice the length of the present one. This observation implies that the natural linewidth for the transition is less than 80 Mc/sec. Hence in the second limiting case maximum hole widths of about 100 Mc/sec might be obtained at high powers from stimulated emission.

A related point of interest arises here which should be noted. Namely, because of the phase-interruption process, more output power is to be expected from a maser transition in which the upper state is optically connected with the ground state than for one which is not. That is, the hole widths in the former case will be larger, even if the effective decay rates and oscillator strengths for the two maser transitions are identical.

11. EFFECTS OF ELASTIC SCATTERING

There exist only two likely sources by which the upper state atoms may migrate over the Doppler distribution during their lifetimes:

(1) large-angle elastic scattering,
(2) small-angle elastic scattering.

We have separated the elastic (atom-atom) scattering process into these two groups both because the nature of the differential scattering cross section readily permits this separation and because the effects of the two processes are in opposite directions.

Process (1) results in violent changes in velocity of the upper state maser atoms and, hence, in changes in the location of the atoms' center frequencies which are comparable to the full width of the Doppler line. This effect therefore tends to restore the gain proportionality condition which holds, for example, over the entire line in the case of natural broadening. Thus, process (1) tends to oppose the burning of holes in the line. Its effects, however, are small in the present case. The total large-angle scattering cross-section for atoms in the upper neon maser state on ground state He atoms is unknown. However the known[17] diffusion rates of the

[17] A. V. Phelps and J. P. Molnar, Phys. Rev. **89**, 1202 (1953).

metastable $He(2\,^3S_1)$ and $Ne(3\,^3P_2)$ levels in their own gases both correspond to total ("hard sphere") large-angle scattering cross sections of about 10^{-15} cm^2 at a pressure of 1 mm Hg, and a temperature of 300°K. Since the upper Ne maser level has a similar configuration ($4\,^1P_1$ is LS notation) to the $Ne(3\,^3P_2)$, it seems unlikely that the total large-angle elastic scattering cross section would be appreciably different in the present case. Hence we expect the total large-angle collision rate for the upper maser level for the pressures in the maser (1 mm Hg of He and 0.1 mm Hg of Ne) to be about 0.5×10^7 sec^{-1}. The latter corresponds to about half the known[12] decay rate of the upper maser level.

Process (1) therefore takes place $\cong$0.5 times per atom on the average and the probability of the atom landing in a hole at low powers may be neglected.

The small-angle collisions, however, will result in the widening of a hole burned in the line. In the real case, the tail of the Van der Waals interaction, the possibility of molecular bonds, etc., will increase the differential scattering cross section enormously at small angles. The magnitudes of these interactions are unknown in the present case. However, one may estimate the size of the effect by assuming a hard-sphere cross section corresponding to typical diffusion rates. From the quantum mechanical treatment of the hard-sphere problem,[18] there is always a cone of small, finite half-angle,

$$\theta\cong h/2\mu_m v_{\rm rel}a, \qquad (35)$$

within which the scattering is nonclassical and within which the total scattering rate is generally about equal to the total large-angle scattering rate. Here, θ is given in the center of mass system, μ_m is the reduced mass, $v_{\rm rel}$ the relative velocity, h is Planck's constant, and a is the molecular diameter. This type of effect is difficult to observe experimentally and would not be included, for example, in the elastic scattering cross section measured in a diffusion experiment. We next assume that the average small-angle scattering event corresponds to scattering in the center-of-mass system through an angle $\theta/2$. This average small-angle collision between the excited neon atom and a ground-state helium atom therefore causes the excited neon atom to change its center frequency by an amount,

$$\delta\nu=\delta v_{\rm Ne}\nu/c=\delta v_{\rm rel}\nu/6c\cong h\nu/24\mu_m ca. \qquad (36)$$

We next choose "a" to correspond to the classical hard-sphere scattering cross section determined from the diffusion data in reference 17 (i.e., $\pi a^2=10^{-15}$ cm^2) and obtain the result

$$\delta\nu\cong 30 \text{ Mc/sec}. \qquad (37)$$

That is, the typical step is about equal to 30 Mc/sec and on the average half the atoms will make one such step in their lifetimes. Hence holes burned in the Doppler line might be enlarged by $\cong$15 Mc/sec due to small-angle scattering.

[18] N. F. Mott and H. S. W. Massey, *The Theory of Atomic Collisions* (Clarendon Press, Oxford, England, 1949), 2nd ed., Chap. II.

12. SEMI-EMPIRICAL REPRESENTATION OF HOLE WIDTHS

From the discussion in Secs. 10 and 11, the hole widths at low powers will approach an unknown constant value falling in the tens of Mc/sec range. From the discussion in Sec. 10, it is apparent that well above threshold the hole widths will increase as the square root of the power in the mode. The power in the mode itself will be approximately proportional to the product of the hole width and the hole depth. Hence at high powers one expects

$$H_i\cong H_0[g(i)-f]/f,$$

where the constant H_0 will also be in the tens of Mc/sec range. Hence we may represent the width of the ith hole by

$$H_i\cong h_0+H_0[g(i)-f]/f \qquad (38)$$

over the entire range in power. The adoption of Eq. (38) permits the evaluation of Eq. (26) with a minimum number of adjustable parameters.

13. COMPARISON OF MODE PULLING ANALYSIS WITH EXPERIMENT

Since the distribution of atoms is peaked at frequency ν_m, there will be a definite order of appearance of the various possible cavity resonances. We shall enumerate these cavity resonance frequencies in order of their appearance:

$$\nu_1,\ \nu_2,\ \nu_3\cdots, \quad \text{where} \quad \cdots\nu_3<\nu_1<\nu_m<\nu_2\cdots. \qquad (39)$$

Condition (39) corresponds to the one illustrated in Fig. 4 and will be assumed below in the cases for which Eq. (26) is evaluated. In making the present comparisons we consider only the Gaussian line for which Eqs. (13), (14), and (15) apply.

Evaluation of Eq. (26) for the first two resonances ($\nu_2>\nu_1$) yields

$$\bar\nu_2-\bar\nu_1\cong\left(\frac{c}{2L}\right)+\frac{\Delta\nu_c}{4(c/2L)}\left[\left(\frac{g_1-f}{f}\right)H_1+\left(\frac{g_2-f}{f}\right)H_2\right]$$
$$-0.56_4\frac{g m}{f}\Delta\nu_c\sin\left(\frac{c/2L}{0.6\Delta\nu_m}\right)\cos\left(\frac{\nu_m-\nu_1-c/4L}{0.3\Delta\nu_m}\right), \qquad (40)$$

where $g(1)$, $g(2)$, and g_m are related through Eq. (14) and it is assumed that $g(2)\geqq f>g(3)$.

The first case of interest corresponds to threshold for the appearance of the $c/2L$ beat and, therefore, the data in Table I. Here, Eq. (40) reduces to

$$\bar\nu_2-\bar\nu_1\cong(c/2L)[1-0.94(\Delta\nu_c/\Delta\nu_m)] \qquad (41)$$

plus terms of order $\Delta\nu_c(c/2L\Delta\nu_m)^3$. The latter amount to frequency fluctuations in the order of 10 kc/sec and are within the limits of error quoted in Table I. Comparison of Eq. (41) with the data in Table I implies that $\Delta\nu_c \cong 1$ Mc/sec, since $\Delta\nu_m \cong 800$ Mc/sec. This value of $\Delta\nu_c$ is in discrepancy with the value ($\cong 0.5$ Mc/sec) obtained from Eq. (3) for the known (0.99) mirror reflectance coefficient and suggests that some additional loss of about 1% per pass is present in the maser other than that introduced by the mirror reflectance coefficient. Fox and Li[19] have made tentative estimates which indicate that a loss of this approximate magnitude might arise from mode mixing at the end plates from flatness irregularities. In support of this contention we note that the present maser obviously has more loss than the first one reported[1] since it oscillated only on the strongest of the five transitions of neon previously observed. Here the main difference between the two masers is that the first maser had Fabry-Perot plates of considerably higher quality. It is conceivable that the reflectance films in the present maser deteriorated by the required amount during bakeout,[8] in which case the contention is not clearly established. It was found, however, that the loss implied by the data in Table I remained constant within the limits of error quoted over six months of continuous operation. In what follows, we assume that the ratio $\Delta\nu_c/\Delta\nu_m \cong 1/800$ has been experimentally determined from the data in Table I.

The second term in Eq. (40) increases with the power and consequently is capable of explaining the anomalous behavior of the $c/2L$ beat reported in Sec. 3. The hole repulsion effect, however, can only explain this anomalous behavior if it dominates over the third term in Eq. (40). The variation of the beat frequency above threshold will obviously be a complicated function of the location of ν_1 in respect to the line center. There are two extreme limiting cases and we only consider these. Because of the large number of exponential terms, it is simplest to solve Eq. (40) numerically. The results quoted below are for the parameters characteristic of the present maser—namely $c/2L \cong 160$ Mc/sec, $\Delta\nu_m \cong 800$ Mc/sec, and $\Delta\nu_c \cong 1$ Mc/sec. In evaluating Eq. (40) we express the total excursion of the beat between threshold for its appearance and threshold for the next cavity resonance. Hence, through the use of Eqs. (14) and (38), only two adjustable parameters h_0 and H_0 appear in the final result.

The first limiting case corresponds to a symmetric placement of the two cavity resonances about the Doppler line. Here the minimum hole repulsion effect should occur since both resonances cross threshold simultaneously and the two hole widths must be the same minimum value at threshold. A numerical solution of Eqs. (14), (38), and (40) for the present maser in this limiting case yields

$$(\bar{\nu}_2-\bar{\nu}_1)_{\max}-(\bar{\nu}_2-\bar{\nu}_1)_{\text{thresh}} \cong 0.75[h_0+0.25H_0-60] \text{ kc/sec}, \quad (42)$$

where the bracketed quantities are expressed in Mc/sec. The fact that an interferometer setting could be found (Sec. 3) for which the $c/2L$ beat was nearly independent of power suggests that the bracketed quantity in Eq. (42) is very close to zero. Hence we expect

$$h_0+0.25H_0 \cong 60 \text{ Mc/sec} \quad (43)$$

for the present maser.

The other limiting case occurs at $\nu_1=\nu_m$. Here the first hole may be enlarged considerably through stimulated emission before the second resonance occurs and the maximum hole repulsion effect should arise. This limit is complicated in the real case by the fact that both the second and third cavity resonances cross threshold simultaneously. Here the two $c/2L$ beats are identical and splitting (see below) does not occur until the fourth and fifth cavity resonances cross threshold. The effect of the third hole will not alter the present estimate substantially, however, and we shall ignore its presence (i.e., the effects of the holes at ν_2 and ν_3 cancel at ν_1, whereas the effects of the holes at ν_1 and ν_3 are additive at ν_2). A numerical evaluation of Eqs. (14), (38), and (40) in this limiting case ($\nu_1=\nu_m$) yields

$$(\bar{\nu}_2-\bar{\nu}_1)_{\max}-(\bar{\nu}_2-\bar{\nu}_1)_{\text{thresh}} \cong 1.3[h_0+0.55H_0-60] \text{ kc/sec} \quad (44)$$

for the parameters in the present maser. The bracketed quantities are again given in Mc/sec. If we assert that the extreme 30 kc/sec shift of the beat reported in Sec. 3 occurs in the second limiting case, Eqs. (43) and (44) imply that $h_0 \cong 40$ Mc/sec and $H_0 \cong 80$ Mc/sec. These values obviously do not represent precise measurements. However, they are sufficiently close to the estimates given in Secs. 10 and 11 to indicate that the hole repulsion effect is the correct explanation for the anomalous power dependence of the beat reported in Sec. 3.

Finally, it is important to note that Eq. (26) also predicts a splitting of the $c/2L$ beat which is compatible with the magnitudes observed. Evaluation of Eq. (26) for three asymmetrically placed holes yields

$$(\bar{\nu}_2-\bar{\nu}_1)-(\bar{\nu}_1-\bar{\nu}_3) \cong \frac{3}{8}\frac{\Delta\nu_c}{(c/2L)}\left[\left(\frac{g_2-f}{f}\right)H_2-\left(\frac{g_3-f}{f}\right)H_3\right] - 1.13\Delta\nu_c\left(\frac{g_m}{f}\right)\left[\sin\left(\frac{c/2L}{0.6\Delta\nu_m}\right)\right]^2 \sin\left(\frac{\nu_m-\nu_1}{0.3\Delta\nu_m}\right), \quad (45)$$

where no terms contained in Eqs. (13) and (26) have been neglected. g_2 and H_2 are the fractional energy gain and hole width at ν_2, respectively; g_3 and H_3 are similarly defined at ν_3; and g_m is the fractional energy gain per pass at the line center. Equation (45) holds only for the case where $g_3 \geqq f$ and the gain at the fourth cavity resonance is below threshold for oscillation. Equation (45) predicts that the beat splitting will vanish for $\nu_m=\nu_1$, since in this case $g_2=g_3$ and $H_2=H_3$.

[19] A. G. Fox and Tingye Li (private communication).

In spite of its formidable appearance, Eq. (45) is a fairly linear function of $(\nu_m-\nu_1)$. The main nonlinear term arises from the power dependent part of the hole width in Eq. (38) and enters Eq. (45) as the difference between the squares of two small quantities. For the same reason, Eq. (45) is not strongly dependent on the parameter H_0 at low powers. The splitting is primarily determined by the last term of Eq. (45). Its maximum value occurs at $\nu_m-\nu_1=\frac{1}{2}(c/2L)$ and at a value for g_m such that the fourth cavity resonance is just below threshold. For the present maser, the maximum value of the second term is a little over 50 kc/sec and hence somewhat in excess of the maximum splitting ($\cong$30 kc/sec) observed. By definition [Eqs. (38) and (39)], the first term in Eq. (45) is positive and, hence subtracts from the nonlinear phase characteristic of the Gaussian. At low powers this subtracted term is mainly determined by the width, h_0 in Eq. (38). For the present maser, the inclusion of h_0 in Eq. (45) would reduce the maximum splitting by about 0.5 kc/sec per Mc/sec of h_0. Hence for the value of h_0 estimated above ($\cong$40 Mc/sec), the first term of Eq. (45) would reduce the maximum splitting from the second term to about that value observed. However, the neglect of the possibility of a slight variation in the mirror reflectance coefficient with polarization may render this close agreement fortuitous. The second term in Eq. (38) will certainly increase with the power and at some point may cancel the nonlinearity responsible for the splitting in Eq. (45). It seems evident from the experimental results (Fig. 3) that this limit is not reached in the present case until more than three modes oscillate.

The evaluation of Eq. (45) for more than three holes is straightforward and not terribly interesting. As the nonlinearity contained in Eq. (45) implies, the splitting increases with the number of modes which go into oscillation. In general, the number of components at $n(c/2L)$ is $(N-n)$ where N is the number of holes $(N\geqq n)$ and these components will not be spaced with perfect evenness. In each case, the absolute frequency of the component increases with the pumping rate if the hole repulsion effect predominates. For a symmetric distribution of holes about the Doppler line, the number of components at $n(c/2L)$ is reduced.

14. CONCLUSION

There exists one perplexing aspect of the hole burning interpretation which should be mentioned. At the highest power levels shown in Fig. 3, the odd-symmetric modes also go into oscillation, as evidenced by the appearance of the 1-Mc/sec satellites in Fig. 2. (These satellites are also split by amounts in the general order of 20 kc/sec.) A question naturally arises regarding the mechanism by which two modes separated by an amount substantially less than the natural linewidth can go into oscillation simultaneously. We note, however, that this question exists regardless of the interpretation of the beat splitting. It seems probable that the answer to this question lies in the different spatial distributions of the two modes.

The main interest in the beat splitting phenomenon, of course, arises from the possibility of using this phenomenon to make an absolute frequency standard in the optical range. As mentioned in Sec. 3, this might be accomplished by the generation of an error frequency in the case of three oscillating modes and the use of this error frequency to adjust the plate separation. The central resonance may then be filtered out with a suitably designed (low quality) Fabry-Perot interferometer. It seems likely from the data in Sec. 3 that this error frequency would be in the general order of about 0.5 kc/sec per Mc/sec separation of the central cavity resonances from the center of the Doppler line and that the sharpness[2,11] of the maser oscillation would not be a prime limitation on the absolute stability obtainable in practice. Throughout the present analysis it has been assumed that the line shape for the transition is symmetric. This assumption will obviously break down at some point, and it is likely that the ultimate limit would be determined by pressure-dependent effects. The presence of other isotopes of neon can introduce such line asymmetries and it would obviously be desirable to isolate one of these isotopes for use in such a standard. From Thomas' numerical calculations and the natural abundance of Ne^{22} in Ne^{20}, the main effect of this line asymmetry would be to introduce a power-dependent frequency shift at the Doppler center in the order of 30 kc/sec. The presence of other isotopes of neon will also result in slight changes in the nonlinear part of the phase characteristic. In this connection it should be mentioned that no significant difference was found in either the mode pulling or beat splitting effects when the present maser was operated on a neon sample containing 99.9% Ne^{20} as opposed to neon samples containing the normal isotopic abundance.

There are two more important sources of practical difficulty in utilizing the beat splitting effect to determine the absolute frequency of the central mode in respect to the Doppler center. The first consists of the possibility that the mirror reflectance coefficient may vary slightly with polarization.[8] If two modes were polarized in one direction and the third orthogonally, this effect could introduce an additional shift in the beat splitting. The effect enters by changing the hole depth and therefore the magnitude of the hole repulsion term. From Eq. (45) the shift in the splitting would amount to about $\frac{3}{8}\Delta\nu_c(2LH/c)\Delta f/f$, where Δf is the extreme variation of the fractional loss between the two orientations and H is the width of the hole for the orthogonally polarized mode. For example, for a hole width of 100 Mc/sec, a 1% variation in the mirror losses between the two extreme orientations would shift the beat splitting by about 2 kc/sec. Second, it has been found by others[9,10] using different observational techniques that the difference frequency between the two $c/2L$ components [Fig. 3(b)] jumps discontinuously to zero at some point below about 1 kc/sec. Under these conditions the $2c/2L$ and $c/2L$ beats are both harmonically related[9] and phase-locked[10] to a very high degree. Hence one completely loses track of the absolute frequencies at this point and it seems apparent that a minimum absolute frequency uncertainty in the order of 2 Mc/sec will exist. This effect is definitely not contained in Eq. (45). Although Eq. (45) does contain a slight nonlinear dependence on $(\nu_m-\nu_1)$, the nonlinear terms are much too small to give, for example, a cubic equation with real roots or a point of inflection near the origin. The effect probably has its origin in nonlinear time-dependent properties of the medium, which, as previously stated, have been neglected for reasons of simplicity.

Neither of these effects would prevent one from using the beat splitting phenomenon to obtain a high degree of relative frequency stabilization when the central mode is slightly detuned from the Doppler center.

ACKNOWLEDGMENTS

The author is particularly indebted to A. J. Rack for the use of his frequency measuring apparatus and help in obtaining the data in Table I. It is also a pleasure to acknowledge the help of D. E. Thomas for his numerical phase shift calculations and the technical assistance of P. Kindlmann. The author is also indebted to Professor W. W. Watson of Yale University for supplying the highly pure Ne^{20} sample used. In addition, the author has benefitted from helpful discussions with Dr. W. L. Faust, Dr. C. G. B. Garrett, Dr. A. Javan, and Dr. J. A. White.

APPENDIX I

The phase shifts associated with a given line shape may be calculated most generally using methods equivalent to the Kramers-Kronig relations.[15] The assumptions on the following derivation are that the process obeys causality (i.e., the amplitude gain function is analytic in the lower half of the complex ω plane), that the gain is small, and that the running wave is amplified linearly as it traverses the cavity. Although the gain saturates in the system during oscillation, we assume that amplification by this saturated gain characteristic is still linear.

We let $K(\omega)$ be the complex, fractional amplitude gain in the medium defined by

$$1+K(\omega)=[1+K_0(\omega)]\exp[-i\phi(\omega)], \tag{I.1}$$

where $K_0(\omega)$ is the fractional amplitude gain per pass and $\phi(\omega)$ is the additional single-pass phase shift resulting from the amplifying medium. Taking the logarithm of (I.1) and expanding in the limits $K(\omega)$, $K_0(\omega)\ll 1$,

$$K(\omega)\cong K_0(\omega)-i\phi(\omega). \tag{I.2}$$

We note at this point that the fractional energy gain per pass, $g_0(\omega)$, is

$$g_0(\omega)\cong 2K_0(\omega). \tag{I.3}$$

We assume $K(\omega)$ vanishes at $\omega=\pm\infty$, $\pm i\infty$, and has no singularities in the lower half-plane. Then

$$\oint\frac{K(\omega)d\omega}{\omega-\omega_0}=0, \tag{I.4}$$

where the integral is taken along the real axis, skirts under the pole at ω_0 and is closed in the lower half of the complex ω plane. Substituting (I.2) and equating real parts,

$$\phi(\omega_0)=-(1/\pi)\int_{-\infty}^{+\infty}\frac{K_0(\omega)d\omega}{\omega-\omega_0}, \tag{I.5}$$

where (I.5) represents the principal part of the integral on the real axis. Equation (I.5) is more convenient for the Lorentzian than the usual form of the Kramers-Kronig relations deduced from it.

We take for the Lorentzian,

$$K_0(\omega)\cong K_m(\Delta\omega)^2[(\Delta\omega)^2+4(\omega_m-\omega)^2]^{-1} \tag{I.6}$$

where the approximation $\omega+\omega_m\cong 2\omega$ has been made. Although $K_0(\omega)$ has a pole in the lower half-plane, $K(\omega)$ does not. We next take,

$$I=\oint\frac{K_0(\omega)d\omega}{\omega-\omega_0}, \tag{I.7}$$

where the path of integration runs along the real axis, skirts under the pole at ω_0 and is closed in the upper half-plane. From (I.5), (I.6), and (I.7),

$$-\pi\phi(\omega_0)+\pi iK_0(\omega_0)=2\pi i\sum(\text{Residues}). \tag{I.8}$$

Hence,

$$\begin{aligned}\phi(\omega_0)&=-2K_0(\omega_0)(\omega_m-\omega_0)/\Delta\omega\\&=-g_0(\omega_0)(\omega_m-\omega_0)/\Delta\omega\end{aligned} \tag{I.9}$$

yielding Eq. (11) of the text.

APPENDIX II

The hole widths may be evaluated through a standard application of time-dependent perturbation theory.[20] We assume that the two maser levels are described by the wave function

$$\Psi(r,t)=A\psi_a+B\psi_b. \tag{II.1}$$

Satisfying

$$(H_0+H')\Psi=i\hbar\partial\Psi/\partial t \tag{II.2}$$

in the presence of the electromagnetic field. H_0 is the original Hamiltonian and H' the time-dependent per-

[20] A similar analysis for levels having different decay rates has been given for the ground state of positronium by V. W. Hughes, S. Marder, and C. S. Wu [Phys. Rev. **106**, 934 (1957), Appendix I]. The approximation, $\gamma_b\gg\gamma_a$, appropriate to the latter reference has, however, not been made here.

turbation. The subscript "a" denotes the upper maser level and "b" the lower. The probability amplitudes A and B are functions of the time. We take the electric field to be linearly polarized in the z direction. Hence,

$$H' = -ezE_0 \cos\omega t = -(ezE_0/2)(e^{i\omega t} + e^{-i\omega t}). \quad \text{(II.3)}$$

Since the perturbation is electric dipole in nature, only off-diagonal matrix elements arise. Neglecting the anti-resonant term and the interaction with all neighboring levels,

$$\begin{aligned} \dot{A} &= (V/i\hbar)e^{-i(\omega-\omega_0)t}B - (\gamma_a/2)A, \\ \dot{B} &= (V/i\hbar)e^{i(\omega-\omega_0)t}A - (\gamma_b/2)B, \end{aligned} \quad \text{(II.4)}$$

where the spontaneous decay rates γ_a and γ_b of the upper and lower maser levels have been introduced phenomenologically. From (II.3)

$$V = -\tfrac{1}{2}eE_0(z)_{a,b}. \quad \text{(II.5)}$$

ω_0 represents the resonant frequency of the atomic transition and ω the frequency of the oscillating field.

We seek those solutions to the coupled Eqs. (II.4) satisfying the initial conditions

$$A = 1 \quad \text{and} \quad B = 0 \quad \text{at} \quad t = 0. \quad \text{(II.6)}$$

Condition (II.6) corresponds to putting the atom in the upper state at $t=0$. The total probability that the atom decays by stimulated emission is then given by

$$P_s = \gamma_b \int_0^\infty |B(t)|^2 dt. \quad \text{(II.7)}$$

Through standard methods, the solution of Eqs. (II.4) for $B(t)$ subject to initial conditions (II.6) is

$$B(t) = \frac{i\hbar}{V} \frac{(\mu_- + \frac{1}{2}\gamma_a)(\mu_+ + \frac{1}{2}\gamma_a)}{(\mu_+ - \mu_-)} [e^{\mu_- t} - e^{\mu_+ t}] e^{i(\omega-\omega_0)t}, \quad \text{(II.8)}$$

where

$$\begin{aligned} \mu_\pm = &-\tfrac{1}{2}[\tfrac{1}{2}(\gamma_a+\gamma_b) + i(\omega-\omega_0)] \\ &\pm \tfrac{1}{2}\{[\tfrac{1}{2}(\gamma_b-\gamma_a) + i(\omega-\omega_0)]^2 - |2V/\hbar|^2\}^{\frac{1}{2}}. \end{aligned} \quad \text{(II.9)}$$

Hence

$$P_s = \gamma_b \left|\frac{\hbar}{V}\right|^2 \left| \frac{(\mu_- + \frac{1}{2}\gamma_a)(\mu_+ + \frac{1}{2}\gamma_a)}{(\mu_+ - \mu_-)} \right|^2 \times \int_0^\infty |e^{\mu_- t} - e^{\mu_+ t}|^2 dt. \quad \text{(II.10)}$$

After some algebraic manipulation, Eq. (II.10) reduces to

$$P_s = \frac{\gamma_b(\gamma_a+\gamma_b)|2V/\hbar|^2}{4(\omega-\omega_0)^2\gamma_a\gamma_b + (\gamma_a\gamma_b + |2V/\hbar|^2)(\gamma_a+\gamma_b)^2}, \quad \text{(II.11)}$$

yielding Eq. (27) of the text.[21] Aside from the minor approximations made in formulating Eqs. (II.4), this expression for the probability function is exact.

[21] *Note added in proof.* A result equivalent to Eq. (II.11) has been obtained from the density matrix method by W. E. Lamb, Jr., and T. M. Sanders, Jr., Phys. Rev. **119**, 1901 (1960).

Theory of an Optical Maser*

WILLIS E. LAMB, JR.
Yale University, New Haven, Connecticut
(Received 13 January 1964)

A theoretical model for the behavior of an optical maser is presented in which the electromagnetic field is treated classically, and the active medium is made up of thermally moving atoms which acquire nonlinear electric dipole moments under the action of the field according to the laws of quantum mechanics. The corresponding macroscopic electric polarization of the medium acts as a source for an electromagnetic field. The self-consistency requirement that a quasistationary field should be sustained by the induced polarization leads to equations which determine the amplitudes and frequencies of multimode oscillation as functions of the various parameters characterizing the maser. Among the results obtained are: threshold conditions, single-mode output as a function of cavity tuning, frequency pulling and pushing, mode competition phenomena including frequency locking, production of combination tones, and population pulsations. A more approximate discussion of maser action using rate equations is also given in which the concept of "hole burning" plays a role.

1. INTRODUCTION

THIS paper gives a theoretical description of the operation of multimode maser oscillators. The type of approach is particularly suitable for gaseous optical masers of the type suggested by Schawlow and Townes,[1] and first realized experimentally by Javan, Bennett, and Herriott,[2] but the equations should also find use in the description of some features of solid-state optical masers.

2. BASIS FOR CALCULATION

We consider a high-Q multimode cavity in which there is a given *classical* electromagnetic field acting on a material medium which consists of a collection of atoms described by the laws of quantum mechanics. No attempt is made to consider noise due to spontaneous emission and thermal, density, or quantum fluctuations. The high degree of spectral purity observed by Javan and co-workers[3] suggests that these should be good approximations.

The effect of the electromagnetic field on the atoms in the cavity is to produce a macroscopic electric polarization $\mathbf{P}(\mathbf{r},t)$ of the medium. This acts as a source for the electromagnetic field in accordance with Maxwell's equations. The conditions for self-consistency (that the field produced should be equal to the field assumed) determine the amplitudes and frequencies of the possible oscillations. The calculations will include nonlinear effects, so that phenomena of frequency pulling and pushing, mode competition, frequency locking, etc., can be described.

The thermal motion of an atom during its natural decay time may carry it several wavelengths through the standing wave pattern of the electromagnetic field. As a result, the atom "sees" Doppler-shifted optical frequencies which depend on its trajectory. This important circumstance considerably influences the behavior of the Javan-Bennett-Herriott maser. When, however, thermal motion is neglected the equations of the paper can be used in a model calculation for an ideal solid-state optical maser.

We will assume that only two atomic states a and b contribute to the maser action. As a related simplification the vector character (polarization) of the electromagnetic field will be ignored. In order to ensure that our analysis should apply, it would be desirable to have the optical configuration favor one plane of polarization, as with windows of the Brewster's angle type. The more complicated problem of a general state of polarization will be dealt with in another paper.

A cavity of the Fabry-Perot type used by Javan, Bennett, and Herriott has, of course, a continuum of modes because it is not enclosed by reflecting walls. However, it follows from work of Fox and Li[4] that there are discrete sets of quasimodes for which the diffractive leakage from the tube is small. The cavity modes of highest Q are the even symmetric ones whose circular frequencies are given by

$$\Omega_n = \pi n c/L\,, \tag{1}$$

where c is the velocity of light, L is the distance between the reflecting plates ($L\sim 100$ cm) and n is a large integer, typically of order 2×10^6. Fox and Li have shown that the modes of next highest Q are those possessing odd radial symmetry, which, for typical geometry differ by about 1 Mc/sec from the former modes. Our discussion will be specifically, but not inevitably, aimed at the modes of highest Q.

---

* This work was supported in part by the U. S. Air Force Office of Scientific Research. The main results of the paper were reported at the Third International Conference on Quantum Electronics, Paris, February, 1963. Lectures on some of the material were given at the 1963 Varenna Summer School.

[1] A. L. Schawlow and C. H. Townes, Phys. Rev. **112**, 1940 (1958).

[2] A. Javan, W. R. Bennett and D. R. Herriott, Phys. Rev. Letters **6**, 106 (1961).

[3] T. S. Jaseja, A. Javan, and C. H. Townes, Phys. Rev. Letters **10**, 165 (1963).

[4] A. G. Fox and T. Li, Bell System Tech. J. **40**, 61 (1961).

Reprinted with permission from *Phys. Rev.*, vol. 134, pp. A1429–A1450, June 15, 1964.

3. ELECTROMAGNETIC FIELD EQUATIONS

We write Maxwell's equations in mks units as

$$\begin{aligned} \operatorname{div}\mathbf{D}=0 \quad \operatorname{curl}\mathbf{E}=-\partial\mathbf{B}/\partial t \\ \operatorname{div}\mathbf{B}=0 \quad \operatorname{curl}\mathbf{H}=\mathbf{J}+\partial\mathbf{D}/\partial t, \end{aligned} \tag{2}$$

where

$$\mathbf{D}=\epsilon_0\mathbf{E}+\mathbf{D},\quad \mathbf{B}=\mu_0\mathbf{H},\quad \mathbf{J}=\sigma\mathbf{E}. \tag{2a}$$

To an approximation whose validity will be discussed in another paper, the array of excited atoms may be regarded as a medium with an electrical state described by a macroscopic polarization $\mathbf{P}(\mathbf{r},t)$ (electric dipole moment density). In order to avoid a complicated boundary value problem, it is convenient to assume the presence of a lossy medium with an Ohmic conductivity σ adjusted to give the desired damping of a normal mode. The electric field then obeys a wave equation

$$\operatorname{curl}\operatorname{curl}\mathbf{E}+\mu_0\sigma\partial\mathbf{E}/\partial t+\mu_0\epsilon_0\partial^2\mathbf{E}/\partial t^2=-\mu_0\partial^2\mathbf{P}/\partial t^2. \tag{3}$$

In the subsequent calculations the main effect of the space dependence of $\mathbf{E}(x,y,z,t)$ comes from the motion of the excited atoms through the field which leads to amplitude modulation of the fields seen by the atoms. The analysis of Fox and Li for the even symmetric modes indicates that the electric field does not vary rapidly across the tube diameter. Accordingly we take only the axial variation of $\mathbf{E}$ into account. Then $\operatorname{curl}\operatorname{curl}\mathbf{E}$ is replaced by $-\partial^2E/\partial z^2$, where z is the axial coordinate, and E is the transverse electric field. For the nth normal mode (unnormalized), we have eigenfunctions

$$U_n(z)=\sin K_n z, \tag{4}$$

with wave number

$$K_n=n\pi/L, \tag{5}$$

where n is a large integer.

In the presence of a given polarization $P(z,t)$, quasi-stationary forced oscillations of the electric field can be expanded in normal mode eigenfunctions

$$E(z,t)=\sum_n A_n(t)U_n(z), \tag{6}$$

where the amplitudes $A_n(t)$ obey a differential equation of a forced, damped simple harmonic oscillator

$$\frac{d^2A_n}{dt^2}+\left(\frac{\sigma}{\epsilon_0}\right)\frac{dA_n}{dt}+\Omega_n^2A_n=-\left(\frac{1}{\epsilon_0}\right)\frac{d^2P_n(t)}{dt^2}, \tag{7}$$

in which $P_n(t)$ is the space Fourier component of $P(z,t)$

$$P_n(t)=\frac{2}{L}\int_0^L dz P(z,t)\sin K_n z. \tag{8}$$

Since $P_n(t)$ will be very nearly monochromatic at an optical frequency[5] (o.f.) ν, we replace its second time derivative by $-\nu^2P_n$ on the right side of Eq. (7). Adjusting the fictional conductivity σ to give the desired Q_n of the nth mode, we write

$$\sigma=\epsilon_0\nu/Q_n. \tag{9}$$

Then $A_n(t)$ obeys

$$\frac{d^2A_n}{dt^2}+\left(\frac{\nu}{Q_n}\right)\frac{dA_n}{dt}+\Omega_n^2A_n=\left(\frac{\nu^2}{\epsilon_0}\right)P_n. \tag{10}$$

In the typical gaseous optical maser, the separation of the principal modes $\Delta\sim150$ Mc/sec is much larger than the cavity mode band width $\nu/Q\sim1$ Mc/sec. Hence we may hope to neglect time Fourier components of $A_n(t)$ and $P_n(t)$ which are at frequencies far from the cavity resonance frequency Ω_n, and write[6]

$$A_n(t)=E_n(t)\cos(\nu_n t+\varphi_n(t)), \tag{11}$$

and

$$P_n(t)=C_n(t)\cos(\nu_n t+\varphi_n(t)) +S_n(t)\sin(\nu_n t+\varphi_n(t)), \tag{12}$$

where the amplitudes $E_n(t)$ and phases $\varphi_n(t)$, as well as the in-phase and quadrature coefficients $C_n(t)$ and $S_n(t)$ are slowly varying functions of t which, together with the frequencies ν_n, are still to be determined. The expressions (11) and (12) are put into Eq. (10) with only the first time derivatives of $E_n(t)$ and $\varphi_n(t)$ retained. Equating the coefficients of $\cos(\nu_n t+\varphi_n)$ and $\sin(\nu_n t+\varphi_n)$ separately to zero, and further neglecting small terms involving $\nu_n\dot{E}_n/Q_n$, $\dot{\varphi}_n\dot{E}_n$ and $\nu_n\dot{\varphi}_nE_n/Q_n$, and recognizing that $\nu_n+\dot{\varphi}_n$ is very close to Ω_n, we find the self-consistency equations

$$(\nu_n+\dot{\varphi}_n-\Omega_n)E_n=-\tfrac{1}{2}(\nu/\epsilon_0)C_n \tag{13}$$

and

$$\dot{E}_n+\tfrac{1}{2}(\nu/Q_n)E_n=-\tfrac{1}{2}(\nu/\epsilon_0)S_n, \tag{14}$$

which serve to determine the amplitudes, frequencies and phases of the o.f. radiation once the polarization state of the medium is known in terms of the $E_n(t)$.

4. POLARIZATION OF THE MEDIUM

The maser action arises from the establishment of a negative temperature distribution for the two excited states a and b of the atoms constituting the medium as shown in Fig. 1. The ground state, far below a and b, is not shown. Consider what happens to an atom which at time t_0 is excited by some process (electron bombardment, collision of the second kind, absorption of resonance radiation, decay from some higher excited

[5] We adopt the convention that all symbols for frequencies should denote circular frequencies. A numerical value, e.g., 150 Mc/sec, however, denotes an ordinary frequency. A decay constant like γ_a which denotes a reciprocal life time $1/\tau_a$ often plays the role of a circular frequency. Numerical values of γ_a will be given as ordinary frequencies.

[6] The representation of an arbitrary function $A_n(t)$ in the form (11) in terms of a variable amplitude $E_n(t)$ and phase $\varphi_n(t)$ is not unique. Despite this, because of the use of the rotating wave approximation it seems possible through Eqs. (13) and (14) to determine both amplitude and phase. [The positive frequency part of (11) is a complex function very closely equal to $A_n^{(+)}(t)=E_n(t)\exp-i(\nu_n t+\varphi_n(t))$ which does have a unique amplitude and phase.]

state, etc.) into the upper maser state a. Let the atom be at position $\mathbf{r}_0$ at t_0, and have velocity $\mathbf{v}$. For the present, we neglect collisions, so that at time $t>t_0$ the atom will be at $\mathbf{r}=\mathbf{r}_0+\mathbf{v}(t-t_0)$. If there is an o.f. electric field $\mathbf{E}(\mathbf{r},t)$ in the cavity, the atom sees a time-dependent field $E(\mathbf{r}_0+\mathbf{v}t-\mathbf{v}t_0, t)$ for $t>t_0$. Associated with this field is a time-dependent perturbation energy whose matrix element is

$$\hbar V(t)=-\wp E(\mathbf{r}_0+\mathbf{v}t-\mathbf{v}t_0, t)\,, \tag{15}$$

where $\wp$ (assumed real) is the matrix element for the electric dipole moment of the atom between states a and b. The perturbation causes the atomic wave function to become a time-dependent linear combination $a(t)\psi_a+b(t)\psi_b$. The quantum-mechanical average value of the electric dipole operator for the atom is $(a^*b+ab^*)\wp$.

To follow the time-dependent wave function (in the subspace of ψ_a and ψ_b), we start from the equations of time-dependent perturbation theory

$$\begin{aligned} i\dot{a}&=W_a a+V(t)b-\tfrac{1}{2}i\gamma_a a\,,\\ i\dot{b}&=W_b b+V(t)a-\tfrac{1}{2}i\gamma_b b\,, \end{aligned} \tag{16}$$

in which the radiative decay of states a and b is described by phenomenological terms containing the decay constants γ_a and γ_b for the two states. Here $\hbar W_a$ and $\hbar W_b$ are the unperturbed energies of states a and b, and the matrix element of the perturbation $V(t)$ is given by Eq. (15).

If the motion of the atom were neglected, and if the maser were working in a single cavity mode, $V(t)$ would be monochromatic, and the rotating wave approximation would allow the Eqs. (16) to be integrated exactly. Even so, there are great algebraic simplifications to be gained by going over to a density matrix description[7] of an ensemble of atoms consisting of all those of a given category which are produced during all times $t_0<t$. A theory of maser action in this case has already been given[8] which is valid when the signals are strong enough to fully saturate the transition $a\leftrightarrow b$. For multimode operation, such an exact solution can no longer be obtained. However, the simpler theory can help with the interpretation of our rather complicated equations, and it will be discussed in Secs. 16–20.

When atomic motion through the electromagnetic field is taken into account an atom does not see a monochromatic perturbation even in single-mode operation. The equations can only be solved in a perturbation expansion of the solution in powers of the $E_n(t)$. It is still advantageous to use the density matrix method, considering first only those atoms characterized by

[7] W. E. Lamb, Jr. and T. M. Sanders, Jr., Phys. Rev. **119**, 1901 (1960), especially pp. 1902–1903; L. R. Wilcox and W. E. Lamb, Jr., *ibid.* **119**, 1915 (1960), especially p. 1928.

[8] W. E. Lamb, Jr., *Quantum Mechanical Amplifiers, in Lectures in Theoretical Physics*, edited by W. E. Brittin and B. W. Downs (Interscience Publishers, Inc., New York, 1960), Vol. II, especially pp. 472–476.

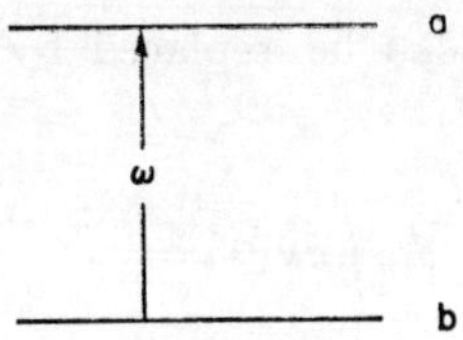

FIG. 1. Two excited energy levels a and b between which the maser action takes place. The levels have a resonance transition frequency $\omega>0$, and are given phenomenological decay constants γ_a and γ_b. The excitation of the states is described by the functions $\lambda_\alpha(r_0,t_0,v)$ which are introduced in Eq. (22).

a, $\mathbf{r}_0$, t_0, $\mathbf{v}$. The density matrix

$$\begin{aligned} \rho(a,\mathbf{r}_0,t_0,\mathbf{v},t)&=\begin{pmatrix} |a|^2 & ab^* \\ a^*b & |b|^2 \end{pmatrix}\\ &=\begin{pmatrix} \rho_{aa} & \rho_{ab} \\ \rho_{ba} & \rho_{bb} \end{pmatrix} \end{aligned} \tag{17}$$

obeys an equation of motion

$$\dot{\rho}=-i[\mathcal{H},\rho]-\tfrac{1}{2}(\Gamma\rho+\rho\Gamma)\,, \tag{18}$$

where Γ is the diagonal matrix

$$\Gamma=\begin{pmatrix} \gamma_a & 0 \\ 0 & \gamma_b \end{pmatrix} \tag{19}$$

and the Hamiltonian matrix $\mathcal{H}$ is

$$\mathcal{H}=\begin{pmatrix} W_a & V(t) \\ V(t) & W_b \end{pmatrix}, \tag{20}$$

with $V(t)$, as given by Eq. (15), having a complicated time dependence because of atomic motion. A solution of Eq. (18) which satisfies the initial conditions

$$\rho(a,\mathbf{r}_0,t_0,\mathbf{v},t_0)=\begin{pmatrix} 1 & 0 \\ 0 & 0 \end{pmatrix} \tag{21}$$

is required. The average electric dipole moment corresponding to this density matrix ρ is $\wp(\rho_{ab}+\rho_{ba})$.

To obtain the macroscopic polarization $P(\mathbf{r},t)$ we have to combine the contributions of all atoms which arrive at $\mathbf{r}$ at time t, no matter when or where they were excited to state a, and also a similar contribution from atoms excited initially to state b. Let $\lambda_\alpha(\mathbf{r}_0,t_0,\mathbf{v})$ be the number of atoms excited to state $\alpha=a, b$ per unit time per unit volume. We have

$$\begin{aligned} P(\mathbf{r},t)=\wp\sum_{\alpha=a,b}\int_{-\infty}^{t}dt_0\int d\mathbf{r}_0\int d\mathbf{v}\lambda_\alpha(\mathbf{r}_0,t_0,\mathbf{v})\\ \times[\rho_{ab}(\alpha,\mathbf{r}_0,t_0,\mathbf{v},t)+\rho_{ba}(\alpha,\mathbf{r}_0,t_0,\mathbf{v},t)]\\ \times\delta(\mathbf{r}-\mathbf{r}_0-\mathbf{v}(t-t_0)). \end{aligned} \tag{22}$$

In practice, $\lambda_\alpha(\mathbf{r}_0,t_0,\mathbf{v})$ will be a slowly varying function of $\mathbf{r}_0$ so that it can be replaced by $\lambda_\alpha(\mathbf{r},t_0,\mathbf{v})$. After integration over $\mathbf{r}_0$

$$P(\mathbf{r},t)=\wp\sum_{\alpha=a,b}\int_{-\infty}^{t}dt_0\int d\mathbf{v} \tag{23}$$
$$\times\lambda_\alpha(\mathbf{r},t_0,\mathbf{v})[\rho_{ab}(\alpha,\,\mathbf{r}-\mathbf{v}(t-t_0),\,t_0,\,\mathbf{v},\,t)+\text{conj.}].$$

Similarly, we will have use for a density matrix describing an ensemble of atoms which arrive at $\mathbf{r}$ with velocity $\mathbf{v}$ at time t regardless of their place $\mathbf{r}_0$, time t_0 or state $\alpha=a,\,b$ of excitation. This will be denoted by

$$\rho(\mathbf{r},\mathbf{v},t)=\sum_{\alpha=a,b}\int_{-\infty}^{t}dt_0\int d\mathbf{r}_0\lambda_\alpha(\mathbf{r}_0,t_0,\mathbf{v})$$
$$\times\rho(\alpha,\mathbf{r}_0,t_0,\mathbf{v},t)\delta(\mathbf{r}-\mathbf{r}_0-\mathbf{v}t+\mathbf{v}t_0). \tag{24}$$

The density matrix resulting from (24) by integration over all velocities will be denoted by $\rho(\mathbf{r},t)$.

5. INTEGRATION OF THE EQUATIONS OF MOTION

The matrix equation of motion for the density matrix $\rho(\alpha,\mathbf{r}_0,t_0,\mathbf{v},t)$ has components

$$\dot\rho_{ab}=-i\omega\rho_{ab}-\gamma_{ab}\rho_{ab}+iV(t)(\rho_{aa}-\rho_{bb}),$$
$$\dot\rho_{aa}=-\gamma_a\rho_{aa}+iV(t)(\rho_{ab}-\rho_{ba}), \tag{25}$$
$$\dot\rho_{bb}=-\gamma_b\rho_{bb}-iV(t)(\rho_{ab}-\rho_{ba}),$$
$$\rho_{ba}=\rho_{ab}^*, \tag{26}$$

where

$$\gamma_{ab}=\tfrac{1}{2}(\gamma_a+\gamma_b) \tag{26a}$$

and

$$\omega=W_a-W_b>0. \tag{27}$$

We consider first the case of excitation to the upper maser state a. At $t=t_0$, $\rho_{aa}=1$ and $\rho_{bb}=\rho_{ab}=\rho_{ba}=0$. The solution to any desired order in the perturbation $V(t)$ can be obtained by iteration. There are contributions to $\rho_{ab}=\rho_{ba}^*$ in first and third order, to ρ_{bb} in second order, and to ρ_{aa} in zeroth and second order. Thus, in zeroth order,

$$\rho_{aa}^{(0)}(a,\mathbf{r}_0,t_0,\mathbf{v},t)=\exp-\gamma_a(t-t_0) \tag{28}$$

and the first-order contribution to ρ_{ab} is

$$\rho_{ab}^{(1)}(a,\mathbf{r}_0,t_0,\mathbf{v},t)$$
$$=i\int_{t_0}^{t}dt'V(t')\exp[(\gamma_{ab}+i\omega)(t'-t)+\gamma_a(t_0-t')], \tag{29}$$

while the third-order contribution is

$$\rho_{ab}^{(3)}(a,\mathbf{r}_0,t_0,\mathbf{v},t)=i\int_{t_0}^{t}dt'V(t')$$
$$\times[\rho_{aa}^{(2)}(a,\mathbf{r}_0,t_0,\mathbf{v},t')-\rho_{bb}^{(2)}(a,\mathbf{r}_0,t_0,\mathbf{v},t')]$$
$$\times\exp(\gamma_{ab}+i\omega)(t'-t), \tag{30}$$

where

$$\rho_{aa}^{(2)}(a,\mathbf{r}_0,t_0,\mathbf{v},t')$$
$$=-\int_{t_0}^{t'}dt''\int_{t_0}^{t''}dt'''V(t'')V(t''')$$
$$\times\{\exp[\gamma_a(t''-t')+(\gamma_{ab}+i\omega)(t'''-t'')$$
$$+\gamma_a(t_0-t''')]+\text{conj.}\}, \tag{31}$$

and similarly

$$\rho_{bb}^{(2)}(a,\mathbf{r}_0,t_0,\mathbf{v},t')$$
$$=+\int_{t_0}^{t'}dt''\int_{t_0}^{t''}dt'''V(t'')V(t''')$$
$$\times\{\exp[\gamma_b(t''-t')+(\gamma_{ab}+i\omega)(t'''-t'')$$
$$+\gamma_a(t_0-t''')]+\text{conj.}\}. \tag{32}$$

6. FIRST-ORDER THEORY

In order to convert the expression (29) for $\rho_{ab}^{(1)}$ into a macroscopic polarization, we must first calculate

$$\rho_{ab}^{(1)}(a,\mathbf{r},\mathbf{v},t)$$
$$=\int_{-\infty}^{t}dt_0\rho_{ab}^{(1)}(a,\,\mathbf{r}_0=\mathbf{r}-\mathbf{v}t+\mathbf{v}t_0,\,t_0,\,\mathbf{v},\,t) \tag{33}$$

as in Eq. (24).

The perturbation $V(t')$ acting at time t' on the atom specified by $\mathbf{r}_0$, t_0, $\mathbf{v}$ is $-(\wp/\hbar)E(\mathbf{r}_0+\mathbf{v}(t'-t_0),\,t')$, but for Eq. (33) we require this for an atom characterized by $\mathbf{r}_0=\mathbf{r}-\mathbf{v}t+\mathbf{v}t_0$, $\mathbf{v}$, t_0, for which the effective perturbation

$$V(t')=-(\wp/\hbar)E(\mathbf{r}-\mathbf{v}(t-t'),\,t') \tag{34}$$

does not depend on t_0. We may then perform the above integration over t_0 if we treat λ_a as a slowly varying function of t_0 and evaluate it at t. For the first-order terms, we have to deal with an expression of the form

$$\int_{-\infty}^{t}dt_0\int_{t_0}^{t}dt'F(t,t')e^{\gamma_a(t_0-t')}.$$

By an interchange[9] of the order of integrations this becomes

$$\int_{-\infty}^{t}dt'\int_{-\infty}^{t'}dt_0F(t,t')e^{\gamma_a(t_0-t')}=(1/\gamma_a)\int_{-\infty}^{t}dt'F(t,t'). \tag{35}$$

Let us assume now that the maser oscillator is running simultaneously in M cavity modes, so that

$$E(\mathbf{r},t)=\sum_{\mu=1}^{M}E_\mu(t)U_\mu(\mathbf{r})\cos(\nu_\mu t+\varphi_\mu(t)), \tag{36}$$

[9] In each double integral an integration is carried over the same triangular area in the t_0, t' plane.

where $E_\mu(t)$ and $\varphi_\mu(t)$ are slowly varying functions of time. We make a rotating wave approximation by keeping only exponential factors like $\exp i(\omega-\nu_\mu)t'$ and neglecting rapidly varying exponentials like $\exp i(\omega+\nu_\mu)t'$. Then $\rho_{ab}^{(1)}(a,\mathbf{r},\mathbf{v},t) =$

$$-\tfrac{1}{2}i\left(\frac{\wp}{\hbar}\right)\left[\frac{\lambda_a(\mathbf{r},v,t)}{\gamma_a}\right]\sum_{\mu=1}^{M}\int_{-\infty}^{t} dt' E_\mu(t') \times U_\mu(\mathbf{r}-\mathbf{v}(t-t'))\exp -i(\nu_\mu t+\varphi_\mu(t')) \times \exp[-\gamma_{ab}+i(\nu_\mu-\omega)](t-t'). \quad (37)$$

We also assume that the amplitudes $E_\mu(t')$ and phases $\varphi_\mu(t')$ do not vary much in a time $1/\gamma_{ab}$, so that they can be evaluated at time t. With a change of variable of integration from t' to $\tau'=t-t'$, we find

$$\rho_{ab}^{(1)}(a,\mathbf{r},\mathbf{v},t)= -\tfrac{1}{2}i\left(\frac{\wp}{\hbar}\right)\left[\frac{\lambda_a(\mathbf{r},\mathbf{v},t)}{\gamma_a}\right]\sum_{\mu=1}^{M} E_\mu(t)\exp -i(\nu_\mu t+\varphi_\mu(t)) \times\int_0^\infty d\tau' U_\mu(\mathbf{r}-\mathbf{v}\tau')\exp -(\gamma_{ab}+i(\omega-\nu_\mu))\tau'. \quad (38)$$

The corresponding contribution to the polarization of the medium is

$$P^{(1)}(a,\mathbf{r},\mathbf{v},t) = -\tfrac{1}{2}i(\wp^2/\hbar)[\lambda_a(\mathbf{r},\mathbf{v},t)/\gamma_a] \times\sum_\mu E_\mu(t)\left\{\exp -i(\nu_\mu t+\varphi_\mu)\int_0^\infty d\tau' U_\mu(\mathbf{r}-\mathbf{v}\tau') \times\exp -(\gamma_{ab}+i(\omega-\nu_\mu))\tau'\right\}+\text{conj.} \quad (39)$$

Let us first assume that the excitation rate density has the form[10]

$$\lambda_\alpha(\mathbf{r},\mathbf{v},t)=W(\mathbf{v})\Lambda_\alpha(\mathbf{r},t) \quad \alpha=a,\,b, \quad (40)$$

where $W(\mathbf{v})$ is the normalized velocity distribution function and $\Lambda_\alpha(\mathbf{r},t)$ is the number of atoms excited to state α per unit volume and time. Because we are assuming a spatial dependence of the electric field only on z, we may change over from a three- to a one-dimensional description. Then the velocity distribution $W(v)$ refers to the z component v of $\mathbf{v}$, and $\mathbf{r}$ is replaced by z.

It will be noted that the quantity $P(a,z,v,t)$ is proportional to $\Lambda_a(z,t)/\gamma_a$. When we now consider the contribution of atoms excited to the lower maser state b there is a complication which we did not meet in the case of a excitation. Spontaneous decay of atoms in the upper maser level a may be one of the excitation mechanisms for state b. This could be plausibly represented by replacing the excitation rate density $\lambda_b(z,v,t)$ by $\lambda_b(z,v,t)+f\gamma_a\rho_{aa}(z,v,t)$, where f is the branching ratio (decay from a to b)/(total decay from a) and λ_b now describes only "external" excitation processes.

In order to reduce somewhat the complexity of the subsequent equations we will now proceed as if f were zero. The effects of cascade excitation $a\to b$ will be discussed by an approximate method in Sec. 20. With this simplification it turns out, as one would expect, that for b excitation $P(b,z,v,t)$ is exactly like (39) except for an over-all sign change and interchange of a and b. Hence, the total polarization $P(z,v,t)=P(a,z,v,t)+P(b,z,v,t)$ is proportional to a quantity

$$N(z,t)=[(\Lambda_a(z,t)/\gamma_a)-(\Lambda_b(z,t)/\gamma_b)], \quad (41)$$

which we will call the "excitation density." This is simply the excess density of active atoms in a steady state in the absence of optical oscillations.

The first-order polarization

$$P^{(1)}(z,t)=\wp\int_{-\infty}^{\infty} dv W(v)[\rho_{ab}^{(1)}(a,z,v,t) +\rho_{ab}^{(1)}(b,z,v,t)+\text{conj.}] \quad (42)$$

is also proportional to $N(z,t)$. For use in Eqs. (13), (14) a spatial Fourier projection on the nth cavity mode is next to be made

$$P_n^{(1)}(t)=(2/L)\int_0^L dz P^{(1)}(z,t)U_n(z). \quad (43)$$

The product $U_n(z)U_\mu(z-v\tau')$ which occurs in (43) may be written as

$$\sin K_n z\,\sin K_\mu(z-v\tau') =\tfrac{1}{2}\cos\{(K_n-K_\mu)z+K_\mu v\tau'\} -\tfrac{1}{2}\cos\{(K_n+K_\mu)z-K_\mu v\tau'\}. \quad (44)$$

The last term will not contribute appreciably to the z integration (43) because the excitation density $N(z,t)$ changes little in an o.f. wavelength. Since the velocity distribution is normally an even function of v only that part of the remainder of (44) which is even in v will contribute to the polarization, i.e.,

$$\tfrac{1}{2}[\cos(K_n-K_\mu)z]\cos Kv\tau',$$

where the subscript μ has been dropped in the last factor since all of the modes considered have very nearly the same wave number $K=\nu/c$.

We find

$$P_n^{(1)}(t)=-\tfrac{1}{2}i(\wp^2/\hbar)\sum_\mu E_\mu\exp -i(\nu_\mu t+\varphi_\mu)N_{n-\mu} \times\int^\infty dv W(v)[\mathfrak{D}(\omega-\nu_\mu+Kv)]+\text{conj.}, \quad (45)$$

where

$$\mathfrak{D}(\omega)=1/(\gamma_{ab}+i\omega) \quad (46)$$

[10] It would be easy to modify the theory to allow the atoms excited to state b to have a different velocity distribution from those excited to state a.

is a convenient abbreviation for a frequently occurring denominator and where

$$N_{n-\mu}(t)=\frac{1}{L}\int_0^L dzN(z,t)\cos\left[(n-\mu)\frac{\pi z}{L}\right] \quad (47)$$

is a spatial Fourier component of the excitation density $N(z,t)$. It should be noted that (45) has a very simple interpretation in terms of Doppler shifts of the atomic transition frequencies by Kv due to the atomic motion. This simplicity will be lost when nonlinear effects are considered.

For the following detailed calculations a Maxwellian distribution

$$W(v)=(u\pi^{1/2})^{-1}\exp-(v^2/u^2) \quad (48)$$

will be assumed. The speed parameter u is related to an effective temperature T by the equation

$$\tfrac{1}{2}mu^2=k_BT, \quad (49)$$

where m is the atomic mass and k_B is the Boltzmann constant. If it should develop that a Maxwellian distribution is not realized in practice, some obvious changes in the later work can be made.

With Eq. (48) the integration over v may profitably be done on (39) before that over τ', and we find

$$P_n^{(1)}(t)=-\tfrac{1}{2}(\wp^2/\hbar Ku)$$
$$\times[\sum_{\mu=1}^{M} E_\mu(t)\exp-i(\nu_\mu t+\varphi_\mu(t))$$
$$\times N_{n-\mu}(t)Z(\nu_\mu-\omega)+\text{conj.}], \quad (50)$$

where $Z(\nu-\omega)$ is an abbreviation for

$$Z(\nu-\omega,\gamma_{ab},Ku)$$
$$=iKu\int_0^\infty d\tau\exp[i(\nu-\omega)\tau-\gamma_{ab}\tau-\tfrac{1}{4}K^2u^2\tau^2], \quad (51)$$

which is a complex function well known in the theory of Doppler broadening.[11] The function Z is, in fact, a function of a single complex variable ζ

$$Z(\zeta)=2i\int_{-\infty}^{i\zeta} dt\exp-(t^2+\zeta^2), \quad (52)$$

where

$$\zeta=\xi+i\eta, \quad (53)$$

with

$$\xi=(\nu-\omega)/Ku, \quad (54)$$

and

$$\eta=\gamma_{ab}/Ku. \quad (55)$$

It is fortunate that extensive tables[12] of the real part Z_r and the imaginary part Z_i of $Z(\xi+i\eta)$ are now available.

$P_n^{(1)}(t)$ is a linear function of the complex electric fields $E_\mu(t)\exp-i\nu_\mu t-i\varphi_\mu(t)$ of the cavity modes and, apart from amplitude modulation arising from a possible slow time variation of the excitation density $N(z,t)$ contains the same frequencies as the cavity field.

To determine amplitude and frequency (or phase) of the oscillations, we write out the contributions of $P_n^{(1)}$ to C_n and S_n of Eqs. (13), (14). These are

$$S_n^{(1)}=-(\wp^2/\hbar Ku)\bar{N}Z_i(\nu_n-\omega)E_n, \quad (56)$$

$$C_n^{(1)}=-(\wp^2/\hbar Ku)\bar{N}Z_r(\nu_n-\omega)E_n, \quad (57)$$

where

$$\bar{N}=N_0(t)=(1/L)\int_0^L dzN(zt), \quad (58)$$

which will be called the "excitation," is the average of the excitation density over the cavity. We have now reverted to the notation of Eq. (51) for the Z function, but to shorten equations have dropped the parameters γ_{ab} and Ku which appear as arguments in (51).

In this approximation, without nonlinear terms, we can only hope to obtain the condition for starting of oscillations and their frequency at threshold. Furthermore, if the conditions are such that several modes can oscillate, they do so independently of each other and hence can be considered separately. The amplitude equation (14) gives

$$\dot{E}_n=-\tfrac{1}{2}(\nu/Q_n)E_\mu-\tfrac{1}{2}(\nu/\epsilon_0)S_n^{(1)} \quad (59)$$

or for a steady state, for which $\dot{E}_n=0$,

$$(\wp^2/\epsilon_0\hbar Ku)\bar{N}Z_i(\nu_n-\omega)=1/Q_n. \quad (60)$$

To first order in $\eta=\gamma_{ab}/Ku$, we have

$$Z(\xi,\eta)\simeq(1-2i\eta\xi)\left[-2\int_0^\xi e^{x^2}dx+i\pi^{1/2}\right]e^{-\xi^2}-2i\eta. \quad (61)$$

Hence for pure Doppler line shape the condition (60) for the onset of oscillations in the nth mode may be written as

$$2\pi^{1/2}[e^2/(4\pi\epsilon_0\hbar u)]$$
$$\times(\wp/e)^2\lambda\bar{N}\exp-(\nu_n-\omega)^2/(Ku)^2=1/Q_n, \quad (62)$$

where $\lambda=2\pi/K$ is the wavelength. In these units,

$$e^2/(4\pi\epsilon_0\hbar c)\approx1/137 \quad (63)$$

is the fine structure constant, so the left-hand side of (62) is the product of this and four other dimensionless factors: c/u, $2\pi^{1/2}$, $\exp-(\nu_n-\omega)^2/(Ku)^2$ and $(\wp/e)^2\lambda\bar{N}$ which is the net number of active atoms in a cylinder of cross sectional area $(\wp/e)^2$ and length λ. As the frequency detuning $\nu_n-\omega$ increases, the excitation $\bar{N}$ required to initiate oscillations increases in proportion to $\exp+(\nu_n-\omega)^2/(Ku)^2$. The frequency of oscillation is

[11] M. Born, *Optik* (Julius Springer-Verlag, Berlin, 1933), pp. 482–486.

[12] B. D. Fried and S. D. Conte, *The Plasma Dispersion Function (Hilbert Transform of the Gaussian)* (Academic Press, Inc., New York, 1961).

determined by Eq. (13) in which we may set $\dot{\varphi}_n=0$ without loss of generality. Using Eq. (57), we find

$$\nu_n=\Omega_n+(\nu/2)(\wp^2\bar{N}/\epsilon_0\hbar Ku)Z_r(\nu_n-\omega). \quad (64)$$

It is convenient to use the threshold condition (60) to express $\bar{N}$ in terms of Q_n. We find

$$\nu_n-\Omega_n=\tfrac{1}{2}(\nu/Q_n)[Z_r(\nu_n-\omega)/Z_i(\nu_n-\omega)] \quad (65)$$

or in the approximation $\gamma_{ab}/Ku\ll 1$

$$\nu_n-\Omega_n=-\left(\frac{1}{\pi^{1/2}}\right)\left(\frac{\nu}{Q_n}\right)\int_0^{\xi_n}dx e^{x^2}. \quad (66)$$

If the integral (66) is expanded to first order in $\xi_n=(\nu_n-\omega)/Ku$ we obtain

$$(\nu_n-\Omega_n)/(\omega-\nu_n)\approx(1/\pi^{1/2})(\nu/QKu)=\mathcal{S}, \quad (67)$$

which implies "linear pulling," i.e., the detuning of ν_n from the cavity frequency Ω_n is proportional to the amount $\omega-\nu_n$ by which the oscillator frequency ν_n is removed from the atomic resonance frequency ω. The right-hand side of (67) ($\mathcal{S}$="stabilization factor"[13]) is about 1/800 for typical values of the parameters used in Sec. 2.

The more accurate expression (65) indicates a "nonlinear pulling" such that the oscillator frequency is nearer to the atomic frequency than it would be for linear pulling. To the next order in $(\Omega_n-\omega)/Ku$, but still for $\gamma_{ab}\ll Ku$, we find

$$(\nu_n-\Omega_n)/(\omega-\nu_n)\simeq[1+\tfrac{1}{3}(\Omega_n-\omega)^2/(Ku)^2]\mathcal{S}. \quad (68)$$

7. THIRD-ORDER TERMS

We now carry out a similar calculation for the third-order quantity $P_n^{(3)}(t)$ using Eqs. (8), (23), (30), (31) and (32). The integration of $\lambda_a(z,v,t_0)\rho_{ab}^{(3)}(a,z,t_0,v,t)$ over times t_0 of excitation involves integrals like

$$\int_{-\infty}^{t}dt_0\int_{t_0}^{t}dt'\int_{t_0}^{t'}dt''\int_{t_0}^{t''}dt'''F(t,t',t'',t''')\exp-\gamma_a(t'''-t_0),$$

which by a repeated interchange of orders of integration can be reduced to

$$(1/\gamma_a)\int_{-\infty}^{t}dt'\int_{-\infty}^{t'}dt''\int_{-\infty}^{t''}dt'''F(t,t',t'',t''').$$

Again, we keep only exponential factors in the time integration which are able to have resonance, and find after changes of variables

$$\tau'=t-t',\quad \tau''=t'-t'',\quad \tau'''=t''-t''' \quad (69)$$

and some algebraic manipulations

$$\begin{aligned}\rho_{ab}^{(3)}(z,v,t)=\tfrac{1}{8}i\wp^3\hbar^{-3}N(z,t)\sum_\mu\sum_\rho\sum_\sigma E_\mu E_\rho E_\sigma\Big\{&[\exp-i(\nu_\mu t+\varphi_\mu)+i(\nu_\rho t+\varphi_\rho)-i(\nu_\sigma t+\varphi_\sigma)]\\ &\times\int_0^\infty d\tau'\int_0^\infty d\tau''\int_0^\infty d\tau'''U_\mu(z-v\tau')U_\rho(z-v\tau'-v\tau'')U_\sigma(z-v\tau'-v\tau''-v\tau''')\\ &\times\{\exp-[(\gamma_{ab}-i\nu_\mu+i\nu_\rho-i\nu_\sigma+i\omega)\tau'+(\gamma_a+i\nu_\rho-i\nu_\sigma)\tau''+(\gamma_{ab}+i\omega-i\nu_\sigma)\tau''']\}\\ &+[\exp-i(\nu_\mu t+\varphi_\mu)-i(\nu_\rho t+\varphi_\rho)+i(\nu_\sigma t+\varphi_\sigma)]\\ &\times\int_0^\infty d\tau'\int_0^\infty d\tau''\int_0^\infty d\tau'''U_\mu(z-v\tau')U_\rho(z-v\tau'-v\tau'')U_\sigma(z-v\tau'-v\tau''-v\tau''')\\ &\times[\exp-[(\gamma_{ab}-i\nu_\mu-i\nu_\rho+i\nu_\sigma+i\omega)\tau'+(\gamma_a-i\nu_\rho+i\nu_\sigma)\tau''+(\gamma_{ab}-i\omega+i\nu_\sigma)\tau''']]\Big\}\\ &+\text{same with } a \text{ and } b \text{ interchanged.}\end{aligned} \quad (70)$$

In calculation of the Fourier projection $P_n^{(3)}(t)$ integrals of the form

$$\left(\frac{2}{L}\right)\int_0^L dzN(zt)U_n(z)U_\mu(z-v\tau')U_\rho(z-v\tau'-v\tau'')U_\sigma(z-v\tau'-v\tau''-v\tau''')$$

appear. The product of the four sine functions can be reduced to

$$\begin{aligned}\tfrac{1}{8}[&\cos(K_\rho-K_\sigma+K_\mu-K_n)z\cos Kv(\tau'''-\tau')+\cos(K_\rho-K_\sigma-K_\mu+K_n)z\cos Kv(\tau'''+\tau')\\ &+\cos(K_\rho+K_\sigma-K_\mu-K_n)z\cos Kv(\tau'+2\tau''+\tau''')]\end{aligned}$$

[13] The term "stabilization factor" ($\approx$cavity band width/atom bandwidth) was previously used in a theory of the ammonia beam maser (Ref. 8, p. 460) where its numerical value was large compared to unity.

apart from rapidly oscillating terms and those odd in v. The integration over v may now be carried out, and we find

$$P_n^{(3)}(t)=\tfrac{1}{32}i\wp^4\hbar^{-3}\sum_\mu\sum_\rho\sum_\sigma E_\mu E_\rho E_\sigma\Big\{[\exp-i(\nu_\mu t+\varphi_\mu)+i(\nu_\rho t+\varphi_\rho)-i(\nu_\sigma t+\varphi_\sigma)]$$
$$\times\int_0^\infty d\tau'\int_0^\infty d\tau''\int_0^\infty d\tau'''[N_{(\rho-\sigma+\mu-n)}\exp-\tfrac{1}{4}(Ku)^2(\tau'''-\tau')^2+N_{(\rho-\sigma-\mu+n)}\exp-\tfrac{1}{4}(Ku)^2(\tau'''+\tau')^2$$
$$+N_{(\rho+\sigma-\mu-n)}\exp-\tfrac{1}{4}(Ku)^2(\tau'+2\tau''+\tau''')^2][\exp-(\gamma_{ab}-i\nu_\mu+i\nu_\rho-i\nu_\sigma+i\omega)\tau'$$
$$-(\gamma_a+i\nu_\rho-i\nu_\sigma)\tau''-(\gamma_{ab}+i\omega-i\nu_\sigma)\tau''']+[\exp-i(\nu_\mu t+\varphi_\mu)-i(\nu_\rho t+\varphi_\rho)+i(\nu_\sigma t+\varphi_\sigma)]$$
$$\times\int_0^\infty d\tau'\int_0^\infty d\tau''\int_0^\infty d\tau'''[N_{(\rho-\sigma+\mu-n)}\exp-\tfrac{1}{4}(Ku)^2(\tau'''-\tau')^2+N_{(\rho-\sigma-\mu+n)}\exp-\tfrac{1}{4}(Ku)^2(\tau'''+\tau')^2$$
$$+N_{(\rho+\sigma-\mu-n)}\exp-\tfrac{1}{4}(Ku)^2(\tau'+2\tau''+\tau''')^2][\exp-(\gamma_{ab}-i\nu_\mu-i\nu_\rho+i\nu_\sigma+i\omega)\tau'-(\gamma_a-i\nu_\rho+i\nu_\sigma)\tau''$$
$$-(\gamma_{ab}-i\omega+i\nu_\sigma)\tau''']\Big\}+\text{same with }\gamma_a\text{ and }\gamma_b\text{ interchanged}+\text{complex conjugate.} \qquad (71)$$

One of the most important characteristics of the third-order polarization is that it has constituents which oscillate at all possible frequencies $\nu_\mu\mp\nu_\rho\pm\nu_\sigma$. The combination of signs appearing here is correlated with use of the rotating wave approximation. Terms with frequencies such as $\nu_\mu+\nu_\rho+\nu_\sigma$ (near the third harmonic) are thereby neglected.

The formidable expression (71) can be simplified in either of two limiting cases: (a) no atomic motion ($u=0$), or (b) "Doppler limit," i.e., Ku much larger than γ_{ab} and various frequency differences such as $\nu_\mu+\nu_\rho-2\nu_\sigma$, etc. In the absence of atomic motion, one finds after some rewriting

$$P_n^{(3)}(t)=\tfrac{1}{32}i\hbar^{-3}\wp^4\sum_\mu\sum_\rho\sum_\sigma E_\mu E_\rho E_\sigma[\exp-i(\nu_\mu t+\varphi_\mu)+i(\nu_\rho t+\varphi_\rho)-i(\nu_\sigma t+\varphi_\sigma)]$$
$$\times[N_{(\rho-\sigma+\mu-n)}+N_{(\rho-\sigma-\mu+n)}+N_{(\rho+\sigma-\mu-n)}]\cdot\mathfrak{D}(\nu_\rho-\nu_\mu-\nu_\sigma+\omega)$$
$$\times[\mathfrak{D}_a(\nu_\rho-\nu_\sigma)+\mathfrak{D}_b(\nu_\rho-\nu_\sigma)][\mathfrak{D}(\omega-\nu_\sigma)+\mathfrak{D}(\nu_\rho-\omega)]+\text{conjugate,} \qquad (72)$$

where $\mathfrak{D}(\omega)$ was defined by Eq. (46) and

$$\mathfrak{D}_\alpha(\omega)=1/(\gamma_\alpha+i\omega),\quad \alpha=a,b. \qquad (73)$$

The "Doppler limit" is appropriate for many possible gaseous optical masers, and for most of the remainder of this paper we will be dealing only with this case. Then

$$\exp-\tfrac{1}{4}(Ku)^2(\tau'''-\tau')^2$$

acts like a delta function of $\tau'''-\tau'$ and the integration over τ''' can be done in the form

$$\int_0^\infty d\tau'\int_0^\infty d\tau'''G(\tau',\tau''')\exp-\tfrac{1}{4}(Ku)^2(\tau'''-\tau')^2\simeq\frac{2\pi^{1/2}}{Ku}\int_0^\infty d\tau'G(\tau',\tau'). \qquad (74)$$

The other Gaussian factors do not have their full peaks in the range of integration, and give contributions which we neglect because they lead to expressions with higher powers of Ku in the denominator. Then, after performing the simple integrations over τ' and τ'', we find

$$P_n^{(3)}(t)=\tfrac{1}{32}i\pi^{1/2}[\wp^4/(\hbar^3Ku)]\sum_\mu\sum_\rho\sum_\sigma E_\mu E_\rho E_\sigma N_{(\rho-\sigma+\mu-n)}\{\mathfrak{D}(\omega-\tfrac{1}{2}\nu_\mu+\tfrac{1}{2}\nu_\rho-\nu_\sigma)[\mathfrak{D}_a(\nu_\rho-\nu_\sigma)+\mathfrak{D}_b(\nu_\rho-\nu_\sigma)]$$
$$\times[\exp-i(\nu_\mu-\nu_\rho+\nu_\sigma)t-i(\varphi_\mu-\varphi_\rho+\varphi_\sigma)]+\mathfrak{D}(-\tfrac{1}{2}\nu_\mu-\tfrac{1}{2}\nu_\rho+\nu_\sigma)[\mathfrak{D}_a(\nu_\sigma-\nu_\rho)+\mathfrak{D}_b(\nu_\sigma-\nu_\rho)]$$
$$\times[\exp[-i(\nu_\mu+\nu_\rho-\nu_\sigma)t-i(\varphi_\mu+\varphi_\rho-\varphi_\sigma)]]\}+\text{complex conjugate.} \qquad (75)$$

By interchanging ρ and σ in the second group of terms, this may be written more compactly as

$$P_n^{(3)}(t)=\tfrac{1}{32}i\pi^{1/2}[\wp^4/(\hbar^3Ku)]\sum_\mu\sum_\rho\sum_\sigma E_\mu E_\rho E_\sigma[\exp[-i(\nu_\mu-\nu_\rho+\nu_\sigma)t-i(\varphi_\mu-\varphi_\rho+\varphi_\sigma)]]$$
$$\times[N_{(\rho-\sigma+\mu-n)}\mathfrak{D}(\omega-\tfrac{1}{2}\nu_\mu+\tfrac{1}{2}\nu_\rho-\nu_\sigma)+N_{(\sigma-\rho+\mu-n)}\mathfrak{D}(-\tfrac{1}{2}\nu_\mu-\tfrac{1}{2}\nu_\sigma+\nu_\rho)][\mathfrak{D}_a(\nu_\rho-\nu_\sigma)+\mathfrak{D}_b(\nu_\rho-\nu_\sigma)]$$
$$+\text{complex conjugate.} \qquad (76)$$

8. SINGLE FREQUENCY OPERATION

For this case, the triple summation over μ, ρ, and σ reduces to a single term

$$P_n^{(3)}(t)=\tfrac{1}{16}i\pi^{1/2}\bar{N}[\wp^4/(\hbar^3\gamma_a\gamma_bKu)] \times\gamma_{ab}[\mathfrak{D}(\omega-\nu_n)+\mathfrak{D}(0)]E_n^{(3)}\exp -i(\nu_nt+\varphi_n) +\text{complex conjugate.} \quad (77)$$

For use in Eqs. (13), (14) we need the in-phase and quadrature components of this

$$C_n^{(3)}(t)=\tfrac{1}{8}\pi^{1/2}\bar{N}[\wp^4/(\hbar^3\gamma_a\gamma_bKu)] \times\gamma_{ab}(\omega-\nu_n)\mathfrak{L}(\omega-\nu_n)E_n^3 \quad (78)$$

and

$$S_n^{(3)}(t)=\tfrac{1}{8}\pi^{1/2}\bar{N}[\wp^4/(\hbar^3\gamma_a\gamma_bKu)] \times[1+\gamma_{ab}^2\mathfrak{L}(\omega-\nu_n)]E_n^3, \quad (79)$$

where the Lorentzian function is denoted by

$$\mathfrak{L}(\omega-\nu)=[\gamma_{ab}^2+(\omega-\nu)^2]^{-1}. \quad (80)$$

The amplitude determining Eq. (13) is then of the form[14]

$$\dot{E}_n=\alpha_nE_n-\beta_nE_n^3, \quad (81)$$

where

$$\alpha_n=-\tfrac{1}{2}(\nu/Q_n)+\tfrac{1}{2}\nu\bar{N}[\wp^2/(\epsilon_0\hbar Ku)]Z_i(\nu_n-\omega), \quad (82)$$

$$\beta_n=(\nu/16)\pi^{1/2}\bar{N}[\wp^4/(\epsilon_0\hbar^3\gamma_a\gamma_bKu)] \times[1+\gamma_{ab}^2\mathfrak{L}(\nu_n-\omega)]. \quad (83)$$

It is useful to employ the starting condition (60) to express the coefficients α_n, β_n (and others which appear later) in terms of a ratio

$$\mathfrak{N}=\bar{N}/\bar{N}_T \quad (84)$$

called the "relative excitation" where $\bar{N}_T$ is the excitation required for threshold oscillations when the cavity frequency Ω_n is tuned to the peak ω of the atomic resonance curve. We find

$$\alpha_n=\tfrac{1}{2}(\nu/Q_n)\{[Z_i(\nu_n-\omega)/Z_i(0)]\mathfrak{N}-1\} \quad (85)$$

and

$$\beta_n=\tfrac{1}{16}\pi^{1/2}(\nu/Q_n)[\mathfrak{N}\wp^2/(\hbar^2\gamma_a\gamma_bZ_i(0))] \times[1+\gamma_{ab}^2\mathfrak{L}(\nu_n-\omega)]. \quad (86)$$

A stable steady state occurs for an intensity of oscillations

$$E_n^2=\alpha_n/\beta_n, \quad (87)$$

which is easily related to the relative excitation with the help of Eqs. (85) and (86).

The frequency determining equation (again $\dot{\varphi}_n$ may be taken zero) is

$$\nu_n=\Omega_n-\tfrac{1}{2}(\nu/(\epsilon_0E_n))[C_n^{(1)}+C_n^{(3)}]. \quad (88)$$

Since the frequency of oscillation ν_n will differ little from the cavity resonance frequency Ω_n, the right side of Eq. (88) may, to a sufficiently good approximation, be evaluated for $\nu_n=\Omega_n$. We may then write

$$\nu_n=\Omega_n+\sigma_n+\rho_nE_n^2, \quad (89)$$

where

$$\sigma_n=\tfrac{1}{2}\nu[\wp^2\bar{N}/(\epsilon_0\hbar Ku)]Z_r(\Omega_n-\omega) \quad (90)$$

and

$$\rho_n=+\tfrac{1}{16}\pi^{1/2}\nu[\wp^4\bar{N}/(\epsilon_0\hbar^3\gamma_a\gamma_bKu)] \times\gamma_{ab}(\Omega_n-\omega)\mathfrak{L}(\Omega_n-\omega) \quad (91)$$

or expressing these coefficients in terms of the relative excitation (84)

$$\sigma_n=\tfrac{1}{2}(\nu/Q_n)\mathfrak{N}Z_r(\Omega_n-\omega)/Z_i(0) \quad (92)$$

and

$$\rho_n=+\tfrac{1}{16}\pi^{1/2}(\nu/Q_n)\mathfrak{N}[\wp^2/(\hbar^2\gamma_a\gamma_bZ_i(0))] \times\gamma_{ab}(\Omega_n-\omega)\mathfrak{L}(\Omega_n-\omega) \quad (93)$$

so that

$$\rho_nE_n^2=\rho_n\alpha_n/\beta_n =\tfrac{1}{2}(\nu/Q_n)\{[Z_i(\Omega_n-\omega)/Z_i(0)]\mathfrak{N}-1\}\gamma_{ab}(\Omega_n-\omega)/ [2\gamma_{ab}^2+(\Omega_n-\omega)^2]. \quad (94)$$

The frequency now depends on the relative excitation (and hence on the power level) as well as on the detuning, i.e., there is frequency "pushing" as well as "pulling."

Equations (89), (92), and (94) indicate that for

$$(\Omega_n-\omega)^2+2\gamma_{ab}^2<-\gamma_{ab}(\Omega_n-\omega)Z_i(\Omega_n-\omega)/ Z_r(\Omega_n-\omega) \quad (95)$$

an increase of excitation should move the frequency in the direction from ω toward Ω_n. For small detuning the right side of (95) is approximately $\tfrac{1}{2}\pi^{1/2}\gamma_{ab}Ku$.

The dependence of power level on excitation and detuning is given by Eq. (87). Using the value of $Z_i(\xi_n,\eta)$ for $\mathfrak{N}=0$, this equation may be written approximately as

$$(\wp E_n)^2/(\hbar^2\gamma_a\gamma_b) =8\left\{\left[\exp-\frac{(\Omega_n-\omega)^2}{(Ku)^2}\right]-\mathfrak{N}^{-1}\right\}\Big/[1+\gamma_{ab}^2\mathfrak{L}(\Omega_n-\omega)]. \quad (96)$$

This expression agrees with the linear approximation (62) in predicting threshold for relative excitation

$$\mathfrak{N}=\bar{N}/\bar{N}_T=\exp(\Omega_n-\omega)^2/(Ku)^2 \quad (97)$$

when there is detuning. Because of its derivation from a third order perturbation theory, Eq. (96) should not be trusted unless it predicts a value of the "saturation

[14] It is easy to see that the coefficient α_n is simply related to the gain (negative absorption) coefficient of the medium at frequency ν_n for small signals. However, the gain coefficient for a strong signal cannot be safely inferred from Eq. (81), since standing rather than traveling waves were assumed in its derivation. A theory of a traveling wave maser along the lines of this paper will be given later.

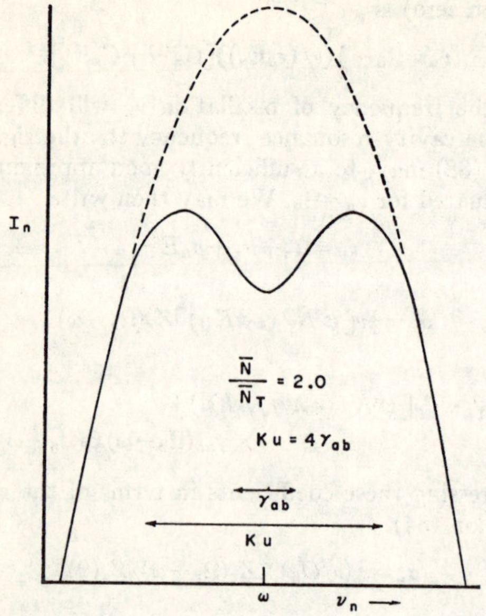

FIG. 2. Relative intensity of oscillation as a function of detuning. The solid curve, drawn for parameters $\bar{N}=2\bar{N}_T$ and $Ku=4\gamma_{ab}$ represents Eq. (96). The dotted curve indicates the Doppler gain profile of the numerator of (96).

parameter"[15]

$$I_n=\tfrac{1}{2}(\wp E_n)^2/(\hbar^2\gamma_a\gamma_b) \tag{98}$$

much less than unity.

The numerator in Eq. (96) has a peak for resonant tuning $\Omega_n=\omega$, but the denominator, which comes from the nonlinear term involving the coefficient β_n also has a peak when $\Omega_n=\omega$. Under certain conditions, the over-all curve of $E_n{}^2=\alpha_n/\beta_n$ versus detuning $\Omega_n-\omega$ should have a flattened peak at resonance, or even a dip between two maxima. The condition for the appearance of two maxima is

$$(\gamma_{ab}/Ku)^2<\tfrac{1}{2}\{1-\exp-[(\Omega_n{}^*-\omega)^2/(Ku)^2]\}, \tag{99}$$

where $\Omega_n{}^*-\omega$ is the detuning required to stop oscillations at the given level of excitation. The double peak[16] (see Fig. 2) should thus be seen somewhat above threshold, i.e., for relative excitation

$$\mathfrak{N}>1/[1-2(\gamma_{ab}/(Ku))^2]. \tag{100}$$

Under this degree of relative excitation, the electric field at central tuning is given by

$$I_n=\tfrac{1}{2}(\wp E_n)^2/(\hbar^2\gamma_a\gamma_b)\approx 4(\gamma_{ab}/(Ku))^2\ll 1 \tag{101}$$

so that the neglect of higher orders of perturbation theory should not be too serious, provided, as we assume, that $\gamma_{ab}\ll Ku$.

[15] The significance of this quantity is shown more clearly in Sec. 18.

[16] This dip has recently been observed. R. A. McFarlane, W. R. Bennett, and W. E. Lamb, Appl. Phys. Letters **2**, 189 (1963); A. Szöke and A. Javan, Phys. Rev. Letters **10**, 521 (1963).

TABLE I. This shows the twenty-seven possible values of the summation indices μ, ρ, σ which appear in Eq. (76) for "three"-mode oscillation. The fourth column gives the corresponding frequencies $\nu_\mu-\nu_\rho+\nu_\sigma$. The last column contains numerical values for the amount by which these frequencies exceed ν_1. We have taken $\nu_2-\nu_1=150$ Mc/sec and $\nu_3-\nu_2=151$ Mc/sec in order to simulate (but greatly to exaggerate) nonlinear pulling effects.

μ	ρ	σ	$\nu_\mu-\nu_\rho+\nu_\sigma$	Typical $\delta\nu$ (Mc/sec)
1	1	1	ν_1	0
1	1	2	ν_2	150
1	1	3	ν_3	301
1	2	1	$2\nu_1-\nu_2$	−150
1	2	2	ν_1	0
1	2	3	$\nu_1+\nu_3-\nu_2$	151
1	3	1	$2\nu_1-\nu_3$	−301
1	3	2	$\nu_1+\nu_2-\nu_3$	−151
1	3	3	ν_1	0
2	1	1	ν_2	150
2	1	2	$2\nu_2-\nu_1$	300
2	1	3	$\nu_2+\nu_3-\nu_1$	451
2	2	1	ν_1	0
2	2	2	ν_2	150
2	2	3	ν_3	301
2	3	1	$\nu_1+\nu_2-\nu_3$	−151
2	3	2	$2\nu_2-\nu_3$	−1
2	3	3	ν_2	150
3	1	1	ν_3	301
3	1	2	$\nu_2+\nu_3-\nu_1$	451
3	1	3	$2\nu_3-\nu_1$	602
3	2	1	$\nu_1+\nu_3-\nu_2$	151
3	2	2	ν_3	301
3	2	3	$2\nu_3-\nu_2$	452
3	3	1	ν_1	0
3	3	2	ν_2	150
3	3	3	ν_3	301

The physical interpretation of the dip is discussed in Secs. 17 and 18 from several points of view.

9. MULTIPLE MODE OPERATION

As the excitation is increased beyond that required for threshold of single frequency oscillation, other frequencies appear in the output of an optical maser oscillator. We wish to use the expression (76) for the nonlinear polarization in the electromagnetic field equations (13), (14) in order to account for the observed phenomena.

The first theory of an oscillator capable of multifrequency operation was given by van der Pol[17] in 1921–22. The necessary nonlinear features were provided by cubic terms in the current-voltage characteristic of a triode vacuum tube. The tank had two *R-L-C* circuits with resonance frequencies Ω_1 and Ω_2. Van der Pol found that steady oscillations could occur only at frequencies near Ω_1 or Ω_2, but that simultaneous steady operation at the two frequencies was impossible. There were hysteresis phenomena, i.e., the choice of steady state of oscillation depended on the past history of the circuit parameters.

Multicavity magnetrons provide very important

[17] B. van der Pol, Phil. Mag. **43**, 700 (1922) and a review article Proc. Inst. Radio Engrs. **22**, 1051 (1934).

TABLE II. This table gives various quantities needed for the evaluation of Eq. (76) for three-frequency oscillation as described in the text.

μ	ρ	σ	n	$\nu_\rho-\nu_\sigma$	$\omega-\frac{1}{2}\nu_\mu+\frac{1}{2}\nu_\rho-\nu_\sigma$	$-\frac{1}{2}\nu_\mu-\frac{1}{2}\nu_\sigma+\nu_\rho$	$\nu_\mu-\nu_\rho+\nu_\sigma$	$\rho-\sigma+\mu-n$	$\sigma-\rho+\mu-n$
1	1	1	1	0	$\omega-\nu_1$	0	ν_1	0	0
1	2	2	1	0	$\omega-\nu_{12}$	$\frac{1}{2}\Delta$	ν_1	0	0
1	3	3	1	0	$\omega-\nu_2$	Δ	ν_1	0	0
2	2	1	1	Δ	$\omega-\nu_1$	$\frac{1}{2}\Delta$	ν_1	2	0
3	3	1	1	2Δ	$\omega-\nu_1$	Δ	ν_1	4	0
2	3	2	1	Δ	$\omega-\nu_{12}$	Δ	$2\nu_2-\nu_3$	2	0
2	2	2	2	0	$\omega-\nu_2$	0	ν_2	0	0
1	1	2	2	$-\Delta$	$\omega-\nu_2$	$-\frac{1}{2}\Delta$	ν_2	-2	0
2	1	1	2	0	$\omega-\nu_{12}$	$-\frac{1}{2}\Delta$	ν_2	0	0
2	3	3	2	0	$\omega-\nu_{23}$	$\frac{1}{2}\Delta$	ν_2	0	0
3	3	2	2	Δ	$\omega-\nu_2$	$\frac{1}{2}\Delta$	ν_2	2	0
1	2	3	2	$-\Delta$	$\omega-\nu_{23}$	"0"	$\nu_1+\nu_3-\nu_2$	-2	0
3	2	1	2	Δ	$\omega-\nu_{12}$	"0"	$\nu_1+\nu_3-\nu_2$	2	0
3	3	3	3	0	$\omega-\nu_3$	0	ν_3	0	0
1	1	3	3	-2Δ	$\omega-\nu_3$	$-\Delta$	ν_3	-4	0
2	2	3	3	$-\Delta$	$\omega-\nu_3$	$-\frac{1}{2}\Delta$	ν_3	-2	0
3	1	1	3	0	$\omega-\nu_2$	$-\Delta$	ν_3	0	0
3	2	2	3	0	$\omega-\nu_{23}$	$-\frac{1}{2}\Delta$	ν_3	0	0
2	1	2	3	$-\Delta$	$\omega-\nu_{23}$	$-\Delta$	$2\nu_2-\nu_1$	-2	0

examples of oscillators capable of multifrequency operation, and here again the normal pattern is that an oscillation existing at one frequency tends to suppress one at another frequency.

There is, of course, a very close connection between van der Pol's work and ours. Where he dealt with a nonlinear triode characteristic for a vacuum tube, we are concerned with the nonlinear response of an assembly of atomic systems which obey the laws of quantum mechanics. A one-to-one correspondence can be set up between the two problems. The Fabry-Perot cavity modes correspond to the resonant constituents of van der Pol's plate circuit. As we will see, however, the effective tube characteristics of the optical medium differ qualitatively from that assumed by van der Pol, and hence the optical maser behaves in a very different fashion with respect to multifrequency operation than the oscillator in van der Pol's original model.

Rather than deal next with the case of two-frequency operation, it will perhaps save space to consider first the more general case of three-frequency operation. Probably most of the interesting phenomena for optical gaseous masers can be understood without dealing explicitly with more than three frequencies. After the more general equations have been obtained, we can easily drop terms and discuss two frequency oscillation as a special and simpler case.

We consider the expression (76) for the third-order polarization, in which the indices μ, ρ, σ, and n can each take on values 1, 2, and 3. It is useful to have the ingredients of the summands in tabular form. The entries in the fourth column of Table I give the frequencies of the various summands in $P_n^{(3)}$ identified by the μ, ρ, and σ values in the first three columns. It will be noted that besides the three frequencies ν_1, ν_2, and ν_3 assumed in the cavity excitation, there are nine additional frequencies present in the polarization of the medium. Hence there must be fields in the cavity at these new frequencies, and the desired self-consistency for three-frequency operation is in jeopardy. However, under certain conditions which will be determined later, the fields at the new frequencies do not produce appreciable effects, even though the frequencies lie close to cavity resonance, so that the calculation can be made as planned.

As is already apparent from the single-frequency case, the oscillation frequencies ν_n are typically very close to the cavity frequencies Ω_n, which are equally spaced and separated by

$$\Delta\approx 150 \text{ Mc/sec.} \tag{102}$$

Hence three of the new frequencies: $2\nu_2-\nu_3$, $\nu_1+\nu_3-\nu_2$, and $2\nu_2-\nu_1$ are very close to the three main frequencies ν_1, ν_2, and ν_3, respectively, and the corresponding terms are carried along in the calculations. The remaining frequencies can be ignored as long as the oscillator is appreciably below threshold for four-frequency operation.

Any pulling of ν_n from Ω_n is typically measured in kc/sec and hence the ν_n are not detuned from Ω_n by an appreciable fraction either of the cavity bandwidths ν/Q_n which are about 1 Mc/sec, or of the radiative decay constants γ_a, γ_b, or γ_{ab} which may be 10 Mc/sec or more. The entries in columns 5, 6, and 7 of Table II occur in the frequency denominators in Eq. (76) and are there to be combined with imaginary numbers $-i\gamma_a$, $-i\gamma_b$, or $-i\gamma_{ab}$. We have neglected the small terms arising from frequency pulling. Thus $\nu_2-\nu_1$ is freely replaced by Δ of Eq. (102), etc. A symbol like ν_{12} denotes a frequency halfway between ν_1 and ν_2, etc.,

$$\nu_{12}=\tfrac{1}{2}(\nu_1+\nu_2), \text{ etc.} \tag{103}$$

The entries in the last two columns of Table II are the

integers characterizing the spatial Fourier components (47) of the excitation density which are needed for the evaluation of the contribution to (76) arising from the indices μ, ρ, σ, and n.

The combinations (μ,ρ,σ) which contribute third order polarizations at frequency ν_1 are (1,1,1), (1,2,2), (1,3,3), (2,2,1), and (3,3,1) while (2,3,2) gives a contribution near ν_1. Each summand has a product of two-resonance denominators. The dominant terms are (1,1,1), (1,2,2), and (1,3,3) since they do not necessarily contain inverse powers of Δ. The other terms have at least one power of Δ in the denominator. All terms will ultimately be expanded to order $1/\Delta^2$.

We find

$$\begin{aligned} P_1^{(3)}(t)=\tfrac{1}{16}i\pi^{1/2}[\wp^4/\hbar^3Ku)][\{&E_1^3\bar{N}(\gamma_{ab}/(\gamma_a\gamma_b))(\mathfrak{D}(\omega-\nu_1)+\mathfrak{D}(0)) \\ &+E_1E_2^2\bar{N}(\gamma_{ab}/(\gamma_a\gamma_b))(\mathfrak{D}(\omega-\nu_{12})+\mathfrak{D}(\Delta/2))+E_1E_3^2\bar{N}(\gamma_{ab}/(\gamma_a\gamma_b))(\mathfrak{D}(\omega-\nu_2)+\mathfrak{D}(\Delta)) \\ &+E_2^2E_1[\tfrac{1}{2}N_2\mathfrak{D}(\omega-\nu_1)+\tfrac{1}{2}\bar{N}\mathfrak{D}(\Delta/2)][\mathfrak{D}_a(\Delta)+\mathfrak{D}_b(\Delta)]+E_3^2E_1[\tfrac{1}{2}N_4\mathfrak{D}(\omega-\nu_1)+\tfrac{1}{2}\bar{N}\mathfrak{D}(\Delta)] \\ &\times[\mathfrak{D}_a(2\Delta)+\mathfrak{D}_b(2\Delta)]\}\exp-i(\nu_1t+\varphi_1)+\tfrac{1}{2}E_2^2E_3[N_2\mathfrak{D}(\omega-\nu_{12})+\bar{N}\mathfrak{D}(\Delta)][\mathfrak{D}_a(\Delta)+\mathfrak{D}_b(\Delta)] \\ &\times\exp-i((2\nu_2-\nu_3)t-(2\varphi_2-\varphi_3))]+\text{complex conjugate.} \end{aligned} \tag{104}$$

The in-phase coefficient $C_1^{(3)}$ is then

$$\begin{aligned} C_1^{(3)}=\tfrac{1}{8}\pi^{1/2}[\wp^4/(\hbar^3Ku)][&E_1^3\bar{N}(\gamma_{ab}/(\gamma_a\gamma_b))(\omega-\nu_1)\mathfrak{L}(\omega-\nu_1)+E_1E_2^2\bar{N}(\gamma_{ab}/(\gamma_a\gamma_b))[(\omega-\nu_{12})\mathfrak{L}(\omega-\nu_{12})+(2/\Delta)] \\ &+E_1E_3^2\bar{N}(\gamma_{ab}/(\gamma_a\gamma_b)[(\omega-\nu_2)\mathfrak{L}(\omega-\nu_2)+(1/\Delta)]+E_2^2E_1N_2\gamma_{ab}(\omega-\nu_1+\Delta)\mathfrak{L}(\omega-\nu_1)/\Delta^2 \\ &+\tfrac{1}{4}E_3^2E_1N_4\gamma_{ab}(\omega-\nu_1+2\Delta)\mathfrak{L}(\omega-\nu_1)/\Delta^2+E_2^2E_3\Delta^{-2}\{N_2\gamma_{ab}(\omega-\nu_{12}+\Delta)\mathfrak{L}(\omega-\nu_{12})\cos\psi \\ &+[N_2(\gamma_{ab}^2-(\omega-\nu_{12})\Delta)\mathfrak{L}(\omega-\nu_{12})-\bar{N}]\sin\psi\}], \end{aligned} \tag{105}$$

where the "relative phase angle" ψ is defined as

$$\psi=(2\nu_2-\nu_1-\nu_3)t+(2\varphi_2-\varphi_1-\varphi_3). \tag{106}$$

The quadrature coefficient $S_1^{(3)}$ is given by

$$\begin{aligned} S_1^{(3)}=\tfrac{1}{8}\pi^{1/2}[\wp^4/(\hbar^3Ku)]\{&E_1^3\bar{N}[\gamma_{ab}^2\mathfrak{L}(\omega-\nu_1)+1]/(\gamma_a\gamma_b)+E_1E_2^2\bar{N}[\gamma_{ab}^2\mathfrak{L}(\omega-\nu_{12})+4\gamma_{ab}^2\Delta^{-2}]/(\gamma_a\gamma_b) \\ &+E_1E_3^2\bar{N}[\gamma_{ab}^2\mathfrak{L}(\omega-\nu_2)+\gamma_{ab}^2\Delta^{-2}]/(\gamma_a\gamma_b)+E_2^2E_1\Delta^{-2}[N_2(\gamma_{ab}^2-(\omega-\nu_1)\Delta)\mathfrak{L}(\omega-\nu_1)-2\bar{N}] \\ &+\tfrac{1}{4}E_3^2E_1\Delta^{-2}[N_4(\gamma_{ab}^2-2(\omega-\nu_1)\Delta)\mathfrak{L}(\omega-\nu_1)-2\bar{N}]+E_2^2E_3\Delta^{-2}[[N_2(\gamma_{ab}^2-(\omega-\nu_{12})\Delta)\mathfrak{L}(\omega-\nu_{12})-\bar{N}]\cos\psi \\ &-N_2\gamma_{ab}(\omega-\nu_{12}+\Delta)\mathfrak{L}(\omega-\nu_{12})\sin\psi]\}. \end{aligned} \tag{107}$$

There are similar expressions for the other coefficients which appear in the self-consistent field equations (13) and (14).

The generalization of Eq. (81) takes the form

$$\begin{aligned} \dot{E}_1&=\alpha_1E_1-\beta_1E_1^3-\theta_{12}E_1E_2^2-\theta_{13}E_1E_3^2-(\eta_{23}\cos\psi+\xi_{23}\sin\psi)E_2^2E_3, \\ \dot{E}_2&=\alpha_2E_2-\beta_2E_2^3-\theta_{21}E_2E_1^2-\theta_{23}E_2E_3^2-(\eta_{13}\cos\psi+\xi_{23}\sin\psi)E_1E_2E_3, \\ \dot{E}_3&=\alpha_3E_3-\beta_3E_3^3-\theta_{31}E_3E_1^2-\theta_{32}E_3E_2^2-(\eta_{21}\cos\psi+\xi_{21}\sin\psi)E_2^2E_1. \end{aligned} \tag{108}$$

The coefficients α_n and β_n were already calculated in the single-frequency case, and are given by Eqs. (82) and (83). The other coefficients are given by

$$\theta_{12}=\tfrac{1}{16}\pi^{1/2}\nu(\wp^4/(\epsilon_0\hbar^3Ku))\{\bar{N}\gamma_{ab}^2(\gamma_a\gamma_b)^{-1}[\mathfrak{L}(\omega-\nu_{12})+4\Delta^{-2}]-2\bar{N}\Delta^{-2}+N_2\Delta^{-2}[\gamma_{ab}^2-(\omega-\nu_1)\Delta]\mathfrak{L}(\omega-\nu_1)\}, \tag{109}$$

$$\theta_{13}=\tfrac{1}{16}\pi^{1/2}\nu(\wp^4/(\epsilon_0\hbar^3Ku))\{\bar{N}\gamma_{ab}^2(\gamma_a\gamma_b)^{-1}[\mathfrak{L}(\omega-\nu_2)+\Delta^{-2}]-\tfrac{1}{2}\bar{N}\Delta^{-2}+\tfrac{1}{4}N_4\Delta^{-2}[\gamma_{ab}^2-2(\omega-\nu_1)\Delta]\mathfrak{L}(\omega-\nu_1)\}, \tag{110}$$

$$\eta_{23}=\tfrac{1}{16}\pi^{1/2}\nu(\wp^4/(\epsilon_0\hbar^3Ku\Delta^2))\{N_2[\gamma_{ab}^2-(\omega-\nu_{12})\Delta]\mathfrak{L}(\omega-\nu_{12})-\bar{N}\}, \tag{111}$$

$$\xi_{23}=-\tfrac{1}{16}\pi^{1/2}\nu\,\wp^4(\epsilon_0\hbar^3Ku\Delta^2)^{-1}N_2\gamma_{ab}(\omega-\nu_{12}+\Delta)\mathfrak{L}(\omega-\nu_{12}), \tag{112}$$

$$\theta_{21}=\tfrac{1}{16}\pi^{1/2}\nu\,\wp^4(\epsilon_0\hbar^3Ku)^{-1}\{\bar{N}\gamma_{ab}^2(\gamma_a\gamma_b)^{-1}[\mathfrak{L}(\omega-\nu_{12})+4\Delta^{-2}]+N_2\Delta^{-2}[\gamma_{ab}^2+(\omega-\nu_2)\Delta]\mathfrak{L}(\omega-\nu_2)-2\bar{N}\Delta^{-2}\}, \tag{113}$$

$$\theta_{23}=\tfrac{1}{16}\pi^{1/2}\nu\,\wp^4(\epsilon_0\hbar^3Ku)^{-1}\{\bar{N}\gamma_{ab}^2(\gamma_a\gamma_b)^{-1}[\mathfrak{L}(\omega-\nu_{23})+4\Delta^{-2}]+N_2\Delta^{-2}[\gamma_{ab}^2-(\omega-\nu_2)\Delta]\mathfrak{L}(\omega-\nu_2)-2\bar{N}\Delta^{-2}\}, \tag{114}$$

$$\xi_{13}=-\tfrac{1}{16}\pi^{1/2}\nu\,\wp^4(\epsilon_0\hbar^3Ku\Delta^2)^{-1}N_2\gamma_{ab}[(\omega-\nu_{23}-\Delta)\mathfrak{L}(\omega-\nu_{23})+(\omega-\nu_{12}+\Delta)\mathfrak{L}(\omega-\nu_{12})], \tag{115}$$

$$\eta_{13}=\tfrac{1}{16}\pi^{1/2}\nu\,\wp^4(\epsilon_0\hbar^3Ku\Delta^2)^{-1}\{N_2[\gamma_{ab}^2+(\omega-\nu_{23})\Delta]\mathfrak{L}(\omega-\nu_{23})+N_2[\gamma_{ab}^2-(\omega-\nu_{12})\Delta]\mathfrak{L}(\omega-\nu_{12})+2\bar{N}\}, \tag{116}$$

$$\theta_{31}=\tfrac{1}{16}\pi^{1/2}\nu\,\wp^4(\epsilon_0\hbar^3Ku)^{-1}\{\bar{N}\gamma_{ab}^2(\gamma_a\gamma_b)^{-1}[\mathfrak{L}(\omega-\nu_2)+\Delta^{-2}]+\tfrac{1}{4}N_4\Delta^{-2}[\gamma_{ab}^2+2(\omega-\nu_3)\Delta]\mathfrak{L}(\omega-\nu_3)-\tfrac{1}{2}\bar{N}\Delta^{-2}\}, \tag{117}$$

$$\theta_{32}=\tfrac{1}{16}\pi^{1/2}\nu\,\wp^4(\epsilon_0\hbar^3Ku)^{-1}\{\bar{N}\gamma_{ab}^2(\gamma_a\gamma_b)^{-1}[\mathfrak{L}(\omega-\nu_{23})+4\Delta^{-2}]+N_2\Delta^{-2}[\gamma_{ab}^2+(\omega-\nu_3)\Delta]\mathfrak{L}(\omega-\nu_3)-2\bar{N}\Delta^{-2}\}, \tag{118}$$

$$\eta_{21}=\tfrac{1}{16}\pi^{1/2}\nu\,\wp^4(\epsilon_0\hbar^3Ku\Delta^2)^{-1}\{N_2[\gamma_{ab}^2+(\omega-\nu_{23})\Delta]\mathfrak{L}(\omega-\nu_{23})-\bar{N}\}, \tag{119}$$

$$\xi_{21}=-\tfrac{1}{16}\pi^{1/2}\nu\,\wp^4(\epsilon_0\hbar^3Ku\Delta^2)^{-1}N_2\gamma_{ab}(\omega-\nu_{23}-\Delta)\mathfrak{L}(\omega-\nu_{23}). \tag{120}$$

The frequency and phase determining equations are of the form

$$\nu_1+\dot{\varphi}_1=\Omega_1+\sigma_1+\rho_1E_1^2+\tau_{12}E_2^2+\tau_{13}E_3^2+E_2^2E_3E_1^{-1}(\eta_{23}\sin\psi-\xi_{23}\cos\psi), \qquad (121)$$

$$\nu_2+\dot{\varphi}_2=\Omega_2+\sigma_2+\rho_2E_2^2+\tau_{21}E_1^2+\tau_{23}E_3^2+E_1E_3(\eta_{13}\sin\psi-\xi_{13}\cos\psi), \qquad (122)$$

$$\nu_3+\dot{\varphi}_3=\Omega_3+\sigma_3+\rho_3E_3^2+\tau_{31}E_1^2+\tau_{32}E_2^2+E_2^2E_1E_3^{-1}(\eta_{21}\sin\psi-\xi_{21}\cos\psi), \qquad (123)$$

where the coefficients σ_n and ρ_n are already given by Eqs. (90) and (91), and the η's and ξ's by Eqs. (111), (112), (115), (116), (119), and (120). The remaining coefficients are

$$\tau_{12}=-\tfrac{1}{16}\pi^{1/2}\nu\wp^4(\epsilon_0\hbar^3Ku)^{-1}\{\bar{N}(\gamma_a\gamma_b)^{-1}[\gamma_{ab}(\omega-\nu_{12})\mathcal{L}(\omega-\nu_{12})+2\gamma_{ab}\Delta^{-1}]+N_2\Delta^{-2}\gamma_{ab}(\omega-\nu_1+\Delta)\mathcal{L}(\omega-\nu_1)\}, \qquad (124)$$

$$\tau_{13}=-\tfrac{1}{16}\pi^{1/2}\nu\wp^4(\epsilon_0\hbar^3Ku)^{-1}\{\bar{N}(\gamma_a\gamma_b)^{-1}[\gamma_{ab}(\omega-\nu_2)\mathcal{L}(\omega-\nu_2)+\gamma_{ab}\Delta^{-1}]+\tfrac{1}{4}N_4\Delta^{-2}\gamma_{ab}(\omega-\nu_1+2\Delta)\mathcal{L}(\omega-\nu_1)\}, \qquad (125)$$

$$\tau_{21}=-\tfrac{1}{16}\pi^{1/2}\nu\wp^4(\epsilon_0\hbar^3Ku)^{-1}\{\bar{N}(\gamma_a\gamma_b)^{-1}[\gamma_{ab}(\omega-\nu_{12})\mathcal{L}(\omega-\nu_{12})-2\gamma_{ab}\Delta^{-1}]+N_2\Delta^{-2}\gamma_{ab}(\omega-\nu_2-\Delta)\mathcal{L}(\omega-\nu_2)\}, \qquad (126)$$

$$\tau_{23}=-\tfrac{1}{16}\pi^{1/2}\nu\wp^4(\epsilon_0\hbar^3Ku)^{-1}\{\bar{N}(\gamma_a\gamma_b)^{-1}[\gamma_{ab}(\omega-\nu_{23})\mathcal{L}(\omega-\nu_{23})+2\gamma_{ab}\Delta^{-1}]+N_2\Delta^{-2}\gamma_{ab}(\omega-\nu_2+\Delta)\mathcal{L}(\omega-\nu_2)\}, \qquad (127)$$

$$\tau_{31}=-\tfrac{1}{16}\pi^{1/2}\nu\wp^4(\epsilon_0\hbar^3Ku)^{-1}\{\bar{N}(\gamma_a\gamma_b)^{-1}[\gamma_{ab}(\omega-\nu_2)\mathcal{L}(\omega-\nu_2)-\gamma_{ab}\Delta^{-1}]+\tfrac{1}{4}N_4\Delta^{-2}\gamma_{ab}(\omega-\nu_3-2\Delta)\mathcal{L}(\omega-\nu_3)\}, \qquad (128)$$

$$\tau_{32}=-\tfrac{1}{16}\pi^{1/2}\nu\wp^4(\epsilon_0\hbar^3Ku)^{-1}\{\bar{N}(\gamma_a\gamma_b)^{-1}[\gamma_{ab}(\omega-\nu_{23})\mathcal{L}(\omega-\nu_{23})-2\gamma_{ab}\Delta^{-1}]+N_2\Delta^{-2}\gamma_{ab}(\omega-\nu_3-\Delta)\mathcal{L}(\omega-\nu_3)\}. \qquad (129)$$

10. TWO-FREQUENCY OPERATION

We may here drop all terms referring to the third frequency ν_3. There are now no "combination tones" in near resonance with Ω_1 and Ω_2. The amplitudes E_1 and E_2 are determined by the differential equations

$$\dot{E}_1=\alpha_1E_1-\beta_1E_1^3-\theta_{12}E_1E_2^2,$$
$$\dot{E}_2=\alpha_2E_2-\theta_{21}E_2E_1^2-\beta_2E_2^3, \qquad (130)$$

where the coefficients α_n, β_n, θ_{12}, and θ_{21} are now given by Eqs. (82), (83), (109), and (113).

Introducing the squared amplitudes

$$X=E_1^2 \quad\text{and}\quad Y=E_2^2, \qquad (131)$$

Eqs. (126) become

$$\dot{X}=2X(\alpha_1-\beta_1X-\theta_{12}Y),$$
$$\dot{Y}=2Y(\alpha_2-\theta_{21}X-\beta_2Y). \qquad (132)$$

The condition for a steady state of oscillation is $\dot{X}=0$, $\dot{Y}=0$ and may represent graphically in an X-Y plane by the point of intersection of the two straight lines

$$L_1:\ \beta_1X+\theta_{12}Y=\alpha_1,$$
$$L_2:\ \theta_{21}X+\beta_2Y=\alpha_2, \qquad (133)$$

if there is one in the first quadrant, together with the single-frequency solutions

$$X=\alpha_1/\beta_1,\quad Y=0 \quad\text{and}\quad X=0,\quad Y=\alpha_2/\beta_2. \qquad (134)$$

The differential equations (132) allow us to follow the temporal behavior of the state X, Y of oscillation in the phase plane X, Y. Through any point in this plane (except stationary points) there passes a curve which indicates the path followed by the representative point (X,Y) on its way to a stable state of oscillation. The parametric equations of the curves are $X=X(t)$, $Y=Y(t)$ with the time t as parameter. From the differential equations it is seen that each curve has a vertical tangent when it crosses the first of the straight lines (133), and a horizontal tangent when it crosses the second straight line. This principle facilitates a very simple, if qualitative, graphical integration of the differential equations (132) for the paths of the phase points in the cases discussed below.

The various possibilities are depicted[18] in Figs. 3–5 where it is assumed that both α_1 and α_2 are positive,

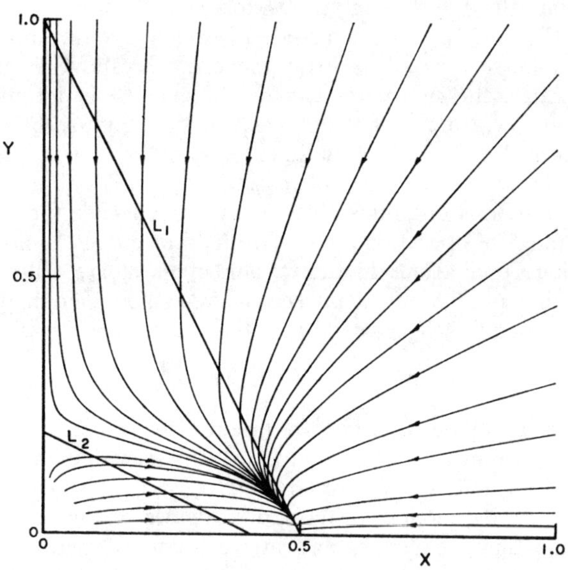

FIG. 3. Phase curves showing the transient behavior of two-mode oscillation. The straight lines L_1 and L_2 of Eq. (133) are taken to have coefficients $\alpha_1=1$, $\alpha_2=0.4$, $\beta_1=\beta_2=2$, $\theta_{12}=\theta_{21}=1$. The slope of a phase curve is zero when it crosses line L_2 and infinite when it crosses line L_1. Although both modes are above threshold, the favored X oscillation is able to quench the Y oscillation.

[18] The phase paths of Figs. 2–4 were kindly integrated on an analog computer by Dr. B. Wise of the Engineering Science Laboratory, Oxford, to whom the author is very indebted.

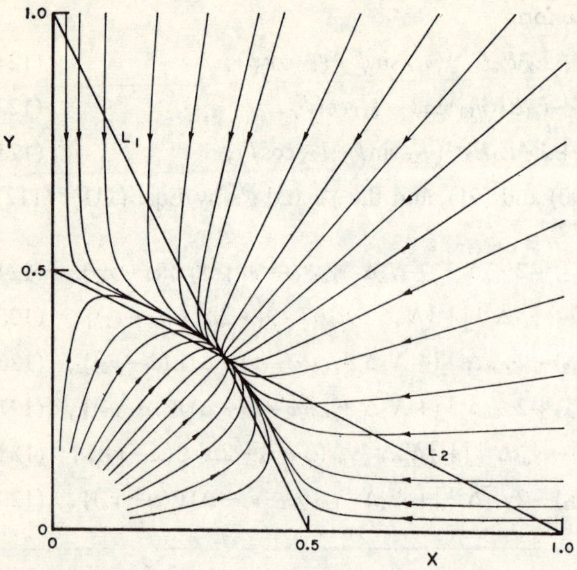

FIG. 4. Diagram similar to Fig. 3, except that the gain parameter for the second mode has been raised to $\alpha_2=1$. Simultaneous oscillations at both frequencies occur at the single stable steady state. Both Figs. 3 and 4 correspond to "weak" coupling.

so that the two modes are individually above threshold. The coefficients β_n are necessarily positive if the optical medium is an active one, and while the mode competition coefficients θ could conceivably have the opposite sign, they have been assumed positive (and equal to each other) in drawing the figures.

Figure 3 applies when mode 1 is well above threshold, but mode 2 is only a little above its threshold, either because the cavity resonance frequency Ω_2 is detuned from the atomic transition frequency ω or possibly because $Q_2<Q_1$. It is clear that the point $(\alpha_1/\beta_1,0)$ represents a stable state of oscillation, while $(0,\alpha_2/\beta_2)$ corresponds to an unstable steady state. Hence there is a range of operation above threshold of modes 1 and 2 where oscillations in the favored mode 1 are able to inhibit oscillations at the second frequency. One might say that the effective gain for the second mode

$$\alpha_2'=\alpha_2-\theta X=\alpha_2-\theta\alpha_1/\beta_1$$

is being made negative by the presence of oscillations at ν_1. From Eqs. (109) and (113) it is seen that the inhibiting effect is enhanced when $\nu_{12}=\omega$, or when the two cavity modes are on opposite sides of the atomic transition frequency, and approximately equally far from it. The physical interpretation of this effect which clearly involves o.f. saturation will be brought out more clearly in Sec. 18.

As the excitation increases, α_2' will eventually become positive, and the relevant diagrams are Figs. 4 and 5. The former applies when $\beta_1\beta_2>\theta^2$ (weak coupling) and the latter when $\beta_1\beta_2<\theta^2$ (strong coupling).

The two cases of weak and strong coupling give very different behaviors. For weak coupling the point of intersection of the two straight lines gives stable steady-state operation, while the single-frequency operating points are unstable. The optical maser oscillates simultaneously at two frequencies under these conditions. For strong coupling, on the other hand, the point of intersection of the two straight lines represents an unstable steady state and would not be realized in practice. All other points in the state diagram evolve into one or the other of the single-frequency operation points. Which of the two is reached depends on the past history of the state of oscillation. In other words, there is hysteresis.

In the Doppler broadened gaseous optical maser Eqs. (83), (109), and (113) indicate that the case of weak coupling is naturally favored, since $\beta_1\beta_2$ tends to be greater than $\theta_{12}\theta_{21}$. Hence, with possible exceptions such as the one discussed in the next section, double frequency operation is preferred. However, when van der Pol's theory of a double resonance feed back triode oscillator is transcribed into our notation, one finds that in his case $\theta=2\beta$. This results from his assumption of a term $(E_1\cos\nu_1t+E_2\cos\nu_2t)^3$ in the triode output current. After discarding terms which have frequencies far from ν_1 and ν_2, this becomes

$$\tfrac{3}{4}(E_1^3+2E_1E_2^2)\cos\nu_1t+\tfrac{3}{4}(E_2^3+2E_2E_1^2)\cos\nu_2t,$$

which leads to the stated relation $\theta=2\beta$. The van der Pol oscillator prefers operation at a single frequency and exhibits hysteresis phenomena, as in the case of strong coupling. Evidently the atomic medium of an optical

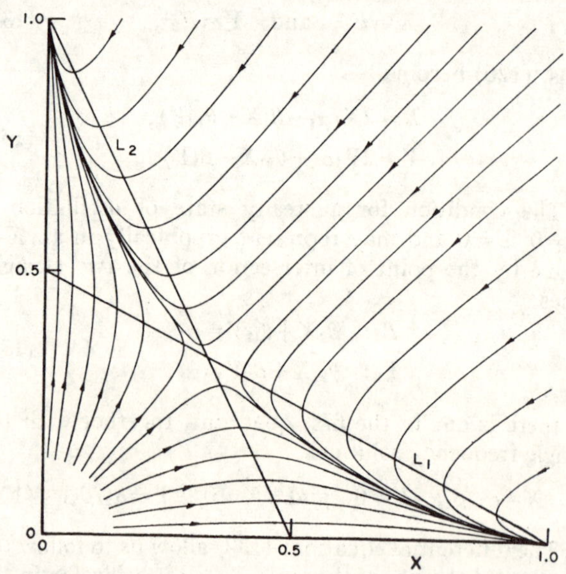

FIG. 5. Phase curves showing the transient behavior of two-mode oscillation when the straight lines L_1 and L_2 of Eq. (133) are taken to have the coefficients $\alpha_1=\alpha_2=1$, $\beta_1=\beta_2=1$, $\theta_{12}=\theta_{21}=2$ (strong coupling). There are two possible stable steady states, each corresponding to single-frequency operation. The particular state reached depends on the initial conditions. Hysteresis phenomena would occur if the parameters characterizing the oscillator were slowly changed.

maser differs from van der Pol's triode oscillator because the nonlinear response of the atoms to frequencies ν_1 and ν_2 has a resonant character not assumed of the triode. It will become clearer from the discussion of Sec. 18 that two groups of atoms with different velocity are driving the two oscillations with only a limited degree of interference. One could easily make a comparable model of the van der Pol type with *two* triodes each with its own tank circuit. A small amount of coupling between the two oscillators would produce intermode effects described by the θ coefficients in the weak-coupling case.

There is also a further case in which the two straight lines coincide, i.e., when $\alpha_1/\alpha_2=\beta_1/\theta=\theta/\beta_2$. In that case, there is a neutral steady state for the representative point lying anywhere on the line within the first quadrant. In practice, the state of operation when this condition is nearly satisfied should be very sensitive to microphonic disturbances.

11. INTENSITIES AND FREQUENCIES IN TWO-MODE OPERATION

The two-mode steady-state solution of (130) is given by

$$E_1^2=(\beta_2\alpha_1-\theta_{12}\alpha_2)/(\beta_1\beta_2-\theta_{12}\theta_{21}),\quad E_2^2=(\beta_1\alpha_2-\theta_{21}\alpha_1)/(\beta_1\beta_2-\theta_{12}\theta_{21}),\tag{135}$$

where, as explained before, the right-hand sides may be evaluated for $\nu_1=\Omega_1$ and $\nu_2=\Omega_2$ without appreciable error. The frequencies are obtained by dropping inapplicable terms from Eqs. (121)–(123). We may set $\dot\varphi_1=\dot\varphi_2$ without loss of generality, and find

$$\nu_1=\Omega_1+\sigma_1+\rho_1E_1^2+\tau_{12}E_2^2,\quad \nu_2=\Omega_2+\sigma_2+\rho_2E_2^2+\tau_{21}E_1^2,\tag{136}$$

where the right-hand sides are to be evaluated for $\nu_n=\Omega_n$, and the E_n^2 are as given in (135).

These equations are fairly complicated, and probably can be used in full generality only for a numerical analysis[19] of very detailed data on optical maser operation. Such a study would be simplified if the values of the E's could be inferred experimentally, since that would effectively reduce the dependence of Eqs. (136) on cavity tuning. For the present we will merely work out the frequencies for the important case of "midtuning" where $\nu_{12}=\frac{1}{2}(\nu_1+\nu_2)=\omega$, $\omega-\nu_1=\frac{1}{2}\Delta$ and $\nu_2-\omega=\frac{1}{2}\Delta$. If we regard Δ as being much greater than the γ's, we then may approximate the coefficients appearing in Eqs. (135)–(136) as follows:

$$\beta_1\simeq\beta_2\simeq\theta_{12}\simeq\theta_{21}\simeq\tfrac{1}{16}\pi^{1/2}\wp^4\bar N(\epsilon_0\hbar^3Ku\gamma_a\gamma_b)^{-1}=\beta,\tag{137}$$

$$\rho_2\approx-\rho_1=2\gamma_{ab}\Delta^{-1}\beta=\rho,\tag{138}$$

$$\tau_{12}\approx-\tau_{21}\approx-\rho,\tag{139}$$

[19] Such an analysis is being made by Dr. R. L. Fork and Dr. M. A. Pollack. The author is very grateful to them for helpful discussions on this and other parts of the manuscript.

and assuming $Q_1=Q_2=Q$, the gain parameters become

$$\alpha_1\approx\alpha_2\simeq\tfrac{1}{2}\nu Q_n^{-1}\{\mathfrak{N}[Z_i(\tfrac{1}{2}\Delta)/Z_i(0)]-1\}=\alpha.\tag{140}$$

The intensities are $E_1^2=E_2^2=\frac{1}{3}\alpha/\beta$, and the frequencies are

$$\nu_1=\Omega_1+\sigma_1-\rho\alpha/\beta,\quad \nu_2=\Omega_2+\sigma_2+\rho\alpha/\beta,\tag{141}$$

so that the beat frequency is

$$\nu_2-\nu_1\simeq\Delta+(\sigma_2-\sigma_1)+4\gamma_{ab}\alpha/\Delta.\tag{142}$$

It will be seen that in the approximations of Eq. (137) we have $\beta_1\beta_2=\theta_{12}\theta_{21}$, and hence in order to decide whether the coupling is "weak" or "strong" it is necessary to evaluate the coefficients β and θ more exactly. In case of exact midtuning, one finds that $\beta^2-\theta^2$ has the same sign as $\bar N+N_2$. Ordinarily this would be positive since $\bar N$ must be positive for oscillations to occur. However, it is possible in principle to arrange to have $\bar N>0$ and $N_2<-\bar N$ by having $N(z)>0$ in the middle two quarters of the tube length, and $N(z)<0$ in the end quarters. In practice the last requirement could be met by adjusting the gas discharge conditions near the ends so that the lower maser atomic level is more populated than the upper one. In He-Ne masers an increase of tube diameter near the ends might be helpful in this respect. Diffusion of Ne metastables to the walls is thereby reduced, and electron excitation of the lower maser level is increased.

12. NORMAL THREE-FREQUENCY OPERATION

Equations (108) and (121)–(123) are fairly complicated, but can be readily used to discuss a number of special cases, as indicated in the following three sections.

In general, unless care is taken to adjust the cavity tuning very accurately, the three frequencies ν_1, ν_2, and ν_3 will be such that the relative phase angle ψ of Eq. (106) is a linear function of the time. Then the last terms in Eqs. (108), (121)–(123) are periodic functions of time, and in some approximation their effects average out. If we neglect these terms, we can get a steady-state solution for the intensities E_1^2, E_2^2, E_3^2 from the system of inhomogeneous linear equations

$$\begin{aligned}\alpha_1&=\beta_1E_1^2+\theta_{12}E_2^2+\theta_{13}E_3^2,\\ \alpha_2&=\beta_2E_2^2+\theta_{21}E_1^2+\theta_{23}E_3^2,\\ \alpha_3&=\beta_3E_3^2+\theta_{31}E_1^2+\theta_{32}E_2^2,\end{aligned}\tag{143}$$

and for the frequencies ν_1, ν_2, ν_3 from

$$\begin{aligned}\nu_1&=\Omega_1+\sigma_1+\rho_1E_1^2+\tau_{12}E_2^2+\tau_{13}E_3^2,\\ \nu_2&=\Omega_2+\sigma_2+\rho_2E_2^2+\tau_{21}E_1^2+\tau_{23}E_3^2,\\ \nu_3&=\Omega_3+\sigma_3+\rho_3E_3^2+\tau_{31}E_1^2+\tau_{32}E_2^2,\end{aligned}\tag{144}$$

again taking $\dot\varphi_1=\dot\varphi_2=\dot\varphi_3=0$. Since the coefficients in Eqs. (143) and (144) are slowly varying functions of frequency, it will suffice first to determine the E_n^2 from

(143) and then to calculate the ν_n's from (144). The equations are a fairly obvious generalization of those for the two frequency case. It will be noted that operation with $E_1^2 \neq 0$, $E_2^2 \neq 0$ can inhibit normal oscillation at ν_3 until, with increasing excitation,

$$\alpha_3' = \alpha_3 - \theta_{31}E_1^2 - \theta_{32}E_2^2 \tag{145}$$

becomes positive.

Further discussion of pulling, pushing, and mode competition for this case will be left to readers requiring a numerical analysis of their data.

13. COMBINATION TONES

As mentioned in Secs. 7, 9, the third-order polarization $P_n^{(3)}(t)$ of the active medium has constituents which oscillate at all possible frequencies of the form $\nu_\mu - \nu_\rho + \nu_\sigma$. Even for "two"-frequency oscillation, there are additional frequencies $2\nu_2 - \nu_1 \equiv \nu_3'$ and $2\nu_1 - \nu_2 \equiv \nu_0'$ in the polarization which are very close to resonance with the principal cavity modes just above and below the two Ω_1 and Ω_2 of main interest. As a consequence of Maxwell's equations, fields at frequencies ν_3' and ν_0' necessarily exist in the cavity and can appear in the output. Well below the threshold for normal three-frequency oscillation, E_3 (and also E_0) will be much smaller than E_1 and E_2. We also assume $\gamma_{ab} \ll \Delta \ll Ku$ and $\nu_2 \simeq \omega$ so that

$$\eta_{21} \simeq -\tfrac{1}{16}\pi^{1/2}\nu\wp^4(\epsilon_0\hbar^3Ku\Delta^2)^{-1}[\bar{N}+2N_2], \qquad \xi_{21} \simeq 0. \tag{146}$$

For a steady state, the third Eq. (108) then gives

$$E_3 \simeq \eta_{21}E_2^2E_1[\alpha_3 - \theta_{31}E_1^2 - \theta_{32}E_2^2]^{-1}\cos\psi. \tag{147}$$

The relative phase angle ψ is determined by Eq. (123) to have $\sin\psi \simeq 0$, whence $|\cos\psi| \simeq 1$, and

$$E_3 \approx -\tfrac{1}{16}\pi^{1/2}\nu\wp^4(\epsilon_0\hbar^3Ku\Delta^2)^{-1} \times[\alpha_3 - \theta_{31}E_1^2 - \theta_{32}E_2^2]^{-1}|\bar{N}+2N_2|E_2^2E_1. \tag{148}$$

This expression is intended to be used under excitation conditions for which the denominator is negative. When the excitation increases, and the factor involving α_3 turns positive, the neglected nonlinear terms in E_3 would have to be taken into account in order to describe the previously discussed normal three-frequency operation.

In order to obtain the combination tone $(2\nu_2 - \nu_1)$ experimentally, one should adjust the cavity tuning so Ω_2 is slightly above the atomic transition frequency ω, thereby making Ω_1 a little nearer resonance than Ω_3. The excitation should be increased until "two"-frequency operation is obtained, but not yet genuine three-frequency operation.

Under these conditions, the θ's in Eq. (148) contain a factor Δ^{-2} and to simplify the discussion they will now be neglected. Equation (148) then has a factor in its denominator

$$\alpha_3 \approx -\tfrac{1}{2}(\nu/Q_3)/G_3, \tag{149}$$

where the gain factor G_3 is given approximately by

$$G_3 \approx [1 - (\bar{N}/\bar{N}_T)(Z_i(\Delta)/Z_i(0))]^{-1}, \tag{150}$$

if one is not too near to threshold for normal three-frequency operation. Equations (148)–(150), within their domain of validity, indicate that E_3 is smaller than E_1 by a factor

$$E_3/E_1 \approx \tfrac{1}{8}\pi^{1/2}\wp^4E_2^2(\epsilon_0\hbar^3Ku\Delta^2)^{-1}Q_3G_3|\bar{N}+2N_2|. \tag{151}$$

Using Eq. (62) for $n=3$, this can be expressed in a convenient form

$$E_3/E_1 = \tfrac{1}{4}[\tfrac{1}{2}(\wp E_2)^2/(\hbar^2\gamma_a\gamma_b)][(\gamma_a\gamma_b)/\Delta^2]G_3 \times|1+(2N_2/\bar{N})|\exp(\Delta/Ku)^2, \tag{152}$$

which shows how the amplitude of the combination tone depends on G_3, and on the saturation parameter as given by Eq. (96) for $n=2$. It should be noted that Eq. (152) could vanish if the spatial distribution of the excitation density is such that $2N_2 + \bar{N} = 0$. (If the excitation is confined to the central region of the Fabry-Perot tube, N_2 and $\bar{N}$, by Eqs. (47), (58) have opposite signs.) An experimental study of the above phenomena might facilitate determination of some of the quantities which enter into our equations but for which direct experimental values are not yet available.

14. FREQUENCY LOCKING PHENOMENA

It has been observed by Javan[20] and by Fork[19] that when the cavity tuning is gradually changed in normal three-frequency operation so that the separation of the beat notes $\nu_2 - \nu_1$ and $\nu_3 - \nu_2$ approaches a small value (typically of order 1 kc/sec), a frequency jump occurs. This phenomenon can be easily understood by reference to Eqs. (121)–(123). For simplicity, we neglect the small frequency pushing associated with the terms involving ρ_{nm} and τ_{nm}, since the nonlinear pulling terms σ_n already give sufficient generality to the frequency relationships. By subtracting the sum of Eqs. (121)–(123) from twice Eq. (122), we find a differential equation for the relative phase angle ψ of Eq. (106) in the form

$$\dot{\psi} = \sigma + A\sin\psi + B\cos\psi, \tag{153}$$

where

$$\sigma = 2\sigma_2 - \sigma_1 - \sigma_3, \tag{154}$$

and A and B are slowly varying quantities which depend on the E_n, ξ_{nm}, and η_{nm}. We evaluate the ξ_{nm} and η_{nm} with the usual approximations $\gamma_{ab} \ll \Delta \ll Ku$. Then

$$\eta_{23} \simeq \eta_{21} \simeq -\tfrac{1}{16}\pi^{1/2}\nu\wp^4(\epsilon_0\hbar^3Ku\Delta^2)^{-1}\{\bar{N}+2N_2\},$$
$$\eta_{13} \simeq \tfrac{1}{8}\pi^{1/2}\nu\wp^4(\epsilon_0\hbar^3Ku\Delta^2)\{\bar{N}-2N_2\}, \tag{155}$$
$$\xi_{23} \simeq \xi_{21} \simeq \xi_{13} \simeq 0.$$

Because of the nearly symmetrical tuning of ν_1 and ν_3

[20] A. Javan, private communications for which the author is very grateful.

we may set $E_1=E_3$ and find

$$A\approx\tfrac{1}{8}\pi^{1/2}\nu\wp^4(\epsilon_0\hbar^3Ku\Delta^2)^{-1}\times\{(\bar{N}+2N_2)E_2^2+2(\bar{N}-2N_2)E_1^2\}, \quad (156)$$

$$B\approx 0.$$

The differential equation

$$\dot{\psi}=\sigma+A\sin\psi+B\cos\psi \quad (157)$$

has an implicit solution

$$t(\psi)=\int_{\psi_0}^{\psi}dx/(\sigma+A\sin x+B\cos x), \quad (158)$$

where ψ_0 is the value of ψ at $t=0$. The character of the function $\psi(t)$ depends critically on whether

$$(A^2+B^2)^{1/2}<|\sigma| \quad \text{or} \quad (A^2+B^2)^{1/2}>|\sigma|.$$

In the first case, the integrand of (158) has no singularities in the range of integration, and one finds that as ψ approaches infinity so does t. Asymptotically, apart from pulsations, one has $\psi\simeq\sigma t+\text{const}$. On insertion of this value in Eqs. (121)–(123) one finds that the frequencies are given by

$$\nu_n\sim\Omega_n+\sigma_n \quad (159)$$

apart from pushing effects and pulsations of phase φ_n, in agreement with the results of Sec. 6 with the approximations made here.

In the second case, $(A^2+B^2)>\sigma^2$, the integral diverges, i.e., $t\rightarrow\infty$ when ψ reaches the value $-\sin^{-1}\times(\sigma/(A^2+B^2)^{1/2}$. In other words, $\psi(t)$ approaches this value asymptotically. The disappearance from ψ of any linear dependence on t forces $\nu_3-\nu_2=\nu_2-\nu_1$ and frequency locking ensues, with a definite relative phase angle $2\varphi_2-\varphi_1-\varphi_3$.

Let us suppose that the maser is in normal three-frequency operation with two distinct beat notes $\nu_3-\nu_2$ and $\nu_2-\nu_1$ near to $\Delta\approx 150$ Mc/sec, i.e., $\sigma=(\nu_3-\nu_2)-(\nu_2-\nu_1)$ is somewhat greater than $(A^2+B^2)^{1/2}$. As the middle cavity frequency Ω_2 is tuned closer to the atomic resonance frequency ω, the separation of beat note frequencies $|\sigma|$ decreases. There should be some pulsations in phase which would increase in amplitude as symmetrical tuning is approached. When $|\sigma|$ reaches $(A^2+B^2)^{1/2}$, a quick transition to the locked state should be made, and only one beat note should be observed. Under the additional simplifying assumption $E_2\gg E_1\approx E_3$, and with use of the starting condition (60) for single-frequency oscillation, the separation of the two beat notes which could be attained just before locking occurs should be given by

$$|\sigma|=\tfrac{1}{8}[(\bar{N}+2N_2)/\bar{N}_T](\wp E_2/\hbar\Delta)^2\nu/Q, \quad (160)$$

which is conveniently expressed as a small fraction of the cavity bandwidth.

It might be pointed out that the above phenomenon is very closely related to one discussed by van der Pol[21] in 1924–27. He considered a self-sustained triode oscillator, capable of oscillation at frequency ν_1. If an external signal at ν is injected into the tank circuit, it may be possible to detect a beat note at $|\nu-\nu_1|$ using a square-law detector. If, however, ν is tuned gradually towards ν_1, a very sudden jump occurs, after which oscillations occur only at ν and the beat note disappears. The width of the "quiet" frequency range depends approximately linearly on the amplitude of the injected signal, when this is small. In the case of the optical maser, we can think that an oscillator at ν_1 is being perturbed by an "external" signal at the combination tone frequency $\nu=2\nu_2-\nu_3$ which arises from the third-order polarization $P_1^{(3)}(t)$ induced in the nonlinear active medium.

15. POPULATION CHANGES AND PULSATIONS

In the absence of o.f. oscillations, the density of atoms in one of the two maser states, say a, can be determined by suitable integrations from $\rho_{aa}^{(0)}(a,z_0,t_0,v,t)$ as given by Eq. (24). When oscillations set in, there are contributions of second order which can be calculated from $\rho_{aa}^{(2)}(\alpha,z_0,t_0,v,t)$ using Eq. (31) for $\alpha=a$ and a similar equation for $\alpha=b$. One has

$$\rho_{aa}(z,t)=\int_{-\infty}^{\infty}dv\int_{-\infty}^{t}dt_0\int dz_0\delta(z-z_0-vt+vt_0)\times\sum_{\alpha=a,b}\lambda_\alpha(z_0,t_0,v)\rho_{aa}(\alpha,z_0,t_0,v,t). \quad (161)$$

It will suffice merely to give the result. One finds, with obvious approximations

$$\rho_{aa}(z,t)=[\Lambda_a(z,t)/\gamma_a]+\{[\Lambda_a(z,t)/\gamma_a]-[\Lambda_b(z,t)/\gamma_b]\}\times\tfrac{1}{8}(\wp^2/Ku)\sum_\mu\sum_\rho\{E_\mu E_\rho\mathfrak{D}_a(\nu_\mu-\nu_\rho)iZ(\omega-\nu_\rho)\times\exp i[(\nu_\mu-\nu_\rho)t+(\varphi_\mu-\varphi_\rho)]+\text{c.c.}\}\times\cos[(n_\mu-n_\rho)\pi z/L] \quad (162)$$

apart from terms with rapid spatial oscillations.

For single-frequency operation, one finds

$$\rho_{aa}(z,t)=[\Lambda_a(z,t)/\gamma_a]-\{[\Lambda_a(z,t)/\gamma_a]-[\Lambda_b(z,t)/\gamma_b]\}\times\tfrac{1}{4}(\wp E_1)^2(\gamma_aKu)^{-1}Z_i(\omega-\nu_1), \quad (163)$$

which contains the lowest order effects of o.f. saturation. In some cases, the density could be monitored by observation of the decay radiation emitted from state a (or b) in transition to some lower level. (Of course, if trapping of resonance radiation were involved, the interpretation would be somewhat complicated.) The change produced by o.f. radiation in $\rho_{aa}(z,t)$ might serve to aid in the determination of parameters like

[21] B. van der Pol, Phil. Mag. **3**, 65 (1927) and the review article cited in Ref. 17.

γ_a, E_1^2, etc., which enter into all our equations, but which might not otherwise be known.

For two-frequency operation of the optical maser, ρ_{aa} has a pulsating constituent at a frequency near Δ besides a dc part. If we assume $\gamma_{ab}\sim\gamma_a\gg\Delta\ll Ku$

$$\begin{aligned}\rho_{aa}(z,t)&\\ =[\Lambda_a(z,t)/\gamma_a]&-\{[\Lambda_a(z,t)/\gamma_a]-[\Lambda_b(z,t)/\gamma_b]\}\\ &\times\tfrac{1}{4}\pi^{1/2}\wp^2(Ku\gamma_a)^{-1}[E_1^2\exp-(\omega-\nu_1)^2/(Ku)^2\\ &+E_2^2\exp-(\omega-\nu_2)^2/(Ku)^2]+\{[\Lambda_a/\gamma_a]-[\Lambda_b/\gamma_b]\}\\ &\times\tfrac{1}{4}\pi^{1/2}\wp^2(Ku\Delta)^{-1}E_1E_2[\exp-(\omega-\nu_1)^2/(Ku)^2\\ &+\exp-(\omega-\nu_2)^2/(Ku)^2]\cos(\pi z/L)\sin(\nu_2-\nu_1)t.\end{aligned}\tag{164}$$

From this expression one sees that the amplitude of the pulsations at frequency near Δ, relative to the dc change in population due to o.f. oscillation, should be

$$2E_1E_2(E_1^2+E_2^2)^{-1}(\gamma_a/\Delta)\cos(\pi z/L),$$

if, for simplicity, we neglect the Gaussian exponentials. Here again, it might be useful to use this phenomenon as a diagnostic tool while undertaking a systematic study of two-frequency operation.

16. CONNECTIONS WITH PREVIOUS CALCULATIONS

The basic paper in this field is, of course, that of Schawlow and Townes[1] who give expressions for threshold equivalent to (62). Townes[22] has also given an equation for linear pulling, as has Javan[23] for nonlinear pulling.

Oscillations of an optical maser involve the propagation of radiation in a nonlinear medium. Several papers have recently appeared which deal with this subject. For various reasons, these do not apply very closely to our particular problem. Thus, Bloembergen *et al.*[24] and Franken and Ward[25] treated harmonic generation which plays a relatively minor role for us. Teng and Statz[26] discussed low-order nonlinearities in a gaseous medium, but, as will be discussed below, their treatment of Doppler broadening is not adequate for our purposes. Also our model for radiation damping is more realistic than theirs which involves just one relaxation time τ, while our equations contain two decay constants γ_a and γ_b. To be sure, the combination $\gamma_{ab}=\frac{1}{2}(\gamma_a+\gamma_b)$ enters most equations, and this might be identified with τ^{-1}.

Among other publications which deal with maser theory are those of Wagner and Birnbaum[27] and of McCumber.[28] These papers consider to some extent the quantum nature of the electromagnetic field. They differ greatly in spirit and content from ours, and we will not attempt to make a comparison here. The work of Haken and Sauermann[29] and of Davis[30] seems much closer to ours; but there are significant differences in the models used, and in the appearance of our equations.

As mentioned in Sec. 4 an earlier calculation[8] applicable to an optical gaseous maser neglected complications arising from the atomic motions and multimode oscillation. It was then possible to work with a density matrix $\rho(\mathbf{r},t)$ characterizing an ensemble[7] of atoms at position $\mathbf{r}$ at time t which were excited at any time $t_0 \lessgtr t$. This obeyed

$$i\dot{\rho}=[\mathcal{H},\rho]-(i/2)(\Gamma\rho+\rho\Gamma)+i\Lambda,\tag{165}$$

which differs from (18) by the term containing a source matrix Λ describing the (slowly varying) rate densities of excitations Λ_a and Λ_b

$$\Lambda=\begin{pmatrix}\Lambda_a & 0\\ 0 & \Lambda_b\end{pmatrix}.\tag{166}$$

(In most applications Λ will be a diagonal matrix.)

It is possible to carry the calculation to higher order in the E_n for multimode oscillation without atomic motion by an iterative procedure in which we begin by neglecting any time dependence in the population difference $\rho_{aa}-\rho_{bb}$. In the rotating wave approximation, one of Eqs. (25) then gives a quasisteady-state solution for $\rho_{ab}(z,t)$

$$\rho_{ab}(z,t)=-\tfrac{1}{2}i(\wp/\hbar)\sum_\mu E_\mu U_\mu(z)\mathfrak{D}(\omega-\nu_\mu)\times(\rho_{aa}-\rho_{bb})\exp-i(\nu_\mu t+\varphi_\mu).\tag{167}$$

Inserting this in another one of Eqs. (25) and again making a rotating wave approximation we find rate equations

$$\begin{aligned}\dot{\rho}_{aa}&=-\gamma_a\rho_{aa}+R(\rho_{bb}-\rho_{aa})+\Lambda_a,\\ \dot{\rho}_{bb}&=-\gamma_b\rho_{bb}+R(\rho_{aa}-\rho_{bb})+\Lambda_b,\end{aligned}\tag{168}$$

where

$$R=\tfrac{1}{4}(\wp/\hbar)^2\sum_\lambda\sum_\mu E_\lambda E_\mu U_\lambda(z)U_\mu(z)\times[\mathfrak{D}(\omega-\nu_\mu)(\exp i(\nu_\lambda-\nu_\mu)t+i(\varphi_\lambda-\varphi_\mu))+\text{c.c.}].\tag{169}$$

The "rate constant" R has pulsations for cases of multifrequency operation, and through Eq. (168) these would lead to pulsations in the populations ρ_{aa} and ρ_{bb} at all beat frequencies $\nu_\lambda-\nu_\mu$. If it were deemed necessary to continue the iteration procedure the pulsating population difference $\rho_{aa}-\rho_{bb}$ could be approximately evalu-

[22] C. H. Townes, *Advances in Quantum Electronics*, edited by J. Singer (Columbia University Press, New York, 1961), pp. 3–11.
[23] A. Javan, E. A. Ballik, and W. L. Bond, J. Opt. Soc. Am. **52**, 96 (1962).
[24] J. A. Armstrong, N. Bloembergen, J. Ducuing, and P. S. Pershan, Phys. Rev. **127**, 1918 (1962).
[25] P. A. Franken and J. F. Ward, Rev. Mod. Phys. **35**, 23 (1963).
[26] C. L. Tang and H. Statz, Phys. Rev. **128**, 1013 (1962).
[27] W. G. Wagner and G. Birnbaum, J. Appl. Phys. **32**, 1185 (1961).
[28] D. E. McCumber, Phys. Rev. **130**, 675 (1963).
[29] H. Haken and H. Sauermann, Z. Physik **173**, 261 (1963); **176**, 47 (1963).
[30] L. W. Davis, Proc. Inst. Elec. Engrs. **51**, 76 (1963).

ated from Eq. (168) and put in Eqs. (25) to obtain an improved Eq. (167).

If we neglect the pulsations in R, we obtain

$$R=\tfrac{1}{2}(\wp/\hbar)^2\gamma_{ab}\sum_{\mu} E_\mu^2[U_\mu(z)]^2\mathcal{L}(\omega-\nu_\mu), \quad (170)$$

which is a plausible generalization of the rate of transitions previously obtained. In a steady state, the population difference implied by (168), (170) is

$$\rho_{aa}-\rho_{bb}=[(\Lambda_a/\gamma_a)-(\Lambda_b/\gamma_b)] \times[1+(2\gamma_{ab}R/(\gamma_a\gamma_b))]^{-1}, \quad (171)$$

which shows clearly the effect of o.f. saturation.

Since the rate constant (170) depends on position z through normal modes $U_n(z)$, the population difference (171) also depends on position, and may be said to have "holes" burned in it. Consequences of this for mode competition can be discussed along the lines followed in Sec. 10. Under most conditions the behavior will correspond to the weak-coupling case, although if the excitation density were such that $-N_2$ were larger than $\bar{N}$ the strong coupling case might be realized.

Combining Eqs. (165), (167), (171), and (41) we obtain a polarization

$$P(z,t)=\wp(\rho_{ab}+\rho_{ab}^*) \quad (172)$$

and the coefficients S_n and C_n which enter the amplitude and frequency determining Eqs. (13) and (14) are

$$S_n=-\frac{2}{L}(\wp^2/\hbar)\gamma_{ab}\mathcal{L}(\omega-\nu_n)E_n \times\int_0^L dz[U_n(z)]^2N(z,t)[1+(2\gamma_{ab}R(z))/\gamma_a\gamma_b)]^{-1} \quad (173)$$

and

$$C_n=[(\omega-\nu_n)/\gamma_{ab}]S_n. \quad (174)$$

These expressions depend nonlinearly on the mode amplitudes E_n because of saturation. However, since we have already neglected the beat frequency pulsations of $\rho_{aa}-\rho_{bb}$ which lead to terms with Δ in the denominator, it might not be consistent to keep any terms in R which are off-resonance by more than about $|\omega-\nu_\mu|\sim(\gamma_{ab}\Delta)^{1/2}$.

We will now consider only single-frequency operation, for which our Eqs. (173)–(174) are essentially exact. Then a single summand $\mu=n$ contributes to R and we find

$$S_n=-\frac{2}{L}\wp^2\hbar^{-1}\gamma_{ab}E_n\int_0^L dz[U_n(z)]^2N(z)[\gamma_{ab}^2+(\omega-\nu_n)^2 +\gamma_{ab}^2(\wp E_n)^2[U_n(z)]^2/(\hbar^2\gamma_a\gamma_b)]^{-1}. \quad (175)$$

The integral may be easily calculated if $N(z)$ is a slowly varying function of position, and we find

$$S_n=-\wp^2\hbar^{-1}\gamma_{ab}\bar{N}\mathcal{L}(\omega-\nu_n)f(w)E_n, \quad (176)$$

where

$$w=[(\wp E_n)^2/(\hbar^2\gamma_a\gamma_b)]\gamma_{ab}^2\mathcal{L}(\omega-\nu_n) \quad (177)$$

and

$$f(w)=(2/w)[1-(1+w)^{-1/2}]. \quad (178)$$

If Eq. (175) for S_n is expanded to third order in E_n, we obtain a result in agreement with that given by Eqs. (56), (72) for the single-frequency case with no atomic motion. It should be noted that whether (175) is expanded or not, the amplitude of oscillation has a maximum for resonant tuning and falls off monotonically with detuning. There is no indication here that the double maximum of Sec. 8 might be a spurious one which arises from the neglect of fifth or higher order terms.

17. DISCUSSION OF DOPPLER BROADENING

The effect of atomic motion upon our equations is rather curious and warrants discussion which, for simplicity, will be given for single-frequency operation. As we have seen in Sec. 16, the optical properties of the medium may be described by a nonlinear susceptibility

$$\chi(\omega-\nu_n, E_n)=P_n/(\epsilon_0E_n). \quad (179)$$

It would perhaps be plausible to hope that the following simple recipe would take atomic motion into account. Because an atom moving with velocity v sees a Doppler shifted frequency, in the laboratory frame of reference it effectively has its resonance frequency shifted by $\omega v/c$. The effective susceptibility ought then to be

$$\chi_{\text{eff}}=\int_{-\infty}^{\infty} dvW(v)\chi(\omega-\omega(v/c)-\nu_n, E_n), \quad (180)$$

which can easily be expressed in terms of the $Z(\zeta)$ function which is discussed in Sec. 6. The effective damping constant becomes

$$\gamma_{ab}[1+(\wp E_n)^2/(\hbar^2\gamma_a\gamma_b)]^{1/2}$$

instead of γ_{ab}, and when this is much smaller than the Doppler parameter Ku, the line shape should be Gaussian with a normal Doppler width.

The above prescription is incorrect except to the first order in E_n if standing waves rather than running waves are involved in any way. This follows from a study of Eq. (70) contributing to the third-order polarization which involves a threefold integration over times τ', τ'', and τ'''. For single-frequency oscillation, the integrand contains

$$U_n(z-v\tau')U_n(z-v\tau'-v\tau'')U_n(z-v\tau'-v\tau''-v\tau''') \times\exp[-i(\omega-\nu_n)\tau']\exp[\pm i(\omega-\nu_n)\tau'''].$$

Each of the U_n's is a standing wave which can be expressed as the sum of two running waves like $\exp\pm i(Kz-v\tau')$, etc. The physical interpretation is that in order to contribute to $P_n^{(3)}$ at time t an atom has to interact three times with the o.f. field: first at $t'''=t-\tau'-\tau''-\tau'''$, then at $t''=t-\tau'-\tau''$, and finally at $t'=t-\tau'$. At each of these interactions the atom has a "choice" of interacting with one or the other of the two running waves. The τ dependence of a typical term of the integrand of (71) is then

$$\exp-\gamma_{ab}(\tau'+\tau''')-\gamma_a\tau'' \exp i(\nu_n-\omega)(\tau'\pm\tau''')\pm iKv\tau' \pm iKv(\tau'+\tau'')\pm iKv(\tau'+\tau''+\tau''').$$

The physical consequence of the appearance of terms involving $\pm Kv$ in the exponent is that each interaction involves a Doppler shift. Since we are interested in the case of a large Doppler width Ku, such a shift can take a given interaction very far from resonance even if $\omega=\nu_n$. This is expressed by the destructive interference of contributions to the integral at various value of τ', τ'', and τ'''. The interference will be least when the accumulated Doppler phase angle

$$\pm Kv\tau'\pm Kv(\tau'+\tau'')\pm Kv(\tau'+\tau''+\tau''')$$

is zero. In order to obtain a nonvanishing spatial Fourier projection (8) for $P_n^{(3)}$, the choices $\pm$ can not be all alike. The six remaining possibilities for the phase angle are

$$\pm Kv[\tau'+(\tau'+\tau'')-(\tau'+\tau''+\tau''')]=\pm Kv(\tau'-\tau'''),$$
$$\pm Kv[\tau'-(\tau'+\tau'')+(\tau'+\tau''+\tau''')]=\pm Kv(\tau'+\tau'''),$$

and

$$\pm Kv[-\tau'+(\tau'+\tau'')+(\tau'+\tau''+\tau''')]=\pm Kv(\tau'+2\tau''+\tau''').$$

As seen in Sec. 7, only the first two possibilities are able to lead to vanishing interference, and then only when $\tau'=\tau'''$.

Physically, one may say that a dominant type of process involves three interactions: first, one with a right (left) running wave at t''', then one with a left (right) running wave at t'', and finally one with a left (right) running wave at t', with the time intervals obeying $t-t'=t''-t'''$ so that the accumulated Doppler phase angle

$$\pm[Kv(t-t')+Kv(t-t'')-Kv(t-t''')]$$

cancels out at time t.

The above cancellation of Doppler interference would not occur if waves running in only one direction were present. The nonlinear terms are much less broadened and weakened by Doppler effect for a standing wave maser oscillator than would be inferred from a study of nonlinear propagation alone. The double peak in the power as a function of tuning met in Sec. 8 can occur only because β_n (saturation) is not as much Doppler broadened as α_n (linear gain profile).

18. HOLE BURNING

In his discussion of maser action Bennett[31] has made use of the notion of "hole burning." Since it aids the physical understanding of the rather complicated equations, we will now show how this phenomenon is described in the present work. As Bennett's treatment does not bring in the population pulsations of Sec. 15, we will base our discussion on the simplified theory of Sec. 16 in which pulsations of population were neglected.

In Sec. 16 the atoms had zero velocity. It is possible to generalize the discussion for the case of atoms having a velocity distribution $W(v)$ at the cost of further fairly plausible assumptions which are no worse than approximations already made. We deal first with those atoms which have a definite velocity v and which were excited at z_0 at time t_0. The perturbation experienced by such an atom is

$$V(t)=-(\wp/\hbar)\sum_\mu E_\mu U_\mu(z_0+v(t-t_0))\cos(\nu_\mu t+\varphi_\mu) \quad (181)$$

so that instead of seeing fields at frequencies ν_μ the moving atom sees fields at twice as many frequencies, i.e., $\nu_\mu\pm Kv$. The rate concept approach of Sec. 16 can not be used since the atoms characterized by different values of z_0 and t_0 experience different perturbations, i.e., the *phases* are not the same for the various members of the ensemble of atoms arriving at z at time t. It is plausible, however, to estimate the effect of saturation on the population difference $\rho_{aa}(v,t)-\rho_{bb}(v,t)$ by using an equation like (171) with a perturbation $V(t)$ given by (181) but with the terms involving z_0 and t_0 omitted. The replacement for the velocity-dependent rate constant $R(v)$ is then plausibly

$$R(v)=\tfrac{1}{8}\gamma_{ab}(\wp/\hbar)^2\sum_\mu E_\mu^2[\mathcal{L}(\nu_\mu-\omega+Kv) +\mathcal{L}(\nu_\mu-\omega-Kv)]. \quad (182)$$

If $v=0$, this reduces to the space average of Eq. (170). The corresponding population difference for atoms having velocity v is then

$$\rho_{aa}(v,t)-\rho_{bb}(v,t) =W(v)[(\Lambda_a/\gamma_a)-(\Lambda_b/\gamma_b)] \times[1+2\gamma_{ab}R(v)/(\gamma_a\gamma_b)]^{-1}. \quad (183)$$

A plot of (183) against v would show the assumed velocity distribution with "holes" burned into it due to o.f. saturation effects. These holes would be appreciable whenever $R(v)$ became comparable to $\gamma_a\gamma_b/(\gamma_a+\gamma_b)$ which could occur near $Kv=\pm(\nu_\mu-\omega)$, so that *two* holes could be burned for each cavity mode in oscillation. At a resonance, where $\nu_n=\omega$, the two holes for the nth mode would merge and reinforce one another. The holes in the velocity distribution would have been

[31] W. R. Bennett, Jr., Appl. Opt. Suppl. 1, 24–61 (1962), especially pp. 58–59. It should be noted that the holes are in first instance burned in the curve of population difference versus velocity, and only indirectly in a curve of gain versus frequency.

seen in Sec. 15 if the integration over velocities had not been carried out so soon.

With the above approximate expression for $\rho_{aa}(v) - \rho_{bb}(v)$ we may use Eqs. (25) to calculate a pseudo-first-order value of $\rho_{ab}(z,v,t)$ and $P(z,v,t)$ in the manner of Sec. 16. The result will be the same as before, but with a velocity dependent reduction factor

$$[1+2\gamma_{ab}R(v)/(\gamma_a\gamma_b)]^{-1}$$

to express the effect of saturation, where $R(v)$ is given by (182). There are similar reductions in the related functions $S_n(v)$ and $C_n(v)$. The coefficients S_n and C_n which enter the equations of self-consistency [(13),(14)] result from integration of these quantities over velocity. Thus

$$S_n = -\wp^2\hbar^{-1}\bar{N}\gamma_{ab}E_n\int_{-\infty}^{\infty} dvW(v)\mathcal{L}(\omega-\nu_n+Kv) \times[1+2\gamma_{ab}R(v)/(\gamma_a\gamma_b)]^{-1}. \quad (184)$$

For single-mode operation, the rate $R(v)$ in the denominator of (184) contains the two terms given by Eq. (182). If an expansion is made to first order in $R(v)$ the integration over v can be done easily for a Maxwellian velocity distribution in the limit of large Doppler width Ku, and one gets results equivalent to those of Sec. 8. The possible dip in output power as a function of cavity tuning in single-mode operation can be interpreted as a consequence of the merging of the two holes at $Kv=\pm(\nu_n-\omega)\to 0$.

For multimode operation a similar calculation can be made with the complete expression (183) for $R(v)$. This will give the dominant terms of the equations of Sec. 9 for a Maxwellian velocity distribution and large Doppler width. However, the frequency locking terms involving ψ will be missing.

For two-mode oscillation, (184) leads to especially strong mode competition attributable to hole burning when $Kv=\nu_2-\omega=\omega-\nu_1$, i.e., the traveling wave along $+z$ for mode 2 and the traveling wave along $-z$ for mode 1 are both in approximate resonance with the atomic transition frequency for an atom having velocity v. This effect can be correlated with the peak in θ for $\omega=\nu_{12}$ mentioned in Sec. 10.

19. APPROXIMATE HIGHER ORDER THEORY FOR SINGLE MODE OPERATION

It would be possible, but quite tedious, to extend the calculations of the text to fifth and higher order for the single-frequency case. The simpler approximate procedure outlined below may serve in the absence of such calculations. It was mentioned above that an expansion of (184) to first order in $R(v)$ reproduces the equations of single-mode operation correct to third order in E_n. If this expansion is not made one may hope to have equations which are valid for stronger signals. The v integration is complicated, and we will content ourselves here with two special cases (a) $Ku\gg|\nu_n-\omega|\gg\gamma_{ab}$ and (b) $\nu_n=\omega$, $Ku\gg\gamma_{ab}$. In the former case, one finds approximately

$$S_n = -\pi^{1/2}\wp^2\bar{N}E_n/(\hbar Ku)[1+\tfrac{1}{4}(\wp E_n)^2/(\hbar^2\gamma_a\gamma_b)]^{1/2}, \quad (185)$$

while at resonance, when $\nu_n=\omega$,

$$S_n = -\pi^{1/2}\wp^2\bar{N}E_n/(\hbar Ku)[1+\tfrac{1}{2}(\wp E_n)^2/(\hbar^2\gamma_a\gamma_b)]^{1/2}, \quad (186)$$

and the merging of the two holes shows up in a simple manner through the doubling of the term expressing the effects of saturation. It will be remembered that a similar doubling of the coefficient β_n took place in Sec. 8 and was responsible for the possible dip in maser output versus cavity tuning. Although the more general behavior of output versus tuning implied by Eqs. (185)–(186) should be qualitatively correct, it must be remembered that rather uncontrolled approximations have been made in their derivation.

20. EXCITATION OF LOWER MASER LEVEL BY SPONTANEOUS DECAY OF UPPER MASER STATE

It was mentioned in Sec. 6 that the lower maser level could, at least in part, be excited by spontaneous emission from the upper. For the present this complication will be treated only in an approximation in which the rate concept is valid. For simplicity we ignore atomic motion, although the work of Secs. 16 and 19 suggests how this could be allowed for approximately. We write rate equations like (168)

$$\dot\rho_{aa} = -\gamma_a\rho_{aa}+R(\rho_{bb}-\rho_{aa})+\Lambda_a,$$
$$\dot\rho_{bb} = -\gamma_b\rho_{bb}+R(\rho_{aa}-\rho_{bb})+\Lambda_b+f\gamma_a\rho_{aa}, \quad (187)$$

where the extra source term $f\gamma\rho_{aa}$ describes the effects of radiative cascade excitation of b from a assuming that a fraction f of the decays from a are to b. The Λ's describe the uncorrelated excitation of the two levels. In a steady state one finds a population density difference

$$\rho_{aa}-\rho_{bb} = [(\Lambda_a/\gamma_a)(1-f(\gamma_a/\gamma_b))-(\Lambda_b/\gamma_b)] \times[1+R\{\gamma_a(1-f)+\gamma_b\}/(\gamma_a\gamma_b)]^{-1}, \quad (188)$$

which should be compared to Eq. (171). It will be seen that the effect of a nonvanishing branching ratio f is merely to change the unsaturated population difference (obtained for $R=0$), and also to modify the value of R for which a given degree of saturation would be obtained. The saturation parameter of Sec. 8 will be modified in an obvious fashion. Thus if $f=0$, a value of rate $R=\frac{1}{2}\gamma_a\gamma_b/\gamma_{ab}$ would cause 50% saturation, while if $f=1$ the value would be $R=\gamma_a$. It should be recalled that the dominant part of the third-order terms $S_n^{(3)}$ and $C_n^{(3)}$ are direct manifestations of saturation phenomena. At the present state of maser art, the decay constants γ_a and γ_b are not well enough known for the effect of a nonvanishing value of f to be easily seen.

When population pulsations are taking place there will be a correlated time-dependent excitation of the lower level by cascade. It is possible that more interesting consequences than those obtained would result, and it is hoped to explore this possibility in a later paper.

21. OTHER SOURCES OF BROADENING

For some kinds of line broadening, especially in certain solid-state optical masers, one could adopt the recipe proposed in Sec. 17, and rejected for the case of Doppler broadening. If the effect of environment could be described by a distribution function for the atomic resonance frequencies ω, an averaged nonlinear susceptibility could be used. This could also be done for the case of isotopic mixtures of the active atoms in gaseous masers.

Although γ_a and γ_b were introduced into our equations to describe spontaneous radiative decay of the states a and b, it is plausible that such phenomenological decay constants might also describe certain kinds of collision broadening. In that case, the γ's would be functions of the pressure.[32] A more detailed discussion of collision broadening for a gaseous optical maser will be given in another paper.

[32] Evidence for such a dependence has recently been obtained by Javan and Szöke, Ref. 16.

Quantum Theory of an Optical Maser.* I. General Theory

Marlan O. Scully† and Willis E. Lamb, Jr.
Department of Physics, Yale University, New Haven, Connecticut
(Received 9 February 1967)

A quantum statistical analysis of an optical maser is presented in generalization of the recent semiclassical theory of Lamb. Equations of motion for the density matrix of the quantized electromagnetic field are derived. These equations describe the irreversible dynamics of the laser radiation in all regions of operation (above, below, and at threshold). Nonlinearities play an essential role in this problem. The diagonal equations of motion for the radiation are found to have an apparent physical interpretation. At steady state, these equations may be solved via detailed-balance considerations to yield the photon statistical distribution $\rho_{n,n}$. The resulting distribution has a variance which is significantly larger than that for coherent light. The off-diagonal elements of the radiation density matrix describe the effects of phase diffusion in general and provide the spectral profile $|E(\omega)|^2$ as a special case. A detailed discussion of the physics involved in this paper is given in the concluding sections. The theory of the laser adds another example to the short list of solved problems in irreversible quantum statistical mechanics.

I. INTRODUCTION

The theory of an optical maser due to Lamb[1] treats the atoms quantum-mechanically while considering the radiation as a classical electromagnetic field. This theory has provided a basis for understanding a wide range of observed laser phenomena and has been extensively tested by Javan and Szöke,[2] Fork and Pollack,[3] and others. Extensions of the theory to allow for the presence of a magnetic field[4,5] or cavity anisotropy[6] have been made by several authors, and there is no doubt that remarkable fits are being obtained with experimental data. The ring laser has been analyzed by Aronowitz,[7] and by Gyorffy and Lamb,[8] again in good agreement with observations. Various forms of modulation can be discussed, as in the work of Harris.[9] The buildup in time of oscillations from a

* This work was supported in part by the National Aeronautics and Space Administration and in part by the U.S. Air Force Office of Scientific Research. The main results of the paper were reported at the International Conference on the Physics of Quantum Electronics, Puerto Rico, July 1965.

† This paper is based on a thesis submitted by M. Scully to Yale University in partial fulfillment of the requirements for the Ph.D degree.

[1] W. E. Lamb, Jr., Phys. Rev. **134**, A1429 (1964).
[2] A. Szöke and A. Javan, Phys. Rev. Letters **10**, 521 (1963).
[3] R. L. Fork and M. A. Pollack, Phys. Rev. **139**, A1408 (1965).
[4] R. L. Fork and M. Sargent, III, Phys. Rev. **139**, A617 (1965).
[5] M. Sargent, III, W. E. Lamb, Jr., and R. L. Fork (to be published).
[6] W. M. Doyle and M. B. White, Phys. Rev. **147**, 359 (1966).
[7] F. Aronowitz, Phys. Rev. **139**, A635 (1965).
[8] B. L. Gyorffy and W. E. Lamb, Jr. (to be published).
[9] S. E. Harris and R. Targ, Appl. Phys. Letters **5**, 202 (1964); S. E. Harris and O. P. McDuff, *ibid.* **5**, 205 (1964).

Reprinted with permission from *Phys. Rev.*, vol. 159, pp. 208–226, July 10, 1967.

very low level has been investigated by Pariser and Marshall,[10] and satisfactory accord with theory is obtained.

In view of the successes of the semiclassical theory, it may be asked why there is need for a better treatment. One reason is that the foregoing theory implies that laser radiation in an ideal steady state is absolutely monochromatic. To be sure, an actual laser has mechanical and statistical disturbances, and these give rise to a finite radiation band width. The intrinsic line width, expressing the effects of thermal noise, vacuum fluctuation fields, and spontaneous emission, is in any case far too small to detect with present techniques. Still, the proper calculation of such effects has provided a challenging problem in nonequilibrium statistical mechanics. Another defect of the semiclassical theory is that oscillations will not grow spontaneously, but require an initial optical-frequency (o.f.) field from which to start. One would like to know how oscillations can develop from a state with no radiation initially present. Since spontaneous radiation must be involved, it is clear that this kind of question requires the quantum theory of radiation.

Still another problem requiring a fully quantum-mechanical theory is to determine the statistical distribution of the energy stored in the laser cavity, i.e., the "photon" statistics. This information is a prerequisite for a proper discussion of the statistical distribution of photoelectrons[11–14] produced by a laser.

A number of papers have appeared recently dealing with a quantum-mechanical laser. The earliest of these replaced the photon emission and annihilation operators by *c* numbers,[15] or prematurely factored[16] the density matrix, and hence are a disguised form of the semiclassical theory. Extensions of the semiclassical theory to include an injected noise signal have been given.[17–19]

We now turn to an enumeration of the fully quantum-mechanical treatments. One of these has been given by the authors[20] and extended in a recent publication.[21] The present paper is a detailed account of that theory and is the first in a series on the quantum theory of the laser. The treatment will closely parallel that of the semiclassical theory. McCumber,[22] Kemmeny,[23] and Korenman[24] have applied a Green's-function technique to the problem. Lax[25] has also given a treatment of the laser spectrum by postulating quantum noise sources determined from general considerations, and in collaboration with Louisell,[26] has subsequently calculated an equation of motion for the density matrix. Willis[27] has extended his earlier treatment, based on methods due to Bogoliubov. The approach of the Haken school has been generalized[28] to include quantum noise sources. The recent results of Fleck[29] are similar to those presented in Ref. 20.

Before developing the quantum theory, it is desirable to review briefly the semiclassical theory. We are interested in the electromagnetic field in a cavity resonator which for optical frequencies can consist of two plane-parallel mirrors. In the semiclassical theory,[1] it was assumed that a known electromagnetic field $E(z, t)$ was present, which consisted of one or more superposed normal modes of oscillation of the cavity as given by

$$E(z,t)=\sum_n E_n(t)\cos[\nu_n t+\varphi_n(t)]\sin(n\pi z/L). \quad (1)$$

The spatial dependence of the normal modes was taken to be as simple as possible. Each mode of the electromagnetic field was specified by an amplitude $E_n(t)$ and a phase angle $\varphi_n(t)$, which were regarded as slowly varying in an optical period. The frequency of each term was denoted by ν_n. The wave equation for the electric field with a driving force term on the right-hand side involving the electric polarization of the medium $P(t)$, and an Ohmic dissipation proportional to a fictitious conductivity σ, was

$$\mu_0\epsilon_0\partial^2\mathbf{E}/\partial t^2+\mu_0\sigma\partial\mathbf{E}/\partial t+\nabla\times(\nabla\times\mathbf{E})=-\mu_0\partial^2\mathbf{P}/\partial t^2. \quad (2)$$

In the solution of the inhomogeneous wave equation the projection on the cavity modes P_n of the electric polarization $P(z,t)$, and their in-phase and out-of-phase amplitudes $C_n(t)$ and $S_n(t)$ played an important role. One had the relation

$$P_n(t)=C_n(t)\cos\{\nu_n t+\varphi_n(t)\}+S_n(t)\sin\{\nu_n t+\varphi_n(t)\}. \quad (3)$$

[10] B. Pariser and T. C. Marshall, Appl. Phys. Letters **6**, 232 (1965); B. Pariser, thesis, Columbia University, 1965 (unpublished).

[11] A summary of experimental results is given in *Proceedings of the International Conference on the Physics of Quantum Electronics, Puerto Rico 1965*, edited by P. Kelley, B. Lax, and P. Tannenwald (McGraw-Hill Book Company, Inc., New York, 1965). See especially: J. A. Armstrong and A. W. Smith, *ibid.*, p. 701; F. Johnson, T. McLean and E. Pike, *ibid.*, p. 706; C. Freed and H. A. Haus, *ibid.*, p. 715.

[12] C. Freed and H. A. Haus, Phys. Rev. Letters **15**, 943 (1965).

[13] A. W. Smith and J. A. Armstrong, Phys. Rev. Letters **19**, 650 (1966).

[14] F. T. Arecchi, A. Berne, and P. Bulamacchi, Phys. Rev. Letters **16**, 32 (1966).

[15] H. Haken and H. Sauermann, Z. Physik **173**, 261 (1963); **176**, 58 (1963).

[16] C. R. Willis, J. Math. Phys. **5**, 1241 (1964); Ref. 11, p. 769.

[17] W. E. Lamb, Jr., in *Quantum Optics and Electronics*; *Lectures Delivered at Les Houches During the 1964 Session of the Summer School of Theoretical Physics, University of Grenoble*, edited by C. DeWitt, A. Blandin, and C. Cohen-Tannoudji (Gordon and Breach Science Publishers, Inc., New York, 1965).

[18] H. Risken, Z. Physik, **186**, 85 (1965).

[19] R. D. Hempstead and M. Lax, Bull. Am. Phys. Soc. **11**, 111 (1966).

[20] M. Scully, W. E. Lamb, Jr., and M. J. Stephen, Ref. 11, p. 759.

[21] M. Scully and W. E. Lamb, Jr., Phys. Rev. Letters **16**, 853 (1966).

[22] D. E. McCumber, Phys. Rev. **130**, 675 (1963).

[23] G. Kemeny, Phys. Rev. **133**, A69 (1964).

[24] V. Korenman, Phys. Rev. Letters **14**, 293 (1965); Ref. 11, p. 748.

[25] M. Lax, Ref. 11, p. 735.

[26] M. Lax and W. H. Louisell (to be published).

[27] C. R. Willis, Phys. Rev. **147**, 406 (1966).

[28] H. Haken, Z. Physik **190**, 327 (1966); H. Sauermann, *ibid.* **189**, 312 (1966); H. Risken, C. Schmid, and W. Weidlich, Phys. Letters **20**, 489 (1966).

[29] J. A. Fleck, Jr., Phys. Rev. **149**, 322 (1966).

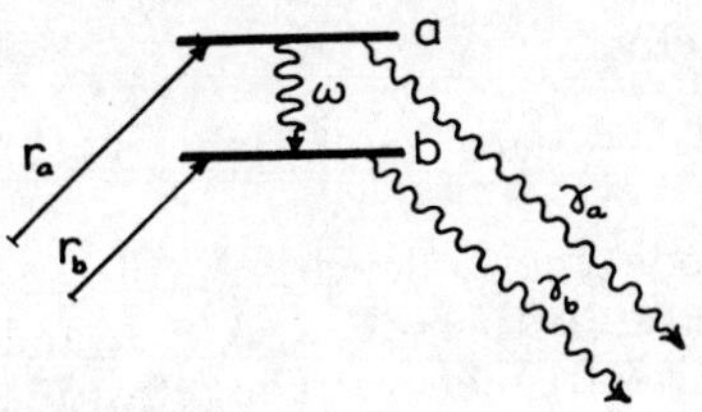

FIG. 1. Maser action takes place between the two excited energy levels a and b separated by a frequency $\omega>0$. These levels are excited at rates r_a and r_b, while the atomic decay constants are given by γ_a and γ_b.

The self-consistency approximation involves calculation of the polarization, i.e., C_n and S_n, of the active medium on the assumption that the electric field E is known, and then substituting that polarization in the right-hand side of the wave equation (2), requiring that the polarization should produce the field which was initially assumed. The result of this requirement is a pair of equations, giving the amplitudes $E_n(t)$ and the frequencies ν_n or phases $\varphi_n(t)$ of each mode of the radiation field:

$$(\nu_n+\dot{\varphi}_n-\Omega_n)E_n=-\tfrac{1}{2}(\nu/\epsilon_0)C_n, \tag{4a}$$

$$\dot{E}_n+\tfrac{1}{2}(\nu/Q_n)E_n=-\tfrac{1}{2}(\nu/\epsilon_0)S_n, \tag{4b}$$

where $\Omega_n=\pi nc/L$ is the cavity resonance frequency and $Q_n=\nu_n\epsilon_0/\sigma_n$ gives the quality factor of the mode.

The next task is to determine the macroscopic driving polarization which is a statistical summation over the microscopic atomic dipoles. The atoms of the active medium are taken to have two excited levels, a and b, separated by a transition frequency ω between which the laser activity is taking place, as in Fig. 1. The levels decay to lower states by radiative decay at rates indicated by γ_a and γ_b. Atoms are excited to these levels by some process such as electron collision from the ground state. Let us imagine that at a time t_0 an atom is brought into state $|a\rangle$ at some point z_0 in the laser cavity. Initially its wave function is ψ_a, but because of the presence of the assumed optical-frequency field, the atom's wave function becomes a linear combination of energy eigenstates, ψ_a and ψ_b. Instead of using wave functions, it is better to work with the elements of a density matrix ρ, which is a more convenient procedure when one wishes to describe a variety of situations in one formalism. The diagonal elements of the 2×2 density matrix are the probabilities of finding the states a and b occupied while the off-diagonal elements are related to the (quantum-mechanically averaged) induced atomic electric-dipole moment at time t of the atom excited at time t_0:

$$p(t,t_0)=\wp[\rho_{a,b}(t,t_0)+\rho_{b,a}(t,t_0)], \tag{5}$$

where $\wp=ex_{ab}$ is the (real) matrix element of the electric-dipole operator connecting states $|a\rangle$ and $|b\rangle$.

The density matrix obeys the differential equation

$$\dot{\rho}=-i[H,\rho]-\tfrac{1}{2}[\Gamma\rho+\rho\Gamma], \tag{6}$$

where

$$\rho=\begin{pmatrix} a^*a & b^*a \\ a^*b & b^*b \end{pmatrix}, \tag{7}$$

$$H=\begin{pmatrix} W_a & V(t) \\ V(t) & W_b \end{pmatrix}, \tag{8}$$

$$\Gamma=\begin{pmatrix} \gamma_a & 0 \\ 0 & \gamma_b \end{pmatrix}, \tag{9}$$

and for the case of stationary atoms

$$\hbar V(t)=-\wp\sum_{n=1}^{M}E_n(t)\sin\left(\frac{n\pi z}{L}\right)\cos[\nu_n t+\varphi_n(t)]. \tag{10}$$

We are interested in a solution $\rho(t,t_0)$ which satisfies a particular initial condition for $t=t_0$, such as

$$\rho(t_0)=\begin{pmatrix} 1 & 0 \\ 0 & 0 \end{pmatrix}. \tag{11}$$

Because of the perturbation $V(t)$, the atom injected at t_0 acquires an electric-dipole moment which decays with a time constant $1/\gamma_{ab}=2/(\gamma_a+\gamma_b)$. In order to calculate adequately the electric polarization, it is necessary to compute the off-diagonal elements of the density matrix to at least third order in the radiative interaction. Having solved for $P(t,t_0)$, we perform a statistical sum over atoms by integration over entrance times t_0.

The differential equation which determines the amplitude E_n as a function of time when only a single mode can oscillate was found to be

$$\dot{E}_n=\alpha_n E_n-\beta_n E_n^3, \tag{12}$$

where the coefficient α_n was given[30] by

$$\alpha_n=-\tfrac{1}{2}(\nu/Q_n)+\tfrac{1}{2}(\nu\wp^2\bar{N}\gamma_{ab}/\hbar\epsilon_0)[(\omega-\nu_n)^2+\gamma_{ab}^2]^{-1} \tag{13}$$

and is the sum of a negative loss term corresponding to the cavity Q and a positive term characterizing the linear pumping. The latter depends on the number density $N(z)$ of excited atoms only through the excitation $\bar{N}$ defined by

$$\bar{N}=\int_0^L N(z)\sin^2\left(\frac{n\pi z}{L}\right)dz \Big/ \int_0^L \sin^2\left(\frac{n\pi z}{L}\right)dz. \tag{14}$$

The parameter β_n is a measure of atomic saturation, which introduces nonlinearities into the problem, and is given[30] by

$$\beta_n=(\tfrac{3}{8}\wp^4\gamma_{ab}^3\nu\bar{N}/\hbar^3\epsilon_0\gamma_a\gamma_b)[(\omega-\nu_n)^2+\gamma_{ab}^2]^{-2}. \tag{15}$$

---

[30] W. E. Lamb, Jr., in *Proceedings of the International School of Physics "Enrico Fermi," Course XXXI*, edited by P. A. Miles (Academic Press Inc., New York, 1964), p. 92, Eqs. (85) and (86).

The steady-state solution of Eq. (12) is clearly

$$E_n^2 = \frac{\alpha_n}{\beta_n} = \frac{[(\text{linear pumping}) - (\text{damping})]}{(\text{nonlinear parameter})}. \quad (16a)$$

Equations (12) and (16a) are basic results of the semiclassical analysis and must, by the correspondence principle, have counterparts in a quantum-mechanical theory of a laser operating in the usual region where huge quantum numbers are involved.

Having set the goal as a quantum treatment paralleling the semiclassical theory, we outline electromagnetic-field quantization in Sec. II, present the model and obtain the equation of motion for the radiation density matrix in Sec. III. In Sec. IV we obtain the steady-state photon statistics $\rho_{n,n}$, while the linewidth analysis is included in Sec. V. Discussion of the physics involved in the paper and a summary will be found in Secs. VI and VII.

II. QUANTIZATION OF THE ELECTROMAGNETIC FIELD

A. Quantum Theory of Radiation

In this section we quantize the radiation field corresponding to a typical laser mode, i.e., a scalar field in a finite one-dimensional cavity. Although there are many textbooks which develop the quantum theory of radiation,[31] they treat the problem for an unbounded region and make use of the vector potential. We are here primarily interested in treating the interaction between the laser radiation and decaying atoms in the electric-dipole approximation. It is unnecessary and risky[32,33] to discuss such a problem using the vector potential, and therefore we prefer to develop the quantum theory of radiation in a form more appropriate for quantum electronics, emphasizing the electric and magnetic fields. Maxwell's equations for a classical free field are

$$\nabla\times\mathbf{H} = \partial\mathbf{D}/\partial t, \quad (16b)$$

$$\nabla\times\mathbf{E} = -\partial\mathbf{B}/\partial t, \quad (16c)$$

$$\nabla\cdot\mathbf{B} = 0, \quad (16d)$$

$$\nabla\cdot\mathbf{E} = 0, \quad (16e)$$

$$\mathbf{B} = \mu_0\mathbf{H}, \qquad \mathbf{D} = \epsilon_0\mathbf{E}. \quad (16f)$$

We take the electric field to be in the x direction and expand in the normal modes of the cavity with an appropriate weighting factor

$$E_x = \sum_s q_s[2\Omega_s^2 M_s/(LA\epsilon_0)]^{1/2} \sin(K_s z), \quad (17)$$

where q_s is the normal mode amplitude with the dimensions of a length, $K_s = s\pi/L$, with $s = 1, 2, 3, \cdots$, and $\Omega_s = s\pi c/L$ the cavity eigenfrequency. The effective transverse area of the optical resonator is denoted by A. The magnetic field in the cavity as implied by Eqs. (17) and (16b) is

$$H_y = \sum_s (\dot{q}_s\epsilon_0/K_s)[2\Omega_s^2 M_s/(LA\epsilon_0)]^{1/2} \cos(K_s z). \quad (18)$$

As is well known, there is an analogy between the dynamical problem of a single mode of the electromagnetic field and that of a mechanical simple harmonic oscillator. We have inserted a quantity M_s into Eqs. (17) and (18) which has the dimensions of a mass in order to emphasize this analogy. The equivalent mechanical oscillator will have a mass M_s and a Cartesian coordinate q_s.

The Hamiltonian for the field

$$H = \tfrac{1}{2}\int d\tau(\epsilon_0 E^2 + \mu_0 H^2) \quad (19)$$

expressed in terms of Eqs. (17) and (18) for E and H becomes

$$H = \tfrac{1}{2}\sum_s [M_s\Omega_s^2 q_s^2 + M_s\dot{q}_s^2], \quad (20)$$

$$H = \tfrac{1}{2}\sum_s [M_s\Omega_s^2 q_s^2 + p_s^2/M_s], \quad (21)$$

where $p_s = M_s\dot{q}_s$ is the canonical momentum of the sth mode. Equation (21) expresses the Hamiltonian for the radiation field as a sum of independent oscillator energies. Each mode of the field is dynamically equivalent to a mechanical harmonic oscillator which is quantized by simply taking over the well-known quantization of the mechanical oscillator

$$[p_s, q_{s'}] = (\hbar/i)\delta_{s,s'}, \quad (22a)$$

$$[q_s, q_{s'}] = [p_s, p_{s'}] = 0. \quad (22b)$$

The nth stationary-state energy of the sth mode of the field is given by

$$E_n = \hbar\Omega_s(n+\tfrac{1}{2}), \quad (23)$$

and the corresponding wave function is[34]

$$\varphi_n(q_s) = (\alpha/\pi^{1/2}2^n n!)^{1/2} H_n(\alpha q_s) \exp(-\tfrac{1}{2}\alpha^2 q_s^2), \quad (24)$$

where $\alpha^2 = (M_s\Omega_s/\hbar)$. It is sometimes convenient to make a canonical transformation to operators a_s and $a_s^\dagger$:

$$a_s = [2M_s\hbar\Omega_s]^{-1/2}(M_s\Omega_s q_s + ip_s), \quad (25a)$$

$$a_s^\dagger = [2M_s\hbar\Omega_s]^{-1/2}(M_s\Omega_s q_s - ip_s). \quad (25b)$$

[31] See, for example, W. Heitler, *The Quantum Theory of Radiation* (Oxford University Press, New York, 1956), 3rd ed.; or W. Louisell, *Radiation and Noise in Quantum Electronics* (McGraw-Hill Book Company, Inc., New York, 1965).

[32] W. E. Lamb, Jr., Phys. Rev. **85**, 259 (1952), especially p. 268.

[33] E. A. Power and S. Zienau, Phil. Trans. Roy. Soc. **251**, 427 (1959).

[34] L. I. Schiff, *Quantum Mechanics* (McGraw-Hill Book Company, Inc., New York, 1955), p. 64.

The Hamiltonian and commutation relations implied are

$$H=\hbar\sum_s(a_s^\dagger a_s+\tfrac{1}{2})\Omega_s, \tag{26}$$

$$[a_s, a_{s'}^\dagger]=\delta_{s,s'}, \tag{27a}$$

$$[a_s, a_{s'}]=[a_s^\dagger, a_{s'}^\dagger]=0. \tag{27b}$$

These operators a_s and $a_s^\dagger$ are the usual annihilation and creation operators for the number states of the sth mode of the radiation field

$$a_s\mid n_s\rangle=n_s^{1/2}\mid n_s-1\rangle, \tag{28a}$$

$$a_s^\dagger\mid n_s\rangle=(n_s+1)^{1/2}\mid n_s+1\rangle. \tag{28b}$$

In terms of these operators, the electric field is

$$E_x=\sum_s \mathcal{E}_s(a_s+a_s^\dagger)\ \sin K_s z, \tag{29}$$

where the quantity

$$\mathcal{E}_s=[\hbar\Omega_s/(LA\epsilon_0)]^{1/2} \tag{30}$$

has the dimension of an electric field.

The states $\mid n\rangle$ are eigenstates of the number operator $a^\dagger a$,

$$a^\dagger a\mid n\rangle=n\mid n\rangle \tag{31}$$

and describe a cavity mode containing exactly n photons. These states have zero average electric field and a mean-square average of

$$\langle[E_x(z)]^2\rangle=\langle n\mid \mathcal{E}_s^2(a^\dagger+a)^2\mid n\rangle\ \sin^2 K_s z \tag{32}$$

$$=2\mathcal{E}_s^2(n+\tfrac{1}{2})\ \sin^2 K_s z. \tag{33}$$

It is the purpose of the next section to investigate more general states of the radiation field, and the electric field calculated from these states. We will be particularly interested in states corresponding to the classical limit of the quantized field.

B. Wave Packets for the Radiation Field

Since the radiation field for a single-cavity mode is dynamically equivalent to the problem of a simple harmonic oscillator, the wave function describing the radiation in the cavity is a linear combination of products of these pure photon eigenstates. For such a state, there would be no definite photon number, but only a distribution of probabilities for finding various numbers of photons if one made an observation of the energy in the cavity. This general state vector for the field is

$$\mid\psi\rangle=\sum_{\{n(s)\}} a_{\{n(s)\}}\mid\{n(s)\}\rangle, \tag{34}$$

where

$$\{n(s)\}=n_1, n_2, \cdots, n_s, \cdots.$$

Concentrating on a single mode of the free field in the q representation,

$$\psi(q, t)=\sum_n a_n(t)\varphi_n(q), \tag{35}$$

$$\psi(q, t)=\sum_n a_n(0)\ \exp(-in\Omega t)\varphi_n(q). \tag{36}$$

This wave packet, in general, has a nonzero average electric field

$$\langle E(z, t)\rangle=\sin(Kz)\sum_n[a_n^*a_{n+1}(n+1)^{1/2}e^{-i\Omega t}+\text{cc}], \tag{37}$$

which has the sinusoidal spatial dependence of a normal mode and a monochromatic time dependence with frequency Ω.

The photon probability distribution is given by $\mid a_n(0)\mid^2$, and the mean photon number is

$$\langle a^\dagger a\rangle=\sum_n na_n^*a_n. \tag{38}$$

The probability amplitudes $a_n(0)$ may be determined from the initial form of the wave function $\psi(q, 0)$ by

$$a_n(0)=\int dq_0\varphi_n(q_0)^*\psi(q_0, 0). \tag{39}$$

We may write the wave function at time t as

$$\psi(q, t)=\int dq_0 G(q, q_0, t)\psi(q_0, 0), \tag{40}$$

i.e., if at time $t=0$, $\psi=\psi(q, 0)$, then the time evolution will be given by folding $\psi(q, 0)$ with the Green's function $G(q, q_0, t)$:

$$G(q, q_0, t)=\sum_n \varphi_n(q_0)^*\varphi_n(q)e^{-in\Omega t}. \tag{41}$$

The physical interpretation of $G(q, q_0, t)$ is that it represents the time development of a wave function which is initially localized as a delta function of q_0. Kennard[35] has given a very ingenious derivation of G, based on the observation that the Green's function is an eigenfunction of the Heisenberg operator $q(-t)$ having the eigenvalue q_0. He found that

$$G(q, q_0, t)=[M\Omega/(2\pi\hbar\mid\sin\Omega t\mid)]^{1/2}$$
$$\times\exp\{iM\Omega[(q^2+q_0^2)\cos\Omega t-2qq_0]/(2\hbar\sin\Omega t)\} \tag{42}$$

develops from a delta function at $t=0$ to a plane wave at $\Omega t=\pi/2$ and back to a delta function at $\Omega t=\pi$, etc. Thus even though the wave packet always returns to its initial state in one period of the oscillator, it has a spread which is a strong function of time. In contrast, however, a wave packet which maintains the same variance while undergoing simple harmonic motion evolves from the ground-state wave function displaced by a distance a:

$$\psi(q, 0)=(\alpha^{1/2}/\pi^{1/4})\ \exp[-\tfrac{1}{2}\alpha^2(q-a)^2]. \tag{43}$$

[35] E. H. Kennard, Z. Physik **44**, 326 (1927).

Then

$$\psi(q,t)=(\alpha^{1/2}/\pi^{1/4})\exp\{-\tfrac{1}{2}i\Omega t-\tfrac{1}{2}\alpha^2[(q-a\cos\Omega t)^2+i(aq\sin\Omega t+\tfrac{1}{2}a^2\sin 2\Omega t)]\},\quad (44)$$

and the probability density is

$$|\psi(q,t)|^2=(\alpha/\pi^{1/2})\exp[-\alpha^2(q-a\cos\Omega t)^2].\quad (45)$$

It may be seen that this packet has the minimum uncertainty product $\Delta q\Delta p=\hbar/2$ allowed by quantum mechanics.

From (17), the average electric field for this wave packet is

$$\langle E\rangle=\sqrt{2}\mathcal{E}\alpha\sin Kz\int_{-\infty}^{\infty}q\,|\psi(q,t)|^2\,dq$$
$$=\sqrt{2}\mathcal{E}\alpha\sin Kz\cos\Omega t.\quad (46)$$

These states provide the closest quantum-mechanical analog for a free classical single-mode field, and are in fact the coherent states $|\alpha\rangle$[36,37]:

$$|\psi\rangle=|\alpha\rangle=\sum_n\{[\alpha\exp(-i\Omega t)]^n/(n!)^{1/2}\}\times\exp(-\tfrac{1}{2}|\alpha|^2)\,|n\rangle.\quad (47)$$

C. Statistical Properties of the Radiation Field

Up to now we have been considering a field which could be represented by a single-state vector $|\psi\rangle$ for which the quantum-mechanical average of an operator Q is

$$\langle Q\rangle=\langle\psi|Q|\psi\rangle.\quad (48)$$

In general, however, we do not know the exact wave function of our system but rather only the probability P_ψ that our system might have this wave function.[38] The ensemble averaged expression for Q is then

$$\langle\langle Q\rangle\rangle_{\text{ensemble}}=\sum_\psi P_\psi\langle\psi|Q|\psi\rangle,\quad (49)$$

which may be written as

$$\langle\langle Q\rangle\rangle_{\text{ensemble}}=\text{Tr}(\rho Q),\quad (50)$$

where

$$\rho=\sum_\psi P_\psi|\psi\rangle\langle\psi|\quad (51)$$

is the weighted projection operator for the states $|\psi\rangle$. This operator ρ represents our state of knowledge or ignorance about the system. In the n representation

[36] For a discussion of the coherent state formalism see R. J. Glauber, in *Quantum Optics and Electronics*; *Lectures Delivered at Les Houches During the 1964 Session of the Summer School of Theoretical Physics, University of Grenoble*, edited by C. DeWitt, A. Blandin, and C. Cohen-Tannoudji (Gordon and Breach Science Publishers, Inc., New York, 1965).

[37] L. Mandel and E. Wolf, Rev. Mod. Phys. **37**, 231 (1965), give an extensive review of the coherence properties of optical fields. See the bibliography of this paper for references to earlier work.

[38] See for example, U. Fano, Rev. Mod. Phys. **29**, 74 (1957); or D. ter Haar, Rept. Progr. Phys. **24**, 305 (1961).

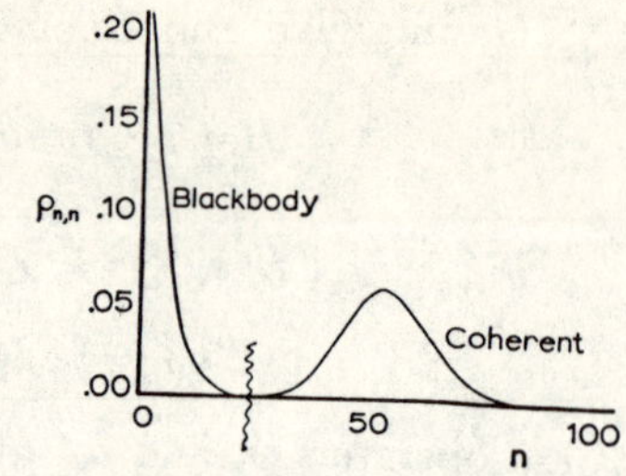

FIG. 2. The photon statistical distribution for single-mode black-body light, Eq. (57), is compared to that for coherent radiation, Eq. (54). The wavy line indicates that two separate curves are shown in the figure.

this operator becomes a matrix $\rho_{n,n'}$ with an infinite number of rows and columns labeled by the integers 1, 2, 3, $\cdots$.

A few pertinent examples of the density matrix for single-mode light are now given. (1) The field might be in a pure number state

$$\rho_{n,n'}=\delta_{n,n'},\quad (52)$$

or (2) a pure coherent state

$$\rho_{n,n'}=\langle n|\alpha\rangle\langle\alpha|n'\rangle,\quad (53)$$

which by (47) is

$$\rho_{n,n'}=\alpha^n\alpha^{*n'}\exp(-|\alpha|^2)/[n!n'!]^{1/2},\quad (54)$$

or (3) a phase-diffused coherent state

$$\rho_{n,n'}=[(\alpha\alpha^*)^n/n!]\exp(-|\alpha|^2)\delta_{n,n'},\quad (55)$$

which has no off-diagonal elements. Neither the pure number state nor the phase-diffused coherent ensemble (nor any ρ diagonal in the n representation) has an average electric field, since the ensemble average field involves $\rho_{n,n+1}$:

$$\langle E\rangle\propto\sum_n[\rho_{n,n+1}(n+1)^{1/2}+\text{cc}].\quad (56)$$

It should be noted that for distributions (3) and (4) the probability for finding n photons $p_n=\rho_{n,n}$ is given by a Poisson distribution characterized by an average n given by $\langle n\rangle=|\alpha|^2$. Another example (4) often met is that of single-mode thermal or black-body radiation of temperature θ:

$$\rho_{n,n'}=\exp(-nh\Omega/k_B\theta)[1-\exp(-h\Omega/k_B\theta)]\delta_{n,n'},\quad (57)$$

which is diagonal in the n representation and therefore contains no phase information, i.e., has zero-ensemble-average electric field.

It will be noted that the probability of finding n photons in the member of the ensemble under consideration, often called the photon statistics, is radically different for black body and for coherent light. Plots of $\rho_{n,n}$ versus n for coherent radiation, Eq. (54), and for incoherent black-body light, Eq. (57) are given in Fig. 2.

III. MODEL AND ANALYSIS

Having developed the quantum theory of radiation in a form suitable for our purposes, we now turn to

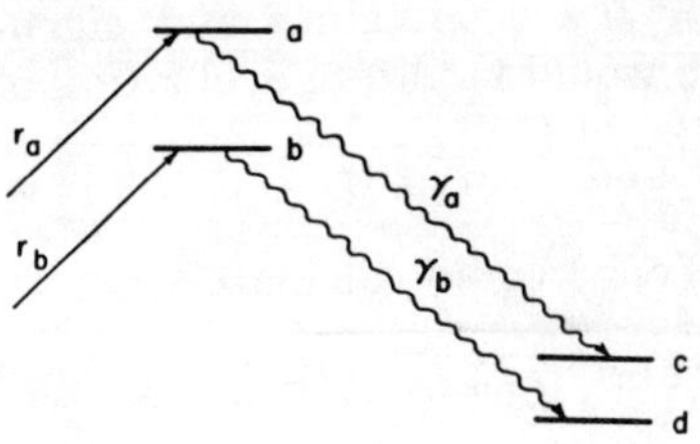

FIG. 3. Atomic-level scheme for atoms. Maser action takes place between levels a and b which are decaying to levels c and d, with decay constants given by γ_a and γ_b, respectively. The corresponding excitations to levels a and b are given by r_a and r_b, respectively.

the fully quantum-mechanical theory of a laser. Both the radiation field and the atomic medium are to be treated according to the laws of quantum mechanics. For simplicity, we consider a gas laser with a single-cavity mode. We neglect the motion of the atoms and spatial variation of the cavity mode. These are nonessential simplifications. The basic idea is the same as in the semiclassical theory. In the earlier work the radiation was described using amplitudes, phases, and frequencies, but now the radiation field has to be characterized in proper quantum-mechanical terminology, i.e., by a density matrix.

To describe laser oscillation, the theory must include a nonlinear active medium and a damping mechanism. To obtain laser pumping action we introduce two-level atoms in their upper state $|a\rangle$ at random times t_0. The more general case of excitation of both the $|a\rangle$ and $|b\rangle$ levels will be dealt with in a later publication. The details of the dissipation mechanism are not very important for the theory of a laser. In the semiclassical theory the damping was represented by Ohmic currents, but it is more convenient for our present purposes to include the dissipation by coupling the electromagnetic field to rapidly decaying, and therefore nonresonant, atoms injected into the cavity in the lower $|\beta\rangle$ of two states $|\alpha\rangle$ and $|\beta\rangle$.

One way of looking at the semiclassical theory is that each atom contributes its mite to the field independently, except insofar as the other atoms have prepared an electromagnetic field with which it interacts. Similarly, in the quantum theory we consider the change in the density matrix for the radiation field due to the injection at time t_0 of a single pumping atom in the upper of the two states $|a\rangle$ and $|b\rangle$ involved in the laser interaction. Working in the n representation, this change is given by

$$\delta\rho_{n,n'}=\rho_{n,n'}(t_0+T)-\rho_{n,n'}(t_0), \tag{58}$$

where T is a time which is long compared with the atomic lifetime, but short compared to the time characterizing the growth or decay of the laser radiation.

The states $|a\rangle$ and $|b\rangle$ of the atom are assumed to decay as in the Wigner–Weisskopf theory of radiation damping. For the state $|a\rangle$, we introduce a state $|c\rangle$ to which the atom decays with the emission of (nonlaser) radiation of type s with a decay constant γ_a. Similarly $|b\rangle$ decays to state $|d\rangle$ with a decay constant γ_b (see Fig. 3).

To obtain $\rho_{n,n'}(t_0+T)$, we must follow the time development of the combined atom-laser field system to time t_0+T and then form the trace of its density matrix over the atomic states

$$\rho_{n,n'}(t_0+T)=\sum_\alpha \rho_{\alpha n,\alpha n'}(t_0+T), \tag{59}$$

where α takes on the values of a, b, c, and d.

Proceeding to calculate $\rho_{n,n'}$ we write the Hamiltonian $\hbar H$, for the interaction of the active atom with the single-mode laser field as

$$H=\nu a^\dagger a+W_a\sigma^\dagger\sigma+W_b\sigma\sigma^\dagger+g(a^\dagger\sigma+a\sigma^\dagger) \tag{60}$$

$$=H_{\rm rad}+H_{\rm atom}+V \tag{61}$$

$$=H_0+V, \tag{62}$$

where ν is the laser frequency to be determined from the theory[39] and $\hbar W_a$ and $\hbar W_b$ are the atomic energies; the raising and lowering operators

$$\sigma^\dagger=\begin{pmatrix}0&1\\0&0\end{pmatrix} \quad\text{and}\quad \sigma=\begin{pmatrix}0&0\\1&0\end{pmatrix} \tag{63}$$

operate on the atomic states

$$|a\rangle=\begin{pmatrix}1\\0\end{pmatrix} \quad\text{and}\quad |b\rangle=\begin{pmatrix}0\\1\end{pmatrix}. \tag{64}$$

The coupling constant is $g=ex_{ab}\mathcal{E}/(\sqrt{2}\hbar)$ which has the dimensions of a frequency, as $\mathcal{E}$ is the electric field (30).

As shown in Appendix I, the Wigner–Weisskopf approximation for our four-level atom interacting with the radiation field yields the following set of equations for the density matrix of the composite system:

$$\dot\rho_{a,n;a,n'}=-i[(H_0+V),\rho]_{a,n;a,n'}-\gamma_a\rho_{a,n;a,n'}, \tag{65a}$$

$$\dot\rho_{b,n+1;b,n'+1}=-i[(H_0+V),\rho]_{b,n+1;b,n'+1}-\gamma_b\rho_{b,n+1;b,n'+1}, \tag{65b}$$

$$\dot\rho_{a,n;b,n'+1}=-i[(H_0+V),\rho]_{a,n;b,n'+1}-\gamma_{ab}\rho_{a,n;b,n'+1}, \tag{65c}$$

$$\dot\rho_{b,n+1;a,n'}=-i[(H_0+V),\rho]_{b,n+1;a,n'}-\gamma_{ab}\rho_{b,n+1;a,n'}, \tag{65d}$$

$$\dot\rho_{c,n;c,n'}=\gamma_a\rho_{a,n;a,n'}, \tag{65e}$$

$$\dot\rho_{d,n+1;d,n'+1}=\gamma_b\rho_{b,n+1;b,n'+1}, \tag{65f}$$

where $\gamma_{ab}=\frac{1}{2}(\gamma_a+\gamma_b)$. We see that Eqs. (65e) and (65f) may be integrated directly to yield

$$\rho_{c,n;c,n'}(t_0+T)=\gamma_a\int_{t_0}^{t_0+T}dt'\rho(t')_{a,n;a,n'}, \tag{66a}$$

$$\rho_{d,n+1;d,n'+1}(t_0+T)=\gamma_b\int_{t_0}^{t_0+T}dt'\rho(t')_{b,n+1;b,n'+1}. \tag{66b}$$

[39] The frequency ν is the frequency of the laser radiation and is determined by the theory. In this paper we are most interested in the case $\omega\neq\nu$; then the laser frequency will be the same as that of the free field. The more general problem of $\omega=\nu$ will be given in a future publication.

Concentrating now on the lasing levels [Eqs. (65a)–(65d)], our equations in expanded form are

$$\dot{\rho}_{a,n;a,n'} = -i[(n-n')\nu - i\gamma_a]\rho_{a,n;a,n'}$$
$$-i[V_{a,n;b,n+1}\rho_{b,n+1;a,n'} - \rho_{a,n;b,n'+1}V_{b,n'+1,a,n'}], \quad (67a)$$

$$\dot{\rho}_{a,n;b,n'+1} = -i[(n-n')\nu + (\omega-\nu) - i\gamma_{ab}]\rho_{a,n;b,n'+1}$$
$$-i[V_{a,n;b,n+1}\rho_{b,n+1;b,n'+1} - \rho_{a,n;a,n'}V_{a,n';b,n'+1}], \quad (67b)$$

$$\dot{\rho}_{b,n+1;a,n'} = -i[(n-n')\nu - (\omega-\nu) - i\gamma_{ab}]\rho_{b,n+1;a,n'}$$
$$-i[V_{b,n+1;a,n}\rho_{a,n;a,n'} - \rho_{b,n+1;b,n'+1}V_{b,n'+1;a,n'}], \quad (67c)$$

$$\dot{\rho}_{b,n+1;b,n'+1} = -i[(n-n')\nu - i\gamma_b]\rho_{b,n+1;b,n'+1}$$
$$-i[V_{b,n+1;a,n}\rho_{a,n;b,n'+1} - \rho_{b,n+1;a,n'}V_{a,n';b,n'+1}]. \quad (67d)$$

The term involving $-i(n-n')\nu$ in these equations will now be transformed away by replacing $\rho_{n,n'}$ by $\rho_{n,n'}\exp[-i(n-n')\nu t]$. It should be kept in mind that subsequently $\rho_{n,n'}$ will be in an interaction picture.

We may write Eqs. (67a)–(67d) as a 2×2 matrix equation

$$\dot{\rho} = -i[C\rho - \rho C'], \quad (68)$$

where

$$\rho = \begin{pmatrix} \rho_{a,n;a,n'} & \rho_{a,n;b,n'+1} \\ \rho_{b,n+1;a,n'} & \rho_{b,n+1;b,n'+1} \end{pmatrix}, \quad (69)$$

$$C = \begin{pmatrix} W_a + n\nu - \tfrac{1}{2}i\gamma_a & V_{a,n;b,n+1} \\ V_{b,n+1;a,n} & W_b + (n+1)\nu - \tfrac{1}{2}i\gamma_b \end{pmatrix}, \quad (70)$$

and the matrix C' is obtained from C by replacing n by n' and taking the Hermitian conjugate.

The solution of (68) is clearly

$$\rho(t) = \exp[-iC(t-t_0)]\rho(t_0)\exp[iC'(t-t_0)]$$
$$= \exp[-iC(t-t_0)]\begin{pmatrix} \rho_{n,n'}(t_0) & 0 \\ 0 & 0 \end{pmatrix}\exp[iC'(t-t_0)]. \quad (71)$$

Next we could solve for the eigenvalues and eigenvectors of C and C' which would facilitate evaluation of

$$\rho_{a,n;a,n'}(t) \quad \text{and} \quad \rho_{b,n;b,n'}(t)$$

and by using (66a) and (66b) could calculate

$$\rho_{c,n;c,n'}(t_0+T) \quad \text{and} \quad \rho_{d,n;d,n'}(t_0+T).$$

Then $\rho_{n,n'}(t_0+T)$ is obtained by contraction with respect to the atomic variables as indicated in Eq. (59):

$$\rho_{n,n'}(t_0+T) = \rho_{a,n;a,n'}(t_0+T) + \rho_{b,n;b,n'}(t_0+T)$$
$$+\rho_{c,n;c,n'}(t_0+T) + \rho_{d,n;d,n'}(t_0+T). \quad (72)$$

Instead of following this approach for obtaining the elements of the density matrix, we adopt another method which has the advantage of side-stepping a considerable portion of the algebraic tedium of the first approach and leads to the same result.

We introduce the notation

$$\int_{t_0}^{t_0+T} \rho_{\beta,\beta'}(t')\,dt' = \sigma_{\beta,\beta'}(t_0+T),$$

$$\beta = 1, 2 \quad \text{and} \quad \beta' = 1, 2,$$

$$\sigma_{1;1} = \sigma_{a,n;a,n'},$$

$$\sigma_{1;2} = \sigma_{a,n;b,n'+1},$$

$$\sigma_{2;1} = \sigma_{b,n+1;a,n'},$$

$$\sigma_{2;2} = \sigma_{b,n+1;b,n'+1}. \quad (73)$$

From Eqs. (66a) and (66b) it is then clear that in the present notation the c- and d-state matrix elements are

$$\rho_{c,n;c,n'}(t_0+T) = \gamma_a\sigma_{1;1}, \quad (74a)$$

$$\rho_{d,n+1;d,n'+1}(t_0+T) = \gamma_b\sigma_{2;2}. \quad (74b)$$

The essence of the simpler approach is the conversion of the differential equations (67) into algebraic equations for σ by integrating both sides from t_0 to t_0+T. We find

$$\rho_{a,n;a,n'}(t_0+T) - \rho_{a,n;a,n'}(t_0) = -\gamma_a\sigma_{11} - i[V_{a,n;b,n+1}\sigma_{21} - \sigma_{12}V_{b,n'+1;a,n'}], \quad (75a)$$

$$\rho_{a,n;b,n'+1}(t_0+T) - \rho_{a,n;b,n'+1}(t_0) = -[i(\omega-\nu) + \gamma_{ab}]\sigma_{12} - i[V_{a,n;b,n+1}\sigma_{22} - \sigma_{11}V_{a,n';b,n'+1}], \quad (75b)$$

$$\rho_{b,n-1;a,n'}(t_0+T) - \rho_{b,n+1;a,n'}(t_0) = -[-i(\omega-\nu) + \gamma_{ab}]\sigma_{21} - i[V_{b,n+1;a,n}\sigma_{11} - \sigma_{22}V_{b,n'+1;a,n'}], \quad (75c)$$

$$\rho_{b,n+1;b,n'+1}(t_0+T) - \rho_{b,n+1;b,n+1}(t_0) = -\gamma_b\sigma_{22} - i[V_{b,n+1;a,n}\sigma_{12} - \sigma_{21}V_{a,n';b,n'+1}]. \quad (75d)$$

As we are interested in times $T \gg 1/\gamma_{ab}$, the first term on the left-hand side of each of Eqs. (75a)–(75d) vanishes. Also we note that

$$\rho_{a,n;a,n'}(t_0) = \rho_{n,n'}(t_0), \quad (76)$$

while the other elements of the density matrix with argument t_0 vanish as the initial excitation is to state $|a\rangle$.

In matrix form, Eqs. (75a)–(75d) are now

$$\begin{pmatrix} -i\gamma_a & -V_{b,n'+1;a,n'} & V_{a,n;b,n+1} & 0 \\ -V_{a,n';b,n'+1} & +(\omega-\nu)-i\gamma_{ab} & 0 & V_{a,n;b,n+1} \\ V_{b,n+1;a,n} & 0 & -(\omega-\nu)-i\gamma_{ab} & -V_{b,n'+1;a,n'} \\ 0 & V_{b,n+1;a,n} & -V_{a,n';n'+1} & -i\gamma_b \end{pmatrix} \begin{pmatrix} \sigma_{11} \\ \sigma_{12} \\ \sigma_{21} \\ \sigma_{22} \end{pmatrix} = \begin{pmatrix} -i\rho_{n,n'}(t_0) \\ 0 \\ 0 \\ 0 \end{pmatrix}. \tag{77}$$

The problem has thus been reduced to solving four simultaneous algebraic equations with four unknowns, which is easily accomplished by matrix techniques. Solving for σ_{11} and σ_{22} we find, with relatively little effort,

$$\gamma_a\sigma_{1;1}=\rho_{nn'}(t_0)-[(n+1)\mathcal{R}_{n,n'}+(n'+1)\mathcal{R}_{n',n}{}^*]\rho_{n,n'}(t_0), \tag{78a}$$

$$\gamma_b\sigma_{2;2}=[\mathcal{R}_{n,n'}+\mathcal{R}_{n',n}{}^*][(n+1)(n'+1)]^{1/2}\rho_{n,n'}(t_0), \tag{78b}$$

where

$$\mathcal{R}_{n,n'}=g^2\frac{\gamma_b(\gamma_{ab}+i\Delta)+g^2(n-n')}{\gamma_a\gamma_b(\gamma_{ab}{}^2+\Delta^2)+2\gamma_{ab}{}^2g^2(n+1+n'+1)+g^2(n'-n)[g^2(n'-n)+i\Delta(\gamma_a-\gamma_b)]} \tag{79}$$

with $\Delta=\omega-\nu$.

We are now in a position to calculate $\rho_{n,n'}(t_0+T)$ as given by Eq. (72). For reasons already mentioned, the first two terms of (72) vanish. Using (74a), and (74b) with $n\rightarrow n-1$ and $n'\rightarrow n'-1$, we have

$$\rho_{n,n'}(t_0+T)=\gamma_a\sigma_{1;1}+\gamma_b\sigma_{2;2}$$
$$(\text{with } n\rightarrow n-1 \text{ and } n'\rightarrow n'-1). \tag{80}$$

Thus we have all the ingredients needed to obtain the change in the radiation-field-density matrix $\delta\rho_{n,n'}$ as given by Eq. (58). From Eqs. (78a), (78b) and (80) we find

$$\delta\rho_{n,n'}=\rho_{n,n'}(t_0+T)-\rho_{n,n'}(t_0)$$
$$=-[(n+1)\mathcal{R}_{n,n'}+(n'+1)\mathcal{R}_{n',n}{}^*]\rho_{n,n'}$$
$$+[\mathcal{R}_{n-1,n'-1}+\mathcal{R}_{n'-1,n-1}{}^*](nn')^{1/2}\rho_{n-1,n'-1}. \tag{81}$$

Just as in the classical theory where we have taken the phase and amplitude of the field to be slowly varying, in the quantum theory the density matrix for the field will not change much due to one atom. Hence we note that for the time interval $t_0<t<t_0+T$, we have $\rho_{n,n'}(t_0)\approx\rho_{n,n'}(t)$. To obtain a macroscopic change in the density matrix, we now multiply (81) by the number of atoms entering the cavity in a time Δt which is long compared to an atomic lifetime but short compared to times characterizing the growth or decay of the radiation field. The number of atoms injected in the upper level in a time Δt is $N_a=r_a\Delta t$, i.e., the rate r_a of injection multiplied by the time Δt. Then the macroscopic change in $\rho_{n,n'}$ due to many atoms acting on the field is

$$\Delta\rho_{n,n'}=r_a\delta\rho_{n,n'}$$
$$=-\Delta t[(n+1)R_{n,n'}+(n'+1)R_{n',n}{}^*]\rho_{n,n'}(t)+\Delta t[R_{n-1,n'-1}+R_{n'-1,n-1}{}^*](nn')^{1/2}\rho_{n-1,n'-1}(t), \tag{82}$$

where $R_{n,n'}=r_a\mathcal{R}_{n,n'}$. The coarse-grained time derivative due to many atoms interacting with the field is then

$$[d\rho_{n,n'}/dt]_{(\text{stimulated and spontaneous emission})}=-[(n+1)R_{n,n'}+(n'+1)R_{n',n}{}^*]\rho_{n,n'}(t)$$
$$+[R_{n-1,n'-1}+R_{n'-1,n-1}{}^*](nn')^{1/2}\rho_{n-1,n'-1}(t). \tag{83}$$

A similar analysis follows for the dissipative interaction, but as mentioned earlier, the details are of secondary interest and we relegate the calculation to Appendix II. We find (to second order in the coupling) from Eq. (II.5)

$$[d\rho_{n,n'}/dt]_{\text{dissipation}}=-\tfrac{1}{2}C(n+n')\rho_{n,n'}$$
$$+C[(n+1)(n'+1)]^{1/2}\rho_{n+1,n'+1}, \tag{84}$$

where the quantity $C=\nu/Q$ is the cavity band width.

Finally, we write the complete equations of motion for the radiation-density matrix (in the interaction picture) as

$$d\rho_{n,n'}/dt=-[(n+1)R_{n,n'}+(n'+1)R_{n',n}{}^*]\rho_{n,n'}$$
$$+[R_{n-1,n'-1}+R_{n'-1,n-1}{}^*](nn')^{1/2}\rho_{n-1,n'-1}$$
$$-\tfrac{1}{2}C(n+n')\rho_{n,n'}+C[(n+1)(n'+1)]^{1/2}\rho_{n+1,n'+1}. \tag{85}$$

Equations (85) are the basic results of this section and provide the quantum equivalent of the classical amplitude and phase equations, with C_n and S_n determined by a self-consistent-field analysis.

IV. DISCUSSION OF EQUATIONS OF MOTION AND PHOTON STATISTICS

It will be noticed that the equations of motion (85) couple only elements of the density matrix having equal degree of off-diagonality $n-n'$, i.e., the coupling is along lines parallel to the main diagonal. Taking advantage of this decoupling, we now investigate the diagonal equations $n=n'$ obtained from (85). For simplicity, in the remainder of this paper we will consider the laser to be tuned to atomic resonance. Detuning and other complications will be discussed in a later paper of this series. These diagonal equations are

$$\dot{\rho}_{n,n}=-A(n+1)[1+(n+1)(B/A)]^{-1}\rho_{n,n}$$
$$+An[1+n(B/A)]^{-1}\rho_{n-1,n-1}$$
$$-Cn\rho_{n,n}+C(n+1)\rho_{n+1,n+1}, \quad (86)$$

where

$$A=2r_a(g^2/\gamma_a\gamma_{ab}), \quad (87)$$

$$B=8r_a(g^2/\gamma_a\gamma_{ab})(g^2/\gamma_a\gamma_b), \quad (88)$$

$$C=\nu/Q. \quad (89)$$

Equations (86) describe the flow of probability for finding n photons in the laser cavity. The separate terms representing the time rates of change of probability have been grouped to make the physical interpretation obvious as depicted in Fig. 4.

These equations for $\rho_{n,n}(t)$ have transient solutions which would describe, for example, the buildup from vacuum to a steady state. We will limit the discussion here to the steady-state solution. By inspection of Fig. 4 it is clear that detailed balance implies that these second-order difference equations reduce to the two equivalent systems of first-order difference equations

$$An[1+(B/A)n]^{-1}\rho_{n-1,n-1}-Cn\rho_{n,n}=0, \quad (90)$$

$$A(n+1)[1+(B/A)(n+1)]^{-1}\rho_{n,n}$$
$$-C(n-1)\rho_{n+1,n+1}=0. \quad (91)$$

The solution of these equations is clearly

$$\rho_{n,n}=\mathfrak{N}\prod_{k=0}^{n}(A/C)[1+(B/A)k]^{-1}, \quad (92)$$

where $\mathfrak{N}$ is a normalization constant.

Let us consider this distribution in three regions of laser operation:

$A>C$ (*above threshold*). The quantity $\rho_{n,n}$ is the product of $n+1$ factors of the form $(A/C)[1+(B/A)k]^{-1}$. For $k<(A/C)(A-C)B^{-1}=n_p$, these factors are each greater than unity, while for $k>n_p$ the factors $(A/C)[1+(B/A)k]^{-1}$ are less than unity; hence $\rho_{n,n}$ increases for n up to n_p and goes monotonically to zero for $n>n_p$. Thus the distribution is peaked at n_p.

$A=C$ (*threshold*). The distribution $\rho_{n,n}$ has a maximum at $n=0$ and decreases in a roughly Gaussian fashion as n increases.

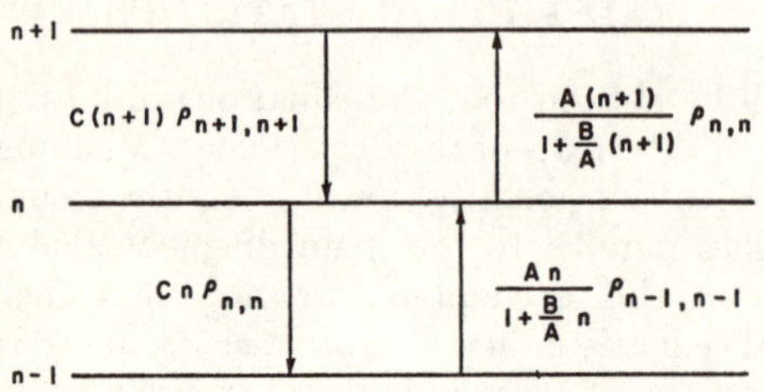

FIG. 4. Flow of probability for finding n photons in the laser cavity due to stimulated emission and damping.

FIG. 5. The laser distribution, Eq. (96), illustrating the three operating regions: (1) 20% below threshold, (2) threshold, and (3) 20% above threshold. Using Eq. (99) the nonlinear parameter B has been chosen to give $\langle n\rangle=50$ at 20% above threshold. The laser distribution (3) should be compared to that for coherent light in Fig. 3.

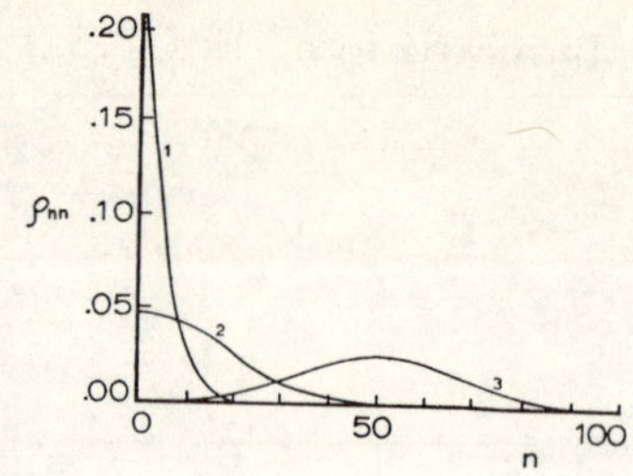

$A<C$ (*below threshold*). Now the distribution falls more rapidly to zero. In this region the nonlinear terms may be ignored and we write

$$\rho_{n,n}=[1-(A/C)](A/C)^n. \quad (93)$$

Hence, below threshold the steady-state solution is essentially that of a black-body cavity

$$\rho_{n,n}=[1-\exp(-\hbar\nu/k_B\theta)]\exp(-n\hbar\nu/k_B\theta), \quad (94)$$

where the effective temperature θ is defined by

$$\exp(-\hbar\nu/k_B\theta)=A/C. \quad (95)$$

The photon distribution in these three regions is displayed in Fig. 5. The steady-state distribution (92) with $\langle n\rangle=10^6$ is compared with a coherent distribution of the same mean value in Fig. 6.

We may write Eq. (92) in a more convenient form as

$$\rho_{nn}=Z^{-1}(A^2/BC)^{n+(A/B)}/[n+(A/B)]!, \quad (96)$$

where the normalization constant Z^{-1} may be expressed in terms of confluent hypergeometric functions

$$Z=\sum_{n=0}^{\infty}\frac{(A^2/BC)^{n+(A/B)}}{[n+(A/B)]!}$$
$$=\left[\frac{(A^2/BC)^{A/B}}{(A/B)!}\right]\left[{}_1F_1\left(1;\frac{A}{B}+1;\frac{A^2}{BC}\right)\right]. \quad (97)$$

Calculating the average value of n, we find

$$\langle n\rangle=Z^{-1}\sum_{n=0}^{\infty}n\frac{(A^2/BC)^{n+(A/B)}}{[n+(A/B)]!}$$
$$=Z^{-1}\sum_{n=1}^{\infty}\left(n+\frac{A}{B}-\frac{A}{B}\right)\frac{(A^2/BC)^{n+A/B}}{(n+A/B)!}$$
$$=Z^{-1}\sum_{n=1}^{\infty}\left[\frac{(A^2/BC)^{n+(A/B)-1}}{(n+A/B-1)!}\right]\frac{A^2}{BC}-\frac{A}{B}(1-\rho_{0,0})$$
$$=[\sum_{n=0}^{\infty}\rho_{n,n}]\frac{A^2}{BC}-\frac{A}{B}(1-\rho_{0,0})$$
$$=(A/C)(A-C)B^{-1}+(A/B)\rho_{0,0}. \quad (98)$$

For a laser appreciably above threshold, the $(A/B)\rho_{0,0}$ term in (98) is clearly insignificant because $\rho_{0,0}\ll 1$, and we have

$$\langle n\rangle=(A/C)(A-C)B^{-1}. \quad (99)$$

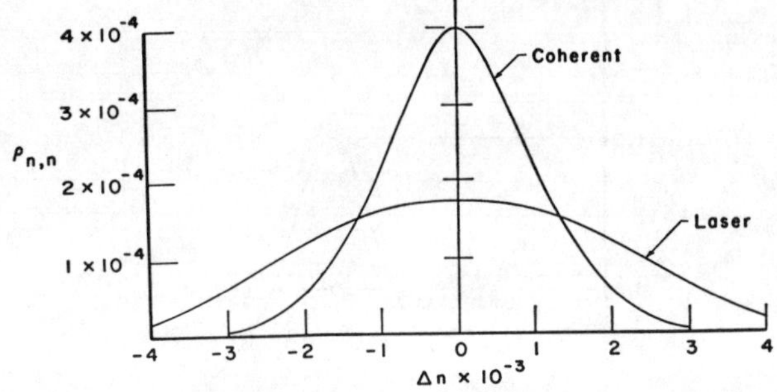

FIG. 6. This figure compares the photon statistics for coherent and laser radiation. The laser is here taken to be 20% above threshold, with the parameter B chosen to give $\langle n\rangle=10^6$.

A similar approximation for the variance of the distribution yields

$$\sigma^2=[A/(A-C)]\langle n\rangle. \quad (100)$$

For a gas laser not too far above threshold, Eqs. (86) may be adequately approximated by retaining only the lowest-order terms in B/A and one finds

$$\dot{\rho}_{n,n}=-[A-B(n+1)](n+1)\rho_{n,n}+[A-Bn]n\rho_{n-1,n-1} -Cn\rho_{n,n}+C(n+1)\rho_{n+1,n+1}. \quad (101)$$

The steady-state solution for (101) is

$$\rho_{n,n}=\mathfrak{N}'\prod_{k=0}^{n}\frac{A-Bk}{C}, \quad (102)$$

where $\mathfrak{N}'$ is the normalization constant. This distribution has a peak at

$$n_p=(A-C)/B. \quad (103)$$

For a sufficiently peaked distribution the average value $\langle n\rangle$ obtained from (103) is

$$\langle n\rangle\approx n_p=\frac{\text{(linear pumping)}-\text{(damping)}}{\text{(nonlinear parameter)}}.$$

We see that the average energy contained in the laser $\langle n\rangle\hbar\nu$ corresponding to this $\langle n\rangle$ has a direct counterpart in Eq. (16a) which expresses the energy $\frac{1}{4}\epsilon_0E^2AL$ of the semiclassical theory.

It will be noted that a consequence of the expansion in (B/A) is that for very large values of n, i.e., $n>A/B$, the distribution $\rho_{n,n}$ goes negative; however, this is well beyond the range of interest, $n=\langle n\rangle\pm O(\langle n\rangle^{1/2})$, and should cause no alarm. Furthermore the difficulty can be avoided by merely letting A/B have an integral value, as this will insure $\rho_{n,n}=0$ for $n>A/B$.

To investigate the electric field of the laser, we must turn to the off-diagonal elements of the density matrix.

V. OFF-DIAGONAL ELEMENTS, CORRELATION TIMES AND SPECTRAL PROFILE

Equations (85) have an infinite number of exponential decaying solutions corresponding to different decay eigenvalues $\mu_s^{(k)}$. These are of the form

$$\rho_{n,n+k}{}^s=\varphi_s(n,k)\exp(-\mu_s^{(k)}t). \quad (104)$$

For the diagonal elements, $k=0$, the lowest eigenvalue $\mu_0^{(0)}=0$ and the corresponding eigenfunction is the steady-state solution (96). For the off-diagonal elements all of the eigenvalues $\mu_s^{(k)}$ are positive. Consequently, the only steady-state solution is

$$\rho_{n,n'}=0, \qquad n\neq n'. \quad (105)$$

It is planned to give a full discussion of these transient solutions (104) in a later paper, but here we will confine our attention to the slowest decay modes for $n\neq n'$. Consequently, for many purposes we may write

$$\rho_{n,n+k}=\varphi_0(n,k)\exp(-\mu_0^{(k)}t). \quad (106)$$

In Appendix III it is shown that the desired eigenfunction for a laser sufficiently above threshold is

$$\varphi_0(n,k)=\left\{\prod_{l=0}^{n}\left[\frac{A-Bl}{C}\right]\prod_{m=0}^{n+k}\left[\frac{A-Bm}{C}\right]\right\}^{1/2} \quad (107)$$

and, to a good approximation, the corresponding eigenvalue is

$$\mu_0^{(k)}=\tfrac{1}{2}k^2D, \quad (108)$$

where

$$D=\tfrac{1}{2}(\nu/Q)\langle n\rangle^{-1}. \quad (109)$$

From (106) and (108) we may then write

$$\rho_{n,n+k}(t)=\rho_{n,n+k}(0)\exp(-\tfrac{1}{2}k^2Dt). \quad (110)$$

The expectation value of the electric field for this density matrix is given by

$$\begin{aligned}E(z,t)&=\mathcal{E}\sin(s\pi z/L)\sum_n(\rho_{n,n+1}(t)(n+1)^{1/2}+\text{c.c.})\\&=\mathcal{E}\sin(s\pi z/L)\sum_n(\rho_{n,n+1}(0)(n+1)^{1/2}e^{-i\nu t}+\text{c.c.})e^{-Dt/2}\\&=E_0\exp[-\tfrac{1}{2}Dt]\cos\nu t. \quad (111)\end{aligned}$$

To obtain the line shape we take the Fourier transform (in the rotating wave approximation) of the average E field.

$$E(\omega)=\int_0^\infty e^{-i\omega t}E_0\exp[-\tfrac{1}{2}Dt]\cos\nu t\,dt \quad (112)$$

$$=E_0[i(\omega-\nu)+\tfrac{1}{2}D]^{-1}, \quad (113)$$

and the spectral profile is

$$|E(\omega)|^2=E_0^2[(\omega-\nu)^2+(\tfrac{1}{2}D)^2]^{-1}. \quad (114)$$

Thus the spectral profile for the laser oscillator is Lorentzian with a width

$$D=\tfrac{1}{4}(\nu/Q)\langle n\rangle^{1/2}, \quad (115)$$

which is the full-width at half-height, in circular frequency units; see Fig. 7. The physical interpretation of this linewidth and the associated decaying electric field will be found in Sec. VI.

A comparison of the present expression for the spectral width D and that derived previously by modifying the semiclassical theory to include noise[17] indicates that we now have twice as much width. From the structure of the calculation, it is apparent that the doubling comes because the present treatment includes the noise due to spontaneous emission of the active atoms as well as thermal and zero-point fluctuations in the cavity walls. The present linewidth is in agreement with the independent results of Korenman[40] and Lax.[41]

VI. DISCUSSION

A. Nature of the Problem

Theoretical physics is most fully developed for treatment of the behavior of isolated conservative dynamical systems. The addition of a given conservative external force field does not present much difficulty, at least in principle. As one attempts to make the discussion of a problem more realistic, the system of interest may be allowed to interact with a thermal reservoir, thereby developing a thermodynamic or statistical mechanical approach. After passage of a sufficiently long time, the system of interest settles down into a state of equilibrium. In cooperative phenomena, the equilibrium state may be one with a sigh degree of long-range spatial order. Such theories predict that statistical fluctuations should occur about the thermodynamic equilibrium state.

An oscillating laser, on the other hand, is clearly not in a state of random fluctuations about thermodynamic equilibrium. It represents an "open" system with a highly organized temporal behavior. The system of interest is in contact with a steady but nonthermal reservoir capable of supplying energy at at a low frequency which is somehow converted into a nearly monochromatic oscillatory behavior at an optical frequency. Such problems push somewhat beyond the range of present-day theoretical physics, and one can make progress only by exploiting some special simplifying feature of the problem. Only much later can one expect to succeed with problems in which the openness permits exchange of both energy and matter, as would be necessary for a basic discussion of a biophysical problem.

B. Model

There are two special features simplifying our model. The first is that the electromagnetic field of a high-Q optical resonator is dynamically equivalent to a system with a single degree of freedom, in this case, a simple harmonic oscillator. The second feature is that under circumstances of interest in laser physics, the density matrix describing the radiation does not change very much during the lifetime of one atom.

Our model for the pumping mechanism is quite realistic for a gaseous optical maser. The excitation of atoms to the laser states $|a\rangle$ or $|b\rangle$ is, as far as the system of interest is concerned, essentially an act of

[40] V. Korenman, Ref. 11, p. 748.

[41] M. Lax, Ref. 11, p. 735.

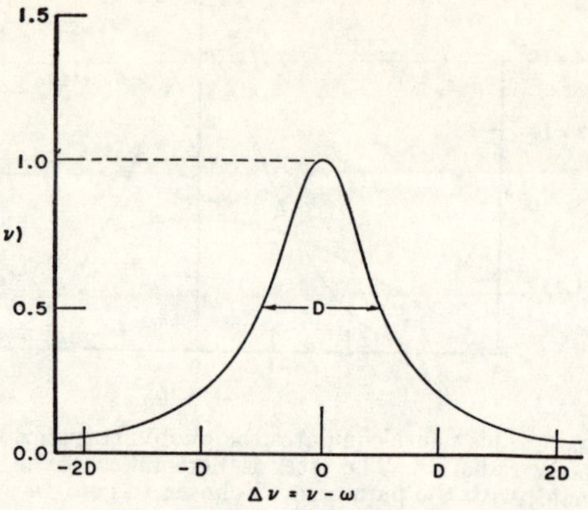

FIG. 7. Spectral profile for a laser oscillator in units of the full-width at half-height D, all frequencies are measured in circular-frequency units.

creation as assumed in the model. In order to simplify the following discussion the case of a excitation will be considered. Assuming that the density matrix describing the radiation field just before the atom is injected is $\rho_{n,n'}(t_0)$, we have calculated the time development of the density matrix for the combined system of atom and field. After several atomic lifetimes, the atom is surely in one of the lower states, c or d, and the only nonvanishing elements of the density matrix are of the form $\rho_{d,n+1;d,n'+1}$ or $\rho_{c,n;c,n'}$. If one then asks for the statistical density matrix describing the radiation field alone, irrespective of whether the atom has ended up in state $|d\rangle$ or $|c\rangle$, i.e., whether the atom did or did not emit a net laser quantum before it decayed, the result is

$$\rho_{n,n'}(t_0+T)=\rho_{c,n;c,n'}(t_0+T)+\rho_{d,n;d,n'}(t_0+T).$$

This process of contraction is an essential feature of the model. A consequence is that even if the radiation field were initially described by a "pure" case density matrix $\rho_{n,n'}(t_0)$, after the injection and decay of one reservoir atom, the density matrix $\rho_{n,n'}(t_0+T)$ would in general be "mixed." One could, of course, in principle, learn more about the combined system by observing whether the atom ended in state $|c\rangle$ or in state $|d\rangle$, but our object is to make a theory describing the system of interest, which is the laser and not the pumping reservoir.

In our theory of a laser, it is not necessary to postulate any noise sources either of a quantum or a classical nature. The noise is automatically produced as a consequence of the contraction process which is basic to the physical problem and follows from the principles of quantum mechanics applied to an open system. There are, however, fluctuations of a shot-effect nature which arise from the injection of pumping and damping atoms. These will be treated in a subsequent paper.

After the change $\delta\rho_{n,n'}$ of the density matrix is calculated for one atom, we pass over to a coarse-grained time derivative due to many atoms. It is thereby implied that $\rho_{n,n'}(t)$ does not change much during the life of any one atom. This may or may not be the case for any given laser. At a steady state, the changes $\delta\rho_{n,n'}$ represent fluctuations which would typically be small fractions of $\rho_{n,n'}$. Our assumption might be less valid in a transient problem. Thus, if initially we had the vacuum radiation field ($\rho_{0,0}=1$, $\rho_{n,n}=0$ for $n\neq 0$), the first atom would change $\rho_{0,0}$ and $\rho_{1,1}$ by small but finite amounts.

The change in $\rho_{n,n'}$ due to other atoms during the life of the atom considered in the calculation of $\rho_{n,n'}$ would involve an approximation in the derivation of Sec. III. A corresponding simplification was made in the semiclassical theory. Thus, in Eq. (37) of Ref. 1, the amplitude $E(t')$ in an integral

$$\int_{-\infty}^{t} dt' E(t') \cdots$$

is replaced by $E(t)$. This assumes that $E(t)$ is slowly varying during an effective atomic lifetime. Except for the earliest stages of buildup from the vacuum state, the validity of our treatment depends on the smallness of the quantity $\{A-(\nu/Q)\}/\gamma_{ab}$, which is 0.01 for the numerical values $A=1.1$ MHz, ($\nu/Q=1.0$ MHz and $\gamma_{ab}=10$ MHz).

C. Approach to Thermal Equilibrium

If we neglect all nonlinear terms involving saturation of the atomic transitions [set $B=0$ in Eq. (91)], our model describes a harmonic-oscillator system of interest in contact with two reservoirs. One of these contains a large number of pumping atoms in the upper laser state $|a\rangle$, and the other contains many damping atoms in the lower state $|\beta\rangle$. One can assign temperatures to each reservoir in the conventional manner. The first reservoir is at $T=-0°$K (very hot) and the second is at $T=0°$K (very cold). Both pumping and damping atoms are separately injected in the optical cavity, and it is assumed that the density matrix for the radiation field is not changed very much by any one atom. In a steady state, the density matrix is

$$\rho_{n,n}\propto (A/C)^n.$$

If $A<C=\nu/Q$, this can be normalized and is a thermodynamic distribution

$$\rho_{n,n}=(A/C)^n[1-(A/C)],$$

corresponding to a temperature θ given by

$$\exp[-(\hbar\nu/k_B\theta)]=A/C\leq 1.$$

The situation is somewhat different from that usually considered, in that the effects of two reservoirs, one hot and one cold, are combined in order to keep the system of interest at an intermediate temperature θ which depends on the strengths A and C of the coupling between the radiation oscillator and the two reservoirs. For the system of interest, the steady-state thermodynamic equilibrium at temperature θ is the same whether one conventional reservoir is involved or the two reservoirs of our model. If the system of interest is not initially in thermal equilibrium, its approach to this state can be determined by solving the equations of motion (91).

This calculation adds another example to a short list of solvable problems where approach to thermal equilibrium is considered in a basic manner. It is similar to the Rayleigh[42] problem of a massive particle sent into a gas of light atoms. Even closer to the laser problem is the generalization of Uhlenbeck and Chang,[43] where a forced simple harmonic oscillator is brought to steady state through collisions with such a gas.

D. The Case of Nonthermal Equilibrium A>C

It is perfectly possible for a two-level system to have a negative temperature if $\rho_{a,a}>\rho_{b,b}$, but it is meaningless for a system, such as a harmonic oscillator, whose energy spectrum has no upper limit to have a negative temperature. Formally, one may try to make θ negative in Eq. (91) with $B=0$ by setting $A>C$, but $\rho_{n,n}$ would then be a steeply increasing function of n, and the photon statistical distribution could not be normalized. To avoid this difficulty, it is necessary to retain the nonlinear B terms in the steady-state solution for $\rho_{n,n}$. Above threshold, the laser photon distribution does rise with increasing n to a peak at $n=n_p$ beyond which saturation effects play an essential role, and bring the distribution down again.

E. Measurement

Quantum electrodynamics is based on a generalization of nonrelativistic quantum mechanics. Despite the analysis given by Bohr and Rosenfeld, it is probably safe to say that the theory of measurement is even less well developed for quantum electrodynamics than for nonrelativistic quantum mechanics.[44] Some of the difficulty, no doubt, arises from the infinite number of degrees of freedom of quantum electrodynamics. In the particular case of a single high-Q cavity mode, however, it seems possible to regard the analogy between a radiation oscillator and a mechanical oscillator as so close that the measurement problems become equivalent. The following discussion is based on this assumption.

The most that can possibly be known about the radiation oscillator at $t=0$ is its wave function, say $\psi(E, 0)$ in the electric field E "coordinate" representation. We assume that this state has been "prepared" somehow. Under the guidance of a definite Hamiltonian, this wave function will evolve into $\psi(E, t)$ at time t. Any Hermitian operator $F(E, -i\hbar\partial/\partial E)$ can, or so the theory contends, be "measured." Each measurement gives as a result one of the eigenvalues F_n of

[42] Lord Rayleigh, Phil. Mag. **32,** 424 (1891).
[43] G. E. Uhlenbeck and C. S. W. Chang, in *Proceedings of the International Synposiun on Transport Processes in Statistical Mechanics,* edited by I. Prigogine (Interscience Publishers, Inc., New York, 1958), pp. 161–168.

the operator F. The probability of finding a particular value F_n when a series of measurements is made on an ensemble of similarly prepared systems is

$$W_n=\left|\int_{-\infty}^{\infty} dE\,\varphi_n(E)^*\psi(E,t)\right|^2,$$

where $\varphi_n(E)$ is the eigenfunction belonging to the eigenvalue F_n of the operator F. The measurements under discussion here are the best permitted. If carried out well, a measurement so disturbs the system of interest that it is pointless to even think of any subsequent measurement of any other operator. The pure case will become a hopeless mixture, even if the system is not physically destroyed. In most physical research, one is not concerned with measurement in this extreme form. Certain scattering experiments are sometimes called measurements, but they do not represent measurement of a Hermitian operator in the strict sense, so we would prefer to call them observations or "bad" measurements. Sometimes, especially when a nearly classical system is under study, one attempts to follow the time development between $t=0$ and $t=t$ by making a series of observations. In our opinion, there is currently no satisfactory theory[44] of "bad" measurements. We do recognize the possibility of "watching" the bob of a pendulum clock swing back and forth. In a similar manner, at least in principle, the temporal oscillations of the intense and highly classical electric field in a laser could be followed by recording on a moving film the deflection of a stream of high-velocity electrons sent across a narrow laser beam.

We have noted in Eq. (109) that the ensemble average of $E(t)$ is a damped oscillating function of time. This damping comes from phase diffusion of the fields for an ensemble of lasers which represents various possible histories of any one laser. An electron-beam probe of any one continuous wave laser would, of course, not show such a dampling, but only a very slight amount of phase irregularity. The average of many similar film records would naturally show the damping phenomenon.

F. Spectrum

In Eq. (112) we have calculated the spectrum associated with the damped oscillating electric field (109) and have found a Lorentzian of full-width at half-maximum just equal to the phase-diffusion constant D. If one had a laser described by a purely diagonal density matrix $\rho_{n,n'}$, the average $\langle E(t)\rangle$ would be zero, and it might seem that a spectrum could not be defined. It is clear that there is no real difficulty here. Any reasonable operational procedure for determining a spectrum would give the desired result. One could, for example, make a Fourier analysis of a very long stretch of the film record mentioned above. The phase information available on the early part of the tracing would, in effect, represent preparation of an ensemble with a nonvanishing off-diagonal density matrix.

[44] W. E. Lamb, Jr. and Y. Aharanov (to be published).

In a subsequent paper, we will work out in detail the theory of a model spectrum analyzer coupled to the laser which does not involve a "bad" measurement of $E(t)$.

G. Photon Statistics

In principle, according to the assumed quantum theory of measurement, one could measure the total amount of energy in the single-mode optical cavity since this is represented by the Hermitian Hamiltonian operator. The result of such a measurement would be an integer multiple n of $\hbar\nu$, apart from the zero-point energy $\frac{1}{2}\hbar\nu$. Each time the measurement was repeated on a similarly prepared system, the n value could change. The statistical distribution of n values after many measurements would be given by the diagonal elements $\rho_{n,n}$ of the density matrix.

In practice, it would not be easy to determine $\rho_{n,n}$ in this manner. As a partial substitute one might count the number of photoelectrons emitted in a certain time interval. In the usual observations, the photoelectron counting is done with the detector located outside the laser cavity and the relationship of the results to $\rho_{n,n}$ is further complicated by diffraction of the radiation escaping from the laser cavity. It would not be very practical, but simpler in principle, to place the photoelectric surface inside the laser cavity. Even here, the photoelectron counting statistics would give a somewhat blurred image of the photon statistical distribution $\rho_{n,n}$. A theory of this process has been given by the authors,[45] and a fuller account is in preparation. The problem of photoelectron counting statistics has also been discussed elsewhere.[46]

H. Absence of Coupling between Elements of the Density Matrix having Different Degrees of "Off Diagonality"

As seen from Eq. (90) the differential equations for $\rho_{n,n'}(t)$ separate into systems of equations connecting elements $\rho_{n,n+k}$ of equal "off-diagonality" $k=n'-n$. For example, if ρ is diagonal initially, it will remain so forever. Similarly, off-diagonal elements like $\rho_{n,n+1}$ evolve completely independently of the diagonal elements $\rho_{n,n}$ and vice versa. This separability is a necessary corollary of the fact that ρ can represent our state of knowledge of any ensemble of lasers. Some restrictions on possible initial values of $\rho_{n,n}$ are imposed by general properties of density matrices. Among these are

$$\sum_n \rho_{n,n}=1 \qquad 0\leq\rho_{n,n}\leq 1, \qquad \text{all } n$$

$$0\leq \text{all eigenvalues of } \|\rho_{n,n'}\| \leq 1.$$

[45] Presented at the Second Rochester Conference on the Quantum Theory of Optical Coherence, June, 1966 (unpublished).
[46] In this context, see especially: L. Mandel, in *Progress in Optics*, edited by E. Wolf (North-Holland, Publishing Company, Amsterdam, 1963), Vol. II; P. L. Kelley and W. H. Kleiner, Phys. Rev. **136**, A316 (1964); R. J. Glauber, Ref. 36.

Since these properties must be satisfied initially, they will continue to be satisfied as the ensemble evolves in time.

I. Symmetry Breaking

In the statistical mechanics of a magnetic substance, the average magnetization at temperature θ is given by an expression like

$$\langle M_z\rangle = \mathrm{Tr}[M_z \rho(\theta)],$$

where $\rho(\theta)$ is the density matrix at equilibrium for the magnetic system.

In the absence of an external magnetic field, the above magnetization is zero on grounds of inversion symmetry. This would seem to rule out the possibility of permanent magnets with a nonzero value of $\langle M_z\rangle$. The resolution of this paradox is well known. The spontaneously magnetized sample is not really in thermodynamic equilibrium, and has to be described by a nonthermodynamic ensemble. Left in contact with a thermal reservoir for a sufficiently long time, the average magnetism of the ensemble would decay to zero, although with a very slow decay rate.

Similar considerations apply to the ensemble average for the quantum laser. General symmetry arguments would lead one falsely to the conclusion that $\langle E(t)\rangle$ should always be zero, but a properly biased ensemble could easily have $\langle E(t)\rangle$ nonzero. The decay time $1/D$ for a typical gas laser is of the order 10^3 sec, which is enormous compared to the period of oscillation 10^{-14} sec.

VII. SUMMARY

This paper develops the quantum theory of a laser oscillator. The radiation field in the cavity resonator is described by a density matrix $\rho_{n,n'}(t)$ in the n representation. A system of differential-difference equations (85) determines the time development of $\rho_{n,n'}(t)$ due to the combined effects of pumping and damping. These are the quantum analog of the amplitude and phase equations (4) in the semiclassical theory, with $S_n(t)$ and $C_n(t)$ determined by a self-consistent field approximation. A steady-state solution for $\rho_{n,n}(\infty)$ is given by Eq. (96), while the off-diagonal elements $\rho_{n,n'}(\infty)=0$ for $n\neq n'$.

In the case of a laser oscillator operating well above threshold, the steady-state photon probability distribution $\rho_{n,n}(\infty)$ is a sharply peaked distribution somewhat broader than a corresponding Poisson distribution. It should be emphasized that the peaked nature of the photon statistical distribution, which is a manifestation of laser coherence, is a result of the nonlinear aspects of the problem. In this case, the off-diagonal elements $\rho_{n,n'}$ $n\neq n'$ decay to their steady state in approximately exponential manner $\exp(-\frac{1}{2}k^2Dt)$ where the decay rate D is given by Eq. (109) and k is the degree of offdiagonality.

The temporal development of the photon statistical distribution $\rho_{n,n}$ has not been discussed in this paper, but has been analyzed by a numerical calculation and the results have been presented in the form of a moving picture.[47] A fuller account of the temporal behavior of $\rho_{n,n'}(t)$ predicted from Eqs. (85) will be presented in a future publication.

APPENDIX I: WIGNER-WEISSKOPF THEORY FOR A FOUR-LEVEL INTERACTING WITH THE LASER FIELD

In the analysis of Sec. III we represented the effect of certain nonlaser modes of the field on our atoms by introducing the radiative damping coefficients γ_a and γ_b. In this appendix we will derive the working relations Eqs. (65) in the Wigner-Weisskopf approximation. We are really interested in the two atomic levels a and b, but in order to consider their radiative decay it is necessary to introduce two more levels c and d. State $|a\rangle$ may decay to $|c\rangle$ with the emission of radiation of frequency ν_s, while $|b\rangle$ goes to $|d\rangle$ with the emission of a photon of frequency ν_σ, as shown in Fig. 8. The concept of an unobserved coordinate or reservoir is nicely illustrated by this example.[48] That the decay radiation is unobserved is clear when we recall that we are looking for $\rho_{nn'}(t)$, which is the trace over all indices except those referring to the laser radiation, i.e.,

$$\rho_{n,n'}(t) = \mathrm{Tr}_\alpha\, \mathrm{Tr}_{\{s,\sigma\}}\rho_{\alpha,\{s,\sigma\},n;\alpha,\{s,\sigma\},n'}, \tag{I.1}$$

where $\alpha=a, b, c, d$ is the atomic index and $\{s,\sigma\}$ denotes the decay radiation.

Let us consider the interaction of a four-level atom with the laser mode of frequency ν and the continuum of decay modes of frequencies ν_s and ν_σ. The Hamiltonian for this system is

$$\begin{aligned}H=&\nu a^\dagger a+\sum_s \nu_s a_s^\dagger a_s+\sum_\sigma \nu_\sigma a_\sigma^\dagger a_\sigma+\sum_{\alpha=a,b,c,d}\epsilon_\alpha A_\alpha^\dagger A_\alpha\\ &+g[a^\dagger A_b^\dagger A_a+aA_a^\dagger A_b]\\ &+\sum_s g_s[a_s^\dagger A_c^\dagger A_a+a_sA_a^\dagger A_c]\\ &+\sum_\sigma g_\sigma[a_\sigma^\dagger A_d^\dagger A_b+a_\sigma A_b^\dagger A_d]\\ =&H_0+H_0^s+H_0^\sigma+\sum_\alpha H_0^\alpha+V+\sum_s V^s+\sum_\sigma V^\sigma,\end{aligned} \tag{I.2}$$

where $a^\dagger$, a; $a_s^\dagger$, a_s; $a_\sigma^\dagger$, a_σ are the emission and ab-

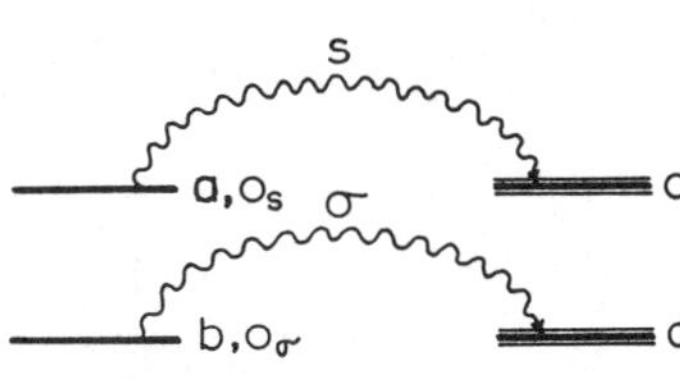

FIG. 8. Level scheme indicating the decay of state $|a, \{0\cdots 0_s\cdots\}\rangle$ to $|c, \{0\cdots 1_s\cdots\}\rangle$ and state $|b, \{0\cdots 0_\sigma\cdots\}\rangle$ to $|d, \{0\cdots 1_\sigma\cdots\}\rangle$.

[47] M. Scully, W. E. Lamb, Jr., and M. Sargent III, in Proceedings of the Fourth International Conference on Quantum Electronics, Phoenix, Arizona, 1966 (unpublished). See also Ref. 45.

[48] M. Lax has given a nice discussion of this point in Appendix A of Phys. Rev. **145**, 110 (1966). Our equations (I.14) correspond to Eqs. (A.10) of that paper.

sorption operators for the laser radiation and those of type s and σ, respectively, while the atomic operators $A_a^\dagger, A_a; A_b^\dagger, A_b; A_c^\dagger, A_c; A_d^\dagger, A_d$ create and annihilate the atom in states a, b, c, and d. The coupling strengths between the fields and the atom are represented by g, g_s, and g_σ.

The equation of motion in the interaction picture for the density matrix of the atom-field system is

$$\dot{\rho}(t)=-i[V(t),\rho]-i[\sum_s V^s(t)+\sum_\sigma V^\sigma(t),\rho]. \quad (I.3)$$

Let us introduce the notation

$$(a,n,0_s)=\alpha, \qquad (b,n+1,0_\sigma)=\beta,$$

$$(c,n,1_s)=\gamma, \qquad (d,n+1,1_\sigma)=\delta, \quad (I.4)$$

with the conventions that a prime on α, β, or δ means that n should be replaced by n' in the definitions of (I.4). We are primarily interested in the quantities $\rho_{\alpha,\alpha'}, \rho_{\beta,\beta'}, \rho_{\beta,\alpha'}, \rho_{\gamma,\gamma'}, \rho_{\delta,\delta'}$, which according to (I.3) obey the differential equations

$$\dot{\rho}_{\alpha,\alpha'}=-i[V,\rho]_{\alpha,\alpha'}-i\sum_s[V_{\alpha,\gamma}{}^s\rho_{\gamma,\alpha'}-\rho_{\alpha,\gamma'}V_{\gamma',\alpha'}{}^s], \quad (I.5a)$$

$$\dot{\rho}_{\beta,\beta'}=-i[V,\rho]_{\beta,\beta'}-i\sum_\sigma[V_{\beta,\delta}{}^\sigma\rho_{\delta,\beta'}-\rho_{\beta,\delta}V_{\delta',\beta'}{}^\sigma], \quad (I.5b)$$

$$\dot{\rho}_{\beta,\alpha'}=-i[V,\rho]_{\beta,\alpha'}-i[\sum_\sigma V_{\beta,\delta}{}^\sigma\rho_{\delta,\alpha'}-\sum_s\rho_{\beta,\gamma'}V_{\gamma',\alpha'}{}^s], \quad (I.5c)$$

$$\dot{\rho}_{\gamma,\gamma'}=-i[V_{\gamma,\alpha}{}^s\,\rho_{\alpha,\gamma'}-\rho_{\gamma,\alpha'}V_{\alpha',\gamma'}{}^s], \quad (I.5d)$$

$$\dot{\rho}_{\delta,\delta'}=-i[V_{\delta,\beta}{}^\sigma\rho_{\beta,\delta'}-\rho_{\delta,\beta'}V_{\beta',\delta'}{}^\sigma]. \quad (I.5e)$$

We next calculate the effects of the decay modes on our four-level atom by solving for $\rho_{\gamma,\alpha'}, \rho_{\delta,\beta'}, \rho_{\delta,\alpha'}, \rho_{\beta,\gamma'}$, etc. For example, $\rho_{\gamma,\alpha'}$ obeys the differential equation

$$\dot{\rho}_{\gamma,\alpha'}=-i[V(t')_{\gamma,\alpha}{}^s\rho(t')_{\alpha,\alpha'}-\rho(t')_{\gamma,\gamma'}V(t')_{\gamma',\alpha'}{}^s]. \quad (I.5f)$$

We then have

$$\rho_{\gamma,\alpha'}=-i\int_{t_0}^t[V(t')_{\gamma,\alpha}{}^s\rho(t')_{\alpha,\alpha'}-\rho(t')_{\gamma,\gamma'}V(t')_{\gamma',\alpha'}{}^s]dt', \quad (I.6a)$$

$$\rho_{\delta,\beta'}=-i\int_{t_0}^t[V_{\delta,\beta}{}^\sigma\rho_{\beta,\beta'}-\rho_{\delta,\delta'}V_{\delta',\beta'}{}^s]dt', \quad (I.6b)$$

$$\rho_{\delta\alpha'}=-i\int_{t_0}^t[V_{\delta,\beta}{}^\sigma\rho_{\beta,\alpha'}-\rho_{\delta,\gamma'}V_{\gamma'\alpha'}{}^s]dt', \quad (I.6c)$$

$$\rho_{\beta\gamma'}=-i\int_{t_0}^t[V_{\beta,\delta}{}^\sigma\rho_{\delta,\gamma'}-\rho_{\beta,\alpha'}V_{\alpha',\gamma'}{}^s]dt'. \quad (I.6d)$$

Since the coupling between the $\{s,\sigma\}$ reservoir and the atom is weak, we assume that for the evaluation of Eqs. (I.6) we may factor the decay radiation density matrix from that of the atom laser, e.g., in Eqs. (I.6),

$$\rho_{\alpha,\alpha'}(t')=\rho_{an0,an'0}(t')\approx\rho_{an,an'}(t')\rho_{0,0}(t'), \quad (I.7)$$

$$\rho_{\gamma,\gamma'}(t')=\rho_{cn1,cn'1}(t')\approx\rho_{cn,cn'}(t')\rho_{1,1}(t'). \quad (I.8)$$

Further, if the number of decay modes is large, then to a good approximation

$$\rho_{0,0}(t')\approx\rho_{0,0}(t_0)=1, \quad (I.9a)$$

$$\rho_{1,1}(t')\approx\rho_{1,1}(t_0)=0. \quad (I.9b)$$

Then Eqs. (I.6) become

$$\rho_{\gamma,\alpha'}(t)=-i\int_{t_0}^t dt'V_{\gamma,\alpha}{}^s(t')\rho_{an,an'}(t'), \quad (I.10a)$$

$$\rho_{\delta,\beta'}(t)=-i\int_{t_0}^t dt'V_{\delta,\beta}{}^\sigma(t')\rho_{bn+1,bn'+1}(t'), \quad (I.10b)$$

$$\rho_{\delta,\alpha'}(t)=-i\int_{t_0}^t dt'V_{\delta,\beta'}{}^\sigma(t')\rho_{bn+1,an'}(t'), \quad (I.10c)$$

$$\rho_{\beta,\gamma'}(t)=-i\int_{t_0}^t dt'V_{\alpha,\gamma'}{}^s(t')\rho_{bn+1,an'}(t'). \quad (I.10d)$$

Substituting Eqs. (I.10) into Eqs. (I.5) and tracing over $\{s\}$ and $\{\sigma\}$ we find

$$\dot{\rho}_{an,an'}=-i[V(t),\rho]_{an,an'}-\mathrm{Tr}_{\{s\sigma\}}\left[\sum_s V_{\alpha,\gamma}{}^s(t)\int_{t_0}^t V_{\gamma,\alpha}{}^s(t')\rho_{an,an'}(t')dt'+\text{etc.}\right], \quad (I.11a)$$

$$\dot{\rho}_{bn+1,bn'+1}=-i[V(t),\rho]_{bn+1,bn'+1}-\mathrm{Tr}_{\{s,\sigma\}}\left[\sum_\sigma V_{\beta,\delta}{}^\sigma(t)\int_{t_0}^t dt'V_{\delta,\beta}{}^\sigma(t')\rho_{bn+1,bn'+1}(t')+\text{etc.}\right], \quad (I.11b)$$

$$\dot{\rho}_{bn+1,an'}=-i[V(t),\rho]_{bn+1,an'}-\mathrm{Tr}_{\{s,\sigma\}}\left[\sum_\sigma V_{\beta,\delta}{}^\sigma(t)\int_{t_0}^t dt'V_{\delta,\beta}{}^\sigma(t')\rho_{bn+1,an'}(t')+\sum_s V_{\gamma',\alpha'}{}^s(t)\int_{t_0}^t dt'V_{\alpha',\gamma'}{}^s\rho_{bn+1,an'}(t')\right], \quad (I.11c)$$

$$\dot{\rho}_{cn,cn'}=\mathrm{Tr}_{\{s,\sigma\}}\left[V_{\gamma,\alpha}{}^s(t)\int_{t_0}^t dt'V_{\alpha',\gamma'}{}^s(t')\rho_{an,an'}(t')+\text{etc.}\right], \quad (I.11d)$$

$$\dot{\rho}_{dn+1,dn'+1}=\mathrm{Tr}_{\{s,\sigma\}}\left[V_{\delta,\beta}{}^\sigma(t)\int_{t_0}^t dt'V_{\beta',\delta'}{}^\sigma(t)\rho_{bn+1,bn'+1}+\text{etc.}\right], \quad (I.11e)$$

where etc. means replace n by n' and take the complex conjugate of the first term. Now we consider the density of modes $W(\nu)$ to be so large that we may replace sums by integrals

$$\sum_s \cdots \to \int_0^\infty d\nu\, W(\nu) \cdots,$$

$$\sum_s \cdots \to \int_0^\infty d\nu\, W(\nu) \cdots.$$

Making the usual approximation of the Wigner–Weisskopf theory in which the matrix elements and the density-of-states factor are evaluated at resonance, thereby neglecting level shifts, Eq. (I.11) become

$$\dot\rho_{an,an'} = -i[V(t),\rho]_{an,an'} - |V_{\alpha,\gamma}|^2 W(\omega(ac)) \int_{t_0}^t dt' \int_0^\infty d\nu\, (\exp\{i[\omega(ac)-\nu](t-t')\} + \text{c.c.})\rho_{an,an'}(t'), \tag{I.12a}$$

$$\dot\rho_{bn+1,bn'+1} = -i[V(t),\rho]_{bn+1,bn'+1} - |V_{\beta,\delta}|^2 W(\omega(bd)) \int_{t_0}^t dt' \int_0^\infty d\nu\, (\exp\{i[\omega(bd)-\nu](t-t')\} + \text{c.c.})\rho_{bn+1,bn'+1}(t'), \tag{I.12b}$$

$$\begin{aligned}\dot\rho_{bn+1,an'} = -i[V(t),\rho]_{bn+1,an'} &- |V_{\beta,\delta}|^2 W(\omega(bd)) \int_{t_0}^t dt' \int_0^\infty d\nu\, \exp\{-i[\omega(bd)-\nu](t-t')\}\rho_{bn+1,an'}(t') \\ &- |V_{\gamma,\alpha}|^2 W(\omega(ac)) \int_{t_0}^t dt' \int_0^\infty d\nu\, \exp\{i[\omega(ac)-\nu](t-t')\}\rho_{bn+1,an'}(t'),\end{aligned} \tag{I.12c}$$

$$\dot\rho_{cn,cn'} = |V_{\alpha,\gamma}|^2 W(\omega(ac)) \int_{t_0}^t dt' \int_0^\infty d\nu\, (\exp\{-i[\omega(ac)-\nu](t-t')\} + \text{c.c.})\rho_{an,an'}(t'), \tag{I.12d}$$

$$\dot\rho_{dn+1,dn'+1} = |V_{\delta,\beta}|^2 W(\omega(bd)) \int_{t_0}^t dt' \int_0^\infty d\nu\, (\exp\{-i[\omega(bd)-\nu](t-t')\} + \text{c.c.})\rho_{bn+1,bn'+1}(t'). \tag{I.12e}$$

In the usual way, we may extend the range of integration to $-\infty$ and use the delta function defined by

$$\int_{-\infty}^\infty d\nu\, \exp[\pm i(\omega-\nu)(t-t')] = 2\pi\delta(t-t'). \tag{I.13}$$

We may then write Eqs. (I.12) as

$$\dot\rho_{an,an'} = -i[H_0+V,\rho]_{an,an'} - \gamma_a\rho_{an,an'}, \tag{I.14a}$$

$$\dot\rho_{bn+1,bn'+1} = -i[H_0+V,\rho]_{bn+1,bn'+1} - \gamma_b\rho_{bn+1,bn'+1}, \tag{I.14b}$$

$$\dot\rho_{bn+1,an'} = -i[H_0+V,\rho]_{bn+1,an} - \gamma_{ab}\rho_{bn+1,an'}, \tag{I.14c}$$

$$\dot\rho_{cn,cn'} = \gamma_a\rho_{an,an'}, \tag{I.14d}$$

$$\dot\rho_{dn+1,dn'+1} = \gamma_b\rho_{bn+1,bn'+1}, \tag{I.14e}$$

where we have transformed back into the Schrödinger picture as it is best suited for the analysis of Sec. III. The decay constants are given by

$$\gamma_a = 2\pi W(\omega(ac))\,|V_{a0,c1}|^2, \tag{I.15}$$

$$\gamma_b = 2\pi W(\omega(bd))\,|V_{b0,d1}|^2, \tag{I.16}$$

$$\gamma_{ab} = \tfrac{1}{2}(\gamma_a+\gamma_b). \tag{I.17}$$

Equations (I.14) describe the interaction of the laser field with the a and b atomic states which are decaying to states c and d with the usual spontaneous radiative decay constants γ_a and γ_b.

APPENDIX II: DAMPING OF THE FIELD

To provide our cavity with a finite Q, we here consider a dissipative interaction, equivalent to the Ohmic losses of the semiclassical theory. One can envision several satisfactory dissipation mechanisms: Interaction with random currents, photon-phonon interaction, interaction with a two-level atomic system, etc. As we have developed a machinery for dealing with the latter type of interaction, we will consider the dissipative subsystem to consist of nonresonant two-level atoms, injected at random times t_0 in the lower of the two states $|\alpha\rangle$ and $|\beta\rangle$. The calculation will then follow along the lines of Sec. III.

For an atom injected in the β state

$$\delta\rho_{nn'}=\gamma_\alpha\sigma_{11}+\gamma_\beta\sigma_{22}[n\to n-1,\, n'\to n'-1]-\rho_{nn'}(t_0), \tag{II.1}$$

$$\gamma_\alpha\sigma_{11}=\frac{2\gamma_\alpha\gamma_{\alpha\beta}g^2((n+1)(n'+1))^{1/2}\rho_{n+1,n'+1}(t_0)}{\gamma_\alpha\gamma_\beta(\gamma_{\alpha\beta}^2+\Delta^2)+2g^2\gamma_{\alpha\beta}^2(n+1+n'+1)+g^2(n'-n)[g^2(n'-n)+i\Delta(\gamma_\alpha-\gamma_\beta)]}, \tag{II.2a}$$

$$\gamma_\beta\sigma_{22}=\frac{-i\gamma_\beta[i\gamma_\alpha(\Delta^2+\gamma_{\alpha\beta}^2)+g^2(n'+1)(\Delta+i\gamma_{\alpha\beta})-g^2(n+1)(\Delta-i\gamma_{\alpha\beta})]\rho_{n+1,n'+1}(t_0)}{\gamma_\alpha\gamma_\beta(\gamma_{\alpha\beta}^2+\Delta^2)+2g^2\gamma_{\alpha\beta}^2(n+1+n'+1)+g^2(n'-n)[g^2(n'-n)+i\Delta(\gamma_\alpha-\gamma_\beta)]}. \tag{II.2b}$$

From Eq. (III.1) and (III.2), we find

$$\delta\rho_{nn'}=-\frac{g^2[\gamma_\alpha\gamma_{\alpha\beta}(n+n')+i\gamma_\alpha\Delta(n'-n)+g^2(n'-n)^2]\rho_{n,n'}(t_0)}{\gamma_\alpha\gamma_\beta(\gamma_{\alpha\gamma}{}^2+\Delta^2)+2\gamma_{\alpha\beta}{}^2g^2(n+n')+g^2(n'-n)[g^2(n'-n)+i\Delta(\gamma_\alpha-\gamma_\beta)]}$$
$$+\frac{2g^2\{\gamma_\alpha\gamma_{\alpha\beta}[(n+1)(n'+1)]^{1/2}\}\rho_{n+1,n'+1}(t_0)}{\gamma_\alpha\gamma_\beta(\gamma_{\alpha\beta}^2+\Delta^2)+2\gamma_{\alpha\beta}^2g^2(n+1+n'+1)+g^2(n'-n)[g^2(n'-n)+i\Delta(\gamma_\alpha-\gamma_\beta)]}. \tag{II.3}$$

Now we replace $\rho_{nn'}(t_0)\to\rho_{nn'}(t)$ and multiply $\delta\rho_{nn'}$ by r_β to obtain the coarse-grained time derivative representing the effects of damping (cavity Q).

Since dissipation, unlike the laser-atom interaction, is a linear process, we will keep only the lowest-order damping terms. We find that the decay of the laser radiation is described by the expression

$$[d\rho_{nn'}/dt]_{\text{damping}}$$
$$\approx-\{g^2(r_\beta/\gamma_\beta)\gamma_{\alpha\gamma}(n+n')[\gamma_{\alpha\beta}^2+\Delta^2]^{-1}\}\rho_{n,n'}(t)$$
$$+\{2g^2(r_\beta/\gamma_\beta)\gamma_{\alpha\beta}[(n+1)(n'+1)]^{1/2}$$
$$\times[\gamma_{\alpha\beta}^2+\Delta^2]^{-1}\}\rho_{n+1,n'+1}(t). \tag{II.4}$$

We define $C=\nu/Q=2r_\beta(g^2/\gamma_\beta)\gamma_{\alpha\beta}[\gamma_{\alpha\beta}^2+\Delta^2]^{-1}$ and write the damping equation in the form appearing in Sec. III.

$$d\rho_{n,n'}/dt=-\tfrac{1}{2}C(n+n')\rho_{n,n'}$$
$$+C[(n+1)(n'+1)]^{1/2}\rho_{n+1,n'+1}. \tag{II.5}$$

APPENDIX III: SOLUTION OF THE OFF-DIAGONAL EQUATIONS

If the laser is far enough above threshold, we expect that the lowest eigenvalue will be small and that the eigenfunction will be similar to that found for the steady-state diagonal equation (102). Guided by these physical considerations we propose to look for solutions of the off-diagonal equations in the form

$$\rho_{n,n+k}(t)=\Phi_n(k,t)=N_k\left\{\prod_{l=0}^{n}\left[\frac{A-Bl}{C}\right]\prod_{m=0}^{n+k}\left[\frac{A-Bm}{C}\right]\right\}^{1/2}\exp(-\mu_0^{(k)}t), \tag{III.1}$$

where N_k is a constant determined by the initial conditions. We will use for $R_{n,n'}$ not the complicated expression corresponding to (79), but an expression to second order in g^2,

$$R_{n,n'}=r_a(g^2/\gamma_a\gamma_{ab})[1-(g^2/\gamma_a\gamma_b)(\gamma_a(n'+1+n'+1)+\gamma_b(n+1+n'+1))/\gamma_{ab}]. \tag{III.2}$$

The use of this approximate form for $R_{n,n'}$ is the analog of the third-order perturbation expansion of the semiclassical theory. Inserting (III.1) and (III.2) into (85) we find, after some algebra,

$$\dot\Phi_n(k,t)=-\tfrac{1}{8}k^2(\gamma_a/\gamma_{ab})B\Phi_n(k,t)-[A-B(n+1+\tfrac{1}{2}k)](n+1+\tfrac{1}{2}k)\Phi_n(k,t)$$
$$+[A-B(n+\tfrac{1}{2}k)][n(n+k)]^{1/2}\Phi_{n-1}(k,t)-C(n+\tfrac{1}{2}k)\Phi_n(k,t)+C[(n+1)(n+1+k)]^{1/2}\Phi_{n+1}(k,t). \tag{III.3}$$

From Eqs. (III.1) we may write

$$\Phi_{n+1}(k,t)=\{[A-B(n+1)]^{1/2}[A-B(n+1+k)]^{1/2}/C\}\Phi_n(k,t), \tag{III.4a}$$

$$\Phi_{n-1}(k,t)=\{[A-Bn]^{-1/2}[A-B(n+k)]^{-1/2}\}C\Phi_n(k,t). \tag{III.4b}$$

Since $n \gg k$, we may write (III.4) to a very good approximation as

$$\Phi_{n+1}(k,t) = [\{(A-B(n+1))/C\} - \tfrac{1}{2}(Bk/C)]\Phi_n(k,t), \tag{III.5a}$$

$$\Phi_{n-1}(k,t) = [\{C/(A-Bn)\} + \tfrac{1}{2}\{BkC/(A-Bn)^2\}]\Phi_n(k,t); \tag{III.5b}$$

likewise the radicals $[n(n+k)]^{1/2}$, etc. appearing in Eq. (III.3) may be approximated by

$$[n(n+k)]^{1/2} \approx n + \tfrac{1}{2}k - \tfrac{1}{8}(k^2/n), \tag{III.6a}$$

$$[(n+1)(n+1+k)]^{1/2} \approx n+1+\tfrac{1}{2}k - \tfrac{1}{8}\{k^2/(n+1)\}. \tag{III.6b}$$

Making use of (III.5) and (III.6), (III.3) becomes

$$\begin{aligned}\dot{\Phi}_n(k,t) = {} & -\tfrac{1}{8}k^2(\gamma_a/\gamma_{ab})B\Phi_n(k,t) - [\tfrac{1}{8}Ck^2\Phi_{n+1}/(n+1) + \tfrac{1}{8}[A-B(n+\tfrac{1}{2}k)]k^2\Phi_{n-1}/n] \\ & -[A-B(n+1+\tfrac{1}{2}k)](n+1+\tfrac{1}{2}k)\Phi_n + C[(A-B(n+1))/C - \tfrac{1}{2}Bk/C](n+1+\tfrac{1}{2}k)\Phi_n \\ & +\{[A-B(n+\tfrac{1}{2}k)](n+\tfrac{1}{2}k)[C/(A-Bn)+\tfrac{1}{2}BkC/(A-Bn)^2]\}\Phi_n - C(n+\tfrac{1}{2}k)\Phi_n.\end{aligned} \tag{III.7}$$

Neglecting terms involving $\langle n\rangle^{-2}$, Eqs. (III.7) become

$$\dot{\Phi}_n(k,t) = -\mu_0^{(k)}\Phi_n = -\{\tfrac{1}{8}Ck^2/\langle n\rangle + \tfrac{1}{8}(A-B\langle n\rangle)k^2/\langle n\rangle + \tfrac{1}{4}k^2[(A-C)/C]B + \tfrac{1}{8}k^2(\gamma_a/\gamma_{ab})B\}\Phi_n. \tag{III.8}$$

Noting that $(A-B\langle n\rangle)$ is C and that

$$\tfrac{1}{8}k^2\gamma_a B/\gamma_{ab} = \tfrac{1}{4}k^2B - \tfrac{1}{8}k^2\gamma_b B/\gamma_{ab}, \tag{III.9}$$

we may write $\mu_0^{(k)}$ as

$$\mu_0^{(k)} = \tfrac{1}{4}(Ck^2/\langle n\rangle) + \tfrac{1}{4}k^2B[(1-\tfrac{1}{2}(\gamma_b/\gamma_{ab}) + (A-C)/C)]. \tag{III.10}$$

The leading term corresponds to

$$D = \tfrac{1}{2}(\nu/Q)\langle n\rangle^{-1}, \tag{III.11}$$

as given in Sec. V.

Quantum Theory of a Simple Maser Oscillator

J. P. Gordon
Bell Telephone Laboratories, Murray Hill, New Jersey
(Received 8 February 1967)

We discuss the quantum theory of a simple model of a maser oscillator, consisting of one radiation-field mode interacting with a large number of stationary three-level atoms. The field and the atoms also interact with separate heat reservoirs which represent dissipation mechanisms and an incoherent pumping mechanism. The model is sufficiently simple that some analytical progress can be made with the nonlinear quantal equations before further approximation is necessary. We start from the quantal equation of motion of the field-atom density operator. We make immediate use of a diagonal coherent-state expansion for the field part of the density operator and a somewhat similar expansion for the atom part. This yields an exact equation of the Fokker-Planck form for a *c*-number weight distribution, which retains all the significance of the original operator equation, and which has the semiclassical equation for the same model as a first, fluctuation-free approximation. We make use of our basic Fokker-Planck equation in a variety of ways. We discuss the reduction of the equation under conditions that the atomic decay constants are large (large atomic linewidth), arriving finally at an equation of motion for a field-only weight function which serves to demonstrate the basic coherence properties of a maser. We derive and discuss the equation of motion of the generalized Wigner density for the maser model. The generalized Wigner density is a smoothed version (a convolution) of our basic weight distribution, and from it we derive an equivalent classical model including noise sources. Finally, we discuss other useful weight distributions and the number representation for the field. The equations we derive in these discussions make contact with the rate equations of Shimoda, Takahasi, and Townes, as well as with the more recent work of Lax and Louisell, Lax, and of Scully and Lamb.

I. INTRODUCTION

The theory of a maser oscillator is an interesting meeting ground for quantal and classical physics. A simple maser represents possibly the most elementary quantal problem involving many particles which obey nonlinear equations. In addition, of course, the maser (laser) is a useful device which is the basis of quantum-electronics research. Hence, it is important to gain as much insight as possible into the behavior of this device, for example, to understand its inherent quantal fluctuations.

Since masers operating at levels near or above the threshold of oscillation normally involve many atoms and many photons, one might expect that there should exist a close equivalent classical model. A tuned circuit oscillator involving a saturable negative resistance is such an equivalent, and part of the effort of this work is to demonstrate precisely the nature of the equivalence. Our equivalent classical model contains appropriate sources of fluctuations (noise).

We discuss a simple model of a maser oscillator, consisting of one radiation field mode interacting with a large number of stationary three-level atoms. The field and the atoms also interact with separate heat reservoirs which represent dissipation mechanisms and an incoherent pumping mechanism. The model is sufficiently simple that some analytical progress can be made with the nonlinear quantal equations before further approximation is necessary.

A considerable amount of progress has already been made in other work on fully quantal treatments of the maser, wherein the radiation field is quantized as well as the atoms. Our model is the same as that of Lax,[1] who has approached the problem by deriving and then working from a set of Langevin-type operator equations involving noncommuting random forces. Lax and Louisell[2] and later Lax[3] have extended this treatment to obtain equations of motion of an "associated classical function" (*c*-number function) which represents a certain (antinormal) ordering of the field density operator. Haken[4] and his co-workers have pursued more or less the same approach, generally treating more complicated models with many field modes, moving atoms, etc. Scully and Lamb[5] have taken a different and independent approach starting from a model nearly equivalent to Lax's. They treat the case of short atomic lifetimes, and work in the photon number (energy) representation for the field. They find the effect on the field density operator of the temporal passage of one atom through the field. They then multiply the result for one atom by the rate of passage of atoms to find the total effect. Two different kinds of atoms represent gain and loss mechanisms.

We have found yet another approach to this problem, which we feel has some advantages over the others. We start (Sec. II) from the quantal equation of motion of the field-atoms density operator. We make immediate use (Sec. III) of a diagonal coherent state expansion for the field part of the density operator and a somewhat similar expansion for the atoms part. This yields an exact equation of the Fokker Planck form for a *c*-number weight distribution, which retains all the

[1] M. Lax, Phys. Rev. **145**, 110 (1966).

[2] M. Lax and W. H. Louisell, J. Quant. Electron. 3, 37 (1967).

[3] M. Lax, Phys. Rev. **157**, 213 (1967); also, *Brandeis Summer Institute Lectures, 1966* (Gordon and Breach Scientific Publishers, Inc., New York, to be published).

[4] V. Arzt, H. Haken, H. Riskin, H. Sauermann, Ch. Schmid, and W. Weidlich, Z. Physik **197**, 207 (1966).

[5] M. Scully and W. E. Lamb Jr., Phys. Rev. Letters **16**, 853 (1966).

Reprinted with permission from *Phys. Rev.*, vol. 161, pp. 367-386, Sept. 10, 1967.

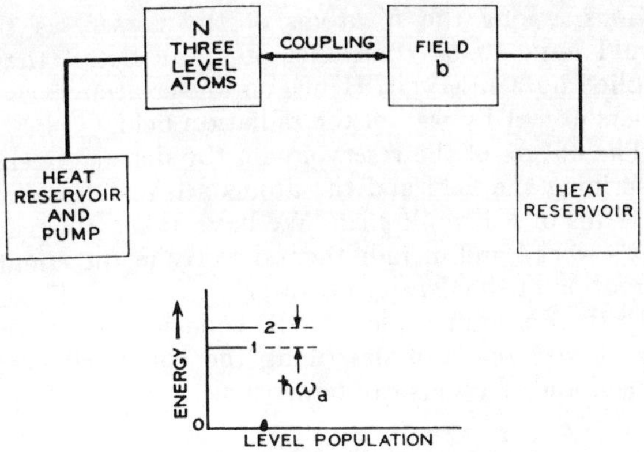

FIG. 1. Block diagram of the maser model.

significance of the original operator equation, and which has the semiclassical equation for the same model as a first, fluctuation free, approximation. The importance of coherent state (minimum uncertainty wave packet) representations has been emphasized by Glauber,[6] who has used them to gain understanding of coherence properties of quantized fields. Further understanding of diagonal coherent state representations has been afforded by the work of Sudarshan[7,8] and Klauder.[8,9]

In the remainder of the paper we make use of our basic Fokker–Planck equation in a variety of ways. In Sec. IV we discuss the reduction of the equation under conditions that the atomic decay constants are large (large atomic linewidth), arriving finally at an equation of motion for a field-only weight distribution which serves to demonstrate the basic coherence properties of a maser. In Sec. V, we derive and discuss the equation of motion of what we call the generalized Wigner density for the maser model. The generalized Wigner density is a smoothed version (a convolution) of our basic weight distribution, and from the diffusion approximation to its equation of motion we derive an equivalent classical model, including noise sources. Finally, in Sec. VI we discuss other useful weight distributions, and the number representation for the field. The equations we derive in these discussions make contact with the rate equations of Shimoda, Takahasi, and Townes[10] as well as with the more recent work of Lax and Louisell,[2] Lax,[3] and Scully and Lamb.[5]

II. THE MODEL AND EQUATION OF MOTION

The model describing the simple maser we consider here is illustrated in Fig. 1. It is identical with one used by Lax.[1] A resonator contains a single important radiation field mode of frequency ω_b described by the uncoupled Hamiltonian

$$\mathbf{H}_b = \hbar\omega_b \mathbf{b}^\dagger \mathbf{b}, \tag{2.1}$$

where $\mathbf{b}$ is the usual annihilation operator and $\mathbf{b}^\dagger$ its adjoint creation operator. The photon number operator $\mathbf{b}^\dagger\mathbf{b}$ has the positive integers and zero as eigenvalues. In contact with the resonator field are a large number N of atoms having three important nondegenerate energy levels labeled 0, 1, 2 in order of increasing energy. The uncoupled Hamiltonian describing the atoms is

$$\mathbf{H}_a = \sum_{j=0}^{2} \epsilon_j \mathbf{N}_j, \tag{2.2}$$

where $\mathbf{N}_j$ is the operator for the number of atoms in the jth level, and ϵ_j is the energy of the jth level. We use the notation of second quantization to describe the atoms, with operators

$$(\mathbf{a}_j)_m \quad \text{and} \quad (\mathbf{a}_j^\dagger)_m$$

being, respectively, annihilation and creation operators for the jth state of the mth atom. The properties of the atomic operators which we shall need are discussed in Appendix A. In terms of these,

$$\mathbf{N}_j = \sum_{m=1}^{N} (\mathbf{a}_j^\dagger \mathbf{a}_j)_m, \tag{2.3}$$

since $(\mathbf{a}_j^\dagger \mathbf{a}_j)_m$ is unity if the mth atom is in the jth state, and zero otherwise.

The coupling between the atoms and the field is described in the rotating wave approximation by the interaction Hamiltonian

$$\mathbf{V} = i\hbar\mu[\mathbf{M}\mathbf{b}^\dagger - \mathbf{M}^\dagger\mathbf{b}], \tag{2.4}$$

where μ is a real coupling constant proportional to the atomic dipole moment associated with the $1\leftrightarrow 2$ transition, and

$$\mathbf{M} = \sum_{m=1}^{N} (\mathbf{a}_1^\dagger \mathbf{a}_2)_m,$$

$$\mathbf{M}^\dagger = \sum_{m=1}^{N} (\mathbf{a}_2^\dagger \mathbf{a}_1)_m. \tag{2.5}$$

The macroscopic polarization of the atoms is proportional to $\mu\mathbf{M}$. Application of a single term of the interaction [expanded according to (2.5)] removes one photon from the field and simultaneously raises one atom from level 1 to level 2, or adds a photon to the field and lowers one atom from level 2 to level 1. Without the rotating wave approximation, terms

$$\mathbf{M}\mathbf{b} \quad \text{and} \quad \mathbf{M}^\dagger\mathbf{b}^\dagger$$

would appear in V. These terms, which do not conserve energy in the individual transitions without a lot of

[6] R. J. Glauber, Phys. Rev. **131**, 2766 (1963); also in *Quantum Optics and Electronics, Les Houches 1964*, edited by C. deWitt, A. Blanden, and C. Cohen-Tannoudji (Gordon and Breach Scientific Publishers, Inc., New York, 1965).
[7] E. C. G. Sudarshan, Phys. Rev. Letters **10**, 277, (1963).
[8] J. R. Klauder and E. C. G. Sudarshan, *Fundamentals of Quantum Optics* (W. A. Benjamin, Inc., New York, to be published).
[9] J. R. Klauder, J. McKenna, and D. G. Currie, J. Math. Phys. **6**, 734 (1965).
[10] K. Shimoda, H. Takahasi, and C. H. Townes, J. Phys. Soc. Japan **12**, 686, 1957.

help from the heat reservoirs, are ignored here from the beginning. This restricts the discussion to the case of sharp resonance, or high Q.

Finally, the model is completed by allowing both the field and the atoms to interact with separate heat reservoirs. The field reservoir simulates the lossy walls of the resonator and the lossy resonator medium. The atom reservior simulates the maser's energy source (the pump) and the various objects which randomly perturb the atoms; i.e., radiation field modes other than the one singled out here, phonon modes, atomic collision, etc. We specifically do not include direct interactions among the N atoms of the maser, as these would have to be considered in more detail than is implied by a reservoir. Hence no maser atom feels the others except by way of the radiation field.

The effects of the reservoirs on the density operator describing the field and the atoms are not describable in terms of a Hamiltonian. We have taken them from Lax's work,[1] and include them directly in the equation of motion of the density operator.

In the Shrödinger picture, the equation of motion of the density operator describing the combined system of field and N atoms can be written

$$\frac{d\rho}{dt}=-\frac{i}{\hbar}[(\mathbf{H}_a+\mathbf{H}_b),\rho]+\mu[(\mathbf{M}\mathbf{b}^\dagger-\mathbf{M}^\dagger\mathbf{b}),\rho]+\sum_{m=1}^{N}\sum_{i,j=0}^{2}w_{ij}\{(\mathbf{a}_i^\dagger\mathbf{a}_j)_m\rho(\mathbf{a}_j^\dagger\mathbf{a}_i)_m-\tfrac{1}{2}(\mathbf{a}_j^\dagger\mathbf{a}_j)_m\rho-\tfrac{1}{2}\rho(\mathbf{a}_j^\dagger\mathbf{a}_j)_m\}$$
$$-\tfrac{1}{2}\gamma\{[\bar{n}\mathbf{b}\mathbf{b}^\dagger+(\bar{n}+1)\mathbf{b}^\dagger\mathbf{b}]\rho+\rho[\bar{n}\mathbf{b}\mathbf{b}^\dagger+(\bar{n}+1)\mathbf{b}^\dagger\mathbf{b}]\}+\gamma\{(\bar{n}+1)\mathbf{b}\rho\mathbf{b}^\dagger+\bar{n}\mathbf{b}^\dagger\rho\mathbf{b}\}. \quad (2.6)$$

In this equation of motion (2.6), the first two terms result from using

$$\mathbf{H}=\mathbf{H}_a+\mathbf{H}_b+\mathbf{V}$$

in the usual equation of motion for the density operator in the Schrödinger picture; i.e.,

$$i\hbar(d\rho/dt)=[\mathbf{H},\rho], \quad (2.7)$$

where the right side is the symbol for the commutator of $\mathbf{H}$ and ρ. The succeeding terms contain the effects of the reservoirs on the atoms and on the field. The positive term in the curly brackets multiplying w_{ij} when $i\neq j$, gives the increase in the population of level i caused by reservoir-induced transitions from level j, with w_{ij} being the probability per unit time and per atom for such a transition. The negative terms, when $i\neq j$, give the decrease in the population of level j caused by reservoir-induced transitions to level i, and also effect the associated decay of the atomic polarization. The terms with $i=j$ do not involve level transitions, and do not affect the level populations. They correspond to reservoir-induced random-phase shifts, and effect additional decay of the polarization.

Finally, γ is the field decay constant caused by its reservoir. In the n representation for the field, one may see that the negative terms involving γ yield a decay of ρ_{nm} proportional to ρ_{nm} while the positive terms yield an increase in ρ_{nm} proportional to $\rho_{n+1,m+1}$ and $\rho_{n-1,m-1}$. The parameter $\bar{n}$ is the mean number of photons in the field when it is allowed to come to thermal equilibrium with its reservoir in the absence of any maser atoms. That is, $\bar{n}=[\exp(\hbar\omega_b/kT)-1]^{-1}$, where T is the resonator temperature.

These reservoir terms constitute a slightly abbreviated version of the more general results of Lax.[1] We have omitted terms whose only effect is to shift the atomic or field frequencies, assuming that any such shifts are incorporated in the definitions of ω_a and ω_b, respectively.

III. THE ANSATZ AND THE FOKKER-PLANCK EQUATION

We will show that (2.6) is consistent with the following expansion of the density operator ρ, namely,

$$\rho(t)=\int^{(6)}P_1(\beta,\beta^*,\mathfrak{X}_j,t)\sigma(\beta,\beta^*,\mathfrak{X}_j)d^{(2)}\beta\, d^{(4)}\mathfrak{X}. \quad (3.1)$$

In (3.1), P_1 is a weight distribution and σ is a simple product density operator for the field and N maser atoms, of the form

$$\sigma(\beta,\beta^*,\mathfrak{X}_j)=|\beta\rangle\langle\beta|\prod_{m=1}^{N}\sigma_m(\mathfrak{X}_j), \quad (3.2)$$

where $|\beta\rangle$ represents a pure coherent state of the field,[6] and

$$\sigma_m(\mathfrak{X}_j)=N^{-1}\{(N-\mathfrak{N}_1-\mathfrak{N}_2)(\mathbf{a}_0^\dagger\mathbf{a}_0)_m+\mathfrak{N}_1(\mathbf{a}_1^\dagger\mathbf{a}_1)_m+\mathfrak{N}_2(\mathbf{a}_2^\dagger\mathbf{a}_2)_m+\mathfrak{M}(\mathbf{a}_2^\dagger\mathbf{a}_1)_m+\mathfrak{M}^*(\mathbf{a}_1^\dagger\mathbf{a}_2)_m\} \quad (3.3)$$

represents a mixed state of the mth atom. We have shortened the writing of (3.1)–(3.3) by using the notation

$$\mathfrak{X}_j(j=1,2,3,4)\equiv(\mathfrak{N}_1,\mathfrak{N}_2,\mathfrak{M},\mathfrak{M}^*),$$
$$d^{(2)}\beta\equiv d(\mathrm{Re}\beta)d(\mathrm{Im}\beta),$$
$$d^{(4)}\mathfrak{X}\equiv d\mathfrak{N}_1 d\mathfrak{N}_2 d(\mathrm{Re}\mathfrak{M})d(\mathrm{Im}\mathfrak{M}). \quad (3.4)$$

We have also used a superscript (6) on the integral sign to represent a sixfold integral. To further simplify later writing we shall use

$$\sigma_b(\beta,\beta^*)\equiv|\beta\rangle\langle\beta|$$

and

$$\sigma_a(\mathfrak{X}_j)\equiv\prod_{m=1}^{N}\sigma_m(\mathfrak{X}_j), \tag{3.5}$$

so that we can write

$$\sigma(\beta,\beta^*,\mathfrak{X}_j)=\sigma_b(\beta,\beta^*)\sigma_a(\mathfrak{X}_j). \tag{3.6}$$

Thus, the subscript b pertains to the field, the subscript a pertains to all the atoms, while the subscript m pertains to the mth atom only.

The usefulness of the expansion (3.1) lies in the fact that the "elementary" density operator σ is of product (a symptom of statistical independence) form, and that in σ each atom has an identical description. The expansion is possible because the density operator equation (2.6) is symmetrical among all the atoms.

The motivation for making the expansion (3.1) was the desire to display the relation between semiclassical and fully quantal treatments of masers before any unessential approximations were made. The expansion of the field in the diagonal coherent state representation seemed appropriate, since a coherent state is the closest the uncertainty principle will allow a quantal description of a field to approach the usual classical description. The accompanying expansion for the atoms part of the density operator was arrived at intuitively to satisfy our expectation that each atom would have an identical description, but that statistical correlations among the atoms would remain within the coherent state field expansion because the present state of each atom depends not only on the present state of the field but also on its dynamically determined past history. For an atoms-field system described by $\sigma(\beta,\beta^*,\mathfrak{X}_j)$ the mean value of the complex field is proportional to β, the mean populations of atom levels 1 and 2 are $\mathfrak{N}_1$ and $\mathfrak{N}_2$, respectively, and the mean macroscopic complex polarization is proportional to $\mathfrak{M}$. A complete statistical relationship between moments of field-atom operators and the parameters of σ is given in Appendix C, but we will not need it at this point. In order to ensure that the eigenvalues of σ_m remain non-negative, the ranges of its parameters are restricted by

$$\begin{aligned}&\mathfrak{N}_1,\mathfrak{N}_2\geq 0,\\ &\mathfrak{N}_1+\mathfrak{N}_2\leq N,\\ &\mathfrak{M}\mathfrak{M}^*\leq\mathfrak{N}_1\mathfrak{N}_2.\end{aligned} \tag{3.7}$$

All of the component density operators of σ are normalized to unit trace, as therefore are σ_a and σ itself. If we trace (3.1) over all the atoms, we obtain

$$\rho_b\equiv \mathrm{Tr}_a\rho=\int^{(6)}P_1(\beta,\beta^*,\mathfrak{X}_j,t)\,|\,\beta\rangle\langle\beta\,|\,d^{(2)}\beta\,d^{(4)}\mathfrak{X}, \tag{3.8}$$

which is the reduced field density operator in the diagonal coherent state expansion. If we trace (3.1) over everything, and assume that ρ is also normalized to unit trace, we obtain

$$1=\int^{(6)}P_1(\beta,\beta^*,\mathfrak{X}_j,t)\,d^{(2)}\beta\,d^{(4)}\mathfrak{X}.$$

Thus P_1 has the form of a probability density; however, there is no *a priori* reason to suppose or expect that P_1 is everywhere positive. In mathematical terminology, P_1 is properly defined as a "distribution"[8] rather than as a function.

For later use we note that if $\mathbf{O}$ is any function of the quantum operators of the problem, then from (3.1) we may write the expression for the mean value of that operator function as

$$\langle\mathbf{O}\rangle\equiv\mathrm{Tr}(\rho\mathbf{O})=\int^{(6)}P_1[\mathrm{Tr}\sigma\mathbf{O}]d^{(2)}\beta\,d^{(4)}\mathfrak{X}. \tag{3.9}$$

To consolidate notation, let us define the angular brackets with subscript 1 by

$$\langle F\rangle_1\equiv\int^{(6)}P_1F\,d^{(2)}\beta\,d^{(4)}\mathfrak{X}, \tag{3.10}$$

where F may be any function of the variables of P_1. Then we can write (3.9) more briefly as

$$\langle\mathbf{O}\rangle=\langle\mathrm{Tr}(\sigma\mathbf{O})\rangle_1. \tag{3.11}$$

While (3.1) is not the most general atom-field density operator, a sensible model starting condition, i.e., the equilibrium state with V removed, may be so represented. If the starting condition is consistent with (3.1) then so will be the subsequent evolution of the density operator, as we shall show.

We shall confine our attention to situations wherein the vast majority of the atoms are continually in the ground-level 0, so that

$$\mathfrak{N}_1,\mathfrak{N}_2,\mathfrak{M},\mathfrak{M}^*\lll N. \tag{3.12}$$

Thus we treat the ground level as an infinite reservoir of atoms; the number of atoms in the two upper levels remain finite. We now show that upon insertion of (3.1) in the density operator equation (2.6), we obtain an equation of the Fokker–Planck form for the evolution of the probability density P_1. This is the basic equation upon which the remainder of our work is built.

To make use of (3.1), we need to know the effect of the various operators in (2.6) on the elementary density operator $\sigma(\beta,\beta^*,\mathfrak{X}_j)$. Using the identity

$$\sigma_b=|\,\beta\rangle\langle\beta\,|=\exp(-\beta^*\beta)\exp(\beta\mathbf{b}^\dagger)\,|\,0\rangle\langle 0\,|\exp(\beta^*\mathbf{b}), \tag{3.13}$$

where $|\,0\rangle$ represents the ground (vacuum) state of the field, we obtain

$$\mathbf{b}^\dagger\sigma_b=(\beta^*+\partial/\partial\beta)\sigma_b, \tag{3.14a}$$

$$\mathbf{b}\sigma_b=\beta\sigma_b. \tag{3.14b}$$

The first (3.14a) of these results may be seen by

performing the indicated differentiation[11] on (3.13). The second follows from (3.13) by virtue of the identities

$$[\mathbf{b}, \exp(\beta \mathbf{b}^\dagger)] = \beta \exp(\beta b^\dagger); \qquad \mathbf{b}\,|\,0\rangle = 0.$$

Alternatively, it is well known that the coherent state $|\beta\rangle$ is an eigenstate of $\mathbf{b}$ with eigenvalue β. The adjoints of (3.14a) and (3.14b) also apply.

Further, we have the identities

$$(a_1^\dagger a_2)_m = N(\partial/\partial\mathfrak{M}^*)\sigma_m, \tag{3.15a}$$

$$(a_j^\dagger a_j)_m = (1 + N(\partial/\partial\mathfrak{N}_j))\sigma_m; \qquad j = 1, 2 \tag{3.15b}$$

along with the adjoint of (3.15a). These follow directly from (3.3); in (3.15b) we have applied the limit (3.12). Using (3.15a) and (3.15b), and the properties of the atomic operators discussed in Appendix A, we find that the macroscopic atomic operators satisfy the relations

$$\mathbf{M}\sigma_a = \{\mathfrak{M}(1 + (\partial/\partial\mathfrak{N}_1)) + \mathfrak{N}_2(\partial/\partial\mathfrak{M}^*)\}\sigma_a, \tag{3.16a}$$

$$\mathbf{M}^\dagger\sigma_a = \{\mathfrak{M}^*(1 + (\partial/\partial\mathfrak{N}_2)) + \mathfrak{N}_1(\partial/\partial\mathfrak{M})\}\sigma_a, \tag{3.16b}$$

$$\mathbf{N}_1\sigma_a = \{\mathfrak{N}_1(1 + (\partial/\partial\mathfrak{N}_1)) + \mathfrak{M}^*(\partial/\partial\mathfrak{M}^*)\}\sigma_a, \tag{3.16c}$$

$$\mathbf{N}_2\sigma_a = \{\mathfrak{N}_2(1 + (\partial/\partial\mathfrak{N}_2)) + \mathfrak{M}(\partial/\partial\mathfrak{M})\}\sigma_a, \tag{3.16d}$$

along with their adjoints. Using (3.14a), (3.14b), and (3.16a)–(3.16d) we can express the first two terms of the right-hand side of (2.6), with σ replacing ρ, in terms of partial differential operators acting on σ. Using (3.14a), (3.14b), and (3.15a)–(3.15b), we can do the same for the reservoir interaction terms.

When we do all this, we find that insertion of (3.1) into (2.6) yields the following equation:

$$\begin{aligned}
\int^{(6)} \frac{\partial P_1}{\partial t}\,\sigma\, d^{(2)}\beta\, d^{(4)}\mathfrak{X} = \int^{(6)} P_1 \Big\{ & [\mu\mathfrak{M} - (\tfrac{1}{2}\gamma + i\omega_b)\beta]\frac{\partial}{\partial\beta} + [\mu\mathfrak{M}^* - (\tfrac{1}{2}\gamma - i\omega_b)\beta]\frac{\partial}{\partial\beta^*} \\
& + [\mu\beta(\mathfrak{N}_2 - \mathfrak{N}_1) - (\Gamma_{12} + i\omega_a)\mathfrak{M}]\frac{\partial}{\partial\mathfrak{M}} + [\mu\beta^*(\mathfrak{N}_2 - \mathfrak{N}_1) - (\Gamma_{12} - i\omega_a)\mathfrak{M}^*]\frac{\partial}{\partial\mathfrak{M}^*} \\
& + [R_1 + w_{12}\mathfrak{N}_2 - \Gamma_1\mathfrak{N}_1 + \mu(\beta^*\mathfrak{M} + \beta\mathfrak{M}^*)]\frac{\partial}{\partial\mathfrak{N}_1} \\
& + [R_2 + w_{21}\mathfrak{N}_1 - \Gamma_2\mathfrak{N}_2 - \mu(\beta^*\mathfrak{M} + \beta\mathfrak{M}^*)]\frac{\partial}{\partial\mathfrak{N}_2} + \bar{n}\gamma\frac{\partial^2}{\partial\beta\partial\beta^*} \\
& + \mu\mathfrak{N}_2\left(\frac{\partial^2}{\partial\beta\partial\mathfrak{M}^*} + \frac{\partial^2}{\partial\beta^*\partial\mathfrak{M}}\right) + \mu\mathfrak{M}\frac{\partial^2}{\partial\mathfrak{N}_1\partial\beta} + \mu\mathfrak{M}^*\frac{\partial^2}{\partial\mathfrak{N}_1\partial\beta^*}\Big\}\,\sigma\, d^{(2)}\beta\, d^{(4)}\mathfrak{X}.
\end{aligned} \tag{3.17}$$

In (3.17) the curly bracketed operator operates on σ. Also, $R_j \equiv Nw_{j0}$ is the rate at which atoms are pumped from the ground state to the jth state; $\Gamma_j \equiv \sum_{i\neq j} w_{ij}$ is the decay constant for the population of the jth state, and $\Gamma_{12} \equiv \frac{1}{2}\sum_j (w_{j1} + w_{j2})$ is the decay constant for the polarization. The next step is term by term partial integration of (3.17), bringing all derivatives away from σ. On performing this under the assumption that the surface integrals vanish, we obtain

$$\begin{aligned}
\int^{(6)} \frac{\partial P_1}{\partial t}\,\sigma\, d^{(2)}\beta\, d^{(4)}\mathfrak{X} = \int^{(6)} d^{(2)}\beta\, d^{(4)}\mathfrak{X}\sigma \Big\{ & -\frac{\partial}{\partial\beta}[\mu\mathfrak{M} - (\tfrac{1}{2}\gamma + i\omega_b)\beta] - \frac{\partial}{\partial\beta^*}[\mu\mathfrak{M}^* - (\tfrac{1}{2}\gamma - i\omega_b)\beta] \\
& - \frac{\partial}{\partial\mathfrak{M}}[\mu\beta(\mathfrak{N}_2 - \mathfrak{N}_1) - (\Gamma_{12} + i\omega_a)\mathfrak{M}] - \frac{\partial}{\partial\mathfrak{M}^*}[\mu\beta^*(\mathfrak{N}_2 - \mathfrak{N}_1) - (\Gamma_{12} - i\omega_a)\mathfrak{M}^*] \\
& - \frac{\partial}{\partial\mathfrak{N}_1}[R_1 + w_{12}\mathfrak{N}_2 - \Gamma_1\mathfrak{N}_1 + \mu(\beta^*\mathfrak{M} + \beta\mathfrak{M}^*)] - \frac{\partial}{\partial\mathfrak{N}_2}[R_2 + w_{21}\mathfrak{N}_1 - \Gamma_2\mathfrak{N}_2 - \mu(\beta^*\mathfrak{M} + \beta\mathfrak{M}^*)] \\
& + \frac{\partial^2}{\partial\beta\partial\beta^*}\bar{n}\gamma + \left(\frac{\partial^2}{\partial\beta\partial\mathfrak{M}^*} + \frac{\partial^2}{\partial\beta^*\partial\mathfrak{M}}\right)\mu\mathfrak{N}_2 + \frac{\partial^2}{\partial\mathfrak{N}_1\partial\beta}\mu\mathfrak{M} + \frac{\partial^2}{\partial\mathfrak{N}_1\partial\beta^*}\mu\mathfrak{M}^*\Big\}\, P_1,
\end{aligned} \tag{3.18}$$

where the long curly bracketed operator operates on P_1.

[11] Throughout this paper we make use of the simplicity of complex derivative notation. Translation to the real and imaginary parts of the complex variable β are made by the identities.

$$\beta \equiv \mathrm{Re}\beta - i\mathrm{Im}\beta,$$

$$\partial/\partial\beta = \tfrac{1}{2}\{[\partial/\partial(\mathrm{Re}\beta)] + i[\partial/\partial(\mathrm{Im}\beta)]\},$$

and their conjugates. Differentiations of functions of β and β^* with respect to β or β^* proceed as though β and β^* were independent variables, as can be proved by making the above translation, and partial integrations may be performed in the same manner. The area element $d^{(2)}\beta$ in the complex β plane always retains its identity as $d(\mathrm{Re}\beta)d(\mathrm{Im}\beta)$. Similar considerations apply to the complex variable $\mathfrak{M}$.

A sufficient condition for the satisfaction of (3.18) is for P_1 to satisfy the Fokker–Planck equation obtained by extracting the integrand; namely

$$\frac{\partial P_1(\beta,\beta^*,\mathfrak{X}_j,t)}{\partial t} = \left\{ -\frac{\partial}{\partial\beta}\left[\mu\mathfrak{M}-(\tfrac{1}{2}\gamma+i\omega_b)\beta\right] - \frac{\partial}{\partial\beta^*}\left[\mu\mathfrak{M}^*-(\tfrac{1}{2}\gamma-i\omega_b)\beta^*\right] \right.$$

$$-\frac{\partial}{\partial\mathfrak{M}}\left[\mu\beta(\mathfrak{N}_2-\mathfrak{N}_1)-(\Gamma_{12}+i\omega_a)\mathfrak{M}\right] - \frac{\partial}{\partial\mathfrak{M}^*}\left[\mu\beta^*(\mathfrak{N}_2-\mathfrak{N}_1)-(\Gamma_{12}-i\omega_a)\mathfrak{M}^*\right]$$

$$-\frac{\partial}{\partial\mathfrak{N}_1}\left[R_1+w_{12}\mathfrak{N}_2-\Gamma_1\mathfrak{N}_1+\mu(\beta^*\mathfrak{M}+\beta\mathfrak{M}^*)\right] - \frac{\partial}{\partial\mathfrak{N}_2}\left[R_2+w_{21}\mathfrak{N}_1-\Gamma_2\mathfrak{N}_2-\mu(\beta^*\mathfrak{M}+\beta\mathfrak{M}^*)\right]$$

$$\left. +\frac{\partial^2}{\partial\beta\partial\beta^*}\bar{n}\gamma + \left(\frac{\partial^2}{\partial\beta\partial\mathfrak{M}^*}+\frac{\partial^2}{\partial\beta^*\partial\mathfrak{M}}\right)\mu\mathfrak{N}_2 + \frac{\partial^2}{\partial\mathfrak{N}_1\partial\beta}\mu\mathfrak{M} + \frac{\partial^2}{\partial\mathfrak{N}_1\partial\beta^*}\mu\mathfrak{M}^* \right\} P_1. \tag{3.19}$$

Equation (3.19) is our basic result. The remainder of this paper consists of an investigation of its properties and utility. Given a solution of (3.19), then $\rho(t)$ as defined by (3.1) satisfies the Schrödinger equation of motion (2.6).

In statistical equations of the Fokker–Planck form, the first-derivative terms on the right-hand side prescribe the mean motions of the variables. They are called drift terms. The second-derivative terms prescribe the fluctuations and are called diffusion terms. In (3.19), each drift coefficient, the square-bracketed term following each first derivative time, is the mean rate of change of the variable in the derivative. To see this mathematically, we can derive from (3.19) the set of equations

$$\frac{d}{dt}\langle\beta\rangle_1 = \langle\mu\mathfrak{M}-(\tfrac{1}{2}\gamma+i\omega_b)\beta\rangle_1,$$

$$\frac{d}{dt}\langle\mathfrak{M}\rangle_1 = \langle\mu\beta(\mathfrak{N}_2-\mathfrak{N}_1)-(\Gamma_{12}+i\omega_a)\mathfrak{M}\rangle_1,$$

$$\frac{d}{dt}\langle\mathfrak{N}_1\rangle_1 = \langle R_1+w_{12}\mathfrak{N}_2-\Gamma_1\mathfrak{N}_1+\mu(\beta^*\mathfrak{M}+\beta\mathfrak{M}^*)\rangle_1,$$

$$\frac{d}{dt}\langle\mathfrak{N}_2\rangle_1 = \langle R_2+w_{21}\mathfrak{N}_1-\Gamma_2\mathfrak{N}_2-\mu(\beta^*\mathfrak{M}+\beta\mathfrak{M}^*)\rangle_1, \tag{3.20}$$

along with the conjugates of the first two. We have used here the notation (3.10).

Equations (3.20) are derived by writing out their left sides according to (3.10), inserting (3.19) for $\partial P_1/\partial t$, and then integrating by parts under our usual assumption that all surface integrals vanish.

So long as we can neglect the diffusion terms in (3.19), then delta function solutions[12] to it are possible. The position of any such delta function moves according to (3.20) with the mean value signs removed. Thus, equations (3.20) without the mean value signs may be considered as the set of dynamic equations which describe the maser when fluctuations are neglected, and indeed they are precisely the equations one would derive for this model problem by the methods of semi classical physics, with β playing the role of the classical field strength.

In contrast to the drift terms which, as we have seen, make quite good sense, the diffusion terms of (3.19) have a curious form. If the complex derivatives are re-expressed in real and imaginary parts,[11] the matrix of the diffusion coefficients is mainly off-diagonal and clearly has some negative eigenvalues. If the diffusion terms were diagonalized by a linear change of variables, the resulting equation would have some negative diagonal diffusion coefficients. Such a situation would not arise in the treatment of any sensible problem derived from classical physics.

Partly because of this problem of the diffusion matrix with negative eigenvalues, no solutions to the complex equation (3.19) are presently known. It would seem that a study of such equations is called for, particularly since (3.19), for example, has been derived from a simple physical model which one might reasonably expect to be well behaved.

To observe the effects of the diffusion terms in (3.19), we must evaluate the time derivatives of second moments. For example, in the same way that Eqs. (3.20) were derived, we find from (3.19) that

$$\frac{d}{dt}\langle\beta^*\beta\rangle_1 = \langle\mu(\beta^*\mathfrak{M}+\beta\mathfrak{M}^*)+\gamma(\bar{n}-\beta^*\beta)\rangle_1, \tag{3.21}$$

where the $\gamma\bar{n}$ on the right is contributed by the $\partial^2/\partial\beta\partial\beta^*$ diffusion term of (3.19). It is clear that $\gamma\bar{n}$ represents the generation of an incoherent or noise field, since no such term appears in the equation for the mean field. To attach physical significance to (3.21), we may derive from (3.11), (3.14), and (3.16) the relations

$$\langle \mathbf{b}^\dagger\mathbf{b}\rangle = \langle\beta^*\beta\rangle_1, \tag{3.22}$$

$$\mu\langle \mathbf{b}^\dagger\mathbf{M}+\mathbf{b}\mathbf{M}^\dagger\rangle = \mu\langle\beta^*\mathfrak{M}+\beta\mathfrak{M}^*\rangle_1. \tag{3.23}$$

[12] Here we are really speaking of solutions in which the weight P_1 is concentrated in a sufficiently small region that we are interested only in its center of gravity and not in any details of its shape. According to the full equation (3.19) such a situation will persist for a certain time depending on the diffusion coefficients.

Hence $\langle\beta^*\beta\rangle_1$ is the mean number of photons in the field, and we see that $\mu\langle\beta^*\mathcal{M}+\beta\mathcal{M}^*\rangle_1$ is the mean rate of transfer of energy from the atoms to the field. Note that this same term (3.23) appears in the mean rate equations (3.20) for $\mathcal{N}_1$ and $\mathcal{N}_2$, as it should.

If we now proceed to examine the equation of motion of $\langle\beta^*\mathcal{M}+\beta\mathcal{M}^*\rangle_1$, we find positive contributions from the $\partial^2/\partial\beta\partial\mathcal{M}^*$ and $\partial^2/\partial\beta^*\partial\mathcal{M}$ diffusion terms, which have the common coefficient $\mu\mathcal{N}_2$. This represents spontaneous emission, and we see that it affects the field as the result of a two-stage process. The remaining diffusion terms do not seem to have any such immediate physical interpretation.

IV. ADIABATIC ELIMINATION OF THE ATOMIC VARIABLES

Pending the availability of solutions to the full equation (3.19), we can examine its behavior in certain limiting situations. In particular, the equations can be reduced in the case that the atomic decay constants Γ are large compared to the field decay constant. Suppose first that $\Gamma_{12}\gg\Gamma_1$, Γ_2, γ. Then the polarization variable $\mathcal{M}$ is "instantaneously" (time scale Γ_{12}^{-1}) rather than dynamically determined by the other variables, and may be eliminated from the equation of motion. Such "adiabatic" approximations are commonly useful in physics.

As a first step we remove the high-frequency motion of P_1. In (3.19), β and $\mathcal{M}$ are tied together in phase, and when conditions for steady oscillation exist, the high-frequency motion of $\langle\beta\rangle_1$ and $\langle\mathcal{M}\rangle_1$ will be nearly as $\exp(-i\omega_0 t)$, where as is well known, or may be shown directly from the mean motions of the variables (3.20),

$$\omega_0=\omega_b+\alpha(\tfrac{1}{2}\gamma)=\omega_a-\alpha\Gamma_{12}, \tag{4.1}$$

where

$$\alpha\equiv(\omega_a-\omega_b)/(\Gamma_{12}+\tfrac{1}{2}\gamma).$$

Hence it is appropriate to change variables from $\mathcal{M}$ and β to $\mathcal{M}'$ and β', where

$$\begin{aligned}\mathcal{M}&=\mathcal{M}'\exp(-i\omega_0 t),\\ \beta&=\beta'\exp(-i\omega_0 t).\end{aligned} \tag{4.2}$$

The required transformation of the partial derivatives is

$$\begin{aligned}\frac{\partial}{\partial\mathcal{M}}&\to\exp(i\omega_0 t)\frac{\partial}{\partial\mathcal{M}'},\\ \frac{\partial}{\partial\beta}&\to\exp(i\omega_0 t)\frac{\partial}{\partial\beta'},\\ \frac{\partial}{\partial t}&\to\frac{\partial}{\partial t}+i\omega_0\left(\frac{\partial}{\partial\mathcal{M}'}\mathcal{M}'+\frac{\partial}{\partial\beta'}\beta'\right)\\ &\qquad -i\omega_0\left(\frac{\partial}{\partial\mathcal{M}'^*}\mathcal{M}'^*+\frac{\partial}{\partial\beta'^*}\beta'^*\right).\end{aligned} \tag{4.3}$$

The conjugates of these equations also apply. Making this change of variables and then omitting the primes for simplicity in writing, we obtain

$$\begin{aligned}\frac{\partial P_1}{\partial t}=\Big\{&-\frac{\partial}{\partial\beta}[\mu\mathcal{M}-\tfrac{1}{2}\gamma(1-i\alpha)\beta]-\frac{\partial}{\partial\beta^*}[\mu\mathcal{M}^*-\tfrac{1}{2}\gamma(1+i\alpha)\beta^*]\\ &-\frac{\partial}{\partial\mathcal{M}}[\mu\beta(\mathcal{N}_2-\mathcal{N}_1)-\Gamma_{12}(1+i\alpha)\mathcal{M}]-\frac{\partial}{\partial\mathcal{M}^*}[\mu\beta^*(\mathcal{N}_2-\mathcal{N}_1)-\Gamma_{12}(1-i\alpha)\mathcal{M}^*]\\ &-\frac{\partial}{\partial\mathcal{N}_1}[R_1+w_{12}\mathcal{N}_2-\Gamma_1\mathcal{N}_1+\mu(\beta^*\mathcal{M}+\beta\mathcal{M}^*)]-\frac{\partial}{\partial\mathcal{N}_2}[R_2+w_{21}\mathcal{N}_1-\Gamma_2\mathcal{N}_2-\mu(\beta^*\mathcal{M}+\beta\mathcal{M}^*)]\\ &+\frac{\partial^2}{\partial\beta\partial\beta^*}\bar{n}\gamma+\left(\frac{\partial^2}{\partial\beta\partial\mathcal{M}^*}+\frac{\partial^2}{\partial\beta^*\partial\mathcal{M}}\right)\mu\mathcal{N}_2+\frac{\partial^2}{\partial\mathcal{N}_1\partial\beta}\mu\mathcal{M}+\frac{\partial^2}{\partial\mathcal{N}_1\partial\beta^*}\mu\mathcal{M}^*\Big\}P_1.\end{aligned} \tag{4.4}$$

The adiabatic elimination of $\mathcal{M}$ and $\mathcal{M}^*$ from the problem now involves the assumption that Γ_{12} is large enough so that $\mathcal{M}$, $\mathcal{M}^*$ come into statistical equilibrium in a time short compared to that required for any of the other variables to change appreciably. A method of adiabatic elimination of variables from Fokker–Planck equations is discussed in Appendix B. In (4.4), it suffices to pick out the terms involving $\partial/\partial\mathcal{M}$ and $\partial/\partial\mathcal{M}^*$ and set them equal to zero. Thus we have

$$\frac{\partial}{\partial\mathcal{M}}\left\{\mu\beta(\mathcal{N}_2-\mathcal{N}_1)-\Gamma_{12}(1+i\alpha)\mathcal{M}-\frac{\partial}{\partial\beta^*}\mu\mathcal{N}_2\right\}P_1=0 \tag{4.5}$$

and its conjugate, which are solved by

$$\mathcal{M}P_1=\frac{\mu}{\Gamma_{12}(1+i\alpha)}\left\{\beta(\mathcal{N}_2-\mathcal{N}_1)-\frac{\partial}{\partial\beta^*}\mathcal{N}_2\right\}P_1 \tag{4.6}$$

and its conjugate. This picking out extracts all terms in (4.4) which are important for time intervals of the order of Γ_{12}^{-1}; the remainder of Eq. (4.4) then holds for time intervals much longer than Γ_{12}^{-1}. We use the adiabatic solution (4.6) to eliminate $\mathcal{M}$ and $\mathcal{M}^*$ from (4.4). In accord with the discussion in Appendix B, the variables $\mathcal{M}$ and $\mathcal{M}^*$ must be placed to the right of all other variables when the substitutions (4.6) and its conjugate are made. This last is important since the products $\mathfrak{M}\beta^*$ and $\mathfrak{M}^*\beta$ occur in (4.4), and for example the adiabatic solution for $\mathfrak{M}$ does not commute with β^*. Bringing all the resulting derivatives to the left to re-establish the Fokker–Planck form, and integrating over the $\mathfrak{M}$ plane, we obtain, as the new equation with $\mathfrak{M}$ and $\mathfrak{M}^*$ eliminated,

$$\frac{\partial P_1(\beta,\beta^*,\mathfrak{N}_1,\mathfrak{N}_2,t)}{\partial t} = \Big\{-\frac{\partial}{\partial\beta}[\pi(\mathfrak{N}_2-\mathfrak{N}_1)-\gamma](\tfrac{1}{2}(1-i\alpha)\beta)-\frac{\partial}{\partial\beta^*}[\pi(\mathfrak{N}_2-\mathfrak{N}_1)-\gamma](\tfrac{1}{2}(1+i\alpha)\beta^*)$$
$$-\frac{\partial}{\partial\mathfrak{N}_1}[R_1+(w_{12}+\pi+\pi\beta\beta^*)\mathfrak{N}_2-(\Gamma_1+\pi\beta\beta^*)\mathfrak{N}_1]-\frac{\partial}{\partial\mathfrak{N}_2}[R_2-(\Gamma_2+\pi+\pi\beta\beta^*)\mathfrak{N}_2+(w_{21}+\pi\beta\beta^*)\mathfrak{N}_1]$$
$$+\frac{\partial^2}{\partial\beta\partial\beta^*}[\gamma\bar{n}+\pi\mathfrak{N}_2]+\frac{\partial^2}{\partial\mathfrak{N}_1\partial\beta}[\pi\beta\mathfrak{N}_2-\tfrac{1}{2}\pi(1-i\alpha)\beta\mathfrak{N}_1]+\frac{\partial^2}{\partial\mathfrak{N}_1\partial\beta^*}[\pi\beta^*\mathfrak{N}_2-\tfrac{1}{2}\pi(1+i\alpha)\beta^*\mathfrak{N}_1]$$
$$+\frac{\partial^2}{\partial\mathfrak{N}_2\partial\beta}[-\tfrac{1}{2}\pi(1+i\alpha)\beta\mathfrak{N}_2]+\frac{\partial^2}{\partial\mathfrak{N}_2\partial\beta^*}[-\tfrac{1}{2}\pi(1-i\alpha)\beta^*\mathfrak{N}_2]-\frac{\partial^3}{\partial\mathfrak{N}_1\partial\beta\partial\beta^*}[\pi\mathfrak{N}_2]\Big\}P_1, \quad (4.7)$$

where

$$\pi\equiv 2\mu^2/\Gamma_{12}(1+\alpha^2)$$

is the decay constant for spontaneous emission of a photon into the laser mode b by an atom in level 2. By contrast, Γ_2 includes spontaneous emission into all other field modes.

In (4.7) the weight distribution P_1 has been integrated over the $\mathfrak{M}$ plane; that is,

$$P_1(\beta,\beta^*,\mathfrak{N}_1,\mathfrak{N}_2,t)\equiv\int^{(2)}P_1(\beta,\beta^*,\mathfrak{N}_1,\mathfrak{N}_2,\mathfrak{M},\mathfrak{M}^*)\,d^{(2)}\mathfrak{M}.$$

By using (4.6) and its conjugate, any moment of the variables including $\mathfrak{M}$ and $\mathfrak{M}^*$ can be re-expressed in terms of the variables other than $\mathfrak{M}$ and $\mathfrak{M}^*$, so that (4.6) and (4.7) together now describe the complete behavior of the maser.

In (4.7), spontaneous emission into the laser mode has become quite explicit, and appears properly in the drift terms which give the mean motion of the atomic populations, and in the field diffusion term $\partial^2/\partial\beta\partial\beta^*$. The origin of all these spontaneous emission terms can be traced to the $\partial^2/\partial\beta\partial\mathfrak{M}^*$ and $\partial^2/\partial\beta^*\partial\mathfrak{M}$ diffusion terms in (3.19). The group of off-diagonal diffusion terms and the third-derivative term give rise to the correlations between level populations and field which are a result of spontaneous emission.

Since intensity[13] and phase fluctuations have different characteristics in an oscillator, it is interesting to transform (4.7) to polar coordinates in the complex β plane. Accordingly, we transform variables in (4.7) from β and β^* to the real variables I and θ, where

$$\beta\equiv I^{1/2}\exp(-i\theta), \quad (4.8)$$

from which we obtain

$$\frac{\partial}{\partial\beta}\beta=\left(\frac{\partial}{\partial\beta^*}\beta^*\right)^*=\frac{\partial}{\partial I}I+\tfrac{1}{2}i\frac{\partial}{\partial\theta},$$

$$\frac{\partial^2}{\partial\beta\partial\beta^*}=\frac{\partial^2}{\partial I^2}I-\frac{\partial}{\partial I}+\frac{\partial^2}{\partial\theta^2}\frac{1}{4I},$$

$$d^{(2)}\beta=\tfrac{1}{2}dI\,d\theta,$$

$$P_1(\beta,\beta^*,\mathfrak{N}_1,\mathfrak{N}_2,t)=2P_1(I,\theta,\mathfrak{N}_1,\mathfrak{N}_2,t).$$

With these substitutions, (4.7) transforms to

$$\frac{\partial P_1(I,\theta,\mathfrak{N}_1,\mathfrak{N}_2,t)}{\partial t} = \Big\{-\frac{\partial}{\partial I}[I\{\pi(\mathfrak{N}_2-\mathfrak{N}_1)-\gamma\}+\gamma\bar{n}+\pi\mathfrak{N}_2]-\frac{\partial}{\partial\theta}[\tfrac{1}{2}\alpha\{\pi(\mathfrak{N}_2-\mathfrak{N}_1)-\gamma\}]$$
$$-\frac{\partial}{\partial\mathfrak{N}_1}[R_1+\{w_{12}+\pi(I+1)\}\mathfrak{N}_2-(\Gamma_1+\pi I)\mathfrak{N}_1]-\frac{\partial}{\partial\mathfrak{N}_2}[R_2-\{\Gamma_2+\pi(I+1)\}\mathfrak{N}_2+(w_{21}+\pi I)\mathfrak{N}_1]$$
$$+\frac{\partial^2}{\partial I^2}[I(\gamma\bar{n}+\pi\mathfrak{N}_2)]+\frac{\partial^2}{\partial\theta^2}\left[\frac{1}{4I}(\gamma\bar{n}+\pi\mathfrak{N}_2)\right]+\frac{\partial^2}{\partial\mathfrak{N}_1\partial I}[\pi(2I+1)\mathfrak{N}_2-\pi I\mathfrak{N}_1]$$
$$+\frac{\partial^2}{\partial\mathfrak{N}_2\partial I}[-\pi I\mathfrak{N}_2]+\frac{\partial^2}{\partial\mathfrak{N}_1\partial\theta}[-\tfrac{1}{2}\pi\alpha\mathfrak{N}_1]+\frac{\partial^2}{\partial\mathfrak{N}_2\partial\theta}[\tfrac{1}{2}\pi\alpha\mathfrak{N}_2]-\frac{\partial^3}{\partial\mathfrak{N}_1\partial I^2}[\pi I\mathfrak{N}_2]-\frac{\partial^3}{\partial\mathfrak{N}_1\partial\theta^2}\left[\frac{\pi\mathfrak{N}}{4I}\right]\Big\}P_1. \quad (4.9)$$

[13] We use intensity (amplitude squared) rather than amplitude as a variable here simply because it is the intensity which appears in the drift and diffusion coefficients. Lax (Ref. 3) has shown that use of intensity leads to "quasilinear" solutions which are fairly accurate even in the region near threshold.

Note that in (4.9) both the intensity and phase diffusion coefficients ($\partial^2/\partial I^2$ and $\partial^2/\partial\theta^2$ terms, respectively) are proportional to the factor $(\gamma\bar{n}+\pi\mathfrak{N}_2)$, which also appears as the spontaneous part of the intensity drift coefficient. Again, there occur a number of off-diagonal diffusion terms whose individual physical significances are not particularly apparent.

Finally, we can eliminate $\mathfrak{N}_1$ and $\mathfrak{N}_2$ from the equation in the case that $\Gamma_1, \Gamma_2 \gg \gamma$, so that the level populations are adiabatically dependent upon the field strength. Because $\mathfrak{N}_2$ and $\mathfrak{N}_1$ appear in the diffusion coefficients, we cannot do this elimination "exactly", as was the case with $\mathfrak{M}$. However, the physically important results are probably contained in the diffusion, or Fokker–Planck, approximation, where the resulting equation is found only up through second derivative terms.

In accordance with the discussion of Appendix B, we accomplish the adiabatic elimination of $\mathfrak{N}_1$ and $\mathfrak{N}_2$ by setting (for $i=1, 2$)

$$\mathfrak{N}_i P_1 = [A_i - (\partial/\partial I)B_i - (\partial/\partial\theta)C_i + \cdots]P_1. \quad (4.10)$$

We substitute (4.10) into (4.9) to eliminate $\mathfrak{N}_1$ and $\mathfrak{N}_2$ from the drift and diffusion coefficients, bring all derivatives to the left, and then solve for the A_i, B_i, and C_i so that all first- and second-derivative terms containing the $\partial/\partial\mathfrak{N}_i$ vanish. We find that by order of magnitude

$$C:B:A = \pi:\pi I:\Gamma.$$

With neglect of w_{12} and w_{21}, and also with neglect of π (but not πI) with respect to Γ_1 and Γ_2, we find the resulting equation

$$\frac{\partial P_1(I,\theta,t)}{\partial t} = \Big\{-\frac{\partial}{\partial I}[\pi(I+1)A_2-\pi IA_1+\gamma(\bar{n}-I)] - \frac{\partial}{\partial\theta}[\tfrac{1}{2}\alpha\{\pi(A_2-A_1)-\gamma\}] + \frac{\partial^2}{\partial I^2}[\gamma\bar{n}I+\mathfrak{D}^{-1}\pi I\Gamma_2\{\Gamma_1A_2-\pi I(A_2-A_1)\}]$$
$$+\frac{\partial^2}{\partial\theta^2}[(4I)^{-1}(\gamma\bar{n}+\pi A_2)+\mathfrak{D}^{-1}\tfrac{1}{2}(\pi\alpha)^2(\Gamma_1A_2+\Gamma_2A_1)] + \frac{\partial^2}{\partial\theta\partial I}[-\mathfrak{D}^{-1}\pi^2\alpha I\Gamma_2(A_2-A_1)]\Big\} P, \quad (4.11)$$

where

$$\mathfrak{D} \equiv \Gamma_1\Gamma_2 + \pi I(\Gamma_1+\Gamma_2),$$

and where A_2 and A_1 are the adiabatic mean values of $\mathfrak{N}_2$ and $\mathfrak{N}_1$, as may be seen from (4.10), and are given by

$$A_2 = \mathfrak{D}^{-1}[\Gamma_1R_1+\pi I(R_1+R_2)],$$
$$A_1 = \mathfrak{D}^{-1}[\Gamma_2R_1+\pi I(R_1+R_2)]. \quad (4.12)$$

Equation (4.11) is in agreement with similar results obtained recently by Lax and by Scully using quite different methods. The fact that we all reach consistent results here lends some extra credence to all of the methods.

The effect of saturation on the diffusion terms of (4.11) is a noteworthy feature. The phase diffusion coefficient, at least on resonance ($\alpha=0$) is only affected in so far as A_2 is affected. The intensity diffusion coefficient, on the other hand, tends to be surpressed further, and for very high power, where

$$\pi I \gg \Gamma_1, \Gamma_2,$$

evaluates approximately to

$$\gamma\bar{n}I + (\Gamma_2/(\Gamma_1+\Gamma_2))R_1.$$

It is in principle possible to have this quantity nearly zero, by having $\bar{n}$, $R_1 \to 0$. In such a case P_1 would come to steady state with a very small variance in I.

In terms of $P_1(I, \theta, t)$, the field density operator is given by

$$\rho_b = \int^{(2)} P_1(I,\theta,t) \mid I^{1/2}\exp[-i(\omega_0t+\theta)]\rangle \times \langle I^{1/2}\exp[-i(\omega_0t+\theta)] \mid dI\, d\theta. \quad (4.13)$$

We may conclude that a maser oscillator, operating under highly saturated conditions, can produce a field which is very nearly in a coherent state. The phase of the field will, as may be shown from (4.11), diffuse slowly away from some initially measured value according to the law

$$(d/dt)\langle\exp(-i\theta)\rangle_1 = -[(4I)^{-1}(\gamma\bar{n}+\pi A_2)]\langle\exp(-i\theta)\rangle_1 \quad (4.14)$$

for the case $\alpha \approx 0$. The result that the drift of θ is zero for $\alpha=0$ is a result of our original choice for ω_0. In (4.14), we have assumed that I and A_2 are sensibly constant in the average taking and have brought them outside of the mean value sign.[13a] Equation (4.14) im-

[13a] *Note added in proof*: If the maser is operating moderately above threshold, it is appropriate to make an adiabatic elimination of the intensity I from (4.11). Straightforward application of the method discussed in Appendix B results in a phase-only diffusion equation, with the diffusion constant given by

$$[4\langle I\rangle_1]^{-1}(\gamma\bar{n}+\pi A_2)(1+\alpha^2),$$

where $A_2 = A_2(\langle I\rangle_1)$. This result generalizes the width of the Lorentzian phase spectrum, multiplying the result given just below in the text by a detuning factor $(1+\alpha^2)$, and specifying that A_2 and I are evaluated with I taken as the adiabatic mean value $\langle I\rangle_1$. This result is consistent with Lax's quasilinear analysis.

plies that the oscillation signal has a Lorentzian power spectrum with a full width at half-power given by

$$\Delta\omega=(\gamma\bar{n}+\pi A_2)/2I.$$

This is a well-known result, which is now justified for arbitrarily high levels of saturation.

One final point is worth mention here. Since we have eliminated $\mathfrak{M}$ and $\mathfrak{M}^*$ from our basic equation first, and then have eliminated $\mathfrak{N}_1$ and $\mathfrak{N}_2$ later, it would seem that the validity of (4.11) should be limited to the regime $\Gamma_{12}\gg\Gamma_1$, $\Gamma_2\gg\gamma$. It turns out otherwise; there is a simplicity in this problem, whose mathematical symptom is that the adiabatic $\mathfrak{M}$ (4.6) has no $\partial/\partial\mathfrak{N}_1$ or $\partial/\partial\mathfrak{N}_2$ terms in it, and whose result is that a simultaneous adiabatic elimination of $\mathfrak{M}$, $\mathfrak{M}^*$, $\mathfrak{N}_1$, and $\mathfrak{N}_2$ in favor of the field variables β and β^* leads precisely to (4.11). Thus (4.11) is in fact valid for any possible relationship of Γ_{12} to Γ_1 and Γ_2.

V. THE GENERALIZED WIGNER DENSITY AND THE APPROXIMATE CLASSICAL MODEL

It is not obvious from our exact basic equation (3.19) for P_1 that there exists a classical model for the maser which approximates the properties of the quantal model including fluctuations. In this section, we show that there is such a classical model, and show quantitatively the relationship between it and the quantal model. To this end we shall use (3.19) to derive the equation of motion of what may be called the generalized Wigner density for the maser.

We symbolize the Wigner density by P_2 and define it from the moment generating function relationship

$$\mathrm{Tr}[\rho(t)\ \exp i(\lambda^*\mathbf{b}+\lambda\mathbf{b}^\dagger+\xi_1\mathbf{N}_1+\xi_2\mathbf{N}_2+\xi_3\mathbf{M}+\xi_4\mathbf{M}^\dagger)]$$
$$\equiv\int^{(6)}P_2(b,b^*,N_1,N_2,M,M^*,t)$$
$$\times\exp i(\lambda^*b+\lambda b^*+\xi_1N_1+\xi_2N_2+\xi_3M+\xi_4M^*)$$
$$\times d^{(2)}b\ dN_1dN_2d^{(2)}M. \tag{5.1}$$

Here, λ, λ^*, and the ξ's are the expansion parameters of the generating functions. If ξ_1 and ξ_2 are real and $\xi_4=\xi_3^*$, then the exponential factor becomes pure imaginary. In (5.1) we have introduced a new set of c-number variables $(b, b^*, N_1, N_2, M, M^*)$, and their associated weight distribution P_2, having the property that any moment of these variables is equal to the *symmetrized* moment of the corresponding operator variables. It is clear from (5.1) that P_2 is the Fourier transform of the symmetrically ordered quantum characteristic function. If the atomic variables are omitted, one can show that P_2 is the ordinary Wigner density[14] for the field. That P_2 should turn out to behave sensibly from a classical viewpoint is not too surprising, as the correspondence between classical variables and the symmetrized form of quantum variables has long been recognized. Nevertheless, it is interesting that the diffusion terms in the Fokker–Planck equation for P_2 turn out to be remarkably simple even for this dynamical nonlinear problem.

In order to derive the equation of motion of P_2, we first need to know its relationship to P_1 whose equation of motion we already have. We find this by inserting the expansion (3.1) in the left side of (5.1). For convenience in writing this, let us use the definition

$$\mathbf{Q}\equiv\exp i(\lambda^*\mathbf{b}+\lambda\mathbf{b}^\dagger+\xi_1\mathbf{N}_1+\xi_2\mathbf{N}_2+\xi_3\mathbf{M}+\xi_4\mathbf{M}^\dagger). \tag{5.2}$$

By virtue of the definitions (5.2) and (3.9), we can write the left side of (5.1) simply as $\langle\mathbf{Q}\rangle$, and from (3.9) we have

$$\langle\mathbf{Q}\rangle=\int^{(6)}P_1(\mathrm{Tr}\sigma\mathbf{Q})d^{(2)}\beta\ d\mathfrak{N}_1d\mathfrak{N}_2d^{(2)}\mathfrak{M}. \tag{5.3}$$

In Appendix C the trace of $\sigma\mathbf{Q}$ is evaluated, and is shown to be given in the limit (3.12) by

$$\mathrm{Tr}\sigma\mathbf{Q}$$
$$=\exp\{i(\nu_1\mathfrak{N}_1+\nu_2\mathfrak{N}_2+\nu_3\mathfrak{M}+\nu_4\mathfrak{M}^*+\lambda^*\beta+\lambda\beta^*)-\tfrac{1}{2}\lambda\lambda^*\}, \tag{5.4a}$$

where

$$i\nu_1\equiv\exp(i\xi_+)[\cos v-i(\xi_-/v)\sin v]-1,$$
$$i\nu_2\equiv\exp(i\xi_+)[\cos v+i(\xi_-/v)\sin v]-1,$$
$$i\nu_3\equiv i\exp(i\xi_+)(\xi_3/v)\sin v,$$
$$i\nu_4\equiv i\exp(i\xi_+)(\xi_4/v)\sin v, \tag{5.4b}$$

and where

$$\xi_+\equiv\tfrac{1}{2}(\xi_2+\xi_1);\qquad \xi_-\equiv\tfrac{1}{2}(\xi_2-\xi_1);\qquad v\equiv(\xi_-^2+\xi_3\xi_4)^{1/2}.$$

Comparison of (5.3) and (5.4) with (5.1) yields the required relation between P_2 and P_1, namely

$$\int^{(6)}P_2\exp\{i(\lambda^*b+\lambda b^*+\xi_1N_1+\xi_2N_2+\xi_3M+\xi_4M^*)\}d^{(2)}b\ dN_1dN_2d^{(2)}M$$
$$=\int^{(6)}P_1\exp\{-\tfrac{1}{2}(\lambda\lambda^*)+i(\lambda^*\beta+\lambda\beta^*+\nu_1\mathfrak{N}_1+\nu_2\mathfrak{N}_2+\nu_3\mathfrak{M}+\nu_4\mathfrak{M}^*)\}d^{(2)}\beta\ d\mathfrak{N}_1d\mathfrak{N}_2d^{(2)}\mathfrak{M}, \tag{5.5}$$

[14] J. E. Moyal, Proc. Cambridge Phil. Soc. **45**, 99 (1949).

where the ν's and the ξ's are related by Eqs. (5.4b). From (5.5) and the equation of motion (3.19) for P_1, we can derive the equation of motion for P_2. It would appear that there is much labor and little insight to be gained by attempting to do this exactly. However, if the sum $\mathfrak{N}_1+\mathfrak{N}_2$ is large compared to unity over the important range of P_1, then Tr $\sigma\mathbf{Q}$ will be appreciable only for small values of the ξ's (particularly in the combinations ξ_+ and v), and it is appropriate to expand (5.4b) in a power series in the ξ's. Keeping only second power terms, we obtain

$$\begin{aligned} i\nu_1&=i\xi_1-\tfrac{1}{2}(\xi_1^2+\xi_3\xi_4),\\ i\nu_2&=i\xi_2-\tfrac{1}{2}(\xi_2^2+\xi_3\xi_4),\\ i\nu_3&=i\xi_3-\tfrac{1}{2}\xi_3(\xi_1+\xi_2),\\ i\nu_4&=i\xi_4-\tfrac{1}{2}\xi_4(\xi_1+\xi_2). \end{aligned} \tag{5.6}$$

The approximation (5.6) yields P_2 as a Gaussian convolution of P_1, as shown in Appendix D, and gives the diffusion approximation to the equation of motion for P_2. The fact that P_2 is a smoothed version of P_1 explains to some extent why the diffusion equation for P_2 is better behaved than is that for P_1. The method of derivation of the diffusion equation for P_2 is given in Appendix E. The result is as follows:

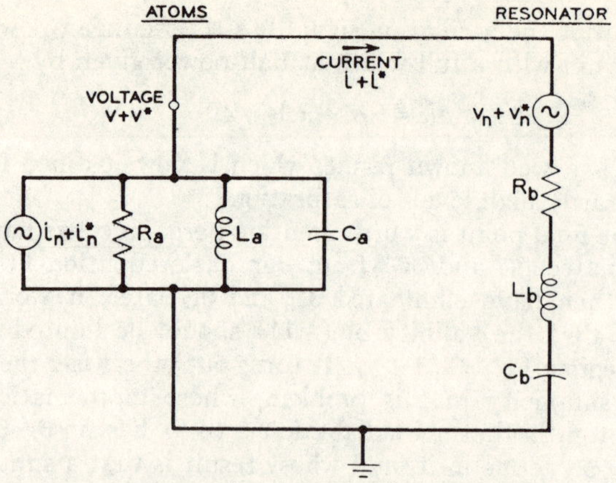

FIG. 2. Circuit equivalent for the maser model.

$$\begin{aligned}
&\frac{\partial P_2(b,b^*,N_1,N_2,M,M^*,t)}{\partial t}\\
&=\Big\{-\frac{\partial}{\partial b}[\mu M-(\tfrac{1}{2}\gamma+i\omega_b)b]-\frac{\partial}{\partial b^*}[\mu M^*-(\tfrac{1}{2}\gamma-i\omega_b)b^*]-\frac{\partial}{\partial M}[\mu b(N_2-N_1)-(\Gamma_{12}+i\omega_a)M]\\
&\quad-\frac{\partial}{\partial M^*}[\mu b^*(N_2-N_1)-(\Gamma_{12}-i\omega_a)M^*]-\frac{\partial}{\partial N_1}[R_1+w_{12}N_2-\Gamma_1N_1+\mu(b^*M+bM^*)]\\
&\quad-\frac{\partial}{\partial N_2}[R_2+w_{21}N_1-\Gamma_2N_2-\mu(b^*M+bM^*)]+\frac{\partial^2}{\partial b\partial b^*}[\gamma(\bar{n}+\tfrac{1}{2})]\\
&\quad+\frac{\partial^2}{\partial M\partial M^*}[\Gamma_{12}(N_1+N_2)+\tfrac{1}{2}(R_1+R_2-\Gamma_1N_1-\Gamma_2N_2+w_{21}N_1+w_{12}N_2)]\\
&\quad+\frac{\partial^2}{\partial N_1^2}[\tfrac{1}{2}(R_1+\Gamma_1N_1+w_{12}N_2)]+\frac{\partial^2}{\partial N_2^2}[\tfrac{1}{2}(R_2+\Gamma_2N_2+w_{21}N_1)]\\
&\quad+\frac{\partial^2}{\partial N_1\partial M}[\tfrac{1}{2}(\Gamma_1-w_{12})M]+\frac{\partial^2}{\partial N_2\partial M^*}[\tfrac{1}{2}(\Gamma_1-w_{12})M^*]\\
&\quad+\frac{\partial^2}{\partial N_2\partial M}[\tfrac{1}{2}(\Gamma_2-w_{21})M]+\frac{\partial^2}{\partial N_2\partial M^*}[\tfrac{1}{2}(\Gamma_2-w_{21})M^*]+\frac{\partial^2}{\partial N_1\partial N_2}[-w_{12}N_2-w_{21}N_1]\Big\}\,P_2.
\end{aligned} \tag{5.7}$$

Equation (5.7), unlike (3.19), looks quite sensible from a classical viewpoint. The diffusion matrix is positive definite and does not couple the atoms with the field. From (5.7) one can derive a simple classical-model equivalent to the maser. The small price one pays is that (5.7) is a good approximation only when the number of atoms in levels one and two is reasonably large. Of course, it is entirely possible to write the exact form of (5.7), but then its advantage of intuitive simplicity would be lost.

Equation (5.7) is equivalent, in the sense of providing equations of motion for various moments of the variables, to the following set of dynamic Langevin equations of motion:

$$\begin{aligned} db/dt&=\mu M-(\tfrac{1}{2}\gamma+i\omega_b)b+F_b(t),\\ dM/dt&=\mu b(N_2-N_1)-(\Gamma_{12}+i\omega_a)M+F_M(t),\\ dN_1/dt&=R_1+w_{12}N_2-\Gamma_1N_1+\mu(b^*M+bM^*)+F_1(t),\\ dN_2/dt&=R_2+w_{12}N_1-\Gamma_2N_2-\mu(b^*M+bM^*)+F_2(t), \end{aligned} \tag{5.8}$$

where the random forces $F(t)$ have zero mean values, broad spectra with respect to the important spectral widths γ and Γ_{12} of our problem, and have the following nonzero correlation functions:

$$
\begin{aligned}
&\langle F_b(t')F_{b*}(t)\rangle=\gamma(\bar{n}+\tfrac{1}{2})\delta_s(t-t')\exp[i\omega_b(t-t')],\\
&\langle F_M(t')F_{M*}(t)\rangle=[\Gamma_{12}(N_1+N_2)+\tfrac{1}{2}(R_1+R_2-\Gamma_1N_1-\Gamma_2N_2+w_{12}N_2+w_{21}N_1)]\delta_s(t-t')\exp[i\omega_a(t-t')],\\
&\langle F_1(t)F_1(t')\rangle=(R_1+\Gamma_1N_1+w_{12}N_2)\delta_s(t-t'),\\
&\langle F_2(t)F_2(t')\rangle=(R_2+\Gamma_2N_2+w_{21}N_1)\delta_s(t-t'),\\
&\langle F_M(t)F_1(t')\rangle=\tfrac{1}{2}(\Gamma_1-w_{12})M(t)\delta_s(t-t'),\\
&\langle F_M(t)F_2(t')\rangle=\tfrac{1}{2}(\Gamma_2-w_{21})M(t)\delta_s(t-t'),\\
&\langle F_1(t)F_2(t')\rangle=-(w_{21}N_1+w_{12}N_2)\delta_s(t-t').
\end{aligned}
\tag{5.9}
$$

In (5.9) the subscript s on the delta function stands for "slow," and indicates consideration of time intervals not too much shorter than the reciprocal bandwidths of the problem; i.e., of the order of γ^{-1} or Γ_{12}^{-1}. The exponential time dependences, and the explicit indication of t in $M(t)$ are included to take care of the high-frequency variation of F_M and F_b. The physical point involved here is that only the spectral region near the resonator and atomic frequencies is important for F_M and F_b, and over this region they have white spectra.

The first two of the dynamic equations (5.8) along with the first two second-moment formulas (5.9) for the random forces are exactly those derivable in the high-Q approximation from the classical equivalent circuit shown in Fig. 2. A series resonant circuit with a noise-voltage source represents the resonator, while a parallel circuit with a noise-current source represents the atoms. The two noise sources are statistically independent. The current in the series circuit is proportional to the field strength, while the voltage across the parallel circuit is proportional to the polarization. If we represent the circuit voltage and current by the sum of positive- and negative-frequency parts

$$(v+v^*),\ (i+i^*);\qquad v,\, i\propto\exp(-i\omega t),$$

then the equivalence relations are

$$
\begin{aligned}
(L_bC_b)^{1/2}&=\omega_b,\\
R_b/L_b&=\gamma,\\
i&=(\hbar\omega\gamma/2R_b)^{1/2}b,\\
v_n&=(2\hbar\omega R_b/\gamma)^{1/2}F_b,\\
(L_aC_a)^{-1/2}&=\omega_a,\\
1/R_aC_a&=2\Gamma_{12},\\
v&=\mu(\hbar\omega 2R_b/\gamma)^{1/2}M,\\
i_n&=[\hbar\omega/R_a\Gamma_{12}(N_1-N_2)]^{1/2}F_M,
\end{aligned}
$$

and

$$R_a=R_b(2\mu^2/\Gamma_{12}\gamma)(N_1-N_2). \tag{5.10}$$

The proportionality between i and b has been so chosen that power and energy relationships come out correctly; e.g., the power delivered to the resonator is

$$vi^*+iv^*=\mu\hbar\omega(Mb^*+M^*b) \tag{5.11}$$

and the energy stored in the resonator is

$$2L_1i^*i=\hbar\omega b^*b. \tag{5.12}$$

The second moments of the random forces show that in the equivalent circuit

$$
\begin{aligned}
\langle v_n(t)v_n^*(t')\rangle&=2R_b\hbar\omega(\bar{n}+\tfrac{1}{2})\delta_s(t-t')\exp[i\omega_b(t-t')]\\
\langle i_n(t)i_n^*(t')\rangle&=\frac{\hbar\omega}{R_a}\,\frac{(N_1+N_2)+(R_1+R_2-w_{01}N_1-w_{02}N_2)/2\Gamma_{12}}{N_1-N_2}\,\delta_s(t-t')\exp[i\omega_a(t-t')].
\end{aligned}
\tag{5.13}
$$

The last two of the dynamic equations (5.8) are the rate equations for the level populations, and the last five of the correlation functions (5.9) are precisely those one should expect as a result of shot noise in the pumping and decay of these level populations. For example, consider the $\langle F_M(t)F_1(t')\rangle$ correlation. A heuristic argument which gives the correct diffusion constant is as follows. The polarization M is proportional to $[N_1N_2]^{1/2}$, hence a fluctuation δN_1 in the decay of N_1 should give rise to a fluctuation δM of M given by

$$\delta M/M=\tfrac{1}{2}\delta N_1/N_1,$$

hence

$$\delta M\delta N_1=(M/2N_1)(\delta N_1)^2.$$

From the $\langle F_1(t)F_1(t')\rangle$ correlation term, the fluctuation $(\delta N_1)^2$ concerned can be seen to be proportional to Γ_1N_1, and hence the correlation term $\langle F_M(t)F_1(t')\rangle$ should contain $(M/2N_1)\Gamma_1N_1=\tfrac{1}{2}\Gamma_1M$, as it indeed does. The w_{12} term in $\langle F_M(t)F_1(t')\rangle$ can be obtained from a similar argument, since a positive fluctuation in the transfer rate $w_{12}N_2$ from level two to level one simultaneously increases N_1 and decreases M, the latter because of the decrease of N_2.

An important physical question to be answered concerns the "number of photons" in the resonator and the fluctuations in that number. This question may be studied through consideration of the moment relation

$$\langle \exp i(\lambda^*\mathbf{b}+\lambda\mathbf{b}^\dagger)\rangle=\langle \exp i(\lambda^*b+\lambda b^*)\rangle_2, \quad (5.14)$$

where on the right-hand side, we are using the angular brackets with subscript 2 to indicate a mean against the probability distribution P_2. Hence this equation is identical with (5.1), with $\xi_k, \nu_k \to 0$. By direct comparison of moments, one can show from (5.14) that

$$\langle \exp(i\chi\mathbf{b}^\dagger\mathbf{b})\rangle=\langle \exp\{i\chi(b^*b-\tfrac{1}{2})+\tfrac{1}{8}\chi^2+\cdots\}\rangle_2, \quad (5.15)$$

where χ is an expansion parameter. For comparison we note that since[15]

$$\langle \beta \mid \exp(i\chi\mathbf{b}^\dagger\mathbf{b}) \mid \beta\rangle=\exp\{[\exp(i\chi)-1]\beta^*\beta\},$$

one finds using (3.11) and (3.2) that

$$\langle \exp(i\chi\mathbf{b}^\dagger\mathbf{b})\rangle=\langle \exp\{[\exp(i\chi)-1]\beta^*\beta\}\rangle_1$$
$$=\langle \exp(i\chi\beta^*\beta-\tfrac{1}{2}\chi^2\beta^*\beta+\cdots)\rangle_1. \quad (5.16)$$

Comparison of (5.15) and (5.16) shows that if $\langle\beta^*\beta\rangle_1$ is greater than $\frac{1}{4}$, then $(b^*b-\frac{1}{2})$ is a significantly better approximation to the photon number than is $\beta^*\beta$. If the mean and variance of the photon number distribution are large compared to unity, then bb^* is a very good approximation to that number. As usual, classical physics is a good approximation to quantal physics in the limit of large quantum numbers.

VI. FURTHER CONSIDERATIONS

There are a number of useful functions (or distributions) other than P_2 which can be derived from our basic weight distribution P_1, and we shall mention a few of these because they shed some further light on our work and, in addition, make further contact with the work of others, notably Lax, and Scully and Lamb, who have treated the same or similar models by quite different methods.

The first of these we label P_{12} because it has the properties of P_1 with respect to the field, and of P_2 with respect to the atoms. To consolidate notation, recall that in (3.6) we abbreviated the atomic variables of P_1 by

$$(\mathfrak{X}_1, \mathfrak{X}_2, \mathfrak{X}_3, \mathfrak{X}_4)\equiv(\mathfrak{N}_1, \mathfrak{N}_2, \mathfrak{M}, \mathfrak{M}^*).$$

Likewise, let us now abbreviate the atomic variables of P_2 by

$$(X_1, X_2, X_3, X_4)\equiv(N_1, N_2, M, M^*),$$
$$d^{(4)}X\equiv dN_1dN_2d(\mathrm{Re}M)d(\mathrm{Im}M). \quad (6.1)$$

[15] This identity follows from reordering the exponential operator into normal ordered form wherein all annihilation operators appear to the right of all creation operators. For an account of such techniques, see W. H. Louisell, *Radiation and Noise in Quantum Electronics* (McGraw-Hill Book Company, Inc., New York, 1964); also Ref. 8.

The function P_{12} is defined from the general relation

$$\mathrm{Tr}_a(\rho\mathbf{Q}_a)\equiv\int^{(6)} P_{12}(\beta, \beta^*, X_j, t)$$
$$\times \exp(i\textstyle\sum\xi_jX_j) \mid \beta\rangle\langle\beta \mid d^{(2)}\beta\, d^{(4)}X, \quad (6.2)$$

where $\mathbf{Q}_a$ is the atomic part of the generating function (5.2), namely,

$$\mathbf{Q}_a\equiv \exp i(\xi_1\mathbf{N}_1+\xi_2\mathbf{N}_2+\xi_3\mathbf{M}+\xi_4\mathbf{M}^\dagger). \quad (6.3)$$

If we expand ρ in our P_1 representation, and make use of (5.14) with $\lambda=\lambda^*=0$, we find the relationship between P_1 and P_{12}, namely,

$$\int^{(4)} P_{12}(\beta, \beta^*, X_j, t) \exp(i\textstyle\sum\xi_jX_j)\, d^{(4)}X$$
$$=\int^{(4)} P_1(\beta, \beta^*, \mathfrak{X}_j, t) \exp(i\textstyle\sum\nu_j\mathfrak{X}_j)d^{(4)}\mathfrak{X}, \quad (6.4)$$

where the ν_j are related to the ξ_j by (5.4b). The passage from P_1 to P_{12} entails the same smoothing with respect to the atomic variables as does the passage from P_1 to P_2. Hence, in the diffusion approximation, the equation of motion for P_{12} can be derived from (3.19) by making the set of substitutions listed in (E5) for the atomic variables only. The resulting equation has many diffusion terms coupling the atoms with the field, and lacks the intuitive simplicity of our P_2 equation. However, like that for P_2, the equation of motion for P_{12} in the diffusion approximation is better behaved than is that for P_1, and it retains the property of directly producing the reduced density operator for the field, as may be seen by setting $\xi_j=0$ in (6.2) and (6.3). The function P_{12} is precisely "the associated classical function" which is being investigated by Lax using Langevin methods.

Other interesting results are obtained by going into the number representation for the field. We do this by taking the n, m field matrix element of (3.1), obtaining

$$\langle n \mid \rho \mid m\rangle=\int^{(4)} P_{n,m}(\mathfrak{X}_j, t)\sigma_a d^{(4)}\mathfrak{X}, \quad (6.5)$$

where the function $P_{n,m}(\mathfrak{X}_j, t)$ is defined by

$$P_{n,m}(\mathfrak{X}_j, t)\equiv\int^{(2)} P_1(\beta, \beta^*, \mathfrak{X}_j, t)\langle n \mid \sigma_b \mid m\rangle d^{(2)}\beta$$
$$=\int^{(2)} P_1 \exp(-\beta\beta^*)\frac{\beta^n\beta^{*m}}{(n!m!)^{1/2}}d^{(2)}\beta. \quad (6.6)$$

The equation of motion for $P_{n,m}(\mathfrak{X}_j, t)$ may be derived from (6.6) and the equation of motion for P_1.

A final function of some appeal is formed by going to the atomic variables X_j [see (6.1)] in the number representation for the field. To do this, we use the defining relation

$$\langle n \mid \mathrm{Tr}_a\rho\mathbf{Q}_a \mid m\rangle\equiv\int^{(4)} P_{n,m}(X_j, t) \exp(i\textstyle\sum\xi_jX_j)\, d^{(4)}X. \quad (6.7)$$

Inserting (6.5) in the left side of (6.7), and again using (5.4), we obtain the relation between $P_{n,m}(X_j,t)$ and $P_{n,m}(\mathfrak{X}_j,t)$, namely,

$$\int^{(4)} P_{n,m}(\mathfrak{X}_j,t)\exp(i\textstyle\sum\nu_j\mathfrak{X}_j)\,d^{(4)}\mathfrak{X}_j$$

$$=\int^{(4)} P_{n,m}(X_j,t)\exp(i\textstyle\sum\xi_jX_j)\,d^{(4)}X. \quad (6.8)$$

Comparing (6.8) with (6.4) we see that the same transformation which takes P_1 to P_{12} takes $P_{n,m}(\mathfrak{X}_j,t)$ to $P_{n,m}(X_j,t)$.

In the course of considering these various functions and their equations of motion, we have found that the equations for P_1-type functions are easier to manipulate (as opposed to solve). Adiabatic eliminations are an example of this. On the other hand, equations for P_2 like functions would appear to have smoother behavior, and in addition have greater intuitive appeal.

To exemplify these remarks, we shall derive the equations of motion for $P_{n,m}(\mathfrak{X}_j,t)$ and $P_{n,m}(X_j,t)$ under conditions appropriate for the adiabatic elimination of the polarization $\mathfrak{M}$. The P_1 equation we shall need is therefore (4.7).

We start with (6.6), transform its right-hand side to the rotating frame using (4.2), and integrate it over the $\mathfrak{M}$ plane, thus obtaining the necessary relation between $P_{n,m}(\mathfrak{N}_1,\mathfrak{N}_2,t)$ and the weight P_1 of (4.7), namely,

$$P_{n,m}(\mathfrak{N}_1,\mathfrak{N}_2,t)=\int^{(2)} P_1(\beta,\beta^*,\mathfrak{N}_1,\mathfrak{N}_2,t)\,\frac{\beta^n\beta^{*m}}{(n!m!)^{1/2}}$$
$$\times\exp[i(m-n)t]\,d^{(2)}\beta, \quad (6.9)$$

where in (6.9) as in (4.7), the primes on the β's are understood. We derive the equation of motion for $P_{n,m}(\mathfrak{N}_1,\mathfrak{N}_2,t)$ by taking the time derivative of (6.9), inserting the equation of motion (4.7) for

$$P_1(\beta,\beta^*,\mathfrak{N}_1,\mathfrak{N}_2,t),$$

integrating by parts with respect to β and β^* to eliminate derivatives with respect to β and β^*, and then using (6.9) again to reexpress the result in terms of the $P_{n,m}$. We thus obtain the following equation of motion for $P_{n,m}$:

$$\frac{\partial P_{n,m}(\mathfrak{N}_1,\mathfrak{N}_2,t)}{\partial t}$$
$$=[(n+1)(m+1)]^{1/2}[\pi\mathfrak{N}_1+\gamma(\bar{n}+1)]P_{n+1,m+1}+(nm)^{1/2}[\pi\mathfrak{N}_2+\gamma\bar{n}]P_{n-1,m-1}$$
$$-[\tfrac{1}{2}\pi(n+m+2)\mathfrak{N}_2+\tfrac{1}{2}\pi(n+m)\mathfrak{N}_1+i(n-m)(\omega_b+\alpha\tfrac{1}{2}\pi(\mathfrak{N}_2-\mathfrak{N}_1))+\gamma\bar{n}(n+m+1)+\tfrac{1}{2}\gamma(n+m)]P_{n,m}$$
$$-(\partial/\partial\mathfrak{N}_1)\{[R_1+w_{12}\mathfrak{N}_2-(\Gamma_1+\tfrac{1}{2}\pi(n+m)-i\alpha\tfrac{1}{2}\pi(n-m))\mathfrak{N}_1]P_{n,m}+(nm)^{1/2}\pi\mathfrak{N}_2P_{n-1,m-1}\}$$
$$-(\partial/\partial\mathfrak{N}_2)\{[R_2+w_{21}\mathfrak{N}_1-(\Gamma_2+\tfrac{1}{2}\pi(n+m+2)+i\alpha\tfrac{1}{2}\pi(n-m))\mathfrak{N}_2]P_{n,m}+[(n+1)(m+1)]^{1/2}\pi\mathfrak{N}_1P_{n+1,m+1}\}. \quad (6.10)$$

An interesting property of (6.10) which was not true in the case of (4.9) is that if we neglect the normally small quantities w_{21} and w_{12}, we can "exactly" carry out the adiabatic elimination of $\mathfrak{N}_1$ and $\mathfrak{N}_2$. When we do this, we reproduce precisely the equation of Scully and Lamb,[5] as we shall demonstrate below. But before doing this we would like to derive from (6.10) the corresponding equation of motion of $P_{n,m}(N_1,N_2,t)$.

To find the equation for $P_{n,m}(N_1,N_2,t)$, we multiply (6.10) by $\exp(i\nu_1\mathfrak{N}_1+i\nu_2\mathfrak{N}_2)$ and integrate over $\mathfrak{N}_1$ and $\mathfrak{N}_2$. We then make use of the identity (6.8) and follow the same method as outlined in Appendix E. Since everything is integrated over the $\mathfrak{M}$ plane, we set $\xi_3=\xi_4=0$ in the relationships (5.4b) between the ν's and the ξ's, obtaining simply

$$i\nu_j=\exp(i\xi_j)-1; \qquad j=1,2. \quad (6.11)$$

Because of the simplicity of (6.11), the passage to $P_{n,m}(N_1,N_2,t)$ can be done exactly in this case. We find that it entails the following set of substitutions into (6.10):

$$\mathfrak{N}_j\to\exp(\partial/\partial N_j)N_j,$$
$$\partial/\partial\mathfrak{N}_j\to(1-\exp(-\partial/\partial N_j)), \qquad j=1,2,$$
$$P_{n,m}(\mathfrak{N}_1,\mathfrak{N}_2,t)\to P_{n,m}(N_1,N_2,t). \quad (6.12)$$

The order of all factors is preserved in these substitutions, so derivatives always come out to the left. After suitably rearranging the resulting equation, we find the following equation of motion for $P_{n,m}(N_1, N_2, t)$

$$(\partial/\partial t)P_{n,m}(N_1, N_2, t)$$
$$=[(n+1)(m+1)]^{1/2}\left[\pi \exp\left(\frac{\partial}{\partial N_1}-\frac{\partial}{\partial N_2}\right)N_1+\gamma(\bar{n}+1)\right]P_{n+1,m+1}+(nm)^{1/2}\left[\pi \exp\left(\frac{\partial}{\partial N_2}-\frac{\partial}{\partial N_1}\right)N_2+\gamma\bar{n}\right]P_{n-1,m-1}$$
$$-\{(n+m+2)\tfrac{1}{2}\pi N_2+(n+m)\tfrac{1}{2}\pi N_1+i(n-m)[\omega_b+\alpha\tfrac{1}{2}\pi(N_2-N_1)]+\gamma\bar{n}(n+m+1)+\tfrac{1}{2}\gamma(n+m)\}P_{n,m}$$
$$+\left[\exp\left(-\frac{\partial}{\partial N_1}\right)-1\right]\left[R_1+w^{12}\exp\left(\frac{\partial}{\partial N_2}\right)N_2-\Gamma_1\exp\left(\frac{\partial}{\partial N_1}\right)N_1\right]P_{n,m}$$
$$+\left[\exp\left(-\frac{\partial}{\partial N_2}\right)-1\right]\left[R_2+w_{21}\exp\left(\frac{\partial}{\partial N_1}\right)N_1-\Gamma_2\exp\left(\frac{\partial}{\partial N_2}\right)N_2\right]P_{n,m}. \quad (6.13)$$

The form of (6.11) makes $P_{n,m}(N_1, N_2, t)$ a Poisson convolution of $P_{n,m}(\mathfrak{N}_1, \mathfrak{N}_2, t)$, and hence N_1 and N_2 in $P_{n,m}(N_1, N_2, t)$ are actually discrete rather than continuous variables. Correspondingly, in view of the identity

$$\{\exp(\pm\partial/\partial N_j)\}f(N_j)=f(N_j\pm1), \quad (6.14)$$

where f represents an arbitrary function (to see this, expand the right-hand side in a Taylor series), we see that (6.13) is in fact a difference equation in N_1 and N_2. Note that (6.13) keeps accurate account of the atomic-population changes. For example, the first term on the right-hand side gives the increase in $P_{n,m}$ due to absorption processes, and the absorption which is due to the maser atoms originates from a state in which N_1 is larger by 1, and N_2 is smaller by 1. The diagonal ($m=n$) terms of (6.13) form precisely the equation one would intuitively derive from energy flow considerations (rate equations). This last is the type of equation proposed and solved in the limit of no saturation by Shimoda, Takahasi, and Townes.[8] Equation (6.13) has also been derived by Lax[3] as an extrapolation of his results obtained by Langevin methods.

As a final exercise we return to (6.10) to effect the adiabatic elimination of $\mathfrak{N}_1$ and $\mathfrak{N}_2$. The procedure here differs somewhat from that of Appendix B, but the principle is the same. We seek instantaneous statistical equations for $\mathfrak{N}_1$ and $\mathfrak{N}_2$ which eliminate derivatives with respect to $\mathfrak{N}_1$ and $\mathfrak{N}_2$ from the equation of motion. In the case of (6.10), this involves setting the two curly bracketed expressions following $\partial/\partial\mathfrak{N}_1$ and $\partial/\partial\mathfrak{N}_2$ equal to zero and solving the two resulting difference equations simultaneously for $\mathfrak{N}_1P_{n,m}$ and $\mathfrak{N}_2P_{n,m}$. The solution is quite simple if conditions are assumed such that we can neglect w_{12} and w_{21} with respect to Γ_1 and Γ_2. In this case we have to solve simultaneously the two equations

$$[R_1-(\Gamma_1+\tfrac{1}{2}\pi(n+m)-i\alpha\tfrac{1}{2}\pi(n-m))\mathfrak{N}_1]P_{n,m}+(nm)^{1/2}\pi\mathfrak{N}_2P_{n-1,m-1}=0$$

and

$$[R_2-(\Gamma_2+\tfrac{1}{2}\pi(n+m)+i\alpha\tfrac{1}{2}\pi(n-m))\mathfrak{N}_2]P_{n-1,m-1}+(nm)^{1/2}\pi\mathfrak{N}_1P_{n,m}=0, \quad (6.15)$$

where in the second equation we have reduced the indices by 1 for convenience. The solutions to these two equations are

$$\mathfrak{N}_1P_{n,m}=D_{n,m}^{-1}\{R_1[\Gamma_2+\tfrac{1}{2}\pi(n+m)+i\alpha\tfrac{1}{2}\pi(n-m)]P_{n,m}+R_2\pi(nm)^{1/2}P_{n-1,m-1}\},$$
$$\mathfrak{N}_2P_{n-1,m-1}=D_{n,m}^{-1}\{R_2[\Gamma_1+\tfrac{1}{2}\pi(n+m)-i\alpha\tfrac{1}{2}\pi(n-m)]P_{n-1,m-1}+R_1\pi(nm)^{1/2}P_{n,m}\}, \quad (6.16)$$

where

$$D_{n,m}=\Gamma_1\Gamma_2+(\Gamma_1+\Gamma_2)\tfrac{1}{2}\pi(n+m)+(1+\alpha^2)[\tfrac{1}{2}\pi(n-m)]^2+i\alpha\tfrac{1}{2}\pi(\Gamma_1-\Gamma_2)(n-m).$$

Inserting (6.16) in (6.10), we find all dependence on $\mathfrak{N}_1$ and $\mathfrak{N}_2$ has disappeared. We may then integrate the $P_{n,m}$ over $\mathfrak{N}_1$ and $\mathfrak{N}_2$, which simply replaces $P_{n,m}$ by the n, m matrix element of the reduced field density operator. Carrying this out and collecting terms, we obtain

$$(d/dt)\langle n|\rho_b|m\rangle$$
$$=[(n+1)(m+1)]^{1/2}[\gamma(\bar{n}+1)+\pi D_{n+1,m+1}^{-1}R_1\Gamma_2]\langle n+1|\rho_b|m+1\rangle+(nm)^{1/2}[\gamma\bar{n}+\pi D_{n,m}^{-1}R_2\Gamma_1]\langle n-1|\rho_b|m-1\rangle$$
$$-\{D_{n+1,m+1}^{-1}R_2[\Gamma_1\tfrac{1}{2}\pi\{(n+m+2)+i\alpha(n-m)\}+(1+\alpha^2)(\tfrac{1}{2}\pi(n-m))^2]$$
$$+D_{n,m}^{-1}R_1[\Gamma_2\tfrac{1}{2}\pi\{(n+m)-i\alpha(n-m)\}+(1+\alpha^2)(\tfrac{1}{2}\pi(n-m))^2]$$
$$+[\gamma\bar{n}(n+m+1)+\tfrac{1}{2}\gamma(n+m)+i\omega_b(n-m)]\}\langle n|\rho_b|m\rangle. \quad (6.17)$$

Specialized to the case considered by Scully and Lamb,[5] i.e., with $\bar{n}=R_1=0$, Eq. (6.17) can be shown to be identical to their result. In our notation, their quantity $R_{n,n'}$ is

$$R_{n,n'}=D_{n+1,n'+1}{}^{-1}R_2\tfrac{1}{2}\pi[\Gamma_1(1+i\alpha)+\tfrac{1}{2}\pi(n-n')(1+\alpha^2)].$$

If one makes the substitutions (6.12) in (6.16), one obtains adiabatic solutions for N_1 and N_2. These solutions, substituted into (6.13), lead also to (6.17). However, it is clearly easier to see how to accomplish the adiabatic elimination of the atomic populations from (6.10) than from (6.13).

ACKNOWLEDGMENTS

The author wishes to thank J. R. Klauder, M. Lax, and W. H. Louisell for many stimulating conversations concerning this work.

APPENDIX A: PROPERTIES OF THE ATOMIC OPERATORS

We will have use for the atomic operators only in pairs such as $(\mathbf{a}_j{}^\dagger \mathbf{a}_k)_m$, which serves to transfer the mth atom from state k to state j. Thus, the total number of atoms is conserved. A convenient basis for matrix representation of such pairs is the set of states with the atom concerned definitely in one particular state. The properties of the atomic operators are specified by

$$(\mathbf{a}_j{}^\dagger \mathbf{a}_k)_m\,|\,p,m\rangle=|\,j,m\rangle\delta_{k,p}, \tag{A1}$$

where δ is the Kronecker delta, and $|\,p,m\rangle$ represents the normalized pth state of the mth atom. Pairs of operators referring to different atoms commute. From (A1), we can derive the useful relations

$$(\mathbf{a}_j{}^\dagger \mathbf{a}_k)_m(\mathbf{a}_p{}^\dagger \mathbf{a}_q)_m=(\mathbf{a}_j{}^\dagger \mathbf{a}_q)_m\delta_{k,p} \tag{A2}$$

and

$$\mathrm{Tr}(\mathbf{a}_j{}^\dagger \mathbf{a}_k)_m=\delta_{j,k}. \tag{A3}$$

Thus, any product of pairs of operators referring to the same atom can be reduced to a single pair. Note that

$$[\sum_{j,k}A_{jk}(\mathbf{a}_j{}^\dagger \mathbf{a}_k)_m][\sum_{p,q}B_{pq}(\mathbf{a}_p{}^\dagger \mathbf{a}_q)_m]=\sum C_{jq}(\mathbf{a}_j{}^\dagger \mathbf{a}_q)_m,$$

where

$$C_{jq}=\sum_k A_{jk}B_{kq}.$$

Hence the pairs of atomic operators may be thought of simply as identifying positions in a matrix.

APPENDIX B: ADIABATIC ELIMINATION

Consider a Fokker–Planck or diffusion equation giving the time evolution of a statistical weight distribution in two variables, say x and y, of the general form

$$\frac{\partial P(x,y,t)}{\partial t}=\left\{-\frac{\partial}{\partial x}A_x-\frac{\partial}{\partial y}A_y+\frac{\partial^2}{\partial x^2}D_{xx}+\frac{\partial^2}{\partial y^2}D_{yy}+\frac{\partial^2}{\partial x\partial y}2D_{xy}\right\}P, \tag{B1}$$

where all partial derivatives on the right-hand side operate on P and where the drift coefficients A_x, A_y and the diffusion coefficients D_{xx}, D_{yy}, D_{xy} may be arbitrary functions of x and y. The drift coefficients have dimensions of velocity in the x–y plane, and prescribe the motion of the mean values of the variables according to

$$(d/dt)\langle x\rangle=\langle A_x\rangle,$$
$$(d/dt)\langle y\rangle=\langle A_y\rangle, \tag{B2}$$

where the angular brackets indicate the statistical mean of the included quantity. More generally, if $f(x,y)$ is any function of x and y, then (B1) leads to and is equivalent to the equation

$$\frac{d}{dt}\langle f(x,y)\rangle=\left\langle\left\{A_x\frac{\partial}{\partial x}+A_y\frac{\partial}{\partial y}+D_{xx}\frac{\partial^2}{\partial x^2}+D_{yy}\frac{\partial^2}{\partial y^2}+2D_{xy}\frac{\partial^2}{\partial x\partial y}\right\}f\right\rangle, \tag{B3}$$

where

$$\langle f(x,y)\rangle\equiv\int^{(2)}f(x,y)P(x,y,t)\,dx\,dy. \tag{B4}$$

Equation (B3) is derived by taking the time derivative of (B4), inserting (B1), and then integrating by parts with the assumption that P is sufficiently well behaved that all surface integrals vanish. In (B3) the derivatives operate on f but not on P. Equations (B2) are special cases of (B3) with $f(x,y)$ set equal to x and y, respectively.

The variable x, say, may be adiabatically eliminated from (B1) if conditions are such that, given any y distribution in some range under consideration, the x distribution has some conditional equilibrium to which it relaxes before the y distribution can change appreciably. Then the x distribution can be assumed to remain adiabatically in its conditional equilibrium and an equation of motion of the y distribution alone obtained.

If we assume that the x distribution is continually in conditional equilibrium then the mean of any function of x and y must be expressible as the mean of some function of y alone. We can formalize this statement in a useful way by requiring satisfaction of a relation of the form

$$\langle xf(x,y)\rangle=\langle af+b_x(\partial f/\partial x)+b_y(\partial f/\partial y)\rangle, \tag{B5}$$

where $f(x,y)$ is again an arbitrary function of x and y, and the three coefficients a, b_x, and b_y are functions of y alone. Special cases of (B5) are

$$\langle x\rangle=\langle a\rangle,$$
$$\langle xy\rangle=\langle ay+b_y\rangle,$$
$$\langle x^2\rangle=\langle ax+b_x\rangle,$$
$$=\langle a^2+b_y(\partial a/\partial y)+b_x\rangle. \tag{B5$'$}$$

In the last example we have had to make use of (B5) twice to eliminate x completely. Thus a is the mean value of x conditional on some particular value of y, b_y provides for correlation between the x and y distributions, while b_x provides for additional variance of the x distribution. Equation (B5) can be used to eliminate x from any function of x and y, because its right-hand side contains one less power of x than does its left. Actually, (B5) is a normally adequate approximation to a more general relation of the same type but including all orders of derivatives.

By partial integration we can transform (B5) to the form

$$\int^{(2)} f(x,y)\left\{x-a+\frac{\partial}{\partial x}b_x+\frac{\partial}{\partial y}b_y\right\}P(x,y,t)\,dx\,dy=0, \tag{B6}$$

where the derivatives operate on P. Because f is an arbitrary function we may extract the integrand, giving an an equivalent to (B5) the relation

$$f(x,y)xP(x,y,t)=f(x,y)\left\{a-\frac{\partial}{\partial x}b_x-\frac{\partial}{\partial y}b_y\right\}P. \tag{B7}$$

We use (B7) to eliminate x from the drift and diffusion coefficients of (B1), and then pull all the resulting derivatives to the left to re-establish the Fokker–Planck form. Generally this process introduces derivatives higher than the second into the result, but if the diffusion approximation is adequate, we can throw away such terms. [If such terms are not negligible, we must go back and extend (B5) to include higher derivatives also.] We then establish values of the three coefficients a, b_x, b_y, and thus establish the adiabatic x distribution, by solving this resulting equation for steady state while ignoring the motion of the y distribution, i.e., ignoring the $\partial/\partial y$ and $\partial^2/\partial y^2$ terms. Hence we set the coefficients of the $\partial/\partial x$, $\partial^2/\partial x^2$, and $\partial^2/\partial x\partial y$ terms equal to zero and solve for a, b_x, and b_y. This eliminates x from everything except P, which we are now free to integrate over x, thus achieving the goal of an equation in y alone.

If the drift coefficients are linear in x while the diffusion coefficients are independent of x, as is so in the elimination of $\mathfrak{M}$ from (4.4) (the extension to more variables is quite straightforward) then our method does not introduce any higher derivatives and so is "exact." As a general example of this sort, suppose that

$$A_x=B-\Gamma x,\qquad A_y=K+Mx, \tag{B8}$$

where B, Γ, K, and M are functions of y alone. The quantity Γ^{-1} is the pertinent relaxation time for the x distribution and must be positive. Then our method results in the solution

$$a=\Gamma^{-1}[B-\Gamma^{-1}(d\Gamma/dy)(2D_{xy}+(M/\Gamma)D_{xx})],$$
$$b_x=D_{xx}/\Gamma,$$
$$b_y=\Gamma^{-1}(2D_{xy}+(M/\Gamma)D_{xx}),$$
$$\frac{\partial P(y,t)}{\partial t}=\left\{-\frac{\partial}{\partial y}\left[K+M\frac{B}{\Gamma}-\frac{d(M/\Gamma)}{dy}\left(2D_{xy}+\frac{M}{\Gamma}D_{xx}\right)\right]+\frac{\partial^2}{\partial y^2}\left(D_{yy}+\frac{M}{\Gamma}2D_{xy}+\frac{M^2}{\Gamma^2}D_{xx}\right)\right\}P,$$

where

$$P(y,t)=\int P(x,y,t)\,dx. \tag{B9}$$

Results obtained by this method correspond precisely with those obtained by careful application of the Langevin method discussed by Lax.[3] We have included this discussion because it was not at first clear how to accomplish the adiabatic elimination directly in a Fokker–Planck equation.

APPENDIX C. EVALUATION OF TRACE (σQ)

We need to evaluate [see (5.3)]

$$\mathrm{Tr}\sigma\mathbf{Q}, \tag{C1}$$

where

$$\mathbf{Q}=\exp i\{\lambda^*\mathbf{b}+\lambda\mathbf{b}^\dagger+\xi_1\mathbf{N}_1+\xi_2\mathbf{N}_2+\xi_3\mathbf{M}+\xi_4\mathbf{M}^\dagger\},\qquad(\xi_4=\xi_3^*); \tag{C2}$$

and σ is the product density operator

$$\sigma=\sigma_b\prod_{m=1}^{N}\sigma_m \tag{C3}$$

[see (3.5), (3.6)]. Since $\mathbf{b}$ and $\mathbf{b}^\dagger$ commute with all the atomic operators, and since pairs of atomic operators pertaining to different atoms also commute, (C1) may be factored; i.e.,

$$\mathrm{Tr}\sigma\mathbf{Q}=[\mathrm{Tr}\sigma_b\mathbf{Q}_b][\mathrm{Tr}\sigma_m\mathbf{Q}_m]^N, \tag{C4}$$

where

$$\mathbf{Q}_b=\exp i(\lambda^*\mathbf{b}+\lambda\mathbf{b}^\dagger) \tag{C5}$$

is the part of $\mathbf{Q}$ referring to the field, and

$$\mathbf{Q}_m=\exp i\{\xi_1(\mathbf{a}_1^\dagger\mathbf{a}_1)_m+\xi_2(\mathbf{a}_2^\dagger\mathbf{a}_2)_m+\xi_3(\mathbf{a}_1^\dagger\mathbf{a}_2)_m+\xi_4(\mathbf{a}_2^\dagger\mathbf{a}_1)_m\} \tag{C6}$$

is the part of $\mathbf{Q}$ referring to the mth atom. In (C4) we have used the fact that Tr $\sigma_m \mathbf{Q}_m$ is independent of the particular atom, i.e., of m.

First, consider the field part: we have

$$\mathrm{Tr}\sigma_b \mathbf{Q}_b = \langle\beta \mid \exp i(\lambda^*\mathbf{b}+\lambda\mathbf{b}^\dagger) \mid \beta\rangle,$$

$$= \exp(-\tfrac{1}{2}\lambda\lambda^*)\langle\beta \mid (\exp i\lambda\mathbf{b}^\dagger)(\exp i\lambda\mathbf{b}) \mid \beta\rangle,$$

$$= \exp(i\lambda^*\beta+i\lambda\beta^*-\tfrac{1}{2}\lambda\lambda^*), \qquad \text{(C7)}$$

where we have first reordered[15] the exponential to take advantage of the fact that $\mid\beta\rangle$ is an eigenstate of $\mathbf{b}$ with eigenvalue β.

The atoms part of $\mathbf{Q}$ is slightly more complicated. By virtue of (A2), we can reduce $\mathbf{Q}_m$ to a linear sum of pairs of atomic operators, plus unity. This may be done efficiently by rewriting (C6) in the form

$$\mathbf{Q}_m = \exp i\{\xi_+\mathbf{n}_+ + \xi_-\mathbf{n}_- + \xi_3\mathbf{a}_1^\dagger\mathbf{a}_2 + \xi_4\mathbf{a}_2^\dagger\mathbf{a}_1\}, \qquad \text{(C8)}$$

where

$$\mathbf{n}_+ \equiv \mathbf{a}_2^\dagger\mathbf{a}_2 + \mathbf{a}_1^\dagger\mathbf{a}_1,$$

$$\mathbf{n}_- \equiv \mathbf{a}_2^\dagger\mathbf{a}_2 - \mathbf{a}_1^\dagger\mathbf{a}_1,$$

$$\xi_+ \equiv \tfrac{1}{2}(\xi_2+\xi_1),$$

$$\xi_- \equiv \tfrac{1}{2}(\xi_2-\xi_1).$$

Now by virtue of the relations [easily derived from (A2)]

$$\mathbf{n}_+\mathbf{O} = \mathbf{O}\mathbf{n}_+ = \mathbf{O}, \qquad \text{(C9)}$$

where $\mathbf{O}$ may be any of the four operators in the exponential of (C8), and

$$(\xi_-\mathbf{n}_- + \xi_3\mathbf{a}_1^\dagger\mathbf{a}_2 + \xi_4\mathbf{a}_2^\dagger\mathbf{a}_1)^2 = v^2(\mathbf{n}_+), \qquad \text{(C10)}$$

where

$$v^2 = \xi_-^2 + \xi_3\xi_4.$$

we can easily reduce $\mathbf{Q}_m$ through the series of self-explanatory steps

$$\begin{aligned}\mathbf{Q}_m &= (\exp i\xi_+\mathbf{n}_+)(\exp i\{\xi_-\mathbf{n}_- + \xi_3\mathbf{a}_1^\dagger\mathbf{a}_2 + \xi_4\mathbf{a}_2^\dagger\mathbf{a}_1\}) \\ &= \{1+[\exp(i\xi_+)-1]\mathbf{n}_+\}[1+(\cos v-1)\mathbf{n}_+ + i(\sin v/v)(\xi_-\mathbf{n}_- + \xi_3\mathbf{a}_1^\dagger\mathbf{a}_2 + \xi_4 a_2^\dagger a_1)] \\ &\quad 1+[\exp(i\xi_+)\cos v-1]\mathbf{n}_+ + i\exp(i\xi_+)(\sin v/v)(\xi_-\mathbf{n}_- + \xi_3\mathbf{a}_1^\dagger\mathbf{a}_2 + \xi_4\mathbf{a}_2^\dagger\mathbf{a}_1).\end{aligned} \qquad \text{(C11)}$$

We can now evaluate Tr $(\sigma_m\mathbf{Q}_m)$, with σ_m given by (3.3). Using (A2) and (A3), we obtain

$$\mathrm{Tr}\sigma_m\mathbf{Q}_m = \{1+[\exp(i\xi_+)\cos v-1](\mathfrak{N}_+/N) + i\exp(i\xi_+)(\sin v/vN)(\xi_-\mathfrak{N}_- + \xi_3\mathfrak{M} + \xi_4\mathfrak{M}^*)\}, \qquad \text{(C12)}$$

where

$$\mathfrak{N}_+ \equiv \mathfrak{N}_2 + \mathfrak{N}_1; \qquad \mathfrak{N}_- \equiv \mathfrak{N}_2 - \mathfrak{N}_1.$$

The atoms part of (C4) is (C12) raised to the Nth power; in the limit of large N [see (3.12)] we obtain

$$[\mathrm{Tr}\sigma_m\mathbf{Q}_m]^N = \exp\{[\exp(i\xi_+)\cos v-1]\mathfrak{N}_+ + i\exp(i\xi_+)(\sin v/v)(\xi_-\mathfrak{N}_- + \xi_3\mathfrak{M} + \xi_4\mathfrak{M}^*)\}. \qquad \text{(C13)}$$

Equations (C13) and (C7) combine according to (C4) to yield the desired result

$$\mathrm{Tr}\sigma\mathbf{Q} = \exp\{-\tfrac{1}{2}(\lambda\lambda^*) + i(\lambda^*\beta + \lambda\beta^* + \textstyle\sum \nu_j\mathfrak{X}_j)\}, \qquad \text{(C14)}$$

where the ν_j are as listed in (5.4b) of the text.

APPENDIX D: EVALUATION OF P_2 FROM P_1 IN THE GAUSSIAN APPROXIMATION

In the approximation, valid for large $\mathfrak{N}_1+\mathfrak{N}_2$, that the ν_j in (5.4) need only be expanded to second power in the ξ_j, we obtain a direct solution for P_2 in terms of P_1. We have from (5.5), the transform relationship

$$\int P_2(b, b^*, X_j, t)\exp[i(\lambda^*b+\lambda b^*+\textstyle\sum\xi_jX_j)]d^{(2)}b\, d^{(4)}X$$

$$= \int P_1(\beta, \beta^*, \mathfrak{X}_j, t)\exp\{[-\tfrac{1}{2}\lambda\lambda^* + i(\lambda^*\beta+\lambda\beta^*+\textstyle\sum\nu_j\mathfrak{X}_j)]\}d^{(2)}\beta\, d^{(4)}\mathfrak{X}, \qquad \text{(D1)}$$

where the ν_j and the ξ_j are related by (5.6), i.e.,

$$i\nu_1 = i\xi_1 - \tfrac{1}{2}(\xi_1^2+\xi_3\xi_4),$$

$$i\nu_2 = i\xi_2 - \tfrac{1}{2}(\xi_2^2+\xi_3\xi_4),$$

$$i\nu_3 = i\xi_3 - \tfrac{1}{2}\xi_3(\xi_1+\xi_2),$$

$$i\nu_4 = i\xi_4 - \tfrac{1}{2}\xi_4(\xi_1+\xi_2).$$

To extract P_2 from (D1) by Fourier transformation we make ξ_1 and ξ_2 real and set $\xi_3=\xi_4^*$ to make the exponential of the left-hand side pure imaginary, then multiply (D1) by the factor

$$(1/4\pi^6)\exp\{-i(\lambda^* b'+\lambda b'^*+\sum \xi_j X_j')\}\, d(\mathrm{Re}\lambda)\, d(\mathrm{Im}\lambda)\, d\xi_1 d\xi_2 d(\mathrm{Re}\xi_4)\, d(\mathrm{Im}\xi_4)$$

and integrate over all the parameters. On the left, this extracts P_2; and so transforms (D1) to

$$P_2(b, b^*, X_j, t)=\int^{(6)} K P_1(\beta, \beta^*, \mathfrak{X}_j, t)\, d^{(2)}\beta\, d^{(4)}\mathfrak{X}, \tag{D2}$$

where the convolution kernel K is given by

$$K=\frac{1}{4\pi^6}\int^{(6)} \exp\{i[\lambda^*(\beta-b)+\lambda(\beta^*-b^*)+\sum \xi_j(\mathfrak{X}_j-X_j)]$$

$$-\tfrac{1}{2}[\lambda\lambda^*+\xi_1^2\mathfrak{N}_1+\xi_2^2\mathfrak{N}_2+\xi_4^*\xi_4(\mathfrak{N}_1+\mathfrak{N}_2)]-\tfrac{1}{2}(\xi_1+\xi_2)(\xi_4^*\mathfrak{M}+\xi_4\mathfrak{M}^*)\}\, d(\mathrm{Re}\lambda)\, d(\mathrm{Im}\lambda)\, d\xi_1 d\xi_2 d(\mathrm{Re}\xi_4)\, d(\mathrm{Im}\xi_4). \tag{D3}$$

Evaluation of the integral of (D3) is facilitated by the following linear transformation which diagonalizes the quadratic terms:

$$\xi_1=\zeta_+[2\mathfrak{N}_2/(\mathfrak{N}_1+\mathfrak{N}_2)]-\tfrac{1}{2}\zeta_-,$$

$$\xi_2=\zeta_+[2\mathfrak{N}_1/(\mathfrak{N}_1+\mathfrak{N}_2)]+\tfrac{1}{2}\zeta_-,$$

$$\xi_4=\zeta_4-\zeta_+[2\mathfrak{M}/(\mathfrak{N}_1+\mathfrak{N}_2)], \tag{D4}$$

along with the conjugate of the last. After this transformation, K factors into the product of six independent integrals, and evaluates to

$$K=\frac{2}{\pi^3(\mathfrak{N}_1+\mathfrak{N}_2)(\mathfrak{N}_1\mathfrak{N}_2-\mathfrak{M}\mathfrak{M}^*)^{1/2}}\exp\left\{-2|\beta-b|^2-\frac{2|\mathfrak{M}-M|^2}{(\mathfrak{N}_1+\mathfrak{N}_2)}-\frac{[(\mathfrak{N}_2-\mathfrak{N}_1)-(N_2-N_1)]^2}{2(\mathfrak{N}_1+\mathfrak{N}_2)}\right.$$

$$\left.-\frac{[\mathfrak{N}_2(\mathfrak{N}_1-N_1)+\mathfrak{N}_1(\mathfrak{N}_2-N_2)-\mathfrak{M}(\mathfrak{M}^*-M^*)-\mathfrak{M}^*(\mathfrak{M}-M)]^2}{2(\mathfrak{N}_1+\mathfrak{N}_2)(\mathfrak{N}_1\mathfrak{N}_2-\mathfrak{M}\mathfrak{M}^*)}\right\}. \tag{D5}$$

APPENDIX E. DERIVATION OF THE "CLASSICAL" FOKKER-PLANCK EQ. (5.7)

We start from the equation of motion (3.19) for P_1, and the relationship between P_1 and P_2 given in (5.5) along with the approximate relations (5.6) between the ν's and the ξ's. We first differentiate the right side of (5.5) with respect to time, and insert the equation of motion (3.19) for $\partial P_1/\partial t$. This yields an equation of motion for $\langle Q\rangle$ [note that (5.5) equates two expressions for $\langle Q\rangle$] of the form

$$\frac{d}{dt}\langle Q\rangle=\int^{(6)} \exp\{-\tfrac{1}{2}(\lambda\lambda^*)+i(\lambda^*\beta+\lambda\beta^*+)\sum \nu_j\mathfrak{X}_j)\}$$

$$\times\mathfrak{F}_1\left(\frac{\partial}{\partial\beta}, \frac{\partial}{\partial\beta^*}, \frac{\partial}{\partial\mathfrak{X}_j}; \beta, \beta^*, \mathfrak{X}_j\right) P_1 d^{(2)}\beta d^{(4)}\mathfrak{X}, \tag{E1}$$

where $\mathfrak{F}_1$ stands for the complete curly bracketed Fokker–Planck operator in the right-hand side of (3.19). The semicolon in $\mathfrak{F}$ indicates that all derivatives are to the left. We integrate (E1) by parts, applying the derivatives to the exponential, where they bring down the parameters λ^*, λ, and the ν_j. Next we express the variables in $\mathfrak{F}_1$ as partial derivatives of the exponential with respect to the parameters. An example of this is

$$\beta^*\to -i(\partial/\partial\lambda+\tfrac{1}{2}\lambda^*).$$

This procedure yields an equation of the form

$$\frac{d}{dt}\langle Q\rangle=\mathfrak{L}_1\left(\lambda, \lambda^*, \nu_j; \frac{\partial}{\partial\lambda}, \frac{\partial}{\partial\lambda^*}, \frac{\partial}{\partial\nu_j}\right)\langle Q\rangle. \tag{E2}$$

In $\mathfrak{L}_1$, all derivatives are to the right, again indicated by the semicolon. Now we use the relations (5.6) to express $\mathcal{L}_1$ in terms of the ξ_j, transforming (E2) to the form

$$\frac{d}{dt}\langle Q\rangle=\mathcal{L}_2\left(\lambda,\lambda^*,\xi_j;\frac{\partial}{\partial\lambda},\frac{\partial}{\partial\lambda^*},\frac{\partial}{\partial\xi_j}\right)\langle Q\rangle. \quad \text{(E3)}$$

The final stage is to express $\langle Q\rangle$ as the left side of (5.5) and to reverse the procedure which led from (E1) to (E2), obtaining an equation of the form

$$\frac{d}{dt}\langle Q\rangle=\int^{(6)}\exp i(\lambda^*b+\lambda b^*+\sum\xi_jX_j)$$

$$\times\mathfrak{F}_2\left(\frac{\partial}{\partial b},\frac{\partial}{\partial b^*},\frac{\partial}{\partial X_j};b,b^*,X_j\right)P_2d^{(2)}bd^{(4)}X, \quad \text{(E4)}$$

where [see also (6.1)] we have used the notation

$$X_j(j=1,2,3,4)\equiv(N_1,N_2,M,M^*)$$

$$d^{(4)}X\equiv dN_1dN_2d(\text{Re}M)d(\text{Im}M).$$

In (E4), the operator $\mathfrak{F}_2$ comes out to be the complete curly bracketed Fokker–Planck operator on the right-hand side of (5.7) after all derivatives higher than the second have been discarded. Extraction of the integrand of (E4) by Fourier transform yields (5.7).

Carrying out the above procedure, we find that passage from $\mathfrak{F}_1$ to $\mathfrak{F}_2$ entails the following set of substitutions:

$$\beta\to b+\frac{1}{2}\frac{\partial}{\partial b^*},$$

$$\frac{\partial}{\partial\beta}\to\frac{\partial}{\partial b},$$

$$\mathfrak{M}\to\left[1+\frac{1}{2}\left(\frac{\partial}{\partial N_2}+\frac{\partial}{\partial N_1}\right)\right]M+\frac{1}{2}\frac{\partial}{\partial M^*}(N_1+N_2),$$

$$\frac{\partial}{\partial\mathfrak{M}}\to\frac{\partial}{\partial M}\left[1-\frac{1}{2}\left(\frac{\partial}{\partial N_2}+\frac{\partial}{\partial N_1}\right)\right],$$

$$\mathfrak{N}_j\to\left(1+\frac{\partial}{\partial N_j}\right)N_j+\frac{1}{2}\left(\frac{\partial}{\partial M}M+\frac{\partial}{\partial M^*}M^*\right),\qquad j=1,2$$

$$\frac{\partial}{\partial\mathfrak{N}_j}\to\frac{\partial}{\partial N_j}-\frac{1}{2}\left(\frac{\partial^2}{\partial N_j^2}+\frac{\partial^2}{\partial M\partial M^*}\right)\qquad j=1,2 \quad \text{(E5)}$$

along with the conjugates of the first four. These replacements are made without disturbing the order of factors. In $\mathfrak{F}_1$, products of field quantities times atomic quantities such as $\beta^*\mathfrak{M}$ and $\beta\mathfrak{N}_1$ occur, but these give no ordering problems since their replacements commute. After the replacements, all derivatives higher than the second are discarded.

Part 4
Laser Amplifiers

TRANSIENTS AND OSCILLATION PULSES IN MASERS

H. STATZ and G. deMARS

Research Division, Raytheon Company, Waltham, Massachusetts

IT HAS been observed by Kikuchi *et al.*[1] that a maser under certain pump conditions, instead of oscillating continuously may either show a "ripple" in the oscillation output or even show discrete pulses of oscillation output with periods in between the pulses where no sign of oscillation can be detected. Similarly, Foner *et al.*[2] detected in their pulsed-field maser that the oscillations may also be discontinuous. Observations of pulsed oscillations have also been made in this laboratory by the authors and also certain transients have been observed which belong to the same class of phenomena. For example, Fig. 1 shows the power output when the maser starts to oscillate after applying the pump or after having saturated the pair of levels between which amplification or oscillation is to take place. The oscillation power output does not increase smoothly in time but rather the oscillation goes off and on and finally approaches a steady-state power output. The power output as a function of time in this initial period resembles a damped vibration. The present measurements were made on a three-level ruby maser operating at 1.3 kmc.

We shall give, in the following, an explanation of the above-described phenomena. The explanation differs from that given by Senitzky.[3] Since we do not require a coherent motion of the spin system, the present treatment applies also to cases in which the time between oscillation pulses is much larger than T_2, the spin-spin relaxation time. However, we require in general three levels and our model is not suitable for a spin $\frac{1}{2}$.

The phenomenon follows qualitatively from accepted equations describing maser action, retaining, however, certain non-

Reprinted with permission from *Quantum Electronics* (New York: Columbia Univ. Press, 1960, pp. 531–537).

linear terms. If we denote the quality factor of the cavity by Q_L including the external coupling and the energy emitting or absorbing properties of the active material, then

$$\frac{dP}{dt} = -\frac{1}{Q_L}\,\omega P \tag{1}$$

where P is the energy stored in the cavity and ω is the frequency at which maser action takes place.

$$\frac{1}{Q_L} = \frac{1}{Q_o} + \frac{1}{Q_e} - Cn = \frac{1}{Q} - Cn. \tag{2}$$

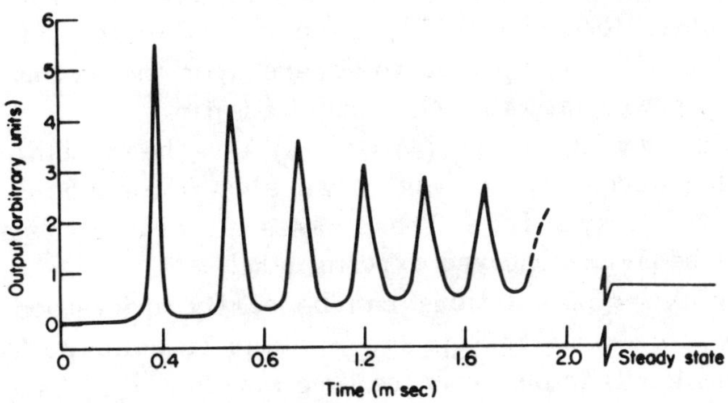

Fig. 1. Experimentally observed energy output from ruby maser

Equation (2) states that the total quality factor depends upon the inverted population $n = n_3 - n_2$. Q_o is the quality factor including wall losses and dielectric absorption in the active material. Q_e describes the energy taken out of the cavity through external coupling. We define $Q^{-1} = Q_o^{-1} + Q_e^{-1}$. C is a constant.

The time rate of change of n is given by

$$\frac{dn}{dt} = -\frac{n - n_1}{\tau} - Dn\,P. \tag{3}$$

In Equation (3) n_1 is the equilibrium value for $n_3 - n_2$ with pump on when there is no electromagnetic energy in the cavity at the signal frequency. τ is the time constant with which a disturbance in n decays. It is normally equal to the spin lattice relaxation time. D is a proportionality constant which determines the rate at which spins flip in an applied rf field. We obtain a set of dimensionless equations from Equations (1), (2), and (3).

$$\frac{d\pi}{dT} = \frac{1}{Q} \pi (\eta - 1) \tag{4}$$

$$\frac{d\eta}{dT} = -\frac{\eta - \alpha}{T_o} + \frac{1 - \alpha}{T_o} \eta \pi. \tag{5}$$

in which $\pi = P/P_o$, where P_o = the steady-state energy stored in the cavity under normal oscillation conditions; $\eta = n/n_o$, where n_o is the amount of level inversion required to overcome the losses characterized by Q^{-1}; $\tau = n_1/n_o$; $T = \omega t$; and $T_o = \omega\tau$. The value of α is expected to depend upon the extent to which the pump power saturates the pump transition.

Solutions to Equations (4) and (5) have been obtained using an analog computer. In Fig. 2 we show three choices of the parameters Q, α, and T_o. These cases are characteristic of the types of behavior observed experimentally.

These types of solutions can be easily understood by considering, for example, Fig. 2a. η starts to build up from some value smaller than one with the time constant T_o. As η crosses the value 1 the maser is basically unstable and oscillations start to build up. $\eta = 1$ corresponds to the case where the losses due to the cavity and the external coupling are just balanced by the emission from the spin system. As the oscillation amplitude becomes rather large it tends to saturate levels 2 and 3, i.e., η tends to decrease. However, the oscillation level π continues to increase until no more net energy is fed into the cavity, i.e., until η has decreased to 1. The electromagnetic energy in the cavity now has its maximum value and continues to flip spins so that η decreases below 1. π now also decreases. At some particular value of $\eta < 1$, the rate at which inverted spins are generated equals that at which spins are flipped. While π further decreases η starts to build up again. At the point where $\eta = 1$, π stops decreasing and for $\eta > 1$ π also increases again.

Qualitatively we can distinguish between two cases. If π has decreased essentially to the noise level in the cavity, before η

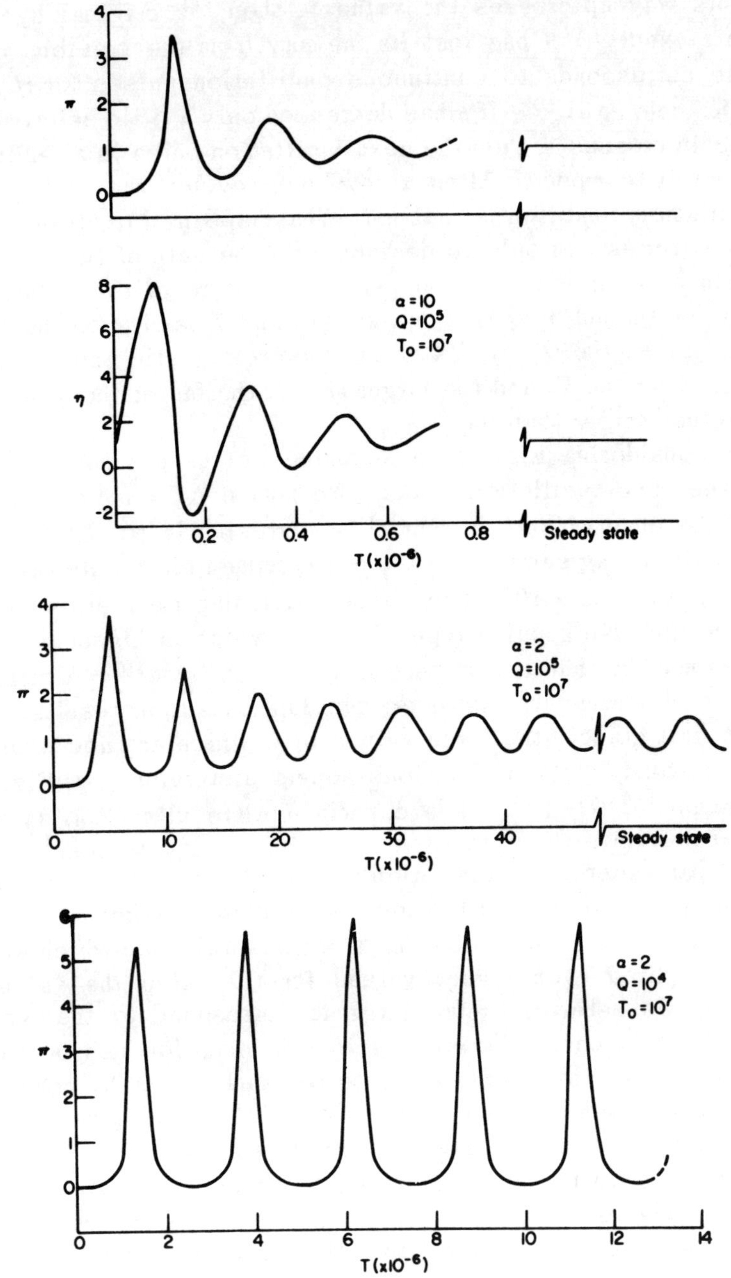

Fig. 2. Theoretical time dependence of electromagnetic energy in maser cavity

on its way up crosses the value 1, then the original cycle repeats exactly. π has lost its memory from the previous cycle. This corresponds to continuous oscillation pulses as reported by Kikuchi *et al.*[1] If π has decreased only a little below 1 when η again crosses 1, then the next oscillation pulse is already considerably damped. After a few such cycles the steady-state oscillation condition is reached. What happens in between these two extremes can only be decided with the help of the exact solution of Equations (4) and (5). The nature of the solution is thus determined by Q on the one side and T_o and α on the other. The smaller the Q, the faster the electromagnetic energy decay; the smaller the T_o and the larger the α, the faster the η recovers to values larger than one.

In considering Fig. 2a for example, let us evaluate the time between two oscillation peaks. We notice $\Delta T = \omega\Delta t = 2 \times 10^6$. For the observations in Fig. 1 ω corresponds to 1.3 kmc. In approximate agreement with the experimental results we thus obtain $\Delta t = 2.5 \times 10^{-4}$ sec. The particular parameters chosen in Fig. 2c give a pulse repetition rate which is higher by about a factor of 20 than that observed by Kikuchi *et al.*[1] A different choice of parameters, however, can duplicate their results.

At first glance, the large values of Q which are necessary to obtain steady-state oscillations appear disturbing. In Fig. 2a, for example, $Q = 10^5$. The experimental results (Fig. 1) were obtained with a Q_e of approximately 1000. Let us consider the other parameters. In the calculations, T_o is 10^7 which corresponds to a τ of 1.2×10^{-3} sec. According to Kikuchi *et al.*,[7] we should have used a value for τ of 2.5×10^{-2} sec which would have required even higher values for Q. Also the value of $\alpha = 10$, we believe, is too large to correspond to the experimental conditions. The answer to this question can be found by considering the effects of cross relaxation[4] or by considering spectral diffusion within one line.[5],[6] In pink ruby which has been used in the present experiments the lines can be shown to be mainly inhomogeneously broadened by dipolar magnetic fields resulting from the Al nuclei. As an oscillation pulse tends to "eat a hole" into the resonance line, cross relaxation supplies inverted spins at the oscillation frequency at a rate much faster than the spin-lattice relaxation mechanism.

At first glance one might think that it would be sufficient to reduce the spin-lattice relaxation time T_o. It is indeed possible to obtain the correct type of solution, i.e. steady-state oscillations are possible with rather small values of Q. When the oscillation approaches its steady-state value, however, the times between the oscillation pulses are too short.

In order to prove that cross relaxation or spin diffusion indeed corrects this deficiency we have constructed a simplified model shown in Fig. 3. We have divided up the line into spins which are resonant more or less exactly at the maser oscillation frequency and spins which are further removed from the oscil-

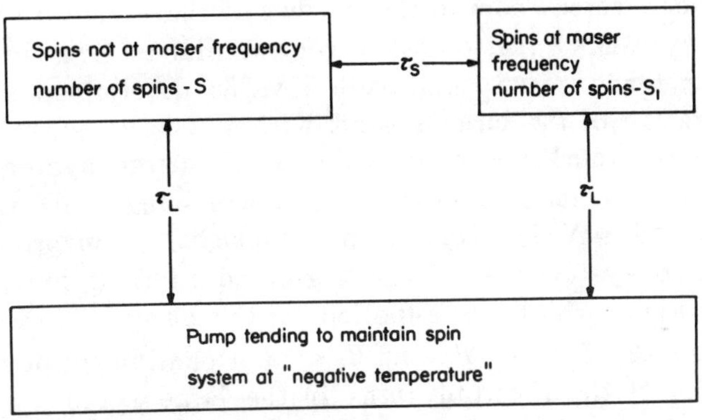

Fig. 3. Block diagram illustrating the equations which include cross relaxation or spectral diffusion

lation frequency. The two spin systems are coupled together by a time constant τ_s which is assumed to be much shorter than the spin-lattice relaxation time. Both systems tend to be maintained at an inverted population characterized by a time constant τ_L which, essentially, will be equal to the spin-lattice relaxation time. The spin system at the maser frequency will be assumed to contain a substantially smaller number of spins than the other system. The appropriate equations describing this system may be written as follows:

$$\frac{dP}{dt} = -\frac{1}{Q_L}\,\omega\,P \tag{6}$$

$$\frac{1}{Q_L} = \frac{1}{Q} - Cn \tag{7}$$

$$\frac{dn}{dt} = -\frac{n - \frac{S_1}{S}N}{\tau_s} - Dn\,P \tag{8}$$

$$\frac{dN}{dt} = -\frac{N - N_2}{\tau_L} + \frac{n - \frac{S_1}{S}N}{\tau_s}. \tag{9}$$

Equations (6) and (7) are identical to Equations (1) and (2). However n refers now to the number of the inverted spins in the small system. Equation (8) is very similar to Equation (3), however, n tends now to approach a value determined by the spin temperature of the large system with a time constant τ_s. If we denote the total number of spins in the large system by S and those in the small system by S_1, then equal spin temperature means $n/S_1 = N/S$, where N is the number of inverted spins in the large system. We have neglected in Equation (8) the inverted spins which are supplied by the pump. We are justified in doing so, if $\tau_s \ll T_L$ and $S_1 \ll S$. Equation (9) describes the behavior of the inverted spins in the large system. In the absence of microwave power at the signal frequency N would approach N_2 with a time constant τ_L. The second term contains the transfer of inverted spins to the small system.

Equations (6-9) have been integrated on an analog computer assuming the following values: $Q = 10^3 - 5 \times 10^3$; $\tau_s = 2 \times 10^{-5}$ sec; $\tau_L = 2.5 \times 10^{-2}$ sec; $S/S_1 = 20$; and $\omega = 2\pi \times 1.3 \times 10^9$ cycles/sec. In reducing the equations again into a dimensionless form analogous to Equations (4) and (5) we have to specify in addition to the above quoted values the quantity N_2/n_o, where n_o is the amount of inversion in the small system required to overcome the losses characterized by Q. We have assumed $N_2/n_o = 40$, which corresponds approximately to a value of $\alpha = 2$ in Equation (5). In the above range of Q values we cover again the type of solutions as in Fig. 2, however, with spacings between the oscillation pulses of a few msec. We have thus indeed shown that cross relaxation or spectral diffusion changes the results in the desired direction.

In the derivation of the equations, it has been assumed that the pump power is being applied at some time and then left on continuously. However similar types of solutions may be obtained under pulsed operating conditions. For example, in the experiments by Foner *et al.*[2] the pump levels have been saturated before the dc magnetic field was pulsed. This saturation of levels is expected to be maintained for some time during the pulse and thus even at the high magnetic field, inverted spins may be supplied at the signal frequency. It is thus possible that the present model may even apply to certain types of pulsed masers.

We wish to thank Dr. L. Rimai for stimulating discussions. We also wish to thank Dr. Foner for an interesting conversation regarding his experiments.

REFERENCES

1. C. Kikuchi, J. Lambe, G. Makhov, and R. W. Terhune, J. Appl. Phys. *30*, 1061 (1959).
2. S. Foner, L. R. Momo, and A. Mayer, Phys. Rev. Lett. *3*, 36 (1959).
3. I. R. Senitzky, Phys. Rev. Lett. *1*, 167 (1958).
4. N. Bloembergen, S. Shapiro, P. S. Pershan, and J. O. Artman, Phys. Rev. *114*, 445 (1959).
5. A. M. Portis, Phys. Rev. *104*, 584 (1956).
6. P. W. Anderson, Phys. Rev. *109*, 1492 (1958).

Pulse Propagation in a Laser Amplifier

JAMES P. WITTKE
RCA Laboratories, Princeton, New Jersey

AND

PETER J. WARTER
Electrical Engineering Department, Princeton University, Princeton, New Jersey
(Received 6 January 1964)

Pulse propagation in maser-type traveling-wave amplifiers with a homogeneously broadened transition is treated by a formalism analogous to the Bloch equations. Phenomenological dephasing (T_2) and recovery (T_1) times are defined, and a linear (nonsaturable) loss mechanism is included. In the numerical calculations, only the case of negligible excitation during the time it takes a pulse to pass is considered. When pulses are allowed to grow until the amplifier is "saturated," steady-state pulses having a unique shape and intensity independent of the initial pulse are found. The parameters of these steady-state pulses depend only on the ratio of the linear loss and gain coefficients. Steady-state pulses have a peak intensity that decreases monotonically from infinity to zero as the linear loss coefficient varies from zero to the gain coefficient, while the pulse width correspondingly varies from zero to infinity, and the pulse energy varies from a *finite* value to zero. Steady-state pulses propagate at a velocity less than that of the small signal velocity in the medium.

1. INTRODUCTION

WITH the advent of traveling-wave maser and laser amplifiers, the question of the response of such devices to pulsed signals has taken on importance. Several authors[1,2] have already treated this problem, approximating the interaction between the (assumed "classical") radiation field and the emissive ions by means of a transition probability for induced emission and absorption. Such an approach is inadequate when considering pulses intense enough to cause appreciable interaction in a time comparable to the relaxation times appropriate to the system. In such a case, the treatment, to be valid, must include the coherent nature of the radiation field–ion interaction.[3] This paper uses such an approach to predict the response of traveling-wave amplifiers of the maser type to pulses. As will be seen, the behavior predicted on this model for intense pulses is, in several important respects, quite different from that given by the earlier treatments.

To handle the corresponding coherent interaction phenomena in the case of nuclear resonance, Bloch[4] postulated a set of equations to describe the dynamics of a nuclear spin system in an external magnetic field when driven by an oscillating magnetic field. He introduced, in a phenomenological way, two relaxation times T_1 and T_2 by means of which he was able to approximate the influence of the surroundings (neighboring ions, nuclei, etc.) on the magnetization of interest. The model we shall employ is a direct adaptation of the Bloch equations to the case of a traveling-wave maser. (The analysis of a system of spin one-half particles by means of the Bloch equations is mathematically identical to that of a quantum-mechanical treatment of any two-level system interacting with an appropriate "classical" radiation field.) The previous treatments[1,2] have used instead an extension of equations developed by Statz and deMars,[5] based on the transition probability approach, to describe the traveling wave maser.

2. FORMULATION OF THE PROBLEM

The model we use is of a homogeneous resonance line with all ions equivalent. A phenomenological relaxation time T_2 is assumed to describe all dephasing interactions in direct analogy to Bloch. A second "relaxation" time T_1 is also assumed. This includes the effects of all the energy states of the emissive ions other than the two connected directly by the radiation field, the energy-exchanging relaxation processes coupling the ions to the surrounding lattice, and even, in our model, the pumping (excitation) processes by means of which the ions are put into, and maintained in, their emissive state. Thus, we assume that the combination of lattice interactions, spontaneous radiative processes, and pumping processes produce, in the absence of a signal, a stationary, initially emissive state in the maser ions, and further, that if the system is disturbed from this steady state, the combination of relaxation and excitation process tends to restore the system to this steady-state condition in a fashion describable by a unique time constant, which we designate as T_1. We further assume that the radiation field is intense enough to be adequately described by a classical field strength E, and, for simplicity, that the transition of interest is electric dipole with moment p.

The equations giving the response of the ions to the radiation field can be obtained using the density matrix

[1] E. E. Schultz-DuBois, Bell Telephone Laboratories, ASTIA document AD232689.
[2] L. M. Frantz and J. S. Nodvik, J. Appl. Phys. **34**, 2346 (1963).
[3] Y. H. Pao, J. Opt. Soc. Am. **52**, 871 (1962).
[4] F. Bloch, Phys. Rev. **70**, 460 (1946).
[5] H. Statz and G. DeMars, in *Quantum Electronics*, edited by C. H. Townes (Columbia University Press, New York, 1960), p. 530.

Reprinted with permission from *J. Appl. Phys.*, vol. 35, pp. 1668–1672, June 1964.

formalism.[6] The basic equations

$$i\hbar \frac{\partial \rho_{ij}}{\partial t} = (E_i - E_j)\rho_{ij} + \sum_k (V_{ik}\rho_{kj} - \rho_{ik}V_{kj}),$$

describe the time dependence of the density matrix, for a system having discrete quantum levels, resulting from the interaction of the field, given by V_{ik}. For a two-state system, this can be written

$$i\hbar(\partial \rho_{11}/\partial t) = (V_{12}\rho_{21} - \rho_{12}V_{21}),$$

and

$$i\hbar(\partial \rho_{12}/\partial t) = -\hbar\omega_0\rho_{12} + V_{12}(1 - 2\rho_{11}),$$

where $E_2 - E_1 = \hbar\omega_0$. The equations for ρ_{22} and ρ_{21} can be obtained from these by using the relations,

$$\rho_{11} + \rho_{22} = 1,$$

and

$$\rho_{12} = \rho_{21}^*.$$

Taking $V_{12} = V_{21} = pE(t)\cos\omega t$, and introducing the phenomenological relaxation terms, gives

$$i\hbar \frac{\partial \rho_{11}}{\partial t} = pE \cos\omega t(\rho_{12}^* - \rho_{12}) - i\hbar\left[\frac{\rho_{11} - \rho(p)}{T_1}\right],$$

and

$$i\hbar \frac{\partial \rho_{12}}{\partial t} = -\hbar\omega_0\rho_{12} + pE \cos\omega t(1 - 2\rho_{11}) - i\hbar\frac{\rho_{12}}{T_2}.$$

Here $\rho(p)$ is the steady-state value of ρ_{11} in the absence of a signal. Since ρ_{11} is real while ρ_{12} is complex, there are three independent equations. However, in the following we shall assume that the applied field is always on resonance, i.e., $\omega = \omega_0$. In this case, and neglecting terms that oscillate at frequency $2\omega_0$, only one component of ρ_{12}, and in particular, the part

$$\phi \equiv \text{Re}[-i\rho_{12}\exp(-i\omega t)],$$

is nonvanishing. There are thus really only two coupled equations to describe the field–ion interaction.

The equations can be put into dimensionless form by the substitutions

$$x \equiv t/T_2,$$
$$\Theta \equiv (pET_2)/\hbar,$$
$$\Phi \equiv 2\phi/[1 - 2\rho(p)],$$
$$\Delta \equiv (1 - 2\rho_{11})/[1 - 2\rho(p)],$$

and

$$\gamma \equiv T_2/T_1.$$

They become:

$$\partial\Delta/\partial x = \Phi\Theta - \gamma(\Delta - 1);$$
$$\partial\Phi/\partial x = -\Phi - \Delta\Theta.$$

[6] See A. A. Vuylsteke, *Elements of Maser Theory* (D. Van Nostrand Company, Inc., Princeton, New Jersey, 1960), Chap. 4.

These equations give the effect of the field on the ions. To obtain a complete solution for the problem, the reaction of the ions on the radiation field should be treated by means of a wave equation with an appropriate (complex) dielectric constant. A considerable mathematical simplification can be achieved, however, if one neglects backward waves and assumes that only a wave moving in the direction of the incident wave is present. A one-dimensional plane wave geometry is assumed. We will makes these assumptions, as have the previous treatments.[1,2]

Conservation of energy can then be expressed by

$$[\partial U(z,t)/\partial t] + v[\partial U(z,t)/\partial z] = P_{em}(z,t) - P_l(z,t).$$

Here a pulse, of energy density

$$U(z,t) = \epsilon E^2(z,t)/8\pi$$

is propagated with a velocity v in the z direction. P_{em} is the power density emitted by the radiating ions, and P_l is a power density lost from the beam due to losses not associated with the emissive ions.

To solve this system of equations for the response of the amplifier to the pulse, we replace the actual distributed amplifier by a series of thin-slab amplifiers, each slab being so thin (Δz) that the reaction of the ions on the field in that slab can be neglected.

A particular form is assumed for the loss mechanism:

$$P_l = \alpha_l v U,$$

where α_l is the loss coefficient. Thus, the loss is assumed linear (nonsaturable). (Such losses seem to be present in many solid-state laser materials, and could arise, for example, from some scattering mechanism such as might be caused by inhomogeneities or by some nonsaturable absorption in the medium.)

The conservation-of-energy equation can also be put into dimensionless form, using, in addition to the earlier substitutions,

$$L \equiv \alpha_l/\alpha_g,$$

and

$$S \equiv \alpha_g \Delta z/2,$$

where $\alpha_g \equiv \alpha_0 \Delta N(0)/N$ is the "gain coefficient" for the emissive medium. [$\Delta N(0)$ is the excess emissive population density, $\Delta N = N_2 - N_1$, before the arrival of the pulse, N is the total ion density, and

$$\alpha_0 = 4\pi p^2 N\omega T_2/\hbar\epsilon$$

is the absorption coefficient for the medium when *all* of the ions are in the lower state, $N_1 = N$; ϵ is the dielectric constant.]

With these definitions, the conservation of energy condition for the slab becomes for a sufficiently thin slab,

$$\Theta_{n+1} = \Theta_n - (\Phi_n + L\Theta_n)S.$$

(Here the time origin in *each* slab is taken as when the pulse first reaches the slab.)

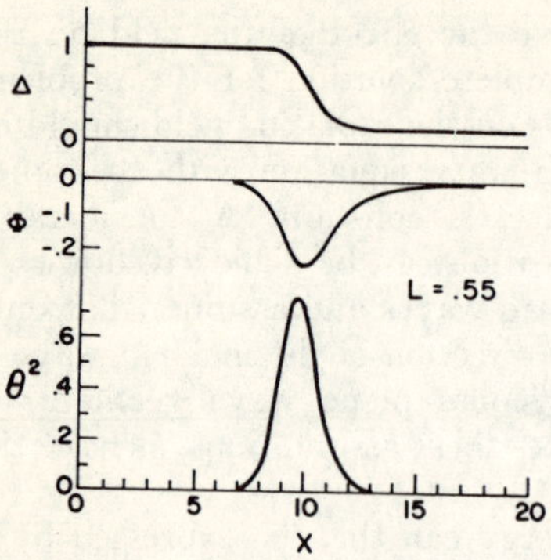

FIG. 1. The steady-state pulse for a loss parameter $L=0.55$. The intensity (θ^2), radiating dipole moment (Φ), and the population inversion (Δ) are shown as functions of the time (x). An arbitrary time origin is shown.

Our determination of amplifier behavior then consists of iterated numerical solutions of the coupled differential equations giving the response of the ions to an applied field, the field at any slab being determined from that at the previous slab by means of the recursion formula with the process continuing for as far along the amplifier as desired.

Note that the equations depend on N and $\rho(p)$ only in the combination $N[1-2\rho(p)]\sim\Delta N$. All that matters is the magnitude of ΔN; it is completely immaterial

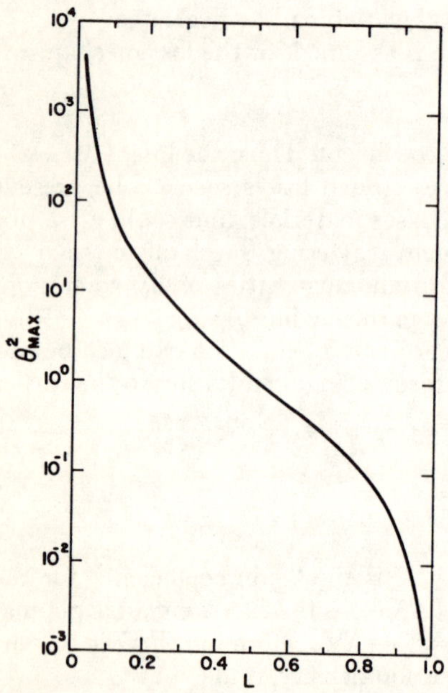

FIG. 2. Dimensionless peak pulse intensity (θ^2) for steady-state pulses as a function of the loss parameter L.

whether a given ΔN is achieved by weakly pumping a system with high ion density N, or strongly pumping $[\rho(p)\rightarrow 0]$ a system with a low ion density. (This, of course, assumes that the system parameters such as α_l, T_1 and T_2 are independent of ion density N.)

We shall restrict this paper to a consideration of cases in which $\gamma=0$. Physically, this means that the pulses of interest are assumed to be so short that no appreciable pumping (or other energy-exchanging processes) can occur during the pulse, which thus leaves the amplifying medium in a nonequilibrium state after it passes. (We have made calculations for several cases in which this assumption does not hold. The behavior of the pulse is then radically different, and more complex. For this reason, we will leave discussion of this situation to a later paper.) This assumption (of weak pumping) leaves the solutions of the set of coupled equations dependent only upon the input pulse and on one factor, the loss parameter L.

It is perhaps worthwhile pointing out that, even in their dimensionless forms, the various quantities appearing in the equations have a direct physical significance. Thus x is the time measured in units of the

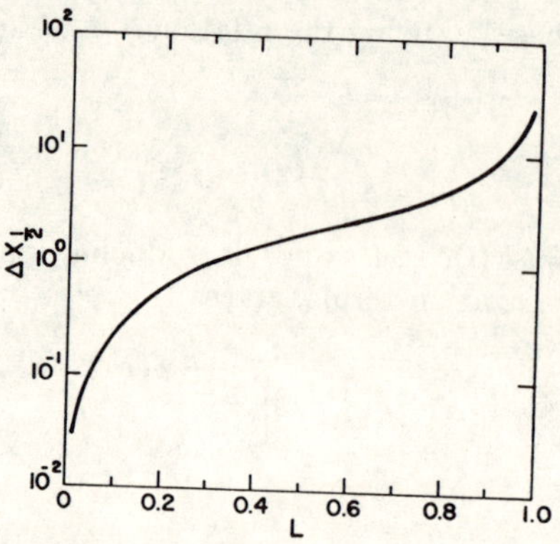

FIG. 3. Dimensionless steady-state pulse duration (width at half-peak intensity) as a function of the loss parameter L.

dephasing time T_2; Δ is the excess (emissive) population, normalized to its original value (before the pulse arrives); Φ is a measure of the component of the (transverse) oscillating dipole moment of the ions that has the proper phase to exchange energy with the radiation field; Θ is the radiation field strength. In a two-state system, in the absence of relaxation phenomena, a resonant field will cause each ion to oscillate between the two states at a (circular) frequency $\Omega=pE/\hbar$. Θ thus measures how far this state-exchanging process proceeds in a dephasing time; γ is the reciprocal energy exchanging "relaxation" time T_1, measured in units of the dephasing time T_2; and L is the ratio of the (scattering) loss coefficient α_l to the "gain coefficient" α_g.

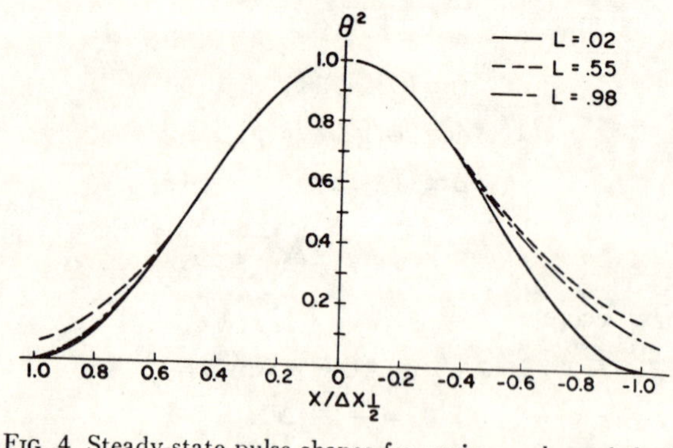

FIG. 4. Steady-state pulse shapes for various values of the loss parameter L. All pulses have been normalized to unity height and width.

3. RESULTS

Steady-State Pulses

One important aspect of the solutions of the equations is the existance of *steady-state pulses*. That is, under the combined effects of the emissive, amplifying interaction and the linear loss (scattering) interaction, for a given values of L, *all* input pulses, as they propagate through the amplifier, approach in peak intensity, pulse duration, and shape, a single, unique steady-state pulse which will propagate through the amplifier without change of shape or intensity. A typical pulse of this type is shown in Fig. 1, where pulse intensity (Θ^2), population inversion (Δ), and radiating dipole moment

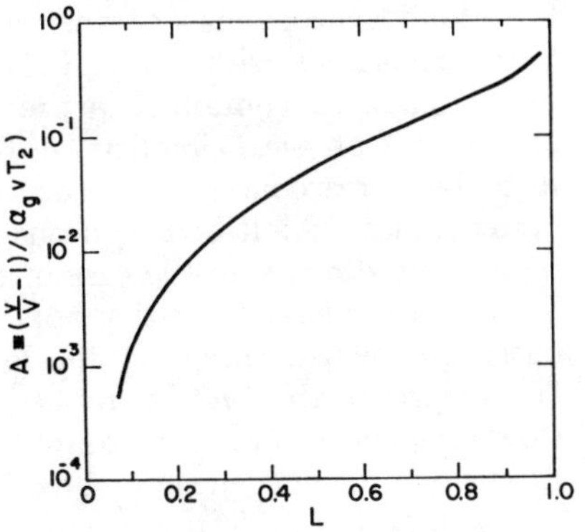

FIG. 5. The "velocity anomaly" for the propagation of steady-state pulses as a function of loss parameter L.

(Φ) are shown as a function of time (x) at one point in the amplifier.

The dependence of the peak pulse intensity for this steady-state pulse upon the loss parameter L is shown in Fig. 2. This shows that the peak intensity is a monotonically decreasing function of L, ranging from infinite intensity when $L=0$ (no linear losses) to zero when $L=1$ ($\alpha_l=\alpha_g$). This is in contrast to the predictions of the transition probability type of theory which does not predict, for finite duration pulses, any limit to the peak intensity for all $\alpha_g>\alpha_l$ and predicts the medium to be transparent, rather than attenuating, in the case $\alpha_l=\alpha_g$.

The pulse duration, defined as the time between half-peak intensity points, is shown in Fig. 3. A monotonic

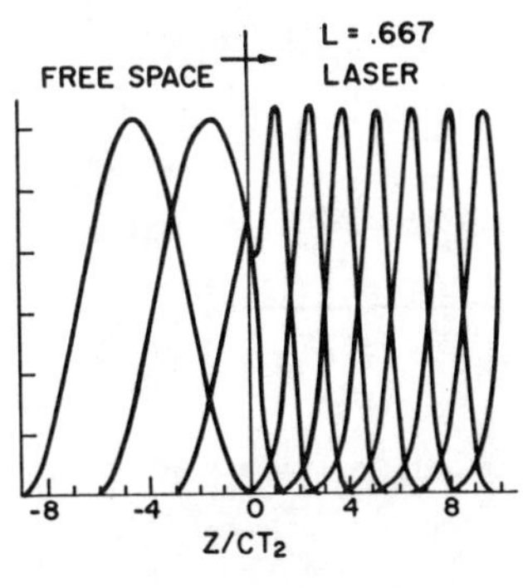

FIG. 6. The time development of a pulse propagating into an amplifier. The pulse shape at a succession of times ($\Delta t=3.0\ T_2$) is shown both outside and within an amplifier. The "velocity anomaly" is clearly seen in the reduction of propagation velocity in the amplifier. (The dielectric constant is assumed to be unity.)

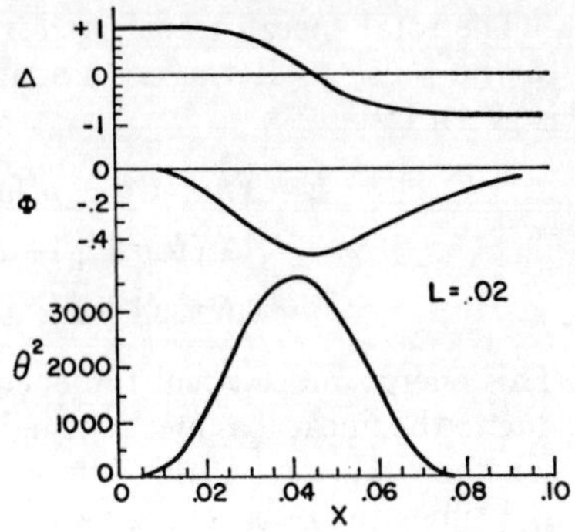

FIG. 7. The steady-state pulse for a very low-loss parameter $L=0.02$. Note that the state population in the amplifier is essentially inverted upon passage of the pulse.

function of L is again seen in which the duration shrinks toward zero as the linear loss decreases and approaches infinity as the loss increases toward $\alpha_e=\alpha_g$.

It should be noted that, as in the case of transition probability theories, when $L>1 (\alpha_l>\alpha_g)$, there is no steady-state pulse but all pulses are attenuated toward zero as they travel through the amplifier.

The shape of the steady-state pulse is more complex to describe. For values of L near 0, it is closely symmetrical about its peak, and the shape, to quite a close approximation, can be expressed by the form

$$I \approx I_0 \sin^2(\pi x/2\Delta x_{\frac{1}{2}}).$$

As L increases, the line becomes broader in the "wings"

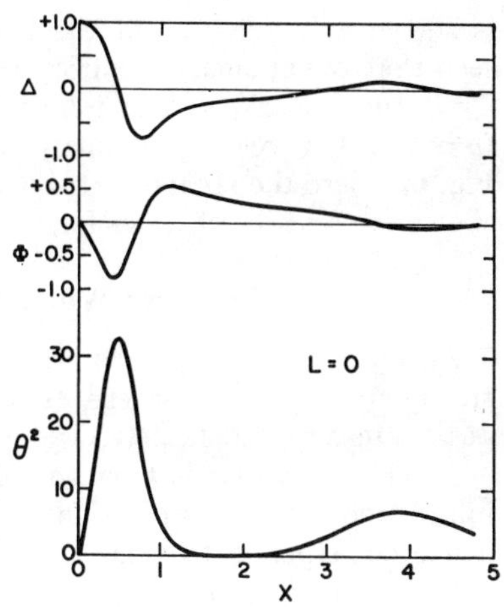

FIG. 8. A growing pulse in the limit of no linear losses $L=0$. There is no steady-state pulse in this case, and the (undamped) train of pulses leaves the energy states saturated rather than inverted.

and then becomes definitely asymmetric, with the leading edge rising more sharply than the trailing edge. Again, however, the top half of the pulse is of approximate $\sin^2$ shape. Examples of steady-state pulses illustrating these characteristics are shown in Fig. 4.

Pulse Energy

A particularly simple relation exists between the final value of Δ (after the pulse has passed) and L;

$$\Delta(\text{final}) = 2L - 1.$$

From this we can get a simple expression for the total energy of a steady-state pulse as follows:

The total energy transferred from the maser medium to the pulse as it traverses a unit length of amplifier is given by

$$E = N\hbar\omega[\rho(p)-\rho(\text{final})],$$
$$=[\Delta N\hbar\omega/2][1-\Delta(\text{final})],$$
$$=\Delta N\hbar\omega(1-L).$$

This energy must equal the energy lost from the pulse due to the linear loss mechanism if the pulse is to remain constant. However, this energy lost is just equal to α_l^{-1} times the pulse energy, or

$$E_{\text{pulse}} = (\Delta N\hbar\omega/\alpha_l)(1-L)$$
$$=\Delta N\hbar\omega(1/\alpha_l - 1/\alpha_g).$$

Note that in the case of no linear loss ($\alpha_l=0$), there is no limit to the pulse energy.

Velocity Anomaly

The steady-state pulses do not propagate down the amplifier at the velocity of light in the medium v. Instead, the velocity V depends on the loss parameter and also on $\alpha_g vT_2$, the gain coefficient for a distance equal to that traveled by light in a time T_2. This dependence is shown in Fig. 5, where the "velocity anomaly" A,

$$A \equiv (v/V-1)/(\alpha_g vT_2),$$

is shown as a function of the loss parameter L. It is seen that the anomaly A varies from zero ($V=v$) when $L=0$ (no linear loss), to infinity ($V=0$) when $L=1$ ($\alpha_e=\alpha_g$). The velocity anomaly can be clearly seen in Fig. 6, where the spatial distribution of pulse energy is shown for a series of times for the case where $L=0.67$.

Behavior When $L=0$

As linear losses are reduced (L approaches 0), the steady-state pulse sharpens (narrows) and grows in peak intensity indefinitely, while Δ(final) approaches -1. This corresponds to a complete inversion of the initial population distribution in the two states, and hence corresponds to what in nuclear magnetic resonance terminology is known as a "180° pulse." This is illustrated in Fig. 7 where the case $L=0.02$ is shown. However, if one solves the limiting case $L=0$ numerically, the system seems to be left in a "saturated" condition, i.e., $\Delta(\text{final})=0$, as shown in Fig. 8. This discrepancy is only apparent, for in the limit $L=0$ there is no steady-state pulse, and the amplifier, with no mechanism to provide damping, generates a train of pulses that leaves the amplifier behind the train saturated, rather than inverted.

4. DISCUSSION

We have presented the results of calculations showing the behavior of a model for a traveling wave maser amplifier when excited by large, monochromatic pulses. The model postulated a dephasing relaxation time T_2 for a homogeneous maser transition, the input pulse to be on resonance with the emissive ions, and a linear loss mechanism contained in the amplifying medium. Despite the physical simplicity of the model, we believe it to be a reasonable representation of actual maser systems that exhibit homogeneous broadening. Thus it may apply directly to the case of ruby optical masers, for which the laser transition appears to be homogeneously broadened, except at the lowest temperatures.[7] It can presumably *not* be applied in a direct way to (low-temperature) ruby microwave masers, whose transitions are inhomogeneously broadened,[8] or to optical maser materials exhibiting inhomogeneous broadening.[9] The neglect of pumping, etc. during the passage of the pulse appears to be a good approximation for most realizable situations (except in the case where $\alpha_l \approx \alpha_g$ and very long steady-state pulses are expected).

Losses that appear to be due to scattering, and are thus likely to be linear, have been directly observed in several solid-state laser materials. Values in ruby are in the range 2%–10% per cm ($\alpha_l \sim 0.02$–0.1 cm^{-1}), while in glass lasers losses are considerably lower ($\alpha_l \approx 0.001$ cm^{-1}). CaF_2 lasers are also known to exhibit such losses.

[7] D. E. McCumber and M. D. Sturge, J. Appl. Phys. **34**, 1682 (1963).
[8] J. Castle (private communication).
[9] Nd-glass lasers probably are severely inhomogeneously broadened, for example.

Propagation of Light Pulses in a Laser Amplifier*

A. Icsevgi† and W. E. Lamb, Jr.
Yale University, New Haven, Connecticut 06520
(Received 1 May 1969)

The problem of a light pulse propagating in a nonlinear laser medium is investigated. The electromagnetic field is treated classically and the active medium consists of thermally moving atoms which have two electronic states with independent decay constants γ_a and γ_b in addition to the decay constant γ_{ab} describing the phase memory. The self-consistency requirement that the field sustained by the polarized medium be equal to the field inducing the polarization leads to coupled equations of motion for the density matrix, and equations of propagation for the electromagnetic field. Although the theory is developed for a Doppler-broadened gaseous medium, it may also be applied to a solid medium with inhomogeneous broadening. A unified treatment is given encompassing a wide range of pulse durations from cw signals to psec pulses. Continuous pumping is allowed, as well as any amount of detuning of the carrier frequency of the pulse from the atomic resonance frequency. The three independent decay constants γ_a, γ_b, and γ_{ab} provide greater flexibility than that obtained by using $1/T_1$ and $1/T_2$. The equations are solved analytically in a few specialized cases and numerically in the general case. Flow charts for accomplishing the numerical integration are given. Among the special problems considered is the apparent paradox of pulses propagating faster than the velocity of light under circumstances described by Basov *et al.* It is shown that this contradiction with relativity arises from the use of an unphysical initial condition.

I. INTRODUCTION

We present a theoretical investigation of the behavior of light pulses traveling in an amplifying laser medium. A semiclassical description of the interaction between matter and radiation will be used, treating the medium quantum mechanically and the radiation field according to Maxwell's theory. The basic ideas are derived from Lamb's theory of optical masers.[1] However, some of his original assumptions and restrictions are relaxed in order to properly apply the theory to the problem at hand. Since there are differences between the problems of a self-sustained oscillator and of an amplifier, it is desirable to develop the theory from first principles.

The medium shall be considered to be a collection of two-level "atoms" (Fig. 1) coupled only through their interaction with the over-all radiation field. If a population inversion between the levels a and b is established, such a medium is capable of amplifying light in a frequency band around the separation of the levels.

In order to carry out necessary statistical summations, it is convenient to represent the state of a two-level atom by means of a 2×2 density matrix ρ. This is related to the wave function description in the follow-

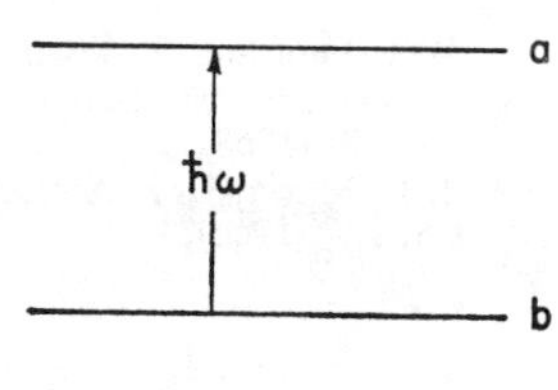

Fig. 1. Energy diagram of the two-level atom. The levels have a resonance transition frequency $\omega>0$.

* Research sponsored by Yale University, the Air Force Office of Scientific Research under AFOSR Grant No. 1324-67, and the NASA under Grant No. NASA-NGR 07-004-035.

† This paper is based on material submitted by A. Icsevgi in partial fulfillment of the requirements for the degree of Doctor of Philosophy at Yale University.

[1] W. E. Lamb, Jr., Phys. Rev. **134**, A1429 (1964).

Reprinted with permission from *Phys. Rev.*, vol. 185, pp. 517–545, Sept. 10, 1969 ing way. Given two time-independent basis functions Ψ_a, Ψ_b, for the states a and b, the wave function for the atom can be written as

$$\Psi(t)=a(t)\Psi_a+b(t)\Psi_b, \tag{1}$$

and the density matrix for such a "pure case" is

$$\rho(t)\equiv\begin{pmatrix}\rho_{aa} & \rho_{ab}\\ \rho_{ba} & \rho_{bb}\end{pmatrix}=\begin{pmatrix}aa^* & ab^*\\ a^*b & bb^*\end{pmatrix}. \tag{2}$$

The off-diagonal elements of ρ will be related to the atomic dipole moment, while the diagonal elements give the probabilities for the atom to be in state a or b.

One wishes to describe the decay of the levels a, b and the damping of the atomic dipole moment. It can be shown, using the Wigner-Weisskopf theory,[2] that for radiative decay this can be achieved phenomenologically by introducing two damping constants γ_a, γ_b such that in the absence of electric field perturbations, the amplitudes a and b decay exponentially as

$$a(t)=a_0e^{-\frac{1}{2}\gamma_a t}, \tag{3a}$$

$$b(t)=b_0e^{-\frac{1}{2}\gamma_b t}. \tag{3b}$$

This implies that ρ_{aa}, ρ_{bb}, ρ_{ab} have decay constants γ_a, γ_b, $\gamma_{ab}=\frac{1}{2}(\gamma_a+\gamma_b)$, respectively. When the contributions to γ_a, γ_b, γ_{ab} of damping mechanisms of nonradiative type, such as phonon interruptions in solids or atomic collisions in gases, are taken into account, the simple relationship between the three constants is destroyed. One finds typically that[3]

$$\gamma_{ab}\geq\tfrac{1}{2}(\gamma_a+\gamma_b). \tag{4}$$

We shall therefore regard γ_a, γ_b, and γ_{ab} as independent. The quantity γ_{ab} represents the decay constant for the dipole moment of the atom, so that $1/\gamma_{ab}$ measures the time of atomic phase memory or coherence. Associated with this damping mechanism is a spectral linewidth γ_{ab} which is sometimes called "homogeneous."

An additional broadening of a different nature is obtained when one considers an ensemble of atoms. In solids the atoms may have slightly different resonance frequencies due to inhomogeneities in the crystal environment. The averaging over resonance frequencies leads to a decay of the polarization or an "inhomogeneous broadening." In gases the averaging over various atomic velocities leads to a similar effect called "Doppler broadening." These complications will be dealt with in a more fundamental way later.

Pulses covering a wide range of duration, from quasi-monochromatic light to recently produced psec pulses,[4] are now available but previous treatments have been concerned either with very broad pulses which permit the use of rate equations,[5] or with ultrashort pulses for which the phase memory time can be considered infinitely long.[6] Basov *et al.*[7] have treated the problem of pulses very broad compared to $1/\gamma_{ab}$ and very sharp with respect to $1/\gamma_a$ and $1/\gamma_b$, but only with homogeneous broadening. Other authors[8–10] have also considered the simplified case of a homogeneously broadened medium but with arbitrary $\gamma_{ab}=1/T_2$ and $\gamma_a=\gamma_b=1/T_1$. In the domain of ultrashort pulses McCall and Hahn have treated the problem of an inhomogeneously broadened attenuator.[11] The recent work of Hopf and Scully[12] deals with an inhomogeneously broadened amplifier and includes the effects of a finite phase memory time, but their analysis applies only to solid-state lasers and they mostly take $\gamma_a=\gamma_b=1/T_1\ll\gamma_{ab}=1/T_2$. It appears that the previous attempts to deal with the problems of pulse propagation in a nonlinear medium have been unnecessarily specialized. The purpose of this paper is to give a unified treatment of the many different aspects of the problem, using one formalism. The discussion is aimed at a gaseous, medium but contains the case of a solid as a special case. It is shown that to lowest order in v/c the inhomogeneous broadening of solids and the Doppler broadening of gases are formally equivalent. This result is a consequence of dealing with running waves instead of standing waves.

The paper contains seven sections. Section II sets up the general formalism, hypotheses, and equations. The subsequent sections correspond to a classification of pulses according to their duration with respect to the atomic phase memory time $1/\gamma_{ab}$ characterizing the medium. Section III deals with monochromatic waves which can be thought of as pulses of infinite duration and which afford exact solutions. Section IV treats long pulses for which rate equations can be derived. Section V considers ultrashort pulses, and Sec. VI analyzes intermediate pulses. Until this section is reached emphasis is placed on analytical methods and solutions, but in Sec. VI we deal with the most general case for which very little can be done analytically, and numerical methods are required. A digital computer program is presented which integrates the set of coupled partial differential equations governing the dynamics of the pulse in the medium. Solutions are obtained and a discussion is given of the effects of varying such param-

[2] V. Weisskopf and E. Wigner, Z. Physik **63**, 54 (1930).

[3] B. L. Gyorffy, M. Borenstein, and W. E. Lamb, Jr., Phys. Rev. **169**, 340 (1968).

[4] A. J. DeMaria, D. A. Stetser, and H. A. Iteynan, Appl. Phys. Letters **8**, 174 (1966).

[5] L. M. Frantz and J. S. Nodvik, J. Appl. Phys. **34**, 2346 (1963).

[6] G. L. Lamb, Phys. Letters **25A**, 181 (1967).

[7] N. G. Basov, R. V. Ambartsumyan, V. S. Zuev, P. G. Kryukov, and V. S. Letokhov, Zh. Eksperim. i Teor. Fiz. **50**, 23 (1966) [English transl.: Soviet Phys.—JETP **23**, 16 (1966)].

[8] J. P. Wittke and P. J. Warter, J. Appl. Phys. **35**, 1668 (1964).

[9] F. T. Arecchi and R. Bonifacio, J. Quant. Electron. **QE-1**, 169 (1965).

[10] C. L. Tang and B. D. Silverman, in *Physics of Quantum Electronics*, edited by P. L. Kelley, B. Lax, and P. E. Tannenwald, (McGraw-Hill Book Co., New York, 1966).

[11] S. L. McCall and E. L. Hahn, Phys. Rev. Letters **18**, 908 (1967).

[12] F. A. Hopf and M. O. Scully, Phys. Rev. **179**, 399 (1969).

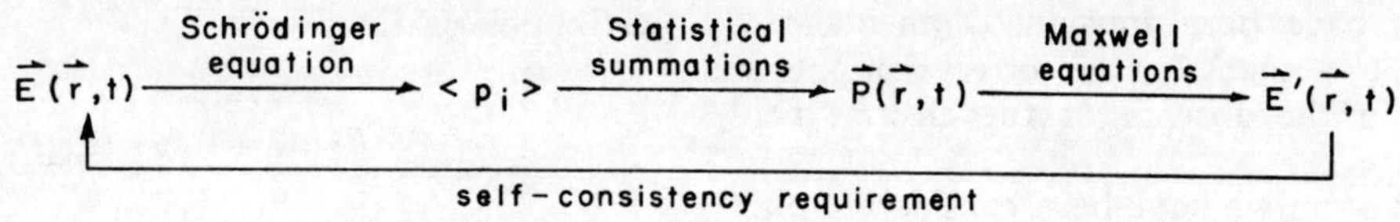

FIG. 2. Schematic basis for the derivation of the equations.

eters as decay constants, Doppler linewidth, gain, loss, pulse shape, duration and energy. The Fortran IV program used is described in Appendix G. Some derivations which would interrupt the continuity are also placed in Appendices. The last section summarizes the results of the calculations.

II. FRAMEWORK FOR ANALYSIS

A. Introduction

The scheme of calculation is well summarized by the diagram of Fig. 2. An assumed electric field $\mathbf{E}(\mathbf{r},t)$ polarizes the atoms of the medium according to the laws of quantum mechanics. The atomic dipole moments $\langle \mathbf{p}_i \rangle$ statistically add up to a macroscopic polarization $\mathbf{P}(\mathbf{r},t)$, which in turn enters Maxwell's equations as a source term and drives the electric field. In order for the scheme to be self-consistent, the electric field $\mathbf{E}'(\mathbf{r},t)$ driven by the polarization $\mathbf{P}(\mathbf{r},t)$ must be exactly equal to the assumed field $\mathbf{E}(\mathbf{r},t)$, i.e.,

$$\mathbf{E}'(\mathbf{r},t)=\mathbf{E}(\mathbf{r},t). \tag{5}$$

The calculation proceeds in three steps: (1) quantum mechanical equations for the density matrix, (2) statistical summations, (3) electromagnetic field equations.

In the remaining parts of this section we shall derive a set of coupled partial differential equations suitable for numerical integration which will be considered valid in all of the specialized cases treated in subsequent sections. In order to derive these equations we shall introduce simplifiying assumptions in two separate stages. First we assume:

(a) The two-level atoms are coupled only through their interaction with the common radiation field $\mathbf{E}(\mathbf{r},t)$.

(b) The radiation field is a scalar $E(z,t)$, representing a uniform plane wave polarized in the x direction, and propagating in the z direction (Fig. 3).

(c) The dipole approximation holds for the interaction of the atoms with the radiation field.

B. Equations of Motion for the Medium

The density matrix for a two-level atom was introduced in Sec. I. In Dirac's notation, the representation-free definition of the density matrix is

$$\rho=|\psi\rangle\langle\psi|. \tag{6}$$

The expectation value for an operator $\mathcal{O}$ is given by

$$\langle\mathcal{O}\rangle=\mathrm{Tr}(\rho\mathcal{O}). \tag{7}$$

The ket $|\Psi\rangle$ obeys Schrödinger's equation

$$i\hbar\frac{\partial}{\partial t}|\Psi\rangle=[H_0+H_{\mathrm{int}}(t)]|\Psi\rangle=H|\Psi\rangle, \tag{8}$$

where H_0 is the Hamiltonian for the unperturbed atom, and

$$H_{\mathrm{int}}=-E\mathcal{P} \tag{9}$$

is the electric field perturbation in the dipole approximation, $\mathcal{P}$ being the dipole moment operator of the atom. Differentiating (6), using (8) and its complex conjugate, we obtain

$$i\hbar\frac{\partial\rho}{\partial t}=[H,\rho]. \tag{10}$$

In the representation for which the basis functions Ψ_a, Ψ_b are eigenfunctions of H_0, we have

$$H_0=\hbar\begin{pmatrix}\omega_a & 0\\ 0 & \omega_b\end{pmatrix} \tag{11}$$

and

$$H_{\mathrm{int}}=\hbar\begin{pmatrix}0 & -(p/\hbar)E(t)\\ -(p/\hbar)E(t) & 0\end{pmatrix}, \tag{12}$$

where the matrix element

$$p=\langle a|\mathcal{P}|b\rangle=\langle b|\mathcal{P}|a\rangle \tag{13}$$

is taken to be real without loss of generality. The diagonal matrix elements of $\mathcal{P}$ vanish by parity consideration. In component form, the equation of motion (10) becomes

$$\dot\rho_{aa}=-\gamma_a\rho_{aa}-i(pE/\hbar)(\rho_{ab}-\rho_{ba}), \tag{14a}$$

$$\dot\rho_{bb}=-\gamma_b\rho_{bb}+i(pE/\hbar)(\rho_{ab}-\rho_{ba}), \tag{14b}$$

$$\dot\rho_{ab}=-(\gamma_{ab}+i\omega)\rho_{ab}-i(pE/\hbar)(\rho_{aa}-\rho_{bb}), \tag{14c}$$

$$\rho_{ba}=\rho_{ab}{}^*, \tag{14d}$$

where the resonance frequency ω is

$$\omega=\omega_a-\omega_b, \tag{15}$$

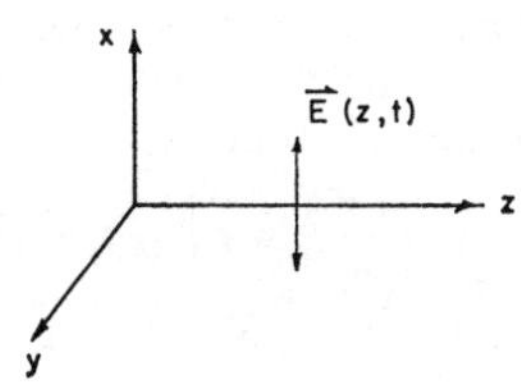

FIG. 3. Linearly polarized plane wave.

and γ_a, γ_b, γ_{ab} have been introduced phenomenologically in order to account for the correct damping of levels a and b in the absence of the electric field perturbation.

The individual atom we have been considering may be specified by the labels

$$\{\alpha,z_0,t_0,v\},$$

where α is the state a or b to which the atom was originally excited, (z_0,t_0) is the space-time point where it was excited and v is the velocity it had. We shall neglect atomic collision effects and hence the atomic velocity v will be a constant for each atom. (For a solid-state laser the label v could be replaced by a frequency label ω since the resonance frequency may vary from one atom to the other due to inhomogeneities in the crystal environment.) The density matrix for this atom has a time dependence so that it should be written as

$$\rho=\rho(\alpha,z_0,t_0,v;\,t). \tag{16}$$

From (7), the expectation value for the atomic dipole moment $\mathcal{P}$ is readily seen to be

$$\langle\mathcal{P}\rangle=p[\rho_{ab}(\alpha,z_0,t_0,v;\,t)+\rho_{ba}]. \tag{17}$$

The macroscopic polarization of the medium at (z,t), $P(z,t)$, will be produced by an ensemble of atoms which arrived at z, at time t, irrespective of their state α, place z_0, time t_0 and velocity v of excitation. The expression (16) should therefore be summed over α, z_0, t_0, v thus leading to a population matrix.

Since we have equations of motion (14) for the individual density matrices, it appears that we have to defer the summation process until $\rho(\alpha,z_0,t_0,v;\,t)$ is determined from those equations. However we will be able to sum over α, z_0, t_0 (but not v) before we integrate the equations. This is achieved in the following way. Consider

$$\rho(v,z,t)=\sum_{\alpha=a}^{b}\int_{-\infty}^{t}dt_0\int dz_0\lambda_\alpha(v,z_0,t_0)\rho(\alpha,z_0,t_0,v;\,t)\times\delta(z-z_0-v(t-t_0)), \tag{18}$$

where $\lambda_\alpha(v,z_0,t_0)$ is the number of atoms excited to state α, per unit time and unit volume. Assuming that this quantity varies slowly with z_0 and t_0, it can be taken outside the integrals and replaced by $\lambda_\alpha(v,z,t)$. The δ function in (18) insures that we are only summing over atoms which will reach the right place z at time t. Hence, $\rho(v,z,t)$ is the population matrix at time t, for a group of atoms of a given velocity v, which will be at z at time t, irrespective of their α, z_0, t_0. It is convenient to define the partial polarization produced by such a group of atoms as

$$P(v,z,t)=p[\rho_{ab}(v,z,t)+\text{c.c.}]. \tag{19}$$

The total polarization $P(z,t)$ is the sum over these partial polarizations

$$P(z,t)=\int dv\,P(v,z,t). \tag{20}$$

The integration over z_0 involved in (18) is straightforward and gives

$$\rho(v,z,t)=\sum_{\alpha=a}^{b}\lambda_\alpha\int_{-\infty}^{t}dt_0\rho(\alpha,\,z-v(t-t_0),\,t_0,\,v;\,t). \tag{21}$$

The set of equations (14) can be written more explicitly as

$$(\partial/\partial t)\rho_{aa}(\alpha,z_0,t_0,v;\,t)=-\gamma_a\rho_{aa}-i(p/\hbar)E(z_0+v(t-t_0),t)(\rho_{ab}-\rho_{ba}), \tag{22a}$$

$$(\partial/\partial t)\rho_{bb}(\alpha,z_0,t_0,v;\,t)=-\gamma_b\rho_{bb}+i(p/\hbar)E(z_0+v(t-t_0),t)(\rho_{ab}-\rho_{ba}), \tag{22b}$$

$$(\partial/\partial t)\rho_{ab}(\alpha,z_0,t_0,v;\,t)=-(\gamma_{ab}+i\omega)\rho_{ab}-i(p/\hbar)E(z_0+v(t-t_0),t)\times(\rho_{aa}-\rho_{bb}). \tag{22c}$$

Let us consider the quantity $(\partial/\partial t)\rho(v,z,t)+v(\partial/\partial z)\times\rho(v,z,t)$. In order to avoid a confusion of symbols, we would like to remind the reader that by $(\partial/\partial t)\times\rho(\alpha,z_0,t_0,v;\,t)$ we mean the derivative of ρ with respect to its fifth variable, evaluated at the given values of the arguments. Similarly, $(\partial/\partial z_0)\rho(\alpha,z_0,t_0,v;\,t)$ will mean the derivative of ρ with respect to its second variable. Introducing the symbol

$$\partial=(\partial/\partial t)+v(\partial/\partial z), \tag{23}$$

and using (21) we have

$$\begin{aligned}\partial\rho(v,z,t)=&\sum_{\alpha=a}^{b}\lambda_\alpha\rho(\alpha,z,t,v;\,t)\\&+\sum_{\alpha=a}^{b}\lambda_\alpha\int_{-\infty}^{t}dt_0\{(\partial/\partial t)\rho(\alpha,\,z-v(t-t_0),\,t_0,\,v;\,t)\\&-v(\partial/\partial z_0)\rho(\alpha,\,z-v(t-t_0),\,t_0,\,v;\,t)\\&+v(\partial/\partial z_0)\rho(\alpha,\,z-v(t-t_0),\,t_0,\,v;\,t)\};\end{aligned} \tag{24}$$

or

$$\partial\rho(v,z,t)=\sum_{\alpha=a}^{b}\lambda_\alpha\rho(\alpha,z,t,v;\,t)+\sum_{\alpha=a}^{b}\lambda_\alpha\int_{-\infty}^{t}dt_0(\partial/\partial t)\rho(\alpha,\,z-v(t-t_0),\,t_0,\,v;\,t). \tag{25}$$

From (22) it is found for example that

$$\begin{aligned}(\partial/\partial t)\rho_{aa}(\alpha,\,z-v(t-t_0),\,t_0,\,v;\,t)=&-\gamma_a\rho_{aa}(\alpha,\,z-v(t-t_0),\,t_0,\,v;\,t)\\&-i(p/\hbar)E(z-v(t-t_0)+v(t-t_0),\,t)\\&\times[\rho_{ab}(\alpha,\,z-v(t-t_0),\,t_0,\,v;\,t)-\text{c.c.}].\end{aligned} \tag{26}$$

On the other hand, we have

$$\rho(\alpha,z,t,v;\,t)=\begin{pmatrix}\delta_{a\alpha} & 0\\ 0 & \delta_{b\alpha}\end{pmatrix}, \tag{27}$$

and

$$\sum_{\alpha=a}^{b}\lambda_\alpha\rho(\alpha,z,t,v;\,t)=\begin{pmatrix}\lambda_a & 0\\ 0 & \lambda_b\end{pmatrix}, \tag{28}$$

so that the *aa* component of Eq. (25) becomes

$$\partial\rho_{aa}(v,z,t)=\lambda_a-\gamma_a\rho_{aa}-i(p/\hbar)E(z,t)\times[\rho_{ab}(v,z,t)-\text{c.c.}], \tag{29}$$

and similarly for the other components. We can write these equations in a more compact form as

$$\partial\rho_{aa}(v,z,t)=\lambda_a-\gamma_a\rho_{aa}-i(p/\hbar)E(z,t)(\rho_{ab}-\rho_{ba}), \tag{30a}$$

$$\partial\rho_{bb}(v,z,t)=\lambda_b-\gamma_b\rho_{bb}+i(p/\hbar)E(z,t)(\rho_{ab}-\rho_{ba}), \tag{30b}$$

$$\partial\rho_{ab}(v,z,t)=-(\gamma_{ab}+i\omega)\rho_{ab}-i(p/\hbar)E(z,t)\times(\rho_{aa}-\rho_{bb}), \tag{30c}$$

$$\rho_{ba}=\rho_{ab}^*. \tag{30d}$$

It is desirable to rewrite these equations in terms of the partial polarization

$$P(v,z,t)=p[\rho_{ab}(v,z,t)+\text{c.c.}]. \tag{31}$$

This can be done with some algebraic manipulation left to Appendix A. The result is

$$[(\partial+\gamma_{ab})^2+\omega^2]P=-2\omega(p^2/\hbar)E(\rho_{aa}-\rho_{bb}), \tag{32a}$$

$$(\partial+\gamma_a)\rho_{aa}=\lambda_a+(\hbar\omega)^{-1}E(\partial+\gamma_{ab})P, \tag{32b}$$

$$(\partial+\gamma_b)\rho_{bb}=\lambda_b-(\hbar\omega)^{-1}E(\partial+\gamma_{ab})P. \tag{32c}$$

C. Electromagnetic Field Equations

The Maxwell's equations appropriate to our case are (in mks units)

$$\nabla\cdot\mathbf{D}=0,\quad \nabla\times\mathbf{E}=-\partial\mathbf{B}/\partial t, \tag{33a}$$

$$\nabla\cdot\mathbf{B}=0,\quad \nabla\times\mathbf{H}=\mathbf{J}+\partial\mathbf{D}/\partial t, \tag{33b}$$

where

$$\mathbf{D}=\epsilon_0\mathbf{E}+\mathbf{P},\quad \mathbf{B}=\mu_0\mathbf{H},\quad \mathbf{J}=\sigma\mathbf{E}. \tag{33c}$$

Here $\mathbf{P}$ is the polarization and $\mathbf{J}$ is the current density. Although the medium is assumed free of real charges and currents, a fictitious Ohmic current $\mathbf{J}$ is introduced in order to account phenomenologically for the linear losses incurred in any absorbing background medium. Taking the curl of the second Eq. (33a) and using (33b) along with $\mathbf{B}=\mu_0\mathbf{H}$, one gets

$$\nabla\times\nabla\times\mathbf{E}+\mu_0\sigma\partial\mathbf{E}/\partial t+c^{-2}\partial^2\mathbf{E}/\partial t^2=-\mu_0\partial^2\mathbf{P}/\partial t^2. \tag{34}$$

In our case

$$\mathbf{E}(\mathbf{r},t)=E(z,t)\hat{x}, \tag{35}$$

and the above equation reduces to

$$-\partial^2E/\partial z^2+\mu_0\sigma\partial E/\partial t+c^{-2}\partial^2E/\partial t^2=-\mu_0\partial^2P/\partial t^2. \tag{36}$$

Coupling this equation with the equations of motion (32) for the medium we get

$$-\partial^2E/\partial z^2+\mu_0\sigma\partial E/\partial t+c^{-2}\partial^2E/\partial t^2=-\mu_0\partial^2P(z,t)/\partial t^2, \tag{37a}$$

$$[(\partial+\gamma_{ab})^2+\omega^2]P(v,z,t)=-2\omega(p^2/\hbar)E(\rho_{aa}-\rho_{bb}), \tag{37b}$$

$$(\partial+\gamma_a)\rho_{aa}(v,z,t)=\lambda_a+(\hbar\omega)^{-1}E(\partial+\gamma_{ab})P(v,z,t), \tag{37c}$$

$$(\partial+\gamma_b)\rho_{bb}(v,z,t)=\lambda_b-(\hbar\omega)^{-1}E(\partial+\gamma_{ab})P(v,z,t), \tag{37d}$$

where $P(z,t)$ and $P(v,z,t)$ are related by Eq. (20). Note that Eqs. (37) were derived under assumptions (a), (b), and (c) of Sec. II A. In the special case $\gamma_a=\gamma_b=\gamma$, the last two equations combine into a single equation for the population inversion density $N=\rho_{aa}-\rho_{bb}$, namely

$$(\partial+\gamma)N=\lambda_a-\lambda_b+2(\hbar\omega)^{-1}E(\partial+\gamma_{ab})P. \tag{38}$$

D. Reduction of the Basic Equations

Equations (37) will now be reduced on the basis of the three additional assumptions (d), (e), and (f)

(d) The optical frequency ω is much larger than the natural linewidth γ_{ab}, i.e.,

$$\omega\gg\gamma_{ab}. \tag{39}$$

(e) The rotating wave approximation holds, i.e. harmonics of the optical frequency will be neglected.

(f) The field can adequately be represented as a running wave with slowly varying amplitude and phase

$$E(z,t)=\mathcal{E}(z,t)\cos[\nu t-Kz+\varphi(z,t)]. \tag{40}$$

In order that the reflected wave may be neglected, we shall have to assume that the properties of the medium may vary only slightly over a wavelength. The frequency ν has a nominal value close to the atomic frequency ω. (Note that ν is a distinct circular frequency and is not equal to $\omega/2\pi$.) For the problem of a laser oscillator inside a cavity, ν represented the frequency of oscillations and was determined by the equations. In our problem ν is a convenient frequency chosen to represent the incoming pulse. A pulse does not have a well-defined frequency and the choice of ν is pretty much arbitrary. Different choices of ν lead to different phase functions $\varphi(z,t)$. We shall therefore set $\nu=\omega$, without loss of generality, when we consider sufficiently short pulses. However, in the case of a nearly monochromatic wave it is convenient to keep ν arbitrary. The wave number K will always be

$$K=\nu/c. \tag{41}$$

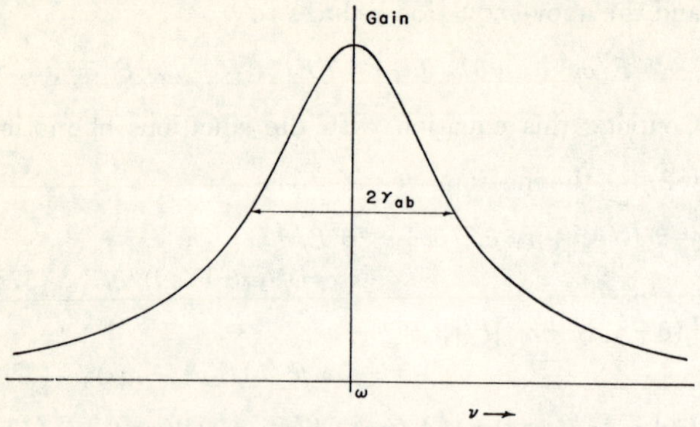

FIG. 4. Lorentzian gain profile with full width at half-maximum $=2\gamma_{ab}$.

The amplitude $\mathcal{E}$ and phase φ shall vary slowly in an optical cycle and wavelength, i.e.,

$$\partial\mathcal{E}/\partial t \ll \nu\mathcal{E}, \quad \partial\mathcal{E}/\partial z \ll K\mathcal{E}, \tag{42a}$$

$$\partial\varphi/\partial t \ll \nu\varphi, \quad \partial\varphi/\partial z \ll K\varphi. \tag{42b}$$

If the field is written as in (40), the response of the medium, neglecting higher harmonics, is of the form

$$P(v,z,t)=C(v,z,t)\cos(\nu t-kz+\varphi) +S(v,z,t)\sin(\nu t-Kz+\varphi), \tag{43}$$

where C and S can be called, respectively, the in-phase and the out-of-phase components of the polarization. When these expressions are substituted into Eqs. (37) and the above approximations are made (see Appendix B) one obtains

$$\partial\mathcal{E}/\partial z+c^{-1}\partial\mathcal{E}/\partial t=-\kappa\mathcal{E}-\tfrac{1}{2}K\langle S/\epsilon_0\rangle_v\,, \tag{44a}$$

$$(\partial\varphi/\partial z+c^{-1}\partial\varphi/\partial t)\mathcal{E}=-\tfrac{1}{2}K\langle C/\epsilon_0\rangle_v\,, \tag{44b}$$

$$\partial S=-\gamma_{ab}S-(\omega-\nu+Kv-\partial\varphi)C -(p^2/\hbar)\mathcal{E}(\rho_{aa}-\rho_{bb})\,, \tag{44c}$$

$$\partial C=-\gamma_{ab}C+(\omega-\nu+Kv-\partial\varphi)S\,, \tag{44d}$$

$$\partial\rho_{aa}=\Lambda_a-\gamma_a\rho_{aa}+\tfrac{1}{2}\mathcal{E}S/\hbar\,, \tag{44e}$$

$$\partial\rho_{bb}=\Lambda_b-\gamma_b\rho_{bb}-\tfrac{1}{2}\mathcal{E}S/\hbar\,, \tag{44f}$$

where

$$\kappa=\tfrac{1}{2}\sigma/\epsilon_0 c \tag{45}$$

is the linear loss coefficient, the symbol $\langle\ \rangle_v$ stands for

$$\langle\cdots\rangle_v=\int(\cdots)W(v)dv\,, \tag{46}$$

the velocity independent source terms Λ_a, Λ_b are such that

$$\lambda_\alpha(v,z,t)=\Lambda_\alpha(z,t)W(v)\,, \tag{47}$$

and $W(v)$ is the velocity distribution function. Atomic motion affects Eqs. (44) in two ways. The Kv terms correspond to the Doppler shift in the frequency seen by the atom and play formally the same role as a $\Delta\omega$ for a solid-state medium. But there are also terms involving $v(\partial/\partial z)$. These formally distinguish the problem of a gaseous amplifier from the problem of a solid-state amplifier for which they would not appear. The physical meaning of the $v(\partial/\partial z)$ terms is that, given an atom of velocity v and its state at some space-time point (z_0,t_0), we can only predict its polarization state at subsequent space time points (z,t) for which $z-z_0=v(t-t_0)$. This is a reasonable result because in our model (constant v) the atom will never leave this path and we would expect to predict its state only at those points which it can reach. Although the $v(\partial/\partial z)$ terms are logically necessary, it can be shown that they are small correction terms that can be neglected. This is seen as follows. If L and $\Delta\tau$ are the length and duration of the pulse related by $L\sim c\Delta\tau$, one has

$$\partial/\partial t\sim 1/\Delta\tau\,, \quad \partial/\partial z\sim 1/L\,, \tag{48}$$

hence,

$$\partial=(\partial/\partial t)+v(\partial/\partial z)\sim\Delta\tau^{-1}+vL^{-1} \sim\Delta\tau^{-1}[1+(v/c)]. \tag{49}$$

Since $v/c\sim 10^{-5}$, the second terms can be neglected and we may drop the $v(\partial/\partial z)$ terms in Eqs. (44). We also use the retarded time, by transforming to new variables

$$\tau=t-z/c\,, \tag{50a}$$

$$z'=z\,, \tag{50b}$$

so that

$$\partial/\partial z+c^{-1}\partial/\partial t=\partial/\partial z'\,, \quad \partial/\partial t=\partial/\partial\tau\,, \tag{51}$$

and Eqs. (44) become

$$\partial\mathcal{E}/\partial z=-\kappa\mathcal{E}-\tfrac{1}{2}K\langle S/\epsilon_0\rangle_v\,, \tag{52a}$$

$$\mathcal{E}\partial\varphi/\partial z=-\tfrac{1}{2}K\langle C/\epsilon_0\rangle_v\,, \tag{52b}$$

$$\partial S/\partial\tau=-\gamma_{ab}S-(\omega-\nu+Kv-\dot\varphi)C -(p^2/\hbar)\mathcal{E}(\rho_{aa}-\rho_{bb})\,, \tag{52c}$$

$$\partial C/\partial\tau=-\gamma_{ab}C+(\omega-\nu+Kv-\dot\varphi)S\,, \tag{52d}$$

$$\partial\rho_{aa}/\partial\tau=\Lambda_a-\gamma_a\rho_{aa}+\tfrac{1}{2}\mathcal{E}S/\hbar\,, \tag{52e}$$

$$\partial\rho_{bb}/\partial\tau=\Lambda_b-\gamma_b\rho_{bb}-\tfrac{1}{2}\mathcal{E}S/\hbar\,, \tag{52f}$$

where the prime of z' has been dropped. In subsequent sections, Eqs. (52) will be used as a starting point to discuss a variety of special cases.

III. MONOCHROMATIC WAVE

A. Fixed Atoms—Homogeneous Broadening

1. Description of the Field and Polarization

The requirement for monochromaticity is met by taking a purely sinusoidal time dependence such as

$$E(z,t)=\mathcal{E}(z)\cos[\nu t-Kz+\varphi(z)]. \tag{53}$$

Neglecting higher-order harmonics, the response of the medium is then also monochromatic:

$$P(z,t)=C(z)\cos(\nu t-Kz+\varphi)+S(z)\sin(\nu t-Kz+\varphi). \quad (54)$$

The relevant quantities $\mathcal{E}$, φ, S, C have no time dependence in this case.

2. *Simplification of the Basic Equations*

The left-hand sides of the last four equations (52) vanish, as do the $\dot{\varphi}$ terms. One then obtains, for the single-velocity $v=0$ case

$$d\mathcal{E}/dz=-\kappa\mathcal{E}-\tfrac{1}{2}KS/\epsilon_0, \quad (55a)$$

$$\mathcal{E}d\varphi/dz=-\tfrac{1}{2}KC/\epsilon_0, \quad (55b)$$

$$\gamma_{ab}C=(\omega-\nu)S, \quad (55c)$$

$$(\omega-\nu)C+\gamma_{ab}S=-(p^2/\hbar)\mathcal{E}(\rho_{aa}-\rho_{bb}), \quad (55d)$$

$$\rho_{aa}=(\Lambda_a/\gamma_a)+\tfrac{1}{2}\mathcal{E}S/(\hbar\gamma_a), \quad (55e)$$

$$\rho_{bb}=(\Lambda_b/\gamma_b)-\tfrac{1}{2}\mathcal{E}S/(\hbar\gamma_b). \quad (55f)$$

The system of equations (55) can easily be combined to give a single equation for the field envelope $\mathcal{E}$. This is achieved in the following way. From (55c) and (55d) one finds

$$S=-(p^2/\hbar\gamma_{ab})\mathfrak{L}(\omega-\nu)\mathcal{E}(\rho_{aa}-\rho_{bb}), \quad (56)$$

where $\mathfrak{L}(\omega-\nu)$ is a dimensionless Lorentzian factor

$$\mathfrak{L}(\omega-\nu)=\gamma_{ab}^2[(\omega-\nu)^2+\gamma_{ab}^2]^{-1}. \quad (57)$$

Then (55e) and (55f) are used to obtain

$$\rho_{aa}-\rho_{bb}=N_0+\mathcal{E}S/(\hbar\gamma), \quad (58)$$

where

$$N_0=(\Lambda_a/\gamma_a)-(\Lambda_b/\gamma_b) \quad (59)$$

is the steady population inversion that would be sustained by the pumping competing with the damping in the absence of electromagnetic fields, and

$$\gamma^{-1}=\tfrac{1}{2}(\gamma_a^{-1}+\gamma_b^{-1}). \quad (60)$$

Substituting (56) in (58) and solving for $\rho_{aa}-\rho_{bb}$, we have

$$\rho_{aa}-\rho_{bb}=N_0[1+(p^2\mathcal{E}^2/\hbar^2\gamma\gamma_{ab})\mathfrak{L}(\omega-\nu)]^{-1}. \quad (61)$$

It is seen from this expression that the population inversion is unchanged to first order by a weak field and that it is saturated ($\rho_{aa}-\rho_{bb}\simeq 0$) by a strong field. The quantity

$$I=p^2\mathcal{E}^2/(\hbar^2\gamma_{ab}\gamma) \quad (62)$$

will be called the "dimensionless intensity." As seen from (61), $I\mathfrak{L}(\omega-\nu)=1$ produces 50% saturation. In terms of I, Eq. (55a) now gives

$$dI/dz=-2\kappa I+2\alpha I[1+I\mathfrak{L}(\omega-\nu)]^{-1}, \quad (63)$$

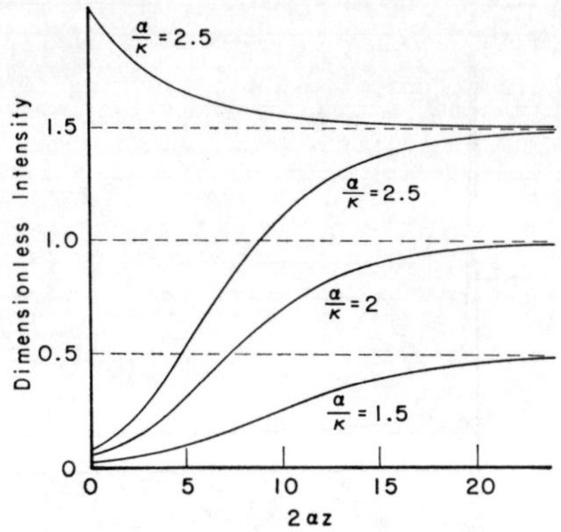

FIG. 5. Amplification of a cw signal for various values of the parameter α/κ. The uppermost curve shows the attenuation of a signal initially stronger than the limiting value. The origin along the z axis must be chosen to correspond to the initial intensity of the signal.

where

$$\alpha=GK\mathfrak{L}(\omega-\nu)=\alpha_0\mathfrak{L}(\omega-\nu) \quad (64)$$

is the small signal gain constant, and

$$G=\tfrac{1}{2}(p^2N_0/\epsilon_0\hbar\gamma_{ab}) \quad (65)$$

is a dimensionless gain parameter.

The condition for a growing solution is obviously $\alpha>\kappa$ and we shall refer to the situation $\alpha=\kappa$ as a threshold. The gain constant α has a Lorentzlan profile with FWHH$=2\gamma_{ab}$ (Fig. 4). We can deduce from (63) the "limiting intensity" I_l by setting

$$dI/dz=0=-2\kappa I_l+2\alpha I_l[1+I_l\mathfrak{L}(\omega-\nu)]^{-1}, \quad (66)$$

hence

$$I_l=[(\alpha-\kappa)/\kappa][\mathfrak{L}(\omega-\nu)]^{-1}. \quad (67)$$

The quantity $(\alpha-\kappa)/\kappa$ is just the dimensionless measure of the amount by which threshold is exceeded. From the Lorentzian factor in (67) we see that the limiting intensity is higher for an off-resonance signal, but the distance required for build-up is increased, due to the Lorentzian profile of the gain.

It is instructive to examine the condition for Eq. (63) to be approximated by the series expansion

$$dI/dz=-2\kappa I+2\alpha I[1-I\mathfrak{L}(\omega-\nu)]=2(\alpha-\kappa)I-2\alpha\mathfrak{L}(\omega-\nu)I^2. \quad (68)$$

A perturbation approach would lead to the above if carried out to third order. Equation (68) yields a limiting intensity

$$I_l=[(\alpha-\kappa)/\alpha][\mathfrak{L}(\omega-\nu)]^{-1} \quad (69)$$

smaller than the exact value (67) by a factor κ/α which is approximately unity near threshold. Hence, for operation around threshold (63) can be approximated by (68).

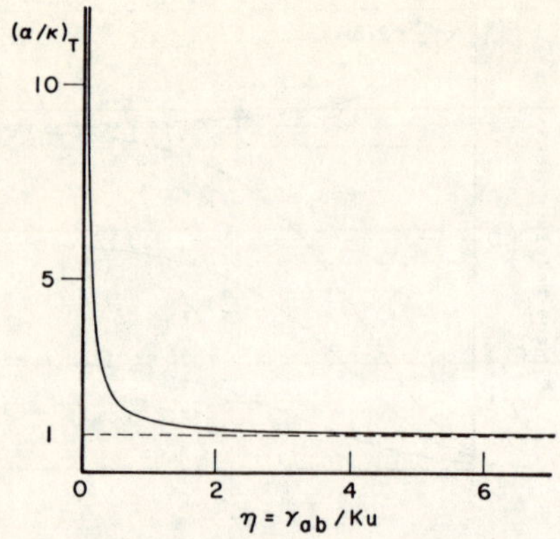

FIG. 6. Threshold value of the parameter α/κ as a function of the dimensionless quantity η.

Equation (63) can be integrated exactly and an implicit solution is

$$I|1-I/I_l|^{\alpha/\kappa}=I_0|1-I_0/I_l|^{\alpha/\kappa}e^{2(\alpha-\kappa)(z-z_0)}, \quad (70)$$

where I_0 is the value of the intensity at position $z=z_0$. A computer illustration of (70) is given in Fig. 5. As can be seen from this figure a small signal is first amplified exponentially with a net gain constant $\alpha-\kappa$, but as the signal grows, saturation effects come into play and a steady state is reached. If an intensity larger than I_l is sent into the medium, it is attenuated down to the same value I_l as is shown by one curve in Fig. 5.

The role of the linear loss κ in the dynamics of this problem is very important since $\kappa=0$ leads to solutions of a different functional form. This can be seen by taking the limit of Eq. (70) as $\kappa\to 0$, or directly by integrating (63) with $\kappa=0$. One then finds

$$Ie^{\mathcal{L}(\omega-\nu)I}=I_0e^{\mathcal{L}(\omega-\nu)I_0}e^{2\alpha(z-z_0)}. \quad (71)$$

Asymptotically, I grows like $2\alpha z$ as $z\to\infty$. This continued rise in intensity is a consequence of neglecting the linear loss mechanism.[13]

B. Moving Atoms—Doppler Broadening (Inhomogeneous Broadening)

For a given value of the velocity v, it is seen from the basic Eqs. (52) that Eqs. (55) must be generalized by formally substituting $\omega-\nu+Kv$ for $\omega-\nu$, the Doppler shift giving an effective resonance frequency equal to $\omega+Kv$. Following the steps given in the preceding paragraph and finally taking an average over the v dependence one finds that Eq. (63) generalizes to

$$dI/dz=-2\kappa I+2\alpha_0 I\langle\mathcal{L}(\omega-\nu+Kv)\times[1+I\mathcal{L}(\omega-\nu+Kv)]^{-1}\rangle_v, \quad (72)$$

where α_0 was defined by Eq. (64). We assume a Maxwellian velocity distribution

$$\langle\cdots\rangle_v=\pi^{-1/2}\int_{-\infty}^{+\infty}e^{-v^2/u^2}(\cdots)d(v/u), \quad (73)$$

and find

$$\langle\mathcal{L}(\omega-\nu+Kv)[1+I\mathcal{L}(\omega-\nu+Kv)]^{-1}\rangle_v =\pi^{-1/2}\gamma_{ab}^2\int_{-\infty}^{+\infty}e^{-v^2/u^2}[(\omega-\nu+Kv)^2+\gamma_{ab}^2(1+I)]^{-1}d(v/u). \quad (74)$$

By introducing parameters

$$\xi=(\omega-\nu)/Ku, \quad (75)$$

$$\eta=\gamma_{ab}/Ku, \quad (76)$$

and

$$x=(\gamma_{ab}/Ku)(1+I)^{1/2}, \quad (77)$$

(74) can be written as

$$\langle\ \rangle_v=\pi^{-1/2}\eta^2\int_{-\infty}^{+\infty}[(t+\xi)^2+x^2]^{-1}e^{-t^2}dt. \quad (78)$$

As shown in Appendix C, this integral can be related to the imaginary part Z_i of the plasma dispersion function

$$Z(\zeta)=i\int_0^{\infty}\exp\{-\tfrac{1}{4}\mu^2-\zeta\mu\}d\mu. \quad (79)$$

As a result (72) becomes

$$dI/dz=-2\kappa I+2\alpha_0\eta^2x^{-1}Z_i(x+i\xi)I. \quad (80)$$

It is also shown in Appendix C that in the special case of an on-resonance signal ($\omega=\nu$), the result can be expressed in terms of the error function defined as

$$\mathrm{erfc}(x)=2\pi^{-1/2}\int_x^{+\infty}e^{-t^2}dt. \quad (81)$$

Equation (80) may be replaced in this case by

$$dI/dz=-2\kappa I+2\pi^{1/2}\alpha_0\eta^2x^{-1}e^{x^2}\mathrm{erfc}(x)\,I. \quad (82)$$

Equations (80) and (82) cannot be integrated analytically. In the following discussion we shall restrict ourselves to the simpler case of Eq. (82) ($\omega=\nu$).

The solution of (82) presents the same qualitative features as (63), namely exponential growth from small values with ultimate saturation. To obtain the threshold condition as a function of η we take the limit of (82) for small I and find

$$(\alpha_0/\kappa)_T=\pi^{-1/2}e^{-\eta^2}[\eta\ \mathrm{erfc}(\eta)]^{-1}. \quad (83)$$

This result is illustrated in Fig. 6. The limiting intensity

[13] W. E. Lamb, Jr., in *Lectures in Theoretical Physics*, edited by W. E. Brittin and B. W. Downs (Interscience Publishers, Inc., New York, 1960), Vol. 2, p. 472.

is obtained from

$$0=-\kappa+\pi^{1/2}\alpha_0\eta^2 x_l^{-1}e^{x_l^2}\text{erfc}(x_l). \tag{84}$$

Given a value of η, this is numerically solved for x_l which in turn gives the limiting intensity I_l through

$$x_l^2=\eta^2(1+I_l). \tag{85}$$

For a special choice of α_0/κ, the result is shown in Fig. 7. Using appropriate expansions of the error function it can be shown that the limiting forms of these results are

$$(\alpha_0/\kappa)_T=1+\eta^{-2}, \tag{86}$$

and

$$I_l=[(\alpha-\kappa)/\kappa]-\eta^{-2}, \tag{87}$$

for $\eta\ll1$ (narrow Doppler line) and

$$(\alpha_0/\kappa)=\pi^{-1/2}\eta^{-1}, \tag{88}$$

and

$$I_l=(\pi\eta^2\alpha_0^2/\kappa^2)-1 \tag{89}$$

for $\eta\ll1$ (broad Doppler line).

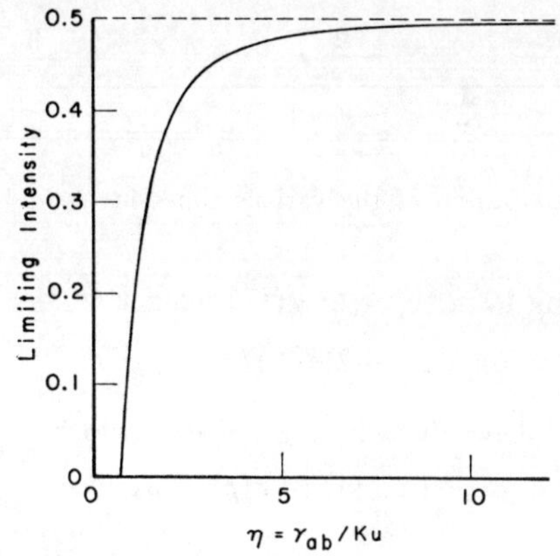

FIG. 7. The limiting intensity I_l as a function of η, for fixed $\alpha/\kappa=1.5$. The asymptotic value in the limit $\eta\to\infty$ (vanishing Doppler width) is $(\alpha-\kappa)/\kappa=0.5$. For η less than η_T the medium is below threshold for the chosen magnitude of α/κ. The value of η_T as a function of α/κ can be obtained from Fig 6.

IV. LONG PULSES

By long pulses we mean that the spectral width of the pulse is small compared to γ_{ab}, which implies that the pulse duration $\Delta\tau_0$ is much longer than the atomic phase memory time $1/\gamma_{ab}$

$$\Delta\tau_0\gg1/\gamma_{ab}. \tag{90}$$

As stated in Sec. I, $\gamma_{ab}>\frac{1}{2}(\gamma_a+\gamma_b)$, so that $1/\gamma_a$ and $1/\gamma_b$ are longer than $1/\gamma_{ab}$. This is seen in Fig. 8 which also shows the two possible ranges of pulse widths 1 and 2 that will be considered in this section.

In Case 1 we shall consider the limit

$$\Delta\tau_0\gg1/\gamma_a, \quad \Delta\tau_0\gg1/\gamma_b. \tag{91}$$

For many gas lasers (but not CO_2), (91) is not a very restrictive condition because $1/\gamma_a$ and $1/\gamma_b$ are not much larger than $1/\gamma_{ab}$ and we have already assumed $\Delta\tau_0\gg1/\gamma_{ab}$. On the other hand, in the CO_2 laser and in most of the solid-state lasers, $1/\gamma_{ab}$ is commonly smaller than $1/\gamma_a$ and $1/\gamma_b$ by several orders of magnitude, and a more complete description of the behavior of "long pulses" should include both Cases 1 and 2. When we treat Case 2 we shall consider the limit

$$\Delta\tau_0\ll1/\gamma_a, \quad \Delta\tau_0\ll1/\gamma_b. \tag{92}$$

In view of the foregoing considerations, Case 1 refers to pulse durations long compared to atomic decay times, and Case 2 to pulse durations short compared to these times.

A. Fixed Atoms (Homogeneous Broadening)

We take the field in the form

$$E(z,t)=\mathcal{E}(z,t)\cos[\nu t-Kz+\varphi(z,t)]. \tag{93}$$

It is again preferable to allow $\nu\neq\omega$ since a long pulse has a pretty well defined frequency. The polarization is then

$$P(z,t)=C(z,t)\cos(\nu t-Kz+\varphi)+S(z,t)\times\sin(\nu t-Kz+\varphi). \tag{94}$$

When $v=0$, Eqs. (52) become

$$\partial\mathcal{E}/dz=-\kappa\mathcal{E}-\tfrac{1}{2}KS/\epsilon_0, \tag{95a}$$

$$\mathcal{E}\,\partial\varphi/\partial z=-\tfrac{1}{2}KC/\epsilon_0, \tag{95b}$$

$$\partial S/\partial\tau=-\gamma_{ab}S-(\omega-\nu-\dot{\varphi})C-(p^2/\hbar)\mathcal{E}(\rho_{aa}-\rho_{bb}), \tag{95c}$$

$$\partial C/\partial\tau=-\gamma_{ab}C+(\omega-\nu-\dot{\varphi})S, \tag{95d}$$

$$\partial\rho_{aa}/\partial\tau=\Lambda_a-\gamma_a\rho_{aa}+\tfrac{1}{2}\mathcal{E}S/\hbar, \tag{95e}$$

$$\partial\rho_{bb}/\partial\tau=\Lambda_b-\gamma_b\rho_{bb}-\tfrac{1}{2}\mathcal{E}S/\hbar. \tag{95f}$$

We now make use of the fact that $\mathcal{E}$, C, S, φ vary slowly in a time $1/\gamma_{ab}$. In this approximation the left-hand sides of (95c) and (95d), as well as the $\dot{\varphi}$ terms can be neglected. Solving these two equations for S we have

$$S=-(p^2/\hbar\gamma_{ab})\mathcal{L}(\omega-\nu)\mathcal{E}(\rho_{aa}-\rho_{bb}). \tag{96}$$

Substituting this expression into (95a), (95e), and (95f) we obtain

$$\partial\mathcal{E}/\partial z=-\kappa\mathcal{E}+\alpha N_0^{-1}(\rho_{aa}-\rho_{bb})\mathcal{E}, \tag{97a}$$

$$\partial\rho_{aa}/\partial\tau=\Lambda_a-\gamma_a\rho_{aa}-\tfrac{1}{2}\gamma_{ab}(p\mathcal{E}/\hbar\gamma_{ab})^2\times\mathcal{L}(\omega-\nu)(\rho_{aa}-\rho_{bb}), \tag{97b}$$

$$\partial\rho_{bb}/\partial\tau=\Lambda_b-\gamma_b\rho_{bb}+\tfrac{1}{2}\gamma_{ab}(p\mathcal{E}/\hbar\gamma_{ab})\times\mathcal{L}(\omega-\nu)(\rho_{aa}-\rho_{bb}), \tag{97c}$$

where α was defined by (64). The last two equations involve $\mathcal{E}^2$ which suggests that we transform (97a) into an equation for the photon flux defined as

$$\mathcal{J}=\tfrac{1}{2}\epsilon_0\mathcal{E}^2c/(\hbar\nu). \tag{98}$$

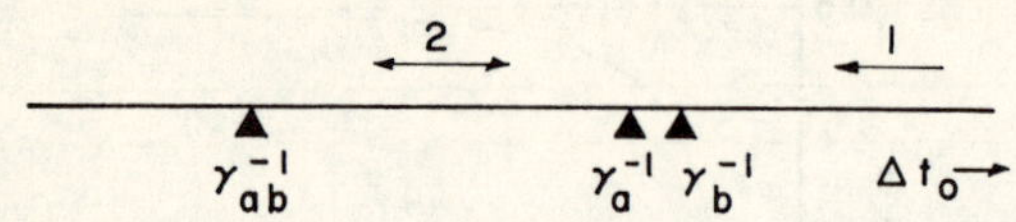

FIG. 8. Visualization of the various times involved in Sec. IV.

Multiplying (97a) by $\epsilon_0 \mathcal{E}^2 c/(\hbar\nu)$ we get

$$\partial \mathcal{I}/\partial z = -2\kappa \mathcal{I} + \sigma(\rho_{aa} - \rho_{bb})\mathcal{I}, \tag{99a}$$

$$\partial \rho_{aa}/\partial \tau + \gamma_a[\rho_{aa} - (\Lambda_a/\gamma_a)] = -\sigma(\rho_{aa} - \rho_{bb})\mathcal{I}, \tag{99b}$$

$$\partial \rho_{bb}/\partial \tau + \gamma_b[\rho_{bb} - (\Lambda_b/\gamma_b)] = \sigma(\rho_{aa} - \rho_{bb})\mathcal{I}, \tag{99c}$$

where

$$\sigma = \sigma_0 \mathcal{L}(\omega - \nu) = 2\alpha N_0^{-1} \tag{100}$$

is the effective cross section for stimulated emission or absorption and

$$\sigma_0 = 2\alpha_0 N_0^{-1} \tag{101}$$

is its resonance value.

Case 1. Pulse duration long compared to atomic decay times. We now assume that the pulse width $\Delta\tau_0$ is long compared to the decay times of the states a and b, i.e., $\Delta\tau_0 \gg 1/\gamma_a$ and $1/\gamma_b$. This means that in (99b) and (99c) we may neglect $\partial\rho_{aa}/\partial\tau$ and $\partial\rho_{bb}/\partial\tau$ compared to $\gamma_a\rho_{aa}$ and $\gamma_b\rho_{bb}$. The equations may then be solved for

$$N = \rho_{aa} - \rho_{bb}. \tag{102}$$

After some algebra we find

$$N = N_0[1 + (2\sigma/\gamma)\mathcal{I}]^{-1}, \tag{103}$$

where N_0 and γ were defined by (59) and (60). We see from (103) that

$$\mathcal{I}_s = \gamma/(2\sigma) \tag{104}$$

is the photon flux that produces 50% saturation. Substituting (103) into (99a) we obtain

$$\partial \mathcal{I}(z,t)/\partial z = -2\kappa\mathcal{I} + \sigma N_0 \mathcal{I}[1 + (2\sigma/\gamma)\mathcal{I}]^{-1}. \tag{105}$$

This equation is formally analogous to Eq. (63), but in the previous section the field intensity was a function of z only, thus leading to the ordinary differential Eq. (63), while in this section we have to deal with the partial differential Eq. (105) because of the additional time dependence of $\mathcal{I}$.

The condition for amplification of a small signal, i.e., threshold, is

$$\sigma N_0 > 2\kappa. \tag{106}$$

The limiting photon flux is readily seen to be

$$\mathcal{I}_l = [(\sigma N_0 - 2\kappa)/(2\kappa)][\gamma/(2\sigma)]. \tag{107}$$

The general solution of (105) is

$$\begin{aligned} \mathcal{I}(z,\tau)\,|1 - [\mathcal{I}(z,\tau)/\mathcal{I}_l]|^{-\sigma N_0/(2\kappa)} &= \mathcal{I}(0,\tau)\,|1 - [\mathcal{I}(0,\tau)/\mathcal{I}_l]|^{-\sigma N_0/(2\kappa)} \\ &\quad \times \exp[(\sigma N_0 - 2\kappa)z], \end{aligned} \tag{108}$$

where $\mathcal{I}(0,\tau)$ is the input pulse at $z=0$. As z increases, this solution approaches the limiting intensity $\mathcal{I}_l$ defined by (107). It can be seen from (108) that the peak of the pulse travels with the velocity of light. Indeed, for any fixed z the peak of the pulse occurs for $\tau = \tau_p$ such that $\mathcal{I}(z,\tau_p)$ is a maximum and it is easily seen that this occurs when $\mathcal{I}(0,\tau)$ has also a maximum. Consequently, $\mathcal{I}(z,\tau)$ and $\mathcal{I}(0,\tau)$ have their peaks at the same retarded time τ_p, for any z. This implies that the peak of the pulse is traveling with the velocity of light c. Figure 9 illustrates the solution (108) obtained by a digital computer.

If one is not too far from threshold, an explicit solution can be found, instead of the implicit solution (108). This is achieved by expanding the initial Eq. (105)

$$\partial \mathcal{I}(z,\tau)/\partial\tau \simeq (\sigma N_0 - 2\kappa)\mathcal{I} - \sigma N_0 \mathcal{I}^2/\mathcal{I}_s, \tag{109}$$

which is readily integrated:

$$\begin{aligned} \mathcal{I}(z,\tau) = \mathcal{I}(0,\tau)\{1 + \mathcal{I}_l^{-1}\mathcal{I}(0,\tau)[\exp\{(\sigma N - 2\kappa)z\} - 1]\}^{-1} \\ \times \exp\{(\sigma N - 2\kappa)z\}, \end{aligned} \tag{110}$$

where

$$\mathcal{I}_l = [(\sigma N_0 - 2\kappa)/(\sigma N_0)][\gamma/(2\sigma)] \tag{111}$$

is the limiting photon flux according to the approximate Eq. (109). We see that this value of the limiting photon flux differs from the exact one (107) by a factor of $\sigma N_0/(2\kappa)$ which is close to unity around threshold.

Case 2. Pulse duration short compared to atomic decay times. $(1/\gamma_{ab} \ll \Delta\tau_0 \ll 1/\gamma_a,\ 1/\gamma_b)$. In this case, the γ_a and γ_b terms are neglected in Eqs. (99), and one obtains

$$\partial \mathcal{I}/\partial z = -2\kappa\mathcal{I} + \sigma\mathcal{I}N, \tag{112a}$$

$$\partial N/\partial\tau = -2\sigma\mathcal{I}N, \tag{112b}$$

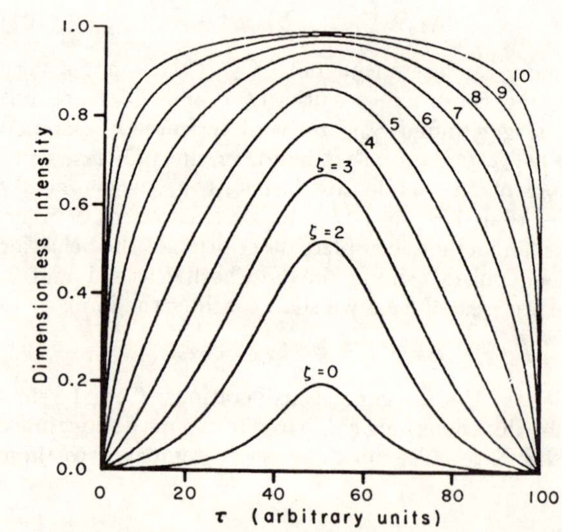

FIG. 9. Evolution of a long pulse as it travels into the medium. The parameter $\sigma N_0/2\kappa$ is fixed at 2.0. The dimensionless intensity $(2\sigma/\gamma)\mathcal{I}$ is plotted versus the retarded time and the successive curves correspond to increasing depths in the medium labelled by values of the dimensionless distance $\zeta = 2\kappa z$.

with

$$N=\rho_{aa}-\rho_{bb}. \quad (113)$$

Integrating (112b) and substituting into (112a) we have

$$\partial \mathcal{I}/\partial z=\left[\sigma N_0 \exp\left\{-2\sigma\int_{-\infty}^{\tau} \mathcal{I}(z,\tau')d\tau'\right\}-2\kappa\right]\mathcal{I}. \quad (114)$$

This equation is of the form: rate of change of $\mathcal{I}$ = (effective gain − loss) × $\mathcal{I}$, with effective gain = gain × saturation factor. We have allowed the signal to be off-resonance by an amount $\omega-\nu$. This detuning reduces the cross section according to (100) which in turn reduces the gain σN_0, but also diminishes the effect of the saturation factor

$$\exp\left(-2\sigma\int_{-\infty}^{\tau} \mathcal{I}d\tau'\right).$$

This result should be expected since off-resonance atoms are harder to saturate.

The evolution of a pulse according to Eq. (114) has been discussed by Basov *et al.*[7] Analytic solutions also have been proposed[5,14,15] for the case $\kappa=0$, but the loss plays an important role in the dynamics of the pulse.

Some insight is gained by first studying the stationary solutions of (114). Such a solution is of the form

$$\mathcal{I}(z,t)=\mathcal{I}(t-z/v)=\mathcal{I}(\bar{\tau}), \quad (115)$$

which represents a pulse moving with velocity v and with unchanging shape. $\mathcal{I}(\bar{\tau})$ obeys

$$(c^{-1}-v^{-1})d\mathcal{I}/d\bar{\tau} = \left\{\sigma N_0 \exp\left[-2\sigma\int_{-\infty}^{\bar{\tau}} \mathcal{I}(\tau')d\tau'\right]-2\kappa\right\}\mathcal{I}. \quad (116)$$

Define the quantity

$$R(\bar{\tau})=\int_{-\infty}^{\bar{\tau}} \mathcal{I}(\tau')d\tau', \quad (117)$$

which is the number of photons in the pulse per units area which have already passed for the given value of $\bar{\tau}$.

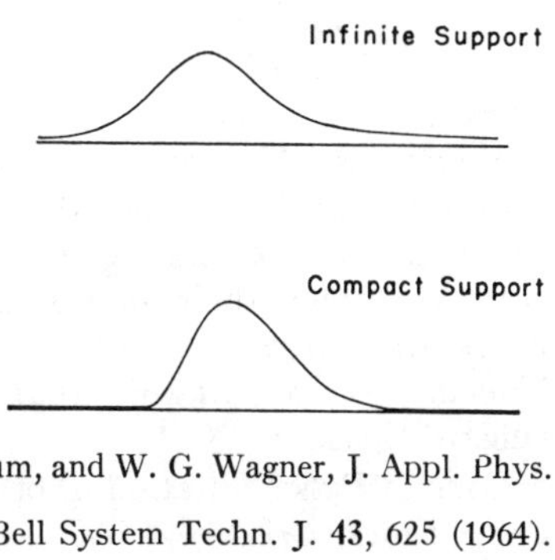

FIG. 10. Illustration of the notion of support of a function.

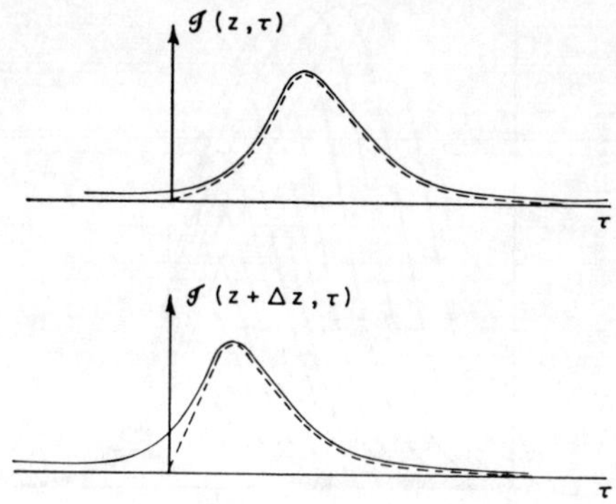

FIG. 11. The upper diagram shows by a solid curve an idealized stationary solution of Eq. (114), for fixed z, and (dotted line) a transient solution approaching the former. The lower figure shows the same situation for a larger value of z. By definition the stationary solution is unchanged in shape but translated to the left since its velocity exceeds c. On the other hand the dotted transient solution is constrained to vanish for $\tau<0$.

Equation (116) may be written in terms of $R(\bar{\tau})$ as

$$(c^{-1}-v^{-1})d^2R/d\bar{\tau}^2=[\sigma N_0 \exp(-2\sigma R)-2\kappa]dR/d\bar{\tau}. \quad (118)$$

Integrating this equation with the initial condition

$$(dR/d\bar{\tau})_{\bar{\tau}=-\infty}=\mathcal{I}(-\infty)=0, \quad (119)$$

we get

$$(c^{-1}-v^{-1})dR/d\bar{\tau}=\tfrac{1}{2}N_0[1-\exp(-2\sigma R)]-2\kappa R. \quad (120)$$

We now show that the velocity v of such a stationary pulse must be greater than the velocity of light c if threshold is exceeded. Consider the limit of (120) for $\bar{\tau}\to-\infty$. Since $R(\bar{\tau})\to 0$ as $\bar{\tau}\to-\infty$, we have

$$(c^{-1}-v^{-1})dR/d\bar{\tau}=(\sigma N_0-2\kappa)R. \quad (121)$$

For physically meaningful solutions $dR/d\bar{\tau}=\mathcal{I}$ and R must be positive, so that above threshold we must have $v>c$, A theory predicting signal velocities larger than c would be in contradiction with special relativity. Our theory is based on Maxwell's equations which are relativistically invariant, and on causal quantum mechanics, therefore we would not expect it to make such predictions.

Although Eq. (114) admits stationary solutions propagating faster than light, it can also be seen from (114) that a physical input pulse will never reach such a stationary form. We proceed to prove this statement. A physical pulse should be represented by a function having nonvanishing values only over a finite interval of the τ axis. Such a function will be said to have "compact support" (see Fig. 10). Special relativity sets a limit to the velocity of the first nonvanishing point of the signal. We see from Eq. (120) that the stationary solutions all have infinite support since (120) is analytic and admits only analytic solutions, while a function with compact support cannot be analytic. On the other hand, the transient equation (114) is support conserving

[14] R. Bellman, G. Birnbaum, and W. G. Wagner, J. Appl. Phys. 34, 780 (1963).

[15] E. O. Schulz-DuBois, Bell System Techn. J. 43, 625 (1964).

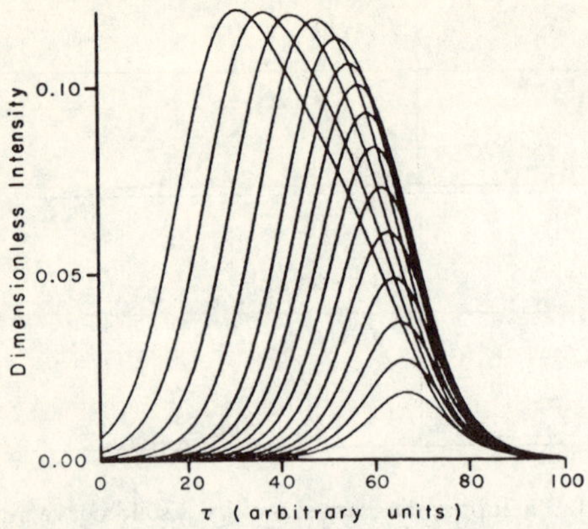

FIG. 12. Evolution of a pulse with infinite tails into a stationary form travelling faster than light, according to Eq. (114), with the following values of the parameters: $\sigma N_0=0.2$ cm^{-1}, $2\kappa=0.03$ cm^{-1}. The dimensionless intensity $\sigma\mathcal{I}T$ is plotted versus τ/T where T is an arbitrary time unit. The successive curves from right to left correspond to increasing distances into the medium, ranging from $z=0$ to $z=60$ cm.

because $\mathcal{I}=0$ implies $\partial\mathcal{I}/\partial z=0$ and the pulse cannot build up from zero. As a result any physical input pulse has compact support, while the stationary solutions have infinite support and Eq. (114) cannot transform the former into the latter. There remains the possibility of an approximate approach to the stationary state in the sense of Fig. 11(a). This type of an approach can only be temporary. Indeed if one considers Fig. 11(a) at a further distance $z+\Delta z$ [Fig. 11(b)], the stationary solution will have translated to the left by $\Delta z/v$. The actual solution is restricted to the same support and in its tendency to follow the stationary solution will distort its shape and therefore inevitably depart from the stationary form.

Basov *et al.*[7] claim that the numerical integration of Eq. (114) displays a quick approach to the stationary state, but they have integrated (114) with an unphysical input pulse extending to infinity at both ends. They have also measured pulse velocities larger than c, but an experimental apparatus can only trace the bulk of the pulse which may temporarily propagate faster than light without contradicting special relativity.

We have integrated Eq. (114) with a digital computer considering two types of input pulses $\mathcal{I}(0,\tau)$. A pulse with exponential tails like

$$\mathcal{I}(0,\tau)=\mathcal{I}_0\,\mathrm{sech}^2[(\tau-\tau_p)/\Delta\tau_0] \tag{122a}$$

reaches effectively a stationary state as shown in Fig. 12.

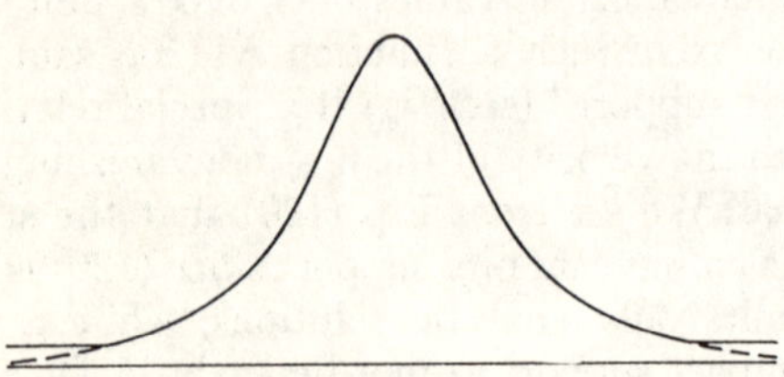

FIG. 13. Truncated hyperbolic secant.

We then alter the above input to give it compact support by simply raising the base line as shown in Fig. 13. The integration with this initial input gives a series of curves properly starting at the origin (see Fig. 14). For a limited time the pulse, except near $\tau=0$, develops towards a stationary form as in the previous case. However this state of affairs cannot go on forever since the pulse must permanently pass through the origin. The numerical integration effectively shows a sudden departure from the quasistationary form with a "piling up" to an increasingly sharper peak near $\tau=0$ (Fig. 15). It is of course understood that the approximations of this section break down when the pulse width is reduced beyond a certain limit. In the next section we will be able to follow the evolution of such a pulse as it becomes too sharp.

In order to demonstrate that the "piling up" is not due to the discontinuities at the junction points $\tau=0$, $\tau=T$, we have also considered input functions of the type

$$\begin{aligned}\mathcal{I}(0,\tau)&=\mathcal{I}_0\exp[-a\tau^{-m}(T-\tau)^{-n}], \quad 0<\tau<T\\ &=0 \text{ otherwise}\end{aligned} \tag{122b}$$

which have continuous derivatives to all orders at the junction points. Figure 16 shows the amplification of such a pulse.

B. Moving Atoms—Doppler Broadening (Inhomogeneous Broadening)

Case 1. Pulse duration long compared to atomic-decay times. By the argument given in III B, Eq. (114) generalizes here to

$$\begin{aligned}\partial\mathcal{I}/\partial z=&-2\kappa\mathcal{I}+\sigma_0N_0\mathcal{I}\langle\mathcal{L}(\omega-\nu+Kv)\\ &\times[1+(2\sigma_0/\gamma)\mathcal{L}(\omega-\nu+Kv)\mathcal{I}]^{-1}\rangle_v.\end{aligned} \tag{123}$$

We have already indicated how to carry out the averaging process involved in this equation. In terms of the parameters ξ, η, and x defined by (75), (76), and

$$x=\eta[1+(2\sigma_0/\gamma)\mathcal{I}]^{1/2}, \tag{124}$$

we obtain

$$\partial\mathcal{I}(z,\tau)/\partial z=[\sigma_0N_0\eta^2x^{-1}Z_i(x+i\xi)-2\kappa]\mathcal{I}, \tag{125}$$

which reduces to

$$\partial\mathcal{I}(z,\tau)/\partial z=[\pi^{1/2}\sigma_0N_0\eta^2x^{-1}e^{x^2}\mathrm{erfc}(x)-2\kappa]\mathcal{I}, \tag{126}$$

when $\xi=(\omega-\nu)/Ku=0$.

Equations (125) and (126) are the counterpart of (80) and (82), but it should be emphasized again that the equations of this section are partial differential equations, while their counterparts in the case of monochromatic waves are ordinary differential equations. The analysis following Eq. (82) applies here with trivial substitutions.

Case 2. Pulse duration short compared to atomic-decay times ($\Delta\tau_0\ll 1/\gamma_a$ and $1/\gamma_b$). Here Eq. (114) must be generalized to include the effects of moving atoms. This is accomplished by replacing $\sigma=\sigma_0\mathcal{L}(\omega-\nu)$ by $\sigma=\sigma_0\mathcal{L}(\omega-\nu+Kv)$ since the atomic frequency is Doppler shifted. An average must then be taken over the v dependence of the right-hand side of (114), which gives

$$\partial \mathcal{I}(z,\tau)/\partial z=[\sigma_0 N_0\langle \mathcal{L}(\omega-\nu+Kv)\times\exp[-2\sigma_0\mathcal{L}(\omega-\nu+Kv)R]\rangle_v-2\kappa]\mathcal{I}, \quad (127)$$

with

$$R=R(z,\tau)=\int_{-\infty}^{\tau}\mathcal{I}(z,\tau')d\tau'. \quad (128)$$

In general, the average in Eq. (127) cannot be evaluated analytically, but if the velocity distribution in Kv is broad compared to γ_{ab}, as it usually is, then in the limit $\gamma_{ab}/Ku\rightarrow 0$, this integral reduces to

$$\langle\ \rangle_v=\pi^{-1/2}\int_{-\infty}^{+\infty}\gamma_{ab}^2[\gamma_{ab}^2+(\omega-\nu+Kv)^2]^{-1}\times\exp\{-2\sigma_0 R\gamma_{ab}^2[\gamma_{ab}^2+(\omega-\nu+Kv)^2]^{-1}\}d(v/u). \quad (129)$$

We first notice that this average is independent of $\omega-\nu$, which is natural since we assumed a flat velocity distribution. Setting $\omega=\nu$ and $y=Kv/\gamma_{ab}$, the average can be written as

$$\langle\ \rangle_v=\pi^{-1/2}(\gamma_{ab}/Ku)\int_{-\infty}^{+\infty}(1+y^2)^{-1}\times\exp[-2\sigma_0 R(1+y^2)^{-1}]dy. \quad (130)$$

By the change of variable

$$y=\cot(\tfrac{1}{2}\theta), \quad (131)$$

this is transformed into

$$\langle\ \rangle_v=\pi^{-1/2}\eta e^{-\sigma_0 R}\int_0^{\pi}e^{\sigma_0 R\cos\theta}d\theta$$
$$=\pi^{1/2}\eta e^{-\sigma_0 R}I_0(\sigma_0 R), \quad (132)$$

where I_0 is the modified Bessel function of order zero. Substituting this result into (127) we get

$$\partial\mathcal{I}/\partial z=\left\{\pi^{1/2}\sigma_0 N_0\eta\exp\left[-\sigma_0\int_{-\infty}^{\tau}\mathcal{I}(z,\tau')d\tau'\right]\times I_0\left[\sigma_0\int_{-\infty}^{\tau}\mathcal{I}(z,\tau')d\tau'\right]-2\kappa\right\}\mathcal{I}. \quad (133)$$

The threshold condition is

$$\pi^{1/2}\sigma_0 N_0\eta=2\kappa, \quad (134)$$

indicating that the required value for $\sigma_0 N_0$ is increased by a factor $\pi^{-1/2}Ku/\gamma_{ab}\gg 1$. Otherwise, Eq. (133) leads

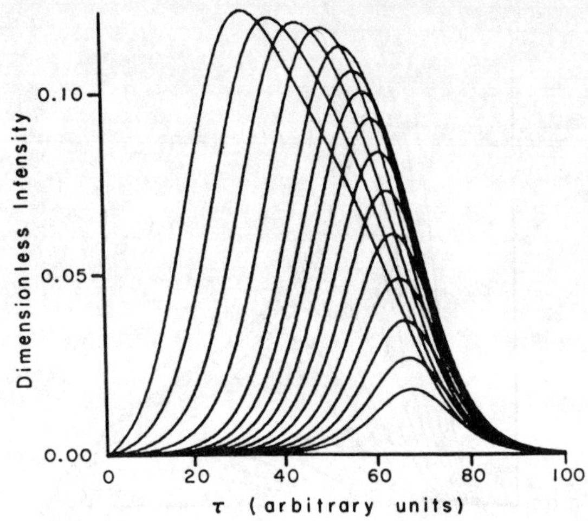

FIG. 14. Integration of Eq. (114) with the same values of the parameters as in Fig. 12, but after truncation of the tails of the input.

to the same qualitative behavior as Eq. (114), with a different saturation factor. Compared to a homogeneously broadened medium, in the case of Doppler broadening (or inhomogeneous broadening), the gain is smaller for a given density N_0. On the other hand, if the same gain is achieved by increasing N_0, the device amplifies better due to a more favorable saturation factor. Figures 16 and 17 represent solutions of (114) and (133) for the same values of the loss and the gain and with the same input pulse.

Equation (133) also has stationary solutions like (114). A stationary solution of (133) satisfies

$$(c^{-1}-v^{-1})d^2R/d\bar{\tau}^2=[\pi^{1/2}\sigma_0 N_0\eta e^{-\sigma_0 R}I_0(\sigma_0 R)-2\kappa]dR/d\bar{\tau} \quad (135)$$

with

$$\mathcal{I}(z,\tau)=\mathcal{I}(t-z/v)=\mathcal{I}(\bar{\tau}) \quad (136)$$

and

$$R(\bar{\tau})=\int_{-\infty}^{\bar{\tau}}\mathcal{I}(\tau')d\tau'. \quad (137)$$

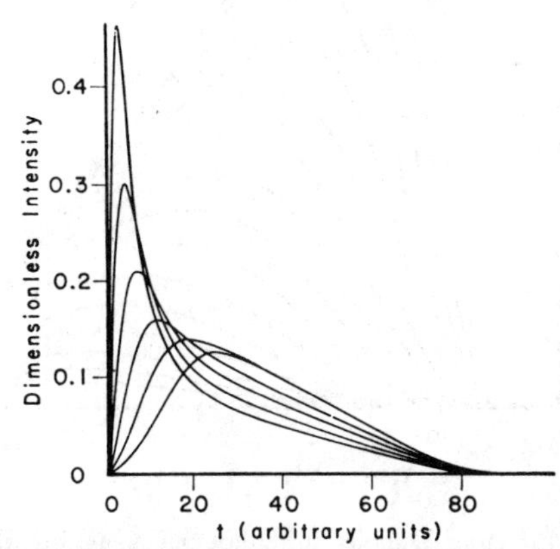

FIG. 15. Extension of Fig. 14 for larger distances into the medium, ranging from $z=60$ to $z=90$ cm by steps of 6 cm.

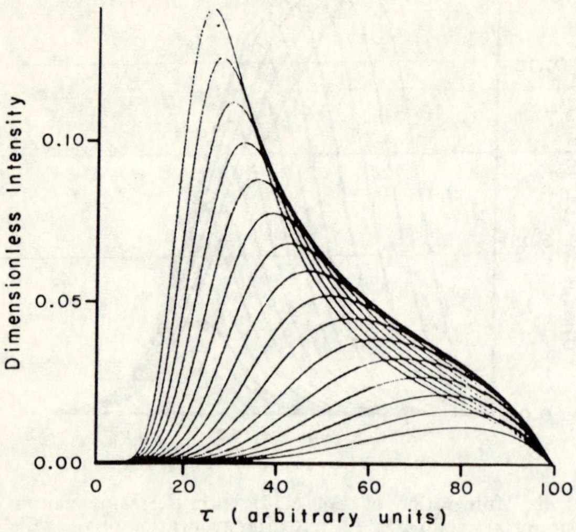

FIG. 16. Integration of Eq. (114) with (122a) as input and with the same values of the parameters as in Fig. 12. The z values range from $z=0$ to $z=45$ cm by steps of 3 cm.

With the initial condition $dR/d\bar{\tau}|_{\bar{\tau}=-\infty}=0$, (135) integrates to

$$(c^{-1}-v^{-1})dR/d\bar{\tau}=\pi^{1/2}\sigma_0 N_0\eta R \times\exp(-\sigma_0 R)[I_0(\sigma_0 R)+I_1(\sigma_0 R)]-2\kappa R, \quad (138a)$$

where I_1 is the modified Bessel function of order one. For $\bar{\tau}\to-\infty$ this equation is approximated by

$$(c^{-1}-v^{-1})dR/d\bar{\tau}\simeq(\pi^{1/2}\sigma_0 N_0\eta-2\kappa)R. \quad (138b)$$

Above threshold $\pi^{1/2}\sigma_0 N_0\eta>2\kappa$, R and $dR/d\bar{\tau}$ are positive for physically meaningful solutions; hence, we must have $v>c$. It is clear from (138a) that such a stationary solution has infinite support and therefore the discussion of IV A Case 2 applies here also.

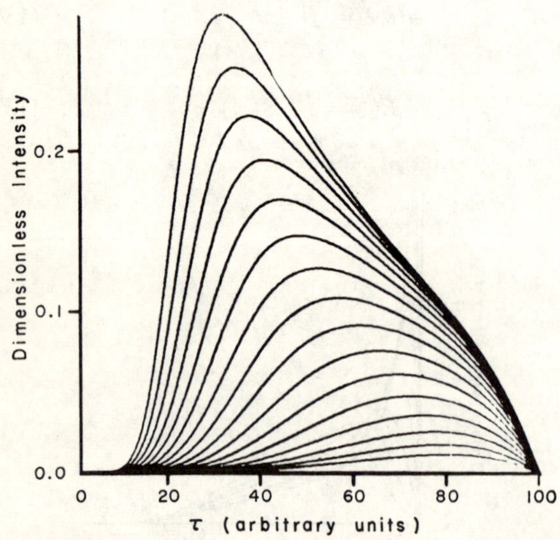

FIG. 17. Effect of Doppler (or inhomogeneous) broadening on pulse propagation. Equation (133) is integrated with the same input pulse, gain and loss parameters and distance range as in Fig. 16. Here we have $Ku\gg\gamma_{ab}$.

V. ULTRASHORT PULSES

In this section we consider pulses much shorter than the atomic phase memory time $1/\gamma_{ab}$, but still much longer than the optical period $2\pi/\nu$. As a consequence the pulse duration is also much shorter than the decay times $1/\gamma_a$, $1/\gamma_b$. The subnsec pulses obtained with gas lasers and the psec pulses produced with solid-state lasers fall under this category.

A. Fixed Atoms

Equations (52) were derived with neglect of variations the field and polarization envelopes and phases in a time $1/\nu$ and a distance $1/K$, and therefore remain valid here. On the other hand, we may now take $\omega=\nu$ without loss of generality since the pulse duration is short enough. Inspection of Eqs. (52) shows that if the input pulse is such that

$$\varphi(0,\tau)\equiv 0, \quad (139)$$

then the phase $\varphi(z,\tau)$, as well as the average in-phase component of the polarization $\langle C\rangle_v$, remains identically zero for any z, provided the velocity distribution is symmetric around $v=0$. Restricting ourselves to "zero phase" input pulses, Eqs. (52) reduce to

$$\partial\mathcal{E}/\partial z=-\kappa\mathcal{E}-\tfrac{1}{2}KS/\epsilon_0, \quad (140a)$$

$$\partial S/\partial\tau=-\gamma_{ab}S-(p^2/\hbar)\mathcal{E}(\rho_{aa}-\rho_{bb}), \quad (140b)$$

$$\partial\rho_{aa}/\partial\tau=\Lambda_a-\gamma_a\rho_{aa}+\tfrac{1}{2}\mathcal{E}S/\hbar, \quad (140c)$$

$$\partial\rho_{bb}/\partial\tau=\Lambda_b-\gamma_b\rho_{bb}-\tfrac{1}{2}\mathcal{E}S/\hbar. \quad (140d)$$

We shall now neglect γ_a, γ_b, and γ_{ab} since the pulse is ultrashort. It was seen that

$$N_0=(\Lambda_a/\gamma_a)-(\Lambda_b/\gamma_b) \quad (141)$$

is the population inversion maintained prior to the arrival of the pulse. If N_0 is to remain finite, in the limit of vanishing γ_a, γ_b we must also take $\Lambda_a=\Lambda_b=0$. Combining the last two equations in (140) we obtain

$$\partial\mathcal{E}/\partial z=-\kappa\mathcal{E}-\tfrac{1}{2}KS/\epsilon_0, \quad (142a)$$

$$\partial S/\partial\tau=-(p^2/\hbar)\mathcal{E}N, \quad (142b)$$

$$\partial N/\partial\tau=\mathcal{E}S/\hbar, \quad (142c)$$

where $N=\rho_{aa}-\rho_{bb}$. The last two equations can easily be integrated, giving

$$S=-pN_0\sin\psi, \quad (143)$$

$$N=N_0\cos\psi, \quad (144)$$

with

$$\psi(z,\tau)=\frac{p}{\hbar}\int_{-\infty}^{\tau}\mathcal{E}(z,\tau')d\tau'. \quad (145)$$

A conservation relation follows as

$$(S/p)^2+N^2=N_0^2. \quad (146)$$

The quantity $\psi(z,\tau)$ is the partial "area" under the pulse envelope at a given space point z. The total area is defined as

$$\theta(z)=\psi(z,+\infty)=\frac{p}{\hbar}\int_{-\infty}^{+\infty}\mathcal{E}(z,\tau')d\tau'. \quad (147)$$

The set of Eqs. (142) is then equivalent to

$$\partial^2\psi/\partial z\partial\tau=-\kappa\partial\psi/\partial\tau+g\sin\psi, \quad (148)$$

where

$$g=\tfrac{1}{2}Kp^2N_0/(\epsilon_0\hbar). \quad (149)$$

The general solution of Eq. (148) is not known. Particular solutions, ignoring the loss term κ, have been constructed by Lamb[6] using the Baeklund transformation. However, the linear loss, although negligible in the problem of an attenuator, plays an important role here.

The evolution of a weak signal according to (148) can be followed analytically, since one has $\psi\ll 1$ and (148) can be approximated by

$$\partial^2\psi/\partial z\partial\tau=-\kappa\partial\psi/\partial\tau+g\psi. \quad (150)$$

Differentiating (150) with respect to τ one finds that the field envelope satisfies the same equation

$$\partial^2\mathcal{E}/\partial z\partial\tau=-\kappa\partial\mathcal{E}/\partial\tau+g\mathcal{E}, \quad (151)$$

which is a linear hyperbolic equation with constant coefficients. Given the boundary condition

$$\mathcal{E}(0,\tau)=\mathcal{E}_0(\tau), \quad (152)$$

(151) has a unique solution

$$\mathcal{E}(z,\tau)=e^{-\kappa z}\int_{-\infty}^{\tau}I_0(\{4gz(\tau-\tau')\}^{1/2})\times[d\mathcal{E}_0(\tau')/d\tau']d\tau'. \quad (153)$$

The area function $\psi(z,\tau)$ is given by a similar expression, since it obeys the same equation

$$\psi(z,t)=e^{-\kappa z}\int_{-\infty}^{\tau}I_0(\{4gz(\tau-\tau')\}^{1/2})\mathcal{E}_0(\tau')d\tau'. \quad (154)$$

For large values of τ this integral diverges. The expressions (153) and (154) are therefore valid only for values of τ such that the area as given by (154) is small.

For a strong signal ($\theta\gtrsim 1$), the solution (153) can be used to determine the behavior of the pulse in the neighborhood of $\tau=0$, since the initial pulse $\mathcal{E}_0(\tau)$ vanishes for $\tau\leq 0$. For small τ we may expand the Bessel function in (154) and get

$$\mathcal{E}(z,\tau)\simeq e^{-\kappa z}\int_0^{\tau}[1+gz(\tau-\tau')][d\mathcal{E}_0(\tau')/d\tau']d\tau'. \quad (155)$$

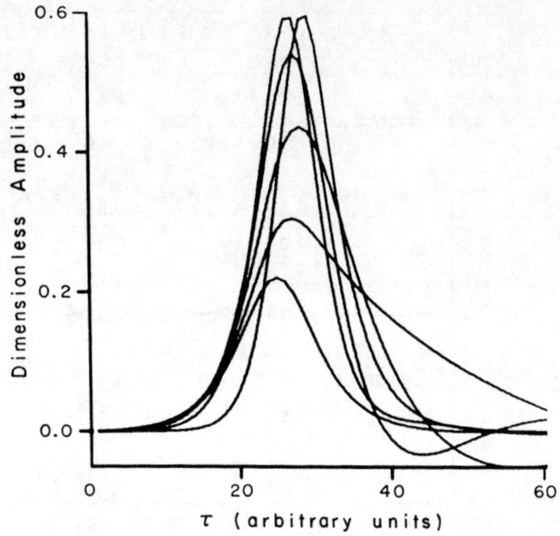

FIG. 18. Propagation of an ultrashort pulse in an unbroadened medium. Equation (148) is integrated with a $\frac{1}{2}\pi$ initial pulse. The dimensionless amplitude $p\mathcal{E}T/\hbar$ is plotted vs τ/T where T is an arbitrary time unit and $gT/\kappa=0.6$. Curves, in order of increasing peak, correspond to distances of 0, 0.4, 1.2, 2.7, 4.0, and 9.5 times the inverse linear loss.

Integrating then by parts we find

$$\mathcal{E}(z,\tau)\underset{\tau\to 0}{\simeq}e^{-\kappa z}\left[\mathcal{E}_0(\tau)+gz\int_0^{\tau}\mathcal{E}_0(\tau')d\tau'\right]. \quad (156)$$

If the first term in the Taylor expansion of $\mathcal{E}_0(\tau)$ is of order n, we may write

$$\mathcal{E}_0(\tau)\simeq\tau^n(n!)^{-1}(d^n\mathcal{E}_0/d\tau^n)_0+\tau^{n+1}[(n+1)!]^{-1}\times(d^{n+1}\mathcal{E}_0/d\tau^{n+1})_0. \quad (157)$$

Substituting into (156) we may carry out the integral

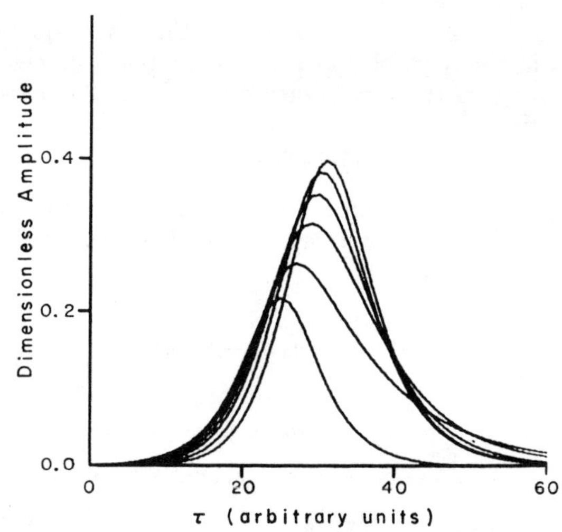

FIG. 19. Effect of nonvanishing γ_{ab} on ultrashort pulse propagation. Equations (171) are integrated with the same input pulse and parameters as in Fig. 13 and with $\gamma_{ab}=0.3$ g/κ. Curves, in order of increasing peak, correspond to distances of 0, 0.8, 1.6, 2.4, 3.2, 4.8 times the inverse linear loss.

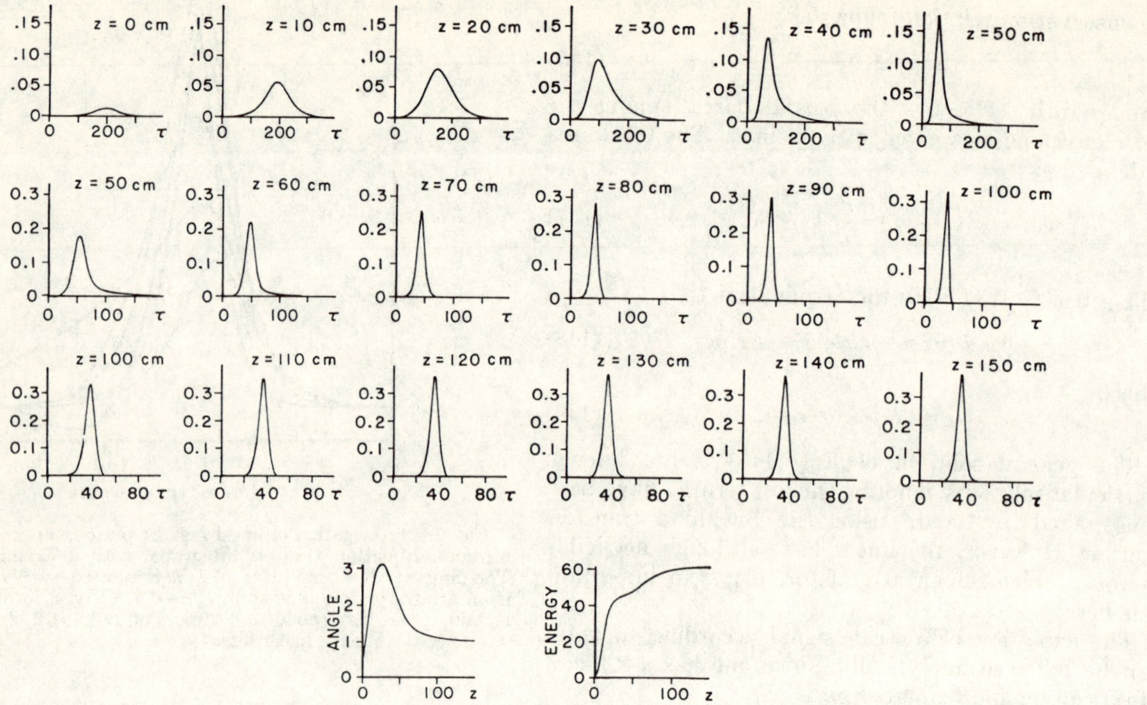

FIG. 20. The series of curves describe the evolution of a $\pi/2$ pulse, initially 100 nsec broad, in a medium with 5 nsec phase memory ($\gamma_{ab}=0.2$ nsec^{-1}). The decay time of the levels a and b is taken to be infinite, and the Doppler broadening is ignored. The quantity $p\mathcal{E}/\hbar$ in nsec^{-1} units is plotted versus τ in nanoseconds. The first curve of each strip duplicates the last curve of the preceeding strip after multiplication of the τ scale by a factor 2. The values of the parameters κ and g are fixed at 0.1 and 0.06. The variation of the angle θ and the energy (in units of the initial energy) as a function of distance is also shown.

and obtain

$$\mathcal{E}(z,\tau) \underset{\tau\to 0}{\simeq} e^{-\kappa z}[1+(gz\tau/n+1)]\mathcal{E}_0(\tau)+O(\tau^{n+2}). \quad (158)$$

We see from this expression that for small τ the effect of the linear loss is dominant since the effect of nonlinear gain is of higher order.

It is possible to find solutions of the exact equation (148) in the form of pulses whose temporal shapes do not change as they travel into the medium. These have the form

$$\psi(z,\tau)=\Psi(\tau-\Delta(z)). \quad (159)$$

The function $\Delta(z)$ represents the amount of time lag associated with z. In terms of the pulse velocity $v(z)$ it can be written as

$$\Delta(z)=\int_0^z [v(z)^{-1}-c^{-1}]dz. \quad (160)$$

Transforming to new variables

$$\tau'=\tau-\Delta(z), \quad (161)$$

$$z'=z, \quad (162)$$

Eq. (148) becomes

$$[\partial^2\psi'(z',\tau')/\partial z'\partial\tau']-[d\Delta(z')/dz'][\partial^2\psi'(z',\tau')/\partial\tau'^2] = -\kappa\partial\psi'/\partial\tau'+g\sin\psi', \quad (163)$$

where

$$\psi'(z',\tau')=\psi(z,\tau). \quad (164)$$

It is easily seen that particular solutions of the form

$$\psi'(z',\tau')=\Psi(\tau') \quad (165)$$

do not exist unless

$$d\Delta(z')/dz'=\text{const}, \quad (166)$$

or

$$v(z)=v. \quad (167)$$

We have thus shown that if a solution with unchanging shape exists, it must travel with a constant velocity. The function $\Psi(\tau)$ must then satisfy

$$(v^{-1}-c^{-1})d^2\Psi/d\tau^2-\kappa d\Psi/d\tau+g\sin\Psi=0. \quad (168)$$

For the physical case $v<c$, (168) is the equation of a pendulum with negative friction which can be thought of as the time reversal image of an ordinary pendulum with positive damping. It may be seen that such a system admits special "limiting solutions" in the form of π pulses. We shall refer to these as the pendulum solutions.

Numerical integration of the transient Eq. (148) shows that an arbitrary initial pulse evolves asymptotically into an unchanging shape which is a hyperbolic secant with total area π, as shown in Fig. 18. It is easily seen from (168) that this corresponds to $v=c$. Equation (168) then becomes

$$\kappa\dot{\Psi}=g\sin\Psi \tag{169}$$

giving

$$(p/\hbar)\mathcal{E}(\tau)=(g/\kappa)\operatorname{sech}[(g/\kappa)(\tau-\tau_0)], \tag{170}$$

where τ_0 is an arbitrary constant of integration. Although theoretically this solution travels with the velocity of light c, a velocity slightly less than c is observed in the numerical integration, as the curves for successive values of z drift slowly towards increasing values of τ (see Fig. 19). This result is accounted for by the fact that we start with an input pulse $\mathcal{E}_0(\tau)$ which is identically zero for $\tau\leq 0$. The transient equation (148) preserves this feature of the pulse for all z, thus preventing the pulse from ever matching exactly one of the solutions represented in (170). However, for increasing τ_0 the mismatch disappears exponentially, thus "pushing" the pulse towards slightly larger values of τ_0. The pendulum solutions with nonzero "mass" ($v\neq c$) proved to be unstable, evolving into the hyperbolic secant form over sufficiently long distances.

For the case of a finite homogeneous broadening γ_{ab}, we shall prove the existence and display the approach to a stationary shape similar to that met above for $\gamma_{ab}=0$. In this case the equations to be considered are

$$\partial\mathcal{E}/\partial z=-\kappa\mathcal{E}-\tfrac{1}{2}KS/\epsilon_0, \tag{171a}$$

$$\partial S/\partial\tau=-\gamma_{ab}S-(p^2/\hbar)\mathcal{E}N, \tag{171b}$$

$$\partial N/\partial\tau=\mathcal{E}S/\hbar. \tag{171c}$$

For a stationary solution traveling with the velocity of light c, we have $\mathcal{E}(z,\tau)=\mathcal{E}(\tau)$, hence $\partial\mathcal{E}/\partial z=0$ and

$$S=-2(\kappa/K)\epsilon_0\mathcal{E}=-pN_0(\kappa/g)(p\mathcal{E}/\hbar) \tag{172}$$

with g given by (149). Substituting (172) into the right-hand side of (171b) and slightly rewriting (171c), we obtain the generalization of (143) and (144)

$$(\partial/\partial\tau)S/pN_0=-(p\mathcal{E}/\hbar)[(N/N_0)-(\gamma_{ab}\kappa/g)], \tag{173a}$$

$$(\partial/\partial\tau)[(N/N_0)-(\gamma_{ab}\kappa/g)]=(p\mathcal{E}/\hbar)(S/pN_0). \tag{173b}$$

Setting as before

$$\psi(z,\tau)=\frac{p}{\hbar}\int_{-\infty}^{\tau}\mathcal{E}(z,\tau')d\tau', \tag{174}$$

we obtain

$$S/pN_0=-[1-(\gamma_{ab}\kappa/g)]\sin\psi, \tag{175}$$

$$(N/N_0)-(\gamma_{ab}\kappa/g)=[1-(\gamma_{ab}\kappa/g)]\cos\psi. \tag{176}$$

From (172) and (175) we have

$$\dot{\psi}=[(g/\kappa)-\gamma_{ab}]\sin\psi, \tag{177}$$

which integrates to

$$(p/\hbar)\mathcal{E}(\tau)=[(g/\kappa)-\gamma_{ab}]\operatorname{sech}\{[(g/\kappa)-\gamma_{ab}]\times(\tau-\tau_0)\}. \tag{178}$$

As before, the pulse shape is a hyperbolic secant and its total area is again π. It now has a smaller peak value and is broader than (170) when $\gamma_{ab}>0$. We have integrated (171) with $\gamma_{ab}=0.1\ g/\kappa$ and $\gamma_{ab}=0.3\ g/\kappa$ and have observed that the solution approaches the steady state (178) [see Fig. 19].

The set of equations (171) can also be used to follow the evolution of pulses of the type considered in paragraph IV A 2. There, the pulse width was assumed to be much shorter than $1/\gamma_a$ and $1/\gamma_b$ but much larger than $1/\gamma_{ab}$, enabling us to set $\gamma_a=\gamma_b=0$ and $\gamma_{ab}=\infty$. Integration of the equations under these conditions showed an increasing sharpening of the pulse leading to a breakdown of the second assumption. We integrated Eqs. (171) with $g/\kappa=0.6$ and $\gamma_{ab}=0.2$ ($T_2=5$ nsec) as in the previous case, but with a much broader initial pulse ($\Delta\tau_0=100$ nsec, $\theta_0=\frac{1}{2}\pi$). The first stage of the evolution confirms the behavior predicted in IV A 2, namely peak velocities temporarily exceeding the velocity of light, but as the pulse gets shorter, it settles down to the stationary form (178) traveling with the velocity of light (see Fig. 20). The peak velocity, if defined by reference to the position of the peak of the field in a diagram of $\mathcal{E}$ as a function of τ, for fixed z, is a misleading concept. Indeed, if $\tau_p(z)$ is the retarded time for which a peak occurs at the space point z, the peak velocity is given by

$$v_p=c[1+c\,d\tau_p/dz]^{-1}, \tag{179}$$

which may become larger than c if $d\tau_p/dz<0$, or even infinite or negative. The fallacy is the following. A pulse of the form $\mathcal{E}(z,t)$ has a spatial shape at any fixed time, and a temporal shape for any fixed z. The two shapes are independent of each other unless they are both unchanging in which case $\mathcal{E}=\mathcal{E}(t-z/v)$. The velocity of the spatial peak of the pulse can be meaningfully defined, but not the velocity of the temporal peak, especially when the pulse is rapidly changing in shape. The meaning of a temporal peak is that a given apparatus is recording a maximum field and it is conceivable that another apparatus, located at a neighboring position, is also recording a maximum at the same instant, thus giving an "infinite velocity".

B. Doppler Broadening

Specializing again for "zero-phase" solutions, Eqs. (52) for $\gamma_a=\gamma_b=\gamma_{ab}=\Lambda_a=\Lambda_b=0$ reduce to

$$\partial\mathcal{E}(z,\tau)/\partial z=-\kappa\mathcal{E}-\tfrac{1}{2}K\langle S/\epsilon_0\rangle_v, \tag{180a}$$

$$\partial S(v,z,\tau)/\partial\tau=-KvC-(p^2/\hbar)\mathcal{E}N, \tag{180b}$$

$$\partial C(v,z,\tau)/\partial\tau=KvS, \tag{180c}$$

$$\partial N(v,z,\tau)/\partial\tau=\mathcal{E}S/\hbar. \tag{180d}$$

A conservation relation in the form

$$(S/p)^2+(C/p)^2+N^2=N_0^2 \tag{181}$$

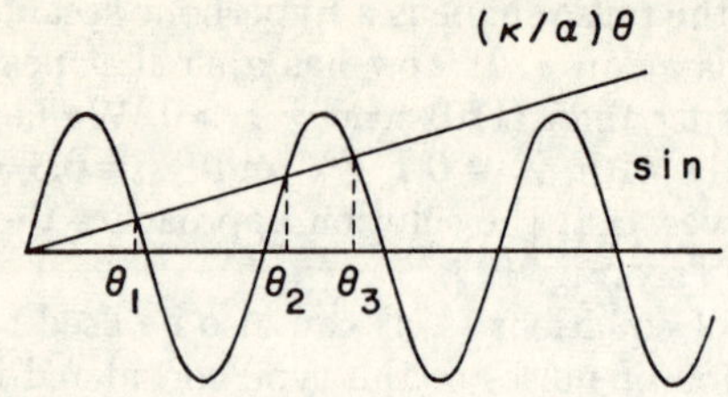

FIG. 21. Graphical determination of the asymptotic values of θ.

is satisfied. As shown in Appendix D, the total area

$$\theta(z)=\frac{p}{\hbar}\int_{-\infty}^{+\infty}\mathcal{E}(z,\tau')d\tau' \tag{182}$$

obeys the "area theorem" of McCall and Hahn[11]

$$d\theta/dz=-\kappa\theta+\alpha\sin\theta, \tag{183}$$

generalized to include the linear loss mechanism. The coefficient α is defined as

$$\alpha=\tfrac{1}{2}\pi(p^2N_0/\epsilon_0\hbar)W(0), \tag{184}$$

where $W(v)$ is the velocity distribution. Note that α is positive for an amplifier and negative for a nonlinear attenuator ($N_0<0$). The area theorem can be thought of as another first integral of the system of equations (180). It expresses the fact that the variation of the area θ depends on θ alone and is not affected by the shape of the field envelope $\mathcal{E}$. Another remarkable property is that this variation is also independent of the detailed velocity distribution $W(v)$. However, it does depend on the value $W(0)$ which usually is related to the width of the distribution.

In the case of fixed atoms the area theorem was not mentioned because the theorem takes a singular form in that limit. Indeed, if we apply the previous result in the limit of an extremely narrow velocity distribution.

$$W(v)\to\delta(v), \tag{185}$$

we have $W(0)=\infty$, hence $\alpha=\infty$, and we find

$$\sin\theta=0,\quad \theta=n\pi. \tag{186}$$

This result is interpreted as follows. Consider an arbitrary input pulse with an arbitrary area $\theta(0)$ at the entry plane $z=0$. At any depth Δz, no matter how small, the area will have reached an integral multiple of π. This unphysical "jump" in θ is a result of the unphysical assumption of no broadening (neither homogeneous nor Doppler), allowing the layer Δz of atoms to ring forever and make a finite contribution to the area.

The solutions $\theta(z)$ of (183) asymptotically approach constant values which are obtained from

$$\sin\theta=(\kappa/\alpha)\theta, \tag{187}$$

giving (see Fig. 21)

$$\theta=\theta_n,\quad (n=1,2,\cdots). \tag{188}$$

If the loss term is small we have

$$\theta_n\simeq n\pi, \tag{189}$$

but this sequence of asymptotic values terminates after approximately $\alpha/\pi\kappa$ values. The solutions of (183) are sketched in Fig. 22. This diagram can be used to follow the evolution of $\theta(z)$ by locating the origin along the z axis to correspond to the initial value of θ. It is seen from Fig. 22 that θ_{2n+1} pulses are stable and θ_{2n} pulses are unstable ($\alpha>0$).

C. Self-Induced Transparency

It can be noted that the area theorem is derived without specifying whether the population is initially inverted or not. For a medium with atoms initially in the ground state (an attenuator), N_0 is negative and the area theorem applies with a negative α

$$d\theta/dz=-\kappa\theta-|\alpha|\sin\theta. \tag{190}$$

Neglecting for simplicity the previously discussed effects of the linear loss terms in this equation, we can sketch the branched solutions as in Fig. 23. For an initial pulse area $\theta_0<\pi$ it is seen that the total area decays to zero. This would be expected on the basis of classical absorption laws. On the other hand if the initial area is

$$\pi<\theta_0<3\pi, \tag{191}$$

the pulse evolves into a 2π pulse, which means that the pulse will not die away over anomalously long distances. This is the "self-induced transparency" effect and has been observed by McCall and Hahn.[11]

Although the area stabilizes at a constant value, it remains to be seen whether the pulse shape can reach a stationary form. In order to investigate this point we shall search for stationary solutions of (180) but we first consider the simple case of fixed atoms by writing Eqs. (171) with $\kappa=0$

$$\partial\mathcal{E}/\partial z=-\tfrac{1}{2}KS/\epsilon_0, \tag{192a}$$

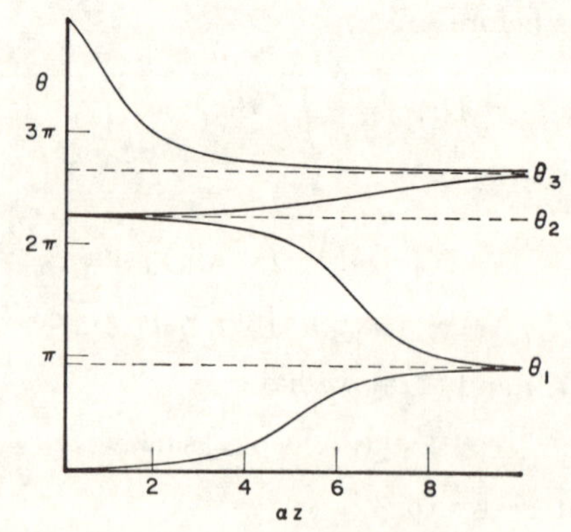

FIG. 22. Branched solutions of the area theorem for an amplifier, Eq. (183), with $\kappa/\alpha=0.1$. The angle θ is plotted vs the dimensionless length αz.

$$\partial S/\partial \tau = -(p^2/\hbar)\mathcal{E}N, \tag{192b}$$

$$\partial N/\partial \tau = \mathcal{E}S/\hbar. \tag{192c}$$

The last two equations are solved by setting

$$S = p|N_0|\sin\psi, \tag{193a}$$

$$N = -|N_0|\cos\psi, \tag{193b}$$

with

$$\dot{\psi} = p\mathcal{E}/\hbar. \tag{194}$$

A stationary solution propagating with velocity v is of the form

$$\mathcal{E}(z,\tau) = \mathcal{E}(t - z/v) = \mathcal{E}(\tau - z(v^{-1} - c^{-1})). \tag{195}$$

After multiplication by $p\mathcal{E}/\hbar = \dot{\psi}$, Eq. (192a) gives

$$(\partial/\partial\tau)(p\mathcal{E}/\hbar)^2 = 2g(v^{-1} - c^{-1})\dot{\psi}\sin\psi, \tag{196}$$

with

$$g = \tfrac{1}{2}Kp^2|N_0|/(\epsilon_0\hbar). \tag{197}$$

Integrating (196), subject to the initial conditions $\mathcal{E}(-\infty) = \psi(-\infty) = 0$, and taking the square root we find

$$p\mathcal{E}/\hbar = \dot{\psi} = (2/\Delta\tau)\sin\tfrac{1}{2}\psi, \tag{198}$$

with

$$\Delta\tau = g^{-1/2}(v^{-1} - c^{-1})^{1/2}. \tag{199}$$

Equation (198) is readily integrated and gives

$$(p/\hbar)\mathcal{E}(z,\tau) = (2/\Delta\tau)\operatorname{sech}\{[t - (z/v)]/\Delta\tau\}, \tag{200}$$

which is a 2π pulse. From (199) the velocity is related to the width through

$$v = c[1 + cg(\Delta\tau)^2]^{-1}. \tag{201}$$

A similar result can be proven for the case of Doppler broadening. The proof is more involved and left to Appendix E. As a result it is found that (200) holds with the following width-velocity relation

$$v = c\{1 + cg(\Delta\tau)^2\pi^{1/2}(Ku\Delta\tau)^{-1}\exp[(Ku\Delta\tau)^{-2}] \times \operatorname{erfc}[(Ku\Delta\tau)^{-1}]\}^{-1}, \tag{202a}$$

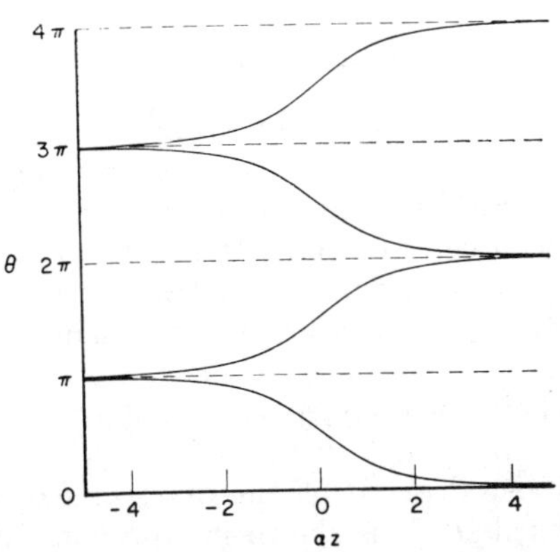

FIG. 23. Branched solutions of the area theorem for an attenuator with $\kappa = 0$.

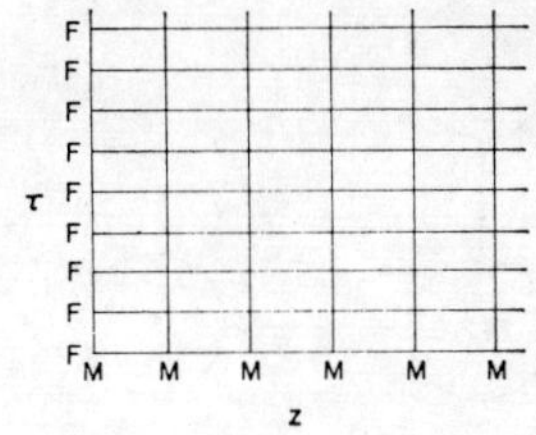

FIG. 24. Rectangular grid used in the numerical integration of Eqs. (226).

which reduces to

$$v = c\{1 + \pi^{1/2}cg\Delta\tau/Ku\}^{-1} \tag{202b}$$

for a broad Doppler line ($Ku\Delta\tau \gg 1$).

VI. GENERAL PULSES

In this section we put aside all the assumptions regarding the relative magnitudes of the pulse width and atomic decay times, and derive a set of equations valid in the most general case compatible with the assumptions of the introductory sections. Computer solutions of these equations are then presented and discussed. Although the terminology is appropriate to the case of gas lasers, the equivalence with the problem of a solid state laser should be borne in mind.

A. Transformation of Equations

Without loss of generality we may set $\nu = \omega$ in the basic equations (52) and obtain

$$\partial\mathcal{E}/\partial z = -\kappa\mathcal{E} - \tfrac{1}{2}K\langle S/\epsilon_0\rangle_v, \tag{203a}$$

$$\mathcal{E}\partial\varphi/\partial z = -\tfrac{1}{2}K\langle C/\epsilon_0\rangle_v, \tag{203b}$$

$$\partial S/\partial\tau = -\gamma_{ab}S - (Kv - \dot{\varphi})C - (p^2/\hbar) \times \mathcal{E}(\rho_{aa} - \rho_{bb}), \tag{203c}$$

$$\partial C/\partial\tau = -\gamma_{ab}C + (Kv - \dot{\varphi})S, \tag{203d}$$

$$\partial\rho_{aa}/\partial\tau = \Lambda_a - \gamma_a\rho_{aa} + \tfrac{1}{2}\mathcal{E}S/\hbar, \tag{203e}$$

$$\partial\rho_{bb}/\partial\tau = \Lambda_b - \gamma_b\rho_{bb} - \tfrac{1}{2}\mathcal{E}S/\hbar. \tag{203f}$$

In the previous discussions the phase equation was neglected by assuming a symmetrical velocity distribution and by restricting the input pulse in such a way that $\varphi(0,\tau) \equiv 0$, thus obtaining the "zero phase" solution $\varphi(z,\tau) \equiv 0$. However it has been shown[12] that if some small $\varphi(0,\tau)$ is injected as input, φ will grow as the wave travels into the medium. It is therefore desirable to introduce this feature into the integration scheme. On the other hand, the phase equation has a singularity for $\mathcal{E} = 0$ which corresponds to the physical fact that the phase is undetermined when the field vanishes. This feature is inconvenient for numerical integration and can be avoided by going over to complex arithmetic in the following way. In Eqs. (203) we make the substitution

$$\mathcal{E}' = \mathcal{E}e^{i\varphi}, \tag{204a}$$

$$\mathcal{P} = -(S + iC)e^{i\varphi}, \tag{204b}$$

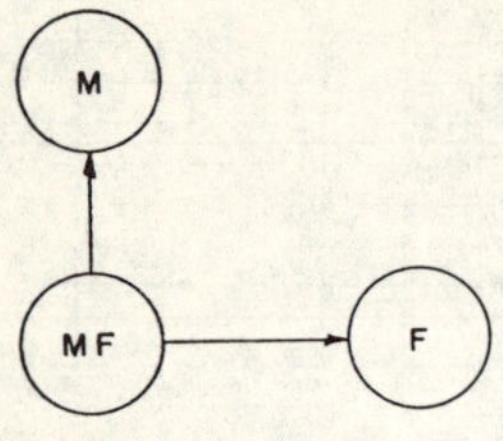

FIG. 25. Schematic representation of the mechanism for propagating the numerical solution.

finding

$$\partial\mathcal{E}'/\partial z = -\kappa\mathcal{E}' + \tfrac{1}{2}K\langle\mathcal{P}/\epsilon_0\rangle_v, \tag{205}$$

and

$$\partial\mathcal{P}/\partial\tau = -(\gamma_{ab} - iKv)\mathcal{P} + (p^2/\hbar)\mathcal{E}'N. \tag{206}$$

We also have

$$\mathcal{E}S = \tfrac{1}{2}(\mathcal{E}'\mathcal{P}^* + \mathcal{E}'^*\mathcal{P}), \tag{207}$$

so that, omitting the prime on the new complex field $\mathcal{E}'$, Eqs. (203) become

$$\partial\mathcal{E}/\partial z = -\kappa\mathcal{E} + \tfrac{1}{2}K\langle\mathcal{P}/\epsilon_0\rangle_v, \tag{208a}$$

$$\partial\mathcal{P}/\partial z = -(\gamma_{ab} - iKv)\mathcal{P} + (p^2/\hbar)\mathcal{E}(\rho_{aa} - \rho_{bb}), \tag{208b}$$

$$\partial\rho_{aa}/\partial\tau = \Lambda_a - \gamma_a\rho_{aa} - \tfrac{1}{4}(\mathcal{E}\mathcal{P}^* + \mathcal{E}^*\mathcal{P})/\hbar, \tag{208c}$$

$$\partial\rho_{bb}/\partial\tau = \Lambda_b - \gamma_b\rho_{bb} + \tfrac{1}{4}(\mathcal{E}\mathcal{P}^* + \mathcal{E}^*\mathcal{P})/\hbar. \tag{208d}$$

In Eqs. (208) $\mathcal{E}$ and $\mathcal{P}$ are complex quantities but ρ_{aa}, ρ_{bb} are real. This complex field $\mathcal{E}$ has a simple interpretation: its modulus $|\mathcal{E}|$ represents the physical field envelope, and its argument φ the physical phase. Equations (208) are free of singularities and can be easily integrated even with a nonzero initial phase.

B. Numerical Integration Procedure

We now discuss the numerical integration of the system of partial differential Eqs. (208) for the propagation of a pulse through the nonlinear amplifying medium.

If at the entry plane $z=0$ of the medium the incoming pulse is a known function $\mathcal{E}(0,\tau)$, then the last three equations are coupled ordinary differential equations for the variables $\mathcal{P}(v;0,\tau)$, $\rho_{aa}(v;0,\tau)$, $\rho_{bb}(v;0,\tau)$. Given the initial values at $\tau=0$ of these variables, they can be determined for any τ. This calculation is done separately for every velocity v, then the polarization is averaged over velocities giving $\mathcal{P}(0,\tau)$. Actually, only a finite number of representative velocities are used. The choice of these will be discussed later.

For any τ, the first equation gives the evolution in the vicinity of z, of $\mathcal{E}(z,\tau)$ as a function of z since the right-hand side is now known. This permits the determination of the field $\mathcal{E}$ as a function of time τ at some small depth $z+\Delta z$ in the medium. Thus the knowledge of the field is propagated step by step into the laser.

The (z,τ) plane is covered with a rectangular grid, the step sizes Hz along z and $H\tau$ along τ being independently chosen (see Fig. 24). Along $\tau=0$ the medium is unreached by the perturbing pulse and therefore is at rest with known values of the populations and zero polarization. This knowledge is represented by "M" (for medium) at the appropriate points. Similarly the field is assumed known at the entry plane $z=0$. This is indicated by a letter "F" in the diagram. Whenever the field $\mathcal{E}$ and the medium properties $\mathcal{P}$, ρ_{aa}, ρ_{bb} are all known at a lattice point, Eq. (208a) provides the derivative of $\mathcal{E}$ along Oz, and the remaining equations give the derivatives of the medium properties along $O\tau$. Schematically this is represented as shown in Fig. 25. Starting from the origin where we have "MF" we can reach every mesh point as indicated by the diagram of Fig. 25. There are many possible ways of covering the entire lattice, but a natural way of doing this, from a physical point of view, is to procede along successive vertical lines, thus determining the temporal shape of the pulse at successive space points, as would be done by a number of fixed measuring apparatuses located at different points.

The total length of the grid along Oz is of course determined by the length of the medium and the extent along $O\tau$ depends on the initial pulse duration and on how far into the tail of the pulse one wishes to investigate. The time step $H\tau$ is usually chosen to fit 25 points under the initial pulse. The space step Hz is chosen to limit the relative variation of the field at each step.

The actual integration of the equations is performed by the following simple predictor-corrector procedure. If

$$y' = dy/dx = f(y) \tag{209}$$

formally represents the set of equations, with y having several components and x standing for either z or τ, and if y_n is the value of the solution at the nth mesh point along Ox, then

$$p_{n+1} = y_{n-1} + 2hy_n' \tag{210}$$

is the predicted value at the $(n+1)$th point, its derivative is

$$p_{n+1}' = f(p_{n+1}), \tag{211}$$

and

$$c_{n+1} = y_n + \tfrac{1}{2}h(p_{n+1}' + y_n') \tag{212}$$

is the corrected value. One then takes $y_{n+1} = c_{n+1}$. It is well known that care must be taken in using such predictor-corrector methods since they are not always stable. An integration formula is said to be stable, if the difference sequence $Y_n - y_n$ between the true solution and the approximate solution remains bounded as $n \to \infty$, and can be made as small as one wishes by a suitable choice of the step size h and of the round-off error. Stability is not an intrinsic property of the formula but also depends on the differential system it is being applied to. It is shown in Appendix F that this formula is stable when applied to the system of Eqs. (208).

TABLE I. This table shows the independent parameters that must be fixed when integrating Eqs. (226).

Name	Gain	Loss	Atomic phase memory	Decay consts.	Doppler linewidth	Pumping ratio	Initial pulse width	Initial pulse angle
Symbol	g	κ	γ_{ab}	γ_a, γ_b	Ku	$(\Lambda_b/\gamma_b)/(\Lambda_a/\gamma_a)$	$\Delta\tau_0$	θ_0
Value	0.06	0.005	0.055	0.1, 0.01	variable	0	variable	variable
Units	$cm^{-1} \times nsec^{-1}$	cm^{-1}	$nsec^{-1}$	$nsec^{-1}$	$nsec^{-1}$	none	nsec	none

It can be noticed that the predictor-corrector formula calls for two backwards points, which means that we need to know the values of the field $\mathcal{E}$ along two vertical lines (see Fig. 24) and the values of the medium properties along two horizontal lines, before we can start using the formula. The state of the medium is known at any time prior to $\tau=0$, therefore we only need to determine the field along the vertical line adjacent to $z=0$. This is done by using

$$p_{n+1}=y_n+hy'_{n-1} \tag{213}$$

instead of the predictor formula (210), and by iterating the corrector formula (212) to convergence, in order to make up for the poor accuracy of (213).

C. Averaging over the Velocities

An important aspect of the program is the averaging of the polarization after each space step. The partial polarization $\mathcal{P}(v,z,\tau)$ is first calculated for each velocity v, along the mesh points of the same vertical line of Fig. 24, then an average is taken which approximates as nearly as possible the integral

$$\pi^{-1/2}\int_{-\infty}^{+\infty} \exp\left(-\frac{v^2}{u^2}\right)\mathcal{P}(v,z,\tau)d\frac{v}{u} \tag{214}$$

by using a finite sum over discrete velocities. For a given number of discrete velocities, this is best achieved by the Hermite-Gauss integration formula of order n

$$\pi^{-1/2}\int_{-\infty}^{+\infty} \exp\left(-\frac{v^2}{u^2}\right)\mathcal{P}(v,z,\tau)d\frac{v}{u} = \sum_{i=1}^{n} \mathcal{P}(v_i,z,\tau)W_i+R_n, \tag{215}$$

where

$$v_i=ux_i, \quad i=1,\cdots,n \tag{216}$$

are proportional to the n zeros x_i of the nth Hermite polynomial $H_n(x)$. The weight factors W_i are given by

$$W_i=2^{n-1}n!n^{-2}[H_{n-1}(x_i)]^{-2}, \tag{217}$$

and the remainder is

$$R_n=[2^{-n}n!/(2n)!]u^{2n}(\partial^{2n}/\partial v^{2n})\mathcal{P}(v,z,\tau). \tag{218}$$

The $2n$th derivative in this last expression must be evaluated for some typical v which we will take to be u. The zeros of the Hermite polynomials, together with the corresponding weight factors, have been tabulated[16] for n up to 64.

If an upper bound is found for $u^{2n}(\partial^{2n}/\partial v^{2n})\mathcal{P}(v,z,\tau)$ one can see from (218) that by choosing n sufficiently large one may achieve any desired accuracy. We proceed to obtain such an upper bound. By integration of (208b) we obtain

$$\mathcal{P}(v,z,\tau)=\frac{p^2}{\hbar}\int_0^\tau d\tau'\mathcal{E}(z,\tau')N(v,z,\tau')\times\exp[-(\gamma_{ab}-iKv)(\tau-\tau')]; \tag{219}$$

hence, if we neglect the variation of N with v,

$$\begin{aligned}|u^{2n}(\partial^{2n}/\partial v^{2n})\mathcal{P}(v,z,\tau)| &= \left|(p^2/\hbar)\int_0^\tau d\tau'\mathcal{E}(z,\tau')N(v,z,\tau')[iKu(\tau-\tau')]^{2n}\times\exp[-(\gamma_{ab}-iKv)(\tau-\tau')]\right| \leq (p^2/\hbar)(\mathcal{E}N)_{\max}\times\int_0^\tau d\tau'[Ku(\tau-\tau')]^{2n}\\ &=(p^2/\hbar Ku)(\mathcal{E}N)_{\max}(Ku\tau)^{2n+1}/(2n+1),\end{aligned} \tag{220}$$

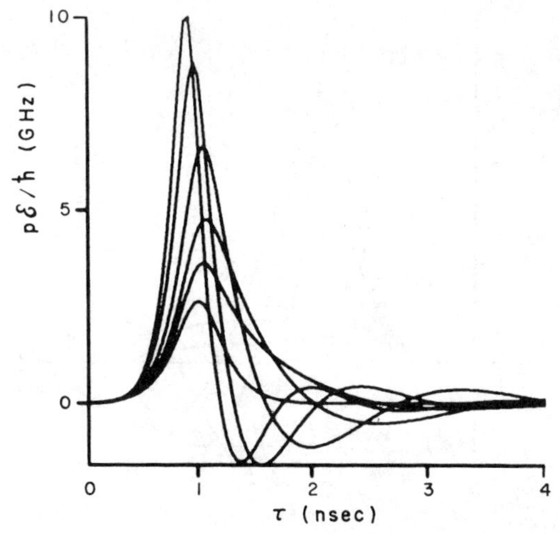

FIG. 26. Standard case. Here, Eqs. (226) are integrated with the following values of the parameters: $\Delta\tau_0=0.5$ nsec, $\theta_0=\frac{1}{2}\pi p$, $Ku=1.4$ GHz, all other parameters being fixed as shown in Table I. Curves, in order of increasing peak, correspond to distances of 0, 0.66, 1.32, 2.64, and 3.96 m into the medium.

[16] T. S. Shao, T. C. Chen, and R. M. Frank, Math. Comp. 18, 598 (1964).

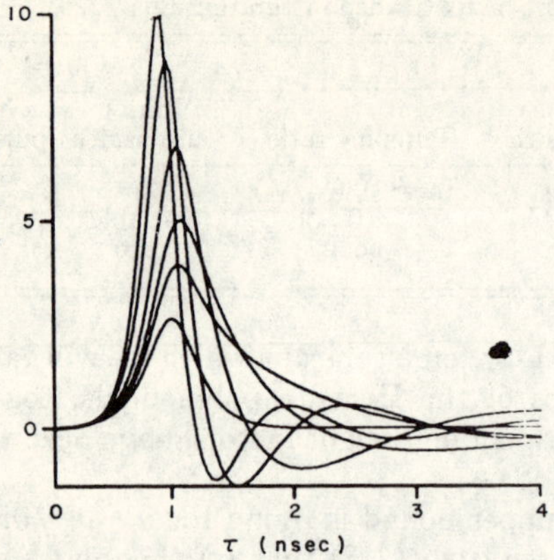

FIG. 27. Fixed atoms. Here, the same situation is represented as in Fig. 26, except for $Ku=0$ instead of $Ku=1.4$ GHz.

where $(\mathcal{E}N)_{\max}$ is the maximum for fixed z of $\mathcal{E}N$ as a function of τ. We find thus

$$R_n \leq (p^2/\hbar Ku)(\mathcal{E}N)_{\max}(Ku\tau)^{2n+1}2^{-n}n!/(2n+1)!. \quad (221)$$

On the other hand we see from (208b) that

$$(p^2/\hbar Ku)(\mathcal{E}N)_{\max} \sim \mathcal{P}(u,z,\tau)_{\max}, \quad (222)$$

so that this quantity may be used as reference in defining a relative error

$$R_n/\mathcal{P}_{\max} \leq 2^{-n}n!(Ku\tau)^{2n+1}/(2n+1)!. \quad (223)$$

The expression (223) is a very satisfactory estimate of the error when $Ku\tau \gg 1$ or $Ku\tau \sim 1$. For $Ku\tau \gg 1$ it tends to overestimate the error. In this case better but more complicated estimates can be found, however (223) will be sufficient for our purposes.

Examination of (223) shows that for $Ku\tau < 1$, one gets excellent accuracy even with a small number of velocities. The relative error (and the number of velocities needed for accurate integration) grows with τ, which means that the effects of Doppler (or inhomogeneous) broadening manifest themselves mainly in the tail of the pulse. In a typical application, the Doppler width may be $Ku=10^9$ Hz and the time of integration $\tau \simeq 5\times 10^{-9}$ sec, so that the number of velocities needed to achieve $R/\mathcal{P}_{\max} < 0.0001$ is 20. In most practical applications we shall restrict ourselves to "zero phase" input pulses. This implies that the real part of $\mathcal{P}$ will be an even function of v, and since the integrating velocities v_i are symmetrically distributed around $v=0$, the actual number of velocities used is reduced by a factor two.

D. Scaling Laws of Equations

In order to determine the number of independent parameters in Eqs. (208), we make the substitutions

$$p\mathcal{E}/\hbar \to \mathcal{E}, \quad \mathcal{P}/pN_0 \to \mathcal{P}, \quad \rho/N_0 \to \rho \quad (224)$$

with

$$N_0 = (\Lambda_a/\gamma_a) - (\Lambda_b/\gamma_b), \quad (225)$$

and Eqs. (208) become

$$\partial\mathcal{E}/\partial z = -\kappa\mathcal{E} + g\langle\mathcal{P}\rangle_v, \quad (226a)$$

$$\partial\mathcal{P}/\partial\tau = -(\gamma_{ab} - iKv)\mathcal{P} + \mathcal{E}(\rho_{aa} - \rho_{bb}), \quad (226b)$$

$$\partial\rho_{aa}/\partial\tau = (\Lambda_a/N_0) - \gamma_a\rho_{aa} - \tfrac{1}{4}(\mathcal{E}\mathcal{P}^* + \mathcal{E}^*\mathcal{P}), \quad (226c)$$

$$\partial\rho_{bb}/\partial\tau = (\Lambda_b/N_0) - \gamma_b\rho_{bb} + \tfrac{1}{4}(\mathcal{E}\mathcal{P}^* + \mathcal{E}^*\mathcal{P}), \quad (226d)$$

with

$$g = GK\gamma_{ab}, \quad (227)$$

where G is the dimensionless coupling constant given by (65). In (226) $\mathcal{P}$ and ρ are dimensionless, while $\mathcal{E}$ has the dimensions of an inverse time. If the gain coefficient $\mathcal{G}$ of the medium is defined as the inverse length over which

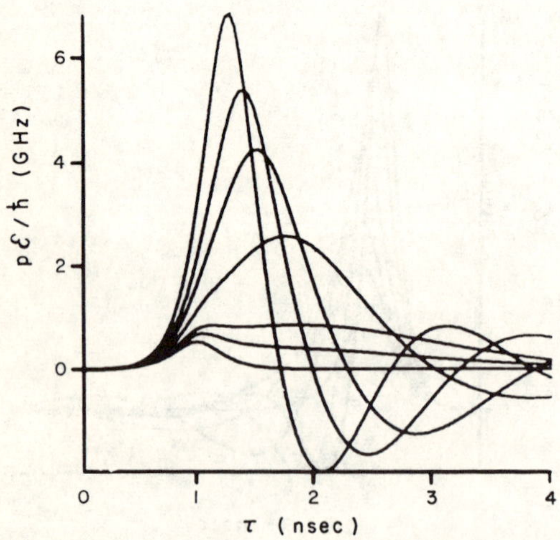

FIG. 28. Weaker pulse. Equations (226) are integrated with the same values of the parameters as in the standard case of Fig. 26, except for $\theta_0=\pi/10$ instead of $\theta_0=\frac{1}{2}\pi$. Curves, in order of increasing peak, correspond to distances of 0, 0.22, 0.38, 0.82, 1.35, 1.80, and 2.56 m into the medium.

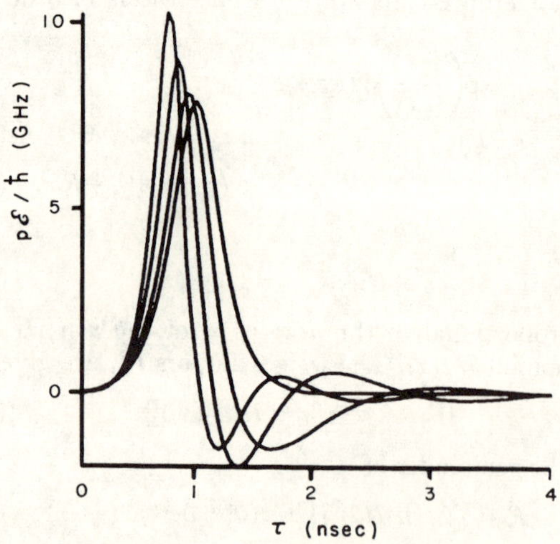

FIG. 29. Stronger pulse. Here the initial pulse angle is $\theta_0=\frac{1}{2}3\pi$. all other parameters being as in the standard case (see Fig. 26), The four curves in order of increasing peak, represent the pulse as seen at distances of 0, 0.56, 1.70, and 3.38 m into the medium.

a weak monochromatic wave is amplified by a factor e, it follows from Sec. III that

$$\mathcal{G}=\pi^{1/2}g/Ku, \quad \text{for } Ku\gg\gamma_{ab}$$
$$=g/\gamma_{ab}, \quad \text{for } \gamma_{ab}\gg Ku. \tag{228}$$

The independent parameters for Eqs. (226) can be taken to be

$$g,\kappa,Ku,\gamma_{ab},\gamma_a,\gamma_b,r=(\Lambda_a/\gamma_a)(\Lambda_b/\gamma_b).$$

E. Numerical Results

We proceed to illustrate the use of Eqs. (226) by numerically integrating them for special choices of the parameters. In Appendix G we describe a FORTRAN IV program for solving these equations. The input pulse is usually assumed to be a hyperbolic secant with truncated tails. The independent parameters of the pulse can then be taken as the width $\Delta\tau_0$ and the angle θ, defined by (147). The values of g and γ_{ab} can be fixed without loss of generality, since they determine the space and time scales. In addition, we also fix, for simplicity, the values of κ, γ_a, γ_b, and r as seen in Table I. Zero-phase input will be assumed in all but one of the examples treated below.

A $\frac{1}{2}\pi$ pulse with 0.5-nsec width traveling in a medium with 1.4-GHz Doppler broadening is taken here as a standard case. The evolution of such a pulse (see Fig. 26) shows a sharpening of the front edge with amplification of the peak power. The trailing edge has an oscillatory behavior with several "undershoots." Figure 27 shows the same pulse traveling in a medium with no Doppler broadening. In this case the history of the pulse is essentially the same except for longer oscillations in the tail. In Fig. 28 the initial pulse is weaker (0.1 π) and in Fig. 29 stronger (1.5 π) compared to the standard case. As would be expected, the weaker the field the more efficient the amplification. Actually, when the initial pulse is too strong (4π) the peak is reduced and the pulse breaks up, as seen in Fig. 30.

We have also considered the effects of varying the pulse width. Figure 31 shows the propagation of a pulse 2.5 nsec broad, other parameters being the same as in the standard case. Larger undershoots are observed. On the other hand, a narrower pulse (0.05 nsec) develops much smaller undershoots (see Fig. 32), undergoes a reduction of the peak, and settles to a stationary form traveling with a velocity less than c. The effect of a larger Doppler linewidth is investigated in this last case.

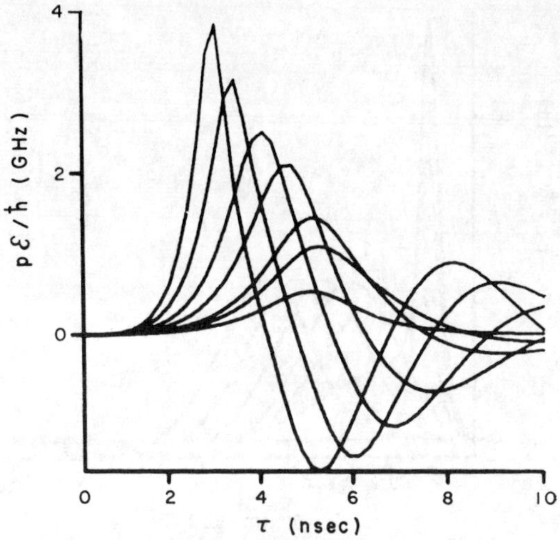

FIG. 31. Broader pulse. Equations (226) are integrated for a pulse initially 2.5 nsec broad, all other parameters having their "standard" values. The increasing peaks belong to pulses located at $z=0$, 0.14, 0.23, 0.46, 0.68, 1.00, and 1.40 m.

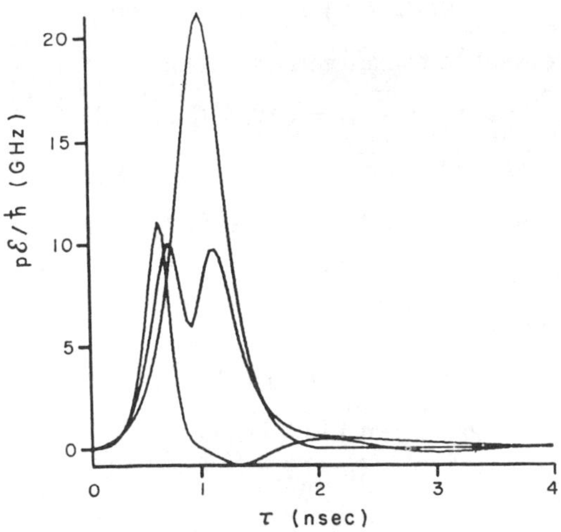

FIG. 30. Pulse breakup. This figure shows how an initially strong pulse ($\theta_0=4\pi$) develops more than one peak. The curve with the highest peak represents the initial pulse, the curve with two peaks corresponds to a distance of 1.30 m and the third curve to a distance of 3.60 m into the medium. Apart from θ_0, all other parameters are given the same values as in the standard case of Fig. 26.

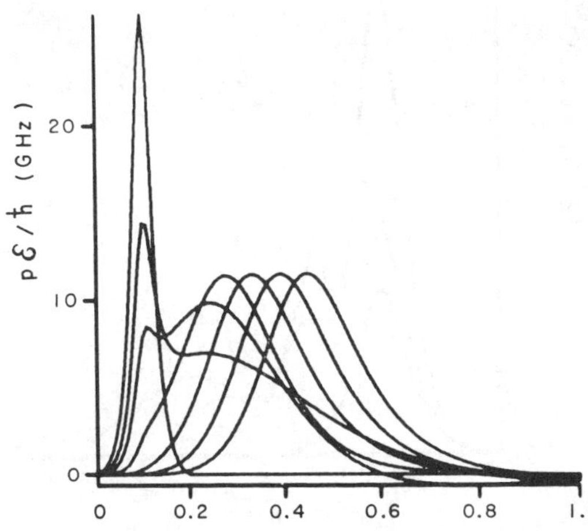

FIG. 32. Sharper pulse. The initial pulse width is taken here to be 0.05 nsec and the other parameters are left unchanged (see Fig. 26). The curve with the highest peak represents the initial pulse. The other peaks from left to right, belong to pulses located at $z=1.75$, 3.50, 7.00, 12.00, and 21.00 m.

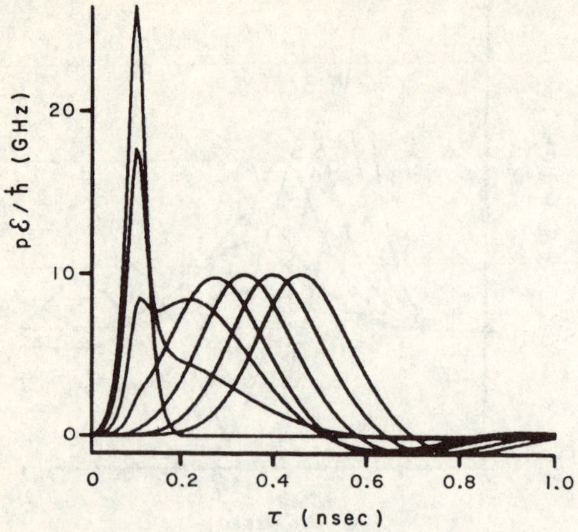

FIG. 33. Broader Doppler line. The calculations of Fig. 32 are repeated here with $Ku=7.0$ GHz instead of $Ku=1.4$ GHz.

Figure 33 describes the same situation as in Fig. 32 except for $Ku=7$ GHz instead of 1.4 GHz. It is seen that the influence of a larger Doppler broadening is not very important. The stationary form is a little broader and the peak is lower.

Finally, we have investigated the effects of a nonzero initial phase $\varphi(0,\tau)$, which for simplicity was taken to be a linear function of time

$$\varphi(0,\tau)=(\omega-\nu)\tau.$$

Such a linear initial phase is equivalent to a detuning of the carrier frequency ν of the pulse from the atomic resonance frequency ω. It was found that a detuning up to Ku has a negligeable effect, but a detuning larger than Ku modifies substantially the evolution of the pulse. This is seen in Fig. 34 which pertains to the standard case with a detuning $(\omega-\nu)=5$ GHz.

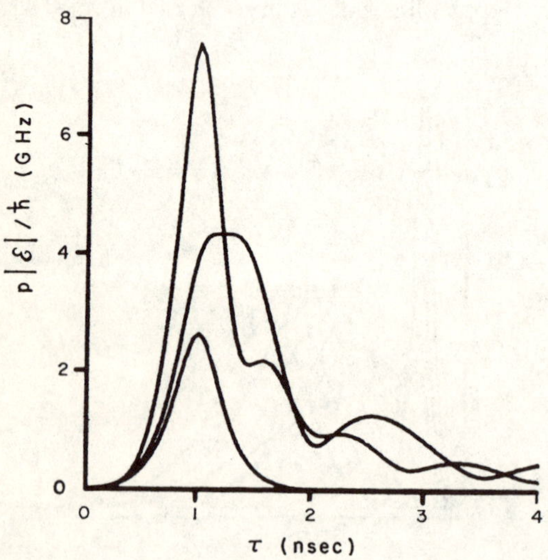

FIG. 34 Effects of detuning. Equations (226) are integrated with the same values of the parameters as in the standard case of Fig. 26, but the carrier frequency of the initial pulse is assumed to be detuned from the atomic resonance frequency by an amount $\omega-\nu=5$ GHz. The three curves, in order of increasing peak, correspond to distances of 0, 1.30, and 3.60 m into the medium.

F. Conclusions

From the calculations we have made it appears that for long enough distances (compared to the inverse gain) the outcoming pulse bears little connection with the input. Over short distances, however, the behavior of the pulse is critically dependent upon its strength as measured by θ and its width. The most general tendency that emerges from the foregoing calculations is for the pulse to develop an oscillating tail. The duration of these oscillations is inversely proportional to the Doppler-width Ku. A strong pulse breaks up into several peaks and undershoots very quickly. On the other hand, a weak pulse is very efficiently amplified and undershoots only after it has gathered sufficient strength. In both cases the total area converges to π. Broad pulses have a tendency to become even broader, while short pulses (compared to $1/\gamma_{ab}$) evolve into quasi-stationary shapes. In the absence of Doppler broadening this stationary shape is usually an hyperbolic secant, but a more complicated form in a Doppler-broadened amplifier. The linear loss κ plays an essential role in the existence of this stationary solution.

We have also checked that as long as the detuning between the carrier frequency of the pulse and the atomic resonance frequency is small compared to Ku, the zero-phase solution remains satisfactory. The amplifier nature of the medium is not affected until this detuning becomes considerably larger than Ku.

APPENDIX A: TRANSFORMATION OF EQS. (30) INTO EQS. (32)

We want to transform Eqs. (30)

$$\partial\rho_{aa}=\lambda_a-\gamma_a\rho_{aa}-i(pE/\hbar)(\rho_{ab}-\rho_{ba}), \qquad (A1)$$

$$\partial\rho_{bb}=\lambda_b-\gamma_b\rho_{bb}+i(pE/\hbar)(\rho_{ab}-\rho_{ba}), \qquad (A2)$$

$$\partial\rho_{ab}=-(\gamma_{ab}+i\omega)\rho_{ab}-i(pE/\hbar)(\rho_{aa}-\rho_{bb}), \qquad (A3)$$

$$\partial\rho_{ba}=-(\gamma_{ab}-i\omega)\rho_{ba}+i(pE/\hbar)(\rho_{aa}-\rho_{bb}), \qquad (A4)$$

where ∂ stands for

$$\partial=(\partial/\partial t)+v(\partial/\partial z), \qquad (A5)$$

to a more convenient form. Define

$$P=p(\rho_{ab}+\rho_{ba}), \qquad (A6)$$

$$Q=p(\rho_{ab}-\rho_{ba}), \qquad (A7)$$

$$N=\rho_{aa}-\rho_{bb}. \qquad (A8)$$

Adding (A3) and (A4) and multiplying by p we get

$$Q=i\omega^{-1}(\partial+\gamma_{ab})P. \qquad (A9)$$

Subtracting (A3) and (A4) multiplying by p and substituting (A9) in the resulting expression we find

$$[(\partial+\gamma_{ab})^2+\omega^2]P=-2\omega(p^2/\hbar)EN. \quad \text{(A10)}$$

On the other hand substituting (A9) into (A3) and (A4) gives

$$\partial\rho_{aa}=\lambda_a-\gamma_a\rho_{aa}+E(\partial+\gamma_{ab})P/\hbar\omega, \quad \text{(A11)}$$

$$\partial\rho_{bb}=\lambda_b-\gamma_b\rho_{bb}-E(\partial+\gamma_{ab})P/\hbar\omega. \quad \text{(A12)}$$

APPENDIX B: SLOWLY VARYING AMPLITUDE AND PHASE APPROXIMATION

If the natural linewidth γ_{ab} is much smaller than the atomic resonance frequency ω, Eqs. (37) may be written as

$$-\partial^2E/\partial z^2+\mu_0\sigma\partial E/\partial t+c^{-2}\partial^2E/\partial t^2=-\mu_0\partial^2P/\partial t^2, \quad \text{(B1)}$$

$$\partial^2P+2\gamma_{ab}\partial P+\omega^2P=-2\omega(p^2/\hbar)E(\rho_{aa}-\rho_{bb}), \quad \text{(B2)}$$

$$\partial\rho_{aa}=\lambda_a-\gamma_a\rho_{aa}+(\hbar\omega)^{-1}E\partial P, \quad \text{(B3)}$$

$$\partial\rho_{bb}=\lambda_b-\gamma_b\rho_{bb}-(\hbar\omega)^{-1}E\partial P, \quad \text{(B4)}$$

where ∂ stands for

$$\partial=(\partial/\partial t)+v(\partial/\partial z). \quad \text{(B5)}$$

Writing the electric field and the polarization as

$$E(z,t)=\mathcal{E}(z,t)\cos\Phi, \quad \text{(B6)}$$

$$P(v,z,t)=C(v,z,t)\cos\Phi+S(v,z,t)\sin\Phi, \quad \text{(B7)}$$

with

$$\Phi=\nu t-Kz+\varphi(z,t), \quad \text{(B8)}$$

we compute the following quantities in the slowly varying amplitude and phase approximation

$$\partial E/\partial t=\dot{\mathcal{E}}\cos\Phi-(\nu+\dot\varphi)\mathcal{E}\sin\Phi, \quad \text{(B9)}$$

$$\partial^2E/\partial t^2=-2\nu\dot{\mathcal{E}}\sin\Phi-(\nu+\dot\varphi)^2\mathcal{E}\cos\Phi, \quad \text{(B10)}$$

$$-\partial^2E/\partial z^2=-2K(\partial\mathcal{E}/\partial z)\sin\Phi +[(\partial\varphi/\partial z)-K]^2\mathcal{E}\cos\Phi, \quad \text{(B11)}$$

$$\partial P=\partial C\cos\Phi-C(\nu-Kv+\partial\varphi)\sin\Phi +\partial S\sin\Phi+S(\nu-Kv+\partial\varphi)\cos\Phi, \quad \text{(B12)}$$

$$\partial^2P=-2\nu\partial C\sin\Phi-C(\nu-Kv+\partial\varphi)^2\cos\Phi +2\nu\partial S\cos\Phi-S(\nu-Kv+\partial\varphi)^2\sin\Phi. \quad \text{(B13)}$$

Substituting into the field Eq. (B1) and equating coefficients of $\sin\Phi$ and $\cos\Phi$ we find

$$\partial\mathcal{E}/\partial z+c^{-1}\partial\mathcal{E}/\partial t=-\kappa\mathcal{E}-\tfrac{1}{2}K\int dv\, S/\epsilon_0, \quad \text{(B14)}$$

$$[\partial\varphi/\partial z+c^{-1}\partial\varphi/\partial t]\mathcal{E}=-\tfrac{1}{2}K\int dv\, C/\epsilon_0, \quad \text{(B15)}$$

where

$$\kappa=\tfrac{1}{2}\sigma/(\epsilon_0 c). \quad \text{(B16)}$$

We next substitute P in the polarization Eq. (B2), equate coefficients of $\sin\Phi$ and $\cos\Phi$ and obtain

$$\partial C=-\gamma_{ab}C+(\omega-\nu+Kv-\partial\varphi)S, \quad \text{(B17)}$$

$$\partial S=-\gamma_{ab}S-(\omega-\nu+Kv-\partial\varphi)C-(p^2/\hbar) \times\mathcal{E}(\rho_{aa}-\rho_{bb}). \quad \text{(B18)}$$

Neglecting rapidly varying terms, the equations for ρ_{aa} and ρ_{bb} are

$$\partial\rho_{aa}=\lambda_a-\gamma_a\rho_{aa}+\tfrac{1}{2}\mathcal{E}S/\hbar, \quad \text{(B19)}$$

$$\partial\rho_{bb}=\lambda_b-\gamma_b\rho_{bb}-\tfrac{1}{2}\mathcal{E}S/\hbar. \quad \text{(B20)}$$

The source terms λ_α in (B3) and (B4) are velocity-dependent according to

$$\lambda_\alpha=\Lambda_\alpha W(v), \quad \text{(B21)}$$

where $W(v)$ is the velocity distribution function. Taking advantage of the homogeneity of the equations, we shall make the following substitutions

$$\lambda_\alpha\to\Lambda_\alpha, \quad \text{(B22)}$$

and

$$\int S dv\to\int SW(v)dv, \quad \int C dv\to\int CW(v)dv. \quad \text{(B23)}$$

The result can then be summarized as

$$\partial\mathcal{E}/\partial z+c^{-1}\partial\mathcal{E}/\partial t=-\kappa\mathcal{E}-\tfrac{1}{2}K\langle S/\epsilon_0\rangle_v, \quad \text{(B24)}$$

$$[\partial\varphi/\partial z+c^{-1}\partial\varphi/\partial t]\mathcal{E}=-\tfrac{1}{2}K\langle C/\epsilon_0\rangle_v, \quad \text{(B25)}$$

$$\partial S=-\gamma_{ab}S-(\omega-\nu+Kv-\partial\varphi)C-(p^2/\hbar) \times\mathcal{E}(\rho_{aa}-\rho_{bb}), \quad \text{(B26)}$$

$$\partial C=-\gamma_{ab}C+(\omega-\nu+Kv-\partial\varphi)S, \quad \text{(B27)}$$

$$\partial\rho_{aa}=\Lambda_a-\gamma_a\rho_{aa}+\tfrac{1}{2}\mathcal{E}S/\hbar, \quad \text{(B28)}$$

$$\partial\rho_{bb}=\Lambda_b-\gamma_b\rho_{bb}-\tfrac{1}{2}\mathcal{E}S/\hbar, \quad \text{(B29)}$$

where

$$\langle\cdots\rangle_v=\int(\cdots)W(v)dv, \quad \text{(B30)}$$

as stated in Eqs. (44).

APPENDIX C: EVALUATION OF INTEGRAL (78)

Consider

$$I=\int_{-\infty}^{+\infty}[(t+\xi)^2+x^2]^{-1}e^{-t^2}dt. \quad \text{(C1)}$$

Using the integral representation

$$[(t+\xi)^2+x^2]^{-1}=$$

$$x^{-1}\int_0^\infty e^{-\mu x}\cos\mu(t+\xi)d\mu\,,\quad (x>0) \tag{C2}$$

we obtain

$$I=x^{-1}\int_{-\infty}^{+\infty} dt\, e^{-t^2}\int_0^\infty d\mu\, e^{-\mu x}\cos\mu(t+\xi)\,, \tag{C3}$$

and by changing the order of integration

$$I=x^{-1}\int_0^\infty d\mu\, e^{-\mu x}\int_{-\infty}^{+\infty} dt\, e^{-t^2}\cos\mu(t+\xi)\,. \tag{C4}$$

Expanding the cosine in terms of exponentials we get

$$I=(2x)^{-1}\left\{\int_0^\infty d\mu\, e^{-\mu x+i\mu\xi}\int_{-\infty}^{+\infty} e^{-t^2+i\mu t}dt+\text{c.c.}\right\}. \tag{C5}$$

Since

$$\int_{-\infty}^{+\infty} e^{-t^2+i\mu t}dt=\pi^{1/2}e^{-\frac{1}{4}\mu^2}, \tag{C6}$$

we obtain

$$I=\pi^{1/2}(2x)^{-1}\int_0^\infty d\mu\, \exp(-\tfrac{1}{4}\mu^2-\mu x)(e^{i\mu\xi}+e^{-i\mu\xi})\,. \tag{C7}$$

With the plasma dispersion function defined as

$$Z(\zeta)=i\int_0^\infty \exp(-\tfrac{1}{4}\mu^2-\zeta\mu)d\mu\,, \tag{C8}$$

we see that

$$I=\pi^{1/2}(2ix)^{-1}\{Z(x+i\xi)+Z(x-i\xi)\}$$
$$=\pi^{1/2}x^{-1}Z_i(x+i\xi)\,. \tag{C9}$$

If $\omega=\nu$, we have $\xi=0$ and the integral I reduces to

$$I=\pi^{1/2}x^{-1}\int_0^\infty d\mu\,\exp(-\tfrac{1}{4}\mu^2-\mu x)$$

$$=2\pi^{1/2}x^{-1}e^{x^2}\int_x^\infty e^{-t^2}dt\,. \tag{C10}$$

The last integral is defined as the error function[17]

$$\text{erfc}(x)=2\pi^{-1/2}\int_x^\infty e^{-t^2}dt\,, \tag{C11}$$

so that

$$I=\pi x^{-1}e^{x^2}\text{erfc}(x)\,. \tag{C12}$$

APPENDIX D: DERIVATION OF AREA THEOREM (183)

We here derive the area theorem

$$d\theta/dz=-\kappa\theta+\alpha\sin\theta\,, \tag{D1}$$

where

$$\theta(z)=\frac{p}{\hbar}\int_{-\infty}^{+\infty}\mathcal{E}(z,\tau')d\tau'\,, \tag{D2}$$

κ is given by Eq. (45) and α is determined below. Starting with Eqs. (180),

$$\partial\mathcal{E}/\partial z=-\kappa\mathcal{E}-\tfrac{1}{2}K\langle S/\epsilon_0\rangle_v\,, \tag{D3}$$

$$\partial S/\partial\tau=-KvC-(p^2/\hbar)\mathcal{E}N\,, \tag{D4}$$

$$\partial C/\partial\tau=KvS\,, \tag{D5}$$

$$\partial N/\partial\tau=\mathcal{E}S/\hbar\,, \tag{D6}$$

we integrate (D3) with respect to τ and use (D2) to find

$$d\theta/dz=-\kappa\theta-\tfrac{1}{2}K[p/(\epsilon_0\hbar)]\left\langle\int_{-\infty}^{+\infty} S(v,z,\tau')d\tau'\right\rangle_v. \tag{D7}$$

From Eq. (D5)

$$S=(Kv)^{-1}\partial C/\partial\tau\,, \tag{D8}$$

hence,

$$d\theta/dz=-\kappa\theta-\tfrac{1}{2}(p/\epsilon_0\hbar)\lim_{\tau\to\infty}\langle C(v,z,\tau)/v\rangle_v\,. \tag{D9}$$

The field $\mathcal{E}(z,\tau)$ is supposed to vanish for very large τ, so let τ_0 be such that

$$\mathcal{E}(z,\tau)=0 \quad\text{for}\quad \tau>\tau_0\,, \tag{D10}$$

then ( ) and ( ) can be integrated to give

$$C(v,z,\tau)=C(v,z,\tau_0)\cos Kv(\tau-\tau_0)+S(v,z,\tau_0)$$
$$\times\sin Kv(\tau-\tau_0)\quad\text{for}\ \tau>\tau_0\,. \tag{D11}$$

We have therefore

$$d\theta/dz=-\kappa\theta-\tfrac{1}{2}[p/(\epsilon_0\hbar)]\lim_{\tau\to\infty}\langle C(v,z,\tau_0)v^{-1}$$
$$\times\cos Kv(\tau-\tau_0)+S(v,z,\tau_0)v^{-1}\sin Kv(\tau-\tau_0)\rangle_v\,. \tag{D12}$$

We may interchange the limiting and averaging processes, and using

$$\lim_{\tau\to\infty} v^{-1}\cos Kv(\tau-\tau_0)=\lim_{\tau\to\infty} v^{-1}\sin Kv(\tau-\tau_0)=\pi\delta(v) \tag{D13}$$

obtain

$$d\theta/dz=-\kappa\theta-\tfrac{1}{2}\pi(p/\epsilon_0\hbar)$$
$$\times\langle\{C(v,z,\tau_0)+S(v,z,\tau_0)\}\delta(v)\rangle_v\,. \tag{D14}$$

---

[17] *Handbook of Mathematical Functions*, edited by M. Abramowitz and I. A. Stegun (Dover Publications, Inc., New York, 1965).

From Sec. V A we know that

$$C(0,z,\tau_0)=0, \tag{D15}$$

$$S(0,z,\tau_0)=-pN_0\sin\psi(z,\tau_0), \tag{D16}$$

hence, if

$$\langle\cdots\rangle_v=\int W(v)(\cdots)dv, \tag{D17}$$

we have

$$d\theta/dz=-\kappa\theta+\alpha\sin\psi(z,\tau_0)=-\kappa\theta+\alpha\sin\theta(z), \tag{D18}$$

with

$$\alpha=\tfrac{1}{2}\pi(p^2N_0/\epsilon_0\hbar)W(0). \tag{D19}$$

APPENDIX E: HYPERBOLIC SECANT SOLUTION FOR NONLINEAR ABSORBER

We seek a solution of (180), for an attenuator, in the form of a pulse traveling with unchanging shape

$$\mathcal{E}(z,\tau)=\mathcal{E}(t-z/v)=\mathcal{E}(\bar{\tau}). \tag{E1}$$

Neglecting the loss term κ, (180a) becomes

$$(c^{-1}-v^{-1})d\mathcal{E}/d\bar{\tau}=-\tfrac{1}{2}K\langle S/\epsilon_0\rangle_v. \tag{E2}$$

The result for the case of homogeneous broadening suggests that we set

$$S(v,z,\tau)=S(0,z,\tau)f(v)=f(v)p|N_0|\sin\psi, \tag{E3}$$

so that

$$(p/\hbar)d\mathcal{E}/d\bar{\tau}=g(v^{-1}-c^{-1})^{-1}\langle f(v)\rangle\sin\psi, \tag{E4}$$

where

$$g=\tfrac{1}{2}Kp^2|N_0|/\epsilon_0\hbar. \tag{E5}$$

Multiply (E4) by $p\mathcal{E}/\hbar=\dot{\psi}$, integrate and obtain

$$(p\mathcal{E}/\hbar)^2=4(\Delta\tau)^{-2}\sin^2\tfrac{1}{2}\psi, \tag{E6}$$

or

$$p\mathcal{E}/\hbar=\dot{\psi}=2(\Delta\tau)^{-1}\sin\tfrac{1}{2}\psi \tag{E7}$$

with

$$\Delta\tau=[(v^{-1}-c^{-1})/g\langle f(v)\rangle]^{1/2}. \tag{E8}$$

Equation (E7) is readily integrated and gives

$$(p/\hbar)\mathcal{E}(\bar{\tau})=2(\Delta\tau)^{-1}\operatorname{sech}(\bar{\tau}/\Delta\tau). \tag{E9}$$

We now determine $f(v)$ by requiring that the remaining equations in (180) be satisfied. The third equation gives

$$\partial C/\partial\tau=\dot{\psi}\partial C/\partial\psi=KvS$$
$$=2(\Delta\tau)^{-1}\sin\tfrac{1}{2}\psi\,\partial C/\partial\psi=p|N_0|Kvf(v)\sin\psi, \tag{E10}$$

hence

$$\partial C/\partial\psi=p|N_0|Kv\Delta\tau f(v)\cos\tfrac{1}{2}\psi, \tag{E11}$$

which integrates to

$$C=2p|N_0|Kv\Delta\tau f(v)\sin(\tfrac{1}{2}\psi). \tag{E12}$$

The fourth equation gives

$$\partial N/\partial\tau=\dot{\psi}\partial N/\partial\psi=\mathcal{E}S/\hbar=\dot{\psi}S/p, \tag{E13}$$

hence

$$\partial N/\partial\psi=|N_0|f(v)\sin\psi, \tag{E14}$$

so that

$$N=|N_0|[2f(v)\sin^2(\tfrac{1}{2}\psi)-1], \tag{E15}$$

where the constant of integration has been chosen to satisfy the initial condition $N=-|N_0|$ and $\psi=0$ at $\tau=-\infty$. Substituting S, C, and N into the conservation relation

$$p^{-2}(S^2+C^2)+N^2=N_0^2, \tag{E16}$$

we find

$$f(v)[\sin^2\psi+4(Kv\Delta\tau)^2\sin^2(\tfrac{1}{2}\psi)+4\sin^4(\tfrac{1}{2}\psi)]$$
$$=4\sin^2(\tfrac{1}{2}\psi), \tag{E17}$$

which is readily solved for $f(v)$

$$f(v)=[1+(Kv\Delta\tau)^2]^{-1}. \tag{E18}$$

The average of this function is

$$\langle f(v)\rangle=\pi^{-1/2}\int_{-\infty}^{+\infty}e^{-v^2/u^2}[1+(Kv\Delta\tau)^2]^{-1}d(v/u)$$
$$=\pi^{1/2}(Ku\Delta\tau)^{-1}\exp[(Ku\Delta\tau)^{-2}]$$
$$\times\operatorname{erfc}\{(Ku\Delta\tau)^{-1}\}, \tag{E19}$$

so that (E8) gives the following velocity-width relation

$$v=c\{1+cg(\Delta\tau)^2\pi^{1/2}(Ku\Delta\tau)^{-1}e^{(Ku\Delta\tau)^{-2}}$$
$$\times\operatorname{erfc}[(Ku\Delta\tau)^{-1}]\}^{-1} \tag{E20}$$

as stated in Eq. (202).

APPENDIX F: STABILITY OF INTEGRATION PROCEDURE

We shall only prove the stability of the step-by-step integration of the polarization and populations along the τ axis. The proof is similar and simpler for the integration of the electric field along the z axis. Since the equations are integrated separately for every velocity, it is sufficient to consider a single representative velocity which may be taken to be $v=0$ without loss of generality. Decomposing $\mathcal{E}$ and $\mathcal{P}$ into their real and imaginary parts

$$\mathcal{E}=\mathcal{C}+i\mathcal{S}, \tag{F1}$$

$$\mathcal{P}=C+iS, \tag{F2}$$

Eqs. (226b), (226c), and (226d) may be written as

$$\partial S/\partial\tau=-\gamma_{ab}S+\mathcal{C}(\rho_{aa}-\rho_{bb}), \tag{F3}$$

$$\partial C/\partial\tau=-\gamma_{ab}C-\mathcal{S}(\rho_{aa}-\rho_{bb}), \tag{F4}$$

$$\partial\rho_{aa}/\partial\tau=(\Lambda_a/N_0)-\gamma_a\rho_{aa}-\tfrac{1}{2}(\mathcal{C}S-\mathcal{S}C), \tag{F5}$$

$$\partial\rho_{bb}/\partial\tau=(\Lambda_b/N_0)-\gamma_b\rho_{bb}+\tfrac{1}{2}(\mathcal{C}S-\mathcal{S}C). \tag{F6}$$

The above equations can formally be written as

$$\dot{Y}^i = f^i(Y^j), \tag{F7}$$

where $Y^i(i=1,2,3,4)$ stands for S, C, ρ_{aa}, ρ_{bb}. If y^i is the approximate solution as opposed to the exact solution Y^i, we define the error

$$\epsilon^i{}_n = Y^i{}_n - y^i{}_n, \tag{F8}$$

where n designates the mesh point.

By definition of the approximate solution we have

$$y^i{}_{n+1} = y^i{}_n + \tfrac{1}{2}h(\dot{y}^i{}_{n+1} + \dot{y}^i{}_n) + R^i{}_n, \tag{F9}$$

where

$$\dot{y}^i{}_n = f^i(y^i{}_n), \tag{F10}$$

and $R^i{}_n$ is the round-off error. On the other hand,

$$Y^i{}_{n+1} = Y^i{}_n + \tfrac{1}{2}h(\dot{Y}^i{}_{n+1} + \dot{Y}^i{}_n) + T^i{}_n, \tag{F11}$$

where $T^i{}_n$ is the truncation error. Subtracting (F9) from (F11) we find

$$\epsilon^i{}_{n+1} = \epsilon^i{}_n + \tfrac{1}{2}h(\dot{\epsilon}^i{}_{n+1} + \dot{\epsilon}^i{}_n) + E^i{}_n, \tag{F12}$$

where

$$\dot{\epsilon}^i{}_n = \dot{Y}^i{}_n - \dot{y}^i{}_n, \tag{F13}$$

and

$$E^i{}_n = T^i{}_n - R^i{}_n. \tag{F14}$$

Equation (F13) can be written as

$$\dot{\epsilon}^i{}_n = f^i(Y^j{}_n) - f^i(y^j{}_n) = \epsilon^j{}_n(\partial f^i/\partial y^j)_n, \tag{F15}$$

where we used the summation convention. Let

$$f^i{}_j = \partial f^i/\partial y^j, \tag{F16}$$

then

$$\epsilon^i{}_{n+1} = \epsilon^i{}_n + \tfrac{1}{2}h(\epsilon^j{}_{n+1} + \epsilon^j{}_n) f^i{}_j + E^i{}_n, \tag{F17}$$

or

$$\epsilon^j{}_{n+1}(\delta^i{}_j - \tfrac{1}{2}hf^i{}_j) = \epsilon^j{}_n(\delta^i{}_j + \tfrac{1}{2}hf^i{}_j) + E^i{}_n. \tag{F18}$$

Using the matrix notation

$$\delta^i{}_j \pm \tfrac{1}{2}hf^i{}_j = \mathbf{1} \pm \tfrac{1}{2}h\mathbf{f}, \tag{F19}$$

Eq. (F18) is solved for $\boldsymbol{\epsilon}_{n+1}$

$$\boldsymbol{\epsilon}_{n+1} = (\mathbf{1} - \tfrac{1}{2}h\mathbf{f})^{-1}(\mathbf{1} + \tfrac{1}{2}h\mathbf{f})\boldsymbol{\epsilon}_n + \mathbf{E}'_n \tag{F20}$$

with

$$\mathbf{E}'_n = (\mathbf{1} - \tfrac{1}{2}h\mathbf{f})^{-1}\mathbf{E}_n. \tag{F21}$$

If $\mathbf{f}$ is treated as constant in n (this idealization is not necessary but greatly simplifies the proof), and if λ^j are the eigenvalues of $\mathbf{f}$, one can find four linear combinations η^j of the ϵ^i's, such that

$$\eta^j{}_{n+1} = (1 - \tfrac{1}{2}h\lambda^j)^{-1}(1 + \tfrac{1}{2}h\lambda^j)\eta^j{}_n + E^{j\prime\prime}{}_n, \tag{F22}$$

where $E^{j\prime\prime}{}_n$ are the same linear combination of the $E^{i\prime}{}_n$ (we shall omit the $''$ and consider $E^j{}_n$ to be the same at every step n). The solution of the difference Eq. (F22) is

$$\eta^j{}_n = C(\rho^j)^n + E^j[1 - (\rho^j)^n][1 - \rho^j]^{-1}, \tag{F23}$$

where C is a constant and ρ^j is given by

$$\rho^j = (1 - \tfrac{1}{2}h\lambda^j)^{-1}(1 + \tfrac{1}{2}h\lambda^j). \tag{F24}$$

If for all j, $|\rho^j| < 1$, i.e., if $\mathrm{Re}\lambda^j < 0$

$$\lim_{n\to\infty} \eta^j{}_n = E^j(1 - \rho^j)^{-1} < +\infty, \tag{F25}$$

therefore $\lim_{n\to\infty} \epsilon^j{}_n$ is also finite for all j since the ϵ's are linear combination of the η's. On the other hand, if $|\rho^j| > 1$ for some j, then at least one $\epsilon^j{}_n$ diverges as $n \to \infty$. Hence, the requirement for stability is

$$|\rho^j| < 1 \quad \text{or} \quad \mathrm{Re}\lambda^j < 0 \quad \text{for all } j. \tag{F26}$$

The limiting errors $E^j(1-\rho^j)^{-1}$ can be made as small as one wishes by choosing h small enough, since E^j is of order h^3.

Now we check that $\mathrm{Re}\lambda^j < 0$. For the system of Eqs. (F3)–(F6), the matrix $f^i{}_j$ is by definition [see Eq. (F16)]

$$f^i{}_j = \begin{bmatrix} -\gamma_{ab} & 0 & \mathcal{C} & -\mathcal{C} \\ 0 & -\gamma_{ab} & -\mathcal{S} & \mathcal{S} \\ -\tfrac{1}{2}\mathcal{C} & \tfrac{1}{2}\mathcal{S} & -\gamma_a & 0 \\ \tfrac{1}{2}\mathcal{C} & -\tfrac{1}{2}\mathcal{S} & 0 & -\gamma_b \end{bmatrix} \tag{F27}$$

and the characteristic equation is

$$(\gamma_{ab} + \lambda)^2[\mathcal{E}^2 + (\gamma_a + \lambda)(\gamma_b + \lambda)] = 0, \tag{F28}$$

where we used $\gamma_{ab} = \tfrac{1}{2}(\gamma_a + \gamma_b)$ for simplicity, and

$$\mathcal{E}^2 = \mathcal{C}^2 + \mathcal{S}^2. \tag{F29}$$

The four roots

$$\lambda_1 = \lambda_2 = -\gamma_{ab}, \tag{F30}$$

$$\lambda_{3,4} = -\gamma_{ab} \pm [\tfrac{1}{4}(\gamma_a - \gamma_b)^2 - \mathcal{E}^2]^{1/2} \tag{F31}$$

are such that $\mathrm{Re}\lambda^j < 0$, and the stability is proven. The fact that $f^i{}_j$ is not constant in n (since the field components $\mathcal{C}$ and $\mathcal{S}$ depend on n) does not change the conclusion. In a more rigorous treatment the power law solution $\eta^j{}_{n+1} \sim (\rho^j)^n$ would be replaced by a product of the form

$$\eta^j{}_n \sim \prod_1^n (\rho^j{}_l) \tag{F32}$$

which also converges to zero as $n \to \infty$, since every factor is less than unity.

APPENDIX G: FORTRAN IV PROGRAM

In this appendix we describe the digital computer program we use to integrate Eqs. (225). The essential steps of the program are shown in flow chart form.

The following notation is used in flow charts

$$\tau_n = nHT, \quad n = -1, 0, 1, 2, \cdots, M \tag{G1}$$

$$z_j = jHS, \quad j = 0, 1, 2 \tag{G2}$$

$$v_k = x_k u, \quad k = 1, 2, 3, \cdots, IS \tag{G3}$$

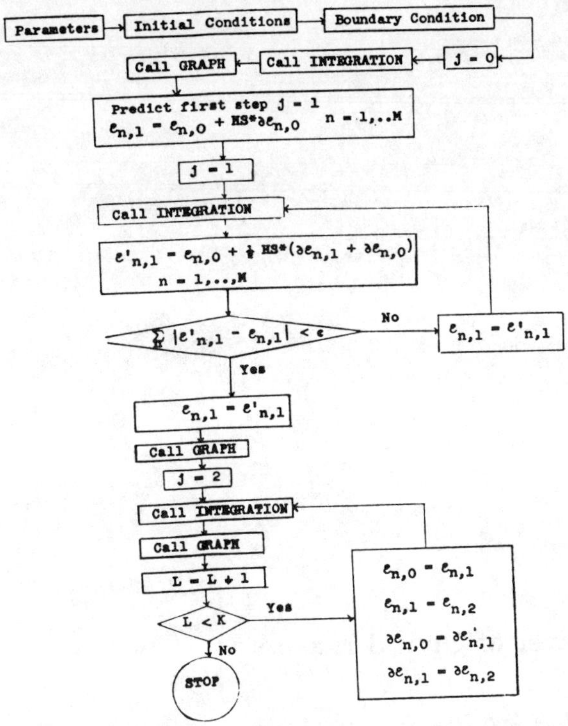

FIG. 35. Flow chart for the MAIN routine. The notation used is explained in Appendix G.

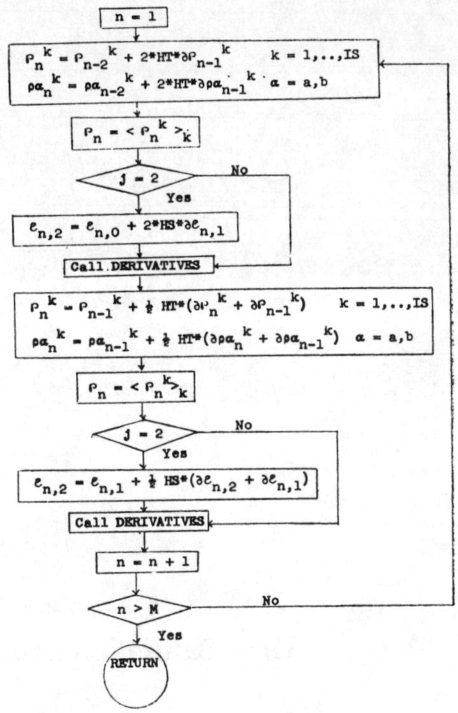

FIG. 36. Flow chart for subroutine INTEGRATION. The notation used is explained in Appendix G.

where HT is the time step size, HS the space step size, M the desired number of time steps in the integration, IS the number of velocities and x_k the zeros of the ISth-order Hermite polynomial,

$$\mathcal{E}_{nj}=\mathcal{E}(\tau_n,z_j)\,, \tag{G4}$$

$$\partial\mathcal{E}_{nj}=\partial\mathcal{E}(\tau_n,z_j)/\partial z\,, \tag{G5}$$

$$\mathcal{P}_{nj}{}^k=\mathcal{P}(v_k,\tau_n,z_j)\,, \tag{G6}$$

$$\partial\mathcal{P}_{nj}{}^k=\partial\mathcal{P}(v_k,\tau_n,z_j)/\partial\tau\,, \tag{G7}$$

$$p\alpha_{nj}{}^k=\rho_{\alpha\alpha}(v_k,\tau_n,z_j)\,, \tag{G8}$$

$$\partial p\alpha_{nj}{}^k=\partial\rho_{\alpha\alpha}(v_k,\tau_n,z_j)/\partial\tau\,, \tag{G9}$$

$$\mathcal{P}_{nj}=\mathcal{P}(\tau_n,z_j)=\langle\mathcal{P}_{nj}{}^k\rangle_k\,. \tag{G10}$$

The input can be classified as follows:

Parameters contain all independent parameters described in Sec. VI-4, the integrating velocities and their weights (see VI-3).

Initial conditions are common to all applications and express the fact that the pulse hits the entry plane ($z=0$) at time $t=\tau=0$ and until then the medium is at rest with zero polarization and given values of the populations. These conditions are built into the program.

Boundary condition the array $\mathcal{E}_{n,0}(n=1,2,\cdots,M)$ represents the incoming pulse and must be read in.

In addition to the MAIN routine, the program includes three subroutines. Subroutine INTEGRATION contains the predictor-corrector formula and is the core of the program. Subroutine DERIVATIVES contains the equations to be integrated, and subroutine GRAPH is devised to give output in the desired format. Flow charts of MAIN and INTEGRATION are seen in Fig. 35 and Fig. 36.

Gain Saturation and Output Power of Optical Masers

W. W. RIGROD
Bell Telephone Laboratories, Incorporated, Murray Hill, New Jersey
(Received 5 April 1963; in final form 8 May 1963)

The nonlinear gain characteristics of optical maser amplifiers at high beam intensities, and the optimum cavity coupling of maser oscillators for maximum output power, are computed for maser media with homogeneous and inhomogeneous line broadening. An approximate expression is derived for the power output of a gas maser oscillating simultaneously at many longitudinal cavity resonances, based on the assumption that the gain saturates independently at each frequency. In each case, the decrease of maser gain with radiation intensity involves an empiric constant, or saturation parameter, which is characteristic of the active medium.

Power and gain measurements at 1.15 μ on three He–Ne maser tubes of different diameter, in a cavity 1.75 m long, are found to satisfy the derived multifrequency power expression, and permit evaluation of the gain–saturation parameter for this gas mixture. The power expression, derived for a single transverse mode, is unexpectedly found to hold for multimode oscillations as well, within the range of measurements. From the measured saturation parameter and the derived expressions, the performance of amplifiers and other oscillators with the same active medium can be predicted.

I. INTRODUCTION

GAIN saturation as a function of steady-state radiation intensity has been analyzed for maser media with homogeneous and inhomogeneous line broadening.[1,2] In the former case, when the gain decreases proportionately over the entire transition line, the gain coefficient g decreases with radiation intensity w according to the relation[1]

$$g=g_0(1+w/w_0)^{-1}, \tag{1}$$

where g_0 is the unsaturated gain coefficient, and w_0 is a saturation parameter, equal to the power density at which the gain falls to $g_0/2$. An example of inhomogeneous line broadening which has been treated analytically is that of the low-pressure gas discharge maser, in which the Doppler-broadened line is much wider than the natural line, and the collision frequency is small compared to the spontaneous decay rates of the excited states.[3,4] In this case, gain saturation for a plane wave of monochromatic radiation is approximately described by[2]

$$g=g_0(1+w/w_0)^{-\frac{1}{2}}, \tag{2}$$

where the symbols have the same meaning as previously, except that here w_0 is the power density at which $g=g_0/\sqrt{2}$. The inhomogeneous line saturates more slowly, principally because its width increases as the beam intensity increases.[3]

Each of these expressions are used below, for appropriate maser media, to study the nonlinear gain characteristics of maser amplifiers at high beam intensities, and to evaluate the optimum cavity coupling of maser oscillators for maximum output power.

Gas–optical masers usually oscillate simultaneously

[1] J. S. Wright and E. O. Schulz du Bois, Solid-State Maser Research, Report No. 5, Contract No. DA-36-039-sc-85357, 20 September 1961 (ASTIA No. Ad 265838).
[2] J. P. Gordon (private communication).

[3] W. R. Bennett, Jr., Phys. Rev. **126**, 580 (1962).
[4] W. R. Bennett, Jr., Appl. Opt. Suppl. **1**, 24 (1962).

Reprinted with permission from *J. Appl. Phys.*, vol. 34, pp. 2602–2609, Sept. 1963.

at a number of longitudinal cavity resonances, in one or more transverse modes. An approximate expression is derived for the multifrequency power-density output of such an oscillator, based on the assumption that the Lorentzian "holes" burnt in the gain curve at each frequency[3] do not overlap appreciably, i.e., that the gain saturates independently at each frequency.

Correlated measurements of the power and unsaturated gain of a number of helium–neon masers at 1.15 μ wavelength, with a cavity about 1.75 m long, have been made in order to test the derived power expression and to evaluate the saturation parameter of the gas mixture for this transition. The measurements are also used to test another formulation,[5] derived for gas masers in which the mode spacing is less than the linewidth, in view of the fact that the precise value of the latter quantity is not known.

II. NET GAIN OF MASER AMPLIFIERS

In a maser amplifier, a beam of radiation flux is directed through the active medium over a long path, with suitable means for avoiding optical feedback. If the intensity w is substantially uniform over the beam area, and the various losses due to scattering and absorption can be lumped into an average absorption coefficient a along the optical path z, the net amplifier gain for an optical path length L can be readily computed from the gain-saturation expressions (1) and (2).

For a medium with *homogeneous* line broadening, (1) applies, and the net gain coefficient is given by

$$g_0(1+w/w_0)^{-1}-a=1/w(dw/dz). \tag{3}$$

With $\beta=w/w_0$, this equation can be written

$$g_0dz=(1+\beta)d\beta/\beta[1-a/g_0(1+\beta)] \tag{4}$$

which, when integrated over a path L, yields

$$(g_0-a)L=\ln(\beta_2/\beta_1)-\frac{g_0}{a}\ln\left[\frac{g_0-a(1+\beta_2)}{g_0-a(1+\beta_1)}\right], \tag{5}$$

where β_1 and β_2 are the normalized input and output power densities, respectively. When the loss parameter $a/g_0=0$, (5) reduces to the simpler form:

$$g_0L=\ln(\beta_2/\beta_1)+\beta_2-\beta_1. \tag{6}$$

These expressions were first derived by Wright and Schulz-DuBois[1] in a study of ruby maser amplifiers. They are plotted in Fig. 1, with gain in decibels along the abscissa, for several values of the normalized input power β_1 and the loss parameter a/g_0.

At small signal levels, $\beta_2-\beta_1 \ll 1$, the first term of (6) is dominant, and the dB gain is approximately proportional to g_0L:

$$10\log_{10}(\beta_2/\beta_1)\approx 4.343 g_0L. \tag{7}$$

[5] A. D. White, E. I. Gordon, and J. D. Rigden, Appl. Phys. Letters 2, 91 (1963).

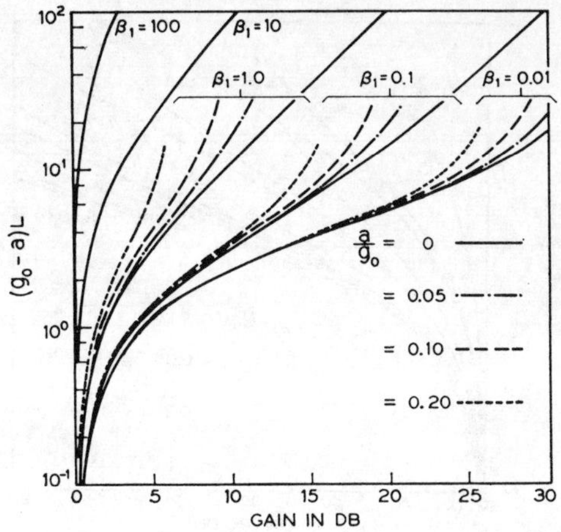

FIG. 1. The ordinates indicate the optical path length L needed to obtain a specified over-all gain in a homogeneous-line maser amplifier with unsaturated gain coefficient g_0, for various levels of input power density $\beta_1=w/w_0$ and the distributed loss parameter a/g_0.

At high signal levels ($\beta_1 \gg 1$), however, the net amplifier gain increases more slowly:

$$g_0L\approx(\beta_2-\beta_1)(1+1/\beta_1), \tag{8}$$

and above about 10 dB, the dB gain increases logarithmically with g_0L.

When the loss parameter a/g_0 is zero, the gain coefficient decreases with increasing beam intensity to a small but finite value. In the presence of scattering or absorption loss, however, the net gain coefficient (g-a) falls to zero at a beam intensity given by

$$(\beta_2)_{\max}=g_0/a-1, \tag{9}$$

and the over-all gain no longer increases with amplifier length beyond this point.

As noted by Bennett,[3] the gas-discharge maser is characterized by homogeneous line broadening at higher pressures, and whenever the Doppler width is not large compared with the natural linewidth. In the low-pressure discharge at optical wavelengths, the line broadening is usually *inhomogeneous*. From (2), the net gain coefficient of a long-path gas-maser amplifier, with the same assumptions as above, would then be given by

$$g_0(1+w/w_0)^{-\frac{1}{2}}-a=1/w(dw/dz) \tag{10}$$

which, when integrated over an optical path L, yields

$$[1-(a/g_0)^2]g_0L=\left\{\ln\left[\frac{(1+\beta)^{\frac{1}{2}}-1}{(1+\beta)^{\frac{1}{2}}+1}\right]+\frac{a}{g_0}\ln\beta-\frac{2g_0}{a}\ln\left[1-\frac{a}{g_0}(1+\beta)^{\frac{1}{2}}\right]\right\}\Bigg|_{\beta_1}^{\beta_2}, \tag{11}$$

where $\beta=w/w_0$ as before. When the loss parameter a/g_0 is zero, this reduces to

$$g_0L=\left\{\ln\left[\frac{(1+\beta)^{\frac{1}{2}}-1}{(1+\beta)^{\frac{1}{2}}+1}\right]+2(1+\beta)^{\frac{1}{2}}\right\}\Bigg|_{\beta_1}^{\beta_2}. \tag{12}$$

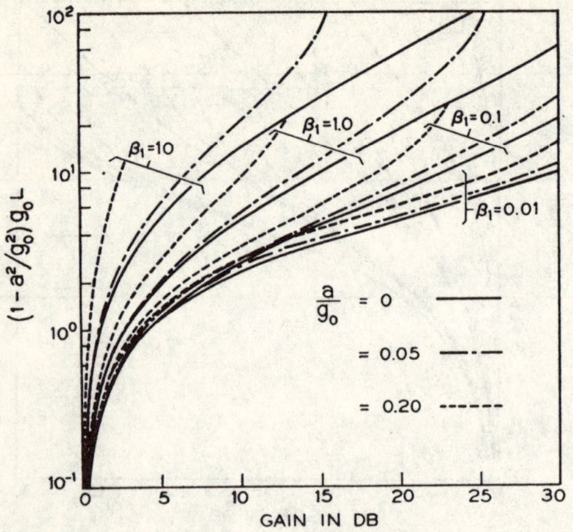

FIG. 2. The curves have the same meaning as in Fig. 1, except that here they apply to a gas-maser amplifier with inhomogeneous line broadening.

The net small-signal gain exponent $(1-a^2/g_0^2)g_0L$ is plotted in Fig. 2 as a function of the dB gain, for several values of a/g_0 and β_1.

At small signal levels, $\beta_2-\beta_1 \ll 1$, the net amplifier gain in dB is proportional to g_0L, exactly as in (6) and (7) for a homogeneous-line maser. At high levels of beam intensity, however,

$$g_0L \approx 2[(1+\beta_2)^{\frac{1}{2}}-(1+\beta_1)^{\frac{1}{2}}], \tag{13}$$

and the dB gain increases only logarithmically with g_0L, although relatively faster than for homogeneous broadening. When distributed loss is present, the maximum possible beam intensity in the amplifier, corresponding to zero gain coefficient in (10), is given by

$$(\beta_2)_{\max} = (g_0/a)^2-1. \tag{14}$$

III. OPTIMUM OUTPUT COUPLING OF MASER OSCILLATORS

Three kinds of maser oscillators are treated in this section: (a) those with a homogeneously broadened line, in which the same group of atoms contribute to gain at all frequencies over the line, (b) single-frequency gas maser oscillators, with inhomogeneously broadened lines, and (c) multifrequency gas maser oscillators. In each case an expression, is derived for the output radiation intensity of the oscillator in terms of its unsaturated gain per pass, loss fraction, and output window transmittances. The optimum transmission fraction for maximum output power can then be readily computed.

A. Homogeneous-Line Masers

In a maser oscillator with homogeneous broadening, both the forward and the backward waves in the optical resonator are amplified by interaction with the *same* group of atoms. Thus, when the power levels in both directions are nearly equal, and the standing wave in the resonator is replaced by an average field intensity, the normalized one-way power density is approximately

$$\beta/2 = w/2w_0, \tag{15}$$

where w_0 is the saturation parameter of the homogeneous line. The condition for steady-state oscillations is

$$(\beta_2/\beta_1)(1-a-t)=1, \tag{16}$$

where (β_2/β_1) is the gain factor, assumed to be the same in both directions, a is the absorption fraction *per pass*, and t the transmission fraction, equal to the average transmittance of both mirrors. For small saturated gain per pass, $(a+t) \ll 1$, this condition reduces to

$$\beta_2/\beta_1 = 1+(\Delta\beta/\beta) \cong 1+a+t. \tag{17}$$

From (3), it follows that oscillations stabilize at an internal power density level w such that

$$gL = g_0L/(1+\beta) \cong a+t. \tag{18}$$

If both mirrors have identical transmittances t, the output power density at each end of the resonator is

$$w_t = \tfrac{1}{2}tw. \tag{19}$$

With the aid of (18), this can be written

$$2w_t/w_0 = t\beta = t[g_0L/(a+t)-1]. \tag{20}$$

The output power has a maximum value at some optimum value of the transmission fraction t_{opt}, ob-

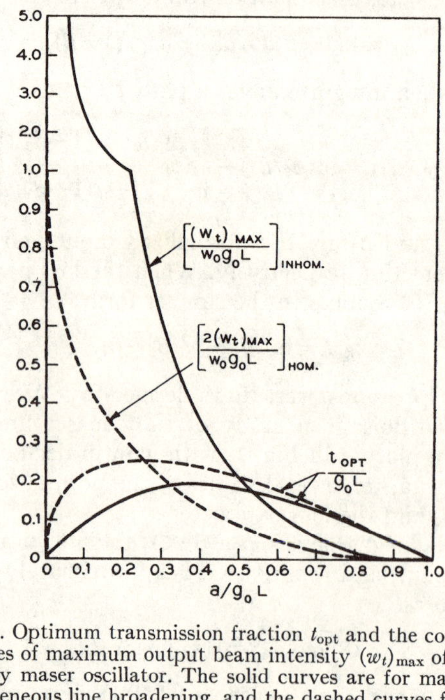

FIG. 3. Optimum transmission fraction t_{opt} and the corresponding values of maximum output beam intensity $(w_t)_{\max}$ of a single-frequency maser oscillator. The solid curves are for masers with inhomogeneous line broadening, and the dashed curves for homogeneous line broadening. (Note scale change at 1 on the ordinate.)

tained by setting dw_t/dt equal to zero:

$$\frac{2(w_t)_{\max}}{w_0}=\frac{t_{\text{opt}}^2}{a}=g_0L\left[1-\left(\frac{a}{g_0L}\right)^{\frac{1}{2}}\right]^2, \tag{21}$$

$$t_{\text{opt}}/g_0L=(a/g_0L)^{\frac{1}{2}}-(a/g_0L). \qquad (22)^6$$

These equations have been plotted as dashed curves in Fig. 3. The optimum cavity coupling is seen to depend on both the loss fraction and the unsaturated gain per pass.

The factor by which the maser output can be increased by optimizing the transmission fraction is

$$\frac{(w_t)_{\max}}{w_{t_0}}=\frac{[1-(a/g_0L)^{\frac{1}{2}}]^2}{[t_0/(a+t_0)-t_0/(g_0L)]}, \tag{23}$$

which is of the order of $(a+t_0)/t_0$ for very large gain fractions.

For a resonator with mirrors of unequal transmittances t_0 and t_n, the output power densities w_{t_0} and w_{t_n} at each end, respectively, are given by

$$w_{t_0}/t_0=w_{t_n}/t_n=\tfrac{1}{2}w_0[g_0L/(a+t)-1], \tag{24}$$

where

$$t=\tfrac{1}{2}(t_0+t_n). \tag{25}$$

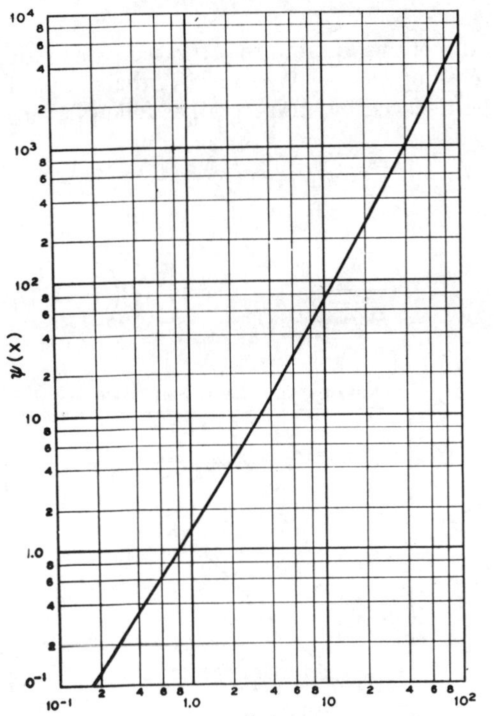

FIG. 4. The function $\Psi(X)$, where $X=g_0L/(a+t)$, the ratio of unsaturated to saturated gain in a multifrequency gas-maser oscillator at the center of the Doppler line.

[6] Equations (21) and (22) were independently derived by R. Kompfner and, in somewhat different notation, by A. Yariv, Proc. IEEE **51**, 4 (1963).

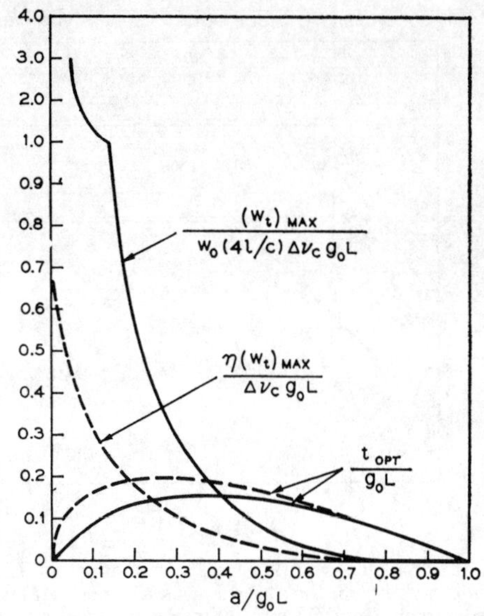

FIG. 5. Optimum transmission fraction t_{opt} and the maximum output beam intensity $(w_t)_{\max}$ of a multifrequency gas-discharge maser. The solid curves are based on $\Psi(X)$ of Eq. (42), and the dashed curves on $\Phi(X)$ of Eq. (48). (Note scale change at 1 on the ordinate.)

Thus, when $t=t_{\text{opt}}$ and $t_n \gg t_0$, the maximum output intensity via the high-transmittance mirror is nearly twice the value given by (21).

B. Single-Frequency Gas Maser Oscillators

In a gas maser oscillator, when the natural linewidth is much smaller than that of the Doppler line, the forward and backward waves for a single resonant frequency usually burns *two* holes in the Gaussian gain curve, which saturate independently.[2,4] Thus the normalized one-way power density of a uniform beam is

$$\beta=w/w_0, \tag{26}$$

where w_0 is the saturation parameter for a single hole, as defined by (2). The saturation parameter for the center frequency of the Doppler curve may differ from this value somewhat, owing to more pronounced dependence of the atom-field interaction on position.[2]

For the usual condition of small saturated gain per pass, the output power density w_t of a maser oscillator can be obtained from (2) and (17) directly:

$$\beta_2/\beta_1\cong 1+g_0L/(1+\beta)^{\frac{1}{2}} \tag{27}$$

$$w_t/w_0=t\beta=t\{[g_0L/(a+t)]^2-1\}. \tag{28}$$

This result is consistent with expression (14) for the maximum beam intensity in a maser amplifier with the same ratio of unsaturated gain to loss per unit length.

The optimum transmission fraction t_{opt} and the maximum output power density w_t, evaluated as before,

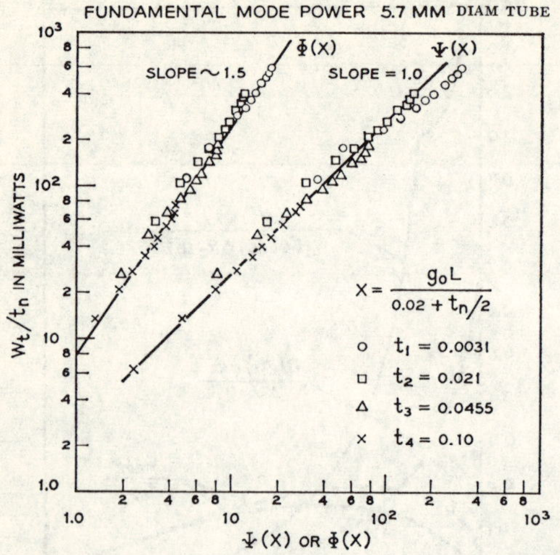

FIG. 6. Internal one-way power in the dominant TEM_{00} mode of a semiconcave resonator at 1.15 μ, as a function of $\Psi(X)$ and $\Phi(X)$, respectively, in a 5.7-mm-diam He–Ne maser tube.

are defined by

$$[(a+t)^3/(a-t)]_{t=t_{opt}}=(g_0L)^2, \tag{29}$$

$$(w_t)_{max}=w_0[2t^2/(a-t)]_{t=t_{opt}}. \tag{30}$$

Their solutions are plotted as solid curves in Fig. 3, normalized in the same way as for homogeneous-line masers. In both cases $(w_t)_{max}$ increases sharply as the loss parameter a/g_0L is reduced, but the effect is most pronounced for inhomogeneous-line masers. Solutions (29)–(30) break down for sufficiently small values of a/g_0L and t/g_0L, however, as the hole widths eventually become comparable to the Doppler linewidth, far above oscillation threshold.

C. Multifrequency Gas Masers

Even when a gas maser is restricted to a single transverse mode of oscillation, it may oscillate simultaneously at several different frequencies. This generally occurs when the frequency spacing of longitudinal cavity resonances $c/2l$ is small compared with the gain bandwidth above threshold, $2\Delta\nu_{max}$. The number of such resonances is approximately

$$N\approx(4l/c)\Delta\nu_{max}, \tag{31}$$

where l is the mirror separation and c the speed of light. Except for the center frequency ν_0, each frequency of oscillation burns two Lorentzian holes in the Gaussian gain curve,

$$g_0(\nu)=g_0(\nu_0)\exp-[(\nu-\nu_0)/\Delta\nu_c)]^2, \tag{32}$$

where

$$\Delta\nu_c=\nu_0/c(2kT/m)^{\frac{1}{2}}=\Delta\nu_D/2(\ln 2)^{\frac{1}{2}}. \tag{33}$$

Here k is Boltzmann's constant, T the absolute temperature, m the atomic mass, and $\Delta\nu_D$ the Doppler width.

When oscillations occur simultaneously at N longitudinal cavity resonances, there may be as many as $2N$ holes in the gain curve, depending on the mirror spacing. However, because of unpredictable nonlinear hole interactions, such as the hole annihilation and repulsion effects discussed by Bennett,[3,4] all that can safely be stated is that there are at least N holes, with a spacing no greater than about $c/2l$, and that the hole widths are greatest near the center of the Doppler line. In any case, the multifrequency output-beam intensity of a gas maser can be computed when it is assumed (1) that the holes do not overlap appreciably, and (2) that, whatever the hole distribution, the gain saturates independently at each frequency.

For small saturated gain per pass, the output power density at each resonant frequency ν_i is given by (28), in which $g_0(\nu_i)L$ is taken to be the same in both directions. From (32), the total output density can then be written

$$w_t=tw_0\sum_i^N\left\{X^2\exp\left[-2\left(\frac{\Delta\nu_i}{\Delta\nu_c}\right)^2\right]-1\right\}, \tag{34}$$

where

$$\Delta\nu_i=\nu_i-\nu_0 \tag{35}$$

and

$$X=g_0(\nu_0)L/(a+t), \tag{36}$$

the ratio of unsaturated to saturated gain at the center of the Doppler line. When $N\gg 1$, the summation (34) can be replaced by an integral, as follows:

$$w_t\approx N[w_t(\nu)]_{av}=Ntw_0[X_{av}{}^2-1], \tag{37}$$

where

$$X_{av}{}^2=\frac{1}{\Delta\nu_{max}}\int_0^{\Delta\nu_{max}}X^2\exp\left[-2\left(\frac{\Delta\nu}{\Delta\nu_c}\right)^2\right]d\nu. \tag{38}$$

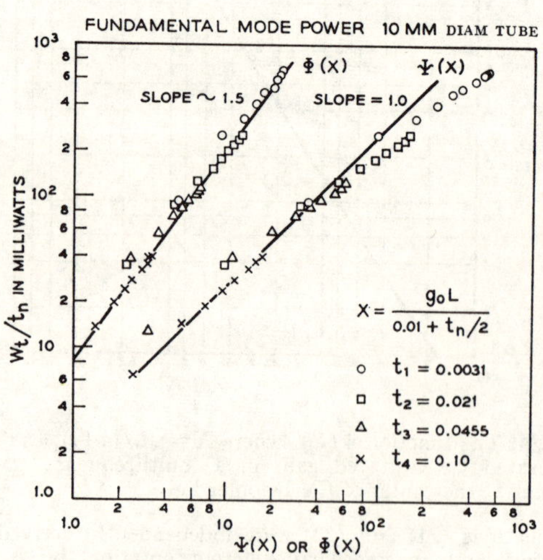

FIG. 7. Same as Fig. 6, for a 10-mm-diam tube.

The gain bandwidth is obtained from (32) and the oscillation-threshold condition,

$$g_0(\nu)L=a+t, \tag{39}$$

yielding

$$\Delta\nu_{max}=\Delta\nu_c(\ln X)^{1/2}, \tag{40}$$

and

$$X_{av}^2=X^2\left[\frac{[(\pi)^{1/2}/2]\,\mathrm{erf}(2\ln X)^{1/2}}{(2\ln X)^{1/2}}\right]. \tag{41}$$

The total output power density (37) can then be written in the form

$$w_t=t[w_0(4l/c)\Delta\nu_c]\Psi(X), \tag{42}$$

where

$$\Psi(X)=\tfrac{1}{2}(\pi/2)^{1/2}X^2\,\mathrm{erf}(2\ln X)^{1/2}-(\ln X)^{1/2} \tag{43}$$

$$=(\ln X)^{1/2}(X_{av}^2-1). \tag{44}$$

The function $\Psi(X)$ is shown in Fig. 4, with $(X-1)$ as abscissa.

The equations defining the optimum transmission fraction t_{opt} and the corresponding value of $(w_t)_{max}$ are analogous to those for single-frequency masers with inhomogeneous broadening, (29) and (30):

$$(a+t)/(a-t)=X_{av}^2, \tag{45}$$

$$(w_t)_{max}=Nw_0[2t^2/(a-t)], \tag{46}$$

where $t\equiv t_{opt}$. The maximum attainable power density, normalized in the form

$$(w_t)_{max}/[w_0(4l/c)\Delta\nu_c g_0L]\equiv(t_{opt}/g_0L)\Psi_{max}(X) \tag{47}$$

and the quantity (t_{opt}/g_0L) have been plotted as solid curves in Fig. 5, as functions of a/g_0L.

An expression different from (42) has been proposed for the multifrequency output power density of a long gas maser, when the mode spacing is less than the linewidth, based on the assumption that the gain coefficient

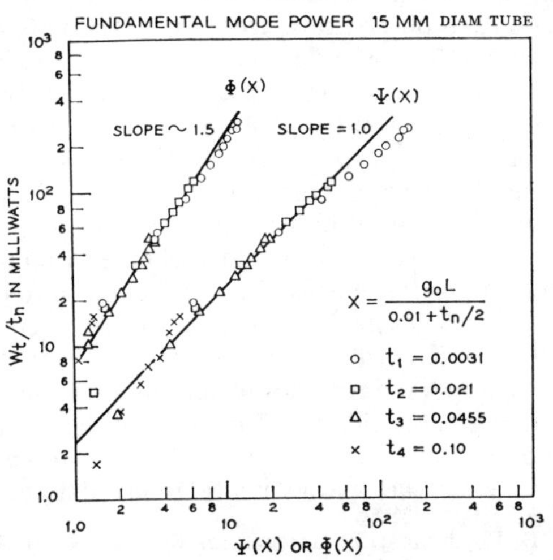

FIG. 8. Same as Fig. 6, for a 15-mm-diam tube.

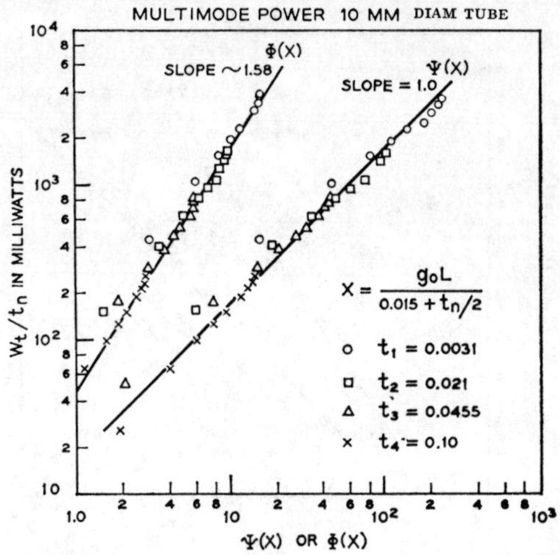

FIG. 9. Internal one-way power at 1.15 μ in multimode pattern of semiconcave resonator, as function of $\Psi(X)$ and $\Phi(X)$, respectively, in a 10-mm-diam tube.

at all frequencies within the Doppler line is uniformly saturated[5]:

$$w_t=t(\Delta\nu_c\cdot\eta^{-1})\Phi(X) \tag{48}$$

where

$$\Phi(X)=\tfrac{1}{2}(\pi)^{1/2}X\,\mathrm{erf}(\ln X)^{1/2}-(\ln X)^{1/2} \tag{49}$$

$$=(\ln X)^{1/2}(X_{av}-1), \tag{50}$$

and η^{-1} corresponds to the saturation parameter per unit bandwidth of a homogeneous line. For this formulation, the optimum transmission fraction and maximum power density are given by the equations:

$$\frac{a+t}{a}=X_{av}=\frac{(\pi)^{1/2}X\,\mathrm{erf}(\ln X)^{1/2}}{2(\ln X)^{1/2}}, \tag{51}$$

$$\eta(w_t)_{max}/\Delta\nu_c g_0L=\frac{t_{opt}^2(\ln X)^{1/2}}{a g_0L}, \tag{52}$$

whose solutions are plotted as dashed curves in Fig. 5.

IV. EXPERIMENTAL MEASUREMENTS

Measurements were undertaken on three He–Ne masers of different diameters with two objects in view: to determine which of the two multifrequency power formulations (42) or (48) was appropriate for the 1.15-μ transition, in a moderately long resonator, and to determine thereby the corresponding saturation parameter w_0 or η^{-1}, respectively.

The functions $\Psi(X)$ and $\Phi(X)$ are proportional to w_t/t, the one-way power density in a maser. For any fixed transverse mode of oscillation, therefore, they should also be proportional to the *total* output power divided by the transmittance of the output window, W_{t_n}/t_n, assuming X to vary with excitation in the same way over the beam area. It should be possible, there-

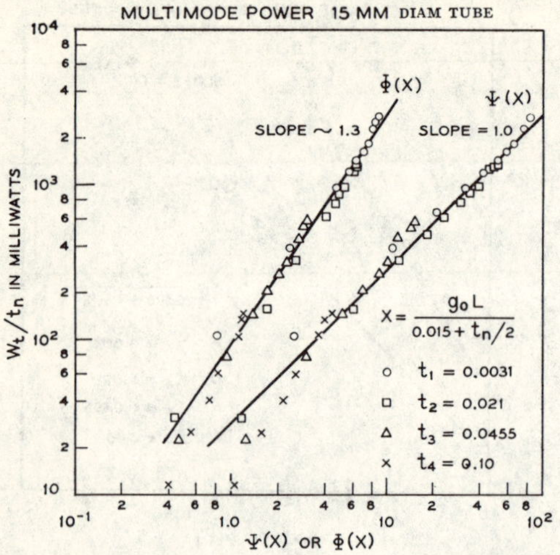

FIG. 10. Same as Fig. 9, for a 15-mm-diam tube.

fore, to test the validity of each function by plotting measured values of W_{t_n}/t_n as a function of both $\Psi(X)$ and $\Phi(X)$, over a wide range of the measured variable X, on log–log coordinate paper. The appropriate function for a given maser should yield a line of unit slope. This procedure would be a more sensitive way of comparing each function with the data than a similar set of plots in Cartesian coordinates.

Three maser tubes with Brewster-angle windows[7] were constructed, having internal diameters of 5.7, 10, and 15 mm, respectively, and a gas fill of 0.1 Torr Ne and 0.7 Torr He. By means of a hot-cathode dc discharge about 1.4 m long, each tube could be excited to various levels of unsaturated gain per pass, each level corresponding to a specific discharge current. The power output of each maser oscillator could then be correlated with its unsaturated gain g_0L, by measuring both quantities with the same discharge current.

Power measurements were made for each tube in a series of external optical resonators, consisting of a low-transmittance ($t_0=0.004$) concave mirror with radius of curvature 5.2 m at one end, and at a distance of about 1.75 m one of four flat output mirrors, with transmittances t_n ranging from 0.0031 to 0.10. The output beam was focused on an Eppley thermopile to measure W_{t_n}, and oscillations at the 3.39-μ transition[8] were suppressed by means of filters inside of the resonator. Power measurements were made with oscillations restricted to the fundamental TEM_{00} mode by means of a centered iris, and where possible with "multimode" oscillations, diffraction-limited only by the bore of each tube.

The small-signal gain measurements were made with a multifrequency linearly polarized probe beam generated by a similar He–Ne maser restricted to the TEM_{00} mode, with fixed and low excitation level. Thus the weak probe beam passed through the tested tubes had substantially the same size and intensity distribution as that generated within each tube when oscillating in that mode during power measurements. From the parameters of this second maser, the peak gain was estimated to be about 1.6 times the measured gain, the latter being a weighted average over the oscillation bandwidth of the source. (It was found by trial that an error in this factor did not affect the slopes of the plotted curves.) Some guidance in estimating the loss fraction entering into X was obtained by measuring the discharge current near oscillation threshold.

The measurements are summarized in Figs. 6–10 in the form of two- and three-decade log–log plots of the one-way power in each maser W_{t_n}/t_n, as a function of $\Psi(X)$ and $\Phi(X)$, respectively. Despite the scatter of plotted points, their average trends may be discerned thanks to two factors: (1) the large range of plotted coordinates attained by varying t_n, and (2) repetition of each set of measurements in five different ways, with three different tubes and two types of resonators. The plots permit three conclusions to be drawn, which are discussed below.

The first observation is that it was not possible to draw straight lines of unit slope through the majority of points for $\Phi(X)$ in any instance, but it was possible to do so for $\Psi(X)$, with varying degrees of success. (The departure of the encircled points from the $\Psi(X)$ lines in Figs. 6–8 for fundamental-mode oscillations may signify that this representation begins to break down for $X>10$ or so.) It follows that, despite the approximations made in deriving (42), it accounts fairly well for the multifrequency power in a He–Ne maser at the 1.15-μ transition, with a frequency spacing between cavity resonances of $c/2l=86$ Mc/sec. As $\Psi(X)$ was derived on the assumption of nonoverlapping holes, and there are at least as many holes as resonances, this result implies that in the tested masers the average hole width was usually less than 86 Mc/sec.

The second observation, from Figs. 9 and 10, is that the multimode power flow is also approximately proportional to the function $\Psi(X)$, indicating that the gain saturates independently at each frequency within the range of these measurements. This result is somewhat puzzling, in view of the relatively close frequency spacing of adjacent-order transverse modes, about 17 Mc/sec. However, it should be borne in mind that different transverse modes of a concave-mirror resonator do not burn holes in the same gain-frequency curve, as they have different transverse distributions and dimensions.[9] The peak intensities of higher-order modes are located at greater distances from the optic axis than for lower-order modes, and tend to deplete the power in the latter.[10] Thus the spatial relations of adjacent-order

[7] W. W. Rigrod, H. Kogelnik, D. J. Brangaccio, and D. R. Herriott, J. Appl. Phys. **33**, 743 (1962).
[8] A. L. Bloom, W. E. Bell, and R. E. Rempel, Appl. Opt. **2**, 317 (1963).
[9] G. D. Boyd and J. P. Gordon, Bell System Tech. J. **40**, 489 (1961).
[10] W. W. Rigrod, Appl. Phys. Letters **2**, 51 (1963).

modes may be more important in determining their relative power than their frequency spacing.

The third observation is that the unit-slope lines of Figs. 6–8 for fundamental-mode oscillations are nearly identical in position, i.e., that the saturation parameter w_0 is independent of tube diameter. It is of some interest to evaluate it, in order to permit prediction of the performance of amplifiers (Fig. 2) and oscillators with the same gas fill. Inasmuch as the gain-saturation expression (2) for a single Lorentzian hole was derived for a uniform plane wave of radiation, there is some question as to the computation of w_0 from measurements with a beam of Gaussian intensity distribution. With this reservation, an order-of-magnitude estimate can be made.

The ratio of peak power density w to the total power W in the TEM_{00} beam is given by

$$w/W = 2/(\pi w_s^2) \cong 0.567/\text{sq mm}, \tag{53}$$

where w_s is the spot radius,[9] taken here to be 1.06 mm, the rms average of its values at the concave and flat mirrors (1.16 and 0.95 mm), respectively. The data of Figs. 6–8 for the three different maser tubes agree roughly on the proportionality constant[11]

$$(W_t/t)/\Psi(X) \cong 1.59 \text{ mW}. \tag{54}$$

With $c/4l = 43.1$ Mc/sec and $\Delta\nu_c = 498$ Mc/sec for Ne at 400°K, the saturation parameter for a single hole is found from (42) to be, in terms of the peak power density,

$$w_0 \cong (0.567)(1.59)(43.1)/(498)$$
$$\cong 0.0594 \text{ mW/sq mm}. \tag{55}$$

In terms of the total power in a Gaussian spot of radius w_s, the total saturation parameter (for a single hole) would be

$$W_0 \cong 0.105(w_s/1.06)^2 \cong 0.093 w_s^2 \text{ mW}, \tag{56}$$

where w_s is the spot radius in mm, provided the unsaturated gain is substantially constant over the spot area.

V. CONCLUSIONS

Gain-saturation relations for maser media with homogeneous[1] and inhomogeneous line broadening have been applied to study the nonlinear gain characteristics of optical maser amplifiers at high beam intensities (Figs. 1 and 2), and the optimum cavity coupling of maser oscillators for maximum output power (Figs. 3 and 5). An approximate expression, Eq. (42), has been derived for the multifrequency power output of a gas optical maser oscillating in the fundamental transverse mode, with the assumption that the Lorentzian "holes" burnt in the Doppler gain-frequency profile do not overlap appreciably, i.e., that the gain saturates independently at each frequency.

Correlated measurements of the internal one-way power and unsaturated gain of helium–neon masers at 1.15 μ in a semiconcave optical resonator 1.75 m long, are found to satisfy Eq. (42) approximately. The measurements disagree with another formulation, Eq. (48), derived for gas masers in which the mode spacing is less than the linewidth, and based on the assumption that the gain saturates to a uniform value over the entire oscillation bandwidth.[5] However, the latter expression may well be more appropriate than Eq. (42) for gas masers at other transition wavelengths, with longer resonators, or higher gas pressures, than those used here.

This result implies that in the tested masers the natural linewidth was less than 86 Mc/sec, the frequency spacing between longitudinal resonances. The power measurements are found to satisfy Eq. (42) even when oscillations occur simultaneously in several transverse modes, whose frequency spacing is relatively small. A possible explanation is suggested, but there is no analysis at present which fully explains the nature of mode competition for active atoms in multimode oscillations. The measurements indirectly confirm Eq. (2), which describes gain saturation in a gas optical maser for monochromatic radiation,[2] and permit the order-of-magnitude evaluation of the saturation parameter w_0 for the gas fill employed.

ACKNOWLEDGMENTS

The writer wishes to thank R. Kompfner for generous permission to publish his calculation of the optimum-cavity coupling for a homogeneous-line maser, which stimulated the present inquiry. He is indebted to J. P. Gordon for informative discussions, and to A. J. Rustako, Jr. for skilled assistance in experimental work. It is a pleasure, moreover, to acknowledge the valuable help given him by D. J. Brangaccio, Mrs. C. A. Lambert, W. L. Bond, Mrs. A. S. Cooper, and Miss D. M. Dodd.

[11] Corrected by a factor of 0.605 to compensate for an error in thermopile calibration.

Saturation Effects in High-Gain Lasers

W. W. RIGROD
Bell Telephone Laboratories, Incorporated, Murray Hill, New Jersey
(Received 17 December 1964)

Earlier calculations of the radiation intensity obtainable from lasers with homogeneous line broadening are generalized to include arbitrarily large loss fractions per pass. The conditions for maximum transmitted or internally dissipated power are derived, as well as the axial distribution of radiation within the active medium. The relevance of these calculations to high-gain gas lasers is discussed.

INTRODUCTION

THE radiation intensity produced by steady-state lasers (optical masers) has been computed previously,[1,2] subject to the restriction of small fractional values of the net loss per pass. With the advent of high-gain lasers which can oscillate even in the presence of large dissipative or coupling losses, however, it is expedient to extend these computations by removing the latter restriction. The analysis presented here treats only the case of homogeneous laser transitions, in which the line shape does not change during gain saturation. The resonator is assumed to support substantially plane electromagnetic waves, with its losses concentrated near the mirrors. Despite these assumptions, the results should furnish a first-order description of the radiation distribution in most forms of high-gain lasers, when the losses exceed several percent.

If we denote the normalized radiation intensities in the $+z$ and $-z$ directions of a laser oscillator by $\beta_+ = w_+/w_0$ and $\beta_- = w_-/w_0$, respectively, where w_0 is the saturation parameter, the saturated gain coefficient $g(z)$ is related to the unsaturated gain coefficient g_0 by[1]

$$g(z) = g_0/(1+\beta_+ + \beta_-). \tag{1}$$

[1] E. O. Schulz du Bois, Bell System Tech. J. **43**, 625 (1964).
[2] W. W. Rigrod, J. Appl. Phys. **34**, 2602 (1963).

As the gain coefficient is isotropic,

$$g(z) = \frac{1}{\beta_+}\frac{d\beta_+}{dz} = -\frac{1}{\beta_-}\frac{d\beta_-}{dz} \tag{2}$$

and we see that

$$\beta_+\beta_- = \text{const.} = C. \tag{3}$$

As shown in Fig. 1, the coordinate z runs between the extremities of the uniform maser medium, which is bounded by the effective reflectances

$$r_1 = 1 - a_1 - t_1 \quad \text{at} \quad z = 0,$$

and

$$r_2 = 1 - a_2 - t_2 \quad \text{at} \quad z = L, \tag{4}$$

where a_1, a_2 are the dissipative losses and t_1, t_2 the respective mirror transmittances.

Oscillations will then stabilize at a level which satisfies the boundary conditions:

$$(\beta_2/\beta_3)r_2 = (\beta_4/\beta_1)r_1 = 1. \tag{5}$$

From (3) we have

$$\beta_1\beta_4 = \beta_2\beta_3 = C \tag{6}$$

and therefore

$$\beta_2/\beta_4 = (r_1/r_2)^{\frac{1}{2}}. \tag{7}$$

Reprinted with permission from *J. Appl. Phys.*, vol. 36, pp. 2487–2490, Aug. 1965.

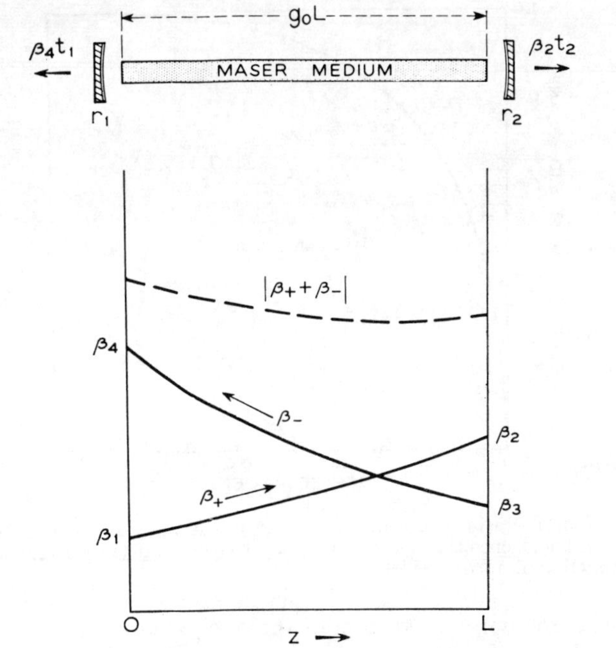

FIG. 1. Schematic diagram showing normalized levels of flux intensity in both directions in an asymmetric laser oscillator; and symbols employed as in the text.

Integrating the gain expression (1) for power flow in the positive direction

$$\frac{1}{\beta_+}\frac{d\beta_+}{dz}=\frac{g_0}{1+\beta_++C/\beta_+} \quad (8)$$

we obtain

$$g_0L=\ln(\beta_2/\beta_1)+\beta_2-\beta_1-C(1/\beta_2-1/\beta_1). \quad (9)$$

The same procedure for gain in the negative direction yields

$$g_0L=\ln(\beta_4/\beta_3)+\beta_4-\beta_3-C(1/\beta_4-1/\beta_3). \quad (10)$$

Adding these two equations, and making use of (5) and (6), we obtain

$$\beta_2=\frac{(r_1)^{\frac{1}{2}}}{[(r_1)^{\frac{1}{2}}+(r_2)^{\frac{1}{2}}][1-(r_1r_2)^{\frac{1}{2}}]}[g_0L+\ln(r_1r_2)^{\frac{1}{2}}] \quad (11)$$

and β_4 is then given by (7).

The total output radiation intensity, via both mirrors, is (w_2+w_4), given by

$$(w_2+w_4)/w_0=\beta_4t_1+\beta_2t_2. \quad (12)$$

When the losses at each mirror are equal,

$$a_1=a_2=a, \quad (13)$$

this reduces to

$$\frac{w_2+w_4}{w_0}=\frac{1-a-(r_1r_2)^{\frac{1}{2}}}{1-(r_1r_2)^{\frac{1}{2}}}[g_0L+\ln(r_1r_2)^{\frac{1}{2}}]. \quad (14)$$

For a symmetrical resonator, defined by

$$r_1r_2=r^2 \quad (15)$$

$$r=1-a-t, \quad (16)$$

the output intensity at each mirror is w_t, given by

$$2w_t/w_0=[(1-a-r)/(1-r)][g_0L+\ln r] \quad (17)$$

$$=[t/(a+t)][g_0L+\ln(1-a-t)]. \quad (18)$$

Thus, when $a_1=a_2$, we can always replace an asymmetrical resonator by an equivalent symmetrical resonator, given by (15), having the same relation between the total output power and the unsaturated gain g_0L. In addition, when one mirror is opaque and lossless, say $t_1=0$ and $r_1=1$, the output intensity is given by

$$\frac{w_t}{w_0}=\beta_2t_2=\frac{t_2}{a_2+t_2}[g_0L+\ln(r_2)^{\frac{1}{2}}] \quad (19)$$

corresponding to the output at each of the two identical mirrors of a symmetrical resonator with twice as much gain. Accordingly, we shall confine our attention henceforth to symmetrical resonators.

The optimum output coupling t_{opt}, for which w_t in (18) is maximized, is defined by

$$t/a=[(1-a-t)/(a+t)][g_0L+\ln(1-a-t)] \quad (20)$$

giving

$$2(w_t)_{\max}/w_0=t^2/a(1-a-t),\ (t=t_{\text{opt}}). \quad (21)$$

For small loss fractions, $(a+t)\ll 1$, these equations reduce to the approximate solutions previously derived[2]:

$$t_{\text{opt}}/a=(g_0L/a)^{\frac{1}{2}}-1 \quad (22)$$

$$2(w_t)_{\max}/w_0=t_{\text{opt}}^2/a. \quad (23)$$

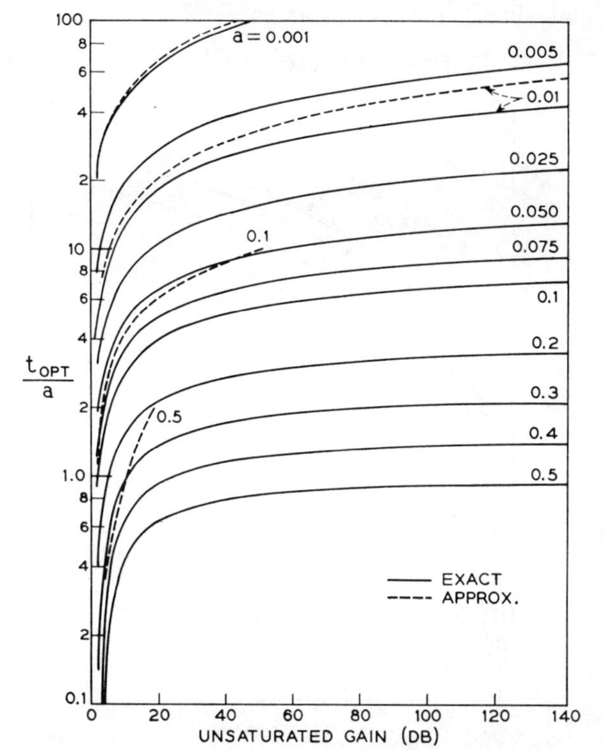

FIG. 2. Optimum coupling coefficient (t_{opt}) for maximum oscillator output power, as a function of the loss fraction (a) and the unsaturated gain in dB, 4.343 g_0L. The solid lines represent the exact solution, and the dashed lines the approximate version.

The exact and approximate solutions for t_{opt} are plotted in Fig. 2 for various values of the unsaturated gain in dB, 4.343 g_0L, and the loss fraction a. The approximate solution is adequate, for high-gain lasers, up to about $a=1\%$. The error in the approximate solution increases rapidly, however, for greater losses, with both a and g_0L. We see that t_{opt}/a decreases rapidly with increasing a, such that $t_{\text{opt}}\approx 1-a$ for $a>0.5$ at high gain values. The curves of Fig. 3 shows the normalized maximum output intensity obtainable with optimum output coupling, as given by (21), for given values of a and 4.343 g_0L. The enhancement of available output power by a decrease in a is greatest for small gain values. At high values of gain and loss ($a>0.5$), the normalized output (21) approaches $(1-a)\ g_0L$. (Gain saturation due to amplified spontaneous emission sets an upper limit to the gain per pass in Figs. 2 and 3.)

A similar procedure may be used to design a resonator for maximum power dissipation within it. In the symmetrical resonator, the normalized radiation intensity reflected from each mirror is

$$\beta_1=\beta_2 r=\beta_2(1-a-t) \tag{24}$$

which can be written in the form:

$$\beta_2-\beta_1=\beta_2 a+\beta_2 t, \tag{25}$$

stating that the power generated per pass $(\beta_2-\beta_1)$ equals the power dissipated in a plus that transmitted via t. The duality between the two forms of loss is shown by comparing the expression for the total (normalized) transmitted power

$$2\beta_2 t=[t/(a+t)][g_0L+\ln(1-a-t)] \tag{26}$$

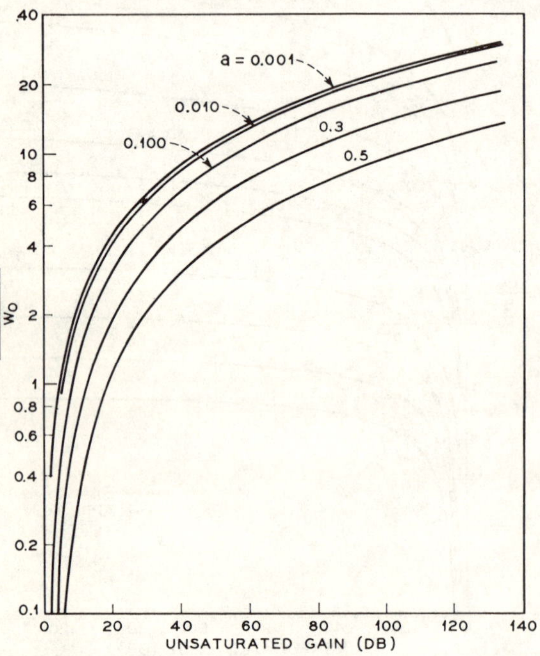

FIG. 3. Normalized values of the maximum output flux intensity, $2(w_t)_{\max}/w_0$, of the homogeneous-line laser oscillator, obtainable with optimum output coupling for given values of a and the single-pass gain 4.343 g_0L(dB).

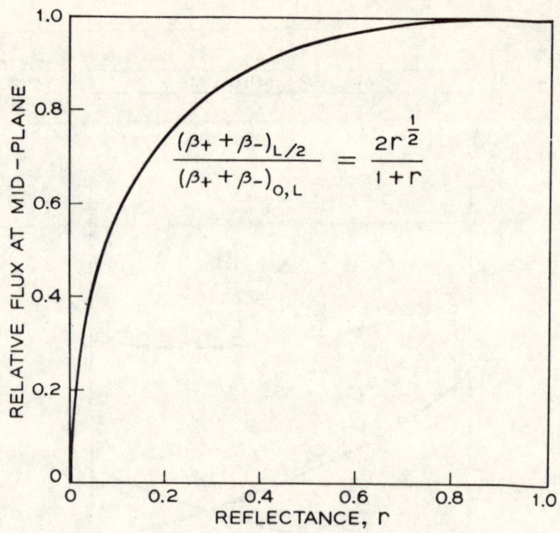

FIG. 4. Ratio of the flux intensity at the midplane ($z=L/2$) of a homogeneous-line maser to that at either end ($z=0,L$), as a function of mirror reflectance.

with the normalized form of the dissipated power:

$$2\beta_2 a=[a/(a+t)][g_0L+\ln(1-a-t)]. \tag{27}$$

For fixed t, there is an optimum value of a for maximum power dissipation in a; and this power is greatest when $t=0$.

It is curious to note that the variation of total (two-way) radiation intensity within the symmetrical laser cavity is independent of the unsaturated gain. When the gain expression (8) is integrated over half the laser length, we find that

$$g_0L/2=\ln(\beta_+/\beta_1)+(\beta_+-C/\beta_+)+(\beta_2-\beta_1). \tag{28}$$

At the midplane $z=L/2$, the flux in both directions is the same,

$$\beta_+=\beta_-=C/\beta_+ \tag{29}$$

yielding

$$g_0L/2=\ln(\beta_+/\beta_-)+(\beta_2-\beta_1). \tag{30}$$

Combining this result with (9), we obtain

$$\ln(\beta_+/\beta_1)=\tfrac{1}{2}\ln(\beta_2/\beta_1)=-\tfrac{1}{2}\ln r. \tag{31}$$

The ratio of the two-way flux intensities at the midplane $z=L/2$, to that at the extremities of the laser, $z=0$ and $z=L$, is then given by

$$\frac{(\beta_++\beta_-)_{L/2}}{(\beta_++\beta_-)_{0,L}}=\frac{2(\beta_1\beta_2)^{\frac{1}{2}}}{\beta_1+\beta_2}=\frac{2(r)^{\frac{1}{2}}}{1+r}, \tag{32}$$

i.e., the ratio of the geometric mean of β_1 and β_2 to their arithmetic mean. This ratio is plotted in Fig. 4 as a function of r, the geometric mean of the mirror reflectances. (The relatively small variation of total flux intensity with distance in a laser has been noted independently by R. N. Zitter.) The expression is not valid, naturally, in the limit of $r=0$, as spontaneous emission noise input would then need to be considered.

The maximum variation in the saturated gain coefficient $g(z)$ along the laser axis is indicated by the ratio

$$\frac{g(L/2)}{g(L)}=\frac{1+\beta_2(1+r)}{1+2\beta_2(r)^{\frac{1}{2}}}=\frac{1+\beta_2}{1+\beta_2[2(r)^{\frac{1}{2}}/(1+r)]} \tag{33}$$

which does not differ appreciably from unity for $r>\frac{1}{2}$.

Gas lasers saturate more or less homogeneously under several conditions: (1) when the half-width of the hole burned in the Gaussian gain profile by monochromatic radiation is comparable to that of the transition itself[3]; (2) when the collision frequency is not small compared to the spontaneous decay rates of the excited states[4]; and (3) in very long resonators, when the mode spacing is appreciably smaller than the hole widths, resulting in nearly uniform gain saturation over the oscillation bandwidth. One or more of these conditions is usually operative in very high-gain gas lasers, except for oscillations not far from threshold.

In the last instance mentioned, the integrated radiation intensity of the laser over the transition can be computed, approximately, by assuming the gain coefficient to saturate homogeneously in each frequency interval within the band.[5] This procedure, employed in Ref. 5 for small-loss lasers, $(a+t)\ll 1$, can be generalized to arbitrarily large loss per pass in the same way as for a true homogeneously broadened laser.

In each frequency interval $d\nu$ under the Gaussian gain profile, the output radiation intensity is given by an expression similar to (18):

$$2\eta w_t(\nu)=[t/(a+t)][g_0(\nu_0)L\exp-[(\nu-\nu_0)/\Delta\nu_c]^2+\ln(1-a-t)], \tag{34}$$

where w_0 has been replaced by η^{-1}, the saturation parameter per unit bandwidth, ν_0 is the center frequency of the transition,

$$\Delta\nu_c=\Delta\nu_D/2(\ln 2)^{\frac{1}{2}}, \tag{35}$$

and $\Delta\nu_D$ is the Doppler width at half-maximum. The oscillation bandwidth, over which $w_t(\nu)\geq 0$, is given by

$$2\Delta\nu_{\max}=2\Delta\nu_c(\ln X)^{\frac{1}{2}}, \tag{36}$$

where

$$X=g_0(\nu_0)L/-\ln(1-a-t). \tag{37}$$

Integration of (34) over the oscillation band yields the total multifrequency output intensity W_t, as follows:

$$2W_t/\eta^{-1}(2\Delta\nu_{\max})=[t/(a+t)]\times[\langle g_0L\rangle+\ln(1-a-t)], \tag{38}$$

where $\langle g_0L\rangle$ is the average value of g_0L over the oscillation bandwidth, given by

$$\langle g_0L\rangle=g_0(\nu_0)L(\pi^{\frac{1}{2}}/2)[\operatorname{erf}(\ln X)^{\frac{1}{2}}/(\ln X)^{\frac{1}{2}}]. \tag{39}$$

Comparison of (38) with (18) shows they are formally identical, except for the replacement of g_0L by its average value, and of w_0 by an equivalent parameter

$$W_0=2\eta^{-1}\Delta\nu_{\max}=\eta^{-1}[2\Delta\nu_c(\ln X)^{\frac{1}{2}}] \tag{40}$$

for the uniformly saturated Gaussian line. The optimum output transmittance t_{opt} is found by setting $dW_t/dt=0$ in (38). After some algebra, we obtain an expression very similar to (20):

$$t/a=[(1-a-t)/(a+t)][\langle g_0L\rangle+\ln(1-a-t)],\quad t=t_{\text{opt}}. \tag{41}$$

The maximum output intensity $(W_t)_{\max}$ at each mirror is given by

$$2(W_t)_{\max}/W_0=t^2/a(1-a-t),\quad t=t_{\text{opt}} \tag{42}$$

in the same form as (21).

In practice, we are usually able to determine the peak gain $g_0(\nu_0)L$ and loss a per pass, but do not know the oscillation bandwidth or the corresponding value of $\langle g_0L\rangle$. However, as previously noted, the approximate solutions for t_{opt}/a, derived for the condition $(a+t)\ll 1$, are adequate for a <0.01 and gains below about 20 dB. In this region, therefore, the solutions for t_{opt}/a and $(W_t)_{\max}$ given in Fig. 5 of Ref. 2 may be used.

For higher values of loss and gain, on the other hand, the curves of Fig. 2 show that t_{opt}/a changes very slowly with g_0L, and for very high values of g_0L approaches the value

$$t_{\text{opt}}/a\approx(1/a)-1. \tag{43}$$

Thus for gains above about 20 dB and $a>0.01$, the peak gain value $g_0(\nu_0)L$ can be used in place of $\langle g_0L\rangle$, with the curves of Fig. 2, to find t_{opt}/a to a quite good approximation. If better accuracy is sought, the values so obtained can be used to compute X, and consequently $\langle g_0L\rangle$ from (39), and with them to redetermine t_{opt}/a. Having found X, we can compute $\Delta\nu_{\max}$ from (35) and (36), evaluate W_0 from (40), and thus find the maximum output intensity via each mirror $(W_t)_{\max}$ from (42) or the curves of Fig. 3. Alternatively, for very high gain per pass, such that $X\gg 1$,

$$\operatorname{erf}(\ln X)^{\frac{1}{2}}\approx 1 \tag{44}$$

and (38) reduces to the expression

$$\frac{2W_t}{\eta^{-1}(2\Delta\nu_c)}\simeq\frac{t}{a+t}\frac{\pi^{\frac{1}{2}}}{2}g_0(\nu_0)L \tag{45}$$

in which $W_t=(W_t)_{\max}$ when $t=t_{\text{opt}}$.

The writer is indebted to Mrs. C. A. Lambert for performing the numerical computations, and to E. I. Gordon for valuable comments.

[3] E. I. Gordon, A. D. White, and J. D. Rigden, *Proceeding of the Symposium on Optical Masers* (Polytechnic Press, Brooklyn, New York, 1963), Vol. 13, p. 309.
[4] W. R. Bennett, Jr., Phys. Rev. **126**, 580 (1962).
[5] A. D. White, E. I. Gordon, and J. D. Rigden, Appl. Phys. Letters **2**, 91 (1963).

Part 5
Internal Modulation and Mode Locking

Mode-Locking of Lasers

PETER W. SMITH, MEMBER, IEEE

Invited Paper

Abstract—This paper is a tutorial and review paper on the subject of mode-locking of lasers. It is intended as an introduction to the subject for the general reader as well as an up-to-date overview for specialists in the field. Emphasis has been placed on giving physical understanding of the phenomena and processes involved, rather than on providing details of specific theories and devices.

I. Introduction

A TYPICAL laser consists of an optical resonator formed by two coaxial mirrors and some laser gain medium within this resonator. The frequency band over which laser oscillation can occur is determined by the frequency region over which the gain of the laser medium exceeds the resonator losses. Often, there are many modes of the optical resonator which fall within this oscillation band, and the laser output consists of radiation at a number of closely spaced frequencies. The total output of such a laser as a function of time will depend on the amplitudes, frequencies, and relative phases of all of these oscillating modes. If there is nothing which fixes these parameters, random fluctuations and nonlinear effects in the laser medium will cause them to change with time, and the output will vary in an uncontrolled way. If the oscillating modes are forced to maintain equal frequency spacings with a fixed phase relationship to each other, the output as a function of time will vary in a well-defined manner. The laser is then said to be "mode-locked" or "phase-locked." The form of this output will depend on which laser modes are oscillating and what phase relationship is maintained. It is possible to obtain: an FM-modulated output, a continuous pulse train, a spacially scanning laser beam, or a "machine gun" output where pulses of light appear periodically at different spacial positions on the laser output mirror.

This paper is intended as a tutorial and review paper on the subject of mode-locking of lasers. It is hoped that it will be of use as an introduction to the subject for the general reader, as well as an up-to-date overview for the specialists in the field.

The paper is organized in the following way. Section II is a historical review of the subject. Section III discusses modes in optical resonators. Section IV treats in detail mode-locking of longitudinal modes of lasers and discusses various techniques for longitudinal mode-locking. Section V treats both transverse mode-locking and the simultaneous locking of longitudinal and transverse laser modes. The paper concludes with a description of some recent experiments using mode-locked lasers. As a review article has recently appeared on the subject of picosecond pulses from mode-locked solid-state lasers [1], this paper will concentrate on the general principles of mode-locking and examples drawn largely from gas laser experiments, but we have attempted to provide a fairly complete set of references covering the entire field.

Manuscript received April 13, 1970; revised July 9, 1970. *This invited paper is one of a series planned on topics of general interest—The Editor.*

The author is a Visiting McKay Lecturer in the Dept. of Electrical Engineering and Computer Sciences, University of California, Berkeley, Calif., on leave from the Bell Telephone Laboratories, Inc., Holmdel, N. J. 07733.

II. Historical Review

Although there were indications of mode-locking in the work of Gürs and Müller [2], [2a] on internal modulation of ruby lasers, and the work of Statz and Tang [3] on He–Ne gas lasers, the first papers to clearly describe this effect were written in 1964 and early 1965 by DiDomencio [4], Hargrove *et al.* [5], and Yariv [8]. Hargrove *et al.* experimentally obtained a continuous train of pulses from a He–Ne laser by mode-locking with an internal acoustic loss modulator. DiDomenico, following a suggestion by E. I. Gordon, showed theoretically that mode-locking could be obtained by internal loss modulation at the resonator mode-spacing frequency. Similar theoretical predictions were made independently by Yariv [8]. Somewhat earlier, Lamb [6], in his now classic paper on the theory of laser operation, had described how the nonlinear properties of the laser medium could cause the modes of a laser to lock with equal frequency spacing. This idea was later discussed by Crowell [7] for the case of many oscillating modes, and he reported experiments with a 6328 Å He–Ne laser demonstrating the "self-locking" of laser modes due solely to the nonlinear behavior of the laser medium. Shortly thereafter, the locking of laser modes with the amplitudes and phases of an FM-modulated wave was discussed by Yariv [8] and Harris and McDuff [9], and demonstrated experimentally by Harris and Targ [10], using a 6328 Å He–Ne laser with an internal phase modulator. Harris later pointed out that internal phase modulation could also be used to produce an output pulse train [11] and this was demonstrated experimentally by Ammann *et al.* [12].

During 1965, experimenters were demonstrating that the techniques of mode-locking could be applied to other laser systems. The argon ion laser [13] and the ruby laser [14] were mode-locked using internal modulation techniques. A most important mode-locking technique was demonstrated by Mocker and Collins [15] who showed that the saturable dye used inside a ruby laser cavity to "Q-switch" the laser (i.e., to cause the laser to emit a short high-power burst of energy) could also be used to obtain mode-locking. Thus the laser output consisted of a number of short intense

Reprinted from *Proc. IEEE*, vol. 58, pp. 1342–1357, Sept. 1970.

pulses of light. These and later experiments demonstrated that Q-switched and mode-locked solid-state lasers could produce much shorter and much higher peak power pulses than those that had been obtained from mode-locked gas lasers. Work was also continuing in 1965 on self-locking of lasers. Statz and Tang [16] published the first of a series of papers they and their collaborators would write on the theory of self-locking. They describe self-locking in terms of "combination tones" (beats) produced by pairs of modes in the nonlinear laser gain medium.

By 1966 there were many workers investigating mode-locking in various laser systems. DeMaria *et al.* reported mode-locking the Nd-doped glass laser with a loss modulator [17], and shortly thereafter, with a saturable dye [18]. The Nd-doped yttrium aluminum garnet (YAG) laser was mode-locked with an internal modulator [19]. It was now becoming clear to experimenters that there was no detector available with a fast enough response time to display the pulses from a mode-locked solid-state laser. Theory showed that the minimum duration of mode-locked pulses was of the order of the reciprocal of the oscillation bandwidth of the laser. From measurements of the bandwidth of laser oscillation it appeared that these pulses could be as short as a few picoseconds (10^{-12} seconds) in duration, yet the fastest detectors had rise times of a fraction of a nanosecond (10^{-9} seconds).

One of the first methods used to estimate these small pulsewidths was measuring the efficiency of the generation of the second harmonic of laser light in a nonlinear crystal. Theory showed that when the incident light is in the form of a pulse train, the average second-harmonic power generated is enhanced over that produced by a continuous wave of the same average power. The amount of this enhancement gives an estimate of the width of the pulses in the pulse train. This method was used by DiDomenico *et al.* [19] to estimate the width of pulses from the mode-locked Nd:YAG laser to be 80 ps (picoseconds) and by Kohn and Pantell [19a] to estimate the width of mode-locked pulses from their ruby laser to be 200 ps.

In late 1966 and early 1967, Maier *et al.* [20], Armstrong [21], and Weber [22] described a new technique for pulsewidth measurement. The laser beam is split with a beam splitter and the two beams are then recombined with a relative delay in a nonlinear crystal. With proper adjustment of polarization and crystal orientation, second-harmonic radiation will be generated only when the pulses in each beam coincide in the nonlinear crystal. With this technique pulsewidth measurements could be obtained in the picosecond region, and Armstrong [21] reported measuring pulsewidths of 6 ps from a Nd: glass laser mode-locked with a saturable dye. Later in the same year Giordmaine *et al.* [23] proposed and demonstrated the technique of pulsewidth measurement using two-photon fluorescence (TPF). In a manner somewhat analagous to the second-harmonic generation method, the beam is split and the two beams made to interfere in a material exhibiting two-photon fluorescence. Enhanced fluorescence is seen at points where pulses coincide in the material. This method is substantially easier to implement experimentally, and workers quickly adopted this technique. Much of this early work had to be reexamined later, however, when several authors [24], [25], [163], [164] pointed out that free running (non-mode-locked) lasers produced TPF results very similar to those of a mode-locked laser! Only by measuring the contrast ratio (the ratio of fluorescence at the peak of the display to the fluorescence level in between the peaks) can one obtain information on the state of mode-locking.

In the meantime work was continuing on mode-locking techniques. To obtain greater pulse powers, the technique of completely coupling a single pulse out of a mode-locked laser was exploited by Michon *et al.* [26] to obtain a pulse of less than 4 ns duration with an energy of 8 mJ from a ruby laser, by Penney and Heynau [27] to obtain a pulse of less than 0.8 ns duration with peak power of 25 MW from a Nd:glass laser, and by Zitter *et al.* [28] to obtain 2 ns, 30-watt pulses from a He-Ne laser. Kachen *et al.* [28a] externally selected and amplified a single mode-locked Nd: glass laser pulse to obtain a 10 to 15 ps pulse with a peak power of 30 GW. A new mode-locking technique was described by Smith [29] who showed that mode-locking could be obtained by moving one laser mirror at a constant velocity. Self-locking was reported for the ruby [16], [26], Nd:glass [78], [78a], and argon ion [67] lasers.

During this period work was also progressing on the theory of mode-locking. Harris [30] published a good review paper on internal modulation theory and techniques. A new theory of self-locking based on the transient response of the laser medium to the incident radiation was proposed by Fox and Smith [31], and further developed by Smith [32]. Statz and his co-workers published a series of papers [33]–[35b] on self-locking, developing their theory based on the nonlinear interaction of laser modes in the gain medium. The theory of mode-locking with saturable absorbers was discussed by Garmire and Yariv [126] and Sacchi *et al.* [128].

In 1968 workers observed mode-locking with several new laser systems. Mode-locking of CO_2 gas lasers [36], [37], the CF_3I photolysis laser [37a], dye lasers [38]–[41], and semiconductor lasers [42] was reported. New techniques of mode-locking were also made available. Fox *et al.* [43] and Chebotayev *et al.* [43a] described the use of a section of laser medium excited to the lower laser state as a saturable absorber to produce mode-locking. Comly *et al.* [44] and Laussade and Yariv [45] described the use of anisotropic molecular liquids within the resonator to cause locking. Huggett [46] used the amplified beats between the laser modes to drive an internal modulator to mode-lock a He–Ne laser.

So far, we have been talking of mode-locking a set of longitudinal resonator modes, characterized by having the same spacial distribution in a plane transverse to the resonator axis, but having a different number of half-wavelengths of light along the axis of the resonator.[1] There are also sets of modes which have the same number of half-wavelengths of light along the resonator axis but have different spacial distributions of energy transverse to this axis. These modes are called transverse modes. (A more complete discussion of

[1] Some authors refer to these as axial modes.

modes in laser resonators will be found in Section III.) In 1968, Auston [47], [48] described how the phase-locking of a set of *transverse* modes of a laser would produce an output beam that would scan back and forth in space as a function of time. Auston [47] and Smith [49] also describe simultaneous locking of longitudinal and transverse laser modes. The light energy can be confined to a small region of space and travels a zigzag path as it bounces back and forth in the laser resonator. These first experiments were performed with a He–Ne laser, but in 1969 transverse mode-locking experiments were reported using Nd:glass [50], [51] and CO_2 [52], [53] laser systems.

Much of the work presently being done has to do with developing higher power, narrower, and better controlled pulses from mode-locked lasers, and using present techniques with new laser systems.

The availability of well-defined light outputs from mode-locked lasers has stimulated the invention of a number of devices that make use of these characteristics. In 1965, Massey *et al.* [54] described the use of an FM demodulator, external to the laser resonator, to produce a single-frequency output from an FM-locked laser. Shortly thereafter, Harris *et al.* [55] described how the single-frequency output of this laser could be stabilized. A somewhat different technique to produce single-frequency output from a mode-locked laser by coupling out only one laser mode was described by Harris and McMurtry [56].

In 1966, Duguay *et al.* [57] described a technique for frequency shifting a train of mode-locked laser pulses by passing them through an external phase modulator driven at the pulse repetition rate. 1967 brought a variety of different uses for mode-locked lasers. Trains of laser pulses were used for testing the pulse response of electrical cables [58] and for measuring the relaxation times of bleachable dyes used for laser mode-locking [59]–[59b]. In 1968, Giordmaine *et al.* [60] described a technique for shortening the duration of laser pulses by giving them a frequency sweep with an external modulator and then passing them through a dispersive structure. This technique was later demonstrated by Duguay and Hansen [61] who compressed pulses from a He–Ne laser by almost a factor of two. Fisher *et al.* [151a] proposed the use of the optical Kerr effect to give the frequency sweep and in this way Laubereau [151b] reported achieving compression of the pulses from a Nd:glass laser by a factor of 10. Optical rectification of picosecond laser pulses in a nonlinear crystal was demonstrated by Brienza *et al.* [61a]. Rentzepis [62] used mode-locked Nd:glass laser pulses to investigate radiationless molecular transitions.

Recently, Duguay and Hansen [63] have constructed a light gate with an open time of approximately 8 ps, using a mode-locked laser pulse to drive a Kerr cell. Perhaps the most dramatic of the recent uses of mode-locked laser pulses, however, are the experiments on thermonuclear neutron emission by groups in the Soviet Union [64] and the United States [159].

Some uses that are presently being investigated include mode-locked lasers as a source for pulse-code modulation communication systems, and fundamental investigations of the interaction of short pulses of radiation with resonant atomic systems.

III. Modes in Laser Resonators

A mode of a resonator can be defined as a self-consistent field configuration. That is, the optical field distribution reproduces itself after one round trip in the resonator. The modes of an open resonator formed by a pair of coaxial plane or spherical mirrors have been studied in great detail [65]. One can identify a set of longitudinal (or axial) modes which all have the same form of spacial energy distribution in a transverse plane, but have different axial distributions corresponding to different numbers of half-wavelengths of light along the axis of the resonator. These longitudinal modes are spaced in frequency by $c/2L$ where c is the velocity of light and L is the optical path length between the mirrors. Fig. 1 illustrates the type of laser output one obtains when a number of longitudinal resonator modes are above threshold for laser operation. Each frequency shown in the top part of the figure corresponds to a different longitudinal mode number, i.e., a different number of half-wavelengths of light in the resonator. For a typical laser this number will be of the order of 10^6.

To obtain laser operation at a single longitudinal mode, it is usually necessary to design a laser resonator sufficiently short so that $c/2L$ is greater than the bandwidth of the gain of the laser material, or to design a complex resonator which has high loss for all modes within the oscillation bandwidth except the favored one.

It is possible to show that for each longitudinal mode number, there exists a set of solutions for the light energy inside a resonator formed by two spherical mirrors which correspond to different energy distributions in a plane transverse to the resonator axis. These solutions are called the transverse modes of the resonator. The cross-sectional amplitude distribution of these modes, for curved mirror resonators with rectangular symmetry, is given closely by [65]

$$A(x, y) = A_{m,n}\left[H_m\left(\frac{\sqrt{2}x}{w}\right)H_n\left(\frac{\sqrt{2}y}{w}\right)\right] \cdot \exp\left(-(x^2 + y^2)/w^2\right) \tag{1}$$

where x and y are the transverse coordinates, and $A_{m,n}$ is a constant whose value depends on the field strength of the mode. w is the radius of the fundamental mode ($m=0, n=0$) at $1/e$ maximum amplitude, $H_a(b)$ is the ath order Hermite polynominal with argument b, and m and n are called the transverse mode numbers.[2] Note that this distribution is independent of the longitudinal mode number. Fig. 2 is a plot of A along the x axis, for $m=0$ to 5. Note that as the transverse mode number increases, the energy is spread further and further from the axis of the resonator. Fig. 3 shows actual mode patterns found with a 6328 Å He–Ne gas laser. In order to obtain oscillation in a single transverse

[2] These modes are sometimes referred to as TEM_{mn} modes by analogy with the modes in waveguides.

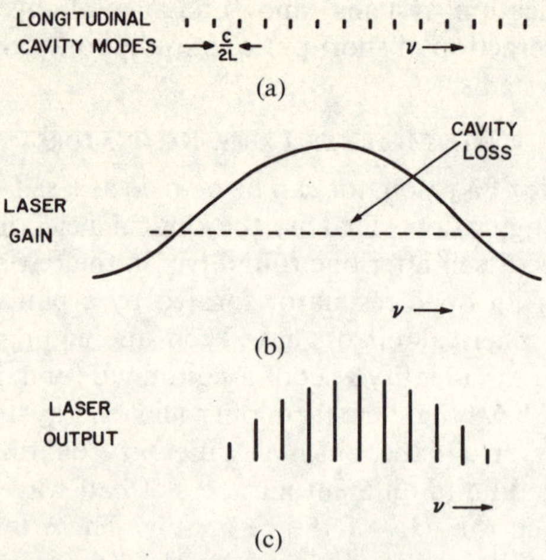

Fig. 1. Laser operation on several longitudinal resonator modes. (a) Longitudinal modes of the laser resonator. (b) Laser gain versus frequency showing the region where the gain exceeds the cavity losses. (c) Laser output: oscillation at resonator mode frequencies for which laser gain exceeds the losses.

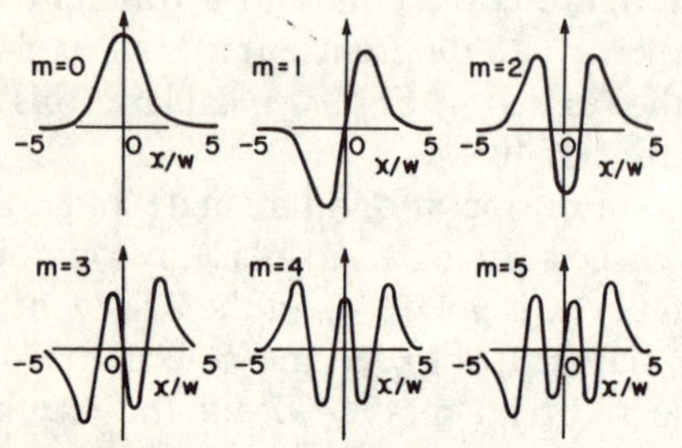

Fig. 2. Plot of field amplitude of transverse modes versus the normalized transverse coordinate x/w (w=spot size of fundamental mode), for $m=1-5$. Reversal of sign of field indicates 180° change of phase.

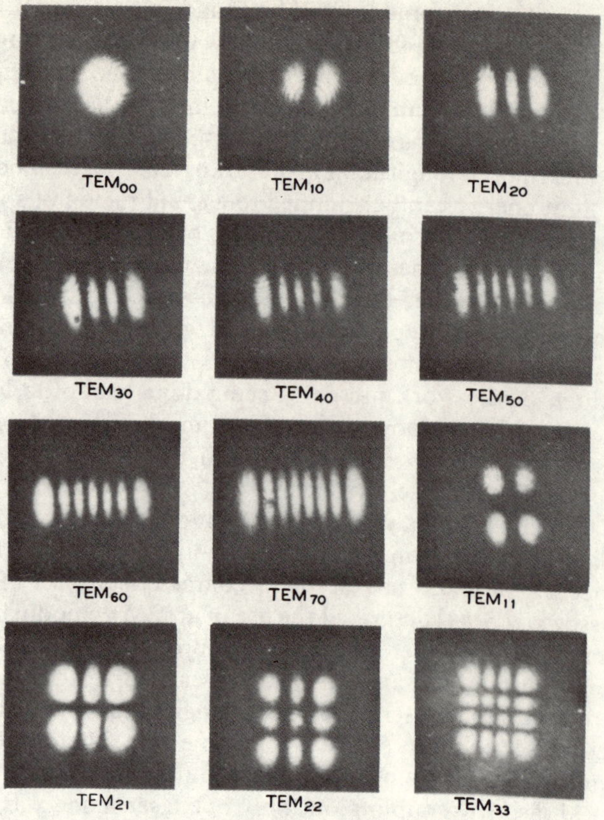

Fig. 3. Mode patterns of a gas laser oscillator with rectangular symmetry [160].

mode, it is necessary to use some device which will give high losses to all transverse modes but the desired one. Since higher order modes spread further from the resonator axis, the easiest way to accomplish single-transverse-mode operation is to insert into the laser cavity a circular aperture whose size is such that the fundamental mode experiences little diffraction loss while higher order modes suffer appreciable attenuation.

For resonators which do not have large diffraction losses, the resonant frequency of a mode can be written [65]

$$\nu = (c/2L)\left[(q+1) + \frac{(m+n+1)}{\pi} \cdot \cos^{-1} \sqrt{(1-L/R_1)(1-L/R_2)}\right] \tag{2}$$

where $q+1$ is the number of half-wavelengths of light along the axis of the resonator, m and n are the transverse mode order numbers, and R_1 and R_2 are the radii of the two mirrors making up the laser resonator. Note the following points. 1) For a given transverse mode (given m and n), longitudinal modes with mode number differing by one are spaced in frequency by $c/2L$. 2) For a given longitudinal mode number (q), transverse modes with the sum of m and n differing by one are spaced in frequency by $(c/2\pi L)\cos^{-1}\sqrt{(1-L/R_1)(1-L/R_2)}$.

Let us consider the output of a laser operating in a number of longitudinal and transverse modes. The total field will be the sum of the individual fields of each of the modes. The optical length L will not be the same for all the modes due to the dispersion of the laser material. In general, both the amplitude and the phase of these modes will vary with time due to random mechanical fluctuations of the laser resonator length and the nonlinear interaction of these modes in the laser medium. The total field will thus vary with time in some uncontrolled way with a characteristic time which is of the order of the inverse of the bandwidth of the oscillating mode frequency spectrum. In the next sections we discuss the results of fixing the frequency spacings and phases of the oscillating modes—mode-locking the laser.

IV. Longitudinal Mode-Locking

Consider a set of oscillating longitudinal modes such as those shown in Fig. 1(c). If we somehow fix the frequency spacing, relative phases, and amplitudes of these modes, the laser output will be a well-defined function of time. Consider the nth mode to have amplitude E_n, angular frequency ω_n, and phase ϕ_n. Then the total laser output field, E_T, can be written

$$E_T = \sum_n E_n e^{i[\omega_n(t-z/c)+\phi_n]} + \text{c.c.} \tag{3}$$

where c.c. represents the complex conjugate and we have assumed the radiation is traveling in the $+z$ direction. If we have equal mode frequency spacing, $\omega_n = \omega_0 + n\Delta$ where

$\Delta=2\pi(c/2L)$ and ω_0 is the optical frequency of the laser output. Thus, we can write

$$E_T = e^{i\omega_0(t-z/c)}\left\{\sum_n E_n e^{i[n\Delta(t-z/c)+\phi_n]}\right\} + \text{c.c.} \qquad (4)$$

This corresponds to a carrier wave of frequency ω_0 whose envelope depends on the values of E_n and ϕ_n. Note, however, that

1) the envelope travels with the velocity of light;
2) the envelope is periodic with period $T=2\pi/\Delta=2L/c$.

For ϕ_n = a constant[3] independent of n, this envelope consists of a single pulse in the period T whose width is approximately the reciprocal of the frequency range over which the E_n's have an appreciable value, i.e., the ratio of the pulse spacing to the pulsewidth is approximately the number of oscillating modes. Within the laser resonator this corresponds to a pulse of light traveling back and forth between the resonator mirrors with the velocity of light. Other selections of amplitudes and phases will result in different envelopes. For any set of mode amplitudes, the narrowest pulse will always result from the phases $\phi_n=0$, for all n. For mode amplitudes and phases corresponding to the sidebands and carrier of an FM-modulated wave, the output intensity will be constant as a function of time.

Fig. 4(a) shows the actual frequency spectrum of the output of a 6328 Å He–Ne laser. The laser is oscillating on a number of longitudinal modes and the amplitudes of these modes fluctuate with time. Fig. 4(b) shows the same laser output when the modes are mode-locked. The oscillation bandwidth has increased and the mode amplitudes are stable.

Fig. 5(a) shows the corresponding output as a function of time. The output consists of a train of narrow pulses separated by the round-trip time of the light in the optical resonator ($2L/c$). An expanded view of one pulse in Fig. 5(b) shows the pulsewidth to be 330 ps. How can a laser be made to operate in this fashion? The remainder of this section discusses techniques for obtaining longitudinal mode-locking.

A. Self-Locking

Under certain conditions, the nonlinear effects of the laser medium itself may cause a fixed phase relationship to be maintained between the oscillating modes. The first detailed discussion of this effect appears to be that of Lamb [6], although earlier papers by Bennett [66] and Statz and Tang [3] mention phase-locking for three-mode laser operation when the central mode is tuned to the center of the atomic gain curve. The first paper to discuss self-phase-locking of a laser oscillating in many oscillating modes was that of Crowell [7], who showed that self-locking the 6328 Å He–Ne laser would result in an output train of pulses of a duration of approximately 1 ns. He found self-locking occurred when oscillation was confined to the fundamental transverse mode and the losses in the optical resonator were "judiciously adjusted."

[3] As origin of time is arbitrary, we shall in subsequent discussion assume that this constant = 0.

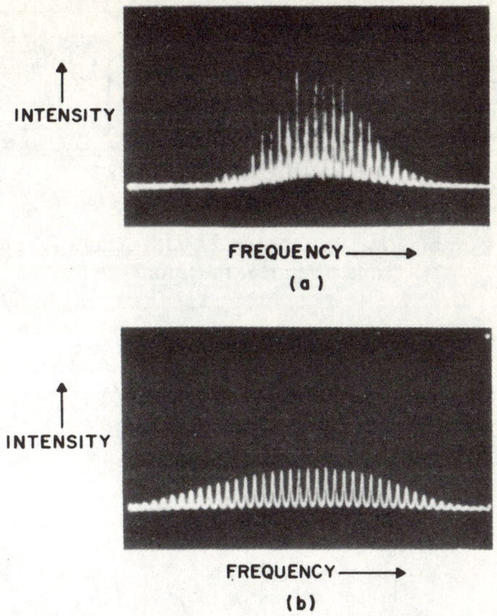

Fig. 4. (a) Output frequency spectrum from a non-mode-locked 6328 Å He–Ne laser. (b) Output frequency spectrum of the same laser when mode-locked [43].

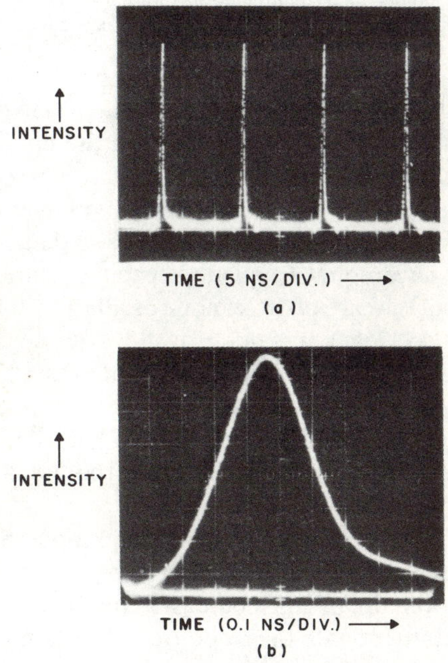

Fig. 5. (a) Output pulse train from mode-locked 6328 Å He–Ne laser. (b) Expanded view of one of the mode-locked pulses showing width of 330 ps. The "tail" is contributed by the diode detector. The laser output pulse is believed to be symmetrical [43].

There have appeared a number of papers on the theory of self-locking [16], [31]–[35b], [68]–[74]. Typically, one of two approaches is used. The first approach uses the concept of combination tones (i.e., the radiation produced by the nonlinear interaction of two or more laser modes with the laser medium). If a combination tone is close in frequency to an oscillating mode, frequency pulling effects will tend to pull the oscillating mode to the same frequency as the combination tone [75]. Fig. 6 shows a simple situation in which three modes at ν_1, ν_2, and ν_3 are oscillating. Due to disper-

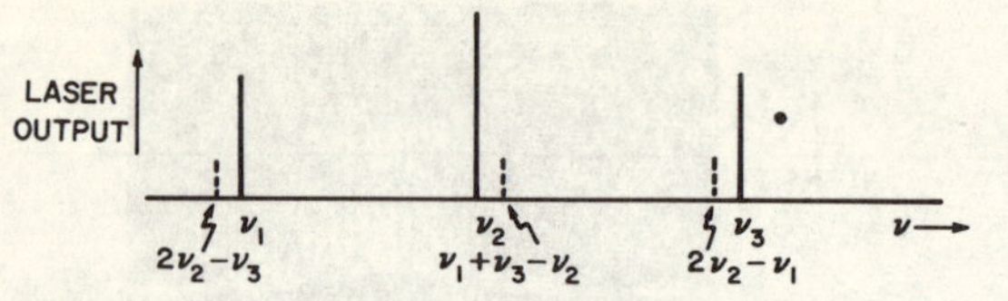

Fig. 6. Combination tones formed by three laser frequencies in a nonlinear medium.

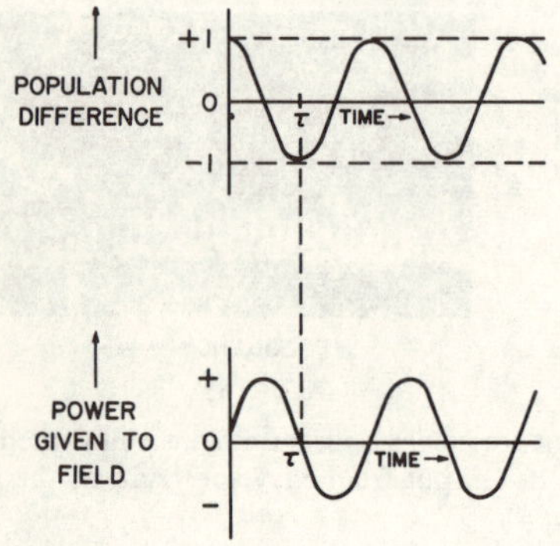

Fig. 7. Plots of normalized population difference and power given to incident field for resonant two-level atomic system for times short compared to atomic decay times.

sion in the gain medium, these modes will not, in general, be exactly equally spaced in frequency. In the nonlinear gain medium, radiation at frequencies $2\nu_2-\nu_3$, $\nu_1+\nu_3-\nu_2$, and $2\nu_2-\nu_1$ is generated. These frequencies will tend to pull the frequencies of the oscillating laser modes so that the oscillating modes have equal frequency spacing. Lamb [6] analyzes in detail the case of three mode oscillation at low power levels (see also [75a]). It is not easy, however, to extend this work to the case of many oscillating modes and high power levels. It has been found necessary to use restrictive assumptions and evaluate special cases [16], [33], [34], [35b], [68]–[69] or to resort to computer solutions of complex equations [71], [74].

Because of the complexity of the equations for a combination tone analysis of a large number of modes in the frequency domain, it may be easier to work in the time domain and investigate the response of the gain medium to a train of pulses of stimulating radiation. A first attempt at this type of analysis was made by Fox and Smith [31], who showed that under certain rather restrictive conditions one would expect the pulses to completely invert the populations of the laser levels. The upper portion of Fig. 7 shows the (normalized) population difference as a function of time for a two-level atomic system under the influence of radiation at the atom's resonant frequency. It is assumed that atomic lifetimes are much longer than the times being considered. +1 indicates all the atoms in the upper laser level and −1 indicates all the atoms in the lower laser level. Although complete saturation would correspond to a population difference of 0, it is seen that after the radiation has interacted with an atomic system with atoms initially in the upper state for a time τ, the population difference becomes inverted, and maximum power is given to the incident field. Thus a pulse of this "size" (called a π or 180° pulse) [76] can interact most efficiently with the atomic gain system. In [31] and [32], it is shown that certain lasers indeed operate in such a way that the self-locked pulses in the laser resonator approximate 180° pulses for the laser medium. The population inversion recovers as a function of time with the time constant of the atomic decay time, and the population difference, as a function of time, varies as shown in Fig. 8. Similar results have also been obtained in [35a] and [74].

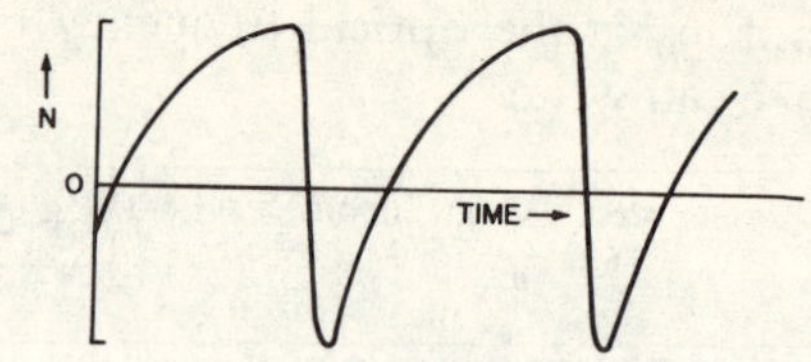

Fig. 8. Variation of population difference as a function of time for a laser with a circulating 180° pulse.

Experimenters have shown that most laser systems can be made to self-lock. The first self-locking experiments were performed with the 6328 Å He–Ne laser [7], [77], but soon self-locking was reported for ruby [16], [26], [77a],[4] argon ion [67], and Nd:glass [78], [78a][4] lasers. Self-locking has also been reported for Nd:YAG lasers [79][4] semiconductor lasers [42] and CO_2 gas lasers [80]. Multiple self-pulsing corresponding to several pulses traveling back and forth in the laser resonator has been observed, notably with the 6328 Å He–Ne laser [32], [77], [82]–[85] where up to six output pulses in one round-trip period have been reported. In general it is not possible to predict the exact conditions for self-pulsing, but the following conditions are usually necessary.

a) The oscillation must be confined to a single transverse mode.
b) The round-trip time for radiation in the resonator must be of the order of or greater than the atomic decay time.
c) The laser must not be operated too much above threshold.

In view of the somewhat uncertain nature of self-locking (self-locked lasers often have unstable outputs), there are some advantages in using a driven cavity perturbation to force the laser to mode-lock. Methods for accomplishing this are described in the following section.

B. Internal Modulation

Hargrove *et al.* [5] first used an internal modulator to produce mode-locking. They used an acoustic loss modulator, driven at the longitudinal mode-spacing frequency ($c/2L$), in the resonator of a 6328 Å He–Ne laser. DiDomenico [4] discussed the theory of operation of a loss modulator driven at a frequency related to the mode-spac-

[4] Recent work has cast doubt on many of these results. Some workers believe that only partial self-locking occurs for solid-state losses (see [81], [162], [163]). There is no doubt that reliable self-locking can be obtained with gas lasers, however.

ing frequency. Shortly thereafter the use of an internal phase modulation (dielectric constant modulation) to lock laser modes with the amplitudes and phases of an FM-modulated wave was discussed by Yariv [8] and Harris and McDuff [9], and demonstrated experimentally by Harris and Targ [10]. It was later shown that internal phase modulation could also be used to produce a train of pulses similar to those produced using a loss modulator [11], [12].

Numerous internal modulation mode-locking experiments have been performed with different laser systems [5], [7], [10]–[14], [17], [19], [35a], [36], [46], [68], [86]–[100] using either acoustic [5], [7], [13], [17], [19], [37a], [86a] or electrooptic [13], [36], [92]–[94], [99], [100] loss modulators to produce output pulses, and using electrooptic phase modulators to produce FM [10]–[12], [89] or pulsing [11], [88], [89], [95] operation. It has been observed that there is a dramatic reduction of the low-frequency noise in the laser output when the laser is mode-locked [96], [99].

Pantell *et al.* [90] have discussed mode-locking a Raman laser using internal modulation techniques. Henneberger and Schulte [91] have described mode-locking a 6328 Å He–Ne laser by vibrating the resonator end mirror at the $c/2L$ rate. The resonator length modulation is equivalent to that produced by an internal phase modulator driven at $c/2L$. Several authors [86], [87], [98] have described placing the internal modulator in a resonator weakly coupled to the main laser resonator in order to reduce the effective insertion loss of the internal modulation device. Gurski [97] describes mode-locking and second-harmonic generation inside the laser resonator using the same nonlinear crystal. Huggett [46] has proposed and demonstrated the rather elegant scheme of using the amplified $c/2L$ beat frequency from the laser output to drive the internal modulator.[5]

A great deal of work has also been done on the theory of mode-locking by internal modulation [8], [30], [68], [92], [99], [101]–[113]. A good review paper on this subject is that of Harris [30]. Although formal results can be obtained, in general the resulting formulas must be solved by computer for the particular case of interest. Fig. 9 shows the results for a mode-locked pulsing operation computed from the theory of McDuff and Harris [104] for a specific case with five free-running oscillating modes. Here α_c is the mode coupling due to loss perturbation which for a small modulator near the end of the laser cavity is approximately one half the average loss per pass introduced by the loss modulator. δ is the peak single-pass phase retardation of a phase modulator.

A simple physical picture of the operation of loss or phase modulators to produce pulse train mode-locking can be given by the following.

Loss Modulation: If the loss is modulated at $c/2L$, one cycle of the modulation frequency corresponds to the time it takes the light to make a round trip in the laser resonator. Thus light incident on a modulator situated at one end of the laser resonator during a certain part of the modulation

[5] T. S. Kinsel has been able to obtain reliable locking of the Nd:YAG laser using this technique (private communication).

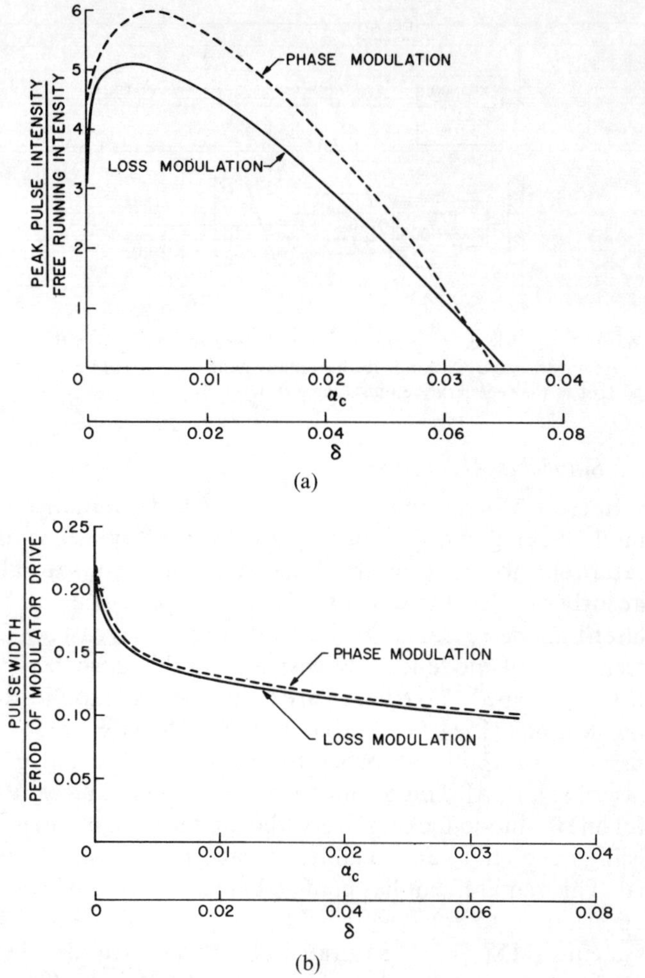

Fig. 9. (a) Pulse peak intensity versus perturbation level for constant detuning. (b) Pulsewidth versus perturbation level for constant detuning [104].

cycle will be again incident at the same point of the next cycle after one round trip in the laser resonator. Light which sees a loss at one time will again see a loss after one round in the resonator. Thus all the light in the resonator will experience loss except that light which passes through the modulator when the modulator loss is zero. Light will tend to build up in narrow pulses in these low-loss time positions. In a general way we can see that these pulses will have a width given by the reciprocal of the gain bandwidth. Pulses wider than this will experience more loss in the modulator. Pulses narrower in time will experience less gain because their frequency spectrum will be wider than the gain bandwidth.

Phase Modulation: Light passing through an electrooptic phase (dielectric constant) modulator will be up- or down-shifted in frequency unless it passes through during the time of the maximum (or minimum) of the dielectric constant cycle. If the modulator is operated at the synchronous rate, this shift will increase every time the light passes through the modulator. Eventually the light frequency will fall outside the gain curve of the laser medium and the light will not experience any gain. Thus the effect of the phase modulator is similar to the loss modulator, and the previous discussion of loss modulation also applies here.

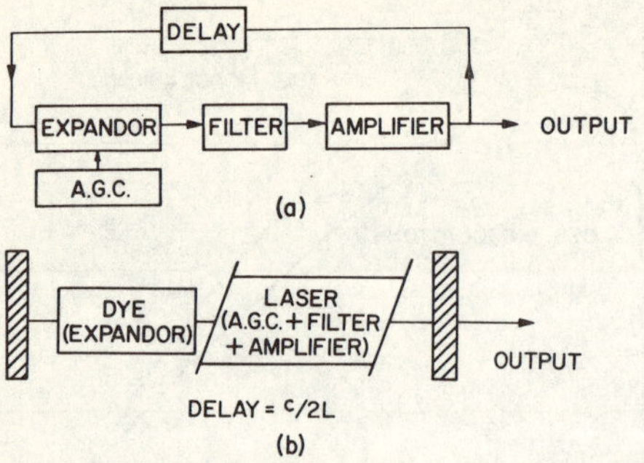

Fig. 10. (a) Block diagram of Cutler's electronic regenerative pulse generator. A.G.C. stands for automatic gain control. (b) Laser equivalent of the regenerative pulse generator.

C. Saturable Absorbers

In 1965, Mocker and Collins [15] reported simultaneously mode-locking and Q-switching a ruby laser by means of a saturable absorber in the laser resonator. (A saturable absorber is a material whose absorption coefficient for the laser light *decreases* as the light intensity is increased.) The technique of mode-locking with a saturable absorber had been described somewhat earlier in a patent application by Rigrod [15a]. DeMaria and Stetser later reported using a saturable absorber to mode-lock the Nd:glass laser [18], [114]. This technique has proved to be a very useful one for mode-locking lasers whose internal light intensity is high enough to appreciably saturate the absorber material. This requirement has confined most saturable absorber applications to liquid [40], [41], [123] and solid-state laser systems [115]–[122]. Saturable absorbers have also been used, however, with gas laser systems. Fox *et al.* [43] and Chebotayev *et al.* [43a] described the use of a pure neon discharge as a saturable absorber for the 6328 Å He–Ne laser. This was demonstrated to operate effectively to mode-lock this laser system. The use of saturable absorbers such as SF_6 to mode-lock the CO_2 gas laser is discussed in [37] and [124].

The theory of the operation of a saturable absorber to produce mode-locking can be discussed [1], [18] in terms of an electronic regenerative pulse generator described by Cutler [125]. Fig. 10(a) shows a block diagram of the system Cutler considered. The expandor is a nonlinear element whose loss decreases as the incident intensity is increased. It was shown [125] that this system would produce a pulse train with pulse separation equal to the group delay time of one round trip through the device and minimum pulsewidth, approximately given by the reciprocal of the filter bandwidth. Fig. 10(b) shows a laser with a saturable absorber in the laser resonator. This system is analogous to the electronic system with the saturable dye acting as an expandor, the laser medium acting as the amplifier and filter, and the round-trip time of light in the laser resonator as the system delay time. It is important for the saturable absorber to have a fast recovery time compared to the resonator round-trip time if it is to act as an expandor in this system. Cutler's analysis shows that if there exists a nonlinear variation of refractive index with frequency within the feedback loop, the pulsewidth will be greater than the reciprocal of the gain bandwidth. For a Gaussian filter function and pulse shape, and assuming a square law saturable absorber, Cutler finds the pulsewidth, Δt, given by

$$\Delta t = \frac{8}{\pi \Delta\omega} \left| \frac{\left(1 + \dfrac{\gamma^2 \Delta\omega^4}{16}\right)}{1 + \left(1 + \dfrac{\gamma^2 \Delta\omega^4}{18}\right)^{1/2}} \right|^{1/2} \tag{5}$$

where $\Delta\omega$ is the bandwidth of the filter and γ is proportional to the curvature of the plot of refractive index versus frequency. Experimental indications of this type of pulse broadening in mode-locked Nd:glass lasers have recently been observed by Treacy [1], [148]–[149].

Although an analysis of Cutler's system is helpful in understanding the operation of a saturable absorber in a mode-locked laser system, the details of the saturation of both the absorber and the laser gain must be considered in a more realistic model. Analysis of saturable absorber mode-locking has been made by several authors [126]–[131a]. In general, the results obtained are in agreement with the predictions of the pulse regenerative amplifier model.

D. Other Methods of Producing Mode-Locking

Several other methods for producing mode-locking of longitudinal modes have been reported. Smith [29] discovered that pulsing mode-locked operation of a 6328 Å He–Ne laser could be obtained by continuously translating one resonator mirror at a constant velocity. A satisfactory explanation for this behavior has not been found. Bambini and Burlamacchi [132] reported a related phenomenon where mode-locking was obtained by low-frequency resonator length modulation. Laussade and Yariv and co-workers have investigated the behavior of an anisotropic molecular liquid inside a laser resonator and have shown both theoretically [45] and experimentally [44], [133] that it may be used to obtain mode-locking. One method of obtaining population inversion and, thus, laser operation in liquid laser systems is to pump them with light from a solid-state laser. Several authors [38], [39], [123], [123a], [134] have reported mode-locked pulse-train output from laser-pumped liquid lasers by mode-locking the pumping laser and adjusting the length of the liquid laser resonator to be integrally related to the length of the pumping laser resonator.

V. Transverse Mode-Locking

Consider a set of transverse modes with the same longitudinal mode number but with transverse mode numbers differing by one. These modes will be approximately equally spaced in frequency (see Section III). Such a set is illustrated in Fig. 11 which shows the set TEM_{0n}, for $n=0, 1, 2, \cdots$. What will occur if we fix the frequency spacings and relative phases of this set of simultaneously oscillating modes? This problem was first investigated by Auston [47]. He showed that a simple result could be obtained if a set of

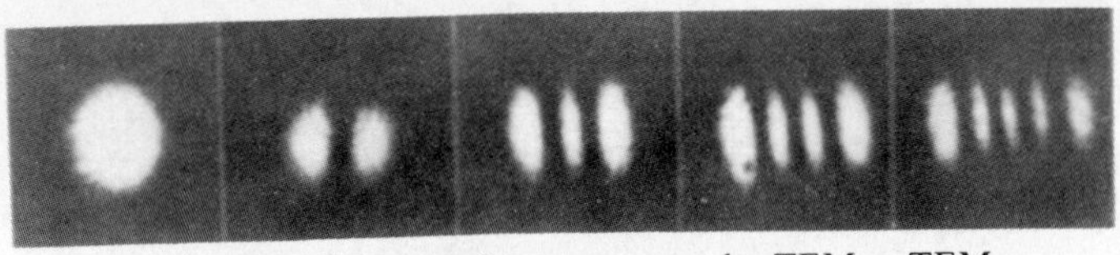

Fig. 11. A linear set of transverse modes TEM_{00}, TEM_{01}, TEM_{02}, TEM_{03}, $\cdots$.

modes is locked with equal frequency spacings and zero phase difference with field amplitudes A_n where

$$|A_n|^2 = \frac{1}{n!}(\bar{n})^n e^{-\bar{n}}. \tag{6}$$

$\bar{n}$ is a parameter which determines the number of oscillating transverse modes. Under these conditions the intensity of the optical field of the laser output will be

$$I(\xi, t) = \frac{1}{\sqrt{\pi}} \exp -(\xi - \xi_0 \cos \Omega t)^2 \tag{7}$$

where ξ = transverse coordinate in the y direction normalized with respect to the fundamental spot size, $\xi_0 = \sqrt{2\bar{n}}$, and Ω = transverse mode frequency spacing. Equation (7) represents a scanning beam of width equal to the fundamental spot size which moves back and forth in the transverse plane with maximum excursion equal to ξ_0. Fig. 12 illustrates this motion over a complete period of $2\pi/\Omega$. Fig. 13 shows how the distribution of A_n's depends on $\bar{n}$. Note that the maximum beam excursion depends on $\sqrt{\bar{n}}$. For large $\bar{n}$, the width of the distribution equals $2\sqrt{2\bar{n}}$, which is simply the total excursion (in units of number of resolvable spots). Thus one obtains the intuitively satisfying result that the maximum number of resolvable spots is roughly equal to the number of oscillating modes. This type of operation was later observed by Auston [48] using a tilting mirror acoustic modulator with a short 6328 Å He–Ne laser. Fig. 14 shows the output versus time of this laser as observed with a photodetector having an aperture in front of it. The coordinate ξ represents the displacement of the aperture from the resonator axis along the direction of beam motion.

Self-locking of transverse laser modes has also been reported by Auston [47], [48] for the 6328 Å He–Ne laser, and by Ito and Inaba [52] for the CO_2 laser.

Simultaneous locking of the modes of a laser oscillating in several longitudinal and transverse modes has also been studied. Two types of behavior have been observed. Smith [49] reports mode-locking all the longitudinal modes corresponding to each transverse mode to form a pulse inside the laser resonator. The pulses corresponding to the different transverse modes do not coincide in time, however, so that the output consists of several pulses in one resonator round-trip period, each one corresponding to a different transverse mode. This type of operation was obtained with a 6328 Å He–Ne laser using a Ne discharge as a saturable absorber [43].

A very different behavior is obtained if a set of transverse modes corresponding to each longitudinal mode number is locked together to form a scanning beam, and each of the sets of longitudinal modes is also locked with zero phase difference to form a pulse in the resonator. In this case the light is confined to a small region of space both in the axial and transverse directions. It can be shown [47] that this "blob" of light bounces back and forth in the laser resonator following the zigzag path to be expected from geometrical optics.

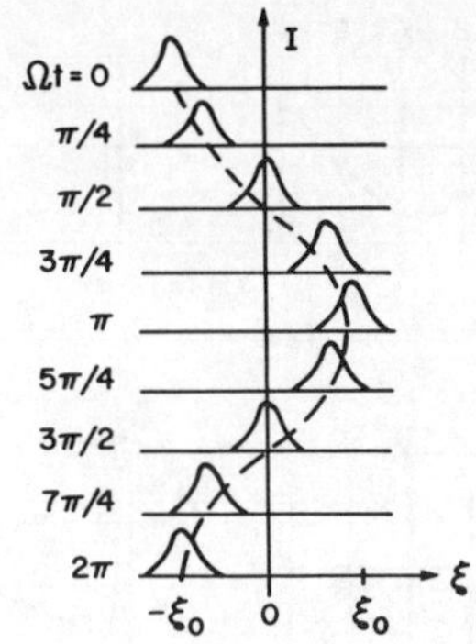

Fig. 12. Plot of distribution of intensity for various times during the scanning period [47].

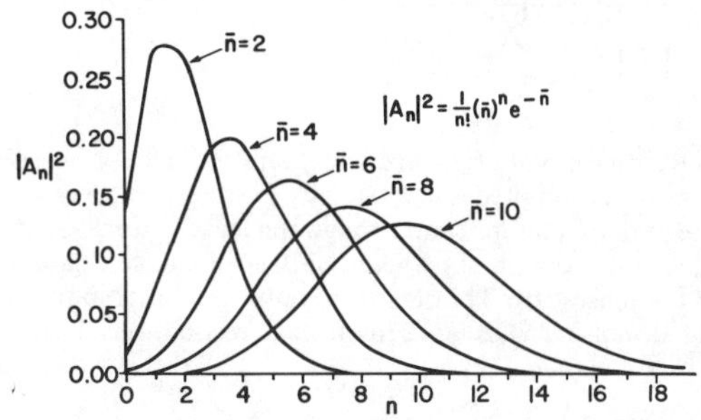

Fig. 13. Distribution of mode amplitudes required to obtain scanning-beam operation for various numbers of oscillating modes [47].

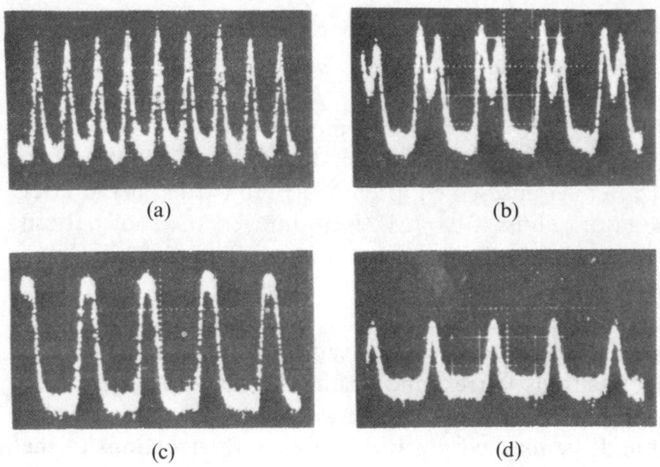

Fig. 14. Output of scanning-beam laser versus time observed through an aperture located at various values of ξ [48]. (a) $\xi = 0$. (b) $\xi = 1.3$. (c) $\xi = 2.6$. (d) $\xi = 3.1$.

The position where this blob of light will hit the laser mirrors can be determined for a resonator consisting of two equal and coaxial mirrors using an elegant description proposed by J. R. Pierce and discussed by Herriott *et al.* [135]. They show that the resonator is equivalent to a series of equally spaced thin lenses and that the beam position at the nth lens can be written

$$x_n = X \sin(n\theta + \alpha) \tag{8}$$

where x_n is the displacement of the beam from the resonator

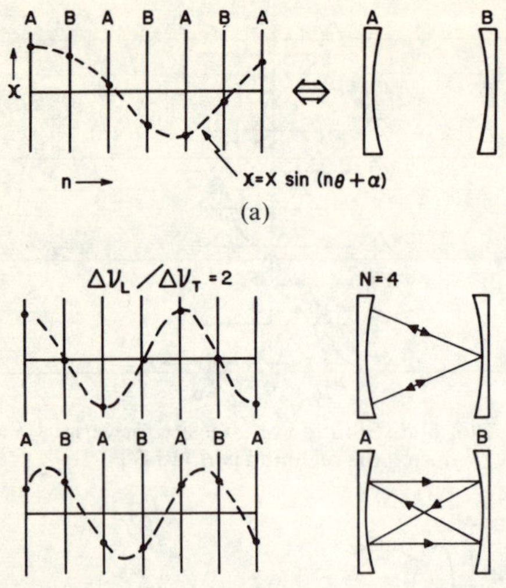

Fig. 15. (a) Resonator with two mirrors, *A* and *B*, replaced by a sequence of thin lenses represented schematically by the positions marked *A* and *B*. The path of a light beam through the lenses is formed by joining with straight lines the points where the dashed sinusoid intersects the planes of the lenses. (b) Thin lens diagrams and corresponding beam paths for the case $\Delta\nu_L/\Delta\nu_T=2$ (confocal resonator). Two possible beam paths are shown [49].

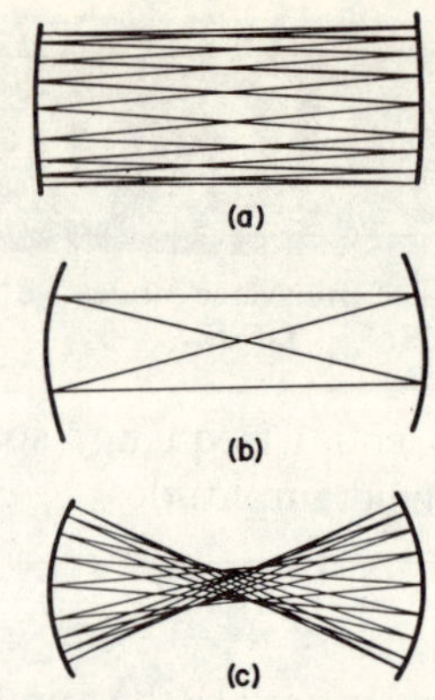

Fig. 16. Locus of "spot" of light. (a) Close-to-plane. (b) Confocal. (c) Close-to-concentric resonators.

axis in the x direction at the nth lens and X and α are arbitrary constants. It can be shown that for our case

$$\theta = \pi\Delta\nu_T/\Delta\nu_L \tag{9}$$

where $\Delta\nu_T$ is the frequency separation between adjacent transverse modes and $\Delta\nu_L(=c/2L)$ is the frequency spacing between adjacent longitudinal modes. Note that if $\theta=2\pi/N$, where N is an integer, the beam displacement will repeat itself after N lenses, i.e., after $N/2$ round trips in the laser resonator. Thus, if $\Delta\nu_L/\Delta\nu_T$ is an integer, the spot pattern on the mirrors will be constant in time. These remarks may perhaps be clarified by referring to Fig. 15. Fig. 15(a) shows a laser resonator replaced by a series of lenses represented schematically by the positions marked *A* and *B*. The "*A*" positions correspond to the mirror *A*, and the "*B*" positions to mirror *B*. The intersections of the dashed sinusoid and the marked positions indicate the locations of the spot on the laser mirrors. Fig. 15(b) shows the special case where $\Delta\nu_L/\Delta\nu_T=2$. Positions of the spot on the laser mirror are now constant with time. Two cases are shown corresponding to two different values of α, and the corresponding beam path in the resonator is indicated. In an experimental situation, the effects of aperturing the beam will determine the value of α and thus the beam path observed. For large integral values of $\Delta\nu_L/\Delta\nu_T$, a pulse of light will appear periodically at different spacial positions on the laser output mirror. This machine gun type of operation may find use in beam scanning systems. Fig. 16 shows the light path for simultaneous locking of longitudinal and a linear set (TEM_{0n}) of transverse modes for (a) close-to-plane parallel, (b) confocal, and (c) close-to-concentric resonators.

Experimental observation of simultaneous mode-locking was reported by Auston [48], who observed self-locking of this type, and Smith [49], who obtained locking with a saturable absorber. Simultaneous locking has also been reported by Bambini and Burlamacchi [132] using low-frequency resonator length modulation and, for a somewhat different situation involving two almost degenerate transverse modes, by Kohiyama *et al.* [136]. All of these experiments were done using the 6328 Å He–Ne laser. Simultaneous self-locking has also been reported for the Nd: glass laser [51] and the CO_2 laser [52], and saturable absorbers have also been used to obtain simultaneous locking with the Nd: glass [50] and CO_2 [53] laser systems.

VI. Current Research Interests

In this section we discuss briefly some current research topics involving mode-locked lasers. The selection of topics is somewhat arbitrary and no attempt is made to be comprehensive. The aim is to give the reader some idea of the range of current work involving mode-locked lasers.

A. Measurement of Picosecond Pulse Durations

The problem of measuring the duration of light pulses of less than several hundred picoseconds duration is not a trivial one. Sampling oscilloscopes are available with rise times of 25 ps but currently available detectors do not have rise times better than 100 to 200 ps, so that pulses less than approximately 200 ps cannot be directly displayed on an oscilloscope screen. Several techniques have been used to estimate the width of picosecond laser pulses. The efficiency of second-harmonic generation in nonlinear crystals is enhanced if the fundamental radiation is in the form of a train of narrow pulses [19], [19a], [140], [145]. The amount of this enhancement can be used to estimate the width of the pulses. DiDomenico *et al.* [19] used this method to estimate a pulsewidth of 80 ps for pulses from a mode-locked Nd: YAG laser. Kohn and Pantell [19a] measured 200 ps pulses from their mode-locked ruby laser.

A better technique became available when Maier *et al.* [20], Armstrong [21], and Weber [22] described a second-harmonic generation intensity correlation technique. In this case no comparison need be made with a non-mode-locked laser output. The technique involves splitting the laser beam with a beam splitter and recombining the beams with a relative delay in a nonlinear crystal. With a proper adjustment of polarizations and crystal orientation, second-harmonic radiation is only generated when the pulses in each beam overlap in the nonlinear crystal. Using this technique, Armstrong [21] measured pulsewidths of 6 ps

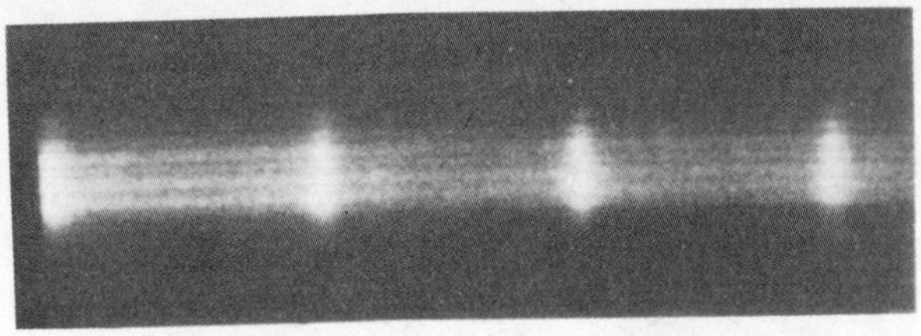

Fig. 17. Two-photon fluorescence pattern observed in rhodamine 6G dye with a mode-locked Nd:glass laser.

from a Q-switched and mode-locked Nd:glass laser, and Glenn and Brienza [141] found that pulsewidths from a Q-switched and mode-locked Nd:glass laser varied from a few ps at the beginning of the pulse train to 15 ps at the end. Krasyuk *et al.* [142] measured pulsewidths of 14 ps from a ruby ring laser. A similar technique which uses a dye exhibiting two photon fluorescence (TPF) as the nonlinear element was proposed by Giordmaine *et al.* [23]. This technique has been widely used because of its apparent simplicity. Fig. 17 shows an experimental TPF display from a mode-locked Nd:glass laser. There is a problem in the interpretation of a fluorescence pattern if the contrast ratio (the ratio of the intensity of fluorescence in the region where pulses overlap to the fluorescence intensity in between these maxima) is not the theoretical maximum [24], [25], [163], [164]. A contrast ratio of 3:1 indicates complete mode-locking with zero phases to produce the narrowest laser pulses possible with a given bandwidth. Completely random phases will give a TPF display with a contrast ratio of 1.5:1, with approximately the same peak width. An FM-locked laser output will produce no peaks (contrast ratio 1:1). There has been much recent work on the interpretation of TPF results. A detailed discussion is beyond the scope of this paper and the reader is referred to Rowe and Li [137], Weber *et al.* [138], Duguay *et al.* [162], and Picard and Schweitzer [165] for recent papers on this subject. Recently, workers have discovered that somewhat more information on pulse shape can be obtained from third-harmonic generation experiments [143]–[144].

Russian workers [139] have described a moving-image camera with an image converter which has a resolution of the order of 10 ps. They show that much more information on pulse structure can be obtained with this technique than can be obtained from TPF experiments. There seems to be no comparable instrument available in this country, however.

B. Frequency-Swept-Pulses and Pulse Compression

Duguay and co-workers showed how mode-locked pulses could be frequency shifted [57], [146], [147] or swept in frequency [60], [61] by passing them through an external phase modulator. Fig. 18 illustrates the operation of this device. Fig. 18(a) shows the pulse train incident on the modulator. Fig. 18(b) shows the variation of the refractive index of the modulator which is being driven at the pulse-repetition frequency. As can be seen from Fig. 18(c), if the pulses pass through the modulator during periods where the refractive index is not to first order changing with time, the average optical frequency will not be shifted but

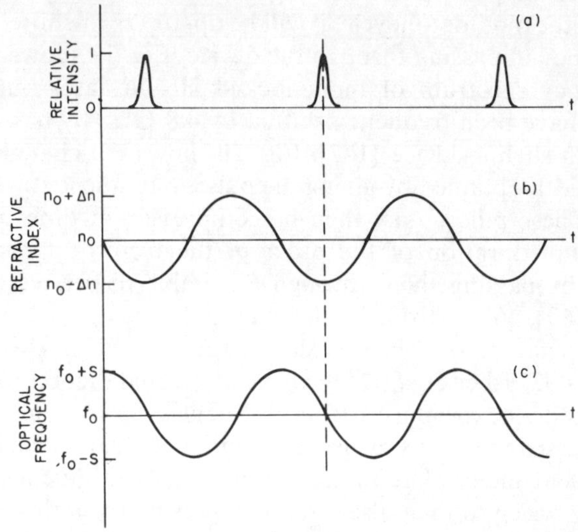

Fig. 18. (a) Laser output versus time. (b) Synchronous variation of modulator's refractive index. (c) Optical frequency of light after passing through modulator [161].

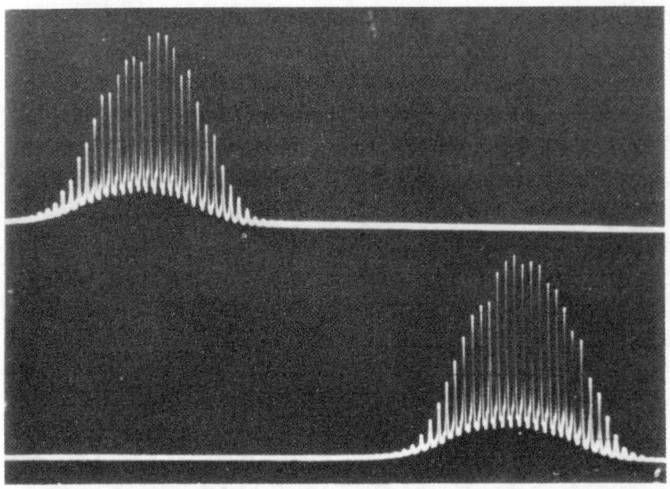

Fig. 19. Frequency spectrum of mode-locked laser pulse train. Upper trace shows unshifted spectrum and lower trace shows frequency-shifted spectrum after pulse train has passed through modulator as described in text [146].

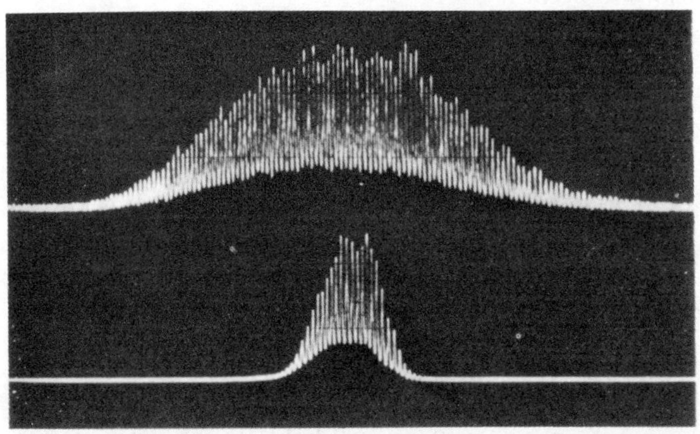

Fig. 20. Frequency spectrum of mode-locked laser pulse train. Lower trace shows spectrum of pulse train entering modulator and upper trace shows increase in width of frequency spectrum after passing through modulator giving frequency sweep to pulses as described in text [60].

pulse. If the pulse passes through the modulator $\pi/2$ later during the modulator cycle, it will be up- or down-shifted in frequency in passing through the device. Fig. 19 shows the frequency spectrum of mode-locked He–Ne laser pulses which have been frequency shifted by 4.8 GHz by passing through such a device [147]. Fig. 20 shows the change in bandwidth obtained by giving the pulses a frequency sweep [60]. These pulses may then be compressed in time to a minimum duration of the order of the reciprocal bandwidth by passing them through a suitable dispersive element [60], [61]. [161]. In this way pulse compression of about a factor of two was demonstrated with a He–Ne laser [61]. Fisher *et al.* [151a] have calculated theoretically that mode-locked pulses from a Nd:glass laser could be compressed to 0.05 ps by using a medium with an intensity-dependent index of refraction (Kerr effect) to give a frequency sweep to the pulses and then passing them through a suitable dispersive element. Laubereau [151b] has reported experimental pulsewidth compression of a factor of 10 using this technique, with CS_2 as the Kerr-effect medium. Hargrove [152] has described the reduction of pulse durations by interferometric combination of frequency shifted pulses.

Recently, Treacy [148]–[149] has reported that the pulses from a Nd:glass laser mode-locked with a saturable absorber can be compressed by passing them through a suitable dispersive element. This indicates that some kind of frequency sweep is occurring within the pulses. A theory of this effect has been given by Cubeddu and Svelto [130], on the assumption that the pulses have a linear frequency sweep caused by the dispersion of the glass matrix in which the Nd atoms are situated. A recent paper by Fisher and Fleck [150] indicates that the frequency sweep may not be a simple linear one. Duguay *et al.* [162] have discussed the effect of the nonlinear index of the glass in producing self-phase modulation of these pulses. So far the Nd:glass laser seems to be the only laser for which such a frequency sweep has been discovered. Smith [151] has shown that no frequency sweep is detectable with the 6328 Å He–Ne laser system when self-locked, or when mode-locked with a Ne absorption cell.

C. Resonant Interaction of Radiation with Matter: Dynamic Polarization Effects

The availability of short intense pulses of coherent radiation has stimulated work on the resonant interaction of radiation with atomic systems. In particular, it is known that pulses of radiation of duration less than the polarization memory time (T_2) of a resonant medium will interact with this medium in a fundamentally different way than will much longer pulses. These effects have been known and studied for some time in the microwave region of the spectrum and recently these ideas and techniques have been applied to the interaction of light pulses with matter. Examples of these studies are photon echo studies [153] and studies of "self-induced transparency" by McCall and Hahn [76].

Fox and Smith[31] and Smith [32] have shown that due to dynamic polarization effects a mode-locked laser pulse may interact with gain medium in such a way that the population difference is reduced below zero after passage of the pulse through the medium, i.e., the laser medium is left in an absorbing state. This type of operation is illustrated in Figs. 7 and 8. Recent studies of the propagation of mode-locked 6328 Å He–Ne laser pulses through a resonant absorber [154], [154a] have also demonstrated the existence of dynamic polarization effects within this system. Mode-locked lasers are a convenient tool for the study of such effects and it is expected that much more work in this area will soon be reported.

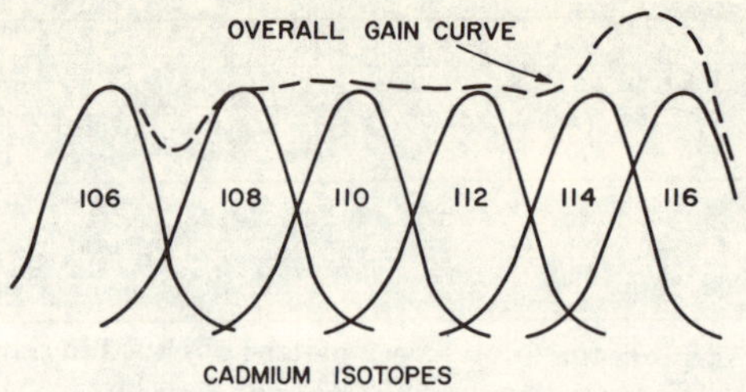

Fig. 21. Gain profile obtained using a mixture of Cd isotopes as indicated. The individual of isotope gain curves have a width of approximately 1200 MHz. Total gain curve width is approximately 8000 MHz.

D. Communications Systems

The ability of a mode-locked laser to emit a train of intense narrow pulses makes it attractive as the source of the carrier for an optical pulse-code modulation system. This type of system has been studied by Denton and Kinsel [155], [155a]. Kinsel [156] describes an optical pulse-code modulation system capable of handling many orders of magnitude more information than currently used microwave systems. When the need for large information capacity transmission systems stimulates their construction, some type of mode-locked laser may well be used as the source of the carrier pulse train.

E. Mode-Locking New Laser Systems

There is still interest in mode-locking new laser systems whose characteristics may be particularly useful for certain experiments. Faxvog *et al.* [157] have recently reported observation of self-locking of the Cd^+ laser operating at 4416 Å. Silfvast and Smith [158] have mode-locked the Cd^+ laser using an internal accoustic loss modulator at both 4416 Å and 3250 Å in the ultraviolet. As the isotope shift for these Cd^+ lines is of the order of the width of the gain curve for a single isotope, a broad gain region can be obtained by using a mixture of isotopes. A broad oscillation spectrum gives the possibility of narrow pulse output. Fig. 21 shows the gain profile obtainable using a mixture of the available Cd^+ isotopes. By using such a mixture Silfvast and Smith [158] obtained an oscillation spectrum of roughly 8000 MHz in width and were able to obtain a continuous train of pulses with pulsewidths of the order of 100 ps.

VII. Concluding Remarks

We have attempted to present a tutorial survey of the subject of mode-locking in lasers with adequate references so that the interested reader may find more detailed information if required. A minimum of mathematics has been used and physical pictures of the processes involved have been presented wherever possible.

References

[1] A. J. DeMaria, W. H. Glenn, Jr., M. J. Brienza, and M. E. Mack, "Picosecond laser pulses," *Proc. IEEE*, vol. 57, pp. 2–25, January 1969.

[2] K. Gürs and R. Müller, "Breitband-Modulation durch Steuerung der Emission eines Optischen Masers (Auskopple-modulation)," *Phys. Lett.*, vol. 5, pp. 179–181, July 1963.

[2a] K. Gürs, "Beats and modulation in optical ruby masers," *Quantum Electronics III*, P. Grivet and N. Bloembergen, Eds. New York: Columbia University Press, 1964, pp. 1113–1119.

[3] H. Statz and C. L. Tang, "Zeeman effect and nonlinear interactions between oscillating laser modes," *Quantum Electronics III*, P. Grivet and N. Bloembergen, Eds. New York: Columbia University Press, 1964, pp. 469–498.

[4] M. DiDomenico, "Small-signal analysis of internal (coupling type) modulation of lasers," *J. Appl. Phys.*, vol. 35, pp. 2870–2876, October 1964.

[5] L. E. Hargrove, R. L. Fork, and M. A. Pollack, "Locking of He-Ne laser modes induced by synchronous intracavity modulation," *Appl. Phys. Lett.*, vol. 5, pp. 4–5, July 1964.

[6] W. E. Lamb, Jr., "Theory of an optical maser," *Phys. Rev.*, vol. 134, pp. A1429–A1450, June 1964.

[7] M. H. Crowell, "Characteristics of mode-coupled lasers," *IEEE J. Quantum Electron.*, vol. QE-1, pp. 12–20, April 1965.

[8] A. Yariv, "Internal modulation in multimode laser oscillators," *J. Appl. Phys.*, vol. 36, pp. 388–391, February 1965.

[9] S. E. Harris and O. P. McDuff, "FM laser oscillation theory," *Appl. Phys. Lett.*, vol. 5, pp. 205–206, November 1964.

[10] S. E. Harris and R. Targ, "FM oscillation of the He-Ne laser," *Appl. Phys. Lett.*, vol. 5, pp. 202–204, November 1964.

[11] S. E. Harris and O. P. McDuff, "Theory of FM laser oscillation," *IEEE J. Quantum Electron.*, vol. QE-1, pp. 245–262, September 1965.

[12] E. O. Ammann, B. J. McMurtry, and M. K. Oshman, "Detailed experiments on helium–neon FM lasers," *IEEE J. Quantum Electron.*, vol.QE-1, pp. 263–272, September 1965.

[13] A. J. DeMaria and D. A. Stetser, "Laser pulse-shaping and mode-locking with acoustic waves," *Appl. Phys. Lett.*, vol. 7, pp. 71–73, August 1, 1965.

[14] T. Deutsch, "Mode-locking effects in an internally modulated ruby laser," *Appl. Phys. Lett.*, vol. 7, pp. 80–82, August 15, 1965.

[15] H. W. Mocker and R. J. Collins, "Mode competition and self-locking effects in a *Q*-switched ruby laser," *Appl. Phys. Lett.*, vol. 7, pp. 270–273, November 15, 1965.

[15a] W. W. Rigrod "Mode-locked laser pulse generator," U. S. Patent 3 492 599, filed September 17, 1965.

[16] H. Statz and C. L. Tang, "Phase locking of modes in lasers," *J. Appl. Phys.*, vol. 36, pp. 3923–3927, December 1965.

[17] A. J. De Maria, C. M. Ferrar, and G. E. Danielson, "Mode locking of a Nd^{3+} doped glass laser," *Appl. Phys. Lett.*, vol. 8, pp. 22–24, January 1, 1966.

[18] A. J. DeMaria, D. A. Stetser, and H. Heynau, "Self mode-locking of lasers with saturable absorbers," *Appl. Phys. Lett.*, vol. 8, pp. 174–176, April 1, 1966.

[19] M. DiDomenico, H. M. Marcos, J. E. Geusic, and R. E. Smith, "Generation of ultrashort optical pulses by mode locking the YAG:Nd laser," *Appl. Phys. Lett.*, vol. 8, pp. 180–182, April 1, 1966.

[19a] R. L. Kohn and R. H. Pantell, "Second harmonic enhancement with an internally-modulated ruby laser," *Appl. Phys. Lett.*, vol. 8, pp. 231–233, May 1966.

[20] M. Maier, W. Kaiser, and J. A. Giordmaine, "Intense light bursts in the stimulated Raman effect," *Phys. Rev. Lett.*, vol. 17, pp. 1275–1277, December 26, 1966.

[21] J. A. Armstrong, "Measurement of picosecond laser pulse widths," *Appl. Phys. Lett.*, vol. 10, pp. 16–18, January 1, 1967.

[22] H. P. Weber, "Method for pulse width measurement of ultrashort light pulses generated by phase-locked lasers using nonlinear optics," *J. Appl. Phys.*, vol. 38, pp. 2231–2234, April 1967.

[23] J. A. Giordmaine, P. M. Rentzepis, S. L. Shapiro, and K. W. Wecht, "Two photon excitation of fluorescence by picosecond light pulses," *Appl. Phys. Lett.*, vol. 11, pp. 216–218, October 1, 1967.

[24] H. P. Weber, "Comments on the pulse width measurement with two-photon excitation of fluorescence," *Phys. Lett.*, vol. 27A, pp. 321–322, July 15, 1968.

[25] J. R. Klauder, M. A. Duguay, J. A. Giordmaine, and S. L. Shapiro, "Correlation effects in the display of picosecond pulses by two-photon techniques," *Appl. Phys. Lett.*, vol. 13, pp. 174–176, September 1968.

[26] M. Michon, J. Ernest, and R. Auffret, "Pulsed transmission mode operation in the case of a mode locking of the modes of a non *Q*-spoiled ruby laser," *Phys. Lett.*, vol. 21, pp. 514–515, June 15, 1966.

[27] A. W. Penney and H. A. Heynau, "PTM single-pulse selection from a mode-locked neodymium 3^+ glass laser using a bleachable dye," *Appl. Phys. Lett.*, vol. 9, pp. 257–258, October 1, 1966.

[28] R. N. Zitter, W. H. Steier, and R. Rosenberg, "Fast pulse dumping and power buildup in a mode-locked He–Ne laser," *IEEE J. Quantum Electron.*, vol. QE-3, pp. 614–617, November 1967.

[28a] G. Kachen, L. Steinmetz, and J. Kysilka, "Selection and amplification of a single mode-locked optical pulse," *Appl. Phys. Lett.*, vol. 13, pp. 229–231, October 1, 1968.

[29] P. W. Smith, "Phase locking of laser modes by continuous cavity length variation," *Appl. Phys. Lett.*, vol. 10, pp. 51–53, January 15, 1967.

[30] S. E. Harris, "Stabilization and modulation of laser oscillators by internal time-varying perturbation," *Proc. IEEE*, vol. 54, pp. 1401–1413, October 1966.

[31] A. G. Fox and P. W. Smith, "Mode-locked laser and the 180° pulse," *Phys. Rev. Lett.*, vol. 18, pp. 826–828, May 15, 1967.

[32] P. W. Smith, "The self-pulsing laser oscillator," *IEEE J. Quantum Electron.*, vol. QE-3, pp. 627–635, November 1967.

[33] H. Statz, "On the conditions for self-locking of modes in lasers," *J. Appl. Phys.*, vol. 38, pp. 4648–4655, November 1967.

[34] H. Statz, G. A. DeMars, and C. L. Tang, "Self-locking of modes in lasers," *J. Appl. Phys.*, vol. 38, pp. 2212–2222, April 1967.

[35] C. L. Tang and H. Statz, "Maximum-emission principle and phase locking in multimode lasers," *J. Appl. Phys.*, vol. 38, pp. 2963–2968, June 1967.

[35a] ——, "Large-signal effects in self-locked lasers," *J. Appl. Phys.*, vol. 39, pp. 31–35, January 1968.

[35b] H. Statz and M. Bass, "Locking in multimode solid-state lasers," *J. Appl. Phys.*, vol. 40, pp. 377–383, January 1969.

[36] D. E. Caddes, L. M. Osterink, and R. Targ, "Mode locking of the CO_2 laser," *Appl. Phys. Lett.*, vol. 12, pp. 74–76, February 1, 1968.

[37] O. R. Wood and S. E. Schwarz, "Passive mode locking of a CO_2 laser," *Appl. Phys. Lett.*, vol. 12, pp. 263–265, April 15, 1968.

[37a] C. M. Ferrar, "*Q*-switching and mode locking of a CF_3I photolysis laser," *Appl. Phys. Lett.*, vol. 12, pp. 381–383, June 1, 1968.

[38] W. H. Glenn, M. J. Brienza, and A. J. DeMaria, "Mode locking of an organic dye laser," *Appl. Phys. Lett.*, vol. 12, pp. 54–56, January 15, 1968.

[39] D. J. Bradley and A. J. F. Durrant, "Generation of ultrashort dye laser pulses by mode locking," *Phys. Lett.*, vol. 27A, pp. 73–74, June 3, 1968.

[40] H. Samuelson and A. Lempicki, "*Q*-switching and mode locking of Nd^{3+}:$SeOCl_2$ liquid laser," *J. Appl. Phys.*, vol. 39, pp. 6115–6116, December 1968.

[41] W. Schmidt and F. P. Schafer, "Self-mode-locking of dye-lasers with saturable absorbers," *Phys. Lett.*, vol. 26A, pp. 558–559, April 22, 1968.

[42] V. N. Morozov, V. V. Nikitin, and A. A. Sheronov, "Self-synchronization of modes in a GaAs semiconducting injection laser," *JETP Lett.*, vol. 7, pp. 256–258, May 1968.

[43] A. G. Fox, S. E. Schwarz, and P. W. Smith, "Use of neon as a nonlinear absorber for mode locking a He-Ne laser," *Appl. Phys. Lett.*, vol. 12, pp. 371–373, June 1, 1968.

[43a] V. P. Chebotayev, I. M. Beterov, and V. N. Lisitsyn, "Selection and self-locking of modes in a He–Ne laser with nonlinear absorption," *IEEE J. Quantum Electron.*, vol. QE-4, pp. 788–790, November 1968.

[44] J. Comly, E. Garmire, J. P. Laussade, and A. Yariv, "Observation of mode locking and ultrashort optical pulses induced by anisotropic molecular liquids," *Appl. Phys. Lett.*, vol. 13, pp. 176–178, February 1, 1968.

[45] J. P. Laussade and A. Yariv, "Mode locking and ultrashort laser pulses by anisotropic molecular liquids," *Appl. Phys. Lett.*, vol. 13, pp. 65–66, July 15, 1968.

[46] G. R. Huggett, "Mode-locking of CW lasers by regenerative RF feedback," *Appl. Phys. Lett.*, vol. 13, pp. 186–187, September 1, 1968.

[47] D. H. Auston, "Transverse mode locking," *IEEE J. Quantum Electron.* (Corresp.), vol. QE-4, pp. 420–422, June 1968.

[48] ——, "Forced and spontaneous phase locking of the transverse

modes of a He–Ne laser," *IEEE J. Quantum Electron.* (Corresp.), vol. QE-4, pp. 471–473, July 1968.

[49] P. W. Smith, "Simultaneous phase-locking of longitudinal and transverse laser modes," *Appl. Phys. Lett.*, vol. 13, pp. 235–237, October 1, 1968.

[50] M. Michon, J. Ernest, and R. Auffert, "Passive transverse mode locking in a confocal Nd^{3+} glass laser," *IEEE J. Quantum Electron.* (Corresp.), vol. QE-5, pp. 125–126, February 1969.

[51] A. A. Mak and V. A. Fromzel, "Observation of self-synchronization of transverse modes in a solid-state laser," *JETP Lett.*, vol. 10, pp. 199–201, October 5, 1969.

[52] H. Ito and H. Inaba, "Self mode locking of the transverse modes in the CO_2 laser oscillator," *Opt. Commun.*, vol. 1, pp. 61–63, May 1969.

[53] V. S. Arakelyan, N. V. Karlov, and A. M. Prokhorov, "Self synchronization of transverse modes of a CO_2 laser," *JETP Lett.*, vol. 10, pp. 178–180, September 20, 1969.

[54] G. A. Massey, M. K. Oshman, and R. Targ, "Generation of single frequency light using the FM laser," *Appl. Phys. Lett.*, vol. 6, pp. 10–11, January 1965.

[55] S. E. Harris, M. K. Oshman, B. J. McMurtry, and E. O. Ammann, "Proposed frequency stabilization of the FM laser," *Appl. Phys. Lett.*, vol. 7, pp. 184–186, October 1965.

[56] S. E. Harris and B. J. McMurtry, "Frequency selective coupling to the FM laser," *Appl. Phys. Lett.*, vol. 7, pp. 265–267, November 15, 1965.

[57] M. A. Duguay, K. B. Jefferts, and L. E. Hargrove, "Optical frequency translation of mode-locked laser pulses," *Appl. Phys. Lett.*, vol. 9, pp. 287–290, October 15, 1966.

[58] H. A. Heynau and A. W. Penney, "An application of mode-locked, laser generated picosecond pulses to the electrical measurement art," *1967 IEEE Int. Conv. Rec.*, vol. 15, pt. 7, pp. 80–86.

[59] J. W. Shelton and J. A. Armstrong, "Measurement of the relaxation time of the Eastman 9740 bleachable dye," *IEEE J. Quantum Electron.* (Corresp.), vol. QE-3, pp. 696–697, December 1967.

[59a] M. E. Mack, "Measurement of nanosecond fluorescence decay time," *J. Appl. Phys.*, vol. 39, pp. 2483–2484, April 1968.

[59b] R. I. Scarlet, J. F. Figueira, and H. Mahr, "Direct measurement of picosecond lifetimes," *Appl. Phys. Lett.*, vol. 13, pp. 71–73, July 15, 1968.

[60] J. A. Giordmaine, M. A. Duguay, and J. W. Hansen, "Compression of optical pulses," *IEEE J. Quantum Electron.*, vol. QE-4, pp. 252–255, May 1968.

[61] M. A. Duguay and J. W. Hansen, "Compression of pulses from a mode-locked He-Ne laser," *Appl. Phys. Lett.*, vol. 14, pp. 14–15, January 1, 1969.

[61a] M. J. Brienza, A. J. DeMaria, and W. H. Glenn, "Optical rectification of mode-locked laser pulses," *Phys. Lett.*, vol. 26A, pp. 390–391, March 1968.

[62] P. M. Rentzepis, "Direct measurements of radiationless transitions in liquids," *Chem. Phys. Lett.*, vol. 2, pp. 117–120, June 1968.

[63] M. A. Duguay and J. W. Hansen, "An ultrafast light gate," *Appl. Phys. Lett.*, vol. 15, pp. 192–194, September 1969.

[64] See, for example, N. G. Basov, P. G. Kriukov, S. D. Zakharov, Y. V. Senatsky, and S. V. Tchekalin, "Experiments on the observation of neutron emission at the focus of high-power laser radiation on a lithium deuteride surface," *IEEE J. Quantum Electron.*, vol. QE-4, pp. 864–867, November 1968.

[65] For an excellent review of this subject see, H. Kogelnik and T. Li, "Laser beams and resonators," *Appl. Opt.*, vol. 5, pp. 1550–1567, October 1966.

[66] W. R. Bennett, Jr., "Hole burning effects in a He-Ne laser," *Phys. Rev.*, vol. 126, pp. 580–593, April 15, 1962.

[67] O. L. Gaddy and E. M. Schaefer, "Self locking of modes in the argon ion laser," *Appl. Phys. Lett.*, vol. 9, pp. 281–282, October 15, 1966.

[68] T. Uchida, "Dynamic behavior of gas lasers," *IEEE J. Quantum Electron.*, vol. QE-3, pp. 7–16, January 1967.

[68a] S. E. Schwarz and P. L. Gordon, "Hamilton's principle and the maximum emission coincidence," *J. Appl. Phys.*, vol. 40, pp. 4441–4447, October 1968.

[69] A. Bambini and P. Burlamacchi, "Stability conditions for mode-locked gas lasers," *IEEE J. Quantum Electron.* (Corresp.), vol. QE-4, pp. 101–102, March 1968.

[70] L. N. Magdich, "Nonstationary phenomena in a laser with interacting modes," *Sov. Phys.—JETP*, vol. 26, pp. 492–494, March 1968.

[71] J. A. Fleck, "Emission of pulse trains by Q-switched lasers," *Phys. Rev. Lett.*, vol. 21, pp. 131–133, July 15, 1968.

[72] D. G. C. Jones, M. D. Sayers, and L. Allen, "Mode self-locking in gas lasers," *J. Phys. A, Proc. Phys. Soc. London (Gen)*, ser. 2, vol. 2, pp. 95–101, January 1969.

[73] D. G. C. Jones, "Self-locking of three modes in the He–Ne laser," *Appl. Phys. Lett.*, vol. 13, pp. 301–302, November 1, 1968.

[74] H. Risken and K. Nummedal, "Self-pulsing in lasers," *J. Appl. Phys.*, vol. 39, pp. 4662–4672, September 1968.

[75] For a discussion of frequency pulling of a laser see, for example, R. H. Pantell, "The laser oscillator with an external signal," *Proc. IEEE*, vol. 53, pp. 474–477, May 1965.

[75a] M. Sargent III, "Mode-locking according to the Lamb theory," *IEEE J. Quantum Electron.*, vol. QE-4, p. 346, May 1968.

[76] For a recent discussion of π and 2π pulses and their interaction with an atomic system see S. L. McCall and E. L. Hahn, "Self-induced transparency," *Phys. Rev.*, vol. 183, pp. 457–485, July 10, 1969.

[77] R. E. McClure, "Mode locking behavior of gas lasers in long cavities," *Appl. Phys. Lett.*, vol. 7, pp. 148–150, September 15, 1965.

[77a] V. V. Korbkin and M. Y. Schelev, "Self-locking of axial emission modes of a ruby laser in the free generation regime," *Sov. Phys.—JETP*, vol. 26, pp. 721–722, April 1968.

[78] M. A. Duguay, S. L. Shapiro, and P. M. Rentzepis, "Spontaneous appearance of picosecond pulses in ruby and Nd:glass lasers," *Phys. Rev. Lett.*, vol. 19, pp. 1014–1016, October 30, 1967.

[78a] S. L. Shapiro, M. A. Duguay, and L. B. Kreuzer, "Picosecond substructure of laser spikes," *Appl. Phys. Lett.*, vol. 12, pp. 36–37, January 15, 1968.

[79] M. Bass and D. Woodward, "Observation of picosecond pulses from Nd:YAG lasers," *Appl. Phys. Lett.*, vol. 12, pp. 275–277, April 15, 1968.

[80] T. J. Bridges and P. K. Cheo, "Spontaneous self-pulsing and cavity dumping in a CO_2 laser with electro-optic Q-switching," *Appl. Phys. Lett.*, vol. 14, pp. 262–264, May 1, 1969.

[81] V. I. Malyshev, A. S. Markin, A. V. Masalov, and A. A. Sychov, "On mode locking in ruby and neodymium lasers operating under free oscillation conditions," *Zh. Eksp. Teor. Fiz.*, vol. 57, pp. 834–840, 1969.

[82] T. Uchida and A. Ueki, "Self locking of gas lasers," *IEEE J. Quantum Electron.*, vol. QE-3, pp. 17–30, January 1967.

[83] F. R. Nash, "Observations of spontaneous phase locking of TEM_{00q} modes at 0.63μ," *IEEE J. Quantum Electron.*, vol. QE-3, pp. 189–196, May 1967.

[84] J. Hirano and T. Kimura, "Generation of high-repetition-rate optical pulses by a He-Ne laser," *Appl. Phys. Lett.*, vol. 12, pp. 196–198, March 1, 1968.

[85] J. Hirano and T. Kimura, "Multiple mode locking of lasers," *IEEE J. Quantum Electron.*, vol. QE-5, pp. 219–224, May 1969.

[86] M. DiDomenico and V. Czarniewski, "Locking of He-Ne laser modes by intracavity acoustic modulation in coupled interferometers," *Appl. Phys. Lett.*, vol. 6, pp. 150–152, April 15, 1965.

[86a] C. M. Ferrar, "Mode-locked flashlamp-pumped coumarin dye laser at 4600 Å," *IEEE J. Quantum Electron.* (Corresp.), vol. QE-5, pp. 550–551, November 1969.

[87] L. C. Foster, M. D. Ewy, and C. B. Crumly, "Laser mode locking by an external Doppler cell," *Appl. Phys. Lett.*, vol. 6, pp. 6–8, January 1, 1965.

[88] M. Michon, J. Ernest, and R. Auffret, "Mode locking of a Q-spoiled Nd^{3+} doped glass laser by intracavity phase modulation," *Phys. Lett.*, vol. 23, pp. 457–458, November 14, 1966.

[89] G. A. Massey, "Laser mode control by internal modulation using the transverse electrooptic effect in quartz," *Appl. Opt.*, vol. 5, pp. 999–1001, June 1966.

[90] R. H. Pantell, H. E. Puthoff, B. G. Huth, and R. L. Kohn, "Mode coupling in an external Raman resonator," *Appl. Phys. Lett.*, vol. 9, pp. 104–106, August 1, 1966.

[91] W. C. Henneberger and H. J. Schulte, "Optical pulses produced by laser length variation," *J. Appl. Phys.* (Communications), vol. 37, p. 2189, April 1966.

[92] R. H. Pantell and R. L. Kohn, "Mode coupling in a ruby laser," *IEEE J. Quantum Electron.*, vol. QE-2, pp. 306–310, August 1966.

[93] K. Gürs, "Modulation and mode locking of the continuous ruby laser," *IEEE J. Quantum Electron.*, vol. QE-3, pp. 175–180, May 1967.

[94] V. Degiorgio and B. Querzola, "Output characteristics of a mode-locked laser," *Nuovo Cimento*, vol. 55B, pp. 272–275, May 11, 1968.

[95] L. M. Osterink and J. D. Foster, "A mode-locked Nd:YAG laser," *J. Appl. Phys.*, vol. 39, pp. 4163–4165, August 1968.

[96] R. Targ and J. M. Yarborough, "Mode-locked quieting of He-Ne and argon lasers," *Appl. Phys. Lett.*, vol. 12, pp. 3–4, January 1, 1968.

[97] T. R. Gurski, "Simultaneous mode-locking and second-harmonic

generation by the same nonlinear crystal," *Appl. Phys. Lett.*, vol. 15, pp. 5–6, July 1, 1969.
[98] L. B. Allen, R. R. Rice, and R. F. Mathews, "Two cavity mode-locking of a He-Ne laser," *Appl. Phys. Lett.*, vol. 15, pp. 416–418, December 15, 1969.
[99] T. Uchida, "Direct modulation of gas lasers," *IEEE J. Quantum Electron.*, vol. QE-1, pp. 336–343, November 1965.
[100] G. W. Hong and J. R. Whinnery, "Switching of phase-locked states in the intracavity phase-modulated He-Ne laser," *IEEE J. Quantum Electron.*, vol. QE-5, pp. 367–376, July 1969.
[101] A. Yariv, "Parametric interactions of optical modes," *IEEE J. Quantum Electron.*, vol. QE-2, pp. 30–37, February 1966.
[102] L. N. Magdich, "Laser mode interaction in the course of *Q*-switching," *Soviet Phys.—JETP*, vol. 24, pp. 11–15, January 1967.
[103] O. P. McDuff, "Internal modulation of lasers," *1967 IEEE Region 3 Conv.*, pp. 121–132.
[104] O. P. McDuff and S. E. Harris, "Nonlinear theory of the internally loss-modulated laser," *IEEE J. Quantum Electron.*, vol. QE-3, pp. 101–111, March 1967.
[105] L. N. Magdich, "Laser mode synchronization by dielectric constant modulation," *Soviet Phys.—JETP*, vol. 25, pp. 223–226, August 1967.
[106] Y. P. Yegorov and A. S. Petrov, "Axial mode locking in internally modulated gas lasers," *Radio Eng. Electron Phys.* (USSR), vol. 12, pp. 1365–1373, August 1967.
[107] H. Haken and M. Pauthier, "Nonlinear theory of multimode action in loss modulated lasers," *IEEE J. Quantum Electron.*, vol. QE-4, pp. 454–459, July 1968.
[108] V. S. Letokhov, "Dynamics of generation of pulsed mode-locking lasers," *Sov. Phys.—JETP*, vol. 27, pp. 746–751, November 1968.
[109] O. P. McDuff and A. L. Pardue, Jr., "Theory of laser mode coupling produced by cavity-length modulation," *IEEE J. Quantum Electron.* (Corresp.), vol. QE-4, pp. 99–101, March 1968.
[110] P. J. Titterton, "Quantum theory of an internally phase-modulated laser—I.," *IEEE J. Quantum Electron.*, vol. QE-4, pp. 85–92, March 1968.
[111] A. E. Siegman and D. J. Kuizenga, "Simple analytic expressions for AM and FM mode-locked pulses in homogeneous lasers," *Appl. Phys. Lett.*, vol. 14, pp. 181–182, March 15, 1969.
[112] J. B. Gunn, "Spectrum and width of mode-locked laser pulses," *IEEE J. Quantum Electron.*, vol. QE-5, pp. 513–516, October 1969.
[113] A. W. Smith, "Effect of host dispersion on mode-locked lasers," *Appl. Phys. Lett.*, vol. 15, pp. 194–196, September 15, 1969.
[114] D. A. Stetser and A. J. DeMaria, "Optical spectra of ultrashort optical pulses generated by mode-locked glass/neodymium lasers," *Appl. Phys. Lett.*, vol. 9, pp. 118–120, August 1, 1966.
[115] A. J. DeMaria, "Ultrashort laser pulses by simultaneously mode-locking and *Q*-switching Nd^{3+} glass lasers," *1967 NEREM* Rec., pp. 34–35, 1967.
[116] A. J. DeMaria, D. A. Stetser, and W. H. Glenn, "Ultrashort light pulses," *Science*, vol. 156, pp. 1557–1568, 1967.
[117] V. I. Malyshev, A. S. Markin, and A. A. Sychev, "Mode self-synchronization in giant pulse of a ruby laser with broad spectrum," *JETP Lett.*, vol. 6, pp. 34–35, July 15, 1967.
[118] A. Schmackpfeffer and H. Weber, "Mode locking and mode competition by saturable absorbers in a ruby laser," *Phys. Lett.*, vol. 24A, pp. 190–191, January 30, 1967.
[119] R. Harrach and G. Kachen, "Pulse trains from mode-locked lasers," *J. Appl. Phys.*, vol. 39, pp. 2482–2483, April 1968.
[120] M. E. Mack, "Mode locking the ruby laser," *IEEE J. Quantum Electron.* (Corresp.), vol. QE-4, pp. 1015–1016, December 1968.
[121] A. R. Clobes and M. J. Brienza, "Passive mode locking of a pulsed Nd:YAG laser," *Appl. Phys. Lett.*, vol. 14, pp. 287–288, May 1, 1969.
[121a] P. C. Magnante, "Mode-locked and bandwidth narrowed Nd: glass laser," *J. Appl. Phys.*, vol. 40, pp. 4437–4440, October 1969.
[122] R. Cubeddu, R. Polloni, C. A. Sacchi, and O. Svelto, "Picosecond pulses, TEM_{00} mode, mode-locked ruby laser," *IEEE J. Quantum Electron.* (Corresp.), vol. QE-5, pp. 470–471, September 1969.
[123] D. J. Bradley, A. J. F. Durrant, F. O'Neill, and B. Sutherland, "Picosecond pulses from mode-locked dye lasers," *Phys. Lett.*, vol. 30A, pp. 535–536, December 29, 1969.
[123a] L. D. Derkachyova, A. I. Krymova, V. I. Malyshev, and A. S. Markin, "Mode locking in polymethylene dye lasers," *Opt. Spect.*, vol. 26, pp. 572–573, June 1969.
[124] J. H. McCoy, "Continuous passive mode locking of a CO_2 laser," *Appl. Phys. Lett.*, vol. 15, pp. 357–360, December 1, 1969.
[125] C. C. Cutler, "The regenerative pulse generator," *Proc. IRE*, vol. 43, pp. 140–148, February 1955.
[126] E. M. Garmire and A. Yariv, "Laser mode-locking with saturable absorbers," *IEEE J. Quantum Electron.*, vol. QE-3, pp. 222–226, June 1967; "Correction," *IEEE J. Quantum Electron.*, vol. QE-3, p. 377, September 1967.
[126a] J. A. Fleck, "Mode-locked pulse generation in passively switched lasers," *Appl. Phys. Lett.*, vol. 12, pp. 178–181, March 1, 1968.
[126b] ——, "Origin of short-pulse emission by passively switched lasers," *J. Appl. Phys.*, vol. 39, pp. 3318–3327, June 1968.
[127] V. S. Letokhov, "Formation of ultrashort pulses of coherent light," *JETP Lett.*, vol. 7, pp. 25–28, January 15, 1968.
[128] C. A. Sacchi, G. Soncini, and O. Svelto, "Self-locking of modes in a passive *Q*-switched laser," *Nuovo Cimento*, vol. 48B, pp. 58–71, March 11, 1967.
[129] V. S. Letokhov, "Generation of ultrashort light pulses in a laser with a nonlinear absorber," *Sov. Phys.—JETP*, vol. 28, pp. 562–568, 1969.
[129a] T. I. Kuznetsova, V. I. Malyshev, and A. S. Markin, "Self-synchronization of axial modes of a laser with saturable filters," *Soviet Phys.—JETP*, vol. 25, pp. 286–291, August 1967.
[130] R. R. Cubeddu and O. Svelto, "Theory of laser self-locking in the presence of host dispersion," *IEEE J. Quantum Electron.*, vol. QE-5, pp. 495–502, October 1969.
[130a] ——, "Effect of dispersion on laser self-locking," *Phys. Lett.*, vol. 29A, pp. 78–79, April 7, 1969.
[131] S. E. Schwarz, "Theory of an optical pulse generator," *IEEE J. Quantum Electron.*, vol. QE-4, pp. 509–514, September 1968.
[131a] V. S. Letokhov and V. N. Marzov, "Generation of ultrashort duration coherent light pulses," *Sov. Phys.—JETP*, vol. 25, pp. 862–866, November 1967.
[132] A. Bambini and P. Burlamacchi, "Phase locking of a multimode gas laser by means of low-frequency cavity-length modulation," *J. Appl. Phys.*, vol. 39, pp. 4864–4865, September 1968.
[133] J. P. Laussade and A. Yariv, "Analysis of mode locking and ultrashort laser pulses with a nonlinear refractive index," *J. Quantum Electron.*, vol. QE-5, pp. 435–441, September 1969.
[134] B. H. Soffer and J. W. Luin, "Continuously tunable picosecond pulse organic dye laser," *J. Appl. Phys.*, vol. 39, pp. 5859–5860, December 1968.
[135] D. Herriott, H. Kogelnik, and R. Kompfner, "Off-axis paths in spherical mirror interferometers," *Appl. Opt.*, vol. 3, pp. 523–526, April 1964.
[136] K. Kohiyama, T. Fujioka, and M. Kobayashi, "Self-locking of transverse higher-order modes in a He-Ne laser," *Proc. IEEE* (Letters), vol. 56, pp. 333–335, March 1968.
[137] H. E. Rowe and T. Li, "Theory of two-photon measurement of laser output," *IEEE J. Quantum Electron.*, vol. QE-6, pp. 49–67, January 1970.
[138] H. P. Weber and R. Dändliker, "Intensity interferometry by two-photon excitation of fluorescence," *IEEE J. Quantum Electron.*, vol. QE-4, pp. 1009–1013, December 1968.
[138a] A. A. Grutter, H. P. Weber, and R. Dändliker, "Imperfectly mode-locked laser emission and its effects on nonlinear optics," *Phys. Rev.*, vol. 185, pp. 629–643, September 10, 1969.
[139] See, for example, A. A. Malyutin and M. Y. Shchelev, "Investigation of the temporal structure of neodymium-laser emission in the mode self-locking regime," *JETP Lett.*, vol. 9, pp. 266–268, April 20, 1969.
[140] M. Bass and K. Andringa, "Reproducible optical second-harmonic generation using a mode-locked laser," *IEEE J. Quantum Electron.*, vol. QE-3, pp. 621–626, November 1967.
[141] W. H. Glenn, Jr., and M. J. Brienza, "Time evolution of picosecond optical pulses," *Appl. Phys. Lett.*, vol. 10, pp. 221–224, April 15 1967.
[142] I. K. Krasyuk, P. P. Pashkin, and A. M. Prokhorov, "Ring ruby laser for ultrashort pulses," *Soviet Phys.—JETP* (Lett.), vol. 7, pp. 89–91, February 1968.
[143] C. C. Wang and E. L. Baardsen, "Optical third harmonic generation using mode-locked and non mode-locked lasers," *Appl. Phys. Lett.*, vol. 15, pp. 396–398, December 15, 1969.
[143a] H. P. Weber and R. Dändliker, "Method for measuring the shape asymmetry of picosecond light pulses," *Phys. Lett.*, vol. 28A, pp. 77–78, November 4, 1968.
[143b] E. I. Blount and J. R. Klauder, "Recovery of laser intensity from

correlation data," *J. Appl. Phys.*, vol. 40, pp. 2874–2875, June 1969.

[144] R. C. Eckardt and C. H. Lee, "Optical third harmonic measurements of subpicosecond light pulses," *Appl. Phys. Lett.*, vol. 15, pp. 425–427, December 15, 1969.

[145] D. M. Thymian and J. A. Carruthers, "Second-harmonic enhancement using a self-locked 0.63-μ He–Ne laser," *IEEE J. Quantum Electron.*, vol. QE-5, pp. 83–86, February 1969.

[146] M. A. Duguay and J. W. Hansen, "Optical frequency shifting of a mode-locked laser beam," *IEEE J. Quantum Electron.*, vol. QE-4, pp. 477–481, August 1968.

[147] C. G. B. Garrett and M. A. Duguay, "Theory of the optical frequency translator," *Appl. Phys. Lett.*, vol. 9, p. 374, November 15, 1966.

[148] E. B. Treacy, "Compression of picosecond light pulses," *Phys. Lett.*, vol. 28A, pp. 34–35, October 1968.

[148a] ——, "Measurement of picosecond pulse substructure using compression techniques," *Appl. Phys. Lett.*, vol. 14, pp. 112–114, February 1, 1969.

[149] ——, "Optical pulse compression with diffraction gratings," *IEEE J. Quantum Electron.*, vol. QE-5, pp. 454–458, September 1969.

[150] R. A. Fisher and J. A. Fleck, Jr., "On the phase characteristics and compression of picosecond pulses," *Appl. Phys. Lett.*, vol. 15, pp. 287–290, November 1, 1969.

[151] P. W. Smith, "On the phase relationship between oscillating modes in a mode-locked 6328 Å laser," to be published.

[151a] R. A. Fisher, P. L. Kelley, and T. K. Gustafson, "Subpicosecond pulse generation using the optical Kerr effect," *Appl. Phys. Lett.*, vol. 14, pp. 140–143, February 15, 1969.

[151b] A. Laubereau, "External frequency modulation and compression of picosecond pulses," *Phys. Lett.*, vol. 29A, pp. 539–540, July 28, 1969.

[152] L. E. Hargrove, "Reduction of mode-locked-laser pulse duration by interferometric combination of frequency-shifted pulses," *J. Opt. Soc. Am.*, vol. 59, pp. 1680–1681, December 1969.

[153] I. D. Abella, N. A. Kurnit, and S. R. Hartmann, "Photon echoes," *Phys. Rev.*, vol. 141, pp. 391–406, January 1966.

[154] A. Frova, M. A. Duguay, C. G. B. Garrett, and S. L. McCall, "Pulse delay effects in the He–Ne laser mode-locked by a Ne absorption cell," *J. Appl. Phys.*, vol. 40, pp. 3969–3972, September 1969.

[154a] P. W. Smith, "Pulse velocity in a resonant absorber," *IEEE J. Quantum Electron.*, vol. QE-6, pp. 416–422, July 1970.

[155] R. T. Denton and T. S. Kinsel, "Terminals for a high-speed optical pulse code modulation communication system: I. 224-Mbit/s single channel," *Proc. IEEE*, vol. 56, pp. 140–145, February 1968.

[155a] T. S. Kinsel and R. T. Denton, "Terminals for a high-speed optical pulse code modulation communication system: II. optical multiplexing and demultiplexing," *Proc. IEEE*, vol. 56, pp. 146–154, February 1968.

[156] T. S. Kinsel, "Light wave of the future: optical PCM," *Electronics*, pp. 123–128, September 16, 1968.

[157] F. R. Faxvog, C. R. Willenbring, and J. A. Carruthers, "Self-pulsing in the He-Cd laser," *Appl. Phys. Lett.*, vol. 16, pp. 8–10, January 1, 1970.

[158] W. T. Silfvast and P. W. Smith, "Mode-locking the He-Cd laser at 4416 Å and 3250 Å," *Appl. Phys. Lett.*, vol. 17, pp. 70–73, July 15, 1970.

[159] G. W. Gobeli, J. C. Bushnell, P. S. Peercy, and E. D. Jones, "Observation of neutrons produced by laser irradiation of lithium deuteride," *Phys. Rev.*, vol. 188, pp. 300–302, December 1969.

[160] H. Kogelnik and W. W. Rigrod, "Visual display of isolated optical-resonator modes," *Proc. IRE* (Corresp.), vol. 50, p. 220, February 1962.

[161] J. W. Hansen, "Optical pulse compression," *Proc. Nat. Electron. Conf.*, vol. 24, pp. 148–151, 1968.

[162] M. A. Duguay, J. W. Hansen, and S. L. Shapiro, "Study of the Nd:glass laser radiation by means of two-photon fluorescence," to be published.

[163] V. S. Letokhov, "Ultrashort fluctuation pulses of light in a laser," *Soviet Phys.—JETP*, vol. 28, pp. 1026–1027, May 1969.

[164] T. I. Kuznetsova, "Concerning the problem of registration of ultrashort light pulses," *Soviet Phys.—JETP*, vol. 28, pp. 1303–1305, June 1969.

[165] R. H. Picard and P. Schweitzer, "Theory of intensity-correlation measurements on imperfectly mode-locked lasers," *Phys. Rev.*, vol. 1, pp. 1803–1819, June 1970.

Theory of FM Laser Oscillation

S. E. HARRIS, MEMBER, IEEE, AND O. P. McDUFF, SENIOR MEMBER, IEEE

Abstract—The paper presents a detailed analysis of FM laser oscillation which includes the effect of arbitrary atomic lineshape, saturation, and mode pulling. Such oscillation is achieved by means of an intracavity phase perturbation, and is a parametric oscillation wherein the laser modes oscillate with FM phases and nearly Bessel function amplitudes. One principal idea is that of the competition between different FM oscillations. The effect of the intracavity phase perturbation is to associate a set of sidebands with each of the previously free-running laser modes. While the free-running laser modes experienced their gain from essentially independent atomic populations, the competing FM oscillations to a large extent see the same atomic population; and in cases of interest the strongest of these oscillations is able to quench the competing weaker oscillations and establish the desired steady state condition. Results of the analysis include the following: threshold and power output, amplitudes and phases of all sidebands, frequency pulling of the entire oscillation, time domain behavior, distortion, supermode conversion efficiency, and effect of mirror motion. Results of numerical application of the theory to a number of specific cases are given.

Manuscript received July 27, 1965; revised August 13, 1965. The work reported in this paper was supported by the Aeronautical Systems Division of the United States Air Force under Contract AF 33(657)-11144. O. P. McDuff was supported by a National Science Foundation Science Faculty Fellowship.

The authors are with the Department of Electrical Engineering, Stanford University, Stanford, Calif.

I. Introduction

THE ATOMIC populations which support most optical maser oscillation are, in general, sufficiently inhomogeneously broadened to allow oscillation in a large number of relatively independent axial modes. To a large extent, the gain of these modes results from their independent interaction with essentially different atomic populations. In gas masers, this is primarily the result of Doppler broadening [1], while in solid state masers it is often the result of what may be termed a spatial broadening [2] wherein, due to their differing spatial variation, the different optical modes interact with different atoms. Modes of the optical resonator which are located sufficiently close to the center of the atomic fluorescence line such that their single pass gain is greater than their single pass loss will oscillate. These modes are driven by spontaneous emission; they saturate nearly independently; and in most instances are, to a good approximation, uncoupled. The output of such multimode masers is not nearly as coherent as is desired for many of the applications of communication and spectroscopy.

Reprinted from *IEEE J. Quantum Electron.*, vol. QE-1, pp. 245–262, Sept. 1965.

In this paper we present a theoretical description of FM laser oscillation. Such oscillation was first demonstrated by Harris and Targ [3], and is a parametric oscillation wherein the atomic population is made to support a single coherent FM oscillation. FM laser oscillation is depicted schematically in Fig. 1, and is achieved by means of an element which allows the path length of the optical resonator to be rapidly varied. Such an element, which we term an intracavity phase perturbation, is driven at a frequency which is approximately, but not exactly, the frequency of the axial mode interval. The resulting laser oscillation consists of a set of modes which have nearly Bessel function amplitudes and FM phases, and which is, in effect, swept over a major portion of the fluorescence line at a sweep frequency which is that of the drive frequency of the phase perturbation.

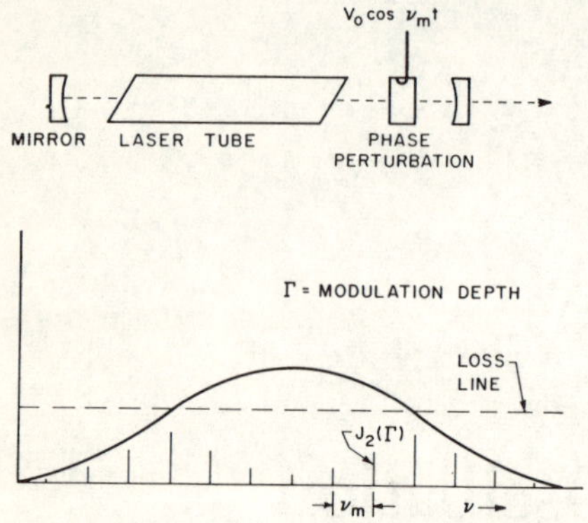

Fig. 1. Schematic of FM laser oscillation.

It will be seen that in order to achieve FM laser oscillation it is necessary for the intracavity phase perturbation to have a strength such that the parametric gain experienced by the laser modes is large compared with their net saturated atomic gains. From a physical point of view, the effect of the intracavity phase perturbation is to associate a set of FM sidebands with each of the previously free-running laser modes, i.e., to cause each of the previously free-running laser modes to become the center frequency or carrier of an FM signal. The resulting multiple FM oscillations then compete for the atomic population in the same sense as did the previously free-running modes. However, while the free-running modes experience their gain from essentially independent atomic populations, the competing FM oscillations, to a large extent, see the same atomic population. For instance, the first upper sideband of an FM oscillation which is centered in the atomic fluorescence line is in the same homogeneous linewidth and therefore sees the same atomic population as does the center frequency of an FM oscillation which is centered one mode above the center of the atomic fluorescence line. The competing FM oscillations are thus much more tightly coupled than were the previous free-running laser modes. In the cases of principal interest, the strongest of the FM oscillations—usually the oscillation whose carrier is at the center of the atomic line—will be able to completely quench the competing weaker oscillations. The result is that shown in the lower part of Fig. 1, wherein the sidesbands of a single coherent oscillation deplete most of the inverted population of the atomic line.

Many of the concepts applicable to free-running laser oscillation may be extended to apply to FM laser oscillation. The threshold of oscillation for a free-running laser mode occurs when gain exceeds loss. For FM laser oscillation, the gain and loss of a particular mode lose their significance, and we consider instead the threshold of the entire oscillation. This threshold will depend on the modulation depth of the oscillation and on a combination of the gains seen by all the FM sidebands. Similarly, instead of considering the pulling of a single free laser mode, we will consider the pulling of the entire FM oscillation.

The paper is divided into two major parts. Sections II through VIII develop the more analytical aspects of the theory, while Sections IX through XII give the results of numerical application of the theory to a number of specific cases and problems.

II. Development of the Basic Equations

We start with a set of equations derived by Lamb [4], termed as self-consistency equations, which describe the effect of an arbitrary optical polarization on the optical electric fields of a high-Q multimode optical resonator. They are as follows:

$$(\nu_n + \dot{\varphi}_n - \Omega_n)E_n = -\frac{1}{2}\left(\frac{\nu}{\epsilon_0}\right)C_n \tag{1a}$$

and

$$\dot{E}_n + \frac{1}{2}\left(\frac{\nu}{Q_n}\right)E_n = -\frac{1}{2}\left(\frac{\nu}{\epsilon_0}\right)S_n. \tag{1b}$$

In the foregoing equations, $E_n(t)$, ν_n, and $\varphi_n(t)$ are the amplitude, frequency, and phase, respectively, of the nth mode; and $C_n(t)$ and $S_n(t)$ are the in-phase and quadrature components of its driving polarization. That is, the total cavity electromagnetic field is given by

$$E(z, t) = \sum_n E_n(t) \cos[\nu_n t + \varphi_n(t)]U_n(z), \tag{2}$$

where $U_n(z) = \sin(n_0 + n)\pi z/L$.[1] The polarization driving the nth mode is obtained from

$$\begin{aligned} P_n(t) &= \frac{2}{L}\int_0^L P(z, t)U_n(z)\,dz \\ &= C_n(t)\cos[\nu_n t + \varphi_n(t)] + S_n(t)\sin[\nu_n t + \varphi_n(t)], \end{aligned} \tag{3}$$

where $P(z, t)$ is the total cavity polarization, and L is the cavity length. The addition of the integer n_0 to the cavity eigenfunction $U_n(z)$ is a departure from the notation of Lamb [4] but will lead to notational simplification in the following work. The integer n_0 is the number of spatial variations of some central mode, which

[1] Except where noted, all sums will be from $-\infty$ to $+\infty$.

we choose to be that mode whose frequency is closest to the center of the atomic fluorescence line. The circular frequency of this central mode is then $\Omega_0 = (n_0\pi c)/L$. Other symbols are defined as follows: $\Omega_n = \Omega_0 + (n\pi c)/L$ = frequency of the nth cavity resonance; $\Delta\Omega$ = frequency interval between axial resonances ($\Delta\Omega = \pi c/L$); $Q_n = Q$ of the nth mode; ν = average optical frequency.[2]

The total cavity polarization $P(z, t)$ consists of a parametric contribution resulting from the intracavity time varying phase perturbation, and of an atomic contribution resulting from the presence of the inverted atomic media. We assume the phase perturbation to have a time varying susceptibility $\Delta\chi'(z, t)$ of the form

$$\Delta\chi'(z, t) = \Delta\chi'(z) \cos \nu_m t, \tag{4}$$

where ν_m is the driving frequency of the perturbation. The parametric contribution to the total polarization $P(z, t)$ is then

$$\begin{aligned} P(z, t) &= \epsilon_0 \Delta\chi'(z, t) E(z, t) \\ &= \epsilon_0 \Delta\chi'(z) \cos \nu_m t E(z, t). \end{aligned} \tag{5}$$

Substituting (5) and (2) into (3), we find the component of $P(z, t)$ which drives the nth mode to be

$$\underset{\text{parametric}}{P_n(t)} = \frac{2\epsilon_0 \cos \nu_m t}{L} \sum_q E_q(t) \cos [\nu_q t + \varphi_q(t)] \cdot \int_0^L \Delta\chi'(z) U_q(z) U_n(z)\, dz. \tag{6}$$

We assume that the driving frequency ν_m is approximately equal to $\Delta\Omega$, and that the cavity Q's are sufficiently high that only the contributions of immediately adjacent modes need be retained. We thus keep only those terms of (6) for which $q = n \pm 1$. We define

$$\delta = \frac{\nu}{c} \int_0^L \Delta\chi'(z) U_{n\pm1}(z) U_n(z)\, dz, \tag{7}$$

which is the coupling coefficient between adjacent modes. The spatial variation of $\Delta\chi'(z)$ will generally be very slow compared with that of the cavity eigenfunction $U_n(z)$; thus by expanding (7) it is seen that δ is given by

$$\delta = \frac{1}{2}\frac{\nu}{c} \int_0^L \Delta\chi'(z) \cos \frac{\pi z}{L}\, dz. \tag{8}$$

Combining (6) and (7), we then have

$$\begin{aligned} \underset{\text{parametric}}{P_n(t)} = \frac{\epsilon_0 \delta c}{\nu L} \{&E_{n+1}(t) \cos [(\nu_{n+1} - \nu_m)t + \varphi_{n+1}(t)] \\ &+ E_{n-1}(t) \cos [(\nu_{n-1} + \nu_m)t + \varphi_{n-1}(t)]\}. \end{aligned} \tag{9}$$

We may arbitrarily choose the frequency of the nth cavity mode ν_n to be any frequency in the vicinity of Ω_n. In that ν_n appears together with the unknown $\dot{\varphi}_n$, any error in its initial choice will be corrected in the solution by the appearance of a constant $\dot{\varphi}_n$. We let

$$\nu_n = \Omega_0 + n\nu_m,$$

and therefore

$$\begin{aligned} \nu_{n+1} - \nu_m &= \Omega_0 + (n+1)\nu_m - \nu_m = \nu_n \\ \nu_{n-1} + \nu_m &= \Omega_0 + (n-1)\nu_m + \nu_m = \nu_n. \end{aligned} \tag{10}$$

We introduce the atomic contribution to the polarization by means of macroscopic quadrature and in-phase components of susceptibility, denoted by χ_n'' and χ_n', respectively. Here χ_n'' and χ_n' depend upon E_n and therefore include the effects of atomic saturation, power dependent mode pulling and pushing, and nonlinear coupling effects. We resolve $P_n(t)$ of (9) into in-phase and quadrature components of the form of (3). Adding the atomic polarizability terms, $C_n(t)$ and $S_n(t)$ then becomes

$$\begin{aligned} C_n(t) = \epsilon_0 \chi_n' E_n + \frac{\epsilon_0 \delta c}{\nu L} [&E_{n+1} \cos (\varphi_{n+1} - \varphi_n) \\ &+ E_{n-1} \cos (\varphi_n - \varphi_{n-1})] \end{aligned} \tag{11a}$$

$$\begin{aligned} S_n(t) = \epsilon_0 \chi_n'' E_n + \frac{\epsilon_0 \delta c}{\nu L} [&-E_{n+1} \sin (\varphi_{n+1} - \varphi_n) \\ &+ E_{n-1} \sin (\varphi_n - \varphi_{n-1})]. \end{aligned} \tag{11b}$$

We define the detuning $\Delta\nu$ to be the frequency difference between the axial mode interval and the driving frequency of the internal phase perturbation, i.e., $\Delta\nu = \Delta\Omega - \nu_m$. We then have

$$\Omega_n - \nu_n = n\Delta\nu, \tag{12}$$

where positive $\Delta\nu$ denotes a driving frequency less than the axial mode interval. Substituting (11) and (12) into (1a) and (1b), we obtain

$$\begin{aligned} [\dot{\varphi}_n - n\Delta\nu + \tfrac{1}{2}\nu\chi_n']E_n = -\frac{\delta c}{2L} [&E_{n+1} \cos (\varphi_{n+1} - \varphi_n) \\ &+ E_{n-1} \cos (\varphi_n - \varphi_{n-1})] \end{aligned} \tag{13a}$$

$$\begin{aligned} \dot{E}_n + \frac{\nu}{2}\left[\frac{1}{Q_n} + \chi_n''\right]E_n = \frac{\delta c}{2L} [&E_{n+1} \sin (\varphi_{n+1} - \varphi_n) \\ &- E_{n-1} \sin (\varphi_n - \varphi_{n-1})], \end{aligned} \tag{13b}$$

which are the fundamental equations of this paper and which, when solved, yield the amplitude, frequency, and phase of the optical modes. Equations (13a) and (13b) are to some extent equivalent to equations that have been given earlier by Gordon and Rigden [5], and by Yariv [6]. Except for notational changes, they are also equivalent to (5a) and (5b) of an earlier paper by Harris and McDuff [7].

III. Discussion of Parameters

For small signal conditions, the quadrature component of the atomic susceptibility χ_n'' is related to the single pass power gain by the relation

$$\frac{\nu L}{c} \chi_n'' = -g_n\Big(1 - \sum_m \beta_{nm} E_m^2\Big), \tag{14}$$

[2] We adopt the convention that all symbols for frequencies shall denote circular frequencies.

where g_n is the unsaturated single pass power gain of the nth mode, and the β_{nm} are saturation parameters which represent the effect of the mth mode on the gain of the nth mode [4].[3] Similarly, the in-phase component of the susceptibility may be expressed as

$$\frac{\nu L \chi_n'}{c} = \sigma_n + \sum_m \tau_{nm} E_m^2, \tag{15}$$

where σ_n is the additional round trip phase retardation which is seen by the nth mode as a result of power independent mode pulling, and the τ_{nm} represent power dependent pulling and pushing effects [4].

The Q of the nth mode may be written

$$\frac{\nu L}{c}\frac{1}{Q_n} = \alpha_n, \tag{16}$$

where α_n is the single pass power loss which is experienced by the nth mode as a result of nonzero output coupling, scattering, diffraction, etc. We note that Q_n is not meant to contain a contribution resulting from a parametric gain or loss; such contributions are accounted for by the right-hand side of (13b).

In practice, the time varying phase perturbation will often be achieved by means of a small perturbing element which has no significant spatial variation in the z direction, i.e., such that $\Delta\chi'(z)$ may be taken as independent of z over the length of the perturbing element. If we let δ_m denote the peak single pass phase retardation of such an element of length a, then it may readily be shown that

$$\delta_m = \frac{\Delta\chi' a\nu}{2c}. \tag{17}$$

If such an element is centered a distance z_0 from an end mirror of an optical cavity which has a total length L, then the coupling coefficient δ, as given by (8), becomes

$$\delta = \frac{\delta_m}{a}\int_{z_0-a/2}^{z_0+a/2} \cos\frac{\pi z}{L}\,dz, \tag{18}$$

which then yields

$$\delta = \frac{L}{a}\frac{2}{\pi}\left(\sin\frac{a}{L}\frac{\pi}{2}\right)\left(\cos\frac{z_0\pi}{L}\right)\delta_m. \tag{19}$$

If the length of the perturbing element is very small compared with the total length of the optical cavity ($a/L \ll 1$), then $\delta \cong \delta_m \cos(\pi z_0)/L$. It is, therefore, desirable for such a perturbing element to be situated as closely as possible to the end of the optical cavity. The coupling coefficient δ will then be very nearly the readily measurable peak single pass phase retardation of the perturbing element. It should be noted that if the perturbing element is not small, then its spatial variation may be of importance. As an extreme case, if the perturbing element were spatially uniform and completely filled the optical cavity, then δ would be zero.

[3] In writing (14) and (15), we have neglected terms which are dependent on the relative phases of the various optical modes. We also note that the notation of this section is not meant to correspond to that of Lamb [4].

IV. Steady State Solution of the Linear Approximation

In the present section we will neglect nonlinearities and take χ_n'' of all modes to be independent of E_n and equal to $-1/Q_n$. That is, we assume an infinity of laser modes, all having a single pass gain equal to their single pass loss. We also assume χ_n' to be zero, and look for solutions with $\dot{E}_n = 0$ and $\dot{\varphi}_n$ constant and independent of n. Equations (13a) and (13b) then become

$$[\dot{\varphi} - n\Delta\nu]E_n = -\frac{\delta c}{2L}[E_{n+1}\cos(\varphi_{n+1}-\varphi_n) + E_{n-1}\cos(\varphi_n - \varphi_{n-1})] \tag{20a}$$

$$0 = \frac{\delta c}{2L}[E_{n+1}\sin(\varphi_{n+1}-\varphi_n) - E_{n-1}\sin(\varphi_n-\varphi_{n-1})]. \tag{20b}$$

Noting the Bessel function identity

$$\frac{2n}{\Gamma}J_n(\Gamma) = J_{n+1}(\Gamma) + J_{n-1}(\Gamma), \tag{21}$$

we see that (20) is satisfied by the set of solutions

$$\begin{aligned}\dot{\varphi}_n &= q\Delta\nu\\ \varphi_{n+1} - \varphi_n &= 0\\ E_n &= J_{n-q}(\Gamma),\end{aligned} \tag{22}$$

where q is an integer, and Γ is given by

$$\Gamma = \frac{c}{L}\frac{1}{\Delta\nu}\delta = \frac{1}{\pi}\frac{\Delta\Omega}{\Delta\nu}\delta = \frac{1}{\pi}\frac{\text{axial mode interval}}{\text{detuning frequency}}\delta. \tag{23}$$

We first consider the $q = 0$ solution, previously given by Harris and McDuff [7], and by Yariv [6], which is a perfect FM oscillation with a modulation depth of Γ and with a center frequency at the zeroth mode. From a varying frequency point of view, the peak-to-peak frequency swing of the oscillation is $2\Gamma\nu_m$. The modulation depth Γ is proportional to the strength of the time varying perturbation and inversely proportional to the detuning frequency $\Delta\nu$. The frequency of the nth cavity mode was originally defined as $\nu_n = \Omega_0 + n\nu_m$. Thus $n = 0$ denotes the only mode of the $q = 0$ solution whose oscillation frequency would, in the absence of the parametric phase perturbation, be a cavity resonance. A schematic of the $q = 0$ solution is shown in the upper part of Fig. 2. Consider next those solutions with $q \neq 0$. $\dot{\varphi} = q\Delta\nu$ denotes a uniform shift of all frequencies from their assumed positions by $q\Delta\nu$. For instance, if $q = 1$, all modes are shifted upward by $\Delta\nu$, and, as shown in the lower part of Fig. 2, the $n = 1$ mode is the only mode which is exactly on a cavity resonance. In this case the mode amplitudes are $E_n = J_{n-1}(\Gamma)$; and the $n = 1$ mode has amplitude $J_0(\Gamma)$ and is the center frequency of the FM oscillation. In a linear theory, any of the previously free-running laser modes may become the carrier of an FM oscillation. The carrier frequency of the qth oscillation is at the qth mode, and is distinguished in that its amplitude is $J_0(\Gamma)$ and in that it is the only sideband of that oscillation

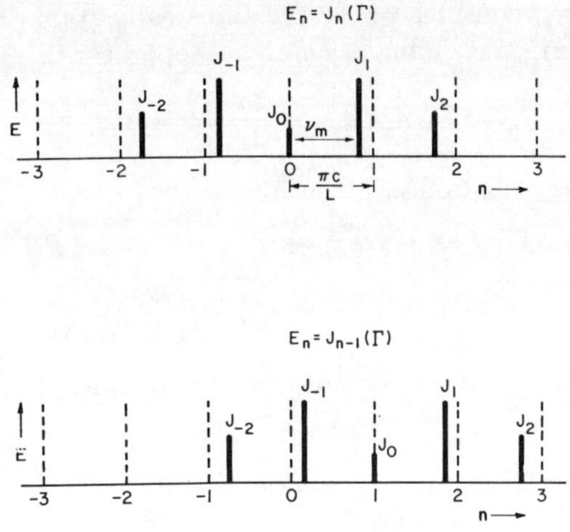

Fig. 2. Schematic of $q = 0$ and $q = 1$ solutions.

whose oscillation frequency is that of an original axial mode.

The total cavity electromagnetic field $E(z, t)$ which corresponds to a particular FM solution is obtained by substituting the solution of (22) into (2). Taking the case of $q = 0$, we have

$$E(z, t) = \sum_n J_n(\Gamma) \cos [(\Omega_0 + n\nu_m)t] \sin \frac{(n_0 + n)\pi z}{L}. \quad (24)$$

By the use of standard trigonometric and Bessel identities, $E(z, t)$ may be put into the closed form

$$E(z, t) = \tfrac{1}{2} \sin \left[\Omega_0 t + \frac{n_0 \pi z}{L} + \Gamma \sin \left(\nu_m t + \frac{\pi z}{L} \right) \right]$$
$$- \tfrac{1}{2} \sin \left[\Omega_0 t - \frac{n_0 \pi z}{L} + \Gamma \sin \left(\nu_m t - \frac{\pi z}{L} \right) \right], \quad (25)$$

which corresponds to a forward and a backward FM traveling wave. We may further write this in standing wave form as

$$E(z, t) = \cos \left[\Omega_0 t + \Gamma \sin \nu_m t \cos \frac{\pi z}{L} \right]$$
$$\cdot \sin \left[\frac{n_0 \pi z}{L} + \Gamma \cos \nu_m t \sin \frac{\pi z}{L} \right]. \quad (26)$$

It is of interest to note that the total cavity electromagnetic field at a particular point of space is in general not frequency modulated. In order to obtain a pure FM signal, it is necessary to couple to either, but not to both, of the FM traveling waves.

When $\Delta\nu = 0$, Γ, as defined by (23), is infinite, and the solutions considered thus far are indeterminant. In this case, (20a) and (20b) have the solution

$$E_n = E_{n+1}$$
$$\varphi_{n+1} - \varphi_n = p\pi \quad (27)$$
$$\dot{\varphi}_n = (-1)^{p+1} \frac{\delta c}{L} = (-1)^{p+1} \frac{\delta}{\pi} \Delta\Omega,$$

where p is an integer.

In the time domain, this solution corresponds to a behavior which consists of a repetitive series of pulses or spikes, with a pulsing frequency which is equal to the driving frequency of the internal phase perturbation. Such pulsing of a laser was first observed by Hargrove, Fork, and Pollack, and was obtained by acoustic modulation of the intracavity loss [8]. Linear analyses of intracavity loss modulation have been given by DiDomenico [9], Yariv [6] and Crowell [10].[4]

V. Dyadic Expansion of Steady State Equations

In the approximation of the previous section, nonlinearities and mode pulling were neglected, and an infinity of laser modes, all with gain equal to loss, were assumed. It was found that the sidebands comprising any of the FM oscillations had exactly Bessel function amplitudes and zero relative phases. That is, any FM oscillation was completely distortion free. When finite atomic linewidth, mode pulling, and atomic saturation are included, this is no longer the case. Relative amplitudes will no longer be exactly Bessel function, and the relative phases will no longer be exactly zero. In this and the following two sections, we develop an effective iterative procedure which will yield the amplitudes and phases of all sidebands of a particular oscillation, and also the center frequency, threshold, and power of the oscillation.

Implicit in the work of this section and of Section VII is the assumption that a stable, steady state solution of (13a) and (13b) exists. As will be seen in Section VIII, this will not always be the case. The general situation is that of a competition between different FM oscillations which are centered at different axial modes and which in the previous section were denoted by the integer q.

In particular, we will consider the $q = 0$ oscillation, i.e., that FM oscillation whose carrier frequency is closest to the center of the atomic line. In latter sections it will be seen that, in most cases of interest, it is this oscillation which will be dominant. Though the formulas will be given explicitly for the $q = 0$ oscillation, their extension to other oscillations will be apparent.

We seek a steady state solution to (13a) and (13b), and thus set $\dot{E}_n$ equal to zero and $\dot{\varphi}_n$ constant and independent of n. We define the quantities Ψ_n and ρ_n as follows:

$$\Psi_n = \frac{\nu L \chi_n'}{c} \quad (28a)$$

$$\rho_n = \frac{\nu L}{c} \left[\frac{1}{Q_n} + \chi_n'' \right]. \quad (28b)$$

Equations (13a) and (13b) may then be written

$$\left[\frac{2L}{\delta c} \dot{\varphi} - \frac{2n}{\Gamma} + \frac{\Psi_n}{\delta} \right] E_n = -[E_{n+1} \cos (\varphi_{n+1} - \varphi_n)$$
$$+ E_{n-1} \cos (\varphi_n - \varphi_{n-1})] \quad (29a)$$

[4] When $\Delta\nu = 0$, (20a) and (20b) also have solutions of the form $E_n = E_{n+1}$, $\varphi_{n+1} - \varphi_n = \pm\pi/2$, $\dot{\varphi}_n = 0$. However, during the course of the numerical analyses which will be described in later sections, solutions of this type were not observed.

$$\frac{\rho_n}{\delta} E_n = [E_{n+1} \sin(\varphi_{n+1} - \varphi_n) - E_{n-1} \sin(\varphi_n - \varphi_{n-1})], \tag{29b}$$

which are a set of simultaneous nonlinear difference equations. The quantity Ψ_n is the additional round trip phase retardation which is seen by the nth mode as a result of the real part of the atomic susceptibility and is described in (15). The quantity ρ_n, described in (14) and (16), is the net saturated single pass power gain of the nth mode. In a free-running laser, all modes oscillate at a power level such that single pass gain equals single pass loss, and therefore ρ_n is zero for all oscillating modes. In the presence of the parametric phase perturbation, this is no longer the case. For those modes which oscillate at a level which is greater than their free-running counterparts, ρ_n is positive; while for those modes which oscillate at a level which is less than their free-running counterparts, ρ_n is negative.

We first consider (29b). We treat the E_n and ρ_n as knowns, and solve for the relative phase angles $(\varphi_n - \varphi_{n-1})$. When considered in this way, (29b) is a linear, first-order, inhomogeneous difference equation with nonconstant coefficients (the E_n). We proceed by construction of a Green's dyadic $G_{n,q}$ such that

$$E_{n+1}G_{n+1,q} - E_{n-1}G_{n,q} = \delta_{n,q}, \tag{30}$$

where $\delta_{n,q}$ is the Kronecker delta [11]. The relative phase angles are then obtained from the relation

$$\sin(\varphi_n - \varphi_{n-1}) = \frac{1}{\delta} \sum_{q=-\infty}^{+\infty} G_{n,q}\rho_q E_q. \tag{31}$$

If we arbitrarily introduce the boundary condition that $G_{n,q} = 0$ at $n = -\infty$, we find the Green's dyadic satisfying (30), and this boundary condition is to be

$$\begin{aligned} G_{n,q} &= \frac{E_q}{E_n E_{n-1}} \qquad n > q \\ &= 0 \qquad n \leq q. \end{aligned} \tag{32}$$

The relative phase angles are then given by

$$\sin(\varphi_n - \varphi_{n-1}) = \frac{1}{E_n E_{n-1}} \frac{1}{\delta} \sum_{q=-\infty}^{n-1} \rho_q E_q^2. \tag{33}$$

Alternately, we could introduce the boundary condition $G_{n,q} = 0$ at $n = +\infty$. In this case the appropriate Green's dyadic is

$$\begin{aligned} G_{n,q} &= -\frac{E_q}{E_n E_{n-1}} \qquad n \leq q \\ &= 0 \qquad n > q, \end{aligned} \tag{34}$$

and the relative phase angles are obtained from

$$\sin(\varphi_n - \varphi_{n-1}) = -\frac{1}{E_n E_{n-1}} \frac{1}{\delta} \sum_{q=n}^{\infty} \rho_q E_q^2. \tag{35}$$

Though the relative phase angles which are predicted by (33) and (35) at first appear to be different, it will be seen that, due to a conservation condition to be proven in the following section, they are in fact identical.

We next consider the first of the steady state equations, i.e., (29a). We define a difference operator $\mathcal{L}$ such that

$$\mathcal{L}E_n = E_{n-1} - \frac{2n}{\Gamma} E_n + E_{n+1} \tag{36}$$

and a perturbation μ_n given by

$$\mu_n = E_{n+1}[1 - \cos(\varphi_{n+1} - \varphi_n)] + E_{n-1}[1 - \cos(\varphi_n - \varphi_{n-1})] - \left[\frac{\Psi_n}{\delta} + \frac{2L}{\delta c}\dot{\varphi}\right]E_n. \tag{37}$$

In terms of these definitions, (29a) may be written

$$\mathcal{L}E_n = \mu_n. \tag{38}$$

By means of the Bessel function identity of (21) and the orthogonality relation

$$\sum_{n=-\infty}^{+\infty} J_{n-q}(\Gamma)J_{n-p}(\Gamma) = \delta_{q,p}, \tag{39}$$

which is a form of Neumann's addition theorem for Bessel functions [12], it is seen that a complete and orthonormal set of eigenvectors $|u_q\rangle$, with corresponding eigenvalues λ_q of the operator $\mathcal{L}$, are given by

$$\begin{aligned} |u_q\rangle &= J_{n+q}(\Gamma) \\ \lambda_q &= q\frac{2}{\Gamma}. \end{aligned} \tag{40}$$

Proceeding by spectral expansion [11], the operator which is inverse to $\mathcal{L}$ may then be written

$$\mathcal{L}^{-1} = \frac{\Gamma}{2} \sum_{\substack{q \\ q\neq 0}} \frac{1}{q} |u_q\rangle\langle u_q|. \tag{41}$$

The solution to (29a), or equivalently (38), is then given by

$$E_n = kJ_n(\Gamma) + \frac{\Gamma}{2} \sum_{\substack{q \\ q\neq 0}} \sum_m \frac{1}{q} J_{n+q}(\Gamma)J_{m+q}(\Gamma)\mu_m. \tag{42}$$

The first term on the right side of this equation is the homogeneous solution of (38), and has an amplitude constant k which is to be determined. The second term is a particular solution which results from the perturbation μ_n.

It is to be noted that (33) or, equivalently, (35), and (42) are all exact, and that no approximations have been made. These equations form the basis of the iterative procedure to be given in Section VII.

VI. Conservation Conditions

We next derive two conservation conditions which are also necessary for the iterative procedure. We return to (13b), multiply through by E_n, and sum both sides over n from $-\infty$ to $+\infty$. We then have

$$\sum_n E_n\dot{E}_n + \frac{\nu}{2}\left[\frac{1}{Q_n} + \chi_n''\right]E_n^2 = \frac{\delta c}{2L} \sum_n [E_n E_{n+1} \sin(\varphi_{n+1} - \varphi_n) - E_n E_{n-1} \sin(\varphi_n - \varphi_{n-1})]. \tag{43}$$

By the change of variable $n = n' + 1$, the second term on the right-hand side of this equation is equal to the first. We thus obtain

$$\sum_{n=-\infty}^{+\infty} E_n\dot{E}_n + \frac{\nu}{2}\left[\frac{1}{Q_n} + \chi_n''\right]E_n^2 = 0. \quad (44)$$

Noting that $E_n\dot{E}_n = (d/dt)(E_n^2/2)$, it is seen that the foregoing equation is a statement of power conservation. That is, the rate of change of total stored energy, plus the net power dissipated or absorbed in all modes, is zero. Though in the absence of the parametric phase perturbation power is conserved by all modes individually,[5] in its presence, power must be conserved jointly. It is noted that the phase perturbation itself does not contribute or absorb any power from the system.

In the steady state, all $\dot{E}_n = 0$, and (44) becomes

$$\sum_n \rho_n E_n^2 = 0, \quad (45)$$

where the ρ_n are the net saturated single pass power gains and are defined in (28b). Equation (45) establishes the equivalence of (33) and (35) of the previous section.

We apply a similar procedure to (13a). Multiplying through by E_n and summing over n from $-\infty$ to $+\infty$, we obtain

$$\sum_{n=-\infty}^{+\infty} [\dot{\varphi}_n - n\Delta\nu + \tfrac{1}{2}\nu\chi_n']E_n^2$$
$$= -\frac{\delta c}{L}\sum_{n=-\infty}^{+\infty} E_nE_{n+1}\cos(\varphi_{n+1} - \varphi_n), \quad (46)$$

which is a reactive conservation condition and which, when the relative steady state amplitudes and phases are known, will determine the frequency of oscillation.

VII Steady State Amplitudes, Phases, Frequency, and Power

As a result of nonzero net saturated gains, and due to the in-phase component of the atomic susceptibility χ_n', actual mode amplitudes and phases will not be exactly those of the ideal FM solution of Section IV. In this section we describe an iterative procedure which allows the calculation to any desired order of the relative mode amplitudes and phases, as well as the center frequency and power of an FM oscillation. The results of the first iteration will reveal the mechanism of the distortion and, are themselves useful for certain applications. The plan of the iterative procedure is to first assume the modes to have Bessel function relative amplitudes. By application of the power conservation condition of (45), the amplitude of the oscillation is determined. Net saturated gains are then obtained, and (33) or, equivalently, (35) is used to find the first-order relative phases. With relative phases known, the contribution to the amplitude perturbation μ_n [(37)], which is a result of the nonzero net saturated gains, may be found. The frequency pulling or pushing of the oscillation $\dot{\varphi}$ is then obtained from the reactive conservation condition of (46). With μ_n now determined, second-order amplitudes may be obtained from (42), and a second iteration may be begun.

In order to effectively illustrate this procedure, it is convenient to assume a specific form for the saturation of the atomic gain.[6] We choose completely inhomogeneous saturation, such that the single pass gain of the nth mode is given by $-g_n(1 - \beta E_n^2)$; i.e., in (14) we take $\beta_{nm} = \beta\delta_{nm}$ where β is a saturation parameter which is the same for all modes. The net saturated gain of the nth mode, as defined by (28b), then becomes

$$\rho_n = \alpha_n - g_n(1 - \beta E_n^2), \quad (47)$$

where α_n is the single pass power loss as given by (16). We let $E_n = kJ_n(\Gamma)$, and substitute from (47) into the power conservation condition of (45). We then have

$$\sum_n \{\alpha_n - g_n[1 - \beta k^2 J_n^2(\Gamma)]\}k^2 J_n^2(\Gamma) = 0. \quad (48)$$

Solving for the oscillation level k^2, we obtain

$$k^2 = \frac{1}{\beta}\frac{\sum_n (g_n - \alpha_n)J_n^2(\Gamma)}{\sum_n g_n J_n^4(\Gamma)}. \quad (49)$$

With the amplitude of all modes determined to first order, the net saturated gains are obtained from (47) with $E_n^2 = k^2 J_n^2(\Gamma)$. First-order relative phase angles are then obtained from (33) or (35). Using (33), we have

$$\sin(\varphi_n - \varphi_{n-1}) = \frac{1}{J_n(\Gamma)J_{n-1}(\Gamma)}\frac{1}{\delta}\sum_{q=-\infty}^{n-1} \rho_q J_q^2(\Gamma). \quad (50)$$

For the case of a symmetrical line shape wherein ρ_n is an even function of n, the evaluation of (50) is simplified by means of the relation

$$\sum_{q=-\infty}^{0} \rho_q J_q^2(\Gamma) = \tfrac{1}{2}\rho_0 J_0^2(\Gamma), \quad (51)$$

which is obtained from the conservation condition (45).

Two points concerning the relative phase angles may be noted. First, nonzero relative phase angles result from an accumulation of nonzero net saturated gains. In general, an FM laser operated at a Γ such that the relative mode amplitudes to at least some extent approximate those of a free-running laser will, at a given δ, have less phase distortion than will an FM laser whose mode amplitudes depart sharply from those of the free-running case. In particular, distortion will be increased when operating at a Γ such that the amplitude of some mode is driven close to zero; or alternately, when operating at a large Γ, such that a number of modes have amplitudes which are considerably larger than the corresponding free-running modes, for instance when Γ is such that the equivalent frequency swing is considerably larger than the spectral width of the free-running laser. In the first case, some particular ρ_n will become large

[5] This statement is strictly true only in the limit of completely inhomogeneous broadening.

[6] We note that this assumption is illustrative and not restrictive; and that the iterative procedure is valid, irrespective of the atomic lineshape and the type of saturation which is assumed.

and negative; in the second case, an accumulation of positive ρ_n's will result.

The second and very important point is that phase distortion may be made arbitrarily small by making the perturbation strength δ increasingly large. It is therefore desirable to obtain a given Γ by working with the largest available δ, and thus from (22) with a correspondingly large detuning.

With relative phases determined, we next find the first-order correction to the center frequency of the FM oscillation. We substitute $E_n = kJ_n(\Gamma)$ into the reactive conservation condition of (46), and note that in the steady state $\dot{\varphi}$ is constant and independent of n. Using the definition of Ψ_n of (28a), we then have

$$\dot{\varphi} = \left(-\frac{c}{2L} \sum_n \Psi_n J_n^2(\Gamma) - \frac{c\delta}{L} \sum_n J_n(\Gamma) J_{n+1}(\Gamma) \cos(\varphi_{n+1} - \varphi_n) \right) k^2 \quad (52)$$

where we have used the relation $\sum_{n=-\infty}^{+\infty} J_n^2(\Gamma) = 1$. The oscillation frequency of the nth axial mode is then $\Omega_0 + n\nu_m + \dot{\varphi}$, and the center frequency of the oscillation is thus $\Omega_0 + \dot{\varphi}$. Equation (52) describes to first order the frequency pulling or pushing of the entire FM oscillation. The first term on the right-hand side gives the contribution to the pushing or pulling which results from the presence of the atomic media, and is in fact simply the weighted average of the pushing or pulling which is associated with each of the laser modes. The second term is a parametric pushing or pulling term. By noting the Bessel identity $\sum_{n=-\infty}^{+\infty} J_n(\Gamma) J_{n+1}(\Gamma) = 0$, it is seen that for an ideal FM signal having relative phase angles which are all zero, the parametric term would be zero. This situation will be approached as δ becomes increasingly large. If Ω_0 is at the center of an exactly symmetrical atomic line, both the atomic and the parametric terms are zero, and the center frequency of the FM oscillation will be Ω_0.

The amplitude perturbation is now completely determined and given by

$$\mu_n = k\left(J_{n+1}(\Gamma)[1 - \cos(\varphi_{n+1} - \varphi_n)] + J_{n-1}(\Gamma)[1 - \cos(\varphi_n - \varphi_{n-1})] - J_n(\Gamma)\left[\frac{\Psi_n}{\delta} + \frac{2L}{\delta c}\dot{\varphi}\right]\right) \quad (53)$$

Substitution of μ_n into (42) yields a set of second-order amplitudes, and the first iteration is complete. The distortion which results from the nonzero net saturated gains is, to first order, phase distortion. At sufficiently large δ, the cosine of the relative phase angles approaches unity, and their contribution to μ_n nearly vanishes. The contribution to μ_n resulting from mode pulling also varies inversely as δ, and thus at large δ the μ_n approach zero; and from (40) the E_n approach $kJ_n(\Gamma)$.

As δ becomes increasingly large, (49) becomes an increasingly exact expression for the level of the FM oscillation. We note that the total power output of the FM oscillation is proportional to the total stored energy E_n^2, which for $E_n = kJ_n(\Gamma)$ equals k^2. From (49), we see that to first order the threshold for FM laser oscillation is given by

$$\sum_n (g_n - \alpha_n) J_n^2(\Gamma) > 0. \quad (54)$$

Equation (54) is the condition for positive power output in an FM laser, and is analogous to the condition $g_n > \alpha_n$ for threshold of a free-running laser. Typically, the cavity loss α_n will be the same for all modes, and (54) becomes

$$\sum_n g_n J_n^2(\Gamma) > \alpha, \quad (55)$$

where α is the single pass power loss of all modes. An FM laser will be above threshold if the sum of the weighted, unsaturated single pass gains is greater than the single pass loss. Note that in an FM laser this threshold condition depends on Γ, which in turn depends on both δ and on the detuning $\Delta\nu$. An FM laser may thus be extinguished by varying either δ or the driving frequency of the perturbation.

Application of the iterative procedure to particular cases, as well as discussion of its convergence, will be given later in the paper.

VIII. Competition and Quenching of FM Oscillations

In the previous section we examined the details of a single steady state FM oscillation, with the implicit assumption that such a stable steady state condition exists. The more general situation is that of a number of competing FM oscillations, with each previously free-running laser mode acting as the center frequency or carrier for a particular oscillation. While the free-running modes were very weakly coupled, the competing FM oscillations, to a large extent, see the same atomic population and are closely coupled. In many cases, the strongest of these oscillations is able to quench the weaker oscillations and thereby establish the desired steady state condition.

The purpose of this section is to develop a simplified set of equations which, to a good approximation, will describe the competition between the FM oscillations. The details of any particular oscillation will not be considered. That is, it will be assumed that δ is sufficiently large that the sidebands which comprise any of the FM oscillations have Bessel function amplitudes and FM phases.

The plan is to first show that a solution consisting of a sum of independent FM oscillations, each with arbitrary amplitude and arbitrary phase, satisfies (13a) and (13b) for an ideal system wherein gain equals loss for all modes. It will, in effect, be shown that the FM oscillations are the normal modes of the lossless system. The competition between oscillations will then be introduced by means of a statement of power conservation. The procedure could perhaps be considered analogous to that used in microwave systems wherein the field configuration of the lossless system is used to take into account small losses.

In the multioscillation situation, each cavity mode contains sidebands from each of the FM oscillations. The variables E_n and φ_n are envelop quantities and become increasingly complicated as the number of FM oscillations is increased. It is, therefore, convenient to change to the new variables X_n and Y_n which are the in-phase and quadrature components, respectively, of a particular mode, as opposed to its amplitude and phase. We let

$$X_n = E_n \cos \varphi_n$$
$$Y_n = E_n \sin \varphi_n, \tag{56}$$

and thus

$$E_n = (X_n^2 + Y_n^2)^{1/2}$$
$$\tan \varphi_n = \frac{Y_n}{X_n}. \tag{57}$$

Taking appropriate partial derivatives of E_n and φ_n and substituting into (13a) and (13b), after some algebra we obtain the completely equivalent equations

$$X_n\dot{Y}_n - Y_n\dot{X}_n + (-n\Delta\nu + \tfrac{1}{2}\nu\chi_n')(X_n^2 + Y_n^2)$$
$$= -\frac{\delta c}{2L}[X_n(X_{n+1} + X_{n-1}) + Y_n(Y_{n+1} + Y_{n-1})] \tag{58a}$$

and

$$X_n\dot{X}_n + Y_n\dot{Y}_n + \frac{\nu}{2}\left[\frac{1}{Q_n} + \chi_n''\right][X_n^2 + Y_n^2]$$
$$= \frac{\delta c}{2L}[X_n(Y_{n+1} + Y_{n-1}) - Y_n(X_{n+1} + X_{n-1})]. \tag{58b}$$

The expected multioscillation solution has mode amplitudes E_n and mode phases φ_n such that

$$E_n(t) \cos [\nu_n t + \varphi_n(t)]$$
$$= \sum_{q=-\infty}^{+\infty} a_q J_{n-q}(\Gamma) \cos [\nu_n t + q\Delta\nu t + \theta_q], \tag{59}$$

where a_q and θ_q are the amplitude and phase, respectively, of the component FM oscillations, and are assumed to be independent of time. As in Section IV, the integer q denotes that mode which is the center frequency of a particular FM oscillation. Expanding (59), we have

$$E_n \cos \varphi_n \cos \nu_n t - E_n \sin \varphi_n \sin \nu_n t$$
$$= \sum_q a_q J_{n-q}(\Gamma) \cos (q\Delta\nu t + \theta_q) \cos \nu_n t$$
$$- \sum_q a_q J_{n-q}(\Gamma) \sin (q\Delta\nu t + \theta_q) \sin \nu_n t. \tag{60}$$

By separately equating in-phase and quadrature components, we see that the expected solution is given by

$$X_n(t) = \sum_q a_q J_{n-q}(\Gamma) \cos (q\Delta\nu t + \theta_q)$$
$$Y_n(t) = \sum_q a_q J_{n-q}(\Gamma) \sin (q\Delta\nu t + \theta_q). \tag{61}$$

It may be shown that, for the ideal lossless case wherein $Q_n = -\chi_n''$ and $\chi_n' = 0$ for all n, the $X_n(t)$ and $Y_n(t)$ of (61) identically satisfy (58a) and (58b). This is true regardless of the amplitude a_q or the phase θ_q of the respective FM oscillations. If nonlinearities are neglected, an FM laser is no better than a free-running laser. As opposed to a set of uncoupled free-running modes, we find a set of uncoupled FM oscillations. These oscillations may be independently excited and might be considered as the normal modes of the lossless system. In the actual situation wherein net saturated gains and χ_n' are not zero, X_n and Y_n will not satisfy (58a) and (58b). That is, the sidebands which comprise each of the FM oscillations will not have exactly Bessel function amplitudes and zero relative phases. However, as δ is increased it is expected that the above solution should become increasingly exact.

We now make use of power conservation to introduce the effects of atomic saturation. To start, the total average stored energy $\mathcal{E}$ of all modes is given by

$$\mathcal{E} = \tfrac{1}{2} \sum_n E_n^2(t)$$
$$= \tfrac{1}{2} \sum_n X_n^2(t) + Y_n^2(t), \tag{62}$$

which in terms of the a_q and θ_q of (61) becomes

$$\mathcal{E} = \tfrac{1}{2} \sum_n \sum_q \sum_p a_q a_p J_{n-q}(\Gamma) J_{n-p}(\Gamma)$$
$$\cdot \cos [(q - p)\, \Delta\nu t + (\theta_q - \theta_p)], \tag{63}$$

where we have used the trigonometric identity for the cosine of a difference of angles. We sum over n, and make use of the orthogonality condition of (39). We then have

$$\mathcal{E} = \tfrac{1}{2} \sum_q \sum_p a_q a_p \cos [(q - p)\, \Delta\nu t + (\theta_q - \theta_p)]\, \delta_{q,p} \tag{64}$$

and thus

$$\mathcal{E} = \tfrac{1}{2} \sum_q a_q^2. \tag{65}$$

We next calculate the total power $\mathcal{P}$ which is absorbed or dissipated by all modes. From the conservation condition (44), $\mathcal{P}$ is given by

$$\mathcal{P} = \frac{\nu}{2} \sum_n \left(\frac{1}{Q_n} + \chi_n''\right) E_n^2. \tag{66}$$

We assume small signal saturation of the form of (14). Using (16), we have

$$\mathcal{P} = \frac{c}{2L}\left[\sum_n (\alpha_n - g_n)E_n^2 + \sum_n \sum_m g_n \beta_{nm} E_m^2 E_n^2\right]. \tag{67}$$

In terms of the X_n and Y_n; $\mathcal{P}$ is given by

$$\mathcal{P} = \frac{c}{2L}\left[\sum_n (\alpha_n - g_n)(X_n^2 + Y_n^2)\right.$$
$$\left. + \sum_n \sum_m g_n \beta_{nm}(X_n^2 X_m^2 + X_m^2 Y_n^2 + X_n^2 Y_m^2 + Y_m^2 Y_n^2)\right]. \tag{68}$$

Consider first the term $X_n^2 X_m^2$. For brevity, we let

$$\xi_q = q\Delta\nu t + \theta_q. \tag{69}$$

Substituting from (61), we have

$$X_n^2X_m^2 = \sum_q \sum_l \sum_s \sum_r a_q a_l a_s a_r J_{n-q}(\Gamma)J_{n-l}(\Gamma) \cdot J_{m-s}(\Gamma)J_{m-r}(\Gamma) \cdot \cos \xi_q \cdot \cos \xi_l \cdot \cos \xi_s \cdot \cos \xi_r . \tag{70}$$

The product of four cosines may be expanded to yield

$$\tfrac{1}{8}[\cos(\xi_q + \xi_l - \xi_s - \xi_r) + \cos(\xi_q - \xi_l + \xi_s - \xi_r) + \cos(\xi_q - \xi_l - \xi_s + \xi_r) + \cos(\xi_q + \xi_l + \xi_s + \xi_r) + \cos(\xi_q + \xi_l + \xi_s - \xi_r) + \cos(\xi_q + \xi_l - \xi_s + \xi_r) + \cos(\xi_q - \xi_l + \xi_s + \xi_r) + \cos(\xi_q - \xi_l - \xi_s - \xi_r)].$$

In summing (70), we will keep only those terms which are independent of the ξ_i, and thereby neglect the effects of a possible phase locking of FM oscillations. Summing over r and s, we find that for $q \neq l$ the first of the above cosine terms contributes the phase independent term

$$\tfrac{1}{4} \sum_q \sum_{\substack{l \\ q \neq l}} a_q^2 a_l^2 J_{n-q}(\Gamma)J_{n-l}(\Gamma)J_{m-q}(\Gamma)J_{m-l}(\Gamma)$$

to the sum of (70), while for $q = l$, it contributes

$$\tfrac{1}{8} \sum_q a_q^4 J_{n-q}^2(\Gamma)J_{m-q}^2(\Gamma).$$

Proceeding similarly, we find contributions from the second and third of the foregoing cosine terms. Examination of each of the last five of those cosine terms shows that, since their arguments do not contain an equal number of plus and minus signs, these terms do not contribute any terms which are independent of the ξ_i. Adding all contributions, we find the phase independent portion of $X_n^2X_m^2$ to be

$$X_n^2X_m^2 = \tfrac{1}{2} \sum_q \sum_l a_q^2 a_l^2 J_{n-q}(\Gamma)J_{n-l}(\Gamma)J_{m-q}(\Gamma)J_{m-l}(\Gamma) + \tfrac{1}{4} \sum_q \sum_l a_q^2 a_l^2 J_{n-q}^2(\Gamma)J_{m-l}^2(\Gamma) - \tfrac{3}{8} \sum_q a_q^4 J_{n-q}^2(\Gamma)J_{m-q}^2(\Gamma). \tag{71a}$$

We next evaluate the contribution of the other terms of (68). Proceeding as before, we find their phase independent contribution to be

$$Y_m^2Y_n^2 = X_n^2X_m^2 \tag{71b}$$

and

$$X_m^2Y_n^2 = \tfrac{1}{4} \sum_q \sum_l a_q^2 a_l^2 J_{n-q}^2(\Gamma)J_{m-l}^2(\Gamma) - \tfrac{1}{8} \sum_q a_q^4 J_{m-q}^2(\Gamma)J_{n-q}^2(\Gamma) \tag{71c}$$

$$X_n^2Y_m^2 = X_m^2Y_n^2 \tag{71d}$$

and

$$X_n^2 + Y_n^2 = \sum_q a_q^2 J_{n-q}^2(\Gamma). \tag{71e}$$

Combining the contributions of (71a)–(71e), we find the total phase independent power which is dissipated or absorbed by all modes to be given by

$$\mathcal{P} = \frac{c}{2L}\Big[\sum_n \sum_q (\alpha_n - g_n)a_q^2 J_{n-q}^2(\Gamma) + \sum_n \sum_m \sum_q \sum_l g_n\beta_{nm} a_q^2 a_l^2 J_{n-q}(\Gamma)J_{n-l}(\Gamma)J_{m-q}(\Gamma)J_{m-l}(\Gamma) + \sum_n \sum_m \sum_q \sum_l g_n\beta_{nm} a_q^2 a_l^2 J_{n-q}^2(\Gamma)J_{m-l}^2(\Gamma) - \sum_n \sum_m \sum_q g_n\beta_{nm} a_q^4 J_{m-q}^2(\Gamma)J_{n-q}^2(\Gamma)\Big]. \tag{72}$$

We may alternately write this as

$$\mathcal{P} = \sum_q \alpha_q a_q^2, \tag{73}$$

where, by comparison with (72), α_q is given by

$$\alpha_q = \frac{c}{2L}\Big[\sum_n (\alpha_n - g_n)J_{n-q}^2(\Gamma) + \sum_n \sum_m \sum_l g_n\beta_{nm} a_l^2 J_{n-q}(\Gamma)J_{n-l}(\Gamma)J_{m-q}(\Gamma)J_{m-l}(\Gamma) + \sum_n \sum_m \sum_l g_n\beta_{nm} a_l^2 J_{n-q}^2(\Gamma)J_{m-l}^2(\Gamma) - \sum_n \sum_m g_n\beta_{nm} a_q^2 J_{m-q}^2(\Gamma)J_{n-q}^2(\Gamma)\Big]. \tag{74}$$

The power which is absorbed by all modes may now be equated to the rate of change of total stored energy. From (65) and (73), we then have

$$\sum_q \frac{d}{dt}\left(\frac{a_q^2}{2}\right) + \alpha_q a_q^2 = 0. \tag{75}$$

Since the a_q are normal modes of the lossless system, it is expected that they should independently satisfy (75). From a more physical point of view, we note that the time varying perturbation couples together only the side bands of a particular FM oscillation; thus each oscillation should independently conserve power. We therefore find the set of equations

$$\frac{da_q}{dt} + \alpha_q a_q = 0, \tag{76}$$

where α_q is given by (74), and which at sufficiently large δ should approximately describe the steady state competition among the FM oscillations. We will refer to (76) as the a Equations.

In effect, α_q is the gain constant of the qth oscillation. In a steady state condition such that only the zeroth oscillation is above threshold, its oscillation level may be determined from the condition $\alpha_0 = 0$. In the more general situation, more than one oscillation is above threshold, and the effective gain of any oscillation is dependent on the amplitude of all other oscillations. We note that the a Equations are not expected to correctly describe the transient condition, and instead should be looked at as a stability test for the steady state solutions of the q simultaneous equations $\alpha_q a_q = 0$. Equation (74) indicates a very close coupling of the FM oscillations. In typical situations, a number of FM oscillations may exist at small Γ. However, as Γ is increased, a point is reached above which the strongest of these oscillations is able to quench all others and establish the desired steady state.

For the special case of completely inhomogeneous saturation, such that the gain of the nth mode is $-g_n(1 - \beta E_n^2)$, i.e., when $\beta_{nm} = \delta_{nm}\beta$, α_q becomes

$$\alpha_q = \frac{c}{2L}\Big[\sum_n (\alpha_n - g_n)J_{n-q}^2(\Gamma) + 2\beta \sum_n \sum_l g_n a_l^2 J_{n-q}^2(\Gamma)J_{n-l}^2(\Gamma) - \beta \sum_n g_n a_q^2 J_{n-q}^4(\Gamma)\Big]. \tag{77}$$

It is of interest to note that in an attempt to determine α_q for the inhomogeneous case of (77) by inspection we probably would have written

$$\alpha_q = \frac{c}{2L} \sum_n [(\alpha_n - g_n) + g_n\beta E_n^2]J_{n-q}^2(\Gamma) \quad \text{(incorrect)}, \tag{78}$$

where, neglecting terms which are phase dependent,

$$E_n^2 = \sum_l a_l^2 J_{n-l}^2(\Gamma). \tag{79}$$

Though the α_q of (78) is very much like that of (77), it differs by a factor of two in describing the saturation of one FM oscillation by another. For the correct α_q of (77), the statement might be made that any FM oscillation saturates another oscillation twice as hard as it saturates itself. This perhaps unexpected factor of two arises due to the power dissipation in the beats of the competing oscillations. It has also been found in simpler problems, such as that of two resonant circuits which are coupled through a nonlinear negative resistance [13].

IX. Variable Modulation Frequency at Constant δ

In previous sections of the paper, we have primarily developed the theory for what may be termed the FM region of operation of an FM laser. However, if either δ or $\Delta\nu$ is sufficiently decreased, the behavior predicted by (13a) and (13b) becomes considerably more complex. In particular, for small detuning, there is a region where the modes have nearly equal amplitudes and is such that the behavior in the time domain consists of a series of spikes or pulses. This type of solution was indicated by (27) of the linear approximation of Section IV, and has been termed as a phase locked solution.

In the numerical analyses of this and the following sections, we assume a Doppler broadened Gaussian atomic line with a homogeneous or natural linewidth which is very small compared with both the axial mode interval and the half-power Doppler width. We include the effects of power independent, but not power dependent mode pulling. Following the notation of Section III, we take

$$g_n = g_0 \frac{Z_i\left(\dfrac{\nu_n - \omega}{Ku}\right)}{Z_i(0)} \tag{80a}$$

and

$$\sigma_n = g_0 \frac{Z_r\left(\dfrac{\nu_n - \omega}{Ku}\right)}{Z_i(0)}, \tag{80b}$$

where g_0 is the unsaturated gain at line center, and Z_r and Z_i are the real and imaginary parts of the plasma dispersion function and are described by Lamb [4]. In the limit of vanishingly small homogeneous linewidth, the functions of (80a) and (80b) are the normalized Gaussian and Hilbert transform of the Gaussian, respectively. In these expressions ω is the frequency of the atomic line center and the parameter Ku is a measure of the Doppler broacening; Ku has units of angular frequency and equals 0.6 times the 3 dB Doppler linewidth. It will be convenient to specify the axial mode interval in units of Ku, and thus make use of the tables of Fried and Conte [14]. For the numerical analyses of this section, we take $g_0 = 0.075$, $\alpha_n = 0.070$, and assume an axial mode interval of 0.1 Ku, corresponding to a ratio of Doppler width to mode spacing of 16.67. These conditions correspond to five free-running laser modes above threshold. This situation is somewhat similar to that of some of the experimental work of Ammann, McMurtry, and Oshman [15], and the results of this section may be compared with their work. Mode amplitudes, phases, and frequencies were obtained by digital computer solution of (13a) and (13b). Solution was accomplished by means of a fourth-order Runge–Kutta method. The equations were programed for twenty-one modes ($n = -10$ to $n = +10$) and were run until a steady state solution to three decimal places was reached. Unless otherwise noted in the following, the $n = 0$ mode was taken at line center and thus has a frequency $\nu_0 = \Omega_0 = \omega$.

In Fig. 3, we show laser mode intensities ($\frac{1}{2}E_n^2$) vs. optical frequency—such as would be observed on a scanning Fabry–Perot interferometer. Figure 3(a) shows the intensities of the modes of the free-running laser, i.e., with $\delta = 0$. In Fig. 3(b), δ is set equal to 0.015, and the frequency of the parametric drive is adjusted such that it is exactly equal to the axial mode interval ($\Delta\nu = 0$). A widening of the optical spectrum and some tendency toward equalization of the mode intensities is observed. Relative phases are found to have values between zero and fifty degrees. Of most interest, a uniform, angular frequency shift of all modes from their free-running positions of $\dot{\varphi} = 0.94\ \delta\ \Delta\Omega/\pi$ is obtained. The direction of this shift is dependent on initial conditions, and is too small to show on the scale of Fig. 3. This solution corresponds to that of the phase locked solution of (27) of the linear approximation of Section IV.

In Figs. 3(c) through 3(f), δ is left constant at 0.015, and the detuning $\Delta\nu$ is increased in steps. At the small detuning of Fig. 3(c) ($\Delta\nu = 0.00035\ \Delta\Omega$), we observe an interesting shift of the envelope of the modes of about 2 $\Delta\Omega$. In addition, there is a uniform angular frequency shift of $\dot{\varphi} = 0.91\ \delta\ \Delta\Omega/\pi$ such as that discussed in connection with Fig. 3(b). Associated with the gross envelope shift is a decrease in laser power as peak relative amplitudes are moved further from the center of the Doppler line. As the detuning is further increased, the envelope shift and decrease in laser power continues, until, as shown in Fig. 3(d), the laser is extinguished.

Figure 3(d) might be considered the beginning of the steady state FM region wherein there is a single FM oscillation with its center frequency at the center of the atomic line, and with a modulation depth Γ which is approximately given by (23). For the detuning of Fig. 3(d), Γ is approximately six, with the result that (55) is not satisfied, and the oscillation is below threshold. In Figs. 3(e) and 3(f), the detuning is further increased, with the result that Γ decreases, relative amplitudes are concentrated closer to the center of the atomic line, and output power increases. In the steady state FM region there is no uniform frequency shift, i.e., $\dot{\varphi}_n = 0$ for all modes. If the detuning is increased past that of Fig. 3(f), we enter the region of multiple FM oscillation that was considered in Section VIII. In this region more than one mode may act as a carrier for an FM oscillation, and a steady state solution in the sense of zero $\dot{E}_n$ does not exist.

Figure 4 shows the time domain behavior which corresponds to the spectrum of Figs. 3(b) and 3(f). If the output signal of the laser is written

$$E(t) = \sum_n E_n \cos [(\Omega_0 + n\nu_m)t + \varphi_n], \tag{81}$$

then the low passed portion (or envelope) of $E^2(t)$, such as would be obtained if the signal were incident on a photodetector, is given by

$$W(t) = \tfrac{1}{2} \sum_s \sum_n E_n E_{n+s} \cos (s\nu_m t + \varphi_{n+s} - \varphi_n). \tag{82}$$

The data for Fig. 4 were obtained by evaluation of (82) for the E_n and φ_n, corresponding to Figs. 3(b) and 3(f). Output intensities are normalized to the total average intensity ($\frac{1}{2} \sum_n E_n^2$) of the free-running laser. We note that the pulsing of the phase locked solution is at the driving frequency of the internal phase perturbation and has peak intensities which are approximately six times the average intensity of the free-running laser. By contrast, the envelope which corresponds to the FM spectrum of Fig. 3(f) is more nearly constant and independent of time. The ripple is entirely even harmonic, and is a result of the distortion of amplitudes and phases from those of an ideal FM signal. As will be seen in the following section, this ripple can be made arbitrarily small, if δ is made sufficiently large. As opposed to the periodic behavior of both the phase locked and FM solutions, we note that the time domain behavior of the envelope of a free-running laser consists of an erratic fluctuation, with peak intensities almost as great as those obtained in the phase locked region.

In Fig. 5 we show output power ($\frac{1}{2} \sum_n E_n^2$) as a function of the normalized detuning. Output power is normalized

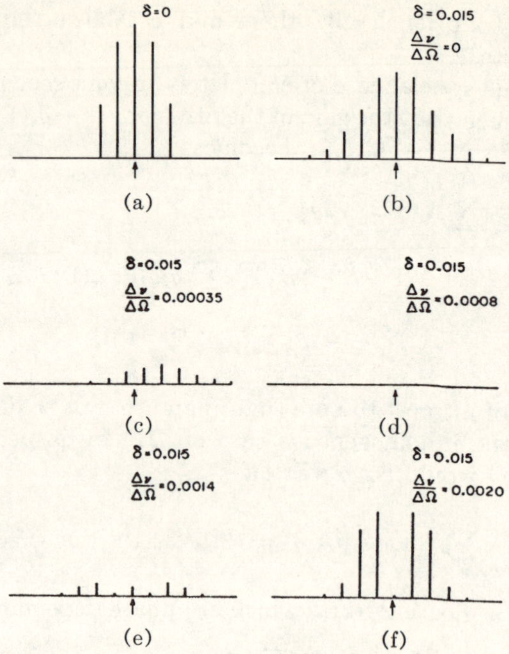

Fig. 3. Laser mode intensities at constant δ and variable detuning: $g_0 = 0.075$; $\alpha_n = 0.070$; $\Delta\Omega = 0.1$ Ku. (Five modes free running.)

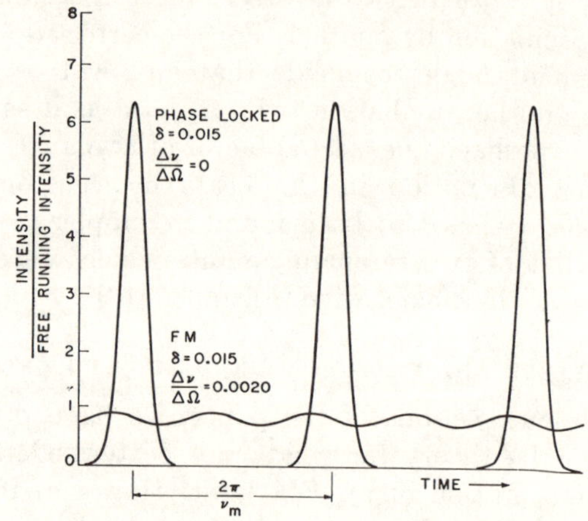

Fig. 4. Output intensity vs. time for phase-locked and FM operation: $g_0 = 0.075$; $\alpha_n = 0.070$; $\Delta\Omega = 0.1$ Ku. (Five modes free running.)

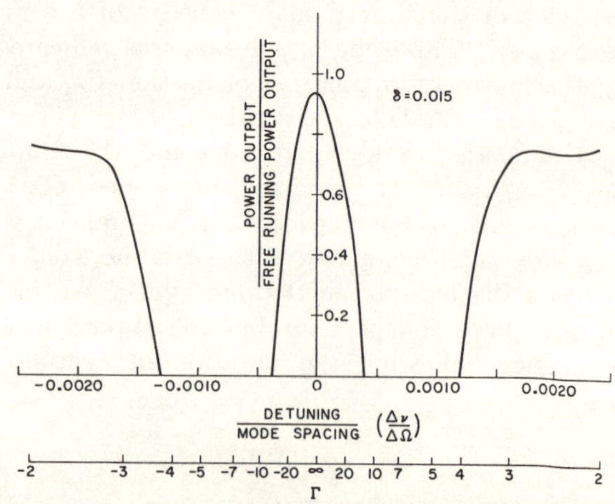

Fig. 5. Output power vs. detuning: $g_0 = 0.075$; $\alpha_n = 0.070$; $\Delta\Omega = 0.1$ Ku. (Five modes free running.)

to that of the free-running laser, and the conditions are again those of Figs. 3 and 4. At zero detuning, output power is about 0.95 that of the free-running laser. As detuning is increased, the mode envelope shifts away from the center of the Doppler line and output power decreases to zero. The oscillation remains below threshold until a detuning corresponding to a Γ of ~ 4. For a further increase of detuning, Γ decreases and the output power rapidly rises. We note that for the case considered, either phase-locked or FM operation may be obtained at nearly full power of the free-running laser.

In Figs. 3 and 4 we considered only positive detunings; that is, we assumed the modulation frequency to be less than the axial mode interval. For negative detunings we obtain very similar results, the one difference being a small asymmetry in frequency. This asymmetry is seen in the power curve of Fig. 5, and was first observed in the experiments of Ammann, McMurtry, and Oshman.[7] It is seen from Fig. 5 that the asymmetry appears in both the phase-locked and FM regions. Peak power in the phase-locked region occurs at a detuning of $\Delta\nu = -0.0004\ \Delta\Omega$ rather than zero. For positive detuning, threshold for the FM region occurs at $0.0012\ \Delta\Omega$, while for negative detuning it occurs at $-0.0013\ \Delta\Omega$. It is found that Γ's obtained with a positive detuning are slightly smaller than those predicted by (23), while Γ's obtained with a negative detuning are slightly larger than those of (23).

These asymmetries may be explained in terms of nonlinear but power independent mode pulling. That is, they are a result of the portion of the plasma dispersion function Z_r which depends nonlinearly upon frequency. Qualitatively, we note that, for an inverted atomic media, the effect of the nonlinear terms in the series expansion of Z_r is to decrease the index of refraction of modes above the center of the atomic line and to increase the index of refraction of modes below the center of the atomic line. These terms thus act to push all modes further from the center of the atomic line than they would otherwise be. The result is that, for negative detuning (modulation frequency greater than the axial mode interval), the average separation between the modulation sidebands and axial modes is somewhat smaller than it would otherwise be; thus, some sort of averaged detuning is somewhat smaller than that which would be obtained in the absence of this nonlinear part of χ_n'. For positive detuning, the opposite situation holds, and an averaged detuning is somewhat larger than that obtained in the absence of mode pulling. This asymmetrical behavior is a distortion and, as predicted by the form of μ_n of (53), will become increasingly small if, at constant Γ, δ becomes increasingly large. From an alternate point of view, if the same Γ is obtained at an increased δ, then $\Delta\nu$ must be increased, and the nonlinear mode pulling is more effectively swamped out.

The behavior in the phase locked region is at first somewhat puzzling; that is, it is physically hard to see how pulsing should result from a time varying phase perturbation. A qualitative explanation is indicated by Fig. 6, which is a plot of $\dot{\varphi}/\Delta\nu$ vs. Γ (or equivalently $\Delta\nu$) at constant δ. Since $\dot{\varphi}$ is a uniform shift of all modes from their assumed positions of (10), the closest integer to $\dot{\varphi}/\Delta\nu$ indicates that mode which is closest to an unperturbed cavity resonance. The curve in Fig. 6 begins at Γ large enough such that the single FM oscillation having its center frequency at the center of the atomic line is already below threshold. As detuning is decreased and thus Γ is increased, a point is reached such that an oscillation with its carrier, i.e., its on-resonance frequency component, at approximately the $+12$ mode, comes above threshold. However, at the given δ the distortion of this oscillation is so large that it bears no resemblance to an FM signal. As $\Delta\nu$ is further decreased, we find that $\dot{\varphi}$ is approximately $\delta\Delta\Omega/\pi$ [the value predicted by (27)], and thus $\dot{\varphi}/\Delta\nu \sim \Gamma$. For large Γ, $J_n(\Gamma)$ has its peak amplitude at approximately $n = \Gamma$. It thus appears that what is happening is that the oscillation which, if undistorted, would have its peak relative amplitude at approximately the center of the atomic line is the one which oscillates—though with a distortion which eliminates most of the FM spectrum.

The foregoing explanation of the phase-locked region leads to the prediction that if δ were increased to the point where the signal were better able to maintain its outer sidebands, its power should decrease; and at a sufficiently large δ, it should fall below threshold. This is indeed the case as is shown in Fig. 7. Here δ is increased over Figs. 3 through 6 by a factor of five, and normalized power vs. normalized detuning is again plotted: We find that the phase-locked solution is completely eliminated. We also note that, due to the larger δ, the asymmetry of Fig. 5 is significantly reduced.

The behavior vs. detuning that has been seen in Figs. 3 through 7 has been somewhat simpler than it would have been had δ not been chosen sufficiently large. At low δ, the region over which the laser is extinguished becomes an unquenched region wherein $\dot{E}_n \neq 0$ and a number of highly distorted FM oscillations are above threshold. (Alternately, this region might be considered as one consisting of multiple phase-locked solutions.) At still lower δ this unquenched region extends into what was previously the steady state FM region. Finally, for very low δ, i.e., δ considerably smaller than the excess gain,[8] the FM solution entirely disappears. There remains a steady state phase-locked solution for very small $\Delta\nu$, and for all other $\Delta\nu$ the situation is unquenched. If spontaneous emission is neglected, then as $\Delta\nu$ approaches zero, the δ which is necessary to obtain a phase-locked

[7] In their paper entitled "Detailed Experiments on Helium-Neon FM Lasers," Ammann, McMurtry, and Oshman have defined their detuning oppositely from that of this paper. That is, their detuning is positive when the modulation frequency is greater than the axial mode interval. Allowing for this difference in definition, the direction of our asymmetry is still opposite to theirs. This is perhaps due to a difference in definition of $\Delta\Omega$.

[8] We define the excess gain as the difference between the unsaturated single pass gain at line center and the single pass loss.

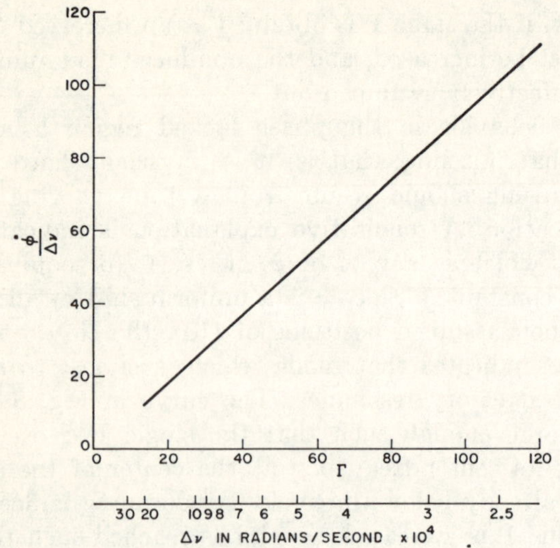

Fig. 6. $\dot{\varphi}/\Delta\nu$ vs. Γ: $g_0 = 0.075$; $\alpha_n = 0.070$; $\Delta\Omega = 0.1$ Ku (Five modes free running.)

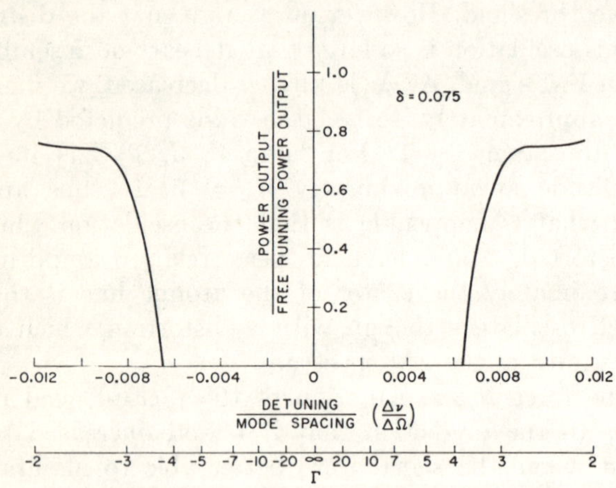

Fig. 7. Output power vs. detuning for larger δ: $g_0 = 0.075$; $\alpha_n = 0.070$; $\Delta\Omega = 0.1$ Ku. (Five modes free running.)

solution also approaches zero. In such small δ cases, the final steady state mode amplitudes are approximately those of the free-running laser, and the principal effect of the perturbation is to cause a locking of the phases of the respective modes. The effect on the phase-locked solution of increasing δ is to cause the mode amplitudes to become more nearly equal and to extend over a larger spectral width than did the modes of the free-running laser. The result is a sharpening of the time domain pulsing and a decrease in output power. The time domain pulses may be made increasingly sharp until the laser is extinguished.

X. Distortion and Super-Mode Conversion

In this section we apply the iterative procedure of Sections V through VII and study some effects of distortion of the ideal FM solution. We assume a somewhat higher gain than in the previous section and take $g_0 = 0.085$, and $\alpha_n = 0.070$. Inhomogeneous saturation, i.e., $\beta_{nm} = \beta\delta_{nm}$ and a ratio of Doppler width to mode spacing of 16.67 are again assumed. These conditions correspond to nine free-running modes in the unperturbed laser.

In Tables I and II we consider mode amplitudes and mode phases, respectively, for a case where $\delta = 0.06$, and the detuning is such that $\Gamma = 4.5$.[9] Column 1 of Table I shows the ideal Bessel function amplitudes which would exist if all net saturated gains were zero and mode pulling were neglected. Column 2 gives the results of digital computer solution of (13a) and (13b). Columns 3, 4, and 5 give the results of the first, second, and third iterations of the procedure of Section VII. In Table II, similar results are given for the relative phase angles. It is noted that these phase angles differ considerably more from their ideal values than do the amplitudes of Table I.

In cases where the distortion is larger than that of the previous example, for instance when δ is considerably lower or Γ is considerably larger, the convergence rate of the iterative procedure will be slower. If the constants of this example are left unchanged, except for reducing both δ and $\Delta\nu$ by a factor of two, it is found that about five iterations are required to reach a comparable steady state. However, even at five iterations, the iterative procedure requires only about 1/20 of the computer time as does solution of (13a) and (13b) to comparable accuracy. During the course of the numerical work, a few cases were found where the numerical procedure failed to converge. These were cases of high distortion and occurred particularly at large Γ.

One manifestation of distortion is the existence of beats when the output of an FM laser is incident on a photodetector. Proceeding from (82) it is seen that the amplitude R_q and phase η_q of the qth beat between the laser modes is given by

$$R_q = (M_q^2 + N_q^2)^{1/2}$$
$$\eta_q = \tan^{-1}\left(\frac{N_q}{M_q}\right), \qquad (83)$$

where

$$M_q = \tfrac{1}{2} \sum_n E_n E_{n+q} \cos(\varphi_{n+q} - \varphi_n)$$
$$N_q = \tfrac{1}{2} \sum_n E_n E_{n+q} \sin(\varphi_{n+q} - \varphi_n). \qquad (84)$$

For the ideal situation where $E_n = J_n(\Gamma)$ and $\varphi_n - \varphi_{n-1} = 0$, by noting (39) we see that all beats for $q \neq 0$ are zero.

Even in the actual case wherein net saturated gains and mode pulling are not zero, it may be shown from the

[9] The presence of a linear term in χ'_n in effect changes the axial mode interval and complicates the specification of the detuning which will lead to a given Γ. The function Z_r of (80b) may be expanded in a power series as follows [14]:

$$Z_r(\zeta) = +2\zeta\left[1 - \frac{2\zeta^2}{3} + \frac{4\zeta^4}{15} - \frac{8\zeta^6}{105} + \cdots\right].$$

If accounted for as part of the detining, the linear term of this series does not lead to distortion. We will thus adopt the procedure of neglecting this term in Ψ_n of (28a), and instead will include its effect in the specification of the detuning. We note that experimentally this problem does not arise in that the linear portion of the mode pulling is automatically included in the measurement of the axial mode interval.

TABLE I

COMPARISON OF MODE AMPLITUDES (E_n)

$g_0 = 0.085$

$\alpha_n = 0.070$

$\left(\frac{\text{Doppler Width}}{\text{Mode Spacing}}\right) = 16.67$

9 Modes Free Running

$\delta = 0.06$

$\Gamma = 4.5$

n	Ideal (Bessel Functions)	Equations (13a) and (13b)	Iterative Procedure 1st Iteration	2nd Iteration	3rd Iteration
0	−0.320	−0.279	−0.282	−0.280	−0.280
1	−0.231	−0.251	−0.254	−0.252	−0.252
2	0.218	0.180	0.180	0.180	0.180
3	0.425	0.410	0.413	0.412	0.412
4	0.348	0.367	0.368	0.369	0.369
5	0.195	0.223	0.222	0.224	0.225
6	0.084	0.107	0.104	0.107	0.107
7	0.030	0.043	0.043	0.042	0.043
8	0.009	0.015	0.014	0.014	0.015
9	0.002	0.005	0.004	0.004	0.005
10	0.001	0.001	0.001	0.001	0.001

TABLE II

COMPARISON OF PHASE ANGLES ($\varphi_n - \varphi_{n-1}$)

$g_0 = 0.085$

$\alpha_n = 0.070$

$\left(\frac{\text{Doppler Width}}{\text{Mode Spacing}}\right) = 16.67$

9 Modes Free Running

$\delta = 0.06$

$\Gamma = 4.5$

n	Ideal	Equations (13a) and (13b)	Iterative Procedure 1st Iteration	2nd Iteration	3rd Iteration
1	0°	5.223°	5.183°	5.194°	5.242°
2	0°	−22.181°	−18.583°	−21.178°	−22.158°
3	0°	17.611°	14.311°	17.645°	17.622°
4	0°	6.298°	4.899°	6.219°	6.352°
5	0°	4.763°	3.730°	4.590°	4.746°
6	0°	5.841°	4.909°	5.611°	5.796°
7	0°	7.317°	6.254°	7.008°	7.265°
8	0°	8.598°	7.290°	8.139°	8.496°
9	0°	9.395°	8.020°	8.919°	9.363°
10	0°	8.977°	8.480°	8.951°	9.310°

formulas of Section VII that, if the center frequency of the FM oscillation is at the center of a symmetrical atomic line, then all odd harmonic beats will be identically zero.

Figure 8 shows the amplitude of the second beat vs. δ at constant Γ. That is, both δ and $\Delta\nu$ are varied such that Γ as given by (23) remains constant. Beat amplitudes are normalized to the total intensity of the FM laser at the given Γ. For reference, the beat which is obtained from (83) by taking the E_n as those of the free-running laser, and assuming that relative phases are zero, is noted in Fig. 8. These conditions are comparable to what has been termed a selflocked beat and which is often observed in practice [10], [15]. Beat amplitudes are shown for $\Gamma = 4.5$ and for $\Gamma = 3.0$. We note that distortion is somewhat larger at the larger Γ, and for either Γ, decreases as δ is increased.

Figure 9 shows the time domain behavior corresponding to certain of the points of Fig. 8. Data for this figure were obtained by evaluation of (82) using E_n and φ_n obtained by five iterations of the procedure of Section VII. It is seen that the ripple may be made increasingly small if δ is made sufficiently large.

As another figure of merit for an FM signal, we consider its supermode conversion efficiency. The super-mode technique is shown schematically in Fig. 10, and was suggested by Massey and demonstrated by Massey, Oshman, and Targ [16]. In this technique the output signal from an FM laser is passed through an external phase modulator which is driven such that its single pass phase retardation is exactly equal to the Γ at which the FM laser is running. By properly adjusting the phase of the external modulator with respect to that of the internal phase perturbation, the resultant light signal can be made to have a modulation depth anywhere between zero and 2Γ. In particular, when the resultant modulation depth is adjusted to zero, then in principal, all of the energy that was previously distributed between all of the sidebands of the FM signal should appear as a single monochromatic optical signal—or "super-mode," as it has been termed by Targ.

We take the light signal emerging from the FM laser to be of the form of (81), and assume the single pass phase retardation of the external modulator to be $-\Gamma \sin \nu_m t$. Each sideband of the FM signal is thus modulated, and after passage through the external modulator, we have

$$E_0(t) = \sum_n E_n \cos [(\Omega_0 + n\nu_m)t + \varphi_n - \Gamma \sin \nu_m t]. \tag{85}$$

By using various trigonometric and Bessel identities it may be shown that $E_0(t)$ may be written

$$E_0(t) = \sum_q F_q \cos (\Omega_0 + q\nu_m)t + \sum_q H_q \sin (\Omega_0 + q\nu_m)t, \tag{86}$$

where F_q and H_q are given by

$$F_q = \sum_n E_{q+n} J_n(\Gamma) \cos \varphi_{q+n}$$
$$H_q = -\sum_n E_{q+n} J_n(\Gamma) \sin \varphi_{q+n}. \tag{87}$$

The power in the desired super-mode is then $F_0^2 + H_0^2$. We define the super-mode conversion efficiency to be the ratio of power in the super-mode to the total power of the incident FM signal, giving

$$\text{super-mode conversion efficiency} = \frac{(F_0^2 + H_0^2)}{\sum_q (F_q^2 + H_q^2)}. \tag{88}$$

Figure 11 shows super-mode conversion efficiency vs. δ at constant Γ for the case of nine free-running modes.

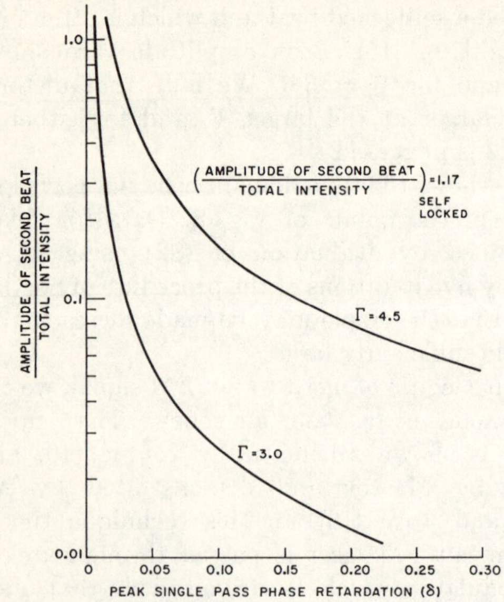

Fig. 8. Amplitude of second beat vs. δ: $g_0 = 0.085$; $\alpha_n = 0.070$; $\Delta\Omega = 0.1$ Ku. (Nine modes free running.)

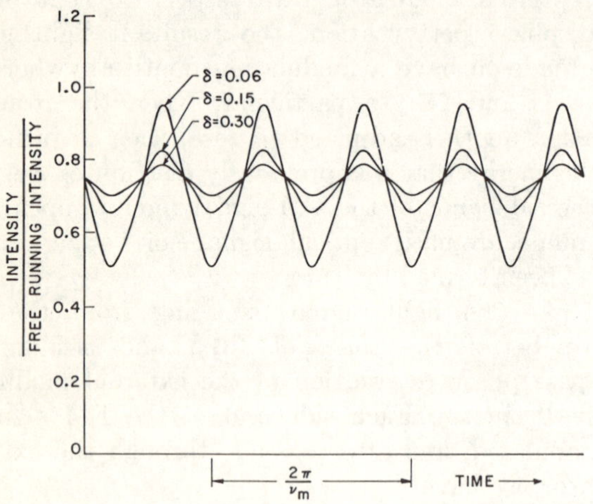

Fig. 9. Output intensity vs. time at different δ: $g_0 = 0.085$; $\alpha_n = 0.070$; $\Delta\Omega = 0.1$ Ku. (Nine modes free running.)

At a $\Gamma = 3.0$, a δ approximately equal to five times the excess gain yields a conversion efficiency of almost 100 percent. At $\Gamma = 4.5$, a somewhat larger δ is required. In obtaining these conversion efficiencies from (88) we have used the Γ as given by (23) and have properly taken into account the linear part of χ_n'. It has been previously noted, however, that because of the nonlinearity of mode pulling vs. frequency, this Γ is not the closest approximation to the existing FM signal. It is thus likely that the efficiencies of Fig. 11 could be somewhat increased by a more optimum choice of the converting Γ.

Examination of the time domain behavior of the converted signal shows that the super-mode process leaves the AM ripple unchanged. At $\delta = .06$, Fig. 9 shows the FM signal to have about a 30 percent AM ripple, while Fig. 11 shows the super-mode conversion efficiency to be about 98 percent. Though at first surprising, the converted signal is found to have a time domain behavior which is identical to that of Fig. 9.

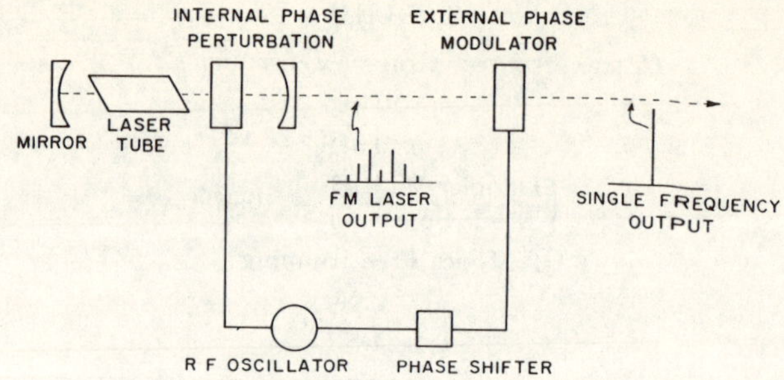

Fig. 10. Schematic of super-mode technique.

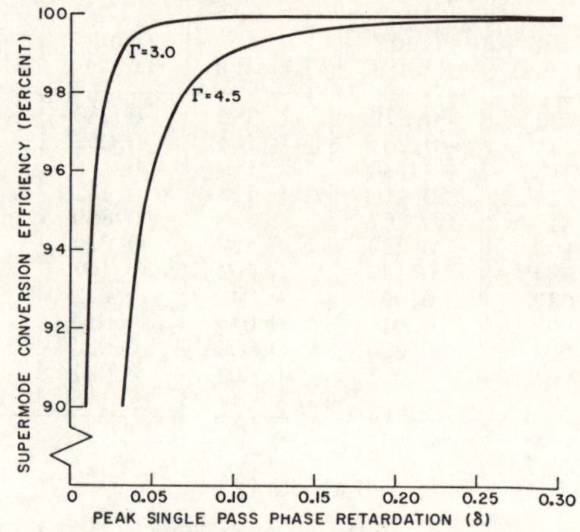

Fig. 11. Super-mode conversion efficiency vs. δ: $g_0 = 0.085$; $\alpha_n = 0.070$; $\Delta\Omega = 0.1$ Ku. (Nine modes free running.)

XI. Application and Validity of the a Equations

In the derivation of the a Equations of Section VIII, two approximations were made. First, it was assumed that δ was sufficiently large that the competing oscillations could be considered to have exactly Bessel function amplitudes. Second, and more important, it was assumed, or perhaps hoped, that phase-locking between different FM oscillations could be neglected without affecting the range of Γ over which only a single steady state FM oscillation would exist. Unfortunately, in certain cases, this second assumption does not appear to be valid.

The thin curve of Fig. 12 shows the application of the a Equations to the case of nine free-running modes of the previous section. Inhomogeneous saturation of all modes is again assumed, and amplitudes of all oscillations are normalized to the total power of the free-running laser. When $\Gamma = 0$, the FM oscillations are identical to the free-running modes. As Γ is slightly increased, the oscillations remain very weakly coupled, and all nine FM oscillations exist simultaneously. For a somewhat larger Γ, certain of these oscillations are quenched, but a condition of multiple oscillation persists. When $\Gamma = 2.3$ or larger, quenching is complete and only the zeroth oscillation remains. Output power first rises as a result of more effective atomic saturation. The slight dip at $\Gamma \sim 3.8$ is attributable to a decrease in amplitude of the one mode, i.e., $J_1(3.832) = 0$. For larger Γ power decreases as relative amplitudes which are further from the center of the Doppler line are increased; and for $\Gamma = 6.7$ the oscillation is below threshold.

The thick curve in Fig. 12 shows normalized output power as obtained from (13a) and (13b), where a δ of 0.06 has been assumed. In regions where this thick line is absent, (13a) and (13b) predict a condition of multiple FM oscillation. Thus, at points where there is only a single thin line and no thick line, (13a) and (13b) are in disagreement with the a Equations.[10]

At Γ's where the a Equations predict multiple FM oscillations, (13a) and (13b) are in agreement. For certain of these cases, the time variation of the nonsteady state E_n which resulted from the digital computer solution of (13a) and (13b) were examined. At $\Gamma = 2.15$, where the a Equations predict that the a_0 and the $a_{\pm 1}$ oscillations should exist simultaneously, that all the E_n were periodic functions of time and had a fundamental frequency variation of $\Delta\nu$ and a higher harmonic variation of $2\Delta\nu$. At $\Gamma = 1.35$, where the a Equations predict that only the $a_{\pm 2}$ oscillations should exist, all the E_n had a perfectly sinusoidal time variation at a frequency of $4\Delta\nu$. These results are consistent with the transformation of (61).

Examination of the time domain behavior of the E_n at $\Gamma = 2.5$, i.e., one of the points of disagreement, showed the zero and also the ± 1 oscillations to exist simultaneously. It was also determined that irrespective of initial conditions, these oscillations existed with a phase relation which, if each were perfect FM, results in partial cancellation of the factor of two which is involved in the cross saturation of FM oscillations, and which was discussed at the end of Section VIII.

It thus appears that the a Equations are over optimistic, and predict a steady state FM oscillation, when perhaps a phase-locked multiple oscillation situation would exist. We note, however, that we have assumed a perfectly symmetrical atomic line with one FM oscillation exactly at its center. This condition is favorable to a phase-locking process, and whether such locking will occur under typical experimental conditions remains to be determined.[11]

As another application of the a Equations, we considered the case of a very narrow high gain line such that only one free-running mode was above threshold. An axial mode interval equal to 1.2 times the Doppler width was assumed, and g_0 and α were taken as 0.56 and 0.056, respectively. The results are shown in Fig. 13, where two interesting effects are observed. First, we find that the FM power may exceed that of the free-running laser by a factor of ~ 3. Second, we find that, for higher Γ, steady state FM oscillations which have their carrier frequency either above or below the center of the atomic line may exist. In such cases, whether the plus or minus oscillation exists is determined by initial conditions.

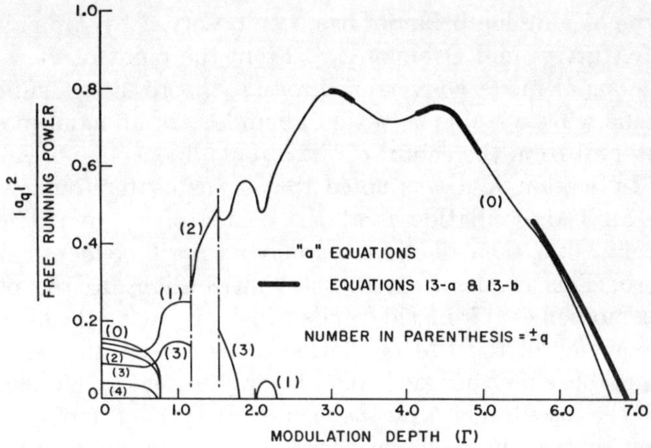

Fig. 12. $|a_q|^2$ vs. Γ: $g_0 = 0.085$; $\alpha_n = 0.070$; $\Delta\Omega = 0.1$ Ku. (Nine modes free running.)

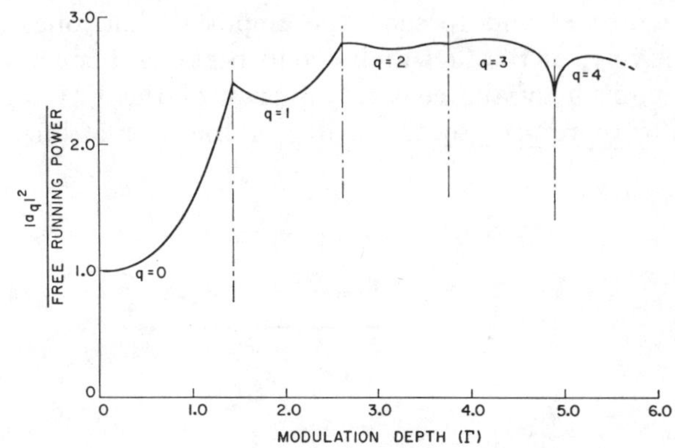

Fig. 13. $|a_q|^2$ vs. Γ for high gain line: $g_0 = 0.56$; $\alpha_n = 0.056$; $\Delta\Omega = 2.0$ Ku. (One mode free running.)

For the case of the high gain line, as well as another case which was considered wherein only two FM oscillations were above threshold, it was found that the results of the a Equations were in complete agreement with those of (13a) and (13b). The agreement in these cases is attributed to the fact that a phase-locking of oscillations requires at least three oscillations—two of which drive the third. It is planned that a further study of phase-locking of FM oscillations will be undertaken.

XII. Effect of Mirror Motion

In previous sections we have always assumed that the center frequency of the $q = 0$ oscillation was at the center of a symmetrical atomic line. Under typical operating conditions, this will seldom be the case. As a result of mirror motion, the center frequencies of all the FM oscillations will drift with respect to the center of the atomic line. It is expected that if the drift from the center of the atomic line is greater than about one-half an axial mode interval, then either the $q = -1$ or $q = +1$ oscillation should become the dominant oscillation. This

[10] Since in all cases the modes display a transient which generally takes the form of a damped or growing beat at harmonics of $\Delta\nu$, it is necessary to make an arbitrary decision as to what is to be considered as a steady state condition. For the present case, a 10 percent variation of the E_n after 100 microseconds was considered as satisfactory. At a proper iteration interval this represents a run time of between 15 and 30 minutes on an IBM 7090, for each data point.

[11] In their experiments on a 6328A He-Ne laser with nine free-running modes above threshold, Ammann, McMurtry, and Oshman, [15], have found that a steady state FM oscillation does exist at certain points, e.g., $\Gamma = 3.9$, where for a running time of 100 μs, (13a) and (13b) predict multiple oscillation. However, this does not resolve the question, in that completely inhomogeneous saturation is not a proper approximation for He-Ne. We have found that if a single pass gain of the form $g_n(1 - \beta E_n^2 - 0.4\,\beta E_{n+1}^2 - 0.4\,\beta E_{n-1}^2)$ is assumed, and all other constants are left unchanged, that (13a) and (13b) will then also predict a steady state oscillation at $\Gamma = 3.9$.

type of jumping behavior has been observed by Ammann, McMurtry, and Oshman [15]. From the point of view of the super-mode conversion process, the resultant super-mode will be stable to within about $\pm\frac{1}{2}$ of an axial mode interval from the center of the atomic line.

In Section X it was noted that if the center frequency of an FM oscillation is at the center of a symmetrical atomic line, then all odd harmonic beats will be identically zero. This results because the contributions to the odd harmonic beats from sidebands which are above the center frequency of the FM oscillation are exactly cancelled by contributions from sidebands below the center frequency of the oscillation. As the center frequency moves off line center, this will no longer be the case. The cancellation of upper and lower contributions is no longer complete, and odd harmonic distortion rapidly increases.

Figures 14 and 15 show the amplitude and phase, respectively, of the first and second beats as a function of the position of the center frequency of the FM oscillation with respect to the center of the atomic line. For these figures, $\delta = 0.15$, $\Gamma = 3.0$, and other constants and the normalization are the same as those of Fig. 8. Data for the figures were obtained by five iterations of the procedure of Section VII. It is seen that the amplitude of the first beat is extremely sensitive to the position of the center frequency of the FM oscillation, while the amplitude of the second beat is nearly independent of this position. In addition, the phase of the first beat changes abruptly as the center of the FM oscillation moves from one side of the atomic line to the other. The behavior of higher odd and even beats is similar to that of the first and second beats, respectively. Harris and Oshman have proposed that these effects could be used as a frequency discriminant for stabilization of an FM laser, and experiments demonstrating the existence of such a discriminant have been performed [17].

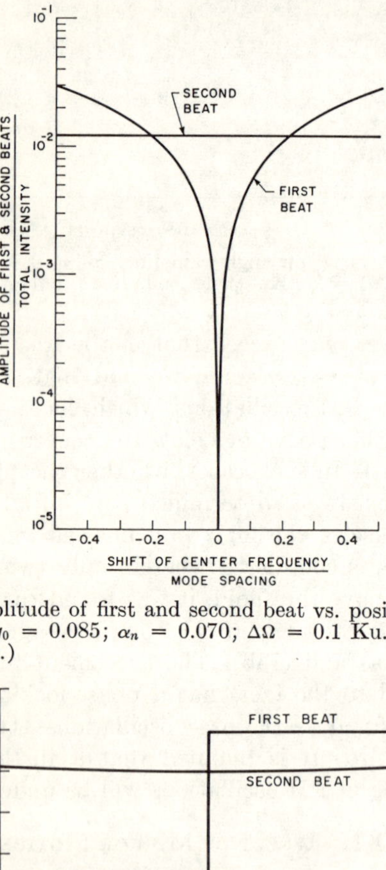

Fig. 14. Amplitude of first and second beat vs. position of center frequency: $g_0 = 0.085$; $\alpha_n = 0.070$; $\Delta\Omega = 0.1$ Ku. (Nine modes free running.)

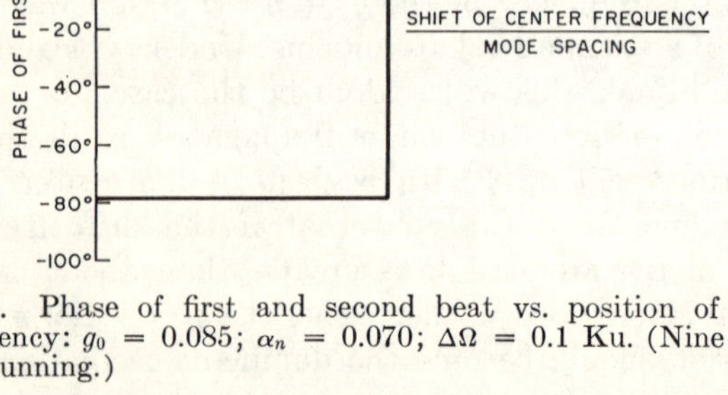

Fig. 15. Phase of first and second beat vs. position of center frequency: $g_0 = 0.085$; $\alpha_n = 0.070$; $\Delta\Omega = 0.1$ Ku. (Nine modes free running.)

Acknowledgment

The authors wish to express their gratitude to E. O. Ammann, B. J. McMurtry, and M. K. Oshman for providing us with the results of their experiments on He-Ne FM lasers, and also for numerous helpful discussions. Knowledge of their experiments increased our confidence in the theory, and in certain instances indicated the direction in which to proceed. We also wish to thank L. Osterink, A. E. Siegman, and Russell Targ for interesting discussions; and gratefully acknowledge the assistance of Mrs. Cora Barry with the numerical computation.

References

[1] W. R. Bennett, Jr., "Hole burning effects in a He-Ne optical maser," *Phys. Rev.*, vol. 126, pp. 580–593, April 1962.

[2] C. L. Tang, H. Statz, and G. de Mars, "Spectral output and spiking behavior of solid-state lasers," *J. Appl. Phys.*, vol. 34, pp. 2289–2295, August 1963.

[3] S. E. Harris and R. Targ, "FM oscillation of the He-Ne laser," *Appl. Phys. Letters*, vol. 5, pp. 202–204, November 1964.

[4] W. E. Lamb, Jr., "Theory of an optical maser," *Phys. Rev.*, vol. 134, pp. A1429–A1450, June 1964.

[5] E. I. Gordon and J. D. Rigden, "The Fabry-Perot electro-optic modulator," *Bell Sys. Tech. J.*, vol. XLII, pp. 155–179, January 1963.

[6] A. Yariv, "Internal modulation in multimode laser oscillators," *J. Appl. Phys.*, vol. 36, pp. 388–391, February 1965.

[7] S. E. Harris and O. P. McDuff, "FM laser oscillation-theory," *Appl. Phys. Letters*, vol. 5, pp. 205–206, November 1964.

[8] L. E. Hargrove, R. L. Fork, and M. A. Pollack, "Locking of He-Ne laser modes induced by synchronous intracavity modulation," *Appl. Phys. Letters*, vol. 5, pp. 4–5, July 1964.

[9] M. DiDomenico, Jr., "Small-signal analysis of internal (coupling type) modulation of lasers," *J. Appl. Phys.*, vol. 35, pp. 2870–2876, October 1964.

[10] M. H. Crowell, "Characteristics of mode-coupled lasers," *IEEE Journal of Quantum Electronics*, vol. QE-1, pp. 12–20, April 1965.

[11] B. Friedman, *Principles and Techniques of Applied Mathematics*. New York: Wiley, 1956.

[12] M. Abramowitz and I. A. Stegun, *Handbook of Mathematical Functions*. National Bureau of Standards, Applied Mathematics Series, vol. 55, 1964, p. 363.

[13] W. J. Cunningham, *Introduction to Nonlinear Analysis*. New York: McGraw-Hill, 1958, p. 313.

[14] B. D. Fried and S. D. Conte, *The Plasma Dispersion Function* (*Hilbert Transform of the Gaussian*). New York: Academic, 1961.

[15] E. O. Ammann, B. J. McMurtry, and M. K. Oshman, "Detailed experiments on helium-neon FM lasers," this issue, pp. 263–272.

[16] G. A. Massey, M. K. Oshman, and R. Targ, "Generation of single frequency light using the FM laser," *Appl. Phys. Letters*, vol. 6, pp. 10–11, January 1965.

[17] S. E. Harris, M. K. Oshman, B. J. McMurtry, and E. O. Ammann, "Proposed frequency stabilization of the FM laser," Appl. Phys. Letters, vol. 7, pp. 184–186, October 1965.

Nonlinear Theory of the Internally Loss-Modulated Laser

ODIS P. McDUFF, SENIOR MEMBER, IEEE AND STEPHEN E. HARRIS, MEMBER, IEEE

Abstract—This paper presents a detailed nonlinear analysis of the internally loss-modulated laser including the effect of arbitrary atomic lineshape, saturation, and mode pulling. Results of the analysis are in part numerical and include a study of the spectral and time domain behavior of the laser output. The results include a determination of the minimum perturbation strength which is necessary to produce phase locking, peak pulse amplitude, and minimum pulsewidth as a function of perturbation strength, a consideration of the detuned case, and a comparison of AM- versus FM-type phase locking. Results are compared with the previously obtained linearized solutions of others.

I. Introduction

PHASE LOCKING of the optical modes of a gas laser by means of internal loss modulation was first reported by Hargrove, Fork, and Pollack.[1] Theoretical studies have been given by DiDomenico,[2] Yariv,[3] and Crowell,[4] who present solutions for a linearized approximation to the problem. Experimental results on helium–neon and argon lasers have been given by Crowell.[4] Deutsch[5] reported phase locking of a ruby laser and Pantell and Kohn[6] have presented a linearized transient study of the ruby laser. Recently, DiDomenico et al.[7] have reported phase locking of a YAlG:Nd laser.

In this paper we present a detailed study of AM-type mode locking which includes the effects of atomic lineshape, saturation, and frequency pulling. The results of this nonlinear study differ appreciably from those of the earlier linear theories. In particular, the modulator drive strength becomes an important parameter in determining the characteristics of the pulsing laser.

The techniques employed and the form of the present paper are similar to those of a previous paper.[9] In the following sections, we first derive and discuss the equations which govern AM-type locking. Conservation conditions are derived and the equations are put into what, in effect, is an integral form, useful in later sections. Numerical results which show the effect of varying the modulator drive strength and frequency are given. It will be seen that at low modulator levels mode phases depart considerably from their ideal values with the result that the laser pulses may have sidelobes of appreciable magnitude. Curves of peak pulse intensity, and pulsewidth versus modulator drive strength are given. Later sections of the paper consider the question of threshold for locking and study the problem of phase locking a laser where only a single mode has gain. It will be seen that, in principle, the pulsewidth in such a laser may be made as small as that of a multimode laser.

Before proceeding we note that periodic pulsing of a laser may also be obtained by means of an internal phase perturbation which is driven at a frequency equal to that of the axial mode interval.[8] Studies of this process have been given by Harris and McDuff[9] and by Ammann, McMurtry, and Oshman.[10] Except for occasional comparisons, this type of locking will not be considered in the present paper.

II. Basic Differential Equations

In this section we summarize the derivation of the coupled-mode differential equations. The procedure is basically the same as that used in an earlier paper on the FM laser.[9]

We start with the self-consistency equations of Lamb[11] which describe the effect of an arbitrary optical polarization upon the electric fields of a high-Q optical resonator. In the present case, the polarization includes a contribution resulting from the inverted atomic media and a parametric contribution resulting from the loss perturbation. We assume the loss perturbation to be represented by a quadrature component of susceptibility,

$$\Delta\chi''(z, t) = \Delta\chi''(z)[1 + \cos \nu_m t], \tag{1}$$

where ν_m is the driving frequency of the perturbation and is approximately equal to $\Delta\Omega$, the frequency separation of the resonances Ω_n of the empty optical cavity.[1] The parametric contribution to total polarization is then

$$\underset{\text{Parametric}}{P(z, t)} = \sum_n \epsilon_0 \, \Delta\chi''(z, t) E_n(t) \sin [\nu_n t + \varphi_n(t)] U_n(z) \tag{2}$$

where

$$U_n(z) = \sin (n_0 + n)\pi z/L,$$

and the total cavity electromagnetic field has been expanded in the form[2]

$$E(z, t) = \sum_n E_n(t) \cos [\nu_n t + \varphi_n(t)] U_n(z).$$

Here $E_n(t)$ and $\varphi_n(t)$ are the slowly time-varying amplitude and phase of the nth mode, ν_n is its frequency, and L is the cavity length. We define

$$\alpha_a = \frac{2\nu}{c} \int_0^L \Delta\chi''(z) U_n^2(z) \, dz \tag{3}$$

Manuscript received July 8, 1966; revised December 1, 1966. The work reported in this paper was sponsored by the National Aeronautics and Space Administration under NASA Grant NGR-05-020-103. O. P. McDuff was supported in part by a National Science Foundation Faculty Fellowship.

O. P. McDuff is with the Department of Electrical Engineering, University of Alabama, University, Ala. He was formerly with Stanford University.
S. E. Harris is with Stanford University, Stanford, Calif.

[1] We adopt the convention that all symbols for frequency shall denote circular frequencies.
[2] Except where noted, all sums will be from $-\infty$ to $+\infty$.

Reprinted from *IEEE J. Quantum Electron.*, vol. QE-3, pp. 101–111, Mar. 1967.

and

$$\alpha_c = \frac{\nu}{c} \int_0^L \Delta\chi''(z) U_{n\pm1}(z) U_n(z). \tag{4}$$

The parametric component of polarization driving the nth mode is given by

$$P_n(t) = \frac{2}{L} \int_0^L P(z, t) U_n(z)\, dz,$$

which becomes upon combining (1) through (4) and assuming that only adjacent mode coupling applies,

$$\begin{aligned} P_n(t) = {} & \frac{\epsilon_0 \alpha_a c}{\nu L} E_n \sin(\nu_n t + \varphi_n) + \frac{\epsilon_0 \alpha_c c}{\nu L} [E_{n+1} \cos(\varphi_{n+1} - \varphi_n) \\ & + E_{n-1} \cos(\varphi_n - \varphi_{n-1})] \sin(\nu_n t + \varphi_n) \\ & + \frac{\epsilon_0 \alpha_c c}{\nu L} [E_{n+1} \sin(\varphi_{n+1} - \varphi_n) \\ & - E_{n-1} \sin(\varphi_n - \varphi_{n-1})] \cos(\nu_n t + \varphi_n). \end{aligned} \tag{5}$$

We include the contribution of the atomic medium to the polarization by macroscopic quadrature and in phase components of susceptibility χ_n'' and χ_n', respectively. The nonlinear effects of the medium are included in the χ_n'' and χ_n'. The self-consistency equations then become

$$\begin{aligned} & [\dot{\varphi}_n - n\, \Delta\nu + \tfrac{1}{2}\nu\chi_n']E_n \\ & = -\frac{\alpha_c c}{2L} [E_{n+1} \sin(\varphi_{n+1} - \varphi_n) - E_{n-1} \sin(\varphi_n - \varphi_{n-1})] \end{aligned} \tag{6}$$

and

$$\begin{aligned} \dot{E}_n + \frac{\nu}{2}\left[\frac{1}{Q_n} + \chi_n''\right]E_n = {} & -\frac{\alpha_a c}{2L} E_n - \frac{\alpha_c c}{2L} \\ & \cdot [E_{n+1} \cos(\varphi_{n+1} - \varphi_n) + E_{n-1} \cos(\varphi_n - \varphi_{n-1})], \end{aligned} \tag{7}$$

where we have taken $\Delta\nu$ to be the detuning of the driving frequency from the axial mode interval $\Delta\Omega$, i.e.,

$$\Omega_n - \nu_n = n\, \Delta\nu. \tag{8}$$

Equations (6) and (7) are the basic differential equations to be considered and when solved give the amplitude, frequency, and phase of the optical modes. They can be combined to give a set of complex coupled-mode equations as presented by DiDomenico,[2] Yariv,[3] and Crowell.[4] The retention of the $\dot{\varphi}_n$ term (which is implicit in Yariv's equations but assumed zero in DiDomenico's and Crowell's) facilitates the inclusion of pulling and pushing of the entire coupled-mode oscillation.

III. Discussion of Parameters

Assuming small gain, we relate the small-signal saturated single-pass power gain to the quadrature component of susceptibility by the relation

$$G_n = -\frac{\nu L}{c} \chi_n'', \tag{9}$$

where G_n is the saturated single-pass fractional power gain of the nth mode and depends nonlinearly upon frequency, excitation, and power level. Similarly, we write

$$\psi_n = \frac{\nu L}{c} \chi_n', \tag{10}$$

where ψ_n is the additional round-trip phase retardation which is seen by the nth mode as a result of the insertion of the atomic medium and also depends nonlinearly upon frequency, excitation, and power level.

Although when possible the equations developed are left in terms of the general expressions G_n and ψ_n, in specific numerical calculations we consider the essence of the problem to be treated by the example of a Doppler-broadened Gaussian atomic line with homogeneous linewidth much smaller than both the axial mode interval and Doppler linewidth. For these cases, we will express the saturated gain as

$$G_n = g_n(1 - \beta E_n^3), \tag{11}$$

where g_n is the unsaturated fractional power gain per pass of the nth mode and β the saturation parameter. For the Doppler-broadened Gaussian line we have then

$$g_n = g_0 \frac{Z_i\left(\dfrac{\nu_n - \omega}{Ku}\right)}{Z_i(0)}. \tag{12}$$

Corresponding to this, we take ψ_n to be

$$\psi_n = g_0 \frac{Z_r\left(\dfrac{\nu_n - \omega}{Ku}\right)}{Z_i(0)}. \tag{13}$$

Here g_0 is the unsaturated line-center gain and Z_r and Z_i are the real and imaginary parts of the plasma dispersion function and are described by Lamb.[11] For vanishingly small homogeneous linewidth, (12) and (13) become g_0 times the normalized Gaussian and the Hilbert transform of the Gaussian, respectively. The parameter ω is the center frequency of the atomic line and Ku equals 0.6 times the half-power Doppler linewidth.

The single-pass power loss α_n is related to the Q of the nth mode by the expression

$$\alpha_n = \frac{\nu L}{c} \frac{1}{Q_n} \tag{14}$$

and includes both dissipative and output coupling loss (mirror transmission). In typical cases α_n is independent of n and we let $\alpha_n = \alpha$.

The time-varying loss perturbation is taken to have $\Delta\chi''$ independent of z over the length l of the perturbing element. By assuming small loss, the power loss per pass $\alpha(t)$ through this perturbing element is readily shown to be

$$\alpha(t) = W(1 + \cos \nu_m t) \tag{15}$$

where W is given by

$$W = \frac{\nu l}{c} \Delta\chi'' \tag{16}$$

and is the average loss. If the perturbing element is located such that its center is a distance z_0 from the end mirror, we obtain the self-coupling term α_a from (3)

$$\alpha_a = \frac{2W}{l} \int_{z_0-l/2}^{z_0+l/2} \sin^2 \frac{(n_0 + n)\pi z}{L} \, dz, \tag{17}$$

or since n_0 is very large,

$$\alpha_a = W. \tag{18}$$

Thus the self-coupling term is simply equal to the average loss introduced by the perturbation. We obtain from (4), (16), and (18) the cross-coupling term

$$\alpha_c = \frac{\alpha_a}{l} \int_{z_0-l/2}^{z_0+l/2} \sin \frac{(n_0 + n)\pi z}{L} \sin \frac{(n_0 + n + 1)\pi z}{L} \, dz \tag{19}$$

which, again noting that n_0 is very large, yields

$$\alpha_c = \frac{L}{l} \frac{\alpha_a}{\pi} \sin \frac{\pi l}{2L} \cos \frac{z_0 \pi}{L}. \tag{20}$$

For the practical case of $l \ll L$ this becomes

$$\alpha_c = \frac{\alpha_a}{2} \cos \frac{z_0 \pi}{L}. \tag{21}$$

The form of (21) is analogous to that obtained by Crowell[4] for loss modulation and Harris and McDuff[9] for phase modulation. It is desirable to locate the perturbing element close to the end of the cavity so that $z_0 \approx 0$ and such that

$$\alpha_c \approx \frac{\alpha_a}{2}. \tag{22}$$

To summarize, we define the loss terms as follows: α_n is the single-pass power loss of the nth mode which results from dissipative and coupling loss not dependent on the internal perturbation; α_a is equal to the average loss per pass introduced by the perturbation; and α_c is the mode coupling term resulting from the internal perturbation. In places where all modes are assumed to have losses independent of n, we will let $\alpha_n = \alpha$. The coupling factor α_c is analogous to δ in the FM laser.[9]

IV. Conservation Conditions

We now derive conservations which result from the basic differential equations. These are similar to those of Harris and McDuff[9] for the phase modulation case and are derived in an identical manner.

Multiplying (7) by E_n, summing over n, and using the constants of the previous section, we obtain

$$\sum_n E_n \dot{E}_n = \frac{c}{2L} [\sum_n (G_n - \alpha_n) E_n^2] - \frac{c}{2L} [\alpha_a \sum_n E_n^2 + 2\alpha_c \sum_n E_{n+1} E_n \cos (\varphi_{n+1} - \varphi_n)]. \tag{23}$$

Noting that $\sum_n E_n \dot{E}_n = d/dt \, [\sum_n E_n^2/2]$, (23) can be interpreted as an expression of conservation of power; i.e., the rate of change of energy stored is equal to the net power generated or absorbed in all modes. We see that, in general, the loss perturbation always absorbs power from the system. However in a hypothetical ideal system wherein all E_n's are equal, all $\varphi_{n+1} - \varphi_n = \pi$, and such that $\alpha_c = \frac{1}{2}\alpha_a$ (modulator at the end of the optical cavity); we see that the parametric term on the right side of (23) is identically equal to zero and contributes no net loss. We shall see later that under various operating conditions the coupled system tries to adjust itself toward this ideal situation. This situation corresponds to the physical picture offered by Crowell[4] of the light pulse going through the perturbing element at that instant of time when its attenuation is zero.

Applying a similar procedure to (6), we obtain the reactive conservation condition

$$\sum_n \dot{\varphi}_n E_n^2 = \sum_n \left(n \, \Delta\nu - \frac{c}{2L} \psi_n\right) E_n^2. \tag{24}$$

Solutions of (6) and (7) which give a nonbeating equilibrium point have $\dot{E}_n = 0$ and all $\dot{\varphi}_n$ equal to some constant. In such cases (24) yields

$$\dot{\varphi} = \frac{\sum_n (n \, \Delta\nu - c/2L \, \psi_n) E_n^2}{\sum_n E_n^2}. \tag{25}$$

The absolute oscillation frequency of the nth mode is $\Omega_n - n \, \Delta\nu + \dot{\varphi}$ and is thus determined when relative mode amplitudes are known.

V. Dyadic Expansion of Steady-State Equations

In this section we develop basic relationships between mode amplitudes and phases for the nonbeating steady-state case. These equations are used later in developing an iterative procedure for determining amplitudes and phases exactly and in calculating the minimum modulator drive that will cause phase locking. The equations are also useful in evaluating the linear approximate solutions obtained by others and in obtaining approximate solutions to the nonlinear problem.

We are interested in solutions of (6) and (7) which have $\dot{E}_n = 0$ and $\dot{\varphi}_n$ equal to a constant independent of n. Equations (6) and (7) become

$$E_{n+1} \sin (\varphi_{n+1} - \varphi_n) - E_{n-1} \sin (\varphi_n - \varphi_{n-1}) = -E_n \left[\dot{\varphi} \frac{2L}{c\alpha_c} - n \, \Delta\nu \frac{2L}{c\alpha_c} + \frac{\psi_n}{\alpha_c} \right] \tag{26}$$

and

$$E_{n+1} \cos (\varphi_{n+1} - \varphi_n) + E_{n-1} \cos (\varphi_n - \varphi_{n-1}) = E_n \left[\frac{G_n - \alpha - \alpha_a}{\alpha_c} \right]. \tag{27}$$

We solve (26) for relative phase angles and treat the rest of the quantities in the equation as known. Proceeding similarly as in the case of phase modulation,[9] we construct a Green's dyadic $\mathcal{G}_{n,q}$ such that

$$E_{n+1} \mathcal{G}_{n+1,q} - E_{n-1} \mathcal{G}_{n,q} = \delta_{n,q}, \tag{28}$$

where $\delta_{n,q}$ is the Kronecker delta, and introduce the boundary condition that $\mathcal{G}_{n,q} = 0$ at $n = +\infty$. We find

$$\mathcal{G}_{n,q} = \frac{-E_n}{E_n E_{n-1}} \qquad n \le q$$
$$= 0 \qquad n > q. \tag{29}$$

From (26), the equation for the relative phase is then

$$\sin(\varphi_n - \varphi_{n-1}) = -\sum_{q=-\infty}^{\infty} \mathcal{G}_{n,q}\left[\dot{\varphi}\frac{2L}{c\alpha_c} - q\,\Delta\nu\frac{2L}{c\alpha_c} + \frac{\psi_q}{\alpha_c}\right]E_q$$
$$\sin(\varphi_n - \varphi_{n-1}) = \frac{1}{E_n E_{n-1}}\frac{1}{\alpha_c}\sum_{q=n}^{\infty}\left[\dot{\varphi}\frac{2L}{c} - q\,\Delta\nu\frac{2L}{c} + \psi_q\right]E_q^2. \tag{30}$$

Similar to the case of phase modulation,[9] there is a complementary form of (30) which sums over q from $-\infty$ to $n - 1$. The equivalence of the two forms is assured by the reactive conservation condition (24). Equation (30) is one of the principal results of this section.

A useful identity is now obtained from (27). First, it is rewritten in the form

$$E_{n+1} - \left(\frac{\alpha + \alpha_a}{\alpha_c}\right)E_n + E_{n-1} = -\frac{G_n}{\alpha_c}E_n + E_{n+1}$$
$$\cdot[1 + \cos(\varphi_{n+1} - \varphi_n)] + E_{n-1}[1 + \cos(\varphi_n - \varphi_{n-1})]. \tag{31}$$

If we define a Green's dyadic

$$\mathcal{G}_{n+1,q} - \left(\frac{\alpha + \alpha_a}{\alpha_c}\right)\mathcal{G}_{n,q} + \mathcal{G}_{n-1,q} = \delta_{n,q} \tag{32}$$

with the boundary conditions $\mathcal{G}_{n,q} = 0$ at $n = \pm\infty$, we obtain the identity

$$E_n = \sum_{q=-\infty}^{\infty}\mathcal{G}_{n,q}\left\{\frac{-G_q}{\alpha_c}E_q + E_{q+1}[1 + \cos(\varphi_{q+1} - \varphi_q)] + E_{q-1}[1 + \cos(\varphi_q - \varphi_{q-1})]\right\}. \tag{33}$$

We choose these boundary conditions because we are interested in a solution with mode amplitudes approaching zero at $|n|$ large. By solving (32) we get

$$\mathcal{G}_{n,q} = \frac{-1}{2\sqrt{\left(\frac{\alpha + \alpha_a}{2\alpha_c}\right)^2 - 1}} \cdot \left[\left(\frac{\alpha + \alpha_a}{2\alpha_c}\right) - \sqrt{\left(\frac{\alpha + \alpha_a}{2\alpha_c}\right)^2 - 1}\right]^{|n-q|}. \tag{34}$$

The identity (33) becomes

$$E_n = \frac{1}{\sqrt{(\alpha + \alpha_a)^2 - 4\alpha_c^2}} \cdot \sum_{q=-\infty}^{\infty}\left[\frac{\alpha + \alpha_a}{2\alpha_c} - \sqrt{\left(\frac{\alpha + \alpha_a}{2\alpha_c}\right)^2 - 1}\right]^{|n-q|}\mu_q \tag{35}$$

with

$$\mu_q = G_q E_q - \alpha_c E_{q+1}[1 + \cos(\varphi_{q+1} - \varphi_q)] - \alpha_c E_{q-1}[1 + \cos(\varphi_q - \varphi_{q-1})]. \tag{36}$$

Equation (35) is the second important result of this section.

To gain a better understanding of (35), consider the case when $\Delta\nu$ and ψ_n are small and/or α_c is large. It will be seen in Section VII, for this case, that the relative phase angles are approximately equal to π. Equation (35) becomes

$$\underset{\text{Approximate}}{E_n} = \frac{1}{\sqrt{(\alpha + \alpha_a)^2 - 4\alpha_c^2}} \cdot \sum_{q=-\infty}^{\infty}\left[\frac{\alpha + \alpha_a}{2\alpha_c} - \sqrt{\left(\frac{\alpha + \alpha_a}{2\alpha_c}\right)^2 - 1}\right]^{|n-q|} G_q E_q. \tag{37}$$

Furthermore, consider the case in which there is gain in only the center mode, i.e., only $G_0 \neq 0$, which in this case yields

$$\underset{\text{One mode}}{E_n} = \frac{G_0 E_0}{\sqrt{(\alpha + \alpha_a)^2 - 4\alpha_c^2}} \left[\left(\frac{\alpha + \alpha_a}{2\alpha_c}\right) - \sqrt{\left(\frac{\alpha + \alpha_a}{2\alpha_c}\right)^2 - 1}\right]^{|n|}. \tag{38}$$

DiDomenico[2] treated the one-mode line in his linearized work and obtained an equation analogous to (38). In this case where losses are assumed independent of n, the linearized treatment gives correct relative amplitudes while the saturation of the center mode sets the scale. One then obtains the following interpretation of (37): the identity is analogous to a homogeneous integral equation in which the nth-mode amplitude is given by a convolution of the single-mode "response" and the product of the saturated gain profile and mode amplitude profile.

VI. Iterative Technique for Solving the Steady-State Problem

If the relative phase angles $\varphi_n - \varphi_{n-1}$ are not exactly π, then small additional terms are present in (36). In this case mode amplitudes and phases may be obtained by an iterative procedure. One begins by making a selection of relative mode amplitudes and phases. The power conservation condition (23) is then used to find the level of oscillation. The mode phases and scaled amplitudes are inserted into (35) and new values $E_n^{(1)}$ are calculated. These new values are used in (25) to calculate the frequency shift $\dot{\varphi}^{(1)}$ of the coupled oscillation. Next the values $E_n^{(1)}$ and $\dot{\varphi}^{(1)}$ are used in (30) to calculate the new relative angles $(\varphi_n - \varphi_{n-1})^{(1)}$. Although not uniquely specified by (30), these angles are taken to be in the second or third quadrant. This agrees with the direct numerical solution of (6) and (7) and, as can be seen from (23), angles in these quadrants cause the loss represented by α_c to subtract from that due to α_a and thereby to reduce the net modulator loss. Finally the new values

$E_n^{(1)}$ and $(\varphi_n - \varphi_{n-1})^{(1)}$ are used in (35) and the cycle is repeated until the process converges.

Unfortunately, this iterative procedure converges relatively slowly. Although (35) includes saturated gain and therefore, in effect, includes the conservation condition (23), the convergence is improved if (23) is used to set the scale each time before repeating the iteration cycle. The iterative procedure is recommended only when α_c is near its optimum value and larger (when $\alpha_c \gtrsim g_0 - \alpha$). In this case, the use of free-running mode amplitudes (with a "tailing off" added on either side) and all relative phase angles equal to π as starting values gives a convergent answer in about ten iterations. This requires about one fourth as much computer time as the direct numerical solution of (6) and (7).

VII. Variation of α_c at Constant Modulator Frequency

In this section we present the results of numerical analysis for a particular set of laser constants. We assume a Doppler-broadened Gaussian atomic line with a homogeneous linewidth much smaller than the axial mode interval and take g_n and ψ_n to be given by (12) and (13). Using the definitions of Section III, we take $g_0 = 0.075$, $\alpha = 0.070$, and assume an axial mode interval of 0.1 Ku. This corresponds to a ratio of Doppler width to mode spacing of 16.67 and gives five free-running modes above threshold. In this and the following section, the modulator is taken to be at the end of the cavity so that the coupling term α_c is equal to one-half the average single-pass loss α_a through the modulator. In specifying the detuning of the modulator drive, we include the linear part of ψ_n in the definition of the mode spacing $\Delta\Omega$.[3] Mode amplitudes, phases, and frequencies were obtained by direct digital computer solution of (6) and (7) using a fourth-order Runge–Kutta method. The equations were programmed for 21 modes ($n = -10$ to $n = +10$) and were run until a steady-state solution to three decimal places was reached. Unless otherwise noted in the following, the $n = 0$ mode was taken to be at line center.

In Fig. 1 we show laser mode amplitudes and the resultant variation of total laser intensity with time at the modulator position. Since the plot is versus $\nu_m t$, it gives directly a picture of spike position relative to phase of the modulator drive. In Fig. 1(a) we show the results for $\alpha_c = 0.00008$ and $\Delta\nu/\Delta\Omega = -0.00003$. As will be seen in Section IX this detuning is approximately the optimum value at low modulator drive levels. The mode amplitudes are essentially the free-running values while the relative phase angles (not shown) vary widely within the range from 90° to 270°. The resultant intensity waveshape has large ripple and a relatively wide spike. This

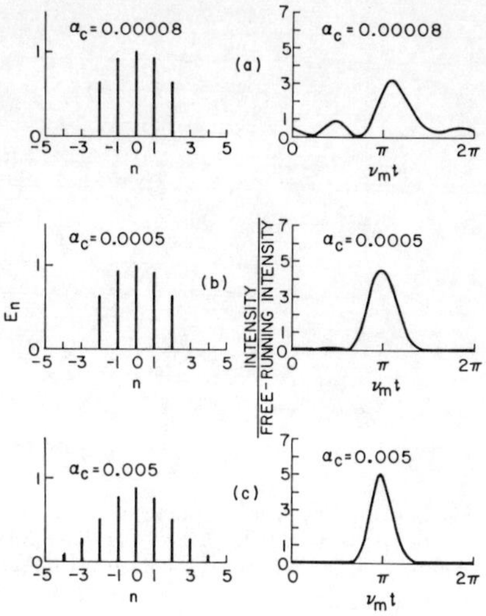

Fig. 1. Laser mode amplitudes and intensity pulseshape at constant detuning and variable α_c: $g_0 = 0.075$; $\alpha = 0.070$; $\Delta\Omega = 0.1\ Ku$; $\Delta\nu = -0.00003\ \Delta\Omega$ (five modes free-running).

operating point is very close to the condition at which phase locking just barely occurs. In Fig. 1(b) we show the results at the same modulator frequency but at a higher drive level, $\alpha_c = 0.0005$. This could be classed as an intermediate region where mode amplitudes are still close to free-running but phase angles have become more uniform. The resultant improvement in spike shape is obvious. Finally, in Fig. 1(c) we show results at very nearly the optimum drive level, $\alpha_c = 0.005$. The mode amplitudes are very close to Gaussian while the angles are close to 180°. This results in a Gaussian-shaped pulse with zero ripple between pulses. A further increase in α_c broadens the linewidth of the Gaussian mode envelope and thereby reduces both the average power output and the half-power pulsewidth.

We find that when the $n = 0$ mode is at line center, the mode amplitudes and relative phase angles are symmetrical about $n = 0$. If the center mode is off line center, at large α_c relatively little happens. The envelope of the mode amplitude profile remains the same. Individual mode amplitudes change as they shift position beneath this envelope. The resultant phase angles become somewhat asymmetrical and, as expected from (25), there is a slight shift in frequency of the coupled oscillation, i.e., all mode frequencies shift from their assumed values by an amount $\dot{\varphi}$. There is only a slight reduction in laser power and spike height, and the overall results are not significant. In contrast, if α_c is near the minimum value that will cause phase locking, as in Fig. 1(a), the effects of the center mode moving off line center are quite severe. Relative phase angles change greatly and a nonsteady-state situation may occur. This case is considered further in Section IX.

Figure 2 shows the variation of the peak pulse intensity as α_c is changed. The detuning and laser constants remain as in Fig. 1. For comparison we show the performance of an identical laser having an internal phase perturbation driven at a frequency equal to its axial mode interval (i.e., in the region termed "phase-locked"[8]–[10]). The

[3] The function Z_r of (13) can be expanded in the power series[12]

$$Z_r(\zeta) = 2\zeta\left(1 - \frac{2\zeta^2}{3} + 4\frac{\zeta^4}{15} - \frac{8\zeta^6}{105} + \cdots\right).$$

We define $\Delta\nu$ with respect to the mode spacing that would be obtained if only the leading term of the expansion were present.

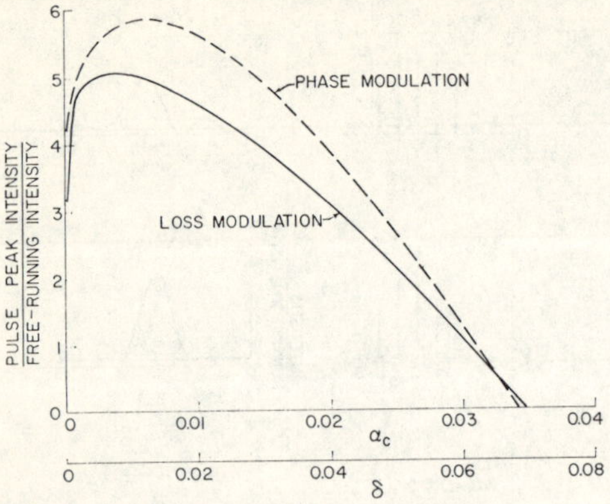

Fig. 2. Pulse peak intensity versus perturbation level at constant detuning: $g_0 = 0.075$; $\alpha = 0.070$; $\Delta\Omega = 0.1\ Ku$; $\Delta\nu = 0$ for phase modulation; $\Delta\nu = -0.00003\ \Delta\Omega$ for loss modulation (five modes free-running).

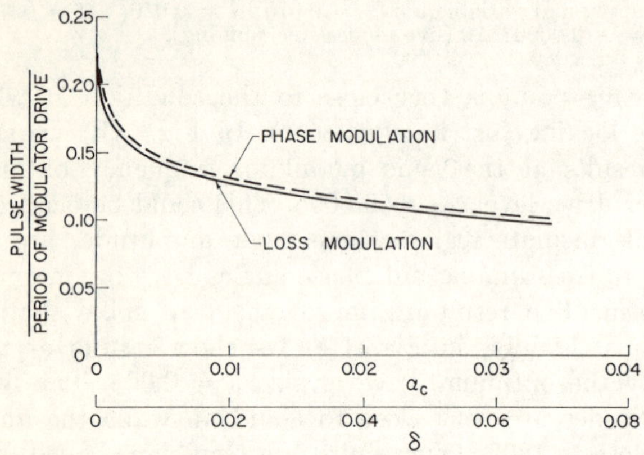

Fig. 3. Pulsewidth versus perturbation level at constant detuning: $g_0 = 0.075$; $\alpha = 0.070$; $\Delta\Omega = 0.1\ Ku$; $\Delta\nu = 0$ for phase modulation; $\Delta\nu = -0.00003\ \Delta\Omega$ for loss modulation (five modes free-running).

quantity δ is the peak single-pass phase retardation of the perturbing element.[4] For both types of modulation we note an initial increase in spike amplitude at low modulation levels as the relative phase angles improve with increasing drive. At high levels of drive, a decrease in spike amplitude occurs with eventual extinguishing of the laser. In loss modulation the decrease at higher drive occurs because of the nonzero modulator loss (proportional to drive level) and because energy is coupled from modes having net gains to modes with net loss. In phase modulation only the latter effect occurs. The optimum peak intensity obtained with loss modulation is about twenty percent below that obtained with phase modulation. Similar results are obtained for other laser constants.

In Fig. 3 we show variation in pulsewidth under the same conditions as Fig. 2. The initial rapid decrease in pulsewidth is due to rapidly improving relative phase angles while the slow continued decrease is a result of the slight broadening of the oscillating linewidth at high drive levels.

[4] The relative alignment of the scales of α_c and δ is such that on a single pass basis the two modulators transfer the same amount of power into adjacent sidebands.

VIII. Variation of Modulator Frequency at Constant α_c

In this section we consider the effects of varying the modulator drive frequency over the range wherein steady-state locking occurs. The height, width, and position of the laser pulses are shown in Figs. 4, 5, and 6, respectively. The results were obtained by direct numerical solution of (6) and (7).

Several interesting points are illustrated by these curves. First, the range of $\Delta\nu$ over which phase locking occurs varies greatly with modulator drive. We will see in Section IX that this range approaches zero as α_c is decreased to its threshold value. Second, there is an optimum frequency which is relatively independent of drive level.[5] As the frequency is changed from this value, the pulse performance is degraded in every respect, i.e., the peak value decreases, the pulse widens slightly, and the phase of the peak varies relative to modulator drive.

These results are consistent with the equations developed in Sections IV and V and can be explained using them. The relative phase angles are dependent on detuning and mode pulling as shown by (30) and depart farther and farther from π as detuning is changed from its optimum value. In a multimode laser, α_c is typically much smaller than the gain G_q of the oscillating modes so that the dominant part of μ_q in (35) and (36) is the G_qE_q term. Thus the *relative* mode amplitudes are approximately independent of phase angles and therefore detuning. The *scale* of mode amplitudes is strongly dependent upon phase angles, however, as is clearly evident from (23). If the mode amplitudes are written in the form $E_n = ke_n$ where $\sum_n e_n^2 = 1$, (23) becomes

$$\sum_n E_n^2 = k^2 = \frac{1}{\beta}\,\frac{\sum_n (g_n - \alpha)e_n^2 - \alpha_a \sum_n e_n^2 - 2\alpha_c \sum_n e_{n+1}e_n \cos(\varphi_{n+1} - \varphi_n)}{\sum_n g_n e_n^4}. \tag{39}$$

Thus the average laser power, which is proportional to $\sum_n E_n^2$, is sensitive to changes in $(\varphi_{n+1} - \varphi_n)$, especially at the larger values of α_c. If $\alpha_a \gtrsim (g_0 - \alpha)$, we see that the laser is extinguished as the relative angles depart greatly from π and, therefore, is extinguished at large detunings. If α_a is smaller than the excess gain, the laser is not extinguished but, as will be seen, ceases to be phase locked at large detunings.

Since the pulsewidth changes little as $\Delta\nu$ is varied, the average power follows very nearly the same shaped curve as does peak power in Fig. 4. For a case corresponding to α_a greater than excess gain, Crowell[4] observed experimentally such a variation in average power as the cavity resonant frequency was changed (equivalent to detuning the modulator drive). He also observed a shift in spike position similar to Fig. 6.

[5] The fact that this frequency differs from zero is a consequence of the definition of axial mode interval. In comparing with experiments, a consistent definition must be used.

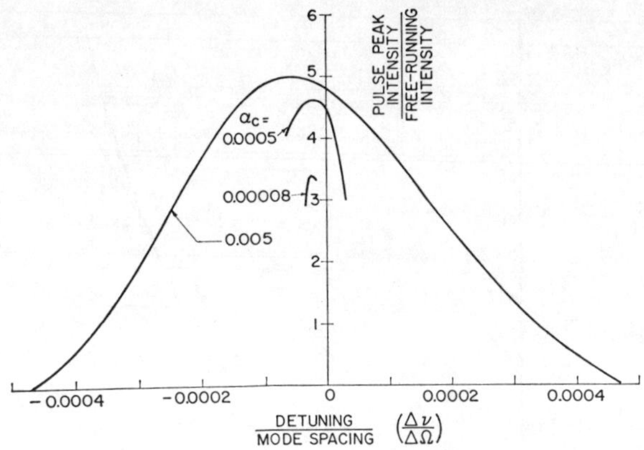

Fig. 4. Pulse peak intensity versus detuning at constant α_c: $g_0 = 0.075$; $\alpha = 0.070$; $\Delta\Omega = 0.1\ Ku$ (five modes free-running).

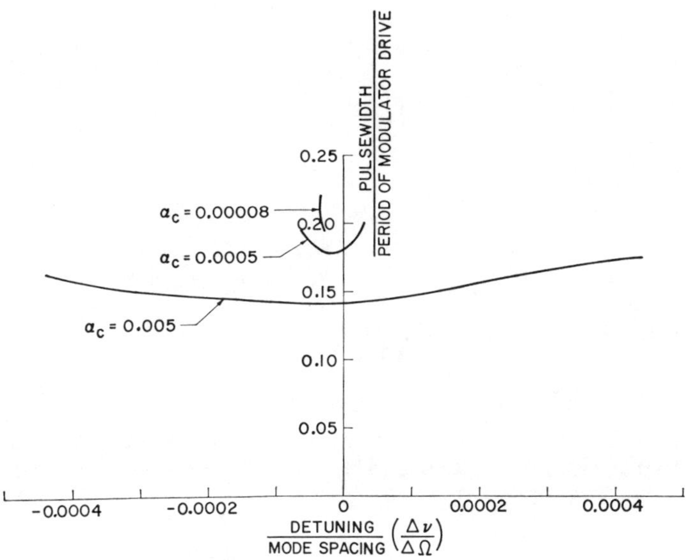

Fig. 5. Pulsewidth versus detuning at constant α_c: $g_0 = 0.075$; $\alpha = 0.070$; $\Delta\Omega = 0.1\ Ku$ (five modes free-running).

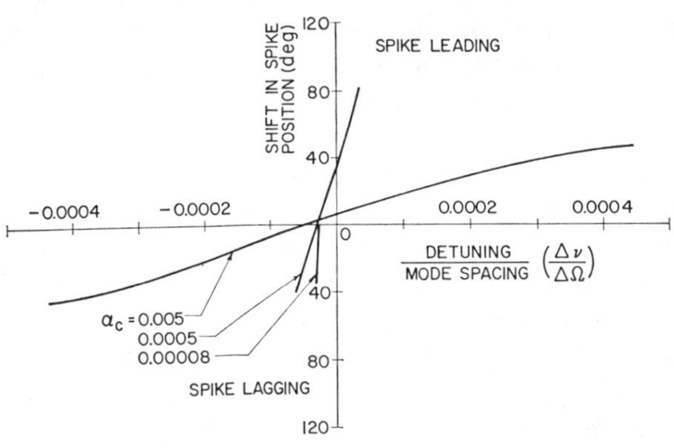

Fig. 6. Pulse relative phase versus detuning at constant α_c: $g_0 = 0.075$; $\alpha = 0.070$; $\Delta\Omega = 0.1\ Ku$ (five modes free-running).

IX. Threshold of Phase Locking

Since in practice the strength of an intracavity loss perturbation may be limited, it is important to know the minimum perturbation that will phase lock a multimode laser. We find that the mode pulling of the active medium specifies such a minimum.

At low modulator drive one expects the parametric terms (right-hand side) in (7) to have negligible effect and the mode amplitudes, therefore, to be approximately equal to their free-running values. This was seen to be the case in Fig. 1(a). By contrast, because the atomic mode pulling term χ_n' depends nonlinearly on n, it is impossible to have a steady-state solution ($\dot{\varphi}$ = constant) to (6) in which the parametric term is neglected. In the absence of the perturbation, the modes are at their free-running frequencies and, therefore, unequally spaced. The perturbation has to pull the modes until their frequency spacing is that of the modulator drive.

Assuming mode amplitudes to be the free-running values, the parametric pulling term then depends upon α_c and the relative phase angles as seen in (6). As α_c is reduced, the sine terms must increase and eventually a limit is reached when one of them becomes ± 1. For a further decrease in α_c a steady-state solution may no longer be obtained, and a beating of mode amplitudes and phases results.

With mode amplitudes known, (30) affords a convenient means to calculate the limiting value of α_c. Each relative phase angle sets a requirement, $\alpha_c \geq (\alpha_c)_n$, where the $(\alpha_c)_n$ are found by letting $\sin(\varphi_n - \varphi_{n-1}) = \pm 1$ in (30); i.e.,

$$(\alpha_c)_n = |Y_n(\Delta\nu)| \tag{40}$$

where we have defined

$$Y_n(\Delta\nu) = \frac{1}{E_n E_{n-1}} \sum_{q=n}^{\infty} \left[\dot{\varphi} \frac{2L}{c} - q\ \Delta\nu \frac{2L}{c} + \psi_q \right] E_q^2. \tag{41}$$

The largest such value of α_c is the minimum that will be sufficient to cause phase locking of all the modes at the given $\Delta\nu$. In Fig. 7 we show a plot of the minimum value of α_c that will cause phase locking versus $\Delta\nu/\Delta\Omega$ for a case having $g_0 = 0.085$, $\alpha = 0.070$, and $\Delta\Omega = 0.1\ Ku$. This case is identical to that considered in Sections VII and VIII except that gain has been increased to give nine modes free-running. The curve is seen to be made up of straight line segments resulting in a sharp minimum at an optimum detuning. We define threshold as this minimum value of α_c.

If the center mode is at line center, there exists a simple procedure for determining the threshold value of α_c and the optimum detuning. The quantities $Y_n(\Delta\nu)$, defined by (41), have a linear dependence upon $\Delta\nu$. If these straight lines are plotted in the $[Y_n, \Delta\nu/\Delta\Omega]$ plane, it is possible to verify that they always fall in the relative positions of Fig. 8; i.e., they intersect below the $\Delta\nu/\Delta\Omega$ axis and the Y_1 line lies above the others in the upper half plane.[6] As noted above, the smallest value of α_c that will phase lock the modes at a given $\Delta\nu$ is equal to the largest $|Y_n|$ at that $\Delta\nu$. From the geometry of Fig. 8, this largest value falls somewhere along the $n = 1$ line. Therefore the threshold point is located at the intersection of the $n = 1$ line and the particular image

[6] Only the $n > 0$ straight lines are shown since $(\varphi_1 - \varphi_0) = (\varphi_0 - \varphi_{-1})$, $(\varphi_2 - \varphi_1) = (\varphi_{-1} - \varphi_{-2})$, etc., when the center mode is at the center of a symmetrical gain profile.

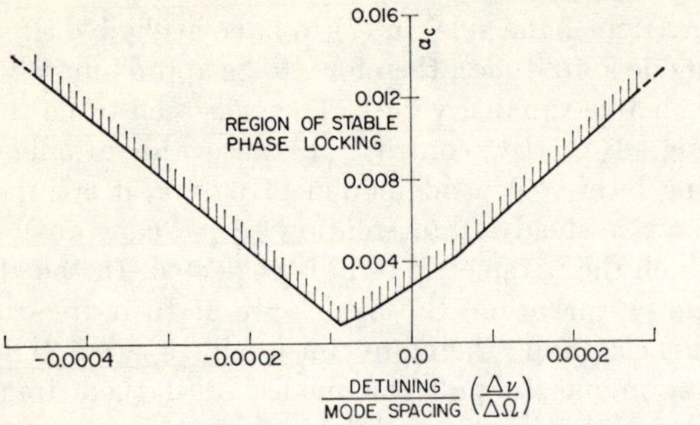

Fig. 7. Value of α_c necessary to cause phase locking versus detuning: $g_0 = 0.85$; $\alpha = 0.070$; $\Delta\Omega = 0.1\ Ku$ (nine modes free-running).

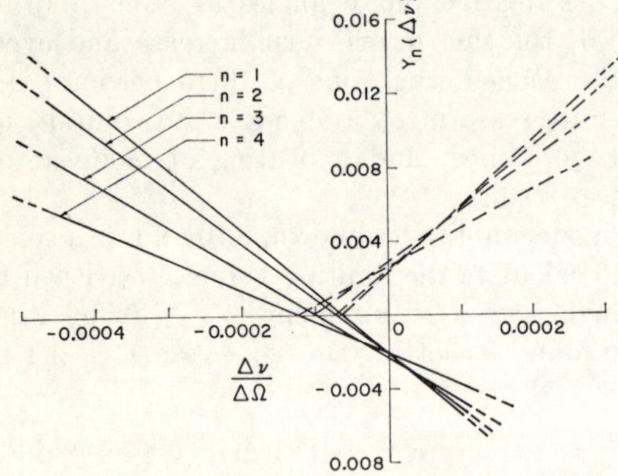

Fig. 8. $Y_n(\Delta\nu)$ versus detuning: $g_0 = 0.085$; $\alpha = 0.070$; $\Delta\Omega = 0.1$ Ku (nine modes free-running).

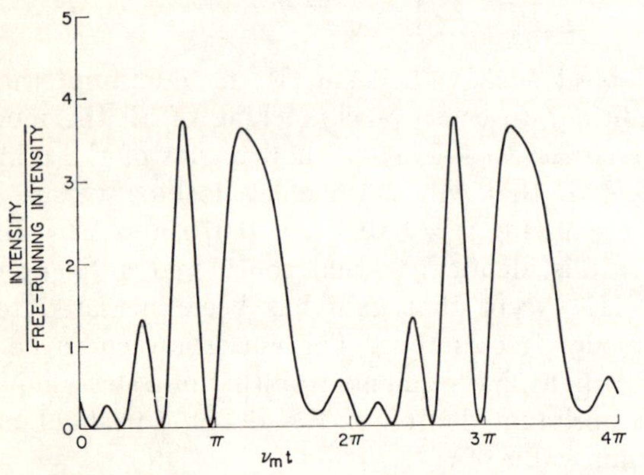

Fig. 9. Output intensity versus time at threshold of locking: $g_0 = 0.085$; $\alpha = 0.070$; $\Delta\Omega = 0.1\ Ku$ (nine modes free-running).

of the other lines (shown dashed) that gives the largest value of Y_1.

At the threshold point in Figs. 7 and 8, $(\varphi_1 - \varphi_0) = \pi/2$ and $(\varphi_4 - \varphi_3) = 3\pi/2$ while the other angles fall between these two extremes (in the second and third quadrants). Since at threshold the relative phase angles vary widely, it is expected that output pulseshapes will be greatly distorted. Figure 9 shows the output intensity resulting from amplitudes and phases calculated at the above threshold point. We note sidelobes which have intensities comparable to the main spike. This might explain the multiple spiking which has been observed experimentally.[4]

The effects of detuning at low perturbations can be

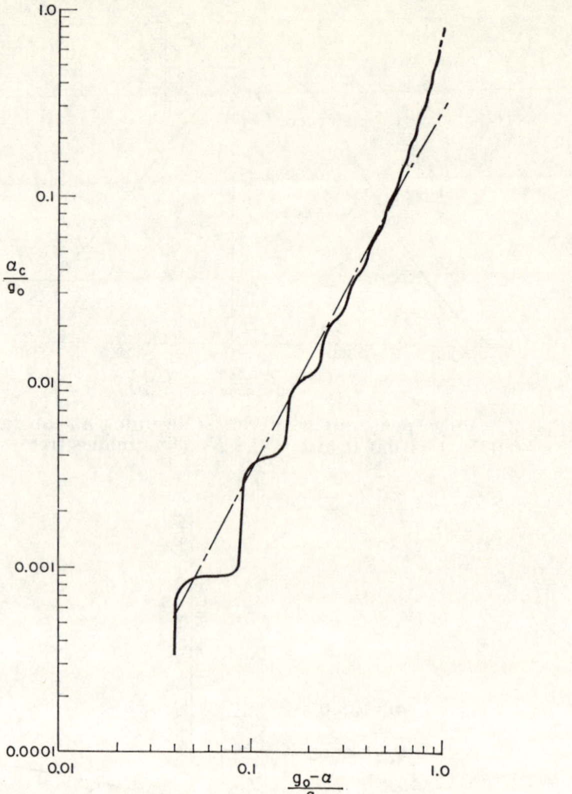

Fig. 10. Variation of threshold value of α_c with excess gain: $\Delta\Omega = 0.1\ Ku$.

understood more clearly by reference to Fig. 8 since, at a given α_c, sin $(\varphi_n - \varphi_{n-1})$ is directly proportional to $Y_n(\Delta\nu)$; i.e., by using (41), then (30) can be written

$$\sin(\varphi_n - \varphi_{n-1}) = \frac{1}{\alpha_c} Y_n(\Delta\nu). \tag{42}$$

At $\alpha_c = 0.004$ then, for example, steady-state locking is to be expected as $\Delta\nu$ is made more and more negative until the uppermost line $(n = 1)$ rises above the point, $Y_n = 0.004$. Here $(\varphi_1 - \varphi_0)$ becomes $\pi/2$ and steady-state locking ceases. Since the lines are widely spaced at this point, the other relative angles are considerably different and the pulse becomes more and more distorted as this nonsteady-state condition is approached. In contrast, as $\Delta\nu$ is *increased* from the value at threshold the angles become more and more nearly equal since the lines are converging. The $n = 3$ line first rises above $Y_n = 0.004$ [therefore, $(\varphi_3 - \varphi_2)$ is the limiting angle] but at that point the other relative angles are approximately equal to $(\varphi_3 - \varphi_2)$. As this nonsteady-state condition is approached, the expected pulseshape is actually improving. Hargrove, Fork, and Pollack[1] noted such a contrast in behavior with detuning as the extremes of the locking range were approached.

In Fig. 10 we show the calculated locking threshold α_c versus excess gain at the same mode spacing, $\Delta\Omega = 0.1\ Ku$, as in the previous figures. The ripple in the curve at low excess gain is a result of the change of the number of free-running modes as the gain is increased. At the left end the curve starts with five modes free-running and each successive hump corresponds to two additional modes coming above the free-running threshold. We also find that the threshold α_c changes as the center mode moves off line center. If we determine the α_c necessary to insure locking *regardless* of the position of the center mode, we find that the space between humps is filled in and the curve becomes very nearly the straight line shown dashed in Fig. 10. At large $(g_0 - \alpha_n)$ the mode amplitudes are affected by the larger α_c and depart from their free-running values. It becomes easier to phase lock the modes than is predicted by assuming free-running amplitudes. This lowers the curve to the dashed line in Fig. 10 at large $(g_0 - \alpha_n)$. In general, we find that the threshold α_c varies approximately as the square of excess gain.

X. Gain in a Single Mode

We have seen that the internal perturbation not only locks the existing free-running modes but also strongly affects their amplitudes and produces nonzero amplitudes where modes were previously below threshold. As an extreme example, this section considers the situation where the atomic gain profile is so narrow that only one laser mode has gain. Since their effects are similar to the multimode case, mode pulling and detuning will not be considered. Mode amplitudes were given by (38) which is rewritten here as

$$E_n = E_0 r^{|n|}, \tag{43}$$

where

$$r = \left[\frac{\alpha + \alpha_a}{2\alpha_c} - \sqrt{\left(\frac{\alpha + \alpha_a}{2\alpha_c}\right)^2 - 1}\right], \tag{44}$$

and E_0 is determined from the power conservation requirement

$$G_0 = \sqrt{(\alpha + \alpha_a)^2 - 4\alpha_c^2}. \tag{45}$$

If inhomogeneous saturation of the form of (11) is assumed, then (45) gives

$$E_0 = \sqrt{\frac{1}{\beta}\left[1 - \frac{1}{g_0}\sqrt{(\alpha + \alpha_a)^2 - 4\alpha_c^2}\right]}. \tag{46}$$

The laser intensity has a time dependence given by[9]

$$W(t) = \tfrac{1}{2} \sum_s \sum_n E_n E_{n+s} \cos(s\nu_m t + \varphi_{n+s} - \varphi_n). \tag{47}$$

For the one active mode line, we substitute (43) into (47), use standard geometric sum formulas and obtain

$$W(t)_{\text{one mode}} = \frac{E_0^2}{2} \sum_s r^{|s|}\left(|s| + \frac{1 + r^2}{1 - r^2}\right) \cos(s\nu_m t + \pi). \tag{48}$$

The resultant output intensity has peak and minimum values given by

$$\begin{aligned} W(t)\Big|_{\text{one mode peak}} &= \frac{E_0^2}{2}\left(\frac{1 + r}{1 - r}\right)^2; \qquad \nu_m t = \pi, 3\pi, 5\pi, \cdots \\ W(t)\Big|_{\text{one mode min}} &= \frac{E_0^2}{2}\left(\frac{1 - r}{1 + r}\right)^2; \qquad \nu_m t = 0, 2\pi, 4\pi, \cdots . \end{aligned} \tag{49}$$

Although the scale of $W(t)$ depends on E_0 and hence upon the gain, the shape of the curve depends only upon r and therefore upon α_c/α (we are still assuming (22) to be true). Normalized plots of laser intensity from (48) are shown in Fig. 11 at several values of α_c/α. At higher modulator drive levels the result is a spiking output which can be as peaked as that of a multimode laser.[7]

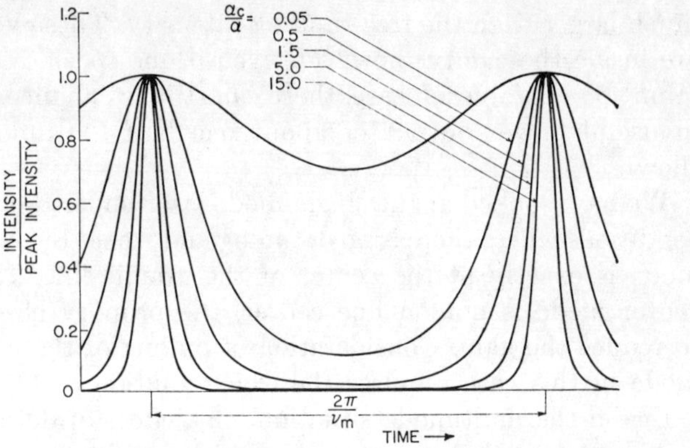

Fig. 11. Normalized intensity versus time for an atomic line having gain in only one mode.

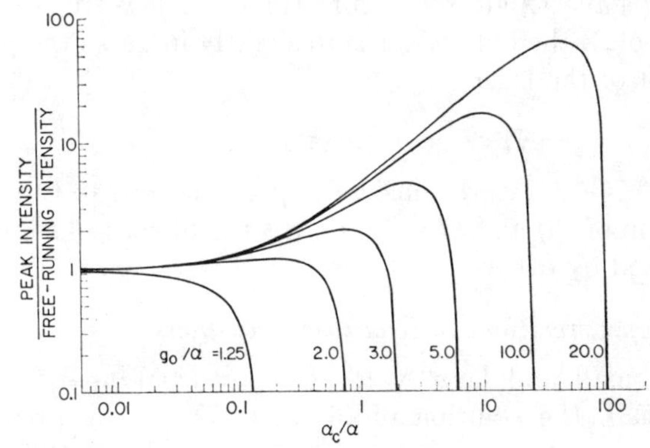

Fig. 12. Pulse peak intensity versus α_c/α for an atomic line having gain in only one mode.

The peak intensity of the spike can be many times the free-running intensity of the single mode. At free running we have

$$g_0(1 - \beta E_0^2) = \alpha. \tag{50}$$

Solving (50) for free-running intensity and combining with (46) and (49), we get

$$\frac{W(t)_{\text{peak}}}{W(t)_{\text{free running}}} = \frac{g_0/\alpha - \sqrt{1 + 4\alpha_c/\alpha}}{g_0/\alpha - 1}\left(\frac{1 + r}{1 - r}\right)^2. \tag{51}$$

Figure 12 shows a plot of (51) versus α_c/α at several values of g_0/α. As shown, there is an optimum modulator drive which produces a spiking output having a peak

[7] In this section we are considering g_0, α_c, and α which are large enough that the small gain approximations of Section III are not always correct. We still take the equations of that section as definitions of these quantities although they may not now be exactly equal to fractional gain and/or loss per pass. For example, the exact expression for gain per pass is

$$\text{Fractional Gain per pass} = \frac{I_{\text{out}} - I_{\text{in}}}{I_{\text{in}}} = e^{G_n} - 1,$$

which becomes approximately equal to G_n only at G_n small.

much larger than the free-running intensity. These values are inside the cavity; however, even if one specifies optimum coupling in each case, there can still be enhancement in useful power output of about one half the amounts shown.

We have looked at the single-mode laser under the ideal conditions where there is no detuning and where the center mode is exactly at the center of the atomic line. If the center mode is not on line center, the primary effect is to reduce the gain. Consideration of pulling of the center mode in this case requires the exact solution of (6) and (7) as in the multimode situation. The effect of detuning is similar to that of the multimode problem having $\alpha_c \gtrsim (g_0 - \alpha_n)$; i.e., detuning the modulator reduces the output until the laser is extinguished. If the $n = \pm 1$ modes have small gain (but still far below free-running threshold), the only effect is to slightly increase the power output of the laser.

XI. Approximate Solutions

In this section we consider approximations to the exact solution of (6) and (7) and discuss the linearized solutions obtained by others.

A. *Approximations to Nonlinear Solutions*

At small and large values of perturbation α_c and at $\Delta\nu$ small, the solution of (6) and (7) can be predicted approximately. In Section IX we considered the small perturbation case. In Table I we show a comparison of the results of an exact solution of (6) and (7) near threshold with the approximate solution obtained by taking free-running mode amplitudes and using (30) to calculate angles. The laser constants are those of the nine-mode case considered in Section IX. The values obtained are seen to be very close to the exact solution.

At larger modulator drive levels we have noted that the relative mode amplitudes are almost perfectly Gaussian. At any particular assumed linewidth of this Gaussian envelope, we can use (30) to calculate relative phase angles and then (39) to calculate the scale factor. If we follow this procedure and determine the oscillating linewidth that maximizes laser intensity, we find at small $\Delta\nu$ that the answers agree closely with the exact solution of (6) and (7). Table II shows the detailed comparison of mode amplitudes and phases at $\Delta\nu = 0$. These results were obtained using laser constants of the five-mode case of Sections VII and VIII. The agreement becomes better over a wider range of $\Delta\nu$ as α_c is increased.

B. *Linearized Solutions*

DiDomenico[2] first obtained an approximate solution having equal mode amplitudes and relative phase angles equal to π. Taking $\Delta\nu$ and ψ_n equal to zero and assuming all G_n equal, we note that (26) and (27) have the solution

$$\begin{aligned} \text{all } E_n &= E_0 \\ \text{all } (\varphi_n - \varphi_{n-1}) &= \theta \qquad (\textit{any value}) \\ \dot{\varphi} &= 0 \end{aligned} \tag{52}$$

TABLE I

Comparison of Exact and Approximate Solutions at Small α_c

$g_0 = 0.085$, $\alpha_n = 0.070$, $\dfrac{\text{Doppler width}}{\text{Mode spacing}} = 16.67$; 9 modes free-running, $\alpha_c = 0.00135$, $\Delta\nu/\Delta\Omega = -0.000088$

n	Exact Solution of (6) and (7) E_n	$(\varphi_n - \varphi_{n-1})$	Approximate Free-Running Amplitudes and (30) E_n	$(\varphi_n - \varphi_{n-1})$
0	0.974	—	1.000	—
1	0.960	134.970°	0.978	138.190°
2	0.897	160.642°	0.907	162.808°
3	0.744	192.654°	0.769	194.864°
4	0.436	222.179°	0.498	221.810°
5	0.084	217.452°	0	—
6	0.008	214.089°	0	—

TABLE II

Comparison of Exact and Approximate Solutions at Large α_c

$g_0 = 0.075$, $\alpha_n = 0.070$, $\dfrac{\text{Doppler width}}{\text{Mode spacing}} = 16.67$; 5 modes free-running, $\alpha_c = 0.005$, $\Delta\nu/\Delta\Omega = 0$

n	Exact Solution of (6) and (7) E_n	$(\varphi_n - \varphi_{n-1})$	Approximate Optimum Gaussian Amplitudes and (30) E_n	$(\varphi_n - \varphi_{n-1})$
0	0.990	—	0.986	—
1	0.894	185.649°	0.874	186.715°
2	0.624	187.605°	0.607	189.556°
3	0.300	190.801°	0.331	194.795°
4	0.096	194.383°	0.142	201.705°
5	0.022	198.244°	0.048	209.460°
6	0.004	202.220°	0.012	217.237°
7	0.000	205.551°	0.003	224.212°

with the value of E_0 determined from the solution of

$$G_n - \alpha = \alpha_a + 2\alpha_c \cos\theta. \tag{53}$$

Here θ can have any value insofar as the linearized equations are concerned. Only by including nonlinearities or by use of a physical argument as per Crowell[4] do we know that θ should be equal to π. We note further that this solution is only an exact solution of the linearized equations when an infinite number of modes have the same gain. For a finite number, an exact solution of the linearized equations gives nonconstant mode amplitudes except for α_c vanishingly small. But then mode pulling ψ_n cannot be neglected as has been shown.

Yariv[3] obtained a solution of the linearized equations

at $\Delta\nu \neq 0$. Taking $\dot{\varphi}$ and ψ_n equal to zero, (26) and (27) yield in this case

$$(\varphi_n - \varphi_{n-1}) = -\pi/2$$

$$G_n - \alpha = \alpha_a \tag{54}$$

$$E_n = kI_n\left(\frac{1}{\pi}\frac{\Delta\Omega}{\Delta\nu}\alpha_c\right),$$

which is Yariv's solution in our notation. Here an infinite number of modes having single-pass gain equal to single-pass loss at the assumed values of E_n is required for the answer to be an exact solution of the linearized equations. Of more importance, the linearized solution fails to simulate the real answer to the nonlinear problem. In the nonlinear case a solution with $(\varphi_n - \varphi_{n-1})$ closer to π oscillates at a higher level and is, therefore, able to negate the solution of (54). Solution (54) causes modulator loss to be high and equal to the average loss, α_a [note (23) or (39)], while in the true solution the average modulator loss is approximately zero. In other terms, the pulse produced by the I_n solution would pass through the modulator when its loss is at the average value rather than zero. In the direct numerical solution of (6) and (7) we have never observed a solution in the form of (54).

Acknowledgment

The authors gratefully acknowledge helpful discussions with L. Osterink and the assistance of Mrs. Cora Barry with the numerical computations.

References

[1] L. E. Hargrove, R. L. Fork, and M. A. Pollack, "Locking of He–Ne laser modes induced by synchronous intracavity modulation," *Appl. Phys. Letters*, vol. 5, pp. 4–5, July 1964.

[2] M. DiDomenico, Jr., "Small-signal analysis of internal (coupling type) modulation of lasers," *J. Appl. Phys.*, vol. 35, pp. 2870–2876, October 1964.

[3] A. Yariv, "Internal modulation in multimode laser oscillators," *J. Appl. Phys.*, vol. 36, pp. 388–391, February 1965.

[4] M. H. Crowell, "Characteristics of mode-coupled lasers," *IEEE J. of Quantum Electronics*, vol. QE-1, pp. 12–20, April 1965.

[5] T. Deutsch, "Mode-locking effects in an internally-modulated ruby laser," *Appl. Phys. Letters*, vol. 7, pp. 80–82, August 1965.

[6] R. H. Pantell and R. L. Kohn, "Mode coupling in a ruby laser," *IEEE J. of Quantum Electronics*, vol. QE-2, pp. 306–310, August 1966.

[7] M. DiDomenico, Jr., J. E. Geusic, H. M. Marcos, and R. G. Smith, "Generation of ultrashort optical pulses by mode locking the YAlG:Nd laser," *Appl. Phys. Letters*, vol. 8, pp. 180–183, April 1966.

[8] S. E. Harris and R. Targ, "FM oscillation of the He–Ne laser," *Appl. Phys. Letters*, vol. 5, pp. 202–204, November 1964.

[9] S. E. Harris and O. P. McDuff, "Theory of FM laser oscillation," *IEEE J. of Quantum Electronics*, vol. QE-1, pp. 245–262, September 1965.

[10] E. O. Ammann, B. J. McMurtry, and M. K. Oshman, "Detailed experiments on helium–neon FM lasers," *IEEE J. of Quantum Electronics*, vol. QE-1, pp. 263–272, September 1965.

[11] W. E. Lamb, Jr., "Theory of an optical maser," *Phys. Rev.* vol. 134, pp. A1429–A1450, June 1964.

[12] B. D. Fried and S. D. Conte, *The Plasma Dispersion Function (Hilbert Transform of the Gaussian).* New York: Academic, 1961.

FM and AM Mode Locking of the Homogeneous Laser—Part I: Theory

DIRK J. KUIZENGA, MEMBER, IEEE, AND A. E. SIEGMAN, FELLOW, IEEE

Abstract—**A new general analysis for mode-locked operation of a homogeneously broadened laser with either internal phase (FM) or amplitude (AM) modulation is presented in this paper. In this analysis, a complex Gaussian pulse is followed through one pass around the laser cavity. Approximations are made to the line shape and modulation characteristics to keep the pulse Gaussian. After one round trip, a self-consistent solution is required. This yields simple analytic expressions for the pulse length, frequency chirp, and bandwidth of the mode-locked pulses. The analysis is further extended to include effects of detuning of the modulator, in which case analytical expressions are obtained for the phase shift of the pulse within the modulation cycle, the shift of the pulse spectrum off line center, the change in pulse length, and the change in power output. Numerical results for a typical Nd:YAG laser are given. In the case of the FM mode-locked laser it is found that there is a frequency chirp on the pulse and that this causes pulse compression and stretching when the modulator is detuned. Etalon effects and dispersion effects are also considered.**

I. Introduction

THE phenomenon of mode locking of a laser by an internal phase or amplitude perturbation to obtain short optical pulses is well known and has been investigated theoretically and experimentally by several authors.

Theoretical studies of the mode-locked laser with an inhomogeneously Doppler-broadened atomic line have been done by DiDomenico [1], Yariv [2], and Crowell [3], all of whom discuss a linearized solution to the problem. More detailed nonlinear calculations for the FM-type mode locking [4] and AM-type mode locking [5] have been presented by Harris and McDuff. In all the above analyses the coupled-mode-equation approach has been used, assuming that the axial modes saturate independently. This has led to a good understanding of mode-locked gas lasers, in particular the He–Ne laser [6] and argon laser [7].

Mode locking has also been observed in solid-state Nd:YAG lasers by DiDomenico *et al.* [8], using an amplitude modulator and by Osterink and Foster [9] using a phase modulator; and in solid-state ruby lasers using an internal amplitude modulator by Deutsch, Pantell, and Kohn [10], [11]. In analyzing these lasers and any other homogeneously broadened lasers, the use of the coupled-mode equations is complicated by the fact that the axial modes do not saturate independently due to the homogeneous broadening, and also by the fact that a very large number of coupled axial modes are usually generated. Haken and Pauthier [12] have suggested one new analytical approach that can be used for the homogeneous AM-type mode locking.

In this paper we present a totally different approach to the homogeneously broadened system [13], working completely in the time domain, rather than the frequency or coupled-mode domain. We assume that there is a short pulse inside the laser cavity, and we follow this pulse once around the laser cavity, through the active medium and the modulator. We then require a self-consistent solution, i.e., no net change in the complete round trip. This approach is very similar to that used some time ago by Cutler [14] to analyze the microwave regenerative pulse generator. A similar approach has recently been used by Gunn [15] to analyze the homogeneously broadened laser with internal amplitude modulation.

In our analysis we assume that the pulse is Gaussian, and we make necessary approximations to the atomic line shape and intracavity modulation functions to keep the pulse Gaussian. One essential approximation is that the bandwidth of the pulse is small compared to the atomic linewidth. Experimental observations in the Nd:YAG laser show that this approximation is quite reasonable.

In this paper we will first consider some properties of Gaussian pulses, the active medium and the FM and AM modulators. The self-consistent solution then leads to a simple expression for the mode-locked pulsewidth, showing the dependence on all the important laser parameters such as linewidth, modulation frequency, depth of modulation, and saturated gain. Detuning of the modulator for the FM and AM cases is also considered as well as modifications to the theory for etalon effects and dispersion.

Where applicable, numerical results for a typical Nd:YAG laser will be presented.

II. Gaussian Pulses

We will consider the most general Gaussian optical pulse given by

$$E(t) = \tfrac{1}{2}E_0 \exp(-\alpha t^2) \exp[j(\omega_p t + \beta t^2)]. \quad (1)$$

The term α determines the Gaussian envelope of the pulse and the term $j\beta t$ is a linear frequency shift during the pulse (chirp). A complex constant γ can be defined as

$$\gamma = \alpha - j\beta, \quad (2)$$

Manuscript received February 18, 1970; revised May 18, 1970. This work was supported by the U. S. Air Force Office of Scientific Research under Contracts AF 49(638)-1525 and F44620-69-C-0017.

The authors are with the Microwave Laboratory and Department of Electrical Engineering, Stanford University, Stanford, Calif. 94305.

Reprinted from *IEEE J. Quantum Electron.*, vol. QE-6, pp. 694–708, Nov. 1970.

so that

$$E(t) = \tfrac{1}{2}E_0 \exp(-\gamma t^2) \exp(j\omega_p t).$$

The Fourier transform of this pulse is given by [16]

$$E(\omega) = (E_0/2)\sqrt{\pi/\gamma} \exp[-(\omega - \omega_p)^2/4\gamma]. \qquad (3)$$

The pulsewidth can be defined in various ways. In this paper we define the pulsewidth (τ_p) as the time between half-intensity points, and from (1) it follows that

$$\tau_p = \sqrt{(2 \ln 2)/\alpha}. \qquad (4)$$

The bandwidth or spectral width of the (Δf_p) is defined as the frequency between half-power points of the pulse spectrum, and from (3) we get[1]

$$\Delta f_p = (1/\pi)\sqrt{2 \ln 2[(\alpha^2 + \beta^2)/\alpha]} \text{ Hz}. \qquad (5)$$

Note how the frequency chirp contributes to the total bandwidth. The pulsewidth–bandwidth product is a parameter often used to characterize pulses, and for the Gaussian pulses used in this analysis, the pulsewidth–bandwidth product is given by

$$\tau_p \cdot \Delta f_p = (2 \ln 2/\pi)\sqrt{1 + (\beta/\alpha)^2}. \qquad (6)$$

Two important special cases are $\beta = 0$ and $|\beta| = \alpha$ and for these two cases $\tau_p \cdot \Delta f_p = 0.440$ and $\tau_p \cdot \Delta f_p = 0.626$. We will see later that these cases apply to AM and FM modulation, respectively.

III. Active Medium

For a laser with a homogeneously broadened line such as the Nd:YAG laser [17], it can be shown (Appendix I) that the amplitude gain is given by

$$g_a(\omega) = \exp g/[1 + 2j(\omega - \omega_a)/\Delta\omega] \qquad (7)$$

where g is the *saturated* amplitude gain through the active medium at line center (ω_a) for one round trip in the cavity. The midband gain coefficient is related to other material constants is shown in Appendix I. It should be noted here that L_c in (89) is twice the length of the active medium for an ordinary Fabry–Perot laser cavity, while L_c is equal to the active medium length for the ring-type laser cavity.

If we now consider the case where the bandwidth of the pulse is much less than the linewidth, we can expand (7) as follows.

$$g_a(\omega) \simeq G \exp\left[-2jg\left(\frac{\omega - \omega_a}{\Delta\omega}\right) - 4g\left(\frac{\omega - \omega_a}{\Delta\omega}\right)^2\right], \qquad (8)$$

where $G = e^g$ and we have assumed that $(\omega - \omega_a)/\Delta\omega < 1$ so that we can neglect higher order terms. The line shape has now become Gaussian, and a Gaussian pulse going through an active medium with this line shape will remain Gaussian.

In (8), we have expanded the line shape about the center frequency ω_a. If we expand the line shape about any other frequency ω_p, we get

$$g_a(\omega) \simeq \exp \frac{g}{1 + 2j\eta}\left[1 - \left(\frac{2j}{1 + 2j\eta}\right)\left(\frac{\omega - \omega_p}{\Delta\omega}\right) - \frac{4}{(1 + 2j\eta)^2}\left(\frac{\omega - \omega_p}{\Delta\omega}\right)^2\right] \qquad (9)$$

where $\omega_a = \omega_p + \nu$, and $\eta = \nu/\Delta\omega$, the shift of the spectrum center frequency ω_p normalized to the atomic linewidth $\Delta\omega$.

We will see later that when we consider detuning of the modulator, the pulse spectrum does indeed shift away from line center, and hence we must consider the line shape given by (9). Note that (8) and (9) have the same form, with g replaced by $g/(1 + 2j\eta)$ and $\Delta\omega$ replaced by $\Delta\omega(1 + 2j\eta)$ in (8).

IV. Modulator

The intracavity modulator may in practice be either a phase modulator or an amplitude modulator.

A. Phase Modulator

The internal phase modulator introduces a sinusoidally varying phase perturbation $\delta(t)$ such that the round-trip transmission through the modulator is given by

$$\exp[-j\delta(t)] = \exp(-j2\delta_c \cos \omega_m t), \qquad (10)$$

where ω_m is the modulation frequency and δ_c is the effective single-pass phase retardation of the modulator. For an ordinary Fabry–Perot laser cavity with an intracavity modulator, it can be shown that [4]

$$\delta_c = \left(\frac{L}{a}\frac{2}{\pi}\sin\frac{a}{L}\frac{\pi}{2}\right)\left(\cos\frac{Z_0\pi}{L}\right)\delta_m, \qquad (11)$$

where L is the length of the cavity, a the length of the modulator crystal, Z_0 the distance of the modulator to a mirror, and δ_m the peak phase retardation through the crystal. Usually $a/L \ll 1$ and $\delta_c \simeq \delta_m \cos(\pi Z_0/L)$. For the ring-type cavity, the pulse passes through the modulator only once per round trip, and we should drop the 2 in (10). Also, $\delta_c \simeq \delta_m$ for the ring cavity.

For the ideal mode-locking case, one can visualize short pulses passing through the modulator consecutively at one or other extremum of the phase variation. If we make the approximation that the pulse is short compared to the modulation period, then the transmission through the modulator is given by

$$\exp[-j\delta(t)] \simeq \exp(\mp j2\delta_c \pm j\delta_c\omega_m^2 t^2). \qquad (12)$$

We note that there can exist two possible solutions for the FM case, one for each extremum of the phase variation. For further reference to these two possibilities we shall call the mode of operation corresponding to the top sign in (12) the positive mode and the other mode the negative mode. Note that the positive mode corresponds to the maximum phase variation and the negative mode to the minimum phase variation.

We can also consider the more general case, when the

[1] We adopt the convention that ω denotes circular frequencies in radians and f denotes frequencies in hertz. We will use ω in the analyses, and change to f in the final results.

pulse goes through the modulator at a phase angle θ from the ideal case. The transmission through the modulator can now be written as

$$\exp[-j\delta(t)] \simeq \exp[\mp j2\delta_c \cos\theta \pm j2\delta_c \sin\theta(\omega_m t) \pm j\delta_c \cos\theta(\omega_m t)^2. \quad (13)$$

For θ positive, the pulse lags behind the modulation signal. We can interpret the terms in the exponent as follows. The first term is an additional phase shift that changes the optical length of the cavity, so that the optical length is now given by

$$L_0 = L \pm (\delta_c/\pi)\lambda_a \cos\theta. \quad (14)$$

The second term is a Doppler frequency shift and the third term gives a linear frequency chirp to the pulse. It is the last term that causes the mode locking.

B. Amplitude Modulator

For the amplitude modulator an idealized modulation characteristic will be assumed, where the amplitude transmission through the modulator is given by

$$a(t) = \exp(-2\delta_l \sin^2 \omega_m t). \quad (15)$$

The ideal mode-locking case is now when the pulse passes through the modulator at the instant of maximum transmission. This occurs twice in every cycle of the modulation signal, and hence the modulation frequency is half that for the phase-modulation case. With the assumption that the pulse is short compared to the modulation period, (15) becomes

$$a(t) \simeq \exp[-2\delta_l(\omega_m t)^2]. \quad (16)$$

For the more general case where the pulse passes through the modulator at a phase angle θ from the ideal case, (15) can be written as

$$a(t) \simeq \exp -[2\delta_l \sin^2\theta + 2\delta_l \sin 2\theta(\omega_m t) + 2\delta_l \cos 2\theta(\omega_m t)^2]. \quad (17)$$

The modulation characteristics of actual amplitude modulators may be considerably different, but it may be assumed that the amplitude transmission can always be written as follows

$$a(t) = \exp -[2\delta_0 + 2\delta_1(\omega_m t) + 2\delta_2(\omega_m t)^2 \cdots], \quad (18)$$

where δ_0, δ_1, and δ_2 depend on the phase angle θ and the depth of modulation. The constants will have to be evaluated for particular modulators such as the acoustic and electrooptic modulators. It can be shown, however, that for $\theta = 0$, $\delta_0 = 0$ and $\delta_1 = 0$.

V. Self-Consistent Solution With No Detuning

We can now consider the pulse going through the active medium and the modulator, and for the approximations we have made in the previous sections, a Gaussian pulse will remain Gaussian and a self-consistent solution becomes possible. The models for analyzing the mode-locked laser are shown in Fig. 1, and we see that the Fabry–Perot and ring cavities are entirely equivalent, except for small differences noted in the previous sections. We will only consider the Fabry–Perot type cavity, but in all cases the results for a ring cavity can be obtained by dividing the depth of modulation by 2.

If $E_1(t)$ is the pulse entering the active medium, then the Fourier transform of the pulse coming out is given by

$$E_2(\omega) = g_a(\omega)E_1(\omega) = \frac{E_0 G}{2}\sqrt{\frac{\pi}{\gamma}} \exp\left[-2jg\left(\frac{\omega-\omega_a}{\Delta\omega}\right) - 4g\left(\frac{\omega-\omega_a}{\Delta\omega}\right)^2 - (\omega-\omega_p)^2/4\gamma\right]. \quad (19)$$

Here we have considered the ideal case where the pulse is on line center ($\omega_p = \omega_a$). This will be the case if the pulse passes through the FM modulator with no Doppler shift or the AM modulator at minimum loss. In this case, (19) can be written as

$$E_2(\omega) = \frac{E_0 G}{2}\sqrt{\frac{\pi}{\gamma}} \cdot \exp[-A(\omega-\omega_a)^2] \exp[-jB(\omega-\omega_a)], \quad (20)$$

where

$$A = 1/(4\gamma) + 4g/\Delta\omega^2 \quad (20a)$$

and

$$B = 2g/\Delta\omega. \quad (20b)$$

Transforming into the time domain, the pulse becomes

$$E_2(t) = (E_0 G/4\sqrt{\gamma A}) \exp[-(t-B)^2/4A] \exp(j\omega_a t). \quad (21)$$

We can now see what the effect of the active medium on the pulses is. Since the pulse spectrum is on line center, the term $4g/\Delta\omega^2$ in (20a) is real and hence $|A| > (1/4\gamma)$, which means the spectral width has been reduced. However, it does not follow that the pulsewidth is increased, because $1/4\gamma$ is in general complex, and for $\beta = \alpha$, the pulsewidth actually remains unchanged, assuming $4g/\Delta\omega^2$ is small. We will consider these conditions in more detail later. The constant B indicates a delay of the pulse envelope.

Now consider the effects of the modulator. The transmission of the modulator can generally be given by $\exp -[\delta_g(\omega_m t)^2]$ where $\delta_g = \mp j\delta_c$ for FM modulator and $\delta_g = 2\delta_l$ for the ideal AM modulator from (12) and (16). The peak of the pulse goes through the modulator at time $t = B$ and hence the pulse coming out of the modulator is given by

$$E_3(t) = E_2(t) \exp[-\delta_g\omega_m^2(t-B)^2]. \quad (22)$$

Finally, the round trip for the pulse is completed by including an additional time delay $2L_0/c$ and an effective reflectivity r of a mirror, to include all losses in the cavity. The pulse after one round trip is then given by

$$E_4(t) = rE_3[t - (2L_0/c)]. \quad (23)$$

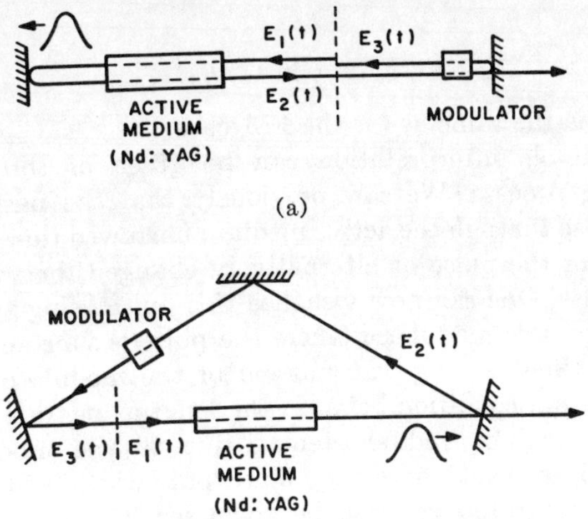

(b)

Fig. 1. Model for mode-locked laser. (a) Fabry–Perot-type cavity. (b) Ring-type cavity.

To obtain a self-consistent solution, the envelope of the pulse must go through the modulator at the same modulation phase every time. Hence, the total round trip time for the pulse is T_m, the modulator period, where $T_m = 2\pi/\omega_m$ for the phase modulator and $T_m = \pi/\omega_m$ for the amplitude modulator.

The self-consistency requirement now becomes

$$E_1(t - T_m)e^{-i\phi} = E_4(t). \tag{24}$$

The phase angle ϕ is included to allow for a possible phase shift (or phase precession [14]) of the optical signal with respect to the pulse envelope. From (1), (21), (22), and (23), the self-consistency condition equation can now be written as

$$E_0/2 \exp\left[-\gamma(t - T_m)^2\right] \exp\left[j\omega_a(t - T_m)\right] \exp(-j\phi)$$

$$= \frac{rE_0G}{4\sqrt{\gamma A}} \exp\left(-[t - B - (2L_0/c)]^2/4A\right)$$

$$\cdot \exp\left(-\delta_g\omega_m^2[t - B - (2L_0/c)]^2\right) \exp\left(j\omega_a[t - (2L_0/c)]\right). \tag{25}$$

From this equation it follows that

$$T_m = 2L_0/c + B \tag{26a}$$

$$\gamma = 1/(4A) + \delta_g\omega_m^2 \tag{26b}$$

$$e^{-i\phi} = (rG/2\sqrt{\gamma A}) \exp(j\omega_a B). \tag{26c}$$

These three equations combined with (20) now essentially solve the problem. From these equations we can obtain the desired analytical expressions for the modulation frequency, the pulsewidth, bandwidth, etc. First, consider (26b). We can substitute for A from (20a), and obtain a second-order equation in γ, which can be solved to give

$$\gamma = \frac{\delta_g\omega_m^2}{2} \pm \frac{1}{2}\sqrt{(\delta_g\omega_m^2)^2 + \frac{\delta_g\omega_m^2\,\Delta\omega^2}{4g}}. \tag{27}$$

The real part of γ must always be positive, and hence we only retain the positive sign in (27).

To see if further approximations are possible, consider the ratio of the second to the first term under the square root sign in (27), namely $\Delta\omega^2/4g\omega_m^2\delta_g$. For lasers with a wide linewidth such as Nd:YAG where $\Delta f \simeq 120$ GHz [17] the modulation frequency will be much less than the linewidth, i.e., $\omega_m \ll \Delta\omega$, and for practical values of g and δ_g we will usually have that

$$\Delta\omega^2/4g\omega_m^2\delta_g \gg 1. \tag{28}$$

With this approximation (27) becomes

$$\gamma \simeq (\omega_m\,\Delta\omega/4)\sqrt{\delta_g/g}. \tag{29}$$

We can now interpret this for the cases of FM and AM intracavity modulation.

A. FM Modulation

For the FM case $\delta_g = \mp j\delta_c$ and for γ_{FM} we now get

$$\gamma_{FM} = \alpha_{FM} - j\beta_{FM} = (1 \mp j)(\omega_m\,\Delta\omega/4)\sqrt{\delta_c/2g}. \tag{30}$$

This equation gives the values for α_{FM} and β_{FM} and from (4) and (5) we can now get the expressions for the pulsewidth and bandwidth[2]

$$\tau_{p0}(\text{FM}) = \frac{\sqrt{2\sqrt{2}\ln 2}}{\pi}\left(\frac{g_0}{\delta_c}\right)^{1/4}\left(\frac{1}{f_m\,\Delta f}\right)^{1/2} \tag{31a}$$

$$\Delta f_{p0}(\text{FM}) = \sqrt{2\sqrt{2}\ln 2}\left(\frac{\delta_c}{g_0}\right)^{1/4}(f_m\,\Delta f)^{1/2}. \tag{31b}$$

We can conclude the following from these two equations.

1) The pulsewidth–bandwidth product

$$\tau_{p0}(\text{FM})\cdot\Delta f_{p0}(\text{FM}) = 2\sqrt{2}\ln 2/\pi = 0.626.$$

2) The pulsewidth is inversely proportional to $(\delta_c)^{1/4}$, and since δ_c is proportional to the square root of P_m, the RF power into the modulator, $\tau_{p0}(\text{FM}) \propto (1/P_m)^{1/8}$ and hence the pulses shorten very slowly with increased modulator drive. This is not the best way to shorten the pulses.

3) The pulse length is proportional to $(f_m\Delta f)^{-1/2}$. It is sometimes assumed that the bandwidth of the pulse is approximately equal to the linewidth and it then follows that $\tau_p \propto 1/\Delta f$. However, the above expression for τ_{p0} shows clearly that this is not the case. Note that increasing the modulation frequency is the better way to shorten the pulses.

The exact modulation frequency can be obtained from (26a). Substituting for the optical length L_0 from (14) and B from (20b), we get for the FM case

$$f_{m0}(\text{FM}) = 1/\left[2L/c \pm 2\left(\frac{\delta_c}{\pi}\right)\frac{\lambda_a}{c} + 2g_0/\Delta\omega\right]. \tag{32}$$

The last two factors are perfectly understandable: the term $2(\delta_c/\pi)\lambda_a/c$ is the extremal "motion of the mirror,"

[2] We have here introduced the subscript 0 to indicate the values of τ_p, Δf_p, g, etc., at zero detuning of the modulator.

looking at phase modulation as a vibrating mirror, and the term $2g_0/\Delta\omega$ is the expected added dispersion or linear delay in the Lorentzian line. These two factors are small compared to $2L/c$, and hence the modulation frequency is approximately equal to $c/2L$.

From (26c) we can get the self-consistent value for g. Substituting for A from (26b), we get

$$rG\sqrt{1 \pm j\delta_c\omega_m/(\alpha - j\beta)} = \exp\left[-j(\phi + \omega_a B)\right].$$

Equating the magnitude parts of this equation, and letting $\beta = \mp\alpha$ and $\Delta f_p = 2\sqrt{\ln 2}\sqrt{\alpha/\pi}$ where Δf_p is the bandwidth of the pulse, we get

$$g_0 = \frac{1}{2}\ln\frac{1}{R} - \frac{1}{4}\ln\left[1 - \frac{16\ln 2\,\delta_c f_m^2}{\Delta f_{p0}^2} + \frac{1}{2}\left(\frac{16\ln 2\,\delta_c f_m^2}{\Delta f_{p0}^2}\right)^2\right] \tag{33}$$

where R is the effective (power) reflection of a mirror and includes all losses. Usually we will have that $\Delta f_p \gg \sqrt{\delta_c} f_m$ and hence we get an approximate value for g_0:

$$g_0 \simeq \tfrac{1}{2}\ln(1/R). \tag{34}$$

This value of g_0 can be used to calculate the pulsewidth and bandwidth, and then (33) can be used to get a better value for g_0. After a few iterations one can obtain the correct values for all the pulse parameters, but the approximate equation will in most cases be within a few percent of the correct value.

Equating the phase part of (26b), we can obtain ϕ, and we can show that $\phi \simeq -2g\omega_a/\Delta\omega$, and since $\omega_a \gg \Delta\omega$, ϕ is some large angle. However, the actual value for ϕ does not affect any of the pulse parameters, and we can neglect it.

B. AM Modulation

For the AM case $\delta_g = 2\delta_l$ and it follows that

$$\gamma_{AM} = \alpha_{AM} = (\omega_m\,\Delta\omega/4)\sqrt{2\delta_l/g}, \tag{35}$$

and $\beta_{AM} = 0$. The pulsewidth and bandwidth are now given by

$$\tau_{p0}(\text{AM}) = \frac{\sqrt{\sqrt{2}\ln 2}}{\pi}\left(\frac{g_0}{\delta_l}\right)^{1/4}\left(\frac{1}{f_m\,\Delta f}\right)^{1/2} \tag{36a}$$

$$\Delta f_{p0}(\text{AM}) = \sqrt{2\sqrt{2}\ln 2}\left(\frac{\delta_l}{g_0}\right)^{1/4}(f_m\,\Delta f)^{1/2}. \tag{36b}$$

These equations are identical to those for the FM case, except that the pulsewidth is shorter by $\sqrt{2}$ for the AM case and hence the pulsewidth–bandwidth product

$$\tau_{p0}(\text{AM})\cdot\Delta f_{p0}(\text{AM}) = 2\ln 2/\pi = 0.440.$$

The exact modulation frequency is given by

$$f_{m0}(\text{AM}) = \tfrac{1}{2}[1/(2L/c + 2g_0/\Delta\omega)]; \tag{37}$$

g_0 is given by

$$g_0 = \frac{1}{2}\ln\left(\frac{1}{R}\right) - \frac{1}{2}\ln\left[1 - 16\ln 2\delta_l\left(\frac{f_m}{\Delta f_{p0}}\right)^2\right].$$

As before, we usually have that $\Delta f_{p0} \ll f_m$, and hence we get

$$g_0 \simeq \tfrac{1}{2}\ln(1/R),$$

which is the same as for the FM case.

A simple interpretation can be given of the mode-locking process. We saw previously that the passage of the pulse through the active medium narrowed the spectral width of the pulse or alternatively changed the width of the pulse. One can now visualize this pulse going through an amplitude modulator where the pulse is shortened due to the time-varying transmission of the modulator. The equilibrium condition between the lengthening due to the active medium and shortening due to the modulator determines what the steady-state pulsewidth will be. A similar interpretation can be given for FM modulation, but it is now easier to visualize the process in the frequency domain. When the pulse passes through the FM modulator, a frequency chirp is put on the pulse. This frequency chirp increases the spectral width of the pulse and an equilibrium state is reached where the increase in spectral width due to the modulator is equal to the narrowing of the spectral width due to the active medium. It is interesting to notice that this equilibrium condition requires a steady-state frequency chirp on the pulse and further that the pulse envelope and frequency chirp contribute equally to the spectral width of the pulse (i.e., $\alpha = \beta$). The interpretation that is usually given for mode locking in an inhomogeneously broadened laser is that the modulator introduces some coupling between adjacent axial modes and that this coupling locks the phases of these modes in such a way as to give short pulses (and hence the terms mode locking or phase locking). This interpretation is not useful for the homogeneously broadened laser, since most of the axial modes are not present in the free-running laser. There are usually only a few axial modes, mostly due to spatial inhomogeneity and hence the term mode locking is somewhat of a misnomer, but we will retain it with a somewhat broadened meaning.

For a typical Nd:YAG laser with 10 percent round-trip loss (i.e., $R = 0.9$), a cavity length of 60 cm, and a linewidth of 120 GHz, the pulse length and bandwidth are given by $\tau_{p0}(\text{FM}) = 39.0\,(1/\delta_c)^{1/4}$ ps and $\Delta f_p(\text{FM}) = 16.1\,(\delta_c)^{1/4}$ GHz and for $\delta_c = 1$ radian, which is easily obtainable, pulses of 39 ps can be generated.

VI. Comparison With Free-Running Laser

There are basically two quantities of interest in comparing the free-running and mode-locked laser. First is the change in output power from the free-running to the mode-locked laser, and second how the axial-mode beats of the free-running laser compare with the modulation frequency for ideal mode locking as discussed in the previous section.

The basic condition for steady-state oscillation in a free-running laser is that the total round-trip gain must be exactly unity, i.e.,[3]

[3] The subscript f indicates parameters for the free-running laser.

$$r_f \exp(g_f/[1 + 2j(\omega - \omega_a)/\Delta\omega]) \exp[-j(\omega/c)2L] = \exp(-jq2\pi). \quad (38)$$

From the magnitude portion of the oscillation condition we get

$$g_f = \tfrac{1}{2}[1 + 4(\omega - \omega_a)^2/\Delta\omega^2] \ln(1/R_f). \quad (39)$$

Usually the free-running axial modes are close to line center and hence $g_f \simeq \frac{1}{2}\ln(1/R_f)$, where R_f is the effective power reflection of a mirror to account for all losses.

The change in output power can be obtained from the discussion in Appendix I and (116). If P_f is the free-running power and P_{ml} is the mode-locked power, then

$$\frac{P_{\mathrm{ml}}}{P_f} = \frac{R_p(g_f/g_0) - 1}{(R_p - 1)} \frac{1}{S(0, \Delta f_{p0})}, \quad (40)$$

where R_p is the normalized pump power to the laser and $S(0, \Delta f_{p0})$ is the saturation function as defined by (113) and in this case we get that

$$S(0, \Delta f_{p0}) = \left[1 - \left(\frac{\Delta f_{p0}}{\Delta f}\right)^2 \frac{1}{2 \ln 2}\right]. \quad (41)$$

From (41), we see that small changes in g_0 will affect the power output, and hence we should use the complete equation for g_0 given by (33). The power output is thus given by

$$\frac{P_{\mathrm{ml}}}{P_f} = \frac{R_p\left\{\dfrac{\ln(1/R_f)}{\ln(1/R) - \frac{1}{2}\ln\left\{1 - 16\ln 2\,\delta_c\left(\dfrac{f_{m0}}{\Delta f_{p0}}\right)^2 + \frac{1}{2}\left[16\ln 2\,\delta_c\left(\dfrac{f_{m0}}{\Delta f_{p0}}\right)^2\right]^2\right\}}\right\} - 1}{(R_p - 1)\left[1 - \dfrac{1}{2\ln 2}\left(\dfrac{\Delta f_{p0}}{\Delta f}\right)^2\right]} \quad (42)$$

where R_p is the normalized pump power, R_f the effective power reflection of the free-running laser, and R the effective power reflection for the mode-locked laser. R_f and R are defined to include all losses in the laser in both cases. Equation (42) will give the small changes in power output when the laser is mode locked. If we assume that the losses in the free-running and mode-locked laser are the same and $\Delta f > \Delta f_{p0}$ and $\Delta f_{p0} > f_{m0}$, then $P_{\mathrm{ml}} = P_f$.

In actuality, however, the free-running laser will always maximize its output power, and particularly if there are small etalon effects in the cavity there will be some axial-mode selection in the free-running laser to minimize the losses. For this reason we may have that $R_f > R$ and there will be a decrease in output power for the mode-locked laser, particularly close to threshold and when δ_c is small.

From the phase portion of the oscillation condition (38), we get

$$\frac{2g_f[(\omega_q - \omega_a)/\Delta\omega]}{1 + \left(\dfrac{\omega_q - \omega_a}{\Delta\omega}\right)^2} + \left(\frac{\omega_q}{c}\right)2L = q2\pi \quad (43)$$

where ω_q is the oscillation frequency of the qth axial mode in the free-running laser. Substituting for g_f from (39), it can readily be shown that the axial-mode beats are given by

$$\Delta f_f = 1/[2L/c + \ln(1/R_f)/\Delta\omega]. \quad (44)$$

Comparing this with the modulation frequency for the FM laser, (32), we get that the axial-mode beat will be exactly between the modulation frequencies for the two modes of the FM laser if $g_0 = \frac{1}{2}\ln(1/R_f)$. This condition can be satisfied if etalon effects and pulling of the modulation frequency due to changes in g_0 [as given by (33)] are neglected. Due to etalon effects, however, $R_f > R$, and the modulation frequency will be lower than the axial-mode beat. Pulling of the modulation frequency due to additional changes in g_0 will further lower the modulation frequency. For the AM mode-locked laser, we will have the same effect as for the FM case.

In the actual free-running laser, the axial-mode beat will not be a single frequency, but due to mode pulling of the etalon effects, and dispersion of the host material and other crystals in the cavity, the axial-mode beats will be spread out.

VII. Self-Consistent Solution With Detuning

Detuning is defined as the frequency shift from the ideal mode-locking frequency (f_{m0}) and is given by

$$\Delta f_m = f_{m0} - f_m. \quad (45)$$

This is consistent with the definition of detuning given by Harris and McDuff [4]. Note that Δf_m is negative for $f_m > f_{m0}$ and vice versa.

The solutions for FM and AM modulation are now considerably different, and we will consider them separately.

A. *FM Modulation*

When the modulation frequency is detuned from the ideal frequency, the pulses pass through the modulator at some phase angle θ away from the extreme phase variation and the transmission through the modulator is given by (13). The pulses now experience some Doppler shift, and in consecutive passes through the modulator, the optical frequency (ω_p) of the pulses is shifted until some equilibrium is reached where the Doppler shift of the modulator is canceled out by an equal and opposite frequency shift from the active medium. The pulse now has a frequency ω_p, such that $\omega_a = \omega_p + \nu$ and ν is the frequency shift of the pulse.

In all cases, we will consider that the detuning is small enough that the laser remains mode locked. For larger detuning, the output of the laser changes to a FM laser type of signal [4], [6], [18], which we will not consider in this paper. FM laser operation has been observed for the Nd: YAG laser, and we will report this in another paper [19].

In Section III., we showed that the line shape could be expanded about any frequency ω_p and that the line shape was the same as before with g replaced by $g/(1 + 2jn)$ and $\Delta\omega$ replaced by $\Delta\omega(1 + 2jn)$ where $\eta = \nu/\Delta\omega$. Hence the pulse coming out of the active medium is given by

$$E_2(t) = E_0G'/(4\sqrt{\gamma A'}) \exp[-(t - B')^2/4A'] \exp(j\omega_p t), \tag{46}$$

where

$$A' = 1/4\gamma + 4g/\Delta\omega^2(1 + 2j\eta)^3 \tag{47a}$$

$$B' = 2g/\Delta\omega(1 + 2j\eta)^2 \tag{47b}$$

$$G' = \exp[g/(1 + 2j\eta)]. \tag{47c}$$

One important difference between the pulse as given by (46) and the pulse without detuning is that B' is now complex. Now consider the term $(t - B')^2/4A'$. If we split B' in its real and imaginary part, substitute it in the above expression and multiply out, we get

$$(t - B')^2/4A' = \left[t - \frac{2g(1 - 4\eta^2)}{\Delta\omega(1 + 4\eta^2)}\right]^2 \Big/ 4A' + \frac{16jg\eta}{\Delta\omega(1 + 4\eta^2)^2} \cdot \left[t - \frac{2g(1 - 4\eta^2)}{\Delta\omega(1 + 4\eta^2)}\right] \Big/ 4A' - \frac{64g^2\eta^2}{\Delta\omega^2(1 + 4\eta^2)^4 4A'}. \tag{48}$$

Now consider the following expansion

$$\frac{16jg\eta}{\Delta\omega(1 + 4\eta^2)^2 4A'} = \frac{K_1}{4A'} + jK_2, \tag{49}$$

where K_1 and K_2 are real. After some algebraic manipulations, it can be shown that

$$K_2 = \frac{16g\nu(\alpha^2 + \beta^2)(1 + 4\eta^2)}{16g(\alpha^2 + \beta^2)(1 - 12\eta^2) + \alpha\,\Delta\omega^2(1 + 4\eta^2)^3} \tag{50a}$$

$$K_1 = K_2\left\{\frac{\beta}{\alpha^2 + \beta^2} - \frac{16g}{\Delta\omega^2}\left[\frac{6\eta - 8\eta^3}{(1 + 4\eta^2)^3}\right]\right\}. \tag{50b}$$

We can now substitute (49) back in (48), complete the square, and thus finally the pulse can be written as

$$E_2(t) = \frac{E_0G'}{4\sqrt{\gamma A'}} \cdot \exp - \left[t - \frac{2g(1 - 4\eta^2)}{\Delta\omega(1 + 4\eta^2)^2} + \frac{K_1}{2}\right]^2 \Big/ 4A' \cdot \exp[j(\omega_0 - K_2)t] \exp\left[K_1^2/16A' + jK_2\left(\frac{2g(1 - 4\eta^2)}{\Delta\omega(1 + 4\eta^2)^2}\right) + \frac{64g^2\eta^2}{\Delta\omega^2(1 + 4\eta^2)^4 4A'}\right]. \tag{51}$$

The expression we have obtained here reveals some very interesting effects of the active medium on the pulses.

1) The delay of the pulse envelope has changed from $2g/\Delta\omega$ to

$$2g(1 - 4\eta^2)/\Delta\omega(1 + 4\eta^2)^2 + K_1/2,$$

and this change will compensate exactly for the change in modulation frequency.

2) A frequency shift K_2 has been introduced and this will exactly cancel out the Doppler shift from the modulator.

3) The last exponential shows the additional attenuation and phase shift that has been introduced.

The self-consistency equation can be obtained as before, and with the transmission through the modulator given by (13), the following conditions will result:

$$T_m = \frac{2L}{c} \mp 2\left(\frac{\delta_c}{\pi}\right)\frac{\lambda_a}{c}\cos\theta + \frac{2g(1 - 4\eta^2)}{\Delta\omega(1 + 4\eta^2)^2} - \frac{K_1}{2} \tag{52a}$$

$$K_2 = \pm 2\delta_c\omega_m \sin\theta \tag{52b}$$

$$\gamma = 1/(4A') \mp j\delta_c \cos\theta\omega_m^2 \tag{52c}$$

$$|rG/2\sqrt{\gamma A'}| \cdot |\exp[K_1^2/16A' + 16g^2\eta^2/(\Delta\omega^2(1 + 4\eta^2)^4 A')]| = 1 \tag{52d}$$

In the last condition, we have only considered the magnitude, since the phase angle ϕ is of no consequence as discussed before.

From (52c) we can solve for γ, and with the same approximation as before we get

$$\gamma \simeq (\omega_m\,\Delta\omega/4)\sqrt{\mp j\delta_c \cos\theta(1 + 2j\eta)^3/g} = (\omega_m\,\Delta\omega/4)\sqrt{\delta_c \cos\theta/g}\,(1 + 4\eta^2)^{3/4}(\cos\psi + j\sin\psi) \tag{53}$$

where

$$\psi = \mp\pi/4 + 3/2\tan^{-1}(2\eta). \tag{54}$$

From (53) we get α_{FM} and β_{FM} and hence the pulsewidth and bandwidth are given by

$$\tau_p(\mathrm{FM}) = \frac{\sqrt{2\ln 2}}{\pi}\left(\frac{g}{\delta_c\cos\theta}\right)^{1/4} \cdot \left(\frac{1}{1 + 4\eta^2}\right)^{3/8}\left(\frac{1}{f_m\,\Delta f\cos\psi}\right)^{1/2} \tag{55}$$

$$\Delta f_p(\mathrm{FM}) = \sqrt{2\ln 2}\left(\frac{\delta_c\cos\theta}{g}\right)^{1/4}(1 + 4\eta^2)^{3/8}\left(\frac{f_m\,\Delta f}{\cos\psi}\right)^{1/2}. \tag{56}$$

Note that the pulsewidth–bandwidth product is now

$$\tau_p\cdot\Delta f_p = \frac{2\ln 2}{\pi\cos\psi}.$$

With the assumption that $\Delta\omega/\omega_m \gg 1$, it can be shown that K_1 and K_2 are given by the following approximate expressions:

$$K_2 \simeq \frac{16g\eta\alpha}{\Delta\omega(1+4\eta^2)^2\cos^2\psi} \tag{57a}$$

$$K_1 \simeq -\frac{16g\eta\tan\psi}{\Delta\omega(1+4\eta^2)^2}. \tag{57b}$$

From (52a) we now get the modulation frequency

$$f_m(\text{FM}) = 1\Big/\left[2L/c \pm 2\left(\frac{\delta_c}{\pi}\right)\frac{\lambda_a}{c}\cos\theta + \frac{2g(1-4\eta^2+4\eta\tan\psi)}{\Delta\omega(1+4\eta^2)^2}\right]. \tag{58}$$

Substituting K_2 in (52b) it can be shown that

$$\frac{\sin\theta}{\sqrt{\cos\theta}} = \pm 2\sqrt{\frac{g}{\delta_c}}\frac{\eta}{(1+4\eta^2)^{5/4}\cos\psi},$$

and from this equation it follows that

$$\cos\theta = \sqrt{4\left(\frac{g}{\delta_c}\right)^2\left[\frac{\eta^4}{(1+4\eta^2)^5\cos^4\psi}\right]+1} - 2\left(\frac{g}{\delta_c}\right)\left(\frac{\eta^2}{(1+4\eta^2)^{5/2}\cos^2\psi}\right). \tag{59}$$

Finally, we can consider (52d), and substituting from (49), this equation can be simplified to

$$|rG/2\sqrt{\gamma A'}|\cdot|\exp K_2^2A'| = 1$$

and g is given by

$$g = (1+4\eta^2)[\tfrac{1}{2}\ln(1/R) + \tfrac{1}{2}\ln 4\gamma A' - K_2^2A'] \tag{60a}$$

and with the same approximations as before, it can be shown that g is approximately given by

$$g \simeq \tfrac{1}{2}(1+4\eta^2)\ln(1/R). \tag{60b}$$

We should note here that the ideal mode-locking frequencies for the positive and negative modes are different. From (58) this frequency difference is

$$4(\delta_c\lambda a/\pi c)f_{m0}^2$$

and for $f_{m0} = 250$ MHz, the frequency difference is 0.28 δ_c kHz, and hence this difference is small, but it is significant because it splits the degeneracy of the two possible modes of the FM mode-locked laser.

We have now obtained all the equations to describe the behavior of the FM mode-locked laser with detuning. In obtaining these equations, several approximations were made in going from the self-consistency conditions given by (52a)–(52d) to the final equations. However, starting with these approximate values of η, g, γ, etc., one can get the exact solutions of (52a)–(52d) by a suitable iterative procedure. This was done for some typical cases of the Nd:YAG laser, and in all cases it was found that the approximations given by (55)–(60) were within a few percent of the exact solutions of the self-consistency conditions.

In particular, we can again consider the case of the Nd:YAG laser with 10 percent total round-trip loss, a 60-cm-long cavity (optical length), and a 120-GHz linewidth. The results are shown in Figs. 2–7. These results clearly illustrate some of the interesting peculiarities of the theory. We will consider these results in some detail and try to obtain some physical insight of what happens in the mode-locked laser.

Fig. 2 shows the phase shift of the pulse with respect to the modulation signal versus detuning. Note the distinct asymmetry in these curves. Fig. 3 shows the frequency shift of the optical frequency of the pulse. We note that the frequency shifts of the positive and negative modes are in opposite directions. This is to be expected, because when the pulse goes through the modulator at a phase angle θ from the extremes of the phase variation, the Doppler frequency shift is in opposite directions for the two modes [see (13)]. We also note the asymmetry of the curves. The reason for this we will see later.

Fig. 4 shows the variation in pulsewidth with detuning for several δ_c. The most surprising observation here is that the pulses continue to get shorter for negative detuning. Fig. 5 shows the variation of the bandwidth with detuning, and we notice that the bandwidth keeps on increasing for positive detuning, even though the pulses get longer as shown in Fig. 4, which is somewhat surprising. Fig. 6 shows the variation of β with detuning, and this figure provides the clue to what is happening in the laser. Consider, say, the positive mode and negative detuning. We notice that β decreases from its value for ideal mode locking. At the same time, we notice from Fig. 4 that the pulses get shorter. What is happening is that the pulses are being compressed as described by Giordmaine *et al.* [18]. They show that when a pulse with a linear frequency chirp is passed through a dispersive medium, and the frequency chirp has the correct sign so that the leading edge of the pulse is retarded, the pulse is compressed. This is precisely what happens in the mode-locked laser. When the modulation signal is detuned, the pulse shifts towards the side of the line. The so-called "anomalous dispersion" of the active medium now provides the dispersion to compress the pulses. It turns out that we get the correct sign for β for pulse compression when the detuning is negative. It can be seen from Fig. 6 that β actually goes through zero. When $\beta \simeq 0$ we get optimum compression, and we actually see from Fig. 4 that the pulse length has a minimum when $\beta \simeq 0$. For large enough negative detuning, β changes sign. It is interesting that the mechanism of mode locking, as described earlier, is now completely changed around. When the pulse now passes through the modulator, the linear frequency chirp induced by the modulator substracts from the frequency chirp on the pulse, and the spectral width of the pulse is decreased. The pulse then passes through the active medium in such a way that the dispersion increases β, and hence the active medium increases the spectral width. The

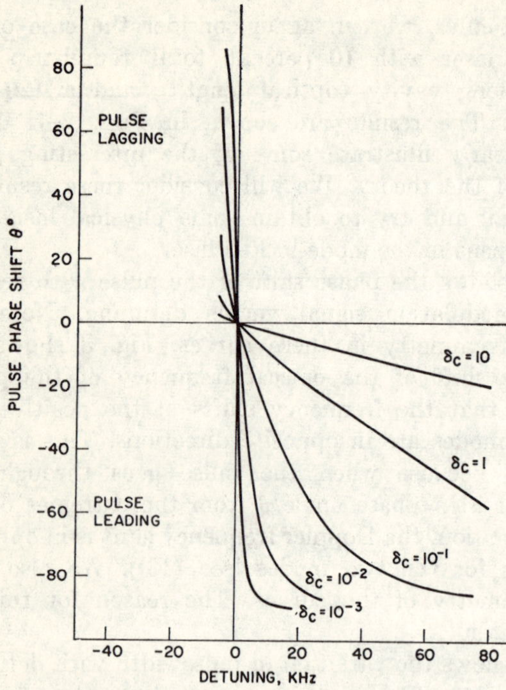

Fig. 2. Phase shift of pulse with respect to modulation signal versus detuning for laser with internal FM modulation. Conditions are for a typical Nd:YAG laser with a 60-cm cavity, 10 percent round-trip loss, and 120-GHz linewidth.

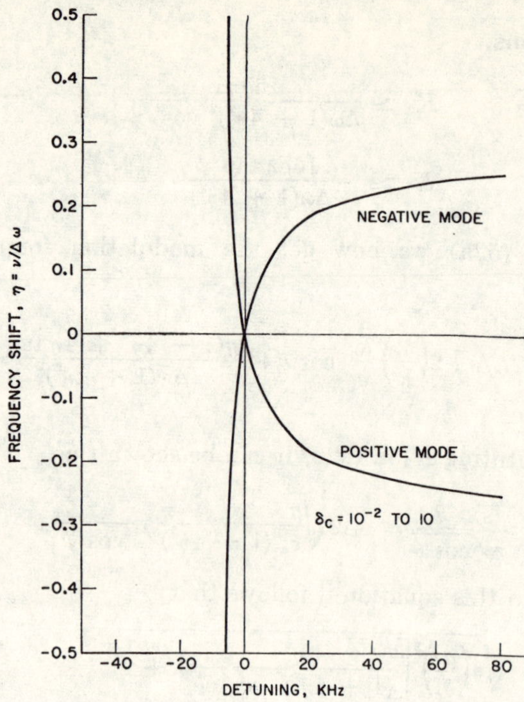

Fig. 3. Frequency shift of pulse off line center versus detuning for laser with internal FM modulation. Same conditions as for Fig. 2. Dependence on δ_c too small to show on the scale of this figure.

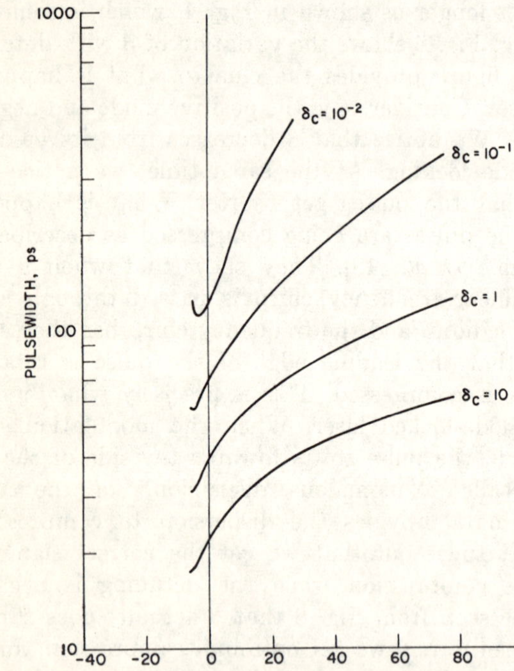

Fig. 4. Pulsewidth versus detuning (FM modulation).

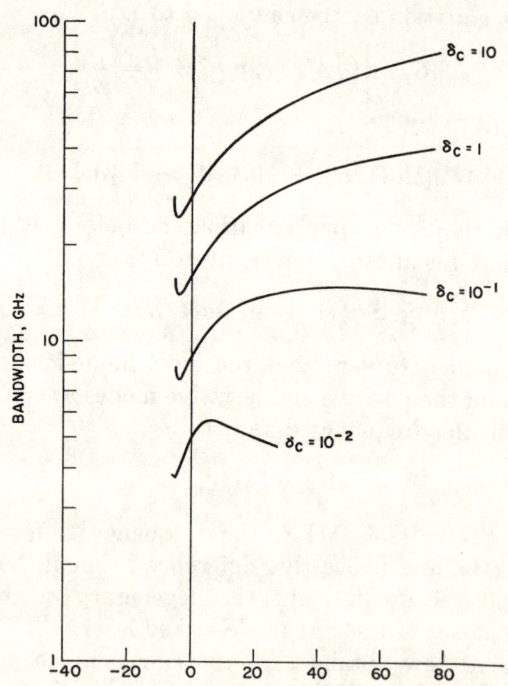

Fig. 5. Bandwidth of pulse versus detuning (FM modulation).

roles of the modulator and active medium have thus been switched around.

For positive detuning, we get pulse stretching, and the pulses keep on getting longer as the detuning increases. From (55) we note that the pulse length approaches infinity as $|\psi| \to 90°$, and from (54) we see that this condition is satisfied when $\eta \to \mp 0.288$ for the positive and negative mode, respectively. When this condition is approached, mode locking becomes impossible. We note from (58), that as $|\psi| \to 90°$ the detuning will also approach infinity, and hence as we keep on increasing the positive detuning, the optical frequency of the pulse will not shift beyond $\eta = \mp 0.288$. For negative detuning, however, the pulse rapidly shifts off line center as shown

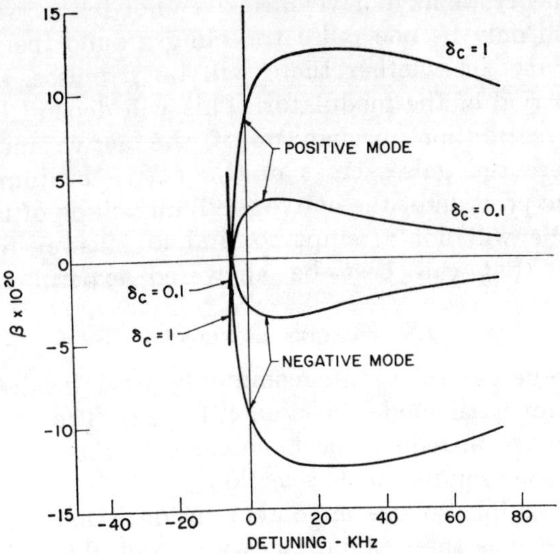

Fig. 6. β (frequency chirp) versus detuning (FM modulation).

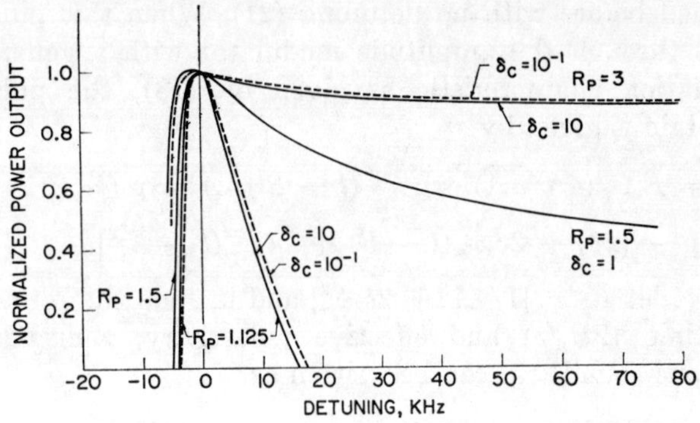

Fig. 7. Power output from laser versus detuning for various depths of modulation and normalized pump powers (FM modulation).

in Fig. 3. For large enough positive or negative detuning the laser will go into the FM-laser mode of operation [18], [19].

We noted previously that the bandwidth kept increasing for positive detuning, even though the pulse length got longer. To explain this, we consider the expression for bandwidth (5),

$$\Delta f_p = \frac{1}{\pi}\sqrt{2\ln 2[(\alpha^2+\beta)^2/\alpha]}.$$

For no detuning, $\alpha = \beta$, and for positive detuning, $\beta > \alpha$ and hence

$$\Delta f_p \simeq \frac{1}{\pi}\sqrt{2\ln 2\beta^2/\alpha} = \tau_p(\beta/\pi).$$

We see that for a pulse with a linear frequency chirp, the bandwidth keeps on increasing even though the pulse gets longer.

We can also obtain the change in output power with detuning from (116) in Appendix I. If $P_{m1}(\Delta f_m)$ is the power output with detuning Δf_m, we obtain

$$\frac{P_{m1}(\Delta f_m)}{P_{m1}(0)} = \frac{[R_p(g_f/g)-1]\left[1-\frac{1}{2\ln 2}(\Delta f_{p0}/\Delta f)^2\right]}{\left[R_p\left(\frac{g_f}{g_0}\right)-1\right]\left\{\left(\frac{1}{1+4\eta^2}\right)\left[1-\frac{1}{2\ln 2(1+4\eta^2)}\left(\frac{\Delta f_p}{\Delta f}\right)^2\right]\right\}}, \tag{61}$$

where R_p is the normalized pump power for the free-running laser, and g_f, g_0, and g are the saturated single-pass gains through the active medium for the free-running laser, mode-locked laser with no detuning, and mode-locked laser with detuning, respectively.

Note that we should use the complete expressions for g and g_0 here. It will be difficult to get the value of g_f due to the complicated nature of the etalon effects, but we can define R_p', where $R_p' = R_p(g_f/g_0)$, and then R_p' is the normalized pump power of the mode-locked laser with no detuning that can be determined. If we assume that $\Delta f \gg \Delta f_p$, and use the approximate expression for g and g_0, (62) simplifies to

$$\frac{P_{m1}(\Delta f_m)}{P_{m1}(0)} = \frac{R_p-(1+4\eta^2)}{R_p-1}. \tag{62}$$

The variation of power output with detuning is shown in Fig. 7. For negative detuning, we found that the pulse rapidly moves off line center, and hence the power output drops rapidly. For positive detuning, the power output changes much slower, particularly far above threshold. It can be seen that the power output does not depend very much on δ_c.

We note that for $\delta_c = 10$ and $R_p = 3$, there is an initial power increase for negative detuning. To understand this, we should note that there are mainly two effects contributing to the change in output power. First, there is the shift off line center that causes a decrease in output power but this effect becomes smaller as we get further above threshold. The second effect is the change in spectral width of the pulse. A narrow spectrum more effectively saturates the line, and hence the power output can increase as the spectrum becomes narrower. For the above-mentioned condition, the laser is far above threshold and the spectral width with no detuning is large. Hence, the second effect can dominate for negative detuning, and give rise to the increase in power output.

B. AM Modulation

Consider now what happens when we detune the modulation frequency with an amplitude modulator. We saw previously that when the pulse frequency shifts off line center, there is a small frequency pulling by the active medium and that a FM modulator could compensate for this frequency change. However, an amplitude modulator can not introduce a frequency shift, and hence we can conclude that with an amplitude modulator, the pulse frequency remains on line center, i.e., $\nu = 0$.

Also, with an amplitude modulator there is no frequency chirp during the pulse, and hence $\beta = 0$. Thus the pulse coming out of the active medium is the same as we obtained before with no detuning (21). When this pulse passes through the amplitude modulator with a general modulation characteristic as given by (18), the pulse we obtain is given by

$$E_3(t) = E_0G/(4\sqrt{\alpha A}) \exp[-(t - B^2/4A] \exp(j\omega_a t) \cdot \exp - [2\delta_0 + 2\delta_1\omega_m(t - B) + 2\delta_2\omega_m^2(t - B)^2]. \quad (63)$$

If we let $K = [1/4A + 2\delta_2\omega_m^2]$ and include the round-trip time $(2L_0/c)$ and effective reflectivity, the pulse after one round trip can be written as

$$E_4(t) = \frac{rE_0G}{4\sqrt{\alpha A}} \exp - K\left(t - B + \frac{\delta_1\omega_m}{K} - \frac{2L_0}{c}\right)^2 \cdot \exp j\omega_a\left(t - \frac{2L_0}{c}\right) \exp[+(\delta_1\omega_m)^2/K] \exp(-2\delta_0). \quad (64)$$

The self-consistency conditions now become

$$K = 1/(4A) + \delta_2\omega_m^2 = \alpha \quad (65a)$$

$$T_m = 2[2L_0/c + B - \delta_1\omega_m/K] \quad (65b)$$

$$rG/(2\sqrt{\alpha A}) \exp[(\delta_1\omega_m)^2/K] \exp - \delta_0 = 1. \quad (65c)$$

From the first equation, we can solve for α, and from this we can get the pulsewidth and bandwidth, which will be the same as the case with no detuning δ_l replaced by δ_2 [(36) and (37)]. For a second equation, we can substitute for K and B, and show that the modulation frequency is given by

$$f_m(\text{AM}) = \frac{1}{2}\left[1 \bigg/ \left(\frac{2L_0}{c} + \frac{2g}{\Delta\omega} - \frac{4\delta_1}{\Delta\omega}\sqrt{\frac{g}{\delta_2}}\right)\right]. \quad (66)$$

For the last equation, we can show that $(\delta_1\omega_m)^2/K \ll g$ assuming $\omega_m/\Delta\omega \ll 1$, and hence we find

$$g \simeq \delta_0 + \tfrac{1}{2}\ln(1/R). \quad (67)$$

For any particular amplitude modulator, we can evaluate δ_0, δ_1, and δ_2 as a function of θ and get the output characteristics of the laser from the above equations. One will usually find that the pulses get longer with detuning, and the output power decreases, and is the same for positive and negative detuning.

VIII. Higher Order Modulation

We have so far only considered the case where the modulator goes through only one cycle per round-trip time of the pulse. It is, however, possible that the modulator goes through several cycles per round-trip time. In general, we can have a modulation frequency p times the fundamental modulation frequency or axial mode spacing. We will call p the order of the modulation. All the previous theory we have developed is still good, with the only modification being that the modulation frequencies we have obtained are multiplied by p. For the particular example of the FM mode-locked Nd:YAG laser we considered, the detuning is multiplied by p.

The advantage in going to higher order modulation is that we can use much higher modulation frequencies and consequently obtain much shorter pulses.

The theory we have developed does not tell us whether there will only be one pulse traveling around inside the laser cavity, or whether there will be p pulses, spaced at the period of the modulator. This will depend on the detailed saturation mechanism of the active medium, and where the pulses cross in the active medium and hence the position of the active medium will be of importance. We will not attempt to find an answer to this question that can best be answered experimentally.

IX. Etalon Effects

Fabry–Perot etalons are commonly used inside laser cavities for axial mode selection [21], [22]. The effect of the etalon is to reduce the bandwidth of the system to produce the required mode selection.

It is possible to use an etalon inside a mode-locked laser to reduce the system bandwidth and obtain longer pulses. This etalon will usually consist of a parallel uncoated glass flat. It is shown in Appendix II that the transmission can be expanded about the peak, to give

$$t_e = \exp\left[j\sqrt{8R}\left(\frac{\omega - \omega_e}{\Delta\omega_e}\right) - 4\left(\frac{\omega - \omega_e}{\Delta\omega_e}\right)^2\right], \quad (68)$$

where $\Delta\omega_e$ is the bandwidth of the etalon and

$$\Delta\omega_e = (c/2h')\sqrt{8/R}\,(1 - R).$$

The one important requirement is that the bandwidth is large compared to the axial mode spacing so that there will be several axial modes under the transmission peak.

If we multiply the gain of the active medium and t_e, we can obtain an effective gain of the system, given by

$$g_e(\omega) = G \exp - \left(\frac{2jg}{\Delta\omega} - \frac{\sqrt{8R}}{\Delta\omega_e}\right)(\omega - \omega_a) - 4\left(\frac{g}{\Delta\omega^2} + \frac{1}{\Delta\omega_e^2}\right)(\omega - \omega_a)^2 \quad (69)$$

and we assume that the transmission peak of the etalon is on line center.

The two effects of the etalon are to change the bandwidth of the system, and to introduce some phase shift $\sqrt{8}\ R/\Delta\omega_e$, which slightly changes the modulation frequency. This is in addition to the change in modulation frequency due to the change in optical length of the cavity when the etalon is introduced.

In particular, we can consider the FM mode-locked laser with no detuning. The pulsewidth is now given by

$$\tau_{p0}(\text{FM}) = \frac{\sqrt{2\sqrt{2}\ln 2}}{\pi}\left(\frac{1}{\delta_c}\right)^{1/4}\left(\frac{1}{f_m}\right)^{1/2}\left(\frac{g_0}{\Delta f^2} + \frac{1}{\Delta f_e^2}\right)^{1/4}. \quad (70)$$

The case of particular interest is where the bandwidth of the etalon is much less than the linewidth, i.e.,

$$1/\Delta f_e^2 > g_0/\Delta f^2.$$

The pulsewidth now becomes:

$$\tau_{p0}(\text{FM}) \simeq \frac{\sqrt{2\sqrt{2}\ln 2}}{\pi}\left(\frac{1}{\delta_c}\right)^{1/4}\left(\frac{1}{f_m\,\Delta f_e}\right)^{1/2}, \quad (71)$$

and we note that the pulsewidth is now entirely determined by the etalon and modulator.

We can consider the same Nd:YAG laser as before with an uncoated quartz etalon of thickness h. The pulsewidths that can be obtained from this system as a function of h are shown in Fig. 8. Note that the bandwidth of the etalon can be written in terms of the index of refraction and the thickness

$$\Delta f_e = (\sqrt{8}c/\pi h)n/(n^2 - 1). \tag{72}$$

Substituting in (71) and evaluating the constants, we get

$$\tau_p = 150\sqrt{h}/(\delta_c)^{1/4} \text{ ps}.$$

From Fig. 8 we note that for etalon thicknesses between 1 mm and 1 cm, we obtain good control of the pulsewidth. For etalon thickness less than 1 mm, the linewidth of the active medium begins to take over, and thus there is no advantage in using an etalon less than 1 mm thick inside the Nd:YAG laser. Another advantage of the etalon is that it usually improves the stability of the mode-locked laser.

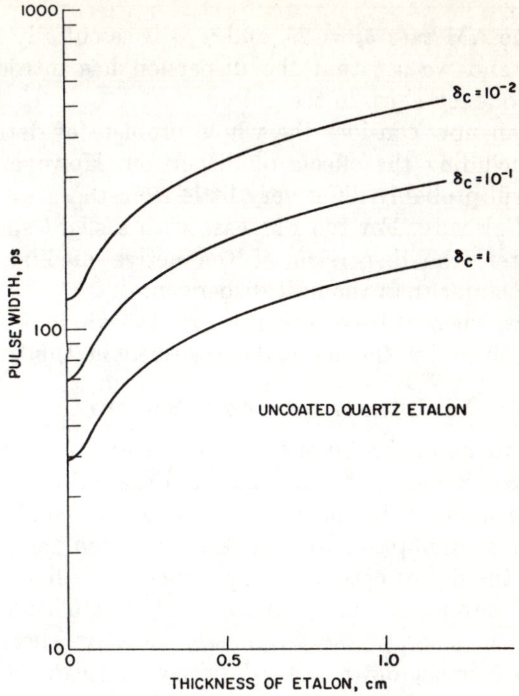

Fig. 8. Pulsewidth in FM mode-locked laser with uncoated quartz etalon.

X. Effects of Host Medium Dispersion and Dispersion of Other Components

The index of refraction of the host medium or any other optical components in the cavity such as the modulator crystal can be written as

$$n = n + n'(\omega - \omega_a) + (n''/2)(\omega - \omega_a)^2 \cdots. \tag{73}$$

For YAG at 1.064 μ, $n' = 12 \times 10^{-17}$ and $n'' \simeq 2 \times 10^{-33}$ [23] and for $LiNbO_3$ (modulator crystal) $n_e' = 3.1 \times 10^{-17}$ and $n_e'' \simeq 0$, $n_0' = 3.8 \times 10^{-17}$ and $n_0'' \simeq 0$ [24].

The total optical length now becomes

$$L_{opt} = L_0 + L_0'(\omega - \omega_a) + L_0''(\omega - \omega_a)^2 \cdots, \tag{74}$$

where

$$L_0 = l_{cav} + (n_1 - 1)l_{c1} + (n_2 - 1)l_{c2} + \cdots$$
$$L_0' = n_1'l_{c1} + n_2 l_{c2} + \cdots$$
$$L_0'' = (n_1''/2)l_{c1} + (n_2''/2)l_{c2} + \cdots.$$

L_0 is just the optical length we considered before. The other two terms now introduce an additional phase shift per round trip given by

$$(\omega/c)[2L_0'(\omega - \omega_a) + 2L_0''(\omega - \omega_a)^2]$$
$$\simeq (\omega_a/c)[2L_0'(\omega - \omega_a) + 2L_0''(\omega - \omega_a)^2].$$

We can add this to the active medium and obtain an effective gain

$$g_e(\omega) = g \exp - \left[2jg'\left(\frac{\omega - \omega_a}{\Delta\omega}\right) - 4g''\left(\frac{\omega - \omega_a}{\Delta\omega}\right)^2\right], \tag{75}$$

where

$$g' = g + (L_0'\omega_a \Delta\omega/c)$$
$$g'' = g + j(L_0''\omega_a \Delta\omega^2/2c).$$

We can now consider the mode-locked laser without detuning. The results we obtained previously are still valid, with g' and g'' replacing g in the appropriate places. In particular, the FM modulation frequency becomes

$$f_{m0}(\text{FM}) = 1\bigg/\left[2L_0/c \pm 2\left(\frac{\delta_c}{\pi}\right)\frac{\lambda_a}{c} + \frac{2g_0}{\Delta\omega} + \frac{2L_0'\omega_a}{c}\right], \tag{76}$$

and a similar result is obtained for AM. This is not a very important change; however, the factor γ as given by (29) becomes

$$\gamma = (\omega_m \Delta\omega/4)\sqrt{\delta_g\bigg/\left(g_0 + j\frac{L_0''\omega_a \Delta\omega^2}{2c}\right)}. \tag{77}$$

We will usually have that

$$g_0 > L_0''\omega_a\Delta\omega^2/2c$$

and γ can be written as

$$\gamma \simeq \frac{\omega_m \Delta\omega}{4}\sqrt{\frac{\delta_g}{g_0}}\left(1 - j\frac{L_0''\omega_a \Delta\omega^2}{4g_0 c}\right). \tag{78}$$

For the FM case, $\delta_g = \mp j\delta_c$ and for γ_{FM} we get

$$\gamma_{FM} = \frac{\omega_m \Delta\omega}{4}\sqrt{\frac{\delta_c}{2g_0}}\left[\left(1 \mp \frac{L_0''\omega_a \Delta\omega^2}{4g_0 c}\right) \mp j\left(1 \pm \frac{L_0''\omega_a \Delta\omega^2}{4g_0 c}\right)\right]. \tag{79}$$

We note that α_{FM} and β_{FM} now have different values for the positive and negative modes and that for the positive mode the dispersion has caused some pulse stretching, while for the negative mode there is some pulse compression. Thus the dispersion is another effect that tends to lift the degeneracy of the two modes.

For the AM case $\delta_g = 2\delta_l$ and γ_{AM} is essentially given by (77) and we see that the dispersion has introduced some frequency chirp in the pulse.

One can now consider the whole problem of detuning again, including the effects of dispersion. However, the results will probably differ very little from those we have obtained already. For the FM case with a slight shift off line center, the dispersion of the active medium will be much larger than the host dispersion.

In cases where it becomes necessary, the whole problem can be solved by the methods described in this paper.

XI. Limitations of the Theory

The two main limitations of the theory are that we assume the line shape is Gaussian, and the pulse is Gaussian with a linear frequency chirp under all conditions.

The first assumption will break down when the bandwidth of the pulses becomes comparable to the linewidth. This will happen as one goes to higher modulation frequencies to generate shorter pulses. Exactly where the theory will break down is hard to say and can best be determined experimentally.

The second assumption may seriously affect some of the results, particularly the behavior of the FM mode-locked laser with detuning. Small distortions in the pulse shape may change the results considerably. However, many of the characteristics described here have been observed in a Nd:YAG laser with a FM modulator, and these results will be presented in Part II of this paper.

Appendix I

Gain and Saturation of Homogeneous Line

In this appendix we consider the basic equations for a homogeneously broadened line, and from these equations we derive an expression for gain of the medium. We also investigate how the line saturates for some general signal. In particular, we consider how the line is saturated by Gaussian pulses with a frequency not necessarily at line center.

The equations for a homogeneously broadened line are given by [25]

$$\ddot{\mathbf{P}} + \frac{2}{T_2}\dot{\mathbf{P}} + \omega_a^2\mathbf{P} = -\frac{2\omega_a L\,|\boldsymbol{\mu}|^2}{3\hbar} N\mathbf{E} \tag{80}$$

$$\dot{N} + \frac{N - N_e}{T_1} = \frac{2}{\hbar\omega_a}\dot{\mathbf{P}}\cdot\mathbf{E}, \tag{81}$$

where $\mathbf{P}$ is the polarization of the medium, $\mathbf{E}$ the electric field of the applied signal, ω_a the center frequency of the line, $|\boldsymbol{\mu}|$ the dipole matrix element, L the Lorentz correction factor to relate local fields to macroscopic fields, and N the population inversion $(N_2 - N_1)$. The full atomic linewidth is given by $\Delta\omega = 2/T_2$. N_e is the population inversion with no applied signal and is proportional to the pump rate. Since $\mathbf{P}$ and $\mathbf{E}$ are polarized in the same direction, we can drop the vector notation.

When a laser is above threshold, the population inversion has a constant value N_0 and a time-dependent component $\Delta N(t)$ that depends on the changes in E and P. We will assume that this component $\Delta N(t)$ is negligible compared to N_0, and hence that N is constant. With this approximation, we neglect changes in the pulse shape of a mode-locked laser due to saturation during the pulse, and also such effects as π pulses [26].

We now consider an electric field $E(t)$ with a Fourier transform $E(\omega)$. Hence we get

$$E(t) = \int_{-\infty}^{\infty} E(\omega)\exp(j\omega t)\,d\omega. \tag{82}$$

Similarly for the polarization

$$P(t) = \int_{-\infty}^{\infty} P(\omega)\exp(j\omega t)\,d\omega. \tag{83}$$

We can substitute in (80), and since N is constant, (80) is linear, and we can get an expression for the susceptibility $\chi(\omega)$, where $P(\omega) = \epsilon_0\chi(\omega)E(\omega)$

$$\chi(\omega) = -\frac{2\omega_a L\,|\boldsymbol{\mu}|^2 N}{3\hbar\epsilon_0}\left[\frac{1}{(\omega_a^2 - \omega^2) + j\,\Delta\omega\,\omega}\right]. \tag{84}$$

We now consider the propagation of an electromagnetic wave through a medium with a susceptibility given by (84). If we neglect losses in the medium, the propagation is governed by Maxwell's equation in the following form

$$\frac{\partial^2 E}{\partial z^2} + \frac{n^2}{c^2}\frac{\partial^2 E}{\partial t^2} = -\mu_0\frac{\partial^2 P}{\partial t^2}, \tag{85}$$

where n is the index of refraction of the medium.

The propagation constant β is defined such that the total field $E(t)$ at point z is given by

$$E(t) = \int_{-\infty}^{\infty} E(\omega)\exp[i(\omega t - \beta z)]\,d\omega. \tag{86}$$

It can be shown [25] that the propagation constant β is given by

$$\beta = \frac{n\omega}{c} + \frac{\chi(\omega)\omega}{2nc}. \tag{87}$$

Considering only the effects of the active medium, i.e., the second term in (87), the gain due to the active medium is given by

$$g_c(\omega) = \exp\left\{-i\left[\frac{\chi(\omega)\omega}{2nc}\right]L_c\right\}, \tag{88}$$

where L_c is the length of the active medium.

If we further make the approximation for $\chi(\omega)$ that $(\omega^2 - \omega_a^2) \simeq 2\omega(\omega - \omega_a)$, it follows that g in (7) is given by

$$g = \left[\frac{L\,|\boldsymbol{\mu}|^2\,\omega_a}{3\hbar\epsilon_0 nc\,\Delta\omega}\right]NL_c. \tag{89}$$

We next consider how the homogeneously broadened line is saturated by a repetitive train of mode-locked pulses. The spectrum of the mode-locked pulses is now no longer narrow compared to the linewidth, and hence we must consider how this changes the expression for the saturation of a homogeneous line obtained by several others [25], [27], [28].

Since we have assumed that N is constant, (81) becomes

$$\frac{N - N_e}{T_1} = \frac{2}{\hbar\omega_a}\overline{\dot{P}(t)E(t)}, \tag{90}$$

and we have taken the average over $\dot{P}(t)E(t)$. When we have a periodic signal, such as a repetitive train of pulses with a periodicity T_m, then (90) becomes

$$\frac{N - N_e}{T_1} = \frac{2}{\hbar\omega_a}\frac{1}{T_m}\int_{-\infty}^{\infty}\dot{P}(t)E(t)\,dt, \tag{91}$$

where $P(t)$ and $E(t)$ are now for a single pulse.

Using the power theorem [16] for Fourier transforms, and noting that $F[\dot{P}(t)] = i\omega F[P(t)]$ where $F[\ ]$ indicates the Fourier transform of the function in parentheses, (90) can be written as

$$\frac{N - N_e}{T_1} = \frac{2}{\hbar\omega_a T_m}\frac{1}{2\pi}\int_{-\infty}^{\infty} i\omega F[P(t)]\cdot F[E(-t)]\,d\omega. \tag{92}$$

Substituting (81) and (82) we get

$$\frac{N - N_e}{T_1} = \frac{2}{\hbar\omega_a T_m}\frac{1}{2\pi}\int_{-\infty}^{\infty} i\omega\chi(\omega)E(\omega)E(-\omega)\,d\omega. \tag{93}$$

From (84) we note that the real part of $\chi(\omega)$, $\chi'(\omega)$ is even in ω, while the imaginary part $\chi''(\omega)$ is odd. Hence (93) can finally be written in the form

$$\frac{N - N_e}{T_1} = -\frac{2}{\hbar\omega_a T_m}\frac{1}{\pi}\int_0^{\infty}\omega\chi''(\omega)E(\omega)E(-\omega)\,d\omega. \tag{94}$$

We can now substitute for $\chi''(\omega)$ and solve for N

$$N = N_e\Big/\left[1 + \frac{4L\,|u|^2\,T_1}{3\hbar^2\epsilon_0\,\Delta\omega T_m}\frac{1}{\pi}\int_0^{\infty}\frac{E(\omega)E(-\omega)\,d\omega}{1 + 4[(\omega - \omega_a)/\Delta\omega]^2}\right] \tag{95}$$

and we have again made the approximation that $(\omega^2 - \omega_a^2) \simeq 2\omega(\omega - \omega_a)$.

The average power density inside the laser cavity is given by

$$I = n\epsilon_0 c\frac{1}{T_m}\int_{-\infty}^{\infty}|E(t)|^2\,dt = \frac{n\epsilon_0 c}{T_m\pi}\int_0^{\infty}E(\omega)E(-\omega)\,d\omega \tag{96}$$

and if we define the saturation power density I_s [25] by

$$I_s = \frac{3n\epsilon_0^2 c\hbar^2\,\Delta\omega}{4L\,|\mu|^2\,T_1}, \tag{97}$$

then (95) can finally be written as

$$N = N_e\Big/\left[1 + \frac{n\epsilon_0 c}{I_s}\frac{1}{T_m\pi}\int_0^{\infty}\frac{E(\omega)E(-\omega)}{1 + 4[(\omega - \omega_a)/\Delta\omega]^2}\right]. \tag{98}$$

We can now consider the special case of a Gaussian pulse given by

$$E(t) = E_0\exp(-\alpha t^2)\cos(\omega_p t + \beta t^2). \tag{99}$$

It can be shown that

$$E(\omega) = \frac{E_0}{2}\sqrt{\frac{\pi}{\gamma}}\exp[-(\omega - \omega_p)^2/4\gamma] + \frac{E_0^*}{2}\sqrt{\frac{\pi}{\gamma^*}}\exp[-(\omega + \omega_p)^2/4\gamma^*] \tag{100}$$

and substituting in (98) and (97), we get

$$N = N_e\Big/\left[1 + \frac{I}{I_s}\sqrt{\frac{\alpha}{2(\alpha^2 + \beta^2)}}\frac{1}{\sqrt{\pi}}\cdot\int_0^{\infty}\frac{\exp[-(\omega - \omega_p)^2\alpha/(\alpha^2 + \beta^2)]}{1 + 4[(\omega - \omega_a)/\Delta\omega]^2}\,d\omega\right]. \tag{101}$$

The integral has no exact analytical solution. However, with the assumption that the spectral width of the pulse is small compared to the full linewidth, we can expand the Lorentzian line shape about ω_p. If we let $\omega_a = \omega_p + \nu$ and $\nu/\Delta\omega = \eta$, it follows that

$$\frac{1}{1 + 4[(\omega - \omega_a)/\Delta\omega]^2} \simeq \frac{1}{(1 + 4\eta^2)}\left[1 + \frac{8\eta}{1 + 4\eta^2}\cdot\left(\frac{\omega - \omega_p}{\Delta\omega}\right) - \frac{4}{1 + 4\eta^2}\left(\frac{\omega - \omega_p}{\Delta\omega}\right)^2\right]. \tag{102}$$

From (111) we now get

$$N \simeq N_e\Big/\left\{1 + \frac{I}{I_s}\left[\left(\frac{1}{1 + 4\eta^2}\right)\left(1 - \left(\frac{\Delta f_p}{\Delta f}\right)^2\cdot\frac{1}{(1 + 4\eta^2)(2\ln 2)}\right)\right]\right\}, \tag{103}$$

where Δf_p is the spectral width of the pulse. The function in the square brackets shows how the line saturates, and we will call this the saturation function $S(\eta, \Delta f_p)$.

From (113) it follows that the laser output power (P) is given by

$$P = P_s\left[\frac{N_e}{N} - 1\right]\frac{1}{S(\eta, \Delta f_p)}. \tag{104}$$

When the laser is free running, $S(\eta, \Delta f_p) = 1$ and N_e/N is equal to the ratio of the pump power to the pump power at threshold R_p. We have shown that the gain through the active medium is proportional to the population inversion N, and if the gain of the free-running laser is g_f, then (114) becomes

$$P = P_s\left[R_p\left(\frac{g_f}{g}\right) - 1\right]\left[\frac{1}{S(\eta, \Delta f_p)}\right]. \tag{105}$$

Finally, the ratio of the output power for any two conditions is given by

$$\frac{P_1}{P_2} = \frac{[R_p(g_f/g_1) - 1]S(\eta_2, \Delta f_{p2})}{[R_p(g_f/g_2) - 1]S(\eta_1, \Delta f_{p1})}. \tag{106}$$

This equation will give the variation in output power for any condition of the mode-locked laser.

Appendix II

Etalon Effects

The amplitude transmission through a lossless Fabry–Perot etalon is given by [29]

$$\tau_e = (1 - R)/(1 - Re^{i\delta}), \tag{107}$$

where

$$\delta = (4\pi/\lambda)nh \cos \theta \tag{108}$$

and R is the reflection of both reflecting surfaces. The effective length of the etalon is given by $h' = nh \cos \theta$. We will consider the transmission of the etalon near a particular transmission peak at ω_e. From (118) we get

$$\delta = \omega_e(2h'/c) + (\omega - \omega_e)(2h'/c). \tag{109}$$

Since ω_e is at the transmission peak, $\omega_e(2\hbar'/c) = 2m\pi$, and hence (117) can be written as

$$\tau_e = \frac{1 - R}{1 - R \exp [i(\omega - \omega_e)(2\hbar'/c)]}. \tag{110}$$

We want to approximate the transmission of the etalon near ω_e by a Gaussian of the form $\exp - [ib(\omega - \omega_e) + a(\omega - \omega_e)^2]$. Expanding the exponential in (120), we get

$$\tau_e \simeq 1 \Bigg/ \left[1 - i\left(\frac{R}{1 - R}\right)\left(\frac{2h'}{c}\right)(\omega - \omega_e) + \frac{1}{2}\left(\frac{R}{1 - R}\right)\left(\frac{2h'}{c}\right)^2(\omega - \omega_e)^2 + \cdots\right]. \tag{111}$$

Expanding the Gaussian

$$\tau_e \simeq 1/[1 + ib(\omega - \omega_e) + [a - (b^2/2)](\omega - \omega_e)^2] + \cdots. \tag{112}$$

Equating coefficients, we get

$$b = -R/(1 - R)(2h'/c)$$
$$a = \tfrac{1}{2} R/(1 - R)^2(2h'/c)^2. \tag{113}$$

Finally, we can write the transmission of the etalon as

$$\tau_e \simeq \exp \left[i\sqrt{8}\, R\left(\frac{\omega - \omega_e}{\Delta\omega_e}\right) - 4\left(\frac{\omega - \omega_e}{\Delta\omega_e}\right)^2\right], \tag{114}$$

where $\Delta\omega_e$ is the effective bandwidth of the etalon, given by

$$\Delta\omega_e = (c/2h)(\sqrt{8/R})(1 - R). \tag{115}$$

References

[1] M. DiDomenico, Jr., "Small signal analysis in internal (coupling type) modulation of lasers," *J. Appl. Phys.*, vol. 35, pp. 2870–2876, October 1964.

[2] A. Yariv, "Internal modulation in multimode laser oscillations," *J. Appl. Phys.*, vol. 36, pp. 388–391, February 1965.

[3] M. H. Crowell, "Characteristics of mode-coupled lasers," *IEEE J. Quantum Electron.*, vol. QE-1, pp. 12–20, April 1965.

[4] S. E. Harris and O. P. McDuff, "Theory of FM laser oscillation," *IEEE J. Quantum Electron.*, vol. QE-1, pp. 245–262, September 1965.

[5] O. P. McDuff and S. E. Harris, "Nonlinear theory of the internally loss-modulated laser," *IEEE J. Quantum Electron.*, vol. QE-3, pp. 101–111, March 1967.

[6] E. O. Ammann, B. J. McMurtry, and M. K. Oshman, "Detailed experiments on He–Ne FM lasers," *IEEE J. Quantum Electron.*, vol. QE-1, pp. 263–272, September 1965.

[7] a. L. M. Osterink and R. Targ, "Single frequency light from an argon FM laser," *Appl. Phys. Lett.*, vol. 10, pp. 115–117, February 1967.
b. L. M. Osterink, "The argon laser with internal phase modulation: Its use in producing single frequency light," Microwave Laboratory, Stanford University, Stanford, Calif., Rept. 1536, May 1967.

[8] M. DiDomenico, Jr., J. E. Geusic, H. M. Marcos, and R. G. Smith, "Generation of ultrashort optical pulses by mode-locking the YAIG:Nd laser," *Appl. Phys. Lett.*, vol. 8, pp. 180–183, April 1966.

[9] L. M. Osterink and J. D. Foster, "A mode-locked Nd:YAG laser," *J. Appl. Phys.*, vol. 39, pp. 4163–4165, August 1968.

10] T. Deutsch, "Mode-locking effects in an internally-modulated ruby laser," *Appl. Phys. Lett.*, vol. 7, pp. 80–82, August 1965.

[11] R. H. Pantell and R. L. Kohn, "Mode coupling in a ruby laser," *IEEE J. Quantum Electron.*, vol. QE-2, pp. 306–310, August 1966.

[12] H. Haken and M. Pauthier, "Nonlinear theory of multimode action in loss modulated lasers," *IEEE J. Quantum Electron.*, vol. QE-4, pp. 454–459, July 1968.

[13] A. E. Siegman and D. J. Kuizenga, "Simple analytic expressions for AM and FM mode-locked pulses in homogeneous lasers," *Appl. Phys. Lett.*, vol. 14, pp. 181–182, March 1969.

[14] C. C. Cutler, "The regenerative pulse generator," *Proc. IRE*, vol. 43, pp. 140–148, February 1955.

[15] J. B. Gunn, "Spectrum and width of mode-locked laser pulses," *IEEE J. Quantum Electron.*, vol. QE-5, pp. 513–516, October 1969.

[16] R. Bracewell, *The Fourier Transform and Its Applications.* New York: McGraw-Hill, 1965.

[17] a. G. E. Geusic, H. M. Marcos, and L. G. Van Uitert, "Laser oscillations in Nd-doped yttrium aluminum, yttrium gallium and gadolinium garnets," *Appl. Phys. Lett.*, vol. 4, pp. 182–184, May 1964.
b. G. E. Geusic, "The continuous Nd:YAG laser," *IEEE J. Quantum Electron.* (Abstracts), vol. QE-3, p. 17, April 1966.

[18] S. E. Harris and R. Targ, "FM oscillations of the He–Ne laser," *Appl. Phys. Lett.*, vol. 5, pp. 202–204, November 1965.

[19] a. D. J. Kuizenga and A. E. Siegman, "FM laser operation of the Nd:YAG laser," this issue, pp. 673–677.
b. ——, "FM-phase-locked and FM-laser operation of Nd:YAG lasers," *J. Opt. Soc. Am.*, vol. 59, p. 506, abstract ThF 21, April 1969.

[20] G. A. Giordmaine, M. A. Duguay, and J. W. Hansen, "Compression of optical pulses," *IEEE J. Quantum Electron.*, vol. QE-4, pp. 252–255, May 1968.

[21] H. Manger and H. Rothe, "Selection of axial modes in optical masers," *Phys. Lett.*, vol. 7, pp. 330–331, December 1963.

[22] H. G. Danielmeyer, "A stabilized efficient single frequency Nd:YAG laser" (to be published).

[23] W. L. Bond, "Measurement of the refractive indices of several crystals," *J. Appl. Phys.*, vol. 36, pp. 1674–1677, May 1965.

[24] G. D. Boyd, W. L. Bond, and H. L. Carter, "Refractive index as a function of temperature in $LiNbO_3$," *J. Appl. Phys.*, vol. 38, pp. 1941–1943, March 1967.

[25] R. H. Pantell and H. E. Puthoff, *Fundamentals of Quantum Electronics.* New York: Wiley, 1969.

[26] a. G. A. Armstrong and E. Courtens, "Exact solution of a π-pulse problem," *IEEE J. Quantum Electron.*, vol. QE-4, pp. 411–419, June 1968.
b. ——, "Pi-pulse propagation in the presence of host dispersion," *IEEE J. Quantum Electron.*, vol. QE-5, pp. 249–259, May 1969.

[27] A. Yariv, *Quantum Electronics.* New York: Wiley, 1967.

[28] A. E. Siegman, *Introduction to Lasers and Masers,* preliminary ed. New York: McGraw-Hill, 1968.

[29] M. Born and E. Wolf, *Principles of Optics.* New York: Pergamon, 1959.

FM and AM Mode Locking of the Homogeneous Laser—Part II: Experimental Results in a Nd:YAG Laser With Internal FM Modulation

DIRK J. KUIZENGA, MEMBER, IEEE, AND A. E. SIEGMAN, FELLOW, IEEE

Abstract—In Part I of this paper [1], a theoretical analysis of the mode-locked homogeneous laser was given. In this part we present experimental results for the Nd:YAG laser with internal phase modulation. $LiNbO_3$ was used as the modulator crystal, and a method to measure the single-pass phase retardation of the modulator accurately at 1.06 μ is described in detail. The pulsewidth and spectral width of the mode-locked laser were measured as a function of depth of modulation, and good agreement with theory was obtained. Etalon effects in the laser were observed, and the results agreed very well with the theory. Mode-locked spectral bandwidths of up to 16 GHz, implying mode-locked pulses as short as 40 ps, were obtained.

I. Introduction

In Part I of this paper [1], we presented a simple theory for the mode-locked homogeneously broadened laser. In this part, we present some experimental results for a mode-locked Nd:YAG laser with internal phase (FM) modulation. Mode locking was obtained in the Nd:YAG laser in the same way as described by Osterink and Foster [2] using $LiNbO_3$ as a phase modulator. Mode locking with an amplitude modulator (AM) was not investigated, although this has been reported by Di Domenico *et al.* [3].

In this paper we first discuss some of the relevant construction details of the Nd:YAG laser and the FM modulator. In particular, the measurement of the depth of modulation (δ_c) of the modulator is considered in detail. The experimental techniques to measure the mode-locked pulses are described. In the experiments, pulses from 40 to 200 ps were obtained, in general with very good agreement between theory and experiment. Experimentally we found that the pulsewidth (τ_p) and bandwidth (Δf_p) of the mode-locked laser were given by $\tau_p = 46/(\delta_c)^{1/4}$ ps and $\Delta f_p = 17.6(\delta_c)^{1/4}$, where δ_c is the single-pass-phase retardation of the FM modulator. For our laser parameters, the theory predicts $\tau_p = 37.9/(\delta_c)^{1/4}$ and $\Delta f_p = 16.5(\delta_c)^{1/4}$. Etalon effects in the cavity were also observed and studied with good agreement between theory and experiment again being obtained.

Manuscript received February 18, 1970; revised May 18, 1970. This work was supported by the U. S. Air Force Office of Scientific Research under Contracts AF 49(638)-1525 and F44620-69-C-0017.

The authors are with the Microwave Laboratory and Department of Electrical Engineering, Stanford University, Stanford, Calif. 94305.

II. Experimental Setup

A schematic of the experimental setup is shown in Fig. 1. We will consider some of the components in more detail.

A. *Nd:YAG Laser*

In our laser the light from the water-cooled linear arc lamps was focused on a water-cooled Nd:YAG rod by a double elliptical pump cavity. Great care was taken to obtain a stable pump cavity and particularly to minimize the vibrations of the Nd:YAG rod due to the water cooling. The Nd:YAG rod was mounted in tapered metal tubes (brass or stainless steel) with solder seals to ensure smooth water flow. The pump cavity was made as small as possible, resulting in a compact, stable laser. Thermal stability was obtained by using a thick invar base on which the pump cavity and mirror holders were mounted.

Two different pump cavities were used in the experiments. In the first cavity, a 5-cm-long by 3-mm-diameter laser rod was pumped with two krypton arc lamps [4] designed for high power output (3 kW). At lower pump levels, these lamps were somewhat unstable due to wandering of the arc at the cathode, causing some variations in the output power of the laser. The second cavity had a 7.5 cm by 3 mm crystal and two xenon lamps. This laser pump cavity was very stable.

The laser optical cavity consisted of a high reflecting flat mirror, and a long-radius (3 or 6 meters) output mirror with 3.5 percent transmission. The laser was linearly polarized by an internal Brewster flat. Although such a Brewster flat is usually a very low loss element, insertion of this element in the Nd:YAG laser caused considerable power loss, due to pump-induced birefringence in the laser rod [5]. Typical power loss was from 30 to 50 percent, but this was pump power dependent, and at higher pump levels the loss in power was worse. Transverse mode control was obtained with an internal iris in the laser cavity, again with considerable loss in laser output power. Due to pump-induced birefringence in the rod, only the small center section of the rod could be used to get TEM_{00} mode power. The maximum linearly polarized TEM_{00} mode power available from our laser was limited to about 1 watt, and did not substantially increase at higher pump levels. Several methods have been proposed to compensate for the

Reprinted from *IEEE J. Quantum Electron.*, vol. QE-6, pp. 709–715, Nov. 1970.

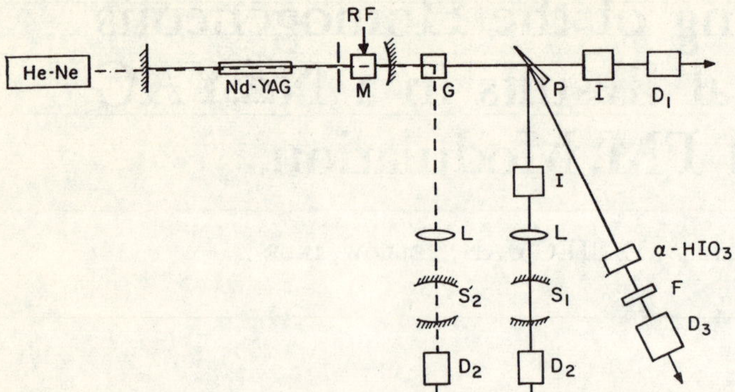

Fig. 1. Schematic of experimental setup.

thermal birefringence to increase the amount of TEM_{00} mode power [6].

The mode-locked laser was very sensitive to any small external reflections back into the laser. All measuring equipment, such as the scanning Fabry–Perot interferometer and the fast pin diode, had to be carefully isolated from the laser with attenuators, beam splitters, and polarization isolators consisting of a polarizer and quarter-wave plate.

B. Modulator

A $LiNbO_3$ crystal was used as a phase modulator. When a voltage is applied along the z axis of the crystal, the indices of refraction for light propagating perpendicular to the z axis and polarized along the z axis or perpendicular to the z axis are given by:

$$n_{\parallel} = n_e - \tfrac{1}{2} r_{33} n_e^3 E_z \qquad (1)$$
$$n_{\perp} = n_0 - \tfrac{1}{2} r_{13} n_0^3 E_z.$$

For $LiNbO_3$ the measured electrooptic coefficients at 6328 Å are [7]

$$r_{33} = 30.8 \times 10^{-10} \text{ cm/V} \qquad (2)$$
$$r_{13} = 8.6 \times 10^{-10} \text{ cm/V}.$$

Since $r_{33} > r_{13}$, the optimum use of $LiNbO_3$ as a phase modulator is with the laser polarized along the z axis of the crystal. In this configuration the peak single-pass-phase retardation is

$$\delta_m = \frac{\pi r_{33} n_e^3 V_z}{\lambda} \left(\frac{l}{d}\right), \qquad (3)$$

where l is the distance the light travels in the crystal, d the thickness of the crystal in the z direction, and V_z the voltage along the z direction. Note that rotating the crystal around the z axis does not change the modulation characteristics of the crystal, except for a possible small change in l.

A 1 cm × 4 mm × 4 mm x-cut antireflection (AR)-coated $LiNbO_3$ crystal was used in the modulator. The crystal was clamped between brass electrodes and resonated with a lumped inductance at 260 MHz and matched into a 50-ohm cable. The modulator had a loaded $Q \simeq 170$. The modulation index δ_m was measured by putting a single-axial-mode He–Ne laser beam through the modulator so that the 6328-Å signal coming out was phase modulated. The Bessel-function sidebands could be measured with a scanning Fabry–Perot (SFP) [8], and hence δ_m at 6328 Å could be measured. Taking into account the change in wavelength and index of refraction, and assuming that r_{33} was independent of wavelength, δ_m could be calculated at 1.06 μ. However, we had sufficient reason to doubt the last assumption particularly from the results of the FM laser [9], and hence δ_m was measured simultaneously at 6328 Å and 1.15 μ by putting both beams through the crystal and measuring δ_m at both wavelengths with two SFP. By polarizing the beams parallel or perpendicular to the z axis, the ratios of all the different electrooptic coefficients could be obtained, as given below:

$$\frac{r_{33}(1.15)}{r_{33}(6328)} = 0.905 \pm 0.012$$
$$\frac{r_{13}(1.15)}{r_{33}(6328)} = 0.236 \pm 0.004 \qquad (4)$$
$$\frac{r_{13}(1.15)}{r_{13}(6328)} = 0.860 \pm 0.01.$$

Note from this that $r_{33}(6328)/r_{13}(6328) = 3.65 \pm 0.11$, which agrees very well with the published value of 3.58 [7]. Making a linear interpolation between 6328 Å and 1.15 μ, the required ratios of r and δ for 1.06 μ and 6328 Å can be obtained. The ratios for the phase retardations are

$$\frac{\delta_{\perp}(1.06)}{\delta_{\parallel}(6328)} = 0.150 \qquad (5)$$
$$\frac{\delta_{\parallel}(1.06)}{\delta_{\parallel}(6328)} = 0.516,$$

where $\delta_{\parallel}$ and $\delta_{\perp}$ are for the laser polarization parallel and perpendicular to the z axis of the $LiNbO_3$ crystal.

It can be shown theoretically that the depth of modulation is proportional to the square root of the RF power into the modulator. This was verified experimentally, and hence the modulator could be calibrated. This gave

$$\delta_{\perp}(1.06) = 0.160 \sqrt{P} \qquad (6)$$
$$\delta_{\parallel}(1.06) = 0.552 \sqrt{P},$$

where P is the RF power in watts.

In the experimental setup, a 6328-Å He–Ne beam was continuously put through the Nd:YAG laser and intracavity modulator, with the 6328-Å output then split off with an appropriate beam splitter and put into a SFP, so that the depth of modulation could be monitored during all experiments. This provided a very accurate method for measuring δ_c.

The radiation at 1.06 μ did induce index inhomogeneities [10] in the $LiNbO_3$ crystal. To eliminate this, the crystal should be heated to 170°C. This was not done, and hence the laser output power was limited to ~300 mW to minimize this damage. At this level it took several hours to damage the crystal.

C. Pulse-Measuring Equipment

The spectrum of the mode-locked laser was observed with a scanning Fabry–Perot (SFP). The free spectral range was initially 20 GHz, and was later increased to 50 GHz. In the latter case, axial modes spaced by 260 MHz could still be resolved, and hence the SFP had a finesse of ≃200. A flat and a 46.9-cm radius of curvature mirror were used, and the laser was carefully matched into the SFP with an appropriate lens [8].

Several techniques have been developed recently to measure short optical pulses [10]–[12], but in this work only two were used, direct measurement with a fast photodiode, and enhancement in optical second-harmonic generation (SHG). A Philco L4501 diode was used with a nominal bandwidth of ~15 GHz, followed by a HP 185B sampling oscilloscope with a rise time of less than 90 ps. This detection scheme was limited by the rise time of the sampling oscilloscope. Experimentally, it was found that the detection limit for pulsewidths was ~120 ps.

It is well known that there is an increase in the average SHG power for a given fundamental input power as the fundamental changes from a steady or constant-power beam to a modulated or time-varying beam with the same average power [14], [15]. When a laser is mode locked, this enhancement in SHG can be clearly observed and is a measure of the pulsewidth. By putting a mode selecting etalon in our laser cavity, either one single axial mode or two equal axial modes could be obtained. The latter is easily obtained with an uncoated quartz flat in the cavity. If we now consider a train of Gaussian pulses as given by [1], eq. (1), the second-harmonic enhancement with the mode-locked pulses becomes

$$E_1 = \sqrt{\frac{2 \ln 2}{\pi}} \left(\frac{T}{\tau_p}\right) = 0.664\left(\frac{T}{\tau_p}\right), \tag{7}$$

relative to the single-axial-mode free-running laser or $E_2 = \frac{2}{3}E_1$ for the mode-locked case relative to two equal axial modes in the free-running laser. Here τ_p is the pulsewidth and T is the time between pulses. Note that for these two cases we need make no assumption about the phase distributions of the axial modes of the free-running laser.

III. Experimental Results

A. Initial Results

In the initial laser setup, the Nd:YAG crystal and $LiNbO_3$ crystal were very carefully aligned with all the AR-coated faces perpendicular to the laser beam. Initially no polarizing element was put in the cavity, and it was found that the laser was linearly polarized along the y axis of the $LiNbO_3$ crystal. The output power of the laser was about 300 mW, and the modulation frequency was 264 MHz.

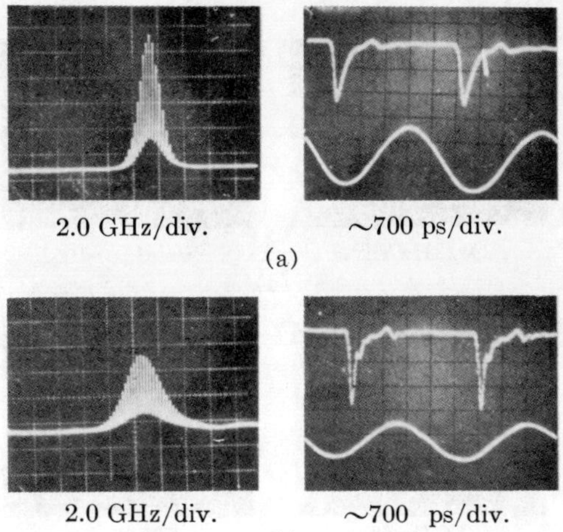

Fig. 2. Typical spectra of FM mode-locked Nd:YAG laser. (a) $\delta_c = 0.046$. (b) $\delta_c = 0.25$.

Some typical results obtained from this mode-locked laser are shown in Fig. 2. It can clearly be seen how the pulses shorten and the spectral width increases as δ_c is increased. The laser output was very stable, but the pulse would jump back and forth between the two possibilities for FM modulation, as shown in Fig. 3. The mechanism of this instability is not very well understood. A detailed investigation of this instability in the FM He–Ne laser was recently reported by Hong and Whinnery [16], but a similar investigation for the Nd:YAG laser has not yet been presented. The theory in [1] shows that these two modes have slightly different modulation frequencies, and one might expect that the laser would jump from one mode to the other as the modulation frequency is changed through zero detuning. However, this is not observed, and the instability is seen for both positive and negative detuning. Double pulsing, with two pulses per modulation cycle, was also observed, but was very unstable and not repeatable.

The depth of modulation δ_c was varied from the smallest value where the laser would just mode lock, to the maximum for the available RF power and polarization of the laser beam in the $LiNbO_3$ crystal. It was found that for δ_c less than $\sim 5 \times 10^{-3}$ (RF power ~1 mW), the pulses became unstable. For δ_c less than $\sim 1.5 \times 10^{-3}$ no more mode-locked pulses could be observed, but the laser spectrum still showed more axial modes than in the free-running laser.

Pulsewidths (τ_p) and bandwidths (Δf_p) were measured for δ_c from 5×10^{-3} to 0.25. The fast photodiode was used to measure the pulsewidth. The results are shown in Fig. 4, with τ_p and Δf_p plotted as a function of δ_c on log–log scales. The solid lines are drawn for slopes $-\frac{1}{4}$ and $+\frac{1}{4}$, and it can be seen that within experimental accuracy $\tau_p \propto (1/\delta_c)^{1/4}$ and $\Delta f_p \propto (\delta_c)^{1/4}$ as the theory predicts ([1], eq. (31) and (32)). The plot of τ_p versus Δf_p in Fig. 5, shows that the experimental points are mostly close to the line for $\tau_p \cdot \Delta f_p = 0.626$, as the theory predicts. This shows indirectly that there is a linear frequency chirp on the pulse such that $\alpha \simeq |\beta|$ ([1], eq. (1)).

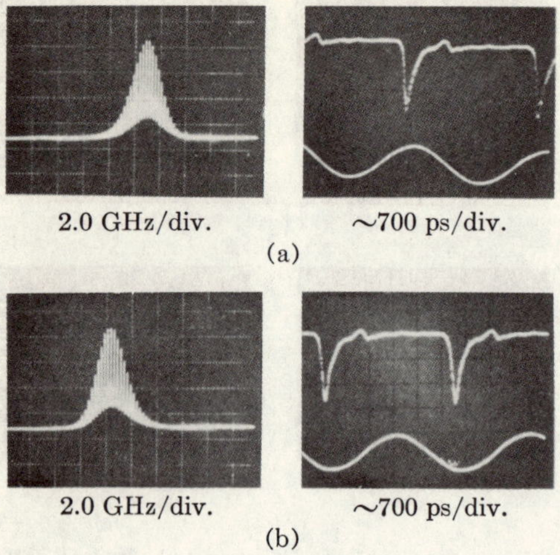

Fig. 3. Alternative pulse positions 180° apart in modulation phase in the FM mode-locked Nd:YAG laser. (a) $\delta_c = 0.133$. (b) $\delta_c = 0.133$.

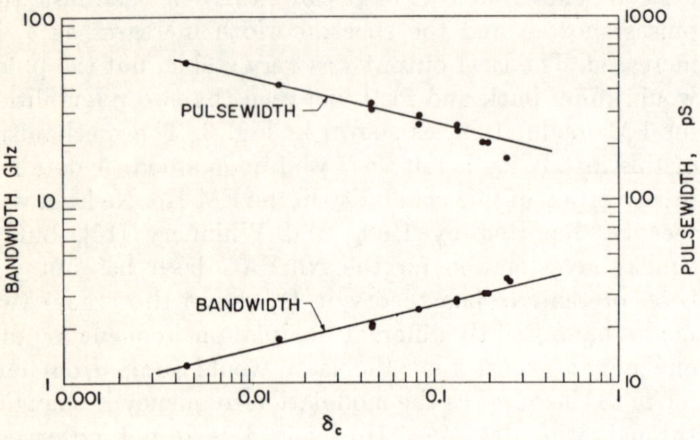

Fig. 4. Pulsewidth and bandwidth versus depth of modulation (δ_c) for laser with internal FM modulation.

In the theory, we assumed that the pulse shape was Gaussian, and consequently, the spectrum was also Gaussian. Figs. 2 and 3 show that the shape of the spectra are Gaussian-like. To test how close the spectra approach a Gaussian, consider the general Gaussian form $I_n = I_{\max} \exp -a(n - n_0)^2$. A plot of $\sqrt{\log (I_{\max}/I_n)}$ versus n is a straight line.

Several spectra over a wide range of δ_c were tested by replotting the SFP data in this way, and in all cases, the experimental points were close to a straight line. A typical result is shown in Fig. 6, which is for the spectrum in Fig. 3(a). We can thus conclude that the assumption of a Gaussian pulse is quite realistic.

We can also compare the experimental pulse length and bandwidth quantitatively with the theoretical predictions. For the measured results of Fig. 4, we get that $\tau_p = 150(1/\delta_c)^{1/4}$ ps and $\Delta f_p = 4.6(\delta_c)^{1/4}$ GHz. But, for our estimated laser parameters of 10 percent round-trip loss, a linewidth of 120 GHz, and axial mode spacing of 264 MHz, the theoretically predicted results from [1] are $\tau_p = 37.9(1/\delta_c)^{1/4}$ ps and $\Delta f_p = 16.5(\delta_c)^{1/4}$. Thus, the

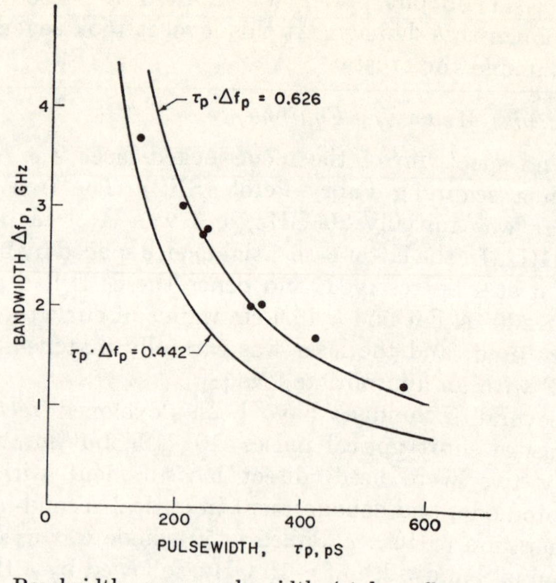

Fig. 5. Bandwidth versus pulsewidth (etalon effects present).

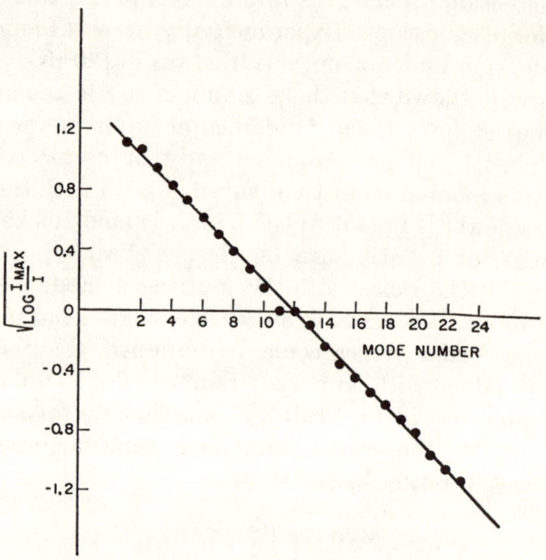

Fig. 6. Test to see if pulse spectrum is Gaussian.

pulses obtained experimentally are considerably longer than predicted by theory. This large discrepancy is due to narrowing of the effective atomic linewidth by etalon effects, as discussed in the next section.

B. Etalon Effects

The existence of mode-selecting etalon effects in our laser was evident from the SFP spectrum of the free-running laser. Fig. 7(a), which is a 10-second time exposure of the free-running laser spectrum, shows distinct clusters of axial modes spaced at ~6.6 GHz, which is approximately the $c/2nL$ for the $LiNbO_3$ crystal. It is thus obvious that the modulator crystal was primarily responsible for the axial mode selection in the free-running laser, even though this crystal was antireflection coated.

In [1] we discussed the effects of a mode-selecting etalon on the characteristics of a mode-locked laser. To

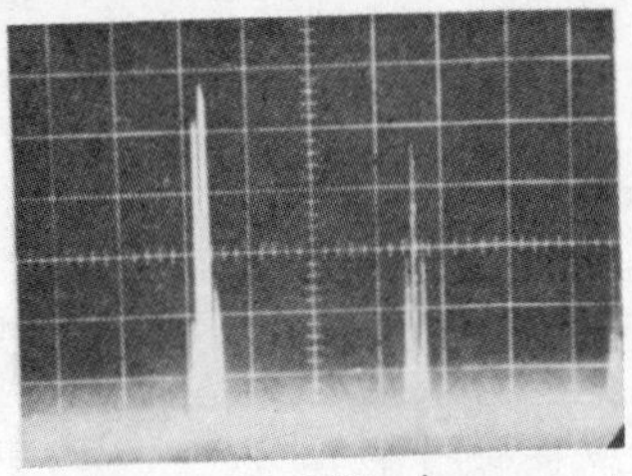

(for 1 cm $LiNbO_3$,

$\Delta f_c = \frac{c}{2nL} = 6.7$ GHz)

→| 6.6 GHz |←

(a)

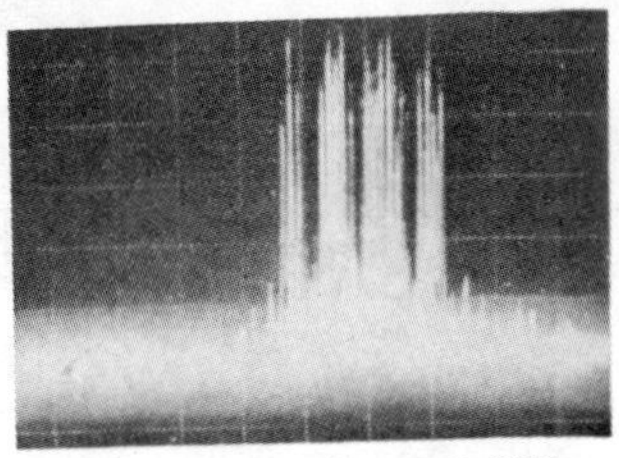

Modulator Crystal Tilted ~ 15°

(for 5 cm YAG, $\Delta f_c = 1.64$ GHz)

→| |← 1.6 GHz

→ 2 GHz/div.

(b)

Fig. 7. Spectrum of the free-running laser (10-second exposure). (a) With etalon effects from modulator crystal. (b) Without etalon effects from modulator, but with some etalon effects from Nd:YAG crystal.

apply this theory here, we have to know the reflectivity of the AR coatings on the modulator crystal. The reflectivities were not measured directly at 1.06 μ, but the wavelength for minimum reflection was found, and from the known indices of refraction and the properties of a single-layer AR coating, the reflectivities of the coatings were calculated. It was found that the coatings were centered at 0.955 and 0.872 μ for the two crystal faces, unfortunately not the same, and not at 1.064 μ. The reflectivities of the coatings at 1.064 μ were 0.5 and 1.4 percent, respectively.

The theory of etalon effects in Part I was done for equal reflecting surfaces, but it can be shown that it is still good with R replaced by the geometric mean of the two reflectivities. In this case, the transmission peak is not 100 percent and this causes somewhat higher losses in the cavity. For the above reflectivities of the coatings, $R = 0.84$ percent, and from [1], eq. (125), we get that the effective linewidth of the laser is $\Delta f_e = 33.1$ GHz. We have that $1/\Delta f_e^2 \gg g_0/\Delta f^2$ ([1], eq. (70)), and hence the pulsewidth is entirely controlled by the etalon. The pulsewidth is given by [1], eq. (71), from which we get $\tau_{p0} = 150/(\delta_c)^{1/4}$, which is exactly what we observed experimentally. While this exact agreement is probably in part fortuitous it still adds credibility to the approach we took in analyzing the etalon effects.

To obtain the shortest possible pulses, we have to eliminate all mode-selection effects, which could best be done with a Brewster-angle modulator crystal. However, the same effect can be achieved by tilting the crystal far enough so that the beams reflected from the two faces of the crystal do not overlap. A tilt of ~15° was sufficient to obtain this effect in a 1-cm crystal. All the light reflected by the crystal faces is lost, and hence this method somewhat increases the loss in the cavity. The crystal is rotated around the z axis so that its modulation properties remain unchanged.

The spectrum of the free-running laser with the modulator crystal tilted is shown in Fig. 7(b). The mode selection due to the modulator crystal has disappeared, but there is still some residual mode selection from the 5-cm-long Nd:YAG crystal. This, however, should not be troublesome. Since a typical mode-locked pulse of 150 ps is only 4.5 cm long, pulses reflected from the two faces are 10 cm apart, and hence do not overlap and can not interfere. The Nd:YAG rod is just tilted enough so that the reflected beams go out of the optical cavity.

C. *Mode-Locked Laser With No Etalon Effects*

When the laser was mode locked after the etalon effects had been removed, it was found that the spectral width of the laser increased considerably. To observe these spectra, the free spectral range of the SFP had to be increased to 50 GHz. Adjacent axial modes could still be observed. With the etalon effects removed, the laser was not very stable anymore, and at that stage a new laser pump cavity with 7.5-cm crystal was tried. The stability of this cavity was far superior to the previous one. However, the same mode-selection effects due to the Nd:YAG rod were observed in the free-running laser. A polarizing Brewster flat was now put in the cavity so that the laser was linearly polarized along the z axis of the $LiNbO_3$ crystal. This increased the maximum available depth of modulation by more than a factor of 3.

Some typical spectra for the improved mode-locked laser are shown in Fig. 8, for zero detuning of the modulator. In Part I of this paper [1] zero detuning was defined such that the pulse went through the modulator at either extremes of the phase variation. However, experimentally, the absolute phase of the pulse with respect to the modulation signal is very difficult to measure. Hence, zero detuning was inferred when the mode-locked laser spectrum was centered on the free-running spectrum. For this condition, the spectra of the two possible modes of operation of the FM mode-locked laser were approximately centered on the same frequency and the power output was a maximum. This confirmed the zero detuning condition.

The modulation index δ_c was varied from a maximum of 1.19 down to 1.75×10^{-3}. For smaller values of δ_c stable mode locking could not be obtained. Note that this depth of modulation required 10 μW of RF power. The minimum depth of modulation was about the same as for the laser with etalon effects. Figs. 8(a) and (b) are spectra of the laser when it was well mode locked. Note that these spectra are smooth, and show no modulation due to etalon effects of the Nd:YAG rod, even though these effects could be seen in the free-running laser. Output power of the laser was again limited to 300 mW TEM_{00} and remained approximately constant over the entire range of δ_c.

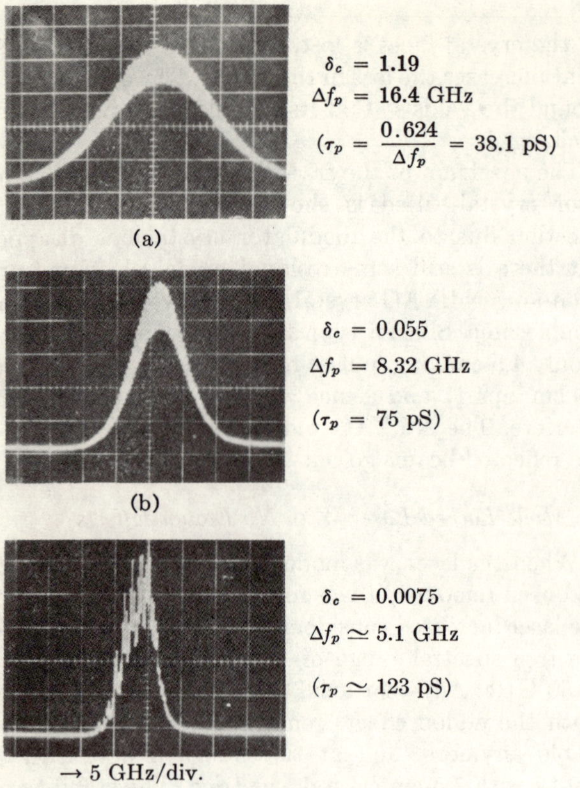

Fig. 8. Typical spectra of the mode-locked Nd:YAG laser with modulator etalon effects eliminated.

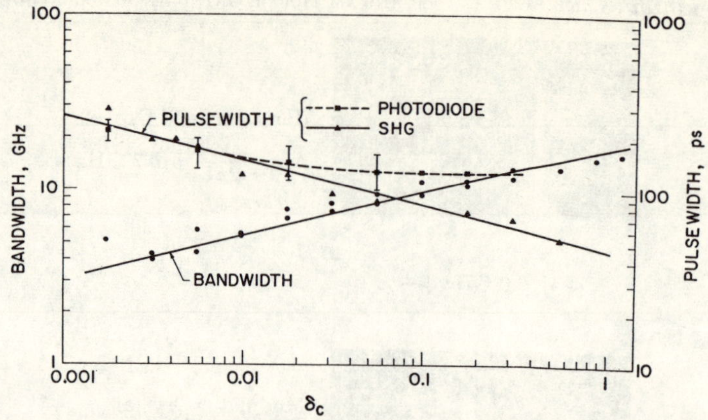

Fig. 9. Pulsewidth and bandwidth versus depth of modulation (δ_c) for laser with internal FM modulation and no etalon effects.

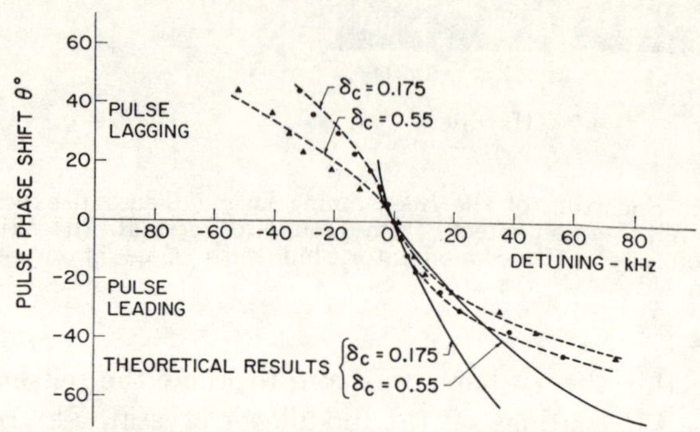

Fig. 10. Phase-shift of pulse with respect to modulation signal versus detuning. Experimental results for FM mode-locked Nd:YAG laser.

The pulsewidths were now far too short to measure directly with the photodiode, and enhancement in SHG was used to measure the pulse length. An α-iodic acid crystal was used for the SHG [17], [18]. The crystal was about 3 mm long, and without focusing the beam very tightly into the crystal, sufficient second harmonic was generated to be detected with a photodiode. Due to fluctuations in the fundamental and SHG, the enhancement in SHG could only be measured to an accuracy of $\sim\pm 20$ percent.

The bandwidth and pulsewidth were measured over the entire range of δ_c, and the results are shown in Fig. 9. The solid lines are drawn for slope $\pm\frac{1}{4}$ and we see that the experimental points are reasonably close to these lines, and hence we conclude that the dependence on δ_c is as predicted by theory. We get that $\tau_p = 46/(\delta_c)^{1/4}$ ps and $\Delta f_p = 17.6(\delta_c)^{1/4}$. The pulsewidth–bandwidth product $\tau_p \cdot \Delta f_p = 0.81$, and even allowing for experimental errors, this is larger than the 0.626 predicted by theory. This result indicates that there is a frequency chirp on the pulse, but the results are not accurate enough to calculate the magnitude of this frequency chirp precisely. The frequency chirp might be measured directly by putting the pulses through an external FM modulator such that the frequency chirp is partly or completely cancelled out [19]. The resultant narrowing of the spectrum would then give an accurate measurement of the frequency chirp. This has not been done experimentally.

The experimental values of $\tau_p = 46/(\delta_c)^{1/4}$ and $\Delta f_p = 17.6(\delta_c)^{1/4}$ are close to the theoretical values of $\tau_p = 37.9/(\delta_c)^{1/4}$ and $\Delta f_p = 16.5(\delta_c)^{1/4}$, assuming a total round-trip loss of 10 percent, a linewidth of 120 GHz and a 264-MHz modulation frequency. The total round-trip loss was not measured directly, but a 10 percent loss with all the components in the cavity was considered a reasonable estimate.

From these results, we can conclude that the theory developed in Part I [1] predicts the behavior of the mode-locked Nd:YAG quite well, both with and without etalon effects in the cavity.

The behavior of the mode-locked laser with detuning of the modulator was not investigated very extensively, and the only characteristic that was measured was the phase shift of the pulses with respect to the modulation signal. This was done by putting the modulation signal and the pulses on the x and y axes of the sampling oscilloscope, and observing the shift of the pulses. Note that the modulation signal should be the signal applied to the crystal, and not the signal going into the modulator. The modulator is a tuned circuit, and adds an additional phase shift with detuning. Experimentally, the modulation signal was obtained by putting a coupling loop close to the conductor connected to a crystal electrode, and hence obtaining the phase of the current in the resonant circuit, which has a constant phase with respect to the voltage on the crystal for small detuning.

The results of the phase measurements are shown in Fig. 10. We see that the comparison between theory and experiment is at best only fair. The experimental results do show some asymmetry with detuning, but not nearly as distinct as the theory predicts. One reason for this may be that the change in velocity of the pulse through the active medium depends very much on the detailed pulse shape and the frequency chirp on the pulse ([1], eq. (51)), and hence the Gaussian pulse with a linear frequency chirp may not be an adequate model to describe the detuning behavior of the laser. In addition to this, relaxation oscillations [2] are observed in the laser at certain positive and negative detunings, which were not considered in the theory, and this may change the behavior of the laser considerably. For very small detuning, the theoretical and experimental results are in reasonable agreement with [1], eq. (58), namely,

$$\left.\frac{d\theta}{df_m}\right|_{\theta=0} \approx \left(\frac{1}{f_{m0}}\right)^2 \frac{\pi\,\Delta f}{\sqrt{2\delta_c\, g}}. \tag{8}$$

This is an important parameter when this phase shift is used to frequency stabilize the laser [2].

Other predicted characteristics of the laser with detuning, such as the shift of the optical spectrum off the atomic line center, were also observed in passing, but not carefully measured. More experimental investigations on the details of the detuning effects are needed.

IV. Conclusion

The most important conclusion from the work presented here is that the simple theory that was derived for the mode-locked Nd:YAG laser in Part I [1] does describe the behavior of this laser very well, at least for no detuning. The simple equations for the pulsewidth can now be used to predict the behavior of the laser, particularly if the modulation frequency were to be increased by shortening the laser, or by modulating at a multiple of the basic axial mode interval. With a modulation frequency of several gigahertz, pulses on the order of 10 ps are predicted. However, the total spectral width now approaches the linewidth, and the simple theory will break down. Thus, we can conclude that much shorter pulses than described here can be generated, but that further theoretical and experimental investigations are needed in this region of operation.

Another conclusion that we can draw from the previous results is that with an appropriate etalon in the laser cavity, the pulsewidths can be controlled over a very wide range, with an upper limit set by the cavity length (so that the pulse remains short compared to the round-trip time) and the lower limit set by the linewidth of the active medium.

Acknowledgment

The authors wish to thank B. Yoshizumi, B. Griffin, and F. Futtere for their technical assistance.

References

[1] D. J. Kuizenga and A. E. Siegman, "FM and AM mode locking of the homogeneous laser—Part I: Theory," this issue, pp. 694–708.
[2] L. M. Osterink and J. D. Foster, "A mode-locked Nd:YAG laser," *J. Appl. Phys.*, vol. 39, pp. 4163–4165, August 1968.
[3] M. DiDomenico, Jr., J. E. Geusic, H. M. Marcos, and R. G. Smith, "Generation of ultrashort optical pulses by mode-locking the YAIG:Nd laser," *Appl. Phys. Lett.*, vol. 8, pp. 180–183, April 1966.
[4] T. B. Read, "The cw pumping of YAG:Nd^{3+} by water-cooled krypton arcs," *Appl. Phys. Lett.*, vol. 9, pp. 342–344, November 1966.
[5] L. M. Osterink and J. D. Foster, "Thermal effects and transverse mode control in a Nd:YAG laser," *Appl. Phys. Lett.*, vol. 12, pp. 128–131, February 1968.
[6] L. M. Osterink, J. D. Foster, G. Massey, and B. Hitz, private communication.
[7] E. H. Turner, "High-frequency electro-optic coefficients of lithium niobate," *Appl. Phys. Lett.*, vol. 8, pp. 303–304, June 1966.
[8] R. L. Fork, D. R. Herriot, and J. Kogelnik, "A scanning spherical mirror interferometer for spectral analysis of laser radiation," *Appl. Opt.*, vol. 3, pp. 1471–1484, December 1964.
[9] a. D. J. Kuizenga and A. E. Siegman, "FM laser operation of the Nd:YAG laser," this issue, pp. 673–677.
b. ——, "FM-phase-locked and FM-laser operation of Nd:YAG lasers," *J. Opt. Soc. Am.*, vol. 59, p. 506, abstract ThF 21, April 1969.
[10] A. Ashkin *et al.*, "Optically-induced refractive index inhomogeneities in $LiNbO_3$ and $LiTaO_3$," *Appl. Phys. Lett.*, vol. 9, pp. 72–74, July 1966.
[11] H. P. Weber, "Method for pulse width measurement of ultrashort light pulses generated by phase-locked lasers using nonlinear optics," *J. Appl. Phys.*, vol. 38, pp. 2231–2234, April 1967.
[12] J. A. Giordmaine, P. M. Rentzepis, S. L. Shapiro, and K. W. Wecht, "Two-photon excitation of fluorescence by picosecond light pulses," *Appl. Phys. Lett.*, vol. 11, pp. 218–220, October 1967.
[13] R. J. Harrach, "Determination of ultrashort pulse widths by two-photon fluorescence patterns," *Appl. Phys. Lett.*, vol. 14, pp. 148–151, March 1969.
[14] J. Ducuing and N. Bloemberger, "Statistical fluctuations in nonlinear optical processes," *Phys. Rev.*, vol. 133A, pp. 1493–1502, March 1964.
[15] a. R. L. Kohn and R. H. Pantell, "Second-harmonic enhancement with an internally modulated ruby laser," *Appl. Phys. Lett.*, vol. 8, pp. 231–233, May 1966.
b. R. L. Kohn, "Mode coupling effects with ruby lasers," Microwave Laboratory, Stanford University, Stanford, Calif., Rept. 1636, April 1968.
[16] G. W. Hong and J. R. Whinnery, "Switching of phase-locked states in the intracavity phase-modulated He–Ne laser," *IEEE J. Quantum Electron.*, vol. QE-5, pp. 367–376, July 1969.
[17] S. K. Kurtz, T. T. Perry, and J. G. Bergman, Jr., "Alpha-iodic acid: A solution-grown crystal for nonlinear optical studies and applications," *Appl. Phys. Lett.*, vol. 12, pp. 186–188, March 1968.
[18] R. W. Wallace, "0.473 μ light source using the second harmonic of the 0.946μ line in Nd^{3+}:YAG" (to be published).
[19] G. A. Giordmaine, M. A. Duguay, and J. W. Hansen, "Compression of optical pulses," *IEEE J. Quantum Electron.*, vol. QE-4, pp. 252–255, May 1968.

Internal Optical Parametric Oscillation

JOEL FALK, MEMBER, IEEE, J. M. YARBOROUGH, MEMBER, IEEE, AND E. O. AMMANN, MEMBER, IEEE

***Abstract*—The theory of optical parametric oscillation internal to the laser cavity is extended to include the dynamics of the population inversion of the laser medium, thus generalizing it to include all laser-oscillator systems. The equations of motion of the oscillator-laser system are solved by digital computer for the case of a *Q*-switched Nd:YAG laser with a $LiNbO_3$ parametric oscillator inside the laser cavity. It is found that this internal optical parametric oscillator operates in a spiking regime, with one or more oscillator pulses per pump pulse. The oscillator pulses inside the nonlinear crystal are often more intense than the laser pulse that would have existed in the absence of parametric oscillation. Oscillator pulse lengths are much shorter than the laser pulse length, with oscillator pulse lengths of typically 5–10 ns compared to laser pulse lengths of 200 ns. The repetition rate of the oscillator pulses is pump-power dependent, with the pulses occurring more frequently as the laser field increases. The theoretical results are compared with experiment, and the analysis is found to provide a good qualitative description of the *Q*-switched parametric oscillator.**

I. Introduction

OPTICAL parametric oscillation, first observed experimentally in 1965, has since been the subject of many theoretical and experimental investigations [1]. Until recently, studies of the optical parametric oscillator concentrated on the situation where the oscillator nonlinear crystal was placed external to the laser pump cavity. Study of the so-called internal optical parametric oscillator (IOPO), where the crystal is situated inside the laser resonator, has been relatively neglected. Oshman and Harris [2] have theoretically treated the CW pumped internal parametric oscillator, but their analysis neglects the temporal variation of the laser population difference and hence is not valid for treatment of pulsed internal parametric oscillation or discussion of a CW internal oscillator that uses a laser pump with an upper-state lifetime comparable to the parametric oscillation build-up period. Preliminary experimental work by Smith and Parker [3] on a CW IOPO verified many aspects of the Oshman–Harris theory. Ammann *et al.* [4] have experimentally studied a repetitively *Q*-switched internal oscillator for which the existing theory does not provide a totally adequate description.

The purpose of this paper is to extend the existing theory to include effects due to the temporal response of the laser population inversion, thus providing a general formulation that covers all possible internal oscillator-laser systems. Equations of motion describing the temporal behavior of a IOPO are developed in Section II. It is shown that these equations, subject to the proper assumptions, reduce to those of Oshman and Harris. We also demonstrate that the equations of motion reduce to the familiar single-mode laser rate equation when the parametric coupling is removed. The internal oscillator equations of Section II are solved by digital computer for several parameter values, and a qualitative general description of the behavior of an internal parametric oscillator is given in Section III. Comparisons of the computer solution with the results of an experimental investigation of a pulsed Nd:YAG, $LiNbO_3$ internal oscillator are made in Section IV.

The behavior predicted by our analysis and substantiated experimentally includes 1) pump-pulse stretching, 2) signal and idler pulses of substantially greater magnitude than those of the pump pulse, and 3) oscillator output occurring as a train of pulses whose repetition rate is dependent on pump electric field strength.

Manuscript received January 28, 1971; revised March 15, 1971. This work was supported by the Air Force Avionics Laboratory, Wright-Patterson Air Force Base, Ohio 45433.

The authors are with the Electro-Optics Organization, GTE Sylvania Inc., Mountain View, Calif. 94040.

Reprinted from *IEEE J. Quantum Electron.*, vol. QE-7, pp. 359–369, July 1971.

II. Equations of Motion of the Internal Optical Parametric Oscillator

A. Derivation of Equations

In this section we derive the equations describing the behavior of the IOPO. The variables of interest are the electric fields of the parametrically coupled waves and the laser population inversion. We consider electric fields at three circular frequencies satisfying the relation

$$\omega_p = \omega_s + \omega_i \tag{1}$$

where the subscripts p, s, and i denote pump, signal, and idler, respectively. We will denote the electric field amplitudes at these frequencies by $E_p(r, t)$, $E_s(r, t)$, and $E_i(r, t)$, and the population inversion per unit volume of the laser medium by N.

The case where the pump, signal, and idler are all resonated will be examined. We will assume that each field is a single spatial eigenmode of its respective cavity. We denote these eigenmodes by $E_{cp}(r)$, $E_{cs}(r)$, and $E_{ci}(r)$ and normalize the modes such that

$$\int_{\mathrm{LC}} E_{cp}^2(r)\, dV = \int_{\mathrm{OC}} E_{cs}^2(r)\, dV = \int_{\mathrm{OC}} E_{ci}^2(r)\, dV = 1 \tag{2}$$

where LC denotes an integration over the laser cavity and OC over the oscillator cavity. Under these conditions, the electric fields obey the equation

$$\ddot{E}_j(r, t) + \frac{1}{\tau_j} \dot{E}_j(r, t) + \omega_j^2 E_j(r, t) = -\frac{1}{\epsilon_j} E_{cj}(r) \int P_j(r, t) E_{cj}(r)\, dV \tag{3}$$

where $j = p, s$, or i, τ_j is the cavity lifetime, ϵ_j the dielectric constant, and $P_j(r, t)$ the total nonlinear polarization at ω_j. This equation is derived from the Maxwell traveling-wave equation and assumes that all fields exist as cavity normal modes. The integral on the right side of (3) is evaluated over the region where its integrand is nonzero.

The nonlinear driving polarizations at the signal and the idler are due to the parametric interaction, while that of the pump has contributions from both parametric and laser effects. We write these polarizations as follows

$$P_p(r, t) = P_p^L(r, t) + P_p^P(r, t) \tag{4}$$

$$P_s(r, t) = P_s^P(r, t) \tag{5}$$

$$P_i(r, t) = P_i^P(r, t) \tag{6}$$

where the superscripts L and P denote laser and parametric contributions, respectively.

The population inversion N is coupled to the pump field and the polarization term $P_p^L(r, t)$ through the laser rate equations [5]

$$\dot{N} + \frac{N - N^e}{T_1} = \frac{2}{\hbar\omega_p} P_p^L(r, t) E_p(r, t) \tag{7}$$

$$\ddot{P}_p^L(r, t) + \frac{2}{T_2} \dot{P}_p^L(r, t) + \omega_p^2 P_p^L(r, t) = -\frac{2\epsilon_0 c n \sigma}{T_2} N E_p(r, t). \tag{8}$$

T_1 is the spontaneous emission time of the upper laser level, N^e the equilibrium population inversion in the presence of pumping, T_2 the transverse relaxation time associated with the linewidth of the laser transition, σ the laser cross section, and n the index of refraction of the laser medium. Equations (3)–(8) are the basic equations from which the equations of motion for IOPO will be derived.

We will consider the oscillator configuration shown schematically in Fig. 1. To simplify the analysis, we will assume that the left-hand oscillator mirror coincides with the laser mirror and both are coated on the nonlinear crystal that extends from $z = 0$ to $z = l_c$. We take the oscillator cavity length to be l, and assume that the laser medium and Q switch completely fill the remainder of the laser cavity. Furthermore, in order to simplify the mathematics, it is assumed that all elements have the same index of refraction at the pump wavelength so that the pump propagation constant is independent of z. This equal-index assumption will lead to a calculated optical cavity length only slightly in error, and therefore should not greatly influence the results of our analysis.

All transverse variations are ignored and the electric fields that are eigenmodes[1] of their respective cavities are written

$$E_{cp}(r) = (2/LA_p)^{1/2} \cos k_p z \tag{9a}$$

$$E_{cs}(r) = (2/lA_s)^{1/2} \cos k_s z \tag{9b}$$

$$E_{ci}(r) = (2/lA_i)^{1/2} \cos k_i z, \tag{9c}$$

where L is the laser cavity length, l the oscillator cavity length, A_j the cross-sectional area of the jth wave, and k_j the propagation constant of that wave. In (9), $k_j = n_j\omega_j/c$, where n_j is the index of refraction at ω_j. The normalization constants have been chosen such that the fields are normalized according to (2).

The parametric contributions to the polarization in (4)–(6) are given by

$$P_p^P(r, t) = \delta E_s(r, t) E_i(r, t) \tag{10a}$$

$$P_s^P(r, t) = \delta E_p(r, t) E_i(r, t) \tag{10b}$$

$$P_i^P(r, t) = \delta E_p(r, t) E_s(r, t) \tag{10c}$$

where δ is the appropriate element of the nonlinear susceptibility tensor [11]. We write all fields and polarizations in complex form

$$E_j(r, t) = \tfrac{1}{2}\mathbf{E}_j(r, t) \exp(i\omega_j t) + \text{c.c.}$$
$$= \tfrac{1}{2}E_{cj}(r)\boldsymbol{\varepsilon}_j'(t) \exp(i\omega_j t) + \text{c.c.} \tag{11a}$$

$$P_j(r, t) = \tfrac{1}{2}\mathbf{P}_j(r, t) \exp(i\omega_j t) + \text{c.c.} \tag{11b}$$

[1] For a discussion of the factors influencing the form of these eigenmodes see [2, eqs. 12] and the subsequent discussion.

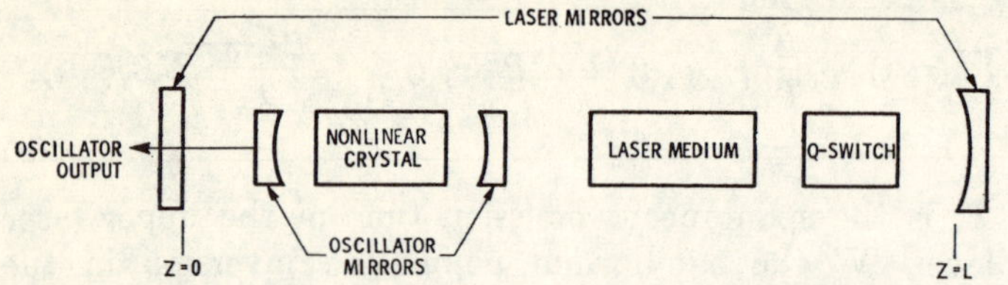

Fig. 1. Schematic of the internal optical parametric oscillator.

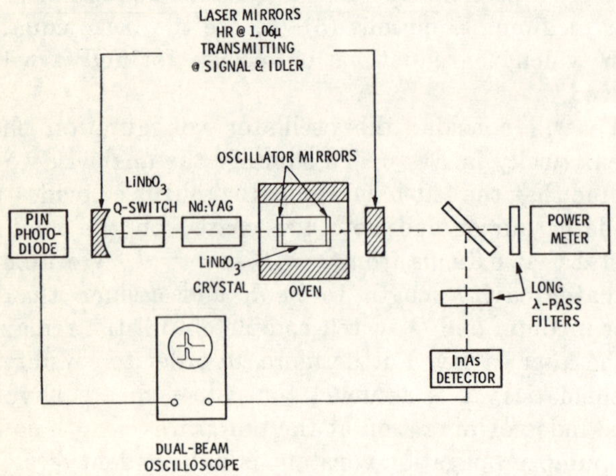

Fig. 2. Internal optical parametric oscillator experimental apparatus.

Substitution of (4)–(6) and (11) into (3) yields

$$\ddot{\varepsilon}_p' + \left(\frac{1}{\tau_p} + i2\omega_p\right)\dot{\varepsilon}_p' + i\,\frac{\omega_p}{\tau_p}\,\varepsilon_p' = -\frac{1}{\epsilon_p}\left\{\int_{LM} E_{cp}(r)[\ddot{\mathbf{P}}_p^L + i2\omega_p\dot{\mathbf{P}}_p^L - \omega_p^2\mathbf{P}_p^L]\,dV + \int_c E_{cp}(r)[\ddot{\mathbf{P}}_p^P + i2\omega_p\dot{\mathbf{P}}_p^P - \omega_p^2\mathbf{P}_p^P]\,dV\right\} \tag{12}$$

$$\ddot{\varepsilon}_k' + \left(\frac{1}{\tau_k} + i2\omega_k\right)\dot{\varepsilon}_k' + i\,\frac{\omega_k}{\tau_k}\,\varepsilon_k' = -\frac{1}{\epsilon_k}\int_c E_{ck}(r)[\ddot{\mathbf{P}}_k^P + i2\omega_k\dot{\mathbf{P}}_k^P - \omega_k^2\mathbf{P}_k^P]\,dV, \qquad k = s, i \tag{13}$$

where $\int_c$ denotes an integration over the nonlinear crystal and $\int_{LM}$ denotes an integration over the laser medium. Assuming that laser and parametric oscillations build up slowly with comparison to their radian frequencies, i.e.,

$$\ddot{\varepsilon}_j'(t) \ll \omega_j\dot{\varepsilon}_j' \qquad \ddot{\mathbf{P}}_j \ll \omega_j\dot{\mathbf{P}}_j \qquad \dot{\mathbf{P}}_j \ll \omega_j\mathbf{P}_j \tag{14}$$

and that, as is the usual case, $\omega_j \gg 1/\tau_j$, we find

$$\dot{\varepsilon}_p' + \frac{1}{2\tau_p}\,\varepsilon_p' = -\frac{i\omega_p\,\delta I_p'}{2\epsilon_p}\,\varepsilon_s'\varepsilon_i' - \frac{i\omega_p}{2\epsilon_p}\int_{LM} E_{cp}(r)\mathbf{P}_p^L\,dV \tag{15}$$

$$\dot{\varepsilon}_s' + \frac{1}{2\tau_s}\,\varepsilon_s' = -\frac{i\omega_s\,\delta I_s'}{2\epsilon_s}\,\varepsilon_p'\varepsilon_i'^* \tag{16}$$

$$\dot{\varepsilon}_i' + \frac{1}{2\tau_i}\,\varepsilon_i' = -\frac{i\omega_i\,\delta I_i'}{2\epsilon_i}\,\varepsilon_p'\varepsilon_s'^* \tag{17}$$

where * denotes a complex conjugate and

$$I_i' \triangleq \int_c E_{cp}(r)E_{cs}(r)E_{ci}(r)\,dV_i. \tag{18}$$

In Appendix I we show that

$$I_i' = \left(\frac{A_i^2}{2LA_pA_sA_i}\right)^{1/2}\left(\frac{l_c}{l}\right)\frac{\sin(\Delta k l_c)}{\Delta k l_c} \tag{19}$$

where $\Delta k \triangleq k_p - k_s - k_i$.

All that remains to complete the field equations is the evaluation of the integral on the right side of (15). Substituting (11a) and (11b) into (8), using (14), and equating components on each side at frequency ω_p yields

$$\dot{\mathbf{P}}_p^L + \frac{1}{T_2}\,\mathbf{P}_p^L = i\,\frac{\epsilon_0 cn\sigma}{\omega_p T_2}\,E_{cp}(r)N\varepsilon_p' \tag{20}$$

We will assume, as is usually the case, that $1/T_2\mathbf{P}_p^L \gg \dot{\mathbf{P}}_p^L$ so that

$$\mathbf{P}_p^L = i\,\frac{\epsilon_0 cn\sigma}{\omega_p}\,E_{cp}(r)N\varepsilon_p'. \tag{21}$$

Subject to this assumption, the right-hand side of the pump field equation may be written

$$\int_{LM} E_{cp}(r)\mathbf{P}_p^L\,dV = i\,\frac{\epsilon_0 cn\sigma}{\omega_p}\left(\frac{V_{LM}}{V_L}\right)N\varepsilon_p' \tag{22}$$

where (V_{LM}/V_L) is a filling factor equal to the ratio of the volume of the pump mode in the laser medium to that in the whole laser cavity. The field equations are now completely specified and we need only to include the population-inversion equation to completely describe the IOPO. Substituting (21) into (7) and equating zero frequency components on each side gives

$$\dot{N} + \frac{N - N^e}{T_1} = -\frac{\epsilon_0 cn\sigma}{\hbar\omega_p}\,E_{cp}^2(r)N\varepsilon_p'^2. \tag{23}$$

Ignoring the spatial variation of N and replacing $E_{cp}^2(r)$ by its spatial average in the laser medium, given by (24)

$$\overline{E_{cp}^2(r)} = \frac{1}{V_{LM}}\int_{LM} E_{cp}^2(r)\,dV = \frac{1}{V_{LM}}\cdot\frac{V_{LM}}{V_L} = \frac{1}{V_L} \tag{24}$$

we rewrite (15)–(17) and (23) using (22) and (24) to give

$$\dot{\varepsilon}_p' + \frac{1}{2\tau_p}\,\varepsilon_p' = \frac{c\sigma}{2n}\left(\frac{V_{LM}}{V_L}\right)N\varepsilon_p' - i\,\frac{\omega_p I_p'}{2\epsilon_p}\,\varepsilon_s'\varepsilon_i' \tag{25}$$

$$\dot{\varepsilon}_s' + \frac{1}{2\tau_s}\,\varepsilon_s' = -i\,\frac{\omega_s I_s'}{2\epsilon_s}\,\varepsilon_p'\varepsilon_i'^* \tag{26}$$

$$\dot{\varepsilon}_i' + \frac{1}{2\tau_i}\,\varepsilon_i' = -i\,\frac{\omega_i I_i'}{2\epsilon_i}\,\varepsilon_p'\varepsilon_s'^* \tag{27}$$

$$\dot{N} + \frac{N - N^e}{T_1} = -\frac{\epsilon_0 cn\sigma}{\hbar\omega_p V_L}\,N\varepsilon_p'^2. \tag{28}$$

Equations (25)–(28) constitute the coupled equations that describe the IOPO. As it is desirable to write these coupled equations in terms of real variables, we define

$$\varepsilon_j' = \varepsilon_i'' \exp[i\phi_i(t)]. \tag{29}$$

Combining (9), (11), and (29), we note that the total

electric fields are given by $E_j(r, t) = (2/V_j)^{1/2}\varepsilon_j'' \cos(k_j z) \cos[\omega_j t + \phi_j(t)]$ where V_j is the volume of the jth mode, and that the total electric field envelopes are then given by

$$\varepsilon_j = (2/V_j)^{1/2}\varepsilon_j''. \tag{30}$$

For convenience, we also define

$$N = N_0 \bar{\mathbf{N}}, \tag{31a}$$

and

$$N_0 = (n/\sigma c \tau_p)(V_L/V_{LM}) \tag{31b}$$

where N_0 is the steady-state value of the laser population difference in the absence of parametric interaction. Substitution of (29)–(31) into (25)–(28) then gives, in terms of real variables, the equations governing the temporal behavior of the IOPO.

$$\dot{\varepsilon}_p + \frac{1}{2\tau_p}\varepsilon_p = \frac{1}{2\tau_p}\bar{\mathbf{N}}\varepsilon_p - \frac{\delta\omega_p I}{4\epsilon_p}\frac{l}{L}\left(\frac{l_c}{l}\right)\varepsilon_s\varepsilon_i \sin\phi \tag{32}$$

$$\dot{\varepsilon}_s + \frac{1}{2\tau_s}\varepsilon_s = \frac{\delta\omega_s I}{4\epsilon_s}\left(\frac{l_c}{l}\right)\varepsilon_p\varepsilon_i \sin\phi \tag{33}$$

$$\dot{\varepsilon}_i + \frac{1}{2\tau_i}\varepsilon_i = \frac{\delta\omega_i I}{4\epsilon_i}\left(\frac{l_c}{l}\right)\varepsilon_p\varepsilon_s \sin\phi \tag{34}$$

$$\dot{\phi} = \frac{\delta I}{4}\left(\frac{l_c}{l}\right)\left[\frac{\omega_p}{\epsilon_p}\frac{l}{L}\frac{\varepsilon_s\varepsilon_i}{\varepsilon_p} - \frac{\omega_s}{\epsilon_s}\frac{\varepsilon_p\varepsilon_i}{\varepsilon_s} - \frac{\omega_i}{\epsilon_i}\frac{\varepsilon_p\varepsilon_s}{\varepsilon_i}\right]\cos\phi \tag{35}$$

$$\dot{\bar{\mathbf{N}}} + \frac{\bar{\mathbf{N}} - \bar{\mathbf{N}}^e}{T_1} = -\frac{\epsilon_0 c n \sigma}{2\hbar\omega_p}\bar{\mathbf{N}}\varepsilon_p^2 \tag{36}$$

where $\phi(t) \triangleq \phi_p(t) - \phi_s(t) - \phi_i(t)$ and $I \triangleq \sin(\Delta k l_c)/\Delta k l_c$. Equations (32)–(36) are solved in Section IV by digital computer. Before proceeding with their solution, we first discuss some significant features of these equations and note several of the limitations of this theory.

B. Discussion of Equations of Motion

Equations (32)–(36) are the coupled equations describing the behavior of the internal oscillator. These equations are very similar to those of Oshman and Harris [2], with the notable exception of dependence upon the population inversion, which is neglected in their treatment. To reduce (32)–(36) to the equations of Oshman and Harris, we neglect $\dot{\bar{\mathbf{N}}}$ in (36) and then solve (36) for $\bar{\mathbf{N}}$, giving

$$\bar{\mathbf{N}} = \frac{\bar{\mathbf{N}}^e}{1 + \dfrac{\epsilon_0 c n \sigma T_1}{2\hbar\omega_p}\varepsilon_p^2}. \tag{37}$$

If we define

$$\beta = (\epsilon_0 c n \sigma T_1)/(2\,\hbar\omega_p) \tag{38}$$

and assume $\beta\varepsilon_p^2 \ll 1$, (37) can be written as

$$\bar{\mathbf{N}} = \bar{\mathbf{N}}^e\,(1 - \beta\varepsilon_p^2). \tag{39}$$

Substitution of (39) into (32) then gives equations identical in form to those of Oshman and Harris.

The IOPO equations can also be reduced to single-mode laser rate equations. Setting the parametric coupling to zero, we can then rewrite (32) as

$$\frac{d}{dt}(\varepsilon_p^2) + \frac{1}{\tau_p}\varepsilon_p^2 = \frac{1}{\tau_p}\bar{\mathbf{N}}\varepsilon_p^2. \tag{40}$$

The population inversion equation (36) remains

$$\dot{\bar{\mathbf{N}}} + \frac{\bar{\mathbf{N}} - \bar{\mathbf{N}}^e}{T_1} = -\frac{\epsilon_0 c n \sigma}{2\hbar\omega_p}\bar{\mathbf{N}}\varepsilon_p^2. \tag{41}$$

Equations (40) and (41) are the rate equations governing a single-mode laser [5, p. 91].

It is evident from the form of (33) and (34) that the IOPO has a threshold since in order for $\dot{\varepsilon}_s$ and $\dot{\varepsilon}_i$ to be positive, the terms on the right-hand side of these equations must exceed the decay terms ε_j/τ_j. Threshold is determined by setting $\dot{\varepsilon}_s = \dot{\varepsilon}_i = 0$ and solving for ε_p, giving

$$\varepsilon_{p,\,\text{thresh}} = \frac{2(l/l_c)n_i n_s \epsilon_0}{\delta I(\tau_s\tau_i\omega_i\omega_s)^{1/2}} \tag{42}$$

where n_i and n_s are the indices of refraction at ω_i and ω_s. For ε_p less than $\varepsilon_{p,\text{thresh}}$ the system just behaves as an ordinary laser. If $\varepsilon_{p,\text{thresh}}$ is exceeded, ε_s and ε_i grow with time, at the expense of ε_p. Since, from (36), $\bar{\mathbf{N}}$ depends on ε_p^2, the population inversion is also affected by the parametric oscillation. In the next section we will consider the detailed dynamics of the IOPO for Q-switched laser operation.

A few comments regarding the relative phase equation (35) are instructive. This equation obviously has a mathematical singularity when ε_p, ε_s, or ε_i goes to zero. Physically, however, none of the fields can vanish, due to the presence of spontaneous emission. In our computer analysis, discussed in Section III, we have limited ε_p, ε_s, and ε_i to be no smaller than the values dictated by spontaneous emission.

There is yet another problem with (35), which is associated with a singularity at $\phi = \pi/2$. We have already noted that minimum threshold occurs for $\phi = \pi/2$. Inspection of (35) shows, however, that at $\phi = \pi/2$ all time derivatives of ϕ are zero, i.e., $\dot{\phi}(\pi/2) = \ddot{\phi}(\pi/2) = \dddot{\phi}(\pi/2) = \cdots = 0$. Thus if the phase is initially $\pi/2$, it can not change from this value. As a practical matter, however, $\pi/2$ is an irrational number that a computer can only approximate. Thus a computer will set $\cos(\pi/2)$ to some small nonzero value and the computed phase will change from $\pi/2$ as ε_p, ε_s, and ε_i change. Fortunately, as we will see in Section III, the computed phase deviates from $\pi/2$ only when ε_s and ε_i approach their spontaneous levels. We note that when ε_s and ε_i are at their spontaneous level all phases exist and ϕ is free to assume any value. In all but one of the computed results shown in the next section, ε_s and ε_i reached their spontaneous levels before ϕ changed significantly. The computer program that solves the equations is thus written so that ϕ is set to $\pi/2$ when the spontaneous levels are reached.

It is interesting to note that the equations governing

the IOPO can also be written in terms of the photon numbers. The equations appear in this form in Appendix II.

C. Limitations of the Theory

In this section we will discuss some of the limitations of the theory implicit in the derivation of the equations of motion. First, we point out that the analysis applies strictly only to homogeneously broadened lasers. This restriction is imposed by the form of (7) and (8), which govern the laser medium. Similar equations could be written for the inhomogeneously broadened case. Practically this restriction causes little difficulty since our computer analysis will only consider Q-switched lasers, and all lasers that can be Q switched at present are homogeneously broadened. In the CW case, or in the case when $\dot{N}$ can be ignored, our equations have been shown to reduce to those of Oshman and Harris [2], which are valid for homogeneously broadened lasers.

It should also be emphasized that the analysis presented in this paper has assumed single longitudinal and transverse modes for the signal, idler, and pump. Comparison of the results of this analysis with experiments requires taking the multimode nature of the pump and oscillator into account. As a first approximation, each mode can be assumed independent, and the analysis can be applied mode by mode. Thus if N pump modes exist, one assumes a power of $1/N$ of the total power per mode.

Although we have expanded the fields here in terms of the normal modes of the cavities, we point out that this is not strictly valid in the case of large output couplings. We will see in Section IV, however, that even when the oscillator is singly resonant the oscillator still behaves qualitatively as the analysis predicts.

III. Solution of Equations of Motion

Some calculated results for internal optical parametric oscillation based on (32)–(36) are given in this section. In order to compare later these results with experiments, we have chosen parameter values typical of the experimental arrangement that employs a Q-switched Nd:YAG laser (1.065-μ transition) as the pump and $LiNbO_3$ as the nonlinear material. The experimental apparatus is shown in Fig. 2 and is identical with that of Ammann *et al.* [4].

To simplify the calculations, we consider the oscillator operating at degeneracy so that $\epsilon_s = \epsilon_i = n^2\epsilon_0$, where n is the appropriate index of refraction, taken in this case to be 2.2. At degeneracy the signal and idler frequencies are equal and are half the pump frequency, so that $\omega_s = \omega_i = \omega_p/2$. A 4-mm-long $LiNbO_3$ crystal oriented at approximately 45° to the c axis is taken as the nonlinear crystal. Mirrors are assumed coated directly on the end faces of the crystal, so that $l = l_c = 4$ mm. We also assume perfect phase matching so that $I = 1$. For this crystal orientation the value of δ is 1.2×10^{-23} CV^{-2}.

For Nd:YAG, the appropriate physical constants are $n = 1.81$, $T_1 = 2.3 \times 10^{-4}$ s [6], and $\sigma = 8.7 \times 10^{-23}$ m^2 [6]. The measured single-pass loss of the laser with the oscillator crystal and Q switch inside is about 10 percent. Most of this loss is due to the Q switch, which polarizes the laser. It is well known that Nd:YAG becomes birefringent when pumped [7], so that an internal polarizer causes a large loss. This problem can be overcome by the use of a Nd:$YAlO_3$ [8] laser pump. The 10 percent loss gives a cavity lifetime for our laser of 11 ns.

Unless otherwise noted, we have assumed the cavity lifetime of the oscillator to be 1.2 ns, which corresponds to 5 percent round-trip loss at the signal and idler wavelength.

We will consider the case where the laser is Q switched. In this situation the initial value of $\bar{\mathbf{N}}$ is $\bar{\mathbf{N}}^e$. We have calculated the behavior of the oscillator for three separate values of $\bar{\mathbf{N}}^e$, arbitrarily taking $\boldsymbol{\varepsilon}_p(t = 0) = 10^4$ V/m. To establish initial values of $\boldsymbol{\varepsilon}_s$ and $\boldsymbol{\varepsilon}_i$, we have assumed, after Byer and Harris [9], that a half-photon per mode drives both the signal and the idler fields. In addition, we have restricted the calculation so that the signal and idler never fall below their spontaneous (initial) values. For the crystal length used here, the appropriate initial value for $\boldsymbol{\varepsilon}_s$ and $\boldsymbol{\varepsilon}_i$ is 0.1 V/m. We have taken $\phi(t = 0) = \pi/2$, which maximizes the parametric gain. We have also carried out calculations with $\phi(t = 0) \neq \pi/2$ and find that for that case, ϕ moves to $\pi/2$ very rapidly.

In Figs. 3–8 we show the computed behavior of the IOPO for three different pumping levels ($\bar{\mathbf{N}}^e = 1.2$, 1.4, and 1.6) and for two round-trip losses (5 and 2 percent). In each case we have plotted $\boldsymbol{\varepsilon}_p^2$, $\boldsymbol{\varepsilon}_s^2$, $\bar{\mathbf{N}}$, and ϕ as functions of time. The dotted curves show the behavior in the absence of parametric coupling (crystal not phase matched). Both $\boldsymbol{\varepsilon}_p^2$ and $\boldsymbol{\varepsilon}_s^2$ have been normalized in each case to the peak value of $\boldsymbol{\varepsilon}_p^2$ in the absence of oscillation. It should be kept in mind that $\boldsymbol{\varepsilon}_p$, $\boldsymbol{\varepsilon}_s$, and $\boldsymbol{\varepsilon}_i$ are electric field strengths *inside* the nonlinear crystal.

In the absence of parametric coupling, the pump field builds up and then decays as the population inversion monotonically decreases. The laser field reaches its peak value at $\bar{\mathbf{N}} = 1.0$, as is the usual behavior for a Q-switched laser. The value of $\bar{\mathbf{N}} = 1.0$ also corresponds to the steady-state value, but this value is not reached on a single laser pulse, such as we have here. Rather, the population inversion overshoots and goes below $\bar{\mathbf{N}} = 1.0$. As $\bar{\mathbf{N}}^e$ is increased, $\boldsymbol{\varepsilon}_{p,\max}^2$ increases and the pulsewidth decreases.

The introduction of parametric coupling produces the waveforms shown in Figs. 3–8 as solid lines. We see that the pulsed IOPO operates in a spiking regime. As the population inversion is increased or the output coupling decreased, the signal spikes occur more frequently and have higher peak power. The spikes are also very narrow, having half-widths of typically 5 ns. When several spikes occur in one laser pulse, the repetition rate is pump-power dependent so that the more intense the pulses, the shorter the interval between them.

In order to better understand the behavior of the IOPO, let us consider in detail the pulse shapes shown in Figs. 3–8. Fig. 3 shows the computed behavior for $\bar{\mathbf{N}}^e = 1.2$

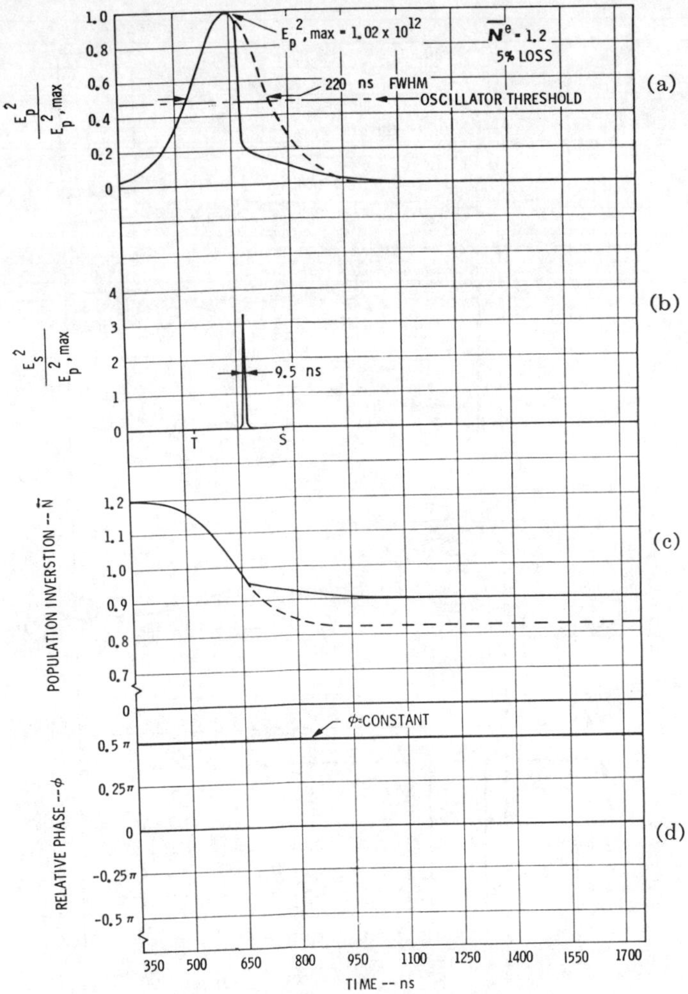

Fig. 3. Computer solutions for IOPO for 5 percent oscillator round-trip loss. Initial normalized inversion $\bar{\mathbf{N}}^e = 1.2$.

with 5 percent round-trip loss at both the signal and idler wavelengths. In the absence of parametric coupling, the laser pulse has a half-width of 220 ns and a peak amplitude of $\mathcal{E}_{p,\max}^2 = 1.02 \times 10^{12}$ V^2/m^2. For a laser mode 1.5 mm in diameter this corresponds to a peak power of about 40 kW. The population inversion monotonically decreases as $\mathcal{E}_p^2$ increases, having its maximum rate of decrease when $\mathcal{E}_p^2$ is at its largest value.

With parametric coupling added, the behavior of the IOPO is considerably more complex. The threshold level for the oscillator has been plotted as a dashed horizontal line in Fig. 3(a). When $\mathcal{E}_p^2$ crosses this level at a time denoted by T, $\mathcal{E}_s$ and $\mathcal{E}_i$ begin to grow and reach their maximum at approximately the peak of the laser pulse. The laser field is quickly depleted by the parametric interaction, as shown. When $\mathcal{E}_p^2$ decreases below the threshold level, the signal field $\mathcal{E}_s^2$ decreases rapidly back to its spontaneous level, which occurs at the time denoted by S in Fig. 3(b). We note that the pump is depleted even after the oscillator falls below threshold, since from (32) the pump is depleted proportionally to $\mathcal{E}_s\mathcal{E}_i$ and these fields are still large immediately after the oscillator falls below threshold. In this case the phase remains at its initial value of $\pi/2$ throughout the pulse. The laser field does not build up again because the population inversion $\bar{\mathbf{N}}$ is less than 1.0 after the oscillator depletes the laser field.

The signal pulse is only 9.5 ns long and its peak power exceeds that of the undepleted laser pulse by a factor of three inside the crystal. The signal power outside the crystal is determined by multiplying by the transmission of the oscillator output mirror. It is important to note that while approximately $\frac{1}{3}$ of the pump pulse is missing, it does not follow that this corresponds to 30 percent energy conversion. The reason that much of the pump pulse is missing is that the oscillator essentially dumps the laser field, and thus the laser lases at a reduced level. Nevertheless, since $\mathcal{E}_{s,\max}^2 > \mathcal{E}_{p,\max}^2$, significant energy conversion may occur. Experimentally we have seen up to 70 percent conversion of the available laser energy.

The population inversion decreases at a much slower rate after the oscillator pulse, since the laser field must begin from a low level and hence uses up population inversion very slowly.

The average signal gain can be calculated by noting that the signal field $\mathcal{E}_s^2$ increased from its spontaneous value of 10^{-2} V^2/m^2 to its peak value of 3.37×10^{12} V^2/m^2 in 140 ns. Since the crystal is 4 mm long, this corresponds to 5250 one-way passes. (Unlike an external oscillator, the internal oscillator has gain in each direction.) The average excess gain is then determined from $g^{5250} = 3.37 \times 10^{14}$, or $g = 1.0064$. The total single-pass loss is 2.5 percent, so the average total single-pass gain is 3.14 percent.

Fig. 4 shows the oscillator behavior when $\bar{\mathbf{N}}^e$ is increased to 1.4 with the round-trip loss still 5 percent. The laser pulsewidth decreases to 115 ns and the peak field increases to $\mathcal{E}_{p,\max}^2 = 3.65 \times 10^{12}$, about 3.6 times as large as before. The oscillator threshold is unchanged, but it is now exceeded earlier in the pump pulse and the pump is depleted before it reaches its maximum value. The first oscillator pulse is narrower than for $\bar{\mathbf{N}}^e = 1.2$ being only 3 ns at half-height, and is ten times as large as the undepleted pump at its peak. Thus increasing $\bar{\mathbf{N}}^e$ from 1.2 to 1.4 has increased the peak laser power by a factor of 3, increased the peak oscillator power by a factor of almost 13, and decreased the oscillator pulsewidth by about a factor of 3.

In this case the oscillator pulse is sufficiently intense that the pump field is driven to its spontaneous level at a time denoted by O in Fig. 4(a). When this occurs, there are then strong fields in the crystal at the signal and idler frequencies and no substantial pump field. The signal and idler, phase matched for mixing to form a field at the pump frequency, drive a pump field whose phase differs by π from the phase of the pump field that initially drove the oscillator. When the signal reaches its spontaneous level, denoted by S_1, its phase is random, i.e., its phase is unlocked from the parametric process and again changes by π so that the signal and idler are again driven by the pump once it crosses threshold.

The pump field begins to increase again since a population inversion still exists ($\bar{\mathbf{N}} > 1$). Thus the pump field

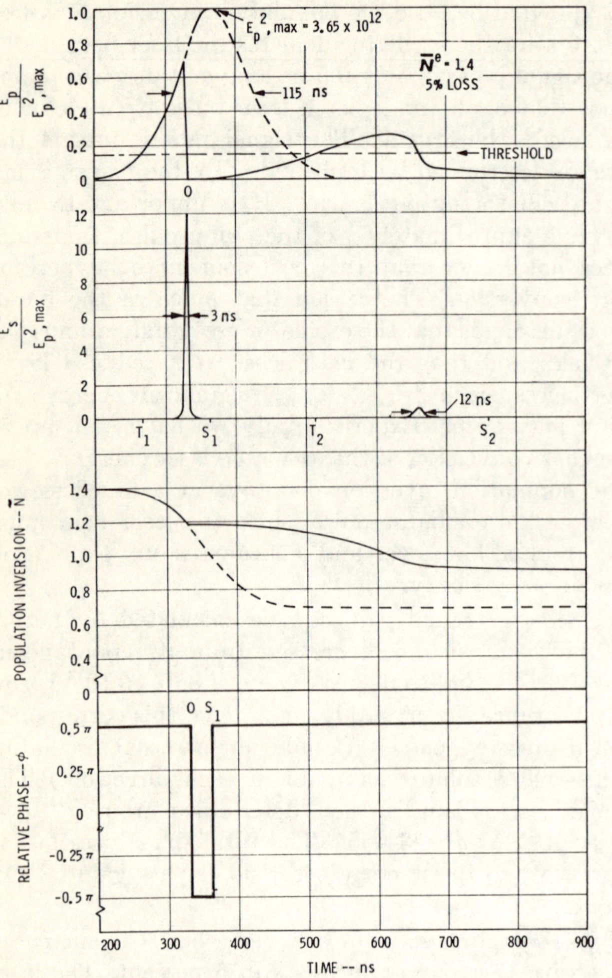

Fig. 4. Computer solutions for IOPO for 5 percent oscillator round-trip loss. Initial normalized inversion $\bar{N}^e = 1.4$.

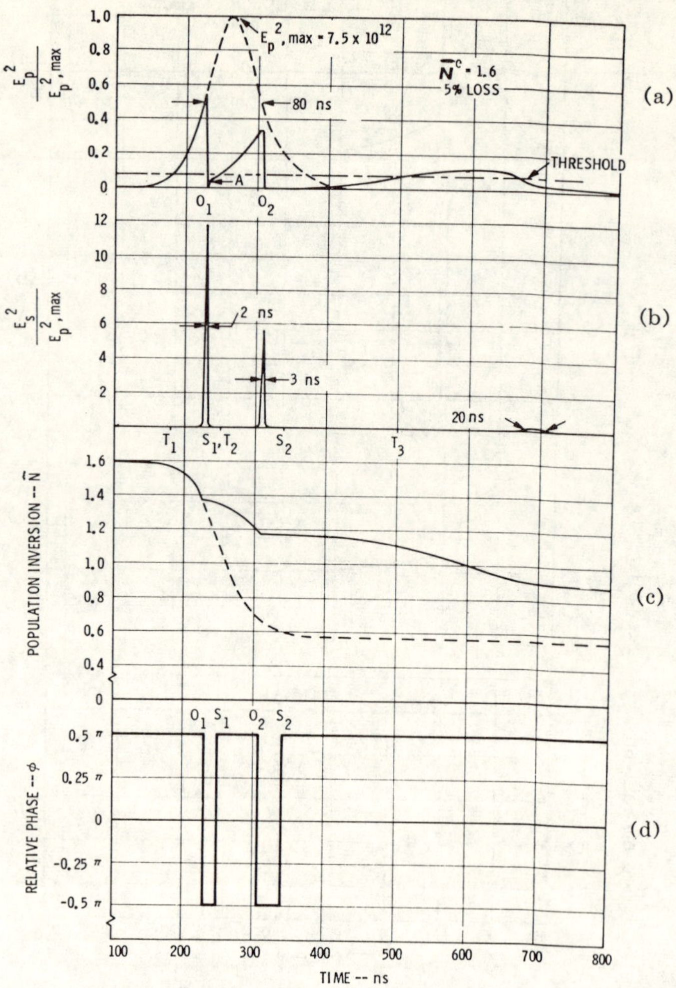

Fig. 5. Computer solutions for IOPO for 5 percent oscillator round-trip loss. Initial normalized inversion $\bar{N}^e = 1.6$.

builds up and the population decreases. At time T_2 the pump again exceeds threshold for the oscillator, so the signal grows slowly and does not reach as high a level as for the first pulse. The pump is again depleted below threshold and the signal and idler again decay to their spontaneous values. The oscillation is not as intense in this case, and ε_p^2 is not depleted to its spontaneous level, so the phase does not change for this pulse.

Note that in this case the duration of the pump pulse is considerable extended, being about four times as long as the undepleted pulse. Such stretching is caused by internal power-dependent losses [10], but the process observed here is somewhat different from others used to date. In this case the oscillator essentially dumps the pump and the pump has to again build up, thus giving the pump a sawtooth-like waveshape. The computed average gain for the first pulse in Fig. 4(a) is 4.0 percent as compared to 3.14 percent for Fig. 3(a).

The effect of increasing the inversion still further is shown in Fig. 5, where the initial inversion has been increased to 1.6. Now three pulses occur rather than two, and the first two are sufficiently strong to deplete the pump to its spontaneous level, denoted by O_1 and O_2 in Fig. 5(a). The sharp rise in the pump from O_1 to A is due to the signal and idler mixing to form pump. There is sufficient energy stored in the signal and idler fields to actually generate a pump field which is 8 percent as large as the undepleted pump field. In this case the pump field is already above threshold by the time the signal reaches its spontaneous level, so that the second signal pulse begins to grow immediately after the first reaches its spontaneous level and the phase changes by π to again favor parametric generation of a new signal and idler. The third pulse is much weaker than the first two and does not deplete the pump to its spontaneous level. The average single-pass gain for the first pulse for $\bar{N}^e = 1.6$ is 4.34 percent.

Figs. 6–8 show the effect of reducing the round-trip loss of the oscillator from 5 to 2 percent. This of course lowers the oscillator threshold, so the signal builds up sooner. In addition, since the losses are less, the peak signal fields reach a higher level. This does not necessarily imply a higher output power, however, since less of the field is transmitted through the oscillator mirror. Figs. 7 and 8 show clearly the effect of the signal and idler mixing to form pump after the pump is depleted to its spontaneous

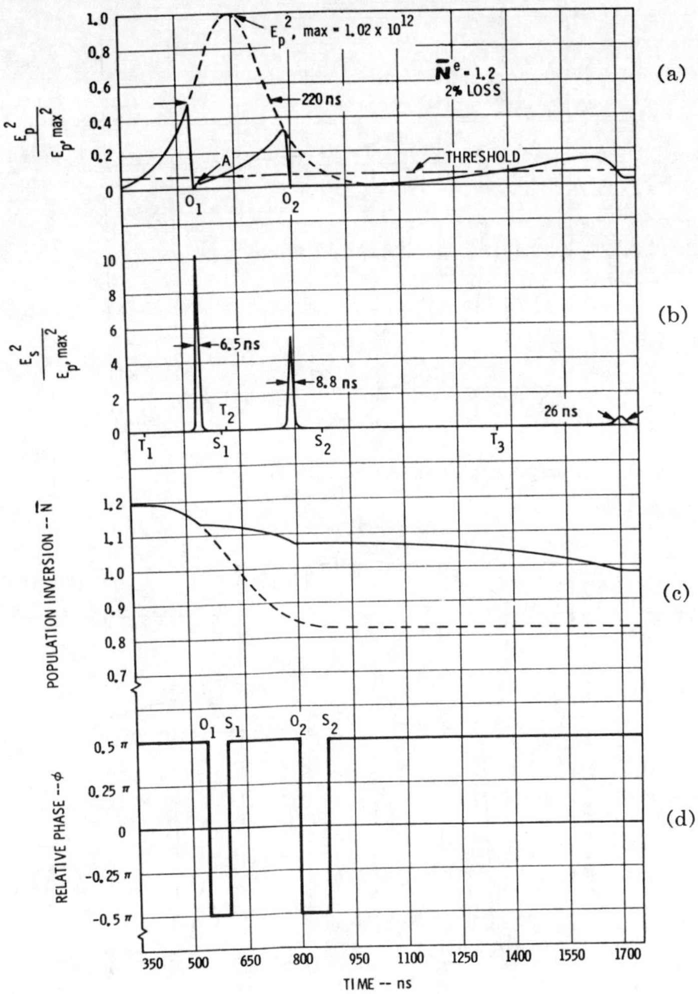

Fig. 6. Computer solutions for IOPO for 2 percent oscillator round-trip loss. Initial normalized inversion $\bar{\mathbf{N}}^e = 1.2$.

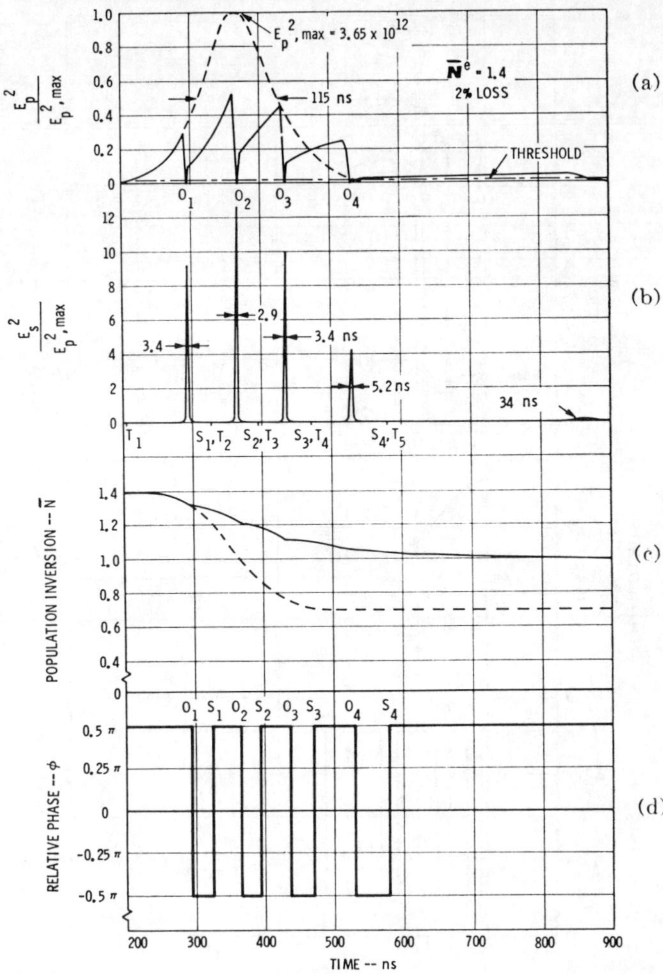

Fig. 7. Computer solutions for IOPO for 2 percent oscillator round-trip loss. Initial normalized inversion $\bar{\mathbf{N}}^e = 1.4$.

level. In fact, the second signal pulse in Fig. 7 drives the pump field to almost 30 percent of its undepleted value. This of course means that the oscillator is undercoupled since much of the signal field was converted back into pump rather than coupled out of the oscillator. For the case of 2 percent round-trip loss of the oscillator, almost all of the signal pulses are strong enough to completely deplete the pump. For the most intense pulse in Fig. 7, the average single-pass gain is 5.95 percent.

Optimum energy transfer to the output signal for a given inversion can be determined by computing the area under all of the signal pulses for different couplings and then multiplying by the mirror transmission. For the cases considered here, assuming that in each case 1 percent of the total loss is due to scattering and diffraction, the output energy from the oscillator and the undepleted laser pulse energy have the ratios shown in Table I.

From this table we see that for $\bar{\mathbf{N}}^e = 1.2$, 2 percent loss (1 percent coupling) gives more output energy than 5 percent loss (4 percent coupling), but for $\bar{\mathbf{N}}^e = 1.6$, 5 percent loss is better. We also see that the signal energy increases slightly faster than the pump energy.

IV. Comparison of Theory With Experiment

In this section we will present experimental results that verify that pulsed IOPOs indeed perform qualitatively in the manner predicted by the computer solutions given in the previous section.

Fig. 9 shows the detected pump and signal pulses from an IOPO operating near degeneracy. The experimental configuration,[2] identical to that used by Ammann *et al.* [4, Fig. 2], used parameter values nearly identical to those used in the computer solutions in the last section. In this figure, a single signal pulse similar to that of Fig. 3 is recorded. For this retraced photograph, the oscillator output coupling was approximately 7 percent. Although the detector used for the signal is not fast enough to accurately measure its pulsewidth, we have determined its duration by frequency doubling the signal and measuring the pulsewidth of the doubled signal. Using this method, the signal pulse length was measured to be 5–10 ns, agreeing well with the computed results.

[2] It should be noted that the detector used to detect the signal outputs shown in [4, Fig. 2] was saturated, so that the pulse lengths reported there are much too long.

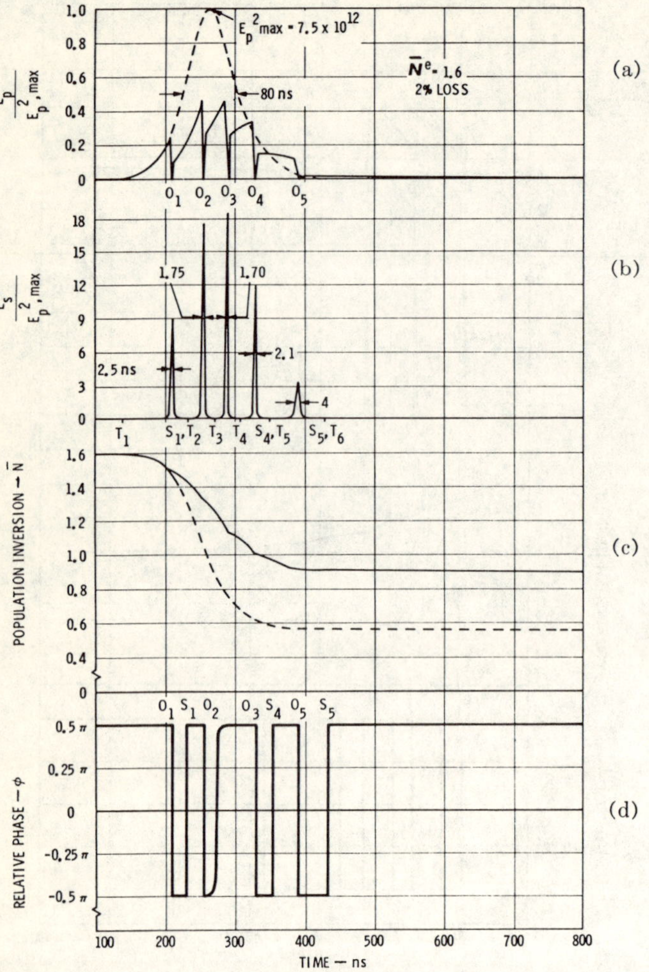

Fig. 8. Computer solutions for IOPO for 2 percent oscillator round-trip loss. Initial normalized inversion $\bar{N}^e = 1.6$.

TABLE I
Output Energy from Oscillator for Various Couplings and Inversions

$\bar{N}^e$	Laser Energy	Signal Energy for 5 Percent Round-Trip Loss (4 percent due to coupling)	Signal Energy for 2 Percent Round-Trip Loss (1 percent due to coupling)
1.2	1	1	1.15
1.4	3.85	3.85	4.21
1.6	7.35	9.55	8.12

In Figs. 10 and 11 the oscillator output is shown as the pump power is increased over that used for Fig. 9. Two and four output pulses, respectively, are observed, much as is shown in Figs. 5 and 7. In addition, as predicted by the computer solutions, the pump pulse is significantly stretched.

Recently we have operated a nondegenerate singly resonant internal oscillator, and we find that the behavior is much the same as for the doubly resonant case. The present analysis is not strictly valid for the singly resonant case since we have assumed no spatial variation in the fields, and there is a spatial variation in the case of the nonresonant field. Nevertheless, the analysis still correctly predicts the qualitative behavior of the singly resonant internal oscillator.

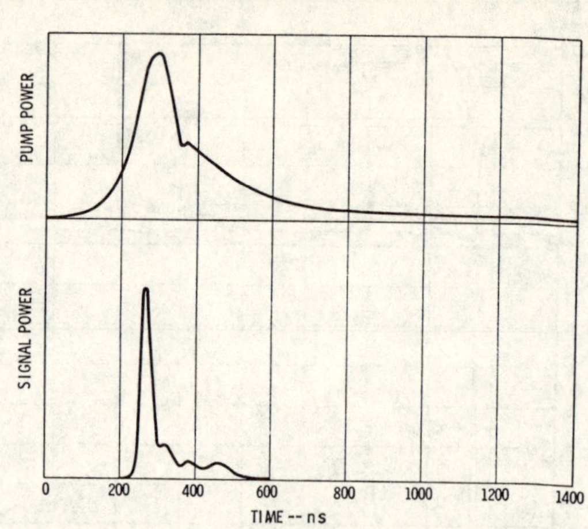

Fig. 9. Experimentally observed pump and signal pulse shapes from IOPO.

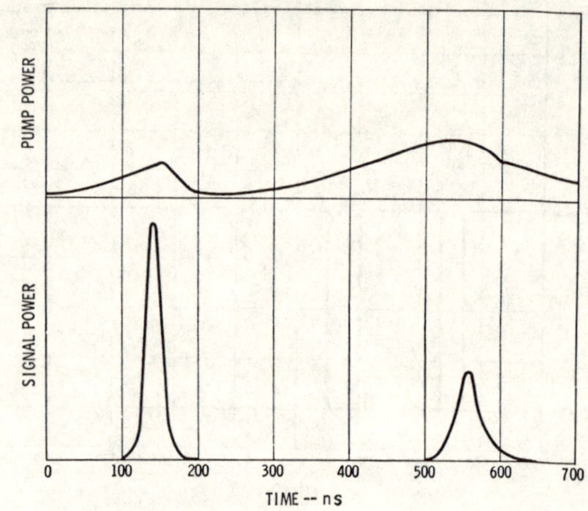

Fig. 10. Experimentally observed pump and signal pulse shapes from IOPO for harder pumping than for Fig. 9.

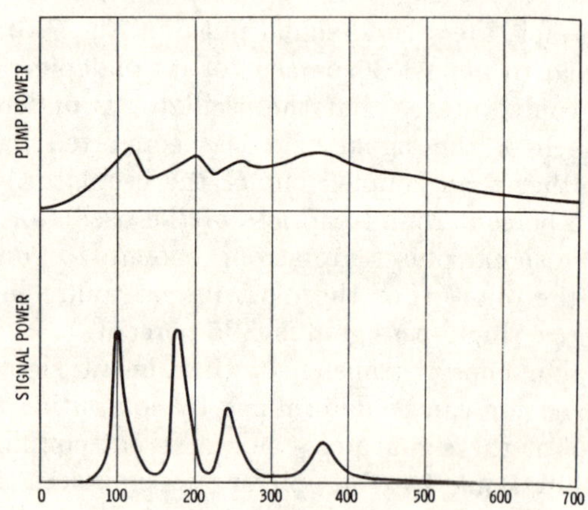

Fig. 11. Experimentally observed pump and signal pulse shapes from IOPO for harder pumping than in Fig. 10.

V. Summary and Conclusions

We have generalized the theory of the internal optical parametric oscillator to include any homogeneously broadened laser–oscillator system. The theory has been shown to reduce to that of Oshman and Harris [2] when the laser population-inversion dynamics are ignored.

The equations of motion have been solved by digital computer for the case of a $LiNbO_3$ parametric oscillator inside a Q-switched Nd:YAG laser. The computer solutions predict that the oscillator output is in a spiking regime, with the parametrically generated fields generally being larger than the pump field that existed inside the nonlinear medium in the absence of parametric oscillation. The repetition rate of the oscillator spikes is pump-power dependent, with the repetition rate increasing as the pump field increases. Thus the repetition rate varies within a given pump pulse.

The theoretical behavior of the IOPO has been verified qualitatively by experiments, which show that the oscillator operates in the spiking regime, with the signal pulses having half-widths of 5–10 ns. The experimentally observed IOPO was very efficient, converting more than 70 percent of the available pump power to signal and idler output.

The short intense output pulses of the IOPO may be of use in other nonlinear optics applications, such as second-harmonic generation.

Appendix I

Derivation of the Phase-Matching Condition

In this Appendix we will evaluate the integral that determines the effects of phase asynchronism on the internal optical parametric oscillator. This integral is given by (18)

$$I_i' \triangleq \int_c E_{cp}(r)E_{cs}(r)E_{ci}(r)\, dV_i. \tag{18}$$

Recall from (9) that

$$E_{cp}(r) = (2/LA_p)^{1/2} \cos k_p z \tag{9a}$$

$$E_{cs}(r) = (2/lA_s)^{1/2} \cos k_s z \tag{9b}$$

$$E_{ci}(r) = (2/lA_i)^{1/2} \cos k_i z. \tag{9c}$$

Substituting (9) into (18) and using appropriate trignometric identities gives

$$I_i' = \left(\frac{1}{2LA_pA_sA_i}\right)^{1/2} \frac{1}{l} \int_0^{A_i} \int_0^{l_c} [\cos(k_p - k_s - k_i)z + \cos(k_p + k_s + k_i)z + \cos(k_p - k_s + k_i)z + \cos(k_p + k_s - k_i)z]\, dz\, dA_i. \tag{43}$$

Since we are ignoring transverse variations, we may take

$$\int_0^{A_i} dA_i = A_i.$$

Evaluating the integrals then gives

$$I_i' = \left(\frac{A_i^2}{2LA_pA_sA_i}\right)^{1/2} \frac{1}{l} \left[\frac{\sin(k_p - k_s - k_i)l_c}{(k_p - k_s - k_i)} + \frac{\sin(k_p + k_s + k_i)l_c}{(k_p + k_s + k_i)} + \frac{\sin(k_p - k_s + k_i)l_c}{(k_p - k_s + k_i)} + \frac{\sin(k_p + k_s - k_i)l_c}{(k_p + k_s - k_i)}\right]. \tag{44}$$

Due to the energy conservation condition, (1), the denominators of each term on the right-hand side of (44) are large except for the first term. We may then omit the other three terms, giving

$$I_i' = \left(\frac{A_i^2}{2LA_pA_sA_i}\right)^{1/2} \left(\frac{l_c}{l}\right) \frac{\sin \Delta k l_c}{\Delta k l_c} \tag{45}$$

where we have written $\Delta k \triangleq k_p - k_s - k_i$.

Appendix II

Expression of the IOPO Equations in Terms of Photon Number

The equations of motion of the IOPO can also be written in terms of photon numbers. This form of the equations will be seen to be easily interpretable in physical terms.

The peak electric field in a resonator mode is related to the number of photons by the relation

$$\boldsymbol{\varepsilon}_j^2 = \frac{4\hbar\omega_j}{\epsilon_j A_j l_j} n_j \tag{46}$$

where $\boldsymbol{\varepsilon}_j$ is the electric field of the jth mode, ω_j the angular frequency, ϵ_j the dielectric constant, A_j the cross-sectional area of the mode, l_j the length of the resonator, and n_j the number of photons in the jth mode. Making use of this relation in (32)–(36) and taking $A_j = A$ then yields the following equations:

$$\dot{n}_p + \frac{1}{\tau_p} n_p = \frac{\bar{\mathbf{N}}}{\tau_p} n_p - \delta I \left(\frac{l_c}{l}\right)\left(\frac{\hbar}{AL}\right)^{1/2} \cdot \left(\frac{\omega_p\omega_s\omega_i}{\epsilon_p\epsilon_s\epsilon_i}\right)^{1/2} \sqrt{n_p n_s n_i} \sin\phi \tag{47}$$

$$\dot{n}_s + \frac{1}{\tau_s} n_s = \delta I \left(\frac{l_c}{l}\right)\left(\frac{\hbar}{AL}\right)^{1/2} \left(\frac{\omega_p\omega_s\omega_i}{\epsilon_p\epsilon_s\epsilon_i}\right)^{1/2} \sqrt{n_p n_s n_i} \sin\phi \tag{48}$$

$$\dot{n}_i + \frac{1}{\tau_i} n_i = \delta I \left(\frac{l_c}{l}\right)\left(\frac{\hbar}{AL}\right)^{1/2} \left(\frac{\omega_p\omega_s\omega_i}{\epsilon_p\epsilon_s\epsilon_i}\right)^{1/2} \sqrt{n_p n_s n_i} \sin\phi \tag{49}$$

$$\dot{\phi} = \delta I \left(\frac{l_c}{l}\right)\left(\frac{\hbar}{AL}\right)^{1/2} \left(\frac{\omega_p\omega_s\omega_i}{\epsilon_p\epsilon_s\epsilon_i}\right)^{1/2} \cdot \left[\sqrt{\frac{n_s n_i}{n_p}} - \sqrt{\frac{n_p n_i}{n_s}} - \sqrt{\frac{n_p n_s}{n_i}}\right] \cos\phi \tag{50}$$

$$\dot{\bar{\mathbf{N}}} + \frac{\bar{\mathbf{N}} - \bar{\mathbf{N}}^e}{T_1} = \frac{2\sigma c}{AL}\left(\frac{\epsilon_0}{\epsilon_p}\right)^{1/2} \bar{\mathbf{N}} n_p. \tag{51}$$

For convenience we define

$$K_1 = \delta I \left(\frac{l_c}{l}\right)\left(\frac{\hbar}{AL}\right)^{1/2} \left(\frac{\omega_p\omega_s\omega_i}{\epsilon_p\epsilon_s\epsilon_i}\right)^{1/2} \tag{52}$$

and

$$K_2 = 2\sigma c/AL(\epsilon_0/\epsilon_p)^{1/2}$$

so that we can rewrite (47)–(51) as

$$\dot{n}_p + \frac{1}{\tau_p} n_p = \frac{1}{\tau_p} \bar{\mathbf{N}} n_p - K_1 \sqrt{n_p n_s n_i} \sin \phi \qquad (53)$$

$$\dot{n}_s + \frac{1}{\tau_s} n_s = K_1 \sqrt{n_p n_s n_i} \sin \phi \qquad (54)$$

$$\dot{n}_i + \frac{1}{\tau_i} n_i = K_1 \sqrt{n_p n_s n_i} \sin \phi \qquad (55)$$

$$\dot{\phi} = K_1 \left[\sqrt{\frac{n_s n_i}{n_p}} - \sqrt{\frac{n_p n_i}{n_s}} - \sqrt{\frac{n_p n_s}{n_i}} \right] \cos \phi \qquad (56)$$

$$\dot{\bar{\mathbf{N}}} + \frac{\bar{\mathbf{N}} - \bar{\mathbf{N}}^e}{T_1} = K_2 \bar{\mathbf{N}} n_p. \qquad (57)$$

Several useful relations can be developed from these equations. Combining (53) and (54) yields

$$\dot{n}_p - \frac{1}{\tau_p} n_p(\bar{\mathbf{N}} - 1) = -\dot{n}_s - \frac{1}{\tau_s} n_s, \qquad (58)$$

which is a photon conservation condition. If we neglect the right-hand side (no parametric coupling), we have just the ordinary rate equation for the number of photons of a single-mode laser.

We can also write (58) as

$$\dot{n}_p + \frac{1}{\tau_p} n_p + \dot{n}_s + \frac{1}{\tau_s} n_s = \frac{1}{\tau_p} \bar{\mathbf{N}} n_p. \qquad (59)$$

This can be interpreted as the population inversion $\boldsymbol{N}$ driving both the pump and signal photon number.

If we assume degeneracy so that $\boldsymbol{\varepsilon}_s = \boldsymbol{\varepsilon}_i$, $\omega_s = \omega_i$, and $n_s = n_i$, and take $\phi = \pi/2$ as is the usual case, we can rewrite (54) as

$$\dot{n}_s + \frac{1}{\tau_s} n_s = K_1 n_s \sqrt{n_p} \qquad (60)$$

or

$$\frac{\dot{n}_s}{n_s} = K_1 \sqrt{n_p} - \frac{1}{\tau_s} = \frac{1}{\tau_s} [K_1 \tau_s \sqrt{n_p} - 1], \qquad (61)$$

which can also be written as

$$\frac{\dot{n}_s}{n_s} = \frac{1}{\tau_s} \left[\left(\frac{n_p}{n_{p,\,\mathrm{thresh}}} \right)^{1/2} - 1 \right] \qquad (62)$$

where $n_{p,\,\mathrm{thresh}} = (1/K_1 \tau_s)^2$.

Equation (62) shows quantitatively how the fractional growth of the signal depends on how far the pump is above threshold.

Acknowledgment

The authors wish to thank S. Barnard for programming the internal oscillator equations and W. H. Lanouette and L. H. Martin for technical assistance in performing the experiments.

References

[1] See S. E. Harris, "Tunable optical parametric oscillators," *Proc. IEEE*, vol. 57, Dec. 1969, pp. 2096–2113, for an extensive bibliography of optical parametric oscillation.

[2] M. K. Oshman and S. E. Harris, "Theory of optical parametric oscillation internal to the laser cavity," *IEEE J. Quantum Electron.*, vol. QE-4, Aug. 1968, pp. 491–502.

[3] R. G. Smith and J. V. Parker, "Experimental observation of and comments on optical parametric oscillation internal to the laser cavity," *J. Appl. Phys.*, vol. 41, 1970, pp. 3401–3408.

[4] E. O. Ammann, J. M. Yarborough, M. K. Oshman, and P. C. Montgomery, "Efficient internal optical parametric oscillation," *Appl. Phys. Lett.*, vol. 16, Apr. 1970, pp. 309–312.

[5] R. H. Pantell and H. E. Puthoff, *Fundamentals of Quantum Electronics.* New York: Wiley, 1969, p. 41.

[6] T. Kushida, H. M. Marcos, and J. E. Geusic, "Laser transition cross section and fluorescence branching ratio for Nd^{3+} in yttrium aluminum garnet," *Phys. Rev.*, vol. 167, Mar. 1968, pp. 289–291.

[7] L. M. Osterink and J. D. Foster, "Thermal effects and transverse mode control in a Nd:YAG laser," *Appl. Phys. Lett.*, vol. 12, Feb. 1968, pp. 128–131.

[8] G. A. Massey, "Criterion for selection of CW laser host materials to increase available power in the fundamental mode," *Appl. Phys. Lett.*, vol. 17, Sept. 1970, pp. 213–215.

[9] R. L. Byer and S. E. Harris, "Power and bandwidth of spontaneous parametric emission," *Phys. Rev.*, vol. 168, Apr. 1968, pp. 1064–1068.

[10] Some of the methods that have been used for pulse stretching are stimulated Rayleigh scattering, two-photon absorption, and second-harmonic generation. For a bibliography of these and other methods for pulse stretching, see J. E. Murray and S. E. Harris, "Pulse lengthening via overcoupled internal second-harmonic generation," *J. Appl. Phys.*, vol. 41, Feb. 1970, pp. 609–613.

[11] G. D. Boyd and D. A. Kleinman, "Parametric interaction of focused Gaussian light beams," *J. Appl. Phys.*, vol. 39, July 1968, pp. 3597–3639.

Theory of Optical Parametric Oscillation Internal to the Laser Cavity

M. KENNETH OSHMAN, MEMBER, IEEE, AND STEPHEN E. HARRIS, MEMBER, IEEE

Abstract—Since the fields inside a laser cavity are much higher than the external fields, an analysis of a parametric oscillator with the nonlinear crystal internal to the laser is performed. Using self-consistency equations as the starting point, the equations of motion of such an oscillator are derived. Depending on various cavity, pumping, and nonlinearity parameters, these lead to several types of oscillation with distinctly different operating characteristics: (1) efficient parametric oscillation similar to that of previous analyses; (2) inefficient parametric oscillation resulting from the fact that the nonlinear interaction drives the phases rather than the amplitudes of the signal, idler, and pump; and (3) a pulsing output from the oscillator with repetitive pulses of the signal and idler. A stability analysis of these various regions shows that they are mutually exclusive and can be experimentally chosen by changing the laser gain, the oscillator output coupling, or the strength of the nonlinear interaction. It is shown that the internal oscillator efficiency rapidly approaches the Manley–Rowe limit, as the available pump power becomes several times greater than that required for threshold. The efficiency of an external oscillator having a triply resonant optical cavity is found to be generally less than that of the corresponding internal oscillator.

I. Introduction

IN THIS PAPER we examine the theory of optical frequency parametric oscillation with the nonlinear crystal placed inside the cavity of the pumping laser. Such an oscillator, proposed originally by Kroll [1], takes advantage of the high pump fields inside the laser cavity, compared to outside, in overcoming threshold and in leading to efficient conversion of pump power to signal and idler power. Several authors have discussed the enhancement of second harmonic generation under similar conditions [2], [3].

With the theory of optical parametric oscillation well known at lower frequencies [4], the extensions to optical frequencies have been considered by a number of authors [1], [5]–[12]. Much of the theoretical apparatus necessary for understanding optical frequency parametric oscillation has been developed to handle the specific problem of harmonic generation although extension to the case of parametric oscillation is straightforward. This includes calculations of the effects of phase matching [7], [13]–[17], focusing [5], [18], [19], double refraction [7], [17], [20], [21], and optical frequency resonators [22], [23]. A number of experiments have been performed involving parametric amplification [8], [24]–[27], and in 1965, Giordmaine and Miller [28] reported the first observation of tunable parametric oscillation. Since that time, several other observations of pulsed parametric oscillation have been reported [29]–[31] and recently Smith *et al.* [32] have observed CW parametric oscillation.

Fig. 1 is a schematic representation of a possible internal parametric oscillator with the pump, signal, and idler simultaneously resonated. In this noncollinear configuration, the interaction region is reduced; however, the problems of multiple wavelength antireflection coatings and mirror coatings characteristic of a collinear configuration are avoided. Two-dimensional parametric amplification of this sort has been observed external to the laser by Akhmanov *et al.* [26]. For analytical simplicity we consider the case of collinear interactions, although the extension to the noncollinear case is straightforward. Such an oscillator might be achieved experimentally using a dichroic beam splitter to divert the signal and idler while allowing the pump to travel through the laser medium.

The analysis leads to the surprising result that an internal parametric oscillator can operate in several previously unpredicted regimes. Briefly, there are three regions of operation: 1) an efficient regime with operating characteristics similar to those of previous analyses; 2) an inefficient regime in which the parametric coupling drives the phases rather than the amplitudes of the oscillations and wherein an interesting shift of signal, idler, and pump frequencies from their normal positions is observed; and 3) a repetitively pulsing regime characterized by short pulses of output power at the signal and idler, accompanied by nearly simultaneous decreases in power at the pump. These results are due basically to the fact that the signal and idler are coupled directly to the saturating gain mechanism of the pumping laser through the nonlinear interaction. The choice of parameters necessary for operation in a particular regime is discussed in terms of the stability conditions for the various types of operation.

After considering the internal oscillator, one is led to question the fundamental difference between it and an oscillator consisting of cavities external to the laser, which again resonate signal, idler, and pump. Again for high-Q cavities, the pump fields could be very high. Experimentally, the differences are immediately apparent. The difficulties of placing the crystal inside the laser cavi-

Manuscript received March 29, 1968; revised May 10, 1968. The work reported in this paper was sponsored by the National Aeronautics and Space Adminstration under NASA Grant NGR-05-020-103. M. K. Oshman's work was partially supported by Sylvania Electronic Systems through the Independent Research Program. Portions of this paper were presented at the 1966 Conference on Electron Device Research.

M. K. Oshman was with the Department of Electrical Engineering, Stanford University. He is now with Sylvania Electronic Systems, Mountain View, Calif.

S. E. Harris is with the Department of Electrical Engineering, Stanford University, Stanford, Calif.

Reprinted from *IEEE J. Quantum Electron.*, vol. QE-4, pp. 491–502, Aug. 1968.

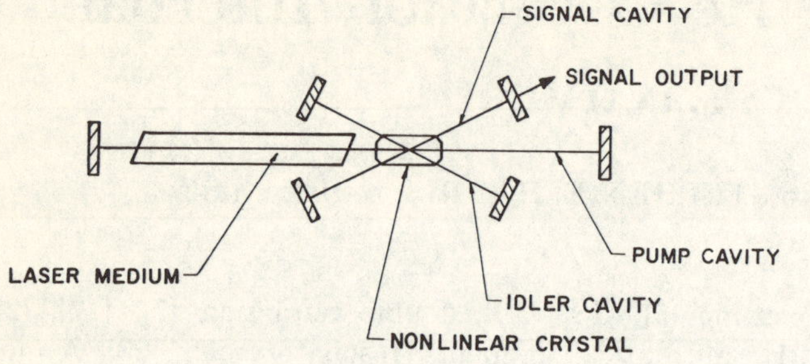

Fig. 1. One geometical arrangement of an internal parametric oscillator.

are large; however, based on other work that uses crystals inside a laser cavity, these seem to be problems that can be overcome. On the other hand, the difficulty of efficiently coupling the pump into an external oscillator is tremendous. Nevertheless, assuming one could optimally couple the pump into such an oscillator, the question remains as to the relative performance of the internal and external oscillator. In particular, as the external oscillator goes above threshold, the pump experiences reflections from the active cavity, thus reducing the efficiency of the oscillator. Similar reflections at microwave frequencies were observed by Ho and Siegman [33]. In Section IV, the efficiency of the external oscillator is compared to that of the internal oscillator, and it is apparent that reflections do render the external oscillator less efficient.

II. Equations of Motion for the Internal Parametric Oscillator

In deriving the equations of motion of the internal parametric oscillator, we use a technique similar to that used by Harris and McDuff [34] to analyze the FM laser. We use the self-consistency equations developed by Lamb [35], which describe the effect of an arbitrary optical polarization on the optical frequency electric fields of a high-Q multi-mode resonator. We direct our attention to three modes of interest with circular frequencies ω_1, ω_2, ω_3 satisfying the relation $\omega_1 + \omega_2 = \omega_3$. In the usual terminology of parametric interactions then ω_1, ω_2, and ω_3 are the signal, idler, and pump frequency, respectively. We shall label all quantities referring to the signal mode, idler, and pump with the subscripts 1, 2, and 3, respectively.

Neglecting transverse variations, the total electric field in the cavity can be written as

$$E(x, t) = \sum_{i=1}^{3} E_i(t) \cos [\omega_i t + \phi_i(t)] u_i(x), \tag{1}$$

where E_i is the amplitude of the ith mode and $u_i(x)$ is the [illegible] variation of the ith mode, which has circular [illegible] and is that mode lying closest in frequency [illegible] malize u_i such that

$$\int_0^L u_i u_j \, dx = \delta_{ij}, \tag{2}$$

[illegible] he cavity extend from $x = 0$ to $x = L$. [illegible] equations then are

$$[\omega_i + \dot{\phi}_i - \Omega_i] E_i = -\frac{1}{2}\left(\frac{\omega_i}{\epsilon_0}\right) C_i \tag{3}$$

$$\dot{E}_i + \frac{1}{2}\left(\frac{\omega_i}{Q_i}\right) E_i = -\frac{1}{2}\left(\frac{\omega_i}{\epsilon_0}\right) S_i. \tag{4}$$

Here Q_i is the Q of the ith mode resulting from the losses due to transmission, scattering, absorption, and diffraction. In deriving these equations, Q_i arises through the definition of a ficticious volume conductivity σ_i, which accounts for these losses. The Q_i and σ_i are related by

$$Q_i = \frac{\omega_i \epsilon_0}{\sigma_i}. \tag{5}$$

We will eventually divide σ_i into two parts, one resulting from volume loss effects and the other resulting from transmission through the mirrors. The C_i and S_i are related to the total polarization $P(x, t)$ through the following equations:

$$\begin{aligned} P_i(t) &= \int_0^L P(x, t) u_i(x) \, dx \\ &= C_i(t) \cos [\omega_i t + \phi_i(t)] \\ &\qquad + S_i(t) \sin [\omega_i t + \phi_i(t)]. \end{aligned} \tag{6}$$

In writing these self-consistency equations, $E_i(t)$ and $\phi_i(t)$ are assumed slowly varying with respect to ω_i.

In considering parametric interactions inside the laser, the polarization at a particular frequency will have two contributions. One primarily affects the pump and results from the presence of the laser medium. The other results from the nonlinear interactions of the electric fields at other frequencies.

The contribution to the polarization due to the laser medium can be introduced as macroscopic quadrature and in-phase components of susceptibility. Then we have

$$C_i(t) = \epsilon_0 \chi_i' E_i(t) \tag{7a}$$

and

$$S_i(t) = \epsilon_0 \chi_i'' E_i(t). \tag{7b}$$

We assume that the signal and idler frequencies are well removed from any transitions of the laser medium, so contributions to them are essentially negligible. For the pump χ_3'' accounts for gain and saturation due to the laser medium; χ_3' results in frequency shifts and mode pulling and pushing effects.

In proceeding to the calculation of the parametric contribution to the polarization, we shall use the specific example of lithium niobate ($LiNbO_3$) as the nonlinear element [36]. This is useful in performing the analysis; however, the technique can easily be generalized to any parametric system. In particular, with $LiNbO_3$, we can assume phase matching is achieved for propagation at 90° to the optic axis [37], as used in the experiments of Giordmaine and Miller [28]. In this way, we can avoid the problems of double refraction, which would needlessly complicate this analysis.

For $LiNbO_3$ oriented with its crystallographic x-axis

along the x-axis of the cavity, the polarization in the crystal is related to the electric fields by

$$P_y(x, t) = 2d_{15}E_yE_z \quad (8)$$

and

$$P_z(x, t) = d_{31}E_y^2. \quad (9)$$

The subscripts y and z refer to fields polarized along the y- and z-axes of the crystal. We shall assume that the signal and idler are ordinary rays polarized along the y-axis and the pump is an extraordinary ray polarized along the z-axis. We should point out that (8) and (9) relate the time-dependent nonlinear polarization to the time-dependent electric fields. Normally, the value of d is defined by relating the Fourier amplitude of the nonlinear polarization to the Fourier amplitudes of the electric fields. Therefore, as discussed by Pershan [38], *the numerical value of d in these equations is a factor of 2 larger than commonly quoted values.*

Using (8) and (9) in conjunction with (1) and (6), we therefore find the following parametric contributions to the polarization (Kleinman's symmetry condition [39] is assumed so that $d_{15} = d_{31}$):

$$C_1(t) = \delta E_2E_3 \cos(\phi_3 - \phi_1 - \phi_2) \quad (10a)$$

$$S_1(t) = -\delta E_2E_3 \sin(\phi_3 - \phi_1 - \phi_2) \quad (10b)$$

$$C_2(t) = \delta E_1E_3 \cos(\phi_3 - \phi_1 - \phi_2) \quad (10c)$$

$$S_2(t) = -\delta E_1E_3 \sin(\phi_3 - \phi_1 - \phi_2) \quad (10d)$$

$$C_3(t) = \delta E_1E_2 \cos(\phi_3 - \phi_1 - \phi_2) \quad (10e)$$

$$S_3(t) = \delta E_1E_2 \sin(\phi_3 - \phi_1 - \phi_2). \quad (10f)$$

Here δ is defined by

$$\delta = d_{15} \int_0^{L'} u_1(x)u_2(x)u_3(x)\, dx, \quad (11)$$

where we have assumed the crystal extends from $x = 0$ to $x = L'$. All contributions at frequencies well removed from the frequencies ω_1, ω_2, and ω_3 are neglected.

In evaluating δ we shall assume that the mirrors of the resonator represent a short-circuit surface for the signal and the idler and an open-circuit surface for the pump. Then the spatial variation of the modes inside the crystal is of the form

$$u_1(x) = \left(\frac{2}{L}\right)^{1/2} \sin k_1x \quad (12a)$$

$$u_2(x) = \left(\frac{2}{L}\right)^{1/2} \sin k_2x \quad (12b)$$

$$u_3(x) = \left(\frac{2}{L}\right)^{1/2} \cos k_3x. \quad (12c)$$

The reason for choosing these boundary conditions is apparent upon inspection of the expression for δ. If all three modes had short-circuit boundaries, then δ would be zero with no parametric interaction. This is the problem met in second harmonic generation where both the fundamental and pump are resonated [40], and results from the fact that a polarization wave is 90° out of phase with the electromagnetic wave it radiates. We define the phase mismatch in terms of the wave vectors as

$$\Delta k = k_3 - k_2 - k_1, \quad (13)$$

and retain only low-frequency terms in (11) with the result

$$\delta = -\left(\frac{2}{L}\right)^{1/2} \frac{L'}{L} d_{15} \frac{\sin \Delta kL'}{\Delta kL}. \quad (14)$$

In practice phase matching can be achieved at 90° to the optic axis by varying the temperature of the crystal [41], [42].

We now combine (6), (7), and (10) with (3) and (4). Since the dielectric constant in the crystal is substantially different from ϵ_0, when using (10) in (3) and (4), we replace ϵ_0 with ϵ, where ϵ is the dielectric constant of the crystal and is approximately the same at ω_1, ω_2, and ω_3. Then one finds

$$\dot{E}_1 = -A_1E_1 + \frac{\omega_1\delta}{2\epsilon} E_2E_3 \sin(\phi_3 - \phi_1 - \phi_2) \quad (15a)$$

$$\dot{E}_2 = -A_2E_2 + \frac{\omega_2\delta}{2\epsilon} E_1E_3 \sin(\phi_3 - \phi_1 - \phi_2) \quad (15b)$$

$$\dot{E}_3 = -A_3E_3 - \frac{\omega_3\delta}{2\epsilon} E_1E_2 \sin(\phi_3 - \phi_1 - \phi_2) \quad (15c)$$

$$\dot{\phi}_3 - \dot{\phi}_2 - \dot{\phi}_1 = -\Omega' + \frac{\delta}{2\epsilon}\left[\frac{\omega_3E_1E_2}{E_3} - \frac{\omega_1E_2E_3}{E_1} - \frac{\omega_2E_1E_3}{E_2}\right] \cdot \cos(\phi_3 - \phi_1 - \phi_2), \quad (15d)$$

where

$$A_1 = \frac{\omega_1}{2}\left[\frac{1}{Q_1} + \chi_1''\right] \quad (16a)$$

$$A_2 = \frac{\omega_2}{2}\left[\frac{1}{Q_2} + \chi_2''\right] \quad (16b)$$

$$A_3 = \frac{\omega_3}{2}\left[\frac{1}{Q_3} + \chi_3''\right] \quad (16c)$$

$$\Omega' = [\omega_3\chi_3' - \omega_2\chi_2' - \omega_1\chi_1']/2. \quad (17)$$

In treating the gain and saturation of the laser, we need to assume some functional form for χ_3''. There have been several different saturation functions suggested for describing the observed effects of various lasers. In this analysis we use a small-signal approximation similar to that derived by Lamb [35]. With this approximation, χ_3'' is related to the single-pass power gain by

$$\frac{\omega_3L}{c}\chi_3'' = -g_0(1 - \beta E_3^2), \quad (18)$$

where g_0 is the single-pass unsaturated power gain and β is a parameter accounting for the effects of saturation.

A more realistic saturation function for inhomogeneously broadened lasers is represented by [43]

$$\frac{\omega_3L}{c}\chi_3'' = -\frac{g_0}{(1 + 2\beta E_3^2)^{1/2}}. \quad (19)$$

The saturation function of (18) produces essentially the same results as (19) under the condition that

$$\left(\frac{g_0}{\alpha_3}\right)^2 - 1 \ll 1,$$

where α_3 is the single-pass power loss. There are a few low-gain lasers, such as Nd : YAG, under which this condition might well be satisfied. Further, we argue that for internal parametric oscillation, the signal and idler drain much of the laser power, thus not permitting the pump to saturate fully. Nevertheless, the reason for using the saturation function of (18) is its simplicity. We have carried out the entire subsequent analysis using (19) and find the same qualitative results. Since the objective here is to point out the general behavior of internal parametric oscillation, we believe the use of the small-signal saturation function is justified. In any particular application, one would have to calculate the following results, using the appropriate saturation function and appropriate parameters.

If we change the time variable so that $\tau = (c/2L)t$, then (15) and (16) become

$$\frac{dE_1}{d\tau} = -\alpha_1 E_1 + \omega_1 \kappa E_2 E_3 \sin(\phi_3 - \phi_1 - \phi_2) \tag{20a}$$

$$\frac{dE_2}{d\tau} = -\alpha_2 E_2 + \omega_2 \kappa E_1 E_3 \sin(\phi_3 - \phi_1 - \phi_2) \tag{20b}$$

$$\frac{dE_3}{d\tau} = -\alpha_3 E_3 + g_0(1 - \beta E_3^2)E_3 - \omega_3 \kappa E_1 E_2 \sin(\phi_3 - \phi_1 - \phi_2) \tag{20c}$$

$$\frac{d}{d\tau}(\phi_3 - \phi_1 - \phi_2) = \Omega + \kappa\left[\frac{\omega_3 E_1 E_2}{E_3} - \frac{\omega_1\omega_2\omega_3}{E_1} - \frac{\omega_2 E_1 E_3}{E_2}\right] \cdot \cos(\phi_3 - \phi_1 - \phi_2), \tag{20d}$$

where

$$\alpha_1 = \frac{L\omega_1}{c}\left[\frac{1}{Q_1} + \chi_1''\right] \tag{21a}$$

$$\alpha_2 = \frac{L\omega_2}{c}\left[\frac{1}{Q_2} + \chi_2''\right] \tag{21b}$$

$$\alpha_3 = \frac{L\omega_3}{c}\left[\frac{1}{Q_3}\right] \tag{21c}$$

$$\Omega = \frac{L}{c}\left[\omega_3\chi_3' - \omega_2\chi_2' - \omega_1\chi_1'\right] \tag{21d}$$

and

$$\kappa = \frac{L\delta}{\epsilon c}. \tag{22}$$

Here α_i is the single-pass power loss for the ith mode. These equations are similar in form to the general equations of motion for parametric interactions described by Siegman [44]. The main difference arises due to the introduction of a saturating gain mechanism. For that reason we believe the form of the subsequent results is generally applicable to other parametric systems coupled directly to the gain mechanism of the pump.

III. Operating Characteristics of the Internal Parametric Oscillator

Using (20a) to (20d) we are in a position to investigate the operating characteristics of the internal parametric oscillator. Previous analyses of parametric oscillators have resulted in the well-known threshold conditions for oscillation and one steady-state region of parametric oscillation. Below threshold, the signal and idler remain zero, whereas above threshold, the pump limits at the threshold level and additional pump power is converted to signal and idler power.

As opposed to this previous solution, the internal parametric oscillator displays three distinct steady-state regions of operation. One region of operation results in the expected behavior, which predicts efficient conversion of pump power to signal power. In addition there is another CW output of the oscillator wherein the phases rather than the amplitudes of the signal and idler are driven by the nonlinear interaction, thereby resulting in less efficient operation. Finally, there is a continuous relaxation oscillation type of solution. In this case the signal and idler spike on and off, resulting in a repetitively pulsing output from the oscillator.

A. Steady-State Solutions for the Internal Parametric Oscillator

Equations (20a) to (20d) produce three distinct steady-state solutions found by setting all derivatives with respect to τ equal to zero. For the following solutions we assume that the mode pulling term Ω is equal to zero. First for the case where there is no parametric interaction ($\kappa = 0$) we have

$$E_1 = E_2 = 0 \tag{23a}$$

$$E_3^2 = \frac{g_0 - \alpha_3}{g_0\beta}. \tag{23b}$$

Equation (23b) predicts the steady-state amplitude of the laser field. For future calculations it is convenient to find the output power of the laser. To do so we divide α_3, the single-pass power loss of the laser, into two parts and let

$$\alpha_3 = \alpha_{3c} + \alpha_{3d}, \tag{24}$$

where α_{3c} is that part of the loss resulting from output coupling and α_{3d} is that part of the loss attributed to all other dissipative mechanisms. Then the output power of the laser P_3 is proportional to $\alpha_{3c}E_3^2$ so

$$P_3 = \gamma\alpha_{3c}E_3^2, \tag{25}$$

where γ is the constant of proportionality. Maximizing P_3 with respect to α_{3c}, the maximum output power of the laser is

$$P_{3,\max} = \gamma\frac{(g_0 - \alpha_{3d})^2}{4g_0\beta}. \tag{26}$$

In the presence of parametric interaction, the first steady-state solution is found by taking $\kappa \neq 0$ and ϕ_3 −

$\phi_1 - \phi_2 = \pi/2$. Then the cosine term of (15d) is zero, the sine terms of (15a) to (15d) are one, and

$$E_1^2 = \frac{1}{\omega_2\omega_3\kappa^2}\left[(g_0 - \alpha_3)\alpha_2 - g_0\beta\frac{\alpha_1\alpha_2^2}{\omega_1\omega_2\kappa^2}\right] \tag{27a}$$

$$E_2^2 = \frac{1}{\omega_1\omega_3\kappa^2}\left[(g_0 - \alpha_3)\alpha_1 - g_0\beta\frac{\alpha_1^2\alpha_2}{\omega_1\omega_2\kappa^2}\right] \tag{27b}$$

$$E_3^2 = \frac{\alpha_1\alpha_2}{\omega_1\omega_2\kappa^2}. \tag{27c}$$

This is the usual efficient solution for parametric oscillation.

$$\frac{g_0 - \alpha_3}{g_0\beta} < \frac{\alpha_1\alpha_2}{2\omega_1\omega_2\kappa^2}\left[1 + 5\sqrt{1 + \frac{24\alpha_1\alpha_2(\alpha_1 - \alpha_2)^2}{\{(\alpha_1 + \alpha_2)(\alpha_1 + \alpha_2 + \alpha_3 - g_0) - 10\alpha_1\alpha_2\}^2}}\right.$$
$$\left. + \frac{(\alpha_1 + \alpha_2)(\alpha_1 + \alpha_2 + \alpha_3 - g_0)}{4}\left[1 - \sqrt{1 + \frac{24\alpha_1\alpha_2(\alpha_1 - \alpha_2)^2}{\{(\alpha_1 + \alpha_2)(\alpha_1 + \alpha_2 + \alpha_3 - g_0) - 10\alpha_1\alpha_2\}^2}}\right]\right]. \tag{32}$$

The final steady-state solution is found by assuming $\phi_3 - \phi_1 - \phi_2 =$ a constant $\neq \pi/2$:

$$E_1^2 = \frac{\omega_1}{\omega_3}\frac{\alpha_1 + \alpha_2}{\alpha_1}E_3^2 \tag{28a}$$

$$E_2^2 = \frac{\omega_2}{\omega_3}\frac{\alpha_1 + \alpha_2}{\alpha_2}E_3^2 \tag{28b}$$

$$E_3^2 = \frac{g_0 - \alpha_1 - \alpha_2 - \alpha_3}{g_0\beta} \tag{28c}$$

$$\sin^2(\phi_3 - \phi_1 - \phi_2) = \frac{g_0\beta\alpha_1\alpha_2}{g_0 - \alpha_1 - \alpha_2 - \alpha_3}\frac{1}{\omega_1\omega_2\kappa^2}. \tag{28d}$$

B. Stability Criteria for Steady-State Solutions

Since there are several steady-state solutions for the internal oscillator, we must find the necessary conditions that a particular solution be stable. There are a number of techniques for approaching this problem, including Liapunov's stability criteria [45] and the Hurwitz test [46], both of which were used. We quote only the results. At the outset we should point out that we find the stability conditions of the various regions to be mutually exclusive. That is, two steady-state solutions are not simultaneously possible.

As is expected, the stability of the free running laser solution is guaranteed so long as parametric oscillation is not possible. That is, below a certain level of pump power, the signal and idler are zero. This condition (for stability of the laser with no parametric oscillation) is just the inverse of the threshold condition for parametric oscillation and is given by

$$E_3^2 = \frac{g_0 - \alpha_3}{g_0\beta} < \frac{\alpha_1\alpha_2}{\omega_1\omega_2\kappa^2}. \tag{29}$$

Stability of the first steady-state parametric oscillation solution (27a) to (27c) requires that three conditions be satisfied. First there is the threshold condition:

$$\frac{g_0 - \alpha_3}{g_0\beta} > \frac{\alpha_1\alpha_2}{\omega_1\omega_2\kappa^2}. \tag{30}$$

The second condition requires that if the single-pass gain of the laser is greater than the sum of the loss to the pump, signal, and idler, i.e.,

$$g_0 > \alpha_1 + \alpha_2 + \alpha_3,$$

then for stability,

$$\frac{g_0 - \alpha_1 - \alpha_2 - \alpha_3}{g_0\beta} < \frac{\alpha_1\alpha_2}{\omega_1\omega_2\kappa^2}. \tag{31}$$

This second condition proves to be the condition that the steady-state solution of (28a) to (28d) not be stable. Finally, there is a third stability condition given by the rather complicated relation (32).

For a better feeling of what this relation entails we assume that the signal loss equals the idler loss, i.e., $\alpha_1 = \alpha_2$, in which case (32) becomes

$$\frac{g_0 - \alpha_3}{g_0\beta} < \frac{3\alpha_1\alpha_2}{\omega_1\omega_2\kappa^2}.$$

This shows that the free running laser power should be less than three times that required for parametric threshold.

This last stability requirement is connected with the third type of operation of the oscillator previously described—the pulsing output. If this condition (32) is not satisfied, then the output of the oscillator consists of a continuous train of pulses. The solution for this region cannot be solved in closed form; however, we have performed a detailed computer solution of this region, which is described in Section III-E.

Finally, for the second steady-state parametric oscillation solution (28a) to (28d) to be stable, it is necessary that

$$g_0 > \alpha_1 + \alpha_2 + \alpha_3 \tag{33}$$

and

$$\frac{g_0 - \alpha_1 - \alpha_2 - \alpha_3}{g_0\beta} > \frac{\alpha_1\alpha_2}{\omega_1\omega_2\kappa^2}. \tag{34}$$

C. Efficient Internal Parametric Oscillation

In describing the characteristics of the various regions of operation of the internal parametric oscillator, it is convenient to define a parameter κ_{th}. This parameter represents that value of κ necessary to overcome threshold in the absence of any output coupling loss to the pump, signal, or idler and is given therefore by the relation

$$\kappa_{\text{th}}^2 = \frac{\alpha_{1d}\alpha_{2d}}{\omega_1\omega_2}\frac{g_0\beta}{g_0 - \alpha_{3d}}, \tag{35}$$

where we have divided α_i into two components such that

$$\alpha_i = \alpha_{ic} + \alpha_{id}. \tag{36}$$

Experimentally, κ is variable by adjusting phase-matching

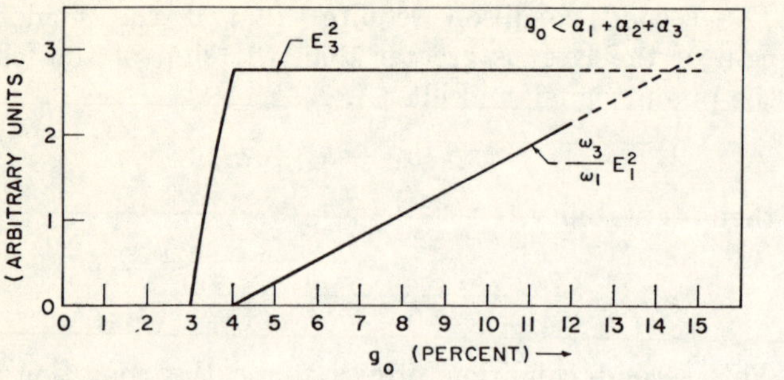

Fig. 2. Pump and signal power inside resonator versus single-pass laser gain in region of efficient parametric oscillation.

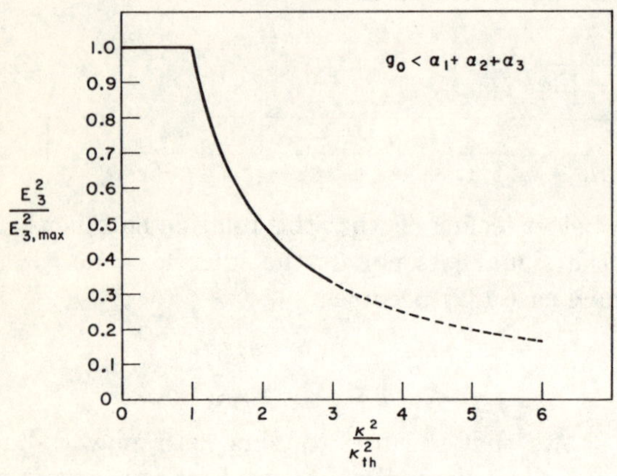

Fig. 3. Pump power inside resonator versus $(\kappa/\kappa_{th})^2$ in region of efficient parametric oscillation.

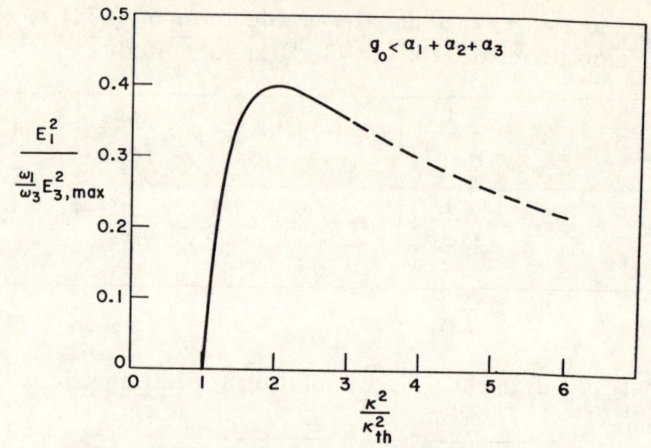

Fig. 4. Signal power inside resonator versus $(\kappa/\kappa_{th})^2$ in region of efficient parametric oscillation.

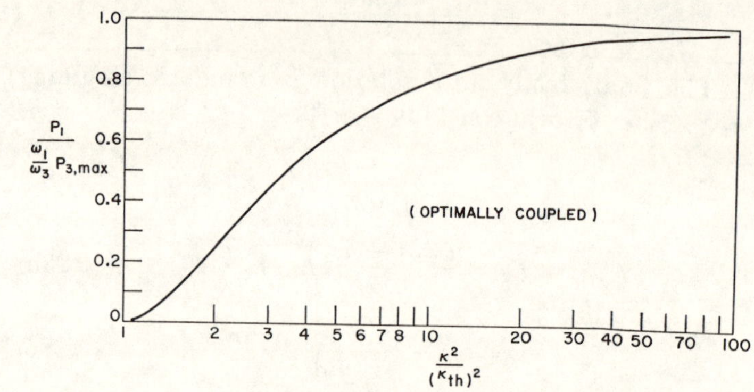

Fig. 5. Optimally coupled signal output power.

conditions, focusing, or crystal nonlinearity. Therefore, the ratio of κ to κ_{th} is a realistic variable and represents the strength of the nonlinear interaction with respect to the strength required for threshold.

Figs. 2 to 4 display some characteristics of the oscillator for the efficient steady-state operation of (27a) to (27c). These figures show the form of the onset of oscillation and the buildup of power for the signal for typical laser operating parameters. In each case the dotted lines indicate the onset of the pulsing output of the oscillator and are continued to show where the steady-state solution would have operated.

In Fig. 2, E_1^2 and E_3^2 are plotted versus laser gain. For the parameters chosen in this case, the laser reaches its oscillation threshold at a gain of 3 percent per pass and increases in power until the parametric oscillator reaches threshold. At that point, the laser limits and further increases in gain result in signal (and idler) power.

In Figs. 3 and 4 the pump and signal power are plotted versus κ^2/κ_{th}^2; $E_{3,max}^2$ is the value that the square of pump electric field amplitude has in the absence of parametric oscillation. As the parametric interaction increases, the pump power continually decreases. Simultaneously, the signal power first increases, goes through a maximum, and begins to decrease. Qualitatively, this can be understood as the result of the fact that the parametric interaction appears to be an increasing loss to the laser. Eventually, the total available power from the gain mechanism goes through a maximum and begins to decrease, much in the same way that increasing output coupling losses to a laser can produce a similar maximum in output power.

In order to determine the efficiency of the oscillator, one must determine the proper output coupling to the signal for given operating conditions. In terms of the ratio κ^2/κ_{th}^2, the efficiency of the oscillator is independent of all other parameters.

The optimum coupling to the signal is found by maximizing the output signal power P_1 with respect to α_{1c}. In a manner analogous to (25), P_1 is given by

$$P_1 = \gamma \alpha_{1c} E_1^2. \tag{37}$$

In terms of the maximum available pump power $P_{3,max}$, the maximization produces, when optimally coupled, the result

$$P_1 = \frac{\omega_1}{\omega_3} P_{3,max} \left[1 - \left(\frac{\kappa_{th}}{\kappa} \right)^2 \right]^2. \tag{38}$$

With perfect conversion, the Manley–Rowe relations limit the signal power to be ω_1/ω_3 times the pump power. Therefore, we see that the internal oscillator satisfies this condition since that is the asymptotic value that the signal power approaches. Fig. 5 is a plot of the relation given by (38). From this graph it is apparent that once the oscillator is above threshold it rapidly becomes very efficient.

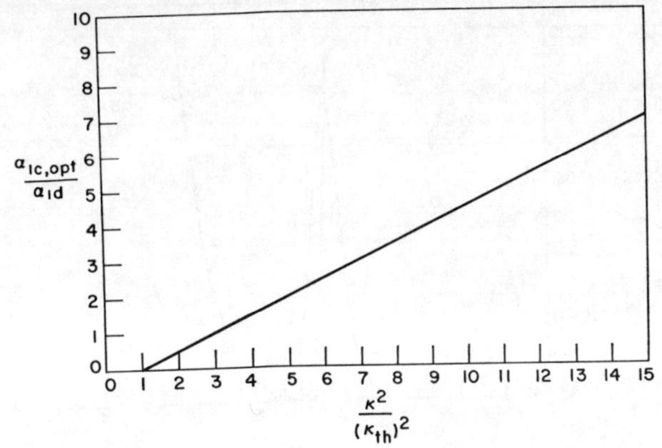

Fig. 6. Ratio of optimum signal coupling loss to signal dissipative loss.

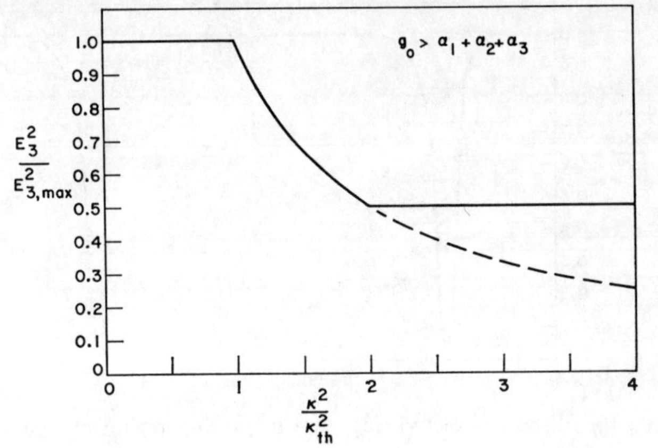

Fig. 8. Pump power versus $(\kappa/\kappa_{th})^2$ showing onset of inefficient region of operation.

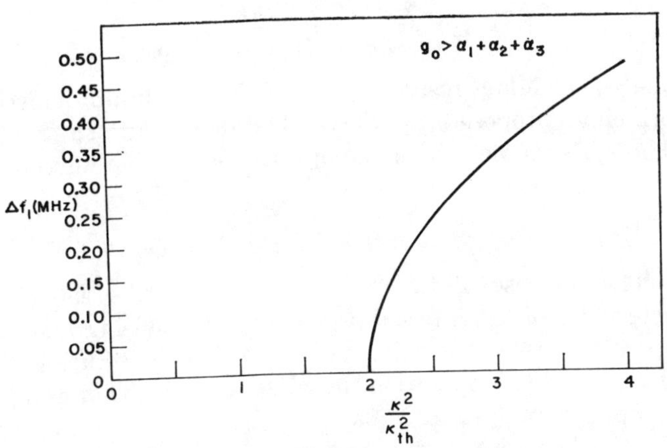

Fig. 7. Frequency shift of signal versus $(\kappa/\kappa_{th})^2$.

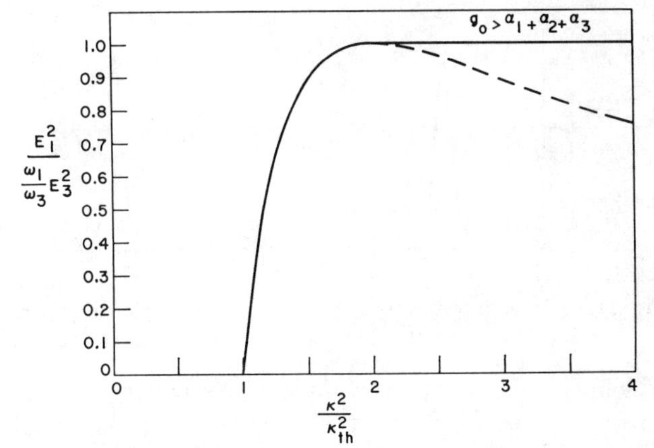

Fig. 9. Signal power versus $(\kappa/\kappa_{th})^2$ showing onset of inefficient region of operation.

Fig. 6 is a plot of the optimum ratio of coupling loss to dissipative loss versus κ^2/κ_{th}^2 and is given by

$$\frac{\alpha_{1c,\text{opt}}}{\alpha_{1d}} = \frac{1}{2}\left[\left(\frac{\kappa^2}{\kappa_{th}^2}\right) - 1\right]. \tag{39}$$

Combining these results with the stability condition given by (32), it is possible to show that should efficient operation be desired, no pulsing output will be experienced so long as optimum coupling to the signal is maintained.

D. *Inefficient Steady-State Region of Operation*

The steady-state region of operation described by (28a) to (28d) represents a case in which the parametric interaction drives the phases of the signal, idler, and pump rather than their amplitudes, thereby resulting in less efficient operation. In general, this is probably a region to be avoided in almost any practical case by appropriately adjusting the operating parameters of the oscillator. Nevertheless, its characteristics are important in understanding the behavior of the internal oscillator.

In (20d) of Section II only one equation was written for all the phases. This equation is the sum of three equations of the form

$$\frac{d\phi_1}{d\tau} = \frac{\omega_1 L}{C}\chi_1' + \frac{\omega_1 \kappa E_2 E_3}{E_1}\cos(\phi_3 - \phi_1 - \phi_2). \tag{40}$$

The fact that $\phi_1 + \phi_2 - \phi_3$ = a constant results from the relations that

$$\dot{\phi}_1 = \Delta\omega_1 \tag{41a}$$

$$\dot{\phi}_2 = \Delta\omega_2 \tag{41b}$$

$$\dot{\phi}_3 = \Delta\omega_3, \tag{41c}$$

where $\Delta\omega_1 + \Delta\omega_2 - \Delta\omega_3 = 0$. That is, $\dot{\phi}_i$ is a constant, representing a shift in frequency of ω_i to $\omega_i + \Delta\omega_i$. These frequency shifts are given by the following equations:

$$\Delta\omega_1 = \alpha_1 \cot(\phi_3 - \phi_1 - \phi_2) \times \frac{c}{2L} \tag{42a}$$

$$\Delta\omega_2 = \alpha_2 \cot(\phi_3 - \phi_1 - \phi_2) \times \frac{c}{2L} \tag{42b}$$

$$\Delta\omega_3 = (\alpha_1 + \alpha_2)\cot(\phi_3 - \phi_1 - \phi_2) \times \frac{c}{2L}, \tag{42c}$$

where

$$\sin^2(\phi_3 - \phi_1 - \phi_2) = \frac{g_0 \beta \alpha_1 \alpha_2}{g_0 - \alpha_1 - \alpha_2 - \alpha_3}\frac{1}{\omega_1 \omega_2 \kappa^2}. \tag{43}$$

The shifts are on the order of a megahertz as shown by Fig. 7 for typical operating parameters.

Figs. 8 and 9 show the behavior of the internal oscillator,

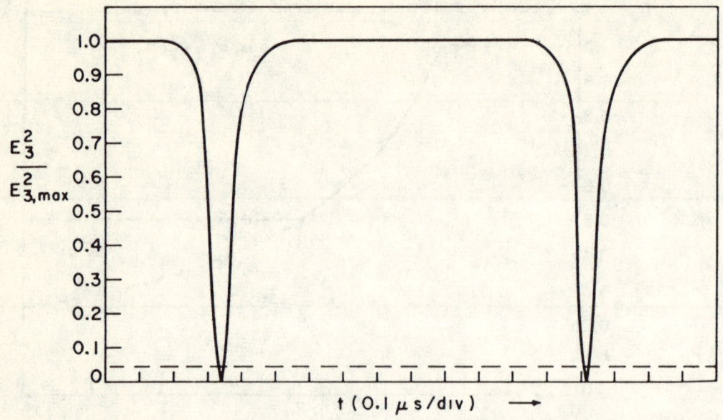

Fig. 10. Pump power versus time in pulsing region of operation.

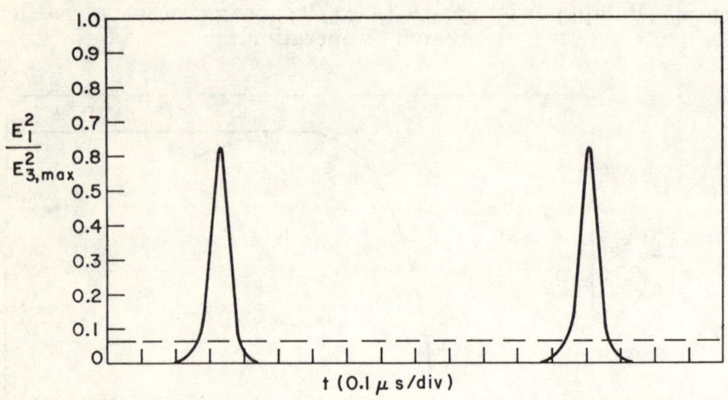

Fig. 11. Signal power versus time in pulsing region of operation.

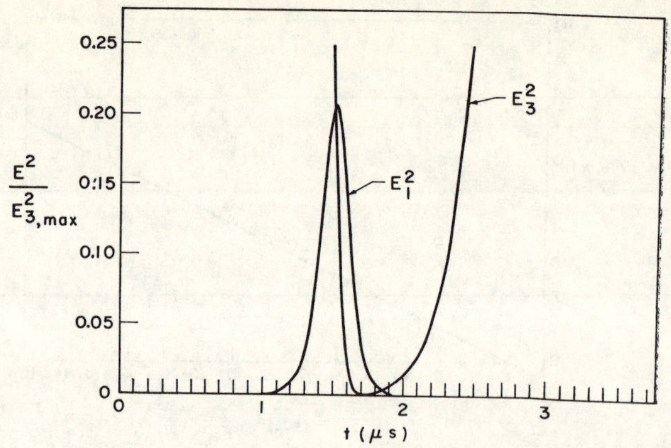

Fig. 12. Details of one pump and signal pulse.

which is able to operate in this inefficient region. The dotted curves represent the behavior had the oscillator remained in the other steady-state region. Once the value of κ^2 is large enough to reach threshold for this second type of steady-state operation, then the pump and signal power no longer change for increasing parametric interaction. The fact that the signal power limits at a maximum in Fig. 9 is a coincidence of the particular parameters chosen for plotting the curve.

It is difficult to decide exactly what is the physical reason for this type of operation. However, we can make some comments that do aid in understanding this behavior.

First we notice that the expression for the pump power (proportional to E_3^2) has very much the same form as the similar expression for pump power in the case of no parametric oscillation. The only difference is that α_3 is replaced by $\alpha_1 + \alpha_2 + \alpha_3$. In addition, a threshold condition for this region of operation is $g_0 > \alpha_1 + \alpha_2 + \alpha_3$, which is similar to the threshold condition for the laser with, once more, the laser loss replaced by the sum of the losses to signal, idler, and pump. For these reasons, one is led to speculate that in this region of operation, the laser appears to have its gain mechanism coupled directly to all three circuits: signal, idler, and pump. Then all three frequencies oscillate with many characteristics similar to a laser rather than a laser pumped parametric oscillator.

Since there are two solutions possible under the condition that $g_0 > \alpha_1 + \alpha_2 + \alpha_3$, the question then arises as to why the system picks one mode of operation as opposed to the other. By investigating the second threshold condition, i.e.,

$$\kappa^2 \geq \frac{\alpha_1\alpha_2}{\omega_1\omega_2} \frac{g_0\beta}{g_0 - \alpha_1 - \alpha_2 - \alpha_3}$$

one then finds a result similar to the maximum-emission principle proposed by Statz, DeMars, and Tang [47]. The systems pick the mode of operation which maximizes the pump power.

E. Pulsing Output of Internal Parametric Oscillator

The purpose of this section is to describe the characteristics of the internal oscillator in its pulsing mode of operation. It has been impossible to find the solutions for this region of operation in closed form. Therefore, a computer analysis of (20a) to (20d) was used to solve for the behavior in this region. The objective has been to outline the qualitative behavior of this spiking regime, and laser parameters were chosen that we believe are representative of existing lasers and materials.

Before proceeding, it is pointed out that we have not used any rate equation approach [48] in developing the equations of motion. Therefore, for some very slow atomic mechanisms (e.g., for use with CO_2 lasers) another equation must be included to account for population changes of the energy levels. Nevertheless, for most lasers, the results of this analysis appear to be consistent with the assumption that no rate equation is necessary.

Figs. 10 and 11 show typical results for the signal and pump fields as a function of time in the pulsing region of operation. The dotted lines represent the corresponding levels had pulsing not begun. The computer analysis has been carried out through many pulses and these appear to be constant in peak height and period. We therefore characterize this region of operation as a repetitively pulsing regime.

Fig. 12 shows an expanded version of one pulse, pointing out the relation between the signal and pump during the course of one pulse. Using this graph we can qualitatively explain the physical reason for this type of operation.

Since the oscillator is well above threshold (because of the stability condition of (32)), the signal and idler build up very rapidly. (As is well known, the buildup time of a parametric oscillator decreases as the margin above

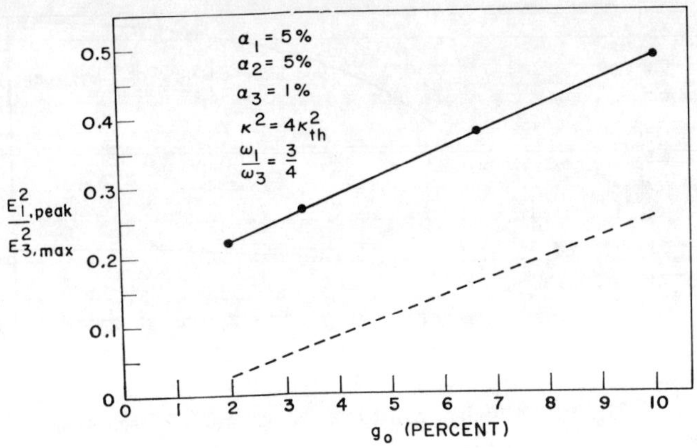

Fig. 13. Peak signal power versus laser gain.

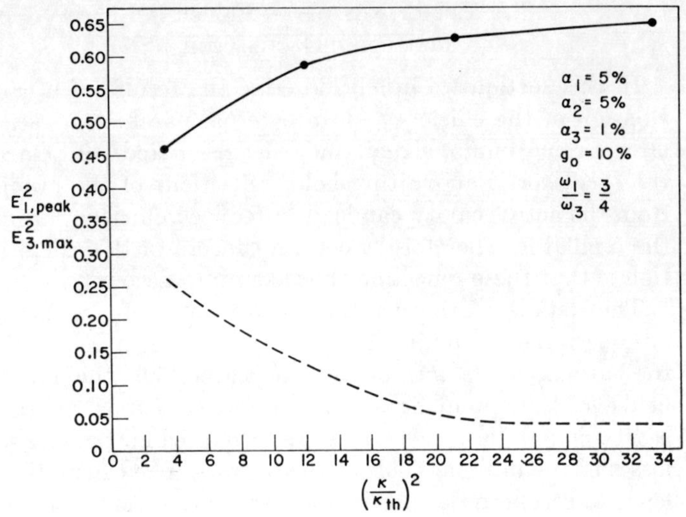

Fig. 16. Peak signal power versus $(\kappa/\kappa_{th})^2$.

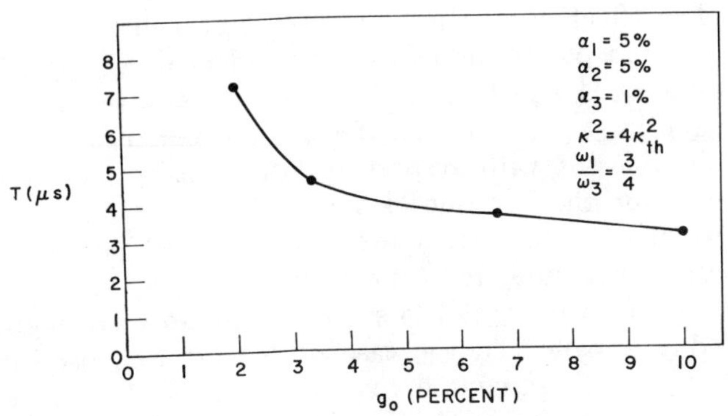

Fig. 14. Pulse period versus laser gain.

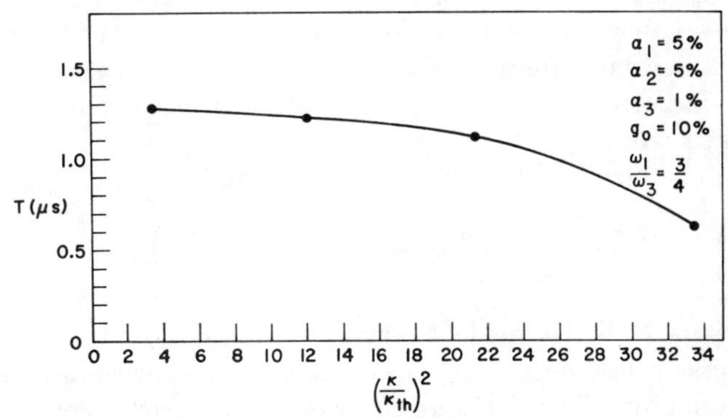

Fig. 17. Pulse period versus $(\kappa/\kappa_{th})^2$.

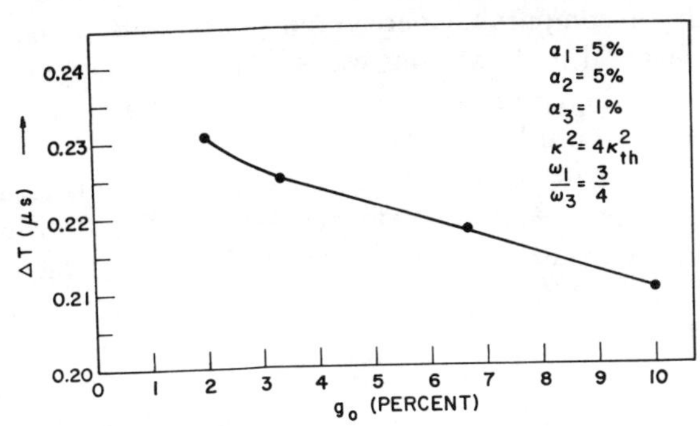

Fig. 15. Pulse width versus laser gain.

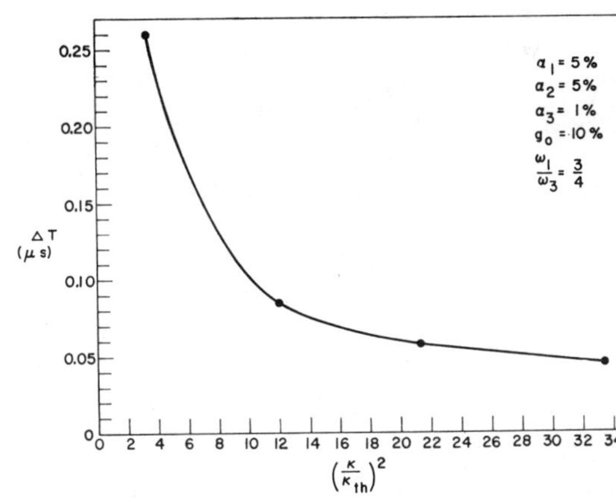

Fig. 18. Pulse width versus $(\kappa/\kappa_{th})^2$.

threshold increases.) The pump is slowed in its decay somewhat by the bandwidth of its resonance. Therefore the signal and idler buildup past the point to which they should have approached, thus draining more power from the pump. As a result, the pump experiences sufficient effective loss to go below threshold. The continuing decrease of the pump is eventually felt by the signal and idler and they then decrease rapidly. With no signal and idler present the pump can once more buildup to its free running value and the process repeats.

Figs. 13 to 15 display the change in pulse parameters as the gain of the laser is changed. The dotted line in Fig. 13 is proportional to the signal power in the steady-state region of operation. Here T is the period between pulses and ΔT is the full width of the pulses between one-half power points. As the gain of the laser is increased, the peak power in the pulses increases and their period and width decrease.

In Figs. 16 to 18 the same characteristics of the pulses are plotted versus κ^2/κ_{th}^2. As κ^2 increases, again the peak power of the pulses increases, and their period and width decrease.

IV. Efficiency of an External Parametric Oscillator

In this section we briefly describe the results of a calculation of the efficiency of an external parametric oscillator having pump, signal, and idler resonated [49]. Once the oscillator is above threshold, reflections of the pump from the active cavity can lead to reduced efficiency from the oscillator. Therefore, a central concern of the calculation is that these reflections be taken into account.

The details of the calculation are not included here, although the technique is summarized. Beginning with the normal mode formulation of Slater [50], the input fields at the pump appear as surface integrals on the cavity boundaries. Reflections are accounted for by energy losses of the internal fields at the mirrors. The calculation leads to second order partial differential equations for the amplitudes of the normal modes at the signal, idler, and pump frequencies. The procedure is very similar to the technique used by Gordon and Rigden [51] in analyzing the Fabry–Perot electrooptic modulator, and the partial differential equations for the mode amplitudes of the parametric oscillator are a straightforward extension of their work.

The result of the calculation of signal power P_1 yields

$$\frac{P_1}{P_3} = \frac{\omega_1}{\omega_3}\frac{\alpha_3 c}{\alpha_3}\left[\frac{\alpha_{1c}}{\sqrt{(\alpha_{1c}+\alpha_{1d})\alpha_{1d}}}\frac{\kappa_{th}}{\kappa} - \frac{\alpha_{1c}}{\alpha_{1d}}\left(\frac{\kappa_{th}}{\kappa}\right)\right]^2, \tag{44}$$

where P_3 is the incident pump power and α_{3c} is the transmission loss necessary for coupling pump power into the resonator. Therefore, α_3 consists of two parts, coupling loss α_{3c} and dissipative loss α_{3d}. In this expression, κ/κ_{th} is reduced from its previously defined value by the ratio α_{3c}/α_3.

Defining

$$\beta = \frac{\alpha_{1c}}{\alpha_{1d}} \tag{45}$$

the expression for P_1/P_3 can be written as

$$\frac{P_1}{P_3} = \frac{\omega_1}{\omega_3}\frac{\alpha_{3c}}{\alpha_3} f(\beta), \tag{46}$$

where

$$f(\beta) = \frac{\kappa_{th}}{\kappa}\frac{\beta}{\sqrt{1+\beta}} - \frac{\beta\kappa_{th}^2}{\kappa^2}. \tag{47}$$

Maximizing $f(\beta)$ yields the coupling to the signal that produces maximum output power. Fig. 19 is a plot of this function maximized with optimum coupling $f(\beta_m)$. On the same scale we have plotted the efficiency of the internal parametric oscillator $(\omega_3 P_1/\omega_1 P_{3,\max})$ also in terms of $(\kappa/\kappa_{th})^2$ for that oscillator.

There could be some objection to displaying the data in this form. However, with a brief explanation it becomes useful in comparing the relative efficiencies of the two oscillators. To the extent that the crystal represents no dissipative loss to the pump (in the form of absorption, scattering, or reflection), the value of κ/κ_{th} is the same

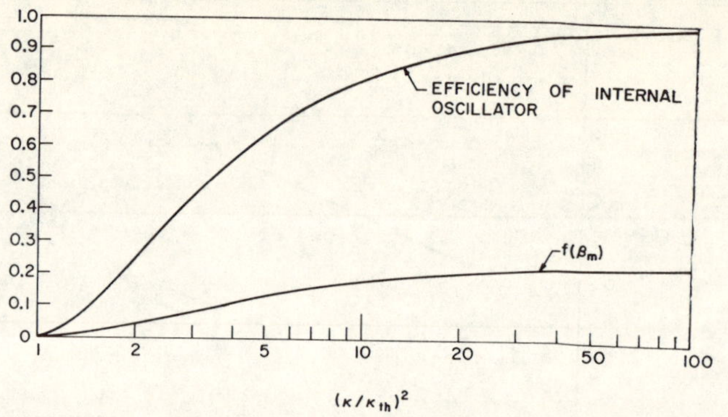

Fig. 19. Efficiency of external parametric oscillator.

in both the internal and external cases. Any real crystal will present some loss, however.

The effect on the internal oscillator will be to lower the pump power through its dependence on losses. The extent of this reduction depends basically on the gain of the laser medium and the relative magnitude of the crystal losses with respect to other dissipative losses. Typically, the insertion of a crystal inside a laser cavity may reduce the laser power by 25 to 50 percent [52] with a comparable reduction in $(\kappa/\kappa_{th})^2$.

The effect of crystal losses at the pump wavelength is readily apparent in the case of the external oscillator. First $(\kappa/\kappa_{th})^2$ is reduced by the factor $(\alpha_{3c}/\alpha_{3c} + \alpha_{3d})^2$ and at a given value of κ/κ_{th}, the output power is reduced by the factor $\alpha_{3c}/\alpha_{3c} + \alpha_{3d}$. It seems reasonable to assume that in many cases the net effect of losses will therefore result in comparable percentage reductions of power in the internal and external cases. Based on this reasoning we can therefore use Fig. 19 to compare the output power of the internal and external oscillators in the case of no losses at the pump and conclude that this is similar to their relative efficiency in the case with losses. However, we caution that this is only a rule of thumb, which could be slightly different for very high-gain or very low-gain lasers.

We can draw some fairly firm conclusions on the relative merits of the internal and external oscillators from these curves. The efficiency of the external oscillator is greatly reduced due to reflections of the pump from the active cavity. As a result, in cases where the crystal losses inserted into the laser do not substantially reduce the laser power (for example to the point that the laser itself is only slightly above threshold), the internal oscillator should produce the higher output power. In addition, the difficulty of coupling the pump into the external oscillator will make the internal oscillator more useful in practice.

V. Summary and Conclusions

Having solved the equations of motion for a parametric oscillator internal to the laser cavity and thereby coupled directly to the saturating gain mechanism of the laser, we have found that such an oscillator can operate in three distinctly different regimes: an efficient regime, a repetitively pulsing regime, and an inefficient regime characterized by frequency shifts of the pump, signal, and idler. The necessary conditions for operation in one particular mode have been derived in terms of the stability criteria for that regime. The regimes are mutually exclusive and may be chosen by varying the laser gain, the oscillator output coupling, or the crystal nonlinearity.

To some extent, each regime may be useful in practice. The pulsing regime offers peak powers slightly greater than the average power of the CW regimes, although the average power is somewhat lower. Some control of repetition rate and length of the pulses is available through changes in the laser gain and crystal nonlinearity. The inefficient regime is generally less useful than the other two regimes, although some applications might make use of the several megahertz frequency shifts of the pump, signal, and idler.

The efficient regime is probably of greatest interest. An important result of the analysis shows that this regime is capable of approaching 100 percent efficiency in conversion of pump power to signal and idler power. By way of comparison, the efficiency of an external oscillator, with pump, signal, and idler resonated, is limited to about 25 percent.

Acknowledgment

It is a pleasure to acknowledge many helpful discussions with B. Byer, A. E. Siegman, E. O. Ammann, and W. B. Tiffany. Mrs. Cora Barry expertly prepared the computer programs.

References

[1] N. M. Kroll, "Parametric amplification in spatially extended media and applications to the design of tuneable oscillators at optical frequencies," *Phys. Rev.*, vol. 127, pp. 1207–1211, August 15, 1962.
[2] J. K. Wright, "Enhancement of second harmonic power generated by a dielectric crystal inside a laser cavity," *Proc. IEEE*, vol. 51, p. 1663, November 1963.
[3] R. G. Smith, K. Nassau, and M. F. Galvin, "Efficient continuous optical second-harmonic generation," *Appl. Phys. Letters*, vol. 7, pp. 256–258, November 15, 1965.
[4] A general treatment of parametric amplifier and oscillator techniques with an extensive bibliography is given by W. H. Louisell, *Coupled Mode and Parametric Electronics*. New York: Wiley, 1960.
[5] R. H. Kingston, "Parametric amplification and oscillation at optical frequencies," *Proc. IRE*, vol. 50, p. 472, April 1962.
[6] S. A. Akhmanov and R. V. Khokhlov, *Zh. Eksperim i Theor. Fiz.*, vol. 43, pp. 351–353, July 1962 (transl.: "Concerning one possibility of amplification of light waves," *Soviet Phys.—JETP*, vol. 16, pp. 252–254, January 1963).
[7] J. A. Armstrong, N. Bloembergen, J. Ducuing, and P. S. Pershan, "Interactions between light waves in a nonlinear dielectric," *Phys. Rev.*, vol. 127, pp. 1918–1939, September 15, 1962.
[8] G. D. Boyd and A. Askin, "Theory of parametric oscillator threshold with single-mode optical masers and observation of amplification in $LiNbO_3$," *Phys. Rev.*, vol. 146, pp. 187–198, June 3, 1966.
[9] V. N. Lugovoy, *Radiotekhn. i Electron.*, vol. 9, p. 596, 1964 (transl.: "A cavity parametric amplifier and oscillator," *Radio Eng. Electron.*, vol. 9, pp. 483–492, April 1964).
[10] S. A. Akhmanov, V. G. Dmitriyev, V. P. Modenov, and V. V. Fadeev, *Radiotekhn. i Electron.*, vol. 10, p. 2157, 1965 (transl.: "Parametric generation in a resonator filled with a nonlinear dielectric," *Radio Eng. Electron.*, vol. 10, pp. 1841–1849, December 1965).
[11] S. A. Akhmanov and R. V. Khokhlov, *Usp. Fiz. Nauk.*, vol. 88, pp. 439–484, March 1966 (transl.: "Parametric amplifiers and generators of light," *Soviet Phys.—Usp.*, vol. 9, pp. 210–222, September–October 1966).
[12] A. Yariv and W. H. Louisell, "Theory of the optical parametric oscillator," *IEEE J. Quantum Electronics*, vol. QE-2, pp. 418–424, September 1966.
[13] J. A. Giordmaine, "Mixing of light beams in crystals," *Phys. Rev. Letters*, vol. 8, pp. 19–20, January 1, 1962.
[14] P. D. Maker, R. W. Terhune, M. Nisenoff, and C. M. Savage, "Effects of dispersion and focusing on the production of optical harmonics," *Phys. Rev. Letters*, vol. 8, pp. 21–22, January 1, 1962.
[15] S. A. Akhmanov, A. I. Kovrigin, R. V. Khokhlov, and O. N. Chunaev, *Zh. Eksper. i Theor. Fiz.*, vol. 45, pp. 1336–1343, November 1963 (transl.: "Coherent interaction length of light waves in a nonlinear dielectric," *Soviet Phys.—JETP*, vol. 18, pp. 919–924, April 1964).
[16] J. E. Midwinter and J. Warner, "The effects of phase matching method and of uniaxial crystal symmetry of the polar distribution of second-order non-linear optical polarization," *Brit. J. Appl. Phys.*, vol. 16, pp. 1135–1142, August 1965.
[17] D. A. Kleinman, "Theory of second harmonic generation of light," *Phys. Rev.*, vol. 128, pp. 1761–1775, November 15, 1962.
[18] J. E. Bjorkholm, "Optical second-harmonic generation using a focused gaussian laser beam," *Phys. Rev.*, vol. 142, pp. 126–136, February 1966.
[19] D. A. Kleinman, A. Ashkin, and G. D. Boyd, "Second-harmonic generation of light by focused laser beams," *Phys. Rev.*, vol. 145, pp. 338–379, May 6, 1966.
[20] N. Bloembergen and P. S. Pershan, "Light waves at the boundary of non-linear media," *Phys. Rev.*, vol. 128, pp. 606–622, October 15, 1962.
[21] G. D. Boyd, A. Ashkin, J. M. Dziedzic, and D. A. Kleinman, "Second-harmonic generation of light with double refraction," *Phys. Rev.*, vol. 137, pp. A1305–A1320, February 15, 1965.
[22] A. Ashkin, G. D. Boyd, and J. M. Dziedzic, "Resonant optical second-harmonic generation and mixing," *IEEE J. Quantum Electronics*, vol. QE-2, pp. 109–124, June 1966.
[23] S. A. Akhmanov, V. G. Dmitriev, and V. P. Modenov, *Radiotekhn. i Electron.*, vol. 10, p. 649, 1965 (transl.: "On the theory of frequency multiplication in a cavity resonator filled with a nonlinear medium," *Radio Eng. Electron.*, vol. 10, pp. 552–559, April 1965).
[24] C. C. Wang and C. W. Racette, "Measurement of parametric gain accompanying optical difference frequency generation," *Appl. Phys. Letters*, vol. 6, pp. 169–171, April 15, 1965.
[25] S. A. Akhmanov, A. I. Kovrigin, A. S. Piskarskas, V. V. Fadeev, and R. V. Khokhlov, *Zh. Eksperim. i Theor. Fiz., Pis'ma Redakt.*, vol. 2, pp. 300–305, 1965 (transl.: "Observation of parametric amplification in the optical range," *JETP Letters*, vol. 2, pp. 191–193, October 1, 1965).
[26] S. A. Akhmanov, A. G. Ershov, V. V. Fadeev, R. V. Khokhlov, O. N. Chunaev, and E. M. Shvom, *Zh. Eksperim. i Theor. Fiz., Pis'ma Redakt.*, vol. 2, pp. 458–463, 1965 (transl.: "Observation of two-dimensional parametric interaction of light waves," *JETP Letters*, vol. 2, pp. 285–288, November 15, 1965).
[27] J. E. Midwinter and J. Warner, "Up-conversion of near infrared to visible radiation in lithium-meta-niobate," *J. Appl. Phys.*, vol. 38, pp. 519–523, February 1967.
[28] J. A. Giordmaine and R. C. Miller, "Tunable coherent parametric oscillation in $LiNbO_3$ at optical frequencies," *Phys. Rev. Letters*, vol. 14, pp. 973–976, June 14, 1965.
[29] S. A. Akhmanov, A. I. Kovrigin, V. A. Kolosov, A. S. Piskarskas, V. V. Fadeev, and R. V. Khokhlov, *Zh. Eksperim. i Theor. Fiz., Pis'ma Redakt.*, vol. 3, pp. 372–378, 1966 (transl.: "Tunable parametric light generator with KDP crystal," *JETP Letters*, vol. 3, pp. 241–245, May 1, 1966).
[30] R. C. Miller and W. A. Nordland, "Tunable $LiNbO_3$ optical oscillator with external mirrors," *Appl. Phys. Letters*, vol. 10, pp. 53–55, January 15, 1967.
[31] L. B. Kreuzer, "Ruby-laser-pumped optical parametric oscillator with electro-optic effect tuning," *Appl. Phys. Letters*, vol. 10, pp. 336–338, June 15, 1967.
[32] R. G. Smith, J. E. Geusic, H. J. Levinstein, J. J. Rubin, S. Singh, and L. G. Van Uitert, "Continuous optical parametric oscillation in $Ba_2NaNb_5O_{15}$," *Appl. Phys. Letters*, vol. 12, pp. 308–310, May 1, 1968.
[33] I. T. Ho and A. E. Siegman, "Passive phase-distortionless parametric limiting with varactor diodes," *IRE Trans. Microwave Theory and Techniques*, vol. MTT-9, pp. 459–472, November 1961.
[34] S. E. Harris and O. P. McDuff, "Theory of FM laser oscillation," *IEEE J. Quantum Electronics*, vol. QE-1, pp. 245–262, September 1965.
[35] W. E. Lamb, Jr., "Theory of an optical maser," *Phys. Rev.*, vol. 134, pp. A1429–A1450, June 15, 1964.
[36] G. D. Boyd, R. C. Miller, K. Nassau, W. L. Bond, and A. Sav-

age, "$LiNbO_3$: An efficient phase matchable nonlinear optical material," *Appl. Phys. Letters*, vol. 5, pp. 234–236, December 1, 1964.

[37] R. C. Miller, G. D. Boyd, and A. Savage, "Nonlinear optical interactions in $LiNbO_3$ without double refraction," *Appl. Phys. Letters*, vol. 6, pp. 77–79, February 15, 1965.

[38] P. S. Pershan, "Non-Linear Optics," in *Progress in Optics V*, E. Wolf, Ed. New York: Wiley, 1966.

[39] D. A. Kleinman, "Nonlinear dielectric polarization in optical media," *Phys. Rev.*, vol. 126, pp. 1977–1979, June 15, 1962.

[40] R. H. Kingston and A. L. McWhorter, "Electromagnetic mode mixing in nonlinear media," *Proc. IEEE*, vol. 53, pp. 4–12, January 1965.

[41] J. Warner, D. S. Robertson, and K. F. Hulme, "The temperature dependence of optical birefringence in lithium niobate," *Phys. Letters*, vol. 20, pp. 163–164, February 1, 1966.

[42] R. C. Miller and A. Savage, "Temperature dependence of the optical properties of ferroelectric $LiNbO_3$ and $LiTaO_3$," *Appl. Phys. Letters*, vol. 9, pp. 169–171, August 15, 1966.

[43] W. W. Rigrod, "Gain saturation and output power of optical masers," *J. Appl. Phys.*, vol. 34, pp. 2602–2609, September 1963.

[44] A. E. Siegman, "Obtaining the equations of motion for parametrically coupled oscillators or waves," *Proc. IEEE*, vol. 54, pp. 756–762, May 1966.

[45] J. La Salle and S. Lefschetz, *Stability by Liapunov's Direct Method.* New York: Academic Press, 1961.

[46] D. F. Tuttle, Jr., *Network Synthesis.* New York: Wiley, 1958.

[47] H. Statz, G. A. DeMars, and C. L. Tang, "Self-locking of modes in lasers," *J. Appl. Phys.*, vol. 38, pp. 2212–2222, April 1967.

[48] A. Yariv, *Quantum Electronics.* New York: Wiley, 1967.

[49] M. K. Oshman, "Studies of optical frequency parametric oscillation," Stanford University, Stanford, Calif., M. L. Rept. 1602, 1967.

[50] J. C. Slater, *Microwave Electronics.* New York: Van Nostrand, 1950.

[51] E. I. Gordon and J. D. Rigden, "The Fabry–Perot electrooptic modulator," *Bell Sys. Tech. J.*, vol. 42, pp. 155–179, January 1963.

[52] R. Targ, L. M. Osterink, and J. M. French, "Frequency stabilization of the FM laser," *Proc. IEEE*, vol. 55, pp. 1185–1192, July 1967.

Part 6
Noise Papers

Noise in a Molecular Amplifier*

M. W. MULLER
Varian Associates, Palo Alto, California and Stanford University, Stanford, California
(Received December 3, 1956)

The quantum theory of noise in lossy electrical circuits is extended to systems which are not in thermal equilibrium. The theory is applied to microwave molecular amplifiers (masers) and predicts a correction to the classically calculated noise figure which can be identified with spontaneous emission from the molecules of the active medium. The correction is small at ordinary temperatures but becomes significant at very low temperatures.

I. INTRODUCTION

THE limiting sensitivity of high-frequency electronic amplifiers is determined primarily by the thermal noise which arises from the uncontrollable motion of charged particles in the dissipative elements of the amplifier, and by shot noise whose source is the finite size of the elementary quantum of electric charge. The invention of the NH_3 "maser,"[1] an amplifier which is based on the interaction of the electromagnetic field with uncharged particles, and its proposed magnetic analogs,[2] provides a means of eliminating shot noise. Moreover, the possibility (or necessity) of employing materials cooled to very low temperatures to interact with radiation in these devices has led to the hope that amplifiers of exceedingly low noise factor might eventually result from this technique.

We propose to examine this idea in the light of the following considerations:

(1). The material which interacts with the radiation in a maser is not in thermal equilibrium[3] and thus caution must be exercised in assigning to it a noise temperature.

(2). Interaction can take place with vacuum fluctuations of the electromagnetic field (spontaneous emission), producing noise which could not be predicted by classical theory.

These considerations have been treated from a phenomenological point of view by Shimoda, Takahasi, and Townes.[4] The present study is based on a theory of noise proposed by Callen and Welton[5] and developed further by Weber[6] and by Ekstein and Rostoker.[7]

II. NOISE THEORY OF NONEQUILIBRIUM SYSTEMS

A circuit capable of supporting a normal mode of frequency ω can be represented by a harmonic oscillator whose Hamiltonian is

$$H=\tfrac{1}{2}(p^2+\omega^2q^2). \quad (1)$$

We consider the circuit coupled to two dissipative media one of which is in thermal equilibrium; the Hamiltonian of the system is[6]

$$H=\tfrac{1}{2}(p^2+\omega^2q^2)+H_{R1}+H_{R2}+\frac{p}{\sqrt{C}}(Q_1+Q_2), \quad (2)$$

where H_{R1}, H_{R2} are the Hamiltonians of the unperturbed media, Q_1, Q_2 are functions of the coordinates and momenta of the particles in the media, and C is the capacity of the circuit.

The total transition probability for the exchange of energy between the circuit and the dissipative media, if initially the circuit is in a state of energy E_F, and the media in states of energies E_{R1} and E_{R2}, is

$$\begin{aligned}W_R=\frac{2\pi}{\hbar}\{&\langle E_F|p/\sqrt{C}|E_F-\hbar\omega\rangle^2\\ &\times[\rho(E_{R1}+\hbar\omega)\langle E_{R1}|Q_1|E_{R1}+\hbar\omega\rangle^2\\ &+\rho(E_{R2}+\hbar\omega)\langle E_{R2}|Q_2|E_{R2}+\hbar\omega\rangle^2]\\ &+\langle E_F|p/\sqrt{C}|E_F+\hbar\omega\rangle^2\\ &\times[\rho(E_{R1}-\hbar\omega)\langle E_{R1}|Q_1|E_{R1}-\hbar\omega\rangle^2\\ &+\rho(E_{R2}-\hbar\omega)\langle E_{R2}|Q_2|E_{R2}-\hbar\omega\rangle^2]\}, \quad (3)\end{aligned}$$

where $\rho(E)$ is the density-in-energy of the states of the media.

Now the information about the energy state of medium 1 is statistical in all cases, while the information about medium 2 may be either statistical or exact. The first of these alternatives arises, for example, when the medium is made active by inverting the state populations from equilibrium, as in some of the proposed paramagnetic masers; the second is characteristic of a process such as is used in the ammonia maser, in which the upper and lower state populations are considered to be known exactly. We will deal with both alternatives here and show that they are substantially equivalent.

The average transition probability due to medium 1 is obtained by averaging over an ensemble of similar systems. This calculation has been carried out by

* Work supported by the U. S. Army Signal Corps.

[1] Gordon, Zeiger, and Townes, Phys. Rev. **99**, 1264 (1955); referred to hereafter as GZT.

[2] A. Honig and J. Combrisson, Phys. Rev. **102**, 917 (1956); G. Feher and E. A. Gere, Phys. Rev. **103**, 501 (1956); N. Bloembergen, Phys. Rev. **104**, 324 (1956).

[3] N. F. Ramsey, Phys. Rev. **103**, 20 (1956).

[4] Shimoda, Takahasi, and Townes (private communication). The author is indebted to Professor Townes for furnishing him with a manuscript of this work before publication.

[5] H. B. Callen and T. A. Welton, Phys. Rev. **83**, 34 (1951).

[6] J. Weber, Phys. Rev. **90**, 977 (1953); **94**, 211, 215 (1954); **101**, 1619, 1620 (1956).

[7] H. Ekstein and N. Rostoker, Phys. Rev. **100**, 1023 (1955).

Reprinted with permission from *Phys. Rev.*, vol. 106, pp. 8–12, Apr. 1, 1957.

Weber[6] but we reproduce it here because we will utilize a similar procedure later.

$$\langle W_{R1}\rangle=\frac{2\pi}{\hbar}\Big\{\langle E_F|p/\sqrt{C}|E_F-\hbar\omega\rangle^2\int_0^\infty \rho(E_{R1}+\hbar\omega)$$
$$\times\langle E_{R1}+\hbar\omega|Q_1|E_{R1}\rangle^2\rho(E_{R1})f(E_{R1})dE_{R1}$$
$$+\langle E_F|p/\sqrt{C}|E_F+\hbar\omega\rangle^2\int_{\hbar\omega}^\infty \rho(E_{R1}-\hbar\omega)$$
$$\times\langle E_{R1}-\hbar\omega|Q_1|E_{R1}\rangle^2\rho(E_{R1})f(E_{R1})dE_{R1}\Big\}$$
$$=\frac{2\pi}{\hbar}\Big\{\langle E_F|p/\sqrt{C}|E_F-\hbar\omega\rangle^2\int_0^\infty \rho(E_{R1}+\hbar\omega)$$
$$\times\langle E_{R1}+\hbar\omega|Q_1|E_{R1}\rangle^2\rho(E_{R1})f(E_{R1})dE_{R1}$$
$$+\langle E_F|p/\sqrt{C}|E_F+\hbar\omega\rangle^2\int_0^\infty \rho(E_{R1})$$
$$\times\langle E_{R1}|Q_1|E_{R1}+\hbar\omega\rangle^2\rho(E_{R1}+\hbar\omega)$$
$$\times[f(E_{R1}+\hbar\omega)/f(E_{R1})]f(E_{R1})dE_{R1}\Big\}. \quad (4)$$

Here $f(E)$ is the statistical weighting factor, and

$$f(E_{R1}+\hbar\omega)/f(E_{R1})=\exp(-\hbar\omega/kT). \quad (5)$$

If we introduce the quantity

$$S_1=\frac{2\pi}{\hbar}\int_0^\infty \rho(E_{R1}+\hbar\omega)\langle E_{R1}|Q_1|E_{R1}+\hbar\omega\rangle^2$$
$$\times\rho(E_{R1})f(E_{R1})dE_{R1}, \quad (6)$$

the transition probability becomes

$$\langle W_{R1}\rangle=S_1[\langle E_F|p/\sqrt{C}|E_F-\hbar\omega\rangle^2$$
$$+\langle E_F|p/\sqrt{C}|E_F+\hbar\omega\rangle^2\exp(-\hbar\omega/kT)]. \quad (7)$$

Now consider the first alternative discussed above for medium 2, i.e., that of an active material produced by state inversion (system 1). Then the calculation used for medium 1 applies to medium 2, with the modification

$$f(E_{R2}+\hbar\omega)/f(E_{R2})=\exp(+\hbar\omega/kT), \quad (8)$$

so that

$$\langle W_{R2}\rangle=S_2[\langle E_F|p/\sqrt{C}|E_F-\hbar\omega\rangle^2$$
$$+\langle E_F|p/\sqrt{C}|E_F+\hbar\omega\rangle^2\exp(+\hbar\omega/kT)]. \quad (9)$$

If the state populations are known exactly (system 2), say n_a and n_b for the upper and lower state respectively, then no averaging is necessary and

$$\langle W_{R2}\rangle=\frac{2C}{\hbar\omega}[\langle E_F|p/\sqrt{C}|E_F-\hbar\omega\rangle^2 b_2'$$
$$+\langle E_F|p/\sqrt{C}|E_F+\hbar\omega\rangle^2 a_2'], \quad (10)$$

where a_2' and b_2' are proportional to the matrix element between the levels, and to n_a and n_b, respectively.

We can now calculate the time-dependent behavior of the circuit by collecting the terms which induce upward and downward transitions. For system 1 the energy U stored in the circuit is governed by the equation

$$\frac{dU}{dt}=\frac{(\hbar\omega)^2}{2C}\Big\{S_1\Big[(n+1)\exp\Big(-\frac{\hbar\omega}{kT}\Big)-n\Big]$$
$$+S_2\Big[(n+1)\exp\Big(+\frac{\hbar\omega}{kT}\Big)-n\Big]\Big\}, \quad (11)$$

where we have put in the values of the harmonic oscillator matrix elements. We define the quantities

$$G_1=\tfrac{1}{2}\hbar\omega S_1[1-\exp(-\hbar\omega/kT)],$$
$$G_2=\tfrac{1}{2}\hbar\omega S_2[1-\exp(+\hbar\omega/kT)]. \quad (12)$$

We can solve Eq. (11) by using the fact that $U=(n+\frac{1}{2})\hbar\omega$, obtaining

$$U=\tfrac{1}{2}\hbar\omega\Big[G_1\frac{1+\exp(-\hbar\omega/kT)}{1-\exp(-\hbar\omega/kT)}+G_2\frac{1+\exp(\hbar\omega/kT)}{1-\exp(\hbar\omega/kT)}\Big]$$
$$\times\Big[\frac{1-\exp[-(G_1+G_2)t/C]}{G_1+G_2}\Big]$$
$$+U_0\exp[-(G_1+G_2)t/C]. \quad (13)$$

This has the proper classical behavior, if we identify the quantities G_1 and G_2 as the classical conductances. We note that $G_2<0$ as would be expected of the conductance of an active element.

If $G_1+G_2>0$, the system is stable and has a steady state energy (after subtracting the zero-point energy) of

$$U(\infty)=\frac{\hbar\omega}{G_1+G_2}\Big[\frac{G_1\exp(-\hbar\omega/kT)}{1-\exp(-\hbar\omega/kT)}$$
$$+\frac{G_2\exp(\hbar\omega/kT)}{1-\exp(\hbar\omega/kT)}\Big]. \quad (14)$$

With the approximation $\hbar\omega\ll kT$, Eq. (14) becomes

$$U(\infty)\cong\Big(\frac{G_1-G_2}{G_1+G_2}\Big)kT, \quad (15)$$

which indicates that at ordinarily encountered temperatures the noise contribution of a negative conductance is given by the Nyquist noise[8] appropriate to its absolute value.

[8] H. Nyquist, Phys. Rev. 35, 110 (1928).

The energy equation of system 2 is

$$\frac{dU}{dt}=\frac{(\hbar\omega)^2}{2C}S_1[(n+1)\exp(-\hbar\omega/kT)-n]+\hbar\omega[(n+1)a_2'-nb_2'], \quad (16)$$

which is solved by

$$U=\tfrac{1}{2}\hbar\omega\left[\frac{G_1}{C}\left(\frac{1+\exp(-\hbar\omega/kT)}{1-\exp(-\hbar\omega/kT)}\right)+a_2'+b_2'\right]\times\left[\frac{1-\exp\{-[(G_1/C)+b_2'-a_2']t\}}{(G_1/C)+b_2'-a_2'}\right]+U_0\exp\{-[(G_1/C)+b_2'-a_2']t\}, \quad (17)$$

and which, in the stable case, has the steady-state energy (less zero-point energy) of

$$U(\infty)=\frac{\hbar\omega a_2'+(G_1/C)\hbar\omega/[\exp(\hbar\omega/kT)-1]}{(G_1/C)+b_2'-a_2'}. \quad (18)$$

The meaning of Eqs. (13) and (17) becomes more apparent if we note that

$$G_k/C=\omega/Q_k \quad (19)$$

is the net probability per unit time of absorption (or emission if G_k, $Q_k<0$) due to the loss represented by Q_k and that b' and a' have the same meanings for the exactly specified active medium of system 2. Thus we can write

$$G_k/C=b_k'-a_k'=\begin{cases} b_k \\ -a_k \end{cases}, \quad (20)$$

where we use b_k for a net absorption probability and a_k for a net emission probability.

If we define an "effective temperature" for the active medium in system 2 by the relation

$$a_2'/b_2'=\exp(\hbar\omega/kT_e), \quad (21)$$

where it is understood that this definition does not imply thermal equilibrium, then from Eqs. (20) and (21)

$$a_2'=\frac{b_2'-a_2'}{\exp(-\hbar\omega/kT_e)-1}=\frac{a_2\exp(\hbar\omega/kT_e)}{\exp(\hbar\omega/kT_e)-1}. \quad (22)$$

With this new notation we can rewrite, for example, Eq. (14):

$$U(\infty)=\frac{\hbar\omega}{b_1-a_2}\left[\frac{b_1}{\exp(\hbar\omega/kT)-1}+\frac{a_2\exp(\hbar\omega/kT)}{\exp(\hbar\omega/kT)-1}\right], \quad (23)$$

and Eq. (18):

$$U(\infty)=\frac{\hbar\omega}{b_1-a_2}\left[\frac{b_1}{\exp(\hbar\omega/kT)-1}+\frac{a_2\exp(\hbar\omega/kT_e)}{\exp(\hbar\omega/kT_e)-1}\right]. \quad (24)$$

Alternatively, by noting that for the medium in thermal equilibrium

$$a_1'/b_1'=\exp(-\hbar\omega/kT), \quad (25)$$

we have, for both Eqs. (14) and (18),

$$U(\infty)=\hbar\omega\left(\frac{a_1'+a_2'}{b_1-a_2}\right). \quad (26)$$

Thus it is quite apparent that systems 1 and 2 are fully equivalent and that there is no further need for separate developments.

A result equivalent to Eq. (26) has been deduced by Shimoda, Takahasi, and Townes[4] on the basis of a stochastic equation for the emission and absorption processes. The present Eq. (26) and the theory leading to it differs from that result in that the noise energy from the active and passive media are treated separately. This feature makes the present theory suitable for the direct calculation of the noise figure of an amplifier, since the concept of noise figure is based on the comparison of excess noise with thermal noise.

III. APPLICATION TO THE REGENERATIVE MASER AMPLIFIER

A simple kind of molecular amplifier is shown in Fig. 1. This regenerative amplifier was discussed originally by GZT on a classical basis. We deal first with the stable amplifier for which $b_1>a_2$.

For the system of Fig. 1,

$$b_1=b_i+b_o+b_c, \quad (27)$$

where the subscripts i, o, c refer to the input coupling, output coupling, and cavity walls, respectively.

In terms of this notation the gain and noise factor of the amplifier as given by GZT appear as

$$\mu=\frac{4b_ib_o}{(b_1-a_2)^2}, \quad (28)$$

$$F=\frac{a_2}{b_i}+\frac{1}{\mu}. \quad (29)\dagger$$

Equation (29) does not take into account any noise generated in the active medium. In order to include this effect, we obtain from the theory of Sec. II the amount of energy stored in the cavity in the steady

† *Note added in proof.*—Eq. (29) does not conform to the conventional definition of noise figure, because it is based on the signal-to-noise ratio in the wave traveling away from the output which contains some (but not all) of the noise generated in the load. The conventional noise figure is $F=a_2/b_i$; if all the noise from the load is assessed against the amplifier, the noise figure is $F=b_1/b_i$. This last equation is based on the signal-to-noise ratio which would be measured immediately following the amplifier and thus is most meaningful in practice. All these equations are equivalent when $\mu\gg1$. I am indebted to the referee and to Professor E. T. Jaynes for this point.

state due to this source:

$$U_2 = \frac{\hbar\omega a_2'}{b_1 - a_2}. \tag{30}$$

Equation (30) is an expression for the total energy integrated over all frequencies. In order to obtain a noise figure we need the spectral density of this energy. The spectral density can be obtained readily by an extension of the theory of Sec. II which has been carried out for media in thermal equilibrium by Weber[6] and Ekstein and Rostoker.[7] Their procedure can be carried out with no essential modification for nonequilibrium systems and leads to the following expression of the spectral density of a noise current generator equivalent to the medium:

$$\begin{aligned}\langle G_I(\omega)\rangle &= \frac{2|G_2|}{\pi}\left[\frac{\hbar\omega}{2} + \frac{\hbar\omega a_2'}{a_2' - b_2'}\right] \\ &= \frac{2|G_2|}{\pi}\left[\frac{\hbar\omega}{2} + \frac{\hbar\omega}{\exp(\hbar\omega/kT_e) - 1}\right].\end{aligned} \tag{31}$$

In Eq. (31) the quantities G_2, a_2', b_2' are of course functions of frequency; their variation is given essentially by the line shape of the transition and thus depends on the details of the amplifier design. It has been shown, however,[1] that if the gain of the amplifier is large, its band width is small compared with the line width, so that over the band width of the amplifier we may consider $\langle G_I(\omega)\rangle$ to be constant.

We can now compute the noise figure of the amplifier by extending the method of GZT:

The amount of thermal noise power incident on the maser cavity in a frequency interval $\Delta\omega$ due to the input signal generator impedance is

$$P_i(\omega)\Delta\omega = \frac{\hbar\omega}{\exp(\hbar\omega/kT) - 1}\left(\frac{\Delta\omega}{2\pi}\right). \tag{32}$$

The net thermal noise power transmitted into the cavity is given by

$$P_{\rm tr}(\omega)\Delta\omega = \left(\frac{\hbar\omega}{\exp(\hbar\omega/kT) - 1}\right)\left(\frac{\Delta\omega}{2\pi}\right) \times \left(\frac{4b_i(b_1 - b_i - a_2)}{(b_1 - a_2)^2 + 4(\omega - \omega_c)^2}\right), \tag{33}$$

where ω_c is the resonant frequency of the cavity in the presence of the active medium. The energy stored in the cavity due to the generator's thermal noise is

$$U_i(\omega)\Delta\omega = \left(\frac{\hbar\omega}{\exp(\hbar\omega/kT) - 1}\right)\left(\frac{\Delta\omega}{2\pi}\right) \times \left(\frac{4b_i}{(b_1 - a_2)^2 + 4(\omega - \omega_c)^2}\right). \tag{34}$$

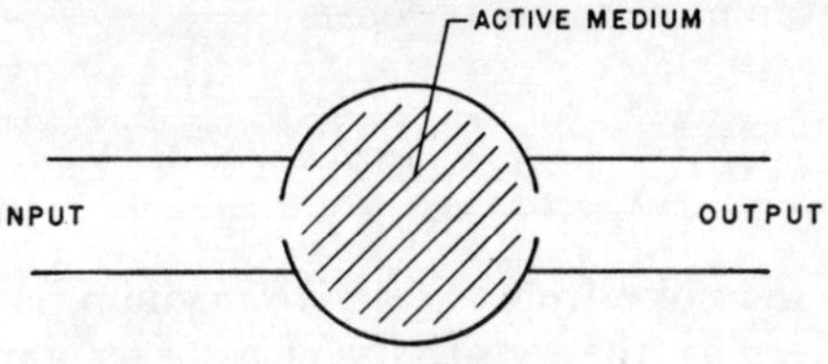

FIG. 1. A cavity maser.

It might be noted that Eq. (34) which for reasons of convenience we have deduced from classical circuit theory, also follows from Eqs. (23) of (24), Eq. (31), and the discussion of band width.

By the same argument the energy stored in the cavity due to thermal noise from the walls is

$$U_c(\omega)\Delta\omega = \left(\frac{\hbar\omega}{\exp(\hbar\omega/kT) - 1}\right)\left(\frac{\Delta\omega}{2\pi}\right) \times \left(\frac{4b_c}{(b_1 - a_2)^2 + 4(\omega - \omega_c)^2}\right), \tag{35}$$

and the energy stored due to noise from the active medium is

$$U_2(\omega)\Delta\omega = \hbar\omega\frac{\Delta\omega}{2\pi}\left(\frac{4a_2'}{(b_1 - a_2)^2 + 4(\omega - \omega_c)^2}\right), \tag{36}$$

so that

$$\int_{-\infty}^{\infty} U_2(\omega)d\omega = U_2. \tag{37}$$

From Eqs. (34), (35), and (36) we find, for the noise power in the output from sources other than the load impedance,

$$\begin{aligned}P_0'(\omega)\Delta\omega = &\left(\frac{\hbar\omega}{\exp(\hbar\omega/kT) - 1}\right)\left(\frac{\Delta\omega}{2\pi}\right) \times \left(\frac{4b_0(b_i + b_c)}{(b_1 - a_2)^2 + 4(\omega - \omega_c)^2}\right) \\ &+ \hbar\omega\frac{\Delta\omega}{2\pi}\left(\frac{4b_0a_2'}{(b_1 - a_2)^2 + 4(\omega - \omega_c)^2}\right).\end{aligned} \tag{38}$$

The total noise power in the output wave guide, including noise power generated in the load and reflected at the output coupling, is

$$\begin{aligned}P_0(\omega)\Delta\omega = &\left(\frac{\hbar\omega}{\exp(\hbar\omega/kT) - 1}\right)\left(\frac{\Delta\omega}{2\pi}\right) \times \left[\frac{4b_0a_2}{(b_1 - a_2)^2 + 4(\omega - \omega_c)^2} - 1\right] \\ &+ \hbar\omega\frac{\Delta\omega}{2\pi}\left(\frac{4b_0a_2'}{(b_1 - a_2)^2 + 4(\omega - \omega_c)^2}\right),\end{aligned} \tag{39}$$

so that the noise figure becomes

$$F=\frac{a_2}{b_i}+\frac{1}{\mu}+\frac{a_2'}{b_i}[\exp(\hbar\omega/kT)-1]. \quad (40)$$

Thus the noise from the active medium has a negligible effect on the sensitivity of a maser amplifier, if the absolute value of effective temperature $|T_e|$ is small compared with the ambient and input temperature T. However it has been proposed that maser amplifiers employing materials at liquid helium temperatures be used, say for radio-astronomical applications. In such applications the noise figure would be based on the temperature of the sky, which is of the order of 1°K. Under these circumstances the correction embodied in Eq. (40) might amount to several decibels at the higher microwave frequencies. When $T_e\sim T$, which will be true under these conditions, and perhaps at ordinary temperatures for system 1, discussed above, the last term on the right-hand side of Eq. (40) will not be negligible. From Eq. (28).

$$b_1-a_2=(4b_ib_0/\mu)^{\frac{1}{2}},$$

and using Eqs. (20), (21), and (27), we obtain

$$\frac{a_2'}{b_i}=\left(\frac{1}{1-\exp(-\hbar\omega/kT_e)}\right)\left[1+\frac{b_0+b_c}{b_i}-\left(\frac{4b_0}{b_i\mu}\right)^{\frac{1}{2}}\right].$$

If we substitute this expression in Eq. (40) and use Eq. (22), we obtain

$$F=\left[1+\frac{b_0+b_c}{b_i}-\left(\frac{4b_0}{b_i\mu}\right)^{\frac{1}{2}}\right]\times\left[1+\exp(\hbar\omega/kT)\left(\frac{1-\exp(-\hbar\omega/kT)}{1-\exp(-\hbar\omega/kT_e)}\right)\right],$$

with the following special cases: If $\hbar\omega\ll kT$, $\hbar\omega\ll kT_e$, $\mu\gg1$,

$$F\cong\left[1+\frac{b_0+b_c}{b_i}\right]\left[1+\frac{T_e}{T}\right].$$

If $T_e=T$ (system 1), $\mu\gg1$,

$$F\cong\left[1+\frac{b_0+b_c}{b_i}\right][1+\exp(\hbar\omega/kT)].$$

From these equations we can conclude that for a regenerative maser it is preferable to obtain the active medium by selective focusing or optical pumping rather than by state inversion.

IV. SUPERREGENERATIVE AMPLIFIER

If in the system of Fig. 1 the quantity of active material is increased until $a_2>b_1$, the amplifier is no longer stable and the energy in the cavity will increase exponentially as predicted by Eq. (13) or (17). If some appropriate means of quenching the oscillation is provided, the system can be used as a super-regenerative amplifier.

It is not possible to predict the noise figure or gain of such an amplifier without knowing the details of its design. It is possible, however, to calculate the excess of noise above its classical value due to emission from the active material. For this purpose we make the following assumptions:

(1) At the start of the exponential rise the cavity is in thermal equilibrium, so that

$$U_0=\hbar\omega/[\exp(\hbar\omega/kT)-1]. \quad (41)$$

(2) The "gain" is large, that is to say, the signal is allowed to build up long enough so that

$$\exp[-(b_1-a_2)t]\gg1. \quad (42)$$

With these assumptions Eq. (17) becomes, after subtracting the zero-point energy,

$$U=\frac{\hbar\omega}{2}\left(\frac{1}{b_1-a_2}\right)\left\{\left[\frac{2b_1}{\exp(\hbar\omega/kT)-1}+\frac{2}{\exp(\hbar\omega/kT)-1}\right]+a_2'\right\}\exp[-(b_1-a_2)t]. \quad (43)$$

The classical thermal noise is due to the term in square brackets in Eq. (42); the excess noise is due to the second term. The contribution arising from U_0 is negligible since $b_1\gg1$, and thus the correction to the classical noise figure is

$$F=F_{cl}\{1+(a_2'/b_1)[\exp(\hbar\omega/kT)-1]\}, \quad (44)$$

where F_{cl} is the noise figure that is obtained when the noise generated by the active medium is neglected.

V. CONCLUSION

It has been shown, by an extension of the quantum theory of noise to systems not in thermal equilibrium, that the active material in a molecular amplifier makes a finite contribution to the noisiness of the system. The applications treated include the stable regenerative maser and the superregenerative maser. The correction due to this noise source is small at the usual microwave frequencies and at ordinary temperatures, but becomes significant at frequencies and temperatures for which $\hbar\omega\sim kT$, a condition which may be met in some proposed devices.

The author is indebted to Dr. W. L. Beaver, Dr. J. C. Helmer, and to Professor M. Chodorow for several stimulating discussions.

Spontaneous Emission and the Noise Figure of Maser Amplifiers

R. V. Pound

Lyman Laboratory, Harvard University

The Nyquist-Johnson formulation of the noise in circuits arising from thermal agitation is applied to circuit elements equivalent to and representing simple resonant absorption, such as occurs in paramagnetic materials or others having r-f line spectra. Such a model for a circuit resistance directly demonstrates the origin of the noise as spontaneous emission by excited particles, in complete analogy to black body radiation in free space. Extension of the conventional theorem to the domain of negative resistances at negative temperatures, as encountered in Maser amplifiers, is valid under circumstances of practical significance. Amplifiers of small noise figure can be made for frequencies ν such that $h\nu/kT_0$, where T_0 is the effective temperature of the signal source, is not large compared to unity. For example, an amplifier of high gain and noise figure less than 2 can be built if use is made of a medium describable by a negative effective temperature smaller, in absolute value, than T_0.

INTRODUCTION

The probability for spontaneous emission of radio-frequency radiation by particles in appropriate low lying excited states is very small. On the other hand such particles can emit electromagnetic radiation strongly when suitably stimulated. Beams or condensed aggregates containing such particles can, thereby, form the basis of practical regeneration of r-f electromagnetic energy. Amplifiers and oscillators based on this principle have been described by Gordon, Zeiger and Townes (*1*) who coined the acronym "Maser" to describe them. Other suggestions have been published by Bloembergen (*2*) and by Bassov and Prokhorov (*3*). Amplification of this kind of very limited magnitude was present on a transient basis in experiments reported to illustrate the existence of negative temperatures, in systems of interacting spins well "insulated" from their lattice (*4*). The usefulness of a method of amplification depends on such quantitative factors as the gain available, the bandwidths, and, especially, the noise figure (*5*). The purpose of this note is to put in view some of the factors affecting the noise figure of these devices; in particular, an aim is to demonstrate the applicability and utility of a wire circuit model for such consideration.

Reprinted with permission from *Ann. Phys.*, vol. 1, pp. 24–32, 1957.

A CIRCUIT DESCRIPTION OF A RESONANT ABSORBER

It is well known that under conditions of thermal equilibrium, the equipartition theorem can be invoked to demonstrate that electrical circuits must possess noise currents and voltages of thermal origin. With each resistance $R(\nu)$, which may be dependent on frequency ν, must be associated a noise voltage generator, for example, such that the mean-squared noise voltage per unit frequency interval is given at temperature T by the Nyquist-Johnson formula

$$\langle e^2 \rangle_{\text{Av}} = 4kTR(\nu). \tag{1}$$

It is well to keep in mind the fact that this representation is the transcription to circuit form of black body radiation and its origin is the spontaneous emission of radiation by particles in excited states. A direct connection between the circuit noise and black body radiation can be made by the well known model of a resistance, matched to the radiation resistance of an antenna which is immersed in black body radiation at temperature T (*5*). From the Rayleigh-Jeans approximation for the energy density in the radiation field, the requirement that there be no net power flow if the resistance is also at temperature T leads to Eq. (1) above. A value for $\langle e^2 \rangle_{\text{Av}}$ valid for all values of $h\nu/kT$ can be written by use in such a model of the Planck radiation law,

$$\langle e^2 \rangle_{\text{Av}} = 4h\nu R\,[\exp\,(h\nu/kT) - 1]^{-1}. \tag{2}$$

Suppose that a cavity is filled with a material that resonates at ν_0, the resonant frequency of the cavity, and further, suppose that the resulting absorption "line" has, per se, a normalized Lorentzian shape, such as is typical of collision broadening or radiation damping,

$$g(\nu) = 2T_2[1 + (2\pi(\nu - \nu_0)T_2)^2]^{-1}, \tag{3}$$

where T_2 has the dimensions of time and is chosen so as to agree with the conventions of the literature of nuclear magnetism. A wire circuit equivalent to the system is that of Fig. 1. The current generators i_S, i_N and the voltage generator

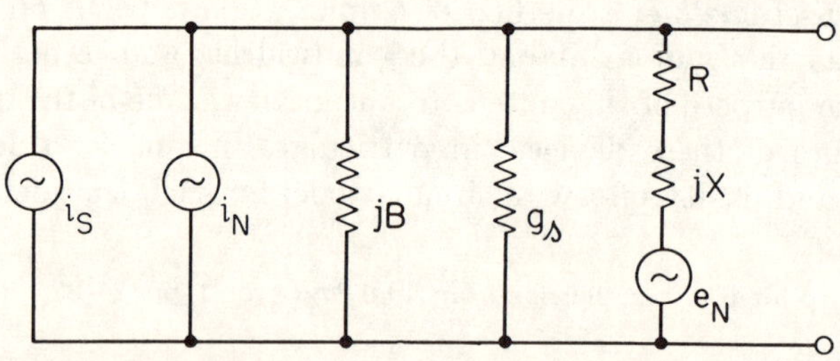

FIG. 1. A wire circuit equivalent to an r-f system including a resonant absorber.

e_N represent, respectively, a signal source and noise sources associated with the two parts of the circuit. The conductance g_s represents, at some point in the circuit, all the losses of the circuit other than the resonant absorbing atomic system. The susceptance jB, in the event that the frequency selectivity of the circuit arising elsewhere than in the absorber is dominated by that of a single cavity mode, is a linear function of frequency

$$B = 2(\nu - \nu_0)Qg_s/\nu_0 , \tag{4}$$

where Q is the conventional quality factor of that mode.

If the absorption at its peak is represented by δ, the ratio of power absorbed by the resonant absorber to $2\pi\nu$ times the energy stored in the circuit, the factor $Q\delta$ represents the ratio of the power absorbed in the absorber to that absorbed by the circuit losses. Therefore, the resistance R, representing the resonant absorption at its peak, is equal to $[g_sQ\delta]^{-1}$. The resonant character of the absorber is correctly simulated if the resistance R is made independent of frequency and the reactance jX is included to represent the dispersion of the medium. The quantity X has the value

$$X = 2\pi(\nu - \nu_0)T_2R \tag{5}$$

for the assumption covered by Eq. (3).

Suppose that the circuit conductance g_s is assumed to be at a temperature so low, compared to that of the absorber R, that the current in g_s caused by the noise current generator i_N, associated with g_s itself, can be ignored in comparison to the current produced by the noise voltage generator e_N associated with R. The power delivered to the circuit losses g_s from the absorber can then be computed. To do this one needs to integrate the mean-square of the current in g_s divided by g_s over all frequencies. One finds

$$P_{g_s} = \langle e^2\rangle_{\text{Av}}\ g_s 2\pi\nu_0 Q\delta^2[4(1 + \delta Q)]^{-1}\{[1 + (2\pi\nu_0T_2/Q)] + (\pi\nu_0T_2/Q)^2\}^{-1/2} \tag{6}$$

In the event that $Q/\pi\nu_0 \ll T_2$, corresponding to a circuit of large width in frequency compared to the width of the resonance of the absorber, the expression becomes

$$P_{g_s} \approx \langle e^2\rangle_{\text{Av}}\, g_s(Q\delta)^2[2(1 + \delta Q)T_2]^{-1}. \tag{7}$$

Putting $\langle e^2\rangle_{\text{Av}}$ from Eq. (2) into this yields

$$P_{g_s} \approx 2h\nu_0(Q\delta)[(1 + \delta Q)T_2]^{-1}\,[\exp\,(h\nu_0/kT) - 1]^{-1}. \tag{8}$$

The factor $(1 + Q\delta)^{-1}$ measures the power gain G_0 of the circuit as a whole when the absorber is added, relative to the circuit without the absorber. The power gain G_0 is defined as the ratio of the power arising from i_s available from the terminals of the entire circuit to that available for the same current generator with δ made zero. For $\delta Q = 1$, for example, $G_0 = \frac{1}{2}$ and then P_{g_s} equals, for $(h\nu/kT) \ll 1$, kTB where B, the effective noise bandwidth of the circuit, is $(1/T_2)$. This bandwidth is twice the width characteristic of the absorber alone because of the critical coupling, $\delta Q = 1$. For $\delta \leq 0$, the power gain $G_0 \geq 1$, and the circuit is an amplifier. The noise power delivered to g_s is

$$P_{g_s} = 2h\nu_0(Q\delta)G_0\{T_2 [\exp (h\nu_0/kT) - 1]\}^{-1} \tag{9}$$

A more detailed examination of the absorption mechanism is required to identify δ. As an example, the resonance concerned might involve only two energy levels per particle, for which

$$\delta = W_{1\to2}(h\nu_0)(N_1 - N_2)/2\pi\nu_0 E \tag{10}$$

where E is the electromagnetic energy stored per unit volume, averaged over the circuit, and $W_{1\to2}$ is the transition probability from the lower state (1) whose population per unit volume is N_1, to the upper (2), whose corresponding population is N_2, induced by the presence of the radiation field. If the circuit were not filled with absorbing material a filling factor could be introduced in the usual manner but, for the purposes of the present discussion, there is no need for such additional complication. The difference $N_1 - N_2$ can be described in terms of the number of particles per unit volume in the upper state, N_2, and the effective temperature, $T_{\text{eff}} = h\nu_0/k \log (N_1/N_2)$, as

$$N_1 - N_2 = N_2 [\exp (h\nu_0/kT_{\text{eff}}) - 1]. \tag{11}$$

These substitutions into the above together with the discussion of $W_{1\to2}$ given below yield

$$P_{g_s} = 2\pi N_2 Q G_0(h\nu_0) \mid (1 \mid \mathbf{M}^+ \mid 2) \mid^2/\hbar \tag{12}$$

To obtain Eq. (12),

$$W_{1\to2} = \mid (1 \mid \mathcal{H}_1 \mid 2) \mid^2 g(\nu)/\hbar^2 \tag{13}$$

has been used for the transition probability, where $\mathcal{H}_1$ represents the interaction of the system with the field, and only a single linearly polarized field component has been allowed, representing a single mode in the resonator. The operator $\mathbf{M}^+$ could be either an electric or magnetic dipole moment operator. It is instructive to recognize, in Eq. (12), that all dependence on the temperature has vanished except that implicit in N_2 and in the factor G_0. The power represented in Eq. (12) is that resulting from *spontaneous emission*, delivered to the circuit with an efficiency that depends on the coupling conditions—as measured by G_0. Normally $G_0 < 1$ and some self-absorption occurs, but for $\delta < 0$, $G_0 > 1$ and there is, in the net, stimulated emission, leading for $\delta Q \leq -1$ to oscillation or self-sustained emission.

To demonstrate the appearance of the spontaneous emission explicitly, one can take the expression for the Einstein coefficient A for spontaneous emission in free space

$$A = 4 \mid (2 \mid \mathbf{M} \mid 1) \mid^2 / 3\hbar \lambda\!\!\!^-{}^3 \tag{14}$$

and recognize that the single mode cavity allows only one component of radiation field. For the present purposes

$$A_x = \mid (2 \mid \mathbf{M}^- \mid 1) \mid^2 / 3\hbar \lambda\!\!\!^-{}^3 = \mid (1 \mid \mathbf{M}^+ \mid 2) \mid^2 / 3\hbar \lambda\!\!\!^-{}^3$$

The cavity, having only a single mode actually at resonance with the system, enhances the spontaneous radiation rate by the factor *(6)*

$$S = 3\lambda^3 Q / 4\pi^2 \text{ (Volume)}. \tag{15}$$

This factor represents the ratio of the energy density in the cavity to that of a single field component in free space, at any temperature, or in the zero point vibrations that may be regarded as inducing the spontaneous emission. It is thereby seen that Eq. (12) can be written

$$P_{g_s} = G_0 S A_x \eta_2 (h\nu_0) \tag{16}$$

where η_2 is the total number of particles in the upper state, equal to ($N_2 \times$ Volume). The restriction to a circuit broad compared to the line width and filled with absorber simplifies the recognition of the factors G_0 and S but, clearly, similar results must be expected irrespective of such restrictions.

With the use of the modified Nyquist-Johnson formula, the power flow is given correctly even for $(h\nu/kT) \gg 1$ provided that the state of the system is correctly described only by N_1/N_2. The circuit model is entirely adequate to represent the "small signal" behavior of the system. The importance of stimulated absorption and emission relative to spontaneous emission together with interference effects among them are automatically taken into account. An important conclusion is that the possibility of a negative sign of T_{eff} does not invalidate the model. As far as algebra alone is concerned, trouble might be expected from the Nyquist-Johnson formula, which must yield a positive mean square, but the common sign of T_{eff} and R satisfies that requirement.

The validity of the noise formulation based on effective temperature is critically dependent on there being statistically no coherence in the states of the system. It is a property of aggregates of particles, constricted within dimensions smaller than the velocity of light c divided by the width of the frequency band $\Delta\nu$ associated with a radiative transition, that states having radiation rates per particle very different from those characteristic of single particles compatible with the representation above, can be produced *(7)*. Clearly, any such coherent states are not properly described in the above manner, or by any temperature, negative or positive. If, for example, a system capable of amplification were initiated by an inversion of a net vector magnetic moment initially in equilibrium at temperature T_c, a small departure from exact inversion will lead to a radiation rate far above that of "thermal" origin. Quantitatively, it is easy to see from a classical model that, for thermal radiation to dominate, the angular departure from exact inversion would have to be less than $kT_c/h\nu N^{1/2}$, when N is the number of particles. This is true because the expectation value of a transverse moment that may be considered the origin of incoherent spontaneous emission would be $\sqrt{N}$ times the moment of an individual particle. For a small angle, the transverse moment due to imperfect inversion would equal the angle times the net moment of the system. On the other hand, any such enhanced radiation damps out through the influence of the forces producing internal equilibrium, as measured by T_2. The damping time is longer than T_2 because of the coupling to the circuit and, in practice, is the net time defined by the net amplifying bandwidth. It is the presence of this same relaxation that renders the assignment of a temperature to the spin system meaningful, however, and, therefore, a close connection is established between the applicability of the above noise calculations and the validity of an assignment of a negative temperature to the Maser system in a sense broader than that signifying only a ratio N_2/N_1.

In practice, the transient amplifier would experience, because of inexact inversion, some "ringing" as an after effect of that sensitizing operation, and it might not prove easy to reduce such ringing below thermal noise in the first few periods of reciprocal bandwidth. In steady state devices (*2*) the presence of strong "pumping" radiation, that mixes one of the states involved in the resonant transitions to be utilized with a third, must be rendered negligible by, similarly, the presence of a short relaxation time T_2 compared to the transition time produced by the pumping signal. For this example, too, the conditions allowing the assignment of a negative temperature to the system composed of two levels would seem to be satisfied.

It is, therefore, possible, subject to these restrictions, to represent not only the transfer characteristics for small signals but the noise properties as well, of both absorption circuits and Maser amplifiers, by the use of equivalent circuits made up of appropriate positive and negative resistances, and reactances, at appropriate positive or negative effective temperatures.

NOISE FIGURE CALCULATIONS

The above conclusions may easily be applied to the calculation of the noise figure of a typical Maser amplifier. Consider the circuit of Fig. 2. The ratio of signal power S_0 to noise power N_0 available in the frequency range between ν and $\nu + d\nu$ from the source is

$$\frac{dS_0}{dN_0} = \frac{i_s^2(\nu)[(\exp(h\nu/kT_0) - 1]d\nu}{4g_0 h\nu d\nu} \tag{17}$$

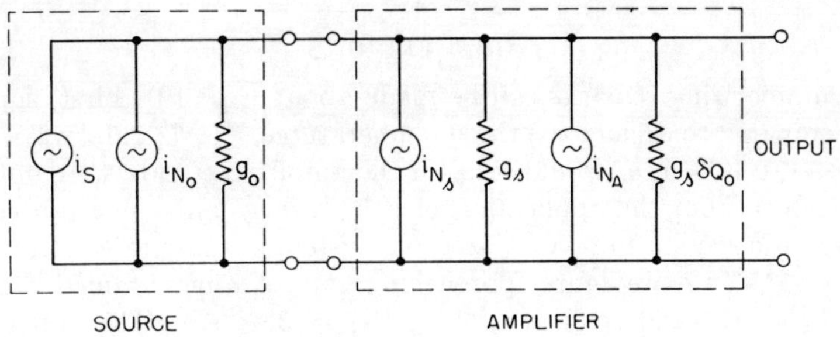

FIG. 2. An equivalent circuit useful for calculation of the noise figure of a Maser amplifier.

where T_0 is the effective temperature of the source g_0 and the ratio may be a function of frequency. The ratio of the signal power to the noise power, in the same frequency range, at the output of the Maser is

$$\frac{dS}{dN} = \frac{i_s^2}{4h\nu}\left\{\frac{g_0}{\exp(h\nu/kT_0) - 1} + \frac{g_s}{\exp(h\nu/kT_c) - 1} + \frac{g_s\delta Q_0}{\exp(-h\nu/\alpha kT_c) - 1}\right\}^{-1} \tag{18}$$

In this expression α is $(-T_{\text{eff}}/T_c)$, T_c is the temperature of the losses of the cavity circuit, and Q_0 is the unloaded Q of the cavity circuit used for the amplifier. The noise figure F, which is, in general, a function of frequency and of the source temperature is, in the same frequency range,

$$F_{T_0}(\nu) = 1 + \frac{[\exp(h\nu/kT_0) - 1]}{[\exp(h\nu/kT_c) + 1]}\left\{(1 - G^{-1})\,x + \left[\frac{Q_{L1}}{P_0 - Q_{L1}}\right](1 + x)\right\} \tag{19}$$

where

$$x = [\exp(h\nu/kT_c) - 1][1 - \exp(-h\nu/\alpha kT_c)]^{-1}$$

and G is the power gain of the Maser, defined as the ratio of the signal power *available* from its output terminals to that *available* from the source itself and Q_{L1} is the Q of the cavity as loaded by the input circuit only, not including the unspecified output load. Note that the gain G defined here differs from G_0, above, because the cavity losses themselves are, here, assigned to the Maser.

In practice a net pass-band will be defined either by this circuit or by those following. In general, the reactive parts of the circuit and the selectivity of the succeeding amplifiers should be included and an effective over-all noise figure calculated by integration over all frequencies. If the bandwidth of the succeeding amplifiers is small, the over-all effective noise figure F_{T_0} can be set equal to $F_{T_0}(\nu)$ above.

CONCLUSIONS

Some summarizing remarks can be made about Eq. (19). First, suppose $h\nu/k$ is small compared to each of the three temperatures, T_0, T_c, and αT_c. Clearly, if $(T_c/T_0) \ll 1$, if $\alpha \sim 1$, as would result if the amplifying state were obtained as a transient effect from the application of a high level pulse of suitable duration (for a magnetic system to reverse M_0) to a system initially in equilibrium at T_c, and if $Q_{L1} \ll Q_0$, a noise figure approaching unity can be obtained. On the other hand, if $T_c = T_0$, and again if $\alpha = 1$, $F_{T_0} = 2 - G^{-1}$. This function ranges from 1 to 2 as G ranges from 1 to ∞.

It might be supposed that an over-all value close to 1 could be obtained for large net G by utilization of a cascade of amplifiers of small incremental gain each or, in the limit, a distributed device amplifying a running wave. Application of the cascade noise-figure formula

$$F_{12} = F_1 + (F_2 - 1)/G_1$$

demonstrates that, if the circuit losses represented by Q_0 are negligible in each stage, identical noise figure is obtained from a cascade of stages of similar T_c and α, with gains G_1, G_2, G_3, $\cdots$ as from a single stage of the same net gain, $G_1G_2G_3 \cdots$.

In circumstances where $(h\nu/kT_0) \gg 1$, Eq. (19) with $T_0 \sim T_c$ and $\alpha \sim 1$ leads to $F_{T_0} \gg 1$. This results not because greater noise is introduced by lowering T_0 but rather because the largeness of $(h\nu/kT_0)$ corresponds to extremely small noise in the signal source. The negative temperatures correspondingly near zero do not have correspondingly little thermal radiation, and, consequently, the large noise figure results from the quietness of the reference. The noise figure is least with the smallest value of α, corresponding to as *small* an absolute negative temperature as possible.

Circuit properties of the Maser are adequately represented by the equivalent negative resistance, so long as signal levels do not produce appreciable changes in the state of the system, in themselves. As is true generally with negative resistance or regenerative devices, the gain-bandwidth product is essentially a constant of the system. A singly resonant circuit that has a gain of ½, for example, without the addition of the radiant system, can be given a gain G with reduction of the bandwidth by the factor $1/2G$, corresponding to the net increase in operating resonant impedance in the presence of given reactive elements. The bandwidth can be increased by further loading of the circuit by, for example, the signal generator and compensating this by a correspondingly larger Maser system.

Attention may be called to a point in the noise properties that seems, from the circuit viewpoint, anomalous. The conclusion that power is delivered to the circuit losses by spontaneous emission processes was so independent of effective temperature that even a system with $N_1 = N_2$, which shows no absorption or no detectable circuit impedance, supplies the normal amount of noise power. In the equivalent circuit, R is infinite as is also $\langle e^2 \rangle_{Av}$. The power transfer found results from the fact that the coupled circuit allows the absorbing medium to relax to thermal equilibrium with the radiation in the circuit. The time constant for such relaxation is short compared to that which would arise from the coupling to radiation fields in free space (*6*). However, such power transfer can continue in accord with this same formula only if the system remains in statistical equilibrium internally, through some interaction between particles. Such an interaction is required to allow the system, on the average, to be described by N_1/N_2. From the interaction with the circuit alone, the particles develop in time with definite correlations that depend on the details of the field distribution of the circuit mode in the medium (*7*). The loss of energy cannot be ascribed to changes in N_1/N_2 alone except after the lapse of times comparable to T_2. Systems with adequately short internal time constants, corresponding to the interparticle interactions, could build up polarization through the interaction with the cavity in such a manner.

In summary, the solid state Maser, because of the feasibility of obtaining small negative effective temperatures, can offer much in improved noise figure for reception of signals from sources of low equivalent temperature, such as may occur in decimeter and centimeter wave bands. Similar analysis can be applied to devices utilizing beams of particles and, in them, the effective temperature is clearly a measure of the degree to which the beam is separated. In addition to the noise considered here and of concern for comparison to small signals, there may also be noise of the type that is proportional to the signal and that results from the fluctuations in the gain of the system. Usually the number of particles participating will be large and, therefore, the fractional fluctuations in output signal will be small.

RECEIVED: January 18, 1957.

REFERENCES

1. J. P. GORDON, H. J. ZEIGER, AND C. H. TOWNES, *Phys. Rev.* **99,** 1264 (1955).
2. N. BLOEMBERGEN, *Phys. Rev.* **104,** 324 (1956).
3. N. G. BASSOV AND A. M. PROKHOROV, *T. Exptl. Theoret. Phys.* (*USSR*) **27,** 431 (1954); *Proc. Acad. Sci.* (*USSR*) **101,** 37 (1945).
4. E. M. PURCELL AND R. V. POUND, *Phys. Rev.* **81,** 279 (1951).
5. J. L. LAWSON AND G. E. UHLENBECK, "Threshold Signals," Radiation Laboratory Series, Vol. 24, Chapter 5. McGraw-Hill, New York, 1950.
6. E. M. PURCELL, *Phys. Rev.* **69,** 681 (A) (1946).
7. R. H. DICKE, *Phys. Rev.* **93,** 99 (1954).

Quantum Noise. IV. Quantum Theory of Noise Sources

MELVIN LAX
Bell Telephone Laboratories, Murray Hill, New Jersey
(Received 16 November 1965)

For a quantum system describable in a Markoffian way, via a set of variables $\mathbf{a}=\{a_1,a_2,\cdots\}$, we show that the Langevin noise sources F_μ in the operator equations of motion $da_\mu/dt=A_\mu(\mathbf{a})+F_\mu$ possess second moments $\langle F_\mu(t)F_\nu(u)\rangle=2\langle D_{\mu\nu}(\mathbf{a},t)\rangle\delta(t\text{-}u)$. The diffusion coefficients $D_{\mu\nu}$ can be determined from a knowledge of the mean equations of motion via the (exact) time-dependent Einstein relation $2\langle D_{\mu\nu}\rangle=-\langle A_\mu a_\nu\rangle-\langle a_\mu A_\nu\rangle+d\langle a_\mu(t)a_\nu(t)\rangle/dt$, where $\langle\ \rangle$ represents a reservoir average. The sources F_μ,F_ν do not commute with one another, and as a result the commutation rules of the a_μ are shown to be preserved in time. The mean motion and diffusion coefficients are calculated for a harmonic oscillator, and for a set of atomic levels. We prove that two dynamically coupled systems (e.g., field and atoms) have uncorrelated Langevin forces if they are coupled to independent reservoirs. Radiation-field–atom coupling adds no new noise sources. We thus obtain simply the maser model including noise sources used in Quantum Noise V. *Direct* calculations of the mean motion and fluctuations in a system coupled to a reservoir yield relationships in agreement with the Einstein relation. For reservoirs violating time reversal, anomalous frequency shifts are found possible that violate the Ritz combination principle since $\Delta\omega_{12}+\Delta\omega_{23}+\Delta\omega_{31}$ need not vanish.

1. INTRODUCTION AND SUMMARY

OUR treatment of quantum noise in this paper and the preceding papers in this series[1] closely parallels a corresponding discussion of noise in classical systems.[2] The first paper (I) in our classical series provides a quasilinear approach to stationary Markoffian random processes. In the quasilinear case, it was easy to obtain the corresponding Langevin theory of noise sources. This work was generalized in (III) and (IV) to include classical nonstationary nonlinear Markoffian processes treated first from a Markoffian point of view and second from a Langevin noise-source point of view. In Paper IV, we emphasized the advantages and flexibility associated with the Langevin noise-source approach.

The present paper extends the noise-source technique to quantum systems. Quantum (and classical) systems experience dissipation and fluctuations through interaction with a reservoir. Our philosophy is that the reservoir can be *completely eliminated* provided that the frequency shifts and dissipation induced by the reser-

[1] A reference to QV is a reference to the author's fifth paper on quantum noise and relaxation. QI: Phys. Rev. **109**, 1921 (1958); QII: Phys. Rev. **129**, 2342 (1963); QIII: J. Phys. Chem. Solids **25**, 487 (1964); QIV: present paper; QV: in *Physics of Quantum Electronics*, edited by P. L. Kelley, B. Lax, and P. E. Tannenwald (McGraw-Hill Book Company, Inc., New York, 1966); QVI: "Moment Treatment of Maser Noise" (unpublished); QVII: "The Rate Equations and Amplitude Fluctuations" (unpublished).

[2] A reference to IV is a reference to the author's fourth paper on classical noise. I: Rev. Mod. Phys. **32**, 25 (1960); II: J. Phys. Chem. Solids **14**, 248 (1960); III: Rev. Mod. Phys. **38**, 359 (1966); IV: Rev. Mod. Phys. (to be published); V: Bull. Am. Phys. Soc. **11**, 111 (1966); VI: *ibid.*

Reprinted with permission from *Phys. Rev.*, vol. 145, pp. 110–129, May 6, 1966.

voir are incorporated into the mean equations of motion, and provided that a suitable operator noise source with the correct moments are added.

Thus, if $\mathbf{a}=\{a_1,a_2,\cdots\}$ is some set of system operators, and

$$d\langle a_\mu\rangle/dt=\langle A_\mu(\mathbf{a})\rangle \quad (1.1)$$

are the correct mean[3] equations of motion including frequency shifts and damping, then we propose that

$$da_\mu/dt=A_\mu(\mathbf{a})+F_\mu(\mathbf{a},t) \quad (1.2)$$

is a valid set of *operator* equations provided that the operators F_μ are endowed with the correct statistical properties. Thus the reservoir has been replaced by the familiar electrical engineer's black box describable by an impedance (the dissipative resistive terms and frequency-shift–producing reactive terms incorporated into A_μ) and an associated noise source F_μ.

The mean equations of motion in more fundamental recent papers on noise in masers are obtained by eliminating the reservoir to second order in perturbation theory. If one calculates, for example, the time rate of change of the occupancy of some state, one obtains the usual sort of transport equation [see Eq. (1.13) with $i=j$] whose coefficients are transition probabilities calculated to second order. One then adopts the *form* of the resulting equations of motion as a model for a maser—but regards the coefficients as experimentally determined, i.e., as the correct, not the second-order transition probabilities. Moreover, the models usually chosen are Markoffian in the sense that the future of all operators (or equivalently of the density matrix) is determined by the present without requiring an integration over past histories.

Although the F's are operators, for most problems we only need to know the reservoir averages over low-order moments and commutators of these operators. We regard our task then, as the determination of the moments of the F_μ, in terms of the *experimental dissipation coefficients, within a Markoffian description.*

The most obvious method of attack is to calculate the reservoir contribution to A_μ to second order in the system-reservoir interaction V [see Appendices A and B]. Then one must calculate the mean moment $\langle F_\mu(t)F_\nu(u)\rangle$ to second order in V (see Appendix B). By comparing the coefficients in these two calculations we arrive at a fluctuation-dissipation relation valid for nonequilibrium situations. We shall show in Sec. 2, that the relation so obtained is indeed *exact.*

To see how to implement this program, we note that Eqs. (1.1) and (1.2) have been so chosen that the first moment of the Langevin forces vanish:

$$\langle F_\mu\rangle=0,\quad A_\mu(\mathbf{a},t)=\langle a_\mu(t+\Delta t)-a_\mu(t)\rangle/\Delta t. \quad (1.3)$$

Thus, by computing the change in a_μ over some suitable short time interval[4] Δt, due to the interaction V, we can determine the reservoir contribution to A_μ.

Next we note, that if the reservoir forces possess a finite correlation time, i.e., $\langle F_\mu(t)F_\nu(u)\rangle\neq 0$ for $|t-u|\sim\tau_c$, the system will acquire a memory of the past and become non-Markoffian. Thus we shall take our moments of the random forces F_μ in the form

$$\langle F_\mu(t)F_\nu(u)\rangle=2\langle D_{\mu\nu}(\mathbf{a},t)\rangle\delta(t-u). \quad (1.4)$$

[A direct proof that $\langle F_\mu(t)F_\nu(u)\rangle=0$ for $t\neq u$ is given in IV (8.12) for the classical case, and in (2.19) for the quantum case.]

Setting $t=s$, $u=s'$ and integrating (1.4) from t to $t+\Delta t$ on both these variables, we obtain[5]

$$2\langle D_{\mu\nu}\rangle=\frac{1}{\Delta t}\int_t^{t+\Delta t}ds\int_t^{t+\Delta t}ds'\langle F_\mu(s)F_\nu(s')\rangle. \quad (1.5)$$

Integrating Eq. (1.2) over an interval Δt, and inserting the results into (1.5), we obtain

$$2\langle D_{\mu\nu}\rangle=\langle[a_\mu(t+\Delta t)-a_\mu(t)][a_\nu(t+\Delta t)-a_\nu(t)]\rangle/\Delta t \quad (1.6)$$

after discarding terms $(A_\mu\Delta t)(A_\nu\Delta t)/\Delta t$ that disappear as $\Delta t\to 0$. Equation (1.6) is reminiscent of the traditional definition of a spatial diffusion coefficient:

$$2D=\langle[x(t)-x(0)]^2\rangle/t.$$

Equation (1.6) is also the basis of our direct perturbation calculation of the diffusion coefficients in Appendix B. We calculate the change Δa_μ in a_μ over the time interval Δt induced by the interaction V, and average $\Delta a_\mu\,\Delta a_\nu$ over the reservoir variables. Since this average depends on the order of the factors, $D_{\mu\nu}$ is not symmetric. Thus F_μ and F_ν *do not commute.*

The *exact* method of Sec. 2 consists in noticing that the equation of motion (2.7) for $\langle a_\mu a_\nu\rangle$ involves $\langle D_{\mu\nu}\rangle$ and emphasizing that this equation can be inverted to solve for $\langle D_{\mu\nu}\rangle$:

$$2\langle D_{\mu\nu}\rangle=-\langle A_\mu a_\nu\rangle-\langle a_\mu A_\nu\rangle+d\langle a_\mu(t)a_\nu(t)\rangle/dt. \quad (1.7)$$

This equation is well known to us as I (5.12). What is new is that we notice that a knowledge of the *mean motion* of all operators provides us with the *fluctuation* moments $D_{\mu\nu}$. In the steady state, the last term of (1.7)

[3] We use single brackets, $\langle\ \rangle$, to denote an average over an ensemble of reservoirs. However, we are dealing with a single system. Thus $\langle a_\mu\rangle$ is still a system operator. A subsequent average over an ensemble of systems will be denoted by double brackets. Thus $\langle\langle a_\mu(t)\rangle\rangle$ is now a number, but it may be time-dependent, if the system is started off in a nonsteady state at time t_0. We will use $\langle\langle a_\mu\rangle\rangle_{ss}$ or $\bar{a}_\mu$ to denote the steady-state system average.

[4] We shall choose this time Δt to be long compared to the reciprocal natural frequency $(\omega_s)^{-1}$ of the system, and short compared to the system relaxation time Γ^{-1}. More precise restrictions will be given in footnote 5 below.

[5] We actually assume that the correlation time τ_c of the Langevin forces is short compared to all system *relaxation* times, Γ^{-1}, but not zero. This correlation time is usually *long* compared to the reciprocals of the natural frequencies $(\omega_s)^{-1}$ of the system. In such a case, the system behaves in an essentially Markoffian way when changes are observed over time intervals Δt that obey $(\omega_s)^{-1}<\tau_c<\Delta t\ll\Gamma^{-1}$. Thus the diffusion coefficients (1.5) or (1.6) are calculated in Appendix B by the use of such time intervals. The motion of $\mathbf{a}(t)$ in $F_\mu(\mathbf{a}(t),t)$ during such time intervals is important as shown in Appendices A and B and in IV, Secs. 5, 6.

can be omitted and (1.7) reduces to the Einstein relation between diffusion coefficients and mobilities [see I (5.18), I (6.14)]. The generalized time-dependent Einstein relation (1.7) is the basis of our *exact* calculations of $D_{\mu\nu}$ for harmonic oscillators in Sec. 3 and nonuniformly spaced multilevel systems in Sec. 4. The resulting exact "fluctuation-dissipation" relations between $D_{\mu\nu}$ and the reservoir contributions to A_μ are in precise agreement with those found by direct use of perturbation theory. The Einstein method, however, guarantees that products of operators propagate properly so that commutation rules are *necessarily preserved in time.*

Note that $\langle D_{\mu\nu}\rangle$ as calculated by (1.7) is not only a system operator,[3] it is, in general, time-dependent. Thus, the way in which the noise sources change during the turn-on of a laser is simply described within the present scheme.

With the noise sources included, our *Eqs. (1.2) are valid operator equations.* In particular, if variables $a_{s+1}, \cdots, a_n$ vary more rapidly than $a_1, a_2, \cdots, a_s$, we can solve for the fast variables in terms of the slow variables, and obtain a set of equations for the usually *much smaller set of slow variables.* The price paid for this is that the slow variable equations contain integrals over history and nonwhite noise sources. If, however, one examines the solutions only at frequencies small compared to the decay constants of the fast variables, then *it is adequate to treat the slow variables as Markoffian and the corresponding noise sources as white.* When this is done, an *enormous practical simplification* has been achieved in the solution of complicated problems such as laser noise.

Since our equations are operator equations, the equations for db/dt and $db^\dagger/dt$ (where b and $b^\dagger$ are destruction and creation operators of a photon field) determine the equation for $db^\dagger b/dt$, the rate of change of the number of photons. By applying the technique just discussed to eliminate all other variables but the upper state population N_2 in a maser and $b^\dagger b$, we show[6] that the *Markoffian* equations for $b^\dagger b$ and N_2 are the familiar rate equations. Moreover, these equations contain just the shot noise sources discussed by McCumber,[7] plus some thermal noise sources important in masers but not lasers. In summary, we have established the rate equations (and their noise sources) on *firm theoretical grounds*, as valid when $b^\dagger b$ and N_2 are the slowest changing variables in the complete set required to describe a maser or laser.[6] This adiabatic approximation can be avoided, however, permitting the extension of the noise calculation to higher frequencies, or to systems with other slowly varying populations or polarizations.[8]

[6] See paper QVII, Ref. 1 (to be published).
[7] D. E. McCumber, Phys. Rev. 141, 306 (1966).
[8] The nonadiabatic treatment (to be published) represents work done jointly with D. E. McCumber.

Summary

Before discussing the details of our computations, it may be worthwhile to *summarize* our principal results:

If b and $b^\dagger$ are the destruction and creation operators associated with a harmonic oscillator, whose commutator is unity, then we find that the appropriate equations including dissipation and fluctuations are

$$db/dt = -(i\omega_0 + \tfrac{1}{2}\gamma)b + f(t); \quad \langle f(t)\rangle = 0, \tag{1.8}$$

$$db^\dagger/dt = (i\omega_0 - \tfrac{1}{2}\gamma)b^\dagger + f^\dagger(t); \quad \langle f^\dagger(t)\rangle = 0. \tag{1.9}$$

The parameter γ is the decay constant of this harmonic oscillator. The nonvanishing moments of the Langevin forces are provided by

$$\langle f^\dagger(t) f(u)\rangle = \gamma\bar{n}\delta(t-u),$$
$$\langle f(u) f^\dagger(t)\rangle = \gamma(\bar{n}+1)\delta(t-u). \tag{1.10}$$

The parameter $\bar{n}$ can be regarded as defined by the second-moment equation

$$d\langle b^\dagger b\rangle/dt = \gamma\bar{n} - \gamma\langle b^\dagger b\rangle; \quad \langle b^\dagger b\rangle \to \bar{n}, \tag{1.11}$$

i.e., the reservoir drives the system occupation number $b^\dagger b$ toward a mean value of $\bar{n}$. A harmonic reservoir at equilibrium would do this if its temperature T_R were given by

$$\bar{n} = [\exp(\hbar\omega_0/kT_R) - 1]^{-1}, \tag{1.12}$$

as shown in Eq. (C11).

Our atomic equations in the absence of any regularities of energy spacing are given as

$$d(a_i^\dagger a_j)/dt = (i\omega_{ij} - \Gamma_{ij})a_i^\dagger a_j + \delta_{ij}\textstyle\sum_k w_{ik} a_k^\dagger a_k + F_{ij}, \tag{1.13}$$

$$\Gamma_{ij} = \Gamma_{ji}; \quad \omega_{ij} = -\omega_{ji}; \quad F_{ji} = (F_{ij})^\dagger,$$

where the w_{ik} is a transition probability and the Γ_{ij} and ω_{ij} are defined in

$$\Gamma_{ij} = \tfrac{1}{2}(\Gamma_i + \Gamma_j) + \Gamma_{ij}^{\text{ph}}; \quad \Gamma_{ii}^{\text{ph}} = 0; \qquad \omega_{ij} = \omega_i - \omega_j + \Delta\omega_{ij}. \tag{1.14}$$

The superscript ph denotes a contribution to the results associated with phase fluctuations. We define

$$\Gamma_j = \sum_{k\neq j} w_{kj}; \quad w_{jj} \equiv 0, \tag{1.15}$$

so that Γ_j represents the total transition rate out of state j. Our random forces F_{ij} have a vanishing first moment and second moments defined in

$$\langle F_{ij}\rangle = 0; \quad \langle F_{ij}(t) F_{kl}(u)\rangle = 2\langle D_{ijkl}\rangle\delta(t-u), \tag{1.16}$$

$$D_{ijkl} = (D_{lkji})^*. \tag{1.17}$$

An explicit and nontrivial expression for the general diffusion constant $\langle D_{ijkl}\rangle$ (in the absence of regularities in energy spacing) is[3]

$$2\langle D_{ijkl}\rangle=\delta_{jk}[\Gamma_{ij}+\Gamma_{kl}-\Gamma_{il}]\langle a_i^\dagger a_l\rangle$$
$$+\delta_{il}\delta_{jk}\sum w_{iq}\langle a_q^\dagger a_q\rangle$$
$$-\delta_{ij}w_{ik}\langle a_k^\dagger a_l\rangle-\delta_{kl}w_{kj}\langle a_i^\dagger a_j\rangle$$
$$-i\delta_{jk}(\omega_{ij}+\omega_{kl}-\omega_{il})\langle a_i^\dagger a_l\rangle. \quad (1.18)$$

The form of Eq. (1.13) was obtained by a perturbation treatment in Appendix A. However, once we grant this form, the diffusion constants of Eq. (1.18) follow without approximation. The parameters Γ_{ij}, w_{ij} can be regarded as experimentally determined. The last term in Eq. (1.18) vanishes in all cases in which ω_{ij} can be decomposed into two parts, one associated with ω_i and another associated with ω_j in the usual subtractive fashion. Whether or not ω_{ij} can be written in such a subtractive form is not a formal but rather a physical question. If the interaction between our system and our reservoir takes the form of

$$V=\hbar\sum a_i^\dagger a_j f_{ij}, \quad (1.19)$$

where f_{ij} represents a set of reservoir operators, then the transition probabilities above are given in

$$w_{mj}=\int_{-\infty}^{\infty} dt\, \exp(-i\omega_{mj}t)\langle f_{jm}(t)f_{mj}(0)\rangle. \quad (1.20)$$

The phase contribution to the damping constant is shown in Appendix A to be given in the form of

$$\Gamma_{ij}^{\text{ph}}=\frac{1}{2}\int_0^\infty dt\langle[f_{ii}(0)-f_{jj}(0),\, f_{ii}(t)-f_{jj}(t)]_+\rangle, \quad (1.21)$$

and the change in ω_{ij} is given in

$$\Delta\omega_{ij}=\Delta\omega_i-\Delta\omega_j+\Delta\omega_{ij}^{\text{ex}}, \quad (1.22)$$

$$\Delta\omega_j=\sum_{m\neq j}\text{Im}\int_0^\infty dt\,\exp(-i\omega_{mj}t)\langle f_{jm}(t)f_{mj}(0)\rangle$$
$$+\tfrac{1}{2}i\int_0^\infty dt\langle[f_{jj}(0),f_{jj}(t)]\rangle, \quad (1.23)$$

$$\Delta\omega_{ij}^{\text{ex}}=\tfrac{1}{2}i\int_0^\infty dt\langle[f_{jj}(0),f_{ii}(t)]\rangle. \quad (1.24)$$

The "extra" term shown in Eq. (1.24) is not ordinarily decomposable in a subtractive way. We shall show in Appendix A that this anomalous frequency-shift term vanisbes in cases in which the reservoir obeys time reversal symmetry and the levels i and j are nondegenerate levels with respect to time reversal. Moreover, we show that this extra term in Eq. (1.24) vanishes whenever the reservoir forces can be decomposed into independent excitations such as phonons or spin waves, provided that these excitation operators belonging to different modes commute with one another at all times. Thus to see an anomalous frequency shift it is desirable to use a reservoir violating time reversal, for example, a ferromagnet or an antiferromagnet. In this case, the extra frequency shift will involve the interactions between the spin waves.

The general formula (1.18) for the diffusion constant D_{ijkl} is not particularly illuminating and it is worthwhile to display explicit results for a number of special cases. From now on if the subscripts of D are indicated by different letters then they are understood to be necessarily different unless otherwise indicated. Our first results are those appropriate to shot noise:

$$2\langle D_{iiii}\rangle=\sum_{q\neq i} w_{iq}\langle a_q^\dagger a_q\rangle+\Gamma_i\langle a_i^\dagger a_i\rangle$$
$$=\text{atomic (rate in+rate out)}, \quad (1.25)$$

$$2\langle D_{iijj}\rangle=-w_{ji}\langle a_i^\dagger a_i\rangle-w_{ij}\langle a_j^\dagger a_j\rangle$$
$$=-(\text{transfer rate}). \quad (1.26)$$

Equation (1.25) describes the typical rate-in plus rate-out contribution to the shot noise source associated with population of level i and Eq. (1.26) contains the sum of the transfer rates from levels i to j. These results have previously been obtained for classical systems in I, Sec. 12 and IV, Sec. 9. The diffusion constant most relevant for off-diagonal elements of the atomic density matrix is given in

$$2\langle D_{ijji}\rangle=(\Gamma_j+2\Gamma_{ij}^{\text{ph}})\langle a_i^\dagger a_i\rangle+\sum w_{iq}\langle a_q^\dagger a_q\rangle, \quad (1.27)$$

which has no simple classical analog. This moment D_{ijji} is valid even in the presence of coupling to a radiation field that induces transitions between levels i and j. In this case the transport equation for the population of level i is given by

$$d\langle a_i^\dagger a_i\rangle/dt=-\Gamma_i\langle a_i^\dagger a_i\rangle+\sum w_{iq}\langle a_q^\dagger a_q\rangle+\langle B_i\rangle, \quad (1.28)$$

$$B_i=\text{radiative rate into } i. \quad (1.29)$$

Assuming that we are in the steady state, in other words setting Eq. (1.28) equal to 0, we can simplify the right-hand side of Eq. (1.27) to obtain the steady-state second moment[3]

$$2\langle\langle D_{ijji}\rangle\rangle_{\text{steady state}}=2\Gamma_{ij}\langle\langle a_i^\dagger a_i\rangle\rangle_{\text{ss}}-\langle\langle B_i\rangle\rangle_{\text{ss}}. \quad (1.30)$$

The subscript ss is to remind us that the steady-state-system ensemble average is understood here. This result is stated without proof in QV (2.7) using $\bar{\sigma}_{ii}$ and $\bar{B}_i$ as abbreviated notations for the ss averages. A more complicated diffusion coefficient is given in

$$2\langle D_{ijjl}\rangle=(\Gamma_j+\Gamma_{ij}^{\text{ph}}+\Gamma_{jl}^{\text{ph}}-\Gamma_{il}^{\text{ph}})\langle a_i^\dagger a_l\rangle$$
$$-i(\omega_{ij}+\omega_{jl}-\omega_{il})\langle a_i^\dagger a_l\rangle, \quad (1.31)$$

which appears to depend on the anomalous frequency shifts. Some diffusion coefficients descriptive of correlated population and phase fluctuation are given by

$$2\langle D_{ijkk}\rangle=-w_{kj}\langle a_i^\dagger a_j\rangle, \quad (k \text{ can equal } i) \quad (1.32)$$

$$2\langle D_{iikl}\rangle=-w_{ik}\langle a_k^\dagger a_l\rangle, \quad (l \text{ can equal } i) \quad (1.33)$$

$$2\langle D_{iiij}\rangle=\Gamma_i\langle a_i^\dagger a_j\rangle, \quad (1.34)$$

$$2\langle D_{ijjj}\rangle=\Gamma_j\langle a_i^\dagger a_j\rangle. \quad (1.35)$$

In Sec. 5, we establish that if two systems that interact with independent reservoirs are coupled together dynamically, no new noise sources are introduced, and no correlations occur between the noise sources associated with different reservoirs. The original noise sources are shown to be slightly modified by the dynamic interaction.

In Sec. 6, we use the results to construct the model of a maser used in QV. In Sec. 7, we obtain a "commutation rule Einstein relationship" and use it to show how the moments must be modified if the population difference in a maser is treated as a number rather than an operator.

Relation to Previous Work

There is, of course, an extensive literature on dissipation in quantum mechanics which we cannot hope to review properly here. This literature can be divided roughly into five categories:

(A) The consideration of a system in interaction with a reservoir, and the (approximate) elimination of the reservoir to obtain effective equations of motion for system operators, or the system density matrix.[9] The disadvantage of this procedure, including our own QIII, is that it provides information only about operators, or fluctuations at one time. To obtain two-time correlations one must use the equilibrium fluctuation-dissipation theorem, as in Sec. 7 of QIII, or one must use our generalization of this theorem in QII to non-equilibrium systems. Only in this way, can Scully, Lamb, and Stephen[9] argue that the decay constant they find is indeed the maser spectral linewidth.

(B) Green's function and moment methods[10] attack two-time correlation functions directly, but generally can be solved only by applying a truncation procedure or a linearization.

(C) A number of papers have been written introducing classical noise sources in a heuristic way.[11]

(D) Quantum noise sources have also been introduced into maser calculations in a heuristic way, or by methods similar to the perturbation techniques adopted here.[12]

(E) Detailed consideration has been given to the harmonic-oscillator[13] and two-level systems.[14] Louisell and Walker[13] have provided an exact solution for the system harmonic oscillator interacting linearly with a bath of harmonic oscillators, for "thermal" initial conditions.[15] Although the Louisell-Walker calculation is exact, it suffers from the usual objections to method (A) above. Schwinger and Senitzky have actually written down harmonic-oscillator equations with quantum noise sources. Senitzky,[13] in our notation,[16] uses the equations

$$\dot{Q}=P,\quad \dot{P}=-\gamma P-\omega_0^2 Q+F(t)\,, \tag{1.36}$$

$$\langle F(t)F(u)\rangle=2\gamma\hbar\omega_0\left[(\bar{n}+\tfrac{1}{2})\delta(t-u)-\frac{i}{2\pi}\mathcal{P}\frac{1}{t-u}\right], \tag{1.37}$$

where we have corrected a sign in the last term involving the principal-valued reciprocal. We show, in Appendix C, that Senitzky's commutation rule

$$\langle[F(t),F(u)]\rangle=-\frac{2i\gamma\hbar\omega_0}{\pi}\mathcal{P}\frac{1}{t-u} \tag{1.38}$$

leads to a commutator

$$\langle[Q,P]\rangle=i\hbar\left[1-\frac{1}{\pi}\left(\frac{\gamma}{\omega_0}\right)+O\left(\frac{\gamma}{\omega_0}\right)^2\right] \tag{1.39}$$

that is close to the desired value when $\gamma\ll\omega_0$, whereas the correct commutation rule [see Eq. (C33) when γ is not frequency-dependent] for the Langevin forces

$$\langle[F(t),F(u)]\rangle=2i\hbar\gamma\delta'(t-u) \tag{1.40}$$

leads to precisely the correct commutation rule for Q and P, as shown in (C37).

In any case, the commutator is odd in $t-u$. Hence its Fourier transform is odd in frequency ω and the spectrum of noise for a *quantum* oscillator can then not take rigorouly the white (flat) form needed to make the treatment Markoffian.

Of course, a noise spectrum need not be exactly white for the Markoffian approximation to be a good

---

[9] Many earlier references are given in QIII. Recent papers with application to masers and lasers include W. Weidlich and F. Haake, Z. Physik **185**, 30 (1965); M. Scully, W. Lamb, and M. Stephen, *Physics of Quantum Electronics*, edited by P. L. Kelly, B. Lax, and P. E. Tannenwald (McGraw-Hill Book Company, Inc., New York, 1966).

[10] D. E. McCumber, Phys. Rev. **130**, 675 (1963); V. Korenman, Phys. Rev. Letters **14**, 293 (1965) and *Physics of Quantum Electronics*, edited by P. L. Kelley, B. Lax, and P. E. Tannenwald (McGraw-Hill Book Company, Inc., New York, 1966). See Sec. 6 of QV; also QVI. C. Willis, J. Math. Phys. **6**, 1984 (1965); R. H. Picard and C. R. Willis, Phys. Rev. **139**, A10 (1965).

[11] K. Shimoda, T. C. Wang, and C. H. Townes, Phys. Rev. **102**, 1308 (1956); W. Lamb, in *Quantum Optics and Electronics* (Gordon and Breach Science Publishers, Inc., New York, 1965). H. A. Haus, IEEE J. Quantum Electron. **1**, 179 (1965). C. Freed and H. A. Haus, Appl. Phys. Letters **6**, 85 (1965). E. I. Gordon, Bell System Tech. J. **43**, 507 (1964). W. G. Wagner and G. Birnbaum, J. Appl. Phys. **32**, 1185 (1961).

[12] H. Haken, Z. Physik **161**, 96 (1964); **182**, 346 (1965). H. Haken and H. Sauermann, *ibid.* **173**, 261 (1963); **176**, 58 (1963). H. J. Pauwels, thesis, Massachusetts Institute of Technology, 1965 (unpublished).

[13] W. H. Louisell and L. R. Walker, Phys. Rev. **137**, B204 (1965); I. R. Senitzky, *ibid.* **119**, 670 (1960); **124**, 642 (1961); J. Schwinger, J. Math. Phys. **2**, 407 (1961).

[14] I. R. Senitzky, Phys. Rev. **131**, 2827 (1963); **134**, A816 (1964); **137**, A1635 (1965).

[15] An extension of Louisell and Walker's work to the initial condition of a definite excitation state has been made by H. Cheng and M. Lax, in *Quantum Theory of the Solid State*, edited by Per-Olav Löwdin (Academic Press Inc., New York, to be published). See also A. E. Glassgold and D. Holliday, Phys. Rev. **139**, A1717 (1965).

[16] Since Senitzky (Ref. 13) uses a harmonic oscillator that couples to a reservoir via its momentum (whereas we use the coordinate), comparison with our notation can be made by the transformations: $Q\to-P$, $P\to Q$, $\beta=\gamma$, $D(t)=F(t)$, and by setting his $4\pi c^2=\omega_0^2$.

one. It need only change little over the width of the resonance. Our method of calculation of the strength of the noise sources seems to (automatically) replace nonwhite sources by white sources whose strengths agree at the resonance frequency. For the harmonic oscillator, a special difficulty arises because there are *two* resonant frequencies ω_0 and $-\omega_0$, and the noise does not agree at these frequencies. However, these frequencies must be well separated $\omega_0 \gg \gamma$, in order that[4,5]

$$\omega_0^{-1} \ll \Delta t \ll \gamma^{-1}, \tag{1.41}$$

which is required for a Markoffian description to exist. If $\gamma \ll \omega_0$ the rotating-wave approximation[17] (RWA) is valid, and the relation $P \propto (b^\dagger + b)$ is essentially a split of P into its positive and negative frequency parts, and $F \propto f^\dagger + f$ is a similar split. Thus $f^\dagger$ should have a white spectrum that corresponds to the spectrum of F at ω_0, and f is similarly determined by the spectrum of F at $-\omega_0$. When we treat b and $b^\dagger$ as our fundamental variables, in the RWA in Sec. 3, these results appear automatically. If, however, we had insisted on using Q, P as variables, without the RWA, we could not have achieved a consistent Markoffian description that preserved the commutation rules.

The two-level analysis of Senitzky[14] does not seem to express the moments of the Langevin forces in terms of transition probabilities and populations so that it is difficult to make a comparison. The work of the Haken school[12] has so far not indicated any population dependence of their noise sources. Their quantum-noise-source treatment also seems to be special to a two-level system.

Our general formula (1.18) for the noise sources in a multilevel system seems therefore to be completely new. The phase or zero-phonon contributions to the decay and frequency shifts in (1.13) are most closely related to the work of McCumber.[18] The general expressions for the phase contributions to the damping and frequency shift (especially the anomalous shift) in a multilevel system seem to be new.

2. EINSTEIN RELATIONS, DIFFUSION COEFFICIENTS, AND EQUIVALENCE TO MOMENT METHODS

If we rewrite our fundamental Langevin equation in the form

$$da_\mu/dt = A_\mu(\mathbf{a},t) + F_\mu(\mathbf{a},t), \tag{2.1}$$

then the assumption that the system has a Markoffian description can be phrased in the form

$$F_\nu(s) \text{ is independent of } a_\mu(t) \quad \text{if} \quad s > t. \tag{2.2}$$

In particular, these operators must commute. If (2.2) were not obeyed, the statistics of the random forces F would depend on past history and the system would acquire memory effects. For present purposes we only need this independent assumption in the much weaker form

$$\langle F_\nu(s) C(\mathbf{a}(t)) \rangle = \langle C(\mathbf{a}(t)) F_\nu(s) \rangle = 0 \quad \text{if} \quad s > t, \tag{2.3}$$

which merely describes a lack of correlation between the random force, at time s, and an arbitrary operator $C(\mathbf{a})$ at an earlier time t. We now write down the algebraic identity

$$a_\mu(t+\Delta t) a_\nu(t+\Delta t) - a_\mu(t) a_\nu(t) = \Delta a_\mu \Delta a_\nu + \Delta a_\mu a_\nu + a_\mu \Delta a_\nu, \tag{2.4}$$

where

$$\Delta a_\mu = a_\mu(t+\Delta t) - a_\mu(t), \tag{2.5}$$

$$\Delta a_\mu \approx A_\mu \Delta t + \int_t^{t+\Delta t} F_\mu(s) ds. \tag{2.6}$$

From Eq. (2.4), we can now write down the equation for the mean motion of the product of two operators:

$$d\langle a_\mu(t) a_\nu(t) \rangle / dt = 2\langle D_{\mu\nu} \rangle + \langle A_\mu a_\nu \rangle + \langle a_\mu A_\nu \rangle. \tag{2.7}$$

In the steady state when the second-order operator does not change with time we obtain the standard Einstein equation

$$2\langle D_{\mu\nu} \rangle_{\text{steady state}} = -\langle A_\mu a_\nu \rangle_{\text{ss}} - \langle a_\mu A_\nu \rangle_{\text{ss}}. \tag{2.8}$$

Aside from the preservation of the order of the operators, Eqs. (2.7) and (2.8) are identical to corresponding classical equations. See, for example, I (14.27) or III (5.10). It is convenient to rewrite Eq. (2.7) in the mnemonic form

$$2\langle D_{\mu\nu} \rangle = -\langle \{da_\mu/dt\} a_\nu \rangle - \langle a_\mu \{da_\nu/dt\} \rangle + \langle d[a_\mu(t) a_\nu(t)]/dt \rangle, \tag{2.9}$$

$$\{da_\mu/dt\} \equiv da_\mu/dt - F_\mu \equiv A_\mu. \tag{2.10}$$

We see, therefore, that $D_{\mu\nu}$ is a measure of the extent to which the usual rules for differentiating a product is violated in a Markoffian system. Equation (2.9) is indeed a useful computational formula because it permits the diffusion coefficients to be calculated in terms of the mean motion of certain system operators. This is in contrast to Eq. (1.6) which requires the direct calculation of fluctuations. Indeed, our presentation in Secs. 3 and 4 will be based on Eq. (2.9). The direct determination of the diffusion constants from the fluctuations moments using Eq. (1.6) is done in Appendix B.

Higher order diffusion coefficients may be determined by a simple generalization of (2.9). But higher moments

---

[17] For a harmonic oscillator, the RWA consists in neglecting the counter-resonant term $\frac{1}{2}\gamma b^\dagger$ in $db/dt = -i\omega_0 b - \frac{1}{2}\gamma(b+b^\dagger) + f$, and a corresponding term $\frac{1}{2}\gamma b$ in $db^\dagger/dt = i\omega_0 b^\dagger - \frac{1}{2}\gamma(b+b^\dagger) + f^\dagger$. The distinction between the free decay frequency $[\omega_0^2 - \frac{1}{4}\gamma^2]^{1/2}$ and ω_0 is thus neglected. The RWA has been extensively exploited in magnetic resonance problems: I. I. Rabi, N. F. Ramsey, and J. Schwinger, Rev. Mod. Phys. **26**, 107 (1954); A. Abragam, *Principles of Nuclear Magnetism* (Oxford University Press, Oxford, England, 1961), Chap. II; F. Bloch and A. J. Siegert, Phys. Rev. **57**, 522 (1940).

[18] D. E. McCumber, Phys. Rev. **135**, A1676 (1964); J. Math. Phys. **5**, 221, 508 (1964).

of the Langevin forces are related in a more complicated way to the diffusion constants as shown in IV.

Equivalence of Langevin and Moment Procedures

We would now like to establish the equivalence of the Langevin method of this paper with the moment procedures adopted in QII. The principal result of paper QII is summarized in the statement that the solution for the mean motion at one time,

$$\langle a_i^\dagger(t)a_j(t)\rangle=\sum O_{qp}{}^{ji}(t,t')\langle a_p{}^\dagger(t')a_q(t')\rangle, \quad (2.11)$$

can be used to obtain the moments containing operators at two times

$$\langle a_i^\dagger(t)a_j(t)a_k{}^\dagger(t')a_l(t')\rangle$$
$$=\sum O_{qp}{}^{ji}(t,t')\langle a_p{}^\dagger(t')a_q(t')a_k{}^\dagger(t')a_l(t')\rangle. \quad (2.12)$$

This two-time result (2.12) possesses the same t dependence as the transient solution (2.11) of the mean motion equations. This is an expression of the Onsager hypothesis concerning the regression of fluctuations, established for a quantum-mechanical system not in equilibrium nor even necessarily in a steady state.

For comparison with the results in this paper, it is necessary to avoid direct use of the Green's function $O_{qp}{}^{ji}$ and replace it by a differential relationship. We note, however, as remarked at the end of paper QII that this principal result (2.12) did not depend on the system being Markoffian. If we assume that the system is Markoffian its future must be predictable from the present, in other words, its time derivative at time t must be expressible in terms of other quantities expressed at time t. In terms of the Green's function the Markoffian requirement takes the form

$$dO_{qp}{}^{ji}(t,t')/dt=\sum B_{ijmn}O_{qp}{}^{nm}(t,t'). \quad (2.13)$$

If we now differentiate Eq. (2.11) and make use of (2.13), we obtain

$$\langle da_i{}^\dagger(t)a_j(t)/dt-\sum B_{ijmn}a_m{}^\dagger(t)a_n(t)\rangle=0, \quad (2.14)$$

which expresses the time derivative of $a_i{}^\dagger a_j$ in terms of similar operators taken at the same time. Applying the same procedure to Eq. (2.12) leads to

$$\langle[da_i{}^\dagger(t)a_j(t)/dt-\sum B_{ijmn}a_m{}^\dagger(t)a_n(t)]$$
$$\times a_k{}^\dagger(t')a_l(t')\rangle=0. \quad (2.15)$$

Since the quantity in brackets in Eq. (2.15) is the difference between the time derivative of the operator $a_i{}^\dagger a_j$ and the corresponding mean motion, it represents the fluctuating random force F_{ij}. Thus, we obtain

$$\langle F_{ij}(t)a_k{}^\dagger(t')a_l(t')\rangle=0, \quad t>t'. \quad (2.16)$$

But Eq. (2.16) is precisely the same as Eq. (2.3), which we took as our expression in the Langevin point of view of the Markoffian nature of our problem. Thus we have established the equivalence of the two procedures. When both are used without further approximation they should yield equivalent results for the mean motion and the noise in a quantum-mechanical system.

Taking the derivative of (2.16) with respect to t', we obtain

$$\langle F_{ij}(t)da_k{}^\dagger(t')a_l(t')/dt'\rangle=0, \quad t>t'. \quad (2.17)$$

Taking the appropriate linear combinations of (2.16) and (2.17), we establish that

$$\langle F_{ij}(t)F_{pq}(t')\rangle=0, \quad t>t'. \quad (2.18)$$

A similar argument proves the same result for $t<t'$, so that we establish

$$\langle F_{ij}(t)F_{pq}(t')\rangle=0, \quad t\neq t'. \quad (2.19)$$

3. THE DAMPED HARMONIC OSCILLATOR

To obtain the diffusion constants using (2.9), it is necessary to obtain the mean equations of motion for operators linear and operators quadratic in the oscillator displacements. In Appendix A, we modify slightly some results of QIII which analyzes a system in interaction with a reservoir and then determines the effective equations of motion for the system after averaging over the reservoir. These results of QIII are sufficiently general to include memory effects. After eliminating these memory effects when the correlation times are short, we arrive at Eq. (A13) for an arbitrary system operator M. Equation (A13) is valid even if only reservoir averages[3] $\langle\ \rangle$ instead of $\langle\langle\ \rangle\rangle$ are taken—see (B16). Thus, we can write

$$\frac{\partial\langle M\rangle}{\partial t}=\langle(M,H)\rangle$$
$$-\frac{1}{\hbar^2}\sum_{ij}\int_0^\infty du\{\langle F_i(u)F_j\rangle\langle[M,Q_i]Q_j(-u)\rangle$$
$$-\langle F_jF_i(u)\rangle\langle Q_j(-u)[M,Q_i]\rangle\}. \quad (3.1)$$

In (3.1), the Q's are system operators and the F's are corresponding reservoir operators that when multiplied together provide the coupling V between the system and reservoir:

$$V=-\sum_i Q_iF_i.$$

The operators $F_i(u)$ and $Q_j(-u)$ are in the interaction representation

$$F_i(u)=\exp(iRu/\hbar)F_i\exp(-iRu/\hbar),$$
$$Q_j(-u)=\exp(-iHu/\hbar)Q_j\exp(iHu/\hbar),$$

where H and R are the system and reservoir Hamiltonians, respectively. [The slightly more general equation (A13) must be used if H depends explicitly on the time.]

For the harmonic-oscillator system, we take

$$H=\hbar\omega_c b^\dagger b; \quad V=i\hbar(b^\dagger g-bg^\dagger). \quad (3.2)$$

With this choice an arbitrary system operator M obeys

the equation of motion

$$\partial\langle M\rangle/\partial t=\langle(M,H)\rangle+\alpha\langle b^\dagger[M,b]\rangle-\beta\langle[M,b]b^\dagger\rangle$$
$$-\alpha^*\langle[M,b^\dagger]b\rangle+\beta^*\langle b[M,b^\dagger]\rangle, \quad (3.3)$$

where the coefficients α and β are integrals over averages of reservoir operators:

$$\alpha=\int_0^\infty du\, e^{-i\omega_c u}\langle gg^\dagger(u)\rangle,$$
$$\beta=\int_0^\infty du\, e^{-i\omega_c u}\langle g^\dagger(u)g\rangle. \quad (3.4)$$

Introducing the interpretations

$$\tfrac{1}{2}\gamma-i\Delta\omega=\alpha-\beta, \quad \omega_0=\omega_c+\Delta\omega, \quad \gamma\bar{n}=2\,\mathrm{Re}\beta \quad (3.5)$$

of the parameters α and β in terms of new real parameters γ, $\bar{n}$, and $\Delta\omega$ which have more direct physical meaning, we obtain the equation of motion for the fairly general operator $(b^\dagger)^r b^s$:

$$\partial\langle(b^\dagger)^r b^s\rangle/\partial t=[i\omega_0(r-s)-\tfrac{1}{2}\gamma(r+s)]\langle(b^\dagger)^r b^s\rangle$$
$$+rs\gamma\bar{n}\langle(b^\dagger)^{r-1}b^{s-1}\rangle. \quad (3.6)$$

These equations are exact in this Markoffian limit when the reservoir consists of a set of harmonic oscillators. (See Appendix D.) In (3.6) ω_0 is the renormalized frequency of the oscillator after the frequency shift $\Delta\omega$ produced by the reservoir has been absorbed. The most important special cases of Eq. (3.6) are given by

$$\partial\langle b\rangle/\partial t=-(i\omega_0+\tfrac{1}{2}\gamma)\langle b\rangle,$$
$$\partial\langle b^\dagger\rangle/\partial t=(i\omega_0-\tfrac{1}{2}\gamma)\langle b^\dagger\rangle, \quad (3.7)$$

$$\partial\langle b^\dagger b\rangle/\partial t=\gamma\bar{n}-\gamma\langle b^\dagger b\rangle. \quad (3.8)$$

We now write out our complete Langevin equations including the noise sources as

$$db/dt=-(i\omega_0+\tfrac{1}{2}\gamma)b+f(t),$$
$$db^\dagger/dt=(i\omega_0-\tfrac{1}{2}\gamma)b^\dagger+f^\dagger(t). \quad (3.9)$$

Our second moment has the typical δ-function form

$$\langle f^\dagger(t)f(u)\rangle=2D_{b^\dagger b}\delta(t-u), \quad (3.10)$$

where the diffusion constant is computed by means of Eq. (2.9) in the form

$$2D_{b^\dagger b}=d\langle b^\dagger b\rangle/dt-\langle\{db^\dagger/dt\}b\rangle-\langle b^\dagger\{db/dt\}\rangle=\gamma\bar{n}. \quad (3.11)$$

Repeating this calculation with the opposite order of the factors yields

$$\langle f(u)f^\dagger(t)\rangle=2D_{bb^\dagger}\delta(t-u)=\gamma(\bar{n}+1)\delta(t-u), \quad (3.12)$$

a second moment clearly distinct from that in (3.11). Indeed, this lack of commutation of the reservoir forces at f and $f^\dagger$ is shown in Appendix C to preserve the commutation rules. We also compute the additional second-moment equations

$$d\langle b^2\rangle/dt=-2i\omega_0\langle b^2\rangle-\gamma\langle b^2\rangle,$$
$$d\langle(b^\dagger)^2\rangle/dt=+2i\omega_0\langle(b^\dagger)^2\rangle-\gamma\langle(b^\dagger)^2\rangle, \quad (3.13)$$

and learn from them that the remaining second moments

$$\langle f(t)f(u)\rangle=\langle f^\dagger(t)f^\dagger(u)\rangle=0 \quad (3.14)$$

are zero.

4. ATOMIC DIFFUSION CONSTANTS

For an arbitrary quantum-mechanical system (which we shall visualize as an atom) whose frequency differences ω_{ij} possess no degeneracies, we shall adopt as our typical equation of motion (A23):

$$d(a_i^\dagger a_j)/dt=(i\omega_{ij}-\Gamma_{ij})a_i^\dagger a_j+\delta_{ij}\sum_{p\neq i}w_{ip}a_p^\dagger a_p+F_{ij} \quad (4.1)$$

obtained in Appendix A by the use of Eq. (3.1). Second moments are defined by[3]

$$\langle F_{ij}(t)F_{kl}(u)\rangle=2\langle D_{ijkl}\rangle\delta(t-u). \quad (4.2)$$

The second moments as usual are calculated by Eq. (2.9) which represents a nonstationary Einstein relationship. For the present case, Eq. (2.9) takes the form

$$2\langle D_{ijkl}\rangle=\frac{d}{dt}\langle a_i^\dagger a_j a_k^\dagger a_l\rangle-\left\langle\left\{\frac{d}{dt}a_i^\dagger a_j\right\}a_k^\dagger a_l\right\rangle$$
$$-\left\langle a_i^\dagger a_j\left\{\frac{d}{dt}a_k^\dagger a_l\right\}\right\rangle, \quad (4.3)$$

where the bracketed symbols defined by Eq. (4.4)

$$\left\{\frac{d}{dt}a_i^\dagger a_j\right\}\equiv\frac{d}{dt}a_i^\dagger a_j-F_{ij} \quad (4.4)$$

simply describe the mean motion of the operators $a_i^\dagger a_j$. In the space of one atom, $\sum a_j^\dagger a_j=1$, we have the identity

$$a_i^\dagger a_j a_k^\dagger a_l=\delta_{jk}a_i^\dagger a_l \quad (4.5)$$

derived in QII. Inserting the bracketed quantities (4.4) into Eq. (4.3), our results immediately simplify into a bilinear expression of the form

$$2\langle D_{ijkl}\rangle=\delta_{jk}\delta_{il}\sum_{p\neq i}w_{ip}\langle a_p^\dagger a_p\rangle$$
$$-\delta_{ij}w_{ik}\langle a_k^\dagger a_l\rangle-\delta_{kl}w_{kj}\langle a_i^\dagger a_j\rangle$$
$$+\langle a_i^\dagger a_l\rangle\delta_{jk}[(\Gamma_{ij}+\Gamma_{kl}-\Gamma_{il})-i(\omega_{ij}+\omega_{kl}-\omega_{il})], \quad (4.6)$$

where the quantities Γ_{ij} and ω_{ij} have the symmetry and antisymmetry properties

$$\Gamma_{ij}=\Gamma_{ji}, \quad \omega_{ij}=-\omega_{ji}. \quad (4.7)$$

We see that our results for the second moments depend on the form of Eq. (4.1) and not how the parameters were obtained. Equations (1.19)–(1.24) display, however, the values of these parameters obtained from the explicit reservoir calculations shown in Appendix A. Equation (4.6) is the general result quoted in Eq. (1.18) and important special cases of the diffusion constants were already presented in Eqs. (1.25)–(1.35).

5. COUPLED SYSTEMS: INDEPENDENT RESERVOIRS

Our systems 1 and 2 are coupled together dynamically via the Hamiltonian H_{12}:

$$H=H_1+H_2+H_{12}, \quad V_1=-\sum Q_jF_j, \quad V_2=-\sum q_jf_j. \tag{5.1}$$

Here V_1 and V_2 are the couplings of systems 1 and 2, respectively, to their corresponding independent reservoirs. The reservoir forces F_j and f_j are definitely uncorrelated since they come from quite independent reservoirs. If M and m are arbitrary operators belonging to the first and second systems, respectively, they obey the Heisenberg equations

$$dM/dt=(M, H_1+H_{12})-\sum (M,Q_j)F_j, \tag{5.2}$$

$$dm/dt=(m, H_2+H_{12})-\sum (m,q_i)f_i. \tag{5.3}$$

The direct use of Eq. (1.6) then yields

$$2\langle D_{mM}\rangle=\langle[m(t+\Delta t)-m(t)] \times[M(t+\Delta t)-M(t)]\rangle/\Delta t \tag{5.4}$$

$$\approx\sum(m,q_i)(M,Q_j)\int_t^{t+\Delta t} ds\int_t^{t+\Delta t} ds' \times\langle f_i(s)F_j(s')\rangle/\Delta t, \tag{5.5}$$

where the first terms in (5.2) and (5.3) have been dropped since they lead to terms of higher order in Δt. The assumed δ-function character of the correlations of the random forces permits the system operators to be evaluated at the initial time t and removed from the integration in Eq. (5.5). The result involves the reservoir forces F and f directly and these are by hypothesis uncorrelated. The vanishing of D_{mM} implies that the effective Langevin forces that enter the equations for dM/dt and dm/dt are necessarily uncorrelated.

The proof we have given, however, is unnecessarily restrictive. The reservoir forces are not in general δ correlated.[5] They merely possess a correlation time that is short compared to all of the typical system relaxation times. In general, however, their correlation times are long compared to the reciprocals of the various oscillation frequencies of the system. Under these circumstances it is not permissible to remove the system operators from underneath the integral sign in Eq. (5.5), and a new proof is needed. The methods of QIII were, in fact, designed to deal with such situations in which the correlation time is short but not zero. The net result of that paper reduced to the Markoffian limit is Eq. (3.1). If we apply (3.1) to calculate the equation of motion of the operator mM, we obtain

$$\begin{aligned}\frac{d\langle mM\rangle}{dt}&=\langle(mM,H)\rangle\\ &-\frac{1}{\hbar^2}\sum\int_0^\infty du\{\langle F_i(u)F_j\rangle\langle[mM,Q_i]Q_j(-u)\rangle\\ &-\langle F_jF_i(u)\rangle\langle Q_j(-u)[mM,Q_i]\rangle\}\\ &-\frac{1}{\hbar^2}\sum\int_0^\infty du\{\langle f_i(u)f_j\rangle\langle[mM,q_i]q_j(-u)\rangle\\ &-\langle f_jf_i(u)\rangle\langle q_j(-u)[mM,q_i]\rangle\}.\end{aligned} \tag{5.6}$$

In Eq. (5.6), correlations between F and f forces have been omitted. However, the time dependence of the system operators is retained underneath the integral sign. If, however, we expand the commutators in Eq. (5.6) and compare them with the corresponding separate equations for the operators M and m, we find that our result has the structure

$$d\langle mM\rangle/dt=\langle m\{dM/dt\}\rangle+\langle\{dm/dt\}M\rangle+0 \tag{5.7}$$

from which we can deduce that the diffusion constant

$$2D_{mM}=0 \tag{5.8}$$

vanishes. The result (5.8) establishes that the Langevin noise sources that enter the equations of two systems that are coupled to independent reservoirs are uncorrelated. While this result is intuitively very reasonable, it is not, in fact, obvious. The Langevin forces are not identical to the reservoir forces which are automatically uncorrelated. Thus, for example, the Langevin force F_{ij} of Eq. (1.13) is not identical to the reservoir operator f_{ij} of Eq. (1.19). The Langevin forces, in effect, involve products of reservoir forces with system operators. Since the time dependence of these system operators must be taken into account, the lack of correlation of the Langevin forces requires the proof just given.

A similar procedure with m and M both taken from the first system yields the same *formal* expression for D_{mM} as if the second system were not present. The interaction operators, however, now include the effect of the dynamic interaction. Unless this interaction is extremely strong, however, its influence during the short correlation time of the reservoirs will be unimportant. This is equivalent, for example, to neglecting the change of atomic state of an atom due to a laser field during the course of its collision with a second atom. Our procedure permits us to include such effects, but we shall omit them in the maser model of the next section.

6. STOCHASTIC MODEL OF A MASER

Our model of a maser is schematically described in Fig. 1. The electromagnetic field is described in terms of a single-cavity mode, although it is easy enough to generalize to the presence of several modes. There are a set of N atoms labeled by the index M. These two systems are coupled by the radiation coupling

$$H_{\text{RAD-ATOM}}=i\hbar\mu\sum_{M=1}^{N}[b^\dagger(a_1^\dagger a_2)^M-b(a_2^\dagger a_1)^M]. \quad (6.1)$$

As shown in Fig. 1, the set of atoms and the radiation field are each coupled to its own reservoir. Indeed, for practical purposes, we can assume that each atom is coupled to its own private reservoir. This is why we have indicated the atomic Langevin forces by a superscript M. Our Langevin equations of motion now take the form

$$db/dt=-(i\omega_c+\tfrac{1}{2}\gamma)b+\mu\sum_M(a_1^\dagger a_2)^M+f, \quad (6.2)$$

$$d(a_1^\dagger a_2)^M/dt=-(\Gamma_{12}{}^M+i\omega^M)(a_1^\dagger a_2)^M+\mu b(a_2^\dagger a_2-a_1^\dagger a_1)^M+F_{12}{}^M, \quad (6.3)$$

$$d(a_2^\dagger a_2)^M/dt=w_{20}(a_0^\dagger a_0)^M+w_{21}(a_1^\dagger a_1)^M-\Gamma_2(a_2^\dagger a_2)^M-B^M+F_{22}{}^M, \quad (6.4)$$

$$d(a_1^\dagger a_1)^M/dt=w_{10}(a_0^\dagger a_0)^M+w_{12}(a_2^\dagger a_2)^M-\Gamma_1(a_1^\dagger a_1)^M+B^M+F_{11}{}^M, \quad (6.5)$$

$$B^M=\mu[b^\dagger(a_1^\dagger a_2)^M+(a_2^\dagger a_1)^M b], \quad (6.6)$$

$$(a_0^\dagger a_0+a_1^\dagger a_1+a_2^\dagger a_2)^M=1, \quad (6.7)$$

$$\omega^M=\omega_2{}^M-\omega_1{}^M, \quad (6.8)$$

where w_{20} and w_{10} are pump terms. We have assumed, for simplicity only, that the interaction between the atoms and the field is equally strong for all atoms. In such a case it is appropriate to introduce averages over the various atom operators:

$$\sigma_{jj}\equiv(1/N)\sum_{M=1}^{N}(a_j^\dagger a_j)^M, \quad (6.9)$$

$$D\equiv\sigma_{22}-\sigma_{11}, \quad (6.10)$$

$$\sigma\equiv(1/N)\sum_{M=1}^{N}(a_1^\dagger a_2)^M\exp(i\omega_0 t). \quad (6.11)$$

Moreover, we shall introduce a new field operator

$$b'=b\exp(i\omega_0 t) \quad (6.12)$$

that has absorbed most of the steady motion in the maser so that b' changes only quite slowly with the time primarily because of the Langevin forces. In the following equations we shall for simplicity drop the prime on b. The average rate of radiation per atom is then given by

$$B=(1/N)\sum B^M=\mu(b^\dagger\sigma+\sigma^\dagger b) \quad (6.13)$$

and the appropriately averaged atomic random force is given by

$$F_{ij}=(1/N)\sum F_{ij}{}^M\exp(i\omega_{ji}t). \quad (6.14)$$

Strictly speaking F_{12} contains a factor $\exp(i\omega_0 t)$ rather than $\exp(i\omega_{21}t)$,but the spectrum of F_{12} can be assumed not to change much over the small difference between these two frequencies. Our coupled atomic and field equations now take the form

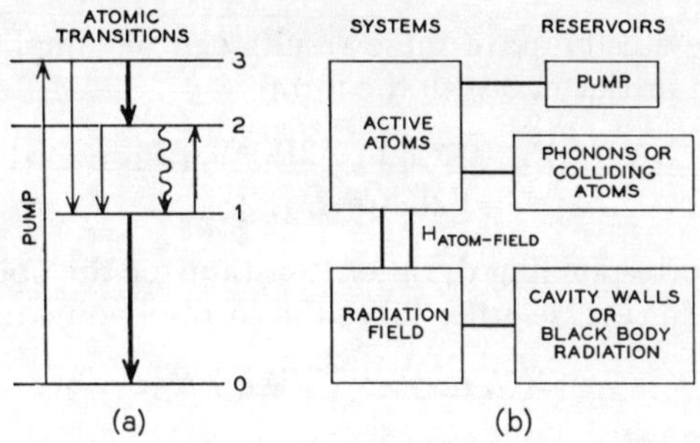

FIG. 1. Model for maser or laser: Radiative transitions (wavy arrow) are induced by the dynamic atom-field coupling. Nonradiative transitions (straight arrows) and quantum noise sources are derivable consequences of the coupling to the reservoirs. We assume in this paper that the transition rate from 3 to 2 is so fast that we are effectively pumping directly into state 2.

$$d\sigma_{22}/dt=w_{20}\sigma_{00}+w_{21}\sigma_{11}-\Gamma_2\sigma_{22}-B+F_{22}, \quad (6.15)$$

$$d\sigma_{11}/dt=w_{10}\sigma_{00}+w_{12}\sigma_{22}-\Gamma_1\sigma_{11}+B+F_{11}, \quad (6.16)$$

$$db/dt=-[\tfrac{1}{2}\gamma+i(\omega_c-\omega_0)]b+N\mu\sigma+F, \quad (6.17)$$

where $F=f\exp(i\omega_0 t)$, as in (C19), and

$$\sigma_{00}=1-\sigma_{11}-\sigma 22.$$

The case of homogeneous broadening can be obtained by making the specialization

$$\omega^M=\omega_a, \quad \Gamma_{12}{}^M=\Gamma. \quad (6.18)$$

In this case Eq. (6.3) reduces to the form

$$d\sigma/dt=-[\Gamma+i(\omega_a-\omega_0)]\sigma+\mu bD+F_{12}. \quad (6.19)$$

The moments of our Langevin forces are now given by

$$\langle F(t)F_{ij}{}^M(u)\rangle=\langle F_{ij}{}^M(u)F(t)\rangle=0, \quad (6.20)$$

$$\langle F_{ij}{}^{M'}(t)F_{kl}{}^M(u)\rangle=0 \quad \text{for} \quad M'\neq M, \quad (6.21)$$

$$\langle F^\dagger(t)F(u)\rangle=\gamma\bar{n}\delta(t-u), \quad (6.22)$$

$$\langle F(u)F^\dagger(t)\rangle=\gamma(\bar{n}+1)\delta(t-u), \quad (6.23)$$

$$\langle F_{ij}{}^M(t)F_{kl}{}^M(u)\rangle=2\langle D_{ijkl}{}^M\rangle\delta(t-u), \quad (6.24)$$

$$\langle F_{ij}(t)F_{kl}(u)\rangle=(1/N)2\langle D_{ijkl}\rangle\delta(t-u), \quad (6.25)$$

$$\langle D_{ijkl}\rangle=(1/N)\sum_M\langle D_{ijkl}{}^M\rangle. \quad (6.26)$$

Here $\bar{n}=\bar{n}(\omega_0)$ as in (C40).

The atomic and field forces are uncoupled as they should be. The factor $1/N$ that appears in Eq. (6.25) is essentially a consequence of the fact that the individual F^M are uncorrelated. The most important diffusion constants for the discussion of phase noise in a maser above threshold are given in

$$2\langle D_{1221}{}^M\rangle=\langle(a_1^\dagger a_1)^M\rangle[\Gamma_2+2\Gamma_{12}{}^{\text{ph}}]+\sum_{p\neq 1}w_{1p}\langle(a_p^\dagger a_p)^M\rangle, \quad (6.27)$$

$$2\langle D_{1221}\rangle=\{\langle\sigma_{11}\rangle[\Gamma_2+2\Gamma_{12}{}^{\text{ph}}]+\sum_{p\neq 1}w_{1p}\langle\sigma_{pp}\rangle\}. \quad (6.28)$$

In the steady state these results can be simplified as shown in Eq. (1.30) to the form[3]

$$2\langle\langle D_{1221}\rangle\rangle_{ss}=\{\bar{\sigma}_{11}[\Gamma_2+2\Gamma_{12}^{ph}]+\Gamma_1\bar{\sigma}_{11}-\bar{B}\}=[2\Gamma\bar{\sigma}_{11}-\bar{B}]. \quad (6.29)$$

The corresponding diffusion constant for the operators taken in reverse order takes the corresponding form

$$2\langle\langle D_{2112}\rangle\rangle_{ss}=[2\Gamma\bar{\sigma}_{22}+\bar{B}]. \quad (6.30)$$

The additional diffusion constants

$$2\langle D_{1111}\rangle=\sum_{p\neq 1} w_{1p}\langle\sigma_{pp}\rangle+\Gamma_1\langle\sigma_{11}\rangle, \quad (6.31)$$

$$2\langle D_{2222}\rangle=\sum_{p\neq 2} w_{2p}\langle\sigma_{pp}\rangle+\Gamma_2\langle\sigma_{22}\rangle, \quad (6.32)$$

$$2\langle D_{1122}\rangle=2\langle D_{2211}\rangle=-[w_{12}\langle\sigma_{22}\rangle+w_{21}\langle\sigma_{11}\rangle], \quad (6.33)$$

$$2\langle D_{1222}\rangle=\Gamma_2\langle\sigma\rangle;\quad 2\langle D_{2212}\rangle=-w_{21}\langle\sigma\rangle, \quad (6.34)$$

$$2\langle D_{2122}\rangle=-w_{21}\langle\sigma^\dagger\rangle;\quad 2\langle D_{2221}\rangle=\Gamma_2\langle\sigma^\dagger\rangle, \quad (6.35)$$

$$D_{ijkl}=(D_{lkji})^* \quad (6.36)$$

can be obtained directly from Eqs. (1.25)–(1.35).

The random forces F_{ij} can by the law of large numbers be taken as Gaussian random variables since they are averages over a large set F_{ij}^M of identically distributed variables. In the optical region the $F(t)$ are produced by the vacuum fluctuations of the electromagnetic field and are clearly Gaussian. In the microwave region, the source of electromagnetic noise will be the cavity walls. Since many atoms contribute to this "black body" radiation, we can again take F as Gaussian.

7. PRESERVATION OF COMMUTATION RULES (AND SECOND MOMENTS)

Since we have computed our diffusion constants by comparing the mean equations of linear operators with the mean equations of motion of quadratic operators, we have necessarily guaranteed that the correct equation of motion is obtained for the product of any two operators. Since this is true for either order in which the product is taken, commutators obey the correct equations of motion. Thus, if the commutation rules are obeyed at an initial instant of time they will necessarily be preserved in time. In spite of this, it is of some interest to display directly what the commutator of the Langevin forces

$$\langle[F_\mu(t),F_\nu(u)]\rangle=2\langle D_{\mu\nu}-D_{\nu\mu}\rangle\delta(t-u) \quad (7.1)$$

depends on. Let us rewrite Eq. (2.7) as

$$d\langle a_\mu a_\nu\rangle/dt=2\langle D_{\mu\nu}\rangle+\langle A_\mu a_\nu\rangle+\langle a_\mu A_\nu\rangle. \quad (7.2)$$

We can next interchange the indices μ and ν and subtract to obtain

$$d\langle[a_\mu,a_\nu]\rangle/dt=2\langle(D_{\mu\nu}-D_{\nu\mu})\rangle+\langle[A_\mu,a_\nu]\rangle+\langle[a_\mu,A_\nu]\rangle. \quad (7.3)$$

Thus, the commutators of our random forces are expressed by

$$2\langle(D_{\mu\nu}-D_{\nu\mu})\rangle=\langle[a_\nu,A_\mu]+[A_\nu,a_\mu]\rangle+d\langle[a_\mu,a_\nu]\rangle/dt \quad (7.4)$$

$$=\langle[a_\nu,\{da_\mu/dt\}]+[\{da_\nu/dt\},a_\mu]\rangle+d\langle[a_\mu,a_\nu]\rangle/dt \quad (7.5)$$

in terms of certain commutators and their time derivatives.

Let us work out one important case for the maser problem as an example. The forces in the population equations commute and therefore we shall not consider them. Instead, let us consider the forces that enter the off-diagonal equations. The appropriate commutator taking account of the factor $1/N$ in (6.25) is

$$2\langle D_{1221}-D_{2112}\rangle/N=\langle[\sigma^\dagger,\{d\sigma/dt\}]+[\{d\sigma^\dagger/dt\},\sigma]\rangle+d\langle[\sigma,\sigma^\dagger]\rangle/dt, \quad (7.6)$$

where the quantity in brackets is given by

$$\{d\sigma/dt\}=[\Gamma+i(\omega_a-\omega_0)]\sigma+\mu bD. \quad (7.7)$$

Thus, we obtain

$$2\langle D_{1221}-D_{2112}\rangle/N=-2\Gamma\langle[\sigma^\dagger,\sigma]\rangle-d\langle[\sigma^\dagger,\sigma]\rangle/dt+\mu\langle b[\sigma^\dagger,D]+[D,\sigma]b^\dagger\rangle. \quad (7.8)$$

The necessary commutators are

$$[\sigma^\dagger,\sigma]=D/N, \quad (7.9)$$

$$[\sigma^\dagger,D]=-2\sigma^\dagger/N;\quad [D,\sigma]=-2\sigma/N. \quad (7.10)$$

Thus, our commutators in the steady state take the simple form of

$$2\langle\langle D_{1221}-D_{2112}\rangle\rangle_{ss}/N=-2(\Gamma/N)\bar{D}-2\bar{B}/N=-(2/N)[\Gamma\bar{D}+\bar{B}]. \quad (7.11)$$

This commutator is precisely what one obtains if one subtracts Eq. (6.30) from Eq. (6.29). Below threshold in paper QV, however, we treat D as a c number. This means that the commutators involving D vanish. Thus if the "dielectric" approximation is made, the last term in Eq. (7.8) must be omitted and the commutator reduces to

$$2\langle\langle D_{1221}-D_{2112}\rangle\rangle_{ss}/N=-2(\Gamma/N)\bar{D} \quad (7.12)$$

as quoted in QV.

Note added in proof. After the completion of this manuscript (and after the results summarized in Secs. 1 and 6 were presented at the 1965 Puerto Rico conference) we have learned that several members of the Haken school have adopted a Markoffian approach closely related to our own. See H. Haken and W. Weidlich [Z. Physik **189**, 1 (1966)]; C. Schmid and H. Risken [*ibid.* **189**, 365 (1966)]. These papers treat the atomic fluctuations and lead to moments in agreement with ours. For the electromagnetic field, the noise sources are not derived by them but are taken from

Senitzky—see, e.g., H. Sauerman, Z. Physik **189**, 312 (1966). Our procedure obtains the field noise sources by the same method as that used for the atomic noise sources, and moreover derives the independence of field and atomic sources.

ACKNOWLEDGMENTS

The author is indebted to Dr. Donald R. Fredkin for extensive correspondence and for early calculations of the mean motion of a harmonic oscillator, Eq. (3.6), and an atomic system, Eq. (A21). Fredkin also suggested retaining the secular terms $k=n$, $l=m$. The remaining secular terms $k=l$, $m=n$ were added later, during the author's investigation of the shape of zero-phonon lines. I also wish to thank J. R. Klauder for critical conversations, and W. H. Louisell and D. E. McCumber for a careful reading of the manuscript.

APPENDIX A: MEAN MOTION BY DENSITY-MATRIX METHODS

As discussed in the text, our aim is to calculate the system density matrix accurate to second order in the coupling. We are not concerned with treating the bath to the same accuracy, therefore we shall not follow QIII precisely, since the latter treats the bath on the same footing as the system. The chief difference is that we shall now set the density matrix of system and bath $\rho=\sigma f_0+\Delta\rho$, where $\sigma=\mathrm{tr}_R\rho$ is the system density matrix and $f_0=\exp(-\beta R)/\mathrm{tr}[\exp(-\beta R)]$ is the unperturbed bath density matrix. (Previously, we had set $\rho=\sigma f+\Delta\rho$.) The pattern of calculation is the same as in QIII, and most results look very similar with slightly different meanings for the symbols. To avoid confusion, and establish notation, we shall outline the key steps in the argument.

We start with a total Hamiltonian $H_T=H+R+V$ decomposable into a system part H, a reservoir part R, and an interaction

$$V=-\sum Q_jF_j, \tag{A1}$$

where the Q's are system operators, and the F's are reservoir operators whose mean values vanish in the decoupled reservoir:

$$\mathrm{tr}_R(F_jf_0)=0,$$

where

$$f_0=\exp(-\beta R)/\mathrm{tr}[\exp(-\beta R)]. \tag{A2}$$

The density matrix ρ of system+reservoir obeys

$$\partial\rho/\partial t=(H+R+V,\rho), \tag{A3}$$

where

$$(A,B)=[A,B]/i\hbar. \tag{A4}$$

The trace of (A3) then yields

$$\partial\sigma/\partial t=(H,\sigma)+\mathrm{tr}_R(V,\Delta\rho). \tag{A5}$$

Subtracting $f_0\partial\sigma/\partial t$ from (A3), $\Delta\rho=\rho-f_0\sigma$ obeys

$$\partial\Delta\rho/\partial t=(H+R,\Delta\rho)+(V,\sigma f_0)+C(\Delta\rho),$$
$$C(\Delta\rho)=(V,\Delta\rho)-f_0\,\mathrm{tr}_R(V,\Delta\rho). \tag{A6}$$

A systematic expansion in V can be set up by first ignoring C and the iterating. Since we wish σ to second order in V, i.e., $\Delta\rho$ to first order, we stop at the first term

$$\Delta\rho\approx\Delta\rho_1=\int_{-\infty}^{t}(V(t',t),\sigma(t,t')f_0)dt',$$
$$\partial\sigma/\partial t=(H,\sigma)+\int_{-\infty}^{t}dt'\,\mathrm{tr}_R(V,(V(t',t),\sigma(t,t')f_0)), \tag{A7}$$

where

$$V(t',t)\equiv U(t,t')VU(t,t')^{-1},$$
$$\sigma(t,t')=U(t,t')\sigma(t')U(t,t')^{-1}=u(t,t')\sigma(t')u(t,t')^{-1}, \tag{A8}$$

where $U(t,t')$ is the operator solution of the unperturbed Schrödinger equation

$$i\hbar dU(t,t')/dt=(H+R)U(t,t'),$$
$$U(t',t')=1,$$
$$U(t,t')=u(t,t')\exp[-iR(t-t')/\hbar],$$
$$i\hbar du(t,t')/dt'=H(t)u(t,t'). \tag{A9}$$

Equation (A7) is very similar to QIII (2.23) with $f(t,t')$ in the latter replaced by f_0, and V obeys $\mathrm{tr}_RVf_0=0$, but the system average of V need not vanish. We have obtained the same result found by Argyres by projection techniques.[19]

Inserting (A1) into (A7), we obtain the analog of QIII (3.6):

$$\frac{\partial\sigma}{\partial t}=(H,\sigma)+\sum_{i,j}\int_{-\infty}^{t}dt'\{\langle\tfrac{1}{2}[F_i(t-t'),F_j]_+\rangle(Q_i,(Q_j(t',t),\sigma(t,t')))$$
$$+\langle(F_i(t-t'),F_j)\rangle(Q_i,\tfrac{1}{2}[Q_j(t',t),\sigma(t,t')]_+)\}. \tag{A10}$$

The trace of (A10) against an arbitrary system operator M yields

$$\partial\langle\langle M\rangle\rangle/\partial t=\langle\langle(M,H)\rangle\rangle$$
$$-\hbar^{-2}\sum_{i,j}\int_{-\infty}^{t}dt'\{\langle\tfrac{1}{2}[F_i(t-t'),F_j]_+\rangle$$
$$\times\langle\langle[[M(t,t'),Q_i(t,t')],Q_j]\rangle\rangle_{t'}$$
$$+\langle[F_i(t-t'),F_j]\rangle\langle\langle\tfrac{1}{2}[[M(t,t'),Q_i(t,t')],Q_j]_+\rangle\rangle_{t'}\}, \tag{A11}$$

where $\langle L\rangle_{t'}=\mathrm{tr}_SL\sigma(t')$. If the reservoir correlation

[19] P. N. Argyres, in *Magnetic and Electric Resonance and Relaxation*, edited by J. Smidt (North-Holland Publishing Company, Inc., Amsterdam, 1963), p. 555.

times are short, the important (t') are close to t. In this small time interval, the system may change rapidly due to its unperturbed motion, but dissipative effects can be assumed small. Thus we can set $\sigma(t,t')$ [which, by (A8) is the density matrix at t obtained from $\sigma(t')$ by propagating with neglect of interactions] approximately equal to $\sigma(t)$:

$$\sigma(t,t')\equiv u(t,t')\sigma(t')u(t,t')^{-1}\approx\sigma(t). \tag{A12}$$

This makes our density matrix equation (A10) Markoffian. Equation (A11) then reduces to the Markoffian form

$$d\langle\langle M\rangle\rangle/dt=\langle\langle(M,H)\rangle\rangle$$
$$-\hbar^{-2}\sum_{i,j}\int_0^\infty du\{\langle F_i(u)F_j\rangle\langle\langle[M,Q_i]Q_j(t-u,\,t)\rangle\rangle$$
$$-\langle F_jF_i(u)\rangle\langle\langle Q_j(t-u,\,t)[M,Q_i]\rangle\rangle\}. \tag{A13}$$

If the system has a time-independent Hamiltonian, $H(t)=H$, then

$$Q_j(t-u,\,t)=\exp(-iHu/\hbar)Q_j\exp(iHu/\hbar)$$
$$\equiv Q_j(-u), \tag{A14}$$

and (A13) reduces to the result (3.1) quoted in the text.

We now wish to apply Eq. (3.1) to the case of an atomic system. The interaction Hamiltonian takes the form

$$V=\hbar\sum a_m{}^\dagger a_n f_{mn}, \tag{A15}$$

and Hermiticity guarantees

$$V^\dagger=V;\quad f_{mn}{}^\dagger=f_{nm}. \tag{A16}$$

Making use of the correspondences

$$Q_i\to a_k{}^\dagger a_l,\quad F_i\to\hbar f_{kl},$$
$$Q_j(-u)\to a_m{}^\dagger a_n\exp(-i\omega_{mn}u),$$
$$F_j\to\hbar f_{mn},\quad M=a_i{}^\dagger a_j, \tag{A17}$$

we obtain our equation for the mean motion of an atomic operator $a_i{}^\dagger a_j$ in the form[20]

$$d\langle a_i{}^\dagger a_j\rangle/dt=i\omega_{ij}\langle a_i{}^\dagger a_j\rangle$$
$$-\sum w_{klmn}{}^+\langle[a_i{}^\dagger a_j,a_k{}^\dagger a_l]a_m{}^\dagger a_n\rangle$$
$$+\sum w_{mnkl}{}^-\langle a_m{}^\dagger a_n[a_i{}^\dagger a_j,a_k{}^\dagger a_l]\rangle, \tag{A18}$$

where the coefficients defined by

$$w_{klmn}{}^+=\int_0^\infty du\exp(-i\omega_{mn}u)\langle f_{kl}(u)f_{mn}(0)\rangle,$$
$$w_{mnkl}{}^-=\int_0^\infty du\exp(-i\omega_{mn}u)\langle f_{mn}(0)f_{kl}(u)\rangle, \tag{A19}$$

obey the Hermiticity property

$$(w_{mnkl}{}^-)^*=w_{lknm}{}^+. \tag{A20}$$

Making use of the identity (4.5), Eq. (A18) can be simplified to the form

$$d\langle a_i{}^\dagger a_j\rangle/dt=i\omega_{ij}\langle a_i{}^\dagger a_j\rangle$$
$$+\sum[-\langle a_i{}^\dagger a_n\rangle w_{jmmn}{}^+ +\langle a_k{}^\dagger a_n\rangle w_{kijn}{}^+$$
$$+\langle a_m{}^\dagger a_l\rangle w_{mijl}{}^- -\langle a_m{}^\dagger a_j\rangle w_{mnni}{}^-]. \tag{A21}$$

We shall now retain only the secular terms, i.e., those on the right-hand side of (A18) or (A21) that vary as $e^{i\omega_{ij}t}$. This is equivalent to retaining those $w_{klmn}{}^\pm$ for which[21]

$$\omega_{kl}+\omega_{mn}=0, \tag{A22a}$$

i.e.,

$$k=n,\quad l=m\quad\text{or}\quad k=l,\quad m=n. \tag{A22b}$$

After removal of the rapid time dependence contained in $a_i{}^\dagger a_j$ these are the only terms which survive a short time average: average over a time Δt short compared to any of the relaxation times but long compared to the reciprocal natural frequencies of the system. An explicit proof of this point is given in Appendix B. The set of conditions (A22) define the only ways in which energy can be conserved if the levels are irregularly spaced. In this Appendix, we henceforth assume that there are no special degeneracies such as would occur for example, in a harmonic oscillator. For this reason we have given a separate treatment of a harmonic oscillator in Sec. 3. Retaining only the secular terms then, Eq. (A21) reduces to the form[20]

$$d\langle a_i{}^\dagger a_j\rangle/dt=(i\omega_{ij}-\Gamma_{ij}{}^c)\langle a_i{}^\dagger a_j\rangle$$
$$+\delta_{ij}\sum_{m\neq i}w_{im}\langle a_m{}^\dagger a_m\rangle, \tag{A23}$$

where the transition probability w_{im} is defined by

$$w_{im}=w_{miim}{}^+ +w_{miim}{}^-$$
$$=\int_{-\infty}^\infty du\exp(-i\omega_{im}u)\langle f_{mi}(u)f_{im}(0)\rangle \tag{A24}$$

and the complex parameter $\Gamma_{ij}{}^c$ is given by

$$\Gamma_{ij}{}^c=-[w_{iijj}{}^+ +w_{iijj}{}^-]+\sum_{\text{all }m}(w_{jmmj}{}^+ +w_{immi}{}^-), \tag{A25}$$

$$\Gamma_{ij}{}^c=\Gamma_{ij}-i\Delta\omega_{ij}, \tag{A26}$$

$$\Gamma_{ij}=\tfrac{1}{2}(\Gamma_i+\Gamma_j)+\Gamma_{ij}{}^{\text{ph}}, \tag{A27}$$

$$\Delta\omega_i=-\text{Im}[\sum_{m\neq j}w_{jmmj}{}^+ +\sum_{m\neq i}w_{immi}{}^-]+\Delta\omega_{ij}{}^{\text{ph}}, \tag{A28}$$

where Γ_i, Γ_j are the decay rates (1.15) and the first term in (A28) is the "Lamb shift" (second-order

[20] For simplicity of notation we change double brackets to single brackets. This is permissible in view of (B16). In this appendix, $\omega_{ij}=\omega_i-\omega_j$ is an unperturbed frequency difference. In the body of the paper, the perturbed frequency $\omega_i-\omega_j+\Delta\omega_{ij}$ is represented by ω_{ij}, for the sake of brevity.

[21] The terms in (A22b) always satisfy (A22a). If the system possesses special regularities of spacing, as in a harmonic oscillator or spin system, then (A22a) permits more secular terms than those explicitly shown in (A22b).

perturbation theory energy shift due to reservoir interactions). The contribution of phase fluctuations or in a solid what might be called zero-phonon contributions[18] are summarized in

$$(\Gamma_{ij}-i\Delta\omega_{ij})^{\mathrm{ph}}=w_{jjjj}^{+}+w_{iiii}^{-}-w_{iijj}^{+}-w_{iijj}^{-}$$
$$=\int_0^\infty du\langle[f_{ii}(0)][f_{ii}(u)-f_{jj}(u)]-[f_{ii}(u)-f_{jj}(u)]f_{jj}(0)\rangle, \quad (A29)$$

$$\Gamma_{ij}^{\mathrm{ph}}=\frac{1}{2}\int_0^\infty du\langle[f_{ii}(0)-f_{jj}(0), f_{ii}(u)-f_{jj}(u)]_+\rangle, \quad (A30)$$

$$-i\Delta\omega_{ij}^{\mathrm{ph}}=\frac{1}{2}\int_0^\infty du\langle[f_{ii}(0)+f_{jj}(0), f_{ii}(u)-f_{jj}(u)]\rangle \quad (A31)$$

$$=-i\Delta\omega_i^{\mathrm{ph}}+i\Delta\omega_j^{\mathrm{ph}}-i\Delta\omega_{ij}^{\mathrm{ex}}, \quad (A32)$$

$$-i\Delta\omega_i^{\mathrm{ph}}=\frac{1}{2}\int_0^\infty du\langle[f_{ii}(0),f_{ii}(u)]\rangle, \quad (A33)$$

$$-i\Delta\omega_{ij}^{\mathrm{ex}}=\frac{1}{2}\int_{-\infty}^\infty du\langle[f_{jj}(0),f_{ii}(u)]\rangle. \quad (A34)$$

The *extra* contribution to the frequency shift described by (A34) is *anomalous* in that it is not expressible as the difference between a frequency shift of level i and a shift of level j. In order to understand these formulas we have applied them to the case where the electronic levels are coupled to lattice vibrations through the interaction (A15) with the reservoir forces defined in terms of the normal phonon coordinates by

$$f_{ii}(u)=(\hbar)^{-1}\textstyle\sum_\mu A_\mu{}^i q_\mu(u). \quad (A35)$$

Neglecting anharmonic interactions between the phonons, the time dependence of these phonon coordinates is given by

$$q_\mu(u)=q_\mu\cos\omega_\mu u+(p_\mu/M\omega_\mu)\sin\omega_\mu u \quad (A36)$$

and the commutator is given by

$$[q_\mu(0),q_\nu(u)]=\delta_{\mu\nu}(i\hbar/M\omega_\mu)\sin\omega_\mu u. \quad (A37)$$

Expressing the time integral

$$\int_0^\infty du\sin\omega_\mu u=\mathcal{P}(1/\omega_\mu) \quad (A38)$$

in terms of the principal-valued reciprocal, the shift in level i due to zero-phonon contributions is given by

$$\Delta\omega_i^{\mathrm{ph}}=-\tfrac{1}{2}\textstyle\sum(A_\mu{}^i)^2/(\hbar M\omega_\mu{}^2) \quad (A39)$$

and the extra anomalous frequency shift for this case vanishes:

$$\Delta\omega_{ij}^{\mathrm{ex}}=0. \quad (A40)$$

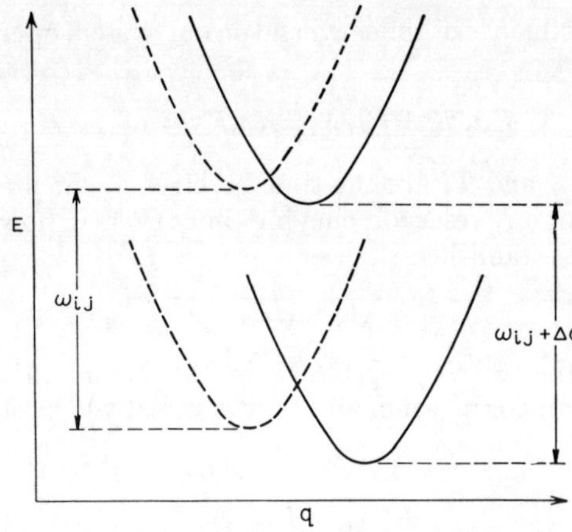

FIG. 2. Configurational coordinate curves of electronic energy E versus normal coordinate q are shown for an electron in two electronic states i and j with $\omega_{ij}=\omega_i-\omega_j>0$. The dashed curves neglect electron-phonon interactions. The solid curves show the shift produced by such interactions. In particular $\Delta\omega_{ij}$ is the change in separation between minima.

The complete energy shift then takes the form

$$\hbar\Delta\omega_{ij}=-\tfrac{1}{2}\textstyle\sum[(A_\mu{}^i)^2-(A_\mu{}^j)^2]/(M\omega_\mu{}^2). \quad (A41)$$

This result was previously obtained in Eq. (6.7) of our paper on the Franck-Condon principle.[22] The frequency difference ω_{ij} is simply the distance between the minima of the two parabolas shown in Fig. 2 and the level shift $\Delta\omega_{ij}$ is the extent to which this separation has been changed by the linear interactions with the lattice. It is to be emphasized that the perturbative equations on which our mean-motion equation (A23) is based have disregarded the effects of multiphonon transitions. Thus, the frequency ω_{ij} refers to what is customarily called the *zero-phonon line*.[18] This line is indeed a transition from the lowest phonon state in one parabola of Fig. 2 to the lowest phonon state in the other parabola. Since we have neglected changes in the curvatures in these parabolas the zero-point phonon energies cancel and the zero-zero difference is simply the difference between the minima of the parabolas. To see why the anomalous frequency shift vanishes in this case we note that if q is any operator whatever, we can make use of stationarity in the form

$$\langle[q(0),q(t)]\rangle=-\langle[q(t),q(0)]\rangle=-\langle[q(0),q(-t)]\rangle. \quad (A42)$$

Thus the integrand of Eq. (A34) is an odd function of the time whereas the integration is taken over an even interval in the time. To get a nonvanishing frequency shift it is necessary therefore to obtain cross terms

$$\langle[q_\mu(u),q_\nu(t)]\rangle\neq0 \quad \text{for} \quad \Delta\omega_{ij}^{\mathrm{ex}}\neq0. \quad (A43)$$

We shall now show that even when such cross terms are available if the reservoir obeys time reversal the anomalous frequency shift will vanish.

[22] M. Lax, J. Chem. Phys. 20, 1752 (1952).

The influence of time reversal on our system operators is given by

$$Ka_j^\dagger K^{-1}=a_{Tj}^\dagger\,,\quad Ka_iK^{-1}=a_{Ti}\,, \tag{A44}$$

where Tj and Ti are the time reverses of the states j and i. For any reservoir obeying time reversal, however, we have established[23]

$$\langle L\rangle=\langle\bar{L}\rangle\,, \tag{A45}$$

where the *barring operation* is the combination of Hermitian conjugation and time reversal which obeys

$$V_{\text{bar}}\equiv\bar{V}\equiv KV^\dagger K^{-1};\quad (AB)_{\text{bar}}=\bar{B}\bar{A}\,;$$
$$B(t)_{\text{bar}}=\bar{B}(-t)\,. \tag{A46}$$

Making use of

$$\bar{V}=\hbar\sum(a_i^\dagger a_j)_{\text{bar}}\bar{f}_{ij}=\hbar\sum a_{Tj}^\dagger a_{Ti}f_{ij}\,, \tag{A47}$$

we find that the reservoir operators are changed under barring into

$$\bar{f}_{ij}=f_{Tj,Ti}\,. \tag{A48}$$

In this way, we derive

$$\langle[f_{jj}(0),f_{ii}(u)]\rangle=\langle[f_{Ti,Ti}(-u),f_{Tj,Tj}(0)]\rangle$$
$$=-\langle[f_{Tj,Tj}(0),f_{Ti,Ti}(-u)]\rangle\,. \tag{A49}$$

It is clear then that the anomalous frequency shift will vanish unless either Ti differs from i or Tj differs from j or $\langle L\rangle$ does not equal $\langle\bar{L}\rangle$, in other words, if the reservoir violates time reversal, as it would, for example, in any magnetic or antiferromagnetic material.

APPENDIX B: DIRECT CALCULATION OF TRANSPORT EQUATION AND DIFFUSION COEFFICIENTS

An arbitrary-system Heisenberg operator M obeys

$$idM/dt=[M,H_T];\quad H_T=H_0+V;\quad \hbar=1\,, \tag{B1}$$

where $H_0=H+R$ is the sum of system and reservoir Hamiltonians and V is the interaction. We shall assume in this appendix that H_T does not depend explicitly on the time, so that $(H_T)_{\text{Heisenberg}}=(H_T)_{\text{Schrödinger}}$ and shall hereafter regard H_0 and V as Schrödinger operators. The transformation

$$M(t)=e^{iH_0(t-t_0)}m(t)e^{-iH_0(t-t_0)} \tag{B2}$$

leads to an operator $m(t)$ that varies much more slowly in time [see the similar transformation IV (5.21)]. We can then treat the drift and diffusion of $m(t)$ as if all correlation times are short compared to this motion. After the drift $\langle\Delta m/\Delta t\rangle$ is included in the equation for dm/dt, the reservoir is then eliminated and replaced by Langevin forces with *zero* correlation time that lead to the *same* diffusion. Then, we can transform back to M equations at the time instant t. Thus

$$id\langle M\rangle/dt=[M,H_0]+e^{iH_0(t-t_0)}i\langle\Delta m/\Delta t\rangle e^{-iH_0(t-t_0)}\,, \tag{B3}$$

$$2\langle D_{MN}\rangle=e^{iH_0(t-t_0)}\langle\Delta m\Delta n\rangle/\Delta t e^{-iH_0(t-t_0)}\,, \tag{B4}$$

where

$$\Delta m=m(t+\Delta t)-m(t) \tag{B5}$$

and $\langle\;\;\rangle$ are reservoir averages. These procedures are not equivalent to calculating $\langle\Delta M/\Delta t\rangle$ and $\langle\Delta M\Delta N\rangle/\Delta t$ (which would unnecessarily smooth the unperturbed motion) but are precisely equivalent to our classical procedures IV (5.24), (5.25). The slow operator m obeys

$$dm/ds=-i[m(s),V(t_0-s)]\,, \tag{B6}$$

where $V(t_0-s)$ is an interaction operator

$$V(t_0-s)\equiv e^{iH_0(t_0-s)}Ve^{-iH_0(t_0-s)}\,. \tag{B7}$$

Equation (B6) can be converted to an integral equation and solved iteratively:

$$\Delta m=-i\int_t^{t+\Delta t}[m(s),V(t_0-s)]ds$$
$$\approx\Delta^{\text{I}}m+\Delta^{\text{II}}m\,, \tag{B8}$$
$$\Delta^{\text{I}}m=-i\int_t^{t+\Delta t}[m(t),V(t_0-s)]ds\,,$$
$$\Delta^{\text{II}}m=-\int_t^{t+\Delta t}ds$$
$$\times\int_t^s ds'[[m(t),V(t_0-s')],V(t_0-s)]\,,$$

$$\langle\Delta m/\Delta t\rangle\approx\langle\Delta^{\text{I}}m+\Delta^{\text{II}}m\rangle/\Delta t=\langle\Delta^{\text{II}}m/\Delta t\rangle\,. \tag{B9}$$

Inserting (B9) result into (B3), we obtain

$$\frac{d\langle M\rangle}{dt}=-i[M,H_0]-\frac{1}{\Delta t}\int_t^{t+\Delta t}ds$$
$$\times\int_t^s ds'\langle[[M(t),V(t-s')],V(t-s)]\rangle\,. \tag{B10}$$

Note that t_0, the arbitrary time at which the representations become identical, has disappeared from (B10). This is also true of our expression for the diffusion constant:

$$2D_{MN}=-\frac{1}{\Delta t}\int_t^{t+\Delta t}ds\int_t^{t+\Delta t}ds'\langle[M(t),V(t-s')]$$
$$\times[N(t),V(t-s)]\rangle \tag{B11}$$

obtained by inserting $\Delta^{\text{I}}m\,\Delta^{\text{I}}n$ into (B4).

Equation (B11) is consistent with (B10) in the sense that the use of (B10) plus the Einstein relation (2.9) leads, after combining two sets of terms, to (B11). If

[23] M. Lax, *Symmetry Principles in Solid-State Physics* (to be published), Chap. 10.

we insert (A1): $V=-\sum Q_iF_i$, use the stationarity of reservoir averages, and restore $\hbar$'s, we obtain

$$\frac{d\langle M\rangle}{dt}=\langle(M,H_0)\rangle-(\hbar^2\Delta t)^{-1}\int_t^{t+\Delta t}ds\int_t^s ds'$$
$$\times\{\sum_{ij}\langle F_i(s-s')F_j(0)\rangle\langle[M(t),Q_i(t-s')]Q_j(t-s)\rangle$$
$$-\langle F_j(0)F_i(s-s')\rangle\langle Q_j(t-s)[M(t),Q_i(t-s')]\rangle\},\quad\text{(B12)}$$

$$2D_{MN}=-(\hbar^2\Delta t)^{-1}\sum_{ij}\int_t^{t+\Delta t}ds$$
$$\times\int_t^{t+\Delta t}ds'\langle F_i(s-s')F_j(0)\rangle$$
$$\times\langle[M,Q_i(t-s')][N,Q_j(t-s)]\rangle.\quad\text{(B13)}$$

Again these results are consistent. Let us integrate first over s and set $s=s'+u$ to obtain

$$\frac{d}{dt}\langle M\rangle=\langle(M,H)\rangle-(\hbar^2\Delta t)^{-1}\int_t^{t+\Delta t}ds'\int_0^{t+\Delta t-s'}du$$
$$\times\sum_{ij}\{\langle F_i(u)F_j(0)\rangle\langle[M,Q_i(t-s')]Q_j(t-s'-u)\rangle$$
$$-\langle F_j(0)F_i(u)\rangle\langle Q_j(t-s'-u)[M,Q_i(t-s')]\rangle\}.\quad\text{(B14)}$$

The integral over $\langle F_i(u)F_j(0)\rangle$ converges when the upper limit is greater than the correlation time τ_c over which $\langle F_i(u)F_j(0)\rangle$ is appreciable. If we choose $\Delta t\gg\tau_c$, we can replace the upper limit $t+\Delta t-s'$ by ∞. With $v=s'-t$, we get

$$\frac{d\langle M\rangle}{dt}=\langle(M,H)\rangle-(\hbar^2\Delta t)^{-1}\int_0^{\Delta t}dv\int_0^{\infty}du$$
$$\times\sum_{ij}\{\langle F_i(u)F_j(0)\rangle\langle[M,Q_i(-v)]Q_j(-v-u)\rangle$$
$$-\langle F_j(0)F_i(u)\rangle\langle Q_j(-v-u)[M,Q_i(-v)]\rangle\}.\quad\text{(B15)}$$

The integral over u represents an integral over the duration of the collision between system and reservoir. The average over v (or s') is an average over the *starting* time of the collision. It is permissible in this average to let $\Delta t\to 0$ to obtain

$$\frac{d\langle M\rangle}{dt}=\langle(M,H)\rangle-\hbar^{-2}\int_0^{\infty}du$$
$$\times\sum_{ij}\{\langle F_i(u)F_j(0)\rangle\langle[M,Q_i]Q_j(-u)\rangle$$
$$-\langle F_j(0)F_i(u)\rangle\langle Q_j(-u)[M,Q_i]\rangle\},\quad\text{(B16)}$$

the result (3.1) derived in (A13). This result, however, retains certain rapidly oscillatory terms whose effect on the long-term motion is of second order. These terms will automatically disappear if Δt is kept large enough. In particular, the system-operator time dependence must be representable as a sum of exponentials

$$Q_i(t)=\sum_\alpha Q_{i\alpha}e^{i\omega_i^\alpha t}.\quad\text{(B17)}$$

The integral over v will involve integrals of the form

$$\frac{1}{\Delta t}\int_0^{\Delta t}dv\exp[i\omega_i^\alpha t+i\omega_j^\beta t].\quad\text{(B18)}$$

For $\omega_i^\alpha+\omega_j^\beta\neq 0$, we shall choose Δt large enough so that

$$(\omega_i^\alpha+\omega_j^\beta)\Delta t\gg 1.\quad\text{(B19)}$$

All such terms can then be neglected unless

$$\omega_i^\alpha+\omega_j^\beta=0.\quad\text{(B20)}$$

For the *secular terms* which obey (B20), the integral over v is insensitive to Δt and one can let $\Delta t\to 0$. Thus, the sole effect of the average over v (or s') is to retain only the secular terms when (B15) is expanded using (B17). Our result can be written in the form

$$\frac{d\langle M\rangle}{dt}=\langle(M,H)\rangle-\hbar^{-2}\sum_{i\alpha,j\beta}\delta(\omega_i^\alpha,-\omega_j^\beta)$$
$$\times\left\{\int_0^\infty e^{-i\omega_j^\beta u}\langle F_i(u)F_j(0)\rangle du\langle[M,Q_{i\alpha}]Q_{j\beta}\rangle\right.$$
$$\left.-\int_0^\infty e^{-i\omega_j^\beta u}\langle F_j(0)F_i(u)\rangle du\langle Q_{j\beta}[M,Q_{i\alpha}]\rangle\right\},\quad\text{(B21)}$$

where the Kronecker delta, $\delta(\omega_i^\alpha,-\omega_j^\beta)$, selects the secular terms.

If we apply a similar procedure to (B13), we get

$$2D_{MN}=-\hbar^{-2}\sum_{\alpha i,\beta j}\delta(\omega_i^\alpha,-\omega_j^\beta)\int_{-\infty}^{\infty}e^{-i\omega_j^\beta u}du$$
$$\times\langle F_i(u)F_j(0)\rangle\langle[M,Q_{i\alpha}][N,Q_{j\beta}]\rangle.\quad\text{(B22)}$$

If the Einstein relation (2.9) is applied to (B21), and the summation indices αi and βj are interchanged in the second term, the same result (B22), is obtained. Thus the Einstein method and the direct method are necessarily in agreement.

The variables Q_j can often be so chosen that

$$Q_j(t)=Q_j\exp(i\omega_j t),\quad\text{(B23)}$$

and the indices α, β are superfluous. For a harmonic oscillator, this suggests the use of b and $b^\dagger$ rather than Q and P. For multilevel "atomic" system the variables $a_i^\dagger a_j$ already obey this requirement. The translation of variables

$$M\to a_i^\dagger a_j,\quad Q_i\to a_k^\dagger a_l,\quad Q_j\to a_m^\dagger a_n,$$
$$F_i\to\hbar f_{kl},\quad F_j\to\hbar f_{mn}\quad\text{(B24)}$$

in (B21) leads to

$$d\langle a_i^\dagger a_j\rangle/dt = i\omega_{ij}\langle a_i^\dagger a_j\rangle - \sum\delta(\omega_{kl}, -\omega_{mn})\{\langle[a_i^\dagger a_j, a_k^\dagger a_l]a_m^\dagger a_n\rangle w_{klmn}^+ - \langle a_m^\dagger a_n[a_i^\dagger a_j, a_k^\dagger a_l]\rangle w_{klmn}^-\}, \quad \text{(B25)}$$

i.e., to Eq. (A18) with the selection of secular terms built in, and the w^+, w^- defined as in (A19).

A similar translation of variables in (B22) leads to

$$2\langle D_{ijkl}\rangle = -\sum \delta(\omega_{mn}, -\omega_{pq})\langle[a_i^\dagger a_j, a_m^\dagger a_n] \times[a_k^\dagger a_l, a_p^\dagger a_q]\rangle w_{mnpq}, \quad \text{(B26)}$$

$$w_{mnpq} = \int_{-\infty}^{\infty} du\, e^{-i\omega_{pq}u}\langle f_{mn}(u) f_{pq}(0)\rangle = w_{mnpq}^+ + w_{mnpq}^-. \quad \text{(B27)}$$

Using (4.5), (B20) can be simplified to

$$2\langle D_{ijkl}\rangle = \sum\langle -a_i^\dagger a_q w_{jklq} + a_i^\dagger a_l w_{jnnk} + a_m^\dagger a_q w_{milq}\delta_{jk} + a_m^\dagger a_l w_{mijk}\rangle. \quad \text{(B28)}$$

In (B28), we assume each factor w_{mnpq} carries with it, the corresponding selection factor $\delta(\omega_{mn}, -\omega_{pq})$. Equations (B25)–(B28) are sufficiently general to *include such regularly spaced systems as harmonic oscillators.* If no regularities occur among the spaces, the only secular terms that remain obey (A22b), i.e., the surviving terms have the form w_{mnnm} and w_{mmnn}.

For the former, one has the identification

$$w_{mnnm} = w_{mnnm}^+ + (w_{mnnm}^+)^* = w_{nm} \quad \text{(B29)}$$

(see A20 and A24) in terms of transition probabilities. Retaining only these secular terms, (B28) reduces to

$$2D_{ijkl} = w_{kj}\langle a_i^\dagger a_j\rangle\delta_{kl} - w_{ik}\langle a_k^\dagger a_l\rangle\delta_{ij} + \sum_{m\neq i} w_{im}\langle a_m^\dagger a_m\rangle\delta_{il}\delta_{jk} + \langle a_i^\dagger a_l\rangle \times[\Gamma_j + w_{jjjj} - w_{iijj} + w_{iill} - w_{jjll}]\delta_{jk}. \quad \text{(B30)}$$

Using Eqs. (A20) and (A29), we can write

$$(\Gamma_{ij} - i\Delta\omega_{ij})^{\text{ph}} = \tfrac{1}{2}[w_{jjjj} + w_{iiii} - 2w_{iijj}]. \quad \text{(B31)}$$

Thus, the coefficient of the last term in Eq. (B30) can be rewritten in the form

$$\Gamma_j + w_{jjjj} - w_{iijj} + w_{iill} - w_{jjll} = \Gamma_{ij} + \Gamma_{jl} - \Gamma_{il} - i(\Delta\omega_{ij} + \Delta\omega_{jl} - \Delta\omega_{il}), \quad \text{(B32)}$$

which brings it into agreement with the result (4.6) of the Einstein method.

APPENDIX C: A MARKOFFIAN AND NON-MARKOFFIAN DISCUSSION OF HARMONIC-OSCILLATOR COMMUTATION RULES

Markoffian Commutation Rules

If we subtract Eq. (3.8) from the corresponding equation for $\langle bb^\dagger\rangle$, we find that

$$d\langle[b,b^\dagger]\rangle/dt = \gamma - \gamma\langle[b,b^\dagger]\rangle \quad \text{(C1)}$$

so that, if the commutation rule $\langle[b,b^\dagger]\rangle = 1$ is obeyed at any time, it will be obeyed forever after, even in the Markoffian approximation.

As a further check on our consistency, let us define[24]

$$\langle b_\omega^\dagger b_\omega\rangle = \int_{-\infty}^{\infty} e^{-i\omega t}\langle b^\dagger(t) b(0)\rangle dt, \qquad \langle b^\dagger(t) b(0)\rangle = \frac{1}{2\pi}\int_{-\infty}^{\infty} e^{i\omega t}\langle b_\omega^\dagger b_\omega\rangle d\omega. \quad \text{(C2)}$$

The transforms of Eqs. (3.9) then yield

$$[\tfrac{1}{2}\gamma + i(\omega_0 - \omega)]b_\omega = f_\omega; \quad [\tfrac{1}{2}\gamma - i(\omega_0 - \omega)]b_\omega^\dagger = f_\omega^\dagger, \quad \text{(C3)}$$

$$\langle b_\omega^\dagger b_\omega\rangle = [(\tfrac{1}{2}\gamma)^2 + (\omega_0 - \omega)^2]^{-1}\langle f_\omega^\dagger f_\omega\rangle, \quad \text{(C4)}$$

$$\langle f_\omega^\dagger f_\omega\rangle = \int_{-\infty}^{\infty} e^{-i\omega t}\langle f^\dagger(t) f(0)\rangle dt = \gamma\bar{n}. \quad \text{(C5)}$$

Set $t=0$ in the second Eq. (C2) and insert (C4) to obtain

$$\langle b^\dagger b\rangle = \frac{1}{2\pi}\int_{-\infty}^{\infty} [(\tfrac{1}{2}\gamma)^2 + (\omega_0 - \omega)^2]^{-1}\gamma\bar{n}\, d\omega = \bar{n}. \quad \text{(C6)}$$

Similarly,

$$\langle bb^\dagger\rangle = \bar{n} + 1 \quad \text{(C7)}$$

and again the commutation rules are preserved.

Relation between Langevin Forces and Reservoir Forces

Because of the linear nature of our system, there is little distinction between the Langevin force f (the extra term on the right-hand side of a dynamic equation), and the reservoir force g that gives rise to the

[24] For any two random operators $A(t)$, $B(t)$, we define

$$A(\omega) = \int_{-\infty}^{\infty} e^{i\omega s} A(s) ds,$$

$$B^\dagger(\omega') = \int_{-\infty}^{\infty} e^{-i\omega' t} B^\dagger(t) dt = [B(\omega')]^\dagger.$$

Assuming stationarity, setting $t = s + u$, and integrating over u, we find for the product average

$$\langle A(\omega) B^\dagger(\omega')\rangle = 2\pi\delta(\omega - \omega')\langle A_\omega B_\omega^\dagger\rangle,$$

where

$$\langle A_\omega B_\omega^\dagger\rangle \equiv \int_{-\infty}^{\infty} du\, e^{-i\omega u}\langle A(0) B^\dagger(u)\rangle.$$

This procedure follows our classical discussion IV, footnote 13, with $\alpha \to B^\dagger$, $\alpha^* \to A$ since in the quantum-mechanical case, it is usually the daggered operators that carry the positive frequencies. Similarly,

$$\langle A^\dagger(\omega) B(\omega')\rangle = 2\pi\delta(\omega - \omega')\langle A_\omega^\dagger B_\omega\rangle,$$

$$\langle A_\omega^\dagger B_\omega\rangle \equiv \int_{-\infty}^{\infty} e^{-i\omega u} du\langle A^\dagger(u) B(0)\rangle.$$

A simple rule to remember is that if we stick to $\exp(-i\omega u)$, it is always the daggered operator that carries the time dependence, and orders of operators are always preserved. One can readily verify, by integration by parts, that it is appropriate to use the rules

$$(dA/dt)_\omega = -i\omega A_\omega; \quad (dA^\dagger/dt)_\omega = i\omega A^\dagger_\omega.$$

Langevin force. The Heisenberg equation

$$db/dt=-i\omega_c b+g(t) \tag{C8}$$

suggests that we set

$$f(t)=g(t)-\langle g(t)\rangle_p;\quad \langle g(t)\rangle_p=-(\tfrac{1}{2}\gamma+i\Delta\omega)b, \tag{C9}$$

where $\langle\ \rangle_p$ is an average over the reservoir as it is perturbed by the system. This interpretation is permissible for non-Markoffian systems. We have, however, made a Markoffian approximation by giving f a flat ("white") frequency spectrum chosen automatically by (3.4) to coincide with the spectrum of g at the frequency ω_c. Thus, for example,

$$\bar{n}=\frac{\mathrm{Re}\beta}{\frac{1}{2}\gamma}=\frac{\mathrm{Re}\int_0^\infty du\, e^{-i\omega_c u}\langle g^\dagger(u)g\rangle}{\mathrm{Re}\int_0^\infty du\, e^{-i\omega_c u}\langle[g,g^\dagger(u)]\rangle} \tag{C10}$$

$$=[\exp(\hbar\omega_c/kT_R)-1]^{-1}. \tag{C11}$$

To obtain the last step, we split $g^\dagger(u)g$ into an anticommutator plus a commutator. We next use the fact that the anticommutator is a real, even function of u, and the commutator is an imaginary, odd function of u to extend the integrals to $-\infty$. If the reservoir is at equilibrium at temperature T_R, we can then use the usual relation QIII (7.7) relating commutators and anticommutators to obtain (C11).

The dissipation coefficient $\frac{1}{2}\gamma$ and frequency shift $\Delta\omega$ regarded as functions of ω_c can be written

$$\tfrac{1}{2}\gamma(\omega_c)=-i\int_0^\infty du\,\sin\omega_c u\langle[g,g^\dagger(u)]\rangle, \tag{C12}$$

$$\Delta\omega(\omega_c)=i\int_0^\infty du\,\cos\omega_c u\langle[g,g^\dagger(u)]\rangle, \tag{C13}$$

so that the Hilbert transform relationship

$$\cos\omega_c u=\frac{1}{\pi}\mathcal{P}\int_{-\infty}^{\infty}\frac{\sin\omega u}{\omega-\omega_c}d\omega \tag{C14}$$

leads to a Kramers-Kronig relation[25]:

$$\Delta\omega(\omega_c)=-\frac{1}{\pi}\mathcal{P}\int_{-\infty}^{\infty}\frac{d\omega}{\omega-\omega_c}\tfrac{1}{2}\gamma(\omega). \tag{C15}$$

For the case $\gamma(\omega)=\gamma=\text{const}$, $\Delta\omega=0$. In any case, $\Delta\omega$ is usually small compared to the range of frequencies over which $\gamma(\omega)$ varies significantly. Since we must assume that the spectrum of g varies slowly near ω_c, we cannot, with the accuracy of the present discussion distinguish between $\gamma(\omega_c)$ and $\gamma(\omega_0)$. We shall conjecture, that a somewhat more self-consistent analysis would require us to absorb $\Delta\omega$ into a new unperturbed frequency ω_0 and, in our Markoffian analysis, evaluate $\gamma=\gamma(\omega_0)$, $\bar{n}=n(\omega_0)$ at the new frequency ω_0.

It is really more accurate to write

$$\langle e^{-i\omega_0 t}f^\dagger(t)e^{i\omega_0 u}f(u)\rangle\approx\gamma\bar{n}\delta(t-u) \tag{C16a}$$

than

$$\langle f^\dagger(t)f(u)\rangle\approx\gamma\bar{n}\delta(t-u), \tag{C16b}$$

since these equations imply

$$\gamma\bar{n}=\int_{-\infty}^{\infty}e^{-i\omega_0 t}\langle f^\dagger(t)f(0)\rangle dt \tag{C17a}$$

or

$$\gamma\bar{n}=\int_{-\infty}^{\infty}\langle f^\dagger(t)f(0)\rangle dt, \tag{C17b}$$

respectively. Within the Markoffian limit of delta-function autocorrelations, there is no distinction between these alternatives, but if the correlation time is long compared to ω_0^{-1}, the first of these alternatives is preferable. We shall therefore define

$$b'=be^{i\omega_0 t},\quad (b')^\dagger=b^\dagger e^{-i\omega_0 t}, \tag{C18}$$

$$F=fe^{i\omega_0 t},\quad F^\dagger=f^\dagger e^{-i\omega_0 t}, \tag{C19}$$

so that our new equations of motion are

$$db'/dt=-\tfrac{1}{2}\gamma b'+F(t),$$
$$d(b')^\dagger/dt=-\tfrac{1}{2}\gamma(b')^\dagger+F^\dagger(t), \tag{C20}$$

with

$$\langle F^\dagger(t)F(u)\rangle\approx\gamma\bar{n}\delta(t-u),$$
$$\langle F(u)F^\dagger(t)\rangle\approx\gamma(\bar{n}+1)\delta(t-u). \tag{C21}$$

The $\delta(t-u)$ could then have a width of the order of the correlation time without changing the results appreciably. These points are somewhat beyond the scope of a Markoffian approach.

Non-Markoffian Discussion of Harmonic Oscillators

As mentioned in the Introduction near Eq. (1.41), this consistent fusion of quantum mechanics with a Markoffian description has been achieved by using a rotating-wave approximation (RWA). The RWA was introduced by the choice of the interaction V in (3.2). The full interaction would have been

$$V=-QG;\qquad G=(2\hbar\omega_c)^{1/2}(g+g^\dagger), \tag{C22}$$

$$Q=(\hbar/2\omega_c)^{1/2}i(b-b^\dagger);\quad P=(\tfrac{1}{2}\hbar\omega_c)^{1/2}(b+b^\dagger) \tag{C23}$$

and would have lead to a decay term of the form $\frac{1}{2}\gamma(b+b^\dagger)$ in both the db/dt and $db^\dagger/dt$ equations, or to the equations

$$dQ/dt=P;\quad dP/dt=-\omega_0^2Q-\gamma P+F(t). \tag{C24}$$

[25] N. N. Bogoliubov and D. V. Shirkov, *Introduction to the Theory of Quantized Fields* (Interscience Publishers, Inc., New York, 1959), Sec. 46. Equation (C15) is really an equation for $\Delta\omega(\omega_c)-\Delta\omega(\infty)$, but we can usually assume $\Delta\omega(\infty)=0$.

A proper comparison with the full interaction procedure can only be made outside the Markoffian framework. The Langevin equations (C24) when combined and written in Fourier form lead to[24]

$$[\omega_0^2-\omega^2-i\omega\gamma]Q_\omega=F_\omega \tag{C25}$$

so that the spectrum of F is given by

$$F_\omega^\dagger F_\omega=DQ_\omega^\dagger Q_\omega;\quad D=[(\omega_0^2-\omega^2)^2+(\omega\gamma)^2], \tag{C26}$$

where

$$\langle Q_\omega^\dagger Q_\omega\rangle=\int_{-\infty}^{\infty}e^{-i\omega t}\langle Q(t)Q(0)\rangle dt \tag{C27}$$

can be found from QIII Sec. 7 to be

$$\langle Q_\omega^\dagger Q_\omega\rangle=2\gamma(\omega)\hbar\omega\bar{n}(\omega)/D(\omega), \tag{C28}$$

where the dissipation coefficient γ in the non-Markoffian case is allowed to be frequency-dependent. In this way, we find

$$\langle F_\omega^\dagger F_\omega\rangle=2\gamma(\omega)\hbar\omega\bar{n}(\omega), \tag{C29}$$

$$\langle F_\omega F_\omega^\dagger\rangle=2\gamma(\omega)\hbar\omega[\bar{n}(\omega)+1], \tag{C30}$$

$$\bar{n}(\omega)=[\exp(\hbar\omega/kT)-1]^{-1}, \tag{C31}$$

where T is the temperature of the reservoir with which the harmonic oscillator interacts. If the reservoir consists of a set of harmonic oscillators, these results are, in fact, exact.[26]

The results (C29), (C30) are not independent since

$$F_\omega^\dagger=F_{-\omega};\quad \gamma(-\omega)=\gamma(\omega),$$
$$\bar{n}(-\omega)=-[\bar{n}(\omega)+1]. \tag{C32}$$

The commutator is given by

$$\langle[F_\omega,F_\omega^\dagger]\rangle=2\hbar\omega\gamma(\omega) \tag{C33}$$

an odd function of ω. *The noise cannot therefore be made white even when $\gamma(\omega)=\gamma=a$ constant.* It is easy to verify in this case that the commutation rules are obeyed by using

$$\langle[Q,P(t)]\rangle=\frac{1}{2\pi}\int_{-\infty}^{\infty}e^{i\omega t}d\omega\langle[Q_\omega,P_\omega^\dagger]\rangle, \tag{C34}$$

$$P_\omega^\dagger=i\omega Q_\omega^\dagger, \tag{C35}$$

$$\langle[Q,P]\rangle=\frac{1}{2\pi}\int d\omega i\omega\langle[F_\omega,F_\omega^\dagger]\rangle/D(\omega). \tag{C36}$$

If we use the commutator (C33) with $\gamma(\omega)=\gamma$; $\omega_0=$ const,

$$\langle[Q,P]\rangle=\frac{i\hbar}{\pi}\int d\omega\frac{\gamma\omega^2}{(\omega^2-\omega_0^2)^2+(\omega\gamma)^2}=i\hbar. \tag{C37}$$

Senitzky[13] has chosen the commutation rule (1.38) in terms of principal-valued reciprocals. Thus,

$$\langle[F_\omega,F_\omega^\dagger]\rangle_{\text{Senitzky}}=2\gamma\hbar\omega_0\omega/|\omega|. \tag{C38}$$

The insertion of (C38) into (C36) then yields

$$\langle[Q,P]\rangle=i\hbar[1-\mu^2]^{-1/2}$$
$$\times\{1-1/\pi\arctan[2\mu(1-\mu^2)^{1/2}]\}, \tag{C39}$$

where $\mu=(\gamma/2\omega_0)$. This result has errors of first order in γ/ω_0 as shown in (1.39).

If we were to regard (C22) as a split of F into its positive- and negative-frequency parts, we would have

$$\langle f_\omega f_\omega^\dagger\rangle=(\omega/\omega_0)\gamma[\bar{n}(\omega)+1],\quad \omega>0$$
$$=0,\qquad \omega<0. \tag{C40}$$

If, furthermore, we were to set $\omega=\omega_0$ in (C40), this would lead to a spectrum that is white except for a jump at $\omega=0$ from the spectrum at $-\omega_0$ to that at ω_0. This is Senitzky's result. Our procedure is equivalent to the choice

$$\langle f_\omega f_\omega^\dagger\rangle=\gamma[\bar{n}(\omega_0)+1],\quad \text{all }\omega,$$
$$\langle f_\omega^\dagger f_\omega\rangle=\gamma\bar{n}(\omega_0),\quad \text{all }\omega, \tag{C41}$$

i.e., two white spectra with no jump. Within the RWA, our procedure leads to exact preservation of the commutation rules as shown in (C1) and (C6).

APPENDIX D: SOLUTION OF THE MARKOFFIAN HARMONIC-OSCILLATOR DENSITY-MATRIX EQUATION

Instead of proving the exactness of (3.6), we shall instead derive an equation for the density matrix σ of a Markoffian harmonic oscillator. We shall then solve the equation for σ and show that the solution for σ is identical to the exact solution found by Louisell and Walker[13] for the case of rotating-wave coupling to a reservoir consisting of a set of harmonic oscillators.

An equation for $\sigma(t)$ can be obtained by combining (A10) and (A12). To avoid having to re-identify the constants, we shall start instead by rewriting (3.3), using the identification (3.5), in the form

$$\partial\langle M\rangle/\partial t=\langle(M,\hbar\omega_0 b^\dagger b)\rangle+\tfrac{1}{2}\gamma\langle b^\dagger[M,b]-[M,b^\dagger]b\rangle$$
$$+\gamma\bar{n}\langle[b,[M,b^\dagger]]\rangle. \tag{D1}$$

We then note that if

$$\partial\sigma/\partial t=\textstyle\sum_i A_i\sigma B_i, \tag{D2}$$

where A_i and B_i are any operators,

$$\partial\langle M\rangle/\partial t=\textstyle\sum_i\langle B_iMA_i\rangle \tag{D3}$$

[26] By using the Heisenberg equations of motion and solving for the reservoir oscillators in terms of the known motion of the system oscillator, one can readily show that the mean equation of motion of our system oscillator, e.g., QIII (4.13) (omitting $j\neq i$ terms) is exact. The spectrum (C28) or QIII (7.11) is then exact since it uses only the correct mean motion and the fluctuation-dissipation theorem. Note, however, that if $\gamma(\omega)\neq$constant we must also regard $\omega_0^2\equiv\omega_c^2+\text{Re}b(\omega)$ of QIII (4.13) as frequency-dependent. The moments of the Langevin forces (C29), (C30) are expressible directly in terms of $\gamma(\omega)$, independently of ω_0.

so that the equation for σ can be obtained by inspecting the equation for $\langle M\rangle$. After some rearrangement, our equation for σ takes the form

$$\partial\sigma/\partial t=(\hbar\omega_0 b^\dagger b,\sigma)+\gamma\sigma+\tfrac{1}{2}\gamma\{b^\dagger[b,\sigma]+[\sigma,b^\dagger]b\} +\gamma(\bar{n}+1)[[b,\sigma],b^\dagger], \quad \text{(D4)}$$

$$\partial\sigma/\partial t=(\hbar\omega_0 b^\dagger b,\sigma)+\gamma\sigma+\tfrac{1}{2}\gamma\{b^\dagger\partial\sigma/\partial b^\dagger+\partial\sigma/\partial b b\} +\gamma(\bar{n}+1)\partial^2\sigma/\partial b\partial b^\dagger, \quad \text{(D 5)}$$

where we use

$$\frac{\partial}{\partial b}\frac{\partial\sigma}{\partial b^\dagger}\equiv[[b,\sigma],b^\dagger]=\frac{\partial}{\partial b^\dagger}\frac{\partial\sigma}{\partial b}\equiv[b,[\sigma,b^\dagger]]. \quad \text{(D6)}$$

The Louisell-Walker solution[13] (when no external driving forces are present) is a function only of $b^\dagger b$. For such a σ, the first term in (D4) or (D5) vanishes. The remaining terms in (D5) have, with malice aforethought, been arranged so that if $\sigma=\sigma^{(n)}$ is in normal order[27] (all $b^\dagger$ operators to the left of all b operators) these terms are already in normal order.

To solve (D5) we use the normal ordering operator $\mathfrak{N}$ discussed by Louisell.[27] If $g(\bar{b},\bar{b}^\dagger)$ is any classical function of the c numbers $\bar{b}$ and $\bar{b}^\dagger$, and if $g^n(\bar{b},\bar{b}^\dagger)$ is the same classical function arranged so that all $(\bar{b}^\dagger)$'s appear to the left of all $\bar{b}$'s, then the operator obtained by replacing $\bar{b}$ by b and $\bar{b}^\dagger$ by $b^\dagger$ is written

$$g^n(b,b^\dagger)\equiv\mathfrak{N}\{g^n(\bar{b},\bar{b}^\dagger)\}. \quad \text{(D7)}$$

Thus, the operator $\mathfrak{N}$ converts a c-number function to an operator (that is already in normal order) by a definite rule.

Let us now assume that a solution to (D5) can be written in the form

$$\sigma^{(n)}(b,b^\dagger)=\mathfrak{N}\{S(z)\};\quad z=\bar{b}^\dagger\bar{b}. \quad \text{(D8)}$$

Since differentiations do not disturb normal order, we have

$$\partial\sigma^{(n)}/\partial b b=\mathfrak{N}\{\partial S(z)/\partial\bar{b}\bar{b}\}. \quad \text{(D9)}$$

[27] For a lucid account of normal ordering and the normal ordering operator $\mathfrak{N}$ see W. H. Louisell, *Radiation and Noise in Quantum Electronics* (McGraw-Hill Book Company, Inc., New York, 1964), Chap. 3.

Underneath the $\mathfrak{N}$, we have only c numbers, and can use the usual rules of differentiation, and can write factors in any order. Thus,

$$\partial\sigma^{(n)}/\partial b b=\mathfrak{N}\{\bar{b}^\dagger\partial S/\partial z\bar{b}\}=\mathfrak{N}\{z\partial S/\partial z\}, \quad \text{(D10)}$$

Eq. (D5) can in this manner be rewritten as

$$\mathfrak{N}\left\{\frac{\partial S(z)}{\partial t}\right\}=\mathfrak{N}\left\{\gamma\frac{\partial}{\partial z}(zS) +\gamma(\bar{n}+1)\frac{\partial}{\partial z}\left[z\left(\frac{\partial S}{\partial z}\right)\right]\right\}, \quad \text{(D11)}$$

from which we obtain the c-number equation

$$\frac{\partial S(z,t)}{\partial t}=\gamma\frac{\partial}{\partial z}(zS)+\gamma(\bar{n}+1)\frac{\partial}{\partial z}\left[z\frac{\partial S}{\partial z}\right]. \quad \text{(D12)}$$

By direct substitution, we can verify that

$$S(z,t)=[y(t)]^{-1}\exp[-z/y(t)] \quad \text{(D13)}$$

is a solution of (D12), provided that y obeys

$$dy/dt=\gamma(\bar{n}+1)-\gamma y \quad \text{(D14)}$$

so that the density matrix is given by

$$\sigma=[y(t)]^{-1}\mathfrak{N}\{\exp[-\bar{b}^\dagger\bar{b}/y(t)]\}. \quad \text{(D15)}$$

The relationship Eq. (3.68) of Louisell[27]

$$\mathfrak{N}\{\exp[(e^x-1)\bar{b}^\dagger\bar{b}]\}=\exp(xb^\dagger b) \quad \text{(D16)}$$

permits this result to be rewritten in the form

$$\sigma=\frac{1}{y(t)}\left[1-\frac{1}{y(t)}\right]^{b^\dagger b} \quad \text{(D17)}$$

from which it is evident that

$$\text{tr}\sigma=1. \quad \text{(D18)}$$

The relations (D14), (D15) define the exact Louisell-Walker[13] solution for the case $\gamma=\text{const}$ (independent of frequency), and $y(t)$ has the interpretation

$$y(t)=\langle b(t)b^\dagger(t)\rangle. \quad \text{(D19)}$$

Author Index

Subject Index

M

N

O

P

Q

R

Editor's Biography

Frank S. Barnes (S'54–M'58–F'70) was born in Pasadena, Calif., on July 31, 1932. He received the B.S.E. degree from Princeton University, Princeton, N.J., in 1954 and the M.S. and Ph.D. degrees from Stanford University, Stanford, Calif., in 1956 and 1958, respectively, all in electrical engineering.

He was a Fulbright Professor at the College of Engineering, Baghdad, Iraq, during the academic year 1957–1958. In 1958 and 1959 he worked as a Research Associate with the Colorado Research Corporation, Broomfield. In 1959 he joined the Electrical Engineering Department at the University of Colorado, Boulder, as an Associate Professor. He became Chairman of the Department in 1963 and full Professor in 1964. He has been carrying out research activity in the fields of masers and molecular beam standards, application of lasers to biology, and millimeter waves.

Dr. Barnes is a member of the Society of Sigma Xi, the American Physical Society, the American Association for the Advancement of Science, and Eta Kappa Nu.